Probability and Mathematical Statistics (Continued)

SEBER • Linear Regression Analysis
SEBER • Multivariate Observations
SEN • Sequential Nonparametrics: Invariance Principles and Statistical Inference
SERFLING • Approximation Theorems of Mathematical Statistics
TJUR • Probability Based on Radon Measures
WILLIAMS • Diffusions, Markov Processes, and Martingales, Volume I: Foundations
ZACKS • Theory of Statistical Inference

Applied Probability and Statistics

ABRAHAM and LEDOLTER • Statistical Methods for Forecasting
AGRESTI • Analysis of Ordinal Categorical Data
AICKIN • Linear Statistical Analysis of Discrete Data
ANDERSON, AUQUIER, HAUCK, OAKES, VANDAELE, and WEISBERG • Statistical Methods for Comparative Studies
ARTHANARI and DODGE • Mathematical Programming in Statistics
BAILEY • The Elements of Stochastic Processes with Applications to the Natural Sciences
BAILEY • Mathematics, Statistics and Systems for Health
BARNETT • Interpreting Multivariate Data
BARNETT and LEWIS • Outliers in Statistical Data
BARTHOLOMEW • Stochastic Models for Social Processes, *Third Edition*
BARTHOLOMEW and FORBES • Statistical Techniques for Manpower Planning
BECK and ARNOLD • Parameter Estimation in Engineering and Science
BELSLEY, KUH, and WELSCH • Regression Diagnostics: Identifying Influential Data and Sources of Collinearity
BHAT • Elements of Applied Stochastic Processes
BLOOMFIELD • Fourier Analysis of Time Series: An Introduction
BOX • R. A. Fisher, The Life of a Scientist
BOX and DRAPER • Evolutionary Operation: A Statistical Method for Process Improvement
BOX, HUNTER, and HUNTER • Statistics for Experimenters: An Introduction to Design, Data Analysis, and Model Building
BROWN and HOLLANDER • Statistics: A Biomedical Introduction
BROWNLEE • Statistical Theory and Methodology in Science and Engineering, *Second Edition*
CHAMBERS • Computational Methods for Data Analysis
CHATTERJEE and PRICE • Regression Analysis by Example
CHOW • Analysis and Control of Dynamic Economic Systems
CHOW • Econometric Analysis by Control Methods
COCHRAN • Sampling Techniques, *Third Edition*
COCHRAN and COX • Experimental Designs, *Second Edition*
CONOVER • Practical Nonparametric Statistics, *Second Edition*
CONOVER and IMAN • Introduction to Modern Business Statistics
CORNELL • Experiments with Mixtures: Designs, Models and The Analysis of Mixture Data
COX • Planning of Experiments
DANIEL • Biostatistics: A Foundation for Analysis in the Health Sciences, *Third Edition*
DANIEL • Applications of Statistics to Industrial Experimentation
DANIEL and WOOD • Fitting Equations to Data: Computer Analysis of Multifactor Data, *Second Edition*
DAVID • Order Statistics, *Second Edition*
DAVISON • Multidimensional Scaling
DEMING • Sample Design in Business Research
DILLON and GOLDSTEIN • Multivariate Analysis: Methods and Applications
DODGE and ROMIG • Sampling Inspection Tables, *Second Edition*
DOWDY and WEARDEN • Statistics for Research

continued on back

MULTIVARIATE OBSERVATIONS

G. A. F. SEBER

Professor of Statistics
University of Auckland
Auckland, New Zealand

JOHN WILEY & SONS

New York • Chichester • Brisbane • Toronto • Singapore

Library of Congress Cataloging in Publication Data:

Seber, G. A. F. (George Arthur Frederick), 1938–
Multivariate observations.

(Wiley series in probability and mathematical statistics. Probability and mathematical statistics)
Includes bibliographical references and index.
1. Multivariate analysis. I. Title. II. Series.

QA278.S39 1984 519.5′35 83-21741
ISBN 0-471-88104-X

Printed in the United States of America

10 9 8 7 6 5 4 3 2 1

Preface

The theory of multivariate analysis had its beginnings in the 1930s and much of the early work was restricted to the multivariate normal distribution. The mathematics was intractable for other distributions—it was hard enough for the normal case—and many of the procedures developed were found to have optimal properties under normality. However, the methods were not welcomed with open arms, as the computations were usually very time-consuming, even with a good desk-top calculator, and the distributions of the various test statistics were not tabulated. For example, likelihood ratio test statistics were developed for a wide range of hypothesis tests, but they all involve finding $d \times d$ determinants, where d is the dimension of the data, and only their asymptotic distributions were known. A great deal of effort was then expended in finding good approximations for the distributions of these test statistics and their power functions. Eigenvalues of various matrices of random variables held a prominent place in the growing industry of multivariate distribution theory.

With the advent of powerful computers, however, the subject began to free itself from its multivariate normal straitjacket and researchers began to tackle problems in a more realistic fashion, without being cramped by lack of computational power. Automated procedures have allowed the graphical exploration of data, which is so necessary for good data analysis, to become a practical possibility and have facilitated the calculation of tables of exact percentage points for a number of test statistics. However, good approximations are still important for the automatic presentation of significance levels in a computer printout. There has also been a growing interest in robust procedures that are close to optimal under normality assumptions but which are better than the usual normal-based procedures when these assumptions are not satisfied.

However, the classical normal-based procedures still have a place for a number of reasons. First, some of these procedures are robust under moderate departures from the usual assumptions. Second, tests and graphical procedures exist for investigating normality. Third, when the normal-based procedure is optimal, it can be used as a guideline for comparing less optimal but more robust procedures. Fourth, many of the "classical" methods are derived using techniques that are important in their own right. Fifth, many of the current statistical computer packages automatically compute various normal-based statistics and it is important that package users know what these statistics are and how they behave in practice.

With the above comments in mind, I have attempted to write a textbook including both data-oriented techniques and a reasonable coverage of classical methods supported by comments about robustness and general practical applicability. Numerous books have been written on multivariate analysis, ranging from the very theoretical to the very applied with no proofs. Why, then, another book on the subject? In the first place, the subject is growing rapidly right across the theoretical–applied spectrum, and regular updating is needed. I have therefore attempted to provide a text that gives a comprehensive up-to-date survey of the subject and is useful as a reference book. This has not been easy, as entire books and monographs on topics covered by single chapters of this book are beginning to appear on the market.

Secondly, there is a current shortage of textbooks suited to graduate majors in statistics that are theoretical and yet give something of the more applied, data-oriented developments. The subject is a rich one, with room for everyone, regardless of their applied or theoretical bias, and it is hoped that this book caters to both viewpoints.

The basic prerequisites for reading this book are a good knowledge of matrix algebra and an acquaintance with the multivariate normal distribution, multiple linear regression, and simple analysis of variance and covariance models. This book could be regarded as a companion volume to Seber [1977].

Chapter 1 discusses the nature of multivariate data and problems of simultaneous inference. Chapter 2 selects from the extensive subject of multivariate distribution theory sufficient ideas to indicate the breadth of the subject and to provide a basis for later proofs. Attention is focused on the multivariate normal distribution, Wishart's distribution, Hotelling's T^2 distribution, together with the multivariate beta distribution and some of its derivatives. Inference for the multivariate normal, both in the one- and two-sample cases, is discussed extensively in Chapter 3, while Chapter 4 surveys graphical and data-oriented techniques. Chapter 5 discusses at length the many practical methods for expressing multivariate data in fewer dimensions in the hope of tracking down clustering or internal structure more effectively. The growing field of discriminant analysis is reviewed in Chapter 6, while Chapter 7 endeavors to summarize the essentials of the almost unmanageable body of literature on cluster analysis. Chapters 8 and 9 develop linear models, with the general theory given in Chapter 8 and applications to simple multivariate analysis of variance and covariance models given in Chapter 9. The final chapter, Chapter 10, is mainly concerned with computational techniques; however, log–linear models and incomplete data are also discussed briefly.

Appendixes A and B summarize a number of useful results in matrix algebra, with projection matrices discussed in Appendix B. Order statistics and probability plotting are considered in Appendix C, and Appendix D is a collection of useful statistical tables. Finally, there is a set of outline solutions for the exercises.

I would like to express my thanks to Dan Coster, Lee Kaiser and Katrina Sharples for collectively reading most of the manuscript. Thanks are also due to Dr. John Anderson and Dr. Peter Lachenbruch for providing some references and material for Chapter 6, and to Dr. Bill Davis for kindly computing some percentage points for Appendixes D14 and D15 ($d = 2$). I am also grateful to John Pemberton for carrying out most of the computing involved in the numerical examples.

I wish to thank the authors, editors, and owners of copyright for permission to reproduce the following published tables and figures: Table 3.1 (Dr. Brian Joiner); Tables 3.2, 4.2, 4.3, 5.2, 5.4, 5.5, 5.7, Appendixes D10, D11, and Fig. 5.2 (copyright by the Royal Statistical Society); Tables 3.4, 5.11, 5.12, 5.13, 6.1, 6.3, 9.12, 9.13, 9.14, 9.15, Figs. 5.3, and 5.4 (copyright by the Biometrics Society); Table 5.8 (copyright by the CSIRO Editorial and Publication Service); Table 7.5, Figs. 4.7, 7.6, 7.7, and 7.8 (copyright by the Institute of Statisticians); Tables 7.6, 7.7, and Appendix D18 (copyright by Elsevier Biomedical Press B.V.); Appendix D12 (copyright by John Wiley & Sons, Ltd.); Figs. 5.10 and 5.14, (copyright by the American Association for the Advancement of Science); Figs. 5.11 and 5.12 (copyright, 1970, by the American Psychological Association, reproduced by permission of the publishers and authors); Appendix D16 (copyright by Gordon and Breach Science Publishers); Appendix D15 (copyright by Marcel Dekker, Inc.); Tables 3.3, 4.4, 6.2, Appendixes D2 to D9 (inclusive), D13, D15, D17, and Fig. 6.3 (copyright by Biometrika Trust); Tables 3.5, 4.1, 4.5, Appendix D1, Figs. 4.3, 4.5, 4.8, 4.9, 4.10, 4.11, and 5.1 (copyright by the American Statistical Association); Appendix D14 (selected values by permission of Dr. K. C. S. Pillai); Appendix D15 (copyright by the Institute of Statistical Mathematics, Tokyo); Tables 5.28 and 9.3 (copyright by The Zoological Society, London); and Tables 5.14–5.21 (Dr. Ivor Francis).

G. A. F. SEBER

Auckland, New Zealand
September 1983

Contents

Notation

Script Letters

$\mathscr{A}$	$\mathscr{C}$	$\mathscr{D}$	$\mathscr{E}$	$\mathscr{N}$	$\mathscr{P}$	$\mathscr{R}$	ℓ
A	C	D	E	N	P	R	l

Greek Alphabet

	Lowercase	Uppercase
Alpha	α	
Beta	β	
Gamma	γ	Γ
Delta	δ	Δ
Epsilon	ε	
Eta	η	
Theta	θ	Θ
Lambda	λ	Λ
Mu	μ	
Nu	ν	
Xi	ξ	
Pi	π	Π
Rho	ρ	
Sigma	σ	Σ
Tau	τ	
Phi	ϕ	Φ
Chi	χ	
Psi	ψ	Ψ
Omega	ω	Ω

Special Symbols

$\square$ = end of theorem or example
∂ = partial derivative sign
$\cap$ = intersection
$\cup$ = union

∞ = infinity
$\propto$ = proportional to
$\mathbf{O}$ = bold-faced "oh" to be distinguished from $\mathbf{0}$, bold-faced zero
$\Leftrightarrow$ = if and only if
$\perp$ = perpendicular (to)
$\otimes$ = Kronecker product
$\in$ = an element of
$\subset$ = included in, a subset of

Multivariate Observations

CHAPTER 1

Preliminaries

1.1 NOTATION

Matrices and vectors are denoted by bold-faced letters $\mathbf{A}$ and $\mathbf{a}$, respectively, and scalars by italics. If $\mathbf{a}$ is an $n \times 1$ column vector with elements $a_1, a_2, \ldots, a_n$, we write $\mathbf{a} = [(a_j)]$, and the *length* or *norm* of $\mathbf{a}$ is denoted by $\|\mathbf{a}\|$. Thus

$$\|\mathbf{a}\| = (\mathbf{a}'\mathbf{a})^{1/2} = \left(a_1^2 + a_2^2 + \cdots + a_n^2\right)^{1/2}.$$

The $n \times 1$ vector with all its elements equal to unity is represented by $\mathbf{1}_n$.

If the $m \times n$ matrix $\mathbf{A}$ has elements a_{ij}, we write $\mathbf{A} = [(a_{ij})]$, and the sum of the diagonal elements, called the *trace* of $\mathbf{A}$, is denoted by $\operatorname{tr}\mathbf{A}$ $(= a_{11} + a_{22} + \cdots + a_{kk}$, where k is the smaller of m and n). The *transpose* of $\mathbf{A}$ is represented by $\mathbf{A}' = [(a'_{ij})]$, where $a'_{ij} = a_{ji}$. A matrix $\mathbf{A}^-$ that satisfies $\mathbf{A}\mathbf{A}^-\mathbf{A} = \mathbf{A}$ is called a *generalized inverse* of $\mathbf{A}$. If $\mathbf{A}$ is square, its determinant is written $|\mathbf{A}|$, and if $\mathbf{A}$ is nonsingular ($|\mathbf{A}| \neq 0$), its inverse is denoted by $\mathbf{A}^{-1}$. The $n \times n$ matrix with diagonal elements $d_1, d_2, \ldots, d_n$ and zeros elsewhere is represented by $\operatorname{diag}(d_1, d_2, \ldots, d_n)$, and when all the d_i's are unity, we have the identity matrix $\mathbf{I}_n$.

Given an $n \times n$ symmetric matrix $\mathbf{A}$ such that $\mathbf{x}'\mathbf{A}\mathbf{x} \geq 0$ for every $\mathbf{x} \neq \mathbf{0}$, we say that $\mathbf{A}$ is positive semidefinite and write $\mathbf{A} \geq \mathbf{O}$, where $\mathbf{O}$ is an $n \times n$ matrix of zeros. We can therefore talk about a partial ordering of symmetric matrices $\mathbf{A}$ and $\mathbf{B}$ and write $\mathbf{B} \geq \mathbf{A}$ when $\mathbf{B} - \mathbf{A} \geq \mathbf{O}$. If $\mathbf{x}'\mathbf{A}\mathbf{x} > 0$ for every $\mathbf{x} \neq \mathbf{0}$, we say that $\mathbf{A}$ is positive definite and write $\mathbf{A} > \mathbf{O}$.

For any $m \times n$ matrix $\mathbf{A}$ the space spanned by the columns of $\mathbf{A}$, called the range space of $\mathbf{A}$, is denoted by $\mathscr{R}[\mathbf{A}]$. The null space or kernel of $\mathbf{A}$ $(= \{\mathbf{x}: \mathbf{A}\mathbf{x} = \mathbf{0}\})$ is denoted by $\mathscr{N}[\mathbf{A}]$. The symbol R^d will represent d-dimensional Euclidean space.

It is common practice to distinguish between random variables and their values by using upper- and lowercase letters, for example, Y and y. However when this notation is applied to vectors and matrices of random variables, there are some problems. First, there is the need to distinguish between a random matrix $[(Y_{ij})]$ and its observed value $\mathbf{Y} = [(y_{ij})]$: This could be achieved by using, for example, script or bold-faced italic letters. Second, a vector of random variables $\mathbf{Y} = (Y_1, Y_2, \ldots, Y_n)'$ looks like a matrix of observed values $\mathbf{Y} = [(y_{ij})]$. Clearly the reader will be more concerned with knowing whether a vector or matrix of data is involved than with the distinction between random variables and their observed values. For this reason I have decided not to make the latter distinction. However, to help the reader distinguish between data and constants, the letters $A, B, \ldots, P$ and their lowercase counterparts will generally refer to constants, whereas $Q, R, \ldots, Z$ and their lowercase counterparts will refer to random variables or their values. There will be some exceptions, as there are not enough letters to go round!

If x and y are random variables, then the symbols $\mathrm{E}[x]$, $\mathrm{var}[x]$, $\mathrm{cov}[x, y]$, and $\mathrm{E}[x|y]$ represent expectation, variance, covariance, and conditional expectation, respectively. Multivariate analogues of these are defined in Section 1.3.

We say that $y \sim N_1(\theta, \sigma^2)$ if y is normally distributed with mean θ and variance σ^2: z has a standard normal distribution if $z \sim N_1(0, 1)$. The t- and chi-square distributions with k degrees of freedom are denoted by t_k and χ_k^2, respectively, and the F-distribution with m and n degrees of freedom is denoted by $F_{m,n}$. The gamma and beta functions are respectively defined as

$$\Gamma(k) = \int_0^\infty y^{k-1} e^{-y}\, dy \qquad (k > 0)$$

and

$$B[a, b] = \frac{\Gamma(a)\Gamma(b)}{\Gamma(a+b)} = \int_0^1 y^{a-1}(1-y)^{b-1}\, dy \qquad (a, b > 0). \quad (1.1)$$

We shall also use the "dot" and "bar" notation representing sum and average, respectively: For example,

$$\bar{y}_{i\cdot} = \sum_{j=1}^{n} y_{ij}/n.$$

In the case of a single subscript, x_j, say, we omit the dot and write $\bar{x}$ for the mean.

It is assumed that the reader is well versed in linear algebra, though many of the standard (and not so standard) results are included in Appendixes A and B at the end of this book. References to these Appendixes are denoted by, for example, A2.3.

1.2 WHAT IS MULTIVARIATE ANALYSIS?

The term *multivariate analysis* means the analysis of many variables, which effectively includes all of statistical sampling. However, this term has traditionally referred to the study of vectors of correlated random variables, for example, n random vectors $\mathbf{x}_1, \mathbf{x}_2, \ldots, \mathbf{x}_n$, each of dimension d (we depart from the more common but less convenient notation of p for dimension). Typically these vectors arise from taking measurements on d variables or characteristics for each of n objects or people. The vectors may or may not come from the same probability distribution and the variables x_{ij} within each $\mathbf{x}_i = (x_{i1}, x_{i2}, \ldots, x_{id})'$ will generally be correlated. The variables may be quantitative (discrete or continuous) or qualitative (ordered or unordered categories). For example, we may wish to record the following $d = 6$ variables for each of n men in a given age bracket: $x_{i1} =$ *height*, $x_{i2} =$ *weight*, $x_{i3} =$ *blood pressure* (continuous quantitative variables); $x_{i4} =$ *number of children* (discrete quantitative variable); $x_{i5} =$ *income category*, that is, low, medium, or high (ordered qualitative variable, also referred to as an ordered multistate or ordinal variable); and $x_{i6} =$ *marital status category*, that is, married, never married, or previously married (unordered qualitative variable, also referred to as a disordered multistate or nominal variable). The qualitative variables are coded: A natural order for the income categories would be $x_{i5} = 1, 2, 3$, respectively, whereas for marital status any order would do, such as $x_{i6} = 1, 2, 3$, respectively, or some permutation thereof. Many qualitative variables are defined on only two categories, for example, sex and presence or absence of a "characteristic" or symptom. Such variables, usually coded $(1, 0)$, are called binary variables. The variable x_{i6} can, for example, be represented by two binary variables x_{i7} and x_{i8} as follows:

$$x_{i7} = \begin{cases} 1 & \text{if married,} \\ 0 & \text{otherwise;} \end{cases}$$

$$x_{i8} = \begin{cases} 1 & \text{if never married,} \\ 0 & \text{otherwise.} \end{cases} \tag{1.2}$$

The three marriage categories are now coded as $(x_{i7}, x_{i8}) = (1, 0), (0, 1), (0, 0)$, respectively. In general, a k-state nominal or ordinal variable can be represented by $k - 1$ binary variables.

It is convenient to express the random variables x_{ij} (or their values) in matrix form $\mathbf{X} = [(x_{ij})]$, namely,

$$\begin{array}{cc} & \text{variable} \\ & \begin{array}{cccc} 1 & 2 & \cdots & d \end{array} \\ \begin{array}{r} \\ \text{object} \\ \text{(person)} \\ \\ \end{array} \begin{array}{c} 1 \\ 2 \\ \vdots \\ n \end{array} & \begin{pmatrix} x_{11} & x_{12} & \cdots & x_{1d} \\ x_{21} & x_{22} & \cdots & x_{2d} \\ \vdots & \vdots & & \vdots \\ x_{n1} & x_{n2} & \cdots & x_{nd} \end{pmatrix}. \end{array}$$

We can also write

$$\mathbf{X} = \begin{pmatrix} \mathbf{x}_1' \\ \mathbf{x}_2' \\ \vdots \\ \mathbf{x}_n' \end{pmatrix} \tag{1.3}$$

$$= \left(\mathbf{x}^{(1)}, \mathbf{x}^{(2)}, \ldots, \mathbf{x}^{(d)}\right), \tag{1.4}$$

where the row representation (1.3) emphasizes the objects, and the column representation (1.4) emphasizes the variables. In our previous example $\mathbf{x}^{(1)}$ is the vector of height measurements for the n men. Some authors use the transpose of $\mathbf{X}$ as their data matrix, but this is inconvenient, for example, when studying linear models. Throughout this book the suffix i generally refers to the ith observation vector, while suffixes j and k usually refer to variables or characteristics.

Although no distinction is made between random variables or their values, it should be noted that sometimes the n objects represent the whole population or a nonrandom subset of the population. In this case the use of random variables is not appropriate.

Many univariate statistical methods generalize quite naturally to higher dimensions, but there are some fundamental difficulties in going above one or two dimensions. First, there is the problem of graphical representation. Hopefully one may be able to transform the d-dimensional points representing the $\mathbf{x}_i$ into points in a two- or at most three-dimensional space without too much change in relative positions. Dimension-reducing techniques are discussed in detail in Chapter 5. Second, the notion of rank ordering that underlies univariate distribution free methods does not readily extend into several dimensions. A great deal of information is lost in transforming vectors into one-dimensional ranked data. Third, there is the problem of choosing d variables from a pool of variables that may be very large: A preliminary and possibly subsequent selection is almost always required.

Most problems in multivariate analysis fall into one of two categories. They are either about the "external" structure of the data (what can we say about the configuration or interpoint distances of the n points in d-dimensional Euclidean space R^d?) or about the "internal" structure of the variables (are the jth and kth variables strongly correlated?). For example, in the previous illustration with six variables and n men, we may wish to look for any natural grouping of the data or any of the n men that are "outlying" or different from the rest. If the sample is large, we could use it to estimate population averages: for example, average height, proportion married, and so on. Alternatively, we may wish to examine the relationship between blood pressure and the other five variables using a regression model.

In the previous example we considered just a single population; instead we may have a sample from each of two populations, say, men and women. Interest now extends to the geometrical relationships between the two swarms of points in R^d. Comparing the two sample vector means reduces to examining the relative positions of the centers of gravity of the two swarms, while comparing covariance structures is related to comparing the overall shapes of the two swarms. If our interest is "internal," then we may wish to see if the regression relationship of blood pressure on the other variables is the same for both populations.

A related problem is that of discrimination. Suppose samples are available from two populations and the origin of a further observation is unknown. On the basis of the d measurements we wish to assign the object to one population or the other. For example, using measures like weight and blood pressure, we may wish to decide whether a male aged 40–45 is at risk with regard to having a heart attack in the next 10 years. The two samples could be observations taken from 10-year case histories of men in the required age bracket: Those who do not have a heart attack form one sample, while those that do make up the other sample. An assignment rule can be estimated from the sample data (see Chapter 6) and used to assign current observations. A doctor may wish to counsel any man who is assigned to the heart attack group.

A natural extension of the previous examples is to the case of more than two populations. We may not even know the number of populations represented by the sample data, so that the problem is to divide the overall configuration into an unknown number of well-defined clusters. A typical problem in this area that has created some controversy is the classification of mental illnesses based on observations of certain quantitative and qualitative variables. How many categories (clusters) do we need and where do we draw the line between normal and abnormal?

An important group of multivariate problems comes under the title of ordination techniques. Instead of having observations on n objects, we have measures of similarity c_{rs}, say, between the rth and sth objects. We wish to represent the objects by a configuration of points in a low-dimensional space so that the distances between the points follow the same pattern as the c_{rs}. Hopefully the configuration will uncover any underlying structure or clustering of the objects. For example, the objects may be ancient graves, with c_{rs} being the proportion of artifacts common to graves r and s.

1.3 EXPECTATION AND COVARIANCE OPERATORS

Let $\mathbf{X} = [(x_{ij})]$ be a matrix of random variables. We define the expectation operator $\mathscr{E}$ for a matrix by

$$\mathscr{E}[\mathbf{X}] = \left[\left(\mathrm{E}[x_{ij}]\right)\right],$$

the matrix of $E[x_{ij}]$. From the linear properties of E,

$$\mathscr{E}[\mathbf{AXB}+\mathbf{C}] = \mathbf{A}\mathscr{E}[\mathbf{X}]\mathbf{B}+\mathbf{C}.$$

The concept of covariance can also be generalized. We define the generalized covariance operator $\mathscr{C}$ of two vectors of random variables $\mathbf{x} = [(x_i)]$ and $\mathbf{y} = [(y_j)]$ (not necessarily of the same dimension) as follows:

$$\begin{aligned}\mathscr{C}[\mathbf{x},\mathbf{y}] &= \left[\left(\operatorname{cov}[x_i, y_j]\right)\right]\\ &= \mathscr{E}\left[(\mathbf{x}-\boldsymbol{\mu}_x)(\mathbf{y}-\boldsymbol{\mu}_y)'\right]\\ &= \mathscr{E}[\mathbf{x}\mathbf{y}'] - \boldsymbol{\mu}_x\boldsymbol{\mu}_y' \qquad (1.5)\end{aligned}$$

(see Exercise 1.1), where $\boldsymbol{\mu}_x = \mathscr{E}[\mathbf{x}]$, the mean of $\mathbf{x}$, and so on. If $\mathbf{x}$ and $\mathbf{y}$ are statistically independent, then $\mathscr{C}[\mathbf{x},\mathbf{y}] = \mathbf{O}$. When $\mathbf{x} = \mathbf{y}$, then $\mathscr{C}[\mathbf{y},\mathbf{y}]$, written as $\mathscr{D}[\mathbf{y}]$, is called the dispersion (variance–covariance) matrix of $\mathbf{y}$. Thus

$$\begin{aligned}\mathscr{D}[\mathbf{y}] &= \left[\left(\operatorname{cov}[y_i, y_j]\right)\right]\\ &= \mathscr{E}\left[(\mathbf{y}-\boldsymbol{\mu}_y)(\mathbf{y}-\boldsymbol{\mu}_y)'\right]. \qquad (1.6)\end{aligned}$$

From Exercise 1.1, we have

$$\mathscr{C}[\mathbf{Ax},\mathbf{By}] = \mathbf{A}\mathscr{C}[\mathbf{x},\mathbf{y}]\mathbf{B}', \qquad (1.7)$$

from which it follows that

$$\mathscr{D}[\mathbf{Ay}] = \mathbf{A}\mathscr{D}[\mathbf{y}]\mathbf{A}'. \qquad (1.8)$$

Suppose $\mathbf{x}_1, \mathbf{x}_2, \ldots, \mathbf{x}_n$ are n statistically independent d-dimensional random vectors with means $\boldsymbol{\mu}_1, \ldots, \boldsymbol{\mu}_n$ and dispersion matrices $\boldsymbol{\Sigma}_1, \ldots, \boldsymbol{\Sigma}_n$, respectively. Then we define a generalized quadratic form

$$\sum_{i=1}^{n}\sum_{j=1}^{n} a_{ij}\mathbf{x}_i\mathbf{x}_j' = \mathbf{X}'\mathbf{A}\mathbf{X},$$

where $\mathbf{X} = (\mathbf{x}_1, \ldots, \mathbf{x}_n)'$, and $\mathbf{A} = [(a_{ij})]$ is symmetric. For $d = 1$ the above expression reduces to a quadratic $\Sigma_i \Sigma_j a_{ij} x_i x_j$.

LEMMA 1.1

$$\mathscr{E}[\mathbf{X}'\mathbf{A}\mathbf{X}] = \sum_{i=1}^{n} a_{ii}\boldsymbol{\Sigma}_i + \mathscr{E}[\mathbf{X}']\mathbf{A}\mathscr{E}[\mathbf{X}].$$

Proof

$$\begin{aligned}
\mathscr{E}[\mathbf{X}'\mathbf{A}\mathbf{X}] &= \mathscr{E}\left[\sum_i \sum_j a_{ij}\{(\mathbf{x}_i - \boldsymbol{\mu}_i)(\mathbf{x}_j - \boldsymbol{\mu}_j)' + \boldsymbol{\mu}_i\mathbf{x}_j' + \mathbf{x}_i\boldsymbol{\mu}_j' - \boldsymbol{\mu}_i\boldsymbol{\mu}_j'\}\right] \\
&= \sum_i \sum_j a_{ij}\mathscr{E}[(\mathbf{x}_i - \boldsymbol{\mu}_i)(\mathbf{x}_j - \boldsymbol{\mu}_j)'] + \sum_i \sum_j a_{ij}\boldsymbol{\mu}_i\boldsymbol{\mu}_j' \\
&= \sum_i \sum_j a_{ij}\mathscr{C}[\mathbf{x}_i, \mathbf{x}_j] + (\boldsymbol{\mu}_1, \ldots, \boldsymbol{\mu}_n)\mathbf{A}(\boldsymbol{\mu}_1, \ldots, \boldsymbol{\mu}_n)' \\
&= \sum_i a_{ii}\boldsymbol{\Sigma}_i + \mathscr{E}[\mathbf{X}']\mathbf{A}\mathscr{E}[\mathbf{X}].
\end{aligned}$$

COROLLARY If $\boldsymbol{\Sigma}_1 = \boldsymbol{\Sigma}_2 = \cdots = \boldsymbol{\Sigma}_n$ ($= \boldsymbol{\Sigma}$, say), then

$$\mathscr{E}[\mathbf{X}'\mathbf{A}\mathbf{X}] = (\operatorname{tr}\mathbf{A})\boldsymbol{\Sigma} + \mathscr{E}[\mathbf{X}']\mathbf{A}\mathscr{E}[\mathbf{X}]. \tag{1.9}$$

□

Finally we mention one other concept that sometimes arises in the multivariate literature. Let $\mathbf{A}$ and $\mathbf{B}$ be $m \times m$ and $n \times n$ matrices, respectively; then

$$\mathbf{A} \otimes \mathbf{B} = \begin{pmatrix} a_{11}\mathbf{B} & a_{12}\mathbf{B} & \cdots & a_{1m}\mathbf{B} \\ a_{21}\mathbf{B} & a_{22}\mathbf{B} & \cdots & a_{2m}\mathbf{B} \\ \vdots & \vdots & & \vdots \\ a_{m1}\mathbf{B} & a_{m2}\mathbf{B} & \cdots & a_{mm}\mathbf{B} \end{pmatrix}$$

is called the Kronecker or direct product of $\mathbf{A}$ and $\mathbf{B}$. Thus if $\mathbf{x}_1, \mathbf{x}_2, \ldots, \mathbf{x}_n$ are independent d-dimensional random vectors with a common dispersion matrix $\boldsymbol{\Sigma}$, then the "rolled out" vector of the $\mathbf{x}_i$'s, namely, $\mathbf{y} = (\mathbf{x}_1', \mathbf{x}_2', \ldots, \mathbf{x}_n')'$, has dispersion matrix

$$\mathscr{D}[\mathbf{y}] = \begin{pmatrix} \boldsymbol{\Sigma} & \mathbf{O} & \cdots & \mathbf{O} \\ \mathbf{O} & \boldsymbol{\Sigma} & \cdots & \mathbf{O} \\ \vdots & \vdots & & \vdots \\ \mathbf{O} & \mathbf{O} & \cdots & \boldsymbol{\Sigma} \end{pmatrix} = \mathbf{I}_n \otimes \boldsymbol{\Sigma}. \tag{1.10}$$

Some authors find it convenient to use $\mathbf{y}$ rather than $\mathbf{X}$ in deriving certain theoretical results. A common notation is $\mathbf{y} = \operatorname{vec}[\mathbf{x}_1, \ldots, \mathbf{x}_n] = \operatorname{vec}[\mathbf{X}']$ (see Henderson and Searle [1979] for a useful review of the properties of vec).

1.4 SAMPLE DATA

Suppose $\mathbf{x}_1, \mathbf{x}_2, \ldots, \mathbf{x}_n$ represent a random sample from a d-dimensional distribution with mean $\boldsymbol{\mu}$ and dispersion matrix $\boldsymbol{\Sigma}$. By analogy with the univariate sample mean $\bar{x}$ and unbiased estimate of variance, $s^2 = \sum_i (x_i - \bar{x})^2/(n-1)$, we consider the multivariate analogues

$$\bar{\mathbf{x}} = \frac{1}{n}\sum_{i=1}^{n} \mathbf{x}_i = \frac{1}{n}\mathbf{X}'\mathbf{1}_n$$

and

$$\mathbf{S} = \frac{1}{n-1}\sum_{i=1}^{n} (\mathbf{x}_i - \bar{\mathbf{x}})(\mathbf{x}_i - \bar{\mathbf{x}})' = \frac{1}{n-1}\mathbf{Q},$$

say. Then

$$\mathscr{E}[\bar{\mathbf{x}}] = \frac{1}{n}\sum_{i=1}^{n} \mathscr{E}[\mathbf{x}_i] = \boldsymbol{\mu}$$

and

$$\mathscr{D}[\bar{\mathbf{x}}] = \frac{1}{n^2}\sum_{i=1}^{n} \mathscr{D}[\mathbf{x}_i] = \frac{1}{n}\boldsymbol{\Sigma},$$

by Exercise 1.3. Writing $\mathbf{y}_i = \mathbf{x}_i - \boldsymbol{\mu}$, then $\bar{\mathbf{y}} = \bar{\mathbf{x}} - \boldsymbol{\mu}$ and

$$\begin{aligned}\sum_{i=1}^{n} (\mathbf{x}_i - \bar{\mathbf{x}})(\mathbf{x}_i - \bar{\mathbf{x}})' &= \sum_{i=1}^{n} (\mathbf{y}_i - \bar{\mathbf{y}})(\mathbf{y}_i - \bar{\mathbf{y}})' \\ &= \sum_{i=1}^{n} \mathbf{y}_i\mathbf{y}_i' - n\bar{\mathbf{y}}\bar{\mathbf{y}}', \end{aligned} \tag{1.11}$$

by Exercise 1.2. Since $\mathbf{y}_1, \mathbf{y}_2, \ldots, \mathbf{y}_n$ have mean zero and dispersion matrix $\boldsymbol{\Sigma}$, $\mathscr{E}[\mathbf{y}_i\mathbf{y}_i'[= \mathscr{D}[\mathbf{y}_i]$ [by (1.6)] and

$$\begin{aligned}\mathscr{E}[\mathbf{Q}] &= \sum_{i=1}^{n} \mathscr{D}[\mathbf{y}_i] - n\mathscr{D}[\bar{\mathbf{y}}] \\ &= n\boldsymbol{\Sigma} - n(\boldsymbol{\Sigma}/n) = (n-1)\boldsymbol{\Sigma}.\end{aligned}$$

Thus $\bar{\mathbf{x}}$ and $\mathbf{S}$ are unbiased estimators of $\boldsymbol{\mu}$ and $\boldsymbol{\Sigma}$, respectively. We note that for continuous random variables, $\mathbf{Q}$ is positive definite with probability 1 under very general conditions, which are usually satisfied (see A5.13).

Since data are frequently "centered," that is, the mean is substracted, we introduce the notation

$$\begin{aligned}\tilde{\mathbf{X}}' &= [\mathbf{x}_1 - \bar{\mathbf{x}}, \ldots, \mathbf{x}_n - \bar{\mathbf{x}}] \\ &= \mathbf{X}' - \bar{\mathbf{x}}(1, 1, \ldots, 1) \\ &= \mathbf{X}'\left(\mathbf{I}_n - \frac{1}{n}\mathbf{1}_n\mathbf{1}_n'\right) \\ &= \mathbf{X}'(\mathbf{I}_n - \mathbf{P}_1), \end{aligned} \tag{1.12}$$

where $\mathbf{P}_1$ and $\mathbf{I}_n - \mathbf{P}_1$ are symmetric idempotent matrices of ranks (traces) 1 and $n - 1$, respectively (see Exercise 1.5 with $\mathbf{K} = \mathbf{1}_n$). With this approach, an alternative proof that $\mathbf{S}$ is unbiased can be given. First,

$$\begin{aligned}\mathbf{Q} &= \tilde{\mathbf{X}}'\tilde{\mathbf{X}} \\ &= \mathbf{X}'(\mathbf{I}_n - \mathbf{P}_1)(\mathbf{I}_n - \mathbf{P}_1)'\mathbf{X} \\ &= \mathbf{X}'(\mathbf{I}_n - \mathbf{P}_1)\mathbf{X}, \end{aligned} \tag{1.13}$$

and $\mathscr{E}[\mathbf{X}'] = (\boldsymbol{\mu}, \ldots, \boldsymbol{\mu}) = \boldsymbol{\mu}\mathbf{1}_n'$. Then, by (1.9),

$$\begin{aligned}\mathscr{E}\left[\mathbf{X}'(\mathbf{I}_n - \mathbf{P}_1)\mathbf{X}\right] &= \boldsymbol{\Sigma}\,\mathrm{tr}[\mathbf{I}_n - \mathbf{P}_1] + \boldsymbol{\mu}\mathbf{1}_n'(\mathbf{I}_n - \mathbf{P}_1)\mathbf{1}_n\boldsymbol{\mu}' \\ &= (n-1)\boldsymbol{\Sigma},\end{aligned}$$

as $(\mathbf{I}_n - \mathbf{P}_1)\mathbf{1}_n = \mathbf{0}$. Hence

$$\mathbf{S} = \frac{\tilde{\mathbf{X}}'\tilde{\mathbf{X}}}{n-1} \tag{1.14}$$

is unbiased. An alternative estimate is the biased estimate $\hat{\boldsymbol{\Sigma}} = \mathbf{Q}/n = (n-1)\mathbf{S}/n$: $\hat{\boldsymbol{\Sigma}}$ is sometimes called the sample covariance matrix, and preferences for $\mathbf{S}$ or $\hat{\boldsymbol{\Sigma}}$ vary among statisticians.

Data are frequently scaled as well as centered. If r_{jk} is the sample correlation coefficient for the jth and kth variables, then

$$\begin{aligned} r_{jk} &= \left(\sum_{i=1}^{n}(x_{ij} - \bar{x}_{\cdot j})(x_{ik} - \bar{x}_{\cdot k})\right) \Big/ \left(\sum_{i=1}^{n}(x_{ij} - \bar{x}_{\cdot j})^2 \sum_{i=1}^{n}(x_{ik} - \bar{x}_{\cdot k})^2\right)^{1/2} \\ &= s_{jk}/(s_{jj}s_{kk})^{1/2}, \end{aligned}$$

where $[(s_{jk})] = \mathbf{S}$. If we write

$$\mathbf{D}_s = \mathrm{diag}(s_{11}, s_{22}, \ldots, s_{dd}), \tag{1.15}$$

then

$$\mathbf{R} = [(r_{jk})]$$
$$= \mathbf{D}_s^{-1/2}\mathbf{S}\mathbf{D}_s^{-1/2}, \tag{1.16}$$

and $\mathbf{R}$ is called the *sample correlation matrix*. Its definition is independent of the factor $n-1$, so that $\mathbf{S}$ or $\hat{\mathbf{\Sigma}}$ can be used. We note that $\mathbf{R}$ is a biased estimate of the *population correlation matrix*

$$\mathbf{P}_\rho = \begin{pmatrix} 1 & \rho_{12} & \cdots & \rho_{1d} \\ \rho_{21} & 1 & \cdots & \rho_{2d} \\ \vdots & \vdots & & \vdots \\ \rho_{d1} & \rho_{d2} & \cdots & 1 \end{pmatrix}, \tag{1.17}$$

where $\rho_{jk} = \sigma_{jk}/(\sigma_{jj}\sigma_{kk})^{1/2}$. We note that

$$\mathbf{P}_\rho = \mathbf{D}_\sigma^{-1/2}\mathbf{\Sigma}\mathbf{D}_\sigma^{-1/2},$$

where $\mathbf{D}_\sigma = \text{diag}(\sigma_{11}, \sigma_{22}, \ldots, \sigma_{dd})$. Methods for simulating correlation matrices are given by Chalmers [1975] and Bendel and Mickey [1978]. In data analysis, with observed values $\mathbf{x}_1, \mathbf{x}_2, \ldots, \mathbf{x}_n$, the scaled and centered data take the form $z_{ij} = (x_{ij} - \bar{x}_{\cdot j})/s_{jj}^{1/2}$ or $z_{ij} = (x_{ij} - \bar{x}_{\cdot j})/\hat{\sigma}_{jj}^{1/2}$. If we set $\mathbf{Z} = [(z_{ij})]$, the columns of $\mathbf{Z}$ have zero sample means and unit sample variances.

Finally, for the estimation of the mode and related contours (isopleths) see Sager [1979].

1.5 MAHALANOBIS DISTANCES AND ANGLES

If $\mathbf{x}$ is a vector of random variables with mean $\boldsymbol{\mu}$ and positive definite dispersion matrix $\mathbf{\Sigma}$, we define the Mahalanobis distance between $\mathbf{x}$ and $\boldsymbol{\mu}$ as

$$\Delta(\mathbf{x}, \boldsymbol{\mu}) = \left\{(\mathbf{x} - \boldsymbol{\mu})'\mathbf{\Sigma}^{-1}(\mathbf{x} - \boldsymbol{\mu})\right\}^{1/2}. \tag{1.18}$$

This distance measure differs from the usual Euclidean distance $\|\mathbf{x} - \boldsymbol{\mu}\|$ in that account is taken of the relative dispersions and correlations of the elements of $\mathbf{x}$. Also, for some distributions like the so-called family of elliptically symmetric distributions, the density function of $\mathbf{x}$ is constant for all $\mathbf{x}$ such that $\Delta(\mathbf{x}, \boldsymbol{\mu}) = c$. We can then interpret $\Delta(\mathbf{x}, \boldsymbol{\mu})$ as a "probabilistic" distance—equal distances imply equal likelihoods.

Given a random sample $\mathbf{x}_1, \mathbf{x}_2, \ldots, \mathbf{x}_n$, we have the following sample versions of (1.18):

$$D(\mathbf{x}, \bar{\mathbf{x}}) = \left\{(\mathbf{x} - \bar{\mathbf{x}})'\mathbf{S}^{-1}(\mathbf{x} - \bar{\mathbf{x}})\right\}^{1/2}$$

and

$$D(\mathbf{x}_r, \mathbf{x}_s) = \left\{(\mathbf{x}_r - \mathbf{x}_s)'\mathbf{S}^{-1}(\mathbf{x}_r - \mathbf{x}_s)\right\}^{1/2},$$

where $\mathbf{S}$ is defined by (1.14). In a similar fashion, we can define the Mahalanobis angle θ between $\mathbf{x}_r$ and $\mathbf{x}_s$, subtended at the origin, by

$$\cos\theta = \frac{\mathbf{x}_r'\mathbf{S}^{-1}\mathbf{x}_s}{D(\mathbf{x}_r, \mathbf{0})\,D(\mathbf{x}_s, \mathbf{0})}.$$

This contrasts with the usual measure

$$\cos\theta = \frac{\mathbf{x}_r'\mathbf{x}_s}{\|\mathbf{x}_r\|\,\|\mathbf{x}_s\|}.$$

In the case of two populations with means $\boldsymbol{\mu}_1$ and $\boldsymbol{\mu}_2$, and common dispersion matrix $\boldsymbol{\Sigma}$, we can define

$$\Delta(\boldsymbol{\mu}_1, \boldsymbol{\mu}_2) = \left\{(\boldsymbol{\mu}_1 - \boldsymbol{\mu}_2)'\boldsymbol{\Sigma}^{-1}(\boldsymbol{\mu}_1 - \boldsymbol{\mu}_2)\right\}^{1/2}$$

as the Mahalanobis distance between the two means. Given a sample of size n_i, with sample mean $\bar{\mathbf{x}}_i$ and unbiased estimate $\mathbf{S}_i$ of dispersion, from the ith population ($i = 1, 2$), we have the sample version

$$D(\bar{\mathbf{x}}_1, \bar{\mathbf{x}}_2) = \left\{(\bar{\mathbf{x}}_1 - \bar{\mathbf{x}}_2)'\mathbf{S}_p^{-1}(\bar{\mathbf{x}}_1 - \bar{\mathbf{x}}_2)\right\}^{1/2},$$

where $\mathbf{S}_p = [(n_1 - 1)\mathbf{S}_1 + (n_2 - 1)\mathbf{S}_2]/(n_1 + n_2 - 2)$ is a pooled unbiased estimate of $\boldsymbol{\Sigma}$.

The concept of Mahalanobis distance, whether in a population or sample form, arises several times in this book, for example, Hotelling's T^2, weights in robust estimation, discriminant analysis, and cluster analysis. For further details see Mardia [1975, 1977]: The second paper deals with the situation where $\boldsymbol{\Sigma}$ or $\mathbf{S}$ is singular.

1.6 SIMULTANEOUS INFERENCE

1.6.1 Simultaneous Tests

In multivariate analysis a variety of techniques have been developed for carrying out simultaneous hypothesis tests and constructing simultaneous confidence intervals. Some of the problems associated with simultaneous inference in the univariate case are described in Seber [1977: Chapter 5] and these are magnified in the multivariate generalizations.

Suppose we are interested in r simultaneous tests and let E_i $(i = 1, 2, \ldots, r)$ be the event that the ith hypothesis is accepted, given that it is true. Then we would like to find the appropriate critical level for each test such that the overall probability that they are simultaneously accepted is $1 - \alpha$, given they are all true, that is,

$$\mathrm{pr}\left[\bigcap_{i=1}^{r} E_i\right] = 1 - \alpha.$$

For example, suppose we are interested in a population parameter $\boldsymbol{\theta}$ that can be partitioned in the form $\boldsymbol{\theta}' = (\boldsymbol{\theta}_1', \boldsymbol{\theta}_2', \ldots, \boldsymbol{\theta}_r')$. Then instead of testing H_0: $\boldsymbol{\theta} = \mathbf{0}$, we may be interested in testing each individual hypothesis $H_{0i}: \boldsymbol{\theta}_i = \mathbf{0}$ $(i = 1, 2, \ldots, r)$. Here $H_0 = \cap_{i=1}^{r} H_{0i}$ and testing H_0 is equivalent to simultaneously testing each H_{0i}. The difference between the simultaneous procedure and the "direct" test of H_0 is that if H_0 is rejected by the simultaneous procedure, we can pinpoint which of the H_{0i} are at fault. If T_i is a test statistic for H_{0i}, then, typically, we require a constant c_α such that

$$\begin{aligned} 1 - \alpha &= \mathrm{pr}\left[\bigcap_{i=1}^{r} (|T_i| \le c_\alpha | H_0 \text{ true})\right] \\ &= \mathrm{pr}\left[\max_{1 \le i \le r} |T_i| \le c_\alpha | H_0 \text{ true}\right]. \end{aligned} \tag{1.19}$$

To find the critical value c_α for the distribution of $\max|T_i|$ we need to know the joint distribution of the $|T_i|$. Usually the T_i contain common random variables (e.g., common s^2 in univariate testing) so that the $|T_i|$ are correlated and follow a multivariate version of one of the univariate distributions like t-, F-, and chi-square distributions. For a survey of this problem see Miller [1977, 1981] and, in particular, Krishnaiah [1979], who calls the simultaneous test procedure outlined above a finite intersection test. Further details are given by Krishnaiah et al. [1980].

A conservative procedure is always available using the Bonferroni inequality

$$\mathrm{pr}\left[\bigcap_{i=1}^{r} E_i\right] \ge 1 - \sum_{i=1}^{r} \mathrm{pr}[\overline{E}_i], \tag{1.20}$$

where $\overline{E}_i$ is the complement of E_i. The lower bound is surprisingly good if the values of $\mathrm{pr}[\overline{E}_i]$ are small (Miller [1977: p. 779]). If we use the critical level α/r for each test, that is, $\mathrm{pr}[E_i] = 1 - (\alpha/r)$, then

$$\mathrm{pr}\left[\bigcap_{i=1}^{r} E_i\right] \ge 1 - r\left(\frac{\alpha}{r}\right) = 1 - \alpha. \tag{1.21}$$

In terms of the above example, we find c_α such that for each i,

$$\mathrm{pr}[|T_i| \le c_\alpha | H_{0i} \text{ true}] = 1 - \frac{\alpha}{r}.$$

Another approximate procedure is frequently available. For example, suppose we are able to find the appropriate critical value d_α for the simultaneous test of a wider class of hypotheses H_{0b} ($b \in B$), where $H_{01}, \ldots, H_{0r}$ are included in this class. Then

$$\mathrm{pr}\left[\bigcap_{i=1}^{r} E_i\right] \ge \mathrm{pr}\left[\bigcap_{b \in B} E_b\right] = 1 - \alpha, \tag{1.22}$$

and using d_α instead of c_α will lead to a conservative test procedure. This approach forms the basis for numerous procedures for finding simultaneous confidence intervals (see Miller [1977, 1981], Seber [1977: Chapter 5]).

Simultaneous hypothesis testing is closely related to a principle of testing that is particularly appropriate for multivariate problems called Roy's union–intersection principle, discussed below.

1.6.2 Union–Intersection Principle

Given (H_0, H), null and alternative hypotheses, respectively, it is sometimes convenient to express H_0 as the intersection of a number of more "primitive" composite hypotheses and H as the union of the same number of corresponding alternatives. Symbolically we have

$$H_0 = \bigcap_a H_{0a}, \qquad H = \bigcup_a H_a, \tag{1.23}$$

where (H_{0a}, H_a) form a natural pair, and $a \in A$, a general set. For example, suppose we wish to test $H_0: \boldsymbol{\theta} = \mathbf{0}$, where $\boldsymbol{\theta}$ is d dimensional; then one choice is $H_{0a}: \mathbf{a}'\boldsymbol{\theta} = 0$ and $H_a: \mathbf{a}'\boldsymbol{\theta} \ne 0$. Since $\boldsymbol{\theta} = \mathbf{0}$ if and only if $\mathbf{a}'\boldsymbol{\theta} = 0$ for all $\mathbf{a}$ in A ($= R^d$), (1.23) is satisfied. In practice, this means that a multivariate test about a vector parameter can be derived in terms of a family of univariate tests. This method of test construction, introduced by Roy [1953] (see also Roy [1957], Roy et al. [1971: Chapter 4], Olkin and Tomsky [1981]), is called the union–intersection principle and is a useful tool for developing multivariate tests.

It should be noted that the decomposition (1.23) is not unique, so that the union–intersection principle can lead to different test procedures. For example, setting $\phi_1 = \theta_1$, $\phi_i = \theta_i - \theta_{i-1}$ ($i = 2, 3, \ldots, d$), we can write $H_{0a}: \phi_a = 0$ ($a = 1, 2, \ldots, d$), $H_0: \boldsymbol{\phi} = \mathbf{0}$, and $H_0 = \cap_a H_{0a}$. Alternatively, we can use the so-called step-down procedure (see Roy et al. [1971]) and consider $H_{01}: \theta_1 = 0$, $H_{0a}: \theta_a - (\theta_1, \theta_2, \ldots, \theta_{a-1})\boldsymbol{\gamma}_{a-1} = 0$ ($a = 2, \ldots, d$), where $\boldsymbol{\gamma}_{a-1}$ is appropriately defined (section 3.3.4). Again $H_0 = \cap_a H_{0a}$, and H_0 is not rejected if each H_{0a} is not rejected. Clearly the choice of decomposition (1.23) will depend on the type of alternatives H_a we regard as important.

1.7 LIKELIHOOD RATIO TESTS

Suppose that a set of random variables has density function $f(\mathbf{y}; \boldsymbol{\theta})$, where $\boldsymbol{\theta} \in \Omega$ and we wish to test $H_0 : \boldsymbol{\theta} \in \omega$, where ω is a subset of Ω. If we write $f(\mathbf{y}; \boldsymbol{\theta})$ as a function of $\boldsymbol{\theta}$, $L(\boldsymbol{\theta})$ say, then the likelihood ratio

$$\ell = \underset{\boldsymbol{\theta} \in \omega}{\text{supremum}}\, L(\boldsymbol{\theta}) \Big/ \underset{\boldsymbol{\theta} \in \Omega}{\text{supremum}}\, L(\boldsymbol{\theta})$$

can be used to test H_0. We reject H_0 at the α level of significance if $\ell < \ell_\alpha$, where $\text{pr}[\ell < \ell_\alpha | H_0 \text{ true}] = \alpha$. This test procedure, called the *likelihood ratio test*, usually has good power properties. Under fairly general conditions, and with large samples, $-2\log \ell$ is approximately distributed as χ^2_ν when H_0 is true. Here ν is, roughly speaking, the difference in dimension (or the number of "free" parameters) of Ω and ω. In some cases ℓ can also be derived using the union–intersection principle, though there are several examples in this book where the two procedures lead to different tests.

EXERCISES 1

1.1 Given random vectors $\mathbf{x}$ and $\mathbf{y}$, prove the following:
   (a) $\mathscr{C}[\mathbf{Ax}, \mathbf{By}] = \mathbf{A}\mathscr{C}[\mathbf{x}, \mathbf{y}]\mathbf{B}'$.
   (b) $\mathscr{C}[\mathbf{x}, \mathbf{y}] = \mathscr{E}[\mathbf{xy}'] - \mathscr{E}[\mathbf{x}](\mathscr{E}[\mathbf{y}])'$.
   (c) $\mathscr{D}[\mathbf{x} - \mathbf{a}] = \mathscr{D}[\mathbf{x}]$.

1.2 If $\mathbf{y}_1, \mathbf{y}_2, \ldots, \mathbf{y}_n$ are d-dimensional vectors, show that

$$\sum_{i=1}^{n} (\mathbf{y}_i - \overline{\mathbf{y}})(\mathbf{y}_i - \overline{\mathbf{y}})' = \sum_{i=1}^{n} \mathbf{y}_i\mathbf{y}_i' - n\overline{\mathbf{y}}\overline{\mathbf{y}}'.$$

1.3 Let $\mathbf{x}_1, \mathbf{x}_2, \ldots, \mathbf{x}_n$ be a random sample from a distribution with dispersion matrix $\boldsymbol{\Sigma}$. Prove the following:
   (a) $\mathscr{D}\left[\sum_{i=1}^{n} a_i\mathbf{x}_i\right] = \left(\sum_{i=1}^{n} a_i^2\right)\boldsymbol{\Sigma}$.
   (b) $\mathscr{C}\left[\sum_{i=1}^{n} a_i\mathbf{x}_i, \sum_{j=1}^{n} b_j\mathbf{x}_j\right] = \mathbf{O}$ if and only if $\sum_{i=1}^{n} a_ib_i = 0$.

1.4 Let $\mathbf{x}$ be a vector of random variables with mean $\boldsymbol{\theta}$ and dispersion matrix $\boldsymbol{\Sigma}$. If $\mathbf{A}$ is symmetric, show that

$$E[\mathbf{x}'\mathbf{A}\mathbf{x}] = \text{tr}[\mathbf{A}\boldsymbol{\Sigma}] + \boldsymbol{\theta}'\mathbf{A}\boldsymbol{\theta}.$$

1.5 Suppose $\mathbf{y} = \mathbf{K}\boldsymbol{\beta} + \mathbf{u}$, where $\mathbf{K}$ is $n \times p$ of rank p, and $\mathscr{E}[\mathbf{u}] = \mathbf{0}$. If $\mathbf{P} = \mathbf{K}(\mathbf{K}'\mathbf{K})^{-1}\mathbf{K}'$, prove the following:
   (a) $\mathbf{P}$ and $\mathbf{I}_n - \mathbf{P}$ are symmetric and idempotent.
   (b) $(\mathbf{I}_n - \mathbf{P})\mathbf{K} = \mathbf{O}$.

(c) $\text{rank}[\mathbf{I}_n - \mathbf{P}] = \text{tr}[\mathbf{I}_n - \mathbf{P}] = n - p$.
(d) $\mathbf{y}'(\mathbf{I}_n - \mathbf{P})\mathbf{y} = \mathbf{u}'(\mathbf{I}_n - \mathbf{P})\mathbf{u}$.

1.6 Let $\mathbf{x}_1, \mathbf{x}_2, \ldots, \mathbf{x}_n$ be a random sample from a d-dimensional distribution, and let

$$g_{rs} = (\mathbf{x}_r - \bar{\mathbf{x}})'\hat{\mathbf{\Sigma}}^{-1}(\mathbf{x}_s - \bar{\mathbf{x}}),$$

where $\hat{\mathbf{\Sigma}} = \tilde{\mathbf{X}}'\tilde{\mathbf{X}}/n$ [see (1.12)]. Show the following:
(a) $E[g_{rr}] = d$.
(b) $E[g_{rs}] = -d/(n-1) \qquad (r \neq s)$.
[Hint: Consider $\sum_{r=1}^{n}(\mathbf{x}_r - \bar{\mathbf{x}})'\hat{\mathbf{\Sigma}}^{-1}(\mathbf{x}_s - \bar{\mathbf{x}})$ for the two cases.]
Mardia [1977]

1.7 Let $\mathbf{x}_1, \mathbf{x}_2, \ldots, \mathbf{x}_n$ be a random sample from a distribution with dispersion matrix $\mathbf{\Sigma}$, and let $\tilde{\mathbf{x}}_r = \mathbf{x}_r - \bar{\mathbf{x}}$. Find $\mathscr{C}[\tilde{\mathbf{x}}_r, \tilde{\mathbf{x}}_s]$ for all r, s.

1.8 The following algorithm provides a method for updating a sample mean and a sample dispersion matrix when a further observation is added, or an observation deleted. Let $\mathbf{x}_1, \mathbf{x}_2, \ldots, \mathbf{x}_n$ be n d-dimensional observations with associated weights w_i $(i = 1, 2, \ldots, n)$. Define

$$W_n = \sum_{i=1}^{n} w_i,$$

$$\bar{\mathbf{x}}_n = \sum_{i=1}^{n} \frac{w_i \mathbf{x}_i}{W_n},$$

$$\mathbf{Q}_n = \sum_{i=1}^{n} w_i(\mathbf{x}_i - \bar{\mathbf{x}}_n)(\mathbf{x}_i - \bar{\mathbf{x}}_n)',$$

$$\mathbf{d}_{n+1} = \mathbf{x}_{n+1} - \bar{\mathbf{x}}_n,$$

$$\mathbf{e}_n = \mathbf{x}_n - \bar{\mathbf{x}}_n.$$

Show that

$$\bar{\mathbf{x}}_{n+1} = \bar{\mathbf{x}}_n + \frac{w_{n+1}\mathbf{d}_{n+1}}{W_{n+1}},$$

$$\mathbf{Q}_{n+1} = \mathbf{Q}_n + \left(w_{n+1} - \frac{w_{n+1}^2}{W_{n+1}}\right)\mathbf{d}_{n+1}\mathbf{d}_{n+1}',$$

$$\bar{\mathbf{x}}_{n-1} = \bar{\mathbf{x}}_n - \frac{w_n \mathbf{e}_n}{W_{n-1}},$$

$$\mathbf{Q}_{n-1} = \mathbf{Q}_n - \left(w_n + \frac{w_n^2}{W_{n-1}}\right)\mathbf{e}_n\mathbf{e}_n'.$$

[Note: If a positive weight corresponds to adding a new observation to the sample, and a negative weight to removing one, then precisely the same algorithm can be used in both cases. Also, by starting with $\bar{\mathbf{x}}_1 = \mathbf{x}_1$ and $\mathbf{Q}_1 = \mathbf{O}$, $\bar{\mathbf{x}}_n$ and $\mathbf{Q}_n$ can be computed in a stepwise fashion as the $\mathbf{x}_i$ are introduced one at a time. This procedure has the advantage of only requiring one pass through the data.] Clarke [1971]

1.9 Given $\mathbf{y} \sim N_d(\mathbf{0}, \mathbf{\Sigma})$ and $\mathbf{A}$ a $d \times d$ symmetric matrix of rank r, show that $\mathbf{y}'\mathbf{A}\mathbf{y} \sim \chi_r^2$ if and only if $\mathbf{A\Sigma A} = \mathbf{A}$. [Hint: Use A6.5.]

CHAPTER 2

Multivariate Distributions

2.1 INTRODUCTION

Most of the univariate continuous distributions have multivariate analogues with the property that their univariate marginal distributions all belong to the same family. For example, there are the multivariate versions of the normal, gamma, beta, t-, F- (see Johnson and Kotz [1972], Ronning [1977], Krishnaiah [1979, 1980], Dawid [1981]) logistic normal (Aitchison and Shen [1980]), lognormal (Press [1972a]), reliability (Block [1977]), and stable distributions (Press [1972b]). A useful family that provides both longer- and shorter-tailed alternatives to the multivariate normal is the family of elliptically symmetric distributions (see Devlin et al. [1976], Chmielewski [1981]): This family is useful for the empirical and theoretical study of robustness to nonnormality (Muirhead [1982]). Another useful family of symmetric distributions is given by Cook and Johnson [1981]. However, we shall consider only those distributions, and some close relations, that are actually used in this text. Our main multivariate distributions are the multivariate normal and multivariate beta. In addition, we consider the Wishart distribution, which is a multivariate generalization of the chi-square distribution, though the Wishart is not described as the multivariate chi square, as not all the marginals are chi square. A univariate distribution called Hotelling's T^2 is also introduced. It is represented by T^2, as it is constructed in much the same way as one constructs the square of a t-statistic.

2.2 MULTIVARIATE NORMAL DISTRIBUTION

We shall consider two definitions of the multivariate normal (MVN) distribution. The first simply defines the distribution in terms of the density function. The second definition is based on a unique property of the MVN distribution, namely, that any linear combination of its elements is univariate normal. Although not needed in what follows, the second definition has two advantages: It can be extended to include the so-called singular MVN and, more importantly, it emphasizes the fact that some multivariate problems can be handled by looking at univariate "reductions."

Definition 1a Let $\mathbf{y} = (y_1, y_2, \ldots, y_d)'$ be a d-dimensional vector of random variables. Then $\mathbf{y}$ is said to have a (nonsingular) MVN distribution if its density function is

$$f(\mathbf{y}) = (2\pi)^{-d/2}|\boldsymbol{\Sigma}|^{-1/2}\exp\left[-\tfrac{1}{2}(\mathbf{y}-\boldsymbol{\theta})'\boldsymbol{\Sigma}^{-1}(\mathbf{y}-\boldsymbol{\theta})\right]$$

$$\left(-\infty < y_j < \infty, j = 1, 2, \ldots, d\right), \quad (2.1)$$

where $\boldsymbol{\Sigma} = [(\sigma_{jk})]$ is positive definite (written $\boldsymbol{\Sigma} > \mathbf{O}$).

It is readily shown (e.g., Seber [1977: Chapter 2]) that $\mathscr{E}[\mathbf{y}] = \boldsymbol{\theta}$ and $\mathscr{D}[\mathbf{y}] = \boldsymbol{\Sigma}$, so that it is convenient to use the notation $\mathbf{y} \sim N_d(\boldsymbol{\theta}, \boldsymbol{\Sigma})$ or $\mathbf{y} \sim N_d$ for short. There are two important special cases: First, $\mathbf{y} - \boldsymbol{\theta} \sim N_d(\mathbf{0}, \boldsymbol{\Sigma})$ and, second, if the y_j are mutually independent with univariate normal distributions $N_1(\theta_j, \sigma^2)$ $(j = 1, 2, \ldots, d)$, then $\mathbf{y} \sim N_d(\boldsymbol{\theta}, \sigma^2\mathbf{I}_d)$. We now list some of the main properties of the MVN distribution for future reference.

THEOREM 2.1 Suppose $\mathbf{y} \sim N_d(\boldsymbol{\theta}, \boldsymbol{\Sigma})$ and let

$$\mathbf{y} = \begin{pmatrix} \mathbf{y}^{(1)} \\ \mathbf{y}^{(2)} \end{pmatrix}, \quad \boldsymbol{\theta} = \begin{pmatrix} \boldsymbol{\theta}^{(1)} \\ \boldsymbol{\theta}^{(2)} \end{pmatrix}, \quad \boldsymbol{\Sigma} = \begin{pmatrix} \boldsymbol{\Sigma}_{11} & \boldsymbol{\Sigma}_{12} \\ \boldsymbol{\Sigma}_{21} & \boldsymbol{\Sigma}_{22} \end{pmatrix},$$

where $\mathbf{y}^{(i)}$ and $\boldsymbol{\theta}^{(i)}$ are $d_i \times 1$ vectors and $\boldsymbol{\Sigma}_{ii}$ is $d_i \times d_i$ $(d_1 + d_2 = d)$.

(i) If $\mathbf{C}$ is a $q \times d$ matrix of rank q, then $\mathbf{Cy} \sim N_q(\mathbf{C}\boldsymbol{\theta}, \mathbf{C}\boldsymbol{\Sigma}\mathbf{C}')$.

(ii) Any subset of the elements of $\mathbf{y}$ has an MVN distribution; in particular $\mathbf{y}^{(1)} \sim N_{d_1}(\boldsymbol{\theta}^{(1)}, \boldsymbol{\Sigma}_{11})$.

(iii) The moment-generating function of $\mathbf{y}$ is given by

$$M(\mathbf{t}) = E[\exp(\mathbf{t}'\mathbf{y})]$$

$$= \exp\left(\mathbf{t}'\boldsymbol{\theta} + \tfrac{1}{2}\mathbf{t}'\boldsymbol{\Sigma}\mathbf{t}\right). \quad (2.2)$$

As the characteristic function of $\mathbf{y}$, $E[\exp(i\mathbf{t}'\mathbf{y})]$, is analytic, it follows that (2.2) uniquely determines (2.1).

(iv) $\mathbf{y}^{(1)}$ and $\mathbf{y}^{(2)}$ are statistically independent if and only if $\mathscr{C}[\mathbf{y}^{(1)}, \mathbf{y}^{(2)}] = \mathbf{O}$. [This result extends to a general partition of $\mathbf{y}$ into, say, k subsets $\mathbf{y}^{(i)}$ $(i = 1, 2, \ldots, k)$; the $\mathbf{y}^{(i)}$ are mutually independent if and only if $\mathscr{C}[\mathbf{y}^{(i)}, \mathbf{y}^{(j)}] = \mathbf{O}$ for all $i, j, i \neq j$].

(v) If $\mathbf{u}_i = \mathbf{A}_i\mathbf{y}$ $(i = 1, 2, \ldots, m)$ and $\mathscr{C}[\mathbf{u}_i, \mathbf{u}_j] = \mathbf{O}$ for all i, j, $i \neq j$, then the $\mathbf{u}_i$ are mutually independent.

(vi) $(\mathbf{y} - \boldsymbol{\theta})'\boldsymbol{\Sigma}^{-1}(\mathbf{y} - \boldsymbol{\theta}) \sim \chi^2_d$.

(vii) $\mathbf{y}'\boldsymbol{\Sigma}^{-1}\mathbf{y}$ has a noncentral chi-square distribution with d degrees of freedom and noncentrality parameter $\delta = \boldsymbol{\theta}'\boldsymbol{\Sigma}^{-1}\boldsymbol{\theta}$.

(viii) The conditional distribution of $\mathbf{y}^{(2)}$, given $\mathbf{y}^{(1)}$, is $N_{d_2}(\boldsymbol{\theta}^{(2)} + \boldsymbol{\Sigma}_{21}\boldsymbol{\Sigma}_{11}^{-1}[\mathbf{y}^{(1)} - \boldsymbol{\theta}^{(1)}], \boldsymbol{\Sigma}_{22\cdot1})$, where

$$\boldsymbol{\Sigma}_{22\cdot1} = \boldsymbol{\Sigma}_{22} - \boldsymbol{\Sigma}_{21}\boldsymbol{\Sigma}_{11}^{-1}\boldsymbol{\Sigma}_{12}.$$

Proof These results are generally well known and are proved in most books that consider the MVN in any detail (e.g., Seber [1977: Theorems 2.1–2.6 and Exercise 2c, no. 2]): As (vii) is not always derived, it is given in Exercise 2.1. □

The MVN distribution has many unique characteristics (Kagan et al. [1973]) and we now use one of these (see Exercise 2.2) to define the MVN, as in Rao [1973; Chapter 8].

Definition 1b $\mathbf{y}$ is said to have a MVN distribution if $\boldsymbol{\ell}'\mathbf{y}$ is univariate normal for all real $\boldsymbol{\ell}$. If $\mathscr{E}[\mathbf{y}] = \boldsymbol{\theta}$, $\mathscr{D}[\mathbf{y}] = \boldsymbol{\Sigma} > \mathbf{O}$, and $\mathbf{y}$ is $d \times 1$, then we write $\mathbf{y} \sim N_d(\boldsymbol{\theta}, \boldsymbol{\Sigma})$. □

One is frequently interested in linear combinations of independent univariate normal variables z_i with common variance σ^2, say $\mathbf{y} = \mathbf{A}\mathbf{z}$, where $\mathscr{D}[\mathbf{y}] = \mathbf{A}\mathscr{D}[\mathbf{z}]\mathbf{A}' = \sigma^2\mathbf{A}\mathbf{A}'$ may be singular (e.g., Exercise 2.5). This situation of singular $\boldsymbol{\Sigma}$ can be included in Definition 1b if we remove the restriction $\boldsymbol{\Sigma} > \mathbf{O}$ and include the degenerate distribution $N_1(b, 0)$ in the class of univariate distributions so that $\boldsymbol{\ell}'\mathbf{y} = b$ is possible (see Exercises 2.4 and 2.5). With this extension of Definition 1b, Theorem 2.1(i)–(v) can be readily proved, with arbitrary rank for $\mathbf{C}$ in (i), for the case of singular $\boldsymbol{\Sigma}$ (see Exercise 2.3): (viii) is also true if we use a generalized inverse $\boldsymbol{\Sigma}_{22}^{-}$ (Rao [1973: pp. 522–523]). When $\boldsymbol{\Sigma}$ is singular, (2.2) still uniquely determines the distribution of probability (usually called the *singular* MVN distribution), though (2.1) no longer exists. In this case there exists at least one nonzero $\boldsymbol{\ell}$ such that $\boldsymbol{\ell}'(\mathbf{y} - \boldsymbol{\theta}) = 0$ (see Exercise 2.4). Furthermore, when $\boldsymbol{\Sigma}$ has rank m, it can be shown that $\mathbf{y}$ is expressible in the form $\mathbf{y} - \boldsymbol{\theta} = \mathbf{B}\mathbf{u}$, where $\mathbf{B}$ is $d \times m$ of rank m and $\mathbf{u}$ has a (nonsingular) MVN distribution with nonsingular $\mathscr{D}[\mathbf{u}]$.

One advantage of Definition 1b is that it can be extended to random variables defined on an infinite-dimensional space. Instead of the vector $\mathbf{y}$ in d-dimensional Euclidean space, we may be interested in a time series $Y = \{Y_t: 0 \leq t < \infty\}$ in a Hilbert space or an infinite sequence $Y = \{Y_1, Y_2, \ldots\}$ in a Banach space. Also related to time series is the development of the so-called complex MVN distribution apparently introduced explicitly by Wooding [1956]. For literature reviews relating to this distribution, see Krishnaiah [1976] and Saxena [1978].

Considerable interest has been expressed in the problem of finding conditions under which the quadratic $(\mathbf{y}-\boldsymbol{\theta})'\mathbf{A}(\mathbf{y}-\boldsymbol{\theta})$ has a chi-square distribution, given that $\mathbf{y} \sim N_d(\boldsymbol{\theta}, \boldsymbol{\Sigma})$, where $\boldsymbol{\Sigma}$ may be singular. For a review, see Khatri [1978].

As far as this book is concerned, it does not matter which definition of the MVN we use for a starting point, as essential results are readily proved either way. However, the second definition avoids explicit mention of the density function (2.1), and it is this idea that suggests a similar method for handling the Wishart distribution discussed below (Rao [1973]).

2.3 WISHART DISTRIBUTION

2.3.1 Definition and Properties

Like most well-known distributions, the Wishart distribution arose out of the sampling distribution of an important sample statistic, in this case $\Sigma_{i=1}^n(\mathbf{y}_i - \bar{\mathbf{y}})(\mathbf{y}_i - \bar{\mathbf{y}})'$, a multivariate analogue of the univariate sum of squares $\Sigma_i(y_i - \bar{y})^2$ (Johnson and Kotz [1972: p. 159]). Two definitions of the Wishart distribution are now given.

Definition 2a Let $\mathbf{W} = [(w_{jk})]$ be a $d \times d$ symmetric matrix of random variables that is positive definite with probability 1, and let $\boldsymbol{\Sigma}$ be a $d \times d$ positive definite matrix. If m is an integer such that $m \geq d$, then $\mathbf{W}$ is said to have a (nonsingular) Wishart distribution with m degrees of freedom if the joint density function of the $\frac{1}{2}d(d+1)$ distinct elements of $\mathbf{W}$ (in, say, the upper triangle) is

$$f(w_{11}, w_{12}, \ldots, w_{dd}) = c^{-1}|\mathbf{W}|^{(m-d-1)/2}\mathrm{etr}\left(-\tfrac{1}{2}\boldsymbol{\Sigma}^{-1}\mathbf{W}\right), \qquad (2.3)$$

where etr represents the operator e^{trace},

$$c = 2^{md/2}|\boldsymbol{\Sigma}|^{m/2}\Gamma_d\left(\tfrac{1}{2}m\right), \qquad (2.4)$$

and

$$\Gamma_d\left(\tfrac{1}{2}m\right) = \pi^{d(d-1)/4}\prod_{j=1}^{d}\Gamma\left[\tfrac{1}{2}(m+1-j)\right] \qquad (2.5)$$

(the so-called multivariate gamma function). We shall write $\mathbf{W} \sim W_d(m, \boldsymbol{\Sigma})$ or $\mathbf{W} \sim W_d$ for short. □

It is shown below (Theorem 2.2, Corollary 1) that for every $\boldsymbol{\ell}$, $\boldsymbol{\ell}'\mathbf{W}\boldsymbol{\ell} \sim \boldsymbol{\ell}'\boldsymbol{\Sigma}\boldsymbol{\ell}\chi_m^2$, a quadratic version of the linear property of the MVN. Unfortunately, the resemblance ends here, as this quadratic property does not uniquely determine (2.3); in fact, it is held by $X\mathbf{W}$ where X has an independent beta distribution (Mitra [1969]). However, as in the case of the MVN, it would be helpful to have a definition of the Wishart distribution that does not involve knowing explicitly the form of the density function (2.3). Such a definition is given below and it has the additional, though somewhat minor, advantage that if we use Definition 1b for the MVN, we can remove the restriction $|\mathbf{\Sigma}| \neq 0$ so that $\mathbf{\Sigma}$ is positive semidefinite, that is, $\mathbf{\Sigma} \geq \mathbf{O}$.

Definition 2b Suppose that $\mathbf{x}_1, \mathbf{x}_2, \ldots, \mathbf{x}_m$ are independently and identically distributed (i.i.d.) as $N_d(\mathbf{0}, \mathbf{\Sigma})$; then

$$\mathbf{W} = \sum_{i=1}^{m} \mathbf{x}_i \mathbf{x}_i' \tag{2.6}$$

is said to have a Wishart distribution with m degrees of freedom. □

If $\mathbf{\Sigma} > \mathbf{O}$ and $m \geq d$, then it can be shown that $\mathbf{W} > \mathbf{O}$ with probability 1 (see A5.13); otherwise, by the definition of $\mathbf{W}$, $\mathbf{W} \geq \mathbf{O}$ and there is a nonzero probability that $|\mathbf{W}| = 0$. In the latter case we say that $\mathbf{W}$ has a singular Wishart distribution and (2.3) does not exist. Of course, (2.6) leads to (2.3) for the nonsingular case (see Exercise 2.16), but we shall use (2.6), as it does not require manipulations of the density function. Although not always necessary, *we shall now assume* $\mathbf{\Sigma} > \mathbf{O}$, unless otherwise stated.

Analogous to Theorem 2.1(i), we have the following.

THEOREM 2.2 If $\mathbf{W} \sim \mathbf{W}_d(m, \mathbf{\Sigma})$ and $\mathbf{C}$ is a $q \times d$ matrix of rank q, then

$$\mathbf{CWC}' \sim W_q(m, \mathbf{C\Sigma C}').$$

Proof Let $\mathbf{x}_1, \mathbf{x}_2, \ldots, \mathbf{x}_m$ be i.i.d. $N_d(\mathbf{0}, \mathbf{\Sigma})$ and let $\mathbf{W}_0 = \Sigma_i \mathbf{x}_i \mathbf{x}_i'$. Then $\mathbf{W}_0 \sim W_d(m, \mathbf{\Sigma})$ so that $\mathbf{CW}_0\mathbf{C}'$ has the same distribution as $\mathbf{CWC}'$. Now

$$\mathbf{CW}_0\mathbf{C}' = \sum_{i=1}^{m} (\mathbf{Cx}_i)(\mathbf{x}_i'\mathbf{C}') = \sum_{i=1}^{m} \mathbf{y}_i \mathbf{y}_i', \tag{2.7}$$

where, by Theorem 2.1(i), the $\mathbf{y}_i$ are i.i.d. $N_q(\mathbf{0}, \mathbf{C\Sigma C}')$. Hence, by (2.6), $\mathbf{CW}_0\mathbf{C}' \sim W_q(m, \mathbf{C\Sigma C}')$. □

(We note that if $m \geq q$ and $\mathbf{C}$ has rank q, then $\mathbf{C\Sigma C}'$ is positive definite (see A5.7) and $\mathbf{CW}_0\mathbf{C}'$ has a nonsingular distribution. However, both assumptions and the assumption of $\mathbf{\Sigma} > \mathbf{O}$ can be dropped if we allow singular normal and Wishart distributions. In future proofs involving the Wishart distribution we shall omit the step of introducing $\mathbf{W}_0$ and simply assume that $\mathbf{W} = \Sigma_i \mathbf{x}_i \mathbf{x}_i'$.)

COROLLARY 1 If ℓ is any nonzero $d \times 1$ vector of constants, then $\ell'\mathbf{W}\ell \sim \sigma_\ell^2\chi_m^2$, where $\sigma_\ell^2 = \ell'\boldsymbol{\Sigma}\ell > 0$ (since $\boldsymbol{\Sigma} > \mathbf{O}$).

Proof Let $\mathbf{C} = \ell$ in the proof of Theorem 2.2. Then, from (2.7), $\ell'\mathbf{W}_0\ell = \sum_{i=1}^m y_i^2$, where the y_i are i.i.d. $N_1(0, \sigma_\ell^2)$. Hence if $\ell \neq \mathbf{0}$, $\ell'\mathbf{W}\ell/\sigma_\ell^2 \sim \chi_m^2$. □

COROLLARY 2 Setting $\ell' = (0,\ldots,0,1,0,\ldots,0)$, we have, from Corollary 1,

$$w_{jj} \sim \sigma_{jj}\chi_m^2. \tag{2.8}$$

In spite of Corollary 2, we do not call the Wishart distribution the multivariate chi-square distribution, as the marginal distribution of w_{jk} ($j \neq k$) is not chi square. We normally reserve the term *multivariate* for the case when all univariate marginals belong to the same family.

2.3.2 *Generalized Quadratics*

If we write

$$\mathbf{X} = \begin{pmatrix} \mathbf{x}_1' \\ \mathbf{x}_2' \\ \vdots \\ \mathbf{x}_m' \end{pmatrix} = [(x_{ij})] = \left(\mathbf{x}^{(1)}, \mathbf{x}^{(2)}, \ldots, \mathbf{x}^{(d)}\right), \tag{2.9}$$

then

$$\mathbf{W} = \sum_{i=1}^m \mathbf{x}_i\mathbf{x}_i' = \mathbf{X}'\mathbf{X}.$$

By regarding this as a multivariate sum of squares (see Section 1.3), it is natural to consider more general expressions of the form $\mathbf{X}'\mathbf{A}\mathbf{X} = \sum_i\sum_j a_{ij}\mathbf{x}_i\mathbf{x}_j'$, where $\mathbf{A} = [(a_{ij})]$ is an $m \times m$ symmetric matrix. In univariate linear model theory it is the nature of $\mathbf{A}$ that determines the distributional properties. Fortunately, these same properties carry through to the multivariate case, and we shall see in Theorem 2.4 later that we can always handle $\mathbf{X}'\mathbf{A}\mathbf{X}$ by considering a univariate version $\ell'\mathbf{X}'\mathbf{A}\mathbf{X}\ell$ ($= \mathbf{y}'\mathbf{A}\mathbf{y}$). But first let us introduce a lemma.

LEMMA 2.3 Suppose $\mathbf{x}_1, \mathbf{x}_2, \ldots, \mathbf{x}_m$ are i.i.d. $N_d(\mathbf{0}, \boldsymbol{\Sigma})$ and let $\mathbf{X} = (\mathbf{x}_1, \mathbf{x}_2, \ldots, \mathbf{x}_m)'$. Then we have the following:

(i) $\mathbf{x}^{(j)} \sim N_m(\mathbf{0}, \sigma_{jj}\mathbf{I}_m)$ [see (2.9)].
(ii) If $\mathbf{a}$ is an $m \times 1$ vector of constants, then

$$\mathbf{X}'\mathbf{a} \sim N_d\left(\mathbf{0}, \|\mathbf{a}\|^2\boldsymbol{\Sigma}\right).$$

(iii) If $\{\mathbf{a}_1, \mathbf{a}_2, \ldots, \mathbf{a}_r\}$, $r \leq m$, is a mutually orthogonal set of $m \times 1$ vectors, then the random vectors $\mathbf{X}'\mathbf{a}_i$ $(i = 1, 2, \ldots, r)$ are mutually independent.

(iv) If $\mathbf{b}$ is a $d \times 1$ vector of constants, then $\mathbf{Xb} \sim N_m(\mathbf{0}, \sigma_b^2 \mathbf{I}_m)$, where $\sigma_b^2 = \mathbf{b}'\mathbf{\Sigma}\mathbf{b}$.

Proof (i) The jth element of $\mathbf{x}_i$ is $N_1(0, \sigma_{jj})$.

(ii) By Exercise 2.6 and (2.9),

$$\mathbf{X}'\mathbf{a} = \sum_{\alpha=1}^{m} a_\alpha \mathbf{x}_\alpha \sim N_d\left(\mathbf{0}, \sum_{\alpha=1}^{m} a_\alpha^2 \mathbf{\Sigma}\right). \tag{2.10}$$

(iii) Let $\mathbf{u}_i = \mathbf{X}'\mathbf{a}_i$ and let $\mathbf{a}_i = (a_{i1}, a_{i2}, \ldots, a_{im})'$. Then

$$\begin{aligned}
\mathscr{C}[\mathbf{u}_i, \mathbf{u}_j] &= \mathscr{C}\left[\sum_\alpha a_{i\alpha}\mathbf{x}_\alpha, \sum_\beta a_{j\beta}\mathbf{x}_\beta\right] \\
&= \sum_\alpha \sum_\beta a_{i\alpha} a_{j\beta} \mathscr{C}[\mathbf{x}_\alpha, \mathbf{x}_\beta] \\
&= \sum_\alpha a_{i\alpha} a_{j\alpha} \mathscr{D}[\mathbf{x}_\alpha] \\
&= \mathbf{a}'_i \mathbf{a}_j \mathbf{\Sigma} \\
&= \mathbf{O}.
\end{aligned}$$

Now

$$\begin{aligned}
\mathbf{u}_i &= (a_{i1}\mathbf{I}_d, \ldots, a_{im}\mathbf{I}_d)\begin{pmatrix} \mathbf{x}_1 \\ \vdots \\ \mathbf{x}_m \end{pmatrix} \\
&= \mathbf{Ax},
\end{aligned}$$

say, where $\mathbf{x} \sim N_{md}(\mathbf{0}, \mathbf{I}_m \otimes \mathbf{\Sigma})$, by Exercise 2.7. Thus, by Theorem 2.1(v), the $\mathbf{u}_i$ are mutually independent.

(iv) Let

$$\mathbf{y} = \mathbf{Xb} = \begin{pmatrix} \mathbf{x}_1'\mathbf{b} \\ \mathbf{x}_2'\mathbf{b} \\ \vdots \\ \mathbf{x}_m'\mathbf{b} \end{pmatrix},$$

then the elements y_i of $\mathbf{y}$ are i.i.d. $N_1(0, \sigma_b^2)$, where [by (1.8)]

$$\sigma_b^2 = \text{var}[\mathbf{x}_i'\mathbf{b}] = \mathscr{D}[\mathbf{b}'\mathbf{x}_i] = \mathbf{b}'\mathbf{\Sigma}\mathbf{b}.$$

COROLLARY Let $\mathbf{a} = \mathbf{1}_m/m$; then $\|\mathbf{a}\|^2 = 1/m$ and, by (ii) above,

$$\bar{\mathbf{x}} = \mathbf{X}'\mathbf{a} \sim N_d(\mathbf{0}, \boldsymbol{\Sigma}/m). \qquad \square$$

We now show that certain properties of quadratic forms extend naturally to the multivariate case. The key results are collected together in the following theorem.

THEOREM 2.4 Let $\mathbf{X}' = (\mathbf{x}_1, \mathbf{x}_2, \ldots, \mathbf{x}_m)$, where the $\mathbf{x}_i$ are i.i.d. $N_d(\mathbf{0}, \boldsymbol{\Sigma})$, and let $\mathbf{y} = \mathbf{X}\boldsymbol{\ell}$, where $\boldsymbol{\ell}(\neq \mathbf{0})$ is a $d \times 1$ vector of constants. Let $\mathbf{A}$ and $\mathbf{B}$ be $m \times m$ symmetric matrices of ranks r and s, respectively, and let $\mathbf{b}$ be an $m \times 1$ vector of constants.

(i) $\mathbf{X}'\mathbf{A}\mathbf{X} \sim W_d(r, \boldsymbol{\Sigma})$ if and only if $\mathbf{y}'\mathbf{A}\mathbf{y} \sim \sigma_\ell^2\chi_r^2$ for any $\boldsymbol{\ell}$, where $\sigma_\ell^2 = \boldsymbol{\ell}'\boldsymbol{\Sigma}\boldsymbol{\ell}$.

(ii) $\mathbf{X}'\mathbf{A}\mathbf{X}$ and $\mathbf{X}'\mathbf{B}\mathbf{X}$ have independent Wishart distributions with r and s degrees of freedom, respectively, if and only if $\mathbf{y}'\mathbf{A}\mathbf{y}/\sigma_\ell^2$ and $\mathbf{y}'\mathbf{B}\mathbf{y}/\sigma_\ell^2$ are independently distributed as chi square with r and s degrees, respectively, for any $\boldsymbol{\ell}$.

(iii) $\mathbf{X}'\mathbf{b}$ and $\mathbf{X}'\mathbf{A}\mathbf{X}$ are independently distributed as N_d and $W_d(r, \boldsymbol{\Sigma})$, respectively, if and only if $\mathbf{y}'\mathbf{b}$ and $\mathbf{y}'\mathbf{A}\mathbf{y}/\sigma_\ell^2$ are independently distributed as N_1 and χ_r^2, respectively, for any $\boldsymbol{\ell}$.

Proof (i) Given $\mathbf{W} = \mathbf{X}'\mathbf{A}\mathbf{X} \sim W_d(r, \boldsymbol{\Sigma})$, then $\mathbf{y}'\mathbf{A}\mathbf{y} = \boldsymbol{\ell}'\mathbf{X}'\mathbf{A}\mathbf{X}\boldsymbol{\ell} = \boldsymbol{\ell}'\mathbf{W}\boldsymbol{\ell} \sim \sigma_\ell^2\chi_r^2$, by Theorem 2.2, Corollary 1. Conversely, suppose that $\mathbf{y}'\mathbf{A}\mathbf{y}/\sigma_\ell^2 \sim \chi_r^2$ for some $\boldsymbol{\ell}$; then, since $\mathbf{y} \sim N_m(\mathbf{0}, \sigma_\ell^2\mathbf{I}_m)$ [by Lemma 2.3(iv)] it follows from A6.5 that $\mathbf{A}$ is indempotent of rank r. Hence, by A6.2, we can write

$$\mathbf{A} = \sum_{i=1}^{r} \mathbf{a}_i\mathbf{a}_i', \tag{2.11}$$

where the $\mathbf{a}_i$ are an orthonormal set of eigenvectors corresponding to the r unit eigenvalues of $\mathbf{A}$. Therefore

$$\mathbf{X}'\mathbf{A}\mathbf{X} = \sum_{i=1}^{r} \mathbf{X}'\mathbf{a}_i\mathbf{a}_i'\mathbf{X} = \sum_{i=1}^{r} \mathbf{u}_i\mathbf{u}_i',$$

where $\mathbf{u}_i = \mathbf{X}'\mathbf{a}_i$. But, by Lemma 2.3(ii) and (iii), the $\mathbf{u}_i$ are i.i.d. $N_d(\mathbf{0}, \boldsymbol{\Sigma})$, since $\|\mathbf{a}_i\|^2 = 1$, so that $\mathbf{X}'\mathbf{A}\mathbf{X} \sim W_d(r, \boldsymbol{\Sigma})$ by Definition 2b.

(ii) Given $\mathbf{X}'\mathbf{A}\mathbf{X}$ and $\mathbf{X}'\mathbf{B}\mathbf{X}$ with independent Wishart distributions, then the quadratics $\mathbf{y}'\mathbf{A}\mathbf{y}$ and $\mathbf{y}'\mathbf{B}\mathbf{y}$ are statistically independent, being functions of the Wishart matrices. Also, by (i), the quadratics are each distributed as σ_ℓ^2 times a chi-square variable. Conversely, suppose that $\mathbf{y}'\mathbf{A}\mathbf{y}/\sigma_\ell^2$ and $\mathbf{y}'\mathbf{B}\mathbf{y}/\sigma_\ell^2$ are independently distributed as χ_r^2 and χ_s^2, respectively for some $\boldsymbol{\ell}$. Then, by A6.5, $\mathbf{A}$ and $\mathbf{B}$ are indempotent matrices of ranks r and s, respectively, and it follows from A6.6 that $\mathbf{AB} = \mathbf{O}$. Thus we can write

$$\mathbf{A} = \sum_{i=1}^{r} \mathbf{a}_i \mathbf{a}'_i \quad \text{and} \quad \mathbf{B} = \sum_{j=1}^{s} \mathbf{b}_j \mathbf{b}'_j,$$

where $\{\mathbf{a}_1, \mathbf{a}_2, \ldots, \mathbf{a}_r\}$ and $\{\mathbf{b}_1, \mathbf{b}_2, \ldots, \mathbf{b}_s\}$ are orthonormal sets of eigenvectors. Moreover, since $\mathbf{A}\mathbf{a}_i = \mathbf{a}_i$ and $\mathbf{B}\mathbf{b}_j = \mathbf{b}_j$, we have

$$\mathbf{a}'_i \mathbf{b}_j = \mathbf{a}'_i \mathbf{A}'\mathbf{B}\mathbf{b}_j = \mathbf{a}'_i \mathbf{A}\mathbf{B}\mathbf{b}_j = 0$$

for all i, j so that the combined set of vectors $\{\mathbf{a}_1, \ldots, \mathbf{a}_r, \mathbf{b}_1, \ldots, \mathbf{b}_s\}$ is orthonormal. Hence, as in the proof of (i),

$$\mathbf{X}'\mathbf{A}\mathbf{X} = \sum_{i=1}^{r} \mathbf{u}_i \mathbf{u}'_i \tag{2.12}$$

and

$$\mathbf{X}'\mathbf{B}\mathbf{X} = \sum_{j=1}^{s} \mathbf{v}_j \mathbf{v}'_j,$$

where $\mathbf{u}_i = \mathbf{X}'\mathbf{a}_i$, $\mathbf{v}_j = \mathbf{X}'\mathbf{b}_j$, and the set $\{\mathbf{u}_1, \ldots, \mathbf{u}_r, \mathbf{v}_1, \ldots, \mathbf{v}_s\}$ are mutually independent [see Lemma 2.3(iii)].

(iii) Suppose that $\mathbf{X}'\mathbf{b}$ and $\mathbf{X}'\mathbf{A}\mathbf{X}$ are independently distributed as N_d and W_d, respectively. Then $\mathbf{y}'\mathbf{b} = \boldsymbol{\ell}'\mathbf{X}'\mathbf{b}$ is N_1 and $\mathbf{y}'\mathbf{A}\mathbf{y} \sim \sigma_\ell^2 \chi_r^2$, as in the proof of (i). Furthermore, $\mathbf{y}'\mathbf{b}$ and $\mathbf{y}'\mathbf{A}\mathbf{y}$ are statistically independent, being functions of independent sets of random variables. Conversely, suppose that $\mathbf{y}'\mathbf{b}$ and $\mathbf{y}'\mathbf{A}\mathbf{y}$ are independently distributed as N_1 and $\sigma_\ell^2 \chi_r^2$, respectively, for some $\boldsymbol{\ell}$. Now we already know from Lemma 2.3(ii) that $\mathbf{X}'\mathbf{b} \sim N_d$. Also, by (i), $\mathbf{X}'\mathbf{A}\mathbf{X} \sim W_d(r, \boldsymbol{\Sigma})$. Since $\mathbf{y} \sim N_m(\mathbf{0}, \sigma_\ell^2 \mathbf{I}_m)$, we have $\mathbf{y}'\mathbf{b} \sim N_1(\mathbf{0}, \sigma_\ell^2 \mathbf{b}'\mathbf{b})$, $\mathbf{y}'\mathbf{b}\mathbf{b}'\mathbf{y}/\mathbf{b}'\mathbf{b} \sim \sigma_\ell^2 \chi_1^2$ and $\mathbf{A}\mathbf{b}\mathbf{b}' = \mathbf{O}$ (A6.6 with $\mathbf{B} = \mathbf{b}\mathbf{b}'$). Postmultiplying by $\mathbf{b}$, we have $\mathbf{A}\mathbf{b} = \mathbf{0}$, so that arguing as in (ii), $\{\mathbf{a}_1, \mathbf{a}_2, \ldots, \mathbf{a}_r, \mathbf{b}\}$ are mutually orthogonal. Hence $\mathbf{u}_1, \mathbf{u}_2, \ldots, \mathbf{u}_r$ and $\mathbf{X}'\mathbf{b}$ are mutually independent, and $\mathbf{X}'\mathbf{b}$ is independent of $\mathbf{X}'\mathbf{A}\mathbf{X}$ [by (2.12)].

COROLLARY 1 $\mathbf{X}'\mathbf{A}\mathbf{X} \sim W_d(r, \boldsymbol{\Sigma})$ if and only if $\mathbf{A}^2 = \mathbf{A}$.

COROLLARY 2 The Wishart variables $\mathbf{X}'\mathbf{A}\mathbf{X}$ and $\mathbf{X}'\mathbf{B}\mathbf{X}$ are independent if and only if $\mathbf{AB} = \mathbf{O}$.

COROLLARY 3 $\mathbf{X}'\mathbf{A}\mathbf{X}$ and $\mathbf{X}'\mathbf{b}$ are independently distributed as W_d and N_d, respectively, if and only if $\mathbf{A}\mathbf{b} = \mathbf{0}$ and $\mathbf{A}^2 = \mathbf{A}$. □

Although we could have stated the preceding corollaries as the main theorem, it is convenient to follow the above approach of Rao [1973] and prove multivariate results using the corresponding univariate theory.

For further properties of the Wishart distribution, the reader is referred to Johnson and Kotz [1972: Chapter 38] and Muirhead [1982: Section 3.2]. A related distribution is the inverse Wishart, the distribution of $\mathbf{W}^{-1}$ (see Press [1972a] and Siskind [1972] for some properties).

2.3.3 Noncentral Wishart Distribution

Let $\mathbf{x}_1, \mathbf{x}_2, \ldots, \mathbf{x}_m$ be independently distributed as $N_d(\boldsymbol{\mu}_i, \boldsymbol{\Sigma})$ $(i = 1, 2, \ldots, m)$. Then, by analogy with the noncentral chi-square distribution (see Exercise 2.1), we can define $\mathbf{W} = \mathbf{X}'\mathbf{X}$ to have the noncentral Wishart distribution, written $W_d(m, \boldsymbol{\Sigma}; \boldsymbol{\Delta})$, with noncentrality matrix $\boldsymbol{\Delta}$ defined to be (see A5.4)

$$\boldsymbol{\Delta} = \sum_{i=1}^{m} (\boldsymbol{\Sigma}^{-1/2}\boldsymbol{\mu}_i)(\boldsymbol{\Sigma}^{-1/2}\boldsymbol{\mu}_i)'$$

$$= \boldsymbol{\Sigma}^{-1/2}\mathbf{M}'\mathbf{M}\boldsymbol{\Sigma}^{-1/2},$$

where $\mathbf{M} = (\boldsymbol{\mu}_1, \boldsymbol{\mu}_2, \ldots, \boldsymbol{\mu}_m)'$. We note that $\mathscr{E}[\mathbf{X}']\mathscr{E}[\mathbf{X}] = \mathbf{M}'\mathbf{M} = \Sigma_i \boldsymbol{\mu}_i \boldsymbol{\mu}_i'$, and it follows from (1.9) with $\mathbf{A} = \mathbf{I}_m$ that

$$\mathscr{E}[\mathbf{W}] = m\boldsymbol{\Sigma} + \mathbf{M}'\mathbf{M}.$$

When $\boldsymbol{\Delta} = \mathbf{O}$, $\mathbf{M} = \mathbf{O}$ (by A4.5), the $\mathbf{x}_i$ are $N_d(\mathbf{0}, \boldsymbol{\Sigma})$, and $W_d(m, \boldsymbol{\Sigma}; \mathbf{O})$ is the same as $W_d(m, \boldsymbol{\Sigma})$.

The noncentral Wishart distribution has a complicated density function, though it can be handled in a formal fashion using zonal polynomials (see Johnson and Kotz [1972: Chapter 48], Muirhead [1982]). As the density function depends on the eigenvalues of $\boldsymbol{\Delta}$, it is common practice in theoretical studies to variously define $\boldsymbol{\Delta}$, $\frac{1}{2}\boldsymbol{\Delta}$, and $\boldsymbol{\Sigma}^{-1}\mathbf{M}'\mathbf{M}$ (see A1.4) as the noncentrality parameter matrix. Some authors use $\mathbf{M}$ and write $\mathbf{W} \sim W_d(m, \boldsymbol{\Sigma}; \mathbf{M})$. Gleser [1976] has given a canonical representation for the noncentral Wishart that is useful for simulation.

Most of the properties of the (central) Wishart carry over to the noncentral case. For example, Theorem 2.4 and its corollaries still hold if the chi-square and Wishart distributions are replaced by their noncentral counterparts. Thus if $\boldsymbol{\Delta} = \boldsymbol{\Sigma}^{-1/2}\mathscr{E}[\mathbf{X}']\mathbf{A}\mathscr{E}[\mathbf{X}]\boldsymbol{\Sigma}^{-1/2}$, $\mathbf{X}'\mathbf{A}\mathbf{X} \sim W_d(m, \boldsymbol{\Sigma}; \boldsymbol{\Delta})$ if and only if $\mathbf{A}^2 = \mathbf{A}$.

A particular case often studied in the literature is the so-called "linear" noncentral distribution in which $\text{rank}[\boldsymbol{\Delta}] = 1$ $(= \text{rank}[\mathbf{M}']$, by A2.2). In this case there is only one linearly independent vector in the set $\{\boldsymbol{\mu}_1, \boldsymbol{\mu}_2, \ldots, \boldsymbol{\mu}_m\}$ so that $\boldsymbol{\mu}_i = k_i\boldsymbol{\mu}$ for all i and certain k_i and $\boldsymbol{\mu}$. Geometrically this is equivalent to saying that the points $\boldsymbol{\mu}_i$ are collinear and lie on a straight line through the origin and the point $\boldsymbol{\mu}$. If $\boldsymbol{\mu}_i = \boldsymbol{\mu}$ $(i = 1, 2, \ldots, m)$, then $\mathbf{M}'\mathbf{M} = m\boldsymbol{\mu}\boldsymbol{\mu}'$ and the rank 1 condition is automatically satisfied.

2.3.4 Eigenvalues of a Wishart Matrix

Suppose that $\mathbf{W} \sim W_d(m, \mathbf{I}_d)$, $m \geq d$, and let $c_1 \geq c_2 \geq \cdots \geq c_d \geq 0$ be the ordered eigenvalues (characteristic roots) of $\mathbf{W}$. The joint distribution of the c_j was shown independently by Fisher [1939], Hsu [1939], and Roy [1939] to have density function

$$f(\mathbf{c}) = k\left[\exp\left(-\frac{1}{2}\sum_{j=1}^{d} c_j\right)\right]\left(\prod_{j=1}^{d} c_j\right)^{(m-d-1)/2} \prod_{i<j}^{d}(c_i - c_j),$$

$$c_1 \geq c_2 \geq \cdots \geq c_d > 0, \quad (2.13)$$

where k is a constant. The marginal distributions of c_1 and c_d are known (see Johnson and Kotz [1972: Section 39.4] for a review), and tables of percentage points are given by Pearson and Hartley [1972: Table 51]. Hanumara and Strain [1980] mention some applications of these percentage points. Formulas for the joint distribution of a few ordered roots, or any pair of roots, and the marginal distribution of any individual root are given by Krishnaiah and Waikar [1971]. The noncentral case is considered by Krishnaiah and Chattopadhyay [1975]. Krishnaiah and Schuurmann [1974] give formulas for the density function of a root divided by the sum of the roots, and the distribution function of c_d/c_1. For further references see Krishnaiah's [1978] review.

For the general case when $\mathbf{W} \sim W_d(m, \boldsymbol{\Sigma})$, various exact and asymptotic results for some or all of the c_i in terms of the eigenvalues $\lambda_1 \geq \lambda_2 \geq \cdots \geq \lambda_d > 0$ of $\boldsymbol{\Sigma}$ are available (Muirhead [1982]). As the method of principal components is a major application of this theory, further references are given in Section 5.2.5. The noncentral Wishart is considered briefly by Johnson and Kotz [1972] and Muirhead [1982] and the moments of its trace are given by Mathai [1980].

2.3.5 Determinant of a Wishart Matrix

Given $\mathbf{W} \sim W_d(m, \boldsymbol{\Sigma})$, $m \geq d$, the problem of estimating $|\boldsymbol{\Sigma}|$, the so-called "generalized variance," has been considered by several authors (e.g., Shorrock and Zidek [1976], Sinha [1976]). It is sometimes used as a measure of variability (see Kowal [1971]). Since $\mathscr{E}[\mathbf{W}] = m\boldsymbol{\Sigma}$ (Exercise 2.10), we have the estimate $|m^{-1}\mathbf{W}|$, where $|\mathbf{W}|$ is distributed as $|\boldsymbol{\Sigma}|$ times the product of d independent chi-square variables with respective degrees of freedom $m, m-1, \ldots, m-d+1$ (Exercise 2.22). Some approximations for the distribution of $|\mathbf{W}|$ are described in Johnson and Kotz [1972: Chapter 39], and approximations for the central and noncentral distributions of $|\mathbf{W}|^{1/d}/m$ are given by Steyn [1978]. The related problem of estimating $|\boldsymbol{\Sigma}_{22\cdot 1}| = |\boldsymbol{\Sigma}_{22} - \boldsymbol{\Sigma}_{21}\boldsymbol{\Sigma}_{11}^{-1}\boldsymbol{\Sigma}_{12}|$ [see Theorem 2.1(viii)] was considered by Tsui et al. [1980].

2.4 HOTELLING'S T^2 DISTRIBUTION

2.4.1 Central Distribution

If $x \sim N_1(\mu, \sigma^2)$, $w \sim \sigma^2\chi^2_m$, and w is statistically independent of x, then

$$T = \frac{(x-\mu)/\sigma}{(w/m\sigma^2)^{1/2}} = \frac{x-\mu}{(w/m)^{1/2}} \sim t_m, \tag{2.14}$$

where t_m is the t-distribution with m degrees of freedom. Therefore

$$T^2 = \frac{m(x-\mu)^2}{w} = m(x-\mu)w^{-1}(x-\mu) \sim F_{1,m},$$

since we have the identity $t_m^2 \equiv F_{1,m}$ between the t- and F-distributions. A natural generalization of the above statistic is the so-called Hotelling's T^2 statistic

$$T^2 = m(\mathbf{x}-\boldsymbol{\mu})'\mathbf{W}^{-1}(\mathbf{x}-\boldsymbol{\mu}),$$

where $\mathbf{x} \sim N_d(\boldsymbol{\mu}, \boldsymbol{\Sigma})$, $\mathbf{W} \sim W_d(m, \boldsymbol{\Sigma})$, $\mathbf{x}$ is statistically independent of $\mathbf{W}$, and both distributions are nonsingular. We shall show in Theorem 2.8 that $T^2 \sim cF_{d,m-d+1}$, but first let us introduce three lemmas.

LEMMA 2.5 Given $\mathbf{u} = (u_1, u_2, \ldots, u_d)' \sim N_d(\boldsymbol{\theta}, \boldsymbol{\Sigma})$, the conditional distribution of u_d, given $u_1, u_2, \ldots, u_{d-1}$, is of the form $N_1(\beta_0 + \Sigma_{j=1}^{d-1}\beta_j u_j, 1/\sigma^{dd})$, where $[(\sigma^{jk})] = \boldsymbol{\Sigma}^{-1}$.

Proof Let $\mathbf{u}^{(1)} = (u_1, \ldots, u_{d-1})'$, $\boldsymbol{\theta}^{(1)} = (\theta_1, \ldots, \theta_{d-1})'$, and

$$\boldsymbol{\Sigma} = \begin{pmatrix} \boldsymbol{\Sigma}_{11} & \boldsymbol{\sigma}_{1d} \\ \boldsymbol{\sigma}'_{d1} & \sigma_{dd} \end{pmatrix}. \tag{2.15}$$

From Theorem 2.1(viii) with $d_2 = 1$, the conditional distribution of u_d, given $\mathbf{u}^{(1)}$, is normal with mean

$$\begin{aligned} E[u_d|\mathbf{u}^{(1)}] &= \theta_d + \boldsymbol{\sigma}'_{d1}\boldsymbol{\Sigma}_{11}^{-1}(\mathbf{u}^{(1)} - \boldsymbol{\theta}^{(1)}) \\ &= \theta_d - \boldsymbol{\sigma}'_{d1}\boldsymbol{\Sigma}_{11}^{-1}\boldsymbol{\theta}^{(1)} + \boldsymbol{\sigma}'_{d1}\boldsymbol{\Sigma}_{11}^{-1}\mathbf{u}^{(1)} \\ &= \beta_0 + \boldsymbol{\beta}'\mathbf{u}^{(1)}, \end{aligned}$$

say. As $\boldsymbol{\Sigma} > \mathbf{O}$, then $\boldsymbol{\Sigma}_{11} > \mathbf{O}$ and (see A3.2)

$$|\boldsymbol{\Sigma}| = |\boldsymbol{\Sigma}_{11}|(\sigma_{dd} - \boldsymbol{\sigma}'_{d1}\boldsymbol{\Sigma}_{11}^{-1}\boldsymbol{\sigma}_{1d}), \tag{2.16}$$

so that

$$\begin{aligned}\operatorname{var}\left[u_d|\mathbf{u}^{(1)}\right] &= \sigma_{dd} - \boldsymbol{\sigma}_{d1}'\boldsymbol{\Sigma}_{11}^{-1}\boldsymbol{\sigma}_{1d} \\ &= \frac{|\boldsymbol{\Sigma}|}{|\boldsymbol{\Sigma}_{11}|} \\ &= \frac{1}{\sigma^{dd}}. \end{aligned} \tag{2.17}$$

LEMMA 2.6 Consider the linear regression model $\mathbf{y} = \mathbf{K}\boldsymbol{\beta} + \boldsymbol{\varepsilon}$, where $\mathbf{K}$ is $m \times p$ of rank p and $\boldsymbol{\varepsilon} \sim N_m(\mathbf{0}, \sigma^2\mathbf{I}_m)$. Let

$$\begin{aligned} Q &= \|\mathbf{y} - \mathbf{K}\hat{\boldsymbol{\beta}}\|^2 \\ &= \underset{\boldsymbol{\beta}}{\text{minimum}}\|\mathbf{y} - \mathbf{K}\boldsymbol{\beta}\|^2; \end{aligned}$$

then $Q \sim \sigma^2\chi^2_{m-p}$ and $Q = 1/w^{dd}$, where $[(w^{jk})] = \mathbf{W}^{-1}$ and

$$\mathbf{W} = \begin{pmatrix} \mathbf{K}' \\ \mathbf{y}' \end{pmatrix}(\mathbf{K}, \mathbf{y}) = \begin{pmatrix} \mathbf{K}'\mathbf{K} & \mathbf{K}'\mathbf{y} \\ \mathbf{y}'\mathbf{K} & \mathbf{y}'\mathbf{y} \end{pmatrix}.$$

Proof From general regression theory (e.g., Seber [1977: Theorem 3.5]; see also A6.5)

$$\begin{aligned} Q &= \mathbf{y}'(\mathbf{I}_m - \mathbf{P})\mathbf{y} \\ &= \mathbf{y}'\mathbf{y} - \mathbf{y}'\mathbf{K}(\mathbf{K}'\mathbf{K})^{-1}\mathbf{K}'\mathbf{y} \sim \sigma^2\chi^2_{m-p}, \end{aligned}$$

since $(\mathbf{I}_m - \mathbf{P})$ is symmetric and idempotent of rank $m - p$ (Exercise 1.5). Also, from (2.16),

$$|\mathbf{W}| = |\mathbf{K}'\mathbf{K}|\left(\mathbf{y}'\mathbf{y} - \mathbf{y}'\mathbf{K}(\mathbf{K}'\mathbf{K})^{-1}\mathbf{K}'\mathbf{y}\right)$$

so that

$$Q = \frac{|\mathbf{W}|}{|\mathbf{K}'\mathbf{K}|} = \frac{1}{w^{dd}}.$$

LEMMA 2.7 Suppose $\mathbf{W} \sim W_d(m, \boldsymbol{\Sigma})$, where $m \geq d$. Then we have the following:

(i) $\sigma^{dd}/w^{dd} \sim \chi^2_{m-d+1}$ and is independent of all the elements w_{jk} of $\mathbf{W}$ $(j, k = 1, 2, \ldots, d-1)$.
(ii) $\boldsymbol{\ell}'\boldsymbol{\Sigma}^{-1}\boldsymbol{\ell}/\boldsymbol{\ell}'\mathbf{W}^{-1}\boldsymbol{\ell} \sim \chi^2_{m-d+1}$ for any fixed $\boldsymbol{\ell}(\neq \mathbf{0})$.

Proof (i) Since $\mathbf{W} \sim W_d(m, \boldsymbol{\Sigma})$, we can argue as in Theorem 2.2 and write $\mathbf{W} = \sum_{i=1}^m \mathbf{x}_i\mathbf{x}_i'$, where the $\mathbf{x}_i$ are i.i.d. $N_d(\mathbf{0}, \boldsymbol{\Sigma})$. If $\mathbf{x}_i = (x_{i1}, x_{i2}, \dots, x_{id})'$ then, from Lemma 2.5 (with $\boldsymbol{\theta} = \mathbf{0}$), the conditional distribution of x_{id} given $x_{i1}, x_{i2}, \dots, x_{i,d-1}$ is $N_1(\sum_{j=1}^{d-1}\beta_j x_{ij}, 1/\sigma^{dd})$. Thus, conditionally, we have a regression model of the form given in Lemma 2.6 (with $p = d - 1$) so that

$$Q = \underset{\boldsymbol{\beta}}{\text{minimum}} \sum_{i=1}^{m} \left(x_{id} - \sum_{j=1}^{d-1} \beta_j x_{ij} \right)^2$$

$$\sim \frac{1}{\sigma^{dd}} \chi^2_{m-(d-1)}.$$

Since the conditional distribution of Q does not involve the conditioning variables x_{ij}, it is also the unconditional distribution, and Q is independent of x_{ij} $(i = 1, 2, \dots, m;\ j = 1, 2, \dots, d-1)$. Thus Q is independent of $w_{jk} = \sum_{i=1}^m x_{ij}x_{ik}$ for all j, k $(j \neq d, k \neq d)$.

Now

$$\mathbf{W} = \mathbf{X}'\mathbf{X} = \begin{pmatrix} \mathbf{x}^{(1)\prime} \\ \vdots \\ \mathbf{x}^{(d)\prime} \end{pmatrix} (\mathbf{x}^{(1)}, \dots, \mathbf{x}^{(d)}),$$

where $\mathbf{x}^{(j)} = (x_{1j}, x_{2j}, \dots, x_{mj})'$. Hence identifying $\mathbf{y}$ with $\mathbf{x}^{(d)}$ and $\mathbf{K}$ with $(\mathbf{x}^{(1)}, \dots, \mathbf{x}^{(d-1)})$ in Lemma 2.6, we see that $Q = 1/w^{dd}$ and $\sigma^{dd}/w^{dd} \sim \chi^2_{m-d+1}$.

(ii) Let $\mathbf{L}$ be any $d \times d$ orthogonal matrix with its bottom row equal to $\|\boldsymbol{\ell}\|^{-1}\boldsymbol{\ell}'$. Then, from Theorem 2.2,

$$\mathbf{LWL}' \sim W_d(m, \mathbf{L\Sigma L}').$$

Now $(\mathbf{LWL}')^{-1} = \mathbf{LW}^{-1}\mathbf{L}'$ and $(\mathbf{L\Sigma L}')^{-1} = \mathbf{L\Sigma}^{-1}\mathbf{L}'$. Applying part (i) to $\mathbf{LWL}'$, we have

$$\frac{\boldsymbol{\ell}'\boldsymbol{\Sigma}^{-1}\boldsymbol{\ell}/\|\boldsymbol{\ell}\|^2}{\boldsymbol{\ell}'\mathbf{W}^{-1}\boldsymbol{\ell}/\|\boldsymbol{\ell}\|^2} = \frac{(\mathbf{L\Sigma}^{-1}\mathbf{L}')_{dd}}{(\mathbf{LW}^{-1}\mathbf{L}')_{dd}}$$

$$= \frac{(\mathbf{L\Sigma L}')^{dd}}{(\mathbf{LWL}')^{dd}} \sim \chi^2_{m-d+1}. \qquad \square$$

Using Lemma 2.7 we can now readily prove the following theorem, which forms the basis of Hotelling's T^2 distribution.

THEOREM 2.8 Let $T^2 = m\mathbf{y}'\mathbf{W}^{-1}\mathbf{y}$, where $\mathbf{y} \sim N_d(\mathbf{0}, \boldsymbol{\Sigma})$, $\mathbf{W} \sim W_d(m, \boldsymbol{\Sigma})$, and $\mathbf{y}$ and $\mathbf{W}$ are statistically independent. (It is assumed that the distributions are nonsingular, i.e., $\boldsymbol{\Sigma} > \mathbf{O}$, and $m \geq d$, so that $\mathbf{W}^{-1}$ exists with probability 1). Then

$$\frac{m-d+1}{d}\frac{T^2}{m} \sim F_{d,m-d+1}. \tag{2.18}$$

Proof

$$\begin{aligned}\frac{T^2}{m} &= \mathbf{y}'\mathbf{W}^{-1}\mathbf{y} \\ &= \frac{\mathbf{y}'\boldsymbol{\Sigma}^{-1}\mathbf{y}}{\mathbf{y}'\boldsymbol{\Sigma}^{-1}\mathbf{y}/\mathbf{y}'\mathbf{W}^{-1}\mathbf{y}} \\ &= \frac{G}{H},\end{aligned}$$

say.

For a given value of $\mathbf{y}$, it follows, from Lemma 2.7(ii), that the conditional distribution of H, given $\mathbf{y} = \ell$, is χ^2_{m-d+1}. Since this conditional distribution does not depend on $\mathbf{y}$, it is also the unconditional distribution, and H is independent of $\mathbf{y}$. Hence $H \sim \chi^2_{m-d+1}$ and is independent of $G = \mathbf{y}'\boldsymbol{\Sigma}^{-1}\mathbf{y}$, a function of $\mathbf{y}$. From Theorem 2.1(vi), $G \sim \chi^2_d$, so that T^2/m is the ratio of two independent chi-square variables. Hence

$$\frac{m-d+1}{d}\frac{T^2}{m} = \frac{G/d}{H/(m-d+1)} \sim F_{d,m-d+1}.$$

COROLLARY Given $\mathbf{x} \sim N_d(\boldsymbol{\mu}, \lambda^{-1}\boldsymbol{\Sigma})$, $\mathbf{W} \sim W_d(m, \boldsymbol{\Sigma})$, and $\mathbf{x}$ independent of $\mathbf{W}$, then

$$\begin{aligned}T^2 &= \lambda m(\mathbf{x}-\boldsymbol{\mu})'\mathbf{W}^{-1}(\mathbf{x}-\boldsymbol{\mu}) \\ &= \lambda(\mathbf{x}-\boldsymbol{\mu})'(\mathbf{W}/m)^{-1}(\mathbf{x}-\boldsymbol{\mu})\end{aligned} \tag{2.19}$$

satisfies (2.18).

Proof Let $\mathbf{y} = \sqrt{\lambda}(\mathbf{x}-\boldsymbol{\mu})$ in Theorem 2.8. □

As the distribution of T^2 depends on two parameters d and m, we shall use the notation $T^2 \sim T^2_{d,m}$ when T^2 is distributed as in (2.18). For future reference we note the equivalence

$$\frac{T^2_{d,m}}{m}\frac{m-d+1}{d} \equiv F_{d,m-d+1}. \tag{2.20}$$

2.4.2 Noncentral Distribution

Given $\mathbf{y} \sim N_d(\boldsymbol{\theta}, \boldsymbol{\Sigma})$, $\mathbf{W} \sim W_d(m, \boldsymbol{\Sigma})$, and $\mathbf{y}$ independent of $\mathbf{W}$, then $T^2 = m\mathbf{y}'\mathbf{W}^{-1}\mathbf{y}$ is said to have a noncentral T^2 distribution with noncentrality parameter $\delta = \boldsymbol{\theta}'\boldsymbol{\Sigma}^{-1}\boldsymbol{\theta}$. To identify this distribution we briefly retrace the steps leading to the derivation for the central case $\delta = 0$ (i.e., $\boldsymbol{\theta} = \mathbf{0}$, as $\boldsymbol{\Sigma}$ is positive definite).

In Theorem 2.8 we considered the ratio G/H and proved that $H \sim \chi^2_{m-d+1}$ and H is independent of G. This part of the proof is valid, irrespective of whether or not $\boldsymbol{\theta} = \mathbf{0}$. From Theorem 2.1(vii), $\mathbf{y}'\boldsymbol{\Sigma}^{-1}\mathbf{y}$ has a noncentral chi-square distribution with d degrees of freedom and noncentrality parameter δ. Thus $(m - d + 1)G/dH$ is the ratio of a noncentral chi-square variable to an independent chi-square variable, each suitably scaled by its degrees of freedom. Hence

$$\frac{T^2}{m}\frac{m-d+1}{d} \sim F_{d,\, m-d+1,\, \delta},$$

where $F_{d,\, m-d+1,\, \delta}$ is the noncentral F-distribution with d and $m - d + 1$ degrees of freedom, respectively, and noncentrality parameter δ. (For derivations and properties of noncentral distributions see Johnson and Kotz [1970] and Muirhead [1982]).

2.5 MULTIVARIATE BETA DISTRIBUTIONS

2.5.1 Derivation

Suppose that $H \sim \sigma^2\chi^2_{m_H}$, $E \sim \sigma^2\chi^2_{m_E}$, and H and E are statistically independent. For example, H could be the "hypothesis" sum of squares and E the "error" or residual sum of squares in an analysis of variance model. Then the density functions of $T = H/E$ and $V = T/(1 + T) = H/(E + H)$ are

$$f(t) = \frac{1}{B(m_H/2, m_E/2)} \frac{t^{m_H/2-1}}{(1+t)^{(m_E+m_H)/2}}, \qquad 0 \le t < \infty,$$

and

$$g(v) = \frac{1}{B(m_H/2, m_E/2)} v^{m_H/2-1}(1-v)^{m_E/2-1}, \qquad 0 \le v \le 1,$$

where

$$B(a, b) = \frac{\Gamma(a)\Gamma(b)}{\Gamma(a+b)}.$$

For convenience we can use the notation $V \sim B_{m_H/2,\, m_E/2}$ and V is said to have a (type 1) beta distribution with $\frac{1}{2}m_H$ and $\frac{1}{2}m_E$ degrees of freedom, respectively. We also note that $m_E T/m_H \sim F_{m_H,\, m_E}$ and T is said to have a type II or inverted beta distribution with $\frac{1}{2}m_H$ and $\frac{1}{2}m_E$ degrees of freedom, respectively. For future reference it is useful to highlight the relationship $V = 1 - (1 + T)^{-1}$ in terms of distributions, namely,

$$1 - B_{a,b} \sim B_{b,a} \sim \left(1 + \frac{2a}{2b}F_{2a,2b}\right)^{-1}. \tag{2.21}$$

The preceding results can be generalized to the case where $\mathbf{H}$ and $\mathbf{E}$ are matrices with independent nonsingular Wishart distributions, namely, $\mathbf{H} \sim W_d(m_H, \boldsymbol{\Sigma})$ and $\mathbf{E} \sim W_d(m_E, \boldsymbol{\Sigma})$ with $m_H, m_E \geq d$. By analogy with the above univariate approach, we could consider $\mathbf{HE}^{-1}$ and $\mathbf{H}(\mathbf{E} + \mathbf{H})^{-1}$, but these matrices are not symmetric and do not lead to useful density functions. However, since $\mathbf{E}$ and $\mathbf{H}$, and hence $\mathbf{E} + \mathbf{H}$, are positive definite with probability 1, we can obtain symmetry by defining the positive definite matrices (see A5.7)

$$\mathbf{T} = \mathbf{E}^{-1/2}\mathbf{H}\mathbf{E}^{-1/2} \tag{2.22}$$

and

$$\mathbf{V} = (\mathbf{E} + \mathbf{H})^{-1/2}\mathbf{H}(\mathbf{E} + \mathbf{H})^{-1/2}, \tag{2.23}$$

where $\mathbf{E}^{1/2}$ and $(\mathbf{E} + \mathbf{H})^{1/2}$ are the symmetric square roots of $\mathbf{E}$ and $\mathbf{E} + \mathbf{H}$, respectively (see A5.4). We now prove the following theorem.

THEOREM 2.9 The joint density function of the $\frac{1}{2}d(d+1)$ distinct elements of $\mathbf{V}$, namely, $g(v_{11}, v_{12}, \ldots, v_{dd})$ [or $g(\mathbf{V})$ for short] is given by

$$g(\mathbf{V}) = \frac{1}{B_d(m_H/2, m_E/2)}|\mathbf{V}|^{(m_H-d-1)/2}|\mathbf{I}_d - \mathbf{V}|^{(m_E-d-1)/2}, \qquad \mathbf{O} < \mathbf{V} < \mathbf{I}_d, \tag{2.24}$$

where

$$B_d(a,b) = \frac{\Gamma_d(a)\Gamma_d(b)}{\Gamma_d(a+b)}, \qquad d \leq 2a, 2b$$

and $\Gamma_d(a)$ is defined by (2.5). It is assumed that $d \leq m_H, m_E$.

Proof Since $\mathbf{H}$ and $\mathbf{E}$ are independent, their joint distribution is simply the product of their marginal densities, namely [see (2.3)],

$$c_H^{-1}c_E^{-1}|\mathbf{H}|^{(m_H-d-1)/2}|\mathbf{E}|^{(m_E-d-1)/2}\operatorname{etr}\left[-\tfrac{1}{2}\boldsymbol{\Sigma}^{-1}(\mathbf{E} + \mathbf{H})\right],$$

where

$$c_i = 2^{m_i d/2}|\boldsymbol{\Sigma}|^{m_i/2}\Gamma_d(m_i/2), \qquad (i = H, E).$$

We now make the transformation from $(\mathbf{H},\mathbf{E})$ to $(\mathbf{V},\mathbf{Z})$, where $\mathbf{V} = (\mathbf{E}+\mathbf{H})^{-1/2}\mathbf{H}(\mathbf{E}+\mathbf{H})^{-1/2}$ and $\mathbf{Z} = \mathbf{E}+\mathbf{H}$. The Jacobian of this transformation is, by A9.2(c), $|\mathbf{Z}|^{(d+1)/2}$, so that the joint density function of the upper triangular elements of $\mathbf{V}$ and $\mathbf{Z}$ is

$$h(\mathbf{V},\mathbf{Z}) = c_H^{-1}c_E^{-1}|\mathbf{Z}^{1/2}\mathbf{V}\mathbf{Z}^{1/2}|^{(m_H-d-1)/2}|\mathbf{Z}-\mathbf{Z}^{1/2}\mathbf{V}\mathbf{Z}^{1/2}|^{(m_E-d-1)/2}$$
$$\times|\mathbf{Z}|^{(d+1)/2}\operatorname{etr}\left(-\tfrac{1}{2}\boldsymbol{\Sigma}^{-1}\mathbf{Z}\right).$$

Using the fact that the integral of the Wishart density is one, we have

$$g(\mathbf{V}) = \int\cdots\int h(\mathbf{V},\mathbf{Z})\,dz_{11}\,dz_{12}\cdots dz_{dd}$$

$$= c_H^{-1}c_E^{-1}|\mathbf{V}|^{(m_H-d-1)/2}|\mathbf{I}_d-\mathbf{V}|^{(m_E-d-1)/2}$$
$$\times\int\cdots\int|\mathbf{Z}|^{(m_H+m_E-d-1)/2}\operatorname{etr}\left(-\tfrac{1}{2}\boldsymbol{\Sigma}^{-1}\mathbf{Z}\right)dz_{11}\cdots dz_{dd}$$
$$= c_H^{-1}c_E^{-1}|\mathbf{V}|^{(m_H-d-1)/2}|\mathbf{I}_d-\mathbf{V}|^{(m_E-d-1)/2}c_{H+E},$$

where c_{H+E}^{-1} is the constant associated with $W_d(m_H+m_E,\boldsymbol{\Sigma})$ [see (2.4)]. It is readily seen that

$$\frac{c_{H+E}}{c_H c_E} = \frac{\Gamma_d([m_H+m_E]/2)}{\Gamma_d(m_H/2)\Gamma_d(m_E/2)} = \frac{1}{B_d(m_H/2,m_E/2)}.$$

Finally, by A5.7, $\mathbf{V} > \mathbf{O}$ and

$$\mathbf{I}_d - \mathbf{V} = (\mathbf{E}+\mathbf{H})^{-1/2}(\mathbf{E}+\mathbf{H}-\mathbf{H})(\mathbf{E}+\mathbf{H})^{-1/2}$$
$$= (\mathbf{E}+\mathbf{H})^{-1/2}\mathbf{E}(\mathbf{E}+\mathbf{H})^{-1/2} > \mathbf{O}, \qquad (2.25)$$

so that using the convention for positive definite matrices, we can write, symbolically, $\mathbf{O} < \mathbf{V} < \mathbf{I}_d$. □

The density function (2.24) does not depend on $\boldsymbol{\Sigma}$, so that Theorem 2.9 still holds if $\boldsymbol{\Sigma} = \mathbf{I}_d$. In this case it can be shown that $\mathbf{T}$ of (2.22) has density function

$$f(\mathbf{T}) = \frac{1}{B_d(m_H/2,m_E/2)}|\mathbf{T}|^{(m_H-d-1)/2}|\mathbf{I}_d+\mathbf{T}|^{-(m_E+m_H)/2}, \qquad \mathbf{T} > \mathbf{O}. \qquad (2.26)$$

However, (2.26) is not true for general $\boldsymbol{\Sigma}$ (Olkin and Rubin [1964]).

In addition to the symmetric square root $(\mathbf{E} + \mathbf{H})^{1/2}$, there are two other "square roots" based on the Cholesky decompositions $\mathbf{E} + \mathbf{H} = \mathbf{L}'\mathbf{L} = \mathbf{U}'\mathbf{U}$, where $\mathbf{L}$ and $\mathbf{U}$ are lower and upper triangular matrices, respectively, with positive diagonal elements (see A5.11). If we put $\mathbf{V} = (\mathbf{L}')^{-1}\mathbf{H}\mathbf{L}^{-1}$ or $\mathbf{V} = (\mathbf{U}')^{-1}\mathbf{H}\mathbf{U}^{-1}$, we find that the density function of $\mathbf{V}$ is still (2.24) in both cases. However, if we use similar representations for $\mathbf{T}$, we find that we get different expressions for $f(\mathbf{T})$, for each of the three square roots. This rather curious result is due to certain independence properties that depend critically on the square root used (Olkin and Rubin [1964]). However the distribution of $\operatorname{tr}\mathbf{T}$ is the same in each case and it is this function of $\mathbf{T}$ that is featured later [see (2.38)].

By analogy with the univariate theory, we say that (2.26) is the density function for a multivariate type II or inverted beta distribution, while $\mathbf{V}$ is said to have a d-dimensional multivariate type I beta distribution with $\frac{1}{2}m_H$ and $\frac{1}{2}m_E$ degrees of freedom, respectively. By simply interchanging the roles of $\mathbf{H}$ and $\mathbf{E}$, we see that $\mathbf{I}_d - \mathbf{V}$ also has a multivariate Type I beta distribution, but with $\frac{1}{2}m_E$ and $\frac{1}{2}m_H$ degrees of freedom.

We now consider the eigenvalues of $\mathbf{V}$.

2.5.2 Multivariate Beta Eigenvalues

Suppose θ is an eigenvalue of $\mathbf{V}$. Then, since $\mathbf{V} > \mathbf{O}$ with probability 1, $\theta > 0$ with probability 1 (by A5.1). We note that θ is a root of

$$\begin{aligned} 0 &= |\mathbf{V} - \theta\mathbf{I}_d| \\ &= |(\mathbf{E}+\mathbf{H})^{-1/2}\mathbf{H}(\mathbf{E}+\mathbf{H})^{-1/2} - \theta\mathbf{I}_d| \\ &= |\mathbf{E}+\mathbf{H}|^{-1}|\mathbf{H} - \theta(\mathbf{E}+\mathbf{H})|, \end{aligned}$$

that is, a root of

$$|\mathbf{H} - \theta(\mathbf{E}+\mathbf{H})| = 0. \tag{2.27}$$

If $\theta \geq 1$, $-(\theta - 1)\mathbf{H} - \theta\mathbf{E} < \mathbf{O}$ with probability 1 and (2.27) holds with probability 0. Hence $\theta < 1$ with probability 1 and we can express (2.27) in the form

$$|\mathbf{H} - \phi\mathbf{E}| = 0, \tag{2.28}$$

where $\phi = \theta/(1-\theta)$. Since $\mathbf{E}$ and $\mathbf{H}$ are independent nonsingular Wishart matrices, both $\mathbf{E} > \mathbf{O}$ and the eigenvalues of $\mathbf{H}$ are distinct, with probability 1 (see A5.13, A2.8). This implies that the eigenvalues of $\mathbf{H}\mathbf{E}^{-1}$, that is, the roots of (2.28), are distinct with probability 1. Hence the eigenvalues of $\mathbf{V}$ are distinct and we can order them in the form $1 > \theta_1 > \theta_2 > \cdots > \theta_d > 0$.

Although in the above theory we have assumed $\mathbf{H} > \mathbf{O}$ and $\mathbf{E} > \mathbf{O}$, we note from (2.23) that if $|\mathbf{H}| = 0$, then $|\mathbf{V}| = 0$ and at least one of the roots of (2.27) is zero, while if $|\mathbf{E}| = 0$, then at last one of the roots is unity [see (2.25)].

From A1.2 we have

$$|\mathbf{V}| = \prod_{j=1}^{d} \theta_j \quad \text{and} \quad |\mathbf{I}_d - \mathbf{V}| = \prod_{j=1}^{d} (1 - \theta_j),$$

so that from (2.24), $g(\mathbf{V}) = g_1(\theta_1, \ldots, \theta_d)$. Hence, by A9.1, the joint density function of the θ_j is

$$\begin{aligned} f(\boldsymbol{\theta}) &= \pi^{d^2/2} g_1(\theta_1, \ldots, \theta_d) \left[\prod_{j<k} (\theta_j - \theta_k) \right] \Big/ \Gamma_d\left(\tfrac{1}{2}d\right) \\ &= a_1^{-1} \left(\prod_{j=1}^{d} \theta_j \right)^{(m_H - d - 1)/2} \left[\prod_{j=1}^{d} (1 - \theta_j) \right]^{(m_E - d - 1)/2} \prod_{j<k}^{d} (\theta_j - \theta_k), \end{aligned} \tag{2.29}$$

where

$$a_1 = \pi^{-d^2/2} B_d\left(\tfrac{1}{2}m_H, \tfrac{1}{2}m_E\right) \Gamma_d\left(\tfrac{1}{2}d\right). \tag{2.30}$$

We note that (2.29) does not depend on $\boldsymbol{\Sigma}$, a result that can be deduced directly from (2.27) (see Exercise 2.21). The distribution given by (2.29) is sometimes called a generalized beta distribution, as some, but not all, of the marginal distributions are beta (Foster [1957, 1958], Foster and Rees [1957]).

The preceding derivation of (2.29), which depends on (2.24), is valid if $\mathbf{H}$ and $\mathbf{E}$ are both positive definite, that is, $\boldsymbol{\Sigma} > \mathbf{O}$ and $m_H, m_E \geq d$. However, if $m_H < d$, then $\mathbf{H}$ is singular and does not have a density function, so that the above method of deriving (2.29) no longer holds. However, if $m_E \geq d$, then $\mathbf{E} + \mathbf{H} > \mathbf{O}$ (as $\mathbf{E} > \mathbf{O}$ and $\mathbf{H} \geq \mathbf{O}$; see A5.10) and $\mathbf{V}$ still exists, but it is now singular with rank m_H. Thus, with probability 1, there are only m_H nonzero roots of $\mathbf{V}$, namely $\theta_1 > \theta_2 > \cdots > \theta_{m_H} > 0$. With a different approach, it can be shown (e.g., Anderson [1958; p. 315]) that the joint density function of these θ_j is

$$f(\boldsymbol{\theta}) = a_2^{-1} \left(\prod_{j=1}^{m_H} \theta_j \right)^{(d - m_H - 1)/2} \left[\prod_{j=1}^{m_H} (1 - \theta_j) \right]^{(m_E - d - 1)/2} \prod_{j<k}^{m_H} (\theta_j - \theta_k), \tag{2.31}$$

where

$$a_2 = \pi^{-m_H^2/2} B_{m_H}\left(\tfrac{1}{2}d, \tfrac{1}{2}[m_E + m_H - d]\right) \Gamma_{m_H}\left(\tfrac{1}{2}m_H\right).$$

We note that (2.31) is of the same form as (2.29), but with the changes $(d, m_H, m_E) \rightarrow (m_H, d, m_E + m_H - d)$. Equations (2.29) and (2.31) were obtained independently by Fisher [1939], Girshick [1939], Hsu [1939], Roy [1939], and Mood [1951] and can be combined into a single equation

$$a^{-1}\left(\prod_{j=1}^{s} \theta_j\right)^{\nu_1}\left[\prod_{j=1}^{s}(1-\theta_j)\right]^{\nu_2} \prod_{j<k}^{s}(\theta_j - \theta_k), \tag{2.32}$$

where $s = \text{minimum}(d, m_H)$, $\nu_1 = \frac{1}{2}(|m_H - d)| - 1)$, $\nu_2 = \frac{1}{2}(m_E - d - 1)$, and

$$a = \pi^{-s^2/2} B_s(\nu_1 + \tfrac{1}{2}[s+1], \nu_2 + \tfrac{1}{2}[s+1]) \Gamma_s(\tfrac{1}{2}s).$$

Owing to various test statistics and associated simultaneous confidence intervals introduced by Roy [1939, 1953], considerable attention has been focused on the distribution of $\theta_{\max}$ $(= \theta_1)$. In fact, Roy [1939] gave an elegant representation for the distribution function of θ_1 in terms of the determinant of a matrix whose elements are incomplete beta functions. Using this representation, Pillai and his co-workers (see Johnson and Kotz [1972; pp. 18–26] or Pillai [1967] for references) constructed tables of the percentage points of $\theta_{\max}$ for $s = 2(1)20$ and selected values of ν_1 and ν_2. Further values for $s = 2, 3, 4$, were given by Foster and Rees [1957] and Foster [1957, 1958], while Heck [1960] has given a number of charts for $s = 2(1)5$. A useful algorithm for the distribution function of $\theta_{\max}$ when ν_2 is an integer is given by Venables [1974]. Chang [1974] gives exact values for $s = 2(1)5$ and found that Pillai's tables were sufficiently accurate for $s \geq 6$. Tables are given in Appendix D14.

If $m_H, m_E \geq d$, the percentage points for $\theta_{\min}$ $(= \theta_d)$ can also be obtained from the tables by noting that the eigenvalues of $\mathbf{I}_d - \mathbf{V}$, that is, the roots of $|\mathbf{E} - \lambda(\mathbf{E} + \mathbf{H})| = 0$, satisfy $\lambda = 1 - \theta$. Thus $\theta_{\min} = 1 - \lambda_{\max}$, where $\lambda_{\max}$ has the same distribution as $\theta_{\max}$, but with m_E and m_H interchanged. Tables for $\theta_{\min}$ are given by Chang [1974: $s = 2(1)5$] and Schuurmann and Waikar [1974: $s = 4(1)10$].

The reader should be aware that there are a variety of notations used in the literature concerning the eigenvalues. Roy's famous maximum root statistic is actually $\phi_{\max}$, the maximum root of $|\mathbf{H} - \phi\mathbf{E}| = 0$, that is, the maximum eigenvalue of $\mathbf{HE}^{-1}$, while Heck's [1960] charts refer to $\theta_{\max}$. Thus we have three sets of roots θ, λ, and ϕ, which are related as follows [see (2.28)]:

$$\phi = \frac{\theta}{1-\theta} = \frac{1-\lambda}{\lambda}, \tag{2.33}$$

and

$$\phi_{\max} = \frac{\theta_{\max}}{1-\theta_{\max}}. \tag{2.34}$$

However, we need only concern ourselves with θ and ϕ.

When $s = 1$, $|\mathbf{H} - \theta(\mathbf{E} + \mathbf{H})| = 0$ has only one root. Since the trace of a matrix is the sum of its eigenvalues (A1.2),

$$\theta_{max} = \text{tr}\left[\mathbf{H}(\mathbf{E} + \mathbf{H})^{-1}\right],$$

and the multivariate distribution of the roots reduces to the univariate beta distribution of θ_{max}, namely,

$$f(\theta) = \frac{1}{B(\nu_1 + 1, \nu_2 + 1)} \theta^{\nu_1}(1 - \theta)^{\nu_2}, \qquad 0 \leq \theta \leq 1. \tag{2.35}$$

The quantile θ_α can be obtained by making the transformation [see (2.21)]

$$F = \frac{2\nu_2 + 2}{2\nu_1 + 2} \frac{\theta_{max}}{1 - \theta_{max}} \sim F_{2\nu_1 + 2, 2\nu_2 + 2} \tag{2.36}$$

and using the upper tail of the F-distribution. Alternatively, we can use $\phi_{max} = \text{tr}[\mathbf{HE}^{-1}]$, and then

$$F = \frac{(\nu_2 + 1)\phi_{max}}{\nu_1 + 1}. \tag{2.37}$$

We note that the eigenvalues ϕ_j of $\mathbf{HE}^{-1}$ are the same as those of $\mathbf{T} = \mathbf{E}^{-1/2}\mathbf{HE}^{-1/2}$, while the eigenvalues θ_j of $\mathbf{H}(\mathbf{E} + \mathbf{H})^{-1}$ are the same as those of $\mathbf{V}$ (A1.4). There is an extensive literature on the exact and large-sample properties of these eigenvalues and the reader is referred to the reviews of Pillai [1976], Krishnaiah [1978: Section 4], and Muirhead [1978, 1982]. In the literature $\mathbf{HE}^{-1}$ and $\mathbf{H}(\mathbf{E} + \mathbf{H})^{-1}$ are commonly called the multivariate F- and beta matrices: The distributions of their traces are considered in Section 2.5.3.

When $\mathbf{H}$ has a noncentral Wishart distribution, the joint distribution of the θ_j is commonly called the noncentral generalized beta distribution and is discussed by Johnson and Kotz [1972: p. 186, with $\mathbf{S}_1 \to \mathbf{H}$ and $\mathbf{S}_2 \to \mathbf{E}$]. Asymptotic expressions for distributions are given by Constantine and Muirhead [1976] for the θ_j, and by Fujikoshi [1977a] for the ϕ_j. There have also been several papers giving asymptotic expansions for functions of eigenvalues using perturbation techniques (Fujikoshi [1978]).

2.5.3 *Two Trace Statistics*

a LAWLEY–HOTELLING STATISTIC

In relation to hypothesis testing, Lawley [1938] and Hotelling [1951] considered using the sum of the eigenvalues of $\mathbf{HE}^{-1}$ when $m_E \geq d$, namely,

$$T_g^2 = m_E \text{tr}[\mathbf{HE}^{-1}] = m_E U^{(s)}, \tag{2.38}$$

where

$$U^{(s)} = \sum_{j=1}^{s} \phi_j = \sum_{j=1}^{s} \frac{\theta_j}{1-\theta_i}, \tag{2.39}$$

and $s = \text{minimum}(d, m_H)$, the number of nonzero eigenvalues of $\mathbf{HE}^{-1}$. This statistic is called the Lawley–Hotelling trace statistic or Hotelling's generalized T^2 statistic: We shall use T_g^2 instead of the usual T_0^2 to avoid confusion in the notation later.

The exact distribution of T_g^2 was obtained by Hotelling [1951] for $d = 2$ and expressions for general d are given by Pillai and Young [1971] and Krishnaiah and Chang [1972]. Exact and approximate 5% and 1% upper tail critical values for various values of d are given by a number of authors, including Davis [1970a, b, 1980a] and Pillai and Young [1971] (see Appendix D15. McKeon [1974] has shown that the distribution of $U^{(s)}$ can be approximated by cF, where $F \sim F_{a,b}$. Here

$$a = dm_H, \qquad b = 4 + (a+2)/(B-1), \qquad c = a(b-2)/b(m_E - d - 1),$$

where $B = (m_E + m_H - d - 1)(m_E - 1)/(m_E - d - 3)(m_E - d)$. This approximation is surprisingly accurate and supersedes previous approximations by Hughes and Saw [1972] and Pillai and Samson [1959].

When $m_E \to \infty$, $T_g^2 \sim \chi^2_{dm_H}$. However, using more terms of the large-sample expansion for the distribution function of T_g^2, more accurate chi-square approximations have been obtained. Similar expansions are available when $\mathbf{H}$ is noncentral Wishart with noncentrality parameter $\boldsymbol{\Delta}$: T_g^2 now has a limiting noncentral chi-square distribution with noncentrality parameter $\operatorname{tr}\boldsymbol{\Delta}$ (see Johnson and Kotz [1972: Chapter 39], Pillai [1976, 1977], Muirhead [1982]).

b PILLAI'S TRACE STATISTIC

Pillai [1953] proposed several other statistics, including

$$V^{(s)} = \operatorname{tr}\left[\mathbf{H}(\mathbf{E}+\mathbf{H})^{-1}\right] = \sum_{j=1}^{s} \theta_j,$$

where $s = \text{minimum}(d, m_H)$. The exact distribution of $V^{(s)}$ was obtained by Pillai and Jayachandran [1970], thus extending the work of Nanda [1950], who obtained the distribution for special cases only. Pillai [1955] obtained a two-moment beta approximation to the distribution that seems moderately accurate, and used a four-moment Pearson curve approximation to construct tables (Pillai [1960], Mijares [1964]). A more accurate approximation and some tables are given by John [1976, 1977]. Using a simplification of an expression obtained by Krishnaiah and Chang [1972], Schuurmann et al. [1975] computed exact percentage points for $s = 2(1)5$ (see Appendix D16).

When $m_E \to \infty$, $m_E V^{(s)} \sim \chi^2_{dm_H}$. Using the Cornish–Fisher approach of Hill and Davis [1968], Y. S. Lee [1971] derived a large-sample approximation for the upper tail values of $m_E V^{(s)}$. He also gave a large-sample approximation for the noncentral distribution of $m_E V^{(s)}$(i.e., when $\mathbf{H}$ is noncentral Wishart). For further references, see Pillai [1976, 1977].

2.5.4 U-*Distribution*

An important function of $\mathbf{V}$ [see (2.23)] in multivariate statistical inference is $U = |\mathbf{I}_d - \mathbf{V}|$. (It is more convenient to use U rather than $|\mathbf{V}|$ in applications). Following Anderson [1958], if $\mathbf{I}_d - \mathbf{V}$ has a multivariate Type I beta distribution with $\frac{1}{2}m_E$ and $\frac{1}{2}m_H$ degrees of freedom, respectively, then we say that U has a U-distribution with degrees of freedom d, m_H, and m_E, and write $U \sim U_{d, m_H, m_E}$. From (2.25),

$$U = \frac{|\mathbf{E}|}{|\mathbf{E} + \mathbf{H}|} \qquad \left(= \prod_{j=1}^{d} (1 - \theta_j) \right), \tag{2.40}$$

and we require $|\mathbf{E}| \neq 0$, that is, $m_E \geq d$, for $U \neq 0$. The statistic U, originally called Λ, was first introduced by Wilks [1932] as the appropriate likelihood ratio statistic for testing a linear hypothesis (see Chapter 8), and has since been studied extensively. The theoretical properties of U are set out in Anderson [1958: Chapter 8] and Kshirsagar [1972: pp. 292–304]. Further references are given by Johnson and Kotz [1972: pp. 202–204]. The main facts about the distribution of U are as follows.

1. The distribution function of U can be expressed as a computable mixture of incomplete beta functions (Tretter and Walster [1975]).

2. The distribution of U_{d, m_H, m_E} is the some as that of $U_{m_H, d, m_E + m_H - d}$. This fact is useful if $m_H < d$ (as in the cases $m_H = 1, 2$ below). Instead of using the triple (d, m_H, m_E), some authors use (d, m_H, n_0), where $n_0 = m_E + m_H$. If we interchange d and m_H so that $d^* = m_H$ and $m_H^* = d$, and set $m_E^* = m_E + m_H - d$, then $n_0^* = m_E^* + m_H^* = m_E + m_H = n_0$. Thus $n_0^* = n_0$ and n_0 remains unchanged when d and m_H are interchanged.

3. The following special cases hold. When $d = 1$,

$$\frac{1 - U_{1, m_H, m_E}}{U_{1, m_H, m_E}} \frac{m_E}{m_H} \sim F_{m_H, m_E} \quad \text{for any } m_H. \tag{2.41}$$

When $m_H = 1$,

$$\frac{1 - U_{d,1, m_E}}{U_{d,1, m_E}} \frac{m_E + 1 - d}{d} \sim F_{d, m_E + 1 - d} \quad \text{for any } d. \tag{2.42}$$

When $d = 2$,

$$\frac{1 - U_{2, m_H, m_E}^{1/2}}{U_{2, m_H, m_E}^{1/2}} \frac{m_E - 1}{m_H} \sim F_{2m_H, 2(m_E - 1)}, \qquad m_H \geq 2. \tag{2.43}$$

When $m_H = 2$,

$$\frac{1 - U_{d, 2, m_E}^{1/2}}{U_{d, 2, m_E}^{1/2}} \frac{m_E + 1 - d}{d} \sim F_{2d, 2(m_E + 1 - d)}, \qquad d \geq 2. \tag{2.44}$$

In Section 2.5.5b the above results are expressed in a more convenient form. The distribution of U has also been derived for the cases d and m_H both less than or equal to 4 (Anderson [1958: pp. 196–202]).

4. By expanding the characteristic function of $-c \log U$ up to terms of order n_0^{-2} and choosing c appropriately, Bartlett [1938a] showed that for large n_0 (i.e., large m_E), $W = -f \log U$ is approximately distributed as $\chi^2_{dm_H}$, where $f = m_E - \frac{1}{2}(d - m_H + 1) = n_0 - \frac{1}{2}(d + m_H + 1)$. This approximation is surprisingly accurate for the usual critical values, in fact, to three decimal places if $d^2 + m_H^2 \leq \frac{1}{3}f$. Rao [1948: pp. 70–71] expanded the characteristic function to terms of order n_0^{-6} and obtained the first three terms of a rapidly converging series. Rao's approximation is in fact a special case of a more general result of Box [1949], who gave asymptotic approximations to functions of general likelihood ratio statistics (see Anderson [1958: Section 8.6]). It is also a modification of an expansion of $-\log U$ given by Wald and Brookner [1941].

Other approximations and asymptotic expansions have been given by Rao [1951] and Roy [1951]. In particular, Rao showed that

$$\frac{1 - U_{d, m_H, m_E}^{1/t}}{U_{d, m_H, m_E}^{1/t}} \frac{ft - g}{dm_H} \quad \text{is approximately} \quad F_{dm_H, ft - g}, \tag{2.45}$$

where

$$t = \left(\frac{d^2 m_H^2 - 4}{d^2 + m_H^2 - 5} \right)^{1/2} \quad \text{and} \quad g = \frac{dm_H - 2}{2}.$$

This approximation to an F-distribution, which is exact when d or m_H is 1 or 2 [see (2.41)–(2.44)], is better than Bartlett's chi-square approximation and is also better than Rao's three-term chi-square expansion when m_E is small (Mudholkar and Trivedi [1980: Table 1]). Although the three-term approximation is slightly better for large m_E, (2.45) seems adequate for practical situations. If $ft - g$ is not an integer, a conservative value for the degrees of freedom is the integral part of $ft - g$. Mudholkar and Trivedi [1980] give a normal approximation that performs well for small and large m_E; however, it requires much more computation than (2.45).

5. Schatzoff [1966a] gave a method for obtaining the exact distribution function of U when d or m_H is even. He also tabled a conversion factor $C_\alpha = C_\alpha(d, m_H, M)$, where $M = m_E - d + 1$, such that we have the exact relation

$$\text{pr}\left[-C_\alpha^{-1} f \log U \leq z\right] = \text{pr}\left[\chi^2_{dm_H} \leq z\right] = 1 - \alpha \tag{2.46}$$

for specified values of α. Pillai and Gupta [1969] obtained the density and distribution functions of U explicitly for $d = 3, 4, 5, 6$ and m_H odd or even, and made substantial additions to Schatzoff's tables. A modified version of these tables, with helpful methods of interpolation, is given by Pearson and Hartley [1972]. Mathai [1971] gave an explicit expression for the distribution of U and filled some gaps. Infinite series are encountered when d and m_H are both odd, though Lee [1972] was able to reduce the problem to the evaluation of certain univariate integrals. He extended the existing tables so that percentage points were available for $d \leq m_H \leq 20$ and $dm_H \leq 144$, except when d or m_H is odd and greater than 10. Using a differential equation approach, Davis [1979] then filled the remaining gaps: A set of tables is given in Appendix D13. In these tables m_H and d are interchangeable, as $M = n_0 - m_H - d + 1$ is unchanged if we interchange m_H and d. We observe that $C_\alpha > 1$ and $C_\alpha \to 1$ as m_E, and therefore M, tends to infinity. Generally C_α is close to 1.

When $\mathbf{H}$ has a noncentral Wishart distribution, U is said to have a noncentral distribution. An asymptotic expansion up to order m_E^{-2} has been given by Sugiura and Fujikoshi [1969] (see also Pillai [1977] and Muirhead [1982]).

2.5.5 *Summary of Special Distributions*

For convenient reference we now summarize some of the previous distribution theory.

a HOTELLING'S T^2

If $\mathbf{x} \sim N_d(\boldsymbol{\mu}, \boldsymbol{\Sigma})$, $\mathbf{W} \sim W_d(m, \boldsymbol{\Sigma})$, and $\mathbf{x}$ and $\mathbf{W}$ are statistically independent, then

$$T^2 = m(\mathbf{x} - \boldsymbol{\mu})'\mathbf{W}^{-1}(\mathbf{x} - \boldsymbol{\mu}) \sim T^2_{d,m} \qquad (m \geq d),$$

where

$$\frac{T^2_{d,m}}{m} \frac{m - d + 1}{d} \equiv F_{d, m-d+1}. \tag{2.47}$$

b U-STATISTIC

Suppose $\mathbf{E} \sim W_d(m_E, \Sigma)$ $(m_E \geq d)$, $\mathbf{H} \sim W_d(m_H, \Sigma)$, $\mathbf{E}$ and $\mathbf{H}$ are statistically independent and $s = \text{minimum}(m_H, d)$; then

$$U = \frac{|\mathbf{E}|}{|\mathbf{E} + \mathbf{H}|} = \prod_{j=1}^{s} (1 - \theta_j) \sim U_{d, m_H, m_E}. \tag{2.48}$$

The basic parameters underlying the joint distribution of the eigenvalues θ_j are s, $\nu_1 = \frac{1}{2}(|m_H - d| - 1)$, and $\nu_2 = \frac{1}{2}(m_E - d - 1)$. If $s = 1$

$$\frac{1 - U}{U} \frac{\nu_2 + 1}{\nu_1 + 1} \sim F_{2\nu_1+2, 2\nu_2+2}, \tag{2.49}$$

while if $s = 2$,

$$\frac{1 - U^{1/2}}{U^{1/2}} \frac{2\nu_2 + 2}{2\nu_1 + 3} \sim F_{4\nu_1+6, 4\nu_2+4}. \tag{2.50}$$

Tables for finding the upper quantile values for $-f \log U$, where $f = m_E - \frac{1}{2}(d - m_H + 1)$ are given in Appendix D13. A good F-approximation is given by (2.45).

c MAXIMUM ROOT STATISTIC

Let $\theta_{\max}$ be the maximum root of $|\mathbf{H} - \theta(\mathbf{E} + \mathbf{H})| = 0$. Then if $\alpha = \text{pr}[\theta_{\max} \geq \theta_\alpha]$, θ_α is obtained by entering s, ν_1, and ν_2 in Appendix D14.

d TRACE STATISTICS

The Lawley–Hotelling trace statistic is

$$T_g^2 = m_E U^{(s)} = m_E \text{tr}[\mathbf{H}\mathbf{E}^{-1}] = m_E \sum_{j=1}^{s} \frac{\theta_j}{1 - \theta_j}, \tag{2.51}$$

and Pillai's trace statistic is

$$V^{(s)} = \text{tr}\left[\mathbf{H}(\mathbf{E} + \mathbf{H})^{-1}\right] = \sum_{j=1}^{s} \theta_j. \tag{2.52}$$

Percentage points and F-approximations are given in Appendix D (D15 and D16).

e EQUIVALENCE OF STATISTICS WHEN $m_H = 1$

When $m_H = 1$, it transpires that all the above statistics are functions of each other. Since $\text{rank}[\mathbf{H}(\mathbf{E} + \mathbf{H})^{-1}] = \text{rank}[\mathbf{H}] = 1$ (by A2.2), $|\mathbf{H} - \theta(\mathbf{E} + \mathbf{H})| = 0$

has only one nonzero root, which we can call $\theta_{\max}$. Thus

$$U^{(1)} = \theta_{\max}/(1 - \theta_{\max}), \tag{2.53}$$

$$V^{(1)} = \theta_{\max}, \tag{2.54}$$

and, from (2.48),

$$U = 1 - \theta_{\max}. \tag{2.55}$$

Also, since rank [**H**] = 1 and **H** has a Wishart distribution, **H** can be expressed in the form $\mathbf{H} = \mathbf{x}\mathbf{x}'$, where $\mathbf{x} \sim N_d(\mathbf{0}, \boldsymbol{\Sigma})$ and **x** is independent of **E**. Then

$$\begin{aligned} T_g^2 &= m_E U^{(1)} \\ &= m_E \text{tr}[\mathbf{H}\mathbf{E}^{-1}] \\ &= m_E \text{tr}[\mathbf{x}'\mathbf{E}^{-1}\mathbf{x}] \qquad \text{(by A1.1)} \\ &= m_E \mathbf{x}'\mathbf{E}^{-1}\mathbf{x} \\ &= T^2 \sim T^2_{d, m_E}. \end{aligned} \tag{2.56}$$

Combining (2.53) with (2.47) leads to

$$\frac{\theta_{\max}}{1 - \theta_{\max}} \frac{m_E - d + 1}{d} \sim F_{d, m_E - d + 1}.$$

This is another way of writing (2.49). Thus T_g^2 reduces to Hotelling's T^2 when $m_H = 1$. The above result can also be established using the following algebraic technique:

$$\begin{aligned} U &= \frac{|\mathbf{E}|}{|\mathbf{E} + \mathbf{x}\mathbf{x}'|} \\ &= |\mathbf{I}_d + \mathbf{E}^{-1}\mathbf{x}\mathbf{x}'|^{-1} \\ &= \left(1 + \text{tr}[\mathbf{E}^{-1}\mathbf{x}\mathbf{x}']\right)^{-1} \qquad \text{(by A2.4)} \\ &= (1 + \mathbf{x}'\mathbf{E}^{-1}\mathbf{x})^{-1} \qquad (2.57) \\ &= \left(1 + \frac{T^2}{m_E}\right)^{-1} \qquad (2.58) \end{aligned}$$

and $\mathbf{x}'\mathbf{E}^{-1}\mathbf{x} = (1 - U)/U = U^{(1)}$.

For future reference we note from (2.49) and (2.21) that U and hence $V^{(1)}$ have a beta distribution, the latter with $\nu_1 + 1 = \frac{1}{2}d$ and $\nu_2 + 1 = \frac{1}{2}(m_E - d + 1)$ degrees of freedom respectively. Finally, we also have

$$\begin{aligned} V^{(1)} &= U^{(1)}/(U^{(1)} + 1) \\ &= \frac{T^2/m_E}{(T^2/m_E) + 1}. \end{aligned} \tag{2.59}$$

2.5.6 *Factorizations of* U

a PRODUCT OF BETA VARIABLES

Anderson [1958] showed that

$$U = \frac{|\mathbf{E}|}{|\mathbf{E} + \mathbf{H}|} = b_1 b_2 \cdots b_d,$$

where the b_k are independently distributed as beta variables with $\frac{1}{2}(m_E - k + 1)$ and $\frac{1}{2}m_H$ degrees of freedom, respectively. The proof that we give follows Rao [1973].

Let $\mathbf{E}_k$ be the leading $k \times k$ submatrix of $\mathbf{E}$ (with $\mathbf{E}_1 = e_{11}$ and $\mathbf{E}_d = \mathbf{E}$), and let

$$\mathbf{E}_k = \begin{pmatrix} \mathbf{E}_{k-1} & \mathbf{e}_{k-1} \\ \mathbf{e}'_{k-1} & e_{kk} \end{pmatrix}.$$

by A5.9, this matrix is nonsingular as $\mathbf{E} > \mathbf{O}$. Then, from A3.2,

$$\begin{aligned} |\mathbf{E}_k| &= \frac{|\mathbf{E}_{k-1}|\,|\mathbf{E}_k|}{|\mathbf{E}_{k-1}|} \\ &= |\mathbf{E}_{k-1}|\left(e_{kk} - \mathbf{e}'_{k-1}\mathbf{E}_{k-1}^{-1}\mathbf{e}_{k-1}\right) \\ &= |\mathbf{E}_{k-1}|\tilde{e}_{kk}, \end{aligned} \tag{2.60}$$

say, and

$$\begin{aligned} |\mathbf{E}| &= |\mathbf{E}_1|\frac{|\mathbf{E}_2|}{|\mathbf{E}_1|} \cdots \frac{|\mathbf{E}_d|}{|\mathbf{E}_{d-1}|} \\ &= e_{11}\tilde{e}_{22} \cdots \tilde{e}_{dd}. \end{aligned} \tag{2.61}$$

We can use the same decomposition for $|\mathbf{E}_H| = |\mathbf{E} + \mathbf{H}|$ and obtain

$$U = \frac{|\mathbf{E}|}{|\mathbf{E}_H|} = \frac{e_{11}\tilde{e}_{22} \cdots \tilde{e}_{dd}}{e_{H11}\tilde{e}_{H22} \cdots \tilde{e}_{Hdd}} = b_1 b_2 \cdots b_d, \tag{2.62}$$

where $b_k = \tilde{e}_{kk}/\tilde{e}_{Hkk}$ and

$$\tilde{e}_{Hkk} = e_{kk} + h_{kk} - (\mathbf{e}_{k-1} + \mathbf{h}_{k-1})'(\mathbf{E}_{k-1} + \mathbf{H}_{k-1})^{-1}(\mathbf{e}_{k-1} + \mathbf{h}_{k-1}). \tag{2.63}$$

Since $\mathbf{E}$ and $\mathbf{H}$ have independent Wishart distributions, we can find $\mathbf{u}_1, \mathbf{u}_2, \ldots, \mathbf{u}_n$ that are i.i.d. $N_d(\mathbf{0}, \mathbf{\Sigma})$ such that, with $m = m_E$ and $n = m_E + m_H$,

$$\mathbf{E} = \sum_{i=1}^{m} \mathbf{u}_i \mathbf{u}_i' \quad \text{and} \quad \mathbf{H} = \sum_{i=m+1}^{n} \mathbf{u}_i \mathbf{u}_i'. \tag{2.64}$$

From Lemma 2.5, in Section 2.4.1, the conditional distribution of u_{ik} (the kth element of $\mathbf{u}_i$), given $\mathbf{u}_{i(k-1)} = (u_{i1}, u_{i2}, \ldots, u_{i,k-1})'$, is

$$N_1\left(\beta_1 u_{i1} + \beta_2 u_{i2} + \cdots + \beta_{k-1} u_{i,k-1}, \sigma_k^2\right), \tag{2.65}$$

where σ_k^2 is suitably defined. Setting

$$\begin{aligned}
\begin{pmatrix} \mathbf{u}_{1(k)}' \\ \mathbf{u}_{2(k)}' \\ \vdots \\ \mathbf{u}_{n(k)}' \end{pmatrix} &= \left(\begin{array}{c|c} \mathbf{u}_{1(k-1)}' & u_{1k} \\ \mathbf{u}_{2(k-1)}' & u_{2k} \\ \vdots & \vdots \\ \mathbf{u}_{n(k-1)}' & u_{nk} \end{array}\right) \\
&= \left(\begin{array}{cccc|c} u_{11} & u_{12} & \cdots & u_{1,k-1} & u_{1k} \\ u_{21} & u_{22} & \cdots & u_{2,k-1} & u_{2k} \\ \vdots & \vdots & & \vdots & \vdots \\ u_{n1} & u_{n2} & \cdots & u_{n,k-1} & u_{nk} \end{array}\right) \\
&= (\mathbf{K} | \mathbf{y}), \quad \text{say}, \\
&= \left(\begin{array}{c|c} \mathbf{K}_1 & \mathbf{y}_1 \\ \mathbf{K}_2 & \mathbf{y}_2 \end{array}\right),
\end{aligned} \tag{2.66}$$

where $\mathbf{K}_1$ and $\mathbf{y}_1$ have m rows, we have

$$\mathbf{E}_k = \begin{pmatrix} \mathbf{K}_1'\mathbf{K}_1 & \mathbf{K}_1'\mathbf{y}_1 \\ \mathbf{y}_1'\mathbf{K}_1 & \mathbf{y}_1'\mathbf{y}_1 \end{pmatrix} = \begin{pmatrix} \mathbf{E}_{k-1} & \mathbf{e}_{K-1} \\ \mathbf{e}_{k-1}' & e_{kk} \end{pmatrix}. \tag{2.67}$$

Hence, arguing as in Lemmas 2.6 and 2.7, we have

$$\begin{aligned}
\tilde{e}_{kk} &= e_{kk} - \mathbf{e}_{k-1}' \mathbf{E}_{k-1}^{-1} \mathbf{e}_{k-1} \\
&= \mathbf{y}_1'\mathbf{y}_1 - \mathbf{y}_1'\mathbf{K}_1\left(\mathbf{K}_1'\mathbf{K}_1\right)^{-1}\mathbf{K}_1'\mathbf{y}_1 \\
&= \min_{\boldsymbol{\beta}} \|\mathbf{y}_1 - \mathbf{K}_1\boldsymbol{\beta}\|^2 \\
&= \min_{\boldsymbol{\beta}} \sum_{i=1}^{m} \left(u_{ik} - \beta_1 u_{i1} - \cdots - \beta_{k-1} u_{i,k-1}\right)^2,
\end{aligned}$$

and $\tilde{e}_{kk}$ is statistically independent of $\mathbf{K}_1$ (and therefore of $\mathbf{K}$, as the $\mathbf{u}_i$ are independent). Using a similar argument, we also have

$$\tilde{e}_{Hkk} = \min_{\boldsymbol{\beta}} \|\mathbf{y} - \mathbf{K}\boldsymbol{\beta}\|^2,$$

which is independent of $\mathbf{K}$, and

$$b_k = \left(\min_{\boldsymbol{\beta}} \sum_{i=1}^{m} \left(u_{ik} - \boldsymbol{\beta}'\mathbf{u}_{i(k-1)}\right)^2 \right) \Big/ \min_{\boldsymbol{\beta}} \sum_{i=1}^{n} \left(u_{ik} - \boldsymbol{\beta}'\mathbf{u}_{i(k-1)}\right)^2. \tag{2.68}$$

From B3.6, the conditional distribution of b_k, given the $\mathbf{u}_{i(k-1)}$ $(i = 1, 2, \dots, n)$, is beta with $\frac{1}{2}(m_E - k + 1)$ and $\frac{1}{2}m_H$ degrees of freedom. As this distribution does not depend on the $\mathbf{u}_{i(k-1)}$, it is also the unconditional distribution and b_k is independent of the $\mathbf{u}_{i(k-1)}$. Hence b_k is independent of $b_1, b_2, \dots, b_{k-1}$, these being functions of the $\mathbf{u}_{i(k-1)}$. The joint density function of the b_k is then

$$\begin{aligned} f(b_1, b_2, \dots, b_d) &= f_1(b_1) f_2(b_2|b_1) \cdots f_d(b_d|b_1, b_2, \dots, b_{d-1}) \\ &= f_1(b_1) f_2(b_2) \cdots f_d(b_d), \end{aligned}$$

and the b_k are mutually independent beta random variables.

We now show that the b_k can be computed using the Cholesky decomposition method of Section 10.1.1a. Let $\mathbf{E} = \mathbf{T}'\mathbf{T}$ and $\mathbf{E}_H = \mathbf{V}'\mathbf{V}$ be the (unique) Cholesky decompositions of $\mathbf{E}$ and $\mathbf{E}_H$. If $\mathbf{T}_k$ is the $k \times k$ leading submatrix of $\mathbf{T}$, then, from the uniqueness of $\mathbf{T}$ [see (10.5)], $\mathbf{E}_k = \mathbf{T}_k'\mathbf{T}_k$ and (see A5.11)

$$|\mathbf{E}_k| = |\mathbf{T}_k|^2 = t_{11}^2 \cdots t_{kk}^2,$$

where t_{kk} is the kth diagonal element of $\mathbf{T}$. Hence

$$\tilde{e}_{kk} = \frac{|\mathbf{E}_k|}{|\mathbf{E}_{k-1}|} = t_{kk}^2 \tag{2.69}$$

and

$$b_k = \frac{t_{kk}^2}{v_{kk}^2}. \tag{2.70}$$

From the properties of the beta distribution [see (2.21)]

$$\begin{aligned} F_k &= \frac{m_E - k + 1}{m_H} \left(\frac{v_{kk}^2 - t_{kk}^2}{t_{kk}^2} \right) \qquad (2.71) \\ &= \frac{m_E - k + 1}{m_H} \left(\frac{1}{b_k} - 1 \right) \\ &\sim F_{m_H, m_E - k + 1}. \qquad (2.72) \end{aligned}$$

Also, writing $\Lambda_k = b_1 b_2 \cdots b_k$, we note that

$$F_k = \frac{m_E - k + 1}{m_H}\left(\frac{\Lambda_{k-1}}{\Lambda_k} - 1\right). \tag{2.73}$$

When $m_H = 1$ we see, from (2.58), that

$$\frac{|\mathbf{E}_k|}{|\mathbf{E}_{Hk}|} = \left(1 + \frac{T_k^2}{m_E}\right)^{-1},$$

where T_k^2 has a Hotelling's T^2 distribution. Hence

$$b_k = \frac{|\mathbf{E}_k|}{|\mathbf{E}_{Hk}|}\frac{|\mathbf{E}_{H,k-1}|}{|\mathbf{E}_{k-1}|} = \frac{1 + T_{k-1}^2/m_E}{1 + T_k^2/m_E}, \tag{2.74}$$

and, from (2.72),

$$F_k = \frac{m_E - k + 1}{1}\frac{T_k^2 - T_{k-1}^2}{m_E + T_{k-1}^2} \sim F_{1, m_E - k + 1}. \tag{2.75}$$

Later in this book we shall consider hypothesis testing so that **H** becomes a "hypothesis" matrix, and **E** an "error" matrix. Under the null hypothesis **H** has a Wishart distribution; otherwise **H** has a noncentral distribution. The statistic F_k (or b_k) can be then used to test the null hypothesis based on k variables, given that the null hypothesis based on $k - 1$ variables is true: Thus F_k provides a test for "additional information" (see Section 9.5.3). For further details of the above decomposition of U see Hawkins [1976].

b PRODUCT OF TWO U-STATISTICS

If

$$\mathbf{E} = \begin{pmatrix} \mathbf{E}_{11} & \mathbf{E}_{12} \\ \mathbf{E}_{21} & \mathbf{E}_{22} \end{pmatrix} \begin{matrix} \}\, d_1 \\ \}\, d_2 \end{matrix},$$

and $\mathbf{E}_H$ is partitioned in a similar fashion, then, by A3.2,

$$U = \frac{|\mathbf{E}|}{|\mathbf{E}_H|} = \frac{|\mathbf{E}_{11}|}{|\mathbf{E}_{H11}|}\frac{|\mathbf{E}_{22} - \mathbf{E}_{21}\mathbf{E}_{11}^{-1}\mathbf{E}_{12}|}{|\mathbf{E}_{H22} - \mathbf{E}_{H21}\mathbf{E}_{H11}^{-1}\mathbf{E}_{H12}|} = b_{(1)}b_{(2)}, \tag{2.76}$$

say. Writing

$$\begin{pmatrix} \mathbf{u}_1' \\ \mathbf{u}_2' \\ \vdots \\ \mathbf{u}_n' \end{pmatrix} = \left(\begin{array}{cccc|c} u_{11} & u_{12} & \cdots & u_{1d_1} & u_{1,d_1+1}, \ldots, u_{1d} \\ u_{21} & u_{22} & \cdots & u_{2d_1} & u_{2,d_1+1}, \ldots, u_{2d} \\ \vdots & \vdots & & \vdots & \vdots \quad \vdots \\ u_{n1} & u_{n2} & \cdots & u_{nd_1} & u_{n,d_1+1}, \ldots, u_{nd} \end{array}\right)$$

$$(d = d_1 + d_2)$$

$$= (\mathbf{K}|\mathbf{Y}), \tag{2.77}$$

say, we can apply B3.7 and show that, conditional on **K**, $b_{(2)} \sim U_{d_2, m_H, m_E - d_1}$. Since the latter distribution does not depend on **K**, it is also the unconditional distribution, and $b_{(2)}$ is independent of **K**. Hence $b_{(2)}$ is independent of $\mathbf{E}_{H11} = \mathbf{K}'\mathbf{K}$ and $\mathbf{E}_{11} = \mathbf{K}_1'\mathbf{K}_1$, and therefore of $b_{(1)} = |\mathbf{E}_{11}|/|\mathbf{E}_{H11}|$: Here $\mathbf{K}_1$ is the first m rows of **K**. Since $\mathbf{E}_{11}$ and $\mathbf{H}_{11}$ have independent d_1-dimensional Wishart distributions (Exercise 2.14), it follows that $b_{(1)} \sim U_{d_1, m_H, m_E}$. Hence, from (2.76), we can write symbolically

$$U_{d, m_H, m_E} \equiv U_{d_1, m_H, m_E} U_{d_2, m_H, m_E - d_1} \qquad (d = d_1 + d_2). \tag{2.78}$$

When $m_H = 1$, we can argue as in (2.74) and obtain

$$b_{(2)} = \frac{U}{b_{(1)}} = \frac{1 + T_{d_1}^2/m_E}{1 + T_d^2/m_E}.$$

From (2.42) with d and m_E replaced by d_2 and $m_E - d_1$, it follows that

$$\frac{m_E - d + 1}{d_2}\left(\frac{1}{b_{(2)}} - 1\right) = \frac{m_E - d + 1}{d_2} \frac{T_d^2 - T_{d_1}^2}{m_E + T_{d_1}^2} \tag{2.79}$$

$$\sim F_{d_2, m_E - d + 1}. \tag{2.80}$$

Also, $b_{(2)}$ is independent of $T_{d_1}^2$, the latter being a function of $b_{(1)}$.

The distribution of (2.79) is considered again in Section 2.6 using a different approach and different notation. There $\mathbf{E} = \mathbf{W}$ and $\mathbf{H} = \mathbf{x}\mathbf{x}'$, where $\mathscr{E}[\mathbf{x}] = \boldsymbol{\mu} \neq \mathbf{0}$, that is, **H** now has a noncentral Wishart distribution. Suppose

$$\mathbf{x} = \begin{pmatrix} \mathbf{x}^{(1)} \\ \mathbf{x}^{(2)} \end{pmatrix}, \qquad \boldsymbol{\mu} = \begin{pmatrix} \boldsymbol{\mu}^{(1)} \\ \boldsymbol{\mu}^{(2)} \end{pmatrix}, \qquad \boldsymbol{\Sigma} = \begin{pmatrix} \boldsymbol{\Sigma}_{11} & \boldsymbol{\Sigma}_{12} \\ \boldsymbol{\Sigma}_{21} & \boldsymbol{\Sigma}_{22} \end{pmatrix},$$

where $\mathbf{x}^{(1)}$ and $\boldsymbol{\mu}^{(1)}$ are $d_1 \times 1$ vectors and $\boldsymbol{\Sigma}_{11}$ is a $d_1 \times d_1$ matrix. Then, from Theorem 2.1(viii) of Section 2.2,

$$\begin{aligned} \mathscr{E}\left[\mathbf{x}^{(2)}|\mathbf{x}^{(1)}\right] &= \boldsymbol{\mu}^{(2)} - \boldsymbol{\Sigma}_{21}\boldsymbol{\Sigma}_{11}^{-1}\boldsymbol{\mu}^{(1)} + \boldsymbol{\Sigma}_{21}\boldsymbol{\Sigma}_{11}^{-1}\mathbf{x}^{(1)} \\ &= \boldsymbol{\beta}_0 + \boldsymbol{\Sigma}_{21}\boldsymbol{\Sigma}_{11}^{-1}\mathbf{x}^{(1)}, \end{aligned}$$

say. For each of the vectors $\mathbf{u}_1, \ldots, \mathbf{u}_m$, the same conditioning leads to $\boldsymbol{\beta}_0 = \mathbf{0}$ so that $\mathbf{x}$ $(= \mathbf{u}_{m+1})$ can be treated on the same footing, provided that

$$\boldsymbol{\mu}^{(2)} - \boldsymbol{\Sigma}_{21}\boldsymbol{\Sigma}_{11}^{-1}\boldsymbol{\mu}^{(1)} = \mathbf{0}. \tag{2.81}$$

Under this condition, the crucial step (2.65) holds for $\mathbf{x}$ as well; thus (2.80) follows once again, provided that (2.81) holds.

A "direct" proof of (2.80), when (2.81) is true, is given in Section 2.6. Although, in the light of the above proof outline, it may be omitted at a first reading, it does provide a useful pedagogical exercise in handling multivariate distributions.

2.6 RAO'S DISTRIBUTION

Before stating the main result of this section, we will prove the following lemma.

LEMMA 2.10 Let

$$\mathbf{W} = \begin{pmatrix} \mathbf{W}_{11} & \mathbf{W}_{12} \\ \mathbf{W}_{21} & \mathbf{W}_{22} \end{pmatrix} \sim W_d(m, \boldsymbol{\Sigma}), \qquad m \geq d, \quad . \tag{2.82}$$

where $\mathbf{W}_{11}$ is $d_1 \times d_1$ $(d_1 < d)$. If $\boldsymbol{\Sigma}$ is partitioned in the same way as $\mathbf{W}$, $\boldsymbol{\Sigma}_{22\cdot 1} = \boldsymbol{\Sigma}_{22} - \boldsymbol{\Sigma}_{21}\boldsymbol{\Sigma}_{11}^{-1}\boldsymbol{\Sigma}_{12}$, and $d_2 = d - d_1$, then

$$\mathbf{W}_{22\cdot 1} = \mathbf{W}_{22} - \mathbf{W}_{21}\mathbf{W}_{11}^{-1}\mathbf{W}_{12} \sim W_{d_2}(m - d_1, \boldsymbol{\Sigma}_{22\cdot 1}),$$

and $\mathbf{W}_{22\cdot 1}$ is statistically independent of $(\mathbf{W}_{11}, \mathbf{W}_{12})$.

Proof Arguing as in Theorem 2.2, we can assume $\mathbf{W} = \sum_{i=1}^m \mathbf{x}_i\mathbf{x}_i'$, where the $\mathbf{x}_i$ are i.i.d $N_d(\mathbf{0}, \boldsymbol{\Sigma})$. Given the partition $\mathbf{x}_i' = (\mathbf{u}_i', \mathbf{v}_i')$, where $\mathbf{u}_i$ is $d_1 \times 1$, we have, corresponding to (2.82),

$$\mathbf{W} = \begin{pmatrix} \sum_i \mathbf{u}_i\mathbf{u}_i' & \sum_i \mathbf{u}_i\mathbf{v}_i' \\ \sum_i \mathbf{v}_i\mathbf{u}_i' & \sum_i \mathbf{v}_i\mathbf{v}_i' \end{pmatrix} = \begin{pmatrix} \mathbf{U}'\mathbf{U} & \mathbf{U}'\mathbf{V} \\ \mathbf{V}'\mathbf{U} & \mathbf{V}'\mathbf{V} \end{pmatrix}, \tag{2.83}$$

say, where $\mathbf{U}' = (\mathbf{u}_1, \ldots, \mathbf{u}_m)$, and so on. Then

$$\begin{aligned} \mathbf{W}_{22\cdot 1} &= \mathbf{V}'\mathbf{V} - \mathbf{V}'\mathbf{U}(\mathbf{U}'\mathbf{U})^{-1}\mathbf{U}'\mathbf{V} \\ &= \mathbf{V}'(\mathbf{I}_m - \mathbf{P})\mathbf{V}, \end{aligned}$$

say, where $\mathbf{I}_m - \mathbf{P} = \mathbf{I}_m - \mathbf{U}(\mathbf{U}'\mathbf{U})^{-1}\mathbf{U}'$ is a symmetric idempotent (projection) matrix of rank $m - d_1$, with probability 1 (see Appendix B1). Let

$$\mathbf{v}_{i\cdot 1} = \mathbf{v}_i - \boldsymbol{\Sigma}_{21}\boldsymbol{\Sigma}_{11}^{-1}\mathbf{u}_i$$

and

$$\mathbf{V}_{2\cdot 1} = (\mathbf{v}_{1.1}, \ldots, \mathbf{v}_{m\cdot 1})' = \mathbf{V} - \mathbf{U}\boldsymbol{\Sigma}_{11}^{-1}\boldsymbol{\Sigma}_{12}.$$

Then, since $\mathbf{PU} = \mathbf{U}$,

$$(\mathbf{I}_m - \mathbf{P})\mathbf{V}_{2\cdot 1} = (\mathbf{I}_m - \mathbf{P})\mathbf{V} \tag{2.84}$$

and

$$\begin{aligned}\mathbf{W}_{22\cdot 1} &= \mathbf{V}_{2\cdot 1}'(\mathbf{I}_m - \mathbf{P})'(\mathbf{I}_m - \mathbf{P})\mathbf{V}_{2\cdot 1}\\ &= \mathbf{V}_{2\cdot 1}'(\mathbf{I}_m - \mathbf{P})\mathbf{V}_{2\cdot 1}.\end{aligned} \tag{2.85}$$

To handle the distribution theory we first of all condition on $\mathbf{U}$. By Theorem 2.1(viii), with $\boldsymbol{\theta} = \mathbf{0}$,

$$\mathbf{v}_i|\mathbf{u}_i \sim N_{d_2}\left(\boldsymbol{\Sigma}_{21}\boldsymbol{\Sigma}_{11}^{-1}\mathbf{u}_i, \boldsymbol{\Sigma}_{22\cdot 1}\right), \tag{2.86}$$

and the rows $\mathbf{v}_{i\cdot 1}'$ of $\mathbf{V}_{2\cdot 1}$ are i.i.d. $N_{d_2}(\mathbf{0}, \boldsymbol{\Sigma}_{22\cdot 1})$. Since $\mathbf{P}$ is now constant in (2.85), it follows from Theorem 2.4, Corollary 1 in Section 2.3.2 that $\mathbf{W}_{22\cdot 1}$ is conditionally $W_{d_2}(m - d_1, \boldsymbol{\Sigma}_{22\cdot 1})$. Since this distribution does not depend on the $\mathbf{u}_i$, $\mathbf{W}_{22\cdot 1}$ is unconditionally $W_{d_2}(m - d_1, \boldsymbol{\Sigma}_{22\cdot 1})$ and is independent of $\mathbf{U}$.

From $\mathbf{P}(\mathbf{I}_m - \mathbf{P}) = \mathbf{O}$ and Exercise 2.18,

$$\mathbf{PV}_{2.1} = \mathbf{PV} - \mathbf{U}\boldsymbol{\Sigma}_{11}^{-1}\boldsymbol{\Sigma}_{12} \tag{2.87}$$

is conditionally independent of $(\mathbf{I}_m - \mathbf{P})\mathbf{V}_{2.1}$, and therefore of $\mathbf{W}_{22\cdot 1}$ [by (2.85)]. We now have the following density function factorization:

$$\begin{aligned}f(\mathbf{W}_{22\cdot 1}, \mathbf{PV}_{2\cdot 1}, \mathbf{U}) &= f(\mathbf{W}_{22\cdot 1}, \mathbf{PV}_{2\cdot 1}|\mathbf{U})f_3(\mathbf{U})\\ &= f_1(\mathbf{W}_{22\cdot 1}|\mathbf{U})f_2(\mathbf{PV}_{2.1}|\mathbf{U})f_3(\mathbf{U})\\ &= f_1(\mathbf{W}_{22.1})f_{23}(\mathbf{PV}_{2\cdot 1}, \mathbf{U}),\end{aligned}$$

and $\mathbf{W}_{22\cdot 1}$ is statistically independent of $(\mathbf{PV}_{2\cdot 1}, \mathbf{U})$. Since $\mathbf{W}_{11} = \mathbf{U}'\mathbf{U}$ and [from (2.87)]

$$\mathbf{W}_{12} = \mathbf{U}'\mathbf{V} = \mathbf{U}'\mathbf{PV} = \mathbf{U}'\mathbf{PV}_{2\cdot 1} + \mathbf{U}'\mathbf{U}\boldsymbol{\Sigma}_{11}^{-1}\boldsymbol{\Sigma}_{12}$$

are functions of $\mathbf{U}$ and $\mathbf{PV}_{2\cdot 1}$, $\mathbf{W}_{22\cdot 1}$ is independent of $(\mathbf{W}_{11}, \mathbf{W}_{12})$.

COROLLARY If $\boldsymbol{\Sigma}_{12} = \mathbf{O}$, $\mathbf{A} = \mathbf{W}_{21}\mathbf{W}_{11}^{-1}\mathbf{W}_{12} \sim W_{d_2}(d_1, \boldsymbol{\Sigma}_{22})$ and $\mathbf{A}$ is independent of $\mathbf{W}_{22\cdot 1}$.

Proof Since $\mathbf{A}$ is a function of $\mathbf{W}_{11}$ and $\mathbf{W}_{12}$ ($= \mathbf{W}_{21}'$), it is independent of $\mathbf{W}_{22\cdot 1}$ by the above lemma. When $\boldsymbol{\Sigma}_{12} = \mathbf{O}$, then $\boldsymbol{\Sigma}_{22\cdot 1} = \boldsymbol{\Sigma}_{22}$, $\mathbf{V}_{2\cdot 1} = \mathbf{V}$, and

$$\mathbf{W}_{21}\mathbf{W}_{11}^{-1}\mathbf{W}_{12} = \mathbf{V}'\mathbf{PV} = \mathbf{V}_{2\cdot 1}'\mathbf{PV}_{2\cdot 1}.$$

Conditional on $\mathbf{U}$, the rows of $\mathbf{V}_{2\cdot 1}$ are i.i.d. $N_{d_2}(\mathbf{0}, \boldsymbol{\Sigma}_{22})$ [see (2.86)]. Since $\mathbf{P}$ is a projection matrix of rank d_1, the conditional distribution of $\mathbf{V}_{2\cdot 1}'\mathbf{P}\mathbf{V}_{2\cdot 1}$ given $\mathbf{U}$ is $W_{d_2}(d_1, \boldsymbol{\Sigma}_{22})$. Since this distribution does not depend on $\mathbf{U}$, it is also the unconditional distribution. □

Suppose that $\mathbf{x} \sim N_d(\boldsymbol{\mu}, \boldsymbol{\Sigma})$, $\mathbf{W} \sim W_d(m, \boldsymbol{\Sigma})$, $\mathbf{x}$ is independent of $\mathbf{W}$, and we partition both $\mathbf{x} = (\mathbf{x}^{(1)\prime}, \mathbf{x}^{(2)\prime})'$ and $\boldsymbol{\mu} = (\boldsymbol{\mu}^{(1)\prime}, \boldsymbol{\mu}^{(2)\prime})'$ into $d_1 \times 1$ and $d_2 \times 1$ vectors, respectively. Later on we shall be interested in testing the hypothesis H_0 that the so-called population Mahalinobis distance squared for $\mathbf{x}$ is the same as that based on $\mathbf{x}^{(1)}$, namely, $\Delta_d^2 = \Delta_{d_1}^2$, where $\Delta_d^2 = \boldsymbol{\mu}'\boldsymbol{\Sigma}^{-1}\boldsymbol{\mu}$ and $\Delta_{d_1}^2 = \boldsymbol{\mu}^{(1)\prime}\boldsymbol{\Sigma}_{11}^{-1}\boldsymbol{\mu}^{(1)}$. From Exercise 2.20, we see that

$$\Delta_d^2 - \Delta_{d_1}^2 = \boldsymbol{\mu}_{2\cdot 1}'\boldsymbol{\Sigma}_{22\cdot 1}^{-1}\boldsymbol{\mu}_{2\cdot 1}, \tag{2.88}$$

where $\boldsymbol{\mu}_{2\cdot 1} = \boldsymbol{\mu}^{(2)} - \boldsymbol{\Sigma}_{21}\boldsymbol{\Sigma}_{11}^{-1}\boldsymbol{\mu}^{(1)}$, and H_0 is true if and only if $\boldsymbol{\mu}_{2\cdot 1} = \mathbf{0}$ (since $\boldsymbol{\Sigma}_{22\cdot 1} > \mathbf{O}$). Sample versions of the above distances are $T_d^2 = m\mathbf{x}'\mathbf{W}^{-1}\mathbf{x}$ and $T_{d_1}^2 = m\mathbf{x}^{(1)\prime}\mathbf{W}_{11}^{-1}\mathbf{x}^{(1)}$, and by the same algebra that led to (2.88) we have

$$T_d^2 - T_{d_1}^2 = m\mathbf{x}_{2\cdot 1}'\mathbf{W}_{22\cdot 1}^{-1}\mathbf{x}_{2\cdot 1}, \tag{2.89}$$

where

$$\mathbf{x}_{2\cdot 1} = \mathbf{x}^{(2)} - \mathbf{W}_{21}\mathbf{W}_{11}^{-1}\mathbf{x}^{(1)}. \tag{2.90}$$

THEOREM 2.11 When $H_0 : \boldsymbol{\mu}^{(2)} - \boldsymbol{\Sigma}_{21}\boldsymbol{\Sigma}_{11}^{-1}\boldsymbol{\mu}^{(1)} = \mathbf{0}$ is true, and $m \geq d$,

$$\frac{T_d^2 - T_{d_1}^2}{m + T_{d_1}^2}\,\frac{m - d + 1}{d_2} \sim F_{d_2, m-d+1} \qquad (d_2 = d - d_1). \tag{2.91}$$

The above statistic is independent of $T_{d_1}^2$, H_0 true or false.

Proof Let $\mathbf{W}$ be partitioned in the form (2.82) and (2.83), and consider the conditional distribution of $T_d^2 - T_{d_1}^2$ given $\mathbf{x}^{(1)}$ and $\mathbf{U}$. From Theorem 2.1(viii) in Section 2.2 we have, conditionally, $\mathbf{x}^{(2)}|\mathbf{x}^{(1)} \sim N_{d_2}(\boldsymbol{\mu}^{(2)} + \boldsymbol{\Sigma}_{21}\boldsymbol{\Sigma}_{11}^{-1}[\mathbf{x}^{(1)} - \boldsymbol{\mu}^{(1)}], \boldsymbol{\Sigma}_{22\cdot 1})$. Looking at (2.90), we note that

$$\begin{aligned}\mathbf{W}_{21}\mathbf{W}_{11}^{-1}\mathbf{x}^{(1)} &= \mathbf{V}'\mathbf{U}\mathbf{W}_{11}^{-1}\mathbf{x}^{(1)}\\ &= \mathbf{V}'\mathbf{a}, \quad \text{say},\\ &= a_1\mathbf{v}_1 + \cdots + a_m\mathbf{v}_m.\end{aligned}$$

Since $\mathbf{x}$ is independent of $\mathbf{W}$, $\mathbf{x}^{(2)}$ is independent of each $\mathbf{v}_i$. Also the $\mathbf{v}_i$ are mutually independent, so that $\mathbf{x}_{2\cdot 1}$ of (2.90) is conditionally MVN, being a linear combination of $m + 1$ independent MVNs (see Exercise 2.6). Now, by (2.86),

$$\mathscr{E}\left[\mathbf{V}'\mathbf{U}\mathbf{W}_{11}^{-1}\mathbf{x}^{(1)}|\mathbf{x}^{(1)}, \mathbf{U}\right] = \boldsymbol{\Sigma}_{21}\boldsymbol{\Sigma}_{11}^{-1}\mathbf{U}'\mathbf{U}\mathbf{W}_{11}^{-1}\mathbf{x}^{(1)} = \boldsymbol{\Sigma}_{21}\boldsymbol{\Sigma}_{11}^{-1}\mathbf{x}^{(1)},$$

and

$$\mathscr{D}[\mathbf{v}_i|\mathbf{U}] = \boldsymbol{\Sigma}_{22\cdot 1}.$$

Hence, from (2.90)

$$\begin{aligned}\mathscr{E}\left[\mathbf{x}_{2\cdot 1}|\mathbf{x}^{(1)},\mathbf{U}\right] &= \boldsymbol{\mu}^{(2)} + \boldsymbol{\Sigma}_{21}\boldsymbol{\Sigma}_{11}^{-1}\left(\mathbf{x}^{(1)} - \boldsymbol{\mu}^{(1)}\right) - \boldsymbol{\Sigma}_{21}\boldsymbol{\Sigma}_{11}^{-1}\mathbf{x}^{(1)} \\ &= \boldsymbol{\mu}_{2\cdot 1} = \mathbf{0}\end{aligned}$$

and

$$\begin{aligned}\mathscr{D}\left[\mathbf{x}_{2\cdot 1}|\mathbf{x}^{(1)},\mathbf{U}\right] &= \boldsymbol{\Sigma}_{22\cdot 1} + \sum_i a_i^2\boldsymbol{\Sigma}_{22\cdot 1} \\ &= \boldsymbol{\Sigma}_{22\cdot 1}(1 + \mathbf{a}'\mathbf{a}),\end{aligned}$$

where

$$\begin{aligned}\mathbf{a}'\mathbf{a} &= \left(\mathbf{U}\mathbf{W}_{11}^{-1}\mathbf{x}^{(1)}\right)'\left(\mathbf{U}\mathbf{W}_{11}^{-1}\mathbf{x}^{(1)}\right) \\ &= \mathbf{x}^{(1)\prime}\mathbf{W}_{11}^{-1}\mathbf{W}_{11}\mathbf{W}_{11}^{-1}\mathbf{x}^{(1)} \\ &= \mathbf{x}^{(1)\prime}\mathbf{W}_{11}^{-1}\mathbf{x}^{(1)} \\ &= T_{d_1}^2/m.\end{aligned} \tag{2.92}$$

Thus, conditionally, $\mathbf{x}_{2\cdot 1} \sim N_{d_2}(\mathbf{0}, \boldsymbol{\Sigma}_{22\cdot 1}[1 + \mathbf{a}'\mathbf{a}])$ and $\mathbf{y} = \mathbf{x}_{2\cdot 1}/(1 + \mathbf{a}'\mathbf{a})^{1/2} \sim N_{d_2}(\mathbf{0}, \boldsymbol{\Sigma}_{22\cdot 1})$. Since this distribution does not depend on $\mathbf{a}'\mathbf{a}$, it is also the unconditional distribution, and $\mathbf{y}$ is independent of $T_{d_1}^2$. By Lemma 2.10, $\mathbf{W}_{22\cdot 1}$ is independent of $(T_{d_1}^2, \mathbf{y})$, as the latter are functions of $\mathbf{W}_{11}$, $\mathbf{W}_{12}$, and $\mathbf{x}$. Hence

$$\begin{aligned}f\left(\mathbf{W}_{22\cdot 1}, \mathbf{y}, T_{d_1}^2\right) &= f_1(\mathbf{W}_{22\cdot 1})f_{23}\left(\mathbf{y}, T_{d_1}^2\right) \\ &= f_1(\mathbf{W}_{22\cdot 1})f_2(\mathbf{y})f_3\left(T_{d_1}^2\right)\end{aligned} \tag{2.93}$$

and $\mathbf{W}_{22\cdot 1}$, $\mathbf{y}$, and $T_{d_1}^2$ are mutually independent. Since $\mathbf{W}_{22\cdot 1} \sim W_{d_2}(m - d_1, \boldsymbol{\Sigma}_{22\cdot 1})$, by Lemma 2.10, it follows from Theorem 2.8 and (2.92) that

$$(m - d_1)\mathbf{y}'\mathbf{W}_{22\cdot 1}^{-1}\mathbf{y} = \frac{(m - d_1)\mathbf{x}_{2\cdot 1}'\mathbf{W}_{22\cdot 1}^{-1}\mathbf{x}_{2\cdot 1}}{1 + \left(T_{d_1}^2/m\right)}$$

has a Hotelling's T^2 distribution with d_2 and $m - d_1$ degrees of freedom. Hence, using (2.89) and (2.20) [the latter with m and d replaced by $m - d_1$ and d_2],

$$\frac{T_d^2 - T_{d_1}^2}{m + T_{d_1}^2}\,\frac{m - d + 1}{d_2} = \mathbf{y}'\mathbf{W}_{22\cdot 1}^{-1}\mathbf{y}\frac{m - d + 1}{d_2} \sim F_{d_2, m-d+1}$$

and is independent of $T_{d_1}^2$ [by (2.93)]. This independence still carries over if H_0 is false, that is, if the conditional mean of $\mathbf{x}_{2\cdot 1}$, namely, $\boldsymbol{\mu}_{2\cdot 1}$, is not zero. □

We note that the statistic in (2.91) is proportional to $(1 - u)/u$, where $u = (1 + T_d^2/m)/(1 + T_{d_1}^2/m)$ is a statistic due to Rao (see Rao [1948)). Also, if $\boldsymbol{\mu}^{(1)} = \mathbf{0}$, then H_0 becomes $\boldsymbol{\mu}^{(2)} = \mathbf{0}$. Other tests of H_0 are considered by Subrahmaniam and Subrahmaniam [1976], and Kariya and Kanazawa (1978]. A generalization of Theorem 2.11 is described in Section 9.5.3.

2.7 MULTIVARIATE SKEWNESS AND KURTOSIS

In addition to the mean $\boldsymbol{\mu}$ and dispersion matrix $\boldsymbol{\Sigma}$ for a d-dimensional random $\mathbf{x}$, it is convenient, particularly for normality and robustness studies, to have multivariate measures of shewness and kurtosis. Various measures are possible, but the following, due to Mardia [1970, 1974, 1975], are particularly useful, namely,

$$\beta_{1,d} = \mathscr{E}\left[\{(\mathbf{x} - \boldsymbol{\mu})'\boldsymbol{\Sigma}^{-1}(\mathbf{y} - \boldsymbol{\mu})\}^3\right] \tag{2.94}$$

and

$$\beta_{2,d} = \mathscr{E}\left[\{(\mathbf{x} - \boldsymbol{\mu})'\boldsymbol{\Sigma}^{-1}(\mathbf{x} - \boldsymbol{\mu})\}^2\right], \tag{2.95}$$

where $\mathbf{y}$ is independent of $\mathbf{x}$ but has the same distribution. When $\mathbf{x}$ is $N_d(\boldsymbol{\mu}, \boldsymbol{\Sigma})$, $\beta_{1,d} = 0$ and $\beta_{2,d} = d(d+2)$ (Exercise 2.26).

The above measures are natural generalizations of the usual univariate measures:

$$\begin{aligned}\sqrt{\beta_1} &= \sqrt{\beta_{1,1}} \\ &= \left\{\mathrm{E}\left[\frac{(x-\mu)^3(y-\mu)^3}{\sigma^6}\right]\right\}^{1/2} \\ &= \frac{\{\mathrm{E}[(x-\mu)^3]\}^{1/2}\{\mathrm{E}[(y-\mu)^3]\}^{1/2}}{\sigma^3} \\ &= \frac{\mu_3}{\sigma^3},\end{aligned} \tag{2.96}$$

and

$$\begin{aligned}\beta_2 &= \beta_{2,1} \\ &= \mathrm{E}\left[\frac{(x-\mu)^4}{\sigma^4}\right] \\ &= \frac{\mu_4}{\sigma^4}.\end{aligned} \tag{2.97}$$

In theoretical studies a more common measure of univariate kurtosis is $\gamma_2 = \beta_2 - 3$.

If $\mathbf{x}_1, \mathbf{x}_2, \ldots, \mathbf{x}_n$ are a random sample from the underlying distribution of $\mathbf{x}$, then natural sample estimates of skewness and kurtosis are

$$b_{1,d} = \frac{1}{n^2} \sum_{h=1}^{n} \sum_{i=1}^{n} \{(\mathbf{x}_h - \overline{\mathbf{x}})'\hat{\mathbf{\Sigma}}^{-1}(\mathbf{x}_i - \overline{\mathbf{x}})\}^3 \tag{2.98}$$

and

$$b_{2,d} = \frac{1}{n} \sum_{i=1}^{n} \{(\mathbf{x}_i - \overline{\mathbf{x}})'\hat{\mathbf{\Sigma}}^{-1}(\mathbf{x}_i - \overline{\mathbf{x}})\}^2, \tag{2.99}$$

where $\hat{\mathbf{\Sigma}} = \mathbf{Q}/n = \Sigma(\mathbf{x}_i - \overline{\mathbf{x}})(\mathbf{x}_i - \overline{\mathbf{x}})'/n$. An algorithm for computing these measures is given by Mardia and Zemrock [1975a]. When the data are normal,

$$\mathrm{E}[b_{1,d}] = 0 \quad \text{and} \quad \mathrm{E}[b_{2,d}] = \frac{d(d+2)(n-1)}{n+1}.$$

These estimates are considered further in Section 4.3.2.

EXERCISES 2

2.1 Suppose that the random variables x_i $(i = 1, 2, \ldots, n)$ are independently distributed as $N_1(\mu_i, \sigma^2)$. Then $\Sigma_{i=1}^n x_i^2/\sigma^2$ is said to have a noncentral chi-square distribution with n degrees of freedom and noncentrality parameter $\delta = \Sigma_{i=1}^n \mu_i^2/\sigma^2$. If $\mathbf{y} \sim N_d(\mathbf{\theta}, \mathbf{\Sigma})$, prove that $\mathbf{y}'\mathbf{\Sigma}^{-1}\mathbf{y}$ has a noncentral chi-square distribution with d degrees of freedom and noncentrality parameter $\mathbf{\theta}'\mathbf{\Sigma}^{-1}\mathbf{\theta}$.

2.2 Using moment-generating functions, show that Definitions 1a and 1b in Section 2.2 are equivalent.

2.3 Given Definition 1b, prove Theorem 2.1(i)–(iv).

2.4 Let $\mathbf{y}$ be a random vector with $\mathscr{E}[\mathbf{y}] = \mathbf{\theta}$ and $\mathscr{D}[\mathbf{y}] = \mathbf{\Sigma}$. If $\mathbf{\Sigma} > \mathbf{O}$, show that there do not exist nonzero constants $\mathbf{a}$ and b such that $\mathbf{a}'\mathbf{y} = b$. If $\mathbf{\Sigma}$ is singular, show that there exists $\mathbf{a}$, $\mathbf{a} \neq \mathbf{0}$, such that $\mathbf{a}'(\mathbf{y} - \mathbf{\theta}) = 0$.

2.5 Consider the regression model $y_i = \beta_0 + \beta_1 x_{i1} + \cdots + \beta_{p-1} x_{i,p-1} + \varepsilon_i$ $(i = 1, 2, \ldots, n)$, or $\mathbf{y} = \mathbf{X}\boldsymbol{\beta} + \boldsymbol{\varepsilon}$, where $\mathbf{X}$ is $n \times p$ of rank p $(< n)$ and $\boldsymbol{\varepsilon} \sim N_n(\mathbf{0}, \sigma^2\mathbf{I}_n)$. Let $\mathbf{e} = (\mathbf{I}_n - \mathbf{P})\mathbf{y}$, where $\mathbf{P} = \mathbf{X}(\mathbf{X}'\mathbf{X})^{-1}\mathbf{X}'$, be the vector of residuals. Show the following:

(a) $\mathscr{D}[\mathbf{e}]$ is singular.

(b) $\mathbf{e}$ has a singular multivariate normal distribution.

(c) $\mathbf{1}_n'\mathbf{e} = 0$.

2.6 If $\mathbf{y}_1, \mathbf{y}_2, \ldots, \mathbf{y}_m$ are independently distributed as $N_d(\boldsymbol{\theta}_i, \boldsymbol{\Sigma}_i)$, $i = 1, 2, \ldots, n$, prove that

$$\sum_{i=1}^{n} a_i \mathbf{y}_i \sim N_d\left(\sum_{i=1}^{n} a_i \boldsymbol{\theta}_i, \sum_{i=1}^{n} a_i^2 \boldsymbol{\Sigma}_i\right)$$

using (a) Definition 1a and moment-generating functions and (b) Definition 1b.

2.7 If $\mathbf{y}_1, \mathbf{y}_2, \ldots, \mathbf{y}_n$ are a random sample from $N_d(\mathbf{0}, \boldsymbol{\Sigma})$, show that $\mathbf{y} = (\mathbf{y}_1', \mathbf{y}_2', \ldots, \mathbf{y}_n')' \sim N_{dn}(\mathbf{0}, \mathbf{I}_n \otimes \boldsymbol{\Sigma})$ [see (1.10)].

2.8 Let $y_1, y_2, \ldots, y_n$ be i.i.d. $N_1(\theta, \sigma^2)$ and define $Q = \Sigma_{i=1}^n (y_i - \bar{y})^2$.

(a) Prove that $\text{cov}[\bar{y}, y_1 - \bar{y}] = 0$ and hence deduce that $\bar{y}$ is statistically independent of Q.

(b) Find a symmetric matrix $\mathbf{A}$ such that $Q = \mathbf{y}'\mathbf{A}\mathbf{y}$, where $\mathbf{y} = [(y_i)]$. Using A6.5, show that $Q/\sigma^2 \sim \chi_{n-1}^2$.

2.9 Using Definition 2a for the Wishart distribution, show that the moment-generating function of a Wishart matrix $\mathbf{W} = [(w_{jk})] \sim W_d(m, \boldsymbol{\Sigma})$ is given by

$$\begin{aligned} M(\mathbf{T}) &= \mathrm{E}\left[\exp\left(\sum_{j \le k}\sum t_{jk} w_{jk}\right)\right] \\ &= \mathrm{E}[\operatorname{etr}(\mathbf{U}\mathbf{W})] \\ &= |\mathbf{I}_d - 2\mathbf{U}\boldsymbol{\Sigma}|^{-m/2}, \end{aligned}$$

where $\mathbf{U}' = \mathbf{U}$, $u_{jj} = t_{jj}$, and $u_{jk} = u_{kj} = \frac{1}{2} t_{jk}$ $(j < k)$. Since this moment-generating function exists in a neighborhood of $\mathbf{T} = \mathbf{O}$, it uniquely determines the joint density function. [Hint: Use A5.8.]

2.10 Prove that $\mathscr{E}[\mathbf{W}] = m\boldsymbol{\Sigma}$, where $\mathbf{W} \sim W_d(m, \boldsymbol{\Sigma})$.

2.11 Using both definitions for the Wishart distribution, show that $W_1(m, \sigma^2)$ is $\sigma^2 \chi_m^2$.

2.12 Use Exercise 2.9 to prove the following for the nonsingular Wishart matrix $\mathbf{W}$:

(a) $\boldsymbol{\ell}'\mathbf{W}\boldsymbol{\ell} \sim \boldsymbol{\ell}'\boldsymbol{\Sigma}\boldsymbol{\ell}\chi_m^2$ for all $\boldsymbol{\ell}$. [Hint: $\boldsymbol{\ell}'\mathbf{W}\boldsymbol{\ell} = \text{tr}[\boldsymbol{\ell}\boldsymbol{\ell}'\mathbf{W}]$.]

(b) Let

$$\mathbf{W} = \begin{pmatrix} \mathbf{W}_{11} & \mathbf{W}_{12} \\ \mathbf{W}_{21} & \mathbf{W}_{22} \end{pmatrix} \quad \text{and} \quad \boldsymbol{\Sigma} = \begin{pmatrix} \boldsymbol{\Sigma}_{11} & \boldsymbol{\Sigma}_{12} \\ \boldsymbol{\Sigma}_{21} & \boldsymbol{\Sigma}_{22} \end{pmatrix},$$

where $\mathbf{W}_{11}$ and $\boldsymbol{\Sigma}_{11}$ are $r \times r$ matrices $(r < d)$. If $\boldsymbol{\Sigma}_{12} = \mathbf{O}$, then $\mathbf{W}_{11}$ and $\mathbf{W}_{22}$ have independent Wishart distributions.

(c) The distribution of $\operatorname{tr} \mathbf{W}$ is the same as the distribution of a linear combination of independent chi-square variables.

(d) If $\mathbf{W}_1$ and $\mathbf{W}_2$ are independent, and $\mathbf{W}_i \sim W_d(m_i, \boldsymbol{\Sigma})$, then

$$\mathbf{W}_1 + \mathbf{W}_2 \sim W_d(m_1 + m_2, \boldsymbol{\Sigma}).$$

2.13 Using Definition 2b, prove Exercise 2.12(d).

2.14 Using Definition 2b, prove that, in general (i.e., $\boldsymbol{\Sigma}_{12} \neq \mathbf{O}$), $\mathbf{W}_{11}$ and $\mathbf{W}_{22}$ have marginal Wishart distributions.

2.15 If $\mathbf{y} \sim N_d(\mathbf{0}, \boldsymbol{\Sigma})$ and $\mathbf{A}$ is any given $d \times d$ symmetric matrix, show that the moment-generating function of $\mathbf{y}'\mathbf{A}\mathbf{y}$ is

$$M(t) = |\mathbf{I}_d - 2t\mathbf{A}\boldsymbol{\Sigma}|^{-1/2}.$$

[Hint: Use A5.8.]

2.16 Given $\mathbf{x}_1, \mathbf{x}_2, \ldots, \mathbf{x}_m$ i.i.d. $N_d(\mathbf{0}, \boldsymbol{\Sigma})$, with $m \geq d$, find the moment-generating function of

$$\mathbf{W} = \sum_{i=1}^{m} \mathbf{x}_i \mathbf{x}_i'$$

and hence show that Definition 2b for the nonsingular Wishart distribution implies Definition 2a [Hint: Apply Exercise 2.15 to tr[$\mathbf{W}\mathbf{U}$] = $\Sigma_i \mathbf{x}_i'\mathbf{U}\mathbf{x}_i$ and use Exercise 2.9.]

2.17 Using the notation of Theorem 2.1(viii), prove that $\mathbf{y}^{(2)} - \boldsymbol{\Sigma}_{21}\boldsymbol{\Sigma}_1^{-1}\mathbf{y}^{(1)}$ and $\mathbf{y}^{(1)}$ are statistically independent.

2.18 Suppose that the rows of $\mathbf{X}$ are i.i.d. $N_d(\mathbf{0}, \boldsymbol{\Sigma})$ and let $\mathbf{Y} = \mathbf{AXB}$ and $\mathbf{Z} = \mathbf{CXD}$, where $\mathbf{A}$, $\mathbf{B}$, $\mathbf{C}$, and $\mathbf{D}$ are conformable matrices of constants. Show that the elements of $\mathbf{Y}$ are independent of the elements of $\mathbf{Z}$ if and only if (a) $\mathbf{B}'\boldsymbol{\Sigma}\mathbf{D} = \mathbf{O}$ and/or (b) $\mathbf{AC}' = \mathbf{O}$ [Hint: Consider cov[y_{ij}, z_{rs}] = 0.]

2.19 Let x and y be continuous random variables such that the probability density function $f(x|y)$ of x given y does not depend on y. Show that x and y are statistically independent and f is the unconditional density function of x.

2.20 If $\boldsymbol{\Sigma}$ is positive definite, show that

$$(\boldsymbol{\mu}_1', \boldsymbol{\mu}_2')\begin{pmatrix} \boldsymbol{\Sigma}_{11} & \boldsymbol{\Sigma}_{12} \\ \boldsymbol{\Sigma}_{21} & \boldsymbol{\Sigma}_{22} \end{pmatrix}^{-1}\begin{pmatrix} \boldsymbol{\mu}_1 \\ \boldsymbol{\mu}_2 \end{pmatrix} - \boldsymbol{\mu}_1'\boldsymbol{\Sigma}_{11}^{-1}\boldsymbol{\mu}_1 = \boldsymbol{\mu}_{2\cdot 1}'\boldsymbol{\Sigma}_{22\cdot 1}^{-1}\boldsymbol{\mu}_{2\cdot 1},$$

where $\boldsymbol{\mu}_{2\cdot 1} = \boldsymbol{\mu}_2 - \boldsymbol{\Sigma}_{21}\boldsymbol{\Sigma}_{11}^{-1}\boldsymbol{\mu}_1$ and $\boldsymbol{\Sigma}_{22\cdot 1} = \boldsymbol{\Sigma}_{22} - \boldsymbol{\Sigma}_{21}\boldsymbol{\Sigma}_{11}^{-1}\boldsymbol{\Sigma}_{12}$. Prove that $\boldsymbol{\Sigma}_{22\cdot 1}$ is positive definite. [Hint: Use A3.1.]

2.21 Let $\mathbf{W}_i \sim W_d(m_i, \boldsymbol{\Sigma})$ $(i = 1, 2)$, where $\mathbf{W}_1$ and $\mathbf{W}_2$ are independent. Show, without finding the distribution, that the distribution of $|\mathbf{W}_1|/|\mathbf{W}_1 + \mathbf{W}_2|$ does not depend on $\boldsymbol{\Sigma}$.

2.22 Given $\mathbf{W} \sim W_d(m, \boldsymbol{\Sigma})$, where $\boldsymbol{\Sigma} > \mathbf{O}$, show that $|\mathbf{W}|/|\boldsymbol{\Sigma}|$ is distributed as the product of d independent chi-square variables with respective degrees of freedom $m, m-1, \ldots, m-d+1$. [Hint: Use the factorization (2.61).]

2.23 Let $\mathbf{W} \sim W_d(m, \boldsymbol{\Sigma})$, where $\boldsymbol{\Sigma} > \mathbf{O}$ and $m \geq d+2$. By giving appropriate values to the vector ℓ in Lemma 2.7 (Section 2.4.1), prove that

$$\mathscr{E}[\mathbf{W}^{-1}] = \boldsymbol{\Sigma}^{-1}/(m-d-1).$$

If $\mathbf{y} \sim N_d(\mathbf{0}, \boldsymbol{\Sigma})$ and $\mathbf{y}$ is independent of $\mathbf{W}$, find $\mathrm{E}[\mathbf{y}'\mathbf{W}^{-1}\mathbf{y}]$. [Hint: If $x \sim \chi_n^2$, then $\mathrm{E}[x^{-1}] = (n-2)^{-1}$.]

2.24 If $\mathbf{x}_1$ and $\mathbf{x}_2$ are independently distributed as $N_d(\mathbf{0}, \boldsymbol{\Sigma})$, for what values of a and b does

$$a\mathbf{x}_1\mathbf{x}_1' + b\mathbf{x}_1\mathbf{x}_2' + b\mathbf{x}_2\mathbf{x}_1' + a\mathbf{x}_2\mathbf{x}_2'$$

have a Wishart distribution?

2.25 Let $\mathbf{X}' = \{\mathbf{x}_1, \mathbf{x}_2, \ldots, \mathbf{x}_n\}$, where the $\mathbf{x}_i$ are i.i.d $N_d(\mathbf{0}, \boldsymbol{\Sigma})$. If $\mathbf{X}'\mathbf{A}_j\mathbf{X} \sim \mathbf{W}_d(m_j, \boldsymbol{\Sigma})$, $j = 1, 2, \ldots, r$, and $\mathbf{A}_j\mathbf{A}_k = \mathbf{O}$ (all j, k; $j \neq k$), show that the $\mathbf{X}'\mathbf{A}_j\mathbf{X}$ are mutually independent.

2.26 Show that when $\mathbf{x} \sim N_d(\mathbf{u}, \boldsymbol{\Sigma})$ in (2.94) and (2.95), we have $\beta_{1,d} = 0$ and $\beta_{2,d} = d(d+2)$.

CHAPTER 3

Inference for the Multivariate Normal

3.1 INTRODUCTION

It has been said that the multivariate normal (MVN) distribution is a myth and that it has dominated research in multivariate analysis for far too long. This criticism is certainly justified, and with more sophisticated computers we are now able to do much more data snooping without the necessity of imposing distributional assumptions or too much structure on the data. However, the MVN still has an important role to play for at least the following reasons. First, we can transform to normality in many situations: Such transformations are discussed in Chapter 4. Second, some procedures based on the MVN assumptions are in fact robust to departures from normality. This usually happens because of the multivariate central limit theorem, which states that under fairly general conditions the asymptotic distribution of a sample vector mean $\bar{\mathbf{x}}$ is MVN. Third, as procedures based on the MVN are frequently optimal in some sense, they provide a yardstick for comparing less than optimal, but more robust, procedures: A "good" procedure is one that is robust to departures from the assumption of multivariate normality, yet is close to being optimal when the assumption is true.

In this chapter we shall initially consider inferences about the mean and dispersion matrix for the MVN based on a sample of size n. Later we shall consider inferences based on samples from two MVN distributions.

3.2 ESTIMATION

3.2.1 Maximum Likelihood Estimation

Suppose $\mathbf{x}_1, \mathbf{x}_2, \ldots, \mathbf{x}_n$ are independently and identically distributed (i.i.d.) as $N_d(\boldsymbol{\mu}, \boldsymbol{\Sigma})$, and $n - 1 \geq d$. We show below that the observed sample mean $\bar{\mathbf{x}}$

and sample dispersion matrix

$$\frac{1}{n}\mathbf{Q} = \frac{1}{n}\sum_{i=1}^{n}(\mathbf{x}_i - \bar{\mathbf{x}})(\mathbf{x}_i - \bar{\mathbf{x}})' \tag{3.1}$$

are the maximum likelihood estimates of $\boldsymbol{\mu}$ and $\boldsymbol{\Sigma}$, respectively.

The likelihood function is the joint distribution of the $\mathbf{x}_i$ expressed as a function of $\boldsymbol{\mu}$ and $\boldsymbol{\Sigma}$, namely,

$$L(\boldsymbol{\mu}, \boldsymbol{\Sigma}) = (2\pi)^{-nd/2}|\boldsymbol{\Sigma}|^{-n/2}\exp\left[-\frac{1}{2}\sum_{i=1}^{n}(\mathbf{x}_i - \boldsymbol{\mu})'\boldsymbol{\Sigma}^{-1}(\mathbf{x}_i - \boldsymbol{\mu})\right]. \tag{3.2}$$

We note that $\boldsymbol{\Sigma}^{-1} > \mathbf{O}$ as $\boldsymbol{\Sigma} > \mathbf{O}$ (by A5.5) and we have (A1.1)

$$\begin{aligned}\log L(\boldsymbol{\mu}, \boldsymbol{\Sigma}) &= c - \tfrac{1}{2}n\log|\boldsymbol{\Sigma}| - \frac{1}{2}\sum_{i=1}^{n}\left[(\mathbf{x}_i - \boldsymbol{\mu})'\boldsymbol{\Sigma}^{-1}(\mathbf{x}_i - \boldsymbol{\mu})\right] \\ &= c - \tfrac{1}{2}n\log|\boldsymbol{\Sigma}| - \frac{1}{2}\sum_{i=1}^{n}\operatorname{tr}\left[(\mathbf{x}_i - \boldsymbol{\mu})'\boldsymbol{\Sigma}^{-1}(\mathbf{x}_i - \boldsymbol{\mu})\right] \\ &= c - \tfrac{1}{2}n\log|\boldsymbol{\Sigma}| - \tfrac{1}{2}\operatorname{tr}\left[\boldsymbol{\Sigma}^{-1}\sum_{i=1}^{n}(\mathbf{x}_i - \boldsymbol{\mu})(\mathbf{x}_i - \boldsymbol{\mu})'\right],\end{aligned} \tag{3.3}$$

where $c = -\tfrac{1}{2}nd\log 2\pi$. Also

$$\begin{aligned}\sum_{i=1}^{n}(\mathbf{x}_i - \boldsymbol{\mu})(\mathbf{x}_i - \boldsymbol{\mu})' &= \sum_{i=1}^{n}(\mathbf{x}_i - \bar{\mathbf{x}} + \bar{\mathbf{x}} - \boldsymbol{\mu})(\mathbf{x}_i - \bar{\mathbf{x}} + \bar{\mathbf{x}} - \boldsymbol{\mu})' \\ &= \sum_{i=1}^{n}(\mathbf{x}_i - \bar{\mathbf{x}})(\bar{\mathbf{x}}_i - \bar{\mathbf{x}})' + n(\bar{\mathbf{x}} - \boldsymbol{\mu})(\bar{\mathbf{x}} - \boldsymbol{\mu})' \\ &= \mathbf{Q} + n(\bar{\mathbf{x}} - \boldsymbol{\mu})(\bar{\mathbf{x}} - \boldsymbol{\mu})'.\end{aligned} \tag{3.4}$$

Now $\operatorname{tr}[\boldsymbol{\Sigma}^{-1}(\bar{\mathbf{x}} - \boldsymbol{\mu})(\bar{\mathbf{x}} - \boldsymbol{\mu})'] = (\bar{\mathbf{x}} - \boldsymbol{\mu})'\boldsymbol{\Sigma}^{-1}(\bar{\mathbf{x}} - \boldsymbol{\mu}) \geq 0$ (since $\boldsymbol{\Sigma}^{-1} > \mathbf{O}$), so that $\log L$ is maximized for any $\boldsymbol{\Sigma} > \mathbf{O}$ when $\boldsymbol{\mu} = \bar{\mathbf{x}}$ ($= \hat{\boldsymbol{\mu}}$, say). Thus

$$\log L(\hat{\boldsymbol{\mu}}, \boldsymbol{\Sigma}) \geq \log L(\boldsymbol{\mu}, \boldsymbol{\Sigma}) \quad \text{for all } \boldsymbol{\Sigma} > \mathbf{O}.$$

The next step is to maximize

$$\log L(\hat{\boldsymbol{\mu}}, \boldsymbol{\Sigma}) = c - \tfrac{1}{2}n\log|\boldsymbol{\Sigma}| - \tfrac{1}{2}\operatorname{tr}[\boldsymbol{\Sigma}^{-1}\mathbf{Q}],$$

$$= c - \tfrac{1}{2}n\left(\log|\boldsymbol{\Sigma}| + \operatorname{tr}\left[\frac{\boldsymbol{\Sigma}^{-1}\mathbf{Q}}{n}\right]\right)$$

subject to $\boldsymbol{\Sigma} > \mathbf{O}$. Since $\mathbf{Q} > \mathbf{O}$ (with probability 1; see A5.13) the solution follows from A7.1, namely,

$$\boldsymbol{\Sigma} = \mathbf{Q}/n \qquad (= \hat{\boldsymbol{\Sigma}}, \text{say}). \tag{3.5}$$

Thus

$$\log L(\hat{\boldsymbol{\mu}}, \hat{\boldsymbol{\Sigma}}) \geq \log L(\hat{\boldsymbol{\mu}}, \boldsymbol{\Sigma}) \geq \log L(\boldsymbol{\mu}, \boldsymbol{\Sigma}),$$

and $\hat{\boldsymbol{\mu}}$ and $\hat{\boldsymbol{\Sigma}}$ are the maximum likelihood estimates of $\boldsymbol{\mu}$ and $\boldsymbol{\Sigma}$. For future reference we note that since

$$\operatorname{tr}[\hat{\boldsymbol{\Sigma}}^{-1}\mathbf{Q}] = \operatorname{tr}[n\mathbf{I}_d] = nd,$$

we have from (3.3)

$$L(\hat{\boldsymbol{\mu}}, \hat{\boldsymbol{\Sigma}}) = (2\pi)^{-nd/2}|\hat{\boldsymbol{\Sigma}}|^{-n/2}e^{-nd/2}. \tag{3.6}$$

The theoretical properties of $\hat{\boldsymbol{\mu}}$ and $\hat{\boldsymbol{\Sigma}}$ [or the unbiased estimate $\mathbf{S} = \mathbf{Q}/(n-1)$], namely, sufficiency, consistency, completeness, efficiency, and Bayesian characteristics, are proved, for example, by Giri [1977: Chapter 5]. The Bayesian estimation of $\boldsymbol{\Sigma}$ is discussed by Chen [1979], while Wang [1980] considers the sequential estimation of $\boldsymbol{\mu}$. An algorithm for handling missing observations is described by Beale and Little [1975] (see Section 10.3).

The use of $\bar{\mathbf{x}}$ as an estimator of $\boldsymbol{\mu}$ has been criticized on the grounds that it is inadmissable for particular loss functions when $d \geq 3$ (James and Stein [1961], Stein [1956, 1965]). Let $\mathbf{t}$ be an estimator of $\boldsymbol{\mu}$ with quadratic loss $(\mathbf{t} - \boldsymbol{\mu})'(\mathbf{t} - \boldsymbol{\mu})$ $(= \|\mathbf{t} - \boldsymbol{\mu}\|^2)$ and risk (expected loss) $E[\|\mathbf{t} - \boldsymbol{\mu}\|^2]$. Then, without assuming normality, it can be shown, under fairly general conditions, that there exists a $\mathbf{t}$ such that, for $d \geq 3$, $\bar{\mathbf{x}}$ is inadmissable (see Brown [1966; Section 3]); that is,

$$E\left[\|\mathbf{t} - \boldsymbol{\mu}\|^2\right] < E\left[\|\bar{\mathbf{x}} - \boldsymbol{\mu}\|^2\right] \tag{3.7}$$

for all $\boldsymbol{\mu}$. James and Stein [1961: p. 369] demonstrated this briefly for the case of diagonal $\boldsymbol{\Sigma}$ and finite fourth moments about the mean. When the underlying distribution is the MVN, $\bar{\mathbf{x}}$ is most efficient and, for each j, the jth element of $\bar{\mathbf{x}}$ is the minimum variance unbiased estimator of μ_j. However, using the loss function $(\mathbf{t} - \boldsymbol{\mu})'\mathbf{A}(\mathbf{t} - \boldsymbol{\mu})$ with $\mathbf{A} = \boldsymbol{\Sigma}^{-1}$, James and Stein [1961: p. 366] showed that for certain positive c,

$$\mathbf{t} = \left(1 - \frac{c}{\bar{\mathbf{x}}'\mathbf{S}^{-1}\bar{\mathbf{x}}}\right)\bar{\mathbf{x}} \tag{3.8}$$

has smaller risk than $\bar{\mathbf{x}}$ for all $\boldsymbol{\mu}$ (see Muirhead [1982: Chapter 4] for a helpful discussion).

Most of the admissibility studies assume normality with $\boldsymbol{\Sigma}$ known (Berger [1976a, 1978, 1980a, 1982], $\boldsymbol{\Sigma} = \mathbf{I}_d$ (Efron and Morris [1975], Faith [1978], Shinozaki [1980]), $\boldsymbol{\Sigma}$ diagonal (Berger and Bock [1976b]), and $\boldsymbol{\Sigma} = \sigma^2\mathbf{I}_d$ (Efron and Morris [1976a], Rao and Shinozaki [1978], Berger [1980a], Fay and Herriott [1979]). Fay and Herriott state that "in our knowledge, the Census Bureau's use is the largest application of James–Stein procedures in a federal statistical program." The cases of general $\boldsymbol{\Sigma}$ (James and Stein [1961: p. 366], Alam [1975], Efron and Morris [1976a], Haff [1980]) and nonnormal distributions (Berger and Bock [1976a], Berger [1980b], Brandwein and Strawderman [1980]) have received much less attention. The admissible estimation of $\boldsymbol{\Sigma}^{-1}$ is discussed by Efron and Morris [1976b] and Haff [1979]. Evidently, under certain circumstances, there are real gains in using a Stein estimator of $\boldsymbol{\mu}$ that has the effect of pulling the mean $\bar{\mathbf{x}}$ towards the origin. However, as noted by Efron and Morris [1975], this new technique has not been as widely used as its proponents might have anticipated. There are several reasons for this. First, an estimator $\mathbf{t}$ that is better than $\bar{\mathbf{x}}$ for one loss function might not be better for a different loss function: Rao and Shinozaki [1978: p. 23] give an example of this due to Professor P. A. P. Moran. Second, if $\mathbf{t} = [(t_j)]$ and $\bar{\mathbf{x}} = [(\bar{x}_j)]$, then, from (3.7),

$$\mathrm{E}\left[\sum_j (t_j - \mu_j)^2\right] < \mathrm{E}\left[\sum_j (\bar{x}_j - \mu_j)^2\right],$$

which does not imply that $\mathrm{E}[(t_j - \mu_j)^2] < \mathrm{E}[(\bar{x}_j - \mu_j)^2]$ for each $j = 1, 2, \ldots, d$. In fact, some of the individual mean square errors of the t_j will be greater than those of the $\bar{x}_j$ so that for some constant vectors $\mathbf{a}$, a linear combination $\mathbf{a}'\mathbf{t}$ will have a larger mean square error than $\mathbf{a}'\bar{\mathbf{x}}$ (Rao and Shinozaki [1978]). Third, an estimator that is admissible under normality may not be admissible for other distributions. Fourth, further work needs to be done on the cases of a general unknown $\boldsymbol{\Sigma}$, an arbitrary $\mathbf{A}$ in the loss function $(\mathbf{t} - \boldsymbol{\mu})'\mathbf{A}(\mathbf{t} - \boldsymbol{\mu})$ (see Gleser [1979] and, in particular, Berger [1980a] for references) and nonquadratic loss functions (e.g., Berger [1978], Brandwein and Strawderman [1980]). Fifth, estimators that are better than $\bar{\mathbf{x}}$ are only significantly better when $\boldsymbol{\mu}$ is in a fairly small region of R^d (Berger [1980a]). Finally, there is a lack of suitable theory for confidence intervals of admissible estimators.

For at least the above reasons it is perhaps not surprising that practictioners have generally ignored the question of inadmissibility and have been more concerned by the lack of robustness demonstrated by $\bar{\mathbf{x}}$ and $\mathbf{S}$ with regard to nonnormality and outlying observations. In Section 4.4 we look at various robust alternatives to these estimators.

3.2.2 *Distribution Theory*

Having found maximum likelihood estimators of $\boldsymbol{\mu}$ and $\boldsymbol{\Sigma}$ for the MVN distribution, we now look at their distributional properties.

THEOREM 3.1 If $\mathbf{x}_1, \mathbf{x}_2, \ldots, \mathbf{x}_n$ are i.i.d. $N_d(\boldsymbol{\mu}, \boldsymbol{\Sigma})$ and $(n-1) \geq d$, then we have the following:

(i) $\bar{\mathbf{x}} \sim N_d(\boldsymbol{\mu}, \boldsymbol{\Sigma}/n)$, $\mathbf{Q} = (n-1)\mathbf{S} \sim W_d(n-1, \boldsymbol{\Sigma})$, and $\bar{\mathbf{x}}$ and $\mathbf{S}$ are statistically independent.

(ii) $T^2 = n(\bar{\mathbf{x}} - \boldsymbol{\mu})'\mathbf{S}^{-1}(\bar{\mathbf{x}} - \boldsymbol{\mu}) \sim T^2_{d,n-1}$, where $T^2_{d,n-1}$ is Hotelling's T^2 distribution with d and $n-1$ degrees of freedom.

Proof (i) Assume for the moment that $\boldsymbol{\mu} = \mathbf{0}$ and let $y_i = \boldsymbol{\ell}'\mathbf{x}_i$ $(i = 1, 2, \ldots, n)$. Then the y_i are i.i.d. $N_1(0, \sigma_\ell^2)$, where $\sigma_\ell^2 = \boldsymbol{\ell}'\boldsymbol{\Sigma}\boldsymbol{\ell}$. From the univariate theory (see Exercise 2.8), $\bar{y} \sim N_1(0, \sigma_\ell^2/n)$, $\Sigma_i(y_i - \bar{y})^2 \sim \sigma_\ell^2\chi^2_{n-1}$, and $\bar{y}$ is statistically independent of $\Sigma_i(y_i - \bar{y})^2$. Now writing $\mathbf{y}' = (y_1, y_2, \ldots, y_n)$, we have $\bar{y} = \mathbf{y}'\mathbf{b}$ $(\mathbf{b} = \mathbf{1}_n/n)$ and $\Sigma_i(y_i - \bar{y})^2 = \mathbf{y}'\mathbf{A}\mathbf{y}$. Hence, by Theorem 2.4(iii) of Section 2.3.2 with $\mathbf{X}' = (\mathbf{x}_1, \mathbf{x}_2, \ldots, \mathbf{x}_n)$, $\bar{\mathbf{x}} = \mathbf{X}'\mathbf{b}$ and $(n-1)\mathbf{S} = \mathbf{X}'\mathbf{A}\mathbf{X}$ are independently distributed as N_d and $W_d(n-1, \boldsymbol{\Sigma})$, respectively. When $\boldsymbol{\mu} \neq \mathbf{0}$, we simply replace $\mathbf{x}_i$ by $\mathbf{x}_i - \boldsymbol{\mu}$: $\mathbf{S}$ is unchanged [see (1.11)] and $\bar{\mathbf{x}}$ is replaced by $\bar{\mathbf{x}} - \boldsymbol{\mu}$. Thus $\bar{\mathbf{x}} \sim N_d(\boldsymbol{\mu}, \boldsymbol{\Sigma}/n)$ (by the Corollary of Lemma 2.3), $\mathbf{S}$ is independent of $\bar{\mathbf{x}} - \boldsymbol{\mu}$ (and therefore of $\bar{\mathbf{x}}$), and $(n-1)\mathbf{S} \sim W_d(n-1, \boldsymbol{\Sigma})$.

(ii) Setting $m = n - 1$, $\mathbf{W}/m = \mathbf{S}$, $\mathbf{x} = \bar{\mathbf{x}}$, and $\lambda = n$ in (2.19) gives the required result. □

A simple mnemonic rule is as follows. Since $\bar{\mathbf{x}} \sim N_d(\boldsymbol{\mu}, \boldsymbol{\Sigma}/n)$, $(\bar{\mathbf{x}} - \boldsymbol{\mu})'(\boldsymbol{\Sigma}/n)^{-1}(\bar{\mathbf{x}} - \boldsymbol{\mu}) \sim \chi^2_d$ [Theorem 2.1(vi) of Section 2.2]. We now replace $\boldsymbol{\Sigma}$ by its unbiased estimate $\mathbf{S}$, and this leads to $n(\bar{\mathbf{x}} - \boldsymbol{\mu})'\mathbf{S}^{-1}(\bar{\mathbf{x}} - \boldsymbol{\mu})$.

3.3 TESTING FOR THE MEAN

3.3.1 *Hotelling's T^2 Test*

Clearly the T^2 statistic in Theorem 3.1 can be used for testing $H_0: \boldsymbol{\mu} = \boldsymbol{\mu}_0$ against $H_1: \boldsymbol{\mu} \neq \boldsymbol{\mu}_0$. Thus $T_0^2 = n(\bar{\mathbf{x}} - \boldsymbol{\mu}_0)'\mathbf{S}^{-1}(\bar{\mathbf{x}} - \boldsymbol{\mu}_0)$ has a $T^2_{d,n-1}$ distribution when H_0 is true, or from (2.20),

$$F_0 = \frac{T_0^2}{n-1}\frac{n-d}{d} \sim F_{d,n-d} \tag{3.9}$$

when H_0 is true. Since we would reject H_0 if $\bar{\mathbf{x}}$ is too "far" from $\boldsymbol{\mu}_0$, we use the upper tail of the F-distribution. In particular, we reject H_0 at the α level of significance if F_0 exceeds $F^{\alpha}_{d,n-d}$, where $\text{pr}[F_{d,n-d} \geq F^{\alpha}_{d,n-d}] = \alpha$. Methods for computing T_0^2 based on the Cholesky decomposition $\mathbf{U}'\mathbf{U}$ of $\mathbf{Q}$ (or $\mathbf{S}$) are described in Section 10.1.3. For example, solving the triangular system $\mathbf{U}'\mathbf{z} = (\bar{\mathbf{x}} - \boldsymbol{\mu}_0)$ for $\mathbf{z}$ gives us

$$\begin{aligned}
\mathbf{z}'\mathbf{z} &= (\bar{\mathbf{x}} - \boldsymbol{\mu}_0)'\mathbf{U}^{-1}(\mathbf{U}')^{-1}(\bar{\mathbf{x}} - \boldsymbol{\mu}_0) \\
&= (\bar{\mathbf{x}} - \boldsymbol{\mu}_0)'\mathbf{Q}^{-1}(\bar{\mathbf{x}} - \boldsymbol{\mu}_0) \\
&= \frac{T_0^2}{n(n-1)}. \qquad (3.10)
\end{aligned}$$

EXAMPLE 3.1 In Table 3.1 we have bivariate observations on the weights and heights of 39 Peruvian Indians. Suppose we wish to test $H_0 : \boldsymbol{\mu} = (63.64, 1615.38)' = \boldsymbol{\mu}_0$ (corresponding to 140 lb and 63 in., respectively). A plot

TABLE 3.1 Weight and Height of 39 Peruvian Indians[a]

Weight (kg)	Height (mm)	Weight (kg)	Height (mm)
71.0	1629	59.5	1513
56.5	1569	61.0	1653
56.0	1561	57.0	1566
61.0	1619	57.5	1580
65.0	1566	74.0	1647
62.0	1639	72.0	1620
53.0	1494	62.5	1637
53.0	1568	68.0	1528
65.0	1540	63.4	1647
57.0	1530	68.0	1605
66.5	1622	69.0	1625
59.1	1486	73.0	1615
64.0	1578	64.0	1640
69.5	1645	65.0	1610
64.0	1648	71.0	1572
56.5	1521	60.2	1534
57.0	1547	55.0	1536
55.0	1505	70.0	1630
57.0	1473	87.0[b]	1542[b]
58.0	1538		

[a]From Ryan et al. [1976], with kind permission of the authors.
[b]Suspected outlier.

of the data (Fig. 3.1) suggests that the last item is an outlier. Other plots (see Examples 4.3 and 4.6 in Section 4.3) also indicate that the normality assumption is tenable, though there is some suggestion of kurtosis. Ignoring the last item (i.e., $n = 38$), we write

$$\bar{\mathbf{x}} = \begin{pmatrix} 62.5316 \\ 1579.8900 \end{pmatrix}, \qquad \bar{\mathbf{x}} - \boldsymbol{\mu}_0 = \begin{pmatrix} -1.1084 \\ -35.4900 \end{pmatrix},$$

$$\mathbf{Q} = \begin{pmatrix} 1331.62 & 7304.83 \\ 7304.83 & 104118 \end{pmatrix}, \qquad \mathbf{U} = \begin{pmatrix} 36.4914 & 200.179 \\ 0 & 253.073 \end{pmatrix}.$$

Solving

$$\begin{pmatrix} 36.4914 & 0 \\ 200.179 & 253.073 \end{pmatrix} \begin{pmatrix} z_1 \\ z_2 \end{pmatrix} = \begin{pmatrix} -1.1084 \\ -35.4900 \end{pmatrix}$$

gives $\mathbf{z}' = (-0.0303749, -0.116191)$. Hence

$$T_0^2 = n(n-1)\mathbf{z}'\mathbf{z} = 20.2798$$

and

$$F_0 = \frac{T_0^2}{n-1}\frac{n-d}{d} = \frac{20.2798}{37}\frac{36}{2} = 9.87.$$

Since $F_{2,30}^{0.001} = 8.77$, $F_0 > F_{2,30}^{0.001} > F_{2,36}^{0.001}$ and we reject H_0 at the 0.1% level of significance. If the outlier is included, $F_0 = 10.50$. □

Since T_0^2 reduces to the square of the usual univariate t-statistic, $n(\bar{x} - \mu_0)^2/s^2$, when $d = 1$, it is not surprising that T_0^2 has a number of properties analogous to those held by the univariate t-ratio. In the first place the t-ratio is independent of change of scale and origin, and T_0^2 is invariant under all nonsingular linear transformations $\mathbf{z} = \mathbf{F}\mathbf{x} + \mathbf{c}$ of the observations and the hypothesis, where $\mathbf{F}$ is an arbitrary $d \times d$ nonsingular matrix and $\mathbf{c}$ is an arbitrary $d \times 1$ vector. For example, $\mathscr{E}[\mathbf{z}] = \boldsymbol{\phi}_0 = \mathbf{F}\boldsymbol{\mu}_0 + \mathbf{c}$ when H_0 is true, and

$$\begin{aligned} n(\bar{\mathbf{z}} - \boldsymbol{\phi}_0)'\mathbf{S}_z^{-1}(\bar{\mathbf{z}} - \boldsymbol{\phi}_0) &= n(\mathbf{F}\bar{\mathbf{x}} + \mathbf{c} - \mathbf{F}\boldsymbol{\mu}_0 - \mathbf{c})'(\mathbf{F}\mathbf{S}\mathbf{F}')^{-1} \\ &\quad \times (\mathbf{F}\bar{\mathbf{x}} + \mathbf{c} - \mathbf{F}\boldsymbol{\mu}_0 - \mathbf{c}) \\ &= n(\bar{\mathbf{x}} - \boldsymbol{\mu}_0)'\mathbf{F}'(\mathbf{F}')^{-1}\mathbf{S}^{-1}\mathbf{F}^{-1}\mathbf{F}(\bar{\mathbf{x}} - \boldsymbol{\mu}_0) \\ &= n(\bar{\mathbf{x}} - \boldsymbol{\mu}_0)'\mathbf{S}^{-1}(\bar{\mathbf{x}} - \boldsymbol{\mu}_0) = T_0^2, \end{aligned}$$

so that the statistic computed from the transformed data is the same as that computed from the original data. It was this invariance property that led

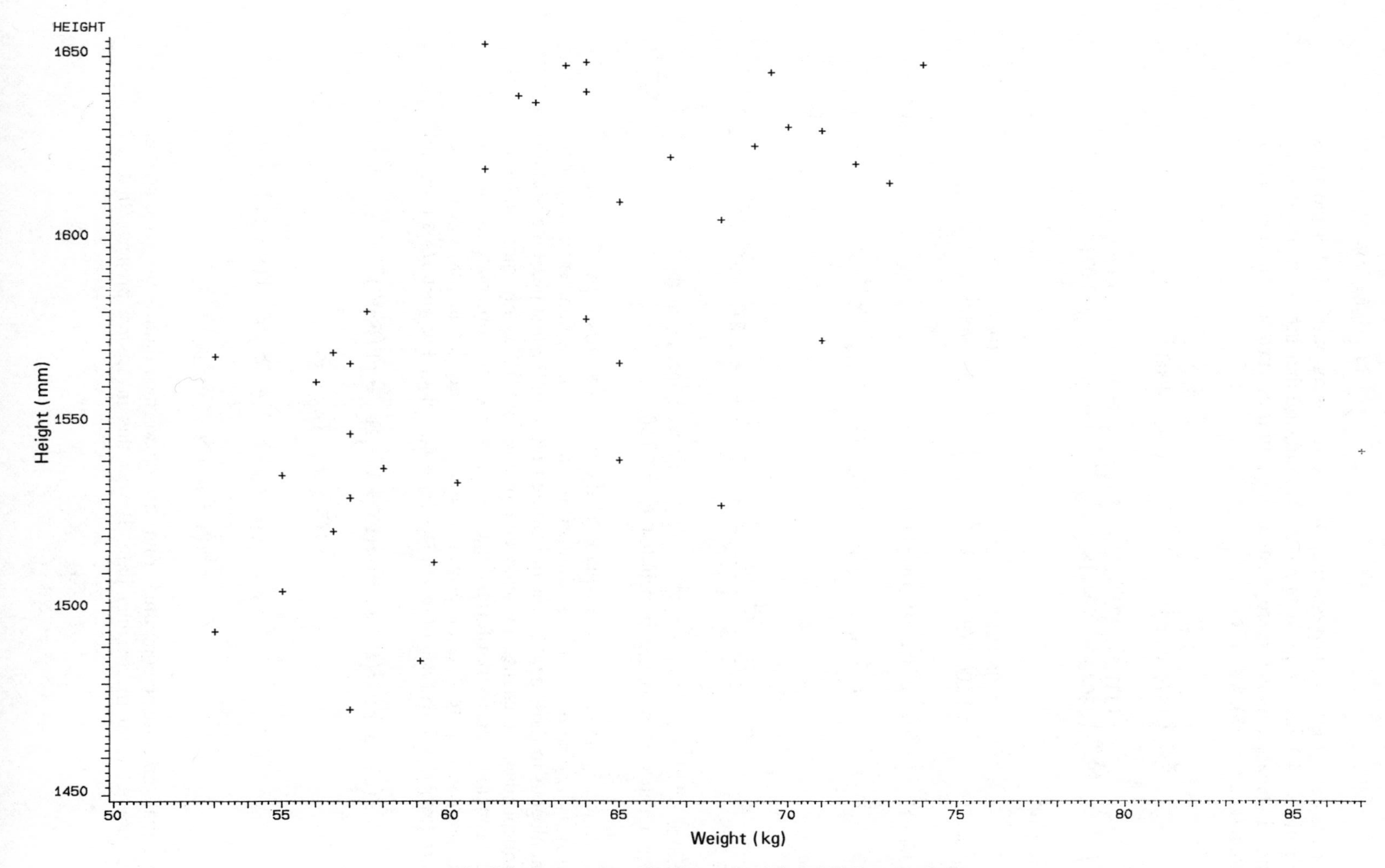

Fig. 3.1 Scatter plot of heights and weights of 39 Peruvian Indians.

Hotelling [1931] to the choice of T_0^2 as a generalization of the t-ratio. In fact, T_0^2 is the uniformly most powerful invariant test; it is also admissable and has certain local minimax properties (see Giri [1977: Chapter 7]). Another property shared by T_0^2 and the t-ratio is given by the following theorem.

THEOREM 3.2 The test statistic T_0^2 is equivalent to the likelihood ratio test statistic for $H_0: \boldsymbol{\mu} = \boldsymbol{\mu}_0$ versus $H_1: \boldsymbol{\mu} \neq \boldsymbol{\mu}_0$.

Proof The likelihood function $L(\boldsymbol{\mu}, \boldsymbol{\Sigma})$ is given by (3.2) and the likelihood ratio statistic is

$$\ell = \frac{\underset{\boldsymbol{\Sigma}}{\text{supremum}}\, L(\boldsymbol{\mu}_0, \boldsymbol{\Sigma})}{\underset{\boldsymbol{\mu}, \boldsymbol{\Sigma}}{\text{supremum}}\, L(\boldsymbol{\mu}, \boldsymbol{\Sigma})}.$$

From (3.6) we have

$$\begin{aligned} \underset{\boldsymbol{\mu}, \boldsymbol{\Sigma}}{\text{supremum}}\, L(\boldsymbol{\mu}, \boldsymbol{\Sigma}) &= L(\bar{\mathbf{x}}, \hat{\boldsymbol{\Sigma}}) \\ &= (2\pi)^{-nd/2} |\hat{\boldsymbol{\Sigma}}|^{-n/2} e^{-nd/2}. \end{aligned}$$

Using the same method of Section 3.2.1 with A7.1 applied to $\log L(\boldsymbol{\mu}_0, \boldsymbol{\Sigma})$, we find that $L(\boldsymbol{\mu}_0, \boldsymbol{\Sigma})$ is maximized at $\boldsymbol{\Sigma} = \hat{\boldsymbol{\Sigma}}_0 = \sum_{i=1}^n (\mathbf{x}_i - \boldsymbol{\mu}_0)(\mathbf{x}_i - \boldsymbol{\mu}_0)'/n$. Hence

$$\begin{aligned} \sup_{\boldsymbol{\Sigma}} L(\boldsymbol{\mu}_0, \boldsymbol{\Sigma}) &= L(\boldsymbol{\mu}_0, \hat{\boldsymbol{\Sigma}}_0) \\ &= (2\pi)^{-nd/2} |\hat{\boldsymbol{\Sigma}}_0|^{-n/2} e^{-nd/2}. \end{aligned}$$

Using the identity (3.4) with $\boldsymbol{\mu} = \boldsymbol{\mu}_0$, we find that

$$\begin{aligned} \ell^{2/n} &= |\hat{\boldsymbol{\Sigma}}| / |\hat{\boldsymbol{\Sigma}}_0| \\ &= \left| \sum_{i=1}^n (\mathbf{x}_i - \bar{\mathbf{x}})(\mathbf{x}_i - \bar{\mathbf{x}})' \right| \Big/ \left| \sum_{i=1}^n (\mathbf{x}_i - \boldsymbol{\mu}_0)(\mathbf{x}_i - \boldsymbol{\mu}_0)' \right| \\ &= |\mathbf{S}| \Big/ \left| \mathbf{S} + \frac{n}{n-1} (\bar{\mathbf{x}} - \boldsymbol{\mu}_0)(\bar{\mathbf{x}} - \boldsymbol{\mu}_0)' \right| \\ &= \left| \mathbf{I}_d + \frac{n}{n-1} \mathbf{S}^{-1} (\bar{\mathbf{x}} - \boldsymbol{\mu}_0)(\bar{\mathbf{x}} - \boldsymbol{\mu}_0)' \right|^{-1}. \end{aligned}$$

Now $\text{rank}[\mathbf{S}^{-1}(\bar{\mathbf{x}} - \boldsymbol{\mu}_0)(\bar{\mathbf{x}} - \boldsymbol{\mu}_0)'] = \text{rank}[(\bar{\mathbf{x}} - \boldsymbol{\mu}_0)(\bar{\mathbf{x}} - \boldsymbol{\mu}_0)'] = \text{rank}[(\bar{\mathbf{x}} - \boldsymbol{\mu}_0)] = 1$ (A2.3) so that by A2.4

$$\begin{aligned}\ell^{2/n} &= \left(1 + \frac{n}{n-1}\text{tr}\left[\mathbf{S}^{-1}(\bar{\mathbf{x}} - \boldsymbol{\mu}_0)(\bar{\mathbf{x}} - \boldsymbol{\mu}_0)'\right]\right)^{-1} \\ &= \left(1 + \frac{n}{n-1}(\bar{\mathbf{x}} - \boldsymbol{\mu}_0)'\mathbf{S}^{-1}(\bar{\mathbf{x}} - \boldsymbol{\mu}_0)\right)^{-1} \\ &= \left\{1 + \frac{T_0^2}{n-1}\right\}^{-1},\end{aligned}$$

which leads to

$$T_0^2 = (n-1)(\ell^{-2/n} - 1). \tag{3.11}$$

By the likelihood ratio principle we reject H_0 if ℓ is too small, that is, if T_0^2 is too large. □

The T_0^2 statistic has one further property that links it closely to the univariate t-ratio. Since $\mathbf{S} > \mathbf{O}$, we have, from A7.6,

$$\begin{aligned}T_0^2 &= n(\bar{\mathbf{x}} - \boldsymbol{\mu}_0)'\mathbf{S}^{-1}(\bar{\mathbf{x}} - \boldsymbol{\mu}_0) \\ &= n\sup_{\ell}\left(\frac{[\ell'(\bar{\mathbf{x}} - \boldsymbol{\mu}_0)]^2}{\ell'\mathbf{S}\ell}\right) \\ &= \sup_{\ell}\left[\frac{n^{1/2}\ell'(\bar{\mathbf{x}} - \boldsymbol{\mu}_0)}{(\ell'\mathbf{S}\ell)^{1/2}}\right]^2 \\ &= \sup_{\ell} t_0^2(\ell),\end{aligned} \tag{3.12}$$

where $t_0(\ell)$ is the usual t-statistic for testing the hypothesis $H_{0\ell}: \ell'\boldsymbol{\mu} = \ell'\boldsymbol{\mu}_0$ (see Section 3.4.4). Hence the T_0^2 test is essentially a univariate t-test based on the "optimal" linear combination of the data $\ell_0'\mathbf{x}_i$, where ℓ_0 is proportional to $\mathbf{S}^{-1}(\bar{\mathbf{x}} - \boldsymbol{\mu}_0)$. We shall see later in Chapter 6 that ℓ_0 is in the direction that, in a certain sense, gives maximum discrimination between $\bar{\mathbf{x}}$ and $\boldsymbol{\mu}_0$.

We note that $H_0 = \cap_{\ell} H_{0\ell}$ so that the union–intersection principle of Section 1.6.2 can be used to construct a test statistic. The acceptance region for $H_{0\ell}$ takes the form $\{\mathbf{X}: t_0^2(\ell) \le k\}$, where $\mathbf{X} = (\mathbf{x}_1, \mathbf{x}_2, \ldots, \mathbf{x}_n)'$. Hence the acceptance region for H_0 is

$$\begin{aligned}\bigcap_{\ell}\{\mathbf{X}: t_0^2(\ell) \le k\} &= \left\{\mathbf{X}: \sup_{\ell} t_0^2(\ell) \le k\right\} \\ &= \{\mathbf{X}: T_0^2 \le k\},\end{aligned} \tag{3.13}$$

and we are led, once again, to the T_0^2 test. In this situation, the likelihood ratio and union–intersection methods lead to the same test statistic. This is not always the case. In Section 1.6.1 we considered the problem of simultaneous testing: Jensen [1972] gives approximate procedures for testing, simultaneously, hypotheses about various subsets of $\boldsymbol{\mu}$.

3.3.2 Power of the Test

We note from Section 2.4.2 that F_0 of (3.9) has a noncentral F-distribution with d and $n-d$ degrees of freedom and noncentrality parameter δ when H_0 is false. To find δ, let $\mathbf{y} = \sqrt{n}(\bar{\mathbf{x}} - \boldsymbol{\mu}_0)$, $\mathbf{W} = (n-1)\mathbf{S}$, and $\boldsymbol{\theta} = \mathscr{E}[\mathbf{y}]$. Then $T_0^2 = (n-1)\mathbf{y}'\mathbf{W}^{-1}\mathbf{y}$ and, from Section 2.4.2, $\delta = \boldsymbol{\theta}'\boldsymbol{\Sigma}^{-1}\boldsymbol{\theta} = n(\boldsymbol{\mu} - \boldsymbol{\mu}_0)'\boldsymbol{\Sigma}^{-1}(\boldsymbol{\mu} - \boldsymbol{\mu}_0)$. If we denote the noncentral F-distribution by $F_{d,n-d,\delta}$, the power of the test for a critical region of size α is

$$P = \text{pr}\left[F_{d,n-d,\delta} \geq F_{d,n-d}^{\alpha}\right]. \tag{3.14}$$

Here P is a function of $\nu_1 = d$, $\nu_2 = n - d$, and δ. It is extensively tabulated by Tiku [1967, 1972] for a range of values of ν_1, ν_2, and the "standardized" noncentrality parameter $\phi = [\delta/(\nu_1 + 1)]^{1/2}$.

In designing an experiment we may wish to choose n to detect a given value of δ with a given power P. Charts are available to help us solve this problem (e.g., Pearson and Hartley [1972: Table 30]). We set the values of α, ν_1, and ϕ and read ν_2 off the appropriate chart.

3.3.3 Robustness of the Test

Departures from the normal distribution are generally measured by the coefficients of skewness and kurtosis (see Section 2.7). For the univariate t-test it has been demonstrated, both theoretically and empirically, that the test is more sensitive to the skewness $\sqrt{\beta_1}$ than to the kurtosis β_2, and we might expect a similar property to hold for T_0^2. Using Monte Carlo simulation, this has been demonstrated by several authors for the bivariate case $d = 2$ (Arnold [1964], Mardia [1970: p. 529], Chase and Bulgren [1971]), and a short summary table reproduced from Mardia [1975: Table 2a] is given here as Table 3.2. Table 3.2 suggests that T_0^2 is reasonably robust with regard to the nominal significance level when the underlying distribution is symmetrical ($\beta_{1,d} = \beta_{1,2} = 0$). However, for skew distributions like the negative exponential, the actual significance level can be many times the nominal level, for example, three times at 5% and six times at 1% (Mardia [1970: Table 3]). These conclusions are supported by the simulation study of Everitt [1979a] for $d > 2$, which shows that any adverse effects tend to increase with d, and by the contamination study of Srivastava and Awan [1982].

Using permutation theory, Mardia [1970] suggested that, for nonnormal data, F_0 of (3.9) is approximately distributed as $F_{ad,a(n-d)}$ when H_0 is true.

TABLE 3.2 Monte Carlo Studies of the One-Sample Hotelling's T^2 for the Bivariate Case ($M = 100 \times$ Maximum of $|\alpha - 0.05|$, α is the Actual Size of the Nominally 5% Test)[a]

Parent Population	Source	Sample Sizes	$\beta_{1,2}$	$\beta_{2,2}$	M (%)
Uniform: independent	Arnold [1964]	4, 6, 8	0	5.6	2
Uniform: Morgenstern	Chase and Bulgren [1971]	5, 10, 20	0	5.60, 5.63	3
Double exponential: independent	Arnold [1964]	8	0	14	1
Negative exponential: independent	Mardia [1970]	4, 6, 8	8	20	10
Negative exponential: Marshall and Olkin	Chase and Bulgren [1971]	5, 10, 20	(8, 10)[b]	(20, 40)	9

[a] From Mardia [1975: Table 2a].
[b] $8 \le \beta_{1,2} \le 10$.

Here

$$a = 1 + \frac{1}{n}\left[\frac{\beta_{2,d} - d(d+2)}{d}\right] + o(n^{-1}), \tag{3.15}$$

and $a \approx 1$ when the data are normal $[\beta_{2,d} = d(d+2)]$. In practice $\beta_{2,d}$ is estimated by $b_{2,d}$ of (2.99). However, Everitt [1979a] showed that this method is inappropriate for skewed distributions like the exponential and lognormal. It is not surprising, as the previous paragraph indicates, that any correction factor like a would be expected to depend on the skewness $\beta_{1,d}$. It transpires that the permutation moments used for deriving a (see Arnold [1964]) do in fact assume symmetry indirectly, so that Mardia's test is recommended for symmetric distributions only. Tests for multivariate skewness and kurtosis are discussed in Section 4.3. A promising robust test of H_0 based on Winsorized ranks is given by Utts and Hettmansperger [1980]. Another robust test is given by Tiku and Singh [1982].

3.3.4 *Step-Down Test Procedure*

Another method for testing $H_0: \boldsymbol{\mu} = \mathbf{0}$, called the step-down procedure, has been proposed (Roy [1958]). Defining $\boldsymbol{\mu}'_{k-1} = (\mu_1, \mu_2, \ldots, \mu_{k-1})$ $(k = 1, 2, \ldots, d;\ \boldsymbol{\mu}_0 = 0)$, we can use the first k elements of each $\mathbf{x}_i$ and test a hypothesis of the form $H_{0k}: \mu_k - \boldsymbol{\gamma}'_{k-1}\boldsymbol{\mu}_{k-1} = 0$ using (see Theorem 2.11 in Section 2.6 with $\boldsymbol{\mu}^{(2)} = \mu_k$ and $\boldsymbol{\mu}^{(1)} = \boldsymbol{\mu}_{k-1}$)

$$F_k = \frac{T^2_{(k)} - T^2_{(k-1)}}{(n-1) + T^2_{(k-1)}} \frac{n-k}{1} \qquad \left(k = 1, 2, \ldots, d;\ T^2_{(0)} = 0\right).$$

Here $T^2_{(k)}$ is the Hotelling's T^2 statistic for testing $\boldsymbol{\mu}_k = \mathbf{0}$: When H_{0k} is true, $F_k \sim F_{1,n-k}$. We note that if $\cap_{k=1}^{r-1} H_{0k}$ is true, then $\boldsymbol{\mu}_{r-1} = \mathbf{0}$ and H_{0r} is true if and only if $\mu_r = 0$ $(r = 1, 2, \ldots, d)$. In particular, $H_0 = \cap_{k=1}^{d} H_{0k}$ so that H_0 can be tested using the sequence of test statistics $F_1, F_2, \ldots, F_d$, in that order. The hypothesis H_0 is not rejected if all d statistics are not significant, while H_0 is rejected if one of the tests is significant. From Theorem 2.11 it follows that F_k is independent of $T^2_{(k-1)}$. Since the elements that make up $T^2_{(k-r)}$ $(r > 1)$ are a subset of those making up $T^2_{(k-1)}$, the proof of Theorem 2.11 can be used to show that F_k is statistically independent of $\{F_1, F_2, \ldots, F_{k-1}\}$. A simple extension of this argument [see (8.79)] shows that when H_0 is true, the F_k are mutually independent and are distributed as $F_{1,n-k}$ $(k = 1, 2, \ldots, d)$. The size α of the overall procedure is then related to the levels α_k of the component tests by

$$1 - \alpha = \prod_{k=1}^{d} (1 - \alpha_i). \tag{3.16}$$

The above procedure is an alternative to Hotelling's T_0^2 for H_0 and it may be used when there is an a priori ordering among the μ_k. Subbaiah and Mudholkar [1978] have compared the powers of the two procedures using simulation and concluded that if the α_k are set equal, there is not a great difference between the powers. As might be expected, if α_1 is made large, the power of the step-down procedure is substantially greater for detecting nonzero μ_1 than the corresponding power of T_0^2. They concluded that "when the variables in a multiresponse experiment are of unequal practical significance, and are ordered accordingly, a step-down analysis seems to yield superior inferences on the earlier variables at the expense of the quality of the inferences on the later variables, as compared with the corresponding inferences obtained using conventional methods such as Hotelling's T^2." The above procedure is generalized in Section 8.8.

3.4 LINEAR CONSTRAINTS ON THE MEAN

3.4.1 Generalization of the Paired Comparison Test

Suppose we wish to test the effect of a new drug on diastolic blood pressure by measuring the blood pressure after administering the drug and a placebo (dummy drug) on two separate occasions to each of n patients. For the ith patient $(i = 1, 2, \ldots, n)$ we have a bivariate random vector $\mathbf{x}_i = (x_{i1}, x_{i2})'$ with mean $\boldsymbol{\mu} = (\mu_1, \mu_2)'$, and to test the effect of the drug we consider $H_0: \mu_1 = \mu_2$. Now if we assume the $\mathbf{x}_i$ are i.i.d. $N_2(\boldsymbol{\mu}, \boldsymbol{\Sigma})$, the $y_i = x_{i1} - x_{i2}$ are i.i.d. $N_1(\mu_y, \sigma_y^2)$, where $\mu_y = \mu_1 - \mu_2$ and $\sigma_y^2 = \sigma_{11} - 2\sigma_{12} + \sigma_{22}$. To test $H_0: \mu_y = 0$ we can use the so-called paired comparison t-statistic (Exercise 3.1)

$n^{1/2}\bar{y}/\{\Sigma_i(y_i - \bar{y})^2/(n-1)\}^{1/2}$. If we are interested in a more general hypothesis of the form $a_1\mu_1 + a_2\mu_2 = c$, that is, $\mathbf{a}'\boldsymbol{\mu} = c$, we simply transform to $y_i = \mathbf{a}'\mathbf{x}_i$ and test $\mu_y = c$.

Suppose now that we wish to compare $d-1$ different drugs (or $d-1$ different strengths of the same drug), so that, with the placebo, we have d treatments applied at different times (say, in random order) to each patient. If $\mathbf{x}_1, \mathbf{x}_2, \ldots, \mathbf{x}_n$ are the $d \times 1$ observation vectors with mean $\boldsymbol{\mu}$, the hypothesis of interest, namely, that of no treatment effects, is $H_0: \mu_1 = \mu_2 = \cdots = \mu_d$ or $\mu_1 - \mu_d = \mu_2 - \mu_d = \cdots = \mu_{d-1} - \mu_d$. To test H_0 we transform to $\mathbf{y}_i = (y_{i1}, \ldots, y_{i,d-1})'$, where $y_{ij} = x_{ij} - x_{id}$ $(j = 1, 2, \ldots, d-1)$, and test $\boldsymbol{\mu}_y = \mathscr{E}[\mathbf{y}_i] = \mathbf{0}$. If there are d drugs and no placebo, we might wish to express H_0 in the form $\mu_1 - \mu_2 = \mu_2 - \mu_3 = \cdots = \mu_{d-1} - \mu_d$ and transform to

$$\begin{aligned} y_{i1} &= x_{i1} - x_{i2}, \\ y_{i2} &= x_{i2} - x_{i3}, \\ &\vdots \\ y_{i,d-1} &= x_{i,d-1} - x_{id}. \end{aligned}$$

Once again we test $\mathscr{E}[\mathbf{y}_i] = \mathbf{0}$.

With the previous motivating example we now consider testing a general linear hypothesis of the form $H_0: \mathbf{A}\boldsymbol{\mu} = \mathbf{b}$, where $\mathbf{A}$ is a known $q \times d$ matrix of rank q and $\mathbf{b}$ is a known $q \times 1$ vector. If we set $\mathbf{y}_i = \mathbf{A}\mathbf{x}_i$, then, by Theorem 2.1(i), the $\mathbf{y}_i$ are i.i.d. $N_q(\boldsymbol{\mu}_y, \boldsymbol{\Sigma}_y)$, where $\boldsymbol{\mu}_y = \mathbf{A}\boldsymbol{\mu}$ and $\boldsymbol{\Sigma}_y = \mathbf{A}\boldsymbol{\Sigma}\mathbf{A}'$. We can now test H_0 by testing $\boldsymbol{\mu}_y = \mathbf{b}$, using

$$\begin{aligned} T_0^2 &= n(\bar{\mathbf{y}} - \mathbf{b})'\mathbf{S}_y^{-1}(\bar{\mathbf{y}} - \mathbf{b}), \\ &= n(\mathbf{A}\bar{\mathbf{x}} - \mathbf{b})'(\mathbf{ASA}')^{-1}(\mathbf{A}\bar{\mathbf{x}} - \mathbf{b}). \end{aligned} \tag{3.17}$$

When H_0 is true, it follows from Section 3.3.1 that $T_0^2 \sim T^2_{q,n-1}$. We now show that H_0 can also be expressed in the form $\mathbf{A}\boldsymbol{\gamma} = \mathbf{0}$. Let $\boldsymbol{\mu}_1$ be any solution of $\mathbf{A}\boldsymbol{\mu} = \mathbf{b}$ and let $\boldsymbol{\gamma} = \boldsymbol{\mu} - \boldsymbol{\mu}_1$ and $\mathbf{z}_i = \mathbf{x}_i - \boldsymbol{\mu}_1$. Then the $\mathbf{z}_i$ are i.i.d. $N_d(\boldsymbol{\gamma}, \boldsymbol{\Sigma})$ and H_0 takes the form $\mathbf{A}\boldsymbol{\gamma} = \mathbf{A}\boldsymbol{\mu} - \mathbf{A}\boldsymbol{\mu}_1 = \mathbf{b} - \mathbf{b} = \mathbf{0}$. We proceed as in (3.17), using $\mathbf{z}_i$ instead of $\mathbf{x}_i$ so that $T_0^2 = n(\mathbf{A}\bar{\mathbf{z}})'(\mathbf{ASA}')^{-1}\mathbf{A}\bar{\mathbf{z}}$.

Having described the general theory of hypothesis testing we now apply the method to several examples.

3.4.2 *Some Examples*

a REPEATED-MEASUREMENT DESIGNS

The blood pressure example of the previous section is a special case of the so-called *repeated measurement* design in which d similar measurements are made on the *same* sampling unit or person. Referring to the general linear hypothesis described above, if we express $H_0: \mu_1 = \mu_2 = \cdots = \mu_d \ (= \mu, \text{ say})$ or $\boldsymbol{\mu} = \mathbf{1}_d\mu$ in the form $\mathbf{A}\boldsymbol{\mu} = \mathbf{0}$, then $\mathbf{A}$ is a $(d-1) \times d$ matrix of rank $d-1$, satisfying $\mathbf{A}\mathbf{1}_d = \mathbf{0}$ (by B2.3). However, $\mathbf{A}$ is not unique. For example, using $H_0 : \mu_1 - \mu_2 = \cdots = \mu_{d-1} - \mu_d = 0$ leads to

$$\mathbf{C}_1\boldsymbol{\mu} = \begin{pmatrix} 1 & -1 & 0 & \cdots & 0 & 0 \\ 0 & 1 & -1 & \cdots & 0 & 0 \\ \vdots & \vdots & \vdots & & \vdots & \vdots \\ 0 & 0 & 0 & \cdots & 1 & -1 \end{pmatrix} \boldsymbol{\mu} = \mathbf{0}, \tag{3.18}$$

while $H_0: \mu_1 - \mu_d = \cdots = \mu_{d-1} - \mu_d = 0$ leads to

$$\mathbf{C}_2\boldsymbol{\mu} = \begin{pmatrix} 1 & 0 & \cdots & 0 & -1 \\ 0 & 1 & \cdots & 0 & -1 \\ \vdots & \vdots & & \vdots & \vdots \\ 0 & 0 & \cdots & 1 & -1 \end{pmatrix} \boldsymbol{\mu} = \mathbf{0}. \tag{3.19}$$

The $(d-1) \times d$ matrix $\mathbf{C}_2$ can be transformed to $\mathbf{C}_1$ by subtracting the second row of $\mathbf{C}_2$ from the first, the third row from the second, and similarly for the rest, so that $\mathbf{C}_1 = \mathbf{F}_1\mathbf{C}_2$, where $\mathbf{F}_1$ is a $(d-1) \times (d-1)$ nonsingular matrix, nonsingular because rank $\mathbf{C}_1 =$ rank $\mathbf{C}_2 = d - 1$ (see A2.2). In fact, if $\mathbf{A}\boldsymbol{\mu} = \mathbf{0}$ is any representation of H_0, where $\mathbf{A}$ is $(d-1) \times d$ of rank $d-1$, then the rows of $\mathbf{A}$ and of $\mathbf{C}_1$ each form a basis for the same vector space, the orthogonal complement of $\mathbf{1}_d$. Hence $\mathbf{A} = \mathbf{F}\mathbf{C}_1$ for some nonsingular $\mathbf{F}$ and, by the invariance of Hotelling's T_0^2 statistic under nonsingular transformations, the test statistic based on $\mathbf{A}\mathbf{x}_i = \mathbf{F}(\mathbf{C}_1\mathbf{x}_i)$ is the same as that based on $\mathbf{y}_i = \mathbf{C}_1\mathbf{x}_i$, the former being (3.17) with $\mathbf{b} = \mathbf{0}$.

The idea of transforming $\mathbf{x}_i$ to $\mathbf{y}_i = \mathbf{C}_1\mathbf{x}_i$ to test H_0 is due to Hsu [1938] and is discussed further by Williams [1970] and Morrison [1972]. From Exercise 3.2 we have the following identity, due to Williams [1970], which can be used if $\mathbf{S}^{-1}$ (or the Cholesky decomposition of $\mathbf{S}$; see Section 10.1.3) is available from other analyses of the data:

$$n\bar{\mathbf{x}}'\mathbf{S}^{-1}\bar{\mathbf{x}} - \frac{n\left(\bar{\mathbf{x}}'\mathbf{S}^{-1}\mathbf{1}_d\right)^2}{\mathbf{1}_d'\mathbf{S}^{-1}\mathbf{1}_d} = n\bar{\mathbf{x}}'\mathbf{A}'(\mathbf{A}\mathbf{S}\mathbf{A}')^{-1}\mathbf{A}\bar{\mathbf{x}}. \tag{3.20}$$

Morrison [1972] also discusses the problem of testing H_0 when (1) $\boldsymbol{\Sigma}$ takes the form

$$\boldsymbol{\Sigma} = \sigma^2 \begin{pmatrix} 1 & \rho & \cdots & \rho & \rho \\ \rho & 1 & \cdots & \rho & \rho \\ \vdots & \vdots & & \vdots & \vdots \\ \rho & \rho & \cdots & \rho & 1 \end{pmatrix} \tag{3.21}$$

(see Exercise 3.3 and Arnold [1979]); (2) $\mathbf{\Sigma}$ is reducible, that is, there exists a $(d-1) \times d$ matrix $\mathbf{A}$ of rank $d-1$ such that $\mathbf{A1}_d = \mathbf{0}$ and $\mathbf{A\Sigma A}'$ is diagonal; (3) $\mathbf{\Sigma}$ has compound symmetry, a block generalization of (3.21); and (4) $\mathbf{\Sigma} = \sigma^2\mathbf{\Sigma}_0$, $\mathbf{\Sigma}_0$ known. The so-called intraclass correlation model leading to (3.21) is sometimes considered in the context of mixed model analysis of variance (e.g., Winer [1962], Scheffé [1959: p. 264]). Morrison [1972] also briefly considers the problem of incomplete data.

The literature on repeated-measurement designs is reviewed by Hedayat and Afsarinejad [1975]. In conclusion, we note that several authors have considered the problem of testing hypotheses about $\boldsymbol{\mu}$ when both null and alternative hypotheses are specified by linear inequalities of the form $\mathbf{b}'\boldsymbol{\mu} \geq 0$ (see Sasabuchi [1980]).

EXAMPLE 3.2 Rao [1948] introduced the now famous example of testing whether the thickness of cork borings on trees was the same in the north, east, south and west directions; thus $d = 4$ and H_0 is $\mu_1 = \mu_2 = \mu_3 = \mu_4$. Observations $\mathbf{x}_i$ on 28 trees are given in Table 3.3, for which

$$\bar{\mathbf{x}} = \begin{pmatrix} 50.5357 \\ 46.1786 \\ 49.6786 \\ 45.1786 \end{pmatrix} \begin{matrix} \mathrm{N} \\ \mathrm{E} \\ \mathrm{S} \\ \mathrm{W} \end{matrix} \quad \text{and} \quad \mathbf{Q}_x = \begin{pmatrix} 7840.96 & 6041.32 & 7787.82 & 6109.32 \\ - & 5938.11 & 6184.61 & 4627.11 \\ - & - & 9450.11 & 7007.61 \\ - & - & - & 6102.11 \end{pmatrix}.$$

TABLE 3.3 Weight of Cork Borings (in Centigrams) in Four Directions for 28 Trees[a]

N	E	S	W	N	E	S	W
72	66	76	77	91	79	100	75
60	53	66	63	56	68	47	50
56	57	64	58	79	65	70	61
41	29	36	38	81	80	68	58
32	32	35	36	78	55	67	60
30	35	34	26	46	38	37	38
39	39	31	27	39	35	34	37
42	43	31	25	32	30	30	32
37	40	31	25	60	50	67	54
33	29	27	36	35	37	48	39
32	30	34	28	39	36	39	31
63	45	74	63	50	34	37	40
54	46	60	52	43	37	39	50
47	51	52	43	48	54	57	43

[a]From Rao [1948: Table 1], by permission of the Biometrika Trustees.

Using the transformation $\mathbf{y}_i = \mathbf{C}_1\mathbf{x}_i$, where $\mathbf{C}_1$ is given by (3.18), we obtain

$$\bar{\mathbf{y}} = \mathbf{C}_1\bar{\mathbf{x}} = \begin{pmatrix} 4.3571 \\ -3.5000 \\ 4.5000 \end{pmatrix},$$

and

$$\mathbf{Q}_y = \mathbf{C}_1\mathbf{Q}_x\mathbf{C}_1' = \begin{pmatrix} 1696.43 & -1500.0 & 121.0 \\ — & 3019.0 & -885.0 \\ — & — & 1537.0 \end{pmatrix}.$$

From (3.17), the test statistic for H_0 is

$$T_0^2 = n(n-1)\bar{\mathbf{y}}'\mathbf{Q}_y^{-1}\bar{\mathbf{y}} = 20.742,$$

and the corresponding F-value, with d now equal to 3, is [see (3.9)]

$$F_0 = \frac{T_0^2}{n-1}\frac{n-d}{d} = \frac{20.742}{27}\frac{25}{3} = 6.40.$$

As $F_0 > F_{3,25}^{0.01} = 4.68$, we reject H_0 at the 1% level of significance so that the bark deposit cannot be considered uniform in the four directions.

The univariate plots for each of the four variables suggest that the marginal distributions are skew. However, in discussing this problem, Mardia [1975] found that the sample multivariate measures of skewness and kurtosis, namely,

$$b_{1,4} = 4.476 \quad \text{and} \quad b_{2,4} = 22.957,$$

were not significant when used as tests for multivariate normality (Section 4.3.2). More appropriately, the coefficients for the transformed data $\mathbf{y}_i$ were

$$b_{1,3} = 1.177 \quad \text{and} \quad b_{2,3} = 13.558,$$

indicating that the $\mathbf{y}_i$ were more like multivariate normal data. Also the T_0^2 test is robust to mild skewness and moderate kurtosis.

As $b_{1,3}$ is small, we can consider the correction (3.15) to the degrees of freedom of F_0, namely,

$$\begin{aligned} a &= 1 + \frac{b_{2,d} - d(d+2)}{nd} \\ &= 1 + \frac{13.558 - 15}{28 \times 3} \\ &= 0.98. \end{aligned}$$

As $a \approx 1$, the correction is unnecessary.

b TESTING SPECIFIED CONTRASTS

In testing whether the μ_i are equal, we saw above that we have a wide choice for the matrix **A**, though $\mathbf{C}_1$ will do in many cases. However, we may be interested in choosing **A** so that attention is focused on certain prescribed contrasts $\mathbf{c}'\boldsymbol{\mu}$ ($\mathbf{c}'\mathbf{1}_d = 0$).

EXAMPLE 3.3 In Example 3.2, Rao was particularly interested in cork differences relating to the contrast north–south versus east–west ($\mu_1 + \mu_3 - \mu_2 - \mu_4$), to which he added the contrasts south versus west ($\mu_3 - \mu_4$) and north versus south ($\mu_1 - \mu_3$). This choice of contrasts leads to the matrix

$$\mathbf{A} = \begin{pmatrix} 1 & -1 & 1 & -1 \\ 0 & 0 & 1 & -1 \\ 1 & 0 & -1 & 0 \end{pmatrix}.$$

Since testing $H_0: \mathbf{A}\boldsymbol{\mu} = \mathbf{0}$ is equivalent to testing $\mu_1 = \mu_2 = \mu_3 = \mu_4$, the value of the test statistic F_0 is the same as that in Example 3.2, namely, 6.402. By the same token, the measures of skewness and kurtosis, $b_{1,3}$ and $b_{2,3}$, are also the same, as **A** is related to $\mathbf{C}_1$ by a nonsingular transformation.

c TEST FOR SYMMETRY

The following is an interesting problem that arises naturally in biology and anthropology. Suppose that $\mathbf{x}_i$ is now a $2d \times 1$ vector in which the first d elements represent measurements or characteristics on the left side of the ith member of the sample, and the second d elements represent the same measurements on the right side. Alternatively, the first d elements could refer to measurements on one of a twin pair and the second d elements refer to the other member. Then a natural hypothesis to test is the symmetry hypothesis $H_0: \mu_i = \mu_{i+d}$ ($i = 1, 2, \ldots, d$) that represents the equality of the left and right sides of an individual or the equality of measurements from twins. In this case H_0 can be expressed in the form $\mathbf{A}\boldsymbol{\mu} = \mathbf{0}$, where $\mathbf{A} = (\mathbf{I}_d, -\mathbf{I}_d)$, so that the above general theory applies, provided that $2d \leq n - 1$. If $\bar{\mathbf{x}}$ and **S** are partitioned in a similar fashion to **A**, namely,

$$\bar{\mathbf{x}} = \begin{pmatrix} \bar{\mathbf{x}}_1 \\ \bar{\mathbf{x}}_2 \end{pmatrix} \begin{matrix} \}d \\ \}d \end{matrix} \quad \text{and} \quad \mathbf{S} = \begin{pmatrix} \overbrace{\mathbf{S}_{11}}^{d} & \overbrace{\mathbf{S}_{12}}^{d} \\ \mathbf{S}_{21} & \mathbf{S}_{22} \end{pmatrix} \begin{matrix} \}d \\ \}d \end{matrix},$$

then, from (3.17),

$$\begin{aligned} T_0^2 &= n(\mathbf{A}\bar{\mathbf{x}})'(\mathbf{A}\mathbf{S}\mathbf{A}')^{-1}\mathbf{A}\bar{\mathbf{x}} \qquad (3.22) \\ &= n(\bar{\mathbf{x}}_1 - \bar{\mathbf{x}}_2)'(\mathbf{S}_{11} - \mathbf{S}_{12} - \mathbf{S}_{21} + \mathbf{S}_{22})^{-1}(\bar{\mathbf{x}}_1 - \bar{\mathbf{x}}_2). \end{aligned}$$

When H_0 is true, $T_0^2 \sim T_{d,n-1}^2$.

d TESTING FOR A POLYNOMIAL GROWTH TREND

Suppose we have a group of n animals all subject to the same conditions, and the size (e.g., length) of each animal is observed at d points in time $t_1, t_2, \ldots, t_d$. Clearly the d lengths, $\mathbf{x}_i$, say, for the ith animal will not be independent, but will be correlated with an unknown covariance structure $\boldsymbol{\Sigma}$. Assuming that each $\mathbf{x}_i$ is $N_d(\boldsymbol{\mu}, \boldsymbol{\Sigma})$, we may wish to test the hypothesis H_0 that the "growth curve" is a polynomial of degree $k - 1$, that is (Rao [1959, 1973]),

$$H_0: \mu_j = \beta_0 + \beta_1 t_j + \cdots + \beta_{k-1} t_j^{k-1}, \qquad j = 1, 2, \ldots, d. \tag{3.23}$$

This hypothesis is a special case of the more general hypothesis $H_0: \boldsymbol{\mu} = \mathbf{K}\boldsymbol{\beta}$, where $\mathbf{K}$ is a $d \times k$ matrix of rank k ($k < d$), or, equivalently $\mathbf{A}\boldsymbol{\mu} = \mathbf{0}$, for some $\mathbf{A}$. The simplest choice for the rows of $\mathbf{A}$ are any $d - k$ linearly independent rows of $\mathbf{I}_d - \mathbf{P}$, where $\mathbf{P} = \mathbf{K}(\mathbf{K}'\mathbf{K})^{-1}\mathbf{K}'$ represents the orthogonal projection of R^d onto $\mathscr{R}[\mathbf{K}]$ (see B2.3). Rao [1959] calls $\mathbf{A}$ the deficiency matrix of $\mathbf{K}'$. The test for H_0 is again given by (3.22) and $T_0^2 \sim T_{d-k, n-1}^2$ when H_0 is true. An alternative method for calculating T_0^2 that does not require $\mathbf{A}$ is given in Section 3.4.3.

e INDEPENDENT ESTIMATE OF Σ

The preceding examples are all special cases of the following problem. Suppose we are given a *single* random vector $\mathbf{x}$ that is $N_d(\boldsymbol{\mu}, \boldsymbol{\Sigma})$ together with an independent matrix $\mathbf{W}$ obtained, say, from previous data, where $\mathbf{W} \sim W_d(m, \boldsymbol{\Sigma})$. If we set $\mathbf{y} = \mathbf{A}\mathbf{x}$, then, from Theorem 2.8, we can test $H_0: \mathbf{A}\boldsymbol{\mu} = \mathbf{0}$ for a $q \times d$ matrix $\mathbf{A}$ of rank q using $T_0^2 = m\mathbf{y}'(\mathbf{A}\mathbf{W}\mathbf{A}')^{-1}\mathbf{y}$. When H_0 is true, $T_0^2 \sim T_{q, m}^2$.

3.4.3 Minimization Technique for the Test Statistic

In Section 3.4.2 we saw how to test the hypothesis $H_0: \boldsymbol{\mu} = \mathbf{K}\boldsymbol{\beta}$ by transforming it to the form $H_0: \mathbf{A}\boldsymbol{\mu} = \mathbf{0}$. Rao [1959, 1973] has given a general method for finding T_0^2 that incorporates either representation of H_0. This method is described by the following theorem, which generalizes Rao [1959: pp. 50–51].

THEOREM 3.3 Let $\mathbf{x}_1, \mathbf{x}_2, \ldots, \mathbf{x}_n$ be i.i.d. $N_d(\boldsymbol{\mu}, \boldsymbol{\Sigma})$, $d \le n - 1$, and let H_0 be the linear hypothesis that $\boldsymbol{\mu} \in V$, where V is a p-dimensional vector subspace of R^d, d-dimensional Euclidean space. Then

$$\underset{\boldsymbol{\mu} \in V}{\text{minimum}}\, n(\bar{\mathbf{x}} - \boldsymbol{\mu})'\mathbf{S}^{-1}(\bar{\mathbf{x}} - \boldsymbol{\mu}) \sim T_{d-p, n-1}^2$$

when H_0 is true.

Proof Let $p_0 = d - p$ and let $\mathbf{P}$ be the symmetric idempotent $d \times d$ matrix representing the orthogonal projection of R^d onto V (see B1.2). We can argue as in Section 3.4.2d and let $\mathbf{A}$ be the $p_0 \times d$ matrix of rank p_0 whose rows are any p_0 linearly independent rows of $\mathbf{I}_d - \mathbf{P}$. Then $V = \mathscr{N}[\mathbf{A}]$, the null space of $\mathbf{A}$, and $H_0: \mathbf{A}\boldsymbol{\mu} = \mathbf{0}$. Since $\mathbf{S}$, and therefore $\mathbf{S}^{-1}$, is positive definite (with probability 1), we can write $\mathbf{S}^{-1} = \mathbf{R}'\mathbf{R}$ (A5.3), where $\mathbf{R}$ is a $d \times d$ nonsingular matrix. If $\mathbf{z} = \mathbf{R}\bar{\mathbf{x}}$ and $\boldsymbol{\theta} = \mathbf{R}\boldsymbol{\mu}$, we wish to minimize

$$(\bar{\mathbf{x}} - \boldsymbol{\mu})'\mathbf{S}^{-1}(\bar{\mathbf{x}} - \boldsymbol{\mu}) = \|\mathbf{z} - \boldsymbol{\theta}\|^2 \tag{3.24}$$

subject to $\mathbf{0} = \mathbf{A}\boldsymbol{\mu} = \mathbf{A}\mathbf{R}^{-1}\mathbf{R}\boldsymbol{\mu} = \mathbf{A}\mathbf{R}^{-1}\boldsymbol{\theta} = \mathbf{A}_1\boldsymbol{\theta}$, say, that is, subject to $\boldsymbol{\theta} \in \mathscr{N}[\mathbf{A}_1] = V_1$, say. This is a standard least squares problem and (3.24) is minimized, subject to $\boldsymbol{\theta} \in V_1$, when $\boldsymbol{\theta} = \hat{\boldsymbol{\theta}} = \mathbf{P}_1\mathbf{z}$ and (B2.2) $\mathbf{P}_1 = \mathbf{I}_d - \mathbf{A}_1'(\mathbf{A}_1\mathbf{A}_1')^{-1}\mathbf{A}_1 = \mathbf{I}_d - \mathbf{P}_2$, say. Hence $\mathbf{z} - \hat{\boldsymbol{\theta}} = \mathbf{P}_2\mathbf{z}$ and

$$\begin{aligned}
\underset{\mathbf{A}\boldsymbol{\mu}=\mathbf{0}}{\text{minimum}}\,(\bar{\mathbf{x}} - \boldsymbol{\mu})'\mathbf{S}^{-1}(\bar{\mathbf{x}} - \boldsymbol{\mu}) &= \|\mathbf{z} - \hat{\boldsymbol{\theta}}\|^2 \\
&= \mathbf{z}'\mathbf{P}_2'\mathbf{P}_2\mathbf{z} \\
&= \mathbf{z}'\mathbf{P}_2\mathbf{z} \\
&= (\mathbf{A}_1\mathbf{z})'(\mathbf{A}_1\mathbf{A}_1')^{-1}\mathbf{A}_1\mathbf{z} \\
&= (\mathbf{A}\bar{\mathbf{x}})'\left[\mathbf{A}\mathbf{R}^{-1}(\mathbf{R}^{-1})'\mathbf{A}'\right]^{-1}\mathbf{A}\bar{\mathbf{x}} \\
&= (\mathbf{A}\bar{\mathbf{x}})'[\mathbf{A}\mathbf{S}\mathbf{A}']^{-1}\mathbf{A}\bar{\mathbf{x}} \\
&= T_0^2/n,
\end{aligned} \tag{3.25}$$

where T_0^2 is the usual Hotelling's T^2 statistic (3.22) for testing $\mathbf{A}\boldsymbol{\mu} = \mathbf{0}$. Thus when H_0 is true, $T_0^2 \sim T^2_{p_0, n-1}$.

COROLLARY If $H_0: \mathbf{A}\boldsymbol{\mu} = \mathbf{b}$, then

$$\underset{\mathbf{A}\boldsymbol{\mu}=\mathbf{b}}{\text{minimum}}\,(\bar{\mathbf{x}} - \boldsymbol{\mu})'\mathbf{S}^{-1}(\bar{\mathbf{x}} - \boldsymbol{\mu}) = \frac{T_0^2}{n}. \tag{3.26}$$

Proof We use the method following (3.17) to "eliminate" $\mathbf{b}$. If $\boldsymbol{\mu}_1$ is a solution of $\mathbf{A}\boldsymbol{\mu} = \mathbf{b}$, $\mathbf{u}_i = \mathbf{x}_i - \boldsymbol{\mu}_1$, and $\boldsymbol{\gamma} = \boldsymbol{\mu} - \boldsymbol{\mu}_1 = \mathscr{E}[\mathbf{u}_i]$, then

$$\begin{aligned}
\min_{\mathbf{A}\boldsymbol{\mu}=\mathbf{b}}\,(\bar{\mathbf{x}} - \boldsymbol{\mu})'\mathbf{S}^{-1}(\bar{\mathbf{x}} - \boldsymbol{\mu}) &= \min_{\mathbf{A}\boldsymbol{\mu}=\mathbf{b}}\,[\bar{\mathbf{u}} - (\boldsymbol{\mu} - \boldsymbol{\mu}_1)]'\mathbf{S}_u^{-1}[\bar{\mathbf{u}} - (\boldsymbol{\mu} - \boldsymbol{\mu}_1)] \\
&= \min_{\mathbf{A}\boldsymbol{\gamma}=\mathbf{0}}\,(\bar{\mathbf{u}} - \boldsymbol{\gamma})'\mathbf{S}_u^{-1}(\bar{\mathbf{u}} - \boldsymbol{\gamma}),
\end{aligned}$$

which is of the same form as (3.25). □

Since T_0^2 is the minimum of $(\bar{\mathbf{x}} - \boldsymbol{\mu})'\mathbf{S}^{-1}(\bar{\mathbf{x}} - \boldsymbol{\mu})$ under H_0, it does not matter how we actually arrive at the minimum. For example, the test statistic for testing $H_0: \mu_1 = \mu_2 = \cdots = \mu_d$ can be derived as in (3.17) or else as in Exercise 3.4. The growth model hypothesis $H_0: \boldsymbol{\mu} = \mathbf{K}\boldsymbol{\beta}$ of (3.23) can also be tackled differently. From Theorem 3.3 we can find T_0^2 by minimizing $n(\bar{\mathbf{x}} - \boldsymbol{\mu})'\mathbf{S}^{-1}(\bar{\mathbf{x}} - \boldsymbol{\mu})$ subject to $\boldsymbol{\mu} = \mathbf{K}\boldsymbol{\beta}$. Now minimizing

$$R = (\bar{\mathbf{x}} - \mathbf{K}\boldsymbol{\beta})'\mathbf{S}^{-1}(\bar{\mathbf{x}} - \mathbf{K}\boldsymbol{\beta})$$

$$= \bar{\mathbf{x}}'\mathbf{S}^{-1}\bar{\mathbf{x}} - 2\boldsymbol{\beta}'\mathbf{K}'\mathbf{S}^{-1}\bar{\mathbf{x}} + \boldsymbol{\beta}'\mathbf{K}'\mathbf{S}^{-1}\mathbf{K}\boldsymbol{\beta}$$

with regard to $\boldsymbol{\beta}$ is a standard problem of generalized least squares (Seber [1977: Section 3.6]). The answer is obtained by differentiating with respect to $\boldsymbol{\beta}$ using A8.1, namely,

$$-2\mathbf{K}'\mathbf{S}^{-1}\bar{\mathbf{x}} + 2\mathbf{K}'\mathbf{S}^{-1}\mathbf{K}\boldsymbol{\beta} = \mathbf{0}, \tag{3.27}$$

which has solution $\boldsymbol{\beta}^* = (\mathbf{K}'\mathbf{S}^{-1}\mathbf{K})^{-1}\mathbf{K}'\mathbf{S}^{-1}\bar{\mathbf{x}}$. Thus

$$T_0^2 = nR_{\min}$$

$$= n(\bar{\mathbf{x}} - \mathbf{K}\boldsymbol{\beta}^*)'\mathbf{S}^{-1}(\bar{\mathbf{x}} - \mathbf{K}\boldsymbol{\beta}^*)$$

$$= n(\bar{\mathbf{x}}'\mathbf{S}^{-1}\bar{\mathbf{x}} - \bar{\mathbf{x}}'\mathbf{S}^{-1}\mathbf{K}\boldsymbol{\beta}^*),$$

and we compare $F_0 = T_0^2(n - d + p)/(n - 1)(d - p)$ with $F^{\alpha}_{d-p,\, n-d+p}$.

EXAMPLE 2.4 The height, in millimeters, of the ramus bone in the jaws of 20 boys was measured at ages 8, $8\frac{1}{2}$, 9, and $9\frac{1}{2}$, and the results from Elston and Grizzle [1962] are given in Table 3.4. A main objective of the study was to establish a standard growth curve for the use of orthodontists, and it is clear from the sample means that a straight line should provide a satisfactory fit for the age range considered. Following Grizzle and Allen [1969], we can use orthogonal polynomials and consider the model

$$\mu_j = \beta_0 + 2\beta_1(t_j - \bar{t}) \qquad (j = 1, 2, 3, 4),$$

where age 8 is chosen as the time origin so that $t_j = j - 1$. Then $\boldsymbol{\mu} = \mathbf{K}\boldsymbol{\beta}$, where

$$\mathbf{K}\boldsymbol{\beta} = \begin{pmatrix} 1 & -3 \\ 1 & -1 \\ 1 & 1 \\ 1 & 3 \end{pmatrix} \begin{pmatrix} \beta_0 \\ \beta_1 \end{pmatrix},$$

TABLE 3.4 Ramus Height of 20 Boys[a]

Individual	Age in Years			
	8	$8\frac{1}{2}$	9	$9\frac{1}{2}$
1	47.8	48.8	49.0	49.7
2	46.4	47.3	47.7	48.4
3	46.3	46.8	47.8	48.5
4	45.1	45.3	46.1	47.2
5	47.6	48.5	48.9	49.3
6	52.5	53.2	53.3	53.7
7	51.2	53.0	54.3	54.5
8	49.8	50.0	50.3	52.7
9	48.1	50.8	52.3	54.4
10	45.0	47.0	47.3	48.3
11	51.2	51.4	51.6	51.9
12	48.5	49.2	53.0	55.5
13	52.1	52.8	53.7	55.0
14	48.2	48.9	49.3	49.8
15	49.6	50.4	51.2	51.8
16	50.7	51.7	52.7	53.3
17	47.2	47.7	48.4	49.5
18	53.3	54.6	55.1	55.3
19	46.2	47.5	48.1	48.4
20	46.3	47.6	51.3	51.8
Mean	48.655	49.625	50.570	51.450
S. D.	2.52	2.54	2.63	2.73

[a]Reproduced from R. C. Elston and J. E. Grizzle [1962]. "Estimation of time-response curves and their confidence bands." *Biometrics*, **18**, 148–159, Table 2. With permission from The Biometric Society.

and $\mathbf{K}$ has orthogonal columns. Now

$$\bar{\mathbf{x}}' = (48.655,\quad 49.625,\quad 50.570,\quad 51.450),$$

$$\mathbf{S} = \begin{pmatrix} 6.330 & 6.189 & 5.777 & 5.548 \\ - & 6.449 & 6.153 & 5.923 \\ - & - & 6.918 & 6.946 \\ - & - & - & 7.465 \end{pmatrix},$$

and

$$\begin{aligned} \boldsymbol{\beta}^* &= (\mathbf{K}'\mathbf{S}^{-1}\mathbf{K})^{-1}\mathbf{K}'\mathbf{S}^{-1}\bar{\mathbf{x}} \\ &= \begin{pmatrix} 50.050 \\ 0.465 \end{pmatrix}. \end{aligned}$$

Hence

$$T_0^2 = n(\bar{\mathbf{x}} - \mathbf{K}\boldsymbol{\beta}^*)'\mathbf{S}^{-1}(\bar{\mathbf{x}} - \mathbf{K}\boldsymbol{\beta}^*)$$

$$= n(\bar{\mathbf{x}}'\mathbf{S}^{-1}\bar{\mathbf{x}} - \bar{\mathbf{x}}'\mathbf{S}^{-1}\mathbf{K}\boldsymbol{\beta}^*)$$

$$= 0.20113,$$

and we compare

$$F_0 = T_0^2 \frac{20 - 4 + 2}{19(4 - 2)} = 0.095$$

with $F_{2,18}^{\alpha}$. Since F_0 is so low, we do not reject the straight line model.

3.4.4 Confidence Intervals

Suppose we wish to construct a confidence interval for a linear function $\mathbf{a}'\boldsymbol{\mu}$, for example, $\mathbf{a}'\boldsymbol{\mu} = (0,\ldots,0,1,0,\ldots,0)\boldsymbol{\mu} = \mu_i$, or a contrast like $\mu_j - \mu_k$. If $\mathbf{a}$ is prespecified, that is, specified before the data are collected, then we can readily construct a t-confidence interval as follows: $\mathbf{a}'\bar{\mathbf{x}} \sim N_1(\mathbf{a}'\boldsymbol{\mu}, \sigma_a^2/n)$, where $\sigma_a^2 = \mathbf{a}'\boldsymbol{\Sigma}\mathbf{a}$, $(n-1)\mathbf{a}'\mathbf{S}\mathbf{a} \sim \sigma_a^2\chi_{n-1}^2$ (by Theorem 2.2, Corollary 1 in Section 2.3.1), $\mathbf{a}'\bar{\mathbf{x}}$ is statistically independent of $\mathbf{a}'\mathbf{S}\mathbf{a}$ (by Theorem 3.1 in Section 3.2.2), and hence

$$T = \frac{\mathbf{a}'\bar{\mathbf{x}} - \mathbf{a}'\boldsymbol{\mu}}{(\mathbf{a}'\mathbf{S}\mathbf{a}/n)^{1/2}} \sim t_{n-1}.$$

A $100(1 - \alpha)\%$ confidence interval for $\mathbf{a}'\boldsymbol{\mu}$ is therefore

$$\mathbf{a}'\bar{\mathbf{x}} \pm t_{n-1}^{\alpha/2}\left(\frac{\mathbf{a}'\mathbf{S}\mathbf{a}}{n}\right)^{1/2}. \tag{3.28}$$

If we are interested in finding confidence intervals for r prespecified linear combinations $\mathbf{a}_i'\boldsymbol{\mu}$ $(i = 1,2,\ldots,r)$, then we run into the usual problems associated with simultaneous inferences (see Section 1.6.1). We shall now consider two methods for constructing simultaneous confidence intervals of the form

$$\mathbf{a}_i'\bar{\mathbf{x}} \pm k_\alpha\left(\frac{\mathbf{a}_i'\mathbf{S}\mathbf{a}_i}{n}\right)^{1/2}, \qquad i = 1,2,\ldots,r. \tag{3.29}$$

The reader is referred to Seber [1977: Chapter 5] for further background details.

The first method consists of using an individual critical level of α/r instead of α so that $k_\alpha = t_{n-1}^{\alpha/2r}$. These so-called Bonferroni t-intervals are conservative in that the overall probability that the r statements are jointly true is at least $1 - \alpha$ [see (1.21)].

A second method of interval construction can be developed along exactly the same lines as Scheffé's S-method for univariate linear models. We may assume, without loss of generality, that the first q vectors of the set $\{\mathbf{a}_1, \mathbf{a}_2, \ldots, \mathbf{a}_r\}$ are linearly independent, and the remaining vectors (if any) are linearly dependent on the first q vectors; thus $q \leq \min(r, d)$. Let $\mathbf{A}' = [\mathbf{a}_1, \mathbf{a}_2, \ldots, \mathbf{a}_q]$, where $\mathbf{A}$ is $q \times d$ of rank q, and let $\boldsymbol{\phi} = \mathbf{A}\boldsymbol{\mu}$ and $\hat{\boldsymbol{\phi}} = \mathbf{A}\overline{\mathbf{x}}$. If $T^2_{q, n-1, \alpha}$ is the upper α quantile value for the $T^2_{q, n-1}$ distribution then, from (2.47),

$$T^2_{q, n-1, \alpha} = \frac{q(n-1)}{n-q} F^{\alpha}_{q, n-q}. \tag{3.30}$$

Using (3.17), we have

$$\begin{aligned}
1 - \alpha &= \text{pr}\left[T^2_{q, n-1} \leq T^2_{q, n-1, \alpha}\right] \\
&= \text{pr}\left[(\mathbf{A}\overline{\mathbf{x}} - \mathbf{A}\boldsymbol{\mu})'(\mathbf{ASA}')^{-1}(\mathbf{A}\overline{\mathbf{x}} - \mathbf{A}\boldsymbol{\mu}) \leq T^2_{q, n-1, \alpha}/n\right] \\
&= \text{pr}\left[(\hat{\boldsymbol{\phi}} - \boldsymbol{\phi})'\mathbf{L}^{-1}(\hat{\boldsymbol{\phi}} - \boldsymbol{\phi}) \leq m\right] \qquad (3.31)\\
&= \text{pr}[\mathbf{d}'\mathbf{L}^{-1}\mathbf{d} \leq m] \\
&= \text{pr}\left[\sup_{\mathbf{h} \neq \mathbf{0}} \left\{\frac{(\mathbf{h}'\mathbf{d})^2}{\mathbf{h}'\mathbf{L}\mathbf{h}}\right\} \leq m\right] \qquad (\text{by } A7.6) \\
&= \text{pr}\left[\frac{(\mathbf{h}'\mathbf{d})^2}{\mathbf{h}'\mathbf{L}\mathbf{h}} \leq m, \text{ all } \mathbf{h}(\neq \mathbf{0})\right] \\
&= \text{pr}\left[|\mathbf{h}'\hat{\boldsymbol{\phi}} - \mathbf{h}'\boldsymbol{\phi}| \leq \left\{T^2_{q, n-1, \alpha}(\mathbf{h}'\mathbf{L}\mathbf{h})/n\right\}^{1/2}, \text{ all } \mathbf{h}\right].
\end{aligned}$$

We can therefore construct a confidence interval for *any* linear function $\mathbf{h}'\boldsymbol{\phi}$, namely,

$$\mathbf{h}'\hat{\boldsymbol{\phi}} \pm \left(T^2_{q, n-1, \alpha}\right)^{1/2}\left(\frac{\mathbf{h}'\mathbf{L}\mathbf{h}}{n}\right)^{1/2}, \tag{3.32}$$

and the overall probability for the whole class of such intervals is exactly $1 - \alpha$. Note that $\mathbf{h}'\mathbf{L}\mathbf{h}/n$ is simply $\text{var}[\mathbf{h}'\hat{\boldsymbol{\phi}}]$ with $\boldsymbol{\Sigma}$ replaced by its unbiased estimator $\mathbf{S}$. Since $\mathbf{h}'\boldsymbol{\phi} = \phi_i$ for suitable $\mathbf{h}$, a confidence interval for every $\mathbf{a}'_i\boldsymbol{\mu} = \phi_i$ $(i = 1, 2, \ldots, q)$ is included in the set of intervals (3.32). In addition, an interval for every ϕ_j $(j = q+1, q+2, \ldots, r)$ is also included in this set owing to the linear dependence of the $\mathbf{a}_j$ $(j = q+1, \ldots, r)$ on the other $\mathbf{a}_i$'s. For example, if $\mathbf{a}_{q+1} = h_1\mathbf{a}_1 + \cdots + h_q\mathbf{a}_q$, then $\phi_{q+1} = \mathbf{a}'_{q+1}\boldsymbol{\mu} = \sum_{i=1}^{q} h_i\phi_i$.

Since the set of intervals

$$\mathbf{a}_i'\bar{\mathbf{x}} \pm \left(T^2_{q,n-1,\alpha}\right)^{1/2}\left(\frac{\mathbf{a}'\mathbf{S}\mathbf{a}}{n}\right)^{1/2}, \qquad i = 1,2,\ldots,r, \tag{3.33}$$

is a subset of (3.32), the above confidence intervals have an overall probability of at least $1 - \alpha$.

In comparing the two sets of conservative confidence limits, we find, for the common situation of $r = q$ (or r not much greater than q), that

$$t_{n-1}^{\alpha/2r} < \left(qF^{\alpha}_{q,n-1}\right)^{1/2} \leq \left(T^2_{q,n-1,\alpha}\right)^{1/2}.$$

The last inequality follows from (3.30). Thus

$$qF^{\alpha}_{q,n-1} \leq \frac{q(n-1)}{n-q}F^{\alpha}_{q,n-q} = T^2_{q,n-1,\alpha}.$$

The Bonferroni intervals are preferred, as they are generally much shorter than the Scheffé intervals. Values of $t_{\nu}^{\alpha/(2r)}$ are tabulated in Appendix D1.

There is a direct link between the Scheffé-type intervals and the test of $H_0 : \mathbf{A}\boldsymbol{\mu} = \mathbf{b}$. If we set $\boldsymbol{\phi} = \mathbf{A}\boldsymbol{\mu} = \mathbf{b}$, we obtain T_0^2 of (3.17), the test statistic for H_0. This statistic T_0^2 is not significant at the α level of significance if and only if $T_0^2 \leq T^2_{q,n-1,\alpha}$, which is true if and only if [see (3.31)] $(\hat{\boldsymbol{\phi}} - \mathbf{b})'\mathbf{L}^{-1}(\hat{\boldsymbol{\phi}} - \mathbf{b}) \leq m$, that is, if and only if $\mathbf{h}'\mathbf{b}$ belongs to the confidence interval (3.32) for *every* $\mathbf{h}$. If T_0^2 is significant, it is due to at least one interval (3.32) not containing $\mathbf{h}'\mathbf{b}$. In this situation we would first look at the intervals for the ϕ_i $(= \mathbf{a}_i'\boldsymbol{\mu})$ and see which ones, if any, did not contain b_i.

Up till now we have been concerned with finding confidence intervals for *prespecified* $\mathbf{a}_i'\boldsymbol{\mu}$ $(i = 1,2,\ldots,r)$. Suppose, however, we wish to choose our confidence intervals on the basis of the data. Then, setting $q = d$, $\mathbf{A} = \mathbf{I}_d$, and $\boldsymbol{\phi} = \boldsymbol{\mu}$ in the above Scheffé intervals, we obtain a set of confidence intervals for all linear functions $\mathbf{h}'\boldsymbol{\mu}$, namely,

$$\mathbf{h}'\bar{\mathbf{x}} \pm \left(T^2_{d,n-1,\alpha}\right)^{1/2}\left(\frac{\mathbf{h}'\mathbf{S}\mathbf{h}}{n}\right)^{1/2}, \tag{3.34}$$

with an overall probability of $1 - \alpha$. No matter how many intervals we select, we have a "protection" of at least $1 - \alpha$. If $H_0 : \boldsymbol{\mu} = \mathbf{b}$ is rejected, then at least one of (3.34) does not contain $\mathbf{h}'\mathbf{b}$ for some $\mathbf{h}$.

In the repeated-measurement design of Section 3.4.2a we were concerned with testing $H_0 : \mu_1 - \mu_d = \cdots = \mu_{d-1} - \mu_d = 0$. If $\phi_i = \mu_i - \mu_d$ $(i = 1,2,\ldots,d-1)$, then we find (Exercise 3.5) that the set of all linear combinations $h_1\phi_1 + \cdots + h_{d-1}\phi_{d-1}$ is identical to the set of all contrasts $c_1\mu_1 + \cdots + c_d\mu_d$ $(\Sigma_i c_i = 0)$. Thus $\boldsymbol{\phi} = \mathbf{0}$ if and only if $\mathbf{h}'\boldsymbol{\phi} = 0$ for every $\mathbf{h}$, if and only if $\mathbf{c}'\boldsymbol{\mu} = 0$ for every contrast. From (3.32) with $q = d - 1$ we can therefore construct a set of confidence intervals for all contrasts $\mathbf{c}'\boldsymbol{\mu}$, namely,

$$\mathbf{c}'\overline{\mathbf{x}} \pm \left(T^2_{d-1,\,n-1,\,\alpha}\right)^{1/2}\left(\frac{\mathbf{c}'\mathbf{S}\mathbf{c}}{n}\right)^{1/2}, \qquad \mathbf{c}'\mathbf{1}_d = 0. \tag{3.35}$$

If T_0^2 is significant, at least one of the above intervals does not contain zero. However, if H_0 is rejected, we would probably be interested in the $r\ [= \frac{1}{2}d(d-1)]$ contrasts $\mu_i - \mu_j\ (i < j)$, of which the $d - 1$ contrasts $\phi_1, \ldots, \phi_{d-1}$ form a basis. In this case the two methods of interval estimation given above lead to intervals of the form

$$\mathbf{c}'\overline{\mathbf{x}} \pm k_\alpha\left(\frac{\mathbf{c}'\mathbf{S}\mathbf{c}}{n}\right)^{1/2}, \tag{3.36}$$

where $k_\alpha = t_{n-1}^{\alpha/2r}$ for the Bonferroni method and $k_\alpha = (T^2_{d-1,\,n-1,\,\alpha})^{1/2}$ for Scheffé's method. For example, if $d = 4$, $n = 21$, and $\alpha = 0.05$, then $r = 6$ and the respective values of k_α are 2.93 and 3.25.

Unfortunately there is a problem associated with the above simultaneous procedures. If contrasts are examined only if T_0^2 is significant, then the correct overall probability is a conditional one, conditional on T_0^2 being significant. For the linear regression model Olshen [1973] showed that the conditional probability is, unfortunately, generally smaller than the unconditional probability, so that we have less "confidence" in the simultaneous conditional procedure than we thought. In the above applications we can also expect the conditional and unconditional probabilities to differ. One way around this problem is to decide to use a simultaneous procedure, irrespective of the significance or otherwise of T_0^2 (Scheffé [1977]).

EXAMPLE 3.5 Consider Example 3.2 (Section 3.4.2) and the cork data of Table 3.3. If we are interested in a confidence interval for each μ_j, then we can use (3.29) with $\mathbf{a}_1' = (1, 0, \ldots, 0)$, and so on, namely,

$$\overline{x}_j \pm k_\alpha\left(\frac{s_{jj}}{n}\right)^{1/2} \qquad (j = 1, 2, 3, 4).$$

With $n = 28$, $r = q = d = 4$, and $\alpha = 0.05$, the two methods give the following values of k_α: From Appendix D1, $t_{27}^{0.05/8} = 2.676$ [see (3.28)] for the Bonferroni method and $(T^2_{4,27,0.05})^{1/2} = \{[4(27)/24]F^{0.05}_{4,24}\}^{1/2} = 3.537$ [see (3.33) and (3.30)] for the Scheffé method. We require the elements of $\overline{\mathbf{x}}$ and the diagonal elements s_{jj} of $\mathbf{S}$, where

$$\overline{\mathbf{x}} = \begin{pmatrix} 50.536 \\ 46.179 \\ 49.679 \\ 45.179 \end{pmatrix} \begin{matrix} \text{N} \\ \text{E} \\ \text{S} \\ \text{W} \end{matrix} \quad \text{and} \quad \mathbf{S} = \begin{pmatrix} 290.406 & 223.753 & 288.438 & 226.271 \\ - & 219.930 & 229.060 & 171.374 \\ - & - & 350.004 & 259.541 \\ - & - & - & 226.004 \end{pmatrix}.$$

Using the Bonferroni intervals, we obtain the following:

$$\mu_1 : 50.54 \pm 8.62,$$

$$\mu_2 : 46.18 \pm 7.50,$$

$$\mu_3 : 49.68 \pm 9.46,$$

$$\mu_4 : 45.18 \pm 7.60.$$

Since the test for $H_0 : \mu_1 = \mu_2 = \mu_3 = \mu_4$ was significant at the 1% level, we can see which contrasts $\mathbf{c}'\overline{\mathbf{x}}$ are responsible for the rejection of H_0. In Example 3.3 (Section 3.4.2) the contrast $\mu_1 - \mu_2 + \mu_3 - \mu_4$ (= N + S − E − W) with $\mathbf{c}' = (1, -1, 1, -1)$ was regarded as a likely candidate. Using (3.35), a 99% confidence interval for this contrast is

$$8.86 \pm \left\{ \frac{3(27)}{25} F_{3,25}^{0.01} \right\}^{1/2} \{128.718/28\}^{1/2},$$

that is, 8.86 ± 8.35 or $[0.51, 17.21]$, which does not contain zero. Corresponding intervals for pairwise contrasts are

$$\mu_1 - \mu_2 : 4.36 \pm 5.83,$$

$$\mu_1 - \mu_3 : 0.86 \pm 5.86,$$

$$\mu_1 - \mu_4 : 5.36 \pm 5.88,$$

$$\mu_3 - \mu_2 : 3.50 \pm 7.78,$$

$$\mu_2 - \mu_4 : 1.00 \pm 7.47,$$

$$\mu_3 - \mu_4 : 4.50 \pm 5.55,$$

and all of these contain zero. Since H_0 is rejected if and only if at least one interval for a contrast does not contain zero, we see that the contrast (N + S − E − W) is responsible for the rejection of H_0.

3.4.5 *Functional Relationships (Errors in Variables Regression)*

Given a random sample $\mathbf{x}_1, \mathbf{x}_2, \ldots, \mathbf{x}_n$ from $N_d(\boldsymbol{\mu}, \boldsymbol{\Sigma})$, suppose that we have the partitions

$$\mathbf{x}_i = \begin{pmatrix} \mathbf{w}_i \\ \mathbf{y}_i \\ \mathbf{z}_i \end{pmatrix} \quad \text{and} \quad \boldsymbol{\mu} = \begin{pmatrix} \boldsymbol{\mu}_w \\ \boldsymbol{\mu}_y \\ \boldsymbol{\mu}_z \end{pmatrix},$$

where $\mathbf{w}_i$, $\mathbf{y}_i$, and $\mathbf{z}_i$ have the same dimension. It is assumed that the jth elements of each mean satisfy

$$\mu_{w_j} = g(\mu_{y_j}, \mu_{z_j}; \boldsymbol{\beta}),$$

where g is some function and $\boldsymbol{\beta}$ is a vector of unknown parameters. For example, if g is linear, we may have the simple model $\mu_{w_j} = \beta_1\mu_{y_j} + \beta_2\mu_{z_j}$, or

$$\boldsymbol{\mu}_w = \beta_1\boldsymbol{\mu}_y + \beta_2\boldsymbol{\mu}_z.$$

The above regression-type model is generally called a *functional relationship*, in this case with replication. A linear relationship exists, but we do not observe it because of errors in the variables. Although we have only considered two vectors $\mathbf{y}$ and $\mathbf{z}$, the extension to more than two is obvious. A procedure for finding the maximum likelihood estimates of $\boldsymbol{\beta}$ and $\boldsymbol{\Sigma}$, together with asymptotic variances and covariances for the estimate of $\boldsymbol{\beta}$, is given by Dolby and Freeman [1975] for a linear or nonlinear g. The reader is referred to their paper and Robinson [1977] for details and references to related models.

Another more general type of model is as follows. Suppose that the $\mathbf{x}_i$ $(i = 1, 2, \ldots, n)$ are independently distributed as $N_d(\boldsymbol{\mu}_{x_i}, \boldsymbol{\Sigma})$, where

$$\mathbf{x}_i = \begin{pmatrix} \mathbf{y}_i \\ \mathbf{z}_i \end{pmatrix} \quad \text{and} \quad \boldsymbol{\mu}_{x_i} = \begin{pmatrix} \boldsymbol{\mu}_{y_i} \\ \boldsymbol{\mu}_{z_i} \end{pmatrix}.$$

A simple linear model is

$$\boldsymbol{\mu}_{y_i} = \boldsymbol{\alpha} + \mathbf{B}\boldsymbol{\mu}_{z_i}, \tag{3.37}$$

where $\boldsymbol{\alpha}$, $\mathbf{B}$, and $\boldsymbol{\mu}_{z_i}$ are unknown parameters (see Gleser [1981] for discussion and references; he assumes $\boldsymbol{\Sigma} = \sigma^2\mathbf{I}_d$). For related topics see Patefield [1981], and Fisher and Hudson [1980].

3.5 INFERENCE FOR THE DISPERSION MATRIX

3.5.1 Introduction

Given a random sample $\mathbf{x}_1, \mathbf{x}_2, \ldots, \mathbf{x}_n$ from $N_d(\boldsymbol{\mu}, \boldsymbol{\Sigma})$, $n - 1 \geq d$, we shall consider testing the following hypotheses about $\boldsymbol{\Sigma}$ in this section.

1. *Blockwise independence.* Let $\boldsymbol{\Sigma}$ be partitioned as follows:

$$\boldsymbol{\Sigma} = \begin{pmatrix} \boldsymbol{\Sigma}_{11} & \boldsymbol{\Sigma}_{12} & \cdots & \boldsymbol{\Sigma}_{1b} \\ \boldsymbol{\Sigma}_{21} & \boldsymbol{\Sigma}_{22} & \cdots & \boldsymbol{\Sigma}_{2b} \\ \vdots & \vdots & & \vdots \\ \boldsymbol{\Sigma}_{b1} & \boldsymbol{\Sigma}_{b2} & \cdots & \boldsymbol{\Sigma}_{bb} \end{pmatrix}, \tag{3.38}$$

where $\boldsymbol{\Sigma}_{rr}$ is $d_r \times d_r$ and $\Sigma_{r=1}^{b} d_r = d$. We shall consider testing $H_0: \boldsymbol{\Sigma}_{rs} = \mathbf{O}$, all r, s $(r \neq s)$: The case $b = 2$ is considered in Section 3.5.2, the general case in Section 3.5.3, and the general case with each $d_r = 1$ in Section 3.5.4.

2. *Sphericity.* In Section 3.5.4 we consider testing the hypotheses

(a) $\boldsymbol{\Sigma} = \sigma^2 \mathbf{I}_d$, σ^2 unknown.
(b) $\boldsymbol{\Sigma} = \mathbf{I}_d$.

The hypotheses $\boldsymbol{\Sigma} = \sigma^2 \boldsymbol{\Sigma}_0$ and $\boldsymbol{\Sigma} = \boldsymbol{\Sigma}_0$ ($\boldsymbol{\Sigma}_0$ known) are also included, as they can be reduced to (a) and (b), respectively, using a suitable transformation of the data.

3. *Equal diagonal blocks.* The hypothesis $\boldsymbol{\Sigma}_{11} = \boldsymbol{\Sigma}_{22} = \cdots = \boldsymbol{\Sigma}_{bb}$ is discussed in Section 3.5.5.

4. *Equal correlations and variances.* In Section 3.5.6 we consider the model (3.21), which is sometimes used in mixed and components of variance models. Here we test $\sigma_{jj} = \sigma^2$ and $\sigma_{jk} = \rho\sigma^2$ $(j, k = 1, 2, \ldots, d;\ j \neq k)$. Likelihood ratio tests and, in one or two cases, union–intersection tests are given for the above hypotheses. For a general review of these tests, together with some tables of significance levels, see Krishnaiah and Lee [1980]. We note that tests for (1) and (2) have also been developed for the case when the $\mathbf{x}_i$ come from a complex multivariate normal distribution (Krishnaiah et al. [1976]). The dispersion matrix of a complex normal plays an important role in the study of spectral density matrices of stationary Gaussian multiple time series.

3.5.2 Blockwise Independence: Two Blocks

Consider the partition

$$\mathbf{x}_i = \begin{pmatrix} \mathbf{y}_i \\ \mathbf{z}_i \end{pmatrix} \begin{matrix} \}d_1 \\ \}d_2 \end{matrix}, \qquad \boldsymbol{\mu} = \begin{pmatrix} \boldsymbol{\mu}_1 \\ \boldsymbol{\mu}_2 \end{pmatrix} \begin{matrix} \}d_1 \\ \}d_2 \end{matrix}, \qquad \boldsymbol{\Sigma} = \begin{pmatrix} \overbrace{\boldsymbol{\Sigma}_{11}}^{d_1} & \overbrace{\boldsymbol{\Sigma}_{12}}^{d_2} \\ \boldsymbol{\Sigma}_{21} & \boldsymbol{\Sigma}_{22} \end{pmatrix} \begin{matrix} \}d_1 \\ \}d_2 \end{matrix}.$$

We wish to test the hypothesis that $\mathbf{y}_i$ is statistically independent of $\mathbf{z}_i$ $(i = 1, 2, \ldots, n)$, that is, test $H_0: \boldsymbol{\Sigma}_{12} = \mathbf{O}$. Two methods for doing this are the likelihood ratio method and the union–intersection principle; Both tests are described below. For future reference we mention the maximum likelihood estimates $\bar{\mathbf{x}}$ and $\mathbf{Q}/n$ of $\boldsymbol{\mu}$ and $\boldsymbol{\Sigma}$, respectively, and these are partitioned appropriately:

$$\bar{\mathbf{x}} = \begin{pmatrix} \bar{\mathbf{y}} \\ \bar{\mathbf{z}} \end{pmatrix} \quad \text{and} \quad \mathbf{Q} = \begin{pmatrix} \mathbf{Q}_{11} & \mathbf{Q}_{12} \\ \mathbf{Q}_{21} & \mathbf{Q}_{22} \end{pmatrix}, \tag{3.39}$$

where $\mathbf{Q} = \Sigma_i(\mathbf{x}_i - \bar{\mathbf{x}})(\mathbf{x}_i - \bar{\mathbf{x}})'$, $\mathbf{Q}_{11} = \Sigma_i(\mathbf{y}_i - \bar{\mathbf{y}})(\mathbf{y}_i - \bar{\mathbf{y}})'$, $\mathbf{Q}_{22} = \Sigma_i(\mathbf{z}_i - \bar{\mathbf{z}})(\mathbf{z}_i - \bar{\mathbf{z}})'$, and $\mathbf{Q}_{12} = \mathbf{Q}_{21}' = \Sigma_i(\mathbf{y}_i - \bar{\mathbf{y}})(\mathbf{z}_i - \bar{\mathbf{z}})'$. We assume $d_1 \geq d_2$.

a LIKELIHOOD RATIO TEST

When H_0 is true, we have two independent random samples $\mathbf{y}_i$ $(i = 1, 2, \ldots, n)$ from $N_{d_1}(\boldsymbol{\mu}_1, \boldsymbol{\Sigma}_{11})$, and $\mathbf{z}_i$ $(i = 1, 2, \ldots, n)$ from $N_{d_2}(\boldsymbol{\mu}_2, \boldsymbol{\Sigma}_{22})$. The overall likelihood for the two samples is then the product of two likelihood functions, one for each sample. Thus using the notation of (3.2), we have, when H_0 is true,

$$L(\boldsymbol{\mu}, \boldsymbol{\Sigma}) = L_1(\boldsymbol{\mu}_1, \boldsymbol{\Sigma}_{11}) L_2(\boldsymbol{\mu}_2, \boldsymbol{\Sigma}_{22}), \tag{3.40}$$

so that the likelihood ratio test for H_0 is

$$\begin{aligned}
\ell &= \sup_{\boldsymbol{\mu}_1, \boldsymbol{\mu}_2, \boldsymbol{\Sigma}_{11}, \boldsymbol{\Sigma}_{22}} L_1(\boldsymbol{\mu}_1, \boldsymbol{\Sigma}_{11}) L_2(\boldsymbol{\mu}_2, \boldsymbol{\Sigma}_{22}) \Big/ \sup_{\boldsymbol{\mu}, \boldsymbol{\Sigma}} L(\boldsymbol{\mu}, \boldsymbol{\Sigma}) \\
&= \frac{L_1(\hat{\boldsymbol{\mu}}_1, \hat{\boldsymbol{\Sigma}}_{11}) L_2(\hat{\boldsymbol{\mu}}_2, \hat{\boldsymbol{\Sigma}}_{22})}{L(\hat{\boldsymbol{\mu}}, \hat{\boldsymbol{\Sigma}})} \\
&= \frac{(2\pi)^{-nd_1/2} |\hat{\boldsymbol{\Sigma}}_{11}|^{-n/2} e^{-nd_1/2} (2\pi)^{-nd_2/2} |\hat{\boldsymbol{\Sigma}}_{22}|^{-n/2} e^{-nd_2/2}}{(2\pi)^{-nd/2} |\hat{\boldsymbol{\Sigma}}|^{-n/2} e^{-nd/2}},
\end{aligned}$$

by (3.6), or

$$\Lambda = \ell^{2/n} = \frac{|\hat{\boldsymbol{\Sigma}}|}{|\hat{\boldsymbol{\Sigma}}_{11}||\hat{\boldsymbol{\Sigma}}_{22}|} = \frac{|\mathbf{Q}|}{|\mathbf{Q}_{11}||\mathbf{Q}_{22}|}, \tag{3.41}$$

where $\hat{\boldsymbol{\Sigma}}_{ii} = \mathbf{Q}_{ii}/n$, $i = 1, 2$. Now, since $\mathbf{Q} > \mathbf{O}$ with probability 1, $\mathbf{Q}_{11}$ is nonsingular (A5.9) and, from A3.2,

$$|\mathbf{Q}| = |\mathbf{Q}_{11}||\mathbf{Q}_{22} - \mathbf{Q}_{21}\mathbf{Q}_{11}^{-1}\mathbf{Q}_{12}|.$$

Setting $\mathbf{E} = \mathbf{Q}_{22} - \mathbf{Q}_{21}\mathbf{Q}_{11}^{-1}\mathbf{Q}_{12}$ $(= \mathbf{Q}_{22\cdot1}$, say) and $\mathbf{H} = \mathbf{Q}_{21}\mathbf{Q}_{11}^{-1}\mathbf{Q}_{12}$, we have from (3.41)

$$\begin{aligned}
\Lambda &= \frac{|\mathbf{Q}_{22\cdot1}|}{|\mathbf{Q}_{22}|} \\
&= \frac{|\mathbf{E}|}{|\mathbf{E} + \mathbf{H}|}.
\end{aligned} \tag{3.42}$$

Since $\mathbf{Q} = (n-1)\mathbf{S} \sim W_d(n-1, \boldsymbol{\Sigma})$ (by Theorem 3.1), we can apply Lemma 2.10 and its Corollary (Section 2.6) to $\mathbf{Q}$. Thus, when $H_0: \boldsymbol{\Sigma}_{12} = \mathbf{O}$ is true, $\boldsymbol{\Sigma}_{22\cdot1} = \boldsymbol{\Sigma}_{22}$ and the random matrices $\mathbf{E}$ and $\mathbf{H}$ are independently distributed as $W_{d_2}(n - 1 - d_1, \boldsymbol{\Sigma}_{22})$ and $W_{d_2}(d_1, \boldsymbol{\Sigma}_{22})$, respectively. It now follows from (2.48), with $m_H = d_1$, $m_E = n - d_1 - 1$, and "d" $= d_2$, that $\Lambda \sim U_{d_2, d_1, n-d_1-1}$

when H_0 is true. Since $U_{d_2, d_1, n-d_1-1}$ has the same distribution as $U_{d_1, d_2, \nu}$, where $\nu = m_H + m_E -$ "d" $= n - d_2 - 1$, we arrive, as expected, at the same test statistic for H_0 if we interchange the roles of $\mathbf{y}_i$ and $\mathbf{z}_i$ and write Λ in the form $|\mathbf{Q}_{11\cdot 2}|/|\mathbf{Q}_{11}|$. We note that H_0 is rejected if the likelihood ratio ℓ is too small or, equivalently, Λ is too small: The lower tail of the U-distribution is therefore appropriate here and tables of critical values are given in Appendix D13. Asymptotic properties of the test statistic are described in Section 3.5.3 with $b = 2$.

From Section 2.5.5b, $\Lambda = \prod_{j=1}^{d_2}(1 - \theta_j)$, where the θ_j are the ordered roots of $|\mathbf{H} - \theta(\mathbf{E} + \mathbf{H})| = 0$. These roots are the eigenvalues of $\mathbf{H}(\mathbf{E} + \mathbf{H})^{-1} = \mathbf{Q}_{21}\mathbf{Q}_{11}^{-1}\mathbf{Q}_{12}\mathbf{Q}_{22}^{-1}$, that is, the roots of

$$|\mathbf{S}_{21}\mathbf{S}_{11}^{-1}\mathbf{S}_{12}\mathbf{S}_{22}^{-1} - \theta\mathbf{I}_{d_2}| = 0, \tag{3.43}$$

where $\mathbf{S}_{ij} = \mathbf{Q}_{ij}/(n-1)$. It does not matter whether we use a divisor of n or $n - 1$, as it cancels out. Since $0 \le \theta_j < 1$, we can write $\theta_j = r_j^2$ ($\theta_j \ne 0$), and the positive square roots r_j are called the sample canonical correlations between the $\mathbf{y}_i$'s and the $\mathbf{z}_i$'s (see Section 5.7). The usual notation adopted is $r_1^2 > r_2^2 > \cdots > r_s^2 > 0$, where $s = \text{minimum}(d_1, d_2)$. These roots can be regarded as estimates of the population canonical correlations, the latter being the ordered positive square roots of the solutions of

$$|\boldsymbol{\Sigma}_{21}\boldsymbol{\Sigma}_{11}^{-1}\boldsymbol{\Sigma}_{12}\boldsymbol{\Sigma}_{22}^{-1} - \rho^2\mathbf{I}_{d_2}| = 0. \tag{3.44}$$

b MAXIMUM ROOT TEST

To apply the union–intersection principle to test $H_0: \boldsymbol{\Sigma}_{12} = \mathbf{O}$, we consider the univariate hypothesis $H_{0ab}: \mathbf{a}'\boldsymbol{\Sigma}_{12}\mathbf{b} = 0$, or, equivalently, $\rho_{ab}^2 = (\mathbf{a}'\boldsymbol{\Sigma}_{12}\mathbf{b})^2/(\mathbf{a}'\boldsymbol{\Sigma}_{11}\mathbf{a})(\mathbf{b}'\boldsymbol{\Sigma}_{22}\mathbf{b}) = 0$, where ρ_{ab} is the correlation between $\mathbf{a}'\mathbf{y}$ and $\mathbf{b}'\mathbf{z}$. An acceptance region for testing H_{0ab} is given by $r_{ab}^2 \le k$, where

$$r_{ab} = \frac{\mathbf{a}'\mathbf{S}_{12}\mathbf{b}}{\{\mathbf{a}'\mathbf{S}_{11}\mathbf{a}\mathbf{b}'\mathbf{S}_{22}\mathbf{b}\}^{1/2}} \tag{3.45}$$

is the sample correlation for the n pairs $(\mathbf{a}'\mathbf{y}_i, \mathbf{b}'\mathbf{z}_i)$, $i = 1, 2, \ldots, n$. Since

$$H_0 = \bigcap_a \bigcap_b H_{0ab},$$

the union–intersection principle (Section 1.6.2) leads to the following acceptance region for testing H_0 (Roy [1953]):

$$\bigcap_a \bigcap_b \{(\mathbf{Y}, \mathbf{Z}): r_{ab}^2 \le k\} = \left\{(\mathbf{Y}, \mathbf{Z}): \sup_{\mathbf{a}, \mathbf{b}} r_{ab}^2 \le k\right\}$$

$$= \{(\mathbf{Y}, \mathbf{Z}): \theta_{\max} \le k\},$$

where, from A7.7 and A1.4, $\theta_{\max}$ is the maximum root of (3.43).

We can also add a third test statistic to Λ and $\theta_{\max}$ for testing H_0, namely, Pillai's trace statistic

$$V^{(s)} = \sum_j \theta_j = \operatorname{tr}\left[\mathbf{H}(\mathbf{E}+\mathbf{H})^{-1}\right] = \operatorname{tr}\left[\mathbf{S}_{21}\mathbf{S}_{11}^{-1}\mathbf{S}_{12}\mathbf{S}_{22}^{-1}\right].$$

For $d = 2$ Pillai and Hsu [1979] showed that these statistics are not seriously affected by slight nonnormality: $V^{(2)}$ is the most robust, followed by Λ.

3.5.3 *Blockwise Independence:* b *Blocks*

Let $\mathbf{x}_1, \mathbf{x}_2, \dots, \mathbf{x}_n$ be a random sample from $N_d(\boldsymbol{\mu}, \boldsymbol{\Sigma})$ and consider the partition

$$\mathbf{x}_i = \begin{pmatrix} \mathbf{x}_{i1} \\ \mathbf{x}_{i2} \\ \vdots \\ \mathbf{x}_{ib} \end{pmatrix}, \quad \boldsymbol{\mu} = \begin{pmatrix} \boldsymbol{\mu}_1 \\ \boldsymbol{\mu}_2 \\ \vdots \\ \boldsymbol{\mu}_b \end{pmatrix}, \quad \boldsymbol{\Sigma} = \begin{pmatrix} \boldsymbol{\Sigma}_{11} & \boldsymbol{\Sigma}_{12} & \cdots & \boldsymbol{\Sigma}_{1b} \\ \boldsymbol{\Sigma}_{21} & \boldsymbol{\Sigma}_{22} & \cdots & \boldsymbol{\Sigma}_{2b} \\ \vdots & \vdots & & \vdots \\ \boldsymbol{\Sigma}_{b1} & \boldsymbol{\Sigma}_{b2} & \cdots & \boldsymbol{\Sigma}_{bb} \end{pmatrix},$$

where $\mathbf{x}_{ir}$ and $\boldsymbol{\mu}_r$ are $d_r \times 1$ vectors and $\boldsymbol{\Sigma}_{rr}$ is $d_r \times d_r$ $(r = 1, 2, \dots, b; \sum_{r=1}^{b} d_r = d)$. Suppose we wish to test the null hypothesis

$$H_0: \boldsymbol{\Sigma} = \begin{pmatrix} \boldsymbol{\Sigma}_{11} & \mathbf{O} & \cdots & \mathbf{O} & \mathbf{O} \\ \mathbf{O} & \boldsymbol{\Sigma}_{22} & \cdots & \mathbf{O} & \mathbf{O} \\ \vdots & \vdots & & \vdots & \vdots \\ \mathbf{O} & \mathbf{O} & \cdots & \mathbf{O} & \boldsymbol{\Sigma}_{bb} \end{pmatrix} = \boldsymbol{\Sigma}_{(b)},$$

say, that the b vectors $\mathbf{x}_{i1}, \mathbf{x}_{i2}, \dots, \mathbf{x}_{ib}$ are mutually independent. From the derivation of (3.41), it is clear that the likelihood ratio statistic ℓ for testing H_0 takes the form

$$\begin{aligned} \Lambda &= \ell^{2/n} \\ &= \frac{|\hat{\boldsymbol{\Sigma}}|}{|\hat{\boldsymbol{\Sigma}}_{11}|\,|\hat{\boldsymbol{\Sigma}}_{22}| \cdots |\hat{\boldsymbol{\Sigma}}_{bb}|} \\ &= \frac{|\mathbf{Q}|}{|\mathbf{Q}_{11}|\,|\mathbf{Q}_{22}| \cdots |\mathbf{Q}_{bb}|}, \end{aligned} \tag{3.46}$$

where $\mathbf{Q} = n\hat{\boldsymbol{\Sigma}} = \sum_i (\mathbf{x}_i - \bar{\mathbf{x}})(\mathbf{x}_i - \bar{\mathbf{x}})'$ and $\mathbf{Q}_{rr} = n\hat{\boldsymbol{\Sigma}}_{rr} = \sum_i (\mathbf{x}_{ir} - \bar{\mathbf{x}}_{\cdot r})(\mathbf{x}_{ir} - \bar{\mathbf{x}}_{\cdot r})'$. Unfortunately, the special method we used to find the distribution of Λ

for the case $b = 2$ in the previous section does not generalize here, so that we must content ourselves with a large-sample test based on the likelihood ratio theory. Using the notation of Section 1.7, the number of parameters in Ω is $d + \frac{1}{2}d(d+1)$, as $\boldsymbol{\mu}$ has d elements and $\boldsymbol{\Sigma}$ has $\frac{1}{2}d(d+1)$ distinct elements in the upper triangle. Similarly, the number of parameters in ω, specified by H_0, is the number of distinct elements in each $\boldsymbol{\mu}_r$ and $\boldsymbol{\Sigma}_{rr}$ $(r = 1, 2, \ldots, b)$, namely,

$$\sum_{r=1}^{b} \left[d_r + \tfrac{1}{2}d_r(d_r+1)\right] = \frac{1}{2}\sum_{r=1}^{b} d_r^2 + \tfrac{3}{2}d.$$

Hence, for large n, $-2\log \ell \,(= -n\log\Lambda)$ is approximately distributed as χ_ν^2 when H_0 is true, where

$$\nu = \tfrac{1}{2}d(d+3) - \frac{1}{2}\left(\sum_r d_r^2 + 3d\right)$$

$$= \frac{1}{2}\left(d^2 - \sum_r d_r^2\right).$$

As the moments of Λ, under H_0, are known exactly, Box [1949] was able to obtain a more accurate large-sample expression for the null distribution, namely (see Anderson [1958: Section 9.5], Muirhead [1982: Section 11.2.4]),

$$\text{pr}[-m\log\Lambda \le z] = \text{pr}\left[\chi_\nu^2 \le z\right] + gm^{-2}\left\{\text{pr}\left[\chi_{\nu+4}^2 \le z\right] - \text{pr}\left[\chi_\nu^2 \le z\right]\right\} + O(m^{-3}), \tag{3.47}$$

where

$$m = n - \frac{3}{2} - \left(d^3 - \sum_r d_r^3\right)\Big/3\left(d^2 - \sum_r d_r^2\right) \tag{3.48}$$

and

$$g = \frac{1}{48}\left(d^4 - \sum_r d_r^4\right) - \frac{5}{96}\left(d^2 - \sum_r d_r^2\right) - \left(d^3 - \sum_r d_r^3\right)^2\Big/72\left(d^2 - \sum_r d_r^2\right). \tag{3.49}$$

Correction factors to make the percentage points of $-m\log\Lambda$ (written as $-2\rho\log\ell$) exactly those of χ_ν^2 are given by Davis and Field [1971] and reproduced by Muirhead [1982: p. 537]. A good Pearson-type approximation is also available, and J. C. Lee et al. [1977: Table 7] use it to provide some accurate percentage points for $-2\log\Lambda$. Muirhead and Waternaux [1980: p. 42] demonstrate, using elliptical distributions, that $-2\log\ell$ can be very sensitive to departures from normality when kurtosis is present, as in "long-tailed" distributions. They provide a correction for $-2\log\ell$ on the assumption of a common kurtosis factor. Properties of the test based on Λ are discussed by Giri [1977: Section 8.3]. The asymptotic nonnull distribution of Λ is given by Olkin and Siotani [1976].

Nagao [1973a, b, 1974] proposed another statistic,

$$L_2 = \tfrac{1}{2}(n-1)\mathrm{tr}\left[\mathbf{Q}\mathbf{Q}_{(b)}^{-1} - \mathbf{I}_d\right]^2, \tag{3.50}$$

where $\mathbf{Q}_{(b)} = \mathrm{diag}(\mathbf{Q}_{11}, \mathbf{Q}_{22}, \ldots, \mathbf{Q}_{bb})$, based on the asymptotic variance of $-2\log\ell$. His statistic reduces to Pillai's trace statistic when $b = 2$ (see Exercise 3.6). He also gave asymptotic expansions for the null and nonnull distributions of $-m\log\Lambda$ and L_2 and showed that their asymptotic relative efficiency was 1.

3.5.4 *Diagonal Dispersion Matrix*

A special case of the hypothesis considered in the previous section is H_{01} that $\boldsymbol{\Sigma}$ is diagonal, that is, $\boldsymbol{\Sigma} = \mathrm{diag}(\sigma_{11}, \ldots, \sigma_{dd})$. Setting $b = d$ and $\mathbf{x}_{ir} = x_{ir}$ $(r = 1, 2, \ldots, d)$, we have from (3.46)

$$\Lambda = \frac{|\hat{\boldsymbol{\Sigma}}|}{\hat{\sigma}_{11}\hat{\sigma}_{22}\cdots\hat{\sigma}_{dd}} = \frac{|Q|}{q_{11}q_{22}\cdots q_{dd}}, \tag{3.51}$$

where $q_{rr} = \Sigma_i(x_{ir} - \bar{x}_{\cdot r})^2 = n\hat{\sigma}_{rr}$ $(r = 1, 2, \ldots, d)$. Since $d_r = 1$, we can use (3.47), with $\nu = \frac{1}{2}d(d-1)$, $m = n - (2d + 11)/6$, and $g = d(d-1)(2d^2 - 2d - 13)/288$. Exact percentage points for the statistic $-m\log\Lambda$, from Mathai and Katiyar [1979a], are given in Appendix D17. A good normal approximation is provided by Mudholkar et al. [1982]. The case $d = 2$ is straightforward, as the hypothesis $\sigma_{12} = 0$ can be tested using the exact t-test of the regression of one variable on the other (see Exercise 3.7).

We note that $r_{jk} = q_{jk}/(q_{jj}q_{kk})^{1/2}$ is the sample correlation for variables labeled j and k, and the correlation matrix $\mathbf{R} = [(r_{jk})]$ is given by

$$\mathbf{R} = \mathrm{diag}\left(q_{11}^{-1/2}, \ldots, q_{dd}^{-1/2}\right)\mathbf{Q}\,\mathrm{diag}\left(q_{11}^{-1/2}, \ldots, q_{dd}^{-1/2}\right). \tag{3.52}$$

Hence, from (3.51), $\Lambda = |\mathbf{R}|$, so that our test statistic for H_{01} is based on the determinant of the correlation matrix. Olkin and Siotani [1976] use this fact to find the asymptotic nonnull distribution of Λ. A union–intersection test of H_{01} based on the concept of maximum eccentricity is given by Schuenemeyer and Bargmann [1978].

In practice we may be interested in testing the hypothesis $H_{02}:\boldsymbol{\Sigma} = \sigma^2\mathbf{I}_d$, where σ^2 is unspecified. The likelihood ratio statistic ℓ_2 for testing H_{02} is now given by (Exercise 3.8)

$$\ell_2^{2/n} = \frac{|\hat{\boldsymbol{\Sigma}}|}{(\operatorname{tr}[\hat{\boldsymbol{\Sigma}}/d])^d} = \frac{|\mathbf{Q}|}{(\operatorname{tr}[\mathbf{Q}/d])^d} = \Lambda_2, \tag{3.53}$$

say. This statistic has a very simple interpretation if we note that testing H_{02} is equivalent to testing that the eigenvalues λ_j of $\boldsymbol{\Sigma}$ are all equal, that is,

$$\begin{aligned} 1 &= \frac{\text{geometric mean of the } \lambda_j}{\text{arithmetic mean of the } \lambda_j} \\ &= \left(\prod_j \lambda_j^{1/d}\right) \Big/ \sum_j \frac{\lambda_j}{d} \\ &= \frac{|\boldsymbol{\Sigma}|^{1/d}}{\operatorname{tr}[\boldsymbol{\Sigma}/d]} \qquad \text{(by A1.2).} \end{aligned}$$

We test this by replacing $\boldsymbol{\Sigma}$ by its maximum likelihood estimate $\hat{\boldsymbol{\Sigma}}$ and seeing if the resulting statistic, $\ell_2^{2/nd}$ is about unity. Using a one-sided test, since the geometric mean cannot exceed the arithmetic mean, we reject H_{02} if ℓ_2 is too small. The ellipsoid $(\mathbf{x}-\boldsymbol{\mu})'\boldsymbol{\Sigma}^{-1}(\mathbf{x}-\boldsymbol{\mu}) = c^2$ reduces to the sphere $(\mathbf{x}-\boldsymbol{\mu})'(\mathbf{x}-\boldsymbol{\mu}) = c^2\sigma^2$ under H_{02} so that H_{02} is called the hypothesis of sphericity. Properties of the above test are discussed, for example, by Muirhead [1982: Section 8.3]. The exact distribution of Λ_2 and a table of percentage points for $d = 4(1)10$ are given by Nagarsenker and Pillai [1973a]. A useful computing formula based on a mixture of incomplete beta functions is given by Gupta [1977], and a good Pearson type I approximation is also available (J. C. Lee et al. [1977]). The large-sample approximation (3.47) for $-m\log\Lambda_2$ still applies, but with $\nu = \frac{1}{2}d(d+1) - 1$, $m = n - 1 - (2d^2 + d + 2)/6d$, and $g = (d+2)(d-1)(d-2)(2d^3 + 6d^2 + 3d + 2)/288d^2$. Venables [1976] shows that Λ_2 can also be derived using a union–intersection argument. Using the family of elliptical distributions, Muirhead and Waternaux [1980] demonstrate that $-2\log\ell_2$ is very sensitive to kurtosis and provide a correction based on the assumption of a common kurtosis factor (see also Muirhead [1982: Section 8.3]).

To test $H_{03}:\boldsymbol{\Sigma} = \mathbf{I}_d$ we use the likelihood ratio (Exercise 3.9)

$$\ell_3 = (e/n)^{dn/2}|\mathbf{Q}|^{n/2}\exp(-\tfrac{1}{2}\operatorname{tr}\mathbf{Q}). \tag{3.54}$$

When H_{03} is true, $-2\log\ell_3$ is asymptotically χ^2_ν, where $\nu = \frac{1}{2}d(d+1)$. The exact distribution and a table of percentage points for $d = 4(1)10$ are given by Nagarsenker and Pillai [1973b]. As might be expected, the test is very sensitive to kurtosis in the parent population (Muirhead and Waternaux [1980: p. 40]). Properties of the test and a modification are given by Muirhead [1982: Section

8.4], who also reproduces percentage points from Davis and Field [1971: $d = 2(1)10$].

The hypotheses that $\boldsymbol{\Sigma} = \sigma^2\boldsymbol{\Sigma}_0$ and $\boldsymbol{\Sigma} = \boldsymbol{\Sigma}_0$, where $\boldsymbol{\Sigma}_0$ is known, can be reduced to H_{02} and H_{03} above, respectively, by making the transformation $\mathbf{y}_i = \boldsymbol{\Sigma}_0^{-1/2}\mathbf{x}_i$. For example, when $\boldsymbol{\Sigma} = \boldsymbol{\Sigma}_0$, $\mathscr{D}[\mathbf{y}_i] = \boldsymbol{\Sigma}_0^{-1/2}\mathscr{D}[\mathbf{x}_i]\boldsymbol{\Sigma}_0^{-1/2} = \boldsymbol{\Sigma}_0^{-1/2}\boldsymbol{\Sigma}_0\boldsymbol{\Sigma}_0^{-1/2} = \mathbf{I}_d$. Nagao [1973a, 1974] has proposed alternative test statistics for these hypotheses based on the asymptotic variance of $-2\log \ell$, where ℓ is the corresponding likelihood ratio. The likelihood ratio test of $H_{04}: \boldsymbol{\mu} = \boldsymbol{\mu}_0$, $\boldsymbol{\Sigma} = \boldsymbol{\Sigma}_0$, is given in Exercise 3.10, and the exact null distribution together with tables are given by Nagarsenker and Pillai [1974: $d = 2(1)6$] (see also Muirhead [1982: Section 8.5]). Some chi-square and F-approximations for a modified likelihood ratio test are given by Korin and Stevens [1973]. Finally, the likelihood ratio test of $H_0: \boldsymbol{\mu} = \boldsymbol{\mu}_0$, $\boldsymbol{\Sigma} = \sigma^2\mathbf{I}_d$ and exact percentage points for $d = 4(1)10$ are given by Singh [1980].

3.5.5 Equal Diagonal Blocks

Using the notation of Section 3.5.3, with $d_r = d_0$ $(r = 1, 2, \ldots, b)$, we now wish to test $H_{05}: \boldsymbol{\Sigma}_{11} = \boldsymbol{\Sigma}_{22} = \cdots = \boldsymbol{\Sigma}_{bb}$, given that the $\boldsymbol{\Sigma}_{rs}$ $(r < s)$ are all equal (to $\boldsymbol{\Sigma}_{12}$). Krishnaiah [1975] gives the following method for doing this, and reviews briefly some of the earlier work relating to the problem (see also Choi and Wette [1972]). Let

$$\overline{\mathbf{x}}_{i\cdot} = \frac{1}{b}(\mathbf{x}_{i1} + \mathbf{x}_{i2} + \cdots + \mathbf{x}_{ib})$$

and

$$\mathbf{y}_i = \begin{pmatrix} \mathbf{y}_{i1} \\ \mathbf{y}_{i2} \\ \vdots \\ \mathbf{y}_{ib} \end{pmatrix} = \begin{pmatrix} \overline{\mathbf{x}}_{i\cdot} \\ \mathbf{x}_{i2} - \overline{\mathbf{x}}_{i\cdot} \\ \vdots \\ \mathbf{x}_{ib} - \overline{\mathbf{x}}_{i\cdot} \end{pmatrix} = \mathbf{C}\mathbf{x}_i, \tag{3.55}$$

say. Since $\mathbf{y}_i = \mathbf{C}\mathbf{x}_i$ for nonsingular $\mathbf{C}$ (Exercise 3.11), the $\mathbf{y}_i$ are i.i.d. $N_d(\mathbf{C}\boldsymbol{\mu}, \mathbf{C}\boldsymbol{\Sigma}\mathbf{C}')$ [Theorem 2.1(i)]. Now for $r \neq 1$,

$$\begin{aligned}
\mathscr{C}[\mathbf{y}_{ir}, \mathbf{y}_{i1}] &= \mathscr{C}[\mathbf{x}_{ir} - \overline{\mathbf{x}}_{i\cdot}, \overline{\mathbf{x}}_{i\cdot}] \\
&= \mathscr{C}[\mathbf{x}_{ir}, \overline{\mathbf{x}}_{i\cdot}] - \mathscr{D}[\overline{\mathbf{x}}_{i\cdot}] \\
&= \frac{1}{b}\sum_{s=1}^{b} \mathscr{C}[\mathbf{x}_{ir}, \mathbf{x}_{is}] - \frac{1}{b^2}\sum_r\sum_s \mathscr{C}[\mathbf{x}_{ir}, \mathbf{x}_{is}] \\
&= \frac{1}{b}[(b-1)\boldsymbol{\Sigma}_{12} + \boldsymbol{\Sigma}_{rr}] - \frac{1}{b^2}\left[b(b-1)\boldsymbol{\Sigma}_{12} + \sum_{r=1}^{b}\boldsymbol{\Sigma}_{rr}\right] \\
&= \frac{1}{b}\boldsymbol{\Sigma}_{rr} - \frac{1}{b^2}\sum_{r=1}^{b}\boldsymbol{\Sigma}_{rr},
\end{aligned}$$

which is zero if H_{05} is true. Setting $\mathbf{y}_{i2}^* = (\mathbf{y}_{i2}', \mathbf{y}_{i3}', \ldots, \mathbf{y}_{ib}')'$, we find that testing H_{05} is equivalent to testing $\mathscr{C}[\mathbf{y}_{i1}, \mathbf{y}_{i2}^*] = \mathbf{O}$, and this can be accomplished using the theory of Section 3.5.2 for testing the independence of two subvectors. We note that for the case $b = 2$ no assumption is needed about the structure of $\mathbf{\Sigma}$, as $\mathbf{\Sigma}$ only has one off-diagonal block $\mathbf{\Sigma}_{12}$.

3.5.6 Equal Correlations and Equal Variances

A hypothesis that sometimes arises in analysis of variance situations is

$$H_{06}: \mathbf{\Sigma} = \sigma^2 \begin{pmatrix} 1 & \rho & \cdots & \rho & \rho \\ \rho & 1 & \cdots & \rho & \rho \\ \vdots & \vdots & & \vdots & \vdots \\ \rho & \rho & \cdots & \rho & 1 \end{pmatrix}. \tag{3.56}$$

If $\mathbf{S} = [(s_{jk})]$, the usual unbiased estimate of $\mathbf{\Sigma}$, then the maximum likelihood estimates of σ^2 and ρ are (Exercise 3.12)

$$\hat{\sigma}^2 = \sum_{j=1}^{d} s_{jj}/d \tag{3.57}$$

and

$$\hat{\sigma}^2\hat{\rho} = \sum\sum_{j \neq k} s_{jk}/d(d-1). \tag{3.58}$$

The likelihood ratio statistic is ℓ_6 (Wilks [1946]; see Exercise 3.13), where

$$\ell_6^{2/n} = \frac{|\mathbf{S}|}{(\hat{\sigma}^2)^d (1-\hat{\rho})^{d-1}[1+(d-1)\hat{\rho}]}. \tag{3.59}$$

Box [1949] showed that

$$-\left(n - 1 - \frac{d(d+1)^2(2d-3)}{6(d-1)(d^2+d-4)}\right)\log \ell_6$$

is asymptotically distributed as chi square with $\frac{1}{2}d(d+1) - 2$ degrees of freedom when H_{06} is true. The exact distribution and percentage points are given by Nagarsenker [1975: $d = 4(1)10$], and a nonparametric test is proposed by Choi [1977]. Finally, the exact distribution and percentage points for jointly testing H_{06} and $\mu_1 = \mu_2 = \cdots = \mu_d$ are available from Mathai and Katiyar [1979b: $d = 4(1)10$].

3.5.7 *Simultaneous Confidence Intervals for Correlations*

Writing $\boldsymbol{\Sigma} = [(\sigma_{jk})]$, we may be interested in constructing simultaneous confidence intervals for the $m = \frac{1}{2}d(d-1)$ correlation coefficients $\rho_{jk} = \sigma_{jk}/(\sigma_{jj}\sigma_{kk})^{1/2}$ $(j < k)$. With the union–intersection principle in mind, we can look at the corresponding univariate problem of finding a confidence interval for $\ell'\boldsymbol{\Sigma}\ell$ $(\ell \neq \mathbf{0})$. Now $\mathbf{Q} = \Sigma_i(\mathbf{x}_i - \overline{\mathbf{x}})(\mathbf{x}_i - \overline{\mathbf{x}})' \sim W_d(n-1, \boldsymbol{\Sigma})$ (by Theorem 3.1) and $\ell'\mathbf{Q}\ell/\ell'\boldsymbol{\Sigma}\ell \sim \chi^2_{n-1}$ (Theorem 2.2, Corollary 1). From A7.5 we have

$$L \leq \frac{\ell'\mathbf{Q}\ell}{\ell'\boldsymbol{\Sigma}\ell} \leq U$$

for all nonzero ℓ if and only if $L \leq \gamma_{\min} < \gamma_{\max} \leq U$, where $\gamma_{\min}$ and $\gamma_{\max}$ are the minimum and maximum eigenvalues, respectively, of $\mathbf{Q}\boldsymbol{\Sigma}^{-1}$. The "optimal" choice of L and U is unknown; however, if we choose L and U such that $\text{pr}[\gamma_{\min} \geq L] = \alpha/2$ and $\text{pr}[\gamma_{\max} \leq U] = \alpha/2$, then

$$\begin{aligned} 1 - \alpha &= \text{pr}\left[L \leq \frac{\ell'\mathbf{Q}\ell}{\ell'\boldsymbol{\Sigma}\ell} \leq U, \text{ for all } \ell \; (\neq \mathbf{0})\right] \\ &= \text{pr}\left[\frac{\ell'\mathbf{Q}\ell}{U} \leq \ell'\boldsymbol{\Sigma}\ell \leq \frac{\ell'\mathbf{Q}\ell}{L}\right]. \end{aligned} \tag{3.60}$$

Also, the eigenvalues of $\mathbf{Q}\boldsymbol{\Sigma}^{-1}$ are the same as those of $\boldsymbol{\Sigma}^{-1/2}\mathbf{Q}\boldsymbol{\Sigma}^{-1/2}$ (A1.4) which, by Theorem 2.2 of Section 2.3.1, is distributed as $W_d(n - 1, \boldsymbol{\Sigma}^{-1/2}\boldsymbol{\Sigma}\boldsymbol{\Sigma}^{-1/2})$, that is, as $W_d(n-1, \mathbf{I}_d)$. As $\boldsymbol{\Sigma}$ is not involved in this latter distribution, $\gamma_{\min}$ and $\gamma_{\max}$ can be tabulated, as in Hanumara and Thompson [1968: $d = 2(1)10$]. For applications of these tables, see Hanumara and Strain [1980]. Further tables, for the choice $L = U^{-1}$, are given by Clemm et al. [1973].

By setting ℓ' equal to $(1,0,\dots,0)$, $(0,1,0,\dots,0)$, and so on, we obtain intervals for the variances $\sigma_{11}, \sigma_{22}, \dots$, namely,

$$\frac{q_{11}}{U} \leq \sigma_{11} \leq \frac{q_{11}}{L} \tag{3.61}$$

and

$$\frac{q_{22}}{U} \leq \sigma_{22} \leq \frac{q_{22}}{L}. \tag{3.62}$$

Unfortunately, intervals for the covariances do not follow so readily. However, setting $\ell' = (1,1,0,\dots,0)$ in (3.60) gives

$$\frac{q_{11} + 2q_{12} + q_{22}}{U} \leq \sigma_{11} + 2\sigma_{12} + \sigma_{22} \leq \frac{q_{11} + 2q_{12} + q_{22}}{L},$$

and, subtracting $\sigma_{11} + \sigma_{22}$ from both sides of the above inequality, we have from (3.61) and (3.62)

$$\frac{2q_{12}}{U} + (q_{11} + q_{22})\left(\frac{1}{U} - \frac{1}{L}\right) \le 2\sigma_{12} \le \frac{2q_{12}}{L} + (q_{11} + q_{22})\left(\frac{1}{L} - \frac{1}{U}\right).$$

We can go one step further and find an interval for ρ_{12} if we divide by $2(\sigma_{11}\sigma_{22})^{1/2}$ and use the square root of the product of (3.61) and (3.62), namely,

$$\frac{L}{U} r_{12} + a_{12}\left(\frac{L}{U} - 1\right) \le \rho_{12} \le \frac{U}{L} r_{12} + a_{12}\left(\frac{U}{L} - 1\right), \qquad (3.63)$$

where $a_{12} = (q_{11} + q_{22})/2(q_{11}q_{22})^{1/2}$ and $r_{12} = q_{12}/(q_{11}q_{22})^{1/2}$.

A different set of intervals is obtained if we "subtract" the interval corresponding to $\ell' = (1, -1, 0, \ldots, 0)$ from that corresponding to $\ell' = (1, 1, 0, \ldots, 0)$. Narrower intervals can be obtained if we use $\ell' = (\ell_1, \ell_2, 0, \ldots, 0)$ and then minimize the width of the corresponding interval with respect to ℓ_1 and ℓ_2 either at the covariance or correlation stage. T. W. Anderson [1965: p. 483] and Aitkin [1969] give two such approaches.

Unfortunately, the above methods give rise to very wide intervals for the ρ_{jk}. This is not surprising, as intervals for the σ_{jj} are also included in the simultaneous set, and the inequalities (3.61) and (3.62) are used in a crude fashion to obtain (3.63). In an attempt to construct other confidence intervals we find that the joint distribution of the r_{jk} is not available in a useful form. Even the asymptotic distribution, which is multivariate normal when the parent population is normal (Aitkin [1969, 1971]), is too complicated to allow the construction of simultaneous confidence intervals from the implicitly defined confidence region

$$(\mathbf{r} - \mathscr{E}[\mathbf{r}])'(\mathscr{D}[\mathbf{r}])^{-1}(\mathbf{r} - \mathscr{E}[\mathbf{r}]) \le k$$

for $\boldsymbol{\rho}$ [see the argument leading from (3.31)]: Here $\mathbf{r}$ and $\boldsymbol{\rho}$ are the sample and population correlation coefficients listed as vectors. However, another approach is available using the marginal distribution of each r_{jk} and Fisher's Z-transformation. We have

$$Z_{jk} = \tfrac{1}{2}\log\left(\frac{1 + r_{jk}}{1 - r_{jk}}\right) = \tanh^{-1} r_{jk} \qquad (= Z[r_{jk}], \text{say}) \qquad (3.64)$$

is approximately $N_1(Z[\rho_{jk}], 1/[n - 3])$. We can therefore construct a large-sample confidence interval for each ρ_{jk} based on this normal distribution and then combine all the intervals using the Bonferroni method [see (1.21)]. Thus for all j, k $(j < k)$, the set of confidence intervals

$$\tanh\left(Z_{jk} - c_\alpha[n - 3]^{-1/2}\right) \le \rho_{jk} \le \tanh\left(Z_{jk} + c_\alpha[n - 3]^{-1/2}\right),$$

where $c_\alpha = z_{\alpha/2m}$ and $m = \frac{1}{2}d(d-1)$, have an approximate overall confidence of at least $100(1-\alpha)\%$. Clearly, if d is large, m and c_α will be large, thus leading to wide intervals. For this reason it is more appropriate to set $\alpha = 0.10$ rather than the customary $\alpha = 0.05$.

The above confidence intervals can be used as a basis for a conservative test procedure, namely, ρ_{jk} is significantly different from zero if its confidence interval does not contain zero, that is, if

$$|r_{jk}| > \tanh\left(c_\alpha[n-3]^{-1/2}\right) = K_\alpha,$$

say. We could calculate K_α and compare its value with each $|r_{jk}|$. However, as we are testing $\rho_{jk} = 0$, we can use various improvements to the normal approximation (3.64) (see Konishi [1978: approximation "S" with $\rho = 0$]). In practice, the researcher may, *before* seeing the data, be interested in only d_1 $(< d)$ of the variables: Then $m = \frac{1}{2}d_1(d_1 - 1)$.

Hills [1969] provides two graphical techniques for examining large correlation matrices. The first method consists of a half-normal plot of the ranked $|Z_{jk}|$. If the ρ_{jk} are all zero, the Z_{jk} have approximately zero means and the probability plot will be approximately linear, provided that the dependence among the Z_{jk} can be ignored and there are enough points, say, at least 50 (see Daniel and Wood [1981]). Any Z_{jk} with a significantly nonzero mean will show up as a point lying above the general linear trend. Hill's second method is a visual clustering technique that endeavors to select clusters of variables which have high positive correlations with each other.

Finally, a word of caution: The above theory leans heavily on the assumption that the $\mathbf{x}_i$ are multivariate normal, and is affected by kurtosis (Muirhead [1982: Section 5.1.4]). We now consider an asymptotic theory that does not require normality.

3.5.8 Large-Sample Inferences

We know that inferences about dispersion matrices are very sensitive to nonnormality of the parent distribution, whereas inferences about means are more robust. A sensible approach, therefore, would be to try and transform the data so that the elements of the dispersion matrix now become the elements of a mean vector. This approach is used for correlation coefficients in the previous section, though normality of the data is assumed. Layard [1972], however, has generalized this approach and does not assume normality. Here asymptotic normality is achieved by the multivariate central limit theorem, and the following notation and theory are based on his paper.

Let $\mathbf{x}_i$ $(i = 1, 2, \ldots, n)$ be i.i.d. with mean $\boldsymbol{\mu}$ and dispersion matrix $\boldsymbol{\Sigma} = [(\sigma_{jk})]$. Define

$$\begin{aligned}\boldsymbol{\sigma}' &= \left(\sigma_{11}, \sigma_{22}, \ldots, \sigma_{dd}, \sigma_{12}, \ldots, \sigma_{1d}, \sigma_{23}, \ldots, \sigma_{2d}, \ldots, \sigma_{d-1,d}\right)\\ &= \left(\{\sigma_{jj}\}, \{\sigma_{jk}\}\right), \qquad (3.65)\end{aligned}$$

say, to be the vector of distinct elements of $\boldsymbol{\Sigma}$. In a similar manner we can define the unbiased estimate of $\boldsymbol{\sigma}'$:

$$\mathbf{s}' = (\{s_{jj}\}, \{s_{jk}\}), \tag{3.66}$$

where $[(s_{jk})] = \mathbf{S} = \Sigma_i (\mathbf{x}_i - \bar{\mathbf{x}})(\mathbf{x}_i - \bar{\mathbf{x}})'/(n-1)$. When $n \to \infty$, $\sqrt{n}(\mathbf{s} - \boldsymbol{\sigma})$ is asymptotically distributed as $N_r(\mathbf{0}, \boldsymbol{\Gamma})$, where $r = \frac{1}{2}d(d+1)$ and $\boldsymbol{\Gamma}$ is described below. Convergence to normality can be speeded up by transforming $\mathbf{s}$ to [see (3.64)]

$$\boldsymbol{\phi}(\mathbf{s}) = \left(\{\log s_{jj}\}, \{\tanh^{-1} r_{jk}\}\right)', \tag{3.67}$$

where $r_{jk} = s_{jk}/(s_{jj}s_{kk})^{1/2}$. Using a Taylor expansion and denoting elements of $\mathbf{s}$ and $\boldsymbol{\phi}$ by the subscripts u and v, respectively, we have

$$\begin{aligned}[\boldsymbol{\phi}(\mathbf{s}) - \boldsymbol{\phi}(\boldsymbol{\sigma})]_v &\approx \sum_{u=1}^{r} (s_u - \sigma_u)\frac{\partial \phi_v(\boldsymbol{\sigma})}{\partial \sigma_u} \\ &= \left(\frac{\partial \phi_v(\boldsymbol{\sigma})}{\partial \sigma_1}, \ldots, \frac{\partial \phi_v(\boldsymbol{\sigma})}{\partial \sigma_r}\right)(\mathbf{s} - \boldsymbol{\sigma}),\end{aligned}$$

or

$$\boldsymbol{\phi}(\mathbf{s}) - \boldsymbol{\phi}(\boldsymbol{\sigma}) \approx \mathbf{A}'(\mathbf{s} - \boldsymbol{\sigma}), \tag{3.68}$$

where $\mathbf{A} = [(a_{uv})] = [(\partial \phi_v(\boldsymbol{\sigma})/\partial \sigma_u)]$. Hence, by A10.2, $\sqrt{n}[\boldsymbol{\phi}(\mathbf{s}) - \boldsymbol{\phi}(\boldsymbol{\sigma})]$ is asymptotically $N_r(\mathbf{0}, \boldsymbol{\Lambda})$, where $\boldsymbol{\Lambda} = \mathbf{A}'\boldsymbol{\Gamma}\mathbf{A}$ and

$$\boldsymbol{\phi}(\boldsymbol{\sigma}) = \left(\{\log \sigma_{jj}\}, \{\tanh^{-1} \rho_{jk}\}\right)'.$$

We now turn our attention to the matrices $\boldsymbol{\Gamma}$ and $\mathbf{A}$. Suppose the general cross moments of the elements of $\mathbf{x}$ are as follows:

$$\mu_{ab\cdots q} = \mathrm{E}\left[(x_1 - \mu_1)^a (x_2 - \mu_2)^b \cdots (x_d - \mu_d)^q\right]. \tag{3.69}$$

For example, $\mu_{20\cdots 0} = \sigma_{11}$, $\mu_{020\cdots 0} = \sigma_{22}$, and $\mu_{110\cdots 0} = \sigma_{12}$. If *as* stands for asymptotic, then it can be shown, either directly (e.g., Seber [1977: pp. 14–16]) or via cumulants (e.g., Muirhead [1982: p. 42] and Cook [1951]), that

$$\begin{aligned}as\,\mathrm{cov}\left[\sqrt{n}\, s_{jk}, \sqrt{n}\, s_{mp}\right] &= \mathrm{E}[(x_j - \mu_j)(x_k - \mu_k)(x_m - \mu_m)(x_p - \mu_p)] \\ &\quad - \mathrm{E}[(x_j - \mu_j)(x_k - \mu_k)]\mathrm{E}[(x_m - \mu_m)(x_p - \mu_p)]\end{aligned} \tag{3.70}$$

for all $j, k, m, p = 1, 2, \ldots, d$, and these are the terms that make up the elements of Γ. For example, the diagonal elements of Γ consist of the d elements

$$as\,\text{var}\left[\sqrt{n}\, s_{jj}\right] = \text{E}\left[(x_j - \mu_j)^4\right] - \left\{\text{E}\left[(x_j - \mu_j)^2\right]\right\}^2$$

$$\left(= \mu_{40\cdots0} - \mu_{20\cdots0}^2, \text{ when } j = 1\right)$$

and the $\frac{1}{2}d(d-1)$ elements (with $j < k$)

$$as\,\text{var}\left[\sqrt{n}\, s_{jk}\right] = \text{E}\left[(x_j - \mu_j)^2 (x_k - \mu_k)^2\right] - \left\{\text{E}\left[(x_j - \mu_j)(x_k - \mu_k)\right]\right\}^2$$

$$\left(= \mu_{220\cdots0} - \mu_{110\cdots0}^2, \text{ when } j = 1, k = 2\right).$$

The reader is referred to Layard [1974] for the case $d = 2$ (see also Exercise 3.14).

The matrix $\mathbf{A}$ in (3.68) is best described by means of an example. If $d = 3$,

$$\boldsymbol{\sigma}' = (\sigma_{11}, \sigma_{22}, \sigma_{33}, \sigma_{12}, \sigma_{13}, \sigma_{23})$$

and

$$\boldsymbol{\phi}'(\boldsymbol{\sigma}) = \left(\log\sigma_{11}, \log\sigma_{22}, \log\sigma_{33}, \tanh^{-1}\rho_{12}, \tanh^{-1}\rho_{13}, \tanh^{-1}\rho_{23}\right).$$

Since $d\tanh^{-1}\rho/d\rho = 1/(1-\rho^2)$,

$$\mathbf{A} = \left[\left(\frac{\partial\phi_v(\boldsymbol{\sigma})}{\partial\sigma_u}\right)\right]$$

$$= \boldsymbol{\Delta}\begin{pmatrix} \mathbf{I}_3 & \mathbf{B} \\ \mathbf{O} & \mathbf{I}_3 \end{pmatrix}, \tag{3.71}$$

where

$$\boldsymbol{\Delta} = \text{diag}\left\{\sigma_{11}^{-1}, \sigma_{22}^{-1}, \sigma_{33}^{-1}, \frac{(\sigma_{11}\sigma_{22})^{-1/2}}{1-\rho_{12}^2}, \frac{(\sigma_{11}\sigma_{33})^{-1/2}}{1-\rho_{13}^2}, \frac{(\sigma_{22}\sigma_{33})^{-1/2}}{1-\rho_{23}^2}\right\}$$

and

$$\mathbf{B} = \begin{pmatrix} -\frac{1}{2}\rho_{12}/(1-\rho_{12}^2), & -\frac{1}{2}\rho_{13}/(1-\rho_{13}^2), & 0 \\ -\frac{1}{2}\rho_{12}/(1-\rho_{12}^2), & 0, & -\frac{1}{2}\rho_{23}/(1-\rho_{23}^2) \\ 0, & -\frac{1}{2}\rho_{13}/(1-\rho_{13}^2), & -\frac{1}{2}\rho_{23}/(1-\rho_{23}^2) \end{pmatrix}. \tag{3.72}$$

For general d, $\mathbf{I}_3$ becomes $\mathbf{I}_d$, the form of $\boldsymbol{\Delta}$ generalizes in an obvious fashion, and $\mathbf{B}$ takes the same form as above with zeros in the ith row in positions corresponding to the ρ_{jk} when neither j nor k equals i.

All the elements of $\boldsymbol{\Gamma}$ and $\mathbf{A}$ are functions of the moments $E[(x_j - \mu_j)(x_k - \mu_k)(x_m - \mu_m)(x_p - \mu_p)]$ and $E[(x_j - \mu_j)(x_k - \mu_k)]$, which are special cases of $\mu_{ab\cdots q}$ [see (3.69)]. We can estimate $\mu_{ab\cdots q}$ by

$$\hat{\mu}_{ab\cdots q} = \frac{1}{n-1}\sum_i (x_{i1} - \bar{x}_{\cdot 1})^a (x_{i2} - \bar{x}_{\cdot 2})^b \cdots (x_{id} - \bar{x}_{\cdot d})^q \tag{3.73}$$

and obtain estimates $\hat{\boldsymbol{\Gamma}}$ and $\hat{\mathbf{A}}$. From a practical viewpoint, therefore, we can assume that $\sqrt{n}\,\boldsymbol{\phi}(\mathbf{s})$ is approximately multivariate normal with mean $\sqrt{n}\,\boldsymbol{\phi}(\boldsymbol{\sigma})$ and "known" dispersion matrix $\hat{\boldsymbol{\Lambda}} = \hat{\mathbf{A}}'\hat{\boldsymbol{\Gamma}}\hat{\mathbf{A}}$, and use $\boldsymbol{\phi}(\mathbf{s})$ to make inferences about $\boldsymbol{\sigma}$ via $\boldsymbol{\phi}(\boldsymbol{\sigma})$. Alternatively, if $n = Nn_0$, we can use Box's [1953] univariate technique (see Scheffé [1959: pp. 83–87]) and split the n observations $\mathbf{x}_i$ into N groups of n_0 observations to obtain N replicates from $N_r(\sqrt{n_0}\,\boldsymbol{\phi}(\boldsymbol{\sigma}), \boldsymbol{\Lambda})$. A third approach is to generate n "pseudoreplicates" using the jackknife technique: The underlying asymptotic distribution is $N_r(\sqrt{n}\,\boldsymbol{\phi}(\boldsymbol{\sigma}), \boldsymbol{\Lambda}_J)$, where $\boldsymbol{\Lambda}_J$ is the unknown dispersion matrix of a pseudoreplicate.

Although Layard's [1972] prime concern was with testing the equality of two dispersion matrices (see Section 3.6.1d), it is clear that the above three methods can be used for making a variety of inferences about $\boldsymbol{\Sigma}$. For example writing $\mathbf{z} = \sqrt{n}\,\boldsymbol{\phi}(\mathbf{s})$ and $\boldsymbol{\theta} = \sqrt{n}\,\boldsymbol{\phi}(\boldsymbol{\sigma})$, we have the linear model

$$\mathbf{z} = \boldsymbol{\theta} + \boldsymbol{\epsilon} \tag{3.74}$$

where $\boldsymbol{\varepsilon} \sim N_r(\mathbf{0}, \mathbf{V})$ and $\mathbf{V}$ $(= \boldsymbol{\Lambda})$ is assumed "known" in the first method. All the previous hypotheses about $\boldsymbol{\Sigma}$ take the form $H_0: \mathbf{A}\boldsymbol{\theta} = \mathbf{0}$. For example, if we wish to test $\sigma_{11} = \sigma_{22} = \cdots = \sigma_{dd}$ or, equivalently, $\log \sigma_{11} = \cdots = \log \sigma_{dd}$, then

$$\mathbf{A} = \left(\begin{array}{cccccc|c} 1 & -1 & 0 & \cdots & 0 & 0 & \\ 0 & 1 & -1 & \cdots & 0 & 0 & \\ \vdots & \vdots & \vdots & & \vdots & \vdots & \mathbf{O} \\ 0 & 0 & 0 & \cdots & 1 & -1 & \end{array}\right)$$

is a $(d-1) \times r$ matrix of rank $d-1$. Alternatively, if we wish to test for mutual independence, that is, $\rho_{jk} = 0$, for all $j, k, j \neq k$, we now have $\mathbf{A} = (\mathbf{O}|\mathbf{I}_{r-d})$. In general, if $\mathbf{A}$ is $q \times r$ of rank q, then, when H_0 is true, we have approximately $\mathbf{Az} \sim N_q(\mathbf{0}, \mathbf{AVA}')$ and $(\mathbf{Az})'(\mathbf{AVA}')^{-1}\mathbf{Az} \sim \chi_q^2$.

The grouping and jackknife methods can also be formulated in a similar fashion. For the grouping method we now have a random sample $\mathbf{z}_1, \mathbf{z}_2, \ldots, \mathbf{z}_N$ from $N_r(\boldsymbol{\theta}, \boldsymbol{\Lambda})$ and we can test $H_0: \mathbf{A}\boldsymbol{\theta} = \mathbf{0}$ using Hotelling's T^2 statistic [see (3.17)]

$$T_0^2 = N(\mathbf{A}\bar{\mathbf{z}})'(\mathbf{AS}_z\mathbf{A}')^{-1}\mathbf{A}\bar{\mathbf{z}},$$

where $\mathbf{S}_z = \sum_i (\mathbf{z}_i - \bar{\mathbf{z}})(\mathbf{z}_i - \bar{\mathbf{z}})'/(N-1)$. When H_0 is true, $T_0^2 \sim T^2_{q, N-1}$.

3.5.9 More General Covariance Structures

A wide class of structural models, called radex models, was introduced by Guttman [1954]. In these models test scores are generated from components that may be viewed as having a special geometrical structure, and hence the more recent name *simplex models*. One such model, called the circumplex by Guttman and studied in detail by Olkin and Press [1969], has a circular symmetric dispersion matrix

$$\boldsymbol{\Sigma}_c = \begin{pmatrix} \sigma_1 & \sigma_2 & \cdots & \sigma_d \\ \sigma_d & \sigma_1 & \cdots & \sigma_{d-1} \\ \vdots & \vdots & & \vdots \\ \sigma_2 & \sigma_3 & \cdots & \sigma_1 \end{pmatrix}, \qquad \sigma_j = \sigma_{d-j+2}\ (j = 2, 3, \ldots, d).$$

A key property of such matrices is that their eigenvectors do not depend on elements of $\boldsymbol{\Sigma}_c$, so that $\boldsymbol{\Sigma}_c$ can be reduced to a diagonal matrix. Olkin [1973] makes use of this property to extend $\boldsymbol{\Sigma}_c$ to the case where each σ_i is replaced by a $k \times k$ matrix $\boldsymbol{\Sigma}_i$, and discusses various hypothesis tests relating to this extended $\boldsymbol{\Sigma}_c$.

We note that $\boldsymbol{\Sigma}_c$ takes the form

$$\boldsymbol{\Sigma}_c = \sigma_1 \mathbf{G}_1 + \sigma_2 \mathbf{G}_2 + \cdots + \sigma_r \mathbf{G}_r, \tag{3.75}$$

where the $\mathbf{G}_i$ are known linearly independent symmetric matrices and the σ_i are unknown parameters. Models with this kind of covariance structure have been studied by several authors (e.g., T. W. Anderson [1969, 1970, 1973], Krishnaiah and Lee [1976], Sinha and Wieand [1979: p. 343], Szatrowski [1980]).

Finally we note that Jöreskog [1970, 1973] has considered questions of inference for a general structure of the form

$$\boldsymbol{\Sigma} = \mathbf{B}(\boldsymbol{\Lambda}\boldsymbol{\Phi}\boldsymbol{\Lambda}' + \boldsymbol{\Psi}^2)\mathbf{B}' + \boldsymbol{\Theta}^2,$$

where $\boldsymbol{\Psi}$ and $\boldsymbol{\Theta}$ are diagonal matrices. Even more general structures are considered by Jöreskog [1981], S. Y. Lee [1979, 1981], and Lee and Bentler [1980].

3.6 COMPARING TWO NORMAL POPULATIONS

3.6.1 Tests for Equal Dispersion Matrices

Although techniques for comparing any number of normal populations are considered in Chapter 9, it is helpful to consider the case of just two populations separately. Suppose we have a random sample $\mathbf{v}_1, \mathbf{v}_2, \ldots, \mathbf{v}_{n_1}$ from

$N_d(\boldsymbol{\mu}_1, \boldsymbol{\Sigma}_1)$ and an independent random sample $\mathbf{w}_1, \mathbf{w}_2, \ldots, \mathbf{w}_{n_2}$ from $N_d(\boldsymbol{\mu}_2, \boldsymbol{\Sigma}_2)$. It is convenient to consider testing $H_0: \boldsymbol{\Sigma}_1 = \boldsymbol{\Sigma}_2$ ($= \boldsymbol{\Sigma}$, say) first, before considering inferences about $\boldsymbol{\mu}_1$ and $\boldsymbol{\mu}_2$. Two basic procedures for testing H_0 are given below, the likelihood ratio method and the union–intersection method. Of the two, the former test procedure is generally more powerful.

a LIKELIHOOD RATIO TEST

If we use the notation of Section 3.2.1, the likelihood function is

$$L_{12}(\boldsymbol{\mu}_1, \boldsymbol{\mu}_2, \boldsymbol{\Sigma}_1, \boldsymbol{\Sigma}_2) = L_1(\boldsymbol{\mu}_1, \boldsymbol{\Sigma}_1) L_2(\boldsymbol{\mu}_2, \boldsymbol{\Sigma}_2),$$

where $L_i(\boldsymbol{\mu}_i, \boldsymbol{\Sigma}_i)$ takes the form (3.2). Maximizing L_{12} is equivalent to simultaneously maximizing each L_i so that L_{12} is maximized at $\hat{\boldsymbol{\mu}}_1 = \bar{\mathbf{v}}$, $\hat{\boldsymbol{\mu}}_2 = \bar{\mathbf{w}}$, $\hat{\boldsymbol{\Sigma}}_1 = \mathbf{Q}_1/n_1 = \sum_i (\mathbf{v}_i - \bar{\mathbf{v}})(\mathbf{v}_i - \bar{\mathbf{v}})'/n_1$ and $\hat{\boldsymbol{\Sigma}}_2 = \mathbf{Q}_2/n_2 = \sum_j (\mathbf{w}_j - \bar{\mathbf{w}})(\mathbf{w}_j - \bar{\mathbf{w}})'/n_2$. From (3.6) the maximum value of L_{12} is

$$\begin{aligned} L_{12}(\hat{\boldsymbol{\mu}}_1, \hat{\boldsymbol{\mu}}_2, \hat{\boldsymbol{\Sigma}}_1, \hat{\boldsymbol{\Sigma}}_2) &= L_1(\hat{\boldsymbol{\mu}}_1, \hat{\boldsymbol{\Sigma}}_1) L_2(\hat{\boldsymbol{\mu}}_2, \hat{\boldsymbol{\Sigma}}_2) \\ &= (2\pi)^{-nd/2} |\hat{\boldsymbol{\Sigma}}_1|^{-n_1/2} |\hat{\boldsymbol{\Sigma}}_2|^{-n_2/2} e^{-nd/2}, \end{aligned}$$

where $n = n_1 + n_2$. Setting $\boldsymbol{\Sigma}_1 = \boldsymbol{\Sigma}_2 = \boldsymbol{\Sigma}$, we now wish to maximize

$$\begin{aligned} \log L_{12}(\boldsymbol{\mu}_1, \boldsymbol{\mu}_2, \boldsymbol{\Sigma}, \boldsymbol{\Sigma}) &= \log L_1(\boldsymbol{\mu}_1, \boldsymbol{\Sigma}) + \log L_2(\boldsymbol{\mu}_2, \boldsymbol{\Sigma}) \\ &= c - \tfrac{1}{2} n \log|\boldsymbol{\Sigma}| - \tfrac{1}{2} \operatorname{tr}\left\{ \boldsymbol{\Sigma}^{-1} \left[\sum_{i=1}^{n_1} (\mathbf{v}_i - \boldsymbol{\mu}_1)(\mathbf{v}_i - \boldsymbol{\mu}_1)' \right.\right. \\ &\qquad \left.\left. + \sum_{j=1}^{n_2} (\mathbf{w}_j - \boldsymbol{\mu}_2)(\mathbf{w}_j - \boldsymbol{\mu}_2)' \right] \right\} \\ &= c - \tfrac{1}{2} n \log|\boldsymbol{\Sigma}| - \tfrac{1}{2} \operatorname{tr}[\boldsymbol{\Sigma}^{-1}\mathbf{Q}] \\ &\qquad + \tfrac{1}{2} \operatorname{tr}\left\{ \boldsymbol{\Sigma}^{-1} \left[n_1(\bar{\mathbf{v}} - \boldsymbol{\mu}_1)(\bar{\mathbf{v}} - \boldsymbol{\mu}_1)' + n_2(\bar{\mathbf{w}} - \boldsymbol{\mu}_2)(\bar{\mathbf{w}} - \boldsymbol{\mu}_2)' \right] \right\} \end{aligned} \tag{3.76}$$

[by (3.3) and (3.4)], where $c = -\tfrac{1}{2} nd \log 2\pi$ and $\mathbf{Q} = \mathbf{Q}_1 + \mathbf{Q}_2$. Since $\operatorname{tr}\{\boldsymbol{\Sigma}^{-1} n_1(\bar{\mathbf{v}} - \boldsymbol{\mu}_1)(\bar{\mathbf{v}} - \boldsymbol{\mu}_1)'\} = n_1(\bar{\mathbf{v}} - \boldsymbol{\mu}_1)'\boldsymbol{\Sigma}^{-1}(\bar{\mathbf{v}} - \boldsymbol{\mu}_1) \geq 0$, and so on, (3.76) is maximized for any $\boldsymbol{\Sigma} > \mathbf{O}$ when $\boldsymbol{\mu}_1 = \bar{\mathbf{v}} = \hat{\boldsymbol{\mu}}_1$, and $\boldsymbol{\mu}_2 = \bar{\mathbf{w}} = \hat{\boldsymbol{\mu}}_2$. Thus

$$\log L_{12}(\hat{\boldsymbol{\mu}}_1, \hat{\boldsymbol{\mu}}_2, \boldsymbol{\Sigma}, \boldsymbol{\Sigma}) \geq \log L_{12}(\boldsymbol{\mu}_1, \boldsymbol{\mu}_2, \boldsymbol{\Sigma}, \boldsymbol{\Sigma}).$$

and the next step is to maximize

$$\log L_{12}(\hat{\boldsymbol{\mu}}_1, \hat{\boldsymbol{\mu}}_2, \boldsymbol{\Sigma}, \boldsymbol{\Sigma}) = c - \tfrac{1}{2} n \left\{ \log|\boldsymbol{\Sigma}| + \operatorname{tr}[\boldsymbol{\Sigma}^{-1}\mathbf{Q}/n] \right\} \tag{3.77}$$

subject to $\boldsymbol{\Sigma} > \mathbf{O}$. Since each $\mathbf{Q}_i > \mathbf{O}$ (with probability 1), $\mathbf{Q} > \mathbf{O}$, and (3.77) is maximized when (A7.1) $\boldsymbol{\Sigma} = \hat{\boldsymbol{\Sigma}} = \mathbf{Q}/n$. Thus

$$\log L_{12}(\hat{\boldsymbol{\mu}}_1, \hat{\boldsymbol{\mu}}_2, \hat{\boldsymbol{\Sigma}}, \hat{\boldsymbol{\Sigma}}) \geq \log L_{12}(\hat{\boldsymbol{\mu}}_1, \hat{\boldsymbol{\mu}}_2, \boldsymbol{\Sigma}, \boldsymbol{\Sigma}) \geq \log L_{12}(\boldsymbol{\mu}_1, \boldsymbol{\mu}_2, \boldsymbol{\Sigma}, \boldsymbol{\Sigma})$$

so that $\hat{\boldsymbol{\mu}}_1$, $\hat{\boldsymbol{\mu}}_2$ and $\hat{\boldsymbol{\Sigma}}$ are the maximum likelihood estimates of $\boldsymbol{\mu}_1$, $\boldsymbol{\mu}_2$, and $\boldsymbol{\Sigma}$ under H_0. Also, from (3.77),

$$\begin{aligned}\log L_{12}(\hat{\boldsymbol{\mu}}_1, \hat{\boldsymbol{\mu}}_2, \hat{\boldsymbol{\Sigma}}, \hat{\boldsymbol{\Sigma}}) &= c - \tfrac{1}{2}n\left\{\log|\hat{\boldsymbol{\Sigma}}| + \operatorname{tr}[\hat{\boldsymbol{\Sigma}}^{-1}\hat{\boldsymbol{\Sigma}}]\right\} \\ &= c - \tfrac{1}{2}n\log|\hat{\boldsymbol{\Sigma}}| - \tfrac{1}{2}nd,\end{aligned}$$

so that

$$L_{12}(\hat{\boldsymbol{\mu}}_1, \hat{\boldsymbol{\mu}}_2, \hat{\boldsymbol{\Sigma}}, \hat{\boldsymbol{\Sigma}}) = (2\pi)^{-nd/2}|\hat{\boldsymbol{\Sigma}}|^{-n/2}e^{-nd/2}.$$

The likelihood ratio statistic is therefore

$$\begin{aligned}\ell &= \frac{L_{12}(\hat{\boldsymbol{\mu}}_1, \hat{\boldsymbol{\mu}}_2, \hat{\boldsymbol{\Sigma}}, \hat{\boldsymbol{\Sigma}})}{L_{12}(\hat{\boldsymbol{\mu}}_1, \hat{\boldsymbol{\mu}}_2, \hat{\boldsymbol{\Sigma}}_1, \hat{\boldsymbol{\Sigma}}_2)} \\ &= \frac{|\hat{\boldsymbol{\Sigma}}|^{-n/2}}{|\hat{\boldsymbol{\Sigma}}_1|^{-n_1/2}|\hat{\boldsymbol{\Sigma}}_2|^{-n_2/2}} \\ &= c_{12}\frac{|\mathbf{Q}_1|^{n_1/2}|\mathbf{Q}_2|^{n_2/2}}{|\mathbf{Q}_1 + \mathbf{Q}_2|^{(n_1+n_2)/2}},\end{aligned} \tag{3.78}$$

where

$$c_{12} = \frac{n^{nd/2}}{n_1^{n_1 d/2} n_2^{n_2 d/2}},$$

a result originally due to Wilks [1932]. Using the usual large-sample theory for the likelihood ratio test, $-2\log \ell$ is asymptotically χ^2_ν, $\nu = \frac{1}{2}d(d+1)$, when H_0 is true. However, there exists a slight modification of ℓ that produces an unbiased test and which leads to a better chi-square approximation and an F-approximation. This modification, called M, is discussed in Section 9.2.6 for the more general case of comparing any number of dispersion matrices. The robustness of these tests is discussed after the union–intersection test below.

We note from A1.2 that

$$\begin{aligned}\ell &= c_{12}\frac{|\mathbf{Q}_1\mathbf{Q}_2^{-1}|^{n_1/2}}{|\mathbf{I}_d + \mathbf{Q}_1\mathbf{Q}_2^{-1}|^{(n_1+n_2)/2}} \\ &= c_{12}\prod_{j=1}^{d}\left\{\phi_j^{n_1/2}(1+\phi_j)^{-(n_1+n_2)/2}\right\}\end{aligned} \tag{3.79}$$

is a function of the characteristic roots ϕ_j of $|\mathbf{Q}_1 - \phi\mathbf{Q}_2| = 0$. Setting $\theta_i = \phi_i/(1+\phi_i)$, we have the roots of $|\mathbf{Q}_1 - \theta(\mathbf{Q}_1 + \mathbf{Q}_2)| = 0$ and ℓ is proportional to $Y = \prod_j \{\theta_j^a (1-\theta_j)^b\}$ with $a = \frac{1}{2}n_1$ and $b = \frac{1}{2}n_2$. Pillai and Nagarsenker [1972] give an expression for the nonnull distribution of Y (i.e., when $\boldsymbol{\Sigma}_1 \neq \boldsymbol{\Sigma}_2$) for general a and b. Pillai and Nagarsenker, and Subrahmaniam [1975: ρ is misprinted at the top of p. 917 but correctly defined later], give asymptotic expansions for certain a and b. Nagarsenker [1978] gives a number of formulas that are useful for computing powers and exact percentage points for significance tests. For further details see Muirhead [1982: Section 8.2].

b UNION–INTERSECTION TEST

Writing $x_i = \ell'\mathbf{v}_i$ and $y_i = \ell'\mathbf{w}_i$, we see that the x_i are i.i.d. $N_1(\ell'\boldsymbol{\mu}_1, \ell'\boldsymbol{\Sigma}_1\ell)$, and the y_i are i.i.d. $N_1(\ell'\boldsymbol{\mu}_2, \ell'\boldsymbol{\Sigma}_2\ell)$. Using the notation above, we also have that the $\ell'\mathbf{Q}_i\ell/\ell'\boldsymbol{\Sigma}_i\ell$, $i = 1, 2$, are independently distributed as $\chi^2_{n_i - 1}$ (see Theorem 2.2, Corollary 1, in Section 2.3.1). Writing $\mathbf{S}_i = \mathbf{Q}_i/(n_i - 1)$, the usual F-test of size α for testing $H_{0\ell}: \ell'\boldsymbol{\Sigma}_1\ell = \ell'\boldsymbol{\Sigma}_2\ell$ has acceptance region of the form $F_1 \leq \ell'\mathbf{S}_1\ell/\ell'\mathbf{S}_2\ell \leq F_2$. Hence, with the union–intersection principle (Roy [1953]), a test of $H_0 = \cap_\ell H_{0\ell}$ has an acceptance region that is the intersection of all the univariate acceptance regions. Using A7.5, this is given by

$$\bigcap_\ell \left\{ (\mathbf{v}_1, \ldots, \mathbf{v}_{n_1}, \mathbf{w}_1, \ldots, \mathbf{w}_{n_2}) : F_1 \leq \frac{\ell'\mathbf{S}_1\ell}{\ell'\mathbf{S}_2\ell} \leq F_2 \right\}$$

$$= \left\{ (\mathbf{V}, \mathbf{W}) : F_1' \leq \inf_\ell \frac{\ell'\mathbf{Q}_1\ell}{\ell'\mathbf{Q}_2\ell} \leq \sup_\ell \frac{\ell'\mathbf{Q}_1\ell}{\ell'\mathbf{Q}_2\ell} \leq F_2' \right\}$$

$$= \{ (\mathbf{V}, \mathbf{W}) : F_1' \leq \phi_{\min} \leq \phi_{\max} \leq F_2' \}$$

$$= \{ (\mathbf{V}, \mathbf{W}) : c_1 \leq \theta_{\min} \leq \theta_{\max} \leq c_2 \}, \tag{3.80}$$

where ϕ $(= \theta/(1-\theta))$ and θ are eigenvalues of $\mathbf{Q}_1\mathbf{Q}_2^{-1}$ and $\mathbf{Q}_1(\mathbf{Q}_1 + \mathbf{Q}_2)^{-1}$, respectively [see (2.33)]. Thus for a test of size α the rejection region is the complement of (3.80), namely, $[\theta_{\min} < c_1] \cup [\theta_{\max} > c_2]$, where c_1 and c_2 satisfy

$$1 - \alpha = \mathrm{pr}[c_1 \leq \theta_{\min} \leq \theta_{\max} \leq c_2 | \boldsymbol{\Sigma}_1 = \boldsymbol{\Sigma}_2].$$

The optimal choice of c_1 and c_2, in the sense of maximizing the power, is not known; however, Schuurmann et al. [1973a] give values of A for the simple choice $c_1 = 1 - A$ and $c_2 = A$, for $d = 2(1)10$ and selected values of $R = \frac{1}{2}(n_1 - d - 1)$, $N = \frac{1}{2}(n_2 - d - 1)$, and α.

Testing $\boldsymbol{\Sigma}_1 = \boldsymbol{\Sigma}_2$ is equivalent to testing that the eigenvalues of $\boldsymbol{\Sigma}_1\boldsymbol{\Sigma}_2^{-1}$ are all unity. Since these eigenvalues are estimated by those of $\mathbf{S}_1\mathbf{S}_2^{-1}$, namely, the ϕ_j multiplied by $(n_2 - 1)/(n_1 - 1)$, several functions such as $\mathrm{tr}[\mathbf{Q}_1\mathbf{Q}_2^{-1}] = \sum_j \phi_j$ have been proposed as test statistics (see Muirhead [1982: Section 8.2.8]).

Unfortunately, like the trace criterion, these functions have the weakness that the effect of large ϕ_j's can be canceled by the presence of small ϕ_j's. Such statistics are more appropriate for testing against one-sided alternatives $\boldsymbol{\Sigma}_1 > \boldsymbol{\Sigma}_2$, that is, against the alternative that the eigenvalues of $\boldsymbol{\Sigma}_1\boldsymbol{\Sigma}_2^{-1}$ are each greater than or equal to unity, with at least one strict inequality. Pillai and Jayachandran [1968] and Chu and Pillai [1979] compare four such tests. However, Pillai and Sudjana [1975] give empirical evidence that the tests are not robust to departures from normality, and that no one statistic is any better than the others with regard to robustness.

Simultaneous confidence intervals for all ratios $\boldsymbol{\ell}'\boldsymbol{\Sigma}_1\boldsymbol{\ell}/\boldsymbol{\ell}'\boldsymbol{\Sigma}_2\boldsymbol{\ell}$, with any non-zero $\boldsymbol{\ell}$, can be constructed using the methods of T. W. Anderson [1965: p. 485]. However, as we found with the single population $\boldsymbol{\Sigma}$ in Section 3.5.7, such intervals are not likely to be useful as they will be too wide.

c ROBUSTNESS OF TESTS

We observed above that the maximum likelihood and union–intersection tests are both based on the roots of $|\mathbf{Q}_1 - \phi\mathbf{Q}_2| = 0$. If $H_0: \boldsymbol{\Sigma}_1 = \boldsymbol{\Sigma}_2$ ($= \boldsymbol{\Sigma}$, say) is true, we can transform the data by premultiplying by $\boldsymbol{\Sigma}^{-1/2}$ (A5.4) without changing the roots ϕ_i as

$$|\boldsymbol{\Sigma}^{-1/2}\mathbf{Q}_1\boldsymbol{\Sigma}^{-1/2} - \phi\boldsymbol{\Sigma}^{-1/2}\mathbf{Q}_2\boldsymbol{\Sigma}^{-1/2}| = 0. \tag{3.81}$$

In studying the null distributions of the two tests with respect to robustness, we can therefore assume that $\boldsymbol{\Sigma} = \mathbf{I}_d$. On this basis, Layard [1972] proved that if n_1 and n_2 tend to infinity in a constant ratio, and both parent distributions have common fourth moments, then $-2\log\ell$ is asymptotically distributed as $\sum_{i=1}^{k} c_i G_i$, where $k = \frac{1}{2}d(d+1)$ and the G_i are independent χ_1^2 variables when H_0 is true. The c_i are functions of correlation coefficients and standardized fourth-order cumulants, and the c_i are all unity when the parent distributions are normal. A similar result was obtained by Muirhead and Waternaux [1980] for a general family of elliptical distributions. For the special case of $d = 2$ and the four elements of $\mathbf{v}_i$ and $\mathbf{w}_j$ all independently and identically distributed with common kurtosis γ_2, Layard [1972] showed that

$$\sum_{i=1}^{3} c_i G_i = (1 + \tfrac{1}{2}\gamma_2)(G_1 + G_2) + G_3. \tag{3.82}$$

For the normal distribution $\gamma_2 = 0$, and $-2\log\ell$ is asymptotically χ_3^2 when H_0 is true. If $\gamma_2 > 0$, the critical values obtained from χ_3^2 will be too small and the significance level greater than the nominal value. If $\gamma_2 < 0$, the opposite situation occurs. Asymptotic nonrobustness was also demonstrated theoretically by Ito [1969: p. 118]. For the above bivariate model, Layard [1972] showed that the asymptotic joint distribution of the two roots (ϕ_1, ϕ_2) $[= (\phi_{\max}, \phi_{\min})]$ also depends on the term $1 + \frac{1}{2}\gamma_2$ in the nonnormal situation so that similar comments about the significance level of the union–intersection test apply.

If the asymptotic distributions of our two test statistics are adversely affected by nonnormality, then we can expect at least a similar nonrobustness for their exact small-sample distributions. For example, Layard [1974], using simulation, examined a modification M of the likelihood ratio statistic ℓ for $d = 2$ and small samples (e.g., $n_1 = n_2 = 25$). Sensitivity to nonzero kurtosis γ_2 is again indicated, and Layard concluded that the significance level is so severely affected by nonnormality that the usefulness of the test is questionable. For a bivariate gamma distribution that is only mildly nonnormal ($\gamma_2 = 0.6$ for each marginal), and with $\rho = 0.9$, the observed significance level was 15.2% instead of the nominal 5%. However, for a heavy-tailed contaminated normal with $\gamma_2 = 5.33$ for the marginals and $\rho = 0.9$, the observed significance level is 40%! Such conclusions are not unexpected, as the nonrobustness of the F-test for comparing two variances is well known. We now consider a robust large-sample approach.

d ROBUST LARGE-SAMPLE TESTS

Layard [1972, 1974] used the large-sample methods described in Section 3.5.8 for testing the equality of two dispersion matrices. Suppose for $i = 1, 2$ we have a random sample of size n_i from a distribution with dispersion matrix $\mathbf{\Sigma}_i$. Then, with the notation of Section 3.5.8, $\boldsymbol{\phi}(\mathbf{s}_i) - \boldsymbol{\phi}(\boldsymbol{\sigma}_i)$ is approximately $N_r(\mathbf{0}, \mathbf{A}'\mathbf{\Gamma}_i\mathbf{A}/n_i)$. Consider H_0, the hypothesis that the two parent distributions are *identical*, apart from a possible shift in location. Then under H_0, $\mathbf{\Sigma}_1 = \mathbf{\Sigma}_2$ [i.e., $\boldsymbol{\phi}(\boldsymbol{\sigma}_1) = \boldsymbol{\phi}(\boldsymbol{\sigma}_2)$] and all the higher moments are equal so that $\mathbf{\Gamma}_1 = \mathbf{\Gamma}_2 = \mathbf{\Gamma}$, say. Hence when H_0 is true, $\boldsymbol{\phi}(\mathbf{s}_1) - \boldsymbol{\phi}(\mathbf{s}_2)$ is approximately $N_r[\mathbf{0}, (1/n_1 + 1/n_2)\mathbf{A}'\mathbf{\Gamma}\mathbf{A}]$, and

$$\frac{n_1 n_2}{n_1 + n_2}[\boldsymbol{\phi}(\mathbf{s}_1) - \boldsymbol{\phi}(\mathbf{s}_2)]'(\mathbf{A}'\hat{\mathbf{\Gamma}}_p\mathbf{A})^{-1}[\boldsymbol{\phi}(\mathbf{s}_1) - \boldsymbol{\phi}(\mathbf{s}_2)]$$

is approximately distributed as χ_r^2, where $\hat{\mathbf{\Gamma}}_p$ is a suitable estimate of $\mathbf{\Gamma}$ based on the pooled data. Layard called this test of H_0 a "standard error" test and recommends using pooled estimates of $\mu_{ab\cdots q}$, namely,

$$\hat{\mu}_{ab\cdots q} = \frac{(n_1 - 1)\hat{\mu}_{ab\cdots q(1)} + (n_2 - 1)\hat{\mu}_{ab\cdots q(2)}}{n_1 + n_2 - 2}$$

in $\hat{\mathbf{\Gamma}}_p$ [see (3.73)]. He also introduced the grouping and jackknife tests of H_0. These two tests, under the linear model formulation of (3.74), amount to testing the equality of the means of two normal populations: This can be accomplished using Hotelling's T^2 test (see Section 3.6.2).

Layard [1974] carried out a simulation study to compare the small-sample robustness of the above three tests with the modified likelihood ratio statistic M of Section 9.2.6 for the case $d = 2$. He concluded that the grouping test maintained its nominal significance level well, but had a much lower power: This property is also shared by its univariate counterpart (Layard [1973]). The standard error and jackknife tests had better powers, and these were close to the power of the M test under normality. However, their significance levels tended to be high and on this count the standard error test seemed preferable, particularly with a contaminated normal distribution. In the latter case the observed significance levels were approximately 7 and 15%, respectively, and there was some evidence that as n_1 and n_2 were increased, the level converged to 5% more rapidly for the standard error test.

3.6.2 *Test for Equal Means Assuming Equal Dispersion Matrices*

a HOTELLING'S T^2 TEST

Given a random sample $\mathbf{v}_1, \mathbf{v}_2, \ldots, \mathbf{v}_{n_1}$ from $N_d(\boldsymbol{\mu}_1, \boldsymbol{\Sigma}_1)$ and an independent random sample $\mathbf{w}_1, \mathbf{w}_2, \ldots, \mathbf{w}_{n_2}$ from $N_d(\boldsymbol{\mu}_2, \boldsymbol{\Sigma}_2)$, we wish to test $H_0: \boldsymbol{\theta} = (\boldsymbol{\mu}_1 - \boldsymbol{\mu}_2) = \mathbf{0}$. We begin by assuming $\boldsymbol{\Sigma}_1 = \boldsymbol{\Sigma}_2$ ($= \boldsymbol{\Sigma}$, say). From Theorem 3.1 in Section 3.2.2, $\bar{\mathbf{v}} \sim N_d(\boldsymbol{\mu}_1, \boldsymbol{\Sigma}/n_1)$, $\mathbf{Q}_1 = (n_1 - 1)\mathbf{S}_1 = \sum_i (\mathbf{v}_i - \bar{\mathbf{v}})(\mathbf{v}_i - \bar{\mathbf{v}})' \sim W_d(n_1 - 1, \boldsymbol{\Sigma})$, and $\bar{\mathbf{v}}$ is statistically independent of $\mathbf{Q}_1$. Similarly, $\bar{\mathbf{w}} \sim N_d(\boldsymbol{\mu}_2, \boldsymbol{\Sigma}/n_2)$, $\mathbf{Q}_2 = (n_2 - 1)\mathbf{S}_2 \sim W_d(n_2 - 1, \boldsymbol{\Sigma})$, and $\bar{\mathbf{w}}$ is statistically independent of $\mathbf{Q}_2$. Since the two samples are independent, the four random variables $\bar{\mathbf{v}}$, $\bar{\mathbf{w}}$, $\mathbf{Q}_1$, and $\mathbf{Q}_2$ are mutually independent. Hence, by Exercises 2.6 and 2.12(d),

$$\bar{\mathbf{v}} - \bar{\mathbf{w}} \sim N_d\left[\boldsymbol{\theta}, \left(\frac{1}{n_1} + \frac{1}{n_2}\right)\boldsymbol{\Sigma}\right], \tag{3.83}$$

$$\mathbf{Q} = \mathbf{Q}_1 + \mathbf{Q}_2 \sim W_d(n_1 + n_2 - 2, \boldsymbol{\Sigma}), \tag{3.84}$$

and $\bar{\mathbf{v}} - \bar{\mathbf{w}}$ is statistically independent of $\mathbf{Q}$. Hence, defining $\mathbf{S}_p = \mathbf{Q}/(n_1 + n_2 - 2)$, we have from (2.19) and (2.20) with $\lambda = n_1 n_2/(n_1 + n_2)$ that

$$\begin{aligned} T^2 &= \frac{n_1 n_2}{n_1 + n_2}(\bar{\mathbf{v}} - \bar{\mathbf{w}} - \boldsymbol{\theta})'\mathbf{S}_p^{-1}(\bar{\mathbf{v}} - \bar{\mathbf{w}} - \boldsymbol{\theta}) \\ &\sim T^2_{d, n_1+n_2-2} \\ &\equiv \frac{d(n_1 + n_2 - 2)}{n_1 + n_2 - d - 1} F_{d, n_1+n_2-d-1}. \end{aligned} \tag{3.85}$$

To test $H_0: \boldsymbol{\theta} = \mathbf{0}$ we calculate

$$T_0^2 = \frac{n_1 n_2}{n_1 + n_2}(\bar{\mathbf{v}} - \bar{\mathbf{w}})'\mathbf{S}_p^{-1}(\bar{\mathbf{v}} - \bar{\mathbf{w}}) \tag{3.86}$$

and reject H_0 at the α level of significance if

$$T_0^2 \geq \frac{d(n_1 + n_2 - 2)}{n_1 + n_2 - d - 1} F^{\alpha}_{d, n_1+n_2-d-1} = T^2_{d, n_1+n_2-2, \alpha}. \tag{3.87}$$

This test statistic is also equivalent to the likelihood ratio test [see (3.11) and Exercise 3.15]. We also note that as $n = n_1 + n_2 \to \infty$,

$$\begin{aligned} T^2_{d, n-2} &= \frac{d(n-2)}{n-d-1} F_{d, n-d-1} \\ &\to dF_{d, \infty} \\ &\equiv \chi^2_d, \end{aligned} \tag{3.88}$$

so that T_0^2 is asymptotically distributed as χ^2_d when H_0 is true.

Using A7.6 and arguing as in (3.12),

$$\begin{aligned} T_0^2 &= \lambda \sup_{\ell} \frac{[\ell'(\bar{\mathbf{v}} - \bar{\mathbf{w}})]^2}{\ell' \mathbf{S}_p \ell} \\ &= \sup_{\ell} \left[\frac{\sqrt{\lambda}\, \ell'(\bar{\mathbf{v}} - \bar{\mathbf{w}})}{(\ell' \mathbf{S}_p \ell)^{1/2}} \right]^2 \\ &= \sup_{\ell} t_0^2(\ell), \end{aligned} \tag{3.89}$$

where $t_0(\ell)$ is the usual t-statistic for testing the hypothesis $H_{0\ell}: \ell'\boldsymbol{\theta} = 0$ based on the two samples $\ell'\mathbf{v}_i$ $(i = 1, 2, \ldots, n_1)$ and $\ell'\mathbf{w}_j$ $(j = 1, 2, \ldots, n_2)$. The supremum occurs when $\ell \propto \mathbf{S}_p^{-1}(\bar{\mathbf{v}} - \bar{\mathbf{w}})$ or, with suitable scaling, $\ell = \mathbf{S}_p^{-1}(\bar{\mathbf{v}} - \bar{\mathbf{w}})$. This particular value of ℓ was introduced by Fisher [1936] as a means of classifying a new observation $\mathbf{x}$ into one of two groups. The function $\ell'\mathbf{x}$ is called Fisher's linear discriminant function (LDF) and it will be discussed further in Chapter 6.

Since $H_0 = \cap_{\ell} H_{0\ell}$, we can duplicate the argument leading to (3.13) and show that T_0^2 also follows from the union–intersection principle. Also, from (3.87),

$$\begin{aligned} 1 - \alpha &= \operatorname{pr}\left[T^2 \leq T^2_{d, n_1+n_2-2, \alpha} \right] \\ &= \operatorname{pr}\left[\lambda(\bar{\mathbf{v}} - \bar{\mathbf{w}} - \boldsymbol{\theta})' \mathbf{S}_p^{-1} (\bar{\mathbf{v}} - \bar{\mathbf{w}} - \boldsymbol{\theta}) \leq T^2_{d, n_1+n_2-2, \alpha} \right], \end{aligned}$$

so that we can construct simultaneous confidence intervals along the lines outlined in Section 3.4.4. For example, a set of simultaneous confidence intervals for the class of all linear functions $\mathbf{h}'\boldsymbol{\theta}$, with overall probability of $1 - \alpha$, is given by [see (3.34)]

$$\mathbf{h}'(\overline{\mathbf{v}} - \overline{\mathbf{w}}) \pm \left(T^2_{d, n_1+n_2-2, \alpha}\right)^{1/2} \left[\left(\frac{1}{n_1} + \frac{1}{n_2}\right)\mathbf{h}'\mathbf{S}_p\mathbf{h}\right]^{1/2}. \tag{3.90}$$

EXAMPLE 3.6 Using individual measurements of cranial length (x_1) and breadth (x_2) on 35 mature female frogs (*Rana esculenta*) and 14 mature male frogs published in a study by Kauri, Reyment [1961] obtained the following results:

$$\overline{\mathbf{v}} = \begin{pmatrix} 22.860 \\ 24.397 \end{pmatrix}, \qquad \overline{\mathbf{w}} = \begin{pmatrix} 21.821 \\ 22.843 \end{pmatrix}$$

$$\mathbf{S}_1 = \begin{pmatrix} 17.683 & 20.292 \\ 20.292 & 24.407 \end{pmatrix}, \qquad \mathbf{S}_2 = \begin{pmatrix} 18.479 & 19.095 \\ 19.095 & 20.755 \end{pmatrix}$$

and

$$\mathbf{S}_p = \frac{(n_1 - 1)\mathbf{S}_1 + (n_2 - 1)\mathbf{S}_2}{n_1 + n_2 - 2} = \begin{pmatrix} 17.903 & 19.959 \\ 19.959 & 23.397 \end{pmatrix}.$$

We note that $\mathbf{S}_1$ and $\mathbf{S}_2$ are very similar so that the assumption of $\boldsymbol{\Sigma}_1 = \boldsymbol{\Sigma}_2$ is tenable. To test the difference between the population means for female and male frogs we can use (3.86) with $d = 2$, $n_1 = 35$, and $n_2 = 14$, that is, $T_0^2 = 1.9693$. From (3.87) we therefore compare

$$F_0 = \frac{n_1 + n_2 - d - 1}{d(n_1 + n_2 - 2)} T_0^2 = \frac{46}{2(47)} T_0^2 = 0.964$$

with $F_{2,46}^{0.05} = 3.2$ and conclude that the test is not significant.

Although the null hypothesis of equal means is not rejected, it is instructive to demonstrate the calculation of confidence intervals. Writing $\boldsymbol{\mu}_1' = (\mu_{F_1}, \mu_{F_2})$ and $\boldsymbol{\mu}_2' = (\mu_{M_1}, \mu_{M_2})$, we can use (3.90) with

$$T^2_{d, n_1+n_2-2, .05} = \frac{2(47)}{46} F_{2,46}^{0.05} = 6.54$$

for constructing the confidence interval for any linear combination of the form $h_1(\mu_{F1} - \mu_{M1}) + h_2(\mu_{F2} - \mu_{M2})$. In particular, setting $\mathbf{h}'$ equal to $(1, 0)$, we have $\mathbf{h}'\mathbf{S}_p\mathbf{h} = s_{11}$, the first diagonal element of $\mathbf{S}_p$, and a confidence interval for $\mu_{F1} - \mu_{M1}$ is

$$22.860 - 21.821 \pm 6.54^{1/2}\left[\left(\tfrac{1}{35} + \tfrac{1}{14}\right)17.903\right]^{1/2}$$

or 1.04 ± 3.42. Similarly, setting $\mathbf{h}' = (0, 1)$, we can find a confidence interval for $\mu_{F_2} - \mu_{M_2}$, namely, 1.55 ± 3.91. The two intervals have a combined "confidence" of at least 95%.

If we had decided before seeing the data that we required these two confidence intervals, then we could use the Bonferroni method with $t_{47}^{0.05/4} = 2.31$ (see Appendix D1) in the above intervals instead of $(T^2_{2,47,0.05})^{1/2} = 6.54^{1/2} = 2.56$. For example, the first interval now becomes 1.04 ± 3.09.

In conclusion we note, from $\mathbf{S}_p$, that an estimate of the correlation between cranial length and breadth is $19.959(17.903 \times 23.397)^{-1/2} = 0.98$, thus suggesting that only one measurement is sufficient. The eigenvalues of $\mathbf{S}_p$ are 40.797 and 0.503, indicating that a single principal component will adequately describe the pooled data (see Section 5.2). The first eigenvector of $\mathbf{S}_p$ is (0.6571, 0.7538).

b EFFECT OF UNEQUAL DISPERSION MATRICES

To examine what happens to our test statistic T_0^2 when $\boldsymbol{\Sigma}_1 \neq \boldsymbol{\Sigma}_2$, we assume that n_1 and n_2 are large enough so that $\mathbf{S}_i = \mathbf{Q}_i/(n_i - 1) \approx \boldsymbol{\Sigma}_i$ $(i = 1, 2)$. Then, from (3.86),

$$\begin{aligned} T_0^2 &= \frac{n_1 n_2}{n_1 + n_2}(\bar{\mathbf{v}} - \bar{\mathbf{w}})'\left(\frac{(n_1 - 1)\mathbf{S}_1 + (n_2 - 1)\mathbf{S}_2}{n_1 + n_2 - 2}\right)^{-1}(\bar{\mathbf{v}} - \bar{\mathbf{w}}) \\ &= \frac{n_1 + n_2 - 2}{n_1 + n_2}(\bar{\mathbf{v}} - \bar{\mathbf{w}})'\left(\frac{(n_1 - 1)}{n_1 n_2}\mathbf{S}_1 + \frac{(n_2 - 1)}{n_1 n_2}\mathbf{S}_2\right)^{-1}(\bar{\mathbf{v}} - \bar{\mathbf{w}}) \\ &\approx (\bar{\mathbf{v}} - \bar{\mathbf{w}})'\left(\frac{1}{n_2}\boldsymbol{\Sigma}_1 + \frac{1}{n_1}\boldsymbol{\Sigma}_2\right)^{-1}(\bar{\mathbf{v}} - \bar{\mathbf{w}}) \\ &= D_0, \end{aligned}$$

say. Since $(\bar{\mathbf{v}} - \bar{\mathbf{w}}) \sim N_d[\boldsymbol{\theta}, (1/n_1)\boldsymbol{\Sigma}_1 + (1/n_2)\boldsymbol{\Sigma}_2]$, it follows from Theorem 2.1(vii) that

$$(\bar{\mathbf{v}} - \bar{\mathbf{w}})'\left(\frac{1}{n_1}\boldsymbol{\Sigma}_1 + \frac{1}{n_2}\boldsymbol{\Sigma}_2\right)^{-1}(\bar{\mathbf{v}} - \mathbf{w}) \tag{3.91}$$

has a noncentral chi-square distribution with d degrees of freedom and noncentrality parameter $\delta = \boldsymbol{\theta}'[(1/n_1)\boldsymbol{\Sigma}_1 + (1/n_2)\boldsymbol{\Sigma}_2]^{-1}\boldsymbol{\theta}$. When $H_0 : \boldsymbol{\theta} = \mathbf{0}$ is true, $\delta = 0$ and we have a central chi-square distribution. We note that when $n_1 = n_2$, D_0 has the same distribution as (3.91), irrespective of the values of $\boldsymbol{\Sigma}_1$ and $\boldsymbol{\Sigma}_2$, thus demonstrating the asymptotic insensitivity of T_0^2 to $\boldsymbol{\Sigma}_1 \neq \boldsymbol{\Sigma}_2$ in this case. Using the approximation $\mathbf{S}_i \approx \boldsymbol{\Sigma}_i$, Ito and Schull [1964: p. 75–78] demonstrated empirically that the effect of $\boldsymbol{\Sigma}_1 \neq \boldsymbol{\Sigma}_2$ on the significance level and the power of the T_0^2 test in minimal if n_1 and n_2 are very large, $n_1 \approx n_2$, and the eigenvalues of $\mathbf{\Sigma}_1\mathbf{\Sigma}_2^{-1}$ lie in the range (0.5, 2). When n_1 and n_2 are markedly different, inequality of $\mathbf{\Sigma}_1$ and $\mathbf{\Sigma}_2$ has a serious effect on the size and power of T_0^2. [It should be noted that the authors used the trace statistic T_g^2 which is equal to T_0^2 as $m_H = 1$; see (2.56)].

c EFFECT OF NONNORMALITY

When $\mathbf{\Sigma}_1 = \mathbf{\Sigma}_2$ we see from (3.88) that for normal data T_0^2 is asymptotically distributed as χ_d^2 when H_0 is true. Even with nonnormal data, the central limit theorem ensures the asymptotic normality of $\bar{\mathbf{v}} - \bar{\mathbf{w}}$ so that D_0, now the same as (3.91), is still asymptotically χ_d^2 when H_0 is true. Hence T_0^2 has the correct asymptotic null distribution when $\mathbf{\Sigma}_1 = \mathbf{\Sigma}_2$, irrespective of the underlying distributions of the $\mathbf{v}_i$ and $\mathbf{w}_j$. This asymptotic robustness might be expected to hold for moderate-sized samples. Such robustness is demonstrated by Everitt [1979a] and his table is reproduced as Table 3.5. For the uniform distribution the observed (true) significance levels are very close to the nominal values. In

TABLE 3.5 Percentage of Samples Exceeding 10%, 5% and 1% Points of the F Statistic Corresponding to Two-Sample T^2 Statistic[a]

	No. of Simulations			Distribution Sampled											
				Multinormal			Uniform			Exponential			Lognormal		
d	N	n_1	n_2	10%	5%	1%	10%	5%	1%	10%	5%	1%	10%	5%	1%
		5	5	10.3	5.1	1.1	10.1	5.7[b]	1.2	8.0[b]	3.6[b]	0.6[b]	7.1[b]	3.0[b]	0.8
		5	10	10.4	4.8	1.0	9.4	5.2	1.3[b]	8.6[b]	4.0[b]	1.2	8.5[b]	3.8[b]	0.7
		5	15	10.2	5.0	1.3[b]	9.6	4.8	0.7[b]	9.3	4.8	1.4[b]	9.3	4.7	1.3[b]
		5	20	10.4	5.2	0.9	9.8	4.8	1.3[b]	10.1	5.6	1.5[b]	9.5	5.2	1.6[b]
2	5,000	10	10	10.1	5.4	1.2	9.8	5.4	1.2	9.4	4.3[b]	0.8	7.3[b]	2.9[b]	0.4[b]
		10	15	10.0	4.8	1.0	9.8	5.2	1.1	9.4	4.5	0.7[b]	8.1[b]	3.2[b]	0.5[b]
		10	20	9.6	4.4	1.0	9.9	5.3	1.5[b]	9.0[b]	4.3[b]	0.7[b]	8.8[b]	4.3[b]	0.8
		15	15	9.6	5.2	0.9	10.7	5.0	1.2	9.3	4.4	0.8	8.1[b]	3.4[b]	0.4[b]
		15	20	9.1[b]	4.3[b]	0.7[b]	10.2	5.0	1.0	9.2	4.4	0.9	8.3[b]	3.5[b]	0.4[b]
		20	20	10.1	5.3	1.0	9.3	5.0	1.0	9.2	4.2[b]	0.8	8.5[b]	3.8[b]	0.3[b]
		10	10	9.5	5.1	0.9	9.6	4.8	0.9	9.6	4.7	0.8	7.2[b]	3.0[b]	0.4[b]
		10	15	9.5	4.9	1.0	9.8	5.0	1.1	9.4	4.5	0.9	7.6[b]	3.4[b]	0.4[b]
4	5,000	10	20	9.2	4.3[b]	1.0	9.7	5.1	1.2	9.2	4.7	0.8	8.3[b]	3.5[b]	0.5[b]
		15	15	9.7	5.1	1.0	9.6	4.9	1.1	9.5	4.6	0.7[b]	7.6[b]	3.0[b]	0.4[b]
		15	20	9.9	5.2	1.1	10.0	5.2	1.2	9.6	4.3[b]	0.9	8.0[b]	3.6[b]	0.4[b]
		20	20	9.5	4.6	1.1	10.7	5.2	1.0	8.5[b]	4.0[b]	0.6[b]	7.4[b]	3.2[b]	0.5[b]
		15	15	9.2	5.1	1.2	10.4	5.7	1.2	7.6[b]	3.0[b]	0.3[b]	8.2[b]	3.2[b]	0.5[b]
6	2,500	15	20	9.0	4.1[b]	0.8	10.1	4.6	1.0	9.0	3.8[b]	0.4[b]	8.4[b]	3.7[b]	0.6
		20	20	9.9	5.4	1.4	9.3	5.2	1.2	9.3	4.6	0.5[b]	8.2[b]	3.2[b]	0.5[b]
8	1,000	20	20	9.7	5.5	1.1	10.4	5.3	1.5	8.2	4.4	0.8	8.2	3.5[b]	0.2[b]

[a]From Everitt [1979a: Table 2].
[b]Indicates that the observed proportion is more than $2\{\alpha(1-\alpha)/N\}^{1/2}$ from the nominal α level.

general, the test is conservative in that nonnormality tends to reduce the true significance level, with the reduction being much greater with a highly skewed distribution like the lognormal. This is in contrast with the one-sample test, where nonnormality tends to increase significance levels. From Table 3.5 we note that the value of d and the equality or inequality of n_1 and n_2 appear to have little effect on the observed significance level. However, if $n_1 = n_2$, we can assume that T_0^2 is robust with regard to both nonnormality and moderate departures from $\boldsymbol{\Sigma}_1 = \boldsymbol{\Sigma}_2$.

Mardia [1971] has given a permutation test based on the statistic $V^{(1)}$ of (2.59), namely,

$$V^{(1)} = \frac{T_0^2/(n-2)}{T_0^2/(n-2)+1}. \tag{3.92}$$

This test is a special case of V^* in Section 9.2.4a with $I = 2$ and $s = 1$. Under the more restrictive null hypothesis H_{01} that the $n = n_1 + n_2$ observations $\mathbf{v}_i$ and $\mathbf{w}_j$ are independently and identically distributed, $V^{(1)}$ has approximately a beta distribution with $\frac{1}{2}ad$ and $\frac{1}{2}a(n-d-1)$ degrees of freedom, respectively, where

$$a = \frac{n-1-2c}{(n-1)(c+1)},$$

$$c = \frac{1}{2}\frac{(n-3)C_1C_2}{n(n-1)},$$

$$C_1 = \frac{n-1}{(n-3)(n-2)}\left\{n(n+1)\left[\frac{1}{n_1}+\frac{1}{n_2}-\frac{4}{n}\right]-2(n-2)\right\}, \tag{3.93}$$

$$C_2 = \frac{(n-1)(n+1)}{d(n-3)(n-d-1)}\left\{b_{2,d}-\frac{(n-1)d(d+2)}{n+1}\right\},$$

$$b_{2,d} = \frac{1}{n}\sum_{i=1}^{n}\left\{(\mathbf{x}_i-\overline{\mathbf{x}})'\left(\frac{\mathbf{Q}}{n}\right)^{-1}(\mathbf{x}_i-\overline{\mathbf{x}})\right\}^2,$$

$\mathbf{x}_i = \mathbf{v}_i$ $(i = 1,2,\ldots,n_1)$ and $\mathbf{x}_{n_1+j} = \mathbf{w}_j$ $(j = 1,2,\ldots,n_2)$. If we use $a = 1$, Mardia [1971] presents evidence to show that the true significance level is not likely to differ by more than 2% from the nominal 5% level. When $n_1 = n_2$, $C_1 = -2(n-1)/(n-3)$, $c = -C_2/n$, and, from (3.93),

$$a^{-1} = 1 - \frac{1}{n}\left\{\frac{n+1}{n-1+2C_2/n}\right\}C_2, \tag{3.94}$$

which is close to one for moderately large samples. In this case we can expect the test to be insensitive to nonnormality. However, the null hypothesis includes $\boldsymbol{\mu}_1 = \boldsymbol{\mu}_2$ and $\boldsymbol{\Sigma}_1 = \boldsymbol{\Sigma}_2$ so that we can expect the test to be affected by $\boldsymbol{\Sigma}_1 \neq \boldsymbol{\Sigma}_2$.

In practice, we would use the statistic $T_0^2(n-d-1)/[d(n-2)]$, which, under the null hypothesis, is approximately distributed as the F-distribution with ad and $a(n-d-1)$ degrees of freedom, respectively [see (3.92) and $B_{a,b}$ of (2.21)]. Tables for handling noninteger degrees of freedom are given by Mardia and Zemroch [1975b] (see also Pearson and Hartley [1972: Table 43]).

Recently four nonparametric tests of H_{01} were given by Friedman and Rafsky [1979]. These tests are multivariate generalizations of the Wald-Wolfowitz and Smirnov two-sample tests and use the minimal spanning tree (see Section 5.9) to order the observations. Friedman and Rafsky [1981] provide two useful P–P plots (Appendix C4): One highlights location differences, the other scale differences.

3.6.3 *Test for Equal Means Assuming Unequal Dispersion Matrices*

If $n_1 = n_2$ ($= n_0$, say), we can reduce the two-sample problem to a one-sample problem by writing $\mathbf{x}_i = \mathbf{v}_i - \mathbf{w}_i$ $(i = 1, 2, \ldots, n_0)$. Then the $\mathbf{x}_i$ are i.i.d. $N_d(\boldsymbol{\theta}, \boldsymbol{\Sigma}_x)$ with $\boldsymbol{\theta} = \boldsymbol{\mu}_1 - \boldsymbol{\mu}_2$ and $\boldsymbol{\Sigma}_x = \boldsymbol{\Sigma}_1 + \boldsymbol{\Sigma}_2$, and we can test $H_0: \boldsymbol{\theta} = \mathbf{0}$ using

$$T_0^2 = n_0 \bar{\mathbf{x}}' \mathbf{S}_x^{-1} \bar{\mathbf{x}}. \tag{3.95}$$

This statistic is distributed as $T^2_{d, n_0 - 1}$ when H_0 is true. However, if $\boldsymbol{\Sigma}_1 = \boldsymbol{\Sigma}_2$, this test is not as powerful as (3.86), which is based on $2(n_0 - 1)$ degrees of freedom. Also, (3.86) is insensitive to unequal $\boldsymbol{\Sigma}_i$ when $n_1 = n_2$, so that the loss of $(n_0 - 1)$ degrees of freedom in using (3.95) will generally not be worth it.

When $n_1 \neq n_2$, the corresponding problem in the univariate case of unequal variances is known as the Behrens–Fisher problem and various solutions have been proposed. One solution, proposed by Scheffé [1943], has been generalized to the multivariate case by Bennett [1951] and extended to more than two populations by Anderson [1963a]. Details of the test are set out in Exercise 3.16. Ito [1969: pp. 91–94] discussed this test in detail and showed that it is most powerful for a limited class of test procedures. However, the test has two disadvantages. First, we would need to choose n_1 of the $\mathbf{w}_j$ observations at random from the n_2 observations in the calculation of $\mathbf{y}_i$ so that different statisticians will obtain different values of T_0^2 in Exercise 3.16. For this reason Scheffé [1970: p. 1503] subsequently rejected the approach. Second, the power of the test will be far from optimal if n_1 is much less than n_2, as we are effectively ignoring $n_2 - n_1$ values of $\mathbf{w}_j$ in calculating $\mathbf{S}_y$. This effect on the power is demonstrated by Ito [1969: p. 97], who compared the test with the most powerful test under the assumption $\boldsymbol{\Sigma}_1 = k\boldsymbol{\Sigma}_2$, with k known (Exercise 3.17). As more powerful tests are given below, we do not recommend Bennett's test.

In the light of (3.91), a natural test statistic for $H_0:\boldsymbol{\theta} = \mathbf{0}$ is

$$(\bar{\mathbf{v}} - \bar{\mathbf{w}})'\left(\frac{1}{n_1}\mathbf{S}_1 + \frac{1}{n_2}\mathbf{S}_2\right)^{-1}(\bar{\mathbf{v}} - \bar{\mathbf{w}}). \tag{3.96}$$

When n_1 and n_2 are very large, this statistic tends to (3.91) and is therefore approximately distributed as χ_d^2 when H_0 is true. The statistic (3.96), a multivariate generalization of Welch's [1947] test statistic, was proposed by James [1954]. Using Taylor expansions, James showed that the upper α critical value of the statistic (3.96) is, to order n_i^{-2},

$$k_\alpha = \chi_d^2(\alpha)\left[A + B\chi_d^2(\alpha)\right], \tag{3.97}$$

where

$$A = 1 + \frac{1}{2d}\sum_{i=1}^{2}\frac{1}{n_i - 1}\left(\operatorname{tr}\left[\frac{\mathbf{S}_T^{-1}\mathbf{S}_i}{n_i}\right]\right)^2,$$

$$B = \frac{1}{d(d+2)}\left[\sum_{i=1}^{2}\frac{1}{n_i - 1}\left\{\frac{1}{2}\left(\operatorname{tr}\left[\frac{\mathbf{S}_T^{-1}\mathbf{S}_i}{n_i}\right]\right)^2 + \operatorname{tr}\left[\frac{\mathbf{S}_T^{-1}\mathbf{S}_i}{n_i}\right]^2\right\}\right],$$

$$\mathbf{S}_T = \frac{1}{n_1}\mathbf{S}_1 + \frac{1}{n_2}\mathbf{S}_2, \tag{3.98}$$

and $\chi_d^2(\alpha)$ is the upper α quantile value of χ_d^2, that is,

$$\operatorname{pr}\left[\chi_d^2 \geq \chi_d^2(\alpha)\right] = \alpha.$$

We would expect (3.96) to be robust to nonnormality, as it has the right asymptotic distribution, irrespective of the underlying distributions of the $\mathbf{v}_i$ and $\mathbf{w}_j$.

A different solution was proposed by Yao [1965], who approximated the null distribution of (3.96) by Hotellings $T_{d,f}^2$, where the degrees of freedom f are now estimated from the data. Using a heuristic argument, Yao showed that

$$\frac{1}{f} = \sum_{i=1}^{n}\frac{1}{n_i - 1}\left\{\frac{(\bar{\mathbf{v}} - \bar{\mathbf{w}})'\mathbf{S}_T^{-1}(\mathbf{S}_i/n_i)\mathbf{S}_T^{-1}(\bar{\mathbf{v}} - \bar{\mathbf{w}})}{(\bar{\mathbf{v}} - \bar{\mathbf{w}})'\mathbf{S}_T^{-1}(\bar{\mathbf{v}} - \bar{\mathbf{w}})}\right\}^2, \tag{3.99}$$

where $\min(n_1 - 1, n_2 - 1) \leq f \leq n_1 + n_2 - 2$. This approach is a generalization of Welch's [1947] approximate degrees of freedom solution to the univariate Behrens–Fisher problem. Using simulation, Yao [1965] and Subrahmaniam and Subrahmaniam [1973, 1975] compared James' expansion method with the above approximate degrees of freedom method. As far as significance levels are concerned, the expansion method gives inflated values that get worse as d increases (e.g., 0.08 instead of the nominal 0.05). However, it also has a slightly higher power, though this would partly reflect the inflated significance level. Subrahmaniam and Subrahmaniam [1973, 1975] also included Bennett's test and demonstrated its general inferiority to the other two tests, particularly when $n_1 \neq n_2$. A very general test procedure for handling problems relating to unequal variances has been given by Johansen [1980]. His procedure leads to a critical value that is similar to k_α of (3.97).

EXAMPLE 3.7 From James [1954] we have the following data:

$$\bar{\mathbf{v}} = \begin{pmatrix} 9.82 \\ 15.06 \end{pmatrix}, \quad \bar{\mathbf{w}} = \begin{pmatrix} 13.05 \\ 22.57 \end{pmatrix}, \quad n_1 = 16, n_2 = 11,$$

$$\mathbf{S}_1 = \begin{pmatrix} 120.0 & -16.3 \\ -16.3 & 17.8 \end{pmatrix} \quad \text{and} \quad \mathbf{S}_2 = \begin{pmatrix} 81.8 & 32.1 \\ 32.1 & 53.8 \end{pmatrix}.$$

From (3.98),

$$\mathbf{S}_T = \begin{pmatrix} 14.936 & 1.899 \\ 1.899 & 6.003 \end{pmatrix}$$

and

$$(\bar{\mathbf{v}} - \bar{\mathbf{w}})'\mathbf{S}_T^{-1}(\bar{\mathbf{v}} - \bar{\mathbf{w}}) = (-3.23, -7.51)\begin{pmatrix} 0.06976 & -0.02207 \\ -0.02207 & 0.17357 \end{pmatrix}\begin{pmatrix} -3.23 \\ -7.51 \end{pmatrix}$$

$$= 9.45.$$

Also

$$\mathbf{B}_1 = \frac{1}{n_1}\mathbf{S}_T^{-1}\mathbf{S}_1 = \begin{pmatrix} 0.5457 & -0.0956 \\ -0.3424 & 0.2155 \end{pmatrix},$$

$$\mathbf{B}_2 = \frac{1}{n_2}\mathbf{S}_T^{-1}\mathbf{S}_2 = \begin{pmatrix} 0.4543 & 0.0956 \\ 0.3424 & 0.7845 \end{pmatrix};$$

$$\operatorname{tr}\mathbf{B}_1^2 = 0.4097, \quad (\operatorname{tr}\mathbf{B}_1)^2 = 0.5794, \quad \operatorname{tr}\mathbf{B}_2^2 = 0.8873, \quad (\operatorname{tr}\mathbf{B}_2)^2 = 1.5346;$$

$$\sum_i \frac{1}{n_i - 1}\operatorname{tr}\mathbf{B}_i^2 = 0.11604, \qquad \sum_i \frac{1}{n_i - 1}(\operatorname{tr}\mathbf{B}_i)^2 = 0.19209;$$

$$A = 1 + \tfrac{1}{4}(0.19209) = 1.0480 \text{ and } B = \tfrac{1}{8}\left[\tfrac{1}{2}(0.19209) + 0.11604\right] = 0.02651.$$

Values of k_α are given in Table 3.6.

For the approximate degrees of freedom method we have (Yao [1965])

$$\frac{1}{f} = \tfrac{1}{15}(0.1657)^2 + \tfrac{1}{10}(0.8343)^2$$

TABLE 3.6 Critical Values for the James (k_α) and Yao (T_α^2) Tests

α	$\chi_2^2(\alpha)$	k_α	T_α^2
0.05	5.991	7.23	8.21
0.025	7.378	9.18	10.70
0.01	9.210	11.90	14.43

and $f = 14.00$. From (2.47)

$$T_{d,f,\alpha}^2 = \frac{df}{f-d+1} F_{d,f-d+1}^{\alpha} \qquad \left(= T_\alpha^2, \text{ say}\right)$$

and this is also tabled in Table 3.6. As expected, the critical values of the two tests are very similar and, in general, the degrees of freedom method is preferred, as it is more conservative. For either test, the value of the test statistic, 9.45, is significant at 5%. When f is not an integer, we can use the tables of Mardia and Zemroch [1975b].

3.6.4 *Profile Analysis: Two Populations*

Suppose that a battery of d psychological tests is administered to a sample of n_1 people from a given population, and let $\boldsymbol{\eta} = (\eta_1, \eta_2, \ldots, \eta_d)'$ be the vector of mean scores for the population. The graph obtained by joining the points $(1, \eta_1), (2, \eta_2), \ldots, (d, \eta_d)$ successively is called the profile of the population. In practice, η_j will be estimated by the sample mean for the jth test ($j = 1, 2, \ldots, d$).

If we have a sample of n_2 people from another population with mean vector $\boldsymbol{\nu} = (\nu_1, \nu_2, \ldots, \nu_d)'$, we would wish to compare the two profiles (j, η_j) and (j, ν_j), $j = 1, 2, \ldots, d$, using the sample data. Suppose, then, that the vectors scores for the first sample are $\mathbf{v}_1, \mathbf{v}_2, \ldots, \mathbf{v}_{n_1}$, and those for the second are $\mathbf{w}_1, \mathbf{w}_2, \ldots, \mathbf{w}_{n_2}$. We shall assume that the $\mathbf{v}_i$'s are i.i.d. $N_d(\boldsymbol{\eta}, \boldsymbol{\Sigma})$ and the $\mathbf{w}_j$'s are i.i.d. $N_d(\boldsymbol{\nu}, \boldsymbol{\Sigma})$. Then the first hypothesis of interest is whether the two profiles are similar, that is, the two graphs are parallel as in Fig. 3.2. Expressed mathematically, we wish to test $H_{01}: \eta_k - \eta_{k-1} = \nu_k - \nu_{k-1}$ ($k = 2, 3, \ldots, d$) or $\mathbf{C}_1\boldsymbol{\eta} = \mathbf{C}_1\boldsymbol{\nu}$, where $\mathbf{C}_1$ is the $(d-1) \times d$ contrast matrix in (3.18). In experimental design terminology H_{01} is the hypothesis of no interaction between the groups and the tests. Now the $\mathbf{C}_1\mathbf{v}_i$ are i.i.d. $N_{d-1}(\mathbf{C}_1\boldsymbol{\eta}, \mathbf{C}_1\boldsymbol{\Sigma}\mathbf{C}_1')$, and the $\mathbf{C}_1\mathbf{w}_j$ are i.i.d. $N_{d-1}(\mathbf{C}_1\boldsymbol{\nu}, \mathbf{C}_1\boldsymbol{\Sigma}\mathbf{C}_1')$, so that testing H_{01} is equivalent to testing whether these two normal populations have the same mean. The theory of Section 3.6.2 applies and the required test statistic is [see (3.86)]

$$T_0^2 = \frac{n_1 n_2}{n_1 + n_2}[\mathbf{C}_1(\bar{\mathbf{v}} - \bar{\mathbf{w}})]'\left(\mathbf{C}_1\mathbf{S}_p\mathbf{C}_1'\right)^{-1}[\mathbf{C}_1(\bar{\mathbf{v}} - \bar{\mathbf{w}})], \qquad (3.100)$$

where

$$(n_1 + n_2 - 2)\mathbf{S}_p = \sum_{i=1}^{n_1} (\mathbf{v}_i - \bar{\mathbf{v}})(\mathbf{v}_i - \bar{\mathbf{v}})' + \sum_{j=1}^{n_2} (\mathbf{w}_j - \bar{\mathbf{w}})(\mathbf{w}_j - \bar{\mathbf{w}})'$$

$$= \mathbf{Q}_1 + \mathbf{Q}_2 = \mathbf{Q}.$$

When H_{01} is true, $T_0^2 \sim T^2_{d-1, n_1+n_2-2}$, that is,

$$F_0 = \frac{T_0^2}{n_1 + n_2 - 2} \frac{n_1 + n_2 - d}{d - 1} \sim F_{d-1, n_1+n_2-d}. \qquad (3.101)$$

In practice, we would plot the sample version of Fig. 3.2 using the sample means instead of the population means.

If we do not reject the hypothesis H_{01} of parallelism, we may now wish to ask the question, Are the population profiles at the same level? Expressed mathematically we wish to test

$$H_{02}: \frac{1}{d}(\eta_1 + \eta_2 + \cdots + \eta_d) = \frac{1}{d}(\nu_1 + \nu_2 + \cdots + \nu_d) \quad \text{or} \quad \mathbf{1}'\boldsymbol{\eta} = \mathbf{1}'\boldsymbol{\nu}.$$

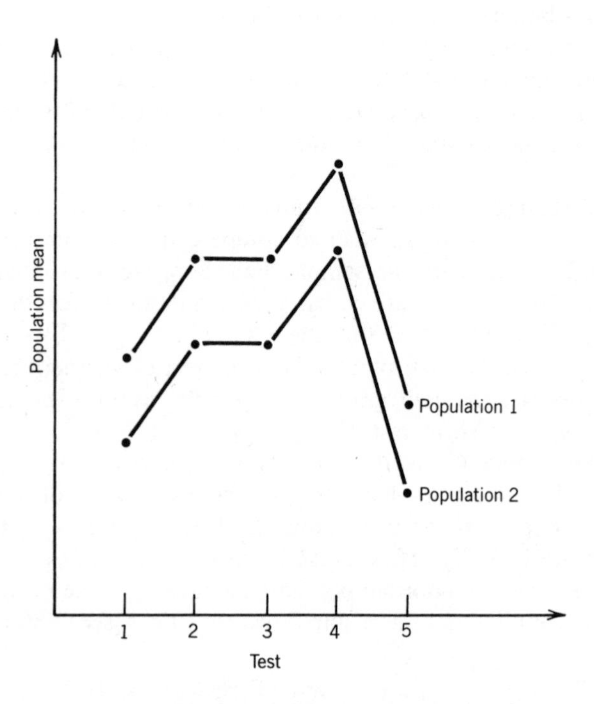

Fig. 3.2 Two parallel profiles

This hypothesis takes the same form as H_{01} so that the test statistic for H_{02} is

$$T_0^2 = \frac{n_1 n_2}{n_1 + n_2} \frac{\left[\mathbf{1}'(\bar{\mathbf{v}} - \bar{\mathbf{w}})\right]^2}{\mathbf{1}'\mathbf{S}_p\mathbf{1}}$$

$$= \left\{ \frac{\mathbf{1}'(\bar{\mathbf{v}} - \bar{\mathbf{w}})}{\left[\mathbf{1}'\mathbf{S}_p\mathbf{1}(1/n_1 + 1/n_2)\right]^{1/2}} \right\}^2 \qquad \left(= t_0^2, \text{ say}\right), \qquad (3.102)$$

which is simply the square of the usual t-statistic t_0 for testing the difference of two means for the data consisting of the column sums $\mathbf{1}'\mathbf{v}_i$ and $\mathbf{1}'\mathbf{w}_j$. When H_{02} is true, $t_0 \sim t_{n_1+n_2-2}$.

The hypothesis H_{02} can be regarded as the hypothesis of no population "main effects" and it can still be true when H_{01} is false, as, for example, in Fig. 3.3. However, H_{02} becomes difficult to interpret when H_{01} is false and, in this case, there is perhaps little point in testing H_{02}. If H_{01} and H_{02} are both true, then the two population profiles coincide.

In the same way we can investigate the hypothesis that there are no test main effects (i.e., effects due to the d variables), that is,

$$H_{03}: \tfrac{1}{2}(\eta_1 + \nu_1) = \tfrac{1}{2}(\eta_2 + \nu_2) = \cdots = \tfrac{1}{2}(\eta_d + \nu_d) \quad \text{or} \quad \mathbf{C}_1(\boldsymbol{\eta} + \boldsymbol{\nu}) = \mathbf{0}.$$

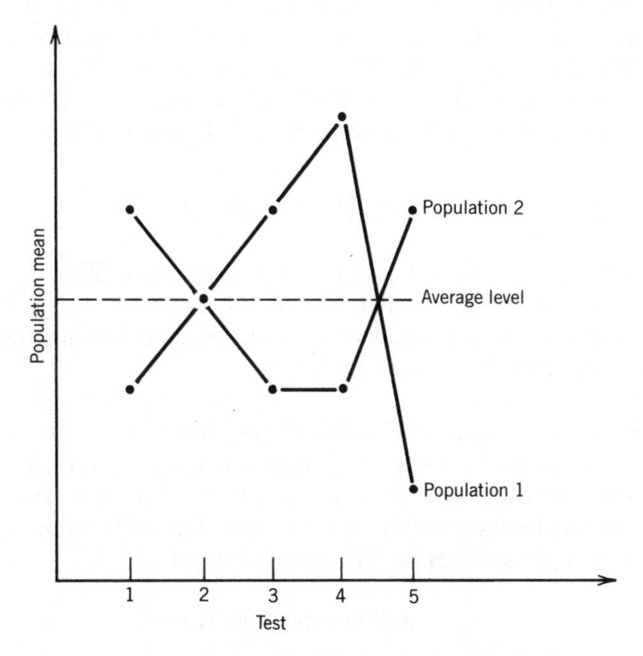

Fig. 3.3 Two profiles with the same average level.

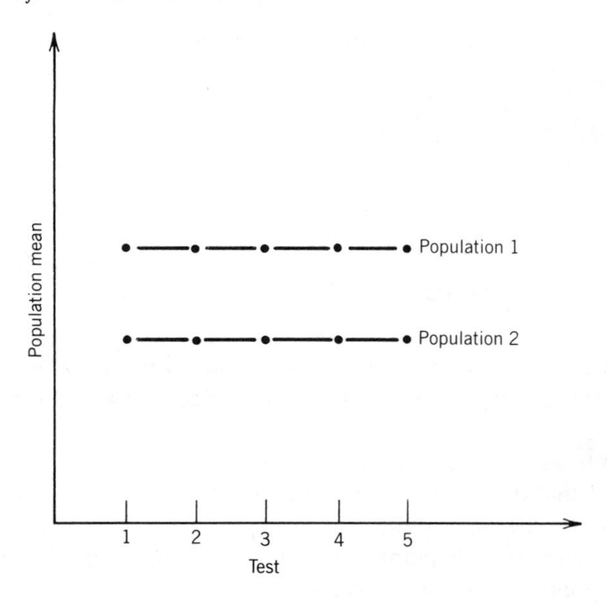

Fig. 3.4 Two parallel profiles with no effects due to the variables.

When H_{01} is true, H_{03} implies that $\eta_1 = \eta_2 = \cdots = \eta_d$ and $\nu_1 = \nu_2 = \cdots = \nu_d$; that is, the two profiles are parallel horizontal lines, as in Fig. 3.4. If H_{02} is also true, the horizontal lines coincide. Now let $\bar{\mathbf{x}} = (n_1\bar{\mathbf{v}} + n_2\bar{\mathbf{w}})/(n_1 + n_2)$ and $\mathbf{Q} = (n_1 + n_2 - 2)\mathbf{S}_p$. Then $\mathbf{C}_1\bar{\mathbf{x}} \sim N_{d-1}(\mathbf{C}_1\mathscr{E}[\bar{\mathbf{x}}], (n_1 + n_2)^{-1}\mathbf{C}_1\mathbf{\Sigma}\mathbf{C}_1')$, $\mathbf{C}_1\mathbf{Q}\mathbf{C}_1' \sim W_{d-1}(n_1 + n_2 - 2, \mathbf{C}_1\mathbf{\Sigma}\mathbf{C}_1')$, and $\mathbf{C}_1\bar{\mathbf{x}}$ and $\mathbf{C}_1\mathbf{Q}\mathbf{C}_1'$ are statistically independent as $\bar{\mathbf{v}}$, $\bar{\mathbf{w}}$, $\mathbf{Q}_1$, and $\mathbf{Q}_2$ are mutually independent. Hence, by (2.19) with $\lambda = n_1 + n_2$, we can test H_{03} given H_{01} (i.e. $\mathbf{C}_1\boldsymbol{\eta} = \mathbf{C}_1\boldsymbol{\nu}$) using

$$T_0^2 = (n_1 + n_2)(\mathbf{C}_1\bar{\mathbf{x}})'\left[\mathbf{C}_1\mathbf{S}_p\mathbf{C}_1'\right]^{-1}\mathbf{C}_1\bar{\mathbf{x}}, \tag{3.103}$$

and $T_0^2 \sim T^2_{d-1,\, n_1+n_2-2}$ when $H_{03} \cap H_{01}$ is true [see (3.101)].

Later, in Section 8.7.2, we shall extend the above theory to the case of profiles for more than two populations. These models are also described as repeated-measurement designs (Timm [1980]).

EXAMPLE 3.8 A sample of 27 children aged about 8–9 years who had an inborn error of metabolism known as transient neonatal tyrosinemia (TNT) were compared with a closely matched sample of 27 normal children (called the control group) by their scores on the Illinois Test of Psycholingual Ability (ITPA). This test gives scores on 10 variables, namely;

$$x_1 = \text{auditory reception score},$$

$$x_2 = \text{visual reception score},$$

x_3 = visual memory,

x_4 = auditory association,

x_5 = auditory memory,

x_6 = visual association,

x_7 = visual closure,

x_8 = verbal expression,

x_9 = grammatic closure,

x_{10} = manual expression.

The data are listed in Table 3.7 and the profiles for the sample means are given in Fig. 3.5. The standard deviation of each mean is $SD/\sqrt{27}$, or roughly 1.35 $(= 7/\sqrt{27})$, so that an inspection of Fig. 3.5 would suggest that the hypothesis of parallelism would not be rejected. The variable x_{10} is omitted from the following analysis.

The sample dispersion matrices $(\mathbf{S}_i = \mathbf{Q}_i/26)$ for the control and TNT groups, respectively, are, to one decimal place,

$$\mathbf{S}_1 = \begin{pmatrix} 49.1 & 16.8 & 2.7 & 27.9 & 10.9 & 30.1 & 6.6 & 64.4 & 23.4 \\ - & 43.4 & 18.2 & 25.8 & 10.7 & 18.4 & 7.0 & 36.9 & 19.9 \\ - & - & 69.4 & 22.5 & 13.5 & 18.8 & 14.4 & 11.2 & 20.6 \\ - & - & - & 77.7 & 32.6 & 39.9 & 17.0 & 56.0 & 32.6 \\ - & - & - & - & 65.4 & 24.9 & 9.7 & 23.6 & 17.5 \\ - & - & - & - & - & 49.7 & 12.8 & 48.1 & 22.0 \\ - & - & - & - & - & - & 18.3 & 14.2 & 9.3 \\ - & - & - & - & - & - & - & 133.9 & 29.7 \\ - & - & - & - & - & - & - & - & 36.3 \end{pmatrix}$$

and

$$\mathbf{S}_2 = \begin{pmatrix} 56.6 & 21.4 & 8.6 & 32.8 & 9.6 & 15.3 & 18.9 & 26.1 & 10.3 \\ - & 65.4 & 8.1 & -1.7 & 17.1 & 16.1 & 14.9 & 11.3 & 16.4 \\ - & - & 30.3 & -0.2 & -0.5 & 3.0 & 5.6 & 10.2 & 10.8 \\ - & - & - & 75.5 & 14.1 & 7.0 & 18.7 & 35.7 & -2.7 \\ - & - & - & - & 55.1 & 7.3 & 14.4 & 27.4 & 6.2 \\ - & - & - & - & - & 42.3 & 5.5 & 4.1 & 10.4 \\ - & - & - & - & - & - & 22.7 & 24.3 & 13.3 \\ - & - & - & - & - & - & - & 82.2 & 19.9 \\ - & - & - & - & - & - & - & - & 44.1 \end{pmatrix}.$$

As these two matrices are not too different and $n_1 = n_2 = 27$, we can assume that our test procedures are reasonably robust with regard to both nonnormality and moderate departures from $\mathbf{\Sigma}_1 = \mathbf{\Sigma}_2$ (see Section 3.6.2c). The pooled dispersion matrix $\mathbf{S}_p$ is the average of $\mathbf{S}_1$ and $\mathbf{S}_2$, and the T_0^2 statistic (3.100) for parallelism takes the value 2.2697. Hence, from (3.101),

$$F_0 = \frac{2.2697}{52}\frac{45}{8} = 0.2455 \sim F_{8,45},$$

and the test statistic is not significant, as $F_{8,45}^{0.05} = 2.2$. Forming the row sums of Table 3.7, we find from (3.102) that $t_0 = 1.88$. Since $t_{52}^{0.05} = 1.68$ and $t_{52}^{0.025} =$

TABLE 3.7 Test Scores in 10 Categories for 54 Children on the Illinois Test of Psycholinguistic Abilities[a]

Control Group	Variables									
	x_1	x_2	x_3	x_4	x_5	x_6	x_7	x_8	x_9	x_{10}
1	40	32	16	20	38	37	28	43	32	30
2	35	30	41	44	39	38	32	36	41	27
3	30	42	48	26	42	34	30	36	43	36
4	22	27	34	19	40	24	29	14	30	31
5	21	38	46	28	48	33	28	22	34	26
6	39	40	47	37	43	40	34	31	49	42
7	39	39	27	42	36	36	34	47	43	38
8	22	23	34	16	33	30	33	20	21	30
9	44	33	43	31	40	40	32	41	40	27
10	34	34	41	41	51	32	41	46	38	38
11	30	43	34	46	50	34	32	45	38	33
12	26	34	32	20	38	20	26	28	28	33
13	44	42	54	48	54	44	34	52	43	44
14	36	39	49	24	49	36	35	50	36	36
15	30	35	32	28	43	32	34	37	39	24
16	18	25	38	24	32	22	30	8	36	23
17	27	28	35	25	42	25	36	16	30	24
18	30	36	42	28	15	26	32	39	28	23
19	33	16	38	35	51	40	33	40	35	40
20	26	37	54	32	36	41	38	27	37	32
21	31	33	33	32	47	37	32	22	36	28
22	29	31	29	26	38	28	21	27	27	22
23	34	29	40	26	33	31	30	39	34	26
24	36	27	34	21	31	27	23	35	36	37
25	42	40	36	31	41	38	35	41	36	31
26	32	38	37	42	44	49	32	43	32	40
27	38	40	40	32	36	41	36	43	41	28
Mean	32.15	33.74	38.30	30.52	40.37	33.89	31.85	34.37	35.67	31.44
SD	7.00	6.58	8.33	8.82	8.09	7.05	4.28	11.57	6.03	6.35

TABLE 3.7 (*continued*)

TNT Group	x_1	x_2	x_3	x_4	x_5	x_6	x_7	x_8	x_9	x_{10}
1	26	38	44	26	37	38	22	28	30	33
2	31	30	26	23	43	35	28	24	27	33
3	28	36	36	16	46	28	33	28	28	39
4	19	43	39	16	46	43	33	28	40	28
5	31	33	26	29	41	33	23	30	23	24
6	35	36	43	32	46	22	37	48	34	43
7	37	35	38	31	44	29	37	41	46	31
8	41	37	36	29	32	30	32	39	48	27
9	35	40	34	49	37	34	34	36	34	40
10	29	19	34	42	32	32	30	40	22	37
11	18	26	27	26	21	22	26	12	27	26
12	38	40	27	30	44	40	29	36	42	38
13	31	23	40	28	37	37	30	45	39	24
14	26	21	31	29	41	23	24	22	28	26
15	33	20	44	34	23	37	28	17	39	35
16	27	36	39	19	28	28	23	38	29	36
17	22	32	35	26	41	38	25	24	38	26
18	37	38	36	47	46	38	36	34	32	40
19	24	40	28	16	36	34	27	24	34	38
20	11	27	36	19	38	22	23	31	41	35
21	24	30	30	26	30	36	30	38	35	32
22	28	14	31	36	48	36	27	38	31	41
23	17	25	31	35	46	25	30	42	36	27
24	31	29	34	28	32	31	24	26	29	31
25	34	38	42	31	43	26	27	27	30	30
26	41	46	40	39	44	46	36	42	41	40
27	23	26	37	21	32	34	22	20	31	18
Mean	28.78	31.78	34.96	28.63	38.30	32.48	28.74	31.78	33.85	32.52
SD	7.53	8.09	5.51	8.69	7.42	6.50	4.76	9.07	6.64	6.42

[a] Data courtesy of Peter Mullins.

2.01, we reject the hypothesis that the profiles are at the same level at the 10% level of significance, but not at the 5% level. Finally, the test statistic (3.103) takes the value $T_0^2 = 129.894$, so that

$$F_0 = \frac{129.894}{52}\frac{45}{8} = 14.05 \sim F_{8,45},$$

which is significant at the 1% level. We conclude that the scores on the d variables are different, which is to be expected from Fig. 3.5.

Summing up, we conclude that the two profiles are very similar. Although there is a positive difference for all the variables (Fig. 3.5), this difference is not

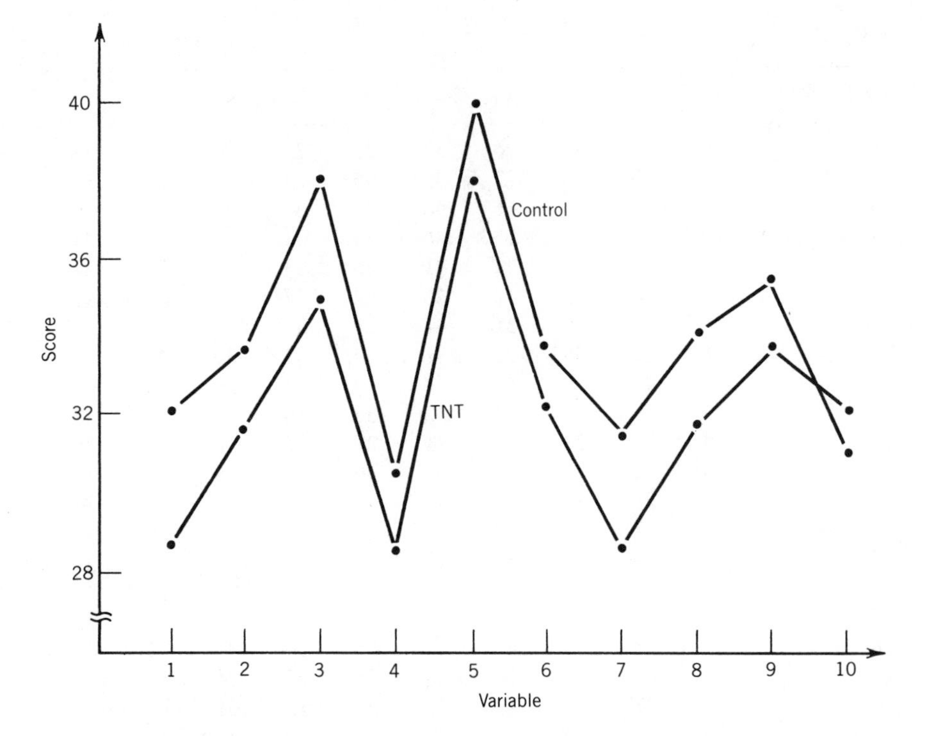

Fig. 3.5 Profiles of average scores in 10 categories for normal children (control) and children with transient neonatal tyrosinemia (TNT).

great when compared to a standard deviation of about 1.4 for each mean. It should be noted that these differences have high positive correlations, which could be a contributing factor to the systematic difference.

EXERCISES 3

3.1 Suppose that $\mathbf{x}_1, \mathbf{x}_2, \dots, \mathbf{x}_n$ are i.i.d. $N_2(\boldsymbol{\mu}, \boldsymbol{\Sigma})$, where $\boldsymbol{\mu}' = (\mu_1, \mu_2)$, and let $y_i = x_{i1} - x_{i2}$ $(i = 1, 2, \dots, n)$. Show that Hotelling's T_0^2 statistic for testing $H_0: \mu_1 = \mu_2$ versus H_1: $\mu_1 \neq \mu_2$ is equivalent to the usual paired comparison t-statistic $\sqrt{n}\,\bar{y}/s_y$.

3.2 Let $\mathbf{x}_1, \mathbf{x}_2, \dots, \mathbf{x}_n$ be a random sample from $N_d(\boldsymbol{\mu}, \boldsymbol{\Sigma})$ and let $\overline{\mathbf{x}}$ and $\mathbf{S}$ be the sample mean and the usual unbiased estimate of $\boldsymbol{\Sigma}$, respectively. Let $\mathbf{A}$ be a $(d-1) \times d$ matrix of rank $d - 1$ such that $\mathbf{A}\mathbf{1}_d = \mathbf{0}$. Using B3.5, verify that

$$n\overline{\mathbf{x}}'\mathbf{S}^{-1}\overline{\mathbf{x}} - \frac{n\left(\overline{\mathbf{x}}'\mathbf{S}^{-1}\mathbf{1}_d\right)^2}{\mathbf{1}_d'\mathbf{S}^{-1}\mathbf{1}_d} = n\overline{\mathbf{x}}'\mathbf{A}'(\mathbf{A}\mathbf{S}\mathbf{A}')^{-1}\mathbf{A}\overline{\mathbf{x}}.$$

Williams [1970]

3.3 Let $\mathbf{x}_1, \mathbf{x}_2, \ldots, \mathbf{x}_n$ be i.i.d. $N_d(\boldsymbol{\mu}, \boldsymbol{\Sigma})$, where $\boldsymbol{\Sigma}$ satisfies (3.21). Let $Q_H = \sum_{i=1}^n \sum_{j=1}^d (\bar{x}_{\cdot j} - \bar{x}_{\cdot\cdot})^2$ and $Q_E = \sum_{i=1}^n \sum_{j=1}^d (x_{ij} - \bar{x}_{i\cdot} - \bar{x}_{\cdot j} + \bar{x}_{\cdot\cdot})^2$, where x_{ij} is the jth element of $\mathbf{x}_i$. Show that $H_0: \mu_1 = \mu_2 = \cdots = \mu_d$ can be tested using $F = (n-1)Q_H/Q_E$, where $F \sim F_{d-1,(n-1)(d-1)}$ when H_0 is true. [Hint: Use Theorem 2.1(iv) and Exercise 1.9.]

Wilks [1946]

3.4 Let $\mathbf{x}_1, \mathbf{x}_2, \ldots, \mathbf{x}_n$ be a random sample from $N_d(\boldsymbol{\mu}, \boldsymbol{\Sigma})$, and let $\bar{\mathbf{x}} = (\bar{x}_1, \bar{x}_2, \ldots, \bar{x}_d)'$ and $\mathbf{S}^{-1} = [(s^{jk})]$. Using Theorem 3.3 in Section 3.4.3, show that Hotelling's T_0^2 statistic for testing $H_0: \mu_1 = \mu_2 = \cdots = \mu_d$ is given by

$$T_0^2 = n \sum_{j=1}^d \sum_{k=1}^d s^{jk} \bar{x}_j \bar{x}_k - n \left[\sum_{j=1}^d \sum_{k=1}^d s^{jk} (\bar{x}_j + \bar{x}_k) \right]^2 \Big/ \left(4 \sum_j \sum_k s^{jk} \right).$$

Verify that this expression is the same as that given by Exercise 3.2.

3.5 Let $\phi_i = \mu_i - \mu_d$ $(i = 1, 2, \ldots, d-1)$. Show that the set of all linear combinations $\sum_{i=1}^{d-1} h_i \phi_i$ is equivalent to the set of all contrasts $\sum_{i=1}^d c_i \mu_i$ $(\sum_{i=1}^d c_i = 0)$.

3.6 Verify that L_2 of (3.50) reduces to $(n-1)\mathrm{tr}[\mathbf{Q}_{12}\mathbf{Q}_{22}^{-1}\mathbf{Q}_{21}\mathbf{Q}_{11}^{-1}]$.

3.7 Suppose $\mathbf{x} \sim N_2(\boldsymbol{\mu}, \boldsymbol{\Sigma})$ where $\sigma_{jj} = \sigma_j^2$ and $\sigma_{12} = \rho\sigma_1\sigma_2$. Verify, from Theorem 2.1(viii), that conditional on x_1, $x_2 = \mu_2 + \rho(\sigma_2/\sigma_1)(x_1 - \mu_1) + \varepsilon$, where $\varepsilon \sim N_1(0, \sigma_2^2[1 - \rho^2])$. If $\mathbf{x}_1, \mathbf{x}_2, \ldots, \mathbf{x}_n$ is a random sample from this model, show that the usual t-statistic for testing the hypothesis H_0 of zero slope $(\rho = 0)$ is a function of r, the sample correlation coefficient of x_1 and x_2. Verify that, given H_0 true, r has density function

$$f(r) = \frac{\Gamma\left(\frac{n-1}{2}\right)}{\Gamma\left(\frac{n}{2} - 1\right)\sqrt{\pi}} (1 - r^2)^{(n-4)/2}, \quad -1 \le r \le 1.$$

3.8 Verify that the likelihood ratio statistic for testing $\boldsymbol{\Sigma} = \sigma^2 \mathbf{I}_d$ based on a random sample from $N_d(\boldsymbol{\mu}, \boldsymbol{\Sigma})$ is given by (3.53).

3.9 Verify (3.54).

3.10 Let $\mathbf{x}_1, \mathbf{x}_2, \ldots, \mathbf{x}_n$ be a random sample from $N_d(\boldsymbol{\mu}, \boldsymbol{\Sigma})$. Show that the likelihood ratio test statistic for $H_{04}: \boldsymbol{\mu} = \boldsymbol{\mu}_0$ and $\boldsymbol{\Sigma} = \boldsymbol{\Sigma}_0$ is given by

$$\ell_4 = (e/n)^{dn/2} |\mathbf{Q}\boldsymbol{\Sigma}_0^{-1}|^{n/2}$$

$$\times \exp\left\{ -\tfrac{1}{2}\left[\mathrm{tr}(\mathbf{Q}\boldsymbol{\Sigma}_0^{-1}) + n(\bar{\mathbf{x}} - \boldsymbol{\mu}_0)'\boldsymbol{\Sigma}_0^{-1}(\bar{\mathbf{x}} - \boldsymbol{\mu}_0) \right] \right\}.$$

3.11 Find the matrix $\mathbf{C}$ of (3.55) and verify that it is nonsingular.

3.12 Let $\mathbf{x}_1, \mathbf{x}_2, \ldots, \mathbf{x}_n$ be i.i.d. $N_d(\boldsymbol{\mu}, \boldsymbol{\Sigma})$, where $\boldsymbol{\Sigma}$ satisfies (3.56). Show that the maximum likelihood estimates of σ^2 and ρ are given by (3.57) and (3.58). [Hint: $\boldsymbol{\Sigma} = \sigma^2(1-\rho)\mathbf{I}_d + \sigma^2\rho\mathbf{1}_d\mathbf{1}_d'$, $\boldsymbol{\Sigma}^{-1}$ takes the same form as $\boldsymbol{\Sigma}$ and $|\boldsymbol{\Sigma}|$ follows from A3.5].

3.13 Verify (3.59).

3.14 Using (3.70), show that when $d = 2$;

$$\boldsymbol{\Gamma} = \begin{pmatrix} \mu_{40} - \mu_{20}^2 & \mu_{22} - \mu_{20}\mu_{02} & \mu_{31} - \mu_{20}\mu_{11} \\ \mu_{22} - \mu_{20}\mu_{02} & \mu_{04} - \mu_{02}^2 & \mu_{13} - \mu_{02}\mu_{11} \\ \mu_{31} - \mu_{20}\mu_{11} & \mu_{13} - \mu_{02}\mu_{11} & \mu_{22} - \mu_{11}^2 \end{pmatrix}.$$

Layard [1974]

3.15 Show that the likelihood ratio statistic for testing the equality of two multivariate normal means, given equal dispersion matrices, is a monotonic function of (3.86).

3.16 Let $\mathbf{v}_1, \mathbf{v}_2, \ldots, \mathbf{v}_{n_1}$ be i.i.d. $N_d(\boldsymbol{\mu}_1, \boldsymbol{\Sigma}_1)$, and let $\mathbf{w}_1, \mathbf{w}_2, \ldots, \mathbf{w}_{n_2}$ be independently i.i.d. $N_d(\boldsymbol{\mu}_2, \boldsymbol{\Sigma}_2)$. Suppose $n_1 < n_2$. Define

$$\mathbf{y}_i = \mathbf{v}_i - \left(\frac{n_1}{n_2}\right)^{1/2}\mathbf{w}_i - (n_1 n_2)^{-1/2}\sum_{\alpha=1}^{n_1}\mathbf{w}_\alpha - n_2^{-1}\sum_{\alpha=1}^{n_2}\mathbf{w}_\alpha, \qquad i = 1, 2, \ldots, n_1.$$

Prove that

$$\mathscr{E}[\mathbf{y}_i] = \boldsymbol{\mu}_1 - \boldsymbol{\mu}_2$$

and

$$\mathscr{C}[\mathbf{y}_h, \mathbf{y}_i] = \delta_{hi}\left(\boldsymbol{\Sigma}_1 + \frac{n_1}{n_2}\boldsymbol{\Sigma}_2\right).$$

Hence show that one possible statistic for testing $\boldsymbol{\mu}_1 = \boldsymbol{\mu}_2$ is $T_0^2 = n_1\bar{\mathbf{y}}'\mathbf{S}_y^{-1}\bar{\mathbf{y}}$. What is the distribution of T_0^2 when $\boldsymbol{\mu}_1 = \boldsymbol{\mu}_2$? Prove that $\bar{\mathbf{y}} = \bar{\mathbf{v}} - \bar{\mathbf{w}}$ and $(n_1 - 1)\mathbf{S}_y = \Sigma_{i=1}^{n_1}(\mathbf{z}_i - \bar{\mathbf{z}})(\mathbf{z}_i - \bar{\mathbf{z}})'$, where $\mathbf{z}_i = \mathbf{v}_i - (n_1/n_2)^{1/2}\mathbf{w}_i$. (Note: This test is not recommended.)

Bennett [1951]

3.17 Given samples from two normal populations as in Exercise 16, suppose that $\boldsymbol{\Sigma}_1 = k\boldsymbol{\Sigma}_2$, where k is known. Derive a Hotelling's T^2 test for testing $\boldsymbol{\mu}_1 = \boldsymbol{\mu}_2$.

3.18 Given a random sample from $N_d(\boldsymbol{\mu}, \boldsymbol{\Sigma})$, obtain the union–intersection test for testing $\boldsymbol{\Sigma} = \boldsymbol{\Sigma}_0$.

CHAPTER 4

Graphical and Data-Oriented Techniques

4.1 MULTIVARIATE GRAPHICAL DISPLAYS

Graphical methods for displaying a set of d-dimensional observations $\mathbf{x}_1, \mathbf{x}_2, \ldots, \mathbf{x}_n$ are still somewhat in their infancy, though there have been some interesting suggestions. Fienberg [1979] gives a readable survey and mentions the following: glyphs and metroglyphs (Anderson [1957]), triangles (Pickett and White [1966]), k-sided polygons (Siegel et al. [1971]) and a variant called STARS (Welsch [1976]), Fourier plots (Andrews [1972], Gnanadesikan [1977: 207–225]), faces (Chernoff [1973]; see also Everitt [1978: Chapter 4]), weather vanes (Bruntz et al. [1974]), and constellations (Wakimoto and Taguri [1978]).

A glyph is a circle of fixed radius with d rays of various lengths representing the values of the d coordinates or variables: These variables may be quantitative or qualitative (e.g., low, medium, high). In the latter case the lowest "level" can be represented by a ray of zero length, as in Fig. 4.1a, where we have $n = 3$ observations and $d = 5$ variables. Reading from left to right, observation 1 has variables 1, 2, and 4 at the medium level, and 3 and 5 at the high level, while observation 4 has variables 1, 3, and 5 at the low level, and 2 and 4 at the medium level. If the glyphs are plotted as points in a two-dimensional scatter plot for two of the variables, then only $d - 2$ rays are required. For example, a four-dimensional plot could be represented by Fig. 4.2, where the lengths of the north and south arrows represent positive and negative values, respectively, of variable 3 (Gower [1967b]). When using an automatic plotting device such as a microfilm plotter, Everitt [1978] suggests the possibility of representing three dimensions by a two-dimensional plot using the intensity of the plotted characteristic to represent the magnitude of the third variable: The size of the plotted points could also be used in this way.

STARS are similar to glyphs, except that the rays are now equally spaced around the circle and the ends of the rays are joined up to form polygons. Fig. 4.1b is a STAR representation of Fig. 4.1a. The method of faces consists of

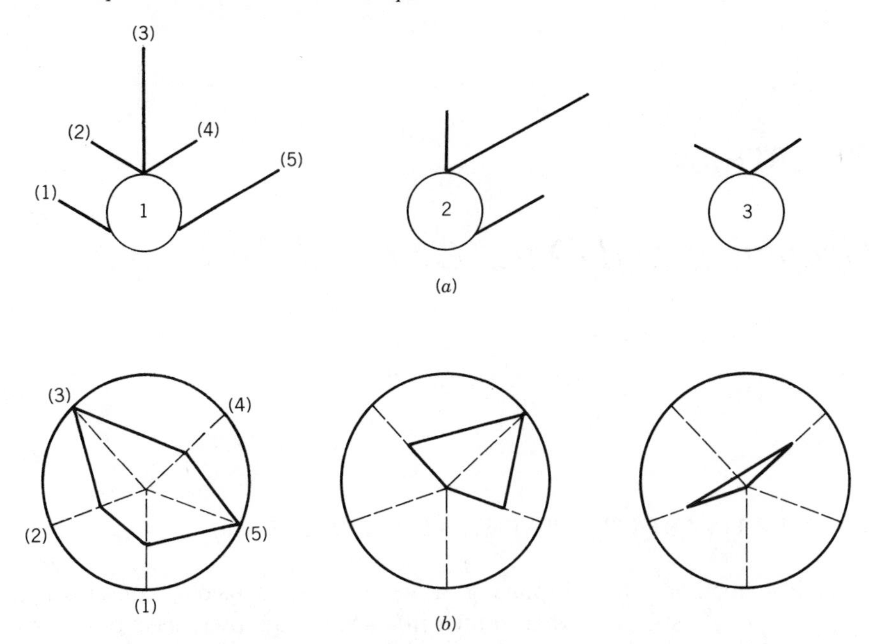

Fig. 4.1 Graphical representation of three four-dimensional points. (*a*) Glyphs. (*b*) STARS.

representing each variable by a different characteristic such as length of nose, shape of face, and size of eyes, as in Fig. 4.3. Unfortunately there is a major problem with the technique. The faces change when the variables are interchanged, so that one may need to try a variety of displays before finding the "best" one (see Chernoff and Rizvi [1975] and Fienberg [1979] for related experiments). Everitt [1978: pp. 87–94] gives a helpful discussion of the method with examples. Further applications are given by Jacob et al. [1976] and by several authors in Wang [1978]. The features in Chernoff's faces, as indicated by Fig. 4.3, are rather crude; but with more sophisticated software, more "attractive" faces can be plotted (Wainer [1981]). The use of asymmetric

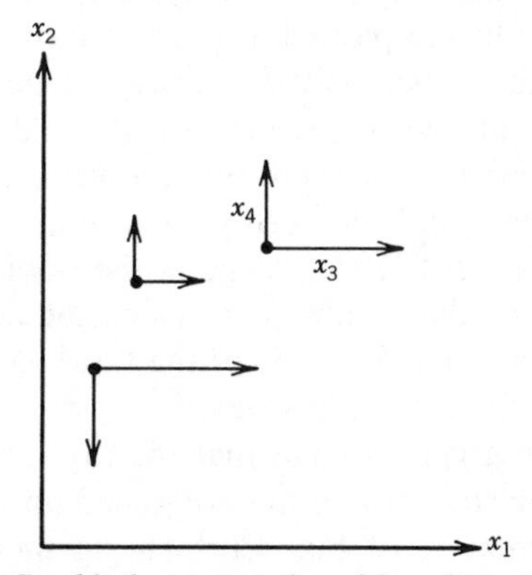

Fig. 4.2 Graphical representation of four-dimensional data.

Fig. 4.3 Chernoff's faces for measurements on permanent first lower premolar of various groups of humans and apes. From Fienberg [1979].

faces can effectively double the number of variables representable (Flury and Riedwyl [1981]).

A helpful two-dimensional representation is the so-called "profile" (Bertin [1967]), already introduced in Section 3.6.4. Each observation is represented by d vertical lines or bars for d-dimensional data, with each line having height proportional to the value of the corresponding variable. The profile refers to the tops of the lines or bars and is usually drawn as a polygonal line, as in Fig. 4.4. There we have profiles for two four-dimensional observations: The profiles are similar, except in the fourth variable. Further details and algorithms are given by Hartigan [1975]. The method can be useful for detecting clusters of similar profiles.

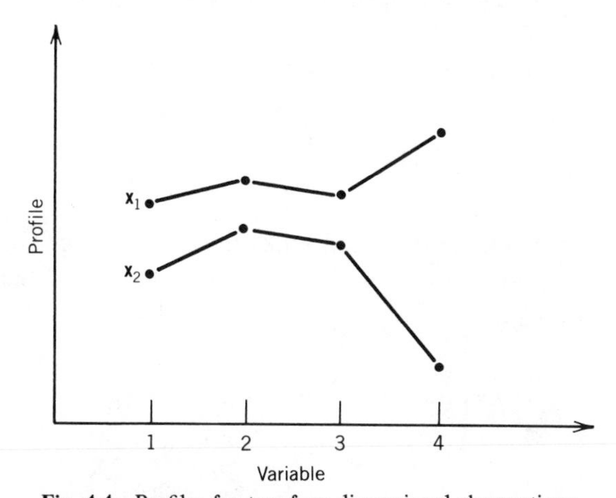

Fig. 4.4 Profiles for two four-dimensional observations.

TABLE 4.1 Percentage of Republican Votes in United States Presidential Elections in 6 Southern States, 1932–1940, 1960–1968[a]

State	1932	1936	1940	1960	1964	1968
Missouri	35	38	48	50	36	45
Maryland	36	37	41	46	35	42
Kentucky	40	40	42	54	36	44
Louisiana	7	11	14	29	57	23
Mississippi	4	3	4	25	87	14
South Carolina	2	1	4	49	59	39

[a] From Kleiner and Hartigan [1981: Table 1].

EXAMPLE 4.1 In Table 4.1 we have data from Kleiner and Hartigan [1981] giving the percentage of Republican votes in the presidential elections in six southern states for six elections. Various graphical displays of this data are given in Fig. 4.5. The inexperienced viewer would probably find the bar profiles the most informative and the faces the least informative. However, it is clear that some practice is needed in interpreting such displays, particularly in the case of faces. □

The Fourier plot method consists of representing a d-dimensional vector $\mathbf{x}' = (x_1, x_2, \ldots, x_d)$ by the finite Fourier series

$$f_{\mathbf{x}}(t) = x_1/\sqrt{2} + x_2 \sin t + x_3 \cos t + x_4 \sin 2t + x_5 \cos 2t + \cdots$$

and plotting $f_{\mathbf{x}}(t)$ for a grid of t-values in the range $-\pi < t < \pi$ (or else replacing t by $2\pi t$ and using the range $0 < t < 1$). An example showing the corresponding curves for five seven-dimensional observations is given in Fig. 4.6. Andrews [1972] gives the following properties for such plots.

1. The function preserves means. We have

$$f_{\bar{\mathbf{x}}}(t) = \frac{1}{n} \sum_{i=1}^{n} f_{\mathbf{x}_i}(t)$$

so that the curve representing the mean looks like an "average" curve.

2. The function preserves distance. A natural measure of distance between two functions is the L_2 norm. From Exercise 4.1 we have

$$\begin{aligned} \int_{-\pi}^{\pi} \left[f_{\mathbf{x}}(t) - f_{\mathbf{y}}(t) \right]^2 dt &= \pi \sum_{i=1}^{d} (x_i - y_i)^2 \\ &= \pi \|\mathbf{x} - \mathbf{y}\|^2, \end{aligned} \tag{4.1}$$

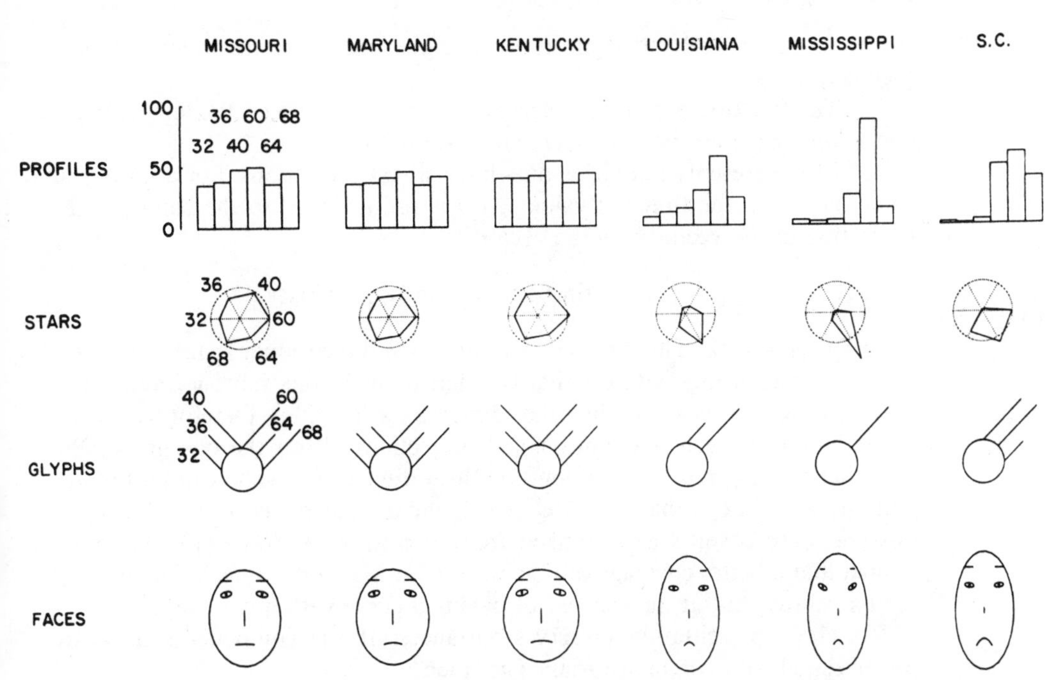

Fig. 4.5 Profiles, STARS, glyphs, and faces for the data in Table 4.1. The circles in the STARS are drawn at 50%. The assignment of the variables to facial features is: 1932—shape of face; 1936—length of nose; 1940—curvature of mouth; 1960—width of mouth; 1964—slant of eyes; 1968—length of eyebrows. From Kleiner and Hartigan [1981].

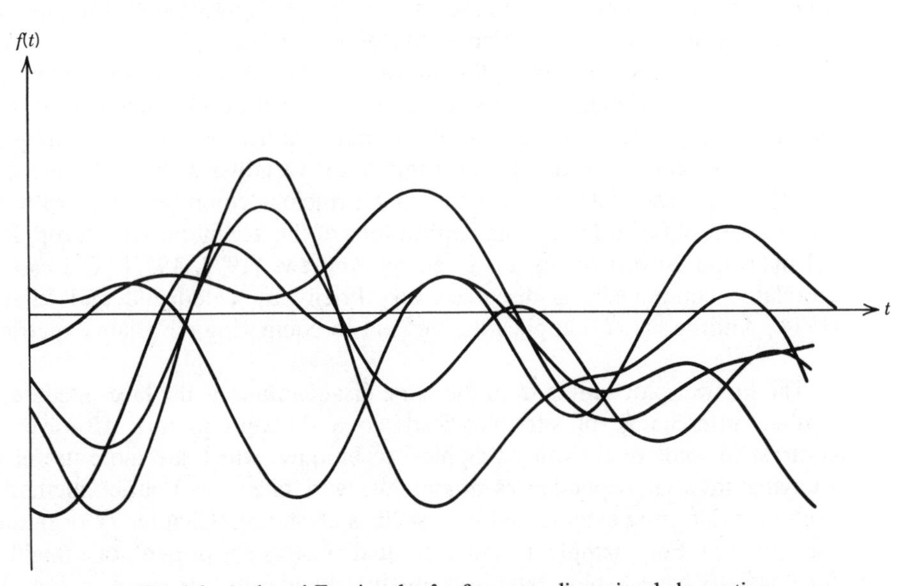

Fig. 4.6 Andrews' Fourier plot for five seven-dimensional observations.

so that points $\mathbf{x}$ and $\mathbf{y}$ that are close together lead to curves which are close together.

3. The function preserves linear relationships. If $\mathbf{y}$ lies on the line joining $\mathbf{x}$ and $\mathbf{z}$, then $f_{\mathbf{y}}(t)$ lies between $f_{\mathbf{x}}(t)$ and $f_{\mathbf{z}}(t)$ for all t.

4. The representation yields one-dimensional projections. For a particular value of $t = t_0$, the function value $f_{\mathbf{x}}(t_0)$ is proportional to the length of the projection of the vector $\mathbf{x}$ on the vector

$$\mathbf{a}_0 = \left(1/\sqrt{2}, \sin t_0, \cos t_0, \sin 2t_0, \cos 2t_0, \ldots\right),$$

since $f_{\mathbf{x}}(t_0) = \mathbf{x}'\mathbf{a}_0$. This projection onto a one-dimensional space may show up clusterings or any data peculiarities that occur in this subspace and which may be otherwise obscured by other dimensions. The plot, therefore, provides a continuum of such one-dimensional projections all on the one graph. We note that $\mathbf{a}_0/\|\mathbf{a}_0\|$ represents a point on the d-dimensional sphere of unit radius and one would hope that, in the course of the plot, as many of these points as possible were covered as t_0 ranged from $-\pi$ to π. Andrews [1972] demonstrated that a better coverage can be achieved using more complex functions of t_0 in $\mathbf{a}_0$ above, but at the expense of having a curve with more "wiggles."

5. The representation preserves variances. If the components of $\mathbf{x}$ are uncorrelated with common variance σ^2, then

$$\text{var}[f_{\mathbf{x}}(t)] = \sigma^2\left(\tfrac{1}{2} + \sin^2 t + \cos^2 t + \sin^2 2t + \cos^2 2t + \cdots\right).$$

If d is odd, this reduces to a constant, $\frac{1}{2}d\sigma^2$; if d is even, the variance lies between $\frac{1}{2}(d-1)\sigma^2$ and $\frac{1}{2}(d+1)\sigma^2$. For all d the variance is therefore either independent of t, or else the relative dependence on t is slight and decreases as d increases. This implies that the variance of $f_{\mathbf{x}}(t)$ is almost constant along the graph. Unfortunately, the components of $\mathbf{x}$ are invariably correlated with unequal variances. However, the above conditions can be approximately satisfied by transforming $\mathbf{x}$ to its vector of standardized principal components (Section 5.2.1). Since low frequencies are more readily seen than high frequencies, it is useful to associate the most important variables with low frequencies. We therefore associate x_1 with the first principal component, x_2 with the second, and so forth. Interesting applications of this technique to anthropological data and British towns are given by Andrews [1972, 1973]. The case of correlated components is discussed very briefly by Goodchild and Vijayan [1974]. Andrews [1973] applies the method to comparing covariance matrices.

The Fourier plot suffers from the same disadvantage as the faces method, in that an interchange of variables leads to a different picture. However, in contrast to some of the other graphical techniques, which are more useful for studying internal dependences among the variables, the Fourier method is useful for detecting external patterns such as clustering (Chapter 7), or outliers (Section 4.5). For example, if well-separated clusters are present, one might be fortunate to have a plot exhibiting distinct bands of curves, as in Fig. 4.7.

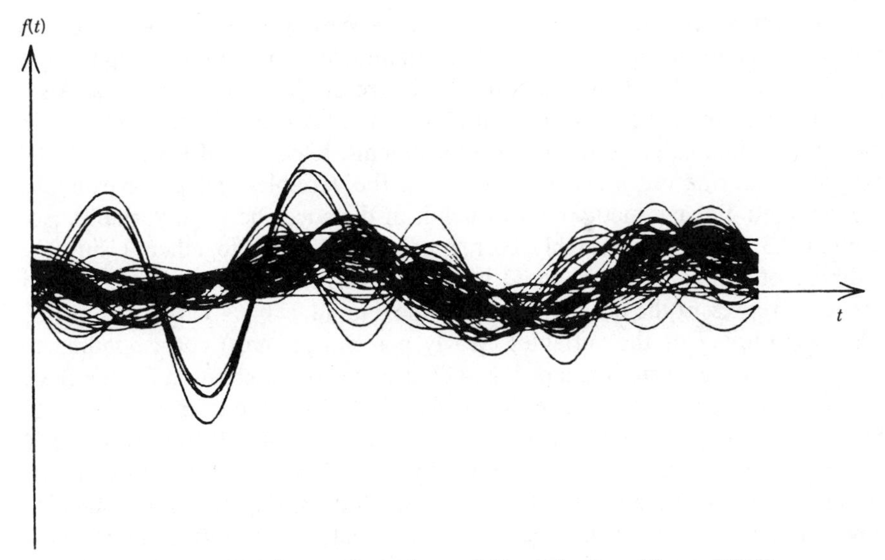

Fig. 4.7 Andrews' curves for a cluster of 86 people. From Morgan [1981].

When $n > 10$ the plot becomes messy and is difficult to study in detail. In this case we would probably separate out interesting looking subsets for closer scrutiny. For larger n, Gnanadesikan [1977: p. 210 ff] suggests using quantile contour plots for a grid of t-values and gives some interesting examples of their use.

EXAMPLE 4.2 Morgan [1981] described briefly a survey of 2622 elderly people from 1975 households in England (see Jolliffe et al. [1982]). For all individuals, 20 variables were selected by the Departments of the Environment and Health and Social Services, and these were then reduced to 10 principal components for each of the age groups 65–74 and 75 +, accounting for more than 80% of the original total variation in each case (see Section 5.2). A k-means method (Section 7.5.3) was used to form 12 clusters of individuals from a subsample of 600 in the 75 + age group. Figure 4.7 illustrates the Andrews curves, computed from the 10 principal component scores, for each of the 86 people in cluster 3. This cluster consisted mainly of single female owner–occupiers, and from the plot there is a suggestion of an outlying group of four individuals. Morgan noted that the "period" of these curves suggests that it is principal component 6 (with coefficient $\sin 3t$) that causes them to separate out. This component identifies employment and these outlying curves correspond to the only four individuals in the cluster who are (part-time) employed. The Andrews curves also clearly identified employed individuals in the other clusters, particularly full-time employed, who have curves with larger amplitudes than part-time employed but share the same period. □

When the number of variables d is large, the above graphical representations become awkward to plot and difficult to assimilate. For example, glyphs with, say, 20 rays would present a confusing picture of short and long rays. Also, any change in the ordering of the variables could lead to a dramatic change in the overall picture. To cope with these problems, Kleiner and Hartigan [1981] proposed carrying out a cluster analysis on the variables, using, for example, the complete linkage method with Euclidean distances between variables [see Section 7.3.1(a)] so that highly correlated variables are together. Using the ordering of the variables imposed by the clustering, they then used trees or castles to represent the observations. For the data in Table 4.1 we have, in Fig. 4.8*a*, a grouping of the variables (yearly percentage rates) corresponding to complete-linkage clustering, and Fig. 4.8*b* is a tree representation for the State of Missouri. We can now add to the displays in Fig. 4.5 the trees and castles of Figs. 4.9 and 4.10, but with the year order given in Fig. 4.8. Evidently, up to 50 variables can be successfully represented by trees, though a lower figure of, say, 30 variables may be appropriate if the branches overlap too much. Castles, however, do not suffer from overlap and can potentially portray many more variables, particularly if the clusters are well defined. The castles can also be tapered for clarity. For example, the above authors gave a further example involving yearly yields of 15 transport companies (called the 15 "variables") for 25 years (called the "objects" or "points"). The companies are clustered in Fig. 4.11*a* and a tapered castle representing yields in 1953 is given in Fig. 4.11*b*.

In the comments that followed Kleiner and Hartigan's paper, Jacob [1981] suggested that the preclustering of variables could be used to advantage in the other methods as well. When "similar" variables are adjacent to one another, glyphs and profiles, for example, convey much more information, as clusters of

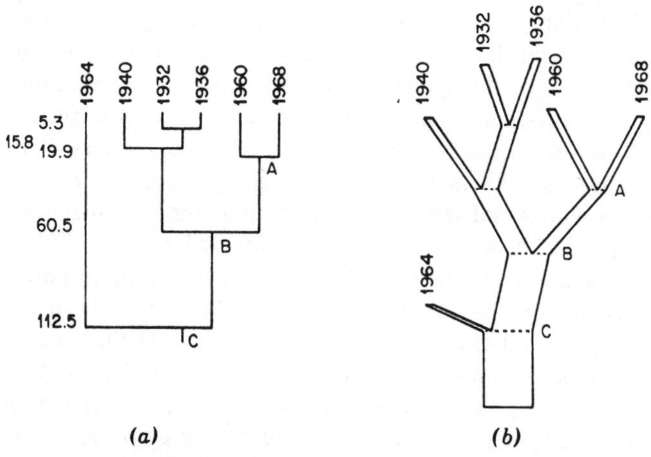

Fig. 4.8 (*a*) Hierarchical clustering by complete linkage for the data in Table 4.1. (*b*) Tree representation of data in Table 4.1 for Missouri. From Kleiner and Hartigan [1981].

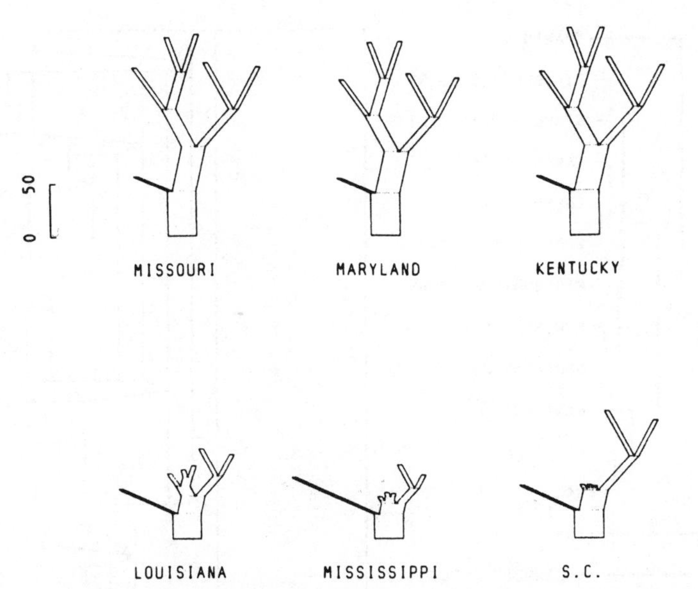

Fig. 4.9 Tree representation of the data in Table 4.1. From Kleiner and Hartigan [1981].

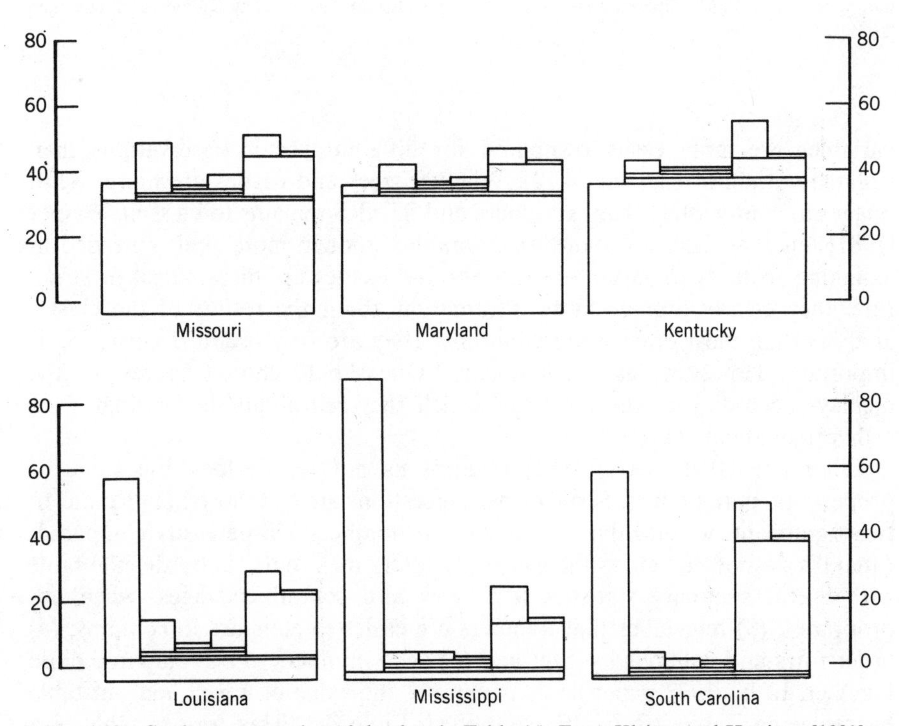

Fig. 4.10 Castle representation of the data in Table 4.1. From Kleiner and Hartigan [1981].

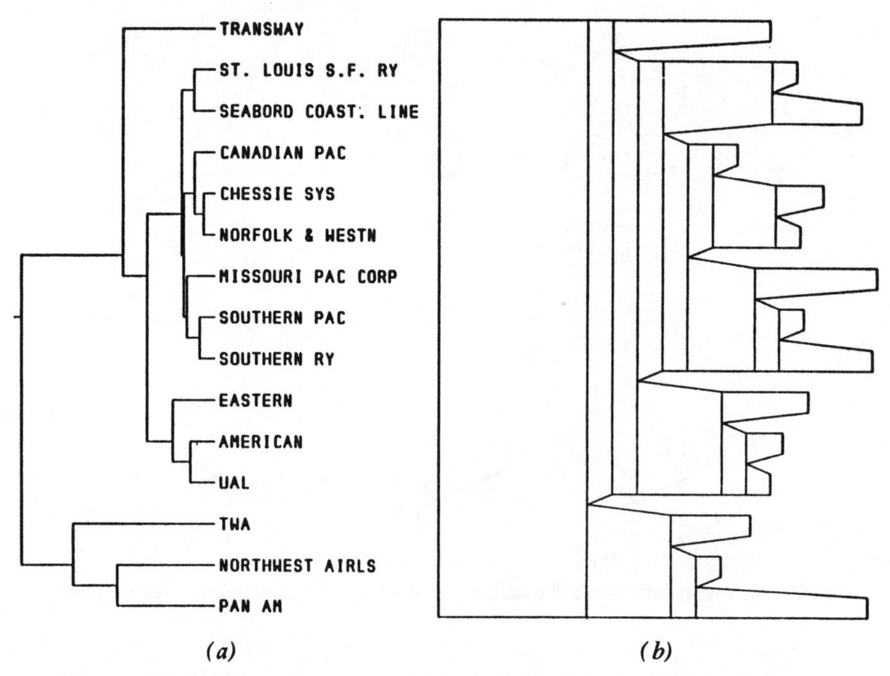

Fig. 4.11 (*a*) Hierarchical clustering for yields of 15 transport companies over 25 years by complete linkage. (*b*) Tapered castle representing yields in 1953. From Kleiner and Hartigan [1981].

variables are more easily compared for different vector observations than isolated variables (e.g., Fig. 4.12). Whether trees and castles are more useful images to contemplate than, say, faces and STARS remains to be seen. Wainer [1981] felt that faces are more "memorable," though more skill is needed in assigning features to variables. As trees and castles are hierarchical in structure, they clearly provide more information about the results of the cluster analysis than most other representations. They are very useful if clustering is important. However, Jacob comments, "One should choose figures for the displays according to the extent to which they tell about the original *data*, rather than about *clusters*."

Ehrenberg [1975, 1977, 1981] reminds us not to overlook the value of properly prepared tables. Some of his suggestions are as follows: (1) Round to two significant or "effective" digits so that numbers can be easily compared; Final 0's do not matter, as the eye easily filters them out; (2) border the table with useful summary statistics (e.g., row and column averages, when appropriate); (3) remember that numbers are easier to compare in columns; (4) order rows and columns by size; and (5) keep numbers to be compared close together. In his 1977 paper he demonstrates the value of a well laid out table by replacing Table 4.2 by the simplified Table 4.3. This former table is a

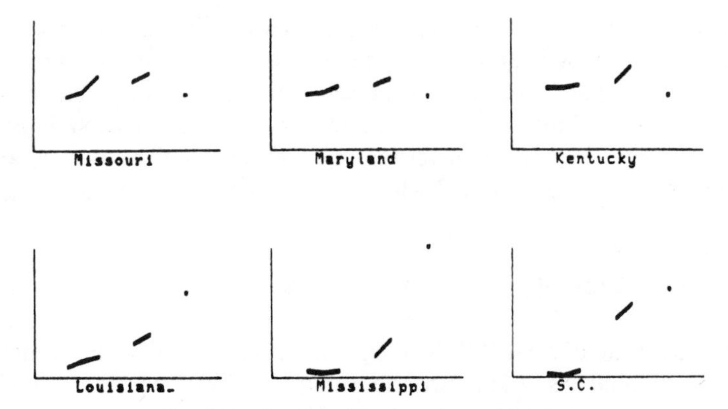

Fig. 4.12 Profiles with clustered coordinates.

10 × 10 correlation matrix where the variables correspond to whether people in a sample of 7000 U.K. adults said they "really liked to watch" a range of 10 television programs like "World of Sport" (WoS). The second table is obtained by rounding to one decimal place, suppressing the unit diagonal elements and zeros before the decimal point, giving both the upper and lower triangles of the matrix, keeping the numbers close together with gaps after about every five rows and columns for easy reading, and reordering the variables. We see at a glance that there are two clusters: The five sports programs have correlations of 0.3 to 0.6; the five current affairs programs have correlations of 0.2 to 0.5; and the correlations between the two groups are about 0.1. Given that the ordering of variables has been chosen, a simple procedure for reordering the correlation matrix is to cut a photocopy of the table into columns and paste them on a blank sheet in the correct order, and then repeat the process on the rows (Chatfield and Collins [1980]).

Finally we note that graphical methods for representing asymmetric matrices have been suggested by several authors (see Constantine and Gower [1978] and

TABLE 4.2 Adults Who "Really Like to Watch"; Correlations to 4 Decimal Places (Programs Ordered Alphabetically with Channel)[a]

		PrB	ThW	Tod	WoS	GrS	LnU	MoD	Pan	RgS	24H
ITV	PrB	1.0000	0.1064	0.0653	0.5054	0.4741	0.0915	0.4732	0.1681	0.3091	0.1242
"	ThW	0.1064	1.0000	0.2701	0.1424	0.1321	0.1885	0.0815	0.3520	0.0637	0.3946
"	Tod	0.0653	0.2701	1.0000	[illegible].0926	0.0704	0.1546	0.0392	0.2004	0.0512	0.2437
"	WoS	0.5054	0.1474	0.0926	1.0000	0.6217	0.0785	0.5806	0.1867	0.2963	0.1403
BBC	GrS	0.4741	0.1321	0.0704	0.6217	1.0000	0.0849	0.5932	0.1813	0.3412	0.1420
"	LnU	0.0915	0.1885	0.1546	0.0785	0.0849	1.0000	0.0487	0.1973	0.0969	0.2661
"	MoD	0.4732	0.0815	0.0392	0.5806	0.5932	0.0487	1.0000	0.1314	0.3267	0.1221
"	Pan	0.1681	0.3520	0.2004	0.1867	0.1813	0.1973	0.1314	1.0000	0.1469	0.5237
"	RgS	0.3091	0.0637	0.0512	0.2963	0.3412	0.0969	0.3261	0.1469	1.0000	0.1212
"	24H	0.1242	0.3946	0.2432	0.1403	0.1420	0.2661	0.1211	0.5237	0.1212	1.0000

[a]From Ehrenberg [1977: Table 4].

Example 2 of Morgan [1981]). Various methods relating to data matrices are given in Section 5.3. Several other graphical tools are also described in Chapter 5, though the emphasis there is on representing a set of d-dimensional vectors by a set of lower-dimensional vectors. Various two-dimensional plots are described in this chapter for investigating distributional properties and for outlier detection. For further methods of display see Barnett [1981].

4.2 TRANSFORMING TO NORMALITY

As the multivariate normal (MVN) distribution has played such a central role in multivariate analysis, it is appropriate that we should consider transformations that help to normalize the data. However, there are some pitfalls associated with the transformations discussed below (see Hernandez and Johnson [1980], Bickel and Doksum [1981], Box and Cox [1982]). Also, the parameters associated with the transformed data may not be as meaningful as those associated with the original data; for example, $\mu_1 - \mu_2$ may be more appropriate than $\log(\mu_1/\mu_2)$. To set the scene we consider the univariate case first.

4.2.1 *Univariate Transformations*

A useful family of transformations is the following:

$$x^{(\lambda)} = \begin{cases} x^{\lambda}, & \lambda \neq 0, \\ \log x, & \lambda = 0 \quad \text{and} \quad x > 0. \end{cases} \tag{4.2}$$

This particular family, studied in detail by Tukey [1957] for $|\lambda| \leq 1$, contains the well-known log, square root, and inverse transformations. To avoid a

TABLE 4.3 The Correlations in Table 4.2 Rounded and Reordered[a]

Programs		WoS	MoD	GrS	PrB	RgS	24H	Pan	ThW	Tod	LnU
World of Sport	ITV		.6	.6	.5	.3	.1	.2	.1	.1	.1
Match of the Day	BBC	.6		.6	.5	.3	.1	.1	.1	0	0
Grandstand	BBC	.6	.6		.5	.3	.1	.2	.1	.1	.1
Prof. Boxing	ITV	.5	.5	.5		.3	.1	.2	.1	.1	.1
Rugby Special	BBC	.3	.3	.3	.3		.1	.1	.1	.1	.1
24 Hours	BBC	.1	.1	.1	.1	.1		.5	.4	.2	.2
Panorama	BBC	.2	.1	.2	.2	.1	.5		.4	.2	.2
This Week	ITV	.1	.1	.1	.1	.1	.4	.4		.3	.2
Today	ITV	.1	0	.1	.1	.1	.2	.2	.3		.2
Line-Up	BBC	.1	0	.1	.1	.1	.2	.2	.2	.2	

[a]From Ehrenberg [1977: Table 5].

discontinuity at $\lambda = 0$, Box and Cox [1964] considered the modification

$$x^{(\lambda)} = \begin{cases} \dfrac{x^{\lambda} - 1}{\lambda}, & \lambda \neq 0, \\ \log x, & \lambda = 0 \quad \text{and} \quad x > 0. \end{cases} \tag{4.3}$$

Using this modification, if we assume that the transformed observations $x_i^{(\lambda)}$ are i.i.d. $N_1(\mu, \sigma^2)$, the likelihood function for the untransformed data is

$$(2\pi\sigma^2)^{-n/2}\left[\exp\left\{-\sum_{i=1}^{n}\frac{\left(x_i^{(\lambda)} - \mu\right)^2}{2\sigma^2}\right\}\right]\left[\prod_{i=1}^{n} x_i^{\lambda-1}\right]. \tag{4.4}$$

Since the last term in square brackets, the Jacobian of the transformation, does not contain μ or σ^2, the maximum likelihood estimates of μ and σ^2 for given λ are

$$\bar{x}(\lambda) = \sum_{i=1}^{n}\frac{x_i^{(\lambda)}}{n} \quad \text{and} \quad \hat{\sigma}_x^2 = \sum_{i=1}^{n}\frac{\left(x_i^{(\lambda)} - \bar{x}^{(\lambda)}\right)^2}{n}. \tag{4.5}$$

If $\dot{x}\ [= (x_1 x_2 \cdots x_n)^{1/n}]$ is the geometric mean of the x_i, then the maximum value of the log likelihood is (apart from a constant)

$$L_{\max}(\lambda) = -\tfrac{1}{2}n\log\hat{\sigma}_x^2 + n\log\dot{x}^{(\lambda-1)} \tag{4.6}$$

$$= -\tfrac{1}{2}n\log\hat{\sigma}_z^2, \tag{4.7}$$

where $z_i^{(\lambda)} = x_i^{(\lambda)}/\dot{x}^{\lambda-1}$. Box and Cox [1964] then suggested choosing $\lambda = \hat{\lambda}$, where $\hat{\lambda}$ maximizes $L_{\max}(\lambda)$. The maximization can be carried out directly using a standard numerical procedure such as solving $dL_{\max}(\lambda)/d\lambda = 0$ iteratively, or by simply plotting $L_{\max}(\lambda)$ against λ. A plot is always useful, as the local behavior of $L_{\max}(\lambda)$ in the neighborhood of $\hat{\lambda}$ can be considered. For example, if $\hat{\lambda} = 0.2$, it may be quite reasonable, for a fairly flat likelihood function, to set $\lambda = 0$, that is, a log transformation. More formally, an approximate $100(1-\alpha)\%$ confidence region for the true value of λ is the set of all λ satisfying

$$L_{\max}(\hat{\lambda}) - L_{\max}(\lambda) \leq \tfrac{1}{2}\chi_{1,\alpha}^2,$$

where $\text{pr}[\chi_1^2 \geq \chi_{1,\alpha}^2] = \alpha$. To test $H_0: \lambda = \lambda_0$ we simply treat $2[L_{\max}(\hat{\lambda}) - L_{\max}(\lambda_0)]$ as being approximately distributed as χ_1^2. Andrews [1971] proposed an "exact" more robust test for H_0, though the empirical study of Atkinson [1973] suggests that the test is less powerful than the likelihood ratio test. Alternatively, robust M-estimators of μ and σ^2 [see (4.32)] can be used instead of (4.5) (Carroll [1980]), and the resulting test for H_0 combines some of the robustness of Andrews' so-called significance method against outliers with some of the power advantages of the likelihood ratio test.

If some of the x_i are negative, we can add a positive constant ξ to all the observations to make them positive. Alternatively, we can include ξ in the above likelihood function, now written $L_{\max}(\xi, \lambda)$ to indicate that x_i is replaced by $x_i + \xi$, and find the maximum likelihood estimates of ξ and λ (Box and Cox [1964]).

John and Draper [1980] provided an alternative family of transformations

$$x^{(\lambda)} = \begin{cases} \text{sign}\left[\dfrac{(|x| + 1)^{\lambda} - 1}{\lambda}\right], & \lambda \neq 0, \\ \text{sign}[\log(|x| + 1)], & \lambda = 0, \end{cases} \tag{4.8}$$

where the sign of $x^{(\lambda)}$ is that associated with the observation x, called the modulus transformation. The power transformation (4.3) is effective in making skewed distributions more symmetrical and, hopefully, more normal. For example, the effect of a logarithmic or square root transformation is to pull in one tail of the distribution. John and Draper [1980] noted that the "modulus transformation, on the other hand, is effective on a distribution that already possesses approximate symmetry about some central point and alters each half of the distribution through the same power transformation in an attempt to make the shape more normal." If all the data are positive, the modulus and power transformations are equivalent. However, an alternative is still provided by the modulus family, since data of the form $x - a$, for some constant a such as a robust estimator of location, can be transformed. The maximum likelihood method described above also applies to the modulus family: Equation (4.7) still holds with $\dot{x}$ in $z^{(\lambda)}$ now being the geometric mean of the $|x_i| + 1$. John and Draper [1980] gave an example where the best power transformation is inadequate (the normal plot is S shaped), while the best modulus transformation gives a linear residual plot. This might have been expected, as the residual plot for the untransformed data was S shaped but symmetric, indicating that a modulus transformation, which treats the tails symmetrically, would be better than a skew-correcting power transformation.

Finally we note that Hinkley [1975, 1977] has given two quick methods of transforming the data to obtain approximate symmetry.

4.2.2 *Multivariate Transformations*

For multivariate data, the above univariate procedures can be applied to each dimension with a separate λ or separate pair (ξ, λ) for each of the d variables. However, Andrews et al. [1971] have given the following multivariate generalization of Box and Cox's technique. We now have a vector of parameters $\boldsymbol{\lambda} = (\lambda_1, \lambda_2, \ldots, \lambda_d)$, one λ_i for each dimension, and the transformed vectors $\mathbf{x}_i^{(\boldsymbol{\lambda})} = (x_{i1}^{(\lambda_1)}, \ldots, x_{id}^{(\lambda_d)})'$ are assumed to be i.i.d. $N_d(\boldsymbol{\mu}, \boldsymbol{\Sigma})$. Corresponding to (4.6), the likelihood function for $\boldsymbol{\lambda}$ is [see (3.6)]

$$L_{\max}(\boldsymbol{\lambda}) = -\tfrac{1}{2} n \log|\hat{\boldsymbol{\Sigma}}| + \sum_{j=1}^{d} (\lambda_j - 1) \sum_{i=1}^{n} \log x_{ij}, \tag{4.9}$$

where x_{ij} is the jth element of $\mathbf{x}_i$ and

$$\hat{\boldsymbol{\Sigma}} = \frac{1}{n} \sum_{i=1}^{n} \left(\mathbf{x}_i^{(\lambda)} - \overline{\mathbf{x}}^{(\lambda)}\right)\left(\mathbf{x}_i^{(\lambda)} - \overline{\mathbf{x}}^{(\lambda)}\right)'. \tag{4.10}$$

We now choose $\boldsymbol{\lambda} = \hat{\boldsymbol{\lambda}}$, where $\hat{\boldsymbol{\lambda}}$ maximizes $L_{\max}(\boldsymbol{\lambda})$. To test $H_0: \boldsymbol{\lambda} = \boldsymbol{\lambda}_0$ we calculate $2[L_{\max}(\hat{\boldsymbol{\lambda}}) - L_{\max}(\boldsymbol{\lambda}_0)]$, which is approximately χ_d^2 when H_0 is true. This statistic can also be used for constructing a confidence region for $\boldsymbol{\lambda}$.

Andrews et al. [1971] give a transformation procedure for improving normality in certain directions. Some interesting plots demonstrating the transformations described above and in the previous section are given in Gnanadesikan [1977: pp. 144–150].

4.3 DISTRIBUTIONAL TESTS AND PLOTS

4.3.1 Investigating Marginal Distributions

In contrast to the univariate case, there is a noticeable dearth of practical alternatives to the MVN distribution. Those distributions that have been used, such as the multivariate *t*-, *F*-, beta, lognormal, and gamma distributions, share with the multivariate normal the perhaps unrealistic property that their univariate marginals all come from the same family. One useful family of distributions that which includes the normal is the family of elliptical distributions (Section 2.1). A first step, therefore, in studying sample data from a multivariate distribution might be to check whether the marginals conform to some known family of univariate distributions. A simple graphical method for doing this is a *Q–Q* or quantile-versus-quantile plot (Wilk and Gnanadesikan [1968]) in which the sample (cumulative) distribution function is compared graphically with a theoretical distribution function *F*. Examples of *F* are the chi-square, gamma, beta, and normal distributions, and details are given in Appendix C. A nice feature of the *Q–Q* plot not shared by its competitor, the so-called *P–P* plot, is that the graph is linear even if the two distributions being compared have different scale or location parameters. If the scales are different, the slope is not unity; if the locations are different, the graph will not pass through the origin.

EXAMPLE 4.3 Normal probability plots (see Appendix C4) were made for the heights and weights of 38 Peruvian Indians (see Table 3.1 on p. 64, with the last observation excluded). These plots are given in Fig. 4.13. In assessing such plots it is helpful, by way of comparison, to have available plots of genuinely normal data for different sample sizes: Such a set of plots is given by Daniel and Wood [1981].

We see that the plot for heights is reasonably linear up to a height of about 1600 mm, and then there is a discontinuity. Looking at the original scatter plot

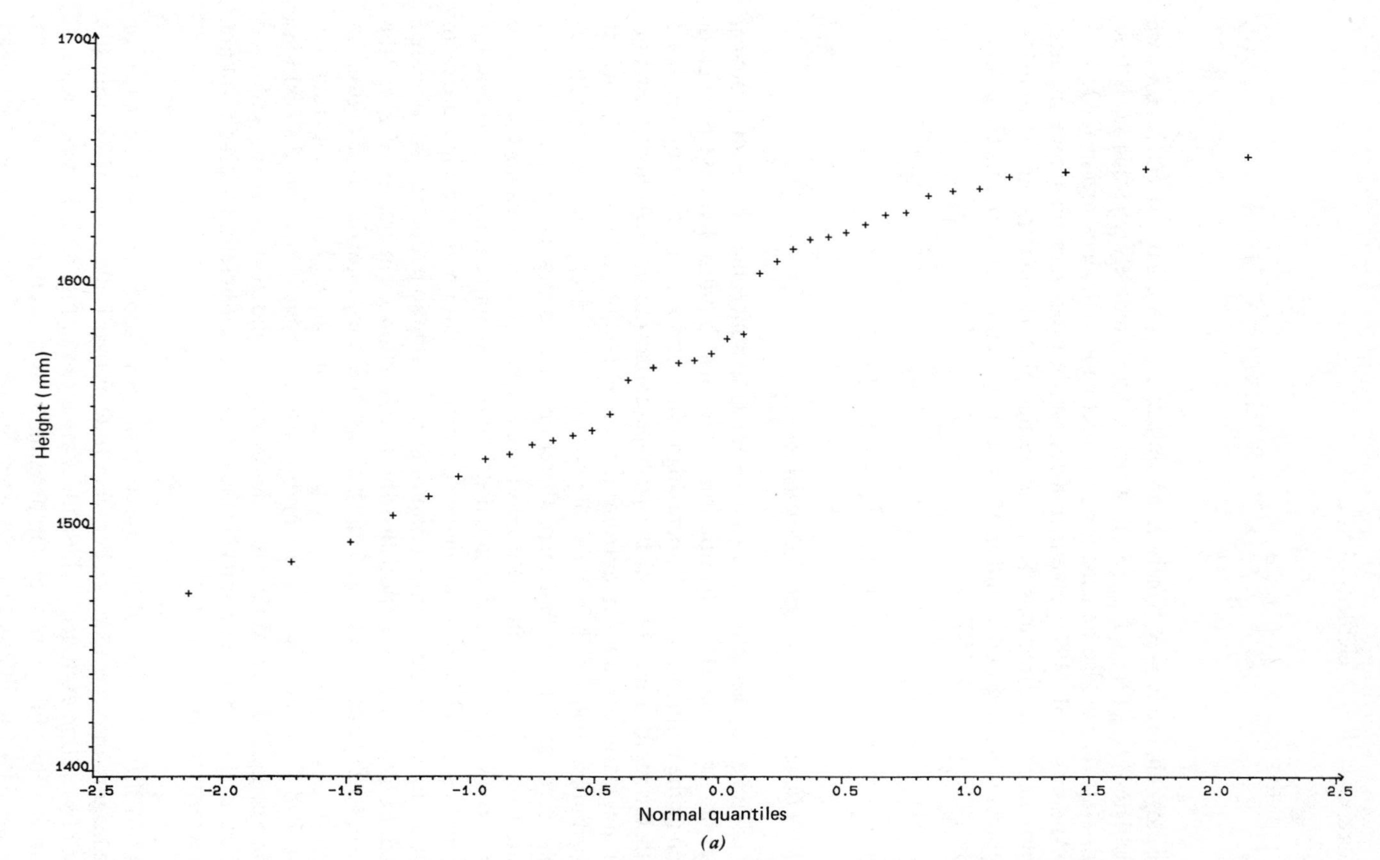

Fig. 4.13 Normal probability plots for the heights and weights of 38 Peruvian Indians given in Table 3.1 (with last observation omitted). (*a*) Height. (*b*) Weight.

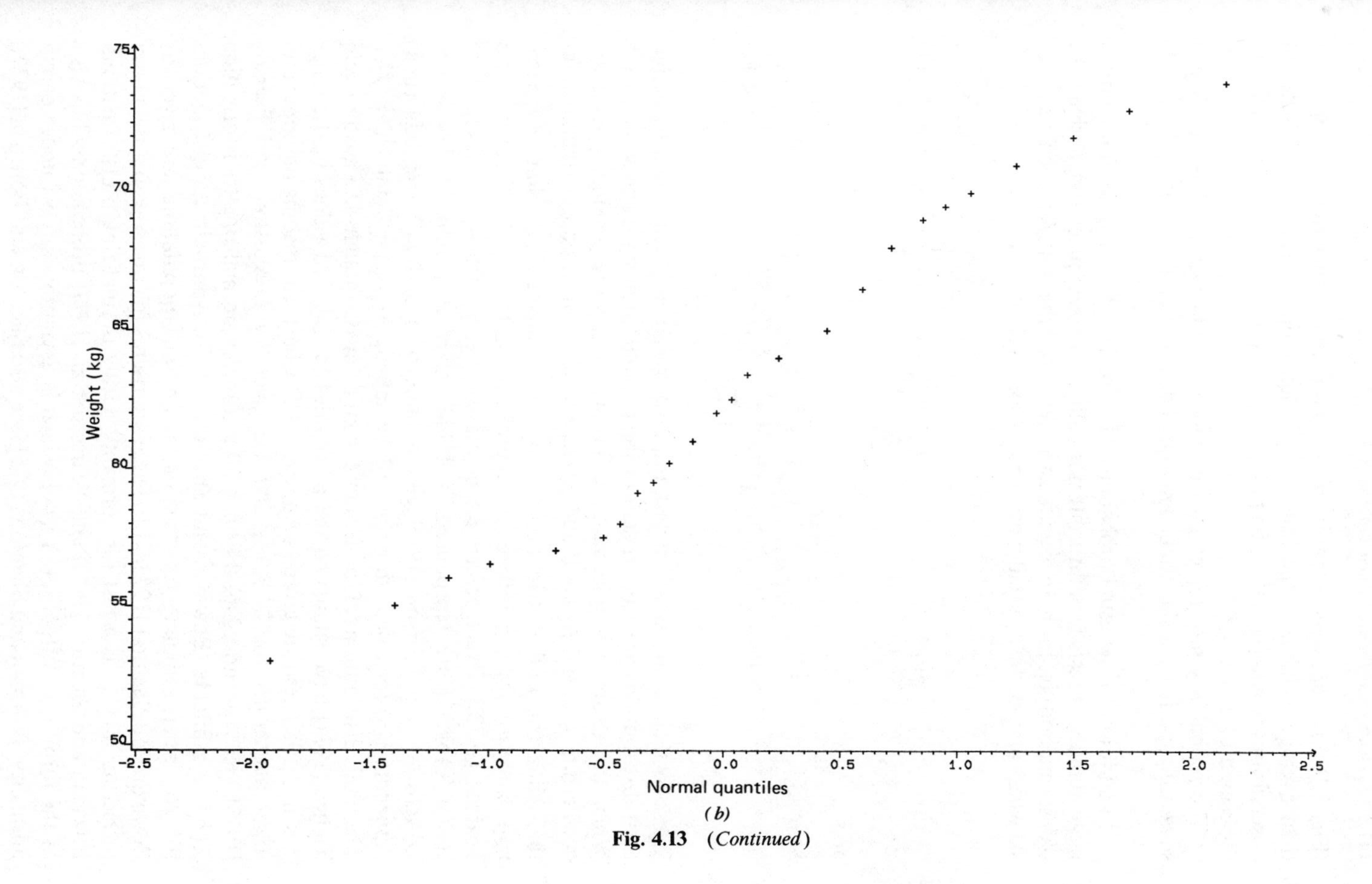

(*b*)
Fig. 4.13 (*Continued*)

(Fig. 3.1) on p. 66, we see that there are about seven taller Indians with lower than average weights in the range 61–64 kg approximately. This cluster of observations would have the effect of raising the probability plot for heights in the region of $0 < x < 0.5$.

The probability plot for weights is somewhat different and looks S shaped suggesting, perhaps, a short-tailed marginal distribution for weight. □

In addition to the usual probability plot for visually assessing normality, a large number of tests for univariate normality have been proposed. Perhaps the oldest method is to base significance tests on the sample coefficients of skewness and kurtosis, usually denoted, respectively, by

$$\sqrt{b_1} = \sqrt{n} \sum_{i=1}^{n} (x_i - \bar{x})^3 \Big/ \left\{ \sum_{i=1}^{n} (x_i - \bar{x})^2 \right\}^{3/2} \tag{4.11}$$

and

$$b_2 = n \sum_{i=1}^{n} (x_i - \bar{x})^4 \Big/ \left\{ \sum_{i=1}^{n} (x_i - \bar{x})^2 \right\}^2 . \tag{4.12}$$

These coefficients are invariant with respect to origin and scale changes, so that their distributions are independent of the population mean μ and variance σ^2. When the underlying population is normal, the corresponding population coefficients $\sqrt{\beta_1}$ and β_2 are 0 and 3, respectively. Various approximations for the symmetric null distribution of $\sqrt{b_1}$ are available, including a normal approximation for $n \geq 150$ and a t-approximation for $n \geq 8$ (D'Agostino and Pearson [1973]). Exact percentage points for $4 \leq n \leq 25$ due to Mulholland [1977: Table 2] are reproduced in Table D3 of Appendix D. For $n \geq 8$, D'Agostino and Pearson [1973] give a table of δ and $1/\lambda$ (reproduced as Appendix D4) for calculating the $N(0,1)$ statistic $X(\sqrt{b_1}) = \delta \sinh^{-1}(b_1^{1/2}/\lambda)$. The null distribution of b_2 is more complicated, though D'Agostino and Tietjen [1971] have simulated the percentiles for selected values in the range $7 \leq n \leq 50$: These are given in Appendix D5. Charts for reading off upper and lower percentiles for $20 \leq n \leq 200$ are given by D'Agostino and Pearson [1973], (see Appendix D6). For $n < 20$ a complex normalizing transformation $X(b_2)$ is available (Bowman and Shenton [1975]). Approximate distributions for $\sqrt{b_1}$ when sampling from various nonnormal populations are given by Bowman and Shenton [1973a, b]: These are useful for investigating the power properties of $\sqrt{b_1}$ as a test for normality. Using normalizing transformations, Bowman and Shenton [1975] and Pearson et al. [1977] propose omnibus tests of the form $Y = X^2(\sqrt{b_1}) + X^2(b_2)$, which is asymptotically χ_2^2 under parent normality. Bowman and Shenton [1975] (see also Shenton and Bowman [1977]) also give useful contour plots for jointly testing $\sqrt{b_1}$ and b_2 when $n > 20$.

Dyer [1974] listed seven further tests of which the Wilk–Shapiro (W) and Anderson–Darling (A_n^2) tests generally provide the most powerful tests for a reasonable class of alternatives (see Dyer [1974], Shapiro et al. [1968], Stephens [1974], Huang and Bolch [1974], and Lin and Mudholkar [1980: Table 1]). Further modifications of some of the tests are given by Green and Hegazy [1976], and large-sample versions of W are given by Shapiro and Francia [1972] and Weisberg and Bingham [1975]. The statistic W is given by

$$W = \left(\sum_{i=1}^{n} a_i^{(n)} x_{(i)} \right)^2 \Big/ \sum_{i=1}^{n} (x_i - \bar{x})^2, \tag{4.13}$$

where $x_{(1)} \le \cdots \le x_{(n)}$ are the ordered observations, and is independent of scale and origin. Since $-a_i^{(n)} = a_{n-i+1}^{(n)}$, we have $a_{k+1}^{(2k+1)} = 0$ and, for $n = 2k$ or $2k + 1$,

$$\sum_{i=1}^{n} a_i^{(n)} x_{(i)} = \sum_{i=1}^{k} a_{n-i+1}^{(n)} \left(x_{(n-i+1)} - x_{(i)} \right). \tag{4.14}$$

The coefficients $a_{n-i+1}^{(n)}$ and percentage points for W are given by Shapiro and Wilk [1965] for $2 \le n \le 50$: These are reproduced in Appendixes D7 and D8. Any departure from normality is detected by a small value of W; an example is given below.

Shapiro and Wilk do not extend their tables beyond $n = 50$ and D'Agostino [1971] gives a number of reasons that suggest that such an extension is not appropriate. However, for $n > 50$, D'Agostino [1971] gives an alternative large-sample test statistic

$$D = \left(\sum_{i=1}^{n} \left[i - \tfrac{1}{2}(n+1) \right] x_{(i)} \right) \Big/ n^{3/2} \left[\sum_{i=1}^{n} (x_i - \bar{x})^2 \right]^{1/2},$$

or its standardized version

$$Y = \frac{\sqrt{n}\left[D - (2\sqrt{\pi})^{-1} \right]}{0.02998598}$$

which, asymptotically, has mean 0 and variance 1 when the x_i are normal. If the x_i are not normal, then $\mathrm{E}[Y]$ tends to differ from zero, with the direction of the difference depending on the parent distribution. D'Agostino [1971, 1972] gave tables of percentiles for n ranging from 10 to 1000 (see Appendix D9) and compared the power of D with W, $\sqrt{b_1}$, and b_2, thus extending the study by Shapiro et al. [1968] (see also Pearson et al. [1977]). From the comparison it appears that W and D are useful all-round tests when the type of deviation from normality is unknown. If the parent distribution is known to be symmetric, b_2 can perform better than W, though their relative performances depend on n and β_2.

The Anderson–Darling statistic is given by

$$A_n^2 = -\frac{1}{n}\left[\sum_{i=1}^{n}(2i-1)\{\log z_i + \log(1 - z_{n+1-i})\} - n\right], \tag{4.15}$$

where

$$z_i = \Phi\left(\frac{x_{(i)} - \bar{x}}{s}\right),$$

$s^2 = \Sigma_i(x_i - \bar{x})^2/(n-1)$ and Φ is the distribution function for $N_1(0,1)$. The null hypothesis of normality is rejected for large values of A_n^2. As there is little difference in the powers of W and A_n^2, there is some computational advantage in using A_n^2, as W requires different coefficients $a_i^{(n)}$ for each n, and percentage points for A_n^2 are now available (Pettitt [1977]). If $\alpha = \text{pr}[A_n^2 \le a_n]$, then, for $n \ge 4$, we have approximately $a_n = a_\infty(1 + c_1 n^{-1} + c_2 n^{-2})$, where a_∞, c_1, and c_2 are given by Table D10. The study by Pettitt confirms that W and A_n^2 have similar powers, though W seems more powerful for the highly skewed distributions, while A_n^2 appears to be more powerful for long-tailed distributions.

A number of tests based on the linearity of the Q–Q probability plot (see Appendix C4) are also available. Two promising tests, called F_1 and F_2, with powers comparable with W are proposed by LaBrecque [1977].

EXAMPLE 4.4 Suppose we wish to test whether the following 11 ordered observations come from a normal distribution: 148, 154, 158, 160, 161, 162, 166, 170, 182, 195, 236. Then $\bar{x} = 172$, $\Sigma_i(x_i - \bar{x})^2 = 6226$, and $s = 24.95$. The ordered values of $(x_{(i)} - \bar{x})/s$ are -0.962, -0.721, -0.561, -0.481, -0.441, -0.401, -0.240, -0.080, 0.401, 0.922, and 2.565. Comparing these values with the standard normal distribution, we see that there is evidence of both skewness and kurtosis. From (4.11) and (4.12) we find that $\sqrt{b_1} = 1.68$ and $b_2 = 4.99$. Using Appendix D3, with $n = 11$, we see that $\sqrt{b_1}$ is significant at the 1% level, as it exceeds the 0.5% value of 1.553 (for a two-tailed test). From Appendix D5, b_2 is just below 5.10, the upper 1% critical value for $n = 11$.

To calculate W of (4.14), we obtain the following from Appendix D7: $a_{11}^{(11)} = 0.5601$, $a_{10}^{(11)} = 0.3315$, $a_9^{(11)} = 0.2260$, $a_8^{(11)} = 0.1429$, and $a_7^{(11)} = 0.0695$. Thus

$$\begin{aligned} W &= \frac{1}{6226}\left\{a_{11}^{(11)}(x_{(11)} - x_{(1)}) + a_{10}^{(11)}(x_{(10)} - x_{(2)}) + \cdots + a_7^{(11)}(x_{(7)} - x_{(5)})\right\}^2 \\ &= \frac{1}{6226}\{0.5601(88) + \cdots + 0.0695(25)\}^2 \\ &= 0.79. \end{aligned}$$

From Appendix D8, this value is just below the 1% critical value of 0.792.

The Anderson–Darling statistic of (4.15) requires values of z_i from the normal tables, for example, $z_1 = \Phi(-0.962) = 0.168$, giving $A_{11}^2 = 1.007$. Using Appendix D10 with $n = 11$ and $p = 0.99$, we obtain

$$a_{11} = 1.0348\left(1 - \frac{1.013}{11} - \frac{0.93}{121}\right) = 0.932.$$

Since $A_{11}^2 > a_{11}$, our test is significant at the 1% level. □

The search for further tests of normality continues. Spiegelhalter [1977, 1980] proposed a Bayesian test statistic S that has good power against certain extreme distributions and which compares very favorably with W. Unfortunately, it requires the computation of five separate statistics; graphs for reading off estimated significance points for $n \leq 100$ are given by Spiegelhalter [1980]. Using the concept of entropy, Vasicek [1976] proposed another test statistic together with a table of 5% significant points.

Lin and Mudholkar [1980] have proposed a test for normality based on the statistic

$$Z = \tfrac{1}{2}\log\left(\frac{1+r}{1-r}\right), \tag{4.16}$$

where r is the correlation of the pairs (x_i, y_i) and

$$y_i = \frac{1}{n}\left\{\sum_{\alpha \neq i} x_\alpha^2 - \left(\sum_{\alpha \neq i} x_\alpha\right)^2 \Big/ (n-1)\right\}^{1/3} \qquad (i = 1, 2, \ldots, n).$$

Given normality of the x_i's, Z is approximately $N_1(0, 3/n)$. More accurately, the upper 100α percentile of the null distribution of Z is

$$z_\alpha = \sigma_n\left[u_\alpha + \tfrac{1}{24}\left(u_\alpha^3 - 3u_\alpha\right)\gamma_{2n}\right], \tag{4.17}$$

where

$$\sigma_n^2 = 3n^{-1} - 7.324n^{-2} + 53.005n^{-3},$$

$$\gamma_{2n} = -11.70n^{-1} + 55.06n^{-2},$$

and $u_\alpha = \Phi^{-1}(\alpha)$, where Φ is the distribution function for the $N_1(0,1)$ distribution. This approximation is adequate for n as small as 5 and the power of this Z-test is fairly close to that of the W-test. Although the Z-test was constructed primarily for detecting nonsymmetric alternatives, it compares reasonably well with other tests for normality against long-tailed symmetric distributions.

Which of the above tests should be used? Some limited power comparisons have been made, with the emphasis largely on the types of alternative distributions. Unfortunately, significance level and sample size both have an effect, and little has been done by way of comparison in this direction. Clearly W ($n \leq 50$), D ($n \geq 10$), and b_2 are useful test statistics, though W is awkward to use and A_n^2 may be preferred. Spiegelhalter's S-statistic looks promising, but only rather crude graphs are published for $n \leq 100$: Further power comparisons are also needed. On the other hand, the Z-statistic, although generally not as powerful as W or S, has the advantage of being readily computerized and is easy to interpret without the need of special tables.

For moderately large samples, Andrews et al. [1972b] have proposed a simple test for nonnormality based on the normalized gaps

$$g_i = \frac{x_{(i+1)} - x_{(i)}}{m_{i+1} - m_i} \qquad (i = 1, 2, \ldots, n-1), \tag{4.18}$$

where $m_i = \mathrm{E}[x_{(i)}]$ under the assumption of normality. If the x_i are i.i.d $N_1(\mu, \sigma^2)$, the g_i are approximately independently and exponentially distributed with scale parameter σ. Gnanadesikan [1977: pp. 165–166] describes three test statistics based on the g_i and the reader is referred to him for details. One can also make a normal probability plot of the cube roots of the ordered g_i [see (4.56)]. Wainer and Schacht [1978] suggest the use of weighted gaps.

The transformation techniques of Box and Cox [1964] described in Section 4.2.1 can be used for testing univariate normality. Gnanadesikan [1977: p. 166] suggests that the pair of parameters (ξ, λ) should be used, as λ appears to be sensitive to skewness, while ξ seems to respond to kurtosis and longer tails. In this case we have the transformation

$$x:\to \begin{cases} \dfrac{(x+\xi)^\lambda - 1}{\lambda}, & \lambda \neq 0, \\ \log(x+\xi), & \lambda = 0, \end{cases}$$

and if the data are already normal, then $\lambda = 1$. We can therefore test for normality by testing $H_0: \lambda = 1$ using the statistic [see (4.7)] $2\{L_{\max}(\hat{\xi}, \hat{\lambda}) - L_{\max}(\hat{\xi}, 1)\}$, which is approximately χ_1^2 when H_0 is true. It should be noted that $L_{\max}(\xi, 1)$ is independent of ξ, so that any value, including $\hat{\xi}$, maximizes $L_{\max}(\xi, 1)$ (see Exercise 4.2).

For further references on univariate testing, see the review of Mardia [1980].

4.3.2 *Tests and Plots for Multivariate Normality*

If we simply investigate marginal distributions, we are ignoring the structure imposed by the dispersion matrix $\mathbf{\Sigma}$. Also, marginal normality does not necessarily imply joint multivariate normality (see Seber [1977: pp. 30–31] for references). In assessing overall multivariate normality it is natural to try and generalize some of the univariate techniques. For example, Mardia [1970, 1974,

1975] introduced the multivariate measures of skewness and kurtosis, $\beta_{1,d}$ and $\beta_{2,d}$, respectively, defined in Section 2.7. When $\mathbf{x}$ is MVN, $\beta_{1,d} = 0$ and $\beta_{2,d} = d(d+2)$, and we can test for these values using the sample estimators

$$b_{1,d} = \frac{1}{n^2} \sum_{h=1}^{n} \sum_{i=1}^{n} g_{hi}^3 \quad \text{and} \quad b_{2,d} = \frac{1}{n} \sum_{i=1}^{n} g_{ii}^2, \tag{4.19}$$

where

$$g_{hi} = (\mathbf{x}_h - \bar{\mathbf{x}})' \left(\frac{\mathbf{Q}}{n} \right)^{-1} (\mathbf{x}_i - \bar{\mathbf{x}}).$$

An algorithm for computing the sample estimators is given by Mardia and Zemrock [1975a]. Mardia [1970] showed that, given multivariate normality, we have asymptotically

$$A = \tfrac{1}{6} n b_{1,d} \sim \chi_f^2, \qquad \text{where } f = \tfrac{1}{6} d(d+1)(d+2), \tag{4.20}$$

and

$$B = \frac{b_{2,d} - d(d+2)}{[8d(d+2)/n]^{1/2}} \sim N_1(0,1). \tag{4.21}$$

Unfortunately, it appears that we must have n impractically large in most cases (at least greater than 50) for these large-sample approximations to be satisfactory. However, Mardia [1974] gives tables of approximate percentiles for $d = 2$ and $n \geq 10$, and provides two alternative large-sample approximations that can be used in conjunction with A and B to give conservative critical values for $d > 2$ and $n \geq 50$.

EXAMPLE 4.5 In Table 3.3 (p. xxx) we have 28 four-dimensional observations for which

$$b_{1,4} = 4.476 \quad \text{and} \quad b_{2,4} = 22.957.$$

From (4.20) and (4.21), with $d = 4$ and $f = 20$, we have

$$A = \tfrac{1}{6}(28)(4.476) = 20.89,$$

which is not significant when compared to χ_{20}^2, and

$$B = \frac{22.957 - 4(6)}{[32(6)/28]^{1/2}} = -0.40,$$

which is not significant when compared to $N_1(0,1)$. Although these distributions are only approximate for $n = 28$, they at least indicate that the data does not show any marked departure from multivariate normality. □

Another approach to the problem is to base tests on the vectors $\mathbf{u}_1$ and $\mathbf{u}_2$ whose elements are the d marginal coefficients of skewness ($\sqrt{b_1}$) and kurtosis (b_2), respectively. However, the distributions of $\mathbf{u}_1$ and $\mathbf{u}_2$ approach multivariate normality very slowly, so that Small [1980] utilized the univariate normalizing transformations of Bowman and Shenton [1975] and considered

$$\mathbf{v}_1 = \delta_1 \sinh^{-1}\left(\frac{\mathbf{u}_1}{\lambda_1}\right), \qquad \mathbf{v}_2 = \gamma_2 \mathbf{1}_d + \delta_2 \sinh^{-1}\left(\frac{[\mathbf{u}_2 - \xi \mathbf{1}_d]}{\lambda_2}\right),$$

where $\sinh^{-1}(\mathbf{a})$ implies that $\sinh^{-1}$ is applied to each element of $\mathbf{a}$. Small [1980] proposed the statistics

$$Q_i = \mathbf{v}_i' \mathbf{V}_i^{-1} \mathbf{v}_i \qquad (i = 1, 2),$$

where the diagonal elements of $\mathbf{V}_1$ and $\mathbf{V}_2$ are unity and the (j,k)th off-diagonal elements are, respectively, r_{jk}^3 and r_{jk}^4, r_{jk} being the sample correlation between the jth and kth variables. Simulated trials by Small indicate that, for $n > 29$ and $2 \le d \le 8$, Q_1 and Q_2 are approximately independently and identically distributed as χ_d^2 when the underlying distribution is MVN. We could also treat $Q_1 + Q_2$ as χ_{2d}^2. The various parameters δ_i, λ_i, and so on, are obtained from Johnson [1965], and details of the computation are given by Small [1980].

Alternative measures of multivariate skewness and kurtosis have been proposed by Malkovich and Afifi [1973] that are essentially the univariate measures for the data $\mathbf{a}'\mathbf{x}_i$, maximized over $\mathbf{a}$. Although such an approach is very appealing, the distributions of the sample estimates seem to be intractable. In a similar vein the authors propose a generalization of the Wilk–Shapiro W-statistic of (4.13).

One of the properties of the MVN is that the regression of one variable on the others is linear. Cox and Small [1978] utilize this property to give a number of tests for multivariate normality. A further test, also based on conditional distributions, is proposed by Hensler et al. [1977].

A univariate procedure of Quesenberry et al. [1976] has been generalized to the multivariate case by Rincón-Gallardo et al. [1979] as follows [with the change in notation $(i, j, k) \to (j, r, d)$]. Define $\bar{\mathbf{x}}_r = \sum_{i=1}^r \mathbf{x}_i / r$ and $\mathbf{Q}_r = \sum_{i=1}^r (\mathbf{x}_i - \bar{\mathbf{x}}_r)(\mathbf{x}_i - \bar{\mathbf{x}}_r)'$. Let $\mathbf{A}_r'\mathbf{A}_r$ be a factorization (e.g., Cholesky decomposition) of $\mathbf{Q}_r^{-1}$ and, for $r = 1, 2, \ldots, n$, define

$$\mathbf{z}_r = \frac{\mathbf{A}_r(\mathbf{x}_r - \bar{\mathbf{x}}_r)}{\left\{[(r-1)/r] - (\mathbf{x}_r - \bar{\mathbf{x}}_r)'\mathbf{Q}_r^{-1}(\mathbf{x}_r - \bar{\mathbf{x}}_r)\right\}^{1/2}}$$

If $\mathbf{z}_r' = (z_{1r},\ldots,z_{dr})$, $r = d+2,\ldots,n$, consider the variables

$$u_{jr} = G_{j+r-d-2}\left(z_{jr}\left\{\frac{j+r-d-2}{1+z_{1r}^2+\cdots+z_{j-1,r}^2}\right\}^{1/2}\right),$$

$$r = d+2, d+3,\ldots,n;\ j = 1,2,\ldots,d, \quad (4.22)$$

where G_ν is the distribution function for student's t_ν distribution. Then, given the $\mathbf{x}_i$ are multivariate normal, the $d(n-d-1)$ random variables u_{jr} are independently and identically distributed with the uniform (rectangular) [0, 1] distribution. Testing for multivariate normality is therefore equivalent to a goodness of fit test of the u_{jr} to the uniform distribution. The authors note that several statistics are available and mention the modified form, U_{MOD}^2, of Watson's U^2 statistic, proposed by Stephens [1970], that has the advantage of having upper 1, 5, and 10 percentage points that are approximately 0.267, 0.187, and 0.152 for $n \geq 10$. The details of computing U_{MOD}^2 are the following. Given $y_1, y_2,\ldots,y_m$ from a uniform [0, 1] distribution, calculate

$$W^2 = \sum_{i=1}^{m}\left(y_i - \frac{2i-1}{m}\right)^2 + \frac{1}{12m},$$

$$U^2 = W^2 - m(\bar{y} - \tfrac{1}{2})^2,$$

and

$$U_{\mathrm{MOD}}^2 = (U^2 - 0.1n^{-1} + 0.1n^{-2})(1.0 + 0.8n^{-1}). \quad (4.23)$$

Another statistic, called the Neyman [1937] smooth statistic, p_4^2, can also be used (Rincón-Gallardo and Quesenberry [1982]). Probability plots can also be carried out. If the u_{jr} are ranked $u_{(1)} \leq u_{(2)} \leq \cdots \leq u_{(m)}$, then $E[u_{(k)}] = k/(m+1)$ and a plot of $u_{(k)}$ versus $k/(m+1)$ should lie approximately on the line through the origin with unit slope. A major weakness of the above approach is that it depends on the order of the sample $\mathbf{x}_1, \mathbf{x}_2,\ldots,\mathbf{x}_n$: Care should be exercised so that the $\mathbf{x}_i$'s are not ordered by the values of one or more of their components. An advantage of the method is that samples can be combined. For example, if we have three samples from parent distributions of the same functional form, but possibly with different parameter values, we can pool the three sets of u-values and test that this common functional form is normal. Also, we can test that the jth variable in the $\mathbf{x}_i$'s is normal by considering the subset $u_{j,d+2}, u_{j,d+3},\ldots,u_{j,n}$ (Rincón-Gallardo and Quesenberry [1982]).

Andrews et al. [1973] suggested using the multivariate transformation technique of Section 4.2.2 for testing multivariate normality. If the data comes from a MVN, no transformation is needed and $\boldsymbol{\lambda} = \mathbf{1}_d$. We can test this using [see (4.9)]

$$2[L_{\max}(\hat{\boldsymbol{\lambda}}) - L_{\max}(\mathbf{1}_d)], \tag{4.24}$$

which is approximately χ_d^2 when $\boldsymbol{\lambda} = \mathbf{1}_d$. If $\boldsymbol{\lambda} \neq \mathbf{1}_d$, we can also test for marginal similarity by testing $\lambda_1 = \lambda_2 = \cdots = \lambda_d$ (Exercise 4.4).

The univariate "gap" methods, using (4.18), are based on the successive interpoint distances $d_i = x_{(i+1)} - x_{(i)}$ for the ranked observations $x_{(i)}$. In the multivariate case there is, unfortunately, no unique method of ordering so that different sets of interpoint distances can be used. For example, Rohlf [1975] suggested using the lengths d_i of the $n - 1$ edge connections in the minimum spanning tree (MST), which is defined as follows. A spanning tree is any set of straight line segments joining various pairs of points $\mathbf{x}_i$ such that there are no closed loops, each point is visited by at least one line, and each point is connected to every other point either directly or through a chain of intermediary points. The length of the tree is the sum of the lengths of its segments so that the MST is the spanning tree of minimum length (see Fig. 4.14). Rohlf [1975] proposed fitting a quantile plot of the ordered d_i^2 to a scaled gamma distribution, as in Appendix C4. The MST has also been found useful in comparing two populations (Friedman and Rafsky [1979]).

A number of graphical methods based on the scaled differences $\mathbf{y}_i = \mathbf{S}^{-1/2}(\mathbf{x}_i - \bar{\mathbf{x}})$ are available. Under the null hypothesis H_0 of multivariate normality, the $\mathbf{z}_i = \boldsymbol{\Sigma}^{-1/2}(\mathbf{x}_i - \boldsymbol{\mu})$ are i.i.d. $N_d(\mathbf{0}, \mathbf{I}_d)$ and $\mathbf{z}_i'\mathbf{z}_i = (\mathbf{x}_i - \boldsymbol{\mu})'\boldsymbol{\Sigma}^{-1}(\mathbf{x}_i - \boldsymbol{\mu}) \sim \chi_d^2$ [Theorem 2.1(vi)]. Clearly the $\mathbf{y}_i$ will have similar properties and the Mahalanobis distances squared (Section 1.5),

$$D_i^2 = \mathbf{y}_i'\mathbf{y}_i = (\mathbf{x}_i - \bar{\mathbf{x}})'\mathbf{S}^{-1}(\mathbf{x}_i - \bar{\mathbf{x}}), \tag{4.25}$$

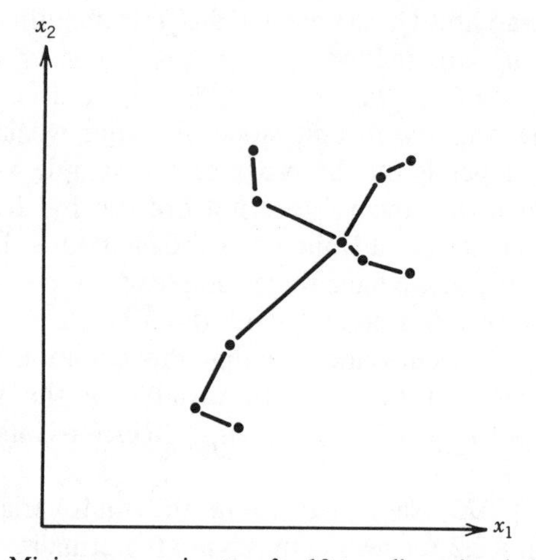

Fig. 4.14 Minimum spanning tree for 10 two-dimensional observations.

will be approximately i.i.d. χ_d^2 under H_0. We note that D_i^2 is unchanged if we work with the correlation matrix $\mathbf{R}$ instead of $\mathbf{S}$, as D_i^2 is invariant under linear transformations of the $\mathbf{x}_i$. Healy [1968] suggested checking for normality by plotting the D_i^2 [or normalized D_i^2 such as $D_i^{2/3}$; see (4.56)] against their expected order statistics. Although Andrews et al. [1973: p. 100] suggested that the above chi-square approximation will be satisfactory for moderate samples ($n > 25$ for the bivariate case), Small [1978] found that much larger values of n are required, particularly as d increases. Now, from Gnanadesikan and Kettenring [1972: p. 113], the exact marginal distribution of $u_i = nD_i^2/(n-1)^2$ is Beta $[a, b]$, where $a = \frac{1}{2}d$ and $b = \frac{1}{2}(n-d-1)$. For sufficiently large n, the correlations between the u_i can be ignored, and the u_i can be treated as independent observations from Beta $[a, b]$. If $u_{(1)} \le u_{(2)} \le \cdots \le u_{(d)}$ are the ranked u_i, $\alpha = \frac{1}{2}(a-1)/a$, and $\beta = \frac{1}{2}(b-1)/b$, then Small [1978] suggested plotting $u_{(i)}$ against u_i^*, where u_i^* is the solution of

$$\int_0^{u_i^*} \frac{\Gamma(\alpha)\Gamma(\beta)}{\Gamma(\alpha+\beta)} y^{\alpha-1}(1-y)^{\beta-1}\,dy = \frac{i-\alpha}{n-\alpha-\beta+1}.$$

An algorithm for doing this is given by Cran et al. [1977]. Any nonnormality in the original observations $\mathbf{x}_i$ will lead to nonlinearity in the plot. Although the largest estimated values u_i^* tend to be too high, Small stated that this is more than offset by the effect of correlation between the u_i that tends to inflate the large $u_{(i)}$ above what might be expected if the u_i were independent.

A different form of scaling can be used that leads to the more general distance measure $(\mathbf{x}_i - \boldsymbol{\mu})'\boldsymbol{\Gamma}(\mathbf{x}_i - \boldsymbol{\mu})$, where $\boldsymbol{\Gamma}$ is an arbitrary positive definite matrix. Arguing as in Appendix C2, we can write

$$(\mathbf{x}_i - \boldsymbol{\mu})'\boldsymbol{\Gamma}(\mathbf{x}_i - \boldsymbol{\mu}) = \sum_{j=1}^{d} \lambda_j w_{ij}^2, \tag{4.26}$$

where the w_{ij}^2 $(j = 1, 2, \ldots, d)$ are i.i.d. χ_1^2. Hence, under H_0, $(\mathbf{x}_i - \boldsymbol{\mu})'\boldsymbol{\Gamma}(\mathbf{x}_i - \boldsymbol{\mu})$ is a linear combination of independent chi-square variables that can be reasonably approximated by a scaled chi-square variable, that is, a gamma variable. Thus if $\boldsymbol{\mu}$ and $\boldsymbol{\Gamma}$ are estimated from the data, a gamma plot can be used for assessing normality. For example, Gnanadesikan and Kettenring [1972] considered the general class of distances $(\mathbf{x}_i - \bar{\mathbf{x}})'\mathbf{S}^b(\mathbf{x}_i - \bar{\mathbf{x}})$. For $b = 1$ we can write, as in (4.26),

$$(\mathbf{x}_i - \bar{\mathbf{x}})'\mathbf{S}(\mathbf{x}_i - \bar{\mathbf{x}}) = \sum_{j=1}^{d} c_j y_{ij}^2, \tag{4.27}$$

where the c_j $(c_1 > c_2 > \cdots > c_d)$ are the eigenvalues of $\mathbf{S}$ and $(y_{i1}, y_{i2}, \ldots, y_{id})$ are the principal components of $\mathbf{x}_i$ (see Section 5.2). In this case, the distance measure is useful for determining which observations have the greatest influence on the orientation and scale of the first few principal components. Similarly, if $b = -1$, the case considered above by Healy [1968],

$$(\mathbf{x}_i - \bar{\mathbf{x}})'\mathbf{S}^{-1}(\mathbf{x}_i - \bar{\mathbf{x}}) = \sum_{j=1}^{d} c_j^{-1} y_{ij}^2,$$

and the emphasis is now on the last few components. Setting $b = 0$, we have (Exercise 4.7)

$$(\mathbf{x}_i - \bar{\mathbf{x}})'(\mathbf{x}_i - \bar{\mathbf{x}}) = \frac{n-1}{n}(\text{tr}[\mathbf{Q}] - \text{tr}[\mathbf{Q}_{-i}]), \tag{4.28}$$

where $\mathbf{Q} = (n-1)\mathbf{S}$ and $\mathbf{Q}_{-i}$ is $\mathbf{Q}$, but with $\mathbf{x}_i$ deleted in the calculations. The measure (4.28) is useful for isolating observations that excessively inflate the overall scale.

EXAMPLE 4.6 We demonstrate the above gamma plot (with $b = -1$) using the height–weight data of 38 Peruvian Indians given in Table 3.1 (excluding the last observation) on p. 64. Applying the method outlined in Appendix C4, we first calculate

$$\begin{aligned} R &= \frac{\text{geometric mean of the } D_i^2}{\text{arithmetic mean of the } D_i^2} \\ &= 0.659486 \end{aligned}$$

and

$$(1 - R)^{-1} = 2.93674,$$

where D_i^2 is given by (4.25). Second, we interpolate linearly in Appendix D2 as follows. For $(1 - R)^{-1}$ equal to 2.9 and 3.0, we obtain the respective values of $\hat{\eta}_1 = 1.32430$ and $\hat{\eta}_2 = 1.37599$. Our required estimate of η is then

$$\begin{aligned} \hat{\eta} &= \hat{\eta}_1 + \frac{0.03674}{3.0 - 2.9}(\hat{\eta}_2 - \hat{\eta}_1) \\ &= 1.343. \end{aligned}$$

Third, we solve

$$p_i = \int_0^{x_i^*} \frac{1}{\Gamma(\hat{\eta})} y^{\hat{\eta}-1} e^{-y}\, dy$$

for x_i^*, where $p_i = (i - \frac{1}{2})/n$, for $i = 1, 2, \ldots, n$. Finally, if $x_{(1)} \le x_{(2)} \le \cdots \le x_{(n)}$ are the ordered D_i^2, we plot the pairs $(x_i^*, x_{(i)})$. This plot is given in Fig. 4.15 and is reasonably linear apart from the curious hump near the upper end. This is no doubt due to those taller Indians with lower than average weight that caused a similar hump in the normal probability plot for the heights [Fig. 4.13*a* in Section 4.3.1].

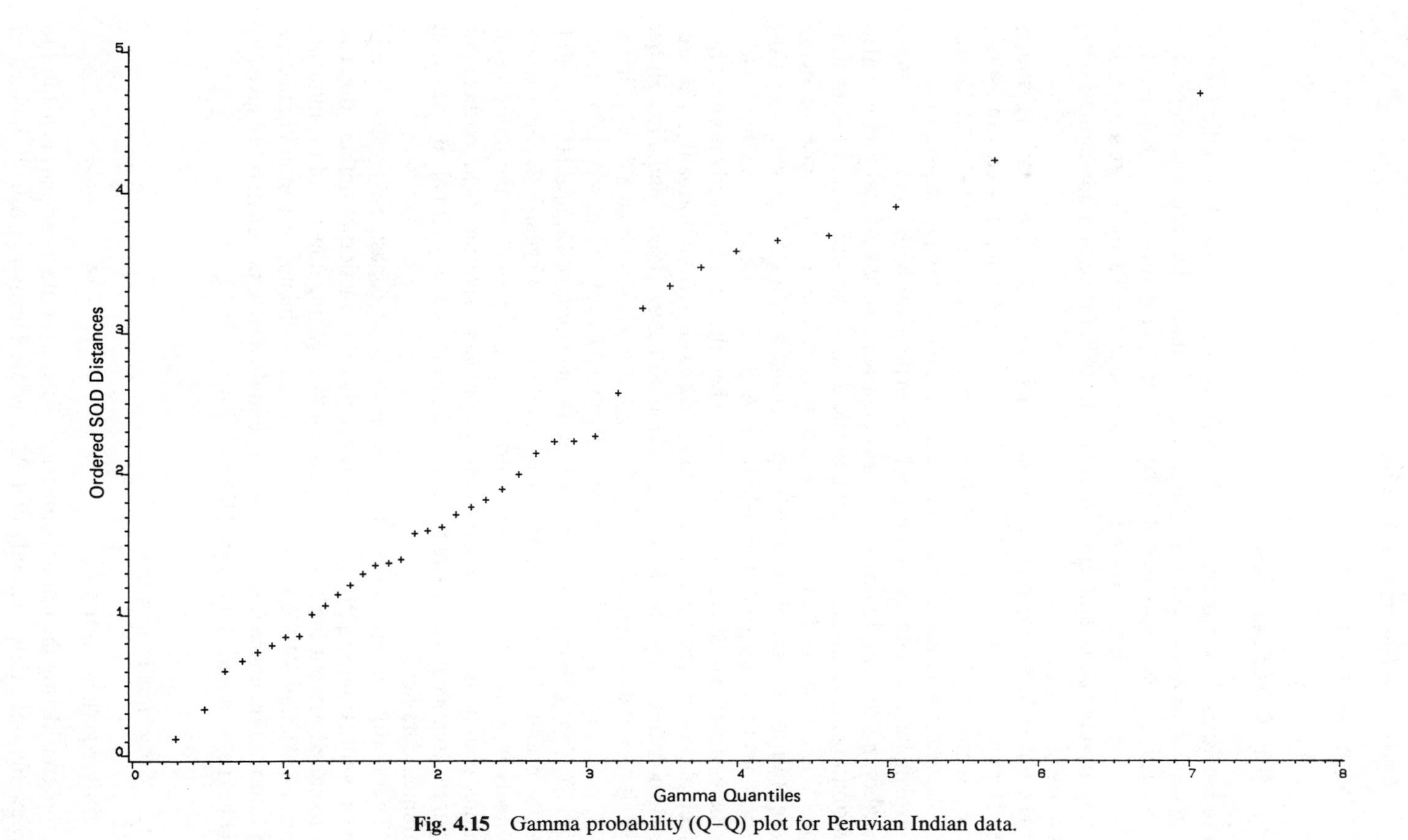

Fig. 4.15 Gamma probability (Q–Q) plot for Peruvian Indian data.

4.4 ROBUST ESTIMATION

4.4.1 Why Robust Estimates?

In recent years there has been a great flurry of activity related to finding robust estimates of location and scale for univariate data. The reader is referred to Huber [1972, 1981] and Bickel [1976] for technical reviews, to Barnett and Lewis [1978: Sections 2.6 and 4.2] and Hogg [1979a, b] for readable, less technical, summaries, and to Hampel [1973, 1978] for helpful nonmathematical overviews.

The purist faces two problems when analyzing data. First, various studies (see Hampel [1973: p. 88, 1978: p. 427]) suggest that practictioners can usually expect something like 0.1–10% of observations (or even more!) to be "dubious," that is, wrong measurements, wrong decimal points, wrongly copied, or simply unexplainable. Second, data are rarely normally distributed and tend to have distributions that are longer in the tails than normal. Occasionally short-tailed distributions arise through rounding of the data or, possibly unknown, truncation. What is needed is a robust estimate, robust in the sense of being insensitive to extreme observations and yet maintaining a high efficiency (low dispersion) for longer-tailed distributions. A good estimator is one that is close to the best, say, an efficiency of 90–95%, when the data is exactly normal, but which also offers good protection against departures from normality. This can be demonstrated by the following example (Tukey [1960], and also Huber [1981]). Given data from $N(\mu, \sigma^2)$, $d_n = \Sigma_i|x_i - \bar{x}|/n$ is about 89% as efficient as $s_n = [\Sigma_i(x_i - \bar{x})^2/n]^{1/2}$; however, if just 0.2% of the observations come from $N(\mu, 9\sigma^2)$, then d_n is more efficient. In the words of Hampel [1973: p. 91], the main aim of robust estimation is "building in safeguards against unsuspectedly large amounts of gross errors, putting a bound on the influence of hidden contamination and questionable outliers, isolating clear outliers for separate treatment (if desired), and still being nearly optimal at the strict parametric model."

Two useful concepts in the theoretical study of robustness are the breakdown point (Hampel [1971]), which is roughly the fraction of outliers that can be tolerated without the estimator potentially breaking down, and the influence function (Hampel [1974]), which quantifies the influence of contamination on an estimate. Influence functions for some multivariate parameters are given by Radhakrishnan and Kshirsagar [1981].

4.4.2 Estimation of Location

a UNIVARIATE METHODS

Before considering the robust estimation of multivariate location it is helpful to review the univariate methods first. Univariate location is usually defined in terms of the population mean or median, and a large number of estimators have been proposed. For example, some 65 of them were compared in a Princeton study under various, mainly symmetrical distributions, ranging from a mildly contaminated normal to the extreme case of the Cauchy distribution; short-tailed and skewed distributions were largely neglected (Andrews, et al. [1972a: see Section 7c for a quick summary]). Although there was not complete agreement among the authors, Huber [1972: pp. 1063–1064] gave a useful summary of the findings and some tentative recommendations. The choice of estimator will depend very much on the degree of nonnormality expected from the data. It is interesting to note from the study how badly the sample mean performed, even with the very mildest contamination. Also, the study supports the principle, not readily accepted by the statistical world, of routinely throwing away extreme observations. Whatever the reader's view of this principle, it is imperative that data should always be checked for outliers, even if one eventually uses the mean of the remaining data. This screening of the data will at least give some protection from extreme outliers and severe contamination.

Relles and Rogers [1977] carried out an unusual and thought-provoking study in which they compared the performance of several outlier-rejecting statisticians with some of the robust procedures described below. They concluded that the statistician is at least 20% less efficient than the best robust estimate and that statisticians should trim off one or two observations more than they think they should. Stigler [1977] applied a number of the methods to well-known sets of scientific data and demonstrated that other aspects of data, such as systematic bias, for example, must be considered in robustness studies. In the discussion following Stigler's paper, Professor D. R. Cox comments, "whether recent work is right to concentrate strongly on robustness to longtailed contamination rather than, for example, robustness to correlations among errors, is not so clear." The problem of unsuspected correlations is highlighted by Hampel [1978: p. 428]. Box [1979: p. 204] raises similar questions in the following comment: "It is currently fashionable to conduct robustness studies in which the normality assumption is relaxed (in favour of heavy tailed distributions or distributions containing outliers) but all other assumptions are retained. Thus it is still assumed that errors are independent, that transformations are correctly specified and so on. This seems to be too naive and narrow a view." These comments are supported by Wegman and Carroll [1977], who demonstrated that current robust estimators are somewhat inadequate for short-tailed distributions or positively correlated data.

We shall now consider four useful classes of robust estimators: the adaptive estimators, the L-estimators (linear function of order statistics), the M-estimators (analogues of maximum likelihood estimators), and the R-estimators (rank test estimators). Recently a new class has been suggested called P-estimators (Johns [1979]): These are analogues of Pitman estimators. The method of cross-validation has also been proposed for constructing robust estimators (Balasooriya and Chan [1981]). An adaptive estimator is one in which the method of estimation is allowed to depend partly on the data (see Hogg [1974] and Forst and Ali [1981] for references). If $x_{(1)} \leq x_{(2)} \leq \cdots \leq x_{(n)}$ are the order statistics for a sample of size n, then an L-estimator takes the form

$$\tilde{\mu} = \sum_{i=1}^{n} a_i x_{(i)}. \tag{4.29}$$

In this class are the sample mean and median, the α-trimmed mean, and the α-Winsorized mean. For the α-trimmed mean a prescribed proportion α of the lower $x_{(i)}$, and the same proportion of the upper $x_{(i)}$ are omitted; the remaining values are averaged. Since the proportion of observations to be trimmed might not be an integer, we set $\alpha n = r + f$, where r is an integer and $0 < f < 1$. Then our estimator is

$$\tilde{\mu} = \frac{\left[(1-f)x_{(r+1)} + x_{(r+2)} + \cdots + x_{(n-r-1)} + (1-f)x_{(n-r)}\right]}{n(1-2\alpha)}. \tag{4.30}$$

For the α-Winsorized mean we simply replace each of the upper r values $x_{(n-r+1)}, x_{(n-r+2)}, \ldots, x_{(n)}$ by $x_{(n-r)}$, and each of the lower r values $x_{(1)}, x_{(2)}, \ldots, x_{(r)}$ by $x_{(r+1)}$, so that our ordered sample is now $\mathbf{y}' = (x_{(r+1)}, \ldots, x_{(r+1)}, x_{(r+2)}, \ldots, x_{(n-r-1)}, x_{(n-r)}, \ldots, x_{(n-r)})$ and

$$\begin{aligned} \tilde{\mu} &= \bar{y} \\ &= \frac{1}{n}\left[r x_{(r+1)} + x_{(r+1)} + x_{(r+2)} + \cdots + x_{(n-r)} + r x_{(n-r)}\right] \end{aligned} \tag{4.31}$$

In general, trimmed means are preferred to Winsorized means. Asymptotic theory for both methods is given by Shorack [1974].

For normal data the α-trimmed mean has an efficiency of roughly $1 - \frac{2}{3}\alpha$, ranging from 1 for a zero-trimmed mean (arithmetic mean) to $\frac{2}{3}$ for a 0.5-trimmed mean (median). The situation is different for nonnormal data and we are faced with the problem of choosing α. A popular choice seems to be $\alpha = 0.1$, which leads to a 10% trim in each tail. Another possibility is to use an adaptive trimmed mean (see Hogg [1974] and the invited comments) in which α is determined from the data using an appropriate measure of nonnormality. Hogg [1967] originally used the kurtosis, but more recently (Hogg [1972]) proposed the measure

$$Q = \frac{\left[\overline{U}(.05) - \overline{L}(.05)\right]}{\left[\overline{U}(.5) - \overline{L}(.5)\right]},$$

where $\overline{U}(\alpha)$ is the average of the largest $n\alpha$th order statistics, with fractional items used if $n\alpha$ is not an integer [as in (4.30)], and $\overline{L}(\alpha)$ is the average of the smallest $n\alpha$th order statistics. Prescott [1978] examined the asymptotic properties of several adaptive trimmed means based on Q. D'Agostino and Lee [1977], and De Wet and Van Wyk [1979], investigated the robustness of several estimators to varying Q or kurtosis in the parent distribution. For highly skewed distributions, asymmetric trimming, with α_1 at one end and α_2 ($\neq \alpha_1$) at the other, may be more appropriate (Hogg [1974], Hertsgaard [1979]).

An M-estimator $\tilde{\mu}$ is obtained by solving an equation of the type

$$\sum_{i=1}^{n} \psi[(x_i - \tilde{\mu})/s] = 0, \tag{4.32}$$

where s is a robust estimator of the population scale parameter for x. Since there is a wide choice of functions ψ, a number of different estimators have been proposed. To give the rational behind the method it is convenient to ignore the scale parameter for the moment and assume that our sample comes from a distribution with probability density function $f(x - \mu)$: We follow Hogg's [1979a, b] discussions.

The logarithm of the likelihood function $L(\mu)$ is

$$\log L(\mu) = \sum_{i=1}^{n} \log f(x_i - \mu) = -\sum_{i=1}^{n} \rho(x_i - \mu),$$

where $\rho(x) = -\log f(x)$. Writing

$$\frac{d \log L(\mu)}{d\mu} = -\sum_{i=1}^{n} \frac{f'(x_i - \mu)}{f(x_i - \mu)} = \sum_{i=1}^{n} \psi(x_i - \mu),$$

where $\rho'(x) = \psi(x)$, the maximum likelihood estimate $\tilde{\mu}$ satisfies

$$\sum_{i=1}^{n} \psi(x_i - \mu) = 0. \tag{4.33}$$

Ignoring any constant in $\log L(\mu)$, we have the following three interesting cases:

(1) Normal: $\rho(x) = x^2/2$, $\psi(x) = x$, $\tilde{\mu} = \bar{x}$.
(2) Double exponential: $\rho(x) = |x|$, $\psi(x) = -1$ for $x < 0$, and $\psi(x) = 1$ for $x > 0$, $\tilde{\mu} =$ *sample median*.
(3) Cauchy: $\rho(x) = \log(1 + x^2)$, $\psi(x) = 2x/(1 + x^2)$, and (4.33) must be solved iteratively.

For theoretical reasons Huber [1964] proposed that we use, for our robust estimator μ, the maximum likelihood estimator for a density function that looks like a normal in the middle and a double exponential in the tails. Thus

$$\rho(x) = \begin{cases} \frac{1}{2}x^2, & |x| \leq a, \\ a|x| - \frac{1}{2}a^2, & |x| > a, \end{cases}$$

and

$$\psi(x) = \begin{cases} -a, & x < -a, \\ x, & |x| \le a, \\ a, & x > a, \end{cases} \tag{4.34}$$

where a is usually chosen to be close or equal to 1.5. With this ψ function, (4.33) must be solved iteratively and the solution is called an M-estimator, M for maximum likelihood. To obtain a scale-invariant version of this estimator we could solve (4.32) using a suitable s. We shall see later that the sample standard deviation is unsatisfactory, as it is not robust. Two useful ad hoc estimators are

$$s_1 = \frac{\text{median}|x_i - \text{median}(x_i)|}{0.6745} \tag{4.35}$$

or MAD (*m*edian of the *a*bsolute *d*eviations), and

$$s_2 = \frac{75\text{th percentile} - 25\text{th percentile}}{2(0.6745)}. \tag{4.36}$$

The divisor 0.6745 is used, as s_i is an estimator of the standard deviation σ when the data are normal. Equation (4.32) can be solved iteratively using, for example, the usual Newton–Raphson method:

$$\tilde{\mu}_{(j+1)} = \tilde{\mu}_{(j)} + s\left\{\sum_i \psi\left[\frac{x_i - \tilde{\mu}_{(j)}}{s}\right]\right\} \Big/ \sum_i \psi'\left[\frac{x_i - \tilde{\mu}_{(j)}}{s}\right], \tag{4.37}$$

where $\tilde{\mu}_{(1)}$ can be taken as the sample median, and the denominator of the second term is simply the number of observations that satisfy $|x_i - \tilde{\mu}_{(j)}|/s \le a$ (Hogg [1979b] also mentions two other methods). Several other ψ functions have been proposed in the literature, including Hampel's ψ, Andrews' sine (wave) function

$$\psi(x) = \begin{cases} \sin\left(\dfrac{x}{a}\right), & |x| \le a\pi, \\ 0, & |x| > a\pi, \end{cases} \tag{4.38}$$

Tukey's biweight function,

$$\psi(x) = \begin{cases} x\left[1 - \left(\dfrac{x}{a}\right)^2\right]^2, & |x| \le a, \\ 0, & |x| > a, \end{cases} \tag{4.39}$$

and, more recently, a promising modification of Hampel's ψ based on the tanh function (Hampel et al. [1981]). The ψ function gives greater weight to observations from the central part of the distribution and less weight, or even zero weight, to observations from the tails; extreme values are pulled toward the origin. The sine and biweight procedures are very similar and are reasonable substitutes for each other. As the ρ functions associated with these two redescending ψ functions are not convex, there could be certain convergence problems in the iterative procedures. Although this situation is not very likely, the procedures should be used with care. In contrast to trimming and Winsorizing, the above ψ functions provide a much smoother down-weighting of extreme values. Adaptive M-estimators can also be used (e.g., Moberg et al. [1978]), and asymmetry of the underlying distribution can also be taken into account (Collins [1976]).

We now briefly consider the class of R-estimators. These are adaptions of various nonparametric statistics for testing hypotheses about a location parameter. Perhaps the simplest is the median, which is also an L-estimator, and it forms the basis of the one-sample sign test for location. Another well-known R-estimator is the so-called Hodges–Lehmann estimator (Hodges and Lehmann [1963]), the median of the averages of all possible pairs of observations.

In spite of the overwhelming evidence, statisticians are still rather slow in using robust procedures along with standard techniques. Unfortunately, old habits die hard and one's activities can so easily be dictated by the available software. Another reason for the delay is perhaps the great number of procedures that are available and the differences of opinion which exist among the proponents of robust techniques. However, there is agreement that a robust method should be used and, as several of the methods work well, we should perhaps concentrate on just one or two methods. Hogg [1979a] suggested using Huber's Equation (4.34) with $a = 1.5$ followed by two or three iterations with the sine ($a = 1.5$) or biweight ($a = 5$) function. He proposed computing both the classical and robust estimates, and looking carefully at those points with low weights if the estimates differed markedly.

A third possible reason for the slow adaption of robust estimates is the absence of a corresponding robust theory for *small*-sample confidence intervals. Although most robust estimates of location are asymptotically normal, we require robust variance estimates of our location estimates (not to be confused with the robust estimation of the population variance). Some progress has been made in this direction using Studentized robust estimates (Shorack [1976], Gross [1976: 25 procedures], Barnett and Lewis [1978: Sections 4.2.4 and 4.3]). For example, the α-trimmed mean provides good protection against mild contamination and it can be Studentized using a Winsorized sample variance (Tukey and McLaughlin [1963], Huber [1970, 1972: pp. 1053–1054]) or the difference between the means of the upper and lower trimmed groups (Prescott [1975]). For M-estimation we can use a Taylor expansion of (4.32) and obtain (Hogg [1979a]; see Exercise 4.8)

$$\tilde{\sigma}_{\psi}^2 = s^2\left\{\frac{1}{n}\sum_{i=1}^{n}\psi^2\left[\frac{x_i - \tilde{\mu}}{s}\right]\right\} \Big/ \left\{\frac{1}{n}\sum_{i=1}^{n}\psi'\left[\frac{x_i - \tilde{\mu}}{s}\right]\right\}^2 \qquad (4.40)$$

as an estimate of the variance of $\sqrt{n}(\tilde{\mu} - \mu)$. Also, $(n-1)^{1/2}(\tilde{\mu} - \mu)/\tilde{\sigma}_\psi$ has approximately a t-distribution with $n - 1$ (or fewer; see Huber [1970]) degrees of freedom. A different method of interval estimation using an M-estimator and percentiles of a certain function is given by Boos [1980]. The effect of asymmetry on the variance estimates is investigated by Carroll [1979], and three estimates of variance are compared by Rocke and Downs [1981].

b MULTIVARIATE METHODS

Having discussed the univariate situation in some detail, we are now perhaps better able to appreciate the problems associated with multivariate robustness. Unfortunately, less progress has been made with the multivariate case and many of the estimators proposed are simply vectors of the corresponding univariate robust estimators. For example, there is $\mathbf{x}_M$, the vector of medians (Mood [1941]); $\mathbf{x}_{HL}$, the vector of Hodges–Lehmann estimators proposed by Bickel [1964]; and $\mathbf{x}_{T(\alpha)}$, the vector of α-trimmed means (Gnanadesikan and Kettenring [1972]). In comparing such vector estimates of location, one is faced with the problem of measuring efficiency. Most measures are based on the eigenvalues of the dispersion matrix of the estimate such as the determinant or trace of the matrix (A1.2). Using the determinant as a measure, Bickel [1964] showed that robustness can be related to the determinant of the population correlation matrix $\mathbf{P}_\rho = [(\rho_{ij})]$. For example, when the absolute value of $|\mathbf{P}_\rho|$ is large, $\mathbf{x}_M$ and $\mathbf{x}_{HL}$ share their corresponding univariate robust properties. However when $|\mathbf{P}_\rho| \approx 0$, the distribution of $\mathbf{x}$ is close to being degenerate (singular) and $\mathbf{x}_M$ and $\mathbf{x}_{HL}$ can have extremely poor efficiencies relative to $\bar{\mathbf{x}}$, even in the presence of heavy contamination. Bickel considers the case $d = 2$ in some detail: $|\mathbf{P}_\rho| \approx 0$ then corresponds to $|\rho| \approx 1$. Gnanadesikan and Kettenring [1972] also considered the case $d = 2$ and compared $\mathbf{x}_M$, $\mathbf{x}_{HL}$, $\mathbf{x}_{T(.1)}$ and $\mathbf{x}_{T(.25)}$ under the assumption of normality. From a limited simulation they concluded that $\mathbf{x}_M$ is the least efficient, $\mathbf{x}_{T(.25)}$ is a substantial improvement, and the best estimators $\mathbf{x}_{HL}$ and $\mathbf{x}_{T(.1)}$ are very similar, being almost as efficient as $\bar{\mathbf{x}}$. Clearly $\mathbf{x}_{T(.1)}$ is a convenient estimator to compute, simply requiring a 10% trim from each end for each variable. Further methods of estimation are considered in Section 4.4.3a below, where procedures for the simultaneous estimation of location and dispersion are also developed.

4.4.3 *Estimation of Dispersion and Covariance*

a UNIVARIATE METHODS

In the univariate case the robust estimation of dispersion is more difficult than it is for location, as there is a conflict between protecting the estimate from outliers and the need to use data in the tails for efficient estimation. Downton [1966] proposed the L-estimator

$$\tilde{\sigma} = \sqrt{\pi} \sum_{i=1}^{n} \frac{(2i - n - 1)x_{(i)}}{n(n-1)} \tag{4.41}$$

of the standard deviation σ of the normal distribution. Healy [1978] proposed a similar estimator based on symmetrically trimmed data. If k [$= (1 - 2\beta)n$] observations, relabeled $y_{(1)} \leq y_{(2)} \leq \cdots \leq y_{(k)}$ ($y_{(i)} = x_{(\beta n + i)}$), are left after trimming βn observations from each end, his estimator of σ is $d(n, k)$, where

$$d(n, k) = b(n, k) \sum_{i=1}^{k} \frac{(2i - k - 1) y_{(i)}}{k(k - \frac{1}{2})}, \tag{4.42}$$

and $b(n, k)$ is the correction for bias. If $0 < \beta < 0.25$ and $n > 10$, $b(n, k)$ is less than 3% below a limiting value b_p that depends only on $p = k/n$, the proportion of the sample left. Healy [1978] tabulated b_p for $p = 0(0.02)0.98$, and his table is reproduced as Table 4.4. He proposed the estimate (4.42) but with $b(n, k)$ replaced by b_p. Prescott [1979] extended the method to the case of asymmetric trimming.

The sensitivity of the usual unbiased estimator, $s^2 = \Sigma_i (x_i - \bar{x})^2/(n - 1)$, to outliers is well known. If the x_i are normal, it requires only a very small amount of contamination to drastically reduce the efficiency of s^2: This is indicated by the example given in Section 4.4.1 and by Table 3 of Healy [1978]. The so-called breakdown point of s^2 is zero, so that just a single outlier can cause s^2 to break down as an estimator of σ^2. To get around this problem we can trim or Winsorize the data (Tukey [1960], Johnson and Leone [1964], Huber [1970]). For example, a β-trimmed estimate of σ^2 is given by

$$\tilde{\sigma}^2 = c(n, k) \sum_{i=1}^{k} \frac{(y_{(i)} - \tilde{\mu})^2}{k - 1}, \tag{4.43}$$

where $\tilde{\mu}$ is a robust estimate of μ such as a trimmed mean, and $c(n, k)$ is a correction for bias. Because of the efficiency loss, it is advisable to use a smaller

TABLE 4.4 The Unbiasing Factor b_p, Where p is the Proportion of the Sample Uncensored[a]

p	0	0·02	0·04	0·06	0·08
0·00	—	119·657	59·826	39·872	29·891
0·10	23·899	19·902	17·045	14·900	13·230
0·20	11·892	10·797	9·882	9·107	8·441
0·30	7·864	7·357	6·909	6·509	6·151
0·40	5·828	5·534	5·266	5·021	4·795
0·50	4·587	4·393	4·213	4·045	3·888
0·60	3·741	3·602	3·470	3·346	3·228
0·70	3·116	3·009	2·907	2·809	2·714
0·80	2·624	2·536	2·450	2·367	2·286
0·90	2·206	2·126	2·046	1·964	1·878

[a] From Healy [1978: Table 1], by permission of the Biometrika Trustees.

proportion β of trimming or Winsorizing for variance estimation than for location estimation. Correction factors like $c(n, k)$ are based on the moments of order statistics of the normal distribution and the assumption that the middle of the sample is sufficiently normal. Johnson and Leone [1964: p. 173] give a table for $n \le 15$.

The estimate MAD of (4.35) is a useful one and, although only 40% efficient for normal data, it is robust to outliers and long-tailed distributions. Hampel [1974] describes it as a "crude but simple and safe scale estimate" that could be used "(i) as a rough but fast scale estimate in cases where no higher accuracy is required; (ii) as a check for more refined computations; (iii) as a basis for the rejection of outliers; and (iv) as a starting point for iterative (and one-step) procedures, especially for many other robust estimators."

A robust estimator of covariance can be obtained from robust variance estimators via the well-known identity

$$\operatorname{cov}[U_1, U_2] = \tfrac{1}{4}\{\operatorname{var}[U_1 + U_2] - \operatorname{var}[U_1 - U_2]\}, \tag{4.44}$$

and this can lead to a robust estimator of ρ_{12}, the correlation between U_1 and U_2. However, such an estimate may not lie in the range $[-1, 1]$ and alternative estimators for ρ_{12} that lie in this range have been suggested by Devlin et al. [1975] using the result

$$\frac{\operatorname{var}[U_1/\sigma_1 + U_2/\sigma_2] - \operatorname{var}[U_1/\sigma_1 - U_2/\sigma_2]}{\operatorname{var}[U_1/\sigma_1 + U_2/\sigma_2] + \operatorname{var}[U_1/\sigma_1 - U_2/\sigma_2]} = \frac{2 + 2\rho_{12} - (2 - 2\rho_{12})}{2 + 2\rho_{12} + (2 - 2\rho_{12})}$$

$$= \rho_{12},$$

where σ_i is the standard deviation of U_i. Therefore, given the sample pairs (x_{i1}, x_{i2}), $i = 1, 2, \ldots, n$ for random variables (X_1, X_2), we proceed as follows: (1) make a robust standardization of the x_{ij} to $z_{ij} = (x_{ij} - \tilde{\mu}_j)/\tilde{\sigma}_j$ $(i = 1, 2, \ldots, n;\ j = 1, 2)$, where $\tilde{\mu}_j$ and $\tilde{\sigma}_j^2$ are robust estimates of the mean and variance of X_j; (2) form the sums $w_{i1} = z_{i1} + z_{i2}$ and $w_{i2} = z_{i1} - z_{i2}$; and (3) calculate robust variance estimates $\tilde{v}_+$ and $\tilde{v}_-$ for the $\{w_{i1}\}$ and $\{w_{i2}\}$, respectively, and obtain

$$\tilde{r} = \frac{\tilde{v}_+ - \tilde{v}_-}{\tilde{v}_+ + \tilde{v}_-} \tag{4.45}$$

as a robust estimate of correlation between X_1 and X_2. Trimmed (or Winsorized) variances can be used, as the scale factor for bias cancels out.

Another method of estimating the correlation is to use bivariate trimming in which observations "outside" the general body of data are trimmed without affecting the general shape of the scatter plot. Two such methods are convex hull trimming (Bebbington [1978]) and ellipsoidal trimming (Titterington [1978]). The use of convex hulls in looking for outliers is described by Barnett [1976a] under the title of partial or *P*-ordering.

In conclusion, we mention one other procedure that involves the simultaneous estimation of μ and σ using a pair of equations like (4.32) (Huber [1964]). With suitable starting values for μ and σ, these equations are solved iteratively. The method seems particularly useful in the multivariate case, as we shall see later.

b MULTIVARIATE METHODS

A robust estimator of the correlation matrix $\mathbf{P}_\rho$ can be obtained by robustly estimating each off-diagonal element using (4.45) for each pair of variables. Devlin et al. [1981] used 5% trimmed means and variances ($\alpha = \beta = 0.05$) and called their estimator $\mathbf{R}^*$(SSD), being based on standardized variables and variances of their sums and differences. However, $\mathbf{R}^*$(SSD) may not be positive definite, and a method of "shrinking" it to achieve positive definiteness (i.e., its eigenvalues positive) is given by Devlin et al. [1975: Equation 10 with $\Delta = 0.25/\sqrt{n}$]. With the Z-transformation (3.64), each element r_{jk}^* is replaced by

$$g\left(r_{jk}^*\right) = \begin{cases} Z^{-1}\left\{Z\left(r_{jk}^*\right) + \Delta\right)\} & \text{if } r_{jk}^* < -Z^{-1}(\Delta), \\ 0 & \text{if } |r_{jk}^*| \le Z^{-1}(\Delta), \\ Z^{-1}\left\{Z\left(r_{jk}^*\right) - \Delta\right)\} & \text{if } r_{jk}^* > Z^{-1}(\Delta). \end{cases} \tag{4.46}$$

The effect of the transformation is to leave large correlations virtually unchanged and to shrink small correlations toward zero. Unfortunately, shrinking tends to introduce bias, particularly with the intermediate correlations. The shrunken estimator is denoted by $\mathbf{R}_+^*$(SSD), and this can be converted to a dispersion estimator, $\mathbf{S}_+^*$(SSD) say, of $\boldsymbol{\Sigma}$ by "rescaling" the correlation matrix using a diagonal matrix of robust (e.g., trimmed) standard deviations, as in (1.16).

An alternative procedure is to use multivariate trimming (MVT) (Gnanadesikan and Kettenring [1972], Gnanadesikan [1977], Devlin et al. [1981]). The process is iterative and requires, at each step, the squared Mahalanobis distances D_i^2, where

$$D_i = \left\{\left(\mathbf{x}_i - \boldsymbol{\mu}^*\right)'\left(\mathbf{S}^*\right)^{-1}\left(\mathbf{x}_i - \boldsymbol{\mu}^*\right)\right\}^{1/2}, \qquad i = 1, 2, \ldots, n, \tag{4.47}$$

where $\boldsymbol{\mu}^*$ and $\mathbf{S}^*$ are the current estimators of $\boldsymbol{\mu}$ and $\boldsymbol{\Sigma}$. A fixed proportion γ ($= 0.1$, say) of the $\mathbf{x}_i$ with the largest D_i are temporarily set aside and $\boldsymbol{\mu}^*$ and $\mathbf{S}^*$ are recomputed using the formulas for $\bar{\mathbf{x}}$ and $\mathbf{S}$, but based on $n(1-\gamma)$ observations. Devlin et al. [1981] suggested starting the process using $\bar{\mathbf{x}}$ and $\mathbf{S}$ in (4.4), and stopping after 25 iterations or when each $Z(r_{jk}^*)$ [see (3.64)] does not change by more than 10^{-3} between successive iterations. The final estimate is denoted by $\mathbf{S}^*$(MVT), which can be converted to $\mathbf{R}^*$(MVT) by scaling with the square roots of the diagonal elements as in (1.16). Unfortunately, as in the univariate case of (4.43), a correction factor for bias is needed so that $\mathbf{S}^*(\text{MVT})$ is an asymptotically unbiased estimate of $\boldsymbol{\Sigma}$ under normality; this correction is not available at present (Gnanadesikan [1977: p. 134]). However, the factor cancels out when forming $\mathbf{R}^*(\text{MVT})$. In order for these estimators to be positive definite (with probability 1), we must have $n(1-\gamma)-1 \geq d$.

Instead of deleting those observations with large values of D_i, we can down-weight them using weights that depend on D_i. This is the basis of the multivariate *M*-estimators proposed by Maronna [1976], Huber [1977a, b], Carroll [1978], Campbell [1980a], and Devlin et al. [1981]. The appropriate equations, which are solved iteratively as in the above MVT method, are

$$\boldsymbol{\mu}^* = \sum_{i=1}^{n} w_1(D_i)\mathbf{x}_i \Big/ \sum_{i=1}^{n} w_1(D_i) \tag{4.48}$$

and

$$\mathbf{S}^* = \frac{1}{n}\sum_{i=1}^{n} w_2(D_i^2)(\mathbf{x}_i - \boldsymbol{\mu}^*)(\mathbf{x}_i - \boldsymbol{\mu}^*)', \tag{4.49}$$

where w_1 and w_2 are appropriate weight functions and D_i is given by (4.47). Under fairly general constraints on w_1 and w_2, Maronna [1976] proved results on the existence and uniqueness of the solutions of (4.48) and (4.49). For elliptically symmetric distributions, namely, those with density function $f(\mathbf{x}) = |\boldsymbol{\Sigma}|^{-1/2}h\{(\mathbf{x}-\boldsymbol{\mu})'\boldsymbol{\Sigma}^{-1}(\mathbf{x}-\boldsymbol{\mu})\}$, he established consistency and asymptotic normality of the solutions. This family of densities is obtained from the family of spherically symmetric (radial) densities of the form $f(\mathbf{x}) = h(\|\mathbf{x}\|)$ by a nonsingular transformation of $\mathbf{x}$, and includes distributions with shorter or longer tails than the MVN. Maronna [1976: p. 53] proposed two types of weight functions for w_1 and w_2. The first results in the maximum likelihood estimators for a d-variate radial t-distribution with f degrees of freedom, and

$$w_1(D) = \frac{d+f}{f+D^2} = w_2(D^2).$$

Devlin et al. [1981] consider only the Cauchy case of $f = 1$ and call the corresponding estimate of $\boldsymbol{\Sigma}$ the maximum likelihood t-estimator, $\mathbf{S}^*(\text{MLT})$. The corresponding correlation matrix, obtained by applying (1.16), is denoted by $\mathbf{R}^*(\text{MLT})$. The second method is based on the rational underlying the use of ψ functions like (4.34) in which observations in the "central part" of the data set (i.e., with small D_i values) are given full weight, while extreme observations (with large D_i values) are down-weighted. The weight functions are

$$w_1(D) = \begin{cases} 1, & D \leq a, \\ \dfrac{a}{D}, & D > a, \end{cases} \tag{4.50}$$

and

$$w_2(D^2) = \{w_1(D)\}^2/c,$$

where c is a "correction" factor for asymptotic unbiasedness under normality; unfortunately c is unknown. As D_i^2 is asymptotically χ_d^2 under normality, Devlin et al. [1981] choose a as the square root of the 90% quantile of χ_d^2 and call the resulting estimator **S***(HUB), as it is based on Huber's ψ function (4.34), that is, $w_1(D) = \psi(D)/D$.

Unfortunately the above M-estimators have a low breakdown point (the proportion of outliers tolerated), namely, no greater than $1/(d+1)$ (Maronna [1976]) or $1/d$ (in the more general framework of Huber [1977b]). This result refers to outliers whose distribution is not far from the assumed parent elliptical symmetry. Devlin et al. [1981] give some empirical evidence that these bounds should be even lower with symmetrical outliers (see Table 4.5). In contrast, the breakdown point for the MVT method is the same as the trimming proportion γ and therefore does not fall off as d increases.

Devlin et al. [1981] compared, using simulation, the estimators **R***(HUB), **R***(SSD), **R***(MVT), **R***(MLT), and **R***(REG), a regression estimator due to Mosteller and Tukey [1977: Chapter 10], and came to the following conclusions. The regression method requires substantial computation and lacks the affine invariance property shared by the others (SSD in special cases only) and **S**; namely, if $\mathbf{y} = \mathbf{Ax}$, then $\mathbf{S}_y = \mathbf{ASA}'$. The MVT, MLT, and HUB are all useful methods, though if extreme outliers are present, robust starting values of $\boldsymbol{\mu}^*$ and $\mathbf{S}^*$ instead of $\bar{\mathbf{x}}$ and $\mathbf{S}$ should be used for the iterative process. The effect of a robust start on the breakdown point is clearly demonstrated in Table 4.5. The MLT method seems the most stable under various conditions, unless d is quite large; MVT is then preferred, having better breakdown

TABLE 4.5 Percentage of Contamination Tolerated[a]

$d = 20$	Type of Contamination	
Estimator	Symmetric	Asymmetric
R	0	0
$\mathbf{R}_+^*$(SSD)	5	5
R*(MVT)	10	4
R*(MVT) with robust start	10	10
R*(HUB)	10	1
R*(HUB) with robust start	≥ 25	1
R*(MLT)	15	2
R*(MLT) with robust start	≥ 25	2

[a]From Devlin et al. [1981: Table 3].
[b]Values tested: 0(1)5, 10(5)25.

properties. If eigenvalues and eigenvectors are extracted, as in principal components, MLT again seems to come out in front. When there are a large number of missing values, $\mathbf{R}_+^*$(SSD) can be useful because of its parsimonious use of the available data; r_{jk}^* is based on the number of complete data pairs for the jth and kth variables only. Unfortunately, the shrinking process of (4.46) brings $\mathbf{R}_+^*$(SSD) closer to $\mathbf{I}_d$ so that estimators of eigenvalues will be biased. This method is therefore not recommended for robust principal component analysis.

Campbell [1980a] proposed using a "redescending" ψ function in $w_1(D) = \psi(D)/D$, namely,

$$w_1(D) = \begin{cases} 1, & D \leq a, \\ \dfrac{a}{D}\exp\left\{\dfrac{-\frac{1}{2}(D-a)^2}{c_2^2}\right\}, & D > a, \end{cases} \tag{4.51}$$

along with

$$w_2(D) = \{w_1(D)\}^2 \qquad \left(= w_1^2(D), \text{ say}\right).$$

Here

$$a = \sqrt{d} + \frac{c_1}{\sqrt{2}}, \tag{4.52}$$

where c_1 and c_2 are chosen by the experimenter. The rationale behind (4.52) is that, under normality, Fisher's square root transformation for the chi-square distribution gives D_i as approximately $N_1(\sqrt{d}, 1/\sqrt{2})$; c_1 is an approximate quantile of $N_1(0,1)$. When $c_1 = \infty$, we arrive at the usual estimators $\overline{\mathbf{x}}$ and $\mathbf{S}$; when $c_2 = \infty$, (4.51) reduces to (4.50). From "extensive practical experience" Campbell suggested $c_1 = 2$ and $c_2 = 1.25$ as suitable values. His equation for $\boldsymbol{\mu}^*$ is (4.48), but his equation for $\mathbf{S}^*$ is

$$\mathbf{S}^* = \sum_{i=1}^{n} w_1^2(D_i)(\mathbf{x}_i - \boldsymbol{\mu}^*)(\mathbf{x}_i - \boldsymbol{\mu}^*)' \Big/ \left[\sum_{i=1}^{n} w_1^2(D_i) - 1\right]. \tag{4.53}$$

From simulation experiments Campbell demonstrated that the robust variance estimates obtained from (4.53) are reasonably efficient under normality in that they are generally within 2–3% of the corresponding elements of $\mathbf{S}$ (which are most efficient under normality). Further studies by Campbell (personal communication) indicate that the values of constants like c_1 and c_2 are critical. With a suitable choice of influence function and associated constants, the estimates are approximately unbiased and, unlike $\mathbf{S}^*$(MVT), correction factors for bias are unnecessary.

4.5 OUTLYING OBSERVATIONS

The concept of an outlier or a "dubious" observation has tantalized statisticians for many years. The key question is, Should we discard one or more observations from a set of data simply because they appear to be "inconsistent" with the rest of the set? We have seen in Section 4.4 that current thinking on the closely related topic of robust estimation supports some form of automatic truncation or modification of the data. However, a closer look at an extreme observation is often warranted, as it may shed light on underlying structures or tell us something about the recording of the data. For example, an observation may deviate sharply from a fitted hypothesized model such as a regression, analysis of variance, or time series model, because the model breaks down at that particular point and not because the observation is spurious. On the other hand, we have already noted in our discussion on robustness that wrong measurements or wrongly recorded data at either the experimental or computational stage are much more common than is generally appreciated, a few percent of wrong values tending to be the norm. As Barnett and Lewis [1978] and Hawkins [1980] give extensive, readable surveys of the vast literature on outliers, we shall simply highlight one or two basic ideas and outline some procedures for multivariate data. Gnanadesikan and Kettenring [1972] give a helpful review of multivariate techniques and this material is reproduced in detail by Gnanadesikan [1977].

To aid our search for outliers we need graphical techniques to provide possible candidates for further investigation, and suitable follow-up tests of "discordancy" of these observations with the rest of the data. With univariate data the observations are readily ranked so that the largest and/or smallest observations come up for scrutiny. When there are several possible outliers, we can test the extreme observations one at a time (consecutive procedures) or as a group (block procedures). The consecutive method, however, may suffer from a masking effect in which the less extreme outliers mask the discordancy of the most extreme observations under investigation.

When we turn to multivariate data, the situation is even more complicated, for a number of reasons. First, for more than two dimensions, there is the problem of presenting the data graphically so as to highlight the presence of any outliers. Second, there is the problem of ordering the data so that we can isolate the extremes or observations that separate themselves from the bulk of the data (see Barnett [1976a]). Third, a multivariate outlier can distort the measures of orientation (correlation) as well as the measures of location and scale. Referring to the bivariate data in Fig. 4.16, observation A will inflate both variances, but will have little effect on the correlation; observation B will reduce the correlation and inflate the variance of x_1, but will have little effect on the variance of x_2; and observation C has little effect on variances but reduces the correlation. From another viewpoint, B and C add what is apparently an insignificant second dimension to data that are essentially one dimensional, lying almost in a straight line. A fourth problem with multivariate outliers is that an outlier can arise because of either a gross error in one of its components or small systematic errors in severable components. Clearly the situation is complex and, as emphasized by Gnanadesikan and Kettenring [1972: p. 109], there is not much point in looking for an omnibus outlier protection procedure. We need an arsenal of methods each designed for a specific purpose.

As a first step in detecting outliers we can look at the d univariate marginal distributions and apply the univariate techniques described by Barnett and Lewis [1978: Chapter 3]. Discordancy tests are provided there for data from the gamma, exponential, truncated exponential, normal, Pareto, Gumbel, Fréchet, Weibull, lognormal, uniform, Poisson, and binomial (but not the beta) distributions. The coefficient of kurtosis, b_2 of (4.12), is a useful statistic for detecting outliers among normally distributed data. If b_2 is significant, the most extreme observation is removed and b_2 retested on the remaining observations. For nonnormal data with a known distribution, a normalizing transformation can sometimes be used as, for example, in the lognormal, Poisson, binomial, and gamma distributions. However, a major weakness of this one-dimensional approach is that outliers like C in Fig. 4.16, which mainly affect the correlation, may not be detected.

In the case of bivariate data, the sample correlation r is very sensitive to outliers and can therefore be used for their detection. For example, when the data are bivariate normal, Gnanadesikan and Kettenring [1972] suggest a normal probability plot of the values

$$Z(r_{-i}) = \tfrac{1}{2}\log\left[\frac{1 + r_{-i}}{1 - r_{-i}}\right], \tag{4.54}$$

where r_{-i} is the sample correlation between the two variables based on the data with $\mathbf{x}_i$ omitted. Devlin et al. [1975] capitalize on the useful concept of the influence curve (Hampel [1974]) and present two graphical methods based on the sample influence function of r. One of their methods leads to the function $(n-1)[Z(r) - Z(r_{-i})]$, which, for a reasonably large sample of normal bivariate data, is approximately distributed as the product of two independent

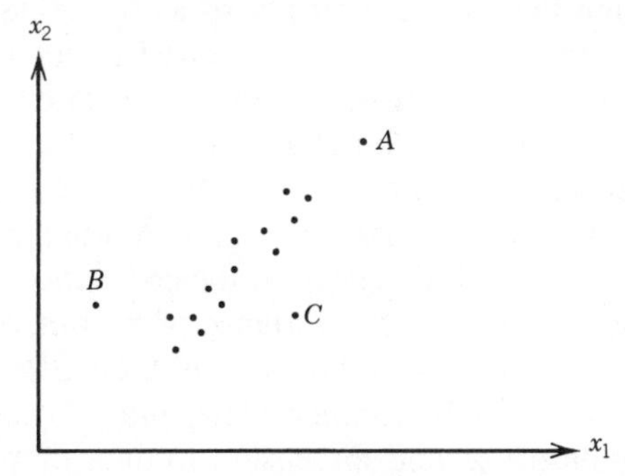

Fig. 4.16 Scatter plot for bivariate data showing three possible outliers A, B, and C.

$N(0,1)$ variables. They propose a further normalizing transformation prior to probability plotting.

For higher-dimensional data we can examine the $\frac{1}{2}d(d-1)$ scatter plots for all the bivariate marginals, provided that d is not too large, and then apply the above bivariate methods for outlier detection. However, although marginal distributions will detect a gross error in one of the variables, they may not show up an outlier with small systematic errors in all its variables. Also, we still have the problem of outliers like B and C in Fig. 4.16, which add spurious dimensions or obscure singularities. Apart from B and C, the data lie almost in a one-dimensional space, the regression line. Going up a dimension we can envisage a set of points that all lie close to a plane except for one or two outliers at some distance from the plane. Generally such outliers can be uncovered by working with robust principal components of the observations (Section 5.2) instead of with the observations: One robust procedure, based on (4.51), is given by Campbell [1980a]. Gnanadesikan and Kettenring [1972: p. 111] note that the first few principal components are sensitive to outliers that inflate variances and covariances (if working with $\hat{\boldsymbol{\Sigma}}$ or $\mathbf{S}$) or correlations (if working with the sample correlation matrix $\mathbf{R}$), and the last few are sensitive to outliers which add spurious dimensions or obscure singularities. If the ith row $\mathbf{y}_i'$ of $\mathbf{Y} = (\mathbf{y}_1, \mathbf{y}_2, \ldots, \mathbf{y}_n)'$ represents the d sample principal components of the observation $\mathbf{x}_i$, then the first and last few principal components are represented by the first and last few columns of $\mathbf{Y}$. In particular, we could look at a scatter plot for the first two columns and use the above bivariate techniques for detecting outliers. One advantage of using principal components is that they are likely to be more normal than the original data, particularly when d is reasonably large, approximate normality being achieved by central limit theorem arguments applied to linear combinations. Another approach is to use one of the hierarchical clustering methods of Section 7.3 such as complete linkage for visually highlighting any cluster(s) of outliers. The multivariate kurtosis $b_{2,d}$ of (4.19) can also be used as an outlier test (Schwager and Margolin [1982]).

We note that all the plotting techniques for assessing multivariate normality in Section 4.3.2 can be used for detecting outliers. In the case of normal or gamma plots, any extreme points could be tested using one of the discordancy tests of Barnett and Lewis [1978: Chapter 3]. Unfortunately all their gamma tests but one assume that the scale parameter λ is unknown, but the shape parameter η is known (see Appendix C2 for definitions). However, the gamma values can be "normalized" using Fisher's transformation $y = x^{1/2}$ or Wilson and Hilferty's transformation $y = x^{1/3}$. In particular, for η not too small,

$$x^{1/2} \text{ is approximately } N_1\left(\left[\lambda\left(\eta - \tfrac{1}{4}\right)\right]^{1/2}, \tfrac{1}{4}\lambda\right) \tag{4.55}$$

and

$$x^{1/3} \text{ is approximately } N_1\left(\left[\lambda\eta\right]^{1/3}\left\{1 - (9\eta)^{-1}\right\}, \frac{1}{9}\left[\frac{\lambda^2}{\eta}\right]^{1/3}\right). \tag{4.56}$$

With either transformation we can apply to the y_i a discordancy test for a sample from a normal distribution with unknown mean and variance: 17 such tests are listed by Barnett and Lewis [1978: pp. 91–92]. They proposed using the transformation $x^{1/2}$, which has the useful property that a change in the shape parameter is reflected by a change in just the normal mean. Such a change can be tested by a so-called "location slippage" test. However, $y = x^{1/3}$ gives a much better normal approximation and Kimber [1979] uses it as a basis for two tests. He selects the two discordancy tests for normal data

$$T = \max\left\{\frac{\bar{y} - y_{(1)}}{s}, \frac{y_{(n)} - \bar{y}}{s}\right\}$$

and

$$b_2 = n\sum_{i=1}^{n}(y_i - \bar{y})^4 \bigg/ \left\{\sum_{i=1}^{n}(y_i - \bar{y})^2\right\}^2,$$

(called K in his paper), where $s^2 = \Sigma_i(y_i - \bar{y})^2/(n-1)$. Kimber also adds a third statistic,

$$Z = \max_{1 \le i \le n}\left(v_i \bigg/ \sum_{\alpha=1}^{n} v_\alpha\right), \tag{4.57}$$

where

$$v_i = -\log u_i - (n-1)\log\left\{\frac{n - u_i}{n-1}\right\}, \qquad u_i = x_i/\bar{x},$$

and the x_i have a gamma distribution. Here Z gives a scale-invariant discordancy test whose null distribution is essentially independent of η, provided that η is not too small (say, $\eta > 4.5$). Large values of T, b_2, and Z signify the presence of a very large or a very small observation in the sample, and significance points for b_2 and Z are given in Appendixes D5 and D6, and Appendix D11, respectively. The critical values for Z are apparently conservative when $\eta < 4.5$, the effect becoming more apparent as η gets smaller. For example, when $n = 20$, the case $\eta = 0.5$ gives a true significance of about 3.7% at the nominal 5%.

Kimber [1979] compared the three tests and came to the following conclusion. None of the three tests are entirely satisfactory over the wide range of situations investigated. However, if the sample size is not too small ($n > 15$, say), Z is the best of the three tests if at most one outlier is involved, while if more than one outlier is likely, b_2 may be used in a consecutive procedure. In the latter case, if the test is significant, we remove the outlier farthest from $\bar{y}$ and repeat the test with $n - 1$ observations. Unfortunately, all the tests have marked weaknesses for small samples, and the use of both Z and b_2 is recommended when $n \leq 15$. There seems to be no great advantage in using T. Although Z appears to be safe to use whatever the value of η, b_2 should be treated with caution if $\eta \leq 1$ is suspected. The transformation $x^{1/3}$ breaks down in this range. The maximum likelihood estimate of η [see C2.4] will throw some light on this question.

If we use Healy's [1968] plot of the D_i^2 [see (4.25)], we can base a formal test for a single outlier on $D_{(n)}^2 = \max_{1 \leq i \leq n} D_i^2$. It transpires (Barnett and Lewis [1978: pp. 210–219]) that if we carry out the so-called "slippage" tests of the mean and dispersion matrix (multivariate versions of models A and B of Ferguson [1961]) using the method of maximum likelihood, both tests lead to $D_{(n)}^2$. This statistic is also equivalent to a statistic proposed by Wilks [1963] (Exercise 4.9), namely,

$$r_{(1)} = 1 - \frac{nD_{(n)}^2}{(n-1)^2}. \tag{4.58}$$

A table of approximate (conservative) significance levels for $D_{(n)}^2$, derived from Wilks' table for $r_{(1)}$ by Barnett and Lewis [1978], is given in Appendix D12. Wilks also gave a test for the presence of two outliers.

An outlier will tend to inflate $\bar{\mathbf{x}}$ and $\mathbf{S}$, and possibly reduce D_i^2, so that it is a good idea to use the robust version of D_i^2, namely, $(\mathbf{x}_i - \boldsymbol{\mu}^*)'(\mathbf{S}^*)^{-1}(\mathbf{x}_i - \boldsymbol{\mu}^*)$, where $\boldsymbol{\mu}^*$ and $\mathbf{S}^*$ are robust estimators of $\boldsymbol{\mu}$ and $\boldsymbol{\Sigma}$. Campbell [1980a] used the procedure given by (4.51) and (4.52) to obtain $\boldsymbol{\mu}^*$ and $\mathbf{S}^*$ (with $c_1 = 2.0$, $c_2 = 1.25$), and recommended the normal probability plot of the "robustified" $D_i^{2/3}$ for detecting outliers. He gave an example, which is typical of robust methods, in which the outliers stood out more clearly with the robust D_i than with the usual D_i. The weight $w_1(D_i)$ will also give some indication of a typical observation and Campbell suggests that a weight less than 0.3 ($c_1 = 2.0$, $c_2 = 1.25$) or 0.6 ($c_1 = 2.0$, $c_2 = \infty$) indicates an outlier.

EXERCISES 4

4.1 Verify Equation (4.1).

4.2 If $L_{\max}(\xi, \lambda)$ is (4.7) with x_i replaced by $x_i + \xi$, verify that $L_{\max}(\xi, 1)$ does not depend on ξ.

4.3 Prove that the coefficients $b_{1,d}$ and $b_{2,d}$ of Equation (4.19) are invariant under the transformation $\mathbf{y} = \mathbf{A}\mathbf{x} + \mathbf{c}$, where $\mathbf{A}$ is nonsingular. Hence verify that these coefficients do not depend on the mean $\boldsymbol{\mu}$ and dispersion matrix $\boldsymbol{\Sigma}$ of $\mathbf{x}$.

4.4 Using the notation following (4.24), derive the likelihood ratio test for the hypothesis $\boldsymbol{\lambda} = \theta \mathbf{1}_d$, where θ is unspecified.

4.5 If $D_i^2 = (\mathbf{x}_i - \bar{\mathbf{x}})'\mathbf{S}^{-1}(\mathbf{x}_i - \bar{\mathbf{x}})$, find $E[D_i^2]$.

4.6 Let $x_1, x_2, \ldots, x_n$ be a random sample from a distribution with coefficients of skewness $\sqrt{\beta_1}$ $(= \mu_3/\sigma^3)$ and kurtosis β_2 $(= \mu_4/\sigma^4)$. Prove that

$$\beta_2 - \beta_1 \geq 1.$$

If $\sqrt{b_1}$ and b_2 are the sample coefficients defined by Equations (4.11) and (4.12), show that

$$b_2 \leq n \quad \text{and} \quad \left|\sqrt{b_1}\right| \leq \sqrt{(n-1)}\,.$$

[Hint: Obtain $b_2 \leq n$ first, then define a random variable U that takes values x_i $(i = 1, 2, \ldots, n)$ with probability $1/n$. What is the population skewness of U?]. This example demonstrates that the sample coefficients may be inadequate estimates of the population coefficients if the latter are large and n is sufficiently small.

Johnson and Lowe [1979]

4.7 Verify Equation (4.28).

4.8 Derive Equation (4.40). Assume $E[\psi\{x - \mu)/\sigma\}] = 0$: This defines μ.

4.9 Let $\mathbf{Q} = \Sigma_{i=1}^n(\mathbf{x}_i - \bar{\mathbf{x}})(\mathbf{x}_i - \bar{\mathbf{x}})'$ and let $\mathbf{Q}_{-i}$ be $\mathbf{Q}$ with $\mathbf{x}_i$ omitted from the data set. If

$$r_{(1)} = \min_{1 \leq i \leq n} \frac{|\mathbf{Q}_{-i}|}{|\mathbf{Q}|},$$

verify Equation (4.58). [Hint: Use Exercise 4.7.]

CHAPTER 5

Dimension Reduction and Ordination

5.1 INTRODUCTION

When taking measurements on people or objects, the researcher will frequently include as many variables as possible to avoid overlooking any that may have future relevance. Unfortunately, when the dimension d is large, a data set may not only be very costly to obtain but it may also be unmanageable and difficult to study without being condensed in some way. In this case several options are available to the researcher.

First, he can try to select a k-subset of the variables so that the "reduced" sample points in k dimensions will reflect certain geometrical or statistical properties of the original d-dimensional points. How this is done will depend on the properties under consideration. For example, in Chapters 6 and 10, methods are given for selecting a subset that has a discriminating power similar to that of the original variables. In Section 5.2.6 a method is given for selecting variables based on a cluster analysis of the variables followed by choosing representative variables from the clusters.

A second approach, considered in this chapter, is to replace a d-dimensional observation by k linear combinations of the variables, where k is much smaller than d. The method of principal components is of this type: Its aim is to choose k so as to explain a reasonable proportion of the total dispersion $\operatorname{tr} \mathbf{S}$, $\mathbf{S}$ being the sample dispersion matrix. A related technique is the biplot that endeavors to describe graphically in two dimensions both relationships among the d-dimensional observations $\mathbf{x}_1, \mathbf{x}_2, \ldots, \mathbf{x}_n$, and relationships among the variables. A special case, the h-plot, gives a two-dimensional picture of the approximate sample variances and correlations of the variables, and it is based on a rank 2 approximation of $\mathbf{S}$. Factor analysis, a method often confused with principal components, attempts to extract a lower-dimensional linear structure from the data that, hopefully, explains the correlations between the variables. Although a popular method in the social sciences, factor analysis has some conceptual and computational problems: It should be used with extreme caution. When attention is focused on comparing one subset of the variables with the subset of remaining variables, the method of canonical correlations can be used to find suitable linear combinations within each subset. However, if one wishes to highlight any grouping of the $\mathbf{x}_i$ observations, but in a lower dimension, then discriminant coordinates (canonical discriminant analysis) can be used. Linear combinations are then chosen to highlight group separation.

A third approach to dimension reduction, also considered in this chapter, is multidimensional scaling (ordination). Here the d-dimensional configuration $\mathbf{x}_1, \mathbf{x}_2, \ldots, \mathbf{x}_n$ is approximated by a configuration $\mathbf{y}_1, \mathbf{y}_2, \ldots, \mathbf{y}_n$ in a lower dimension in such a way that the second configuration is reasonably "close" to the first and distances (metric scaling) or the rank ordering of distances (nonmetric scaling) between points is preserved as much as possible. This method can be used even when the first configuration is not available and the data consist only of measures of similarity or dissimilarity between the n objects. We now begin our review of the above methods by discussing principal components.

5.2 PRINCIPAL COMPONENTS

5.2.1 *Definition*

Let $\mathbf{x}$ be a random d-vector with mean $\boldsymbol{\mu}$ and dispersion matrix $\boldsymbol{\Sigma}$. Let $\mathbf{T} = (\mathbf{t}_1, \mathbf{t}_2, \ldots, \mathbf{t}_d)$ be an orthogonal matrix such that

$$\mathbf{T}'\boldsymbol{\Sigma}\mathbf{T} = \boldsymbol{\Lambda} = \text{diag}(\lambda_1, \lambda_2, \ldots, \lambda_d), \tag{5.1}$$

where $\lambda_1 \geq \lambda_2 \geq \cdots \geq \lambda_d \geq 0$ are the eigenvalues of $\boldsymbol{\Sigma}$ (see A1.3), and let $\mathbf{y} = [(y_j)] = \mathbf{T}'(\mathbf{x} - \boldsymbol{\mu})$. Then $y_j = \mathbf{t}_j'(\mathbf{x} - \boldsymbol{\mu})$ $(j = 1, 2, \ldots, d)$ is called the jth *principal component* of $\mathbf{x}$, and $z_j = \lambda_j^{-1/2} y_j$ is called the jth *standardized principal component* of $\mathbf{x}$. These components have a number of optimal properties described below, and they provide a useful tool for investigating a wide range of data analysis problems. Since the eigenvector $\mathbf{t}_j$ has unit length, we note that y_j is the orthogonal projection of $\mathbf{x} - \boldsymbol{\mu}$ in direction $\mathbf{t}_j$.

LEMMA 5.1 The y_j are uncorrelated and var $y_j = \lambda_j$.

Proof $\mathscr{D}[\mathbf{y}] = \mathscr{D}[\mathbf{T}'(\mathbf{x} - \boldsymbol{\mu})] = \mathbf{T}'\mathscr{D}[\mathbf{x}]\mathbf{T} = \mathbf{T}'\boldsymbol{\Sigma}\mathbf{T} = \boldsymbol{\Lambda}$. □

COROLLARY $\mathscr{D}[\mathbf{z}] = \mathscr{D}[\boldsymbol{\Lambda}^{-1/2}\mathbf{y}] = \boldsymbol{\Lambda}^{-1/2}\boldsymbol{\Lambda}\boldsymbol{\Lambda}^{-1/2} = \mathbf{I}_d$.

5.2.2 *Dimension Reduction Properties*

One method of dimension reduction consists of transforming $\mathbf{x}$ to a lower-dimensional k-vector $\mathbf{y}_{(k)}$ such that, for a suitable choice of the $d \times k$ matrix $\mathbf{A}$, $\mathbf{A}\mathbf{y}_{(k)}$ is "close" to $\mathbf{x}$. Given $\mathscr{D}[\mathbf{y}_{(k)}] = \boldsymbol{\Gamma} > \mathbf{O}$, we can use the more convenient representation $\mathbf{A}\mathbf{y}_{(k)} = (\mathbf{A}\boldsymbol{\Gamma}^{1/2})(\boldsymbol{\Gamma}^{-1/2}\mathbf{y}_{(k)}) = \mathbf{A}_0\mathbf{z}_{(k)}$, where $\mathscr{D}[\mathbf{z}_{(k)}] = \mathbf{I}_k$. We shall assume, for the moment, that $\boldsymbol{\mu} = \mathbf{0}$ and $\mathscr{E}[\mathbf{z}_{(k)}] = \mathbf{0}$.

A reasonable measure of the "nearness" of $\mathbf{x}$ and $\mathbf{A}_0\mathbf{z}_{(k)}$ would be a suitable function f of $\boldsymbol{\Delta} = \mathscr{D}[\mathbf{x} - \mathbf{A}_0\mathbf{z}_{(k)}]$, where $f(\boldsymbol{\Delta})$ measures the "size" of $\boldsymbol{\Delta}$. Before choosing f, we note first of all that f is defined on the set of $d \times d$ positive semidefinite matrices, $\mathscr{P}$, say; thus $\mathbf{C} \in \mathscr{P}$ if and only if $\mathbf{C} \geq \mathbf{O}$. We can therefore define the usual partial ordering on $\mathscr{P}$, namely, if $\mathbf{C} \in \mathscr{P}$ and $\mathbf{D} \in \mathscr{P}$, then $\mathbf{C} \geq \mathbf{D}$ if and only if $\mathbf{C} - \mathbf{D} \in \mathscr{P}$. Clearly a reasonable requirement of f is that it be strictly increasing, that is, $f(\mathbf{C}) > f(\mathbf{D})$ if $\mathbf{C} \geq \mathbf{D}$ and $\mathbf{C} \neq \mathbf{D}$. Furthermore, if we simply rotate the vector $\mathbf{x} - \mathbf{A}_0\mathbf{z}$ by premultiplying by an orthogonal matrix, we would not expect to change the "nearness" of $\mathbf{A}_0\mathbf{z}_{(k)}$ to $\mathbf{x}$. Therefore a second property we would expect of f is that it is invariant under orthogonal transformations, namely, $f(\mathbf{T}'\mathbf{C}\mathbf{T}) = f(\mathbf{C})$ for all orthogonal $\mathbf{T}$.

The above two properties of f were suggested by Okamoto and Kanazawa [1968], and the development below is based mainly on their paper. They noted that the functions $f(\mathbf{C}) = \operatorname{tr}\mathbf{C}$ and the Frobenius norm $\|\mathbf{C}\| = \{\operatorname{tr}[\mathbf{C}\mathbf{C}']\}^{1/2} = \{\Sigma_i\Sigma_j c_{ij}^2\}^{1/2}$ satisfy these two properties; we can also add $|\mathbf{C}|$ to the list (Exercise 5.1). They stated the following lemma, which we now prove.

LEMMA 5.2 Let f be defined on $\mathscr{P}$. For any $\mathbf{C} \in \mathscr{P}$, let $\mu_1[\mathbf{C}] \geq \mu_2[\mathbf{C}] \geq \cdots \geq \mu_d[\mathbf{C}] \geq 0$ be the eigenvalues of $\mathbf{C}$ in decreasing order of magnitude. A necessary and sufficient condition for f to be strictly increasing and invariant under orthogonal transformations is that $f(\mathbf{C}) = g(\mu_1[\mathbf{C}], \mu_2[\mathbf{C}], \ldots, \mu_d[\mathbf{C}])$ for some g that is strictly increasing in each argument. [In essence, this lemma states that if f satisfies the two properties, then minimizing $f(\mathbf{C})$ with respect to $\mathbf{C}$ is equivalent to simultaneously minimizing the eigenvalues of $\mathbf{C}$.]

Proof *Necessity*. There exists orthogonal $\mathbf{T}$ such that

$$\mathbf{T}'\mathbf{C}\mathbf{T} = \operatorname{diag}(\mu_1[\mathbf{C}], \mu_2[\mathbf{C}], \ldots, \mu_d[\mathbf{C}]) = \mathbf{M},$$

say. Since f is invariant under $\mathbf{T}$,

$$f(\mathbf{C}) = f(\mathbf{T}'\mathbf{C}\mathbf{T}) = f(\mathbf{M}) = g(\mu_1[\mathbf{C}], \mu_2[\mathbf{C}], \ldots, \mu_d[\mathbf{C}])$$

for some g. Let $\mathbf{L}$ be the same matrix as $\mathbf{M}$, but with δ ($\delta > 0$) added to the jth diagonal element. Since $\mathbf{L} \geq \mathbf{M}$, $\mathbf{L} \neq \mathbf{M}$, and f is strictly increasing, then

$$\begin{aligned} g(\mu_1[\mathbf{C}], \ldots, \mu_j[\mathbf{C}] + \delta, \ldots, \mu_d[\mathbf{C}]) &= f(\mathbf{L}) \\ &> f(\mathbf{M}) \\ &= g(\mu_1[\mathbf{C}], \ldots, \mu_j[\mathbf{C}], \ldots, \mu_d[\mathbf{C}]), \end{aligned}$$

and g is a strictly increasing function of $\mu_j[\mathbf{C}]$.

Sufficiency. Since $0 = |\mathbf{S}'\mathbf{C}\mathbf{S} - \mu\mathbf{I}_d| = |\mathbf{S}'||\mathbf{C} - \mu\mathbf{I}_d||\mathbf{S}|$ if and only if $|\mathbf{C} - \mu\mathbf{I}_d| = 0$ for any orthogonal $\mathbf{S}$, then $\mathbf{C}$ and $\mathbf{S}'\mathbf{C}\mathbf{S}$ have the same eigenvalues. Since f depends only on the eigenvalues of $\mathbf{C}$, $f(\mathbf{S}'\mathbf{C}\mathbf{S}) = f(\mathbf{C})$ and f is invariant under orthogonal transformations. To prove that f is strictly increasing, let $\mathbf{C} \geq \mathbf{D}$ with $\mathbf{C} \neq \mathbf{D}$. Then $\mu_j(\mathbf{C}) > \mu_j(\mathbf{D})$ for at least one j (see A7.9) and

$$
\begin{aligned}
f(\mathbf{C}) &= g(\mu_1[\mathbf{C}], \mu_2[\mathbf{C}], \ldots, \mu_d[\mathbf{C}]) \\
&> g(\mu_1[\mathbf{D}], \mu_2[\mathbf{D}], \ldots, \mu_d[\mathbf{D}]) \\
&= f(\mathbf{D}).
\end{aligned}
$$

□

Having given the properties of f, we now wish to find $\mathbf{A}_0 z_{(k)}$ (or, equivalently, $\mathbf{A}\mathbf{y}_{(k)}$) such that $f(\boldsymbol{\Delta})$ is minimized. Setting $\mathbf{B} = \mathscr{C}[\mathbf{x}, \mathbf{z}_{(k)}]$,

$$
\begin{aligned}
\boldsymbol{\Delta} &= \mathscr{D}\left[\mathbf{x} - \mathbf{A}_0\mathbf{z}_{(k)}\right] \\
&= \mathscr{C}\left[\mathbf{x} - \mathbf{A}_0\mathbf{z}_{(k)}, \mathbf{x} - \mathbf{A}_0\mathbf{z}_{(k)}\right] \\
&= \mathscr{D}[\mathbf{x}] - \mathbf{A}_0\mathscr{C}[\mathbf{z}_{(k)}, \mathbf{x}] - \mathscr{C}[\mathbf{x}, \mathbf{z}_{(k)}]\mathbf{A}_0' + \mathbf{A}_0\mathscr{D}[\mathbf{z}_{(k)}]\mathbf{A}_0' \qquad [\text{by } (1.7)] \\
&= \boldsymbol{\Sigma} - \mathbf{A}_0\mathbf{B}' - \mathbf{B}\mathbf{A}_0' + \mathbf{A}_0\mathbf{A}_0' \\
&= (\boldsymbol{\Sigma} - \mathbf{B}\mathbf{B}') + (\mathbf{A}_0 - \mathbf{B})(\mathbf{A}_0 - \mathbf{B})' \\
&\geq \boldsymbol{\Sigma} - \mathbf{B}\mathbf{B}', \qquad (5.2)
\end{aligned}
$$

as $(\mathbf{A}_0 - \mathbf{B})(\mathbf{A}_0 - \mathbf{B})' \geq \mathbf{O}$ (see A4.4). Also $\boldsymbol{\Sigma} - \mathbf{B}\mathbf{B}' = \mathscr{D}[\mathbf{x} - \mathbf{B}\mathbf{z}_{(k)}] \geq \mathbf{O}$. Now the rank of the $d \times k$ matrix $\mathbf{B}$ is at most k, so that from A7.10 we have

$$
\mu_i[\boldsymbol{\Sigma} - \mathbf{B}\mathbf{B}'] \geq \begin{cases} \mu_{k+i}[\boldsymbol{\Sigma}] = \lambda_{k+i} & (i = 1, 2, \ldots, d-k), \\ 0 & (i = d-k+1, \ldots, d). \end{cases}
$$

Hence, by Lemma 5.2,

$$
\begin{aligned}
f(\boldsymbol{\Delta}) &\geq f(\boldsymbol{\Sigma} - \mathbf{B}\mathbf{B}') \\
&= g(\mu_1[\boldsymbol{\Sigma} - \mathbf{B}\mathbf{B}'], \mu_2[\boldsymbol{\Sigma} - \mathbf{B}\mathbf{B}'], \ldots, \mu_d[\boldsymbol{\Sigma} - \mathbf{B}\mathbf{B}']) \\
&\geq g(\lambda_{k+1}, \lambda_{k+2}, \ldots, \lambda_d, 0, \ldots, 0), \qquad (5.3)
\end{aligned}
$$

and we have a lower bound on $f(\boldsymbol{\Delta})$. In the following theorem we now show that this lower bound is attainable.

THEOREM 5.3 Let $\mathbf{T} = (\mathbf{T}_1, \mathbf{T}_2)$, where $\mathbf{T}_1 = (\mathbf{t}_1, \mathbf{t}_2, \ldots, \mathbf{t}_k)$, be defined as in (5.1). If f satisfies Lemma 5.2 above, then

$$f(\mathbf{\Delta}) = f\left(\mathscr{D}\left[\mathbf{x} - \mathbf{A}\mathbf{y}_{(k)}\right]\right)$$

is minimized when $\mathbf{A}\mathbf{y}_{(k)} = \mathbf{A}_0 \mathbf{z}_{(k)} = \mathbf{T}_1\mathbf{T}_1'\mathbf{x}$.

Proof Given $\mathbf{A}\mathbf{y}_{(k)} = \mathbf{T}_1\mathbf{T}_1'\mathbf{x} = \mathbf{P}\mathbf{x}$, say, then $\mathbf{x} - \mathbf{A}\mathbf{y}_{(k)} = (\mathbf{I} - \mathbf{P})\mathbf{x} = \mathbf{Q}\mathbf{x}$, say. From (1.8), $\mathbf{\Delta} = \mathscr{D}[\mathbf{Q}\mathbf{x}] = \mathbf{Q}\mathbf{\Sigma}\mathbf{Q}$. Now $\mathbf{\Sigma} = \mathbf{T}\mathbf{\Lambda}\mathbf{T}'$ so that

$$\begin{aligned}
\mathbf{P}\mathbf{\Sigma} &= \mathbf{T}_1\mathbf{T}_1'(\mathbf{T}_1, \mathbf{T}_2)\begin{pmatrix} \mathbf{\Lambda}_1 & \mathbf{O} \\ \mathbf{O} & \mathbf{\Lambda}_2 \end{pmatrix}\begin{pmatrix} \mathbf{T}_1' \\ \mathbf{T}_2' \end{pmatrix} \\
&= (\mathbf{T}_1, \mathbf{O})\begin{pmatrix} \mathbf{\Lambda}_1 & \mathbf{O} \\ \mathbf{O} & \mathbf{\Lambda}_2 \end{pmatrix}\begin{pmatrix} \mathbf{T}_1' \\ \mathbf{T}_2' \end{pmatrix} \\
&= \mathbf{T}_1\mathbf{\Lambda}_1\mathbf{T}_1', \qquad (5.4)
\end{aligned}$$

where $\mathbf{\Lambda}_1$ is $k \times k$. Since $\mathbf{Q} = \mathbf{T}\mathbf{T}' - \mathbf{T}_1\mathbf{T}_1' = \mathbf{T}_2\mathbf{T}_2'$, we have, using a similar argument to that leading to (5.4), that $\mathbf{Q}\mathbf{\Sigma} = \mathbf{T}_2\mathbf{\Lambda}_2\mathbf{T}_2'$. Also, from $\mathbf{T}_2'\mathbf{T}_1 = \mathbf{O}$, we have $\mathbf{Q}\mathbf{\Sigma}\mathbf{P} = \mathbf{O}$ so that

$$\mathbf{\Delta} = \mathbf{Q}\mathbf{\Sigma}\mathbf{Q} = \mathbf{Q}\mathbf{\Sigma} = \mathbf{T}_2\mathbf{\Lambda}_2\mathbf{T}_2' = \mathbf{T}\begin{pmatrix} \mathbf{O} & \mathbf{O} \\ \mathbf{O} & \mathbf{\Lambda}_2 \end{pmatrix}\mathbf{T}'. \qquad (5.5)$$

Finally

$$\begin{aligned}
f(\mathbf{\Delta}) &= f(\mathbf{T}'\mathbf{\Delta}\mathbf{T}) \\
&= f\left(\begin{bmatrix} \mathbf{O} & \mathbf{O} \\ \mathbf{O} & \mathbf{\Lambda}_2 \end{bmatrix}\right) \qquad [\text{by } (5.5)] \\
&= g(\mu_1[\mathbf{\Lambda}_2], \ldots, \mu_{d-k}[\mathbf{\Lambda}_2], 0, \ldots, 0) \\
&= g(\lambda_{k+1}, \ldots, \lambda_d, 0, \ldots, 0), \qquad (5.6)
\end{aligned}$$

which is the lower bound (5.3). Thus $f(\mathbf{\Delta})$ achieves its minimum value when $\mathbf{A}\mathbf{y}_{(k)} = \mathbf{P}\mathbf{x}$. □

In the above theorem we have shown that $\mathbf{A}\mathbf{y}_{(k)} = \mathbf{P}\mathbf{x}$ gives *a* solution of the minimization problem. However, Okamoto and Kanazawa [1968] proved the more general result that (5.6) is true if and only if $\mathbf{A}\mathbf{y}_{(k)} = \mathbf{P}\mathbf{x}$. Since $\mathbf{P}$ represents the orthogonal projection onto $\mathscr{R}[\mathbf{T}_1]$ (see B1.9), the solution is unique if and only if $\lambda_k \neq \lambda_{k+1}$ (by A1.3). Special cases of the above results were obtained by Kramer and Mathews [1956], Rao [1964a, 1973], Darroch [1965: $f = trace$] and Brillinger [1975: pp. 106–107, 339, 342].

Since $\mathbf{Q\Sigma P} = \mathbf{O}$,

$$\mathbf{\Sigma} = \{\mathbf{P\Sigma} + (\mathbf{I} - \mathbf{P})\mathbf{\Sigma}\}\{\mathbf{P} + \mathbf{I} - \mathbf{P}\}$$

$$= \mathbf{P\Sigma P} + \mathbf{Q\Sigma Q}, \tag{5.7}$$

which we can regard as an "orthogonal decomposition" of $\mathbf{\Sigma}$ as $\mathbf{PQ} = \mathbf{O}$. Thus $\mathbf{\Delta} = \mathbf{Q\Sigma Q}$ minimizes $f(\mathbf{\Delta})$ by suitably "projecting" $\mathbf{\Sigma}$ onto the space of positive semidefinite matrices of rank less than or equal to k so as to minimize the "distance" $f(\mathbf{\Sigma} - \mathbf{P\Sigma P})$ between $\mathbf{\Sigma}$ and its "projection" $\mathbf{P\Sigma P}$. Alternatively, $\mathbf{Px}$ is the orthogonal projection of $\mathbf{x}$ onto $\mathscr{R}[\mathbf{P}]$, where $\mathbf{P}$ is chosen so that the functional $f(\mathscr{D}[(\mathbf{I} - \mathbf{P})\mathbf{x}])$ of the perpendicular vector $(\mathbf{I} - \mathbf{P})\mathbf{x}$ is minimized.

Uptil now we have assumed that $\mathscr{E}[\mathbf{x}] = \mathbf{0}$. If $\mathscr{E}[\mathbf{x}] = \boldsymbol{\mu}$, then we simply replace $\mathbf{x}$ by $\mathbf{x} - \boldsymbol{\mu}$ in the above derivations. The best approximation for $\mathbf{x} - \boldsymbol{\mu}$ is therefore $\mathbf{T}_1\mathbf{T}_1'(\mathbf{x} - \boldsymbol{\mu}) = \mathbf{T}_1\mathbf{y}_{(k)}$, where $\mathbf{y}_{(k)} = (y_1, y_2, \ldots, y_k)'$ are the first k principal components defined in Section 5.2.1. The corresponding standardized vector $\mathbf{z}_{(k)}$ gives the first k standardized principal components (see Lemma 5.1).

Since $\mathbf{T}_1\mathbf{y}_{(k)}$ is an "approximation" for $\mathbf{x} - \boldsymbol{\mu}$, a measure of fit is the sum of squares of the residuals, namely,

$$Q_1 = (\mathbf{x} - \boldsymbol{\mu} - \mathbf{T}_1\mathbf{y}_{(k)})'(\mathbf{x} - \boldsymbol{\mu} - \mathbf{T}_1\mathbf{y}_{(k)})$$

$$= \sum_{j=k+1}^{d} y_j^2 \qquad \text{(by Exercise 5.8)}. \tag{5.8}$$

If $\theta_a = \sum_{j=k+1}^{d}\lambda_j^a = \operatorname{trace}\mathbf{\Sigma}^a - \sum_{j=1}^{k}\lambda_j^a$, $a = 1, 2, 3$, $h_0 = 1 - (2\theta_1\theta_3/3\theta_2^2)$, and $\mathbf{x} \sim N_d(\boldsymbol{\mu}, \mathbf{\Sigma})$, then Jackson and Mudholkar [1979] showed that $(Q_1/\theta_1)^{h_0}$ is approximately $N_1(1 + \theta_1^{-2}\theta_2 h_0[h_0 - 1], 2\theta_1^{-2}\theta_2 h_0^2)$. They recommend the use of Q_1 as a suitable statistic for multivariate quality control where good large-sample estimates of $\boldsymbol{\mu}$ and $\mathbf{\Sigma}$ are available. A significant value of Q_1 could suggest that $\mathbf{x}$ may be an outlier. Another measure of fit is (Rao [1964a])

$$Q_2 = \sum_{j=k+1}^{d} z_j^2 \tag{5.9}$$

$$= (\mathbf{x} - \boldsymbol{\mu})'\mathbf{\Sigma}^{-1}(\mathbf{x} - \boldsymbol{\mu}) - \sum_{j=1}^{k} z_j^2 \tag{5.10}$$

based on the standardized principal components z_j. Although $Q_2 \sim \chi_{d-k}^2$ under normality, it requires the additional computation of either $\mathbf{\Sigma}^{-1}$ or the smallest eigenvalues $\lambda_{k+1}, \lambda_{k+2}, \ldots, \lambda_d$; both of these could have some numerical problems if $\mathbf{\Sigma}$ is near singularity. On the other hand, Q_1 can be readily computed from the residual $\mathbf{x} - \boldsymbol{\mu} - \mathbf{T}_1\mathbf{y}_{(k)}$. If Q_1 is significant, then the elements of the residual will shed some light on the adequacy of $\mathbf{y}_{(k)}$. For further related references in quality control see Hawkins [1974] and Jackson and Hearne [1979].

5.2.3 *Further Properties*

Principal components have a number of interesting properties that have been used in the past to motivate their definition. One such motivation is given by the following theorem.

THEOREM 5.4 The principal components $y_j = \mathbf{t}_j'(\mathbf{x} - \boldsymbol{\mu})$, $j = 1, 2, \ldots, d$, have the following properties:

(i) For any vector $\mathbf{a}_1$ of unit length, $\text{var}[\mathbf{a}_1'\mathbf{x}]$ takes its maximum value λ_1 when $\mathbf{a}_1 = \mathbf{t}_1$.
(ii) For any vector $\mathbf{a}_j$ of unit length such that $\mathbf{a}_j'\mathbf{t}_i = 0$ $(i = 1, 2, \ldots, j-1)$, $\text{var}[\mathbf{a}_j'\mathbf{x}]$ takes its maximum value λ_j when $\mathbf{a}_j = \mathbf{t}_j$.
(iii) $\Sigma_{j=1}^d \text{var}\, y_j = \Sigma_{j=1}^d \text{var}\, x_j = \text{tr}\, \boldsymbol{\Sigma}$.

Proof (i) $\text{var}[\mathbf{a}_1'\mathbf{x}] = \mathscr{D}[\mathbf{a}_1'\mathbf{x}] = \mathbf{a}_1'\boldsymbol{\Sigma}\mathbf{a}_1$ and the result follows from A7.4.

(ii) Since $\mathbf{t}_1, \mathbf{t}_2, \ldots, \mathbf{t}_d$ are mutually orthogonal, they form a basis for R^d. We are given $\mathbf{a}_j \perp \{\mathbf{t}_1, \mathbf{t}_2, \ldots, \mathbf{t}_{j-1}\}$ so that $\mathbf{a}_j$ can be expressed in the form

$$\mathbf{a}_j = c_j\mathbf{t}_j + c_{j+1}\mathbf{t}_{j+1} + \cdots + c_d\mathbf{t}_d.$$

Hence $1 = \mathbf{a}_j'\mathbf{a}_j = \Sigma_{\alpha=j}^d c_\alpha^2 \mathbf{t}_\alpha'\mathbf{t}_\alpha = \Sigma_{\alpha=j}^d c_\alpha^2$ and

$$\begin{aligned}
\text{var}\left[\mathbf{a}_j'\mathbf{x}\right] &= \mathbf{a}_j'\boldsymbol{\Sigma}\mathbf{a}_j \\
&= \mathbf{a}_j' \sum_{\alpha=j}^d (c_\alpha \boldsymbol{\Sigma}\mathbf{t}_\alpha) \\
&= \mathbf{a}_j' \sum_{\alpha=j}^d c_\alpha \lambda_\alpha \mathbf{t}_\alpha \\
&= \left(\sum_{\alpha=j}^d c_\alpha \mathbf{t}_\alpha\right)' \left(\sum_{\alpha=j}^d c_\alpha \lambda_\alpha \mathbf{t}_\alpha\right) \\
&= \sum_{\alpha=j}^d c_\alpha^2 \lambda_\alpha \\
&\geq \lambda_j \sum_{\alpha=j}^d c_\alpha^2 \\
&= \lambda_j,
\end{aligned}$$

with equality if $c_j = 1$, $c_{j+1} = \cdots = c_d = 0$. Thus $\text{var}[\mathbf{a}_j'\mathbf{x}]$ is maximized when $\mathbf{a}_j = \mathbf{t}_j$.

(iii) $$\sum_{j=1}^{d} \text{var}\, y_j = \text{tr}\, \boldsymbol{\Lambda} = \text{tr}[\boldsymbol{\Lambda}\mathbf{T}'\mathbf{T}] = \text{tr}[\mathbf{T}\boldsymbol{\Lambda}\mathbf{T}'] = \text{tr}\, \boldsymbol{\Sigma}. \qquad \square$$

Properties (i) and (ii) can also be described as follows. Component y_1 is the normalized linear combination of the elements of $\mathbf{x} - \boldsymbol{\mu}$ with maximum variance λ_1. Now $\mathbf{a}'\mathbf{x}$ is uncorrelated with y_1, that is,

$$\begin{aligned} \mathscr{C}\left[\mathbf{a}'\mathbf{x}, \mathbf{t}_1'(\mathbf{x} - \boldsymbol{\mu})\right] &= \mathscr{C}\left[\mathbf{a}'\mathbf{x}, \mathbf{t}_1'\mathbf{x}\right] \\ &= \mathbf{a}'\boldsymbol{\Sigma}\mathbf{t}_1 \qquad [\text{by (1.7)}] \\ &= \mathbf{a}'\lambda\mathbf{t}_1 \\ &= 0 \end{aligned}$$

if and only if $\mathbf{a} \perp \mathbf{t}_1$. Thus, from (ii), y_2 is the normalized linear combination uncorrelated with y_1 with maximum variance ($= \lambda_2$), and y_j is the normalized linear combination uncorrelated with $y_1, y_2, \ldots, y_{j-1}$ with maximum variance λ_j. This property of the y_j was used by Hotelling [1933] to define principal components. They were also described geometrically by Pearson [1901] in a different context as follows.

Consider a plane through the origin with equation $\mathbf{a}'\mathbf{u} = 0$, where $\mathbf{a}'\mathbf{a} = 1$, with $\mathbf{a}$ a unit vector normal to the plane. The distance of a random point $\mathbf{x} - \boldsymbol{\mu}$ from the plane is the projection of $\mathbf{x} - \boldsymbol{\mu}$ in direction $\mathbf{a}$, namely, $d_1 = \mathbf{a}'(\mathbf{x} - \boldsymbol{\mu})$. If we look for the plane that minimizes $E[d_1^2] = \text{var}\, d_1 = \mathbf{a}'\boldsymbol{\Sigma}\mathbf{a}$, we obtain $\mathbf{a} = \mathbf{t}_d$ as its normal (the unit vector perpendicular to the plane). We next find a plane perpendicular to the first plane such that $E[d_1^2]$ is a minimum: This leads to $\mathbf{t}_{d-1}$, and so on.

Instead of using a plane, Kendall [1975] considers a line through the point $\boldsymbol{\xi}$ with direction cosines $\mathbf{a} = (a_1, a_2, \ldots, a_d)'$, namely,

$$\frac{u_1 - \xi_1}{a_1} = \frac{u_2 - \xi_2}{a_2} = \cdots = \frac{u_d - \xi_d}{a_d} \qquad (= \gamma, \text{say}), \tag{5.11}$$

or

$$\mathbf{u} = \gamma\mathbf{a} + \boldsymbol{\xi} \qquad (\|\mathbf{a}\| = 1). \tag{5.12}$$

The foot of the perpendicular from $\mathbf{x}$ to the line is given by $\mathbf{p} = \gamma\mathbf{a} + \boldsymbol{\xi}$, where γ satisfies $\mathbf{a}'(\mathbf{x} - \mathbf{p}) = 0$, that is, $\gamma = \mathbf{a}'(\mathbf{x} - \boldsymbol{\xi}) = \mathbf{a}'\tilde{\mathbf{x}}$, say. Hence the square of the distance of $\mathbf{x}$ from the line is

$$\begin{aligned} D_1^2 &= \|\mathbf{x} - \mathbf{p}\|^2 \\ &= \|\mathbf{x} - \gamma\mathbf{a} - \boldsymbol{\xi}\|^2 \\ &= \|\tilde{\mathbf{x}} - (\mathbf{a}'\tilde{\mathbf{x}})\mathbf{a}\|^2 \\ &= \tilde{\mathbf{x}}'\tilde{\mathbf{x}} - 2(\mathbf{a}'\tilde{\mathbf{x}})^2 + (\mathbf{a}'\tilde{\mathbf{x}})^2\mathbf{a}'\mathbf{a} \\ &= \tilde{\mathbf{x}}'\tilde{\mathbf{x}} - (\mathbf{a}'\tilde{\mathbf{x}})^2. \end{aligned} \tag{5.13}$$

If $\boldsymbol{\theta} = \mathscr{E}[\tilde{\mathbf{x}}] = \boldsymbol{\mu} - \boldsymbol{\xi}$, then (by Exercise 1.4)

$$\begin{aligned} \mathrm{E}\left[D_1^2\right] &= \operatorname{tr}\boldsymbol{\Sigma} + \boldsymbol{\theta}'\boldsymbol{\theta} - \left\{\operatorname{var}[\mathbf{a}'\tilde{\mathbf{x}}] + (\mathbf{a}'\boldsymbol{\theta})^2\right\} \\ &= \operatorname{tr}\boldsymbol{\Sigma} - \mathbf{a}'\boldsymbol{\Sigma}\mathbf{a} + \left\|\boldsymbol{\theta} - (\mathbf{a}'\boldsymbol{\theta})^2\mathbf{a}\right\|^2, \end{aligned}$$

the last step following from (5.13) with $\tilde{\mathbf{x}}$ replaced by $\boldsymbol{\theta}$. Now minimizing $\mathrm{E}[D_1^2]$ is equivalent to setting $\boldsymbol{\theta} = \mathbf{0}$ and maximizing $\mathbf{a}'\boldsymbol{\Sigma}\mathbf{a}$: the maximum occurs at $\mathbf{a} = \mathbf{t}_1$ [by Theorem 5.4(i)]. Therefore a line through $\boldsymbol{\mu}$ with direction $\mathbf{t}_1$ minimizes the expected squared distance of $\mathbf{x}$ from the line. Alternatively, if we shift the origin to $\boldsymbol{\mu}$, then a line through the origin with direction $\mathbf{t}_1$ minimizes the expected squared distance of $\mathbf{x} - \boldsymbol{\mu}$ from the line. Similarly, $\mathbf{t}_2$ is the direction of a second line through the origin, perpendicular to the first, with the same property, and so forth.

Property (iii) of Theorem 5.4 provides us with a crude but simple technique for deciding what k should be. The ratio $\lambda_j/\Sigma_{\alpha=1}^d\lambda_\alpha = \operatorname{var} y_j/\operatorname{tr}\boldsymbol{\Sigma}$ measures the contribution of y_j to $\operatorname{tr}\boldsymbol{\Sigma}$, the "total variation" of $\mathbf{x}$. We can therefore take out successive components $y_1, y_2, y_3, \ldots$ and stop at y_k when $\Sigma_{j=1}^k\lambda_j/\operatorname{tr}\boldsymbol{\Sigma}$ is close enough to unity. This question is raised later under sample principal components.

If $\lambda_{k+1} = \lambda_{k+2} = \cdots = \lambda_d = 0$, then $y_{k+1}, y_{k+2}, \ldots, y_d$ have zero variances and $\mathbf{t}_j'(\mathbf{x} - \boldsymbol{\mu}) = 0$ $(j = k+1, k+2, \ldots, d)$. In this case, $d - k$ of the x_j variables are linearly dependent on the others and $\mathbf{x}$ lies in a k-dimensional subspace of R^d.

If we standardize the x_j and work with $x_j^* = (x_j - \mu_j)/\sigma_{jj}^{1/2}$, then $\mathscr{D}[\mathbf{x}^*] = \mathbf{P}_\rho$, the correlation matrix of $\mathbf{x}$ [see (1.17)]. We can now extract a new set of principal components, $\mathbf{y}^* = \mathbf{L}'\mathbf{x}^*$, say, where $\mathbf{L}$ is orthogonal and $\mathbf{L}'\mathbf{P}_\rho\mathbf{L}$ is diagonal. However, $\mathbf{y}^*$ will differ from $\mathbf{y}$, as the eigenvectors of $\boldsymbol{\Sigma}$ are, in general, not the same as those of $\mathbf{P}_\rho$. The question of whether to use $\mathbf{y}$ or $\mathbf{y}^*$ is discussed in the next section.

In addition to the above properties, principal components also enjoy various optimal properties in the context of regression (Obenchain [1972], Chen [1974]) and restricted least squares (Fomby et al. [1978]). However, apart from these optimal properties, we sometimes find that the components have a natural physical meaning. For example, in the process of measuring various characteristics of an organism, the first component may provide an overall measure of size, and the second an overall measure of shape. In educational data the first component may provide an overall measure of learning ability, the second a measure of motivation, and so on.

5.2.4 Sample Principal Components

In practice, $\boldsymbol{\mu}$ and $\boldsymbol{\Sigma}$ are not known and have to be estimated from a sample $\mathbf{x}_1, \mathbf{x}_2, \ldots, \mathbf{x}_n$. Let $\hat{\boldsymbol{\mu}} = \overline{\mathbf{x}}$ and let $\hat{\boldsymbol{\Sigma}} = \Sigma_i(\mathbf{x}_i - \overline{\mathbf{x}})(\mathbf{x}_i - \overline{\mathbf{x}})'/n$ with eigenvalues $\hat{\lambda}_1 \geq \hat{\lambda}_2 \geq \cdots \geq \hat{\lambda}_d$ and an orthogonal matrix $\hat{\mathbf{T}} = (\hat{\mathbf{t}}_1, \hat{\mathbf{t}}_2, \ldots, \hat{\mathbf{t}}_d)$ of corresponding eigenvectors. For each observation $\mathbf{x}_i$ we can define a vector of sample (estimated) principal components $\mathbf{y}_i = \hat{\mathbf{T}}'(\mathbf{x}_i - \overline{\mathbf{x}})$ giving us a new data matrix

$$\mathbf{Y}' = (\mathbf{y}_1, \mathbf{y}_2, \ldots, \mathbf{y}_n) = \hat{\mathbf{T}}'(\mathbf{x}_1 - \overline{\mathbf{x}}, \mathbf{x}_2 - \overline{\mathbf{x}}, \ldots, \mathbf{x}_n - \overline{\mathbf{x}}). \tag{5.14}$$

Many authors prefer to use the unbiased estimator $\mathbf{S}$ of $\boldsymbol{\Sigma}$ instead of $\hat{\boldsymbol{\Sigma}}$ in defining the sample components. In this case

$$\mathbf{S}\hat{\mathbf{t}}_j = \frac{n}{n-1}\hat{\boldsymbol{\Sigma}}\hat{\mathbf{t}}_j = \left(\frac{n}{n-1}\hat{\lambda}_j\right)\hat{\mathbf{t}}_j, \tag{5.15}$$

and the eigenvalues of $\mathbf{S}$ are $n\hat{\lambda}_j/(n-1)$. If we use $\hat{\lambda}_j/\text{tr}\,\hat{\boldsymbol{\Sigma}}$ for assessing the relative magnitude of an eigenvalue, then the scale factor $n/(n-1)$ cancels out and the two approaches are identical in this respect.

As in Section 5.2.2, we might expect the first k sample principal components of a vector observation to give us the "best" summary of the observation in k dimensions. Support for this viewpoint is provided by the following theorem.

THEOREM 5.5 Let $\mathbf{A}$ be a $d \times k$ matrix and let $\mathbf{y}_{i(k)} = \mathbf{g}(\mathbf{x}_i - \overline{\mathbf{x}})$, $i = 1, 2, \ldots, n$, for any (measurable) function $\mathbf{g}: R^d \to R^k$ such that $\overline{\mathbf{y}}_{\cdot(k)} = \mathbf{0}$. If f is strictly increasing and invariant under orthogonal transformations (i.e., satisfies Lemma 5.2), then

$$f\left\{\frac{1}{n}\sum_{i=1}^{n}\left(\mathbf{x}_i - \overline{\mathbf{x}} - \mathbf{A}\mathbf{y}_{i(k)}\right)\left(\mathbf{x}_i - \overline{\mathbf{x}} - \mathbf{A}\mathbf{y}_{i(k)}\right)'\right\} \tag{5.16}$$

is minimized with respect to $\mathbf{A}$ and $\mathbf{g}$ by choosing

$$\mathbf{A} = (\hat{\mathbf{t}}_1, \hat{\mathbf{t}}_2, \ldots, \hat{\mathbf{t}}_k) = \hat{\mathbf{T}}_1 \quad \text{and} \quad \mathbf{g}(\mathbf{x} - \overline{\mathbf{x}}) = \hat{\mathbf{T}}_1'(\mathbf{x} - \overline{\mathbf{x}}).$$

Proof Following Okamato [1969], let $\mathbf{v}$ be the random vector taking the value $\mathbf{x}_i$ $(i = 1, 2, \ldots, n)$ with probability n^{-1}, $\mathbf{w} = \mathbf{v} - \overline{\mathbf{x}}$, and $\mathbf{y}_{(k)} = \mathbf{g}(\mathbf{w})$. Then

$\boldsymbol{\eta} = \mathscr{E}[\mathbf{v}] = \sum_{i=1}^{n}\mathbf{x}_i(1/n) = \bar{\mathbf{x}}$, $\mathscr{E}[\mathbf{w}] = \mathbf{0}$, $\mathscr{E}[\mathbf{y}_{(k)}] = \bar{\mathbf{y}}_{\cdot(k)} = \mathbf{0}$, and

$$\begin{aligned}
\mathscr{D}[\mathbf{w}] &= \mathscr{D}[\mathbf{v}] \\
&= \mathscr{E}\left[(\mathbf{v}-\boldsymbol{\eta})(\mathbf{v}-\boldsymbol{\eta})'\right] \\
&= \frac{1}{n}\sum_{i=1}^{n}(\mathbf{x}_i-\boldsymbol{\eta})(\mathbf{x}_i-\boldsymbol{\eta})' \\
&= \frac{1}{n}\sum_{i=1}^{n}(\mathbf{x}_i-\bar{\mathbf{x}})(\mathbf{x}_i-\bar{\mathbf{x}})' \\
&= \hat{\boldsymbol{\Sigma}}.
\end{aligned}$$

Hence (5.16) is equal to

$$f\left(\mathscr{E}\left[(\mathbf{w}-\mathbf{A}\mathbf{y}_{(k)})(\mathbf{w}-\mathbf{A}\mathbf{y}_{(k)})'\right]\right) = f\left(\mathscr{D}\left[\mathbf{w}-\mathbf{A}\mathbf{y}_{(k)}\right]\right),$$

which, by Theorem 5.3, is minimized for all random vectors $\mathbf{y}_{(k)}$ such that $\mathscr{E}[\mathbf{y}_{(k)}] = \mathbf{0}$ (and therefore for all such $\mathbf{y}_{(k)}$ that are measurable functions of $\mathbf{w}$) when $\mathbf{A}\mathbf{y}_{(k)} = \hat{\mathbf{T}}_1\hat{\mathbf{T}}_1'\mathbf{w}$. This last equation holds if we set $\mathbf{A} = \hat{\mathbf{T}}_1$ and $\mathbf{g}(\mathbf{w}) = \hat{\mathbf{T}}_1'\mathbf{w}$. □

An interesting consequence of the above theorem is that by using $\mathbf{v}$, $\bar{\mathbf{x}}$, and $\hat{\boldsymbol{\Sigma}}$ instead of $\mathbf{x}$, $\boldsymbol{\mu}$, and $\boldsymbol{\Sigma}$ there is essentially no difference between a "population" or a "sample" approach to principal components. For example, we have the following sample analogue of Theorem 5.4(i). Given any vector $\mathbf{a}_1$ of unit length, the sample variance

$$\begin{aligned}
\operatorname{var}\left[\mathbf{a}_1'\mathbf{v}\right] &= \mathbf{a}_1'\mathscr{D}[\mathbf{v}]\mathbf{a}_1 \\
&= \mathbf{a}_1'\hat{\boldsymbol{\Sigma}}\mathbf{a}_1 \\
&= \frac{1}{n}\sum_{i=1}^{n}\left[\mathbf{a}_1'(\mathbf{x}_i-\bar{\mathbf{x}})\right]^2
\end{aligned}$$

takes its maximum value of $\hat{\lambda}_1$ when $\mathbf{a}_1 = \hat{\mathbf{t}}_1$. Part (ii) of Theorem 5.4 carries over in a similar fashion, and part (iii) becomes

$$\begin{aligned}
\operatorname{tr}\hat{\boldsymbol{\Sigma}}_y &= \operatorname{tr}\left[\frac{1}{n}\mathbf{Y}'\mathbf{Y}\right] \qquad [\text{from (5.14) and } \bar{\mathbf{y}} = \mathbf{0}] \\
&= \operatorname{tr}\left[\hat{\mathbf{T}}'\hat{\boldsymbol{\Sigma}}_x\hat{\mathbf{T}}\right] \\
&= \operatorname{tr}\left[\hat{\boldsymbol{\Sigma}}_x\hat{\mathbf{T}}\hat{\mathbf{T}}'\right] \\
&= \operatorname{tr}\hat{\boldsymbol{\Sigma}}_x \qquad (= \hat{\lambda}_1 + \cdots + \hat{\lambda}_d).
\end{aligned} \tag{5.17}$$

In reference to the geometrical properties, a line through $\bar{\mathbf{x}}$ with direction $\hat{\mathbf{t}}_1$ minimizes the expected squared distance of $\mathbf{v}$ from the line, that is, minimizes the average squared distance of the points $\mathbf{x}_1, \mathbf{x}_2, \ldots, \mathbf{x}_n$ from the line.

Table 5.1 (or its transpose) gives a common format for setting out the results of a principal component analysis. For an observation $\mathbf{x} = (x_1, x_2, \ldots, x_d)'$, the jth principal component is given by

$$y_j = \hat{\mathbf{t}}_j'(\mathbf{x} - \bar{\mathbf{x}})$$

$$= \hat{t}_{1j}(x_1 - \bar{x}_1) + \hat{t}_{2j}(x_2 - \bar{x}_2) + \cdots + \hat{t}_{dj}(x_d - \bar{x}_d).$$

In general, the $\hat{\lambda}_j$ will be distinct (see A2.8) so that the corresponding eigenvectors $\hat{\mathbf{t}}_j$ will be unique, apart from a sign. Uniqueness can be achieved by, for example, assigning a positive sign to the element of $\hat{\mathbf{t}}_j$ with the largest absolute value. The number k of components selected will depend on the relative magnitudes of the eigenvalues. If there is a clear gap between the large and small eigenvalues λ_j, then, for n much greater than d, this will tend to show up in the sample estimates $\hat{\lambda}_j$, thus giving a clear choice for k. For example, if the $\hat{\lambda}_j$ are

$$(20.1|0.8, 0.6, 0.4|0.02, 0.01, 0.008)$$

or, expressed as a cummulative percentage,

$$(89.15|93.87, 97.41, 99.78|99.89, 99.95, 100),$$

the possible stopping points are indicated by vertical bars, that is, $k = 1$ or 4. Alternatively, we might stop at k if $\hat{\lambda}_{k+1}/G \le \delta_1$ or $\Sigma_{j=k+1}^{d}\hat{\lambda}_j/G \le \delta_2$ $(G = \hat{\lambda}_1 + \cdots + \hat{\lambda}_d)$, where δ_1 and δ_2 are to be chosen. Unfortunately, a joint assessment of the $\hat{\lambda}_j$ is not easy because of sampling variability. We shall see later that for normal data the coefficient of variation of $\hat{\lambda}_j$ is asymptotically $(2/n)^{1/2} + O(n^{-1})$: For $n = 20$ this is around 30%. Gnanadesikan and Wilk [1969: Example 13] calculated the sample eigenvalues for 200 observations

TABLE 5.1 Principal Component Analysis

	Eigenvector				
Variable	$\hat{\mathbf{t}}_1$	$\hat{\mathbf{t}}_2$	$\cdots$	$\hat{\mathbf{t}}_d$	
x_1	$\hat{t}_{11}$	$\hat{t}_{12}$	$\cdots$	$\hat{t}_{1d}$	
x_2	$\hat{t}_{21}$	$\hat{t}_{22}$	$\cdots$	$\hat{t}_{2d}$	
$\cdots$	$\cdots$	$\cdots$	$\cdots$	$\cdots$	
x_d	$\hat{t}_{d1}$	$\hat{t}_{d2}$	$\cdots$	$\hat{t}_{dd}$	
Eigenvalue	$\hat{\lambda}_1$	$\hat{\lambda}_2$	$\cdots$	$\hat{\lambda}_d$	Sum $= G$
Percent of sum G	$100\,\hat{\lambda}_1/G$	$100\,\hat{\lambda}_2/G$	$\cdots$	$100\,\hat{\lambda}_d/G$	
Cumulative %	$100\,\hat{\lambda}_1/G$	$100(\hat{\lambda}_1 + \hat{\lambda}_2)/G$	$\cdots$	100	

simulated from $N_{49}(\mathbf{0}, \mathbf{I}_{49})$ and found that the eigenvalues ranged from 0.3 to 2.13, even though the population eigenvalues are all unity. This range is much wider than that suggested by the above formula, which gives a coefficient of variation of about 10%. A cross-validatory method for choosing k has been proposed by Eastment and Krzanowski [1982].

The problem of sampling variability is particularly crucial when d/n is not small. However, Wachter [1975, 1976a, b, 1980] has given a graphical technique that shows some promise when d and n are both large (tentatively greater than 12). Wachter defined a sample distribution function $\hat{F}_n$ that has a step of size $1/d$ at each $\hat{\lambda}_j$ and called it the "positive random spectrum" of data matrix $\tilde{\mathbf{X}} = (\mathbf{x}_1 - \bar{\mathbf{x}}, \mathbf{x}_2 - \bar{\mathbf{x}}, \ldots, \mathbf{x}_n - \bar{\mathbf{x}})'$. He showed that, irrespective of the distributional assumptions, the $\hat{\lambda}_j$ are asymptotically independent and $\hat{F}_n$ tends to some function F that has the properties of a distribution function, these properties depending on the "spread" or spectrum of the λ_j. Wachter's technique consists of performing a Q–Q plot of $\hat{F}_n$ versus F, or, equivalently, a plot of $\hat{\lambda}_j$ versus $F^{-1}\{j/(d+1)\}$, for different λ_j spectra. For example, a "two-atom" model proposes that the λ_j are all equal to one of two values. The idea is to find a model to match the data.

Exact linear relationships among random variables seldom exist in practice, as all quantitative random variables are measured with error. We can expect $\boldsymbol{\Sigma}$ to have nonzero eigenvalues, though in some cases it may have a few very small eigenvalues. If $n - 1 \geq d$, this means that $\hat{\boldsymbol{\Sigma}}$ is positive definite with probability 1 and the $\hat{\lambda}_j$ will be all positive. Yet if λ_d is very small, we can expect $\hat{\lambda}_d$ to be small also, and it is tempting to test $\lambda_d = 0$ using the distribution of $\hat{\lambda}_d$, even though $\lambda_d \neq 0$. It is, however, a question of whether λ_d is small enough to be ignored rather than whether $\lambda_d = 0$. Clearly the main emphasis of principal components is a descriptive one and formal tests of significance are of lesser importance.

There is one practical problem when successive components are being extracted. It may happen that for some r the eigenvalues $\hat{\lambda}_{r+1}, \hat{\lambda}_{r+2}, \ldots, \hat{\lambda}_d$ are not too different, thus suggesting the hypothesis $H_0: \lambda_{r+1} = \lambda_{r+2} = \cdots = \lambda_d$. If H_0 is true, there is no sense in maximizing the variance var$[\mathbf{a}'\mathbf{v}]$ in any particular direction given by $\mathbf{a}$, subject to $\mathbf{a} \perp \{\hat{\mathbf{t}}_1, \hat{\mathbf{t}}_2, \ldots, \hat{\mathbf{t}}_r\}$. A test for H_0 is given in the next section.

We have already mentioned one serious disadvantage of principal components. They are not scale invariant and the results depend on the units of measurement. If we use standardized data $(x_{ij} - \bar{x}_{\cdot j})/[\Sigma_i(x_{ij} - \bar{x}_{\cdot j})^2]^{1/2}$, then $\hat{\boldsymbol{\Sigma}}$ and $\mathbf{S}$ reduce to the sample correlation matrix $\mathbf{R} = [(r_{jk})]$, where $r_{jj} = 1$ and

$$r_{jk} = \sum_i (x_{ij} - \bar{x}_{\cdot j})(x_{ik} - \bar{x}_{\cdot k}) \Big/ \left\{ \sum_i (x_{ij} - \bar{x}_{\cdot j})^2 \sum_i (x_{ik} - \bar{x}_{\cdot k})^2 \right\}^{1/2} \quad (j \neq k). \quad (5.18)$$

The eigenvalues and eigenvectors of $\mathbf{R}$ will differ from those of $\mathbf{\Sigma}$, and we may end up with a different number k of relevant principal components. Kendall [1975: p. 22] gives an example in which the eigenvalues of $\hat{\mathbf{\Sigma}}$ are $(86.640, 7.094, 0.471, 0.258)$, with the first eigenvalue accounting for $100 \times 86.64/94.463 = 92\%$ of the total, and those of $\mathbf{R}$ are $(1.676, 1.146, 0.960, 0.218)$, with the first eigenvalue now accounting for only 42% of the total. The corresponding eigenvectors also have different elements with regard to both size and sign. Some practitioners maintain that $\mathbf{R}$ should always be used instead of $\hat{\mathbf{\Sigma}}$ or $\mathbf{S}$, as $\mathbf{R}$ does not depend on the scales used for the original variables. This approach would seem reasonable in psychological and educational studies where the scales may be arbitrary and the data little more than ranks. However, the distribution theory associated with $\mathbf{R}$ is much more complex and there are some problems in interpreting the coefficients given by the eigenvectors. Gnanadesikan [1977: p. 12] concludes that there "does not seem to be any *general* elementary rationale to motivate the choice of scaling of the variables as a preliminary to principal components analysis on the resulting covariance matrix." About all that can be said is that similar variables should be measured in the same units where possible.

We note that the sum of the eigenvalues of $\mathbf{R}$ is d, the trace of $\mathbf{R}$, so that the average is 1. The arbitrary rule, sometimes used in computer packages, of including only those eigenvalues greater than unity is often inappropriate. If we obtained eigenvalues $(1.1, 1.0, 0.9)$, we would not wish to exclude the component corresponding to 0.9.

It is well known that estimates of variances, covariances, and correlations can be very sensitive to outliers, so we can expect principal components to have the same sensitivity. For instance, Devlin et al. [1981] mention an example of annual observations on 14 economic variables for 29 chemical companies, in which the first two principal components are given in Fig. 5.1. These two components have the required zero sample correlation, but if the outlier is excluded, the correlation coefficient of the remaining 28 points is 0.99. Admittedly, this is an extreme example, but the message is clear: Robust estimates of $\boldsymbol{\mu}$, $\mathbf{\Sigma}$, and $\mathbf{P}_\rho$ are advisable (Section 4.4.3b). One computational procedure for extracting robust components is given by Campbell [1980a].

EXAMPLE 5.1 Jeffers [1967] described a study carried out on the strength of wooden props used in mining. A number of variables were measured on each pitprop and the prop was compressed in a vertical position until failure occurred. The variables measured were the following:

Topdiam(x_1) = the top diameter of the prop, in inches;

Length(x_2) = the length of the prop, in inches;

Moist(x_3) = the moisture content of the prop, expressed as a percentage of dry weight;

Testsg(x_4) = the specific gravity of the timber at the time of the test;

Ovensg(x_5) = the oven-dry specific gravity of the timber;

Ringtop(x_6) = the number of annual rings at the top of the prop;

Ringbut(x_7) = the number of annual rings at the base of the prop;

Bowmax(x_8) = the maximum bow, in inches;

Bowdist(x_9) = the distance of the point of maximum bow from the top of the prop, in inches;

Whorls(x_{10}) = the number of knot whorls;

Clear(x_{11}) = the length of clear prop from the top of the prop, in inches;

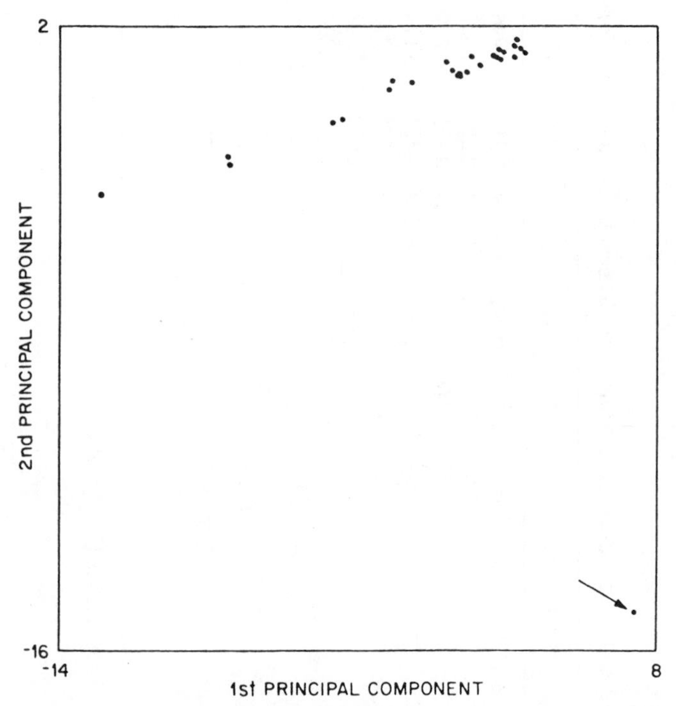

Fig. 5.1 First two principal components for chemical companies data. From Devlin et al. [1981: Fig. 1].

TABLE 5.2 Correlation Matrix for Physical Properties of Pit Props (Upper Triangle)[a]

Variable	x_1	x_2	x_3	x_4	x_5	x_6	x_7	x_8	x_9	x_{10}	x_{11}	x_{12}	x_{13}
x_1	1.000	0.954	0.364	0.342	−0.129	0.313	0.496	0.424	0.592	0.545	0.084	−0.019	0.134
x_2	—	1.000	0.297	0.284	−0.118	0.291	0.503	0.419	0.648	0.569	0.076	−0.036	0.144
x_3	—	—	1.000	0.882	−0.148	0.153	−0.029	−0.054	0.125	−0.081	0.162	0.220	0.126
x_4	—	—	—	1.000	0.220	0.381	0.174	−0.059	0.137	−0.014	0.097	0.169	0.015
x_5	—	—	—	—	1.000	0.364	0.296	0.004	−0.039	0.037	−0.091	−0.145	−0.208
x_6	—	—	—	—	—	1.000	0.813	0.090	0.211	0.274	−0.036	0.024	−0.329
x_7	—	—	—	—	—	—	1.000	0.372	0.465	0.679	−0.113	−0.232	−0.424
x_8	—	—	—	—	—	—	—	1.000	0.482	0.557	0.061	−0.357	−0.202
x_9	—	—	—	—	—	—	—	—	1.000	0.526	0.085	−0.127	−0.076
x_{10}	—	—	—	—	—	—	—	—	—	1.000	−0.319	−0.368	−0.291
x_{11}	—	—	—	—	—	—	—	—	—	—	1.000	0.029	0.007
x_{12}	—	—	—	—	—	—	—	—	—	—	—	1.000	0.184
x_{13}	—	—	—	—	—	—	—	—	—	—	—	—	1.000

[a]From Jeffers [1967: Table 2].

Knots(x_{12}) = the average number of knots per whorl;

Diaknot(x_{13}) = the average diameter of the knots, in inches.

Although one of the aims of the study was to relate the maximum compressive strength to the variables, an initial attempt was made to reduce the dimension of the problem using principal components. Table 5.2 gives the upper triangle of the correlation matrix for the 13 variables measured on $n = 180$ props cut from Corsican pine. The eigenvalues of this matrix are given in Table 5.3 and the first step is to decide on the number of components k. Bearing in mind the sampling variability of the eigenvalues, it is not easy to choose a cutting point. The rather arbitrary rule of considering components with eigenvalues greater than unity would lead to $k = 4$, thus accounting for 74% of the sum of the eigenvalues. This rule tends to give too few components and does not seem appropriate here. A figure approaching 90% would often be regarded as adequate and Jeffers settled for $k = 6$, giving a total contribution of about 87%.

The corresponding eigenvectors for the first six components are given in Table 5.4 and, instead of being scaled to have unit length, they have been scaled so that the largest element in each vector is unity. Jeffers suggested that these elements may be interpretated as the relative weights given to the variables in each component, and that important variables are those with high negative or positive weights (say, greater than 0.7). For example, the first component gives high positive weights to the top diameter, the length, the number of rings at the top and base, the bow, and the number of whorls, and may be interpreted as a general index of the size of the prop. The second

TABLE 5.3 Eigenvalues of the Correlation Matrix in Table 5.2

Component	Eigenvalue	Percent of Sum	Cumulative Percent
1	4.219	32.4	32.4
2	2.378	18.3	50.7
3	1.878	14.4	65.2
4	1.109	8.5	73.7
5	0.910	7.0	80.7
6	0.815	6.3	87.0
7	0.576	4.4	91.4
8	0.440	3.4	94.8
9	0.353	2.7	97.5
10	0.191	1.5	99.0
11	0.051	0.4	99.4
12	0.041	0.3	99.7
13	0.039	0.3	100.0
Sum	13.000		

TABLE 5.4 Eigenvectors for First Six Components of Pitprop Data[a]

Variable	*Eigenvector for component* 1	2	3	4	5	6
TOPDIAM	0·96	0·40	−0·43	−0·11	0·14	0·19
LENGTH	1·00	0·34	−0·49	−0·13	0·19	0·26
MOIST	0·31	1·00	0·29	0·10	−0·58	−0·44
TESTSG	0·43	0·84	0·73	0·07	−0·59	−0·09
OVENSG	0·14	−0·31	1·00	0·06	−0·29	1·00
RINGTOP	0·70	−0·26	0·99	−0·08	0·53	0·08
RINGBUT	0·99	−0·35	0·53	−0·81	0·36	0·00
BOWMAX	0·72	−0·35	−0·51	0·36	−0·31	−0·09
BOWDIST	0·88	0·32	−0·43	0·12	0·18	0·05
WHORLS	0·93	−0·46	−0·25	−0·26	−0·26	−0·28
CLEAR	−0·03	0·38	−0·15	1·00	0·57	0·28
KNOTS	−0·28	0·63	0·19	−0·37	1·00	−0·27
DIAKNOT	−0·27	0·57	−0·68	−0·38	−0·13	1·00

[a]From Jeffers [1967: Table 4].

component, giving high weight only to moisture content and the specific gravity of green timber, is a measure of the degree of seasoning. The third component is a measure of the rate of growth and strength of the timber, while the fourth component is a contrast between the length of the clear prop from the top and the number of rings at the base. The fifth component is a direct measure of the number of knots per whorl, and the sixth component is a combined index of the average diameter of the knots and the strength of the timber, that is, a combined strength index.

EXAMPLE 5.2 A further study described by Jeffers [1967] concerned the degree of variation in 40 individual alate aldelgids (winged aphids) caught in a light trap. Interest centered on the number of distinct taxa present in the trapping area. As the aphids are difficult to identify with any certainty by conventional taxonomic keys, principal component analysis was suggested as a possible aid to the identification of distinct taxa. A total of 19 variables were measured, including body length (Length), body width (Width), forewing length (Forwing), using a microscope. Body length and width and all leg measurements were recorded at a magnification of $\times 10$, wing length at $\times 3$, and all other measurements at $\times 40$. Table 5.5 gives the upper half of the correlation matrix, and the very high correlations indicate a high degree of collinearity among the variables. In effect, very few basic dimensions of the individuals had been measured and Jeffers suggested that new variables, uncorrelated with the ones measured, should be sought. This is indicated by Table 5.6, which shows that the first two components account for 85% of the sum of the eigenvalues, and the first four components account for 92%. From the corresponding eigenvectors in Table 5.7, scaled so that the largest element is unity, the first component is a general index of size, with approximately

TABLE 5.5 Correlation Matrix for Winged Aphid Variables (Lower Triangle)[a]

x_1	x_2	x_3	x_4	x_5	x_6	x_7	x_8	x_9	x_{10}
1.000									
0.934	1.000								
0.927	0.941	1.000							
0.909	0.944	0.933	1.000						
0.524	0.487	0.543	0.499	1.000					
0.799	0.821	0.856	0.833	0.703	1.000				
0.854	0.865	0.886	0.889	0.719	0.923	1.000			
0.789	0.834	0.846	0.885	0.253	0.699	0.751	1.000		
0.835	0.863	0.862	0.850	0.462	0.752	0.793	0.745	1.000	
0.845	0.878	0.863	0.881	0.567	0.836	0.913	0.787	0.805	1.000
−0.458	−0.496	−0.522	−0.488	−0.174	−0.317	−0.383	−0.497	−0.356	−0.371
0.917	0.942	0.940	0.945	0.516	0.846	0.907	0.861	0.848	0.902
0.939	0.961	0.956	0.952	0.494	0.849	0.914	0.876	0.877	0.901
0.953	0.954	0.946	0.949	0.452	0.823	0.886	0.878	0.883	0.891
0.895	0.899	0.882	0.908	0.551	0.831	0.891	0.794	0.818	0.848
0.691	0.652	0.694	0.623	0.815	0.812	0.855	0.410	0.620	0.712
0.327	0.305	0.356	0.272	0.746	0.553	0.567	0.067	0.300	0.384
−0.676	−0.712	−0.667	−0.736	−0.233	−0.504	−0.502	−0.758	−0.666	−0.629
0.702	0.729	0.746	0.777	0.285	0.499	0.592	0.793	0.671	0.668

x_{11}	x_{12}	x_{13}	x_{14}	x_{15}	x_{16}	x_{17}	x_{18}	x_{19}
1.000								
−0.465	1.000							
−0.447	0.981	1.000						
−0.439	0.971	0.991	1.000					
−0.405	0.908	0.920	0.921	1.000				
−0.198	0.725	0.714	0.676	0.720	1.000			
−0.032	0.396	0.360	0.298	0.378	0.781	1.000		
0.492	−0.657	−0.655	−0.687	−0.633	−0.186	0.169	1.000	
−0.425	0.696	0.724	0.731	0.694	0.287	−0.026	−0.775	1.000

[a] From Jeffers [1967: Table 9].

equal weights given to most of the variables. The second component is almost entirely a measure of ovipositor spines, and the third and fourth components are measures of the number of antennal spines and of the number of spiracles, respectively.

As the first two components account for such a high percentage of the variation, they can be plotted for each insect, as in Fig. 5.2. The plot suggests the presence of four major groups, recognizable by differences in the general size of the organism and the number of ovipositor spines. It would appear that not all the variables are necessary: Only one of the main group of variables such as the length of tibia III, together with the number ovipositor spines, the number of antennal spines, and the number of spiracles, need be retained.

TABLE 5.6 Eigenvalues for the 19 Components of the Correlation Matrix in Table 5.5

Component	Eigenvalue	Percent of Sum	Cumulative Percent
1	13.861	73.0	73.0
2	2.370	12.5	85.4
3	0.748	3.9	89.4
4	0.502	2.6	92.0
5	0.278	1.4	93.5
6	0.266	1.4	94.9
7	0.193	1.0	95.9
8	0.157	0.8	96.7
9	0.140	0.7	97.4
10	0.123	0.6	98.1
11	0.092	0.4	98.6
12	0.074	0.4	99.0
13	0.060	0.3	99.3
14	0.042	0.2	99.5
15	0.036	0.2	99.7
16	0.024	0.1	99.8
17	0.020	0.1	99.9
18	0.011	0.1	100.0
19	0.003	0.0	100.0
	19.000		

TABLE 5.7 Eigenvectors for First Four Components of Aphid Variables[a]

Variable	*Eigenvectors for component* 1	2	3	4
LENGTH	0·96	−0·06	0·03	−0·12
WIDTH	0·98	−0·12	0·01	−0·16
FORWING	0·99	−0·06	−0·06	−0·11
HINWING	0·98	−0·16	0·03	−0·00
SPIRAC	0·61	0·74	−0·20	1·00
ANTSEG 1	0·91	0·33	0·04	0·02
ANTSEG 2	0·96	0·30	0·00	−0·04
ANTSEG 3	0·88	−0·43	0·06	−0·18
ANTSEG 4	0·90	−0·08	0·18	−0·01
ANTSEG 5	0·94	0·05	0·11	0·03
ANTSPIN	−0·49	0·37	1·00	0·27
TARSUS 3	0·99	−0·02	0·03	−0·29
TIBIA 3	1·00	−0·05	0·09	−0·31
FEMUR 3	0·99	−0·12	0·12	−0·31
ROSTRUM	0·96	0·02	0·08	−0·06
OVIPOS	0·76	0·73	−0·03	−0·09
OVSPIN	0·41	1·00	−0·16	−0·06
FOLD	−0·71	0·64	0·04	−0·80
HOOKS	0·76	−0·52	0·06	0·72

[a] From Jeffers [1967: Table 11].

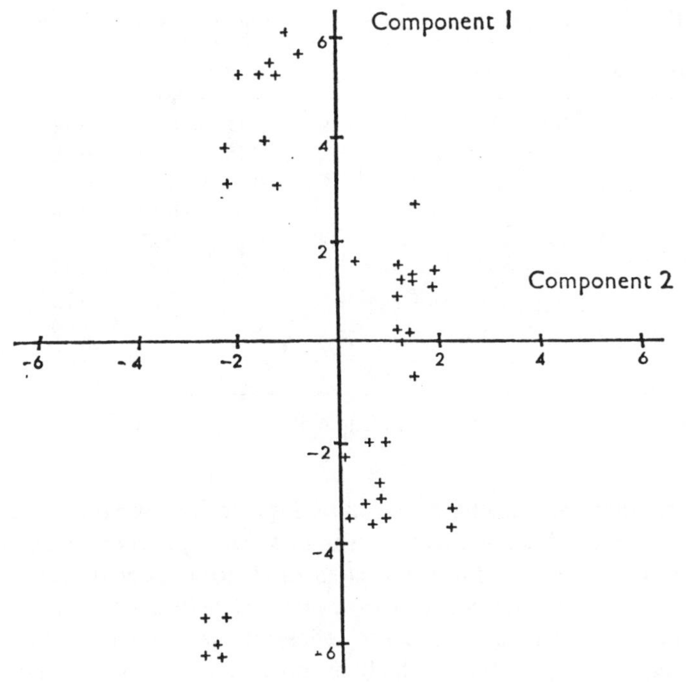

Fig. 5.2 Plotted values of the first two components for individual insects. From Jeffers [1967: Fig. 1].

EXAMPLE 5.3 Duziński and Arnold [1973] described an interesting study that compared the diets of sheep and cattle grazing together on sown pasture. Each animal was fitted with an esophageal fistula so that diets could be sampled at regular intervals. Pastures were sampled on the same day as the diets were sampled by cutting the grass on a number of quadrats and then separating the material into categories. The response or y-variables were the differences between cattle and sheep in the percentages of various substances in the fistulated material. The pasture or x-variables were transformations of total dry matter (x_1), green dry matter (x_2), the percentage of edible green (x_3), green grass leaf (x_4), dry grass leaf (x_5), green clover (x_6), dry clover (x_7), stem (x_8), and inert matter (x_9). The transformations were basically logarithmic in an endeavor to normalize the data. As a first step in the study, an attempt was made to reduce the dimension of the pasture data using principal components. The correlation matrix for 18 pasture samples is given in Table 5.8, and the eigenvalues are listed in Table 5.9. The authors chose the first four principal components, which contributed 91% of the total variation, and the corresponding eigenvectors are given in Table 5.10. As with the previous two examples, we have scaled the eigenvectors so that the largest element in the vector is unity. The authors actually scaled each eigenvector so that its squared norm was equal to the eigenvalue ($\hat{\mathbf{t}}_i'\hat{\mathbf{t}}_i = \hat{\lambda}_i$).

TABLE 5.8 Correlation Matrix for Pasture Variables (Upper Triangle)[a]

Variable	x_1	x_2	x_3	x_4	x_5	x_6	x_7	x_8	x_9
x_1	1.00	0.11	0.04	0.11	0.42	0.11	0.22	0.34	−0.51
x_2	—	1.00	0.86	0.98	−0.11	0.76	−0.36	−0.48	0.13
x_3	—	—	1.00	0.84	−0.33	0.80	−0.57	−0.71	−0.11
x_4	—	—	—	1.00	−0.13	0.64	−0.39	−0.48	0.12
x_5	—	—	—	—	1.00	−0.17	0.21	0.39	−0.06
x_6	—	—	—	—	—	1.00	−0.24	−0.43	0.06
x_7	—	—	—	—	—	—	1.00	0.72	0.30
x_8	—	—	—	—	—	—	—	1.00	0.19
x_9	—	—	—	—	—	—	—	—	1.00

[a]From Dudziński and Arnold [1973: Table 3].

The first component is characterized by high positive weights for green dry matter, percentage of edible green, green grass leaf, and green clover, and high negative weights for dry clover and stem (and, to a lesser degree, dry grass leaf). This principal component gives a contrast between the green and dry components of the pasture and hence represents an index of "greenness." A positive value of y_1 indicates a high proportion of green material, while a negative value indicates that the pasture is largely dry. The second principal component has a high positive weight for total dry matter and a negative weight for inert matter, thus representing a measure of "useful bulk." The third component, with high weights for inert and, to a lesser degree, dry clover, stem, and green dry matter, is used as a measure of "total bulk." The fourth principal component represents a contrast between dry grass leaf and dry clover and measures the type of dry material available to animals. A positive

TABLE 5.9 Eigenvalues of the Correlation Matrix in Table 5.8

Component	Eigenvalue	Percent of Sum	Cumulative Percent
1	4.250	47.2	47.2
2	1.771	19.7	66.9
3	1.460	16.2	83.1
4	0.731	8.1	91.2
5	0.397	4.4	95.7
6	0.230	2.6	98.2
7	0.109	1.2	99.4
8	0.050	0.6	100.0
9	0.003	0.0	100.0
Sum	9.001		

value of y_4 implies that the amount of dry grass leaf is dominant, while a negative value implies that dry clover is dominant.

5.2.5 *Inference for Sample Components*

Suppose that the observations $\mathbf{x}_1, \mathbf{x}_2, \ldots, \mathbf{x}_n$ are a random sample from $N_d(\boldsymbol{\mu}, \boldsymbol{\Sigma})$; then $\bar{\mathbf{x}}$ and $\hat{\boldsymbol{\Sigma}}$ are the maximum likelihood estimators of $\boldsymbol{\mu}$ and $\boldsymbol{\Sigma}$. If $\boldsymbol{\Sigma}$ is positive definite with distinct eigenvalues λ_j, then $\boldsymbol{\Sigma} = \mathbf{T}\boldsymbol{\Lambda}\mathbf{T}'$, where $\mathbf{T}$ is unique (with an appropriate convention for the signs of the eigenvectors $\mathbf{t}_i$; see A1.3). The uniqueness of $\mathbf{T}$ implies that we have a one-to-one relationship between the $\frac{1}{2}d(d+1)$ distinct elements of $\boldsymbol{\Sigma}$ (in the upper triangle) and the $d + \frac{1}{2}d(d-1) = \frac{1}{2}d(d+1)$ "free" elements of $\boldsymbol{\Lambda}$ and $\mathbf{T}$. Since $\mathbf{T}'\mathbf{T} = \mathbf{I}_d$, we can choose $d-1$ elements of $\mathbf{t}_1$, $d-2$ elements of $\mathbf{t}_2$, and so on, or $1 + 2 + \cdots + d - 1 = \frac{1}{2}d(d-1)$ elements of $\mathbf{T}$ altogether. The maximum likelihood estimate $\hat{\boldsymbol{\Sigma}} = \hat{\mathbf{T}}\hat{\boldsymbol{\Lambda}}\hat{\mathbf{T}}'$ of $\boldsymbol{\Sigma}$ is also positive definite with distinct eigenvalues (with probability 1), so that, by A7.2, $\hat{\lambda}_j$ and $\hat{\mathbf{t}}_j$ are the maximum likelihood estimates of λ_j and $\mathbf{t}_j$ and have the usual properties such as asymptotic normality. Similarly, the sample principle components will be maximum likelihood estimates of the population components.

Under normality, $n\hat{\boldsymbol{\Sigma}} \sim W_d(n-1, \boldsymbol{\Sigma})$ (see Theorem 3.1) and the joint distribution of the $\hat{\lambda}_j$ can be found (James [1960]): Sugiyama [1966] gave the distribution of the largest eigenvalue and corresponding eigenvector. However, these distributions are complicated and a number of asymptotic expansions have been given for distinct λ_j and some λ_j equal (see the reviews of Muirhead [1978, 1982] and Tyler [1981]). For distinct eigenvalues, Girshick [1939] showed that the set of random variables $\hat{\lambda}_j$, $\hat{t}_{jk}$ $(j, k = 1, 2, \ldots, d)$ is asymptotically multivariate normal with $\{\hat{\lambda}_1, \hat{\lambda}_2, \ldots, \hat{\lambda}_d\}$ asymptotically independent of

TABLE 5.10 Eigenvectors for First Four Components of Pasture Data

Variable		Eigenvector 1	2	3	4
Total dry matter	(x_1)	−0.07	1.00	−0.18	−0.31
Green dry matter	(x_2)	0.92	0.27	0.35	0.10
Percent edible green	(x_3)	1.00	0.10	−0.03	−0.14
Green grass leaf	(x_4)	0.90	0.24	0.31	0.14
Dry grass leaf	(x_5)	−0.36	0.67	0.09	1.00
Green clover	(x_6)	0.81	0.24	0.29	−0.30
Dry clover	(x_7)	−0.65	0.26	0.63	−0.55
Stem	(x_8)	−0.79	0.37	0.43	−0.19
Inert	(x_9)	−0.04	−0.43	1.00	0.27

$\{\hat{\mathbf{t}}_1, \hat{\mathbf{t}}_2, \ldots, \hat{\mathbf{t}}_d\}$. The asymptotic moments are given by

$$E[\hat{\lambda}_j] = \lambda_j + \sum_{\alpha \neq j} \frac{\lambda_j \lambda_\alpha (\lambda_j - \lambda_\alpha)^{-1}}{n} + O(n^{-2}), \tag{5.19}$$

$$\text{var}[\hat{\lambda}_j] = \frac{2\lambda_j^2}{n}\left[1 - \frac{1}{n} \sum_{\alpha \neq j} \lambda_\alpha^2 (\lambda_j - \lambda_\alpha)^{-2}\right] + O(n^{-3}), \tag{5.20}$$

$$\text{cov}[\hat{\lambda}_j, \hat{\lambda}_k] = O(n^{-2}) \qquad (j \neq k),$$

$$\mathscr{E}[\hat{\mathbf{t}}_j] = \mathbf{t}_j + O_d(n^{-1}),$$

$$\mathscr{D}[\hat{\mathbf{t}}_j] = \frac{1}{n} \sum_{\alpha \neq j} \lambda_j \lambda_\alpha (\lambda_j - \lambda_\alpha)^{-2} \mathbf{t}_\alpha \mathbf{t}'_\alpha + O_d(n^{-2}),$$

$$\mathscr{C}[\hat{\mathbf{t}}_j, \hat{\mathbf{t}}_k] = -\frac{1}{n} \lambda_j \lambda_k (\lambda_j - \lambda_k)^{-2} \mathbf{t}_j \mathbf{t}'_k + O_d(n^{-2}) \qquad (j \neq k).$$

Since the covariances of the $\hat{\lambda}_j$ are asymptotically zero, we have that the standardized random variables $(n/2)^{1/2}(\hat{\lambda}_j - \lambda_j)/\lambda_j$ are asymptotically i.i.d. $N_1(0,1)$. We note that if $\lambda_{j+1} \approx \lambda_j$, then $\mathscr{D}[\hat{\mathbf{t}}_j]$ will be large, indicating instability in the estimate $\hat{\mathbf{t}}_j$. Using a Taylor expansion and (5.20), we also have that the random variables

$$\sqrt{\frac{n}{2}}\,(\log \hat{\lambda}_j - \log \lambda_j)$$

are asymptotically i.i.d. $N_1(0,1)$; however, both normal approximations are not very accurate (Konishi and Sugiyama [1981]).

When some of the λ_j are equal, the above theory does not hold. In the first place there is some arbitrariness in defining eigenvectors for equal eigenvalues (see A1.3), so that we do not have a one-to-one correspondence between $\mathbf{\Sigma}$ and the appropriate elements of $\mathbf{\Lambda}$ and $\mathbf{T}$: A7.2 can no longer be invoked. Secondly, the sample eigenvalues are distinct (with probability 1) so that we will have m estimates of an eigenvalue with multiplicity m. However, the average of these estimates turns out to be the maximum likelihood estimate of the repeated eigenvalue, and the corresponding sample eigenvectors are (non-unique) maximum likelihood estimates of the population eigenvectors. The likelihood ratio test for testing $H_0: \lambda_{a+1} = \lambda_{a+2} = \cdots = \lambda_{a+b}$ is given by

$$-2\log \ell = -n \log\left\{\prod_{j=a+1}^{a+b} \frac{\hat{\lambda}_j}{\bar{\lambda}}\right\}, \tag{5.21}$$

where $\bar{\lambda} = (\sum_{j=a+1}^{a+b} \hat{\lambda}_j)/b$. When H_0 is true, $-2 \log \ell$ is approximately χ_ν^2, with $\nu = \frac{1}{2}(b-1)(b+2)$. If $a + b = d$, so that we are testing for the equality of the last b eigenvalues, then a better chi-square approximation is obtained if n in (5.21) is replaced by

$$n - 1 - a - \frac{2b^2 + b + 2}{6b} + \sum_{j=1}^{a} \left[\frac{\bar{\lambda}}{\hat{\lambda}_j - \bar{\lambda}} \right]^2.$$

This correction was suggested by Lawley [1956] and "confirmed" by James [1969]. The statistic (5.21) and extensions of the above asymptotic theory to the case of multiple eigenvalues were given by Anderson [1963b]; further results are given by Fujikoshi [1978]. Anderson also gave a test for the hypothesis that several consecutive eigenvalues are equal to a given nonzero constant. Asymptotic expansions for this test and (5.21) are given by Fujikoshi [1977b] for the null case, and Sugiura [1976b] for the nonnull case of distinct eigenvalues. The statistic (5.21) can also be applied to the correlation matrix $\mathbf{R}$ instead of $\hat{\boldsymbol{\Sigma}}$, but its asymptotic distribution is no longer chi square (Anderson [1963b: p. 136].

The above asymptotic theory assumes that the underlying distribution is MVN. Unfortunately, the large-sample joint distribution of the $\hat{\lambda}_j$ is not robust to departures from normality, being mainly affected by the nonzero fourth cumulants and cross-cumulants of the parent population. Waternaux [1976] showed that for distinct λ_j, the limiting joint distribution of the $\sqrt{n}(\hat{\lambda}_j - \lambda_j)$ is $N_d(\mathbf{0}, \boldsymbol{\Omega})$, where $\boldsymbol{\Omega} = [(\omega_{jk})]$, $\omega_{jj} = k_4^j + 2\lambda_j^2$ and $\omega_{jk} = k_{22}^{jk}$ $(j \neq k)$. Here k_4^j is the fourth cumulant of the jth element of $\mathbf{x}$ and k_{22}^{jk} is the bivariate cumulant of order 4 of the joint distribution of the jth and kth elements. Using a Fisher–Cornish expression for $\hat{\lambda}_j$, Waternaux also obtained an approximation for the percentiles of the distribution of $\hat{\lambda}_j$. From a simulation study for $d = 3$, she demonstrated that, for nonnormal data, large samples ($n > 100$) are required for the asymptotic means and variances to be adequate. Further terms in the asymptotic joint distribution of the $\sqrt{n}(\hat{\lambda}_j - \lambda_j)$ and an asymptotic expression for the distribution of

$$R_q = \frac{\hat{\lambda}_1 + \cdots + \hat{\lambda}_q}{\hat{\lambda}_1 + \cdots + \hat{\lambda}_d} \tag{5.22}$$

are given by Fujikoshi [1980] (see also Exercise 5.9). He also confirms the nonrobustness to departures from normality in both situations and gives asymptotic expansions for the distribution of certain functions of the $\hat{\lambda}_j$ in the nonnormal case. Davis [1977] shows how the limiting distribution of (5.21) is affected by nonnormality and examines briefly the effect of nonnormality on the theory given by Anderson [1963b] for the multiple eigenvalue case. Tyler [1981] gives a general asymptotic inference for eigenvectors.

5.2.6 Preliminary Selection of Variables

In many experimental situations there is a strong temptation to measure every conceivable characteristic of a sample unit in order that "nothing will be missed," for example, height, weight, sex, marital status, income, and size of family. The dimension of the vector observations **x** can therefore grow rapidly, even though n is usually limited because of cost or shortage of sampling units. Every experimenter is therefore faced with the problem of which characteristics should be included, particularly if the data are going to be used extensively to study relationships between the sample units. Before carrying out a principal component analysis (PCA), it may therefore be appropriate to reduce the number of variables to a smaller subset. The aim would be to discard variables that add little to the resulting PCA. Jolliffe [1972, 1973] described eight rejection methods and applied five of them to artificial and real data sets. His first, and slowest, technique (A2) is a stepwise method that rejects at each stage the variable having the largest multiple correlation with the remaining variables (see Section 5.7.1). The process stops when the multiple correlation first falls below some prescribed level (e.g., 0.15). Two techniques (B2 and B4) involving PCA of the full set of variables first are also included and they associate a variable with each of the principal components. The variables associated with the last few components are rejected, the number of rejections being equal to the number of eigenvalues less than a certain value (e.g., 0.70 for the PCA of the correlation matrix). However, having to carry out an initial PCA of the full set does, in some situations, defeat the purpose of trying to reduce the set.

The two final methods (Cl and C2) are much faster and use agglomerative clustering methods for variables based on the single-linkage and group average linkage methods, respectively (Section 7.3.1). The fusion of clusters continues until all the between-cluster similarities first drop below 0.45 for C2 and 0.55 for C_1. If g clusters have been found at this stage, then, using an appropriate criterion (e.g., one of the innermost variables), one variable is selected from each cluster for retention in the final subset of g variables. The principle behind this technique is that variables in a given cluster are similar, so that a suitably chosen candidate will represent the whole cluster. Although there is not a great deal of difference between C_1 and C_2, C_2 seems preferred. This is supported by the demonstrated superiority of average linkage over single linkage in detecting clusters (Section 7.3.2b).

5.2.7 General Applications

An important application of principal components analysis (PCA) is the screening for outliers and redundant dimensions by plotting the first or last few components (Section 4.5). Here robust estimates of $\boldsymbol{\mu}$ and $\boldsymbol{\Sigma}$ should generally be used, as $\bar{\mathbf{x}}$ and $\mathbf{S}$ (or $\hat{\boldsymbol{\Sigma}}$) can be seriously affected by outliers. A normal probability plot of $Q_1^{h_0}$ for each $\mathbf{x}_i$ can also be used [see (5.8)].

Principal components have also been used to analyze vector time series by approximating the series by a filtered version of itself, but restraining the filter to have reduced rank (Brillinger [1975: Chapter 9]). Alternatively, the structure of a series can be studied by reducing the dimensions of the input and output data and estimating the reduced-dimension transfer function (Priestly et al. [1973]). Methods of robust estimation have also been developed for time series (e.g., Martin [1979]).

Frequently the observations $\mathbf{x}_1, \mathbf{x}_2, \ldots, \mathbf{x}_n$ are not a random sample but may come from different populations or groups, and may even represent the whole population. In this case it is not appropriate to assume an underlying sampling theory, but simple regard PCA as a dimension-reducing technique. For example, suppose that several groups have the same variables measured on them, and interest centers on discovering how similar the groups are with respect to their overall features. An intuitive procedure is to apply a separate PCA to each group and then assess the similarity of the groups by comparing the first few components. The angular measures suggested by Krzanowski [1979a] may be useful for comparing the reduced groups. Alternatively, if the group structure is unknown, PCA can be applied to the n observations and the first few components plotted to help indicate possible clusters in the data.

For the group situation, a number of models have been proposed under the title "fixed (-effects) models of PCA" (see Okamoto [1976] for references). From $\mathbf{y}_i = \hat{\mathbf{T}}'(\mathbf{x}_i - \overline{\mathbf{x}})$ we have $\mathbf{x}_i = \overline{\mathbf{x}} + \hat{\mathbf{T}}\mathbf{y}_i$. Extracting the jth elements and letting $\hat{\mathbf{t}}'_j$ equal the jth row of $\hat{\mathbf{T}}$ gives us

$$\begin{aligned} x_{ij} &= \overline{x}_j + \hat{\mathbf{t}}'_j \mathbf{y}_i \\ &= \overline{x}_j + \sum_{r=1}^{k} y_{ir} \hat{t}_{jr} + \varepsilon_{ij}, \end{aligned} \tag{5.23}$$

where $\varepsilon_{ij} = \sum_{r=k+1}^{d} y_{ir} \hat{t}_{jr}$. By analogy, we have a fixed-effects model

$$x_{ij} = \beta_j + \sum_{r=1}^{k} p_{ir} m_{jr} + \varepsilon_{ij} \tag{5.24}$$

with an appropriate orthogonal structure imposed on the m_{jr} and the p_{ir}, mimicking the orthogonality of the first k columns of $\hat{\mathbf{T}}$ and the zero sample covariances of the y_{ir}. If we use the standardized components $y_{ir} = \hat{\lambda}_r^{1/2} z_{ir}$ in (5.23), we have the analogous representation

$$\sum_{r=1}^{k} p_{ir} m_{jr} = \sum_{r=1}^{k} \delta_r l_{ir} m_{jr}, \tag{5.25}$$

which is of the form (10.20) given by a singular value decomposition. One method of estimation for (5.24) consists first of ignoring (5.25) and forming the usual analysis of variance matrix of residuals $\mathbf{R} = [(x_{ij} - \hat{\beta}_j)]$. Then we obtain the singular value decomposition of $\mathbf{R}$ and use the first k eigenvalues and corresponding eigenvectors to give estimates of δ_r, l_{ir}, and m_{jr}. If $k = d$, there is nothing left over for estimating the "error" ε_{ij}. The smallest eigenvalues are therefore used to estimate the ε_{ij}. This method can be applied to models more general than (5.24). For example, Mandel [1971, 1972] initially proposed using the technique for the model

$$x_{ij} = \mu + \alpha_i + \beta_j + \gamma_{ij} + \varepsilon_{ij},$$

where the interaction γ_{ij} takes the form (5.25) and

$$\mathbf{R} = \left[(x_{ij} - \hat{\mu} - \hat{\alpha}_i - \hat{\beta}_j)\right] = \left[(x_{ij} - \bar{x}_{i\cdot} - \bar{x}_{\cdot j} + \bar{x}_{\cdot\cdot})\right].$$

For further details and references relating to the above model, see Yochmowitz and Cornell [1978], Krzanowski [1979b], Cornelius [1980], Krishnaiah and Yochmowitz [1980], and Marasinghe and Johnson [1981]. A cross-validation method for estimating k in the model (5.24) is given by Wold [1978]. Bradu and Gabriel [1978] show how the biplot of Section 5.3 can be used for choosing the appropriate two-way model.

Another application of PCA is in regression analysis. Consider the regression model

$$\begin{aligned} y_i &= \beta_0 + \beta_1 x_{i1} + \cdots + \beta_{p-1} x_{i,p-1} + \varepsilon_i \qquad (i = 1, 2, \ldots, n), \\ &= \alpha + \beta_1 (x_{i1} - \bar{x}_{\cdot 1}) + \cdots + \beta_{p-1}(x_{i,p-1} - \bar{x}_{\cdot,p-1}) + \varepsilon_i. \end{aligned}$$

This can be expressed in the form

$$\mathbf{y} = \alpha \mathbf{1}_n + \tilde{\mathbf{X}}\boldsymbol{\beta} + \boldsymbol{\varepsilon},$$

where $\tilde{\mathbf{X}} = (\mathbf{x}_1 - \bar{\mathbf{x}}, \mathbf{x}_2 - \bar{\mathbf{x}}, \ldots, \mathbf{x}_n - \bar{\mathbf{x}})'$. If

$$\tilde{\mathbf{X}}'\tilde{\mathbf{X}} = \sum_{i=1}^{n} (\mathbf{x}_i - \bar{\mathbf{x}})(\mathbf{x}_i - \bar{\mathbf{x}})'$$

has an eigenvalue $\hat{\lambda}$ close to zero, with corresponding eigenvector $\hat{\mathbf{t}}$, then

$$\tilde{\mathbf{X}}'\tilde{\mathbf{X}}\hat{\mathbf{t}} = \hat{\lambda}\hat{\mathbf{t}} \approx \mathbf{0},$$

and premultiplying by $\hat{\mathbf{t}}'$ gives us $\tilde{\mathbf{X}}\hat{\mathbf{t}} \approx \mathbf{0}$; that is, multicollinearity is present. If, in general, we have $d - k$ linear constraints $\tilde{\mathbf{X}}\hat{\mathbf{T}}_2 \approx \mathbf{O}$, where $\hat{\mathbf{T}} = (\hat{\mathbf{T}}_1, \hat{\mathbf{T}}_2)$,

then

$$\begin{aligned}\tilde{\mathbf{X}}\boldsymbol{\beta} &= \tilde{\mathbf{X}}\hat{\mathbf{T}}(\hat{\mathbf{T}}'\boldsymbol{\beta}) \\ &\approx (\tilde{\mathbf{X}}\hat{\mathbf{T}}_1, \mathbf{O})(\hat{\mathbf{T}}'\boldsymbol{\beta}) \\ &= (\tilde{\mathbf{X}}\hat{\mathbf{T}}_1)(\hat{\mathbf{T}}_1'\boldsymbol{\beta}) \\ &= \mathbf{Z}\boldsymbol{\phi}, \end{aligned} \tag{5.26}$$

where the regression model is now expressed in terms of the first k principal components instead of the original variables [see (5.14) transposed]. However, care is needed, as y may be highly correlated with one of the omitted components (see Jolliffe [1982] for examples].

The idea of using PCA for analyzing multicollinearity and finding a meaningful regression subset was suggested by Jeffers [1967] and Cox [1968: p. 272], and is discussed in detail by several authors (see Mansfield et al. [1977], Hill et al. [1977], and White and Gunst [1979] for references). The method is also referred to as latent root regression. A useful practical discussion on the role of PCA in detecting multicollinearity in regression models is given by Chatterjee and Price [1977: Chapter 7] and Mandel [1982]. An interesting example is given by Box et al. [1973]. However the technique seems of limited usefulness as a method of finding a regression subset (see Draper and Smith [1981: p. 332 and p. 337].

5.2.8 Generalized Principal Components Analysis

If the d-dimensional observations $\mathbf{x}_1, \mathbf{x}_2, \ldots, \mathbf{x}_n$ all approximately satisfy $d - k$ linear constraints, then the observations lie approximately in a k-dimensional subspace. Using PCA, we can estimate the constraints, namely,

$$\hat{\mathbf{t}}_j'(\mathbf{x} - \bar{\mathbf{x}}) \approx 0 \qquad (j = k+1, k+2, \ldots, d), \tag{5.27}$$

and use the first k principal components as the coordinates in the k-dimensional subspace. However, the observations may, instead, satisfy nonlinear constraints so that a reduced nonlinear coordinate system is more appropriate. For example, suppose $d = 2$ and each $\mathbf{x} = (x_1, x_2)'$ satisfies

$$x_1^2 - 2x_1x_2 + x_1 - x_2 + 3 = 0.$$

If we know that any constraint would be quadratic, we could consider

$$\mathbf{x}^* = \left(x_1^2, x_2^2, x_1x_2, x_1, x_2\right)'$$

and focus our attention on

$$\mathbf{a}'\mathbf{x}^* = a_1x_1^2 + a_2x_2^2 + a_3x_1x_2 + a_4x_1 + a_5x_2,$$

a general quadratic in x_1 and x_2. Working with the n vectors $\mathbf{x}_i^*$, we can calculate $\bar{\mathbf{x}}^*$ and $\mathbf{S}^*$, and extract eigenvalues λ_j^* and eigenvectors $\mathbf{t}_j^*$, say. For this "starred" system the principal component transformation is $\mathbf{y}^* = \mathbf{T}^{*\prime}(\mathbf{x}^* - \bar{\mathbf{x}}^*)$, and we would expect $\lambda_5^* = 0$. Also

$$\begin{aligned} y_5^* &= \mathbf{t}_5^{*\prime}(\mathbf{x}^* - \bar{\mathbf{x}}^*) \\ &= t_{15}^*(x_1^* - \bar{x}_1^*) + t_{25}^*(x_2^* - \bar{x}_2^*) + \cdots + t_{55}^*(x_5^* - \bar{x}_5^*) \\ &= t_{15}^*x_1^* + t_{25}^*x_2^* + \cdots + t_{55}^*x_5^* - c, \qquad \text{say}, \\ &= b(x_1^2 - 2x_1x_2 + x_1 - x_2 + 3) \\ &= 0, \end{aligned} \tag{5.28}$$

since y_5^* has sample mean zero and sample variance λ_5^* zero. The scale factor b arises as $\|\mathbf{t}_5^*\| = 1$. Thus any quadratic constraints approximately satisfied by the data will correspond to small λ_j^*.

If the type of nonlinear constraint is unknown, we may wish to consider a cubic polynomial so that

$$\mathbf{x}^* = (x_1^3, x_2^3, x_1^2x_2, x_1x_2^2, x_1^2, x_2^2, x_1, x_2)'.$$

However, the dimension of $\mathbf{x}^*$ grows rapidly with the dimension d of $\mathbf{x}$ and the degree of the polynomial. For example, with quadratic constraints, $\mathbf{x}^*$ has dimension $2d + \frac{1}{2}d(d-1)$ and we require n to exceed this if the method is to work (e.g., $n > 27$ for $d = 6$).

This nonlinear technique was proposed by Gnanadesikan and Wilk [1966, 1969] and considered independently by Van de Geer [1968]. It can be motivated along the lines of Theorem 5.4. The reader is referred to Gnanadesikan [1977: p. 53] for further details and examples.

5.3 BIPLOTS AND h-PLOTS

The biplot is a procedure introduced by Gabriel [1971] that graphically describes both relationships among the d-dimensional observations $\mathbf{x}_1, \mathbf{x}_2, \ldots, \mathbf{x}_n$ and relationships among the variables. It is based on the standard result, demonstrated below, that any $n \times m$ matrix $\mathbf{B} = [(b_{ij})]$ of rank r can be factorized (nonuniquely) as

$$\mathbf{B} = \mathbf{G}\mathbf{H}', \tag{5.29}$$

where $\mathbf{G}$ and $\mathbf{H}$ are $n \times r$ and $m \times r$ matrices, respectively, of rank r. Thus

$$b_{ij} = \mathbf{g}_i'\mathbf{h}_j, \tag{5.30}$$

where $\mathbf{g}_i'$ and $\mathbf{h}_j'$ are the rows of $\mathbf{G}$ and $\mathbf{H}$, respectively, and we have a representation of the b_{ij} in terms of r-dimensional vectors. This factorization assigns each $\mathbf{g}_i$ to a row of $\mathbf{B}$ and each $\mathbf{h}_j$ to a column of $\mathbf{B}$. If $r = 2$, we can plot the $n + m$ vectors in two dimensions and obtain what Gabriel [1971] calls the biplot. For $r > 2$ it may be possible to approximate $\mathbf{B}$ satisfactorily by a rank 2 matrix $\mathbf{B}_{(2)}$, and the corresponding biplot may shed light on $\mathbf{B}$ itself.

From (5.29) we can write $\mathbf{B} = (\mathbf{G}\mathbf{R}')(\mathbf{H}\mathbf{R}^{-1})'$ for any nonsingular $\mathbf{R}$, and this nonuniqueness of the factorization can lead to rather different looking biplots (e.g., Exercise 5.10). Apart from an orthogonal transformation (rotation or reflection), which does not change the relations between vectors, the factorization can be made unique by imposing a particular metric on the columns of $\mathbf{G}$ or $\mathbf{H}$. Clearly, we would wish to choose a factorization in which the $\mathbf{G}$ and $\mathbf{H}$ have meaningful properties. A natural approach is to use the singular value decomposition (see Section 10.1.5)

$$\mathbf{B} = \mathbf{L}_r\boldsymbol{\Delta}_r\mathbf{M}_r' = \sum_{i=1}^{r} \delta_i \boldsymbol{l}_i \mathbf{m}_i', \tag{5.31}$$

where $\mathbf{L}_r$ is an $n \times r$ matrix of rank r with orthogonal columns $\boldsymbol{l}_i$ (i.e., $\mathbf{L}_r'\mathbf{L}_r = \mathbf{I}_r$), $\mathbf{M}_r$ is an $m \times r$ matrix of rank r with orthogonal columns $\mathbf{m}_i$ (i.e., $\mathbf{M}_r'\mathbf{M}_r = \mathbf{I}_r$), $\boldsymbol{\Delta}_r = \text{diag}(\delta_1, \delta_2, \ldots, \delta_r)$, and $\delta_1 \geq \delta_2 \geq \cdots \geq \delta_r > 0$ are the positive singular values of $\mathbf{B}$ (i.e., the positive square roots of the nonzero eigenvalues of $\mathbf{B}'\mathbf{B}$). Dropping the matrix subscript r, we see from (5.31) that

$$\mathbf{B}'\mathbf{B}\mathbf{M} = \mathbf{M}\boldsymbol{\Delta}\mathbf{L}'\mathbf{L}\boldsymbol{\Delta}\mathbf{M}'\mathbf{M} = \mathbf{M}\boldsymbol{\Delta}^2 \tag{5.32}$$

or

$$\mathbf{B}'\mathbf{B}\mathbf{m}_i = \delta_i^2\mathbf{m}_i, \tag{5.33}$$

so that the columns of $\mathbf{M}$ are eigenvectors of $\mathbf{B}'\mathbf{B}$. Writing

$$\mathbf{B} = \mathbf{L}(\boldsymbol{\Delta}\mathbf{M}') = \mathbf{G}\mathbf{H}',$$

say, we have verified the existence of the factorization introduced in (5.29). For this particular choice of $\mathbf{G}$ and $\mathbf{H}$,

$$\mathbf{H}\mathbf{H}' = \mathbf{M}\mathbf{\Delta}^2\mathbf{M}' = \sum_{i=1}^{r} \delta_i^2 \mathbf{m}_i \mathbf{m}_i' = \mathbf{B}'\mathbf{B}, \tag{5.34}$$

$$\mathbf{G}\mathbf{G}' = \mathbf{L}\mathbf{L}' \tag{5.35}$$

$$= (\mathbf{B}\mathbf{M}\mathbf{\Delta}^{-1})(\mathbf{B}\mathbf{M}\mathbf{\Delta}^{-1})' \qquad [\text{by } (5.31)]$$

$$= \mathbf{B}(\mathbf{B}'\mathbf{B})^{-}\mathbf{B}' \qquad (\text{by Exercise 5.11}). \tag{5.36}$$

The factorization (5.29) is now unique up to an orthogonal transformation, as we have imposed constraints $\mathbf{G}'\mathbf{G} = \mathbf{L}'\mathbf{L} = \mathbf{I}_r$ (see Exercise 5.12).

If we want to find a rank s $(s < r)$ approximation to $\mathbf{B}$, then applying a Frobenius norm criterion leads to the approximation (see A7.3)

$$\mathbf{B}_{(s)} = \sum_{i=1}^{s} \delta_i \boldsymbol{l}_i \mathbf{m}_i',$$

where $\mathbf{B}_{(s)}$ is the value of the $n \times m$ matrix $\mathbf{C}$ of rank s that minimizes

$$\|\mathbf{B} - \mathbf{C}\|^2 = \sum_{i=1}^{n} \sum_{j=1}^{m} \left(b_{ij} - c_{ij}\right)^2.$$

The lack of fit can be measured by (see Exercise 5.13)

$$\left\|\mathbf{B} - \mathbf{B}_{(s)}\right\|^2 = \delta_{s+1}^2 + \delta_{s+2}^2 + \cdots + \delta_r^2, \tag{5.37}$$

and the goodness of fit by

$$\rho_s^{(2)} = 1 - \left(\left\|\mathbf{B} - \mathbf{B}_{(s)}\right\|^2 / \|\mathbf{B}\|^2\right)$$

$$= \sum_{i=1}^{s} \delta_i^2 \Big/ \sum_{i=1}^{r} \delta_i^2. \tag{5.38}$$

In addition to minimizing a certain norm, we now establish another useful property of $\mathbf{B}_{(s)}$.

LEMMA 5.6 If the rows of $\mathbf{B}$ sum to zero (i.e., $\mathbf{B}'\mathbf{1}_n = \mathbf{0}$), then, in terms of the Frobenius norm, $\mathbf{B}_{(s)}$ is the rank s matrix whose column differences best approximate the column differences of $\mathbf{B}$.

Proof Let $\mathbf{C}$ be an $n \times m$ matrix such that $\mathbf{C}'\mathbf{1}_n = \mathbf{0}$, and let $\mathbf{b}_j$ and $\mathbf{c}_j$ be columns of $\mathbf{B}$ and $\mathbf{C}$. If $\mathbf{u}_j = \mathbf{b}_j - \mathbf{c}_j$, then

$$\begin{aligned}\sum_{j=1}^{m}\sum_{k=1}^{m}\|\mathbf{b}_j - \mathbf{b}_k - (\mathbf{c}_j - \mathbf{c}_k)\|^2 &= \sum_{j=1}^{m}\sum_{k=1}^{m}\|\mathbf{u}_j - \mathbf{u}_k\|^2 \\ &= 2m\sum_{j=1}^{m}\|\mathbf{u}_j - \bar{\mathbf{u}}\|^2 \qquad \text{(Exercise 5.14)} \\ &= 2m\sum_{j=1}^{m}\|\mathbf{u}_j\|^2,\end{aligned}$$

as $\mathbf{B}'\mathbf{1}_n = \mathbf{C}'\mathbf{1}_n = \mathbf{0}$ implies that $\bar{\mathbf{u}} = \mathbf{0}$. Now

$$\sum_{j=1}^{m}\|\mathbf{u}_j\|^2 = \sum_{i=1}^{n}\sum_{j=1}^{m}\left(b_{ij} - c_{ij}\right)^2 = \|\mathbf{B} - \mathbf{C}\|^2,$$

which is minimized when $\mathbf{C} = \mathbf{B}_{(s)}$. (The same result applies to row differences if $\mathbf{B}\mathbf{1}_m = \mathbf{0}$.) □

Gabriel [1971] applied the above theory to the data matrix $\mathbf{B} = \tilde{\mathbf{X}} = (\tilde{\mathbf{x}}_1, \tilde{\mathbf{x}}_2, \ldots, \tilde{\mathbf{x}}_n)'$, where $\tilde{\mathbf{x}}_i = \mathbf{x}_i - \bar{\mathbf{x}}$ and $m = r = d$. Then $\mathbf{B}'\mathbf{1}_n = \mathbf{0}$ and

$$\mathbf{B}'\mathbf{B} = \sum_{i=1}^{n}\tilde{\mathbf{x}}_i\tilde{\mathbf{x}}_i' = \mathbf{Q} = (n-1)\mathbf{S}, \tag{5.39}$$

where $\mathbf{S}$ is the unbiased estimate of the dispersion matrix of the $\mathbf{x}_i$. If $\delta_1^2 \geq \delta_2^2 \geq \cdots \geq \delta_d^2$ are the ordered eigenvalues of $\mathbf{Q}$, then from (5.35) and (5.34), $\mathbf{G} = (\boldsymbol{l}_1, \boldsymbol{l}_2, \ldots, \boldsymbol{l}_d)$ and $\mathbf{H} = (\delta_1\mathbf{m}_1, \delta_2\mathbf{m}_2, \ldots, \delta_d\mathbf{m}_d)$, where $\mathbf{G}$ is $n \times d$ and $\mathbf{H}$ is $d \times d$. Also, from (5.34) and (5.36), $\mathbf{H}\mathbf{H}' = \mathbf{Q}$ and $\mathbf{G}\mathbf{G}' = \tilde{\mathbf{X}}\mathbf{S}^{-1}\tilde{\mathbf{X}}/(n-1)$. We can get rid of the factor of $(n-1)^{-1}$ from $\mathbf{G}\mathbf{G}'$ by rescaling, namely,

$$\mathbf{G} = (n-1)^{1/2}(\boldsymbol{l}_1, \boldsymbol{l}_2, \ldots, \boldsymbol{l}_d), \qquad \mathbf{H} = (n-1)^{-1/2}(\delta_1\mathbf{m}_1, \delta_2\mathbf{m}_2, \ldots, \delta_d\mathbf{m}_d). \tag{5.40}$$

Then (5.29) still holds with

$$\mathbf{H}\mathbf{H}' = \mathbf{S} \quad \text{and} \quad \mathbf{G}\mathbf{G}' = \tilde{\mathbf{X}}\mathbf{S}^{-1}\tilde{\mathbf{X}}', \tag{5.41}$$

and the rows $\mathbf{g}_i'$ and $\mathbf{h}_j'$ of $\mathbf{G}$ and $\mathbf{H}$, respectively [see (5.30)] have the following properties:

(1) From (5.41), $\mathbf{g}'_\alpha \mathbf{g}_\beta = \tilde{\mathbf{x}}'_\alpha \mathbf{S}^{-1} \tilde{\mathbf{x}}_\beta$, and the distance between the points represented by $\mathbf{g}_\alpha$ and $\mathbf{g}_\beta$, namely,

$$\|\mathbf{g}_\alpha - \mathbf{g}_\beta\| = \left\{\mathbf{g}'_\alpha \mathbf{g}_\alpha + \mathbf{g}'_\beta \mathbf{g}_\beta - 2\mathbf{g}'_\alpha \mathbf{g}_\beta\right\}^{1/2}$$

$$= \left\{(\tilde{\mathbf{x}}_\alpha - \tilde{\mathbf{x}}_\beta)' \mathbf{S}^{-1} (\tilde{\mathbf{x}}_\alpha - \tilde{\mathbf{x}}_\beta)\right\}^{1/2}$$

$$= \left\{(\mathbf{x}_\alpha - \mathbf{x}_\beta)' \mathbf{S}^{-1} (\mathbf{x}_\alpha - \mathbf{x}_\beta)\right\}^{1/2}$$

is the Mahalanobis distance between the observations $\mathbf{x}_\alpha$ and $\mathbf{x}_\beta$ (Section 1.5).

(2) The sample covariance between variables j and k is

$$s_{jk} = \mathbf{h}'_j \mathbf{h}_k,$$

and the sample variances are $\|\mathbf{h}_j\|^2$.

(3) The correlation between variables j and k is the cosine of the angle between $\mathbf{h}_j$ and $\mathbf{h}_k$ ($= \mathbf{h}'_j \mathbf{h}_k / \|\mathbf{h}_j\| \, \|\mathbf{h}_k\|$).

(4) We note that

$$\|\mathbf{h}_j - \mathbf{h}_k\|^2 = \mathbf{h}'_j \mathbf{h}_j + \mathbf{h}'_k \mathbf{h}_k - 2\mathbf{h}'_j \mathbf{h}_k$$

$$= s_{jj} + s_{kk} - 2s_{jk}$$

is the sample variance of the difference between variables j and k.

If we approximate $\mathbf{B}$ by $\mathbf{B}_{(2)}$, then we can use the corresponding biplot of the $n + d$ vectors $\mathbf{g}_i$ and $\mathbf{h}_j$ to visually provide approximate information on distances between observations, and on sample variances, covariances, and correlations between variables. The approximation $\mathbf{B}_{(2)}$ will be satisfactory if $\rho_2^{(2)}$ of (5.38) is close to unity.

In conclusion, we see that if we are given $\mathbf{S}$, we can find the two largest eigenvalues λ_1 and λ_2, where $\lambda_i = \delta_i^2/(n-1)$, with corresponding unit eigenvectors $\mathbf{m}_1$ and $\mathbf{m}_2$ [see (5.33)], using a principal components program. We then form the matrix

$$\mathbf{H}_{(2)} = \left(\sqrt{\lambda_1}\, \mathbf{m}_1, \sqrt{\lambda_2}\, \mathbf{m}_2\right), \tag{5.42}$$

where the d rows $\mathbf{h}_j$ of $\mathbf{H}_{(2)}$ have, approximately, properties (2)–(4) above. Corsten and Gabriel [1976] call the plot of just the $\mathbf{h}_j$ the h-plot of $\mathbf{S}$, and demonstrated how sample covariances matrices can compared by looking at their h-plots. Instead of using $\|\mathbf{B} - \mathbf{B}_{(2)}\|^2$ as a measure of lack of fit, the authors proposed using $\|\mathbf{S} - \mathbf{H}_{(2)}\mathbf{H}'_{(2)}\|^2$. This leads to the goodness of fit measure

$$\rho_2 = \left(\lambda_1^2 + \lambda_2^2\right) / \sum_{j=1}^{d} \lambda_j^2$$

for h-plotting, where $\Sigma_j \lambda_j^2 = \mathrm{tr}\mathbf{S}^2 = \Sigma_j \Sigma_k s_{jk}^2$.

The general problem of least squares approximation of matrices by "additive" and "multiplicative" models is considered by Gabriel [1978] and Gabriel and Zamir [1979: weighted least squares] while matrix approximation in general is considered by Rao [1980]. Bradu and Gabriel [1978] give an application of biplots to two-way models. A technique, similar in spirit to the biplot, is correspondence analysis (Hill [1974], Mardia et al. [1979], Gordon [1981], GENSTAT statistical package).

EXAMPLE 5.4 Corsten and Gabriel [1976] demonstrated their h-plot method on daily rainfall data in eight regions of Israel: three northern regions (coast, interior, and east), three corresponding central regions, a narrow buffer region between north and center, and a small region in the south. Each observation was a rainy day's precipitation averaged over several rain gauges in one region.

TABLE 5.11 Variances, Covariances, and Correlations of Daily Rainfall (mm) in Israel—945 Rainy Days, Preexperimental, 1949–1960 (Correlations Below Main Diagonal)[a]

	Nc	Ni	Ne	B	Cc	Ci	Ce	S
Nc	86.68	80.49	75.50	61.16	80.57	62.93	47.29	39.38
Ni	.8804	96.43	90.29	61.43	88.35	75.90	63.01	45.64
Ne	.7966	.9032	103.63	61.34	93.63	81.24	67.44	47.59
B	.7802	.7429	.7156	70.90	86.35	64.44	41.45	31.40
Cc	.6591	.6852	.7005	.7811	172.40	126.88	96.76	73.52
Ci	.6317	.7224	.7458	.7153	.9031	114.49	101.74	74.38
Ce	.4229	.5343	.5516	.4099	.6136	.7917	144.24	109.64
S	.3714	.4080	.4104	.3274	.4916	.6103	.8015	129.73

Regions: N-North, B-buffer, C-center, S-South; c-coast, i-interior, e-East.

[a]Reproduced from L. C. A. Corsten and K. R. Gabriel [1976]. *Biometrics*, **32**, 851–863, Table P. With permission from The Biometric Society.

TABLE 5.12 Variances, Covariances, and Correlations of Daily Rainfall (mm) in Israel—208 Rainy Days, North-Seeded in 1961–1967 Experiments (Correlations Below Main Diagonal)[a, b]

	Nc	Ni	Ne	B	Cc	Ci	Ce	S
Nc	133.17	145.30	90.70	116.91	70.73	75.97	44.00	26.82
Ni	.9184	187.96	123.21	142.93	91.51	100.77	63.77	32.27
Ne	.7955	.9096	97.61	92.01	58.55	65.32	42.77	21.66
B	.7805	.8032	.7175	168.48	91.62	90.07	43.84	23.25
Cc	.5980	.6512	.5782	.6886	105.06	113.33	72.43	52.29
Ci	.5622	.6277	.5646	.5926	.9442	137.12	87.93	62.26
Ce	.3786	.4619	.4299	.3354	.7017	.7457	101.40	82.16
S	.2317	.2347	.2186	.1786	.5086	.5301	.8135	100.60

Regions: See Table 5.11

[a]Reproduced from L. C. A. Corsten and K. R. Gabriel (1976). Graphical exploration in comparing variance matrices. *Biometrics*, **32**, 851–863, Table N. With permission from The Biometric Society. [b]*Regions*: See Table 5.11.

TABLE 5.13 Coordinates for *h*-Plots of Rainfall Data[a]

		North-Seeded		Pre-experimental	
North	coast	10.10	−3.72	7.39	−3.81
	interior	12.70	−3.97	8.40	−3.35
	east	8.42	−2.71	8.71	+3.09
Buffer		11.10	−3.86	6.70	−3.31
Center	coast	8.82	3.29	11.80	−1.54
	interior	9.78	4.63	10.08	0.73
	east	6.68	6.69	9.61	6.28
South		4.60	7.50	7.78	7.18
Goodness-of-fit		98.79%		98.73%	
Sample Size		208		845	
λ_1		696.54		639.79	
λ_2		185.12		140.23	
λ_3		63.79		62.36	
$\Sigma_i \lambda_i$		1031.40		918.49	
$\Sigma\lambda^2$		525809.17		434494.42	

[a]Reproduced from L. C. A. Corsten and K. R. Gabriel (1976). Graphical exploration in comparing variance matrices. *Biometrics*, **32**, 851–863, Table of coordinates. With permission from The Biometric Society.

The regions were used for a rainfall stimulation experiment in 1961–1967, and on each experimental day cloud seeding was randomly assigned either to the north or to the center and carried out upwind of those regions. Prior to the experiment, "natural" rainfall data was collected over the same regions from 1949 to 1960. The sample covariances and correlations for two of the cases, pre-experimental and north seeded, are presented in Tables 5.11 and 5.12. By way of illustration, the sample dispersion matrix in Table 5.11 has eigenvalues $\lambda_1 = 639.79$, $\lambda_2 = 140.23$, $\lambda_3 = 62.36$, and so on. The matrix $\mathbf{H}_{(2)}$ of (5.42) is given by the last two columns in the upper part of Table 5.13, and the rows of $\mathbf{H}_{(2)}$, namely, $(7.39, -3.81)$, and so forth, form the coordinates for the arrows of the *h*-plot (Fig. 5.3). The goodness of fit measure, $100\rho_2$, is 98.73%, and such a high value indicates an excellent fit of $\mathbf{H}_{(2)}$ to $\mathbf{H}$. In particular, $\|\mathbf{h}_j\|$ and $\|\mathbf{h}_j - \mathbf{h}_k\|$ for $\mathbf{H}_{(2)}$ will be close to their counterparts for $\mathbf{H}$, the latter being upper bounds. The lengths of the rays in the *h*-plots and the cosines of the angles between the rays will therefore be close to the respective standard derivations of the variables and the correlation coefficients. The axes of the plots are used only for constructing the plot and not for interpreting them. Corsten and Gabriel [1976] came to the following conclusions with regard to the pre-experimental and north-seeded plots.

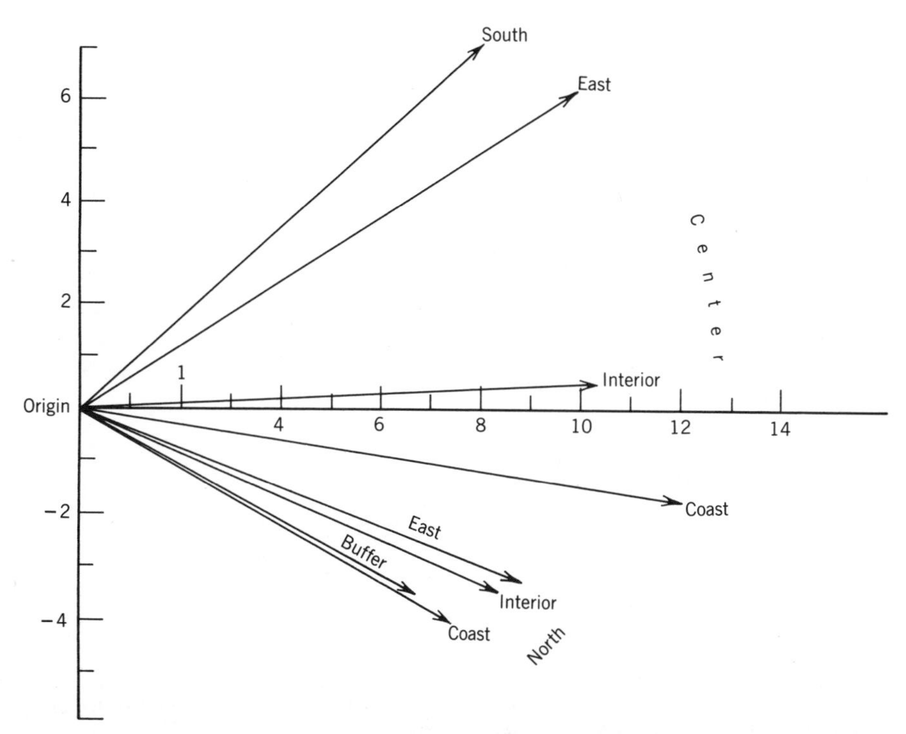

Fig. 5.3 The *h*-plot of variances of rainfall on pre-experimental days. Redrawn from Corsten and Gabriel [1976]. With permission from The Biometric Society.

Figure 5.3 shows that "natural" variability in rainfall was much the same in all regions, except that the northern and buffer regions had slightly smaller standard derivations. The correlations show a clear geographical pattern: a highly correlated cluster of northern regions and the buffer; little correlation of this cluster with the dry regions of the south and center–east; and the central coast and interior are in between, being more highly correlated with the north than with the south. For the north-seeded days, Fig. 5.4 shows that the center–coast and center–interior are less highly correlated with the north and buffer than under "natural" conditions. Furthermore, north and buffer rainfall variability appear to have increased (except in the east) under north seeding, and the seeded areas are more highly correlated. This suggests that seeding not only increased average rainfall, but also increased variability and reduced the correlation with unseeded regions.

The authors suggested a number of test procedures based on the *h*-plot and the reader is referred to their paper for further details.

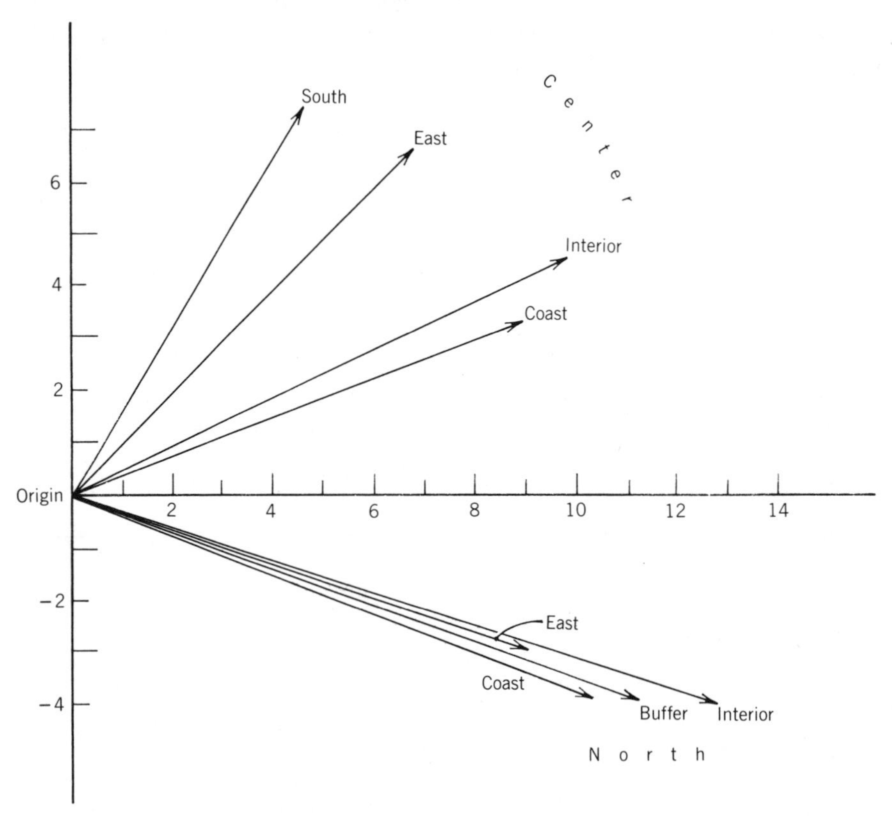

Fig. 5.4 The *h*-plot of variances of rainfall on north-seeded days. Redrawn from Corsten and Gabriel [1976]. With permission from the Biometric Society.

5.4 FACTOR ANALYSIS

5.4.1 Underlying Model

Let $\mathbf{x} = (x_1, x_2, \ldots, x_d)'$ be a d-dimensional random vector with mean $\boldsymbol{\mu}$ and dispersion matrix $\boldsymbol{\Sigma}_x$. Suppose we believe that, apart from a random fluctuation or "error" term, the elements of $\mathbf{x}$ can be explained in terms of a smaller number m of unknown random variables $(f_1, f_2, \ldots, f_m) = \mathbf{f}'$ called (*common*) *factors*. In the simplest model we assume that the factors affect $\mathbf{x}$ linearly so that we consider a linear model of the form

$$\mathbf{x} = \boldsymbol{\mu} + \boldsymbol{\Gamma}\mathbf{f} + \boldsymbol{\varepsilon}, \tag{5.43}$$

where $\boldsymbol{\varepsilon}$ is the "error" term and everything on the right-hand side of (5.43) is unknown. The elements of $\boldsymbol{\varepsilon}$ are usually called the *specific* or *unique factors*. We also make the reasonable assumptions that $\mathbf{f}$ and $\boldsymbol{\varepsilon}$ have zero means and are uncorrelated. The (j, k)th element of the $d \times m$ constant matrix $\boldsymbol{\Gamma}$, γ_{jk}, say, is called the weight or *factor loading* of x_j on the factor f_k as

$$x_j = \mu_j + \sum_{k=1}^{m} \gamma_{jk} f_k. \tag{5.44}$$

For identifiability we require that each column of $\boldsymbol{\Gamma}$ have at least two nonzero elements; otherwise $\mathbf{f}$ and $\boldsymbol{\varepsilon}$ are "confounded" (see Exercise 5.15). It is further assumed that the factors account for all the correlation structure so that the elements of the residual or error vector $\boldsymbol{\varepsilon} = \mathbf{x} - \boldsymbol{\mu} - \boldsymbol{\Gamma}\mathbf{f}$ are uncorrelated, that is, $\boldsymbol{\Psi} = \mathscr{D}[\boldsymbol{\varepsilon}]$ is a diagonal matrix, $\text{diag}(\psi_1^2 \psi_2^2, \ldots, \psi_d^2)$. We can therefore write

$$\begin{aligned}
\boldsymbol{\Sigma}_x &= \mathscr{D}[\boldsymbol{\Gamma}\mathbf{f} + \boldsymbol{\varepsilon}] \\
&= \mathscr{D}[\boldsymbol{\Gamma}\mathbf{f}] + \mathscr{D}[\boldsymbol{\varepsilon}] \\
&= \boldsymbol{\Gamma}\mathscr{D}[\mathbf{f}]\boldsymbol{\Gamma}' + \boldsymbol{\Psi} \\
&= \boldsymbol{\Gamma}\boldsymbol{\Sigma}_f\boldsymbol{\Gamma}' + \boldsymbol{\Psi},
\end{aligned} \tag{5.45}$$

say. However, we also have an equivalent representation

$$\begin{aligned}
\mathbf{x} &= \boldsymbol{\mu} + \left(\boldsymbol{\Gamma}\boldsymbol{\Sigma}_f^{1/2}\right)\left(\boldsymbol{\Sigma}_f^{-1/2}\mathbf{f}\right) + \boldsymbol{\varepsilon} \\
&= \boldsymbol{\mu} + \tilde{\boldsymbol{\Gamma}}\tilde{\mathbf{f}} + \boldsymbol{\varepsilon},
\end{aligned}$$

with

$$\begin{aligned}
\boldsymbol{\Sigma}_x &= \left(\boldsymbol{\Gamma}\boldsymbol{\Sigma}_f^{1/2}\right)\left(\boldsymbol{\Gamma}\boldsymbol{\Sigma}_f^{1/2}\right)' + \boldsymbol{\Psi} \\
&= \tilde{\boldsymbol{\Gamma}}\tilde{\boldsymbol{\Gamma}}' + \boldsymbol{\Psi},
\end{aligned}$$

and $\mathscr{D}[\tilde{\mathbf{f}}] = \mathbf{\Sigma}_f^{-1/2}\mathscr{D}[\mathbf{f}]\mathbf{\Sigma}_f^{-1/2} = \mathbf{I}_m$. In our original model, (5.43), we can therefore assume that the factors are standardized and uncorrelated, that is, $\mathbf{\Sigma}_f = \mathbf{I}_m$. We now have

$$\mathbf{\Sigma}_x = \mathbf{\Gamma}\mathbf{\Gamma}' + \mathbf{\Psi} \tag{5.46}$$

$$= \mathbf{\Pi} + \mathbf{\Psi}, \tag{5.47}$$

say.

Given a random sample of observations $\mathbf{x}_1, \mathbf{x}_2, \ldots, \mathbf{x}_n$, the basic problem is to decide whether $\mathbf{\Sigma}_x$ can be expressed in the form (5.46) for a reasonably small value of m, and to estimate the elements of $\mathbf{\Gamma}$ and $\mathbf{\Psi}$. However, some care is needed in developing estimation procedures, as the model is undetermined. In the first instance we will have, from the sample data, estimates of the $\frac{1}{2}d(d+1)$ distinct elements of the upper triangle of $\mathbf{\Sigma}_x$, but on the right-hand side of (5.46) we have $dm + d$ parameters, dm for $\mathbf{\Gamma}$ and d for $\mathbf{\Psi}$. The solution will be indeterminate unless $\frac{1}{2}d(d+1) - d(m+1) \geq 0$, or $d > 2m$. Even if this condition is satisfied, $\mathbf{\Gamma}$ is not unique. If $\mathbf{L}$ is any $m \times m$ orthogonal matrix, then

$$\mathbf{\Gamma}\mathbf{f} = (\mathbf{\Gamma}\mathbf{L})(\mathbf{L}'\mathbf{f}) = \mathbf{\Gamma}^*\mathbf{f}^*, \tag{5.48}$$

say, where $\mathscr{D}[\mathbf{f}^*] = \mathbf{L}'\mathscr{D}[\mathbf{f}]\mathbf{L} = \mathbf{L}'\mathbf{L} = \mathbf{I}_m$. Therefore any orthogonal rotation of $\mathbf{f}$ will do just as well as $\mathbf{f}$ itself.

There is also a further source of nonuniqueness in the model that is not usually recognized (Francis [1973]). If the $d \times m$ matrix $\mathbf{\Gamma}$ is of rank r $(r < m)$, then $\mathbf{\Gamma}\mathbf{\Gamma}'$ is positive semidefinite of rank r and there exists the factorization (see A4.3) $\mathbf{\Gamma}\mathbf{\Gamma}' = \mathbf{\Gamma}_r\mathbf{\Gamma}_r'$, where $\mathbf{\Gamma}_r$ is $d \times r$ of rank r. In this case the model

$$\mathbf{x} = \boldsymbol{\mu} + \mathbf{\Gamma}_r\mathbf{h} + \boldsymbol{\varepsilon},$$

where $\mathbf{h}$ is a vector of r factors, has the same dispersion as (5.46). Another representation for $\mathbf{\Sigma}_x$, which leads to a different $\mathbf{\Psi}$, is given in Exercise 5.16.

The above problem of uniqueness is generally resolved by choosing an orthogonal rotation $\mathbf{L}$ such that the final $\mathbf{\Gamma}$ [called $\mathbf{\Gamma}^*$ in (5.48)] satisfies the condition that $\mathbf{\Gamma}'\mathbf{\Psi}^{-1}\mathbf{\Gamma}$ is diagonal with positive diagonal elements [see later, following (5.54)]. This restriction requires $\mathbf{\Gamma}$ to be of full rank m. Given a "valid" $\mathbf{\Psi}$, namely, one with all positive diagonal elements, then for this particular $\mathbf{\Psi}$ it can be shown that the above restriction yields a unique $\mathbf{\Gamma}$, provided that a suitable convention is adopted with the sign of each column of $\mathbf{\Gamma}$. However given $\mathbf{\Sigma}_x$, such a $\mathbf{\Psi}$ may not exist for a given m; and if it does exist, it may not be unique (see Lawley and Maxwell [1971: Section 2.3]). Uniqueness can also be obtained by specifying a priori the values of a sufficient number of parameters, the so-called "restricted" model (Lawley and Maxwell [1971: Chapter 7]). One rigorous approach to the problem of uniqueness is given by Williams [1978].

Once Γ is estimated, $\mathbf{f}$ is frequently rotated, as in (5.48), to obtain more "meaningful" factors, that is, obtain a more meaningful structure in Γ^*. Suppose, for example, that the estimated factor loadings for the jth component of $\mathbf{x}$, given by the jth row of the estimate $\hat{\Gamma}$ [see (5.44)], consist of a few moderate values and a large number of small values. Then it may be considered worth "rotating" $\hat{\Gamma}$ so that a few values become relatively large and the rest become zero or negligible. This can be achieved by maximizing certain functions of the squares of the rotated loadings. The most popular method is *varimax* (Kaiser [1958], Lawley and Maxwell [1971: Section 6.3]), which seeks to maximize the sum, over the columns, of the "variance" of the squared loadings in a column of the rotated $\hat{\Gamma}$. A less satisfactory technique is *quartimax*, which seeks to maximize the corresponding sum of variances for the rows.

With an orthogonal rotation, uncorrelated factors remain uncorrelated. However, if we make a general nonsingular linear transformation $\mathbf{g} = \mathbf{R}\mathbf{f}$, we have the model $\Gamma\mathbf{f} = (\Gamma\mathbf{R}^{-1})\mathbf{g}$, and the new factors $\mathbf{g}$ are now correlated with dispersion matrix $\mathscr{D}[\mathbf{g}] = \mathbf{R}\mathbf{R}'$. Such transformations are called *oblique rotations* and are frequently used by practictioners to achieve "better" models. One such rotation is *biquartimin*, which minimizes the sum of the "covariances" of pairs of columns of the squared loadings of $\Gamma\mathbf{R}^{-1}$. For a general reference on rotations see Williams [1979].

The two techniques of principal component analysis (PCA) and factor analysis (FA) are often confused, particularly in the behavioral sciences, and it is perhaps appropriate to make some comparisons at this stage. If $\mathbf{y}$ is the set of d principal components of $\mathbf{x}$, then $\mathscr{D}[\mathbf{y}] = \Lambda = \text{diag}(\lambda_1, \lambda_2, \ldots, \lambda_d)$ and we can write (see Section 5.2.1 for notation)

$$\begin{aligned}
\mathbf{x} - \boldsymbol{\mu} &= \mathbf{T}\mathbf{y} \\
&= (\mathbf{T}_1, \mathbf{T}_2)\begin{pmatrix} \mathbf{y}_1 \\ \mathbf{y}_2 \end{pmatrix}, \qquad \text{say,} \\
&= \mathbf{T}_1\mathbf{y}_1 + \mathbf{T}_2\mathbf{y}_2 \\
&= \left(\mathbf{T}_1\Lambda_1^{1/2}\right)\left(\Lambda_1^{-1/2}\mathbf{y}_1\right) + \mathbf{T}_2\mathbf{y}_2 \\
&= \mathbf{B}_1\mathbf{g}_1 + \boldsymbol{\eta},
\end{aligned} \tag{5.49}$$

say, where the elements of $\mathbf{y}_1$ are the first m principal components, the elements of $\mathbf{g}_1$ are the first m standardized principal components, and $\Lambda_1 = \text{diag}(\lambda_1, \lambda_2, \ldots, \lambda_m)$. This model looks like the FA model (5.43) as $\mathscr{D}[\mathbf{g}_1] = \Lambda_1^{-1/2}\mathscr{D}[\mathbf{y}_1]\Lambda_1^{-1/2} = \mathbf{I}_m$ and $\mathscr{C}[\mathbf{g}_1, \boldsymbol{\eta}] = \Lambda_1^{-1/2}\mathscr{C}[\mathbf{y}_1, \mathbf{y}_2]\mathbf{T}_2' = \mathbf{O}$. However, if $\Lambda_2 = \text{diag}(\lambda_{m+1}, \lambda_{m+2}, \ldots, \lambda_d)$, then $\mathscr{D}[\boldsymbol{\eta}] = \mathbf{T}_2\Lambda_2\mathbf{T}_2'$ is not diagonal, so that $\mathbf{g}_1$ does not "explain" all the correlation structure in $\mathbf{x}$. There are no assumptions underlying PCA; we simply rotate $\mathbf{x}$ to $\mathbf{y}$ with the hope that some elements of $\mathbf{y}$ have small variances. If the eigenvalues of $\boldsymbol{\Sigma}_x$ are distinct, the rotation that achieves the ordering of the eigenvalues is unique (with a suitable convention for the sign of an eigenvector). However, FA makes the strong assumption that there is an underlying linear model that accounts completely for the correlation structure. Moreover, the underlying factors can be arbitrarily rotated without changing the form of the model. Because of the underlying linear structure, FA regression has been proposed as a method of linear prediction instead of the usual least squares regression for situations when the regressors ("independent" variables) are subject to multicollinearity and measurement errors (see Isogawa and Okamoto [1980] for references).

In conclusion, we note that the above approach is not suitable for categorical data. For models and references relating to this topic and the related subject of latent structure analysis, see Bartholomew [1980] and Andersen [1982].

5.4.2 *Estimation Procedures*

Before discussing several methods of estimation, we introduce some terminology commonly used in the factor analysis literature. If $\boldsymbol{\Gamma} = [(\gamma_{jk})]$, then the diagonal elements of (5.46) give us

$$\begin{aligned} \sigma_{jj} &= \sum_{k=1}^{m} \gamma_{jk}^2 + \psi_j^2 \\ &= h_j^2 + \psi_j^2, \end{aligned} \tag{5.50}$$

say, where h_j^2 is called the *communality* or *common variance* and ψ_j^2 is called the *residual variance* or *unique variance* of x_j. Obvious estimates of $\boldsymbol{\mu}$ and $\boldsymbol{\Sigma}$ are $\overline{\mathbf{x}}$ and $\mathbf{S}$, and we now consider several methods of estimating $\boldsymbol{\Gamma}$, $\boldsymbol{\Psi}$, and $\mathbf{f}$ using $\mathbf{S}$ as a starting point.

a METHOD OF MAXIMUM LIKELIHOOD

Suppose that the sample observations $\mathbf{x}_i$ $(i = 1, 2, \dots, n)$ come from $N_d(\boldsymbol{\mu}, \boldsymbol{\Sigma})$. Then $(n-1)\mathbf{S} \sim W_d(n-1, \boldsymbol{\Sigma})$ and, from (2.3), the log-likelihood function of $\boldsymbol{\Gamma}$ and $\boldsymbol{\psi}$ is

$$\log L(\boldsymbol{\Gamma}, \boldsymbol{\Psi}) = c_1 - \tfrac{1}{2}(n-1)\log|\boldsymbol{\Sigma}| - \tfrac{1}{2}\operatorname{tr}\left[\boldsymbol{\Sigma}^{-1}(n-1)\mathbf{S}\right] \tag{5.51}$$

$$= c_1 - \tfrac{1}{2}(n-1)\left\{\log|\boldsymbol{\Gamma}\boldsymbol{\Gamma}' + \boldsymbol{\Psi}| + \operatorname{tr}\left[(\boldsymbol{\Gamma}\boldsymbol{\Gamma}' + \boldsymbol{\Psi})^{-1}\mathbf{S}\right]\right\}. \tag{5.52}$$

It can be shown that maximizing the above expression with respect to the elements of $\boldsymbol{\Gamma}$ and $\boldsymbol{\Psi}$ leads to the following equations for the maximum likelihood estimates $\hat{\boldsymbol{\Gamma}}$ and $\hat{\boldsymbol{\Psi}}$ (Lawley and Maxwell [1971] with their "n" equal to $n-1$):

$$\mathbf{S}\hat{\boldsymbol{\Psi}}^{-1}\hat{\boldsymbol{\Gamma}} = \hat{\boldsymbol{\Gamma}}\left(\mathbf{I}_m + \hat{\boldsymbol{\Gamma}}'\hat{\boldsymbol{\Psi}}^{-1}\hat{\boldsymbol{\Gamma}}\right) \tag{5.53}$$

and

$$\hat{\boldsymbol{\Psi}} = \operatorname{diag}(\mathbf{S} - \hat{\boldsymbol{\Gamma}}\hat{\boldsymbol{\Gamma}}'). \tag{5.54}$$

However, as already noted, these equations do not have a unique solution, as $\boldsymbol{\Gamma}$ can be arbitrarily rotated. It is therefore usual to impose the restriction that

$$\hat{\mathbf{J}} = \hat{\boldsymbol{\Gamma}}'\hat{\boldsymbol{\Psi}}^{-1}\hat{\boldsymbol{\Gamma}} \tag{5.55}$$

is diagonal. This particular constraint can be justified by the following Bayesian-type argument (Bartholomew [1981]). Assuming $\mathbf{f} \sim N_m(\mathbf{0}, \mathbf{I}_m)$ and the conditional distribution of $\mathbf{x} - \boldsymbol{\mu}$, given $\mathbf{f}$, is $N_d(\boldsymbol{\Gamma}\mathbf{f}, \boldsymbol{\Psi})$, then the conditional distribution of $\mathbf{f}$, given $\mathbf{x}$, is (see Exercise 5.17) $N_m(\boldsymbol{\Gamma}'\boldsymbol{\Sigma}^{-1}(\mathbf{x} - \boldsymbol{\mu}), [\boldsymbol{\Gamma}'\boldsymbol{\Psi}^{-1}\boldsymbol{\Gamma} + \mathbf{I}_m]^{-1})$. If $\boldsymbol{\Gamma}'\boldsymbol{\Psi}^{-1}\boldsymbol{\Gamma}$ is diagonal, then we have chosen the particular rotation that leads to the elements of $\mathbf{f}$ being conditionally independent.

Equations (5.53)–(5.55) must be solved iteratively, and unfortunately there are convergence problems. In fact, prior to 1967 practictioners had so much difficulty with solving the equations that the maximum likelihood method was rarely used and another method, principal factor analysis (see below), became popular. However, there was a major breakthrough when Jöreskog [1967] and Lawley [1967] developed a suitable algorithm based on numerically maximizing $\log L(\boldsymbol{\Gamma}, \boldsymbol{\Psi})$ of (5.52) (see Lawley and Maxwell [1971], Jöreskog [1977], and S. Y. Lee [1979] for details, and Williams [1979] for a historical survey). The process begins with an initial estimate $\boldsymbol{\Psi}_0$ from which $\boldsymbol{\Gamma}_0$ is calculated (Exercise 5.18). The function $\log L(\boldsymbol{\Gamma}_0, \boldsymbol{\Psi})$ is then maximized with respect to $\boldsymbol{\Psi}$ to obtain $\boldsymbol{\Psi}_1$ and corresponding $\boldsymbol{\Gamma}_1$; $\log L(\boldsymbol{\Gamma}_1, \boldsymbol{\Psi})$ is then maximized with respect to $\boldsymbol{\Psi}$, and so on. In practice, the likelihood function may not have a true maximum subject to the constraint $\boldsymbol{\Psi} > \mathbf{O}$. This may be due to the fact that the underlying model is inadequate or simply due to sampling variation. In such cases one or more elements of $\boldsymbol{\Psi}$ tend to zero in the course of the iteration and become negative if allowed to do so. To overcome this problem, the maximization is carried out within the region R_δ for which $\psi_j^2 \geq \delta$ for all j, where δ is an arbitrary small positive number (e.g., 0.005). The solution is said to be improper if it occurs on the boundary of R_δ, and proper otherwise. If $\psi_j^2 = \delta$ for some j, it stays at that value until the iterations stop. Improper solutions occur with surprising frequency and the cause cannot be determined because of the constraints that $\hat{\boldsymbol{\Pi}}$ of (5.47) is positive semidefinite and $\hat{\boldsymbol{\Psi}}$ is positive definite. As the maximum likelihood theory has not been proved valid for boundary points, it has been suggested that we assume $\hat{\boldsymbol{\Pi}}$ to be just symmetric (see Van Driel [1978] for a discussion and references).

One satisfactory feature of the maximum likelihood method is that it is independent of the scales of measurement. To see this we note, from Exercise 5.18, that the computational procedure depends only on finding the eigenvalues and eigenvectors of $\tilde{\mathbf{S}} = \mathbf{\Psi}^{-1/2}\mathbf{S}\mathbf{\Psi}^{-1/2}$. If we make the scale transformation $\mathbf{y}_i = \mathbf{D}\mathbf{x}_i$, where the diagonal elements of the diagonal matrix $\mathbf{D}$ all have the same sign, then, since all the following matrices except $\mathbf{S}$ are diagonal,

$$\begin{aligned}\tilde{\mathbf{S}}_y &= (\mathbf{D}\mathbf{\Psi}\mathbf{D}')^{-1/2}\mathbf{D}\mathbf{S}\mathbf{D}'(\mathbf{D}\mathbf{\Psi}\mathbf{D}')^{-1/2} \\ &= \mathbf{\Psi}^{-1/2}\mathbf{D}^{-1}\mathbf{D}\mathbf{S}\mathbf{D}\mathbf{D}^{-1}\mathbf{\Psi}^{-1/2} \\ &= \mathbf{\Psi}^{-1/2}\mathbf{S}\mathbf{\Psi}^{-1/2} = \tilde{\mathbf{S}}_x.\end{aligned}$$

Given a proper solution for k factors, we can now carry out a goodness of fit test of the hypothesis $H_0: \mathbf{\Sigma} = \mathbf{\Gamma}\mathbf{\Gamma}' + \mathbf{\Psi}$ $(m = k)$ versus the alternative that $\mathbf{\Sigma}$ is unrestricted using the likelihood ratio statistic ℓ. Evaluating (5.51) at $\mathbf{\Sigma} = \mathbf{S}$ and $\mathbf{\Sigma} = \hat{\mathbf{\Sigma}} = \hat{\mathbf{\Gamma}}\hat{\mathbf{\Gamma}}' + \hat{\mathbf{\Psi}}$, we have

$$\begin{aligned}-2\log\ell &= (n-1)\{\log|\hat{\mathbf{\Sigma}}| + \operatorname{tr}[\mathbf{S}\hat{\mathbf{\Sigma}}^{-1}] - \log|\mathbf{S}| - d\} \qquad (5.56) \\ &= (n-1)\left\{\log\left(\frac{|\hat{\mathbf{\Sigma}}|}{|\mathbf{S}|}\right)\right\} \qquad \text{(by Exercise 5.19).}\end{aligned}$$

Following Bartlett's [1951] recommendation, when H_0 is true,

$$\left(n - \frac{2d+11}{6} - \frac{2k}{3}\right)\log\frac{|\hat{\mathbf{\Sigma}}|}{|\mathbf{S}|} \tag{5.57}$$

is approximately χ^2_ν, where $\nu = \frac{1}{2}[(d-k)^2 - (d+k)]$. Because of this test procedure, Jöreskog's algorithm actually minimizes

$$\log|\mathbf{\Sigma}| + \operatorname{tr}[\mathbf{S}\mathbf{\Sigma}^{-1}] - \log|\mathbf{S}| - d \tag{5.58}$$

instead of, equivalently, maximizing $\log L(\mathbf{\Gamma}, \mathbf{\Psi})$ of (5.52); $-2\log\ell$ of (5.56) is then $n-1$ times the minimum of (5.58). Lawley and Maxwell [1971: p. 36] suggest that the chi-square test can be trusted only if $n - d > 50$, though the simulations of Geweke and Singleton [1980] suggest that this is probably too pessimistic. The latter authors suggest, for example, that $n = 10$, $d = 1$ and $n = 25$, $d = 2$ are possible thresholds.

It should be noted that the above likelihood ratio test is only valid if $\mathbf{\Gamma}$ has full rank. We can therefore expect problems with $-2\log\ell$ if k is too large. For the same reason we cannot use the usual asymptotic theory to test $H_0: m = k$ versus $H_1: m = k_1$ $(> k)$. Parameters in the k_1 factor model will be underidentified and their estimators inconsistent when the null hypothesis is true (see Geweke and Singleton [1980]).

b PRINCIPAL FACTOR ANALYSIS

Because of the early difficulties with the maximum likelihood method, and the lack of availability of the Jöreskog–Lawley algorithm, the following method is still very popular.

We note that $\mathbf{\Sigma} - \mathbf{\Psi} = \mathbf{\Gamma}\mathbf{\Gamma}'$ is positive semidefinite of rank m. Hence there exists an orthogonal matrix $\mathbf{M}$ such that

$$\mathbf{M}'(\mathbf{\Sigma} - \mathbf{\Psi})\mathbf{M} = \text{diag}(\phi_1, \phi_2, \ldots, \phi_m, 0, \ldots, 0) = \mathbf{\Phi},$$

where the ϕ_j are the positive eigenvalues. If the columns of $\mathbf{M}_1$ are the first m eigenvectors, and $\mathbf{\Phi}_1 = \text{diag}(\phi_1, \phi_2, \ldots, \phi_m)$, then we can write

$$\mathbf{\Sigma} - \mathbf{\Psi} = \mathbf{M}\mathbf{\Phi}\mathbf{M}' = \mathbf{M}_1\mathbf{\Phi}_1\mathbf{M}_1' = \left(\mathbf{M}_1\mathbf{\Phi}_1^{1/2}\right)\left(\mathbf{M}_1\mathbf{\Phi}^{1/2}\right)',$$

and $\mathbf{\Gamma} = \mathbf{M}_1\mathbf{\Phi}_1^{1/2}$ is a solution (with certain optimal properties; see Exercise 5.3 with $\mathbf{A} = \mathbf{\Sigma} - \mathbf{\Psi}$). Therefore, given a suitable estimate $\hat{\mathbf{\Psi}}_1$ of $\mathbf{\Psi}$, we can take out the first m standardized principal components of $\mathbf{S} - \hat{\mathbf{\Psi}}_1$ and obtain an estimate $\hat{\mathbf{M}}_1\hat{\mathbf{\Phi}}_1^{1/2}$ of $\mathbf{\Gamma}$. This technique, called principal factor analysis, was suggested by Thompson [1934].

Several suggestions for $\hat{\mathbf{\Psi}}_1 = \text{diag}(\hat{\psi}_1^2, \hat{\psi}_2^2, \ldots, \hat{\psi}_d^2)$ are given in the literature, the most common being $\hat{\psi}_j^2 = 1/s^{jj}$, where $[(s^{jk})] = \mathbf{S}^{-1}$. This is equivalent to estimating the communality h_j^2 by [see (5.50)]

$$\hat{h}_j^2 = s_{jj} - \hat{\psi}_j^2 = s_{jj} - \left(1/s^{jj}\right) = s_{jj}R_j^2,$$

where R_j is the sample (multiple) correlation between x_j and the fitted regression of x_j on the remaining x-variables (see Seber [1977: p. 333] for a proof).

The key step in the above method is the choice of m. Since $\mathbf{\Sigma} - \mathbf{\Psi}$ has $d - m$ zero eigenvalues, we can expect some of the eigenvalues of $\mathbf{S} - \hat{\mathbf{\Psi}}$ (which is not necessarily positive semidefinite) to be negative. In practice, we could continue to extract positive eigenvalues until $\Sigma_{j=1}^{m}\hat{\phi}_j$ was close to $\Sigma_{j=1}^{d}\hat{\phi}_j = \text{tr}[\mathbf{S} - \hat{\mathbf{\Psi}}_1] = \Sigma_{j=1}^{d}\hat{h}_j^2$, the estimated total communality. Because of the presence of negative $\hat{\phi}_j$, m will then not exceed the number of positive $\hat{\phi}_j$. Once m is chosen, we can iterate the process for $i = 1, 2, 3, \ldots,$ until stability, using

$$\hat{\mathbf{\Psi}}_{i+1} = \text{diag}\left(\mathbf{S} - \hat{\mathbf{\Gamma}}_i\hat{\mathbf{\Gamma}}_i'\right) \tag{5.59}$$

and extracting the m components of $\mathbf{S} - \hat{\mathbf{\Psi}}_{i+1}$ to obtain $\hat{\mathbf{\Gamma}}_{i+1}$. We could also start the process with $\hat{\mathbf{\Psi}}_1 = \mathbf{O}$ and initially extract the components of $\mathbf{S}$. In this case the choice of m depends on the relative magnitudes of the eigenvalues of $\mathbf{S}$, though such an approach will not be satisfactory unless the elements of $\mathbf{\Psi}$ are small compared to the elements of $\mathbf{\Gamma}$; otherwise the eigenvalues will reflect the elements of $\mathbf{\Psi}$ as much as those of $\mathbf{\Gamma}$. Because of the optimal properties of principal components, the above iterative procedure is equivalent to minimizing $\mathrm{tr}[(\mathbf{S} - \boldsymbol{\Sigma})^2] = \Sigma_i\Sigma_j(s_{ij} - \sigma_{ij})^2$ subject to $\boldsymbol{\Sigma} = \boldsymbol{\Gamma}\boldsymbol{\Gamma}' + \boldsymbol{\Psi}$ (see Exercise 5.3). Algorithms for doing this are the unweighted least squares of Jöreskog [1977] and the minres method of Harman [1977] (see also Kennedy and Gentle [1980: pp. 567–574] for a summary).

As in PCA, the question arises to as to whether the dispersion or correlation matrix should be used. Unfortunately, the latter is often used with little or no justification. In the correlation approach the FA model is based on the standardized variables $z_j = (x_j - \mu_j)/\sigma_{jj}$ so that we have the scaled version of (5.46), namely,

$$\begin{aligned}\mathbf{P}_\rho &= \mathbf{D}_\sigma^{-1/2}\boldsymbol{\Sigma}\mathbf{D}_\sigma^{-1/2}\\ &= \left(\mathbf{D}_\sigma^{-1/2}\boldsymbol{\Gamma}\right)\left(\mathbf{D}_\sigma^{-1/2}\boldsymbol{\Gamma}\right)' + \mathbf{D}_\sigma^{-1/2}\boldsymbol{\Psi}\mathbf{D}_\sigma^{-1/2}\\ &= \boldsymbol{\Delta}\boldsymbol{\Delta}' + \boldsymbol{\Theta}, \qquad (5.60)\end{aligned}$$

say, where $\mathbf{P}_\rho = [(\rho_{jk})]$ is the population correlation matrix and $\mathbf{D}_\sigma = \mathrm{diag}(\sigma_{11}^{1/2}, \sigma_{22}^{1/2}, \ldots, \sigma_{dd}^{1/2})$. If we equate diagonal elements of (5.60), and note (5.50),

$$1 = \rho_{jj} = \sigma_{jj}^{-1}h_j^2 + \sigma_{jj}^{-1}\psi_j^2 = H_j^2 + \theta_j^2,$$

say. Here $\tilde{\mathbf{P}}_\rho = \mathbf{P}_\rho - \boldsymbol{\Theta}$ is called the *reduced correlation matrix*. Given estimates $\hat{H}_j^2$ of the communalities of the z_j, $\tilde{\mathbf{P}}_\rho$ can be estimated by the sample version $\tilde{\mathbf{R}}$, which is simply the sample correlation matrix $\mathbf{R} = [(r_{jk})]$ with its unit diagonal elements replaced by the $\hat{H}_j^2$.

A variety of estimates for H_j^2 have been suggested in the literature. The most common method is to use the lower bound (Harris [1978]) $\hat{H}_j^2 = R_j^2 = 1 - 1/r^{jj}$, where $[(r^{jk})] = \mathbf{R}^{-1}$ (see Seber [1977: bottom of p. 335]). However, Gnanadesikan [1977: p. 21] mentions that for d not too small (say, $d > 10$), the different choices of $\hat{H}_j^2$ that he lists, including R_j^2, generally lead to similar results. One can, of course, iterate, and algorithms are given by Jöreskog [1977] and Harman [1977].

C ESTIMATING FACTOR SCORES

In conclusion, we discuss methods of estimating the factor scores $\mathbf{f}$ for each observation $\mathbf{x}_i$. Clearly $\mathbf{f}$ cannot be estimated in the usual fashion, as it is a random vector and the $m + n$ unknown random variables $\mathbf{f}$ and $\boldsymbol{\varepsilon}$ outnumber the n observations. However, treating $\mathbf{f}$ as fixed, we have

$$\mathbf{y} = \mathbf{x} - \boldsymbol{\mu} \sim N_d(\boldsymbol{\Gamma}\mathbf{f}, \boldsymbol{\Psi}), \qquad (5.61)$$

and we can consider the generalized least squares estimate of $\mathbf{f}$ in this model, namely (Bartlett [1938b]), $\mathbf{f}^* = (\boldsymbol{\Gamma}'\boldsymbol{\Psi}\boldsymbol{\Gamma})^{-1}\boldsymbol{\Gamma}'\boldsymbol{\Psi}^{-1}\mathbf{y}$, which is the linear unbiased estimate with minimum variance properties. If we replace parameters by their maximum likelihood estimators and note that $\hat{\mathbf{J}} = \hat{\boldsymbol{\Gamma}}'\hat{\boldsymbol{\Psi}}^{-1}\hat{\boldsymbol{\Gamma}}$ is assumed to be diagonal, then we have

$$\mathbf{f}_i^* = \hat{\mathbf{J}}^{-1}\hat{\boldsymbol{\Gamma}}'\hat{\boldsymbol{\Psi}}^{-1}(\mathbf{x}_i - \bar{\mathbf{x}}) \tag{5.62}$$

as the factor scores for $\mathbf{x}_i$.

An alternative approach is to choose the linear estimate $\hat{\mathbf{f}} = \mathbf{A}\mathbf{y}$ that minimizes the mean square error

$$\begin{aligned}
E\left[\|\hat{\mathbf{f}} - \mathbf{f}\|^2\right] &= E\left[\|\mathbf{A}\mathbf{y} - \mathbf{f}\|^2\right] \\
&= E\left[\|\mathbf{A}\boldsymbol{\Gamma}\mathbf{f} + \mathbf{A}\boldsymbol{\varepsilon} - \mathbf{f}\|^2\right] \\
&= E\left[\mathbf{f}'(\mathbf{A}\boldsymbol{\Gamma} - \mathbf{I}_m)'(\mathbf{A}\boldsymbol{\Gamma} - \mathbf{I}_m)\mathbf{f} + 2\boldsymbol{\varepsilon}'\mathbf{A}'(\mathbf{A}\boldsymbol{\Gamma} - \mathbf{I}_m)\mathbf{f} + \boldsymbol{\varepsilon}'\mathbf{A}'\mathbf{A}\boldsymbol{\varepsilon}\right] \\
&= \operatorname{tr}\left[(\mathbf{A}\boldsymbol{\Gamma} - \mathbf{I}_m)'(\mathbf{A}\boldsymbol{\Gamma} - \mathbf{I}_m)\right] + \operatorname{tr}[\mathbf{A}'\mathbf{A}\boldsymbol{\Psi}] \qquad \text{(by Exercise 1.4)} \\
&= \operatorname{tr}[\boldsymbol{\Gamma}'\mathbf{A}'\mathbf{A}\boldsymbol{\Gamma} + \mathbf{A}'\mathbf{A}\boldsymbol{\Psi} - \mathbf{A}\boldsymbol{\Gamma} - \boldsymbol{\Gamma}'\mathbf{A}' + \mathbf{I}_m] \\
&= \operatorname{tr}[\mathbf{A}'\mathbf{A}\boldsymbol{\Sigma}] - 2\operatorname{tr}[\mathbf{A}\boldsymbol{\Gamma}] + m \qquad \text{(by A1.1)},
\end{aligned} \tag{5.63}$$

making use of the fact that $\boldsymbol{\varepsilon}$ and $\mathbf{f}$ are independent and have zero means. Using the rules of matrix differentiation in A8.2, we differentiate (5.63) with respect to $\mathbf{A}$ and obtain $2\mathbf{A}\boldsymbol{\Sigma} - 2\boldsymbol{\Gamma}' = \mathbf{O}$. Hence $\mathbf{A} = \boldsymbol{\Gamma}'\boldsymbol{\Sigma}^{-1}$ and (Thomson [1951])

$$\hat{\mathbf{f}} = \boldsymbol{\Gamma}'\boldsymbol{\Sigma}^{-1}\mathbf{y}.$$

Using the identities in A3.4, we can write

$$\boldsymbol{\Gamma}'\boldsymbol{\Sigma}^{-1} = \boldsymbol{\Gamma}'(\boldsymbol{\Gamma}\boldsymbol{\Gamma}' + \boldsymbol{\Psi})^{-1} = \left(\mathbf{I}_m + \boldsymbol{\Gamma}'\boldsymbol{\Psi}^{-1}\boldsymbol{\Gamma}\right)^{-1}\boldsymbol{\Gamma}'\boldsymbol{\Psi}^{-1}$$

and

$$\hat{\mathbf{f}} = \left(\mathbf{I}_m + \boldsymbol{\Gamma}'\boldsymbol{\Psi}^{-1}\boldsymbol{\Gamma}\right)^{-1}\boldsymbol{\Gamma}'\boldsymbol{\Psi}^{-1}\mathbf{y}, \tag{5.64}$$

which is a ridge estimator of $\mathbf{f}$ for the conditional model (5.61). Hence $\hat{\mathbf{f}}$ is conditionally biased as

$$\begin{aligned}
\mathscr{E}[\hat{\mathbf{f}}|\mathbf{f}] &= (\mathbf{I}_m + \mathbf{J})^{-1}\boldsymbol{\Gamma}'\boldsymbol{\Psi}^{-1}\boldsymbol{\Gamma}\mathbf{f} \\
&= (\mathbf{I}_m + \mathbf{J})^{-1}\mathbf{J}\mathbf{f} \\
&\neq \mathbf{f}.
\end{aligned}$$

However, as with ridge estimators, the elements of $\hat{\mathbf{f}}$ will have a smaller mean square error than those of $\mathbf{f}^*$. Comparing $\mathbf{f}^* = \mathbf{J}^{-1}\mathbf{\Gamma}'\mathbf{\Psi}^{-1}\mathbf{y}$ with $\hat{\mathbf{f}} = (\mathbf{I}_m + \mathbf{J})^{-1}\mathbf{\Gamma}'\mathbf{\Psi}^{-1}\mathbf{y}$, we see that the elements of $\hat{\mathbf{f}}$ are simply scalar multiples of those of $\mathbf{f}^*$. The choice of estimators for the individual factor scores, namely, $\mathbf{f}_i^*$ of (5.62) and the corresponding $\hat{\mathbf{f}}_i$, is up to the experimenter, as no general preference has been given (Lawley and Maxwell [1971: p. 111]).

If $m = 2$, the $\mathbf{f}_i^*$ or $\hat{\mathbf{f}}_i$ $(i = 1, 2, \ldots, n)$ can be plotted as a scatter diagram. Such plots have been used as a method for detecting clusters, though this is not recommended. It is preferable to detect the clusters first and then, if necessary, apply FA to each cluster. Factor analysis can only be applied to homogeneous data that fit a particular linear model.

5.4.3 *Some Difficulties*

Factor analysis goes back to Spearman [1904], though early developments are associated with Thurstone and Thomson. However, in spite of its long history, it has a thin coverage in the statistical literature and is regarded by many statisticians as lacking statistical respectability. Part of the controversy that began around 1950 (see Kendall and Babington Smith [1950] and the ensuing discussion), and which still exists today, is due to a lack of agreement regarding the meaning of FA and a general confusion between FA and PCA. The situation is not helped by the use of principal factor analysis. In spite of early attempts to clarify the differences (e.g., Kendall and Lawley [1956]), several widely used texts still display this confusion.

One of the main controversies regarding FA is whether or not the factors have any real existence and have causal rather than just statistical implications. Some researchers in psychology maintain the reality of the factors by giving them psychological names. However, the question of existence need not be established before a model can be used, though care is needed in interpreting such a model. Francis [1973, 1974] has raised a number of serious difficulties associated with FA and the following is based on his work.

Suppose, for the purpose of discussion, that $\boldsymbol{\mu} = \mathbf{0}$ and we have three two-dimensional observations $\mathbf{x}_1, \mathbf{x}_2, \mathbf{x}_3$ on a factor model with a single factor. Then, from (5.43),

$$\mathbf{x}_i = \boldsymbol{\gamma} f_i + \boldsymbol{\varepsilon}_i \qquad (i = 1, 2, 3),$$

where f_i and $\boldsymbol{\varepsilon}_i$ are uncorrelated and unobservable, and the elements of $\boldsymbol{\varepsilon}_i$ are uncorrelated. Writing $\mathbf{f} = (f_1, f_2, f_3)'$ and $(\mathbf{x}_1, \mathbf{x}_2, \mathbf{x}_3)' = (\mathbf{x}^{(1)}, \mathbf{x}^{(2)})$, and so on, we have

$$\mathbf{x}^{(j)} = \mathbf{f}\gamma_j + \boldsymbol{\varepsilon}^{(j)} \qquad (j = 1, 2). \tag{5.65}$$

Thus in three-dimensional space we have two observable correlated vectors $\mathbf{x}^{(1)}$ and $\mathbf{x}^{(2)}$, and three unobservable uncorrelated vectors $\mathbf{f}$, $\boldsymbol{\varepsilon}^{(1)}$ and $\boldsymbol{\varepsilon}^{(2)}$. By judiciously choosing the γ_j, we can construct a model in which each $\boldsymbol{\varepsilon}^{(j)}$ is perpendicular to $\mathbf{f}$ (Fig. 5.5). We are then faced with the problem of trying to discover the relationship between the x's and $\mathbf{f}$ without being able to observe $\mathbf{f}$. To make matters worse, the link between the x's and $\mathbf{f}$ is via the uncorrelated vectors $\boldsymbol{\varepsilon}^{(1)}$ and $\boldsymbol{\varepsilon}^{(2)}$, which are perpendicular to $\mathbf{f}$!

Clearly the above approach to the factor model is inadequate and several attempts to provide a more suitable framework have been made (see McDonald and Burr [1967] for a survey of earlier models). Williams [1978, 1979] bases the factor model on a type of second-order stochastic process in which the d variables in $\mathbf{x}$ are a selection from a sequence of random variables. Bartholomew [1981] gives a helpful framework via conditional distributions and a posterior analysis. A useful historical survey of the problem of factor indeterminacy is given by Steiger [1979].

There is a more serious difficulty associated with FA and this relates to the choice of m, the number of factors. Unfortunately, there have been too few simulation studies like that of Francis [1973] in which the methods have been tested out on data from known factor models. The models with known Γ and $\mathbf{C} = \boldsymbol{\Psi}^{1/2}$ used by Francis are given in Table 5.14. The test procedure used was to choose a model and then ask a program to either retrieve these parameters from the known dispersion matrix, or to estimate these parameters using a

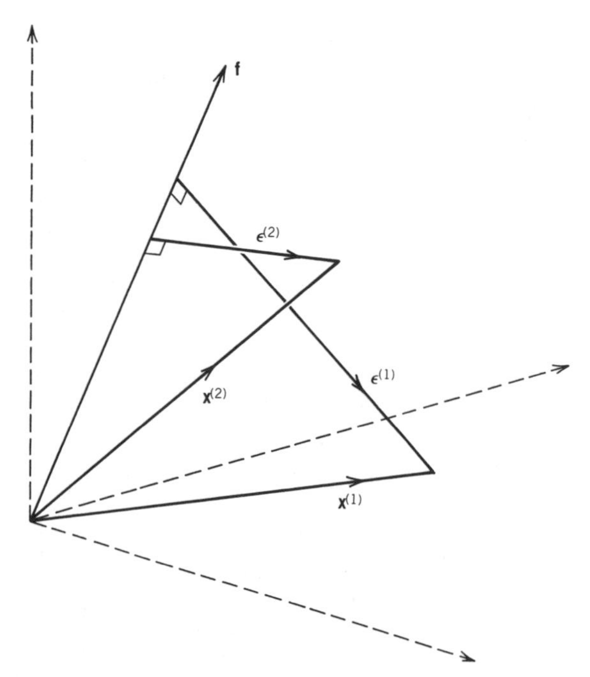

Fig. 5.5 Three-dimensional representation of a one-factor model with three two-dimensional observations. Redrawn from Francis [1973].

TABLE 5.14 True Parameters of 12 Factor Models: $\mathbf{\Gamma}$ is the matrix of factor loadings, and $\mathbf{C} = \mathbf{\Psi}^{1/2}$ the diagonal matrix of the square roots of residual variances[a]

Model	$\mathbf{\Gamma}$			diag $\mathbf{C}$	Model	$\mathbf{\Gamma}$			diag $\mathbf{C}$
I	0	3.1		1.55	IV	9	0		5
	0	3.3		1.65		8	$\sqrt{2}$		5
	0	3.5		1.75		7	$2\sqrt{2}$		5
	0	3.7		1.85		6	$3\sqrt{2}$		5
	0	3.9		1.95		5	$4\sqrt{2}$		5
	4.1	0		2.05		4	$5\sqrt{2}$		5
	4.3	0		2.15		3	$6\sqrt{2}$		5
	4.5	0		2.25		2	$7\sqrt{2}$		5
	4.7	0		2.35		1	$8\sqrt{2}$		5
	4.9	0		2.45		0	$9\sqrt{2}$		5
II	1	−1		2	IV(a)	9	0		5
	2	−1		2		9	1		5
	3	−1		2		9	2		5
	4	−1		4		9	3		5
	5	−1		4		9	4		5
	5	1		4		9	5		5
	4	1		4		9	6		5
	3	1		2		9	7		5
	2	1		2		9	8		5
	1	1		2		9	9		5
III	1	0	0	1	IV(b)	5.0	7.5		5
	1	−1	0	1		4.0	8.0		5
	1	1	0	1		3.3	8.6		5
	1	−1	−1	1		2.5	9.2		5
	1	1	1	1		1.7	9.7		5
	−1	−1	1	1		.8	10.3		5
	−1	1	−1	1		.0	10.8		5
	−1	−1	0	1		−.8	11.4		5
	−1	1	0	1		−1.7	11.9		5
	−1	0	0	1		−2.5	12.5		5
V	10	7	4	15	VIII	1	1	0	1
	10	7	4	15		1	1	0	1
	10	7	4	15		1	1	0	1
	10	7	4	15		1	1	0	1
	10	7	0	15		1	1	0	1
	10	7	0	20		1	0	1	1
	10	7	0	20		1	0	1	1
	10	0	0	20		1	0	1	1
	10	0	0	20		1	0	1	1
	10	0	0	20		1	0	1	1

TABLE 5.14 (*Continued*)

Model	Γ			diag C		Model	Γ			diag C
				(a)	(b)					
VI	10	7	0	20	10	IX	10	10	1	11
	9	6	0	20	10		10	9	2	12
	8	5	0	20	10		10	8	3	13
	7	4	0	30	15		10	7	4	14
	6	3	1	30	15		10	6	5	15
	5	2	1	30	15		10	5	6	16
	4	1	6	30	15		10	4	7	17
	3	0	6	20	10		10	3	8	18
	2	3	11	20	10		10	2	9	19
	1	5	11	10	5		10	1	10	20
VII	10	0	0	15		X	10	7	0	1
	10	0	0	15			10	7	0	3
	10	8	0	15			10	7	0	5
	10	8	0	15			10	7	0	9
	10	8	6	15			5	1	1	1
	0	8	6	20			5	1	1	10
	0	8	6	20			5	1	1	20
	0	0	6	20			1	4	10	1
	0	0	6	20			1	4	10	10
	0	0	0	20			1	4	10	20

[a]From Francis [1973, 1974].

simulated random sample from the model. Each model has two or three factors, $d = 10$ variables, and in most cases the sample size was 50, though several larger samples were taken. The specific factors $\boldsymbol{\varepsilon}$ were generated by the relation $\boldsymbol{\varepsilon} = \mathbf{C}\mathbf{z}$, where $\mathbf{C}$ is a $d \times d$ diagonal matrix with positive diagonal elements and the elements of $\mathbf{z}$ are i.i.d. $N_1(0,1)$. Thus $\mathbf{C}^2 = \boldsymbol{\Psi}$. After the factors were taken out, they were rotated using a varimax (orthogonal) rotation and an oblique rotation. In comparing the fitted $\hat{\boldsymbol{\Gamma}}$ with the true $\boldsymbol{\Gamma}$ it should be noted that the corresponding columns could be of different signs, as it depends on the sign convention adopted by the program. We now summarize Dr. Francis's conclusions, and I am grateful to him for permission to reproduce his findings.

a PRINCIPAL FACTOR ANALYSIS

Five principal factor analysis (PFA) programs were considered from BMD, OSIRIS, and SSPS packages, and Francis selected BMDX72 for carrying out the analysis, as it was the best with regard to accuracy, options, and documentation. This program also has the option for PFA of the dispersion matrix, while the others were restricted to the correlation matrix. The dispersion matrix was mainly used in the study, and PFA was applied to the sample dispersion matrices in Table 5.15 for models I–IV.

In model I all the eigenvalues of $\mathbf{S}$ were greater than 1.0, so that the program, as a default, extracted half the number possible, namely, 5. The actual number of true factors was 2, but we would not know this in practice. The first two columns of $\hat{\Gamma}$ (Table 5.16) are similar to Γ, but the other columns of $\hat{\Gamma}$ are not negligible, although their elements are smaller. This model is ideally suited to varimax rotation and from Table 5.16b we see that the first two columns of $\hat{\Gamma}$ improve slightly on rotation, but the other three columns are changed considerably, so that several of the loadings compared in size with those in the first two columns. As Francis [1974] comments, "there is little doubt that an imaginative user could come up with a plausible explanation for the high estimated loadings on the non-existent factors." If just two columns of $\hat{\Gamma}$ are rotated then varimax recovers the loadings well (Table 5.16c).

The oblique rotations were unsatisfactory (Table 5.16d and e). Although biquartimin does not disturb the first two factors very much, the three fictitious factors now have large loadings that could be taken as indicating important factors. Direct quartimin, another oblique rotation that is supposed to lead to "simple" loadings, destroyed the simple structure of the first two columns of $\hat{\Gamma}$.

Similar comments apply to model II. An interesting feature of this model are the high estimated loadings associated with the fourth to eighth variables in Table 5.17a. The same thing happened, particularly with regard to the nonexistent third and fourth factors, when two more samples were analyzed and when PFA was performed on the true dispersion matrix. The reason for this is that PFA is trying to represent the large elements in $\boldsymbol{\Psi}$ for these four variables by two extra factors. Francis [1973] concluded that "it is not possible to justify the use of factor analysis for finding underlying structure simply because repeated applications of the technique on independent sets of data give rise to similar loadings." The varimax criterion was not well suited to this model, and its application (Table 5.17b) destroyed any similarity with the true structure. Using the correlation matrix led to estimates of the first two factor loadings that were reasonably close to the true loadings if the variables were scaled by their true standard deviations. However, the other columns of $\hat{\Gamma}$ had several large loadings and, once again, orthogonal rotation destroyed any similarity with the true model (Table 5.17c and d).

For model III the good estimates in the first three columns of Table 5.18a are completely lost after any kind of rotation, irrespective of the criterion used or the number of factors rotated; only the result of varimax is displayed in Table 5.18b.

While in models I–III the factors $\mathbf{f}$ are uncorrelated, in model IV they are correlated with correlation $1/\sqrt{2}$. The true loadings of these correlated factors f_1 and f_2 are given in Table 5.14, model IV; the loadings on (f_1', f_2'), an uncorrelated pair of factors defined by $f_1' = f_1$ and $f_2' = \sqrt{2}f_2 - f_1$, which is an oblique rotation of (f_1, f_2), are given in Table 5.14, model IVa; and the loadings on (f_1'', f_2''), are uncorrelated, where $f_1'' = (5f_1 - 3\sqrt{2}f_2)/\sqrt{13}$ and

TABLE 5.15 Sample Covariance Matrices from the First Four Models[a]

	1	2	3	4	5	6	7	8	9	10
					(a) Model I					
1	10.934									
2	8.104	10.709								
3	10,468	8.623	14.528							
4	8,541	10.155	9,629	14.846						
5	11.998	10.494	12.625	11.063	19.832					
6	-0.047	0.025	0.901	1.892	-0.092	15,804				
7	2.385	2.765	3,480	3.880	2.915	12.921	17.580			
8	-0.626	-0.166	0.867	1.466	0.684	15.001	15.426	24.436		
9	-0.168	1.990	1.433	3.292	0.323	13.498	15.365	16.602	21.326	
10	-1.749	-1.139	-0.453	1.873	-2.958	12.491	13.367	15.814	15.385	22.136
					(b) Model II					
1	6.538									
2	4.306	10.726								
3	4.119	6.424	11.158							
4	8.362	13.593	14.434	46.048						
5	7.712	15.703	17.170	35.662	50.866					
6	6.362	12.561	13.009	29.966	32.377	46.468				
7	6.692	9.276	10.110	26.415	22.693	31.466	38.688			
8	3.714	7.595	7.864	16.852	19.896	23.380	17.020	17.505		
9	1.948	5.766	5.695	9.869	11.463	14.367	11.470	9.120	11.323	
10	0.083	1.500	2.336	4.778	4.006	7.754	7.099	3.990	3.929	7.311
					(c) Model III					
1	2.019									
2	1.176	3.068								
3	1.000	-0.358	3.174							
4	1.193	2.612	-0.011	4.186						
5	0.868	-0.259	2.067	-1.143	4.353					
6	-1.314	-0.360	-1.976	-1.194	-1.262	3,983				
7	-0.812	-1.445	0.191	-0.759	-1.034	-0.371	3.370			
8	-0.840	-0.198	-1.861	-0.745	-1.572	2.263	0,119	2,618		
9	-1.235	-2.171	0.508	-2.321	0.161	0.281	1.733	0.087	2.835	
10	-1,065	-1.108	-0.993	-1.110	-0.864	1,117	1.153	0.901	1.194	1.987
					(d) Model IV					
1	82.75									
2	65.42	84.53								
3	62.97	72.65	87,83							
4	69.26	73.96	75.14	101.07						
5	80.51	83.78	81.72	87.21	124.91					
6	71.01	76.29	74.66	82.88	103.16	114.02				
7	73.60	82.38	86.23	95.95	114.34	111.62	152.98			
8	63.87	82.41	81.17	89.27	109.95	112.18	131.56	146.59		
9	54.38	63.39	72.70	85.01	99.63	101.09	118.65	120.62	143.34	
10	75.79	80.76	84.86	102.93	122.04	124.28	142.69	147.02	140.07	192.02

[a] From Francis [1973].

TABLE 5.16 Factor Loadings for Model I[a]

Variable	Factor 1	2	3	4	5	Factor 1	2	3	4	5
	(a) Before rotation					(b) After orthogonal rotation of five factors				
1	.42	3.00	-.21	.37	-.57	-.19	3.08	.04	-.27	.27
2	.60	2.81	.61	-.72	.48	-.03	2.40	-.23	-.65	1.78
3	.80	3.27	-.15	.63	-.93	-.00	3.50	.42	-.43	.07
4	1.11	3.05	1.40	-.28	1.17	.28	2.53	.54	-.28	2.67
5	.62	4.01	-1.03	.34	.12	.27	4.08	-.74	.26	.60
6	3.44	-.58	-.54	-.26	-.25	3.29	.08	.70	-1.11	-.26
7	3.81	.17	-.27	-.51	-.87	3.25	.85	.93	-1.86	-.20
8	4.35	-.80	-1.53	.56	1.29	4.85	.07	.46	.42	-.02
9	4.13	-.46	.50	-1.44	-.34	3.50	-.11	.93	-2.45	.75
10	3.87	-1.21	1.65	1.61	-.19	3.01	-.46	3.51	-.45	.22
	(c) After orthogonal rotation of two factors					(d) After oblique rotation of five factors (biquartimin)				
1	-.12	3.03				-.18	3.32	.83	8.75	6.54
2	.09	2.88				.39	2.98	-1.72	-11.14	-9.28
3	.21	3.36				.15	3.71	2.26	14.75	11.23
4	.56	3.20				.44	3.35	1.10	-18.76	-19.46
5	-.10	4.06				.08	4.44	-1.66	-.74	-1.48
6	3.49	.04				5.26	-.07	.07	1.35	4.73
7	3.72	.84				5.74	.76	.36	8.05	11.72
8	4.43	-.02				6.23	-.04	-.08	-15.39	-12.78
9	4.15	.27				6.58	-.02	-1.23	-3.97	3.11
10	4.03	-.51				4.14	-.48	11.25	9.35	1.93
	(e) After oblique rotation of five factors (direct quartimin)									
1	.09	2.87	.39	.06	-.43					
2	.87	.88	2.25	-.61	-.33					
3	.21	3.54	.12	.55	-.67					
4	.05	.40	3.46	.50	.01					
5	-.39	3.35	.92	-.88	.98					
6	2.19	.36	-.58	.35	1.35					
7	3.19	1.17	-.61	.44	.42					
8	.29	-.09	-.02	.36	4.46					
9	4.23	-.56	.49	.10	.17					
10	.09	-.00	.16	4.43	.30					

[a] From Francis [1973].

$f_2'' = (f_1 + 2\sqrt{2}f_2)/\sqrt{13}$, are given in Table 5.14, model IVb. Although PFA will find estimated loadings for some pair of uncorrelated factors, it would be unlikely to choose any particular pair such as (f_1', f_2'). However, it did apparently choose a pair rather like (f_1'', f_2''), as the first two columns of Table 5.19a are like the interchanged columns of model IVb. Unfortunately, the oblique rotation of the five columns using biquartimin is a disaster. However, if just two factors are rotated, we see from Table 5.19d, that, apart from sign and an interchange of columns, the estimated loadings are close to those of model IV. It is therefore possible that an oblique rotation of PFA loadings may give

TABLE 5.17 Factor Loadings for Model II[a]

Variable	Factor 1	2	3	4	5	Factor 1	2	3	4	5
	(a) Before rotation					(b) After orthogonal rotation of five factors				
1	1.30	.33	.55	.60	.26	.98	-.09	.81	-.88	.36
2	2.33	.72	-.27	.34	1.06	1.94	.61	1.04	-.50	1.35
3	2.48	.80	-.24	.56	.82	2.17	.55	1.00	-.68	1.16
4	5.85	1.67	2.62	-1.41	.30	3.26	1.63	5.53	-1.43	.15
5	6.17	2.90	-1.41	1.04	-.82	6.57	1.71	1.81	-.88	.22
6	6.08	-2.02	-1.39	-1.40	-.75	2.81	5.55	1.66	-1.98	.23
7	4.99	-2.78	1.65	1.70	-.49	1.60	2.52	1.73	-5.16	.20
8	3.48	-.79	-1.14	-.38	.15	1.98	2.84	.72	-1.01	.79
9	2.20	-.84	-.86	.16	1.85	.94	1.63	.41	-.87	2.29
10	1.06	-1.00	-.11	-.19	.84	-.08	1.12	.37	-.70	.99
	(c) Before rotation (from correlation matrix)					(d) After orthogonal rotation of five factors (from correlation matrix)				
1	.56	-.54	.52	.04	.26	.23	.18	.93	-.04	.08
2	.76	-.29	-.17	.13	.27	.57	.15	.40	-.01	.52
3	.78	-.21	-.14	.33	-.06	.76	.15	.29	.20	.27
4	.82	-.19	.05	.05	-.34	.71	.48	.28	.14	-.01
5	.83	-.22	-.27	.09	-.24	.83	.35	.13	.02	.22
6	.87	.20	.00	-.26	-.09	.39	.75	.11	.19	.33
7	.79	.20	.35	-.27	-.10	.19	.78	.32	.30	.13
8	.84	.17	-.13	-.28	.01	.38	.70	.06	.09	.45
9	.72	.34	-.21	-.03	.46	.19	.35	.07	.28	.81
10	.43	.70	.25	.46	-.04	.08	.20	-.03	.95	.15

[a]From Francis [1973].

TABLE 5.18 Factor Loadings for Model III[a]

Variable	Factor 1	2	3	4	5	Factor 1	2	3	4	5
	(a) Before rotation					(b) After orthogonal rotation of five factors				
1	1.08	.04	-.05	.30	.07	.80	-.37	-.23	-.46	.47
2	1.17	-1.04	-.08	.23	.39	1.57	.07	-.40	.12	.09
3	.83	1.30	.16	.68	-.40	-.11	-.74	.18	-1.37	.74
4	1.38	-1.10	.80	.17	.10	1.80	-.50	-.10	-.13	-.53
5	.78	1.40	-1.12	.07	.64	-.27	-.37	-.28	-.26	1.98
6	-1.27	-.94	-.91	.65	-.07	-.32	1.86	-.15	.25	-.34
7	-.87	.55	1.32	.43	.47	-.44	-.20	1.67	-.08	-.41
8	-1.01	-.89	-.34	.29	.07	-.12	1.24	.07	.44	-.52
9	-1.12	1.01	.29	.08	-.11	-1.33	.00	.77	-.19	-.02
10	-1.06	-.03	.22	-.12	.54	-.54	.41	.73	.67	-.20

[a]From Francis [1973].

TABLE 5.19 Factor Loadings for Model IV[a]

	Factor					Factor				
Variable	1	2	3	4	5	1	2	3	4	5
	(a) Before rotation					(b) After orthogonal rotation of five factors				
1	6.96	4.59	−2.30	1.38	.42	3.79	6.83	2.92	−.66	−.12
2	7.67	3.93	.81	−1.15	−.40	4.19	4.72	4.33	−4.20	.47
3	7.83	3.24	2.57	.09	−.04	4.37	3.55	5.22	−3.83	2.19
4	8.73	2.36	1.64	2.73	−1.88	6.18	2.98	6.71	−1.33	1.30
5	10.28	1.86	−1.70	−.62	2.41	7.37	6.91	1.86	−2.85	2.11
6	9.98	.13	−1.43	−.75	.99	8.20	4.96	1.56	−2.67	1.36
7	11.48	−.96	.86	−2.31	−.89	9.83	2.82	2.57	−5.20	1.17
8	11.27	−2.28	.47	−2.53	−.82	10.37	2.16	1.50	−4.88	1.10
9	10.43	−4.10	2.17	2.11	2.75	9.93	1.28	1.63	−1.07	6.18
10	12.67	−3.86	−2.43	1.71	−2.12	13.28	2.58	2.40	−.16	.20
	(c) After oblique rotation of five factors (biquartimin)					(d) After oblique rotation of two factors				
1	7.03	−9.03	−2.46	1.22	−2.64	−.30	8.15			
2	7.75	−4.48	1.26	−4.38	−3.50	−1.48	7.65			
3	7.92	−5.58	2.73	−1.26	−2.52	−2.31	6.90			
4	8.83	−12.33	3.40	−6.45	−16.00	−3.84	6.22			
5	10.39	.48	−3.60	9.58	13.57	−5.46	6.23			
6	10.09	1.01	−2.20	4.07	6.85	−7.00	4.08			
7	11.61	2.61	1.62	−5.01	−1.03	−9.19	3.38			
8	11.40	4.81	1.13	−4.28	.62	−10.37	1.75			
9	10.55	3.01	−.08	16.55	16.33	−11.62	−.73			
10	12.81	−3.52	−.60	−5.17	−11.26	−12.98	.44			

[a] From Francis [1973].

good estimates of the true loadings, but this can only happen under very stringent conditions, including knowledge of the true number of factors and the existence of a suitable method of rotation. The biquartimin, which minimizes the sample covariance of the squared loadings for the pair of columns, is ideal for model IV.

Although the BMD program estimated the communalities by the diagonal elements of **S**, Francis used several other estimators, including the iterative procedure, and came to similar conclusions as those above.

b MAXIMUM LIKELIHOOD FACTOR ANALYSIS

Francis [1973] considered three such programs, SPSS, NIH (a program for the National Institutes of Health, U.S.A.) and UFABY3, a program of Jöreskog documented by Jöreskog and Van Thillo [1971] that supercedes an earlier version for estimating the parameters in an unrestricted model. A further six models were included in the study: Only models I–III satisfied the requirement that $\mathbf{J}$ be diagonal, and models VIII and IX had rank 2. A summary of the results from all the models is given in Table 5.20, with further details in Table 5.21.

In model I, $m = 2$ was "accepted" and varimax produced a good estimate of $\boldsymbol{\Gamma}$ (see Table 5.14 with a sign change in one column and an interchange of two columns). When a third factor was fitted, an improper solution was obtained.

In model II, $m = 1$ was accepted. However, when two factors were fitted, reasonable estimates of $\boldsymbol{\Gamma}$ and $\mathbf{C}$ were obtained. Rotation led to poorer estimates of the loadings.

In model III, $m = 3$ was accepted, but the estimates of $\boldsymbol{\Gamma}$ were not as good as those of the PFA estimates of Table 5.18a. After rotation, all resemblance to $\boldsymbol{\Gamma}$ is lost. When a further factor was introduced, an improper solution was obtained.

In model IV, $m = 2$ was accepted. Here $\hat{\boldsymbol{\Gamma}}$ roughly approximated $\boldsymbol{\Gamma}$ for model IVb before rotation, and IVa after rotation. The elements of diag $\hat{\mathbf{C}}$ ranged from 3.1 to 5.1.

In model V, $m = 1$ was accepted and, in the two cases of 50 samples, the fitting of three factors (the true number) led to an improper solution. For 250 samples $\hat{\mathbf{C}}$ was a good estimate of $\mathbf{C}$, but $\hat{\boldsymbol{\Gamma}}$ was nothing like $\boldsymbol{\Gamma}$ before or after rotation. Even if the true $\boldsymbol{\Sigma}$ were used (corresponding to infinite sample size), which leads to recovering $\mathbf{C}$ exactly, $\hat{\boldsymbol{\Gamma}}$ was nothing like $\boldsymbol{\Gamma}$ before or after rotation. There were similar problems with models VI, VII, and X.

Since the dispersion matrix of a three-factor model with rank $\boldsymbol{\Gamma} = 2$ cannot be distinguished from the dispersion matrix of some two-factor model, we would expect models VIII and IX to behave like a two-factor model. When $\boldsymbol{\Sigma}$ ($n = \infty$) was used in model IX, $\mathbf{C}$ was recovered exactly for $m = 2$, but there was an improper solution from $\mathbf{S}$ for $n = 50$. When $\boldsymbol{\Sigma}$ was used in model VIII, the program refused to fit a second factor. However, for $n = 50$ and 250, reasonable estimates of $\hat{\mathbf{C}}$ and $\hat{\boldsymbol{\Gamma}}$ were obtained: The first column of $\hat{\boldsymbol{\Gamma}}$ was close to the first column of $\boldsymbol{\Gamma}$, and the second column of $\hat{\boldsymbol{\Gamma}}$ was close to a linear combination of columns 2 and 3 of $\boldsymbol{\Gamma}$.

c CONCLUSIONS

If the correct number of factors is known, then PFA may provide a reasonable estimate of $\boldsymbol{\Gamma}$ that may or may not improve with orthogonal rotation. However, a satisfactory method for determining m does not seem to be available, and fictitious factors are all too readily generated. For this reason the maximum likelihood method seems preferable, though the goodness of fit test for m may not lead to the correct value of m. The test appears ready to accept a low value of m as soon as the large variances in $\boldsymbol{\Psi}$ have been explained (e.g., models V, VIa). If the normality assumption is tenable, then an asymptotic sampling theory for the estimates is available (Lawley and Maxwell [1971: Chapter 5]).

TABLE 5.20 Summary of Results from UFABY3[a]

Model	Sample Size	"Accepted" at m	α	m	Score[b]	Improper at m	Illustrated in Table 5.21
I	50	2	.39	2	2	3	x
	50	2	.69	2	3	3	
II	50	1	.03	1	2	3	
		2	.72				x
	50	1	.09	1	2	3	
III	50	3	.44	3	2	4	x
IV	50	2	.39	2	2	3	
V	∞			3	5	4	x
	50	1	.82	1	2	3	
	50	1	.60	1	3	2	
	250	1	.62	1	4	4	
VI(a)	∞			3	4	4	
	50	1	.55	1	0	1,3	
	50	1	.09	1	2	3	
VI(b)	∞			3	4	5	x
	50	2	.15	2	2	3	
	50	2	.07	2	0	2	
VII	∞			3	5	6	x
	50	1	.80	1	2	3	
	50	1	.39	1	3	3	
VIII	∞			1	2		
	50	2	.85	2	2	4	x
	250	2	.58	2	3	3	
IX	∞			2	5	4	x
	50	1	.06	1	2	2	
	250	2	.17	2	3	3	
X	∞			3	4		

[a]From Francis [1973].
[b]Scale of accuracy of $\hat{\mathbf{C}} = \hat{\mathbf{\Psi}}^{1/2} = \text{diag}(\hat{\psi}_1, \ldots, \hat{\psi}_d)$:

Score	Meaning
0	improper: at least one $\psi_j = \delta$.
1	very poor.
2	all $\hat{\psi}_j$ are within 50% of ψ_j.
3	all $\hat{\psi}_j$ are within 20% of ψ_j.
4	all $\hat{\psi}_j$ are within 10% of ψ_j.
5	all $\hat{\psi}_j$ are exact.

TABLE 5.21 UFABY3 Estimated Factor Loadings[a]

Model	Before rotation $\hat{\boldsymbol{\Gamma}}$		diag $\hat{\mathbf{C}}$	After varimax rotation $\hat{\boldsymbol{\Gamma}}$	
I	2.5	−1.7	1.2	3.0	0.1
	2.4	−1.4	1.8	2.7	−0.1
	2.9	−1.6	1.9	3.3	−0.2
	2.8	−1.1	2.4	2.9	−0.5
	3.2	−2.1	2.2	3.8	0.1
	1.9	2.8	2.0	0.1	−3.4
	2.7	2.7	1.7	0.9	−3.7
	2.2	3.5	2.6	0.0	−4.2
	2.3	3.2	2.2	0.2	−4.0
	1.6	3.4	2.8	−0.4	−3.7

Model	Before rotation $\hat{\boldsymbol{\Gamma}}$		diag $\hat{\mathbf{C}}$	Model	Before rotation $\hat{\boldsymbol{\Gamma}}$			diag $\hat{\mathbf{C}}$
II	1.3	0.8	2.1	III	0.8	0.6	−0.2	1.0
	2.3	0.9	2.1		1.4	−0.3	−0.2	0.9
	2.5	1.0	2.0		0.1	1.5	0.2	1.0
	5.3	1.6	3.9		1.8	−0.2	0.6	0.6
	5.9	2.3	3.3		−0.1	1.4	−0.9	1.2
	6.1	−1.7	2.5		−0.6	−1.3	−0.6	1.2
	4.7	−1.3	3.8		−0.8	0.0	1.2	1.1
	3.6	−0.7	2.0		−0.4	−1.2	−0.3	0.9
	2.3	−0.6	2.4		−1.4	0.3	0.5	0.7
	1.1	−0.9	2.3		−0.8	−0.6	0.4	0.9

Model	Before rotation $\hat{\boldsymbol{\Gamma}}$			After rotation $\hat{\boldsymbol{\Gamma}}$		
V	12.7	−1.6	−1.0	11.6	5.5	1.0
	12.7	−1.6	−1.0	11.6	5.5	1.0
	12.7	−1.6	−1.0	11.6	5.5	1.0
	12.7	−1.6	−1.0	11.6	5.5	1.0
	11.9	0.2	2.5	9.7	5.9	4.5
	11.9	0.2	2.5	9.7	5.9	4.5
	11.9	0.2	2.5	9.7	5.9	4.5
	8.5	5.3	−0.8	4.3	8.9	1.2
	8.5	5.3	−0.8	4.3	8.9	1.2
	8.5	5.3	−0.8	4.3	8.9	1.2

TABLE 5.21 (*Continued*)

Model	Before rotation $\hat{\Gamma}$			diag $\hat{C}$	After varimax rotation $\hat{\Gamma}$		
VI(b)	−7.0	10.0	−0.7	10.0	2.0	12.0	0.7
	−6.2	8.9	−0.3	10.0	1.8	10.7	0.4
	−5.3	7.8	−0.0	10.0	1.5	9.3	0.0
	−4.5	6.9	0.3	15.0	1.3	8.0	−0.3
	−4.5	5.0	0.9	15.0	2.0	6.4	−0.7
	−3.6	3.9	1.2	15.0	1.8	5.1	−1.0
	−6.8	0.1	2.6	15.0	6.5	2.9	−1.6
	−5.9	−1.0	2.9	10.0	6.2	1.6	−1.9
	−11.0	−3.6	1.0	9.9	11.5	1.5	0.9
	−11.4	−3.6	−0.8	5.2	11.6	1.7	2.8

Model	Before rotation $\hat{\Gamma}$			After varimax rotation $\hat{\Gamma}$		
VII	8.2	5.2	−2.3	9.9	−1.1	0.4
	8.2	5.2	−2.3	9.9	−1.1	0.4
	12.5	0.6	2.6	10.8	6.6	−1.8
	12.5	0.6	2.6	10.8	6.6	−1.8
	13.6	−3.2	−1.9	10.8	8.2	4.0
VII	5.4	−8.4	0.5	0.9	9.3	3.7
	5.4	−8.4	0.5	0.9	9.3	3.7
	1.1	−3.8	−4.5	−0.0	1.6	5.8
	1.1	−3.8	−4.5	−0.0	1.6	5.8
	0.0	0.0	0.0	0.0	−0.0	−0.0

Model	Before rotation $\hat{\Gamma}$		diag $\hat{C}$	Model	Before rotation $\hat{\Gamma}$		After varimax rotation $\hat{\Gamma}$	
VIII	1.3	−0.4	0.8	IX	13.4	−4.6	13.5	4.5
	1.3	−0.9	1.0		13.2	−3.2	12.5	5.5
	1.1	−0.8	0.8		13.0	−1.8	11.5	6.5
	1.4	−0.8	0.8		12.8	−0.4	10.5	7.5
	0.9	−0.6	1.0		12.7	1.0	9.5	8.5
	0.9	0.9	1.0		12.5	2.4	8.5	9.5
	0.9	1.2	1.0		12.3	3.8	7.5	10.5
	1.0	1.1	0.9		12.1	5.2	6.5	11.5
	0.9	1.1	0.9		11.9	6.6	5.5	12.5
	0.9	1.0	1.0		11.7	8.0	4.5	13.5

[a] From Francis [1973].

If normality cannot be assumed, then the maximum likelihood method can be justified on other grounds (Howe [1955]; see also Morrison [1976]); however, the goodness of fit test is no longer valid. Although Jöreskog has written excellent programs, improper or almost improper solutions come up too frequently. When an improper solution occurs, one of the loadings of the variable with zero-estimated ψ_j becomes very large, and the others very small, so that the whole matrix $\hat{\Gamma}$ is disturbed.

Unless the true model has a "good" structure, like model I, for example, an orthogonal rotation may be worse than useless if the wrong value of m is used. Although oblique rotations are popular with some practitioners in the social sciences (e.g., Cattell and Khanna [1977]), there is a general lack of hard empirical evidence to support this practice. After discussing oblique methods of rotation, Comrey [1973] states in his book (p. 175); "More experience with these methods or perhaps new methods will be required before satisfactory oblique analytic rotation becomes a practical reality, if in fact it ever does." Admittedly, there are circumstances (e.g., m known, the linear model true and "suitable" for the particular rotation used) where the experimenter may be lucky, as in Table 5.19d; however, such a situation seems unlikely to occur.

In conclusion, it must be stated that if FA is carried out, then the results must be interpreted with extreme caution. Even if the postulated model is true —and this is a very strong assumption— the chance of its recovery by present methods does not seem very great. However, proponents of FA and rotational methods (e.g., Cattell and Khanna [1977]) claim that FA is particularly useful as a method of creating hypotheses suggesting underlying causes. A method called Q-mode factor analysis has been used as a cluster technique, though it has been criticized (see Everitt [1979b] for references).

5.5 MULTIDIMENSIONAL SCALING

5.5.1 Classical (Metric) Solution

Principal component analysis is presented in Section 5.2.2 as a linear reduction technique that can be used to replace a set of d-dimensional vectors $\mathbf{x}_1, \mathbf{x}_2, \ldots, \mathbf{x}_n$ by a set of k-dimensional vectors $\mathbf{y}_1, \mathbf{y}_2, \ldots, \mathbf{y}_n$, where k is usually much smaller than d. If $k = 2$, the $\mathbf{y}_i$ can be plotted on a scatter diagram that can then be studied for patterns or clustering. We now consider another reduction technique called multidimensional scaling in which we calculate interpoint distances $\delta_{rs} = \|\mathbf{x}_r - \mathbf{x}_s\|$ and then try to find a set of k-dimensional vectors $\mathbf{y}_i$ with interpoint distances $d_{rs} = \|\mathbf{y}_r - \mathbf{y}_s\|$ such that $d_{rs} \approx \delta_{rs}$ for all r, s. In many applications the $\mathbf{x}_i$ are not given, and δ_{rs} is simply a measure of the proximity of the objects r and s. The proximity measure need not be the Euclidean distance and in fact may not be a distance measure at all. Sometimes the proximity is a distance, but measured with error, so that the problem is to reconstruct the original configuration from approximate interpoint distances only. Proximities are usually described as similarities or dissimilarities (see Chapter 7), depending on the information we have about the objects.

A proximity δ_{rs} is called a *dissimilarity* if $\delta_{rr} = 0$, $\delta_{rs} \geq 0$, and $\delta_{rs} = \delta_{sr}$ for all $r, s = 1, 2, \ldots, n$; the matrix $\mathbf{D} = [(\delta_{rs})]$ is called a *dissimilarity matrix*. We say that $\mathbf{D}$ is Euclidean if there exists a p-dimensional configuration $\mathbf{y}_1, \mathbf{y}_2, \ldots, \mathbf{y}_n$ for some p such that $d_{rs} = \delta_{rs}$. The following theorem (Gower [1966], Mardia et al. [1979]) tells us when $\mathbf{D}$ is Euclidean.

THEOREM 5.7 Let $\mathbf{A} = [(a_{rs})]$, where $a_{rs} = -\frac{1}{2}\delta_{rs}^2$. Define $b_{rs} = a_{rs} - \bar{a}_{r\cdot} - \bar{a}_{\cdot s} + \bar{a}_{\cdot\cdot}$ so that

$$\mathbf{B} = [(b_{rs})] = (\mathbf{I}_n - n^{-1}\mathbf{1}_n\mathbf{1}_n')\mathbf{A}(\mathbf{I}_n - n^{-1}\mathbf{1}_n\mathbf{1}_n'). \tag{5.66}$$

Then $\mathbf{D}$ is Euclidean if and only if $\mathbf{B}$ is positive semidefinite.

Proof Equation (5.66) is left as an exercise (Exercise 5.22). Given $\mathbf{D}$ is Euclidean, there exists a configuration $\mathbf{x}_1, \mathbf{x}_2, \ldots, \mathbf{x}_n$ such that

$$-2a_{rs} = \delta_{rs}^2 = \|\mathbf{x}_r - \mathbf{x}_s\|^2 = \mathbf{x}_r'\mathbf{x}_r + \mathbf{x}_s'\mathbf{x}_s - 2\mathbf{x}_r'\mathbf{x}_s.$$

Also

$$-2\bar{a}_{r\cdot} = \mathbf{x}_r'\mathbf{x}_r + \frac{1}{n}\sum_i \mathbf{x}_i'\mathbf{x}_i - 2\mathbf{x}_r'\bar{\mathbf{x}},$$

$$-2\bar{a}_{\cdot s} = \mathbf{x}_s'\mathbf{x}_s + \frac{1}{n}\sum_i \mathbf{x}_i'\mathbf{x}_i - 2\bar{\mathbf{x}}'\mathbf{x}_s,$$

$$2\bar{a}_{\cdot\cdot} = 2\frac{1}{n}\sum_i \mathbf{x}_i'\mathbf{x}_i - 2\bar{\mathbf{x}}'\bar{\mathbf{x}}.$$

Here, by substituting and canceling, we obtain

$$\begin{aligned} b_{rs} &= \mathbf{x}_r'\mathbf{x}_s - \mathbf{x}_r'\bar{\mathbf{x}} - \bar{\mathbf{x}}'\mathbf{x}_s + \bar{\mathbf{x}}'\bar{\mathbf{x}} \\ &= (\mathbf{x}_r - \bar{\mathbf{x}})'(\mathbf{x}_s - \bar{\mathbf{x}}) \end{aligned} \tag{5.67}$$

and

$$\mathbf{B} = \tilde{\mathbf{X}}\tilde{\mathbf{X}}' \geq \mathbf{O} \qquad \text{(by A4.4)} \tag{5.68}$$

where

$$\tilde{\mathbf{X}}' = (\mathbf{x}_1 - \bar{\mathbf{x}}, \mathbf{x}_2 - \bar{\mathbf{x}}, \ldots, \mathbf{x}_n - \bar{\mathbf{x}}). \tag{5.69}$$

Conversely, suppose $\mathbf{B} \geq \mathbf{O}$ of rank p. There exists an orthogonal matrix $\mathbf{V} = (\mathbf{v}_1, \mathbf{v}_2, \ldots, \mathbf{v}_n)$ such that (see A1.3)

$$\mathbf{V}'\mathbf{B}\mathbf{V} = \begin{pmatrix} \boldsymbol{\Gamma} & \mathbf{O} \\ \mathbf{O} & \mathbf{O} \end{pmatrix}, \tag{5.70}$$

where $\boldsymbol{\Gamma} = \text{diag}(\gamma_1, \gamma_2, \ldots, \gamma_p)$ and $\gamma_1 \geq \gamma_2 \geq \cdots \geq \gamma_p$ are the positive eigenvalues of $\mathbf{B}$. Let $\mathbf{V}_1 = (\mathbf{v}_1, \mathbf{v}_2, \ldots, \mathbf{v}_p)$. Then, from (5.70),

$$\begin{aligned}
\mathbf{B} &= \mathbf{V}\begin{pmatrix} \boldsymbol{\Gamma} & \mathbf{O} \\ \mathbf{O} & \mathbf{O} \end{pmatrix}\mathbf{V}' \\
&= \mathbf{V}_1\boldsymbol{\Gamma}\mathbf{V}_1' \\
&= (\mathbf{V}_1\boldsymbol{\Gamma}^{1/2})(\mathbf{V}_1\boldsymbol{\Gamma}^{1/2})' \\
&= \mathbf{Y}\mathbf{Y}',
\end{aligned}$$

where

$$\begin{aligned}
\mathbf{Y} &= \left(\sqrt{\gamma_1}\,\mathbf{v}_1, \sqrt{\gamma_2}\,\mathbf{v}_2, \ldots, \sqrt{\gamma_p}\,\mathbf{v}_p\right) \\
&= (\mathbf{y}^{(1)}, \mathbf{y}^{(2)}, \ldots, \mathbf{y}^{(p)}) \\
&= \begin{pmatrix} \mathbf{y}_1' \\ \mathbf{y}_2' \\ \vdots \\ \mathbf{y}_n' \end{pmatrix}.
\end{aligned} \tag{5.71}$$

We note that $\|\mathbf{y}^{(j)}\|^2 = \gamma_j\|\mathbf{v}_j\|^2 = \gamma_j$. Since $b_{rs} = \mathbf{y}_r'\mathbf{y}_s$,

$$\begin{aligned}
\|\mathbf{y}_r - \mathbf{y}_s\|^2 &= \mathbf{y}_r'\mathbf{y}_r + \mathbf{y}_s'\mathbf{y}_s - 2\mathbf{y}_r'\mathbf{y}_s \\
&= b_{rr} + b_{ss} - b_{rs} - b_{sr} \\
&= a_{rr} + a_{ss} - 2a_{rs} \qquad (5.72) \\
&= -2a_{rs} \qquad (\text{since } a_{rr} = a_{ss} = 0) \\
&= \delta_{rs}^2,
\end{aligned}$$

and the $\mathbf{y}_i$ give the required configuration. Hence $\mathbf{D}$ is Euclidean. □

The above theorem gives a method for constructing a configuration $\{\mathbf{y}_i\}$ now commonly known as the "classical" method of multidimensional scaling.

The basic idea was introduced by Richardson [1938] and popularized by Torgerson [1952, 1958], who introduced the term *multidimensional scaling*. Gower [1966, 1967a] further clarified the technique under the name of *principal coordinate analysis* and showed that it is closely connected with principal component analysis. When $\mathbf{D}$ is Euclidean, we shall call the configuration $\{\mathbf{y}_i\}$ the classical solution.

We note that the classical solution is not unique, as a shift of origin and a rotation or a reflection will not change interpoint distances. For example, if $\mathbf{L}$ is a $p \times p$ orthogonal matrix, then

$$\|\mathbf{L}(\mathbf{y}_r - \mathbf{y}_s)\|^2 = (\mathbf{y}_r - \mathbf{y}_s)'\mathbf{L}'\mathbf{L}(\mathbf{y}_r - \mathbf{y}_s) = \|\mathbf{y}_r - \mathbf{y}_s\|^2,$$

and the $\mathbf{L}\mathbf{y}_i$ provide another solution. We shall see below that the classical approach has a number of advantages, including some optimal properties. It also places the mean $\bar{\mathbf{y}}$ at the origin, as $(\mathbf{I}_n - n^{-1}\mathbf{1}_n\mathbf{1}_n')\mathbf{1}_n = \mathbf{0}$ implies $\mathbf{B}\mathbf{1}_n = \mathbf{0}$ and

$$n^2\bar{\mathbf{y}}'\bar{\mathbf{y}} = (\mathbf{Y}'\mathbf{1}_n)'(\mathbf{Y}'\mathbf{1}_n) = \mathbf{1}_n'\mathbf{B}\mathbf{1}_n = \mathbf{0},$$

that is, $\bar{\mathbf{y}} = \mathbf{0}$.

Sibson [1979] has provided an approximate error analysis for the classical procedure by applying a perturbation theory to procrustes analysis (Section 5.6). He concluded that the procedure is robust against errors that leave the dissimilarities δ_{rs} approximately linearly related to the $\|\mathbf{x}_r - \mathbf{x}_s\|$. Kruskal [1977b] also notes; "An empirical fact is that classical scaling is amazingly robust. The solution is surprisingly little affected by random error or monotonic distortion in the δ_{ij}." Further support for robustness comes from the theoretical study by Mardia [1978], who shows that a perturbation of the form $\delta_{rs}^2 - 2a$ does not alter the solution much when a is small.

We note that

$$\begin{aligned}\delta_{rs}^2 &= \|\mathbf{y}_r - \mathbf{y}_s\|^2 \\ &= \sum_{j=1}^{p} (y_{rj} - y_{sj})^2.\end{aligned}$$

If γ_j is small, or the elements of $\mathbf{y}^{(j)} = \sqrt{\gamma_j}\,\mathbf{v}_j$ are not too different (y_{rj} and y_{sj} being the rth and sth elements, respectively), then $(y_{rj} - y_{sj})^2$ will be small for all r, s. Therefore if γ_1 and γ_2 are much larger than the other γ's, and the elements of $\mathbf{y}^{(j)}$ ($j = 1, 2$) are reasonably different, then most of the contribution to each δ_{rs}^2 will come from the first two elements of $\mathbf{y}_r$ and $\mathbf{y}_s$. Provided that

$$\sum_{j=1}^{2} (y_{rj} - y_{sj})^2 \approx \delta_{rs}^2,$$

we then have a useful two-dimensional representation.

Sometimes the dissimilarity matrix is not Euclidean, so that some of the eigenvalues of $\mathbf{B}$ are negative. In this case we cannot write $\mathbf{\Gamma} = \mathbf{\Gamma}^{1/2}\mathbf{\Gamma}^{1/2}$, so that Theorem 5.7 does not hold; however, if the first k eigenvalues are comparatively large and positive, and the remaining positive or negative eigenvalues are near zero, then the rows of $\mathbf{Y}_k = (\mathbf{y}^{(1)}, \mathbf{y}^{(2)}, \ldots, \mathbf{y}^{(k)})$ will give a "reasonable" configuration in k dimensions. Further research is needed to see how well the method tolerates negative eigenvalues.

The classical procedure can also be applied to similarities c_{rs}, where $c_{rr} = 1$, $c_{rs} = c_{sr}$, and $0 \leq c_{rs} \leq 1$. We simply convert the similarities into dissimilarities using the transformation

$$\delta_{rs} = (2 - 2c_{rs})^{1/2} \tag{5.73}$$

and set $a_{rs} = c_{rs}$ in Theorem 5.7. If $\mathbf{A} \geq \mathbf{O}$, then $\mathbf{B} \geq \mathbf{O}$, and $\mathbf{D}$ is Euclidean. Hence, from (5.72),

$$\begin{aligned} \|\mathbf{y}_r - \mathbf{y}_s\|^2 &= a_{rr} + a_{ss} - 2a_{rs} \\ &= 2 - 2c_{rs} \\ &= \delta_{rs}^2, \end{aligned}$$

and the $\mathbf{y}_i$ provide a solution once again.

We can summarize the classical scaling procedure as follows:

(1) Form the matrix $\mathbf{A}$, where $a_{rs} = -\frac{1}{2}\delta_{rs}^2$ for dissimilarities and $a_{rs} = c_{rs}$ for similarities.
(2) Obtain $\mathbf{B}$ from $\mathbf{A}$ (called "double centering" by psychometricians).
(3) Extract the k largest positive eigenvalues $\gamma_1 > \gamma_2 > \cdots > \gamma_k$ of $\mathbf{B}$ with corresponding eigenvectors $\mathbf{Y}_k = (\mathbf{y}^{(1)}, \mathbf{y}^{(2)}, \ldots, \mathbf{y}^{(k)})$ that are normalized by $\|\mathbf{y}^{(j)}\|^2 = \gamma_j$ $(j = 1, 2, \ldots, k)$.
(4) The n rows of $\mathbf{Y}_k$ are called the principal coordinates in k dimensions.

In the above discussion we described the method as primarily one that provides a geometrical representation of proximity data. This process is sometimes called ordination. However, if we are given the original observations $\mathbf{x}_i$, then the emphasis is more on dimension reducing. In this situation we might ask how the method compares with principal components. We note from (5.68) that $\mathbf{B} = \tilde{\mathbf{X}}\tilde{\mathbf{X}}'$ and, in the context of principal component analysis, the estimate $\hat{\mathbf{\Sigma}}$ of the dispersion matrix $\mathbf{\Sigma}$ for the $\mathbf{x}_i$ satisfies $n\hat{\mathbf{\Sigma}} = \tilde{\mathbf{X}}'\tilde{\mathbf{X}}$. Hence $\mathbf{B}$ and $n\hat{\mathbf{\Sigma}}$ have the same distinct positive eigenvalues (A1.4) and the same rank (A2.3). If $\mathbf{\Sigma} > \mathbf{O}$ and $n - 1 \geq d$, this common rank is d (with probability 1). The jth principal component of $\mathbf{x}_i$ is $\hat{\mathbf{t}}_j'(\mathbf{x}_i - \overline{\mathbf{x}})$, where $\|\hat{\mathbf{t}}_j\| = 1$ and (see Section 5.2.4)

$$\begin{aligned} \tilde{\mathbf{X}}'\tilde{\mathbf{X}}\hat{\mathbf{t}}_j &= n\hat{\lambda}_j\hat{\mathbf{t}}_j \\ &= \gamma_j\hat{\mathbf{t}}_j \qquad (j = 1, 2, \ldots, d). \end{aligned}$$

Premultiplying by $\tilde{\mathbf{X}}$ leads to $\mathbf{B}\tilde{\mathbf{X}}\hat{\mathbf{t}}_j = \gamma_j \tilde{\mathbf{X}}\hat{\mathbf{t}}_j$, so that $\tilde{\mathbf{X}}\hat{\mathbf{t}}_j$ is an eigenvector of $\mathbf{B}$, normalized by $\hat{\mathbf{t}}_j'\tilde{\mathbf{X}}'\tilde{\mathbf{X}}\hat{\mathbf{t}}_j = \gamma_j \hat{\mathbf{t}}_j'\hat{\mathbf{t}}_j = \gamma_j$, corresponding to the eigenvalue γ_j. Since, apart from a scale factor, the eigenvector corresponding to an eigenvalue of multiplicity 1 is unique (see A1.3), we can choose the sign of $\hat{\mathbf{t}}_j$ such that $\mathbf{y}^{(j)} = \tilde{\mathbf{X}}\hat{\mathbf{t}}_j$. Then

$$\mathbf{Y}_k' = \left(\mathbf{y}^{(1)}, \mathbf{y}^{(2)}, \ldots, \mathbf{y}^{(k)}\right)' \qquad (1 \le k \le d)$$

$$= \begin{pmatrix} \hat{\mathbf{t}}_1' \\ \vdots \\ \hat{\mathbf{t}}_k' \end{pmatrix} \tilde{\mathbf{X}}'$$

and

$$\mathbf{y}_i = \begin{pmatrix} \hat{\mathbf{t}}_1'(\mathbf{x}_i - \bar{\mathbf{x}}) \\ \hat{\mathbf{t}}_2'(\mathbf{x}_i - \bar{\mathbf{x}}) \\ \vdots \\ \hat{\mathbf{t}}_k'(\mathbf{x}_i - \bar{\mathbf{x}}) \end{pmatrix},$$

the first k principal components of $\mathbf{x}_i$. Therefore principal components and principal coordinates are the same in this situation.

For the more general case where we are simply given a Euclidean dissimilarity matrix $\mathbf{D}$ or a positive semidefinite similarity matrix $\mathbf{A}$, the principal coordinates will be the principal components for *some* configuration with $d_{rs}^2 = \delta_{rs}^2$, and will therefore share some of the optimal properties of the sample principal components. When $\mathbf{D}$ is not Euclidean, principal coordinates have the following optimal property. Let $\tilde{\mathbf{X}}_f$ be an $n \times k$ matrix whose rows are an approximating configuration and have mean zero. If $\mathbf{B}_f = \tilde{\mathbf{X}}_f \tilde{\mathbf{X}}_f'$, then $\|\mathbf{B} - \mathbf{B}_f\|^2 = \text{tr}[(\mathbf{B} - \mathbf{B}_f)^2]$ is minimized when $\tilde{\mathbf{X}}_f = \mathbf{Y}_k$ (Mardia [1978], Mardia et al. [1979: Section 14.4]). Mardia [1978] has proposed two coefficients for measuring how close the $d_{rs}(= \|\mathbf{y}_r - \mathbf{y}_s\|)$ for the k-dimensional principal coordinates are to the δ_{rs}. These are

$$\alpha_{1,k} = \left(\sum_{i=1}^{k} \gamma_i \Big/ \sum_{i=1}^{n} |\gamma_i| \right) \times 100\% \tag{5.74}$$

and

$$\alpha_{2,k} = \left(\sum_{i=1}^{k} \gamma_i^2 \Big/ \sum_{i=1}^{n} \gamma_i^2 \right) \times 100\%, \tag{5.75}$$

where $\gamma_1 \ge \gamma_2 \ge \cdots \ge \gamma_n$ are the eigenvalues of $\mathbf{B}$ and $\gamma_k > 0$. The second was also proposed by Saito [1978] and is related to the problem of finding a lower-rank approximation for $\mathbf{B}$ (see A7.3).

Examples of the use of principal coordinates are given by Hills [1969], Blackith and Reyment [1971], and Everitt [1978], and the application to binary data is discussed by Banfield and Gower [1980]. Psychometricians frequently call $\mathbf{B} = \tilde{\mathbf{X}}\tilde{\mathbf{X}}'$ the **Q**-matrix, and $\boldsymbol{\Sigma} = \tilde{\mathbf{X}}'\tilde{\mathbf{X}}/n$ (or more usually the correlation matrix) the **R**-matrix. Thus principal coordinate analysis is a "Q-method," and principal components an "R-method."

Finally, we note that it has been common practice in the past to fit the model $d_{rs} \approx \delta_{rs} + a$, where a is chosen so that the $\delta_{rs} + a$ are more "distance-like." Several methods of estimating a are available (see Saito [1978]), and the problem is usually referred to as the "additive constant" problem.

5.5.2 Nonmetric Scaling

In Section 5.5.1 we considered the problem of finding a set of k-dimensional vectors $\mathbf{y}_i$ whose interpoint distances d_{rs} were as close as possible to a set of dissimilatiries δ_{rs}. If, however, the configuration $\{\mathbf{y}_i\}$ is to be used geometrically for cluster and pattern detection, then an expansion or contraction of the configuration would provide an equally useful picture. Some distortion could also be allowed, provided that the rank ordering of the d_{rs} was the same as that of the δ_{rs}. We could therefore widen our search for configurations and look for one such that

$$d_{rs} \approx f(\delta_{rs}) \tag{5.76}$$

for some unknown monotonic increasing function f satisfying

$$\delta_{r_1 s_1} < \delta_{r_2 s_2} \Leftrightarrow f(\delta_{r_1 s_1}) \le f(\delta_{r_2 s_2}).$$

Our aim, then, would be to find a configuration in k dimensions such that the plot of d_{rs} versus δ_{rs} is monotonically increasing (or approximately so). For example, if $n = 3$ and $\delta_{23} < \delta_{13} < \delta_{12}$, we require three points $\mathbf{y}_i$ such that $d_{23} \le d_{13} \le d_{12}$. A method for doing this developed by Shepard [1962a, b] and Kruskal [1964a, b] (see Kruskal [1977b], Kruskal and Wish [1978], Carroll and Arabie [1980] and De Leeuw and Heiser [1980]), called nonmetric multidimensional scaling, is described below. The term *nonmetric* is used, as only the rank order of the δ_{rs} is important, whereas classical scaling is "metric," being based on the magnitudes of the δ_{rs}.

Given a trial $\{\mathbf{y}_i\}$ configuration, the plot of d_{rs} versus δ_{rs} $(r < s)$ will not generally be monotonic as in Fig. 5.6, but will be partly sawtoothed as in Fig. 5.7; point B lies below point A rather than above it or on the same level. For this situation we would like some measure of departure from monotonicity that we can use to make adjustments to the $\mathbf{y}_i$ to improve the degree of monotonicity. Kruskal [1964a] suggested fitting the nearest monotonic curve by monotone least squares, shown dotted in Fig. 5.7, and measuring the difference between this curve and the solid curve using a scaled sum of squared vertical dif-

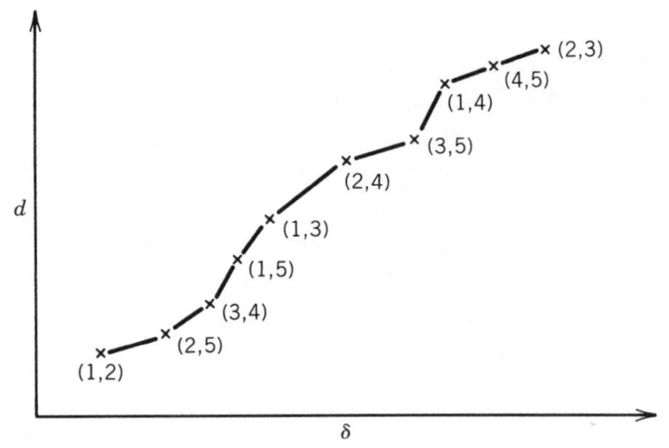

Fig. 5.6 Scatter plot of distance versus dissimilarity where the monotonicity constraint is satisfied.

ferences. Therefore, given $\delta_{r_1s_1} < \delta_{r_2s_2} < \cdots < \delta_{r_ms_m}$ [$r_j < s_j$ and $m = \frac{1}{2}n(n-1)$], we find positive numbers $\hat{d}_{rs}$ to minimize

$$S^2 = \sum_{r<s}\sum (d_{rs} - \hat{d}_{rs})^2 \Big/ \sum_{r<s}\sum d_{rs}^2 \tag{5.77}$$

subject to the monotone constraint

$$\hat{d}_{r_1s_1} \leq \hat{d}_{r_2s_2} \leq \cdots \leq \hat{d}_{r_ms_m}.$$

The minimum value of S, the positive square root of S^2, is called the STRESS

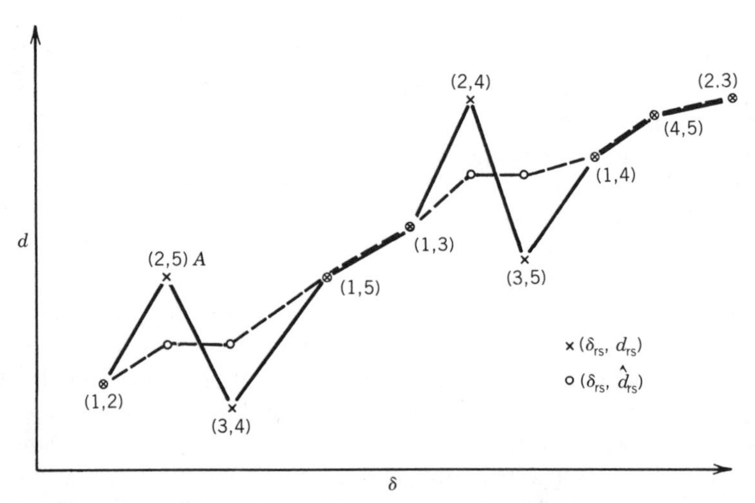

Fig. 5.7 Scatter plot of distance versus dissimilarity where the monotonic constraint is not satisfied. The dotted curve is the monotonic curve of best fit.

S^* and is a measure of the departure from monotonicity. It has the desirable property of being invariant under the similarity group of shift of origin (translation), orthogonal transformations (rotations and reflections), and, with the particular choice of denominator in (5.77), uniform scaling (isotropic contractions and dilations) of the $\mathbf{y}_i$. We note that the fitted $\hat{d}_{rs}$ are *not* distances and there does not necessarily exist a k-dimensional configuration whose interpoint distances are $\hat{d}_{rs}$. They are simply a set of numbers that are used as a reference for measuring the monotonicity of the d_{rs}. In terms of (5.76), we have found a monotonic transformation f, given by $\hat{d}_{rs} = f(\delta_{rs})$, that leads to a minimum S^2.

Having calculated the $\hat{d}_{rs}$, we now "improve" the k-dimensional representation by choosing the nk elements of the $\mathbf{y}_i$ to minimize S^* with respect to the d_{rs}. For this second configuration we find a new set of $\hat{d}_{rs}$, and this in turn leads to a third configuration. The procedure is continued until S^* converges to its minimum over all k-dimensional configurations. Clearly some dissimilarities δ_{rs} are "closer" to k-dimensional curved subspaces than others, and Kruskal [1964a] gives some empirical guidelines, reproduced in Table 5.22, as to how well the configuration fits in terms of monotonicity.

A major problem in using the above method is the choice of k. Kruskal [1964a] recommended beginning with a low value of k, say, $k = 1$, and plotting the minimum stress against k until the minimum stress became acceptably small. Obviously the stress is zero if $k \geq n - 1$, since perfect monotonicity can be achieved with n points in $n - 1$ dimensions. However, we would hope that $k = 1, 2$ or possibly 3 so that the $\mathbf{y}_i$ can be plotted. For example, in Fig. 5.8*b* we would choose $k = 3$ on the basis of Table 5.22. A recent program for performing both metric and nonmetric scaling called KYST (pronounced "kissed"), from the initials of Kruskal, Young, Shepard, and Torgerson, was developed by Kruskal, Young, and Seery to combine the best features of the two previous programs TORSCA and M-D-SCAL. In essence, the Torgerson classical scaling procedure and Young's preliminary iterative procedure were integrated into Kruskal's M-D-SCAL program, the latter extending the work of Shepard both conceptually and computationally. The essential elements of

TABLE 5.22 Guidelines for Determining Appropriateness of Fitted Configuration Obtained by Multidimensional Scaling

Minimum Stress (%)	Goodness of Fit
20	Poor
10	Fair
5	Good
$2\frac{1}{2}$	Excellent
0	Perfect

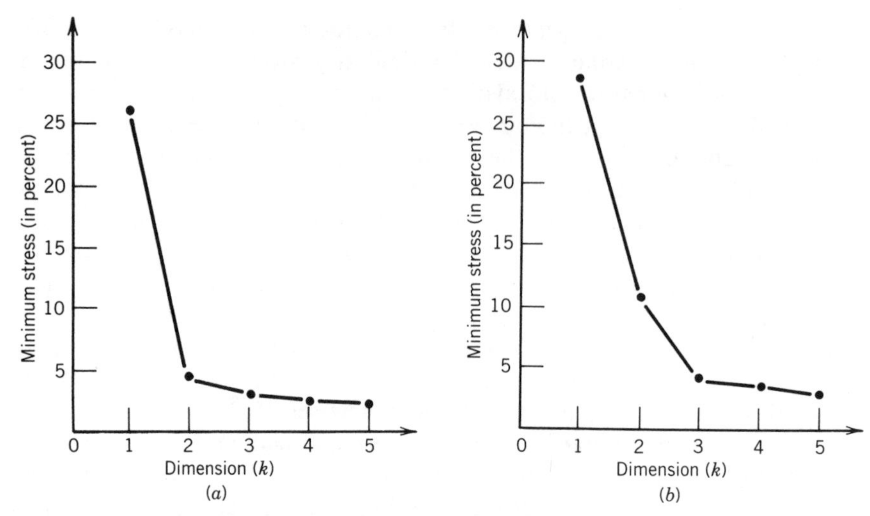

Fig. 5.8 Plot of minimum stress versus dimension. (*a*) Elbow at $k = 2$. (*b*) Elbow at $k = 3$.

KYST are described by Kruskal [1977b], and helpful background material is provided by Kruskal and Wish [1978]. The program begins with the maximum value of k to be entertained and computes an initial configuration. This could be a set of points chosen at random from some distribution, a set of evenly spaced points in R^k, or (preferably) the classical metric solution computed from the δ_{rs}. With a monotonic regression algorithm for computing the $\hat{d}_{rs}$, the k-dimensional configuration with minimum stress is computed iteratively using a gradient method. One coordinate is then dropped to form the initial configuration for a $(k-1)$-dimensional fit. The program has the option of rotating the k-dimensional solution to principal coordinates (in this case the same as principal components) and the last coordinate would then be the one to drop. The procedure continues until the specified minimum value of k is reached (usually $k = 1$). A few ties among the δ_{rs} have little effect on the procedure and missing values can be filled in with appropriate averages of nonmissing values. Similarities c_{rs} are converted into dissimilarities by a suitable transformation, for example, $\delta_{rs} = a - c_{rs}$.

Various procedures are available for handling asymmetric δ_{rs} caused primarily by unreliability or experimental error, the simplest being to take the average $\frac{1}{2}(\delta_{rs} + \delta_{sr})$. For example, in the famous Morse signal example of Shepard [1963] where signals r and s are heard in succession, we can take δ_{rs} as the proportion of listeners who say that the signals are different. The order makes a difference, so that $\delta_{rs} \neq \delta_{sr}$ (see Example 5.5 below).

There are difficulties with nonmetric scaling when the number of observations is too small relative to the number of dimensions. Kruskal and Wish [1978] recommend that $n - 1 \geq 4k$, with empirical support for this rule existing mostly for $k \leq 3$. They also give practical hints for coping with the problems of incomplete convergence (maximum number of iterations reached) or a local rather than a global minimum. They suggest a plot of d_{ij} versus δ_{ij} should be examined for each solution. A table of residuals $d_{ij} - \hat{d}_{ij}$ can also be useful. A number of rules of thumb, like Table 5.22, are available for choosing k. Usually the appropriate dimension is indicated by an elbow in the plot of minimum stress versus dimension, provided that the stress at the elbow is below about 0.10. For example, we would choose $k = 2$ in Fig. 5.8a and $k = 3$ in Fig. 5.8b. However, caution is needed when there is an apparent elbow at $k = 2$, as a very large stress at $k = 1$ can have limited accuracy. On the other hand, $k = 1$ is suggested if the stress there is below 0.15, provided that $n \geq 10$. An alternative, statistical, approach to the problem is to compile tables of plots using simulated data for which the true dimension is known, but with increasing amounts of "noise" (error) added (Spence [1972], Spence and Graef [1974], Kruskal and Wish [1978: p. 54 and their Appendix C]). With increasing error, the elbow gets rounded and may disappear altogether; the plot is also raised. We could try and match up our own plot with one of those tabled, though if the error level is too high, we may conclude that there is no point in trying to choose a dimension k and that multidimensional scaling is inappropriate. Spence [1979] obtained the following empirical formula:

$$\text{STRESS} = -524.25 + 33.80k - 2.54n - 307.26\log k + 588.35(\log n)^{1/2},$$

which fits simulated (pseudo-) random data well for $k = 1(1)5$ and $10 \leq n \leq 60$. He suggested that if the values of STRESS for a data set and different k are well below the values given by the above formula, say, only a third or half as large, then the data are satisfactory for scaling, otherwise the data may have little structure and the results of the scaling should be treated with caution. For rough formal tests, the standard deviation associated with the above formula is about 0.01.

We note that STRESS is just one measure of fit, and other measures have been proposed. One modification consists of replacing d_{rs} by $d_{rs} - \bar{d}_{..}$ in STRESS; another, called SSTRESS (see Takahane et al. [1977]), replaces d_{rs} and $\hat{d}_{rs}$ by their squares. Sammon [1969], in his technique called nonlinear mapping, chose a configuration with interpoint distances d_{rs} to minimize the so-called mapping error

$$E = \left(\sum_{r<s}\sum \frac{(\delta_{rs} - d_{rs})^2}{\delta_{rs}} \right) \Big/ \sum_{r<s}\sum \delta_{rs}. \tag{5.78}$$

Sammon uses a "steepest descent" procedure for minimizing E over the nk elements of the $\mathbf{y}_i$, while Chang and Lee [1973] suggest a relaxation method. These authors and Howarth [1973] give several examples.

Once we have found a configuration $\{\mathbf{y}_i\}$, we can rotate and translate it without changing the d_{rs}. To overcome this indeterminancy, it is often advantageous to shift the centroid to the origin (i.e., subtract $\bar{\mathbf{y}}$ from each $\mathbf{y}_i$) and rotate to a principal components solution, particularly when comparing different configurations.

EXAMPLE 5.5 Shepard [1963, 1980] applied multidimensional scaling to Morse code data collected by Rothkopf [1957]. About 150 observers who did not know Morse code listened to a pair of signals produced at a fixed rapid rate by machine and separated by a quiet period of 1.4 secs, and were required to state whether the two signals they heard were the same or different. A total of 36 signals were used, 26 letters of the alphabet and the numbers 0–9. Each signal consists of a sequence of dots and dashes, e.g. ·--- for letter *J* and ····· for *5*. The (r, s)th element of the 36×36 similarity matrix is the proportion of observers who responded "same" to signal r followed by signal s. The matrix is roughly symmetric and applying nonmetric multidimensional scaling gives the plot of stress versus dimension in Fig. 5.9. Choosing $k = 2$ leads to Fig. 5.10, which shows Shepard's interpretation of the two dimensions. The vertical scale could also be labeled "length of signal" rather than "number of components," though, as noted by Shepard, it is not possible to distinguish between them with the data given. □

The method of multidimensional scaling deals with only one matrix of proximities. Suppose, however, that several such matrices are available for the same objects or stimuli, perhaps one for each person involved with the experiment. If we can regard the matrices as true replicates, then we could simply average over the subjects and obtain a single proximity matrix. However, if there are large systematic differences between the matrices, we may wish to analyze the data as a three-way array, the so-called three-way multidimensional scaling. One such model, called "individual differences scaling," or INDSCAL, was developed by Carroll and Chang [1970]. Suppose we have N

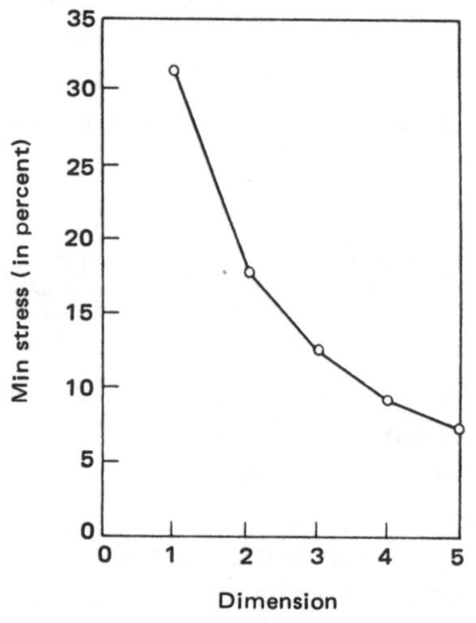

Fig. 5.9 Plot of stress versus number of dimensions for Morse code example.

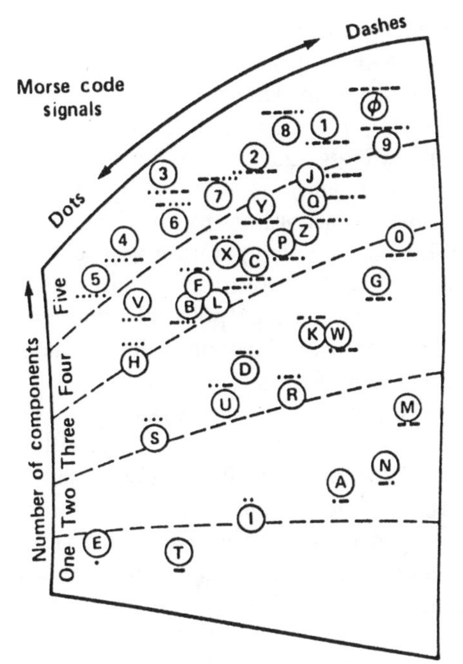

Fig. 5.10 Shepard's interpretation of Morse code configuration. From R. N. Shepard [1980]. *Science*, **210**, 390–398. Copyright 1980 by the American Association for the Advancement of Science.

subjects and n stimuli and let δ_{rsv} be the dissimilarity for the stimuli r and s and the vth person ($r, s = 1, 2, \ldots, n$; $v = 1, 2, \ldots, N$). We then find weights w_{vj}, one for each person in combination with each variable, and a configuration $\{\mathbf{y}_i\}$ such that

$$d_{rsv} = \sum_{j=1}^{k} w_{vj}(y_{rj} - y_{sj})^2 \approx \delta_{rsv} \tag{5.79}$$

for some k. This model is discussed by Kruskal [1977b], Kruskal and Wish [1978: Section 4], and Young et al. [1978, 1980]. The latter authors use an algorithm called ALSCAL (see *Psychometrika*, Vol. 43, 1978, for further references). We note that the solution cannot be rotated because of the weights.

The problem of multidimensional scaling with restrictions is considered by De Leeuw and Heiser [1980].

EXAMPLE 5.6 In April 1968, 18 students taking part in a pilot experiment were each asked to give similarity ratings, on a scale from 1 (extremely dissimilar) to 9 (extremely similar), of all $\binom{12}{2} = 66$ pairs of nations drawn from a set of 12 nations. After this they were asked to state anonymously their opinions regarding the action that the United States should take in Vietnam. They were then classified as "doves," "moderates," or "hawks," according to whether they advocated (1) withdrawal of troops from Vietnam, (2) concessions to bring about a negotiated peace, or (3) continuation or escalation of the current U.S. military involvement. Using the similarity matrix of ratings from each student, Wish et al. [1970] applied INDSCAL to these matrices rather than multidimensional scaling to the average matrix. They argued that INDSCAL would allow for the dimensions obtained by the scaling to be weighted differently by each student according to their backgrounds and political orientation. A three-dimensional configuration was selected and scores on the first two dimensions for the 12 countries are given in Fig. 5.11. The first dimension was identified as measuring "political alignment and ideology," and the second dimension as measuring "economic development."

In Fig. 5.12 we have a plot of the students' weights on these two dimensions with their labels as doves, moderates, or hawks. A diagonal line through the origin separates the doves from the hawks and suggests that political alignment is the more important factor for hawks in judging the similarity of nations. On the other hand, economic development is the more important factor, both in absolute and relative terms, for the doves. The reader is referred to Wish et al. [1970] for further details and a more comprehensive example involving 21 nations and 90 students from 15 different countries. □

Little has been done in the way of inference for multidimensional scaling (e.g., Ramsay [1978: confidence regions for points]). Levine [1978] has studied the values of STRESS, with d_{rs} replaced by $d_{rs} - \bar{d}_{..}$, for random configura-

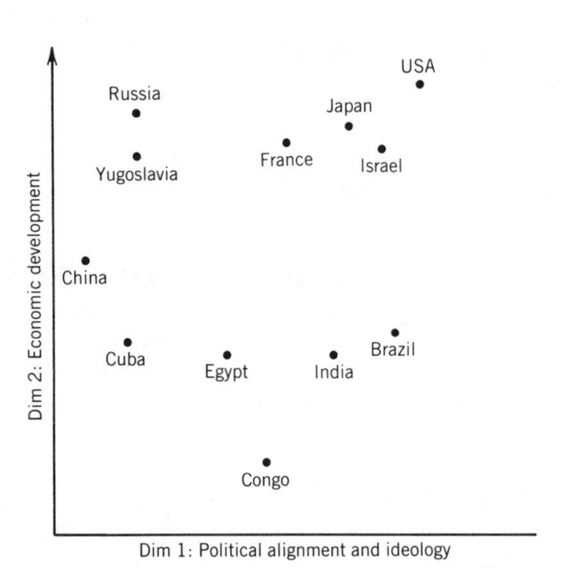

Fig. 5.11 Dimensions one and two of three-dimensional INDSCAL configuration for 12 nations. From Wish et al. [1970]. Copyright 1970 by the American Psychological Association. Reprinted by permission of the publisher and author.

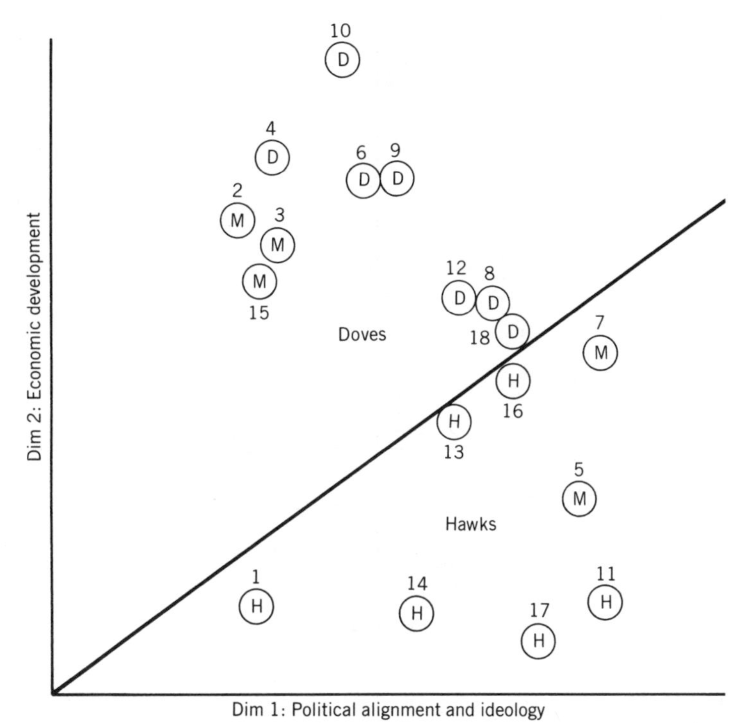

Fig. 5.12 Plot of students' weights on dimensions one and two of three-dimensional INDSCAL configuration for 12 nations. From Wish et al. [1970]. Copyright 1970 by the American Psychological Association. Reprinted by permission of the publisher and the author.

tions with the aim of providing information on testing the null hypothesis of randomness. When fewer than 10 points are considered, low values of STRESS can often be obtained, even for random configurations.

We note that multidimensional scaling can be used for sniffing out one-dimensional structure in the data. The most common application is seriation, where the aim is to determine the chronological ordering of the objects. A famous example of this comes from archeology, where the similarity of two graves is measured by the number of types of pottery they have in common. The argument is that similar graves will be close chronologically. Sometimes associated with this kind of problem is the so-called "horseshoe" phenomenon (Kendall [1971]) pictured in Fig. 5.13. In this case, the two-dimensional scaling solution is found to have a U shape, and an "ordering" along the curve can be detected by joining up points with dissimilarities less than a threshold value. The two-dimensional configuration is almost a one-dimensional configuration that has been bent into a horseshoe shape. Only one curvilinear dimension is needed to give a reasonable description of the configuration, though a second coordinate, giving the perpendicular distance from the nearest point of the curved line of the horseshoe, would improve the description.

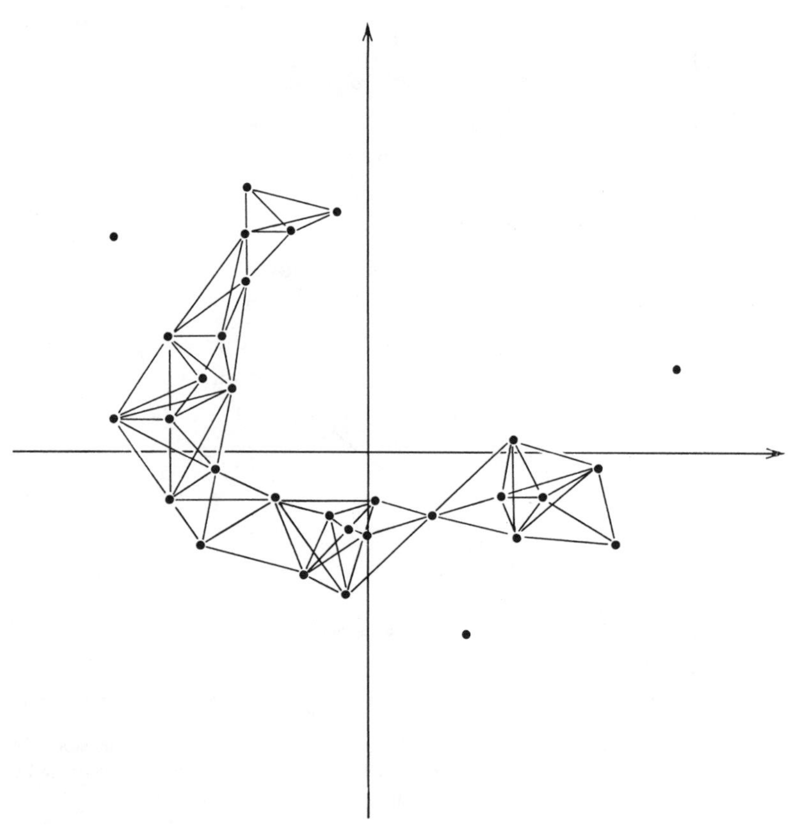

Fig. 5.13 Horseshow phenomenon with points linked that have dissimilarities less than a threshold value.

The horseshoe phenomenon can occur in situations where the distance between two objects can be measured accurately when they are close together, but not when they are far apart. Moderate or large distances lead to similar measures, with the effect of pulling the more distant objects closer together. For example, graves close together in time will have similar types of pottery, but those sufficiently separated in time will have no types in common. The same applies to the so-called color-matching problem where individuals are asked to compare colors. Those close to each other on the spectrum will be recognized as being similar, whereas those further apart will be all treated the same. Again a horseshoe curve results giving the so-called color circle (Fig. 5.14); the one-dimensional coordinate is the wavelength. It is important to draw in the links as shown in Fig. 5.13, particularly when the curve is almost closed and when it is more of a wide band rather than a curve; otherwise the one-dimensional structure may be overlooked. Kendall [1971] gives a number of procedures to help "unbend" the horseshoe so that the gap is clearer.

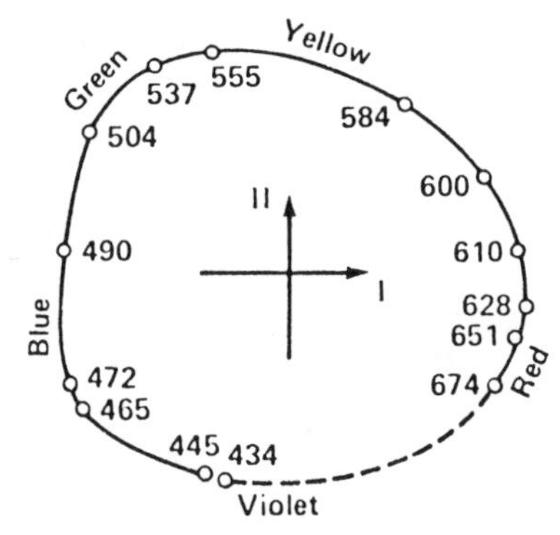

Fig. 5.14 Multidimensional scaling solution for 14 colors. From R. N. Shepard [1980], *Science*, **210**, 390–398. Copyright 1980 by the American Association for the Advancement of Science.

One of the signs that the data is basically one dimensional is the existence of a permutation of the object labels that converts the similarity matrix into a matrix which approximates a so-called Robinson matrix (Robinson [1951]; see also the papers by Gelfand, Kendall, Landau, and de la Vega in Hodson et al. [1971]). In this matrix, the similarity coefficient always decreases when we move from the main diagonal to the left or to the right. For example, the matrix in Table 5.23 can be changed into that of Table 5.24, which is close to a Robinson matrix, by reordering the variables as shown. The relevance of this type of matrix structure is perhaps seen more clearly by the following example (Spaulding [1971]). Suppose we have five collections of cutting tools labeled A, B, C, D, and E, and the tools are made of three materials, stone, bronze, and iron. If a tool of a particular material is present in a collection, we score 1; otherwise we score 0 (see Table 5.25). The pattern is clear: No 1's are separated by 0's in any column or row, and the cluster of 1's moves steadily upward from left to right. The five collections are therefore readily ordered, and they cannot be so simply ordered with any other significantly different row and column permutations. The ordering supports the view that the period of the use of stone did not overlap with that of the use of iron, so that we have five classes

TABLE 5.23 Similarity Matrix for Eight Variables

	1	2	3	4	5	6	7	8
1	1.00	.03	.01	.20	.02	.33	.35	.06
2	0.3	1.00	.98	.54	.96	.02	.15	.57
3	.01	.98	1.00	.54	.98	.01	.13	.58
4	.20	.54	.54	1.00	.55	.25	.41	.86
5	.02	.98	.98	.55	1.00	.02	.15	.60
6	.33	.02	.01	.25	.02	1.00	.51	.14
7	.35	.15	.13	.41	.15	.51	1.00	.33
8	.06	.57	.58	.86	.60	.14	.33	1.00

TABLE 5.24 Approximate Robinson Matrix Obtained by Reordering Variables in Table 5.23

	1	6	7	4	8	5	2	3
1	1.00	.033	.035	.20	.06	.02	.03	.01
6	.33	1.00	.52	.25	.14	.02	.02	.01
7	.35	.52	1.00	.41	.33	.15	.15	.13
4	.20	.25	.41	1.00	.86	.55	.54	.54
8	.06	.14	.33	.86	1.00	.60	.57	.58
5	.02	.02	.15	.55	.60	1.00	.98	.98
2	.03	.02	.15	.54	.57	.98	1.00	.98
3	.01	.01	.13	.54	.58	.98	.98	1.00

TABLE 5.25 Incidence Matrix for Five Collections

	Stone	Bronze	Iron
A	0	0	1
B	0	1	1
C	0	1	0
D	1	1	0
E	1	0	0

of cutting tools: (1) stone only, (2) stone and bronze, (3) bronze only, (4) bronze and iron, and (5) iron only. Using Jaccard's similarity coefficient $\alpha/(\alpha + \beta + \gamma)$ (see Table 7.1), we obtain the Robinson matrix in Table 5.26. For further references on the subject of seriation, see Hodson et al. [1971].

In conclusion, we mention one important practical problem associated with large dissimilarity matrices. If the underlying experiment consists of a number of subjects each comparing $\binom{n}{2} = \frac{1}{2}n(n-1)$ pairs of n stimuli, then there is a practical limit on n above which even the most dedicated subject is exhausted,

TABLE 5.26 Robinson Similarity Matrix for Data in Table 5.25

	A	B	C	D	E
A	1	$\frac{1}{2}$	0	0	0
B	$\frac{1}{2}$	1	$\frac{1}{2}$	$\frac{1}{2}$	0
C	0	$\frac{1}{2}$	1	$\frac{1}{2}$	0
D	0	$\frac{1}{2}$	$\frac{1}{2}$	1	$\frac{1}{2}$
E	0	0	0	$\frac{1}{2}$	1

not to mention the possible unreliability of the resulting data. For example, if $n = 40$, then 780 pairs need to be assessed. Clearly, for quite moderate values of n, the complete pairwise assessment is out of the question and incomplete designs need to be considered. For a discussion and a survey of this topic see Spence and Domoney [1974] and Spence [1982].

5.6 PROCRUSTES ANALYSIS (MATCHING CONFIGURATIONS)

We have seen that in some situations multidimensional scaling of Section 5.5 can be regarded as a technique for trying to match one set of n points in d-dimensional space by another set in a lower-dimensional space. A related technique, commonly known as procrustes analysis (Hurley and Cattell [1962: p. 260]), refers to the problem of matching two configurations of n points in d-dimensional space where there is a preassigned correspondence between the points of one configuration and the points of the other. The origins of the technique go back at least to Mosier [1939] and, although originally developed for use in factor analysis, it is now seen as relating closely to multidimensional scaling. The method is reviewed by Sibson [1978] and the following discussion is based on his paper. Further algebraic results are given by Ten Berge [1977].

Let $\mathbf{X}' = (\mathbf{x}_1, \mathbf{x}_2, \ldots, \mathbf{x}_n)$ and $\mathbf{Y}' = (\mathbf{y}_1, \mathbf{y}_2, \ldots, \mathbf{y}_n)$ be two d-dimensional configurations of n points. If the $\mathbf{y}_i$ are in fact k dimensional ($k < d$), then we can add a column of $d - k$ zeros to each $\mathbf{y}_i$. One measure of goodness of fit of $\mathbf{Y}$ to $\mathbf{X}$ is

$$D(\mathbf{X}, \mathbf{Y}) = \sum_{i=1}^{n} \|\mathbf{x}_i - \mathbf{y}_i\|^2 = \mathrm{tr}\left[(\mathbf{X} - \mathbf{Y})(\mathbf{X} - \mathbf{Y})'\right]. \tag{5.80}$$

Fixing the $\mathbf{x}_i$, we wish to move $\mathbf{y}_i$ relative to $\mathbf{x}_i$ through rotation, reflection, and translation, that is, by the linear transformation

$$\mathbf{T}'\mathbf{y}_i + \mathbf{c} \qquad (i = 1, 2, \ldots, n), \tag{5.81}$$

where $\mathbf{T}$ is an orthogonal matrix, so that (5.80) is minimized. Carrying out the translation first, we see that $\sum_i \|\mathbf{x}_i - \mathbf{y}_i - \mathbf{c}\|^2$ is minimized when $\mathbf{c} = \overline{\mathbf{x}} - \overline{\mathbf{y}}$ (see Exercise 7.20). If the data are centered at the origin so that $\overline{\mathbf{x}} = \overline{\mathbf{y}} = \mathbf{0}$, then $\mathbf{X} = \tilde{\mathbf{X}} = (\mathbf{x}_1 - \overline{\mathbf{x}}, \mathbf{x}_2 - \overline{\mathbf{x}}, \ldots, \mathbf{x}_n - \overline{\mathbf{x}})'$, $\mathbf{Y} = \tilde{\mathbf{Y}}$, and $\mathbf{c} = \mathbf{0}$. The problem of matching now reduces to minimizing

$$\begin{aligned} D(\tilde{\mathbf{X}}, \tilde{\mathbf{Y}}\mathbf{T}) &= \sum_{i=1}^{n} \|\tilde{\mathbf{x}}_i - \mathbf{T}'\tilde{\mathbf{y}}_i\|^2 \\ &= \mathrm{tr}\left[(\tilde{\mathbf{X}} - \tilde{\mathbf{Y}}\mathbf{T})(\tilde{\mathbf{X}} - \tilde{\mathbf{Y}}\mathbf{T})'\right] \\ &= \mathrm{tr}[\tilde{\mathbf{X}}\tilde{\mathbf{X}}'] + \mathrm{tr}[\tilde{\mathbf{Y}}\tilde{\mathbf{Y}}'] - 2\,\mathrm{tr}[\mathbf{T}\tilde{\mathbf{X}}'\tilde{\mathbf{Y}}] \qquad \text{(by A1.1)} \end{aligned}$$

with respect to orthogonal $\mathbf{T}$. This is achieved using the following theorem.

THEOREM 5.8 Let $\mathbf{A}$ be a $d \times d$ matrix with singular value decomposition $\mathbf{A} = \mathbf{L\Delta M}'$ where $\mathbf{L}$ and $\mathbf{M}$ are orthogonal matrices and $\mathbf{\Delta} = \mathrm{diag}(\delta_1, \delta_2, \ldots, \delta_d)$, where each $\delta_j \geq 0$ [see (10.19)]. Then for all orthogonal $\mathbf{T}$,

$$\mathrm{tr}[\mathbf{TA}] \leq \mathrm{tr}\left[(\mathbf{A}'\mathbf{A})^{1/2}\right] \tag{5.82}$$

with equality if $\mathbf{T} = \hat{\mathbf{T}} = \mathbf{ML}'$.

Proof

$$\mathrm{tr}[\mathbf{TA}] = \mathrm{tr}[\mathbf{TL\Delta M}'] = \mathrm{tr}[\mathbf{M}'\mathbf{TL\Delta}] = \mathrm{tr}[\mathbf{N\Delta}],$$

where $\mathbf{N} = \mathbf{M}'\mathbf{TL}$ is an orthogonal matrix, being the product of orthogonal matrices. Now the elements of an orthogonal matrix cannot exceed 1, so that

$$\begin{aligned}
\mathrm{tr}[\mathbf{N\Delta}] &= \sum_{j=1}^{d} n_{jj}\delta_j \\
&\leq \sum_{j=1}^{d} \delta_j \\
&= \mathrm{tr}\,\mathbf{\Delta} \\
&= \mathrm{tr}\left[(\mathbf{\Delta}'\mathbf{\Delta})^{1/2}\right] \\
&= \mathrm{tr}\left[(\mathbf{M}'\mathbf{A}'\mathbf{LL}'\mathbf{AM})^{1/2}\right] \\
&= \mathrm{tr}\left[(\mathbf{M}'\mathbf{A}'\mathbf{AM})^{1/2}\right] \\
&= \mathrm{tr}\left[(\mathbf{AMM}'\mathbf{A}')^{1/2}\right] \qquad \text{(by A4.7)} \\
&= \mathrm{tr}\left[(\mathbf{AA}')^{1/2}\right] \\
&= \mathrm{tr}\left[(\mathbf{A}'\mathbf{A})^{1/2}\right].
\end{aligned}$$

We have equality when $\mathbf{T} = \hat{\mathbf{T}}$, as

$$\begin{aligned}
\hat{\mathbf{T}}\mathbf{A} &= \mathbf{ML}'\mathbf{L\Delta M}' \\
&= \mathbf{M\Delta M}' \\
&= (\mathbf{M\Delta M}'\mathbf{M\Delta M}')^{1/2} \\
&= (\mathbf{M\Delta}^2\mathbf{M}')^{1/2} \\
&= (\mathbf{M\Delta L}'\mathbf{L\Delta M}')^{1/2} \\
&= (\mathbf{A}'\mathbf{A})^{1/2}.
\end{aligned}$$

□

Now $D(\tilde{\mathbf{X}}, \tilde{\mathbf{Y}}\mathbf{T})$ is minimized when $\mathrm{tr}[\mathbf{T}\tilde{\mathbf{X}}'\tilde{\mathbf{Y}}]$ is maximized. Setting $\mathbf{A} = \tilde{\mathbf{X}}'\tilde{\mathbf{Y}}$, we see from the above theorem that a solution is given by $\hat{\mathbf{T}} = \mathbf{ML}'$, and

$$\begin{aligned} D_{\min}^{(1)} &= D(\tilde{\mathbf{X}}, \tilde{\mathbf{Y}}\hat{\mathbf{T}}) \\ &= \mathrm{tr}[\tilde{\mathbf{X}}\tilde{\mathbf{X}}'] + \mathrm{tr}[\tilde{\mathbf{Y}}\tilde{\mathbf{Y}}'] - 2\,\mathrm{tr}\left[(\tilde{\mathbf{Y}}'\tilde{\mathbf{X}}\tilde{\mathbf{X}}'\tilde{\mathbf{Y}})^{1/2}\right]. \end{aligned} \tag{5.83}$$

If $\mathbf{A}$ is nonsingular, then $\mathbf{TA} = (\mathbf{A}'\mathbf{A})^{1/2}$ has a unique solution

$$\hat{\mathbf{T}} = (\mathbf{A}'\mathbf{A})^{-1/2}\mathbf{A}' = (\tilde{\mathbf{Y}}'\tilde{\mathbf{X}}\tilde{\mathbf{X}}'\tilde{\mathbf{Y}})^{-1/2}(\tilde{\mathbf{Y}}'\tilde{\mathbf{X}}).$$

The rotation that takes $\tilde{\mathbf{Y}}$ to $\tilde{\mathbf{Y}}\hat{\mathbf{T}}$ is called the procrustes rotation of $\tilde{\mathbf{Y}}$ relative to $\tilde{\mathbf{X}}$. Since $\mathrm{tr}[(\tilde{\mathbf{Y}}'\tilde{\mathbf{X}}\tilde{\mathbf{X}}'\tilde{\mathbf{Y}})^{1/2}] = \mathrm{tr}[(\tilde{\mathbf{X}}'\tilde{\mathbf{Y}}\tilde{\mathbf{Y}}'\tilde{\mathbf{X}})^{1/2}]$, we see from (5.83) that the roles of $\tilde{\mathbf{X}}$ and $\tilde{\mathbf{Y}}$ can be interchanged, as expected.

In some situations the scales of the two configurations are different, so that (5.81) is now replaced by $b\mathbf{T}'\mathbf{y}_i + \mathbf{c}$ $(b > 0)$. Proceeding as before, we now wish to minimize

$$D(\tilde{\mathbf{X}}, b\tilde{\mathbf{Y}}\mathbf{T}) = \mathrm{tr}[\tilde{\mathbf{X}}\tilde{\mathbf{X}}'] + b^2\mathrm{tr}[\tilde{\mathbf{Y}}\tilde{\mathbf{Y}}'] - 2b\,\mathrm{tr}[\mathbf{T}\tilde{\mathbf{X}}'\tilde{\mathbf{Y}}] \tag{5.84}$$

with respect to b and orthogonal $\mathbf{T}$. This can be done in two stages. If we fix $b > 0$, (5.84) is minimized with respect to $\mathbf{T}$ when $\mathbf{T} = \hat{\mathbf{T}}$. Then

$$D(\tilde{\mathbf{X}}, b\tilde{\mathbf{Y}}\hat{\mathbf{T}}) = \mathrm{tr}[\tilde{\mathbf{X}}\tilde{\mathbf{X}}'] + b^2\mathrm{tr}[\tilde{\mathbf{Y}}\tilde{\mathbf{Y}}'] - 2b\,\mathrm{tr}\left[(\tilde{\mathbf{Y}}'\tilde{\mathbf{X}}\tilde{\mathbf{X}}'\tilde{\mathbf{Y}})^{1/2}\right]. \tag{5.85}$$

If we differentiate with respect to b, the above is minimized for all b when

$$b = \hat{b} = \mathrm{tr}\left[(\tilde{\mathbf{Y}}'\tilde{\mathbf{X}}\tilde{\mathbf{X}}'\tilde{\mathbf{Y}})^{1/2}\right]/\mathrm{tr}[\tilde{\mathbf{Y}}\tilde{\mathbf{Y}}'].$$

Since $\hat{b} > 0$, it also minimizes (5.85) for the restricted case of $b > 0$. Gower [1971b] noted that

$$D_{\min}^{(2)} = D(\tilde{\mathbf{X}}, \hat{b}\tilde{\mathbf{Y}}\hat{\mathbf{T}}) = \mathrm{tr}\left[(\tilde{\mathbf{X}}\tilde{\mathbf{X}}')\right] - \frac{\left\{\mathrm{tr}\left[(\tilde{\mathbf{Y}}'\tilde{\mathbf{X}}\tilde{\mathbf{X}}'\tilde{\mathbf{Y}})^{1/2}\right]\right\}^2}{\mathrm{tr}[\tilde{\mathbf{Y}}\tilde{\mathbf{Y}}']} \tag{5.86}$$

is not symmetric in $\tilde{\mathbf{X}}$ and $\tilde{\mathbf{Y}}$, so that the optimal scaling of $\tilde{\mathbf{Y}}\hat{\mathbf{T}}$ to $\tilde{\mathbf{X}}$ is not the inverse of the optimal scaling from $\tilde{\mathbf{X}}$ to $\tilde{\mathbf{Y}}\hat{\mathbf{T}}$. However, symmetry could be obtained if we followed Gower's suggestion and scaled the configurations so that $\mathrm{tr}[\tilde{\mathbf{X}}\tilde{\mathbf{X}}'] = \mathrm{tr}[\tilde{\mathbf{Y}}\tilde{\mathbf{Y}}']$ $(= 1$, for example).

If we wished to compare two configurations $\tilde{\mathbf{Y}}$ and $\tilde{\mathbf{Z}}$, say, with the same $\tilde{\mathbf{X}}$, then $D_{\min}^{(r)}$ $(r = 1, 2)$ could be used as a measure of fit: The smaller the value, the better the fit. For example, $\tilde{\mathbf{Y}}$ might be the classical scaling solution and $\tilde{\mathbf{Z}}$ the nonmetric solution. However, if we wished to compare $\tilde{\mathbf{Y}}$ with $\tilde{\mathbf{X}}$ and $\tilde{\mathbf{V}}$

with $\tilde{\mathbf{U}}$, then some standardization of $D_{\min}^{(r)}$ is needed to remove the effect of the number of points and the scale of $\tilde{\mathbf{X}}$ (or $\tilde{\mathbf{U}}$). Sibson [1978] suggested

$$\alpha_r = \frac{D_{\min}^{(r)}}{\mathrm{tr}[\tilde{\mathbf{X}}\tilde{\mathbf{X}}']}$$

or

$$\beta_r = \frac{D_{\min}^{(r)}}{\{\mathrm{tr}[\tilde{\mathbf{X}}\tilde{\mathbf{X}}']\mathrm{tr}[\tilde{\mathbf{Y}}\tilde{\mathbf{Y}}']\}^{1/2}}.$$

He noted that β_1 and

$$\alpha_2 = 1 - \frac{\{\mathrm{tr}[(\tilde{\mathbf{Y}}'\tilde{\mathbf{X}}\tilde{\mathbf{X}}'\tilde{\mathbf{Y}})^{1/2}]\}^2}{\{\mathrm{tr}[\tilde{\mathbf{X}}\tilde{\mathbf{X}}']\mathrm{tr}[\tilde{\mathbf{Y}}\tilde{\mathbf{Y}}']\}}$$

are symmetric in $\tilde{\mathbf{X}}$ and $\tilde{\mathbf{Y}}$, which may be preferred. A related coefficient introduced by Robert and Escoufier [1976] (see also Escoufier and Robert [1980]) is

$$RV[\tilde{\mathbf{X}}, \tilde{\mathbf{Y}}] = \frac{\mathrm{tr}[\tilde{\mathbf{Y}}'\tilde{\mathbf{X}}\tilde{\mathbf{X}}'\tilde{\mathbf{Y}}]}{\{\mathrm{tr}[(\tilde{\mathbf{X}}\tilde{\mathbf{X}}')^2]\mathrm{tr}[(\tilde{\mathbf{Y}}\tilde{\mathbf{Y}}')^2]\}^{1/2}}. \qquad (5.87)$$

This coefficient is a measure of similarity between $\tilde{\mathbf{X}}$ and $\tilde{\mathbf{Y}}$, and the authors call it the *RV*-coefficient. If we wish to reduce $\tilde{\mathbf{X}}$ to $\tilde{\mathbf{X}}\mathbf{L}$ and $\tilde{\mathbf{Y}}$ to $\tilde{\mathbf{Y}}\mathbf{M}$, then one criterion for choosing $\mathbf{L}$ and $\mathbf{M}$ would be to minimize $RV[\tilde{\mathbf{X}}\mathbf{L}, \tilde{\mathbf{Y}}\mathbf{M}]$. The authors showed, by suitable choices of $\mathbf{L}$, $\mathbf{M}$, $\tilde{\mathbf{X}}$, and $\tilde{\mathbf{Y}}$ (e.g., $\tilde{\mathbf{Y}} = \tilde{\mathbf{X}}$), that the methods of principal components, multiple regression, canonical correlations (see Section 5.7), and discriminant coordinates (see Section 5.8) arise as special cases of the minimization. They also used the *RV*-coefficient to provide an algorithm for selecting the "best" subset of variables.

An alternative, more robust method than the procrustes least squares approach has been proposed by Siegel and Benson [1982] and called the repeated median algorithm.

5.7 CANONICAL CORRELATIONS AND VARIATES

5.7.1 Population Correlations

Suppose we have have a d-dimensional vector $\mathbf{z} = (z_1, z_2, \ldots, z_d)'$ with zero mean and dispersion matrix $\mathscr{D}[\mathbf{z}] = \mathbf{\Sigma} = [(\sigma_{jk})]$. We can measure the linear association between z_j and z_k by their correlation. If, however, we require a measure of linear association between z_1 and the vector $\mathbf{y} = (z_2, z_3, \dots, z_d)'$, then we can use $\rho^2_{1(23\cdots d)}$, the square of the multiple correlation between z_1 and $\mathbf{y}$. This quantity is defined to be the maximum squared correlation between z_1 and any linear combination $\boldsymbol{\beta}'\mathbf{y}$ and is given by

$$\rho^2_{1(23\cdots d)} = \frac{\boldsymbol{\sigma}'_{21}\boldsymbol{\Sigma}_{22}^{-1}\boldsymbol{\sigma}_{21}}{\sigma_{11}},$$

where the parameters come from the partition

$$\boldsymbol{\Sigma} = \begin{pmatrix} \sigma_{11} & \boldsymbol{\sigma}'_{21} \\ \boldsymbol{\sigma}_{21} & \boldsymbol{\Sigma}_{22} \end{pmatrix}.$$

This maximum occurs when (see Exercise 5.25)

$$\boldsymbol{\beta} = \boldsymbol{\Sigma}_{22}^{-1}\boldsymbol{\sigma}_{21}. \tag{5.88}$$

Taking the generalization one step further, suppose we partition $\mathbf{z}' = (\mathbf{x}', \mathbf{y}')$, where $\mathbf{x}$ and $\mathbf{y}$ are d_1- and d_2- $(= d - d_1)$ dimensional vectors, respectively, and we wish to construct a measure of linear association between $\mathbf{x}$ and $\mathbf{y}$. Then one such measure is the maximum value of ρ^2, the squared correlation between arbitrary linear combinations $\boldsymbol{\alpha}'\mathbf{x}$ and $\boldsymbol{\beta}'\mathbf{y}$. Partitioning

$$\boldsymbol{\Sigma} = \begin{pmatrix} \boldsymbol{\Sigma}_{11} & \boldsymbol{\Sigma}_{12} \\ \boldsymbol{\Sigma}_{21} & \boldsymbol{\Sigma}_{22} \end{pmatrix},$$

where $\boldsymbol{\Sigma}_{ii}$ is $d_i \times d_i$ and $\boldsymbol{\Sigma}_{12} = \boldsymbol{\Sigma}'_{21}$, we have

$$\begin{aligned} \rho^2 &= \frac{\left(\text{cov}[\boldsymbol{\alpha}'\mathbf{x}, \boldsymbol{\beta}'\mathbf{y}]\right)^2}{\text{var}[\boldsymbol{\alpha}'\mathbf{x}]\text{var}[\boldsymbol{\beta}'\mathbf{y}]} \\ &= \frac{\left(\boldsymbol{\alpha}'\boldsymbol{\Sigma}_{12}\boldsymbol{\beta}\right)^2}{\left(\boldsymbol{\alpha}'\boldsymbol{\Sigma}_{11}\boldsymbol{\alpha}\right)\left(\boldsymbol{\beta}'\boldsymbol{\Sigma}_{22}\boldsymbol{\beta}\right)}. \end{aligned} \tag{5.89}$$

From A7.7, the maximum value of ρ^2 is ρ_1^2, the largest eigenvalue of $\boldsymbol{\Sigma}_{11}^{-1}\boldsymbol{\Sigma}_{12}\boldsymbol{\Sigma}_{22}^{-1}\boldsymbol{\Sigma}_{21}$ (or of $\boldsymbol{\Sigma}_{22}^{-1}\boldsymbol{\Sigma}_{21}\boldsymbol{\Sigma}_{11}^{-1}\boldsymbol{\Sigma}_{12}$). The maximum occurs when $\boldsymbol{\alpha} = \mathbf{a}_1$, the eigenvector of $\boldsymbol{\Sigma}_{11}^{-1}\boldsymbol{\Sigma}_{12}\boldsymbol{\Sigma}_{22}^{-1}\boldsymbol{\Sigma}_{21}$ corresponding to ρ_1^2, and $\boldsymbol{\beta} = \mathbf{b}_1$, the corresponding eigenvector of $\boldsymbol{\Sigma}_{22}^{-1}\boldsymbol{\Sigma}_{21}\boldsymbol{\Sigma}_{11}^{-1}\boldsymbol{\Sigma}_{12}$. The maximization of (5.89) can be carried out in two stages (see Exercise 5.30).

The positive square root $\sqrt{\rho_1^2}$ is called the *first canonical correlation* between $\mathbf{x}$ and $\mathbf{y}$: $u_1 = \mathbf{a}'_1\mathbf{x}$ and $v_1 = \mathbf{b}'_1\mathbf{y}$ are called the *first canonical variables* (Hotelling [1936]). Since ρ_1^2 is independent of scale, we can scale $\mathbf{a}_1$ and $\mathbf{b}_1$ such that $\mathbf{a}'_1\boldsymbol{\Sigma}_{11}\mathbf{a}_1 = 1$ and $\mathbf{b}'_1\boldsymbol{\Sigma}_{22}\mathbf{b}_1 = 1$. In this case u_1 and v_1 have unit variances.

The above procedure may be regarded as a dimension-reducing technique in which $\mathbf{x}$ and $\mathbf{y}$ are reduced to u_1 and v_1 such that ρ^2 is minimized. However, u_1, for example, may not be an "adequate" reduction of $\mathbf{x}$, and a possible extension of the procedure would be to reduce $\mathbf{x}$ and $\mathbf{y}$ to r-dimensional vectors $\mathbf{u} = (u_1, u_2, \ldots, u_r)'$ and $\mathbf{v} = (v_1, v_2, \ldots, v_r)'$, respectively, such that (1) the elements of $\mathbf{u}$ are uncorrelated, (2) the elements of $\mathbf{v}$ are uncorrelated, and (3) the squares of the correlations between u_j and v_j $(j = 1, 2, \ldots, r)$ are collectively maximized in some sense. One approach is given by the following theorem, but let us first introduce some notation. Let

$$\mathbf{\Sigma}_{ii} = \mathbf{R}'_{ii}\mathbf{R}_{ii} \qquad (i = 1, 2) \tag{5.90}$$

be the Cholesky decomposition of $\mathbf{\Sigma}_{ii}$ (see A5.11), where $\mathbf{R}_{ii}$ is upper triangular. Define

$$\mathbf{C} = \left(\mathbf{R}'_{11}\right)^{-1}\mathbf{\Sigma}_{12}\mathbf{R}_{22}^{-1}; \tag{5.91}$$

then

$$\mathbf{\Sigma}_{11}^{-1}\mathbf{\Sigma}_{12}\mathbf{\Sigma}_{22}^{-1}\mathbf{\Sigma}_{21} = \mathbf{R}_{11}^{-1}\mathbf{C}\mathbf{C}'\mathbf{R}_{11} \tag{5.92}$$

and

$$\mathbf{\Sigma}_{22}^{-1}\mathbf{\Sigma}_{21}\mathbf{\Sigma}_{11}^{-1}\mathbf{\Sigma}_{12} = \mathbf{R}_{22}^{-1}\mathbf{C}'\mathbf{C}\mathbf{R}_{22}. \tag{5.93}$$

THEOREM 5.9 Let $1 > \rho_1^2 \geq \rho_2^2 \geq \cdots \geq \rho_m^2 > 0$, where $m = \text{rank } \mathbf{\Sigma}_{12}$, be the m nonzero eigenvalues of $\mathbf{\Sigma}_{11}^{-1}\mathbf{\Sigma}_{12}\mathbf{\Sigma}_{22}^{-1}\mathbf{\Sigma}_{21}$ (and of $\mathbf{\Sigma}_{22}^{-1}\mathbf{\Sigma}_{21}\mathbf{\Sigma}_{11}^{-1}\mathbf{\Sigma}_{12}$). Let $\mathbf{a}_1, \mathbf{a}_2, \ldots, \mathbf{a}_m$ and $\mathbf{b}_1, \mathbf{b}_2, \ldots, \mathbf{b}_m$ be corresponding eigenvectors of $\mathbf{\Sigma}_{11}^{-1}\mathbf{\Sigma}_{12}\mathbf{\Sigma}_{22}^{-1}\mathbf{\Sigma}_{21}$ and $\mathbf{\Sigma}_{22}^{-1}\mathbf{\Sigma}_{21}\mathbf{\Sigma}_{11}^{-1}\mathbf{\Sigma}_{12}$, respectively. Suppose $\boldsymbol{\alpha}$ and $\boldsymbol{\beta}$ are arbitrary vectors such that for $r \leq m - 1$, $\boldsymbol{\alpha}'\mathbf{x}$ is uncorrelated with each $\mathbf{a}'_j\mathbf{x}$ $(j = 1, 2, \ldots, r)$ and $\boldsymbol{\beta}'\mathbf{y}$ is uncorrelated with each $\mathbf{b}'_j\mathbf{y}$ $(j = 1, 2, \ldots, r)$. Then we have the following:

(i) The maximum squared correlation between $\boldsymbol{\alpha}'\mathbf{x}$ and $\boldsymbol{\beta}'\mathbf{y}$ is given by ρ_{r+1}^2 and it occurs when $\boldsymbol{\alpha} = \mathbf{a}_{r+1}$ and $\boldsymbol{\beta} = \mathbf{b}_{r+1}$.

(ii) $\text{cov}[\mathbf{a}'_j\mathbf{x}, \mathbf{a}'_k\mathbf{x}] = 0, j \neq k$, and $\text{cov}[\mathbf{b}'_j\mathbf{y}, \mathbf{b}'_k\mathbf{y}] = 0, j \neq \text{k}$.

Proof We observe from (5.91) that rank $\mathbf{C}$ = rank $\mathbf{\Sigma}_{12}$, and that (5.92), (5.93), $\mathbf{C}\mathbf{C}'$, and $\mathbf{C}'\mathbf{C}$ all have the same positive eigenvalues $\rho_1^2, \rho_2^2, \ldots, \rho_m^2$ (by A1.4). From

$$\mathbf{\Sigma}_{11}^{-1}\mathbf{\Sigma}_{12}\mathbf{\Sigma}_{22}^{-1}\mathbf{\Sigma}_{21}\mathbf{a}_j = \rho_j^2\mathbf{a}_j,$$

we have

$$\mathbf{R}_{11}\mathbf{\Sigma}_{11}^{-1}\mathbf{\Sigma}_{12}\mathbf{\Sigma}_{22}^{-1}\mathbf{\Sigma}_{21}\mathbf{R}_{11}^{-1}\mathbf{R}_{11}\mathbf{a}_j = \rho_j^2\mathbf{R}_{11}\mathbf{a}_j$$

or, using (5.92),

$$\mathbf{CC}'(\mathbf{R}_{11}\mathbf{a}_j) = (\mathbf{R}_{11}\mathbf{a}_j)\rho_j^2. \tag{5.94}$$

Hence $\mathbf{a}_{0j} = \mathbf{R}_{11}\mathbf{a}_j$ is an eigenvector of $\mathbf{CC}'$ and, using similar algebra, $\mathbf{b}_{0j} = \mathbf{R}_{22}\mathbf{b}_j$ is an eigenvector of $\mathbf{C}'\mathbf{C}$, corresponding to ρ_j^2.

(i) Let $\mathbf{A}' = (\mathbf{a}_1, \mathbf{a}_2, \ldots, \mathbf{a}_r)$, $\mathbf{A}'_0 = (\mathbf{a}_{01}, \mathbf{a}_{02}, \ldots, \mathbf{a}_{0r})$ and let $\mathbf{B}'$ and $\mathbf{B}'_0$ be similarly defined. We are given

$$\begin{aligned}
\mathbf{0} &= \mathscr{C}[\mathbf{Ax}, \boldsymbol{\alpha}'\mathbf{x}] \\
&= \mathbf{A}\mathscr{D}[\mathbf{x}]\boldsymbol{\alpha} \\
&= \mathbf{A}\boldsymbol{\Sigma}_{11}\boldsymbol{\alpha} \\
&= (\mathbf{AR}'_{11})(\mathbf{R}_{11}\boldsymbol{\alpha}) \\
&= \mathbf{A}_0\boldsymbol{\alpha}_0,
\end{aligned}$$

say, where $\boldsymbol{\alpha}_0 = \mathbf{R}_{11}\boldsymbol{\alpha}$. Using a similar argument, we also have $\mathbf{B}_0\boldsymbol{\beta}_0 = \mathbf{0}$, where $\boldsymbol{\beta}_0 = \mathbf{R}_{22}\boldsymbol{\beta}$. Then

$$\begin{aligned}
\rho_{\max}^2 &= \underset{\mathbf{A}\boldsymbol{\Sigma}_{11}\boldsymbol{\alpha}=\mathbf{0},\, \mathbf{B}\boldsymbol{\Sigma}_{22}\boldsymbol{\beta}=\mathbf{0}}{\text{supremum}} \left\{ \frac{(\boldsymbol{\alpha}'\boldsymbol{\Sigma}_{12}\boldsymbol{\beta})^2}{(\boldsymbol{\alpha}'\boldsymbol{\Sigma}_{11}\boldsymbol{\alpha})(\boldsymbol{\beta}'\boldsymbol{\Sigma}_{22}\boldsymbol{\beta})} \right\} \\
&= \underset{\mathbf{A}_0\boldsymbol{\alpha}_0=\mathbf{0},\, \mathbf{B}_0\boldsymbol{\beta}_0=\mathbf{0}}{\text{supremum}} \left\{ \frac{(\boldsymbol{\alpha}'_0\mathbf{C}\boldsymbol{\beta}_0)^2}{(\boldsymbol{\alpha}'_0\boldsymbol{\alpha}_0)(\boldsymbol{\beta}'_0\boldsymbol{\beta}_0)} \right\} \qquad (5.95) \\
&= \rho_{r+1}^2
\end{aligned}$$

which occurs when $\boldsymbol{\alpha}_0 = \mathbf{a}_{0,r+1}$ and $\boldsymbol{\beta}_0 = \mathbf{b}_{0,r+1}$, that is, when $\boldsymbol{\alpha} = \mathbf{a}_{r+1}$ and $\boldsymbol{\beta} = \mathbf{b}_{r+1}$.

(ii) Since $\mathbf{CC}'$ is symmetric, we can choose its eigenvectors so that they are mutually orthogonal (see A1.3). Hence

$$\begin{aligned}
\mathscr{C}[\mathbf{a}'_j\mathbf{x}, \mathbf{a}'_k\mathbf{x}] &= \mathbf{a}'_j\boldsymbol{\Sigma}_{11}\mathbf{a}_k \\
&= \mathbf{a}'_{0j}\mathbf{a}_{0k} \\
&= 0 \qquad (j \neq k).
\end{aligned}$$

A similar result holds for the columns of $\mathbf{B}'_0$. □

The positive square root $\sqrt{\rho_j^2}$ is called the jth canonical correlation, and $u_j = \mathbf{a}_j'\mathbf{x}$ and $v_j = \mathbf{b}_j'\mathbf{y}$ the jth canonical variables. The latter will be unique (apart from signs) if the canonical correlations are distinct (see A1.3). Theorem 5.9 tells us that the canonical variables have the following property. First, u_1 and v_1 are chosen so that they have maximum squared correlation ρ_1^2. Second, u_2 and v_2 are chosen so that u_2 is uncorrelated with u_1, v_2 is uncorrelated with v_1, and u_2 and v_2 have maximum squared correlation ρ_2^2. Third, u_3 and v_3 are chosen so that u_3 is uncorrelated with both u_1 and u_2, v_3 is uncorrelated with v_1 and v_2, and u_3 and v_3 have maximum squared correlation ρ_3^2, and so on. Since the eigenvectors $\mathbf{a}_{0j}$ and $\mathbf{b}_{0j}$ can be scaled to have unit length, we have $\text{var}\, u_j = \mathbf{a}_j'\boldsymbol{\Sigma}_{11}\mathbf{a}_j = \mathbf{a}_{0j}'\mathbf{a}_{0j} = 1$ and, similarly, $\text{var}\, v_j = 1$. For this scaling the canonical variates all have unit variances. Thus, writing $\mathbf{u} = \mathbf{A}\mathbf{x}$, for example, we have $\mathscr{D}[\mathbf{u}] = \mathbf{A}\boldsymbol{\Sigma}_{11}\mathbf{A}' = \mathbf{I}_r$.

If the $d_1 \times d_2$ matrix $\boldsymbol{\Sigma}_{12}$ has full rank, and $d_1 < d_2$, then $m = d_1$. All the eigenvalues of $\boldsymbol{\Sigma}_{11}^{-1}\boldsymbol{\Sigma}_{12}\boldsymbol{\Sigma}_{22}^{-1}\boldsymbol{\Sigma}_{21}$ are then positive, while $\boldsymbol{\Sigma}_{22}^{-1}\boldsymbol{\Sigma}_{21}\boldsymbol{\Sigma}_{11}^{-1}\boldsymbol{\Sigma}_{12}$ had d_1 positive eigenvalues and $d_2 - d_1$ zero eigenvalues. However, the rank of $\boldsymbol{\Sigma}_{12}$ can vary, even though $\boldsymbol{\Sigma} > \mathbf{O}$. For example, there may be constraints on $\boldsymbol{\Sigma}_{12}$ such as $\boldsymbol{\Sigma}_{12} = \mathbf{O}$ (rank zero) or $\boldsymbol{\Sigma}_{12} = \sigma^2\rho\mathbf{1}_{d_1}\mathbf{1}_{d_2}'$ (rank 1).

Although the above approach to canonical correlations is the more traditional one, another method of derivation that also has considerable appeal can be based on the concept of prediction. For example, if we wished to predict z_1 from $\mathbf{y} = (z_2, z_3, \ldots, z_d)'$, then we could use the linear predictor $\boldsymbol{\beta}'\mathbf{y}$ such that the variance of the error is minimized. Now

$$\begin{aligned}
\mathrm{E}\left[(z_1 - \boldsymbol{\beta}'\mathbf{y})^2\right] &= \mathrm{E}\left[z_1^2 - 2\boldsymbol{\beta}'\mathbf{y}z_1 + \boldsymbol{\beta}'\mathbf{y}\mathbf{y}'\boldsymbol{\beta}\right] \\
&= \text{var}\, z_1 - 2\boldsymbol{\beta}'\mathscr{C}[\mathbf{y}, z_1] + \boldsymbol{\beta}'\mathscr{D}[\mathbf{y}]\boldsymbol{\beta} \\
&= \sigma_1^2 - 2\boldsymbol{\beta}'\boldsymbol{\sigma}_{21} + \boldsymbol{\beta}'\boldsymbol{\Sigma}_{22}\boldsymbol{\beta}. \qquad (5.96)
\end{aligned}$$

Differentiating (5.96) with respect to $\boldsymbol{\beta}$ using A8.1 gives the solution $\boldsymbol{\beta} = \boldsymbol{\Sigma}_{22}^{-1}\boldsymbol{\sigma}_{21}$, the same as (5.88). Extending this concept to $\mathbf{z}' = (\mathbf{x}', \mathbf{y}')$, we wish to find $2r$ linear combinations $\mathbf{u} = \mathbf{A}\mathbf{x}$ and $\mathbf{v} = \mathbf{B}\mathbf{y}$, where $\mathbf{A}$ and $\mathbf{B}$ each have r linearly independent rows satisfying $\mathbf{A}\boldsymbol{\Sigma}_{11}\mathbf{A}' = \mathbf{I}_r$ and $\mathbf{B}\boldsymbol{\Sigma}_{22}\mathbf{B}' = \mathbf{I}_r$, such that $\mathbf{u}$ and $\mathbf{v}$ are "close" to each other. If we use $\mathrm{E}[\|\mathbf{u} - \mathbf{v}\|^2]$ as a measure of closeness, then we find that it is minimized when $\mathbf{u}$ and $\mathbf{v}$ are the canonical variables (see Brillinger [1975: Theorem 10.2.2]). For further derivations of a similar nature see Izenman [1975] and Yohai and Garcia Ben [1980].

Uptil now we have assumed that $\boldsymbol{\mu}_z = \mathscr{E}[\mathbf{z}] = \mathbf{0}$. If this is not the case, we simply replace $\mathbf{z}$ by $\mathbf{z} - \boldsymbol{\mu}_z$ in the above theory; for example, $\mathbf{u} = \mathbf{A}(\mathbf{x} - \boldsymbol{\mu}_x)$.

5.7.2 *Sample Canonical Correlations*

The above theory can now be readily extended to a sample of observations $\mathbf{z}_1, \mathbf{z}_2, \ldots, \mathbf{z}_n$, using the technique of Theorem 5.5 in Section 5.2.4. We define a random variable $\mathbf{w}$ such that $\mathbf{w} = \mathbf{z}_i$ $(i = 1, 2, \ldots, n)$ with probability $1/n$. Then $\boldsymbol{\mu}_w = \mathscr{E}[\mathbf{w}] = \bar{\mathbf{z}}$ and $\mathscr{D}[\mathbf{w}] = \hat{\boldsymbol{\Sigma}} = \sum_{i=1}^{n}(\mathbf{z}_i - \bar{\mathbf{z}})(\mathbf{z}_i - \bar{\mathbf{z}})'/n$ so that we can handle the sampling situation by applying the population theory of the previous section to $\mathbf{w} - \boldsymbol{\mu}_w$. Writing $\bar{\mathbf{z}}' = (\bar{\mathbf{x}}', \bar{\mathbf{y}}')$, we can partition $\hat{\boldsymbol{\Sigma}}$ in the same way as $\boldsymbol{\Sigma}$, namely,

$$n\hat{\boldsymbol{\Sigma}} = \begin{pmatrix} \sum_{i=1}^{n}(\mathbf{x}_i - \bar{\mathbf{x}})(\mathbf{x}_i - \bar{\mathbf{x}})', & \sum_{i=1}^{n}(\mathbf{x}_i - \bar{\mathbf{x}})(\mathbf{y}_i - \bar{\mathbf{y}})' \\ \sum_{i=1}^{n}(\mathbf{y}_i - \bar{\mathbf{y}})(\mathbf{x}_i - \bar{\mathbf{x}})', & \sum_{i=1}^{n}(\mathbf{y}_i - \bar{\mathbf{y}})(\mathbf{y}_i - \bar{\mathbf{y}})' \end{pmatrix}$$

$$= \begin{pmatrix} \tilde{\mathbf{X}}'\tilde{\mathbf{X}}, & \tilde{\mathbf{X}}'\tilde{\mathbf{Y}} \\ \tilde{\mathbf{Y}}'\tilde{\mathbf{X}}, & \tilde{\mathbf{Y}}'\tilde{\mathbf{Y}} \end{pmatrix}$$

$$= \begin{pmatrix} \mathbf{Q}_{11}, & \mathbf{Q}_{12} \\ \mathbf{Q}_{21}, & \mathbf{Q}_{22} \end{pmatrix},$$

say. We assume that $d_1 \le d_2$; then, given $\boldsymbol{\Sigma} > \mathbf{O}$ and $n - 1 \ge d$, we know that, with probability 1, $n\hat{\boldsymbol{\Sigma}}$ is positive definite and there are no constraints on $\mathbf{Q}_{12}$, that is, rank $\mathbf{Q}_{12} = d_1$. Let $r_1^2 > r_2^2 > \cdots > r_{d_1}^2 > 0$ be the eigenvalues of $\mathbf{Q}_{11}^{-1}\mathbf{Q}_{12}\mathbf{Q}_{22}^{-1}\mathbf{Q}_{21}$, with corresponding eigenvectors $\hat{\mathbf{a}}_1, \hat{\mathbf{a}}_2, \ldots, \hat{\mathbf{a}}_{d_1}$, and define $u_{ij} = \hat{\mathbf{a}}_j'(\mathbf{x}_i - \bar{\mathbf{x}})$, where $\hat{\mathbf{a}}_j$ is scaled so that

$$1 = \hat{\mathbf{a}}_j'\hat{\boldsymbol{\Sigma}}_{11}\hat{\mathbf{a}}_j = \hat{\mathbf{a}}_j'\frac{1}{n}\sum_i(\mathbf{x}_i - \bar{\mathbf{x}})(\mathbf{x}_i - \bar{\mathbf{x}})'\hat{\mathbf{a}}_j = \frac{1}{n}\sum_{i=1} u_{ij}^2.$$

Then $\sqrt{r_j^2}$ is called the jth sample canonical correlation. These correlations are distinct with probability 1 (otherwise any equality implies some constraints on the original sample vectors $\mathbf{z}_i$). We call u_{ij} the jth sample canonical variable of $\mathbf{x}_i$. In a similar fashion we define $v_{ij} = \hat{\mathbf{b}}_j'(\mathbf{y}_i - \bar{\mathbf{y}})$ to be the jth sample canonical variable of $\mathbf{y}_i$, where $\hat{\mathbf{b}}_1, \hat{\mathbf{b}}_2, \ldots, \hat{\mathbf{b}}_{d_1}$ are the corresponding eigenvectors of $\mathbf{Q}_{22}^{-1}\mathbf{Q}_{21}\mathbf{Q}_{11}^{-1}\mathbf{Q}_{12}$. The u_{ij} and v_{ij} are also called the scores of the ith individual on the jth canonical variables. We note that r_j^2 is the square of the correlation of the (u_{ij}, v_{ij}) $(i = 1, 2, \ldots, n)$. The sample canonical variables have the same optimal properties as those described in the previous section, except that population variances and covariances are replaced by their sample counterparts.

For computational purposes, we observe that we can use $\mathbf{Q}_{ab}$, $\hat{\boldsymbol{\Sigma}}_{ab} = \mathbf{Q}_{ab}/n$, or $\mathbf{S}_{ab} = \mathbf{Q}_{ab}/(n-1)$ $(a, b = 1, 2)$ in calculating the required eigenvalues and eigenvectors as the factors n and $n - 1$ cancel out. Also, given the Cholesky decomposition $\mathbf{Q}_{ii} = \mathbf{T}_{ii}'\mathbf{T}_{ii}$, where $\mathbf{T}_{ii}$ is upper triangular, we note from the sample versions of (5.91) and (5.94) that $\hat{\mathbf{a}}_{0j} = \mathbf{T}_{11}\hat{\mathbf{a}}_j$ is an eigenvector of $\hat{\mathbf{C}}\hat{\mathbf{C}}'$, where

$$\hat{\mathbf{C}} = (\mathbf{T}_{11}')^{-1}\mathbf{Q}_{12}\mathbf{T}_{22}^{-1}. \tag{5.97}$$

Similarly, $\hat{\mathbf{b}}_{0j} = \mathbf{T}_{22}\hat{\mathbf{b}}_j$ is an eigenvector of $\hat{\mathbf{C}}'\hat{\mathbf{C}}$. Details of the computation are given in Section 10.1.5d, where we shall also require

$$\mathbf{U} = [(u_{ij})] = \begin{pmatrix} \mathbf{x}_1' - \overline{\mathbf{x}}' \\ \vdots \\ \mathbf{x}_n' - \overline{\mathbf{x}}' \end{pmatrix} (\hat{\mathbf{a}}_1, \hat{\mathbf{a}}_2, \ldots, \hat{\mathbf{a}}_{d_1}) = \tilde{\mathbf{X}}\hat{\mathbf{A}}' \tag{5.98}$$

and

$$\mathbf{V} = [(v_{ij})] = \tilde{\mathbf{Y}}\hat{\mathbf{B}}'. \tag{5.99}$$

If any variable is qualitative with k states, it can be replaced by $k-1$ zero–one binary variables [see (1.2)] and the analysis carried out as above. An application of canonical analysis to mixed data is given by Lewis [1970] (see Mardia et al. [1979: Section 10.4]).

Kendall [1975] noted that "the difficulties of interpretation are such that not many examples of convincing applications of canonical correlation analysis appear in the literature." On the credit side, Gnanadesikan [1977: pp. 282–284] suggests that valuable insight can be obtained from two- and three-dimensional displays of the canonical variables (or their principal components), particularly in the identification of outliers that may be inducing an artificial linear relationship. Normal probability plots and univariate outlier procedures can also be applied to the canonical variables, with separate plots for each set.

EXAMPLE 5.7 In Table 5.27 we have a well-known set of data from Frets [1921] that has been discussed by various authors (e.g., Rao [1952], Anderson [1958], Mardia et al. [1979]). The data are summarized as follow:

$$\overline{\mathbf{z}}' = (\overline{\mathbf{x}}', \overline{\mathbf{y}}') = (185.72, 151.12, 183.84, 149.24)$$

and

$$\begin{pmatrix} \mathbf{Q}_{11} & \mathbf{Q}_{12} \\ \mathbf{Q}_{21} & \mathbf{Q}_{22} \end{pmatrix} = \left(\begin{array}{cc|cc} 2287.04 & 1268.84 & 1671.88 & 1106.68 \\ 1268.84 & 1304.64 & 1231.48 & 841.28 \\ \hline 1671.88 & 1231.48 & 2419.36 & 1356.96 \\ 1106.68 & 841.28 & 1356.96 & 1080.56 \end{array}\right).$$

The rows of $\mathbf{Q}_{12}$ appear to be approximately linearly dependent, and this is confirmed by transforming to the correlation matrix

$$\mathbf{R}_{12} = \begin{pmatrix} 0.7108 & 0.7040 \\ 0.6932 & 0.7086 \end{pmatrix}.$$

Since these elements are approximately equal, the rank of $\mathbf{R}_{12}$ (and therefore of $\mathbf{Q}_{12}$) is nearly 1. In fact, the eigenvalues of $\mathbf{Q}_{11}^{-1}\mathbf{Q}_{12}\mathbf{Q}_{22}^{-1}\mathbf{Q}_{21}$, which are also the

TABLE 5.27 Head Measurements on the First and Second Adult Sons in 25 Families[a]

Head Length, First Son	Head Breadth, First Son	Head Length, Second Son	Head Breadth, Second Son
x_1	x_2	y_1	y_2
191	155	179	145
195	149	201	152
181	148	185	149
183	153	188	149
176	144	171	142
208	157	192	152
189	150	190	149
197	159	189	152
188	152	197	159
192	150	187	151
179	158	186	148
183	147	174	147
174	150	185	152
190	159	195	157
188	151	187	158
163	137	161	130
195	155	183	158
186	153	173	148
181	145	182	146
175	140	165	137
192	154	185	152
174	143	178	147
176	139	176	143
197	167	200	158
190	163	187	150

[a]From Frets [1921].

same as those of $\mathbf{R}_{11}^{-1}\mathbf{R}_{12}\mathbf{R}_{22}^{-1}\mathbf{R}_{21}$ (see Exercise 5.27) are (0.6218, 0.0029). Taking square roots, $r_1 = 0.7885$ and $r_2 = 0.0537$.

The corresponding eigenvectors are

$$\hat{\mathbf{a}}_1 = \begin{pmatrix} 0.0577 \\ 0.0722 \end{pmatrix}, \qquad \hat{\mathbf{a}}_2 = \begin{pmatrix} 0.1429 \\ -0.1909 \end{pmatrix},$$

$$\hat{\mathbf{b}}_1 = \begin{pmatrix} 0.0512 \\ 0.0819 \end{pmatrix}, \qquad \hat{\mathbf{b}}_2 = \begin{pmatrix} 0.1796 \\ -0.2673 \end{pmatrix}.$$

Hence the first canonical variables are

$$\begin{aligned} u_1 &= \hat{\mathbf{a}}_1'(\mathbf{x} - \bar{\mathbf{x}}) \\ &= 0.0577(x_1 - 185.72) + 0.0707(x_2 - 151.12) \end{aligned}$$

and

$$v_1 = \mathbf{b}_1'(\mathbf{y} - \bar{\mathbf{y}})$$

$$= 0.0512(y_1 - 183.84) + 0.0819(y_2 - 149.24).$$

Since the second canonical correlation r_2 is close to zero, we can compare the two head dimensions of first and second sons by concentrating on just the first canonical variates. However, in this example it is helpful to standardize the four original variables by dividing each by its sample standard deviation (SD) to get

$$u_1 = 0.552x_1^* + 0.522x_2^*$$

and

$$v_1 = 0.504y_1^* + 0.538y_2^*,$$

where $x_1^* = (x_1 - \bar{x}_1)/SD$, and so on. Now u_1 and v_1 can be interpreted as "girth" measurements being approximately proportional to the sum of the standardized length and breadth for each brother. We also find that the second canonical variables, namely,

$$u_2 = 1.366x_1^* - 1.378x_2^*$$

and

$$v_2 = 1.769y_1^* - 1.759y_2^*,$$

can be interpreted as "shape" measurements, being approximately proportional to the differences between the standardized length and breadth for the two brothers. Clearly there is little correlation between the head "shapes" of first and second brothers. We note that the eigenvectors (0.552, 0.522)′, and so on, associated with the standardized variables x_i^* and y_j^* can be obtained directly from the eigenanalysis of $\mathbf{R}_{11}^{-1}\mathbf{R}_{12}\mathbf{R}_{22}^{-1}\mathbf{R}_{21}$ instead of $\mathbf{Q}_{11}^{-1}\mathbf{Q}_{12}\mathbf{Q}_{22}^{-1}\mathbf{Q}_{21}$. Some statistical packages do this.

Finally, in Fig. 5.15 we have a plot of u_1 versus v_1 for each of the 25 pairs of brothers. There do not appear to be any outliers.

5.7.3 *Inference*

If the sample $\mathbf{z}_1, \mathbf{z}_2, \ldots, \mathbf{z}_n$ comes from $N_d(\boldsymbol{\mu}, \boldsymbol{\Sigma})$, then the exact joint distribution of the sample canonical correlations, which depends only on the population canonical correlations, has been given by Constantine [1963] and James [1964] (see also Muirhead [1982: Section 11.3]). However, the distribution is

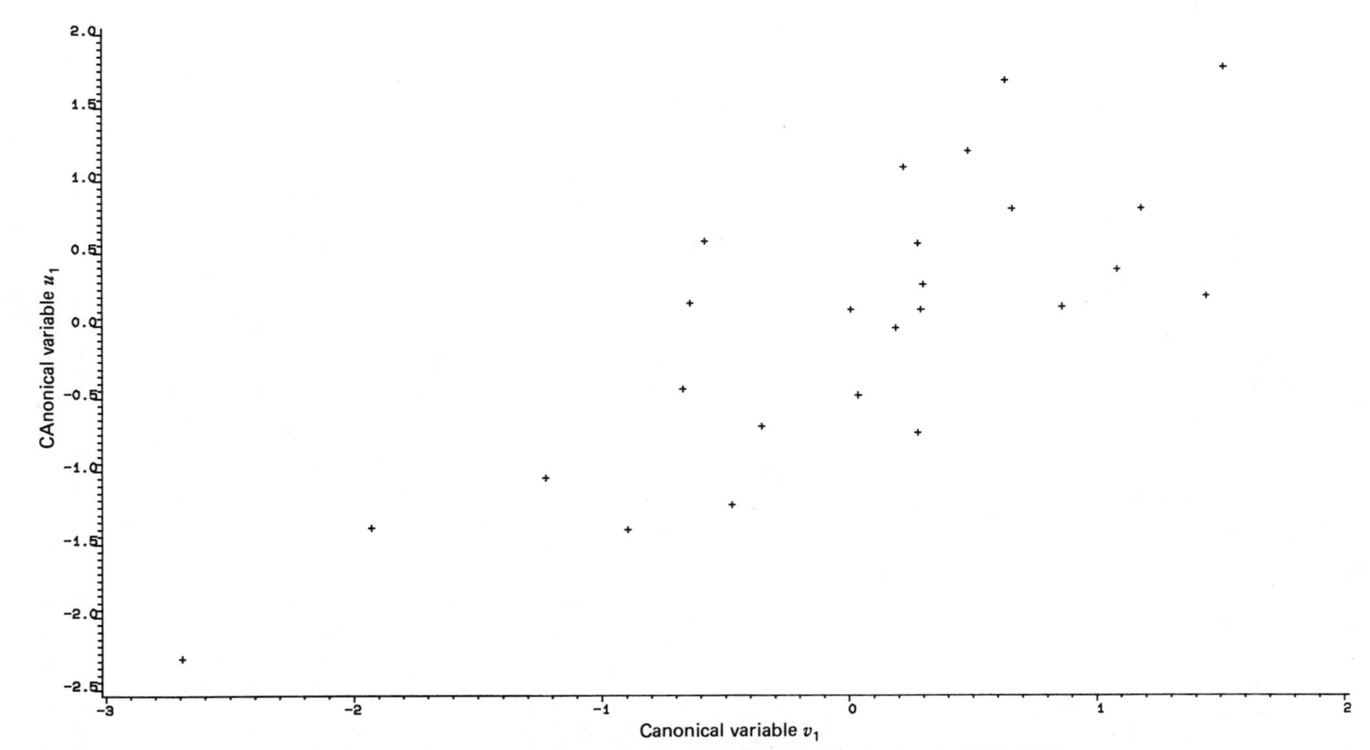

Fig. 5.15 Scatter plot of first two canonical variables for the data in Table 5.27.

intractable and an asymptotic theory has been developed. Since $\hat{\boldsymbol{\Sigma}}$ is the maximum likelihood estimate of $\boldsymbol{\Sigma}$, then, provided that the population canonical correlations are distinct, r_j^2, $\hat{\mathbf{a}}_j$, and $\hat{\mathbf{b}}_j$ are the maximum likelihood estimates of ρ_j^2, $\mathbf{a}_j$, and $\mathbf{b}_j$ (given an appropriate sign convention for the eigenvectors). Large-sample maximum likelihood theory then applies and Hsu [1941] showed that as $n \to \infty$, the $\sqrt{n}(r_j^2 - \rho_j^2)$ $(j = 1, 2, \ldots, m)$ are asymptotically independently distributed as $N_1(0, 4\rho_j^2[1 - \rho_j^2]^2)$. The asymptotic theory for the sample eigenvalues and eigenvectors (using a different scaling) is summarized by Brillinger [1975: Theorem 10.2.3], who suggested that it is probably more sensible to estimate variances using the jackknife method. Using the transformation (3.64), we note from Glynn and Muirhead [1978] that

$$E\left[\tanh^{-1} r_j\right] = \tanh^{-1}\rho_j + O(n^{-2})$$

and

$$\operatorname{var}\left[\tanh^{-1} r_j\right] = n^{-1} + O(n^{-2}).$$

Further references on asymptotic expansions are given by Fujikoshi [1977a], Muirhead and Waternaux [1980], and Muirhead [1982].

One of the most commonly used tests in canonical correlation analysis is the test that $d_1 - k$ of the smallest correlations are zero, that is, $H_{0k}: \rho_{k+1} = \rho_{k+2} = \cdots = \rho_{d_1} = 0$ $(\rho_k > 0)$. When H_{0k} is true, the canonical variables u_{ij}, v_{ij} $(j = k+1, \ldots, d_1)$ have no predictive value for comparing $\mathbf{x}_i$ and $\mathbf{y}_i$, so that the relationship between $\mathbf{x}_i$ and $\mathbf{y}_i$ can be summarized by means of the first k canonical variables. If k is small, this represents a substantial reduction in dimensionality. The likelihood ratio test for testing H_{0k} leads to the statistic

$$\ell = \prod_{j=k+1}^{d_1} \left(1 - r_j^2\right),$$

and $-2n \log \ell$ is asymptotically χ_ν^2, where $\nu = (d_1 - k)(d_2 - k)$, when H_{0k} is true (Bartlett [1938a, 1947]). A special case of this procedure, namely, H_0: $\rho_1 = \rho_2 = \cdots = \rho_{d_1} = 0$ or $\boldsymbol{\Sigma}_{12} = \mathbf{O}$ is discussed in Section 3.5.2. The chi-square approximation can be improved if we use Lawley's modification

$$-\left(n - 1 - k - \tfrac{1}{2}(d_1 + d_2 + 1) + \sum_{j=1}^{k} r_j^{-2}\right)\log \ell \qquad (5.100)$$

(see Glynn and Muirhead [1978]). To determine k we can test the sequence of hypotheses H_0, H_{01}, $H_{02}, \ldots$, until we obtain a nonsignificant test for H_{0r}. We then choose $k = r$. Fujikoshi and Veitch [1979] propose several other procedures using extensions of Mallows' [1973] C_p statistic and Akaike's [1973] information criterion. A large-sample expansion of the nonnull distribution of (5.100) is given by Fujikoshi [1977b]. It should be noted that Harris [1975: p. 144] claims that (5.100) is not approximately chi square and that such tests should be abandoned.

Unfortunately, as with other tests based on the sample dispersion matrix, $-n \log \ell$ is so nonrobust to departures from normality, particularly to the presence of outliers and to long-tailed distributions, that there is some doubt whether it should be used at all. Muirhead and Waternaux [1980] obtained the asymptotic joint distribution of the r_j^2 in the nonnull case and showed that the $\sqrt{n}(r_j^2 - \rho_j^2)$ are still asymptotically normal with mean 0, but the asymptotic variance–covariance matrix now contains additional terms depending on fourth-order cumulants of the parent population. The limiting distribution of $-n \log \ell$ is now a weighted sum of χ_1^2 variables, the weights being complicated functions of the cumulants. In one special case, when the parent population is elliptical, there is substantial algebraic simplification and a correction factor can be calculated. Asymptotic inference for eigenvectors is given by Tyler [1981].

In conclusion, we note that canonical correlation analysis has been applied to time series (Brillinger [1975]), reduced rank regression with random regressors (Izenman [1975]), and the analysis of systems of linear equations (Hooper [1959], Hannan [1967]).

5.7.4 More Than Two Sets of Variables

Uptil now we have considered two sets of variables represented by $\mathbf{z}' = (\mathbf{x}', \mathbf{y}')$. If we transform $\mathbf{x}_0 = (\mathbf{R}'_{11})^{-1}\mathbf{x}$ and $\mathbf{y}_0 = (\mathbf{R}'_{22})^{-1}\mathbf{y}$, where $\boldsymbol{\Sigma}_{ii} = \mathbf{R}'_{ii}\mathbf{R}_{ii}$, then

$$\mathscr{D}[\mathbf{x}_0] = \left(\mathbf{R}'_{11}\right)^{-1}\boldsymbol{\Sigma}_{11}\mathbf{R}_{11}^{-1} = \mathbf{I}_{d_1},$$

$\mathscr{D}[\mathbf{y}_0] = \mathbf{I}_{d_2}$ and $\mathscr{C}[\mathbf{x}_0, \mathbf{y}_0] = \mathbf{R}_{11}'^{-1}\boldsymbol{\Sigma}_{12}\mathbf{R}_{22}^{-1} = \mathbf{C}$. Therefore, if $\mathbf{z}'_0 = (\mathbf{x}'_0, \mathbf{y}'_0)$, we have

$$\mathscr{D}[\mathbf{z}_0] = \begin{pmatrix} \mathbf{I}_{d_1} & \mathbf{C} \\ \mathbf{C}' & \mathbf{I}_{d_2} \end{pmatrix} = \boldsymbol{\Sigma}_0,$$

say. The jth canonical variables are given by (see Theorem 5.9)

$$u_j = \mathbf{a}'_j\mathbf{x} = \left(\mathbf{R}_{11}^{-1}\mathbf{a}_{0j}\right)'\mathbf{R}'_{11}\mathbf{x}_0 = \mathbf{a}'_{0j}\mathbf{x}_0$$

and

$$v_j = \mathbf{b}'_j\mathbf{y} = \mathbf{b}'_{0j}\mathbf{y}_0,$$

where $\|\mathbf{a}_{0j}\| = \|\mathbf{b}_{0j}\| = 1$. Since the correlation of $\boldsymbol{\alpha}'\mathbf{x}$ and $\boldsymbol{\beta}'\mathbf{y}$ reduces to the expression in curly brackets in (5.95), the first variables u_1 and v_1 are obtained by maximizing $\boldsymbol{\alpha}_0'\mathbf{C}\boldsymbol{\beta}_0$ subject to $\|\boldsymbol{\alpha}_0\| = \|\boldsymbol{\beta}_0\| = 1$. If

$$\boldsymbol{\Phi} = \begin{pmatrix} \boldsymbol{\alpha}_0' & \mathbf{0}' \\ \mathbf{0}' & \boldsymbol{\beta}_0' \end{pmatrix} \begin{pmatrix} \mathbf{I}_{d_1} & \mathbf{C} \\ \mathbf{C}' & \mathbf{I}_{d_2} \end{pmatrix} \begin{pmatrix} \boldsymbol{\alpha}_0 & \mathbf{0} \\ \mathbf{0} & \boldsymbol{\beta}_0 \end{pmatrix}$$

$$= \begin{pmatrix} 1 & \phi_{12} \\ \phi_{21} & 1 \end{pmatrix},$$

where $\phi_{12} = \boldsymbol{\alpha}_0'\mathbf{C}\boldsymbol{\beta}_0 = \phi_{21}$, then u_1 and v_1 are obtained by maximizing ϕ_{12} or, equivalently, a suitable function of $\boldsymbol{\Phi}$ such as $\mathbf{1}'\boldsymbol{\Phi}\mathbf{1} = 2 + 2\phi_{12}$ or $\operatorname{tr} \boldsymbol{\Phi}^2 = \Sigma_i \Sigma_j \phi_{ij}^2 = 2 + 2\phi_{12}^2$. The next canonical variables are found by the same procedure, but with further constraints on $\boldsymbol{\alpha}_0$ and $\boldsymbol{\beta}_0$ [u_1 and u_2 are uncorrelated and v_1 and v_2 are uncorrelated, so that $\mathbf{a}_{01}'\boldsymbol{\alpha}_0 = 0$ and $\mathbf{b}_{01}'\boldsymbol{\beta}_0 = 0$; see (5.95)]. Using the above approach, Kettenring [1971] generalized the procedure to three or more sets of variables. For example, if $\mathbf{z}' = (\mathbf{w}', \mathbf{x}', \mathbf{y}')$ and $\mathbf{w}_0 = (\mathbf{R}_{33}')^{-1}\mathbf{w}$, then

$$\mathscr{D}[\mathbf{z}_0] = \begin{pmatrix} \mathbf{I}_{d_1} & \mathbf{C}_{12} & \mathbf{C}_{13} \\ \mathbf{C}_{12}' & \mathbf{I}_{d_2} & \mathbf{C}_{23} \\ \mathbf{C}_{13}' & \mathbf{C}_{23}' & \mathbf{I}_{d_3} \end{pmatrix} = \boldsymbol{\Sigma}_0,$$

where $\mathbf{C}_{ab} = (\mathbf{R}_{aa}')^{-1}\boldsymbol{\Sigma}_{ab}\mathbf{R}_{bb}^{-1}$. Then, writing

$$\mathbf{D}_0 = \begin{pmatrix} \boldsymbol{\alpha}_0' & \mathbf{0}' & \mathbf{0}' \\ \mathbf{0}' & \boldsymbol{\beta}_0' & \mathbf{0}' \\ \mathbf{0}' & \mathbf{0}' & \boldsymbol{\gamma}_0' \end{pmatrix},$$

we have

$$\boldsymbol{\Phi} = \mathbf{D}_0\boldsymbol{\Sigma}_0\mathbf{D}_0' = \begin{pmatrix} 1 & \phi_{12} & \phi_{13} \\ \phi_{21} & 1 & \phi_{23} \\ \phi_{31} & \phi_{32} & 1 \end{pmatrix}, \tag{5.101}$$

where $\phi_{ij} = \phi_{ji}$. The first canonical variables t_1, u_1, and v_1 for the three groups, respectively, are now found by minimizing some function of the elements of $\boldsymbol{\Phi}$ subject to $\|\boldsymbol{\alpha}_0\| = \|\boldsymbol{\beta}_0\| = \|\boldsymbol{\gamma}_0\| = 1$. In the sample case $\boldsymbol{\Sigma}_0$ and $\mathbf{C}_{ab}$ are replaced by their sample equivalents. Kettenring [1971] considered five functions of $\boldsymbol{\Phi}$, including the ones mentioned above, and the reader is referred to his paper and Gnanadesikan [1977] for further details and examples.

5.8 DISCRIMINANT COORDINATES

Although discriminant analysis is not discussed until Chapter 6, it is appropriate to mention in this chapter a dimension-reducing technique that is useful for examining any clustering effects in the data. Suppose we have n d-dimensional observations in which n_i belong to group i $(i = 1, 2, \ldots, g;$ $n = \sum_i n_i)$. Let $\mathbf{x}_{ij}$ be the jth observation in group i, and define

$$\overline{\mathbf{x}}_{i\cdot} = \frac{1}{n_i} \sum_{j=1}^{n_i} \mathbf{x}_{ij} \quad \text{and} \quad \overline{\mathbf{x}}_{\cdot\cdot} = \frac{1}{n} \sum_{i=1}^{g} \sum_{j=1}^{n_i} \mathbf{x}_{ij}.$$

Then we can define two matrices, the *within-groups* matrix

$$\begin{aligned} \mathbf{W} &= \sum_{i=1}^{g} \sum_{j=1}^{n_i} (\mathbf{x}_{ij} - \overline{\mathbf{x}}_{i\cdot})(\mathbf{x}_{ij} - \overline{\mathbf{x}}_{i\cdot})' \\ &= \sum_{i=1}^{g} (n_i - 1)\mathbf{S}_i, \end{aligned}$$

say, and the *between-groups* matrix

$$\begin{aligned} \mathbf{B} &= \sum_{i=1}^{g} \sum_{j=1}^{n_i} (\overline{\mathbf{x}}_{i\cdot} - \overline{\mathbf{x}}_{\cdot\cdot})(\overline{\mathbf{x}}_{i\cdot} - \overline{\mathbf{x}}_{\cdot\cdot})' \\ &= \sum_{i=1}^{g} n_i (\overline{\mathbf{x}}_{i\cdot} - \overline{\mathbf{x}}_{\cdot\cdot})(\overline{\mathbf{x}}_{i\cdot} - \overline{\mathbf{x}}_{\cdot\cdot})' \cdot \\ &= (n - g)\mathbf{S}, \end{aligned}$$

say. These are multivariate analogues of the usual analysis of variance sums of squares and are discussed further in Section 9.2. Clearly the degree of clustering within the groups will be determined by the relative "magnitudes" of **B** and **W**. However, as a first step, we could reduce the multivariate observations $\mathbf{x}_{ij}$ to univariate observations $z_{ij} = \mathbf{c}'\mathbf{x}_{ij}$ and obtain the usual sums of squares

$$\sum_i \sum_j (z_{ij} - \bar{z}_{i\cdot})^2 = \mathbf{c}'\mathbf{W}\mathbf{c}$$

and

$$\sum_i \sum_j (\bar{z}_{i\cdot} - \bar{z}_{\cdot\cdot})^2 = \mathbf{c}'\mathbf{B}\mathbf{c}.$$

How do we choose **c**? If we were testing for equal group population means for the z_{ij} observations, we could use the ratio

$$F_c = \frac{\mathbf{c}'\mathbf{B}\mathbf{c}}{\mathbf{c}'\mathbf{W}\mathbf{c}}$$

and reject the hypothesis of equal means if F_c is too large. The magnitude of F_c reflects the degree of group clustering in the z_{ij} and one criteria would be to choose **c** so as to maximize F_c. To do this we assume that at least one within-group sample dispersion matrix $\mathbf{S}_i$ is positive definite (with probability 1). Then **W**, a sum of positive semidefinite matrices, is also positive definite (A5.10) with Cholesky decomposition $\mathbf{W} = \mathbf{T}'\mathbf{T}$ (A5.11). Setting $\mathbf{b} = \mathbf{Tc}$,

$$\begin{aligned} F_c &= \frac{\mathbf{b}'(\mathbf{T}')^{-1}\mathbf{B}\mathbf{T}^{-1}\mathbf{b}}{\mathbf{b}'\mathbf{b}} \\ &= \frac{\mathbf{b}'\mathbf{A}\mathbf{b}}{\mathbf{b}'\mathbf{b}}, \qquad \text{say}, \\ &= \mathbf{a}'\mathbf{A}\mathbf{a}, \end{aligned}$$

where $\mathbf{a} = \mathbf{b}/\|\mathbf{b}\|$, that is, $\|\mathbf{a}\| = 1$. Maximizing F_c subject to $\mathbf{c} \neq \mathbf{0}$ is equivalent to maximizing $\mathbf{a}'\mathbf{A}\mathbf{a}$ subject to $\mathbf{a}'\mathbf{a} = 1$. From A7.4, the maximum occurs when $\mathbf{a} = \mathbf{a}_1$, the unit eigenvector corresponding to the largest eigenvalue λ_1 ($= \mathbf{a}_1'\mathbf{A}\mathbf{a}_1$) of **A**.

We now ask what other directions **c** might give us useful separation of the groups? A reasonable procedure might be to choose a second **a** that is perpendicular to $\mathbf{a}_1$ such that $\mathbf{a}'\mathbf{A}\mathbf{a}$ is again minimized subject to $\|\mathbf{a}\| = 1$. This problem is formally equivalent to extracting principal components so that from Theorem 5.4 (with $\boldsymbol{\Sigma} \to \mathbf{A}$) in Section 5.2.3, the solution is $\mathbf{a}_2$, the unit eigenvector corresponding to λ_2 ($= \mathbf{a}_2'\mathbf{A}\mathbf{a}_2$), the second largest eigenvalue of **A**. We can therefore find d orthogonal directions given by $\mathbf{a}_r$ $(r = 1, 2, \ldots, d)$ that correspond to the eigenvalues $\lambda_1 > \lambda_2 > \cdots > \lambda_d > 0$ of **A**. We note that

$$(\mathbf{T}')^{-1}\mathbf{B}\mathbf{T}^{-1}\mathbf{a}_r = \mathbf{A}\mathbf{a}_r = \lambda_r \mathbf{a}_r,$$

and multiplying on the left by $\mathbf{T}^{-1}(n-g)^{1/2}$, we have

$$\mathbf{W}^{-1}\mathbf{B}\mathbf{c}_r = \lambda_r \mathbf{c}_r.$$

Hence $\mathbf{W}^{-1}\mathbf{B}$ has eigenvalues λ_r and suitably scaled eigenvectors $\mathbf{c}_r$ $(r = 1, 2, \ldots, d)$, where $\mathbf{c}_r$ satisfies $\mathbf{a}_r = \mathbf{T}\mathbf{c}_r(n-g)^{-1/2}$, the $\mathbf{a}_r$ being orthonormal. Let $\mathbf{C}' = (\mathbf{c}_1, \mathbf{c}_2, \ldots, \mathbf{c}_k)$ $(k \le d)$ and consider the transformation $\mathbf{z}_{ij} = \mathbf{C}\mathbf{x}_{ij}$. Then, following Gnanadesikan [1977], we shall call the k elements of $\mathbf{z}_{ij}$ the first k *discriminant coordinates*. The more common term *canonical variates* is reserved for canonical correlation analysis in Section 5.7. Since these coordinates are determined so as to emphasize group separation, but with decreasing effectiveness, we have to decide on k. Clearly some trial and error is inevitable, though the relative magnitudes of the λ_r will give some indication. For $k = 2$ or 3, plots of the $\mathbf{z}_{ij}$ will be helpful for studying the degree and nature of the group separations and for highlighting outliers. In studying these plots we use the usual Euclidean distance for assessing interpoint distances as the elements of $\mathbf{z}_{ij}$ are "standardized." To see this we first recall that $\mathbf{W} = \mathbf{T}'\mathbf{T}$ so that

$$\mathbf{c}_r'\mathbf{S}\mathbf{c}_s = (n-g)^{-1}\mathbf{c}_r'\mathbf{T}'\mathbf{T}\mathbf{c}_s = \mathbf{a}_r'\mathbf{a}_s = \delta_{rs},$$

or $\mathbf{CSC}' = \mathbf{I}_k$. Assuming equal group dispersion matrices, then $\mathscr{C}[\mathbf{c}_r'\mathbf{x}_{ij}, \mathbf{c}_s'\mathbf{x}_{ij}] = \mathbf{c}_r'\boldsymbol{\Sigma}\mathbf{c}_s$ and, replacing $\boldsymbol{\Sigma}$ by $\mathbf{S}$, we see that the $\mathbf{z}_{ij}$ have zero sample covariances and unit sample variances. Furthermore, if the samples are large enough for the central limit theorem to apply, and $\mathbf{S} \approx \boldsymbol{\Sigma}$, then $\bar{\mathbf{z}}_{i\cdot} \sim N_k(\boldsymbol{\mu}_i, \mathbf{I}_k n_i^{-1})$ approximately, for some unknown $\boldsymbol{\mu}_i$. A $100(1-\alpha)\%$ circular confidence region for $\boldsymbol{\mu}_i$ is then approximately given by

$$n_i(\bar{\mathbf{z}}_{i\cdot} - \boldsymbol{\mu}_i)'(\bar{\mathbf{z}}_{i\cdot} - \boldsymbol{\mu}_i) \leq \chi_2^2(\alpha).$$

A computational procedure for finding the λ_r and $\mathbf{c}_r$ is given in Section 10.1.4c, and a geometrical interpretation of the above procedure is given by Campbell and Atchley [1981]. A ridge-type adjustment to $\mathbf{W}$ is proposed by Campbell [1980b], and robust M-estimation (see Section 4.4.3b) of the discriminant coordinates is developed by Campbell [1982] using a functional relationship model.

Some authors work with the matrices $\mathbf{S}_B = \mathbf{B}/(g-1)$ and $\mathbf{S} = \mathbf{W}/(n-g)$ instead of $\mathbf{B}$ and $\mathbf{W}$. The only change is that the eigenvalues λ_r above are now all multiplied by $(n-g)/(g-1)$; the $\mathbf{c}_r$ are unchanged, as we still require $\mathbf{CSC}' = \mathbf{I}_k$.

EXAMPLE 5.8 Reeve [1941] published a series of trivariate measurements on skulls of six collections (four subspecies) of anteaters. The subspecies, locality of the collection, and numbers of skull are given in Table 5.28. The features chosen for measurement were (1) the basal length, excluding the premaxilla; (2) the occipitonasal length; and (3) the greatest length of the nasals. The data used for analysis consisted of the common logarithms of the original observations measured in millimeters. The appropriate "within" and "between" matrices are

$$\mathbf{B} = \begin{pmatrix} 0.0200211 & 0.0174448 & 0.0130811 \\ - & 0.0158517 & 0.0150665 \\ - & - & 0.0306818 \end{pmatrix}$$

and

$$\mathbf{W} = \begin{pmatrix} 0.0136309 & 0.0127691 & 0.0164379 \\ - & 0.0129227 & 0.0171355 \\ - & - & 0.0361519 \end{pmatrix}.$$

TABLE 5.28 Trivariate Measurements on Anteater Skulls Found in Six Locations[a]

Subspecies	Locality	No. of skulls	Mean vector
instabilis	Sta. Marta, Colombia	21	2·054 2·066 1·621
chapadensis	Minas Geraes, Brazil	6	2·097 2·100 1·625
chapadensis	Matto Grosso, Brazil	9	2·091 2·095 1·624
chapadensis	Sta. Cruz, Bolivia	3	2·099 2·102 1·643
chiriquensis	Panama	4	2·092 2·110 1·703
mexicana	Mexico	5	2·099 2·107 1·671

[a]From Reeve [1941], courtesy of The Zoological Society of London.

The eigenvalues λ_r of $\mathbf{W}^{-1}\mathbf{B}$ are (2.400, 0.905, 0.052) and, as λ_3 is small, we can choose $k = 2$. Now the first two scaled eigenvectors are

$$\mathbf{c}_1' = (108.918, -33.823, -35.497)$$

and

$$\mathbf{c}_2' = (-40.608, 59.952, 22.821),$$

and these can be used to calculate the individual discriminant coordinates $\mathbf{z}_{ij}$ for each skull. However, following Seal [1964: p. 136], we can plot the discriminant coordinates $\mathbf{C}\bar{\mathbf{x}}_{i\cdot}$ $(i = 1, 2, \ldots, 6)$ for the mean of each of the six collections, as in Fig. 5.16. These coordinates are (96.3, 77.4), (99.7, 77.8), (99.2, 77.8), (99.2, 78.2), (96.1, 80.4), and (98.0, 79.2). The radius of the 95% confidence circle for $\boldsymbol{\mu}_i$ is $\{\chi_2^2(0.05)/n_i\}^{1/2} = 2.448/\sqrt{n_i}$, and these are also shown in Fig. 5.16. These circles are different from those used by Seal [1964].

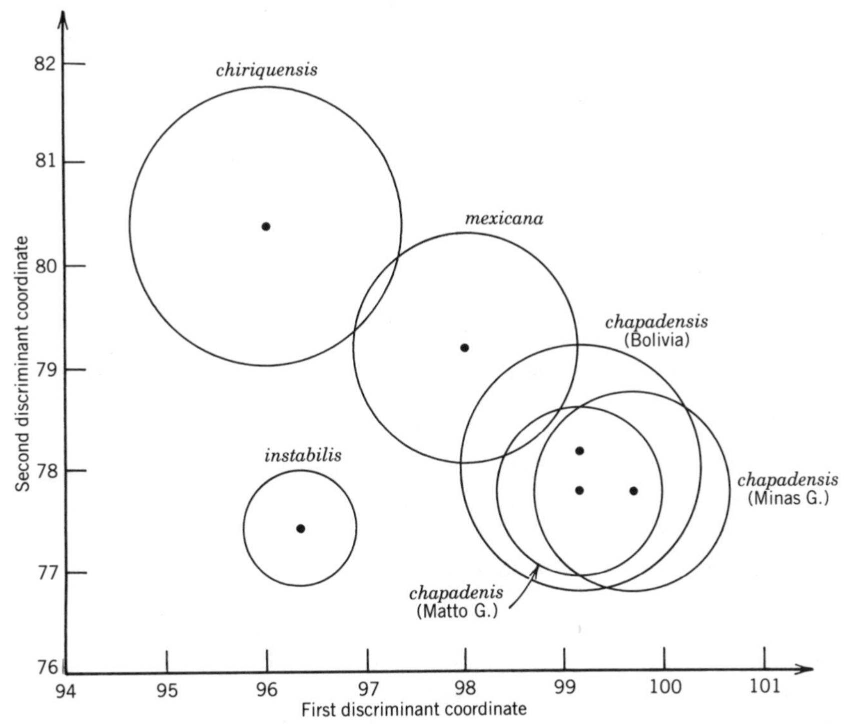

Fig. 5.16 Scatter plot of the first two discriminant coordinates for the sample means of six collections of anteater skulls: a 95% confidence circle for each population mean is also drawn.

5.9 ASSESSING TWO-DIMENSIONAL REPRESENTATIONS

In applying one of the previous dimension-reducing techniques, we would hope to obtain a useful reduction to two dimensions so that the scatter diagram of the reduced configuration is "similar" to the original configuration. As in multidimensional scaling, we can measure similarity by comparing the two-dimensional interpoint distances d_{rs} with the original distances or dissimilarities δ_{rs}. Various measures of "distortion" have been proposed (see Cormack [1971: Table 3], Orlóci [1978: pp. 264–276], including the so-called cophenetic correlation, the correlation between the $\binom{n}{2}$ pairs (d_{rs}, δ_{rs}) (see Rohlf [1974a] and references therein), the rank correlation (Johnson [1967]), the Goodman–Kruskal γ statistic (Hubert [1974]), and Hartigan's [1967] measure $\Sigma_{r<s}|d_{rs} - \delta_{rs}|$. These measures are discussed further in Section 7.3.2a.

One useful graphical method for examining the effectiveness of a reduction is the minimum spanning tree (MST) described in the paragraph following (4.24). Given the lengths δ_{rs} of all $\binom{n}{2}$ possible segments between pairs of points in the original configuration, we can define the MST as the spanning

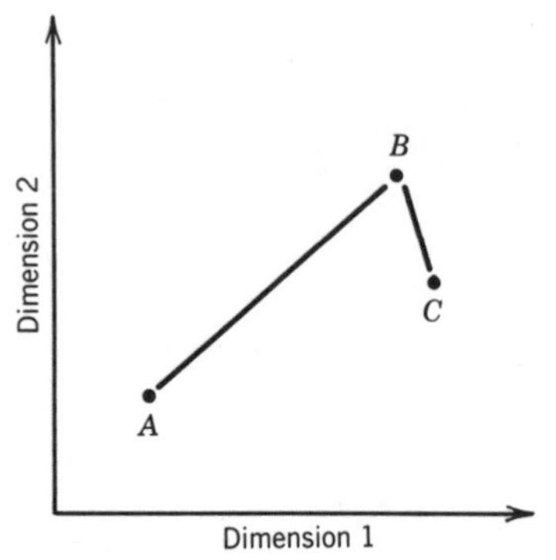

Fig. 5.17 Links in the minimum spanning tree transferred to a two-dimensional reduction of the data.

tree of minimum length. The links in the MST can be transferred to the scatter diagram, provided that there is a one-to-one correspondence between the initial and reduced configurations. For example, in Fig. 5.17 we have the MST for three points. There is some distortion in the two-dimensional reduction, as C appears closer to A than B, whereas in the original configuration B is closer to A. When distortion of this type occurs, then caution must be exercised in making inferences about possible clustering from the scatter diagram. Suitable algorithms for finding the MST are available (e.g., Gower and Ross [1969]).

5.10 A BRIEF COMPARISON OF METHODS

The methods of principal components, principal coordinates, and h-plotting all involve the extraction of eigenvalues and eigenvectors. We have already noted that principal components have a number of useful optimal properties that aid interpretation and which can be used for detecting outliers and redundant dimensions. However, principal coordinate analysis is more general, being the same as principal components for vector data and Euclidean distances, yet requiring only the matrix of similarities or dissimilarities. In particular, metrics other than the Euclidean metric can be used for measuring interpoint distances. If the method does not give an adequate reduction, that is, a low enough dimension, Gower [1966] suggests using the principal coordinate solution as a starting point for the iterative algorithm of nonmetric multidimensional scaling. Kruskal [1977b] uses this starting point in his multidimensional scaling program KYST and notes that sometimes the iterative procedure does not change this initial configuration very much. Random start procedures have also been used, though there is some controversy over their usefulness (see Psychometrika, Vol. 43, 1978, pp. 111–119). In general, fewer dimensions will be needed to reproduce ordinal information than to reproduce metric information, so that nonmetric multidimensional scaling is likely to be more useful for complex nonlinear-type data. Multidimensional scaling is discussed further in Section 7.8 in relation to cluster analysis.

Banfield and Gower [1980] discuss the ordination of multivariate binary data with special reference to principal coordinate analysis. By plotting certain additional points, they show how to (1) check on the suitability of the ordination, (2) provide a useful screening test before any cluster analysis, and (3) recover appropriate values of the entries of the 2×2 tables from which the ordination was derived.

Theoretical and empirical evidence has been given that shows that factor analysis can be completely misleading if the number of factors is unknown. The "reality" of the derived factors is controversial, and, all in all, the method does not inspire much confidence in spite of its support from social scientists. Discriminant coordinates is a useful tool for examining external structure, such as the clustering of groups and discriminant analysis (Chapter 6), while canonical analysis is useful for investigating internal structure among the variables.

EXERCISES 5

5.1 Let f be a real-valued function defined on the $n \times n$ positive definite matrices $\mathbf{C}$. Verify that f is invariant under orthogonal transformations [i.e., $f(\mathbf{T}'\mathbf{C}\mathbf{T}) = f(\mathbf{C})$ for orthogonal $\mathbf{T}$] and is strictly increasing in the eigenvalues of $\mathbf{C}$ for the functions (a) $\operatorname{tr}\mathbf{C}$, (b) $|\mathbf{C}|$, and (c) $\|\mathbf{C}\| = \{\operatorname{tr}[\mathbf{C}\mathbf{C}']\}^{1/2}$. [Hint: Use A7.9.]

5.2 Given $n \times n$ positive definite matrices $\mathbf{A}$ and $\mathbf{B}$ such that $\mathbf{A} - \mathbf{B}$ is positive semidefinite, show that the eigenvalues of $\mathbf{A}\mathbf{B}^{-1}$ are all greater than or equal to unity. Deduce that $|\mathbf{A}| \geq |\mathbf{B}|$. [Hint: Use A5.12.]

5.3 Let $\mathbf{A}$ be a $d \times d$ positive definite matrix. Then there exists an orthogonal matrix $\mathbf{T}$ such that $\mathbf{T}'\mathbf{A}\mathbf{T} = \mathbf{\Lambda}$, where $\lambda_1 \geq \lambda_2 \geq \cdots \geq \lambda_d > 0$ are the eigenvalues of $\mathbf{A}$. If $\mathbf{B}$ is a $d \times k$ matrix of rank k $(k < d)$, and f is a function satisfying the two conditions of Lemma 5.2 in Section 5.2.2, show that $f(\mathbf{A} - \mathbf{B}\mathbf{B}')$ is minimized when $\mathbf{B} = \mathbf{T}_1\mathbf{\Lambda}_1^{1/2}$, where $\mathbf{T}_1$ is the matrix consisting of the first k columns of $\mathbf{T}$, and $\mathbf{\Lambda}_1 = \operatorname{diag}(\lambda_1,\ldots,\lambda_k)$.

5.4 Given $\mathbf{x} = (x_1, x_2)'$, where x_1 and x_2 are random variables with zero means, unit variances, and correlation ρ, find the principal components of $\mathbf{x}$.

5.5 Let $\mathbf{x}$ be a d-dimensional vector of random variables with mean $\mathbf{0}$ and dispersion matrix $\mathbf{\Sigma} = [(\sigma_{jk})]$, and let $y_j = \mathbf{t}_j'\mathbf{x}$ be the jth principal component of $\mathbf{x}$ with variance λ_j. Show that the correlation between x_i and y_j is $t_{ij}(\lambda_j/\sigma_{ii})^{1/2}$, where t_{ij} is the ith element of $\mathbf{t}_j$.

5.6 Suppose $z_0, z_1,\ldots,z_d$ are independently and identically distributed with mean 0 and variance σ^2. Let $x_i = z_0 + z_i$, $i = 1,2,\ldots,d$. Verify that there is a principal component of $\mathbf{x} = (x_1, x_2,\ldots,x_d)'$ that is proportional to

$\bar{x}$. What is its variance? Deduce that this is the first principal component, that is, the component with maximum variance.

5.7 Let $\mathbf{x}$ have mean zero and dispersion matrix

$$\boldsymbol{\Sigma} = \begin{pmatrix} \alpha+1 & 1 & 1 \\ 1 & \alpha+1 & 1 \\ 1 & 1 & \alpha+1 \end{pmatrix}.$$

Verify that $\boldsymbol{\Sigma} - \mathbf{I}_3\alpha$ has rank 1 and a nonzero eigenvalue of 3. Hence find the eigenvalues of $\boldsymbol{\Sigma}$ and the first principal component of $\mathbf{x}$.

5.8 Verify Equation (5.8).

5.9 Let R_q be defined as in Equation (5.22) and let

$$\rho_q = \frac{\lambda_1 + \lambda_2 + \cdots + \lambda_q}{\lambda_1 + \lambda_2 + \cdots + \lambda_d}.$$

Show that for normal data $\sqrt{n}\,(R_q - \rho_q)$ is asymptotically $N_1(0, v)$, where

$$v = \frac{2\left\{(1-\rho_q)^2\left(\lambda_1^2 + \lambda_2^2 + \cdots + \lambda_q^2\right) + \rho_q^2\left(\lambda_{q+1}^2 + \lambda_{q+2}^2 + \cdots + \lambda_d^2\right)\right\}}{\{\mathrm{tr}\,\boldsymbol{\Sigma}\}^2}.$$

[Hint: Use (5.19), (5.20), and A10.2].

Kshirsagar [1972: p. 454]

5.10 Draw the biplots for the two factorizations

$$\begin{pmatrix} 2 & 2 & -4 \\ 2 & 1 & -3 \\ 0 & -\frac{3}{2} & \frac{3}{2} \\ -1 & -\frac{1}{2} & \frac{3}{2} \end{pmatrix} = \begin{pmatrix} 2 & 2 \\ 2 & 1 \\ 0 & -\frac{3}{2} \\ -1 & -\frac{1}{2} \end{pmatrix} \begin{pmatrix} 1 & 0 & -1 \\ 0 & 1 & -1 \end{pmatrix}$$

$$= \begin{pmatrix} 2 & -4 \\ 0 & -1 \\ -3 & \frac{9}{2} \\ 0 & \frac{1}{2} \end{pmatrix} \begin{pmatrix} -3 & -1 & 4 \\ -2 & -1 & 3 \end{pmatrix}.$$

Gabriel [1971]

5.11 Verify Equation (5.36).

5.12 Show that $\mathbf{G}$ in (5.29) is unique, up to an orthogonal transformation, if $\mathbf{G}'\mathbf{G} = \mathbf{I}_r$. [Hint: Suppose that there are two solutions $\mathbf{G}_i$ $(i = 1, 2)$. Using B1.7 prove that $\mathbf{B}(\mathbf{B}'\mathbf{B})^-\mathbf{B}' = \mathbf{G}_i\mathbf{H}_i'(\mathbf{H}_i\mathbf{H}_i')^-\mathbf{H}_i\mathbf{G}_i' = \mathbf{G}_i\mathbf{G}_i'$. Now use B1.9.]

5.13 Verify Equation (5.37).

5.14 Prove that for $\mathbf{x}_1, \mathbf{x}_2, \ldots, \mathbf{x}_n$,

$$\sum_{\substack{r=1 \\ r<s}}^{n} \sum_{s=1}^{n} \|\mathbf{x}_r - \mathbf{x}_s\|^2 \Big/ \binom{n}{2} = \frac{2}{n-1} \sum_{i=1}^{n} \|\mathbf{x}_i - \overline{\mathbf{x}}\|^2.$$

5.15 Show that there is an identifiability problem in the model (5.44) if the first column of $\boldsymbol{\Gamma}$ has only one nonzero element.

5.16 Let $\boldsymbol{\Gamma}$ be a $d \times m$ matrix of rank r and let $\mathbf{Q}$ be an $m \times (m - r)$ matrix for which $\boldsymbol{\Gamma}\mathbf{Q} = \mathbf{O}$ and $\mathbf{Q}'\mathbf{Q} = \mathbf{I}_{m-r}$. If $\mathbf{M}$ is any $d \times (m - r)$ matrix with mutually orthogonal rows, show that

$$\boldsymbol{\Gamma}\boldsymbol{\Gamma}' + \boldsymbol{\Psi} = (\boldsymbol{\Gamma} + \mathbf{M}\mathbf{Q}')(\boldsymbol{\Gamma} + \mathbf{M}\mathbf{Q}')' + \boldsymbol{\Psi} - \mathbf{M}\mathbf{M}'.$$

5.17 Given that $\mathbf{f} \sim N_m(\mathbf{0}, \mathbf{I}_m)$ and $\mathbf{x}|\mathbf{f} \sim N_m(\boldsymbol{\mu} + \boldsymbol{\Gamma}\mathbf{f}, \boldsymbol{\Psi})$, show that $\mathbf{f}|\mathbf{x} \sim N_m(\boldsymbol{\Gamma}'\boldsymbol{\Sigma}^{-1}(\mathbf{x} - \boldsymbol{\mu}), [\boldsymbol{\Gamma}'\boldsymbol{\Psi}\boldsymbol{\Gamma} + \mathbf{I}_m]^{-1})$. [Hint: Use A2.4 and show that $\boldsymbol{\Gamma}'\boldsymbol{\Sigma}^{-1} = (\mathbf{I}_m + \boldsymbol{\Gamma}'\boldsymbol{\Psi}^{-1}\boldsymbol{\Gamma})^{-1}\boldsymbol{\Gamma}'\boldsymbol{\Psi}^{-1}$.]

5.18 Let $\boldsymbol{\Theta} = \text{diag}(\theta_1, \theta_2, \ldots, \theta_m)$, where $\theta_1 \geq \theta_2 \geq \cdots \geq \theta_m$ are the m largest eigenvalues of $\hat{\boldsymbol{\Psi}}^{-1/2}\mathbf{S}\hat{\boldsymbol{\Psi}}^{-1/2}$, and let $\boldsymbol{\Omega} = (\boldsymbol{\omega}_1, \boldsymbol{\omega}_2, \ldots, \boldsymbol{\omega}_m)$ be the $d \times m$ matrix whose columns are the corresponding orthonormal eigenvectors. Verify that

$$\hat{\boldsymbol{\Gamma}} = \hat{\boldsymbol{\Psi}}^{1/2}\boldsymbol{\Omega}(\boldsymbol{\Theta} - \mathbf{I}_m)^{1/2}$$

satisfies Equations (5.53) and (5.55).

5.19 Given $\hat{\boldsymbol{\Sigma}} = \hat{\boldsymbol{\Gamma}}\hat{\boldsymbol{\Gamma}}' + \hat{\boldsymbol{\Psi}}$, where $\hat{\boldsymbol{\Gamma}}$ and $\hat{\boldsymbol{\Psi}}$ are the respective maximum likelihood estimates of $\boldsymbol{\Gamma}$ and $\boldsymbol{\Psi}$ (see Section 5.4.2a), show that $\text{tr}[\mathbf{S}\hat{\boldsymbol{\Sigma}}^{-1}] = d$.

5.20 If $\hat{\mathbf{f}}$ is defined by Equation (5.64), show that

$$\mathscr{E}\left[(\hat{\mathbf{f}} - \mathbf{f})(\hat{\mathbf{f}} - \mathbf{f})'\right] = (\mathbf{I}_m + \mathbf{J})^{-1}.$$

5.21 Consider the factor analysis model

$$\mathbf{x} - \boldsymbol{\mu} = \boldsymbol{\Gamma}\mathbf{f} + \boldsymbol{\varepsilon} = \boldsymbol{\gamma}_1 f_1 + \boldsymbol{\gamma}_2 f_2 + \boldsymbol{\varepsilon},$$

where f_1 and f_2 are correlated random variables with zero means and unit variances. Let βf_1 be the conditional regression of f_2 on f_1, and let $f_{2\cdot1} = f_2 - \beta f_1$. Show that f_1 and $f_{2\cdot1}$ are uncorrelated, and express the model in the form

$$\mathbf{x} - \boldsymbol{\mu} = \boldsymbol{\theta}_1 f_1 + \boldsymbol{\theta}_2 f_{2\cdot1} + \boldsymbol{\varepsilon}.$$

If estimates of $\boldsymbol{\theta}_1$ and $\boldsymbol{\theta}_2$ are available, can we estimate $\boldsymbol{\gamma}_1$? [This is the problem of trying to find an oblique rotation that sends $(f_1, f_{2\cdot1})$ into (f_1, f_2).]

Francis [1973]

5.22 Verify Equation (5.66).

5.23 Given the dissimilarity matrix

$$\mathbf{D} = \begin{pmatrix} 0 & 2 & 2 & 2\sqrt{2} \\ 2 & 0 & 2\sqrt{2} & 2 \\ 2 & 2\sqrt{2} & 0 & 2 \\ 2\sqrt{2} & 2 & 2 & 0 \end{pmatrix},$$

find a two-dimensional configuration, centered at the origin, whose interpoint distances are given by $\mathbf{D}$.

5.24 In Theorem 5.8 [see (5.82)] show that $\mathbf{T}$ is a maximizing orthogonal matrix if and only if $\mathbf{TA}$ is positive semidefinite.

Ten Berge [1977]

5.25 Verify (5.88).

5.26 Given $\mathscr{D}[(\mathbf{x}', \mathbf{y}')'] = \mathbf{\Sigma} > \mathbf{O}$, show that $\rho_1^2 < 1$, where ρ_1 is the first canonical correlation between $\mathbf{x}$ and $\mathbf{y}$. [Hint: Use A3.2.]

5.27 Show that the population canonical correlations are the same, irrespective of whether we use the dispersion matrix $\mathbf{\Sigma}$ or the correlation matrix.

5.28 Given the population canonical variables u_i and v_j, show that $\text{cov}[u_i, v_j] = 0$ $(i \neq j)$. [Hint: Show that $\mathbf{a}_i = \mathbf{\Sigma}_{11}^{-1}\mathbf{\Sigma}_{12}\mathbf{b}_i$.]

5.29 Use the method of Lagrange multipliers to verify that ρ_1^2 is the maximum value of $(\boldsymbol{\alpha}'\mathbf{\Sigma}_{12}\boldsymbol{\beta})^2$ subject to $\boldsymbol{\alpha}'\mathbf{\Sigma}_{11}\boldsymbol{\alpha} = 1$ and $\boldsymbol{\beta}'\mathbf{\Sigma}_{22}\boldsymbol{\beta} = 1$. [Hint: Use A8.1.]

5.30 Given $\mathbf{z}' = (\mathbf{x}', \mathbf{y}')$, use Equation (5.88) to find the square of the multiple correlation between $\mathbf{c}'\mathbf{x}$ and $\mathbf{y}$. Show that the maximum value of this with respect to $\mathbf{c}$ is ρ_1^2, the square of the first canonical correlation.

Cramer [1973]

5.31 Let $\mathbf{z}' = (\mathbf{x}', \mathbf{y}')$, where $\mathbf{x}$ and $\mathbf{y}$ are both two dimensional, and suppose that

$$\mathscr{D}[\mathbf{z}] = \sigma^2 \begin{pmatrix} 1 & a & b & b \\ a & 1 & b & b \\ b & b & 1 & c \\ b & b & c & 1 \end{pmatrix},$$

where $|a|$, $|b|$, and $|c|$ are all less than unity. Find the first canonical correlation and the corresponding canonical variables.

CHAPTER 6

Discriminant Analysis

6.1 INTRODUCTION

The problem we consider in this chapter is the following. Given that an object (or person) is known to come from one of g distinct groups G_i $(i = 1, 2, \ldots, g)$ in a population $\mathscr{P}$, we wish to assign the object to one of these groups on the basis of d measured characteristics $\mathbf{x}$ associated with the object. We would like our assignment rule to be optimal in some sense such as minimizing the number or cost of any errors of misclassification that we might make on the average. The literature on this subject of discrimination is extensive, as seen from the bibliographies of Cacoullos [1973], Lachenbruch [1975a], Huberty [1975], Goldstein and Dillon [1978], and Lachenbruch and Goldstein [1979]. The subject also goes under the title of classification, though this term is usually reserved for cluster analysis (see Chapter 7). Another subtitle is "statistical pattern recognition with supervised learning," where the groups are pattern types, signal sources, or characters, and the reader is referred to Dolby [1970] for a readable introduction to statistical character recognition. The literature on pattern recognition, statistical or otherwise, is very extensive: Fu [1977] and Zadeh [1977] list about 22 books between them! A nonstatistical method of discrimination is the use of diagnostic keys (Payne and Preece [1980], Jolliffe et al. [1982]).

A perusal of the 579 references in the bibliography of Lachenbruch [1975a] will indicate the wide-ranging applications of discriminant analysis. We consider just three examples to give the flavor:

(1) A number of dated literary works by an author are available and these can be put into g chronological groups. An undated work turns up and we wish to classify it chronologically. A major problem is to decide what style characteristics should be measured, particularly those that change with time.

(2) A student in a class either passes or fails an examination at the end of the year and therefore belongs to one of $g = 2$ groups. However, one student misses the exam owing to illness and the instructor wishes to decide whether to pass or fail the student on the basis of $\mathbf{x}$, a vector of scores on tests and assignments done during the year.

(3) On the basis of preliminary tests, a doctor wishes to decide whether a patient is free of a certain disease or whether the patient should be referred to a medical clinic for more comprehensive tests.

When developing an allocation rule, there are four cases that can be distinguished, depending on the information or sample data we have on each group. The distribution of $\mathbf{x}$ is (1) completely known, (2) known except for some unknown parameters, (3) partially known, and (4) completely unknown. The difference between (1) and (2), in practice, is usually in the size of the samples available from each group. For very large samples we may effectively ignore the sampling variation in the parameter estimates. Model (1) provides a convenient starting point for the discussion of allocation rules and we begin here with the case of just two groups. It should be emphasized that we consider groups that are qualitatively distinct. There is, however, an important class of problems where the groups are quantitatively distinct and we simply wish to split a single populations into two groups on the basis of some variable (e.g., high or low intelligence). This latter problem, called probit discrimination, is discussed by Albert and Anderson [1981].

6.2 TWO GROUPS: KNOWN DISTRIBUTIONS

6.2.1 *Misclassification Errors*

Suppose we have a population $\mathscr{P}$ with a proportion π_1 in group G_1 and the remaining proportion π_2 $(= 1 - \pi_1)$ in group G_2. For $i = 1, 2$, let $f_i(\mathbf{x})$ be the probability or probability density function of $\mathbf{x}$ if $\mathbf{x}$ comes from G_i. Consider the following classification rule for an identified object from $\mathscr{P}$ with an observed $\mathbf{x}$ (sometimes called the "feature" vector). Assign to G_i if $\mathbf{x}$ belongs to R_i, where R_1 and R_2 are mutually exclusive and $R_1 \cup R_2 = R$, the entire sample space for $\mathscr{P}$. We can make one of two errors: Assign $\mathbf{x}$ to G_2 when it belongs to G_1 or assign $\mathbf{x}$ to G_1 when it belongs to G_2. The respective probabilities of making these errors are

$$P(2|1) = \int_{R_2} f_1(\mathbf{x})\, d\mathbf{x} \quad \text{and} \quad P(1|2) = \int_{R_1} f_2(\mathbf{x})\, d\mathbf{x}, \qquad (6.1)$$

where a single integral sign is used to represent multiple integration or summation over d dimensions. Then the total probability of misclassification is

$$\begin{aligned} P(\mathbf{R},\mathbf{f}) &= \sum_{i=1}^{2} \text{pr}[\text{wrongly assign } \mathbf{x} \text{ to } G_i] \\ &= \sum_{i=1}^{2} \text{pr}\big[\text{assign } \mathbf{x} \text{ to } G_i | \mathbf{x} \in G_j\big]\text{pr}\big[\mathbf{x} \in G_j\big] \qquad (j = 1 \text{ or } 2,\ j \neq i) \\ &= P(1|2)\pi_2 + P(2|1)\pi_1, \end{aligned} \tag{6.2}$$

where $\mathbf{R} = (R_1, R_2)'$ and $\mathbf{f} = (f_1, f_2)'$, a convenient misuse of the vector notation (Lachenbruch [1975a]). The following lemma is central to our discussion.

LEMMA 6.1 The integral $\int_{R_1} g(\mathbf{x})\, d\mathbf{x}$ is minimized with respect to R_1 when $R_1 = R_{01} = \{\mathbf{x}\colon g(\mathbf{x}) < 0\}$.

Proof Let $R_{02} = \{\mathbf{x}\colon g(\mathbf{x}) \geq 0\}$. For general R_1 we can write $R_1 = (R_1 \cap R_{01}) \cup (R_1 \cap R_{02})$ so that $g(\mathbf{x}) < 0$ in $R_{01} - R_1 \cap R_{01}$ $(\subset R_{01})$ and $g(\mathbf{x}) \geq 0$ in $R_1 - R_1 \cap R_{01}$ $(\subset R_{02})$. Hence

$$\begin{aligned} \int_{R_1} g(\mathbf{x})\, d\mathbf{x} &= \int_{R_1 \cap R_{01}} g(\mathbf{x})\, d\mathbf{x} + \int_{R_1 - R_1 \cap R_{01}} g(\mathbf{x})\, d\mathbf{x} \\ &\geq \int_{R_1 \cap R_{01}} g(\mathbf{x})\, d\mathbf{x} \\ &= \int_{R_{01}} g(\mathbf{x})\, d\mathbf{x} - \int_{R_{01} - R_1 \cap R_{01}} g(\mathbf{x})\, d\mathbf{x} \\ &\geq \int_{R_{01}} g(\mathbf{x})\, d\mathbf{x}, \end{aligned}$$

which completes the proof. Furthermore, let $B = \{\mathbf{x}\colon g(\mathbf{x}) = 0\}$ and let A be a subset of B. Then the integral or summation of $g(\mathbf{x})$ over $R_{01} \cup A$ is the same as that over R_{01}. Hence R_{01} is not unique, as the boundary points B may be arbitrarily assigned to R_{01} or R_{02}. (If $\mathbf{x}$ is continuous, then $\text{pr}[\mathbf{x} \in B] = 0$.)

6.2.2 *Some Allocation Principles*

a MINIMIZE TOTAL PROBABILITY OF MISCLASSIFICATION

From (6.2) and (6.1),

$$\begin{aligned} P(\mathbf{R},\mathbf{f}) &= \pi_1\left(1 - \int_{R_1} f_1(x)\, d\mathbf{x}\right) + \pi_2 \int_{R_1} f_2(\mathbf{x})\, d\mathbf{x} \\ &= \pi_1 + \int_{R_1} \left[\pi_2 f_2(\mathbf{x}) - \pi_1 f_1(\mathbf{x})\right] d\mathbf{x}. \end{aligned} \tag{6.3}$$

By Lemma 6.1, this is minimized if we choose $R_1 = R_{01}$, where $R_{01} = \{\mathbf{x}: \pi_2 f_2(\mathbf{x}) - \pi_1 f_1(\mathbf{x}) < 0\}$. This classification rule, due to Welch [1939], that minimizes the total probability of misclassification is the following: Assign $\mathbf{x}$ to G_1 if

$$\frac{f_1(\mathbf{x})}{f_2(\mathbf{x})} > \frac{\pi_2}{\pi_1}, \tag{6.4}$$

and to G_2 otherwise. As we noted above, the assignment on the boundary $f_1(\mathbf{x})/f_2(\mathbf{x}) = \pi_2/\pi_1$ can be arbitrary, for example, assign to G_1 with probability $\frac{1}{2}$. Although for continuous $\mathbf{x}$ the probability that $\mathbf{x}$ falls on the boundary is zero, the class interval represented by the recorded (rounded) value of $\mathbf{x}$ can overlap the boundary with a nonzero probability.

EXAMPLE 6.1 Given that $f_i(\mathbf{x})$ is the density function for $N_d(\boldsymbol{\mu}_i, \boldsymbol{\Sigma}_i)$, and $\boldsymbol{\Sigma}_1 = \boldsymbol{\Sigma}_2 = \boldsymbol{\Sigma}$, we shall derive the optimal classification rule based on (6.4) and the corresponding probabilities of misclassification. Now

$$f_i(\mathbf{x}) = (2\pi)^{-d/2}|\boldsymbol{\Sigma}|^{-1/2}\exp\left[-\tfrac{1}{2}(\mathbf{x} - \boldsymbol{\mu}_i)'\boldsymbol{\Sigma}^{-1}(\mathbf{x} - \boldsymbol{\mu}_i)\right]$$

and

$$\begin{aligned} f_1(\mathbf{x})/f_2(\mathbf{x}) &= \exp\left[-\tfrac{1}{2}(\mathbf{x} - \boldsymbol{\mu}_1)'\boldsymbol{\Sigma}^{-1}(\mathbf{x} - \boldsymbol{\mu}_1) + \tfrac{1}{2}(\mathbf{x} - \boldsymbol{\mu}_2)'\boldsymbol{\Sigma}^{-1}(\mathbf{x} - \boldsymbol{\mu}_2)\right] \\ &= \exp\left[(\boldsymbol{\mu}_1 - \boldsymbol{\mu}_2)'\boldsymbol{\Sigma}^{-1}\mathbf{x} - \tfrac{1}{2}(\boldsymbol{\mu}_1 - \boldsymbol{\mu}_2)'\boldsymbol{\Sigma}^{-1}(\boldsymbol{\mu}_1 + \boldsymbol{\mu}_2)\right]. \end{aligned} \tag{6.5}$$

Taking logarithms, (6.4) is as follows: Assign to G_1 if

$$D(\mathbf{x}) = \boldsymbol{\lambda}'\left[\mathbf{x} - \tfrac{1}{2}(\boldsymbol{\mu}_1 + \boldsymbol{\mu}_2)\right] > \log(\pi_2/\pi_1), \tag{6.6}$$

where $\boldsymbol{\lambda} = \boldsymbol{\Sigma}^{-1}(\boldsymbol{\mu}_1 - \boldsymbol{\mu}_2)$. The equation $D(\mathbf{x}) = \log(\pi_2/\pi_1)$ defines a hyperplane for separating the two groups.

To find the probabilities of misclassification, define

$$\Delta^2 = (\boldsymbol{\mu}_1 - \boldsymbol{\mu}_2)'\boldsymbol{\Sigma}^{-1}(\boldsymbol{\mu}_1 - \boldsymbol{\mu}_2) = \boldsymbol{\lambda}'(\boldsymbol{\mu}_1 - \boldsymbol{\mu}_2), \tag{6.7}$$

the so-called Mahalanobis distance squared between $\boldsymbol{\mu}_1$ and $\boldsymbol{\mu}_2$. Then

$$\begin{aligned} \mathrm{E}\left[D(\mathbf{x})|\mathbf{x} \in G_i\right] &= D(\boldsymbol{\mu}_i) \\ &= \boldsymbol{\lambda}'\left[\boldsymbol{\mu}_i - \tfrac{1}{2}(\boldsymbol{\mu}_1 + \boldsymbol{\mu}_2)\right] \\ &= \tfrac{1}{2}(-1)^{i+1}\Delta^2 \qquad [\text{by } (6.7)] \end{aligned} \tag{6.8}$$

and

$$\text{var}[D(\mathbf{x})|\mathbf{x} \in G_i] = \mathscr{D}[\boldsymbol{\lambda}'\mathbf{x}] = \boldsymbol{\lambda}'\boldsymbol{\Sigma}\boldsymbol{\lambda} = \Delta^2. \tag{6.9}$$

Since $D(\mathbf{x})$ is normally distributed,

$$\begin{aligned} P(2|1) &= \text{pr}\left[D(\mathbf{x}) \le \log(\pi_2/\pi_1)|\mathbf{x} \in G_1\right] \\ &= \text{pr}\left[\frac{D(\mathbf{x}) - \Delta^2/2}{\Delta} \le \frac{\log(\pi_2/\pi_1) - \Delta^2/2}{\Delta}\middle|\mathbf{x} \in G_1\right] \\ &= \text{pr}\left[Z \le \left\{\log\left(\frac{\pi_2}{\pi_1}\right) - \tfrac{1}{2}\Delta^2\right\}/\Delta\right] \\ &= \Phi\left(\left\{\log(\pi_2/\pi_1) - \tfrac{1}{2}\Delta^2\right\}/\Delta\right), \end{aligned} \tag{6.10}$$

where $Z \sim N_1(0,1)$ with distribution function Φ, and Δ is the positive square root of Δ^2. To find $P(1|2)$ we simply interchange the subscripts 1 and 2: Thus

$$\begin{aligned} P(1|2) &= \Phi\left(\left\{\log(\pi_1/\pi_2) - \tfrac{1}{2}\Delta^2\right\}/\Delta\right) \\ &= \Phi\left(-\left\{\log(\pi_2/\pi_1) + \tfrac{1}{2}\Delta^2\right\}/\Delta\right). \end{aligned} \tag{6.11}$$

If $\pi_1 = \pi_2 = \frac{1}{2}$, then $P(2|1) = P(1|2) = \Phi(-\frac{1}{2}\Delta)$, and we assign $\mathbf{x}$ to G_1 if [see (6.6)]

$$\boldsymbol{\lambda}'\mathbf{x} > \tfrac{1}{2}(\boldsymbol{\mu}_1 - \boldsymbol{\mu}_2)'\boldsymbol{\Sigma}^{-1}(\boldsymbol{\mu}_1 + \boldsymbol{\mu}_2) = \tfrac{1}{2}(\boldsymbol{\lambda}'\boldsymbol{\mu}_1 + \boldsymbol{\lambda}'\boldsymbol{\mu}_2), \tag{6.12}$$

that is, if $\boldsymbol{\lambda}'\mathbf{x}$ is closer to $\boldsymbol{\lambda}'\boldsymbol{\mu}_1$ than $\boldsymbol{\lambda}'\boldsymbol{\mu}_2$ (see Exercise 6.1). The rule (6.12) goes back to Fisher [1936], though his approach was distribution free. We look for the linear combination $\boldsymbol{\lambda}'\mathbf{x}$ that maximizes the ratio of the difference of its means in the two groups to its standard deviation. From A7.6 the square of this ratio $[\boldsymbol{\lambda}'(\boldsymbol{\mu}_1 - \boldsymbol{\mu}_2)]^2/\boldsymbol{\lambda}'\boldsymbol{\Sigma}\boldsymbol{\lambda}$ is a maximum when $\boldsymbol{\lambda} \propto \boldsymbol{\Sigma}^{-1}(\boldsymbol{\mu}_1 - \boldsymbol{\mu}_2)$ or, with suitable scaling, $\boldsymbol{\lambda} = \boldsymbol{\Sigma}^{-1}(\boldsymbol{\mu}_1 - \boldsymbol{\mu}_2)$, thus leading to (6.12).

EXAMPLE 6.2 Suppose that $\mathbf{x}$ is MVN as in Example 6.1, but $\boldsymbol{\Sigma}_1 \ne \boldsymbol{\Sigma}_2$. Then

$$\begin{aligned} Q(\mathbf{x}) &= \log[f_1(\mathbf{x})/f_2(\mathbf{x})] \\ &= \tfrac{1}{2}\log(|\boldsymbol{\Sigma}_2|/|\boldsymbol{\Sigma}_1|) - \tfrac{1}{2}(\mathbf{x} - \boldsymbol{\mu}_1)'\boldsymbol{\Sigma}_1^{-1}(\mathbf{x} - \boldsymbol{\mu}_1) + \tfrac{1}{2}(\mathbf{x} - \boldsymbol{\mu}_2)'\boldsymbol{\Sigma}_2^{-1}(\mathbf{x} - \boldsymbol{\mu}_2) \\ &= c_0 - \tfrac{1}{2}\left[\mathbf{x}'(\boldsymbol{\Sigma}_1^{-1} - \boldsymbol{\Sigma}_2^{-1})\mathbf{x} - 2\mathbf{x}'(\boldsymbol{\Sigma}_1^{-1}\boldsymbol{\mu}_1 - \boldsymbol{\Sigma}_2^{-1}\boldsymbol{\mu}_2)\right], \end{aligned} \tag{6.13}$$

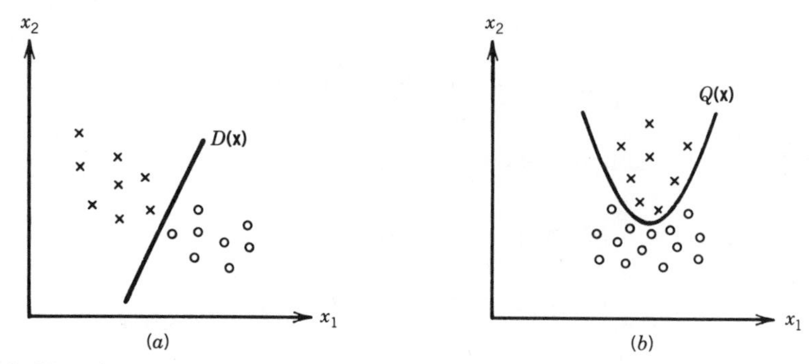

Fig. 6.1 Use of (*a*) linear discrimination and (*b*) quadratic discrimination to separate two swarms of points.

and our optimal rule is as follows: Assign to G_1 if $Q(\mathbf{x}) > \log(\pi_2/\pi_1)$. In contrast to $D(\mathbf{x})$, which is linear in $\mathbf{x}$, $Q(\mathbf{x})$ is quadratic. For example, in Fig. 6.1*a*, $\boldsymbol{\Sigma}_1 = \boldsymbol{\Sigma}_2$, so that the two swarms of points, which are similar in shape, can be adequately separated by a straight line. However, in Fig. 6.1*b*, $\boldsymbol{\Sigma}_1 \neq \boldsymbol{\Sigma}_2$ so that one swarm is long and thin vertically while the other is circular. Here a curved discriminant function gives adequate separation.

The special case of $\boldsymbol{\mu}_1 = \boldsymbol{\mu}_2$ is considered by Bartlett and Please [1963], Geisser [1973], Desu and Geisser [1973], Lachenbruch [1975b], and Schwemer and Mickey [1980].

EXAMPLE 6.3 Suppose that $\mathbf{x} = (x_1, x_2)'$ is a pair of independent Bernoulli variables each taking the value 1 or 0. Let $x_j = 1$ with probability p_{ij}, and $x_j = 0$ with probability $1 - p_{ij}$, if $\mathbf{x}$ comes from G_i $(i, j = 1, 2)$. We will show that the optimal classification rule leads to a linear discriminant function.

Now

$$\frac{f_1(x)}{f_2(x)} = \frac{p_{11}^{x_1}(1-p_{11})^{1-x_1} p_{12}^{x_2}(1-p_{12})^{1-x_2}}{p_{21}^{x_1}(1-p_{21})^{1-x_1} p_{22}^{x_2}(1-p_{22})^{1-x_2}}$$

$$= \frac{(1-p_{11})(1-p_{12})}{(1-p_{21})(1-p_{22})}\left(\frac{p_{11}}{1-p_{11}}\,\frac{1-p_{21}}{p_{21}}\right)^{x_1}\left(\frac{p_{12}}{1-p_{12}}\,\frac{1-p_{22}}{p_{22}}\right)^{x_2}$$

and

$$\log\left[\frac{f_1(\mathbf{x})}{f_2(\mathbf{x})}\right] = \beta_0 + \beta_1 x_1 + \beta_2 x_2, \tag{6.14}$$

where

$$\beta_0 = \log\left[\frac{(1-p_{11})(1-p_{12})}{(1-p_{21})(1-p_{22})}\right], \qquad \text{etc.}$$

The rule is as follows: Assign to G_1 if $\beta_0 + \beta_1 x_1 + \beta_2 x_2 > \log(\pi_2/\pi_1)$, where the β_j are functions of the four parameters p_{ij}.

The generalization to a d-dimensional vector $\mathbf{x}$ of independent Bernouilli variables is obvious. The rule is now the following: Assign to G_1 if

$$\beta_0 + \beta_1 x_1 + \cdots + \beta_d x_d > \log(\pi_2/\pi_1), \tag{6.15}$$

where the β_j are functions of $2d$ parameters p_{ij} $(i = 1, 2; j = 1, 2, \ldots, d)$.

b LIKELIHOOD RATIO METHOD

When π_1 is unknown, an intuitive rule is to choose G_i that maximizes the likelihood function for the sample $\mathbf{x}$. This leads to the following rule: Assign to G_1 if $f_1(\mathbf{x})/f_2(\mathbf{x}) > 1$, and this is a special case of the previous rule with $\pi_1 = \pi_2 = \frac{1}{2}$.

c MINIMIZE TOTAL COST OF MISCLASSIFICATION

Let $C(2|1)$ and $C(1|2)$ be the costs of misclassifying a member of G_1 and G_2, respectively. Then, from (6.2), the total expected cost of misclassification is

$$C_T = C(2|1)P(2|1)\pi_1 + C(1|2)P(1|2)\pi_2.$$

The rule (6.4) depends only on Lemma 6.1 and not on the condition $\pi_1 + \pi_2 = 1$. Hence, replacing π_i by $C(j|i)\pi_i$, it follows that C_T is minimized when R_1 is chosen so that $C(1|2)\pi_2 f_2(\mathbf{x}) < C(2|1)\pi_1 f_1(\mathbf{x})$. Our rule now becomes as follows: Assign to G_1 if

$$\frac{f_1(x)}{f_2(x)} > \frac{\pi_2 C(1|2)}{\pi_1 C(2|1)}.$$

This rule reduces to (6.4) if the costs are equal.

The above problem is a special case of general decision theory and the above procedure is in fact a Bayes procedure (see Anderson [1958: Chapter 6] for a helpful discussion). It is sometimes called the optimal Bayes procedure, as it minimizes C_T.

d MAXIMIZE THE POSTERIOR PROBABILITY

Suppose for the moment that $\mathbf{x}$ is a discrete random vector. Then, by Bayes' theorem, the posterior probability of G_i is

$$\begin{aligned} q_i(\mathbf{x}_0) &= \text{pr}[\mathbf{x} \in G_i | \mathbf{x} = \mathbf{x}_0] \\ &= \frac{\text{pr}[\mathbf{x} = \mathbf{x}_0 | \mathbf{x} \in G_i]\text{pr}[\mathbf{x} \in G_i]}{\sum_{j=1}^{2} \text{pr}[\mathbf{x} = \mathbf{x}_0 | \mathbf{x} \in G_j]\text{pr}[\mathbf{x} \in G_j]} \\ &= \frac{f_i(\mathbf{x}_0)\pi_i}{f_1(\mathbf{x}_0)\pi_1 + f_2(\mathbf{x}_0)\pi_2}. \end{aligned} \tag{6.16}$$

If $\mathbf{x}$ is continuous, then we have the approximation

$$\text{pr}[\mathbf{x}_0 \le \mathbf{x} \le \mathbf{x}_0 + d\mathbf{x}_0 | \mathbf{x} \in G_i] = \int_{\mathbf{x}_0}^{\mathbf{x}_0 + d\mathbf{x}_0} f_i(\mathbf{x})\, d\mathbf{x}$$

$$\approx f_i(\mathbf{x}_0)\, d\mathbf{x}_0.$$

Using this approximation and letting $d\mathbf{x}_0 \to \mathbf{0}$, we see, intuitively, that (6.16) will still hold so that (6.16) is the posterior probability for G_i, $\mathbf{x}$ discrete or continuous.

A reasonable classification criterion would be to assign $\mathbf{x}$ to the group with the larger posterior probability, that is, assign $\mathbf{x}$ to G_1 if

$$q_1(\mathbf{x}) > q_2(\mathbf{x}), \tag{6.17}$$

which is the same as $f_1(\mathbf{x})/f_2(\mathbf{x}) > \pi_2/\pi_1$.

e MINIMAX ALLOCATION

A rule that minimizes the total probability of misclassification may not do so well for an individual group. This is particularly the case if one of the π_i is small, as in the detection of a rare disease. To avoid a possible "imbalance" we can use a so-called *minimax* rule that allocates $\mathbf{x}$ so as to minimize the maximum individual probability of misclassification, that is, which minimizes the greater of $P(1|2)$ and $P(2|1)$. Now, for $0 \le \alpha \le 1$,

$$\max\{P(1|2), P(2|1)\} \ge (1-\alpha)P(2|1) + \alpha P(1|2), \tag{6.18}$$

and, by (6.4), the right-hand side of (6.18) is minimized when $R_1 = R_{01} = \{\mathbf{x}: f_1(\mathbf{x})/f_2(\mathbf{x}) > \alpha/(1-\alpha) = c\}$. If we choose c (i.e., $\alpha = \alpha_0$, say) so that the misclassification probabilities for R_{01} are equal, namely, $P_0(1|2) = P_0(2|1)$, then

$$(1-\alpha_0)P(2|1) + \alpha_0 P(1|2) \ge (1-\alpha_0)P_0(2|1) + \alpha_0 P_0(1|2)$$

$$= (1 - \alpha_0 + \alpha_0)P_0(2|1)$$

$$= P_0(2|1),$$

and, from (6.18),

$$\max\{P(1|2), P(2|1)\} \ge P_0(2|1) = \max\{P_0(1|2), P_0(2|1)\}.$$

The minimax rule is therefore the following: Assign $\mathbf{x}$ to G_1 if $f_1(\mathbf{x})/f_2(\mathbf{x}) > c$, where c satisfies $P_0(1|2) = P_0(2|1)$.

EXAMPLE 6.4 If the two distributions are normal with common covariance matrix (see Example 6.1), then the minimax procedure is as follows: Assign to G_1 if

$$D(\mathbf{x}) > \log c,$$

where $D(\mathbf{x})$ is given by (6.6), and [see (6.10) and (6.11)]

$$\Phi\left(\frac{\{\log c - \frac{1}{2}\Delta^2\}}{\Delta}\right) = \Phi\left(\frac{\{-\log c - \frac{1}{2}\Delta^2\}}{\Delta}\right).$$

The above equation has solution $\log c = 0$ or $c = 1$, so that the minimax rule is the same as the likelihood ratio method. Both procedures do not require knowledge of π_1.

f SUMMARY

In general, all the rules except the minimax rule are equivalent to (6.4), so that, initially, we shall adopt this method. However, the method of maximizing the posterior probability, although equivalent to (6.4), is more common in the literature for two reasons: It generalizes more naturally to the case of more than two groups, and it provides a good framework for the logistic discrimination method discussed in Section 6.4.

6.3 TWO GROUPS: KNOWN DISTRIBUTIONS WITH UNKNOWN PARAMETERS

6.3.1 General Methods

It is now convenient to use the notation $f_i(\mathbf{x}|\boldsymbol{\theta}_i)$ instead of $f_i(\mathbf{x})$, where $\boldsymbol{\theta}_i$ is the vector of parameters for group G_i. In Section 6.2 we saw that when $\boldsymbol{\theta}_i$ is known, all the classification rules take the following form: Assign to G_1 if $f_1(\mathbf{x}|\boldsymbol{\theta}_1)/f_2(\mathbf{x}|\boldsymbol{\theta}_2) > c$, for an appropriate value of c. However, if $\boldsymbol{\theta}_i$ is unknown and we have sample data (commonly called "training" data) from each group, say, $\mathbf{x}_{11}, \mathbf{x}_{12}, \ldots, \mathbf{x}_{1n_1}$ from G_1 and $\mathbf{x}_{21}, \mathbf{x}_{22}, \ldots, \mathbf{x}_{2n_2}$ from G_2, then we can replace $\boldsymbol{\theta} = (\boldsymbol{\theta}_1', \boldsymbol{\theta}_2')'$ by $\hat{\boldsymbol{\theta}} = \hat{\boldsymbol{\theta}}(\mathbf{z})$, an efficient estimate of $\boldsymbol{\theta}$ (such as the maximum likelihood estimate) based on the sample data $\mathbf{z} = (\mathbf{x}_{11}', \ldots, \mathbf{x}_{1n_1}', \mathbf{x}_{21}', \ldots, \mathbf{x}_{2n_2}')'$. The optimal region $\mathbf{R}_0$ is now estimated by $\hat{\mathbf{R}}_0 = (\hat{R}_{01}, \hat{R}_{02})'$, where $\hat{R}_{01} = \{\mathbf{x}: f_1(\mathbf{x}|\hat{\boldsymbol{\theta}}_1)/f_2(\mathbf{x}|\hat{\boldsymbol{\theta}}_2) > c\}$. We would like $\hat{R}_{01}$ to be close to R_{01}, and this will be the case for sufficiently large samples. As a rule of thumb, Lachenbruch and Goldstein [1979] suggest that n_i should exceed three times the number of parameters in $\boldsymbol{\theta}_i$: This number is adequate for groups that are "well separated," but can be increased for groups that are close together.

In assessing a given allocation rule, there are various associated error probabilities, usually called error rates, that need to be considered (see

Hills [1966] for a theoretical treatment). We use Lachenbruch's [1975a: p. 30] definitions:

(1) The optimum error rates

$$\begin{aligned} e_{i,\text{opt}} &= P(j|i) \\ &= \int_{R_{0j}} f_i(\mathbf{x}|\boldsymbol{\theta}_i)\,d\mathbf{x} \end{aligned} \tag{6.19}$$

and [see (6.2)]

$$\begin{aligned} e_{\text{opt}} &= \pi_1 e_{1,\text{opt}} + \pi_2 e_{2,\text{opt}} \\ &= P(\mathbf{R}_0, \mathbf{f}). \end{aligned} \tag{6.20}$$

(2) The actual error rates

$$e_{i,\text{act}} = \int_{\hat{R}_{0j}} f_i(\mathbf{x}|\boldsymbol{\theta}_i)\,d\mathbf{x} \tag{6.21}$$

and

$$\begin{aligned} e_{\text{act}} &= \pi_1 e_{1,\text{act}} + \pi_2 e_{2,\text{act}} \\ &= P(\hat{\mathbf{R}}_0, \mathbf{f}). \end{aligned} \tag{6.22}$$

If a large number of observations are classified using the partition $\hat{\mathbf{R}}_0$, then about $100e_{\text{act}}\%$ will be misclassified. We note that $e_{\text{opt}} \le e_{\text{act}}$, by the definition of e_{opt}.

(3) The expected actual error rates $\text{E}[e_{i,\text{act}}]$ and

$$\text{E}[e_{\text{act}}] = \pi_1 \text{E}[e_{1,\text{act}}] + \pi_2 \text{E}[e_{2,\text{act}}]. \tag{6.23}$$

We also have the following intuitive estimates:

(1) The plug-in estimates

$$\hat{e}_{i,\text{act}} = \int_{\hat{R}_{0j}} f_i(\mathbf{x}|\hat{\boldsymbol{\theta}}_i)\,d\mathbf{x} \tag{6.24}$$

and

$$\hat{e}_{\text{act}} = P(\hat{\mathbf{R}}_0, \hat{\mathbf{f}}), \tag{6.25}$$

obtained by using the estimates of the unknown parameters in f_1 and f_2. Some authors call these estimates the apparent error rates (e.g., Hills [1966], Glick [1973], Goldstein and Dillon [1978]).

(2) The apparent error rates

$$e_{i,\text{app}} = m_i/n_i, \qquad e_{\text{app}} = \pi_1 e_{1,\text{app}} + \pi_2 e_{2,\text{app}}, \tag{6.26}$$

where m_i of the n_i observations from group G_i are incorrectly classified by the assignment rule. If π_1 and π_2 are unknown, and the $n = n_1 + n_2$ observations are a random sample from the combined group population $\mathscr{P}$, we have the intuitive estimates $\hat{\pi}_i = n_i/(n_1 + n_2)$ and

$$\begin{aligned} \hat{e}_{\text{app}} &= \hat{\pi}_1 e_{1,\text{app}} + \hat{\pi}_2 e_{2,\text{app}} \\ &= \frac{n_1}{n_1 + n_2}\frac{m_1}{n_1} + \frac{n_2}{n_1 + n_2}\frac{m_2}{n_2} \\ &= \frac{m_1 + m_2}{n_1 + n_2}. \end{aligned} \tag{6.27}$$

This estimate is also called the (overall) apparent error rate by some authors, and the resubstitution method by others, for example, Habbema and Hermans [1977]. It is used, for example, in the SPSS and BMD07M programs.

(3) The leaving-one-out or cross-validation method proposed by Lachenbruch and Mickey [1968]. This technique consists of determining the allocation rule using the sample data minus one observation, and then using the rule to classify the omitted observation. Doing this for each of the n_i observations from G_i gives

$$e_{i,c} = \frac{a_i}{n_i}, \tag{6.28}$$

the proportion from G_i that are misclassified. We can then define

$$e_c = \pi_1 e_{1,c} + \pi_2 e_{2,c}, \tag{6.29}$$

which can be estimated, as in (6.27), by

$$\hat{e}_c = \frac{a_1 + a_2}{n_1 + n_2}. \tag{6.30}$$

Because the omission of one observation forms the basis of jackknifing, the above estimates are sometimes incorrectly called jackknife estimates.

(4) The bootstrap estimate proposed by Efron [1979, 1981] (see also McLachlan [1980a]). As $e_{i,\text{app}}$ is biased, Efron suggested estimating its bias using a so-called "bootstrap" technique. In this technique a new sample of size n_i $(i = 1, 2)$ is taken *with replacement* from the original sample of n_i observations. The new sample is not just a random permutation of the original sample as sampling is now with replacement. From these two new samples we obtain a new classification rule using the same method as before. Suppose that m_i^* of the new sample and m_i^{**} of the original sample are misclassified by the new rule, and let $d_i = (m_i^{**} - m_i^*)/n_i$. Then $\text{E}[d_i]$ is approximated by $\bar{d}_i$, obtained by averaging d_i over a large number of repeated realizations (say, 100) of new samples. The bootstrap estimates are then defined as [see (6.26)]

$$e_{i,\text{boot}} = (m_i/n_i) + \bar{d}_i \tag{6.31}$$

and

$$e_{\text{boot}} = \pi_1 e_{1,\text{boot}} + \pi_2 e_{2,\text{boot}}. \tag{6.32}$$

It appears that the estimate of the bias of $e_{i,\text{app}}$, $\bar{d}_i$, has good efficiency for the important case of groups that are close together (McLachlan [1980a]).

The plug-in estimate $\hat{e}_{\text{act}}$ not only relies heavily on the correct specification of the f_i, but it also has poor small-sample properties. Setting $r_i(\mathbf{x}) = f_i(\mathbf{x}|\boldsymbol{\theta}_i)\pi_i$, Hills [1966] proves the following lemma for the rule based on

$$\begin{aligned} R_{01} &= \left\{\mathbf{x}\colon \frac{f_1(\mathbf{x}|\boldsymbol{\theta}_1)}{f_2(\mathbf{x}|\boldsymbol{\theta}_2)} > \frac{\pi_2}{\pi_1}\right\} \\ &= \{\mathbf{x}\colon r_1(\mathbf{x}) > r_2(\mathbf{x})\} \\ &= \left\{\mathbf{x}\colon r_1(\mathbf{x}) = \max_{1 \le i \le 2} r_i(\mathbf{x})\right\} \end{aligned}$$

(see also Glick [1972: p. 118], though he uses nonerror rates rather than error rates).

LEMMA 6.2 Let $\hat{r}_i(\mathbf{x}) = f_i(\mathbf{x}|\hat{\boldsymbol{\theta}}_i)\pi_i$ be a pointwise unbiased estimate of $r_i(\mathbf{x})$, that is, $\text{E}_{\theta}[\hat{r}_i(\mathbf{x})] = r_i(\mathbf{x})$ for almost all $\mathbf{x}$. Then, referring to (6.22) and (6.25), we have

$$\text{E}[\hat{e}_{\text{act}}] \le e_{\text{opt}} < \text{E}[e_{\text{act}}]. \tag{6.33}$$

Proof If $\hat{r}(\mathbf{x}) = \max_{1 \le i \le 2} \hat{r}_i(\mathbf{x})$, then $\hat{R}_{01} = \{\mathbf{x} : \hat{r}(\mathbf{x}) = \hat{r}_1(\mathbf{x})\}$ and $\hat{R}_{02} = R - \hat{R}_{01} = \{\mathbf{x} : \hat{r}(\mathbf{x}) = \hat{r}_2(\mathbf{x})\}$. Since the expected value of the maximum of several random variables is not less than the maximum of the expected values, we have

$$\begin{aligned} \mathrm{E}_{\hat{\theta}}[\hat{r}(\mathbf{x})] &\geq \max_{1 \leq i \leq 2} \mathrm{E}_{\hat{\theta}}[\hat{r}_i(\mathbf{x})] \\ &= \max_{1 \leq i \leq 2} r_i(\mathbf{x}) \qquad (6.34) \\ &= r(\mathbf{x}), \qquad (6.35) \end{aligned}$$

say. Now

$$\mathrm{E}[\hat{e}_{\text{act}}] = \mathrm{E}_{\hat{\theta}}\left\{\int_{\hat{R}_{02}} \pi_1 f_1(\mathbf{x}|\hat{\boldsymbol{\theta}}_1)\, d\mathbf{x} + \int_{\hat{R}_{01}} \pi_2 f_2(\mathbf{x}|\hat{\boldsymbol{\theta}}_2)\, d\mathbf{x}\right\},$$

and using

$$\int_{\hat{R}_{02}} f_1(\mathbf{x}|\hat{\boldsymbol{\theta}}_1)\, d\mathbf{x} = 1 - \int_{\hat{R}_{01}} f_1(\mathbf{x}|\hat{\boldsymbol{\theta}}_1)\, d\mathbf{x}, \qquad \text{etc.,}$$

we have

$$\begin{aligned} 1 - \mathrm{E}[\hat{e}_{\text{act}}] &= \mathrm{E}_{\hat{\theta}}\left\{\int_{\hat{R}_{01}} \hat{r}_1(\mathbf{x})\, d\mathbf{x} + \int_{\hat{R}_{02}} \hat{r}_2(\mathbf{x})\, d\mathbf{x}\right\} \\ &= \mathrm{E}_{\hat{\theta}}\left\{\int_{\hat{R}_{01}} \hat{r}(\mathbf{x})\, d\mathbf{x} + \int_{\hat{R}_{02}} \hat{r}(\mathbf{x})\, d\mathbf{x}\right\} \\ &= \mathrm{E}_{\hat{\theta}}\left\{\int_R \hat{r}(\mathbf{x})\, d\mathbf{x}\right\}, \qquad (R = \hat{R}_{01} \cup \hat{R}_{02}) \\ &= \int_R \mathrm{E}_{\hat{\theta}}[\hat{r}(\mathbf{x})]\, d\mathbf{x} \qquad \text{(by Fubini's theorem)} \\ &\geq \int_R r(\mathbf{x})\, d\mathbf{x} \qquad \text{[by (6.35)]} \\ &= \int_{R_{01}} r_1(\mathbf{x})\, d\mathbf{x} + \int_{R_{02}} r_2(\mathbf{x})\, d\mathbf{x} \\ &= \pi_1 + \pi_2 - e_{\text{opt}} \\ &= 1 - e_{\text{opt}}. \end{aligned}$$

Hence $\mathrm{E}[\hat{e}_{\text{act}}] \leq e_{\text{opt}}$. Also $e_{\text{opt}} < e_{\text{act}}$ (with probability 1, since, in general, $\hat{\boldsymbol{\theta}} \neq \boldsymbol{\theta}$), so that $e_{\text{opt}} < \mathrm{E}_{\hat{\theta}}[e_{\text{act}}] = \mathrm{E}[e_{\text{act}}]$.

COROLLARY The above lemma still holds if $\hat{r}_i(\mathbf{x}) = f_i(\mathbf{x}|\hat{\boldsymbol{\theta}}_i)\hat{\pi}_i$ and $E[\hat{r}_i(\mathbf{x})] \geq r_i(\mathbf{x})$. The main change in the proof occurs in (6.34) where $=$ is replaced by $\geq$. □

We note that with a suitable estimate $\hat{\boldsymbol{\theta}}_i$, $f_i(\mathbf{x}|\hat{\boldsymbol{\theta}}_i)$ will be an asymptotically unbiased estimate of $f_i(\mathbf{x}|\boldsymbol{\theta}_i)$ so that Lemma 6.2 will generally hold for large samples. If $\hat{\pi}_i$ is asymptotically unbiased, then the Corollary of Lemma 6.2 will also hold asymptotically. We can therefore expect $\hat{e}_{\text{act}}$ to be an underestimate of e_{act} (and of $E[e_{\text{act}}]$), and the evidence indicates that the bias can be severe; $\hat{e}_{\text{act}}$ is therefore unsatisfactory. Unfortunately, $e_{i,\text{app}}$ and e_{app} also tend to be underestimates (see McLachlan [1976]), though they do have the advantage of being easy to calculate and do not require any distributional assumptions. In general, they should be used only when n_1 and n_2 are reasonably large: Lachenbruch (personal communication) recommends that n_i exceed twice the number of parameters in $\boldsymbol{\theta}_i$. For smaller samples, $e_{i,c}$ and $e_{i,\text{boot}}$ are preferred, as they are both almost unbiased estimates of $e_{i,\text{act}}$. Although its variance is large, $e_{i,c}$ has been popular in the literature, particularly as convenient computing formulas are available for normal distributions (see Exercise 6.6). However, Efron [1979] demonstrated in the normal case that $e_{i,\text{boot}}$ can have a standard deviation one-third that of $e_{i,c}$; further comparisons by McLachlan [1980a] indicate that in the normal case $e_{i,\text{boot}}$ has a good efficiency compared with an efficient parametric estimate when the groups are close together. Additional references on the general subject of error estimation are given in the reviews of Toussaint [1974], Lachenbruch [1975a], and Glick [1978].

Instead of estimating $f_i(\mathbf{x}|\boldsymbol{\theta}_i)$ by $f_i(\mathbf{x}|\hat{\boldsymbol{\theta}}_i)$, where $\hat{\boldsymbol{\theta}}_i = \hat{\boldsymbol{\theta}}_i(\mathbf{z})$ is an estimate based on the sample data, an alternative approach called the *predictive* method is to use

$$h_i(\mathbf{x}|\mathbf{z}) = \int_{\Theta} f_i(\mathbf{x}|\boldsymbol{\theta}_i) g_i(\boldsymbol{\theta}_i|\mathbf{z})\, d\boldsymbol{\theta}_i. \tag{6.36}$$

Here g_i can be regarded as either some weighting function based on the sample data or a full Bayesian posterior density function for $\boldsymbol{\theta}_i$ based on the prior density $g_{2i}(\boldsymbol{\theta}_i)$ and the data $\mathbf{z}$, that is,

$$g_i(\boldsymbol{\theta}_i|\mathbf{z}) \propto g_{1i}(\mathbf{z}|\boldsymbol{\theta}_i) g_{2i}(\boldsymbol{\theta}_i).$$

The expression (6.36) arises in Bayesian statistical prediction theory as the predictive density for a "future" observation from G_i as assessed on the basis of the data $\mathbf{z}$ (see Aitchison et al. [1977] for references).

Aitchison and Kay [1975] introduced the useful concept of the atypicality index (see also Aitchison et al. [1977] and Moran and Murphy [1979] for further comments). If $\hat{f}_i$ is the estimated or predictive density, an observation $\mathbf{y}$ is said to be more typical of group i than $\mathbf{x}$ if $\hat{f}_i(\mathbf{y}) > \hat{f}_i(\mathbf{x})$. The atypicality index of $\mathbf{x}$ with regard to group i is then defined as

$$I_i(\mathbf{x}) = \int_{\mathbf{y} \in S} \hat{f}_i(\mathbf{y})\, d\mathbf{y}, \tag{6.37}$$

where S is the set of all $\mathbf{y}$ more typical than $\mathbf{x}$. Atypicality indices are more readily calculated for discrete data: The integral in (6.37) is then replaced by a sum (Aitchison and Aitken [1976]). If $I_1(\mathbf{x})$ and $I_2(\mathbf{x})$ are both near 1, $\mathbf{x}$ may be an outlier and not belong to either G_1 or G_2.

6.3.2 *Normal Populations*

a LINEAR DISCRIMINANT FUNCTION

Suppose that $f_i(\mathbf{x}|\boldsymbol{\theta}_i)$ is the density function for $N_d(\boldsymbol{\mu}_i, \boldsymbol{\Sigma}_i)$, where $\boldsymbol{\Sigma}_1 = \boldsymbol{\Sigma}_2$ $(= \boldsymbol{\Sigma})$, and let $\mathbf{x}_{i1}, \mathbf{x}_{i2}, \ldots, \mathbf{x}_{in_i}$ be a random sample from G_i $(i = 1, 2)$. We can estimate $\boldsymbol{\mu}_i$ by $\overline{\mathbf{x}}_{i\cdot} = \Sigma_j \mathbf{x}_{ij}/n_i$ and $\boldsymbol{\Sigma}$ by the pooled estimator

$$\mathbf{S}_p = \frac{(n_1 - 1)\mathbf{S}_1 + (n_2 - 1)\mathbf{S}_2}{n_1 + n_2 - 2} = \frac{\mathbf{Q}_p}{n_1 + n_2 - 2}, \tag{6.38}$$

where

$$\mathbf{S}_i = \sum_{j=1}^{n_i} (\mathbf{x}_{ij} - \overline{\mathbf{x}}_{i\cdot})(\mathbf{x}_{ij} - \overline{\mathbf{x}}_{i\cdot})'/(n_i - 1).$$

Referring to (6.6), our estimated rule is as follows. Assign to G_1 if

$$D_s(\mathbf{x}) > \log(\pi_2/\pi_1) \qquad (= \log c), \tag{6.39}$$

where

$$\begin{aligned} D_s(\mathbf{x}) &= \hat{\boldsymbol{\lambda}}'\left[\mathbf{x} - \tfrac{1}{2}(\overline{\mathbf{x}}_{1\cdot} + \overline{\mathbf{x}}_{2\cdot})\right] \\ &= (\overline{\mathbf{x}}_{1\cdot} - \overline{\mathbf{x}}_{2\cdot})'\mathbf{S}_p^{-1}\left[\mathbf{x} - \tfrac{1}{2}(\overline{\mathbf{x}}_{1\cdot} + \overline{\mathbf{x}}_{2\cdot})\right], \end{aligned}$$

and

$$\hat{\boldsymbol{\lambda}} = \mathbf{S}_p^{-1}(\overline{\mathbf{x}}_{1\cdot} - \overline{\mathbf{x}}_{2\cdot}).$$

We shall call $D_s(\mathbf{x})$ the linear discriminant function (LDF), the sample estimate of $D(\mathbf{x})$ in (6.6). Although it is traditional to use $\mathbf{S}_p$, we could also use $\mathbf{Q}_p/(n_1 + n_2)$, the maximum likelihood estimator of the common dispersion matrix $\boldsymbol{\Sigma}$. Di Pillo [1976, 1977, 1979] proposed $\mathbf{S}_p + k\mathbf{I}_d$, a ridge-type estimator, instead of $\mathbf{S}_p$ as the former can lead to a smaller e_{act}. Other ridge-type adjustments have been suggested by Campbell [1980a]. If we have a random sample of size n $(= n_1 + n_2)$ from the combined groups, so-called mixture sampling, then the n_i are random and $\hat{\pi}_i = n_i/n$ is the maximum likelihood estimate of π_i.

If $\pi_1 = \pi_2$, (6.39) reduces to the sample equivalent of (6.12), namely, assign $\mathbf{x}$ to G_1 if

$$\hat{\boldsymbol{\lambda}}'\mathbf{x} > \tfrac{1}{2}(\hat{\boldsymbol{\lambda}}'\overline{\mathbf{x}}_{1\cdot} + \hat{\boldsymbol{\lambda}}'\overline{\mathbf{x}}_{2\cdot}). \tag{6.40}$$

This method was first proposed by Fisher [1936], who used it to classify two species of iris on the basis of four measurements: sepal length, sepal width, petal length, and petal width. As the classification works very well, the iris data have been quoted by many authors. However, it turns out that a single measurement, petal length, does just as well. Also, the assumption $\boldsymbol{\Sigma}_1 = \boldsymbol{\Sigma}_2$ is violated.

Using the rule (6.39), we have from Exercise 6.7 that

$$e_{\text{act}} = \pi_1 \Phi\left(\{\log c - D_s(\boldsymbol{\mu}_1)\}/\sigma\right) + \pi_2 \Phi\left(\{-\log c + D_s(\boldsymbol{\mu}_2)\}/\sigma\right), \quad (6.41)$$

where Φ is the distribution function for $N_1(0,1)$, and $\sigma = (\hat{\boldsymbol{\lambda}}'\boldsymbol{\Sigma}\hat{\boldsymbol{\lambda}})^{1/2}$. Replacing $\boldsymbol{\mu}_i$ and $\boldsymbol{\Sigma}$ by their estimates leads to (see Exercise 6.7)

$$\hat{e}_{\text{act}} = \pi_1 \Phi\left(\{\log c - \tfrac{1}{2}D^2\}/D\right) + \pi_2 \Phi\left(\{-\log c - \tfrac{1}{2}D^2\}/D\right), \quad (6.42)$$

where D is the positive square root of

$$D^2 = \hat{\boldsymbol{\lambda}}'\mathbf{S}_p\hat{\boldsymbol{\lambda}} = (\bar{\mathbf{x}}_{1\cdot} - \bar{\mathbf{x}}_{2\cdot})'\mathbf{S}_p^{-1}(\bar{\mathbf{x}}_{1\cdot} - \bar{\mathbf{x}}_{2\cdot}), \quad (6.43)$$

the well-known Mahalanobis "D-squared" measure of distance between two samples. However, we have already noted that $\hat{e}_{\text{act}}$ is not a good estimator of e_{act}. The general estimators e_{app} and e_c mentioned above [(6.26) and (6.29)] can be improved upon when the data are normal (Lachenbruch [1975a: Chapter 2], McLachlan [1976, 1980a], Schervish [1981]). For example, using $\tilde{D}$ instead of D in (6.42), where

$$\tilde{D}^2 = \frac{n_1 + n_2 - d - 3}{n_1 + n_2 - 2} D^2,$$

gives a better estimate of e_{act} than e_c. Also, McLachlan [1976] gave a formula for the asymptotic bias of $e_{i,\text{app}}$ that can be used to improve on $e_{i,\text{app}}$. However, these methods are appropriate for normal data only. We note that some packages, for example, SPSS and BMD07M, calculate plug-in estimates of the posterior probabilities, namely [see (6.16)],

$$\hat{q}_i(\mathbf{x}) = \frac{\pi_i f_i(\mathbf{x}|\hat{\boldsymbol{\theta}}_i)}{\{\pi_1 f_1(\mathbf{x}|\hat{\boldsymbol{\theta}}_1) + \pi_2 f_2(\mathbf{x}|\hat{\boldsymbol{\theta}}_2)\}},$$

for the sample data.

Campbell [1978] considers the problem of detecting outliers in the sample data. If D^2_{-m} is the value of D^2 calculated from the data but with observation m omitted, then Campbell suggests a Q–Q plot of the values of $D^2 - D^2_{-m}$ for each group versus the quantiles of a gamma distribution with unknown shape parameter (see Appendix C4). Campbell also suggests a similar plot using $\hat{\boldsymbol{\lambda}}'\hat{\boldsymbol{\lambda}}$ instead of D^2.

A sequential method of discrimination based on the LDF is discussed by Srivastava [1973].

EXAMPLE 6.5 In Table 6.1 we have four measurements on two species of flea beetles, namely,

x_1 = the distance of the transverse groove from the posterior border of the prothorax (in microns),

x_2 = the length of the elytra (in 0.01 mm),

x_3 = the length of the second antennal joint (in microns),

x_4 = the length of the third antennal joint (in microns).

Measurements were made on 19 specimens of male *Haltica oleracea* L. and on

TABLE 6.1 Four Measurements on Samples of Two Species of Flea-Beetles [a]

	Haltica oleracea					*Haltica carduorum*			
No.	x_1	x_2	x_3	x_4	No.	x_1	x_2	x_3	x_4
1	189	245	137	163	1	181	305	184	209
2	192	260	132	217	2	158	237	133	188
3	217	276	141	192	3	184	300	166	231
4	221	299	142	213	4	171	273	162	213
5	171	239	128	158	5	181	297	163	224
6	192	262	147	173	6	181	308	160	223
7	213	278	136	201	7	177	301	166	221
8	192	255	128	185	8	198	308	141	197
9	170	244	128	192	9	180	286	146	214
10	201	276	146	186	10	177	299	171	192
11	195	242	128	192	11	176	317	166	213
12	205	263	147	192	12	192	312	166	209
13	180	252	121	167	13	176	285	141	200
14	192	283	138	183	14	169	287	162	214
15	200	294	138	188	15	164	265	147	192
16	192	277	150	177	16	181	308	157	204
17	200	287	136	173	17	192	276	154	209
18	181	255	146	183	18	181	278	149	235
19	192	287	141	198	19	175	271	140	192
					20	197	303	170	205

[a] Reproduced from A. A. Lubischew (1962). "On the use of discriminant functions in taxonomy." *Biometrics*, **18**, 455–477, Table 2. With permission from The Biometric Society.

20 specimens of male *Haltica carduorum* Guer. Rounding off to two decimal places for ease of presentation, we obtain the two sample means

$$\bar{\mathbf{x}}_{1\cdot} = \begin{pmatrix} 194.47 \\ 267.05 \\ 137.37 \\ 185.95 \end{pmatrix}, \qquad \bar{\mathbf{x}}_{2\cdot} = \begin{pmatrix} 179.55 \\ 290.80 \\ 157.20 \\ 209.25 \end{pmatrix},$$

and the respective sample dispersion matrices are

$$\mathbf{S}_1 = \begin{pmatrix} 187.60 & 176.86 & 48.37 & 113.58 \\ - & 345.39 & 75.98 & 118.78 \\ - & - & 66.36 & 16.24 \\ - & - & - & 239.94 \end{pmatrix}$$

and

$$\mathbf{S}_2 = \begin{pmatrix} 101.84 & 128.06 & 36.99 & 32.59 \\ - & 389.01 & 165.36 & 94.37 \\ - & - & 167.54 & 66.53 \\ - & - & - & 177.88 \end{pmatrix}.$$

It seems reasonable to pool $\mathbf{S}_1$ and $\mathbf{S}_2$ and obtain

$$\mathbf{S}_p = \frac{(n_1 - 1)\mathbf{S}_1 + (n_2 - 1)\mathbf{S}_2}{n_1 + n_2 - 2}$$

$$= \begin{pmatrix} 143.56 & 151.80 & 42.53 & 71.99 \\ - & 367.79 & 121.88 & 106.24 \\ - & - & 118.31 & 42.06 \\ - & - & - & 208.07 \end{pmatrix}.$$

Then

$$\hat{\boldsymbol{\lambda}} = \mathbf{S}_p^{-1}(\bar{\mathbf{x}}_{1\cdot} - \bar{\mathbf{x}}_{2\cdot}) = \begin{pmatrix} 0.345 \\ -0.130 \\ -0.106 \\ -0.143 \end{pmatrix},$$

$$-\tfrac{1}{2}(\bar{\mathbf{x}}_{1\cdot} - \bar{\mathbf{x}}_{2\cdot})'\mathbf{S}_p^{-1}(\bar{\mathbf{x}}_{1\cdot} + \bar{\mathbf{x}}_{2\cdot}) = -\tfrac{1}{2}\left(\bar{\mathbf{x}}'_{1\cdot}\mathbf{S}_p^{-1}\bar{\mathbf{x}}_{1\cdot} - \bar{\mathbf{x}}'_{2\cdot}\mathbf{S}_p^{-1}\bar{\mathbf{x}}_{2\cdot}\right)$$

$$= 15.805,$$

and $D_s(\mathbf{x})$ of (6.39) is given by

$$D_s(\mathbf{x}) = 15.805 + 0.345x_1 - 0.130x_2 - 0.106x_3 - 0.143x_4.$$

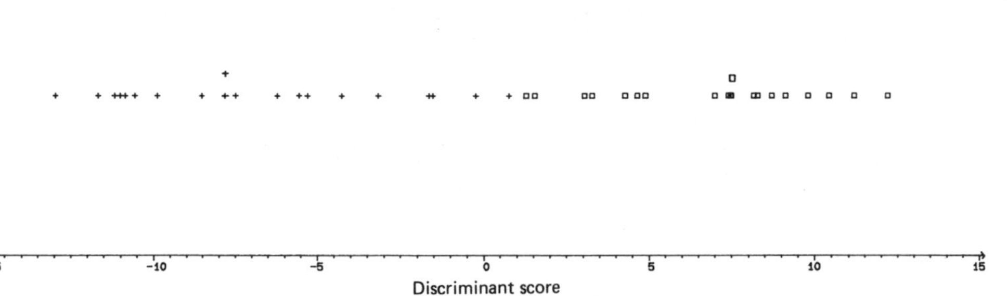

Fig. 6.2 LDF scores for two species of flea beetles, *H. carduorum* (+) and *H. oleracea* (□).

Assuming $\pi_1 = \pi_2$, we assign $\mathbf{x}$ to *H. oleracea* if $D_s(\mathbf{x}) > 0$. Applying this rule to the data in Table 6.1, we find that only one observation, namely, number 8 in group 2, is classified incorrectly. Values of $D_s(\mathbf{x})$ are plotted in Fig. 6.2. We conclude that the four variables chosen seem to provide an adequate LDF for classifying a flea beetle from one of the two species.

b QUADRATIC DISCRIMINANT FUNCTION

When $\boldsymbol{\Sigma}_1 \neq \boldsymbol{\Sigma}_2$, a quadratic discriminant is appropriate. Replacing $\boldsymbol{\mu}_i$ by $\overline{\mathbf{x}}_{i\cdot}$ and $\boldsymbol{\Sigma}_i$ by $\mathbf{S}_i$ in (6.13) gives us the sample estimate

$$Q_s(\mathbf{x}) = \tfrac{1}{2}\log\left(\frac{|\mathbf{S}_2|}{|\mathbf{S}_1|}\right) - \tfrac{1}{2}(\mathbf{x} - \overline{\mathbf{x}}_{1\cdot})'\mathbf{S}_1^{-1}(\mathbf{x} - \overline{\mathbf{x}}_{1\cdot}) + \tfrac{1}{2}(\mathbf{x} - \overline{\mathbf{x}}_{2\cdot})'\mathbf{S}_2^{-1}(\mathbf{x} - \overline{\mathbf{x}}_{2\cdot}),$$

which we shall call the quadratic discriminant function (QDF). We note that quadratic discriminants can also arise from nonnormal distributions (e.g., Cooper [1963]).

c ROBUSTNESS OF LDF AND QDF

The classification rule (6.39) based on the LDF is derived on the basis of the following assumptions: (1) The distributions are normal; (2) $\boldsymbol{\Sigma}_1 = \boldsymbol{\Sigma}_2$; and (3) the $n_1 + n_2$ sample observations are correctly classified. We might ask by how much does the rule depart from optimality when one or more of these assumptions no longer hold? One survey of this topic is given by Krzanowski [1977]. In relation to (1), Lachenbruch [1975a: Chapter 3] summarized the results of a number of studies on discrete data and concluded that "the general indications seem to be that the linear discriminant function performs fairly well on discrete data of various types." This is supported by the example of Titterington et al. [1981]. However, Moore [1973] demonstrated that for binary data the LDF gives poor results for populations whose log likelihood undergoes a "reversal." A reversal occurs when $L(\mathbf{x}) = \log[f_1(\mathbf{x}|\boldsymbol{\theta}_1)/f_2(\mathbf{x}|\boldsymbol{\theta}_2)]$, the true log likelihood, is not a monotonic function of $x_1 + x_2 + \cdots + x_d$. To illustrate this, suppose that the values $(0,0)$ and $(1,1)$ of any two binary variables occur more frequently in one group (G_1), while the values of (0, 1) and (1, 0) occur more frequently in the other (G_2). For example, in the study of infant maturity we could define (Yerushalmy et al. [1965])

$$x_1 = \begin{cases} 0 & \text{if birth weight is low,} \\ 1 & \text{if birth weight is high,} \end{cases}$$

and

$$x_2 = \begin{cases} 0 & \text{if gestation length is short,} \\ 1 & \text{if gestation length is long.} \end{cases}$$

A baby would be classified as "normal" if it had low birth weight and short gestation, or high birth weight and long gestation; otherwise the baby would be classified as "abnormal." The optimal rule would be the following: Assign to G_1 if $L(\mathbf{x}) > c$, and this would tend to assign babies correctly. However, for any linear rule Assign to G_1 if

$$a(\mathbf{x}) = \beta_0 + \sum_{j=1}^{d} \beta_j x_j > c, \tag{6.44}$$

we have

$$a(1,1) = a(0,1) + a(1,0) - a(0,0),$$

where $a(x_1, x_2)$ is an abbreviation of $a(x_1, x_2, \ldots, x_d)$. Hence

$$a(1,1) > \max\{a(0,1), a(1,0)\} \tag{6.45}$$

implies that

$$a(0,0) < \min\{a(0,1), a(1,0)\} \tag{6.46}$$

If (1, 1) is assigned to G_1, and both (0, 1) and (1, 0) assigned to G_2, then $a(1,1) > c$, $\min\{a(0,1), a(1,0)\} < c$, and $a(0,0) < c$. Thus (0, 0) cannot be assigned to G_1. The essence of the problem is that we are trying to use a linear function $a(\mathbf{x})$ that is essentially a monotonically increasing function of $x_1 + x_2$, to approximate $L(\mathbf{x})$, which is not monotonic.

One reason for the breakdown of the LDF in the above example is that x_1 and x_2 are positively correlated in G_1, but negatively correlated in G_2, so that the assumption of equal dispersion matrices is violated. However, Moore [1973] showed that reversals can occur even when there are moderate positive correlations between the variables. Dillon and Goldstein [1978] extended

Moore's study and again demonstrated the poor performance of the LDF because of reversals when there are large positive correlations, even when the dispersion matrices are equal. Krzanowski [1977] considered the case of a mixture of binary and normal random variables and showed that the LDF performs poorly if there is a moderate positive correlation among all the binary variables or the correlations between the discrete and continuous variables differ markedly for the two groups.

When we turn to continuous data, limited studies (e.g., Lachenbruch et al. [1973], Crawley [1979], Chinganda and Subrahmaniam [1979]) indicate that the LDF is not robust to nonnormal distributions like the very skew lognormal: There are large total error rates and individual actual error rates are widely different. However, a small amount of skewness can be tolerated (Subrahmaniam and Chinganda [1978]), and the LDF seems to be moderately robust to longer-tailed symmetric distributions and mixtures of normals (Ashikaga and Chang [1981]).

When $\mathbf{\Sigma}_1 \neq \mathbf{\Sigma}_2$, we saw above that the optimal rule is quadratic rather than linear. Although the QDF may tolerate mild nonnormality (Clarke et al. [1979]), it is unsatisfactory for binary data (see Moore [1973: p. 403]). Since the LDF still might be satisfactory for small departures from $\mathbf{\Sigma}_1 = \mathbf{\Sigma}_2$, Marks and Dunn [1974] compared the asymptotic and small-sample performance of the QDF, LDF, and the "best" linear discriminant function for normal data under both proportional and nonproportional differences in the $\mathbf{\Sigma}_i$. Their study, which extends that of Gilbert [1969], showed that for small samples (say, $n_1, n_2 < 25$), the QDF performs worse than a linear discriminant when $\mathbf{\Sigma}_1 \approx \mathbf{\Sigma}_2$ and d is moderately large ($d > 6$). However, the performance of the QDF relative to the LDF improves as the covariance differences increase. This investigation was taken further by Wahl and Kronmal [1977], who compared performances for moderate to large sample sizes. The "best" linear discriminant was omitted, as it offers little advantage in the situation when the LDF does better than the QDF and is not as good as the QDF when the covariance differences are greater. From the two studies it is clear that sample size is a critical factor in choosing between the LDF and QDF with normal data, and the following broad recommendations can be made:

(1) For small covariance differences and small d ($d \leq 6$) there is generally little to choose between the LDF and QDF.

(2) For small samples ($n_1, n_2 < 25$) and the covariance differences and/or d large, the LDF is preferred. However, when both covariance differences and d are large, the misclassification probabilities $P(i|j)$ may be too large for practical use.

(3) For large covariance differences and $d > 6$, QDF is much better than LDF, provided that the samples sizes are sufficiently large. Recommended sample sizes are $d = 4$, $n_1 = n_2 = 25$, and 25 additional observations for every two dimensions, for example, $d = 6, 8, 10$ and $n_1 = n_2 = 50, 75, 100$. For more than 100 observations the asymptotic results, which favor the QDF, are reached fairly quickly.

The deterioration of the QDF as d increases is also supported by Van Ness [1979] and is due to the fact that $\mathbf{S}_i$ is not a reliable estimate of $\boldsymbol{\Sigma}_i$ when d becomes a moderate fraction of n_i. The poor small-sample performance of the QDF, even when $\boldsymbol{\Sigma}_1 \neq \boldsymbol{\Sigma}_2$, is also supported by the simulation studies of Aitchison et al. [1977: p. 23] and Remme et al. [1980]. In summary, we conclude that for normal data the LDF performs satisfactorily, provided that the dispersion matrices are not too different; and the QDF is very poor for small samples, but is satisfactory for large covariance differences, provided that the samples are large enough.

Other departures from the assumptions of normality and $\boldsymbol{\Sigma}_1 = \boldsymbol{\Sigma}_2$ have been studied from a contamination viewpoint. A change partway through an investigation can lead to a shift in the mean (location contamination) or a change in the covariance structure (scale contamination) for a portion of the sampled observations. Some effects of scale contamination have been studied by Ahmed and Lachenbruch [1975, 1977], who showed that a large-sample LDF is little affected by mild scale contamination, particularly if $\pi_1 = \pi_2$. Broffit et al. [1981] showed that location contamination had little effect on the LDF. However, it would seem preferable to use robust estimates of $\boldsymbol{\mu}_i$ $(i = 1, 2)$ and $\boldsymbol{\Sigma}$ in the calculation of the LDF if there is likely to be contamination or outliers. Such an LDF seems to perform as well as the usual LDF when there is no contamination. Broffitt et al. [1980] considered using robust estimators for the QDF to try and mitigate some of the effects of nonnormality on the QDF; however, they found that any gains were not sufficient to warrant the extra computational effort. A ridge regression approach using a biased estimate $\mathbf{S} + k\mathbf{I}_d$ of $\boldsymbol{\Sigma}$ instead of $\mathbf{S}$ has been suggested by Di Pillo [1976] to help offset the increase in error rates through using estimates rather than true parameters values.

The assumption that the initial samples are correctly classified may not always hold, and the effects of initially misclassifying a portion of the samples on the LDF and QDF have been studied by McLachlan [1972] and Lachenbruch [1966, 1974, 1979]. It transpires that if the initial misclassification rates for the two samples are about the same and all the other assumptions are satisfied, then the misclassification probabilities $P(i|j)$ for the LDF are little affected, irrespective of whether the initial misclassification is random or nonrandom. Unfortunately the individual apparent error rates $e_{i,\text{app}}$ are grossly distorted and totally unreliable for any sample sizes. The situation with regard to the QDF is even worse, as the $P(i|j)$ are adversely affected even if the sample misclassification rates are about the same. The effect appears to get worse with increasing covariance differences and increasing initial misclassification rates. For this reason Lachenbruch [1979] recommends that if initial misclassification is suspected, then all the sample observations should be carefully checked and reassigned if necessary.

d MISSING VALUES

In a regression model with missing response (Y) values, the usual methods for estimating these values (see Seber [1977: Section 10.2]) are equivalent to "predicting" the missing values on the basis of the complete data. A similar approach can be used for handling missing x_j values in discriminant analysis. For example, Chan and Dunn [1974] compared a number of procedures for estimating missing values, including a method D in which each x_j is treated as a response and the resulting regression on the other x's (regressors) using complete data only is used to estimate (predict) any missing values of x_j. Chan et al. [1976] also compared several methods and recommended a modification of D, called D^*, in which one uses regressor data with all the gaps filled by substituting sample means. As these methods do not produce consistent estimates of the LDF coefficients, Little [1978] considered two further consistent modifications that he called D_1 and D_2: D_1 uses a consistent estimator of $\boldsymbol{\Sigma}$, and D_2 is an iterative procedure which gives maximum likelihood estimators under normality assumptions.

e PREDICTIVE DISCRIMINANT

Using the predictive density function $h_i(\mathbf{x}|\mathbf{z})$ of (6.36) instead of the plug-in estimate $f_i(\mathbf{x}|\hat{\boldsymbol{\theta}}_i)$, we have the following allocation rule: Assign to G_1 if

$$\frac{h_1(\mathbf{x}|\mathbf{z})}{h_2(\mathbf{x}|\mathbf{z})} > c. \tag{6.47}$$

If $\boldsymbol{\Sigma}_1 \neq \boldsymbol{\Sigma}_2$ and $\nu_i = n_i - 1$, then using the conventional "improper" or "vague" prior for $\boldsymbol{\theta}$, the vector of the distinct elements of $\boldsymbol{\mu}_1$, $\boldsymbol{\mu}_2$, $\boldsymbol{\Sigma}_1^{-1}$, and $\boldsymbol{\Sigma}_2^{-1}$, leads to (Aitchison et al. [1977])

$$h_i(\mathbf{x}|\mathbf{z}) = T_d\left(\mathbf{x}, \nu_i|\overline{\mathbf{x}}_{i\cdot}, \{1 + n_i^{-1}\}\mathbf{S}_i\right), \tag{6.48}$$

where

$$T_d(\mathbf{x}, \nu|\mathbf{b}, \mathbf{C}) = \frac{\Gamma(\nu+1)/2}{\pi^{d/2}\Gamma(\nu-d+1)/2}|\nu\mathbf{C}|^{-1/2}$$
$$\times\left[1 + (\mathbf{x}-\mathbf{b})'(\nu\mathbf{C})^{-1}(\mathbf{x}-\mathbf{b})\right]^{-(\nu+1)/2} \tag{6.49}$$

is the density function of a d-dimensional Student's distribution with ν degrees of freedom. Substituting (6.48) in (6.47) leads to a ratio that we shall call the PU—the predictive discriminant for unequal dispersion matrices.

If $\boldsymbol{\Sigma}_1 = \boldsymbol{\Sigma}_2 = \boldsymbol{\Sigma}$, the corresponding predictive density is

$$h_i(\mathbf{x}|\mathbf{z}) = T_d\left(\mathbf{x}, \nu|\overline{\mathbf{x}}_{i\cdot}, \{1 + n_i^{-1}\}\mathbf{S}_p\right),$$

where $\nu = n_1 + n_2 - 2$ and $\mathbf{S}_p$ is given by (6.38). The rule (6.47) now takes the form

$$R(\mathbf{x}) = \frac{1 + [n_2/(n_2+1)](\mathbf{x}-\overline{\mathbf{x}}_{2\cdot})'\mathbf{Q}_p^{-1}(\mathbf{x}-\overline{\mathbf{x}}_{2\cdot})}{1 + [n_1/(n_1+1)](\mathbf{x}-\overline{\mathbf{x}}_{1\cdot})'\mathbf{Q}_p^{-1}(\mathbf{x}-\overline{\mathbf{x}}_{1\cdot})}$$
$$> c^{2/(\nu+1)}\left[\frac{n_2(n_1+1)}{n_1(n_2+1)}\right]^{d/(\nu+1)}, \tag{6.50}$$

and we shall call the ratio $R(\mathbf{x})$ the PE—the predictive discriminant for equal dispersion matrices. Aitchison et al. [1977] gave an example demonstrating that the plug-in (estimative) and predictive rules have very different properties and compared the two rules by considering how close $\log[f_1(\mathbf{x}|\hat{\boldsymbol{\theta}}_1)/f_2(\mathbf{x}|\hat{\boldsymbol{\theta}}_2)]$ and $\log[h_1(\mathbf{x}|\mathbf{z})/h_2(\mathbf{x}|\mathbf{z})]$ are to $LO(\text{true}) = \log[f_1(\mathbf{x}|\boldsymbol{\theta}_1)/f_2(\mathbf{x}|\boldsymbol{\theta}_2)]$, the so-called true log-odds. They concluded from theoretical considerations using the Kullback and Liebler [1951] information measure, and a simulation study using small samples ($n_1, n_2 \leq 20, d \leq 9$), that the predictive method gave a closer estimate. More specifically, if $\boldsymbol{\Sigma}_1 \neq \boldsymbol{\Sigma}_2$, the PU is moderately superior to the LDF and only slightly better than the PE; while if $\boldsymbol{\Sigma}_1 = \boldsymbol{\Sigma}_2$, the PE is by far the best and the PU and LDF are comparable. In both the cases the QDF is decidely the worst; the poor small-sample properties of the QDF have already been mentioned above. We note that for large samples the estimative and predictive rules are equivalent, as the multivariate Student's distribution tends to the MVN (see Exercise 6.9). Further evidence in favor of the PE is given by McLachlan [1979].

However, Moran and Murphy [1979] pointed out that the LDF and QDF are primarily associated with allocation rules rather than with the estimation of log-odds. They argue that it is not appropriate to use these discriminant functions as representative of the frequentist approach to the estimation of log-odds without at least introducing adjustments for bias. When this is done, the performance of the estimative (plug-in) methods is much closer to that of the predictive. The unbiased LDF (Exercise 6.10) can be expressed in the form

$$\text{ULDF} = \frac{1}{2}\left(\frac{n_1 + n_2 - d - 3}{n_1 + n_2 - 2}\right)\{Q_{22}(\mathbf{x}) - Q_{11}(\mathbf{x})\} + \tfrac{1}{2}d\left(\frac{1}{n_1} - \frac{1}{n_2}\right), \tag{6.51}$$

where $Q_{ii}(\mathbf{x}) = (\mathbf{x} - \bar{\mathbf{x}}_{i\cdot})'\mathbf{S}_p^{-1}(\mathbf{x} - \bar{\mathbf{x}}_{i\cdot})$. This may be compared with

$$\text{LDF} = \tfrac{1}{2}\{Q_{22}(\mathbf{x}) - Q_{11}(\mathbf{x})\}. \tag{6.52}$$

We assign $\mathbf{x}$ to G_1 if the value of the discriminant function exceeds $\log c$. If $n_1 = n_2$ and $c = 1$, then, using the ULDF, we assign $\mathbf{x}$ to G_1 if $Q_{11}(\mathbf{x}) < Q_{22}(\mathbf{x})$, that is, if

$$(\mathbf{x} - \bar{\mathbf{x}}_{1\cdot})'\mathbf{Q}_p^{-1}(\mathbf{x} - \bar{\mathbf{x}}_{1\cdot}) < (\mathbf{x} - \bar{\mathbf{x}}_{2\cdot})'\mathbf{Q}_p^{-1}(\mathbf{x} - \bar{\mathbf{x}}_{2\cdot}),$$

which is the same as $R(\mathbf{x}) > 1$ in (6.50). In this case the correction for bias is unnecessary, as the ULDF, LDF, and PE allocation rules are all identical.

Moran and Murphy [1979] suggested an alternative "frequentist" approach based on a generalized likelihood ratio test discussed by Anderson [1958: p. 141]. If we assume that $\mathbf{x} \sim N_d(\boldsymbol{\mu}, \boldsymbol{\Sigma})$ and the $\mathbf{x}_{ij}$ ($j = 1, 2, \ldots, n_i$) are i.i.d. $N_d(\boldsymbol{\mu}_i, \boldsymbol{\Sigma}_i)$, then the likelihood ratio statistic for testing $H_0: (\boldsymbol{\mu}, \boldsymbol{\Sigma}) = (\boldsymbol{\mu}_1, \boldsymbol{\Sigma}_1)$ versus $H_1: (\boldsymbol{\mu}, \boldsymbol{\Sigma}) = (\boldsymbol{\mu}_2, \boldsymbol{\Sigma}_2)$ can be used as a discriminant function. When

$\boldsymbol{\Sigma}_1 = \boldsymbol{\Sigma}_2$, the corresponding likelihood ratio rule is the following: Assign to G_1 if $R(\mathbf{x}) > (\pi_2/\pi_1)^{2/(\nu+3)}$, which is of the same form as (6.50). If we ignore costs, then $c = \pi_2/\pi_1$, and these two rules are almost the same when $n_1 = n_2$, and identical if $\pi_1 = \pi_2$ as well. When $n_1 \neq n_2$ and $\pi_1 = \pi_2$, the likelihood ratio rule becomes Assign to G_1 if $R(\mathbf{x}) > 1$, that is, if

$$Z = -\frac{n_1}{n_1+1}Q_{11}(\mathbf{x}) + \frac{n_2}{n_2+1}Q_{22}(\mathbf{x}) > 0.$$

The distribution of Z has been studied by John [1960, 1963] and Memon and Okamoto [1971]. The latter authors obtained an asymptotic distribution for Z and good approximations for the individual actual error rates $e_{i,\text{act}}$ of (6.21) using a method similar to that used by Okamoto [1963] for the LDF. It transpires that Z has certain minimax properties (Das Gupta [1965]) with respect to the $e_{i,\text{act}}$ (see Section 6.2.2e) in contrast to the LDF that can have widely different error rates (Moran [1975]). Also, $e_{1,\text{act}} + e_{2,\text{act}}$ is smaller for Z than for the LDF over a wide range of values of d and $\Delta = \{(\boldsymbol{\mu}_1 - \boldsymbol{\mu}_2)'\boldsymbol{\Sigma}^{-1}(\boldsymbol{\mu}_1 - \boldsymbol{\mu}_2)\}^{1/2}$ (Siotani and Wang [1977: Table 4.2]).

The predictive estimate $h_i(\mathbf{x}|\mathbf{z})$, derived using Bayesian methods, has also been derived using a non-Bayesian argument, so that the estimate can be given a frequency interpretation and misclassification probabilities can be calculated (Han [1979]). However, all the theory in this section is very much tied to the normal distribution and robustness properties need investigation.

6.3.3 *Multivariate Discrete Distributions*

a INDEPENDENT BINARY VARIABLES

Consider Example 6.3 in Section 6.2.2, where $\mathbf{x}' = (x_1, x_2)$ is a vector of independent binary variables, each taking the value 1 or 0, with $\text{pr}[x_j = 1] = p_{ij}$ when $\mathbf{x} \in G_i$ $(i = 1, 2;\ j = 1, 2)$. Then $\mathbf{x}'$ takes the four values (1, 1), (1, 0), (0, 1) and (0, 0) with respective probabilities $\theta_{(i)1} = p_{i1}p_{i2}$, $\theta_{(i)2} = p_{i1}(1 - p_{i2})$, $\theta_{(i)3} = (1 - p_{i1})p_{i2}$, and $\theta_{(i)4} = (1 - p_{i1})(1 - p_{i2})$. Therefore, given $\mathbf{x} \in G_i$, $\mathbf{x}$ falls in the multinomial cell k with probability $\theta_{(i)k}$ $(k = 1, 2, 3, 4)$, and the probability function for $\mathbf{x}$ is

$$f_i(\mathbf{x}) = p_{i1}^{x_1}(1 - p_{i1})^{1-x_1}p_{i2}^{x_2}(1 - p_{i2})^{1-x_2}. \tag{6.53}$$

If an observation $\mathbf{x}$ to be classified falls into cell j, then the optimal rule (6.4) is Classify as G_1 if

$$\frac{f_1(\mathbf{x})}{f_2(\mathbf{x})} = \frac{\theta_{(1)j}}{\theta_{(2)j}} > \frac{\pi_2}{\pi_1}. \tag{6.54}$$

The parameters $\theta_{(i)k}$ have to be estimated from the sample data. Suppose, then, that we have a random sample of n observations from the population $\mathscr{P}$, and of the n_i that come from G_i suppose that $n_{(i)k}$ fall into cell k, that is, $n_i = \Sigma_k n_{(i)k}$. Since

$$\text{pr}[\mathbf{x} \text{ in cell } k \text{ and } \mathbf{x} \in G_i] = \text{pr}[\mathbf{x} \text{ in cell } k|\mathbf{x} \in G_i]\text{pr}[\mathbf{x} \in G_i]$$
$$= \theta_{(i)k}\pi_i, \tag{6.55}$$

the likelihood function is

$$L(\{p_{ij}, \pi_j\}) = \prod_{i=1}^{2} \{\text{pr}[n_{(i)k} \text{ cell frequencies and data from } G_i]\}$$
$$= \prod_{i=1}^{2}\prod_{k=1}^{4} \{\theta_{(i)k}\pi_i\}^{n_{(i)k}}$$
$$= \prod_{i=1}^{2}\left\{\pi_i^{n_i}\prod_{k=1}^{4}\theta_{(i)k}^{n_{(i)k}}\right\}, \tag{6.56}$$

where the $\theta_{(i)k}$ are replaced by their functions of the p_{ij}. Maximizing (6.56) gives us the maximum likelihood estimates $\hat{p}_{ij}$ (= proportion of n_i observations with $x_j = 1$) and $\hat{\pi}_i = n_i/n$ (see Exercise 6.11): These can be plugged into (6.54).

Instead of using (6.54), we can, of course, use the LDF of (6.6) and rely on its robustness to discrete data and mild inequality of dispersion matrices. Moore [1973] showed that for his simulated data the actual error rates for the LDF and the above plug-in method, which he calls the first-order procedure, were very similar. The rule (6.54) can also be expressed in the form (6.14), namely,

$$\beta_0 + \beta_1 x_1 + \beta_2 x_2 > \log(\pi_2/\pi_1), \tag{6.57}$$

where the β_j are now estimated directly using the method of logistic discrimination described in Section 6.4. A cross-validatory technique has also been proposed by Mabbett et al. [1980].

The above methods clearly extend to the case of d independent binary variables. The numbers of parameters to be estimated for each of the methods are (excluding the π_i): $2d$ probabilities p_{ij} for the multinomial approach, $2d + \frac{1}{2}d(d+1)$ elements of $\boldsymbol{\mu}_1$, $\boldsymbol{\mu}_2$, and $\boldsymbol{\Sigma}$ for the LDF, and $d + 1$ β's for the logistic method.

b CORRELATED BINARY VARIABLES

If $\mathbf{x}' = (x_1, x_2)$, as in the example above, but the x_i are now correlated binary variables, then we can no longer express the $\theta_{(i)k}$ as functions of the p_{ij}'s. For instance, $\theta_{(i)1} \neq p_{i1}p_{i2}$, as x_1 and x_2 are not independent. However, the multinomial model (6.56) still applies and the maximum likelihood estimate of

$\theta_{(i)k}$ is $\hat{\theta}_{(i)k} = n_{(i)k}/n_i$; $\hat{\pi}_i$ is still n_i/n. Substituting these estimates in (6.54) gives the simple rule Assign to G_1 if

$$n_{(1)j} > n_{(2)j}. \tag{6.58}$$

The extension to a d-dimensional $\mathbf{x}$ is straightforward with $2^d - 1$ independent cell probabilities to be estimated for each group.

Although, for rule (6.58), e_{act} and $E[\hat{e}_{\text{act}}]$ both converge rapidly to e_{opt} (e_{act} in a probabilistic sense; see Glick [1973], who uses nonerror rather than error rates), the method is generally unsatisfactory, as very large samples are usually required to provide satisfactory estimates of all the cell probabilities. For example, if $d = 10$ then $2^d - 1 = 1023$ and, in practice, many of the cells would be empty or have few observations. We note, in passing, that Goldstein and Wolf [1977] give an expression for the "bias" $E[\hat{e}_{\text{act}}] - e_{\text{opt}}$ [see (6.25)]. To overcome this problem of sparseness, the number of parameters must be reduced by imposing further constraints on the multinomial model. One method is to express $f(\mathbf{x})$ in terms of the independent binary model and a multiplicative "adjustment" for dependence. For example, Bahadur [1961] based the adjustment factor on the correlation structure: The number of parameters is reduced by assuming various "higher-order" covariances to be zero. Two other adjustments, using orthogonal polynomials, were proposed by Martin and Bradley [1972] and Ott and Kronmal [1976]. Although each of the three methods has its own particular strengths (see Goldstein and Dillon [1978] for a discussion), none of the three allows for a satisfactory degree of "control" in the construction of models for the cell probabilities $\theta_{(i)k}$ (Lachenbruch and Goldstein [1979]). A similar type of adjustment, but based on $\log \theta_{(i)k}$, was proposed by Brunk and Pierce [1974] and analyzed using Bayesian methods. However, an alternative approach is to use the models described in Section 10.2. For example, if $d = 3$, we can extend (10.28) to give

$$\begin{aligned}
\log f(\mathbf{x}) &= \log \operatorname{pr}[X_1 = x_1, X_2 = x_2, X_3 = x_3] \\
&= u + u_1(1 - 2x_1) + u_2(1 - 2x_2) \\
&\quad + u_3(1 - 2x_3) + u_{12}(1 - 2x_1)(1 - 2x_2) \\
&\quad + u_{23}(1 - 2x_2)(1 - 2x_3) + u_{31}(1 - 2x_3)(1 - 2x_1) \\
&\quad + u_{123}(1 - 2x_1)(1 - 2x_2)(1 - 2x_3) \\
&= (u + u_1 + u_2 + u_3 + u_{12} + u_{23} + u_{31} + u_{123}) \\
&\quad + (-2u_1 - 2u_{12} - 2u_{31} - 2u_{123})x_1 \\
&\quad + \cdots + (4u_{12} + 4u_{123})x_1x_2 + \cdots + (-8u_{123})x_1x_{23} \\
&= \beta_0' + \beta_1'x_1 + \cdots + \beta_{12}'x_1x_2 + \cdots + \beta_{123}'x_1x_2x_3.
\end{aligned} \tag{6.59}$$

If we can assume $\beta'_{12} = \beta'_{23} = \beta'_{31} = \beta'_{123} = 0$, or that these parameters are the same for both $f_1(\mathbf{x})$ and $f_2(\mathbf{x})$, then

$$\log[f_1(\mathbf{x})/f_2(\mathbf{x})] = \beta_0 + \beta_1 x_1 + \beta_2 x_2 + \beta_3 x_3, \tag{6.60}$$

where β_j is the difference between the corresponding β_j''s. For general d, the same linear model applies if second- and higher-order interactions are the same for both probability functions f_i. The β_j can be estimated using the methods of Section 6.4.

Instead of trying to reduce the number of parameters, another approach to the sparseness problem is to use a nearest neighbor method (Section 6.5.2a) in which $f_i(\mathbf{x})$ is estimated by the proportion of the n_i observations with values close to or equal to $\mathbf{x}$. Gardner and Barker [1975] give an interesting diagnostic example in which the simple allocation rule $\Sigma_{j=1}^{d} x_j > c$, that is, assign to the "sick" group if the number of positive symptoms exceeds a certain value, compares satisfactorily with more sophisticated methods.

c CORRELATED DISCRETE VARIABLES

Suppose $\mathbf{x}' = (x_1, x_2)$, where x_1 and x_2 are discrete random variables taking s_1 and s_2 values, respectively. Then the multinomial model (6.56) still applies, except there are now $s_1 s_2$ cell probabilities instead of 4. The problem of sparseness is now even greater and all the methods above for reducing the number of parameters apply here. Of course, any discrete random variable can be converted to a set of binary variables. For example, if $y = 1, 2, 3$, we can define

$$x_1 = \begin{cases} 1, & y = 1, \\ 0, & \text{otherwise}, \end{cases}$$

$$x_2 = \begin{cases} 1, & y = 2, \\ 0, & \text{otherwise}, \end{cases} \tag{6.61}$$

with $x_1 = 0$ and $x_2 = 0$ corresponding to $y = 3$. However, this procedure is more appropriate for unordered categories, for example, married, never married, or previously married. Another method for comparing the two multinomial distributions, based on a measure of distance between two distributions, has been proposed by Dillon and Goldstein [1978] (see Section 6.5.2c).

6.3.4 *Multivariate Discrete–Continuous Distributions*

Suppose $\mathbf{x}' = (\mathbf{x}^{(1)\prime}, \mathbf{x}^{(2)\prime})$, where $\mathbf{x}^{(1)}$ is continuous and $\mathbf{x}^{(2)}$ discrete. Even though most problems in medical discriminations fall into this mixed category, it is surprising that this situation has received comparatively little attention in the literature (see Krzanowski [1975, 1980], Tu and Han [1982] for references). However, Krzanowski [1975, 1976, 1977] has developed a model that assumes that $\mathbf{x}^{(2)}$ is a d_2-dimensional vector of binary random variables generating $S = 2^{d_2}$ states; and $\mathbf{x}^{(1)}$ is a d_1-dimensional MVN vector with mean $\boldsymbol{\mu}^{(s)}$ depending on the state s of $\mathbf{x}^{(2)}$ ($s = 1, 2, \dots, S$) and constant dispersion matrix $\boldsymbol{\Sigma}$. Then

$$f(\mathbf{x}) = g(\mathbf{x}^{(1)}|\mathbf{x}^{(2)})h(\mathbf{x}^{(2)}),$$

and the allocation rule is Assign to G_1 if

$$\frac{g_1(\mathbf{x}^{(1)}|\mathbf{x}^{(2)})h_1(\mathbf{x}^{(2)})}{g_2(\mathbf{x}^{(1)}|\mathbf{x}^{(2)})h_2(\mathbf{x}^{(2)})} > \frac{\pi_2}{\pi_1}. \tag{6.62}$$

Since g is the density function for the MVN, the logarithm of (6.62) reduces to [see (6.5)]

$$\left(\boldsymbol{\mu}_1^{(s)} - \boldsymbol{\mu}_2^{(s)}\right)'\boldsymbol{\Sigma}^{-1}\left\{\mathbf{x}^{(1)} - \tfrac{1}{2}\left(\boldsymbol{\mu}_1^{(s)} + \boldsymbol{\mu}_2^{(s)}\right)\right\} > \log\left\{\frac{h_2(\mathbf{x}^{(2)})\pi_2}{h_1(\mathbf{x}^{(2)})\pi_1}\right\},$$

$$\left(\mathbf{x}^{(2)} \text{ belongs to state } s,\ s = 1, 2, \dots, S\right), \tag{6.63}$$

Thus for each s there is a separate classification rule. However, very large data sets would be needed to estimate all the quantities $\boldsymbol{\mu}_i^{(s)}$, $h_i(\mathbf{x}^{(2)})$, and $\boldsymbol{\Sigma}$. A zero frequency for state s would mean that $\boldsymbol{\mu}_i^{(s)}$ could not be estimated without additional constraints on the model. Krzanowski [1975] suggested using a log-linear model like (6.60) to model the probability $h_i(\mathbf{x}^{(2)})$, and a similar linear expression to model $\boldsymbol{\mu}_i^{(s)}$ in terms of main effects and first-order interactions, higher-order interactions being assumed zero. For $d = 2$ we have, as in (6.59),

$$\boldsymbol{\mu}_i^{(s)} = \boldsymbol{\alpha}_i + \boldsymbol{\beta}_{i1}x_1^{(2)} + \boldsymbol{\beta}_{i2}x_2^{(2)} + \boldsymbol{\beta}_{i12}x_1^{(2)}x_2^{(2)},$$

where $\mathbf{x}^{(2)} = (x_1^{(2)}, x_2^{(2)})'$ is in state s. Lachenbruch and Goldstein [1979] noted a number of extensions of the above procedure to discrete random variables taking more than two values, and to normal variables with unequal dispersion matrices. Krzanowski [1980] extended the model to the case of categorical (unordered multistate) variables with more than two states. He (Krzanowski [1979c]) also described some linear transformations of the data, analogous to those used in canonical correlation analysis, for simplifying the underlying correlation structure and reducing the number of variables.

Another approach to the problem of mixed discrete and continuous data, which does not require normality assumptions, is the logistic model described below.

6.4 TWO GROUPS: LOGISTIC DISCRIMINANT

6.4.1 General Model

Since the classification rules developed in this chapter depend only on the ratio of the density functions, we can simply model the ratio without specifying the individual densities $f_i(\mathbf{x})$. The so-called logistic model assumes that

$$\log\left[\frac{f_1(\mathbf{x})}{f_2(\mathbf{x})}\right] = \alpha + \boldsymbol{\beta}'\mathbf{x}, \tag{6.64}$$

and the allocation rule of (6.4) becomes Assign to G_1 if $\alpha + \boldsymbol{\beta}'\mathbf{x} > \log(\pi_2/\pi_1)$, that is, if

$$\alpha_0 + \boldsymbol{\beta}'\mathbf{x} > 0, \tag{6.65}$$

where

$$\alpha_0 = \alpha + \log(\pi_1/\pi_2). \tag{6.66}$$

With this model the posterior probabilities of (6.16) take the simple form

$$\begin{aligned} q_1(\mathbf{x}) &= \text{pr}[\mathbf{x} \in G_1 | \mathbf{x}] \\ &= \frac{f_1(\mathbf{x})\pi_1}{f_1(\mathbf{x})\pi_1 + f_2(\mathbf{x})\pi_2} \\ &= \frac{\exp[\alpha + \log(\pi_1/\pi_2) + \boldsymbol{\beta}'\mathbf{x}]}{\exp[\alpha + \log(\pi_1/\pi_2) + \boldsymbol{\beta}'\mathbf{x}] + 1} \end{aligned} \tag{6.67}$$

and

$$q_2(\mathbf{x}) = 1 - q_1(\mathbf{x}).$$

Hence

$$\log\left\{\frac{q_1(\mathbf{x})}{q_2(\mathbf{x})}\right\} = \log\left\{\frac{f_1(\mathbf{x})\pi_1}{f_2(\mathbf{x})\pi_2}\right\} = \alpha_0 + \boldsymbol{\beta}'\mathbf{x}. \tag{6.68}$$

This model for the "posterior odds" was suggested by Cox [1966] and Day and Kerridge [1967] as a basis for discrimination. There is a growing body of literature on logistic discrimination and this is reviewed by Lachenbruch [1975a] and Anderson [1982]. The discussion in this section is largely based on Anderson's paper (see also Anderson and Blair [1982]).

One advantage of the above logistic approach is that we only need to estimate the $d + 1$ parameters α_0 and $\boldsymbol{\beta}$ from the sample data, without having to specify f. By contrast, the previous methods require not only the specification of f_i, $f_i(\mathbf{x}|\boldsymbol{\theta}_i)$, say, but also the estimation of many more unknown parameters $\boldsymbol{\theta}_i$ $(i = 1, 2)$. Another advantage is that the family of distributions satisfying (6.64) is quite wide. These include the following:

(1) MVN distributions with equal dispersion matrices. That is, from (6.6) we have

$$\alpha + \boldsymbol{\beta}'\mathbf{x} = -\tfrac{1}{2}\boldsymbol{\lambda}'(\boldsymbol{\mu}_1 + \boldsymbol{\mu}_2) + \boldsymbol{\lambda}'\mathbf{x}, \tag{6.69}$$

where $\boldsymbol{\lambda} = \boldsymbol{\Sigma}^{-1}(\boldsymbol{\mu}_1 - \boldsymbol{\mu}_2)$.

(2) Multivariate independent dichotomous variables [see (6.14)].

(3) Multivariate discrete distributions following the log-linear model with the same interactions in each group [see (6.60)].

(4) Mixtures of (1) and (2) that are not necessarily independent.

(5) Certain truncated versions of the above.

(6) Versions of the above with $\mathbf{x}$ replaced by some vector function of $\mathbf{x}$ in (6.64).

Finally, the logistic model is particularly useful for handling diagnostic data (Dawid [1976]).

6.4.2 Sampling Designs

There are three common sampling designs that yield data suitable for estimating $\boldsymbol{\beta}$: (1) mixture sampling in which a sample of $n = n_1 + n_2$ members is randomly selected from the total population $\mathscr{P}$ so that the n_i are random (Day and Kerridge [1967]); (2) separate sampling in which for $i = 1, 2$ a sample of fixed size n_i is selected from G_i (Anderson [1972], Prentice and Pyke [1979]); and (3) conditional sampling in which, for $j = 1, 2, \ldots, m$, $n(\mathbf{x}_j)$ members of $\mathscr{P}$ are selected at random from all members of $\mathscr{P}$ with $\mathbf{x} = \mathbf{x}_j$ (e.g., in bioassay where the two groups refer to "response" and "no response"). In all three cases we adopt the notation used for (3) and assume that the n observations take m different values $\mathbf{x}_j$ $(j = 1, 2, \ldots, m)$ with frequency $n_i(\mathbf{x}_j)$ in group G_i.

a CONDITIONAL SAMPLING

Here $n(\mathbf{x}_j) = n_1(\mathbf{x}_j) + n_2(\mathbf{x}_j)$ is fixed and $n_i(\mathbf{x}_j)$ is random. The likelihood function for this design is

$$\begin{aligned} L_C &= \prod_{j=1}^{m} \{\mathrm{pr}[\mathbf{x} \in G_1 | \mathbf{x} = \mathbf{x}_j]\}^{n_1(\mathbf{x}_j)} \{\mathrm{pr}[\mathbf{x} \in G_2 | \mathbf{x} = \mathbf{x}_j]\}^{n_2(\mathbf{x}_j)} \\ &= \prod_{i=1}^{2} \prod_{j=1}^{m} \{q_i(\mathbf{x}_j)\}^{n_i(\mathbf{x}_j)}. \end{aligned} \tag{6.70}$$

From (6.67) we see that L_C is a function of α_0 and $\boldsymbol{\beta}$ that can be maximized to obtain maximum likelihood estimates.

b MIXTURE SAMPLING

In this case n is fixed and the $\mathbf{x}_j$ and $n(\mathbf{x}_j)$ are now random variables. Let $L(\mathbf{x}, G_i) = \pi_i f_i(\mathbf{x})$ and

$$f(\mathbf{x}) = \pi_1 f_1(\mathbf{x}) + \pi_2 f_2(\mathbf{x}).$$

Then the likelihood function is

$$\begin{aligned} L_M &= \prod_{i=1}^{2} \prod_{j=1}^{m} \{L(\mathbf{x}_j, G_i)\}^{n_i(\mathbf{x}_j)} \\ &= \left[\prod_{i=1}^{2} \prod_{j=1}^{m} \left\{\frac{L(\mathbf{x}_j, G_i)}{f(\mathbf{x}_j)}\right\}^{n_i(\mathbf{x}_j)}\right]\left[\prod_{j=1}^{m} \{f(\mathbf{x}_j)\}^{n(\mathbf{x}_j)}\right] \\ &= \left[\prod_{i=1}^{2} \prod_{j=1}^{m} \{q_i(\mathbf{x}_j)\}^{n_i(\mathbf{x}_j)}\right] L, \qquad \text{say}, \\ &= L_C L. \end{aligned} \tag{6.71}$$

Since $f_i(\mathbf{x})$ is unspecified, no assumptions are made about the form of $f(\mathbf{x})$, the density function for the above sampling scheme. We can therefore assume that $f(\mathbf{x})$ contains no useful information about α_0 and $\boldsymbol{\beta}$. Even if we knew something about $f_i(\mathbf{x})$, the extra information about α_0 and $\boldsymbol{\beta}$ in L would be small compared with that contained in L_C. Therefore, as in conditional sampling, maximum likelihood estimates are again found by maximizing L_C.

In practice, we frequently have $m = n$, that is, all the observations are different and $n_i(\mathbf{x}_j)$ is 0 or 1.

c SEPARATE SAMPLING

Separate sampling is generally the most common sampling situation and the likelihood function is

$$\begin{aligned} L_S &= \prod_{i=1}^{2} \prod_{j=1}^{m} \{L(\mathbf{x}_j | G_i)\}^{n_i(\mathbf{x}_j)} \\ &= \prod_{i=1}^{2} \prod_{j=1}^{m} \{f_i(\mathbf{x}_j)\}^{n_i(\mathbf{x}_j)} \end{aligned} \tag{6.72}$$

$$\begin{aligned} &= \prod_{i=1}^{2} \prod_{j=1}^{m} \left\{\frac{L(\mathbf{x}_j, G_i)}{\pi_i}\right\}^{n_i(\mathbf{x}_j)} \\ &= L_M \pi_1^{-n_1} \pi_2^{-n_2}. \end{aligned} \tag{6.73}$$

If π_1 and π_2 are known, then this model is equivalent to (6.71). However, if π_1 and π_2 are unknown, we proceed as follows (Anderson and Blair [1982]). We assume that $\mathbf{x}$ is discrete so that the values of f may be taken as multinomial probabilities (Section 6.3.3).

From (6.64) we have

$$f_1(\mathbf{x}) = f_2(\mathbf{x})\exp(\alpha + \boldsymbol{\beta}'\mathbf{x})$$

and, from (6.72),

$$L_S = \prod_{j=1}^{m} p_{\mathbf{x}_j}^{n(\mathbf{x}_j)}\exp\{n_1(\mathbf{x}_j)[\alpha + \boldsymbol{\beta}'\mathbf{x}_j]\}, \qquad (6.74)$$

where $p_{\mathbf{x}} = f_2(\mathbf{x})$. The problem is to maximize L_S subject to the constraints that f_2 and f_1 are probability functions, namely,

$$\sum_{\mathbf{x}} p_{\mathbf{x}} = 1 \qquad (6.75)$$

and

$$\sum_{\mathbf{x}} p_{\mathbf{x}}\exp(\alpha + \boldsymbol{\beta}'\mathbf{x}) = 1. \qquad (6.76)$$

Using Lagrange multipliers, it can be shown that (Exercise 6.12) the answer is given by estimating $p_{\mathbf{x}}$ by

$$\hat{p}_{\mathbf{x}} = \frac{n(\mathbf{x})}{n_1\exp(\alpha + \boldsymbol{\beta}'\mathbf{x}) + n_2},$$

substituting $\hat{p}_{\mathbf{x}}$ into L_S of (6.74), and then maximizing the resulting expression

$$L'_S = L'_C n_1^{-n_1} n_2^{-n_2} \prod_{j=1}^{m} [n(\mathbf{x}_j)]^{n(\mathbf{x}_j)},$$

where

$$L'_C = \prod_{i=1}^{2}\prod_{j=1}^{m}\{\tilde{q}_i(\mathbf{x}_j)\}^{n_i(\mathbf{x}_j)}, \qquad (6.77)$$

$$\begin{aligned}\tilde{q}_1(\mathbf{x}) &= \frac{n_1\exp(\alpha + \boldsymbol{\beta}'\mathbf{x})}{n_1\exp(\alpha + \boldsymbol{\beta}'\mathbf{x}) + n_2}\\ &= \frac{\exp[\alpha + \log(n_1/n_2) + \boldsymbol{\beta}'\mathbf{x}]}{\exp[\alpha + \log(n_1/n_2) + \boldsymbol{\beta}'\mathbf{x}] + 1}\end{aligned}$$

and

$$\tilde{q}_2(\mathbf{x}) = 1 - \tilde{q}_1(\mathbf{x}).$$

We note that $\tilde{q}_1(\mathbf{x})$ is the same as $q_1(\mathbf{x})$ of (6.67), except that $\alpha_0 = \alpha + \log(\pi_1/\pi_2)$ is replaced by $\alpha + \log(n_1/n_2)$. Hence, with a correct interpretation of α_0, maximizing L'_C is equivalent to maximizing L_C. We have therefore reduced the problem of maximizing L_S subject to (6.75) and (6.76) to maximizing L_C again. Prentice and Pyke [1979] claim that the restriction to discrete variables can be dropped, as the above estimates will still have satisfactory properties for continuous variables. However, Anderson and Blair [1982] show that for continuous variables the estimates are no longer technically maximum likelihood and suggest an alternative method called penalized maximum likelihood estimation.

6.4.3 *Computations*

We have shown above that the appropriate likelihood function for all three sampling schemes and discrete or continuous data is

$$L_C = \prod_{i=1}^{2} \prod_{j=1}^{m} \{q_i(\mathbf{x}_j)\}^{n_i(\mathbf{x}_j)}, \tag{6.78}$$

where

$$q_1(\mathbf{x}) = \frac{\exp(\beta_0 + \boldsymbol{\beta}'\mathbf{x})}{1 + \exp(\beta_0 + \boldsymbol{\beta}'\mathbf{x})}$$

and

$$q_2(\mathbf{x}) = \frac{1}{1 + \exp(\beta_0 + \boldsymbol{\beta}'\mathbf{x})}.$$

The role of β_0 is not the same for the three sampling schemes. In conditional and mixture sampling, $\beta_0 = \alpha_0$ $[= \alpha + \log(\pi_1/\pi_2)]$. However, in separate sampling, $\beta_0 = \alpha + \log(n_1/n_2) = \alpha_0 + \log(n_1\pi_2/n_2\pi_1)$ and α_0 cannot be estimated from the estimate of β_0 unless π_1 and π_2 are known or have independent estimates.

Now

$$\begin{aligned}\log L_C &= \sum_{i=1}^{2} \sum_{j=1}^{m} n_i(\mathbf{x}_j) \log q_i(\mathbf{x}_j) \\ &= \sum_j \left\{ n_1(\mathbf{x}_j)\left[\beta_0 + \boldsymbol{\beta}'\mathbf{x}_j\right] + n(\mathbf{x}_j) \log q_2(\mathbf{x}_j) \right\},\end{aligned}$$

and the maximum likelihood equations are

$$\frac{\partial \log L_C}{\partial \beta_k} = \sum_j \{n_1(\mathbf{x}_j) - n(\mathbf{x}_j)q_1(\mathbf{x}_j)\}x_{jk} = 0 \qquad (k = 0,1,\dots,d),$$

where x_{jk} is the kth element of $\mathbf{x}_j$ ($x_{j0} \equiv 1$). We also have

$$\frac{-\partial^2 \log L_C}{\partial \beta_k \, \partial \beta_l} = \sum_j n(\mathbf{x}_j)q_1(\mathbf{x}_j)q_2(\mathbf{x}_j)x_{jk}x_{jl} \qquad (k, l = 0,1,\dots,d)$$

$$= a_{kl}, \tag{6.79}$$

say. Day and Kerridge [1967] and Anderson [1972] originally suggested using the Newton–Raphson procedure for maximizing L_C, though the procedure may not converge (Jones [1975]). However, Anderson [1982] noted that the quasi-Newton methods (see Gill and Murray [1972]) have desirable properties with regard to starting values, speed of convergence, and, given sufficient iterations (tentatively, no less than $d + 1$), asymptotic variance estimation. The starting values $\beta_0 = 0$ and $\boldsymbol{\beta} = \mathbf{0}$ proposed by Anderson [1972] work well in practice, and convergence to the maximum is rapid as L_C has a unique maximum for finite $\boldsymbol{\beta}$, except under two easily recognized circumstances (Albert [1978], Anderson [1982]): (1) complete separation in which all the sample points from G_1 and G_2 lie on opposite sides, respectively, of a hyperplane and (2) zero marginal proportions occurring with discrete data. If there is complete separation, suppose that the m sample values are labeled so that $\mathbf{x}_1, \mathbf{x}_2, \dots, \mathbf{x}_{m_1}$ belong to G_1 and $\mathbf{x}_{m_1+1}, \dots, \mathbf{x}_m$ belong to G_2. Now a family of parallel hyperplanes perpendicular to the unit vector $\boldsymbol{\gamma}$ can be represented by $\boldsymbol{\gamma}'\mathbf{x} = c$, where c, the perpendicular distance from the origin to a plane, varies. With complete separation we can find a separating hyperplane $h(\mathbf{x}) = \boldsymbol{\gamma}'\mathbf{x} - c_0 = 0$ such that all the points from G_1 lie on one side [$h(\mathbf{x}_j) > 0$, $j = 1, 2, \dots, m_1$] and all the points from G_2 lie on the other side [$h(\mathbf{x}_j) < 0$, $j = m_1 + 1, \dots, m$]. Taking $\beta_0 = -kc_0$ and $\boldsymbol{\beta} = k\boldsymbol{\gamma}$ in (6.78), we have

$$L_C = \prod_{j=1}^{m_1} \left[1 + \exp\left(-\beta_0 - \boldsymbol{\beta}'\mathbf{x}_j\right)\right]^{-n_1(\mathbf{x}_j)} \prod_{j=m_1+1}^{m} \left[1 + \exp\left(\beta_0 + \boldsymbol{\beta}'\mathbf{x}_j\right)\right]^{-n_2(\mathbf{x}_j)}$$

$$= \prod_{j=1}^{m} \left\{1 + \exp\left[-kh(\mathbf{x}_j)\right]\right\}^{-n_1(\mathbf{x}_j)} \prod_{j=m_1+1}^{m} \left\{1 + \exp\left[kh(\mathbf{x}_j)\right]\right\}^{-n_2(\mathbf{x}_j)}$$

$$\to 1, \qquad \text{as } k \to \infty. \tag{6.80}$$

There are other limiting values of β_0 and $\boldsymbol{\beta}$ that give the same upper bound of 1 for L_C, so that L_C does not have its maximum at a unique point. Although this nonunique solution indicates that β_0 and $\boldsymbol{\beta}$ cannot be estimated with much precision, Anderson [1972] noted that any separating hyperplane should give a reasonably good discrimination rule, particularly if the n_i are large, since all the given sample observations are correctly allocated. Fortunately, as noted by Day and Kerridge [1967: p. 320], the iterative procedures described above for maximizing L_C must at some stage produce a solution $\beta_0^{(r)}$, $\boldsymbol{\beta}^{(r)}$ such that $g_r(\mathbf{x}) = \beta_0^{(r)} + \boldsymbol{\beta}^{(r)\prime}\mathbf{x}$ is a separating hyperplane. To see this, suppose, for example, that $\mathbf{x}_1$ lies on the G_2 side of the plane $h(\mathbf{x}) = 0$. Then $h(\mathbf{x}_1) < 0$,

$$\{1 + \exp[-kh(\mathbf{x}_1)]\}^{-1} \le \tfrac{1}{2}$$

and, since all the factors $q_i(\mathbf{x}_j)$ in L_C are less than unity, $L_C \le \frac{1}{2}$. Hence if there is at least one point on the wrong side of the plane, then $L_C \le \frac{1}{2}$; $h(\mathbf{x}) = 0$ will be separating if $L_C > \frac{1}{2}$. Since $L_C = 1$ is the maximum value, it follows that any convergent process for maximizing the likelihood must eventually produce a discriminant $g_r(\mathbf{x})$ that separates the two groups. The first discriminant to achieve this will have a likelihood L_C that is greater than that of any other discriminant which does not completely separate the data. It is easy to check whether $g_r(\mathbf{x})$ is separating or not, as the values $g_r(\mathbf{x}_j)$ are required at each step of the iteration.

The problem of zero marginal proportions can be illustrated as follows (J. A. Anderson [1974]). Suppose x_1, the first element of $\mathbf{x}$, is dichotomous, taking values 0 or 1. Suppose further that in the sample data $x_1 = 0$ for all $\mathbf{x}$ from G_1 and $x_1 \ne 0$ for at least one $\mathbf{x}$ from G_2. Then the sample proportion for $x_1 = 1$ in G_1 is zero. With this data we have

$$\text{in } G_1: \beta_0 + \boldsymbol{\beta}'\mathbf{x} = \beta_0 + \beta_2 x_2 + \cdots + \beta_d x_d,$$

$$\text{in } G_2: \beta_0 + \boldsymbol{\beta}'\mathbf{x} = \beta_0 + \beta_1 x_1 + \cdots + \beta_d x_d.$$

Thus β_1 only appears in those terms of L_C for which $x_1 = 1$ in G_2, that is, in the terms $\{\exp(\beta_0 + \boldsymbol{\beta}'\mathbf{x}) + 1\}^{-1}$ with $x_1 = 1$. Without affecting any other terms of L_C, we can make β_1 arbitrarily large and negative so that the terms with β_1 tend to 1. The maximum likelihood estimate of β_1 is therefore $-\infty$. This is unsatisfactory, as we do not wish to estimate $\text{pr}[\mathbf{x} \in G_1 | \mathbf{x}]$ as zero (implying certainty about group membership) when $x_1 = 1$, and yet we wish to retain x_1 as a predictor, as it contains useful information about group membership. An ad hoc method for getting around this problem is described by J. A. Anderson [1974, 1982].

We noted that the iterative maximization of L_C is applicable to all three sampling procedures when it comes to finding the maximum likelihood estimates $\hat{\beta}_0$ and $\hat{\boldsymbol{\beta}}$. For mixture and conditional sampling, the asymptotic variances of the estimates are given by the diagonal elements of $\mathbf{A}^{-1}$, where $\mathbf{A} = [(a_{kl})]$ [see (6.79)] is the observation information matrix evaluated at the maximum likelihood point. In the case of separate sampling there are the additional constraints (6.75) and (6.76). However, Anderson [1972] showed that $\mathbf{A}^{-1}$ can still be used, except that the (1,1) element, the asymptotic variance of $\hat{\beta}_0$, is now in error by $O(n^{-1})$.

Although the maximum likelihood estimates are asymptotically unbiased, they can have substantial small-sample biases that will affect the plug-in estimate of the linear discriminant (6.65). Estimated corrections for bias are given by Anderson and Richardson [1979], and, on the basis of their simulation studies, they recommend a wider use of these corrections. McLachlan [1980b] also investigated the corrections and showed that they can be applied to separate as well as mixture sampling.

EXAMPLE 6.6 (Anderson [1972]) People who suffer from rheumatoid arthritis can also contract the eye disease *keratoconjunctivitis sicca* (KCS). Although the disease can be diagnosed reliably by an ophthalmic specialist, such a service is not routinely available for screening all rheumatoid patients. It was felt that logistic discrimination based on a straightforward checklist of symptoms could be used to enable medical staff of a rheumatic center who are not ophthalmic specialists to decide which rheumatoid arthritic patients should be referred to the eye hospital.

The diagnostic check list consisted of 10 symptoms of the presence or absence type, with $x_j = 1$ if the jth symptom is present and $x_j = 0$ if it is absent ($j = 1, 2, \ldots, 10$). The sample or training data, called set I, consisted of 37 rheumatoid patients without KCS (i.e., "normals" or group 1) and 40 with KCS (group 2). The estimated linear discriminant function is

$$\begin{aligned} z_1 &= \hat{\beta}_0 + \hat{\boldsymbol{\beta}}'\mathbf{x} \\ &= \hat{\beta}_0 + \hat{\beta}_1 x_1 + \cdots + \hat{\beta}_{10} x_{10} \\ &= 4.0 - 4.4x_1 - 2.1x_2 - 1.1x_3 - 4.7x_4 - 3.5x_5 \\ &\quad -0.8x_6 + 0.8x_7 - 2.4x_8 + 1.8x_9 - 0.9x_{10}. \end{aligned}$$

The z_1 scores of all the patients were calculated using this equation and are shown in Fig. 6.3*a*. We note that our assignment rule for this separate sampling experiment is Assign to the normal group if

$$z_1 > \log\left(\frac{n_1\pi_2}{n_2\pi_1}\right) = \log\left(\frac{0.925\pi_2}{\pi_1}\right) \qquad (= c_1, \text{ say}).$$

As doubts about diagnosis occur when a patient's z_1 score is small, it was decided to give a patient a queried diagnosis if his score was in the range -2 to 2. This range for c_1 corresponds to the range (0.15, 7.99) for π_2/π_1 and leads to the following diagnostic system:

$$z_1 \geq 2 : \text{call patient normal,}$$

$$-2 < z_1 < 2 : \text{query diagnosis,}$$

$$z_1 \leq -2 : \text{diagnose KCS.}$$

To test the method, the 10 symptoms were then observed on a further set of 41 patients, series II, which included 17 normals and 24 cases of KCS. The z_1 scores were again calculated for these patients, using the same discriminant function, and are shown in Fig. 6.3*b*. We see that the scores for the two series are quite comparable.

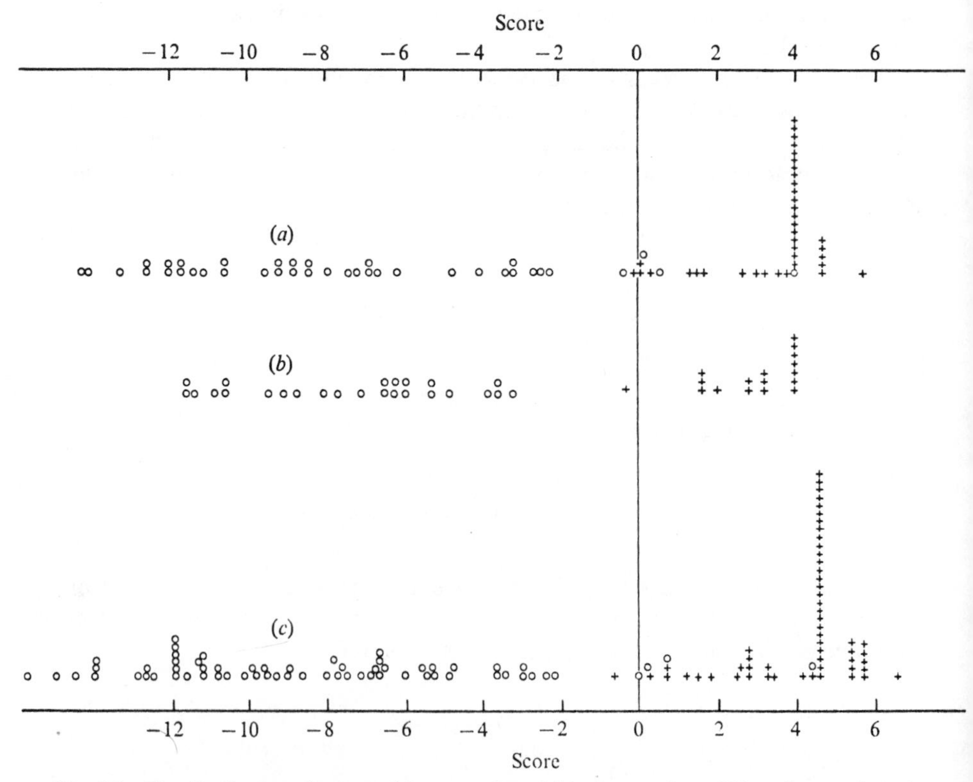

Fig. 6.3 The distribution of scores of keratoconjunctivitis sicca patients (O) and normals (+). (*a*) Series I patients estimated from series I. (*b*) Series II patients estimated from series I. (*c*) Series I + II patients estimated from series I + II. From Anderson [1972], by permission of the Biometrika Trustees.

For applying the diagnostic method to further patients, the two sets of data, series I and II, were combined to give the linear discriminant

$$z_2 = 4.7 - 5.2x_1 - 3.0x_2 - 1.3x_3 - 5.4x_4 - 4.0x_5$$

$$+1.1x_6 + 0.8x_7 - 1.9x_8 + 2.1x_9 - 2.0x_{10}.$$

The scores of all the patients in series I and II were calculated using z_2, and these are shown in Fig. 6.3*c*. A summary of all the results illustrated in Fig. 6.3 are given in Table 6.2, and we conclude from these that the results of diagnosing series I and II patients using z_2 are very similar to the results from using z_1. Also, the functions for z_1 and z_2 are very alike, so that, overall, the proposed diagnostic procedure is stable and repeatable. In addition, the error and query rates are at acceptable levels.

6.4.4 LGD Versus LDF

In Section 6.3.2c we saw that Fisher's linear discriminant function (LDF) has certain robust properties and we now ask how the logistic discriminant (LGD), $\hat{\beta}_0 + \hat{\boldsymbol{\beta}}'\mathbf{x}$, compares. Clearly, if the underlying distributions are normal with equal dispersion matrices, then the LDF will be asymptotically more efficient, as it is based on a "full" rather than a conditional likelihood procedure. This was demonstrated by Efron [1975], who gave the following table of the asymptotic relative efficiency (ARE) of the LGD with respect to the LDF for different values of $\Delta = [(\boldsymbol{\mu}_1 - \boldsymbol{\mu}_2)'\boldsymbol{\Sigma}^{-1}(\boldsymbol{\mu}_1 - \boldsymbol{\mu}_2)]^{1/2}$ and $\pi_1 = \pi_2$ (the case

TABLE 6.2 Evaluation of the Logistic Method of the Diagnosis of Kera-Conjunctivitis Sicca in Rheumatoid Arthritis[a]

	Discriminator estimated from Series I					
	Kerato-conjunctivitis sicca			No *kerato-conjunctivitis sicca*		
	Correct	Query	Wrong	Correct	Query	Wrong
Series I	36	3	1*	30	7	0
Series II	24	0	0	13	4	0
	Discriminator estimated from Series I and II					
	Kerato-conjunctivitis sicca			No *kerato-conjunctivitis sicca*		
	Correct	Query	Wrong	Correct	Query	Wrong
Series I+II	60	3	1*	47	7	0

* This patient had no symptoms.

[a]From Anderson [1972: Table 3], by permission of the Biometrika Trustees.

most favorable to the LGD):

Δ	0	0.5	1	1.5	2	2.5	3	3.5	4
ARE	1.000	1.000	0.995	0.968	0.899	0.786	0.641	0.486	0.343

He noted that just when good discrimination becomes possible ($2.5 \le \Delta \le 3.5$), the ARE of the logistic procedure falls off sharply; there is a similar decline in a measure describing the asymptotic relative accuracy of the actual error rates. O'Neill [1980] extended the asymptotic results of Efron from normal distributions to an exponential family using plug-in maximum likelihood estimates in the resulting linear discriminants, and demonstrated a similar loss of asymptotic efficiency. However, this is not the whole story. In the first place, the situation is different for finite samples, as the maximum likelihood estimates for both procedures are biased. Secondly, if the wrong distribution is assumed, then the maximum likelihood estimates will be inconsistent and the resulting estimated LDF will also be inconsistent. Thirdly, for independent binary data, different population means implies different dispersion matrices, and the relevance of the LDF approach is then an open question. Press and Wilson [1978] presented two studies involving nonnormal data and found that the apparent error rates, e_{app} of (6.26), were smaller for the LGD than the LDF, thus supporting the results of Halperin et al. [1971]. The differences were not great and the authors felt that the two methods would not generally give markedly different results. Krzanowski [1975: Table 3] found that both methods yielded almost identical results for several sets of mixed binary–continuous data. In a simulation study carried out at Auckland University, Crawley [1979] showed that (1) for normal distributions with equal dispersion matrices, the performance of the LGD, in terms of apparent error rates, is very comparable with the LDF; (2) for normal data with unequal dispersion matrices, the LGD is slightly better than the LDF, and the difference in the apparent error rates $e_{i,\text{app}}$ ($i = 1, 2$) for the LGD is much less, particularly with increasing covariance differences; and (3) for nonnormal data the LGD is much better than the LDF with respect to e_{app} and the difference in the $e_{i,\text{app}}$. Crawley concluded that the LGD is preferred when the distributions are clearly nonnormal or the dispersion matrices are clearly unequal. The results also suggest that when using the LDF, widely different rates of misclassification are an indication that the LGD is more appropriate. Further support for the LGD comes from Byth and McLachlan [1980], and O'Hara et al. [1982].

6.4.5 *Predictive Logistic Model*

Writing $\boldsymbol{\theta}' = (\alpha_0, \boldsymbol{\beta}')$ and $\mathbf{y}' = (1, \mathbf{x}')$, we have, from (6.66) and (6.67),

$$q_1(\mathbf{x}|\boldsymbol{\theta}) = \frac{\exp(\boldsymbol{\theta}'\mathbf{y})}{\exp(\boldsymbol{\theta}'\mathbf{y}) + 1}. \tag{6.81}$$

Suppose $\mathbf{z}$ represents the sample data and $\hat{\boldsymbol{\theta}} = \hat{\boldsymbol{\theta}}(\mathbf{z})$ is the maximum likelihood estimate of $\boldsymbol{\theta}$ with estimated approximate dispersion matrix $\hat{\mathbf{V}}$. Then, with the Bayesian version of the maximum likelihood theory, $\boldsymbol{\theta}$ is approximately $N_{d+1}(\hat{\boldsymbol{\theta}}, \hat{\mathbf{V}})$. With this prior distribution for $\boldsymbol{\theta}$, we can proceed as in (6.36) and obtain

$$q_1(\mathbf{x}|\mathbf{z}) = \int_{\Theta} q_1(\mathbf{x}|\boldsymbol{\theta}) g(\boldsymbol{\theta}|\mathbf{z})\, d\boldsymbol{\theta},$$

where g is the density function for $N_{d+1}(\hat{\boldsymbol{\theta}}, \hat{\mathbf{V}})$. Using an approximation $q_1(\mathbf{x}|\boldsymbol{\theta}) \approx \Phi(\boldsymbol{\theta}'\mathbf{y}/\sqrt{b})$ due to Aitchison and Begg [1976], with $b = 2.942$ and Φ the distribution function for the $N_1(0,1)$ distribution, Aitken [1978] showed that

$$q_1(\mathbf{x}|\mathbf{z}) \approx \Phi\left(\hat{\boldsymbol{\theta}}'\mathbf{y}\{b + \mathbf{y}'\hat{\mathbf{V}}\mathbf{y}\}^{-1/2}\right).$$

Other values of b can also be used (e.g., Lauder [1978: $b = k^2$]). Instead of the LGD

$$\log\left\{\frac{q_1(\mathbf{x}|\hat{\boldsymbol{\theta}})}{1 - q_1(\mathbf{x}|\hat{\boldsymbol{\theta}})}\right\} = \hat{\boldsymbol{\theta}}'\mathbf{y} = \hat{\alpha}_0 + \hat{\boldsymbol{\beta}}'\mathbf{x},$$

we can use the predictive logistic discriminant function $\log\{q_1(\mathbf{x}|\mathbf{z})/[1 - q_1(\mathbf{x}|\mathbf{z})]\}$. This method was compared with several others for the case of multivariate binary data by Aitken [1978], and it has been used for handling imprecise data (Aitchison and Begg [1976], Lauder [1978]).

6.4.6 *Quadratic Discrimination*

For the case of two normal distributions $N_d(\boldsymbol{\mu}_i, \boldsymbol{\Sigma}_i)$ $(i = 1, 2)$ we see from (6.13) that

$$\begin{aligned}\log[q_1(\mathbf{x})/q_2(\mathbf{x})] &= \log(\pi_1/\pi_2) + c_0 + \mathbf{x}'(\boldsymbol{\Sigma}_1^{-1}\boldsymbol{\mu}_1 - \boldsymbol{\Sigma}_2^{-1}\boldsymbol{\mu}_2) \\ &\quad + \tfrac{1}{2}\mathbf{x}'(\boldsymbol{\Sigma}_1^{-1} - \boldsymbol{\Sigma}_2^{-1})\mathbf{x} \\ &= \delta_0 + \boldsymbol{\delta}'\mathbf{x} + \mathbf{x}'\boldsymbol{\Gamma}\mathbf{x},\end{aligned}$$

say, where $\boldsymbol{\Gamma} = [(\gamma_{rs})]$ is symmetric. The above function is linear in the coefficients δ_0, $\boldsymbol{\delta}$, and γ_{rs} $(r \le s)$ so that it can be written in the form $\beta_0 + \boldsymbol{\beta}'\mathbf{y}$ with $\mathbf{y} = (x_1, x_2, \dots, x_d, x_1^2, x_1x_2, x_2^2, \dots)'$ and $1 + d + \frac{1}{2}d(d+1)$ β's to be estimated. In principle, these parameters can be estimated as before, using logistic discrimination, but in practice there are too many parameters to estimate iteratively for $d > 4$. Anderson [1975] suggested several approximations for $\mathbf{x}'\boldsymbol{\Gamma}\mathbf{x}$, the simplest being the rank 1 approximation $\boldsymbol{\Gamma} \approx \lambda_1 \mathbf{t}_1 \mathbf{t}_1'$, where λ_1 is the largest eigenvalue of $\boldsymbol{\Gamma}$ and $\mathbf{t}_1$ the corresponding unit eigenvector (A7.3). In this case the logistic model is

$$q_1(\mathbf{x}) = 1 - q_2(\mathbf{x}) = \frac{e^q}{1 + e^q},$$

where $q = \delta_0 + \boldsymbol{\delta}'\mathbf{x} + \lambda_1(\mathbf{t}_1'\mathbf{x})^2$. Since $\mathbf{t}_1'\mathbf{t}_1 = 1$, the number of parameters has now been reduced to $1 + 2d$. Although q is nonlinear in $\mathbf{t}_1$, the parameters can be estimated once again by maximizing L_C of (6.78) using another quasi-Newton iterative procedure.

6.5 TWO GROUPS: UNKNOWN DISTRIBUTIONS

6.5.1 *Kernel Method*

If the form of $f_i(\mathbf{x})$ is unknown, then it must be estimated directly from the sample data $\mathbf{x}_{ij}$ ($i = 1, 2$; $j = 1, 2, \ldots, n_i$). We note from Glick [1972] that if $\hat{f}_i(\mathbf{x})$ and $\hat{\pi}_i$ are consistent estimates of $f_i(\mathbf{x})$ and π_i, respectively, and $\int[\hat{\pi}_1 \hat{f}_1(\mathbf{x}) + \hat{\pi}_2 \hat{f}_2(\mathbf{x})]\, d\mathbf{x} \to 1$ as $n_i \to \infty$ ($i = 1, 2$), then e_{act} and $\hat{e}_{\text{act}}$ both tend to e_{opt} (in a probabilistic sense; see Section 6.3.1 for definitions). Methods are described below that, under fairly general conditions, give consistent estimates of $f_i(\mathbf{x})$.

To simplify the discussion we shall drop the suffix i denoting group membership and consider the problem of estimating $f(\mathbf{x})$ from a sample $\mathbf{x}_1, \mathbf{x}_2, \ldots, \mathbf{x}_n$. Recently Habbema et al. [1974a], Hermans and Habbema [1975], and Aitchison and Aitken [1976] have used the potential function (kernel) method of multivariate density estimation (Cacoullos [1966], Murthy [1966], Breiman et al. [1977], and the review of Fryer [1977]; see also Susarla and Walter [1981] for a more general approach). Here $f(\mathbf{x})$ is estimated by

$$g(\mathbf{x}|\lambda) = \frac{1}{n} \sum_{j=1}^{n} K(\mathbf{x}|\mathbf{x}_j, \lambda),$$

where $K(\mathbf{y}|\mathbf{z}, \lambda)$ is a kernel probability or probability density function on $\mathbf{y}$ with mode at $\mathbf{z}$, and a smoothing parameter λ whose value depends on the group G_i.

a CONTINUOUS DATA

A popular choice for K is the MVN distribution, as it is unimodal and has convenient scale properties. One kernel density is therefore

$$K(\mathbf{y}|\mathbf{z}, \lambda) = (2\pi\lambda^2)^{-d/2}|\mathbf{S}|^{-1/2}\exp\left\{\frac{1}{2\lambda^2}(\mathbf{y} - \mathbf{z})'\mathbf{S}^{-1}(\mathbf{y} - \mathbf{z})\right\}, \quad (6.82)$$

where $\mathbf{S} = \Sigma_j(\mathbf{x}_j - \overline{\mathbf{x}})(\mathbf{x}_j - \overline{\mathbf{x}})'/(n-1)$. Habbema et al. [1974a] used diag(**S**), the matrix of diagonal elements, rather than **S** itself. If we write $\mathbf{y} = (y_1, y_2, \dots, y_d)'$, $\mathbf{x}_j = (x_{j1}, x_{j2}, \dots, x_{jd})'$, and so on, their kernel is

$$K(\mathbf{y}|\mathbf{z}, \lambda) = (2\pi\lambda^2)^{-d/2}\left(\prod_{k=1}^{d} s_k\right)^{-1} \exp\left\{-\frac{1}{2}\sum_{k=1}^{d}\left(\frac{y_k - z_k}{\lambda s_k}\right)^2\right\}, \qquad (6.83)$$

where $s_k^2 = \Sigma_{j=1}^n (x_{jk} - \overline{x}_{\cdot k})^2/(n-1)$. Sometimes it is convenient to work in terms of the scaled data $\tilde{y}_k = y_k/s_k$, $\tilde{z}_k = z_k/s_k$, with corresponding kernel

$$\tilde{K}(\tilde{\mathbf{y}}|\tilde{\mathbf{z}}, \lambda) = (2\pi\lambda^2)^{-d/2}\exp\left\{-\frac{1}{2\lambda^2}(\tilde{\mathbf{y}} - \tilde{\mathbf{z}})'(\tilde{\mathbf{y}} - \tilde{\mathbf{z}})\right\}. \qquad (6.84)$$

If λ is estimated by $\hat{\lambda}$, say, $f(\mathbf{x})$ can be estimated by

$$\begin{aligned}\hat{f}(\mathbf{x}) &= g(\mathbf{x}|\hat{\lambda}) \\ &= \frac{1}{n}\sum_{j=1}^{n} K(\mathbf{x}|\mathbf{x}_j, \hat{\lambda}),\end{aligned}$$

which will be consistent at all continuity points $\mathbf{x}$ if $\hat{\lambda} \to 0$ sufficiently slowly as $n \to \infty$ [see Murthy [1966] with $B_{kn} = (s_k\hat{\lambda})^{-1}$ and $(B_{1n}B_{2n}\cdots B_{dn})/n \to 0$]. When $\mathbf{x}$ is one of the observations, say, $\mathbf{x}_r$, we can estimate $f(\mathbf{x}_r)$ by

$$\hat{f}^{(r)}(\mathbf{x}_r) = \frac{1}{n-1}\sum_{j=1,\, j\neq r}^{n} K(\mathbf{x}_r|\mathbf{x}_j, \hat{\lambda}). \qquad (6.85)$$

As the maximum likelihood estimate of λ for the likelihood $\Pi_j g(\mathbf{x}_j|\lambda)$ turns out to be zero, Habbema et al. [1974a] suggested a "leaving-one-out" modification of the maximum likelihood method. Working with the scaled data, they maximized

$$h(\lambda) = \prod_{r=1}^{n} h^{(r)}(\tilde{\mathbf{x}}_r) \qquad (\lambda > 0)$$

with respect to λ, where

$$h^{(r)}(\tilde{\mathbf{x}}_r) = \frac{1}{n-1}\sum_{j=1,\, j\neq r}^{n} \tilde{K}(\tilde{\mathbf{x}}_r|\tilde{\mathbf{x}}_j, \lambda).$$

The above method can be used for each group G_i to obtain estimates $\hat{f}_i(\mathbf{x})$ $(i = 1, 2)$ and the rule Assign to G_1 if $\hat{f}_1(\mathbf{x})/\hat{f}_2(\mathbf{x}) > \pi_2/\pi_1$. Habbema et al. [1974a] used a forward selection procedure for choosing a subset of the x-variables (see Section 6.10), though it is more suited to small-scale problems, as it is rather time-consuming on the computer. For this reason the authors recommended using the estimate of λ based on the full set of variables at each stage. The above kernel method has been used by Habbema [1976] for classifying human chromosomes; a useful review of this latter topic is given by Habbema [1979]. Hermans and Habbema [1975] concluded for the two medical examples they investigated that the kernel and normal based LDF procedures gave similar results for normal data; otherwise the kernel method was better. Although a MVN kernel has been used above, it should be clear that we are not making any assumptions about $f(\mathbf{x})$ itself: Any other unimodel density could be used as a kernel.

b BINARY DATA

For multivariate binary data, Aitchison and Aitken [1976] suggested using the kernel

$$K(\mathbf{y}|\mathbf{z},\lambda) = \lambda^{d-D(\mathbf{y},\mathbf{z})}(1-\lambda)^{D(\mathbf{y},\mathbf{z})} \tag{6.86}$$

where $\frac{1}{2} \le \lambda \le 1$ and $D(\mathbf{y},\mathbf{z}) = \|\mathbf{y}-\mathbf{z}\|^2$. With binary data the dissimilarity coefficient $D(\mathbf{y},\mathbf{z})$ is simply the number of disagreements in corresponding elements of $\mathbf{y}$ and $\mathbf{z}$. The "extreme" kernels are

$$K(\mathbf{y}|\mathbf{z},\tfrac{1}{2}) = (\tfrac{1}{2})^d$$

and

$$K(\mathbf{y}|\mathbf{z},1) = \begin{cases} 1, & \mathbf{y}=\mathbf{z}, \\ 0, & \mathbf{y}\neq\mathbf{z}. \end{cases}$$

which represent estimating $f(\mathbf{x})$ by a uniform distribution and the relative frequency $n(\mathbf{x})/n$, respectively. As in the continuous case, the maximum likelihood estimate of λ based on $\prod_j g(\mathbf{x}_j|\lambda)$ is unsatisfactory ($\hat{\lambda}=1$) and we can use the leaving-one-out modification based on the likelihood $\prod_r f^{(r)}(\mathbf{x}_r)$ [see (6.85)]. The authors noted that, using the transformation $w_k = |y_k - z_k|$ $(k=1,2,\ldots,d)$ from $\mathbf{y}$ to $\mathbf{w}$ with fixed $\mathbf{z}$, we can write

$$K(\mathbf{y}|\mathbf{z},\lambda) = \prod_{k=1}^{d} \lambda^{1-w_k}(1-\lambda)^{w_k}, \tag{6.87}$$

which is easier to handle. They gave an outline proof that $\hat{f}(\mathbf{x})$ [$= g(\mathbf{x}|\hat{\lambda})$] is a consistent estimator of $f(\mathbf{x})$ for all $\mathbf{x}$, and noted that atypicality indices can be readily calculated [see (6.37) with a summation instead of an integral]. Studies by Aitken [1978] and Remme et al. [1980] show that the kernel method compares favorably with other methods. For a further discussion on kernel-based estimates, see Titterington [1980] and Titterington et al. [1981]. The problem of missing data is considered by Titterington [1977].

Unfortunately the leaving-one-out estimate of λ can behave erratically, even for large samples, and it is strongly influenced by the presence of empty or nearly empty multinomial cells. Hall [1981] demonstrated the problem theoretically and by example, and proposed an alternative estimate of λ.

c CONTINUOUS AND DISCRETE DATA

If $\mathbf{y}' = (\mathbf{y}^{(1)\prime}, \mathbf{y}^{(2)\prime})$ is a mixture of continuous variables $\mathbf{y}^{(1)}$ and binary variables $\mathbf{y}^{(2)}$, Aitchison and Aitken [1976] suggested using a product of kernel functions, say,

$$K(\mathbf{y}|\mathbf{z}, \lambda_1, \lambda_2) = K_1(\mathbf{y}^{(1)}|\mathbf{z}^{(1)}, \lambda_1) K_2(\mathbf{y}^{(2)}|\mathbf{z}^{(2)}, \lambda_2),$$

where K_1 is the normal kernel (6.82) and K_2 is the binomial kernel (6.86). Then $f(\mathbf{x})$ can be estimated by

$$\hat{f}(\mathbf{x}) = \frac{1}{n} \sum_{j=1}^{n} K_1(\mathbf{x}^{(1)}|\mathbf{x}_j^{(1)}, \hat{\lambda}_1) K_2(\mathbf{x}^{(2)}|\mathbf{x}_j^{(2)}, \hat{\lambda}_2),$$

where $\hat{\lambda}_1$ and $\hat{\lambda}_2$ are, for example, the leaving-one-out maximum likelihood estimates. The factorization of the kernel in no way implies independence of the continuous and binary components.

Discrete data with more than two categories can either be converted to binary data [see (6.61)], or else modeled using an extension of the binomial kernel method: Aitchison and Aitken [1976] suggested two such extensions. Kernel methods can also be adapted to handle missing data and unclassified observations, the so-called mixture problem (Murray and Titterington [1978]). A "variable" (adaptive) kernel method that seems to work better for skewed distributions (e.g., lognormal) has been proposed (Breiman et al. [1977], Habbema et al. [1978], Remme et al. [1980: p. 103].

6.5.2 *Other Nonparametric Methods*

a NEAREST NEIGHBOR TECHNIQUES

The first nonparametric method of classification was the nearest neighbor rule of Fix and Hodges [1951]. Suppose we have observations $\mathbf{x}_{ij}$, $j = 1, 2, \ldots, n_i$, from G_i $(i = 1, 2)$. If $\mathbf{x}$ is to be assigned, we order the $n_1 + n_2$ observations using a distance function $D(\mathbf{x}, \mathbf{x}_{ij})$. We choose some integer K and let K_i be the number of observations from G_i in the K closest observations to $\mathbf{x}$. Then we assign $\mathbf{x}$ to G_1 if

$$\frac{K_1}{n_1} \bigg/ \frac{K_2}{n_2} > \frac{\pi_2}{\pi_1}. \tag{6.88}$$

For further references see Lachenbruch [1975a: p. 57].

Hills [1967] applied a similar technique to binary data using the metric $D(\mathbf{x},\mathbf{y}) = \|\mathbf{x} - \mathbf{y}\|^2$, the number of disagreements between $\mathbf{x}$ and $\mathbf{y}$. If $D(\mathbf{x},\mathbf{y}) = h$, then $\mathbf{y}$ is called a near neighbor to $\mathbf{x}$ of order h. For example, 1100, 1010, and 0000 are all near neighbors to 1000, of order 1. We can use the assignment rule (6.88), but with K_i equal to the number of sample observations from G_i such that $D(\mathbf{x},\mathbf{x}_{ij}) \le h$. In this case K_i/n_i is not a consistent estimator of $f_i(\mathbf{x})$ unless $h = 0$, the multinomial situation described in Section 6.3.3b. A difficulty with this method is the choice of h so that $K_i \neq 0$ for a given $\mathbf{x}$. A nearest neighbor approach for estimating scale factors in the kernel method has been proposed by Breiman et al. [1977].

b PARTITIONING METHODS

A method similar in spirit to the nearest neighbor technique is the use of "statistically equivalent blocks" (Anderson [1966], Gessaman [1970], Gessaman and Gessaman [1972]). The sample from G_1 is ordered on the basis of the first variable and is then divided into g approximately equal groups. Within each group the observations are ordered on the basis of the second variable and the group is divided into g approximately equal groups. The process is repeated for $k \le d$ of the variables so that there are g^k groups or "blocks" with approximately the same number of observations in each block. Suppose that an observation $\mathbf{x}$ falls into block j and there are n_{ij} observations from G_i in this block. Then we assign to G_1 if

$$\frac{n_{1j}/n_1}{n_{2j}/n_2} > \frac{\pi_2}{\pi_1}. \tag{6.89}$$

The ordering can also be carried out on the first k principal components of each observation. A different procedure is generally obtained if the sample from G_2 is used to obtain the blocks.

Another partitioning method that shows promise because of its generality and asymptotic optimality has been proposed by Gordon and Olshen [1978].

c DISTANCE METHODS

Nearest neighbor methods are one way of measuring the closeness of $\mathbf{x}$ to G_i. Another approach is to calculate the distance between the sample distribution functions of the two groups for two cases, the first including $\mathbf{x}$ in G_1, and the second including $\mathbf{x}$ in G_2. We assign $\mathbf{x}$ to the group that gives the greater distance. For example, Dillon and Goldstein [1978] used a distance method proposed by Matusita (see Matusita [1964] for references) for handling multinomial data. The problem with distance methods is to find a suitable distance function.

d RANK PROCEDURES

Given a discriminant function $D(\mathbf{x})$, it is possible to determine an assignment rule based only on ranked values of D. Suppose that D treats the observations from the same group symmetrically, that is, D is invariant under permutations of $\mathbf{x}_{11}, \mathbf{x}_{12}, \ldots, \mathbf{x}_{1n_1}$ from G_1 and permutations of $\mathbf{x}_{21}, \mathbf{x}_{22}, \ldots, \mathbf{x}_{2n_2}$ from G_2. We can assume that D gives larger values for observations from G_1 than from G_2. Then assuming, for the moment, that $\mathbf{x} \in G_1$, we can compute D, say $D_1(\cdot)$, using the $n_1 + 1$ observations from G_1 and the n_2 observations from G_2. Let $R_1(\mathbf{x})$ be the rank of $D_1(\mathbf{x})$ among $D_1(\mathbf{x}_{11}), \ldots, D_1(\mathbf{x}_{1n_1}), D_1(\mathbf{x})$, ranked from the smallest to the largest. Then Broffitt et al. [1976] showed that for $\mathbf{x}$ continuous, $R_1(\mathbf{x})$ will have a uniform distribution over the integers $1, 2, \ldots, n_1 + 1$. A small rank would indicate that $\mathbf{x}$ looks more like a member of G_2. The probability of obtaining a value at least as large as $R_1(\mathbf{x})$ is $P_1(\mathbf{x}) = R_1(\mathbf{x})/(n_1 + 1)$. We can repeat the process by assuming $\mathbf{x}$ now comes from G_2 and calculating a corresponding discriminant $D_2(\cdot)$. Let $R_2(\mathbf{x})$ be the rank of $-D_2(\mathbf{x})$ among $-D_2(\mathbf{x}_{21}), \ldots, -D_2(\mathbf{x}_{2n_2}), -D_2(\mathbf{x})$. A small rank makes $\mathbf{x}$ look more like a member of G_1. An assignment rule is therefore

$$
\begin{aligned}
&\text{Assign } \mathbf{x} \text{ to } G_1 \text{ if } P_1(\mathbf{x}) > P_2(\mathbf{x}), \\
&\text{Assign } \mathbf{x} \text{ to } G_2 \text{ if } P_2(\mathbf{x}) > P_1(\mathbf{x}), \qquad (6.90) \\
&\text{Use a nonrank procedure if } P_1(\mathbf{x}) = P_2(\mathbf{x}).
\end{aligned}
$$

Clearly *any* discriminant D can be used, provided that it is symmetric and $\mathbf{x}$ is continuous. A nice feature of the method is that the probabilities of misclassification $P(i|j)$ will be approximately equal, as they are asymptotically equal. The above discussion is based on Randles et al. [1978a, b] and the reader is referred to their papers for further details and extensions. The ranking procedure is considered further by Beckman and Johnson [1981].

Instead of forming the discriminant function and then using ranks, Conover and Iman [1980, 1981] proposed ranking the data first and then using the LDF or QDF of Section 6.3.2. The rank transformation consists of ranking the kth components, x_{ijk}, say, of all the observations $\mathbf{x}_{ij}$ from the smallest, with rank 1, to the largest, with rank $n = n_1 + n_2$. Each component is ranked separately for $k = 1, 2, \ldots, d$ and we replace x_{ijk} by its rank number. A similar transformation is applied to $\mathbf{x}$, the observation to be assigned. Each component x_k of $\mathbf{x}$ is replaced by a rank (actually a number obtained by linear interpolation between two adjacent ranks) representing its position in the n values of x_{ijk}. We now treat the vectors of ranks from each group as though they were multivariate normal and compute a linear or quadratic discriminant function. We assign $\mathbf{x}$, now replaced by a vector of ranks, to G_1 or G_2 on the basis of the discriminant function value.

Conover and Iman [1980] compared their ranking methods, called RLDF and RQDF, with several nonparametric techniques, including the partition and kernel procedures. They concluded that if the data are MVN, very little is lost by using the RLDF and RQDF methods instead of the LDF and QDF methods. When the data were nonnormal, the ranking methods were superior and they compared favorably with the other nonparametric methods.

A histogram method, also based on order statistics, is proposed by Chi and Van Ryzin [1977].

e SEQUENTIAL DISCRIMINATION ON THE VARIABLES

A sequential approach to classification is as follows. After each step we decide to either allocate our current **x** to one of the G_i or introduce another variable into **x** (Kendall and Stuart [1966], Kendall [1975]). Thus if $R = \log[f_2(x_1, x_2,\ldots,x_r)/f_1(x_1, x_2,\ldots,x_r)]$, our rule is

$$\begin{aligned} &\text{If } R < a, \text{ assign to } G_1; \\ &\text{If } R > b, \text{ assign to } G_2; \\ &\text{If } a \le R \le b, \text{ recalculate } R \text{ using } (x_1, x_2,\ldots,x_{r+1}). \end{aligned} \tag{6.91}$$

First we order the x_1 values of all the $n = n_1 + n_2$ sample points and choose a_1 and b_1 such that all observations with $x_1 < a$ belong to G_1, and all observations with $x_1 > b$ belong to G_2. For those observations with $a_1 \le x_1 \le b_1$ we go through the same procedure with x_2, choosing a_2 and b_2 such that the observations with $x_2 < a$ belong to G_1 and those with $x_2 > b_2$ belong to G_2. For the observations with $a_1 \le x_1 \le b_1$ and $a_2 \le x_2 \le b_2$ we proceed to x_3 and continue the process until all n observations are assigned to their correct groups, or we run out of variables. A natural ordering for the variables is to order the x_k according to the degree of overlap between the two groups, x_1 having the smallest overlap and therefore being the best one-variable discriminator.

Although the above technique is distribution free and easy to understand, it can leave a substantial proportion of the sample observations unclassified, so that there is a nonzero probability that a new observation **x** may not be classified. For example, Kendall [1975] applied the method to Fisher's famous iris data (50 four-dimensional observations from each of two species) and found that 13 of the 100 observations were unclassified. However, Richards [1972] suggested a refinement and an extension that led to a further classification of 10 of the 13 observations. The refinement consists of looking at all the variables at each stage, except the one used at the previous stage, and using the best one-variable discriminator. For example, x_1 might also be the best discriminator in terms of the smallest overlap instead of x_3 at stage 3. His extension consists of using two-variable discriminators based on the bivariate frequency distribution to classify observations not classified by the one-variable method. A related procedure for discrete or categorical data is given by Sturt [1981].

A general theory of sequential discrimination is given by Hora [1980].

6.6 UTILIZING UNCLASSIFIED OBSERVATIONS

In some situations unclassified observations (say, n_3 of them) are also available and the question arises as to whether these should be included in estimating a discriminant function. For example, Conn's syndrome manifests itself in two forms: either as a tumor (adenoma), which is treated by surgery, or as a certain condition (bilateral hyperplasia) for which there is a drug therapy. Since the only way to confirm a diagnosis is to operate to see if a tumor exists, those cases treated by drug therapy will remain unclassified.

An interesting (univariate) example from fisheries biology has been given by Hosmer [1973a, b]. The sex of the halibut can only be determined by dissection. However, it is much easier to measure the length of a fish, which, for a given age class, is related to sex. For example, 11-year-old males are, on the average, about 19 cm longer than 11-year-old females, though the length distributions overlap. The International Halibut Commission has two different sources of data: sex, age, and length data from research cruises, and age and length data only from commercial catches, as the fish have been cleaned by the time the boats have returned to port. For each age class, the length distribution for the combined sexes is closely approximated by a mixture of two normal distributions, and the problem is to estimate the various parameters (especially π_1, the proportion of males) as efficiently as possible using the classified data from research cruises and the larger body of unclassified data from commercial catches. Here the emphasis is more on estimation than discrimination, and Hosmer [1973a] gives iterative methods for three sampling schemes. For further details see Hosmer and Dick [1977].

In two papers McLachlan [1975, 1977a] considered the problem of unclassified observations for the case of MVN data with equal dispersion matrices, $\pi_1 = \pi_2$, and n_3 ($= m$ in his notation) large and small, respectively. As in the above fisheries example, a classified observation is often much more expensive to make than an unclassified one, so that a large sample of unclassified observations may contribute as much to the estimation of the LDF as a small sample of classified observations. For the case of small n_3, McLachlan [1977a] showed that the LDF based solely on classified data can be improved upon [i.e., $E[e_{act}]$ of (6.23) can be lowered] by using the LDF to classify the unclassified data and then constructing a new LDF based on an appropriate weighting of classified and unclassified data. O'Neill [1978] considered the more general case of unknown π_i and gave an iterative maximum likelihood method for estimating the LDF based on mixture sampling with part of the data classified. He concluded that if the populations are separate enough, namely, $2.5 \leq \Delta \leq 4$, then the information in an unclassified observation varies from about one-fifth to two-thirds that of a classified observation. Ganesalingam and McLachlan [1978] came to a similar conclusion in a univariate study: For $3 \leq \Delta \leq 4$ the ratio is about $\frac{1}{4}$ to $\frac{2}{3}$. Since the ratio of costs of classified to unclassified observations is frequently much greater than 5:1, the LDF could be estimated using a large number of inexpensive, unclassified observations rather than a small, correctly classified sample. However, whatever data are available, the message is clear: If there is a "reasonable" separation of the classified data, then the unclassified data should be used as well. This viewpoint is also supported by Titterington [1976], who considered the same problem, but for the predictive discriminant rather than the LDF. Ganesalingam and McLachlan [1981] discussed the estimation of π_1 and gave further references on the topic.

The above problem is also considered by Anderson [1979] for the less restrictive logistic model. Suppose we have n_i observations from G_i ($i = 1, 2$) and $n_3 = n - n_1 - n_2$ unclassified observations with corresponding density functions $f_i(\mathbf{x})$ and $f_3(\mathbf{x}) = \pi_1 f_1(\mathbf{x}) + \pi_2 f_2(\mathbf{x})$. Then, for the separate sampling scheme, the likelihood function is [see (6.72)]

$$L_{\text{SEP}} = \prod_{i=1}^{3} \prod_{j=1}^{m} \{ f_i(\mathbf{x}_j) \}^{n_i(\mathbf{x}_j)}, \tag{6.92}$$

where $f_1(\mathbf{x})/f_2(\mathbf{x}) = \exp(\alpha + \boldsymbol{\beta}'\mathbf{x})$. J. A. Anderson [1979] showed that for discrete data, maximizing L_{SEP} with respect to α, $\boldsymbol{\beta}$, and π_1 $(= 1 - \pi_2)$ subject to the constraints $\Sigma_{\mathbf{x}} f_i(\mathbf{x}) = 1$ ($i = 1, 2, 3$) is equivalent to maximizing

$$L^*_{\text{SEP}} = \prod_{j=1}^{m} \left\{ \frac{1}{n_1^*} q_1^*(\mathbf{x}_j) \right\}^{n_1(\mathbf{x}_j)} \left\{ \frac{1}{n_2^*} q_2^*(\mathbf{x}_j) \right\}^{n_2(\mathbf{x}_j)} \left\{ \frac{\pi_1}{n_1^*} q_1^*(\mathbf{x}_j) + \frac{\pi_2}{n_2} q_2^*(\mathbf{x}_j) \right\}^{n_3(\mathbf{x}_j)},$$

where

$$q_1^*(\mathbf{x}) = \frac{\exp\left[\alpha + \log\left(n_1^*/n_2^*\right) + \boldsymbol{\beta}'\mathbf{x}\right]}{\exp\left[\alpha + \log\left(n_1^*/n_2^*\right) + \boldsymbol{\beta}'\mathbf{x}\right] + 1},$$

$$q_2^*(\mathbf{x}) = 1 - q_1^*(\mathbf{x}),$$

and

$$n_i^* = n_i + \pi_i n_3 \qquad (i = 1, 2).$$

Once again a quasi-Newton procedure can be used. When $n_3 = 0$, L^*_{SEP} is proportional to L'_C of (6.77). J. A. Anderson [1979: p. 21] argued that the above method can also be applied to continuous data (or continuous and discrete data), but with a slight loss of efficiency.

6.7 ALL OBSERVATIONS UNCLASSIFIED (METHOD OF MIXTURES)

In the previous section we considered the possibility of having some of the sample observations unclassified. We now go a step further and suppose that we have mixture sampling but with all n observations unclassified. This mixture problem in which one endeavors to estimate the underlying parameters of the mixing distribution has a long history, though much of the research has been confined to mixtures of univariate normals (see the reviews of Macdonald [1975] and Odell and Basu [1976], and the paper by James [1978]) or multivariate normals (e.g., Day [1969], Wolfe [1970]). In the latter case it is assumed that we have a sample of n observations from the mixture density function

$$f(\mathbf{x}) = \sum_{i=1}^{2} \pi_i f_i(\mathbf{x}|\boldsymbol{\mu}_i, \boldsymbol{\Sigma}_i), \tag{6.93}$$

where the π_i are the unknown mixing proportions and f_i is the density function for $N_d(\boldsymbol{\mu}_i, \boldsymbol{\Sigma}_i)(i = 1, 2)$. Maximum likelihood estimates of π_1, $\boldsymbol{\mu}_i$, and $\boldsymbol{\Sigma}_i$ are obtained by maximizing $\prod_{j=1}^{n} f(\mathbf{x}_j)$. These estimates give us the plug-in estimates $\hat{f}_i(\mathbf{x})$ and $\hat{f}(\mathbf{x})$. From (6.17) we have the rule Assign $\mathbf{x}$ to the group with the larger plug-in estimate of the posterior probability

$$\hat{q}_i(\mathbf{x}) = \frac{\hat{\pi}_i \hat{f}_i(\mathbf{x})}{\hat{f}(\mathbf{x})}.$$

Parameter estimation for the case $\boldsymbol{\Sigma}_1 = \boldsymbol{\Sigma}_2$ is discussed further by Tang and Chang [1972] and Ganesalingam and McLachlan [1979].

For the general case of g rather two groups, Wolfe [1970] gave two estimation programs, NORMIX and NORMAP, where the latter assumes $\boldsymbol{\Sigma}_1 = \boldsymbol{\Sigma}_2 = \cdots = \boldsymbol{\Sigma}_g$. Unfortunately such methods suffer from the problem that there may be more than one solution to the maximum likelihood equations. In fact, Day [1969] pointed out that for $n < 100$, d moderate, and distributions not widely separated, there invariably appear to be several solutions. Evaluating the likelihood at each local maximum may not be feasible with the large-scale problems that can be met in practice. Everitt [1974: p. 85] also found that NORMAP failed to converge for a number of simple examples. As might be expected, both procedures are very sensitive to outliers, and some attempt at screening these out before analysis should be made. Hartigan [1975: Chapter 5] also discusses some of the difficulties and gives a general algorithm.

The method of mixtures is discussed further in Section 7.5.4, where the emphasis is on dividing the n observations into two clusters rather than on estimating an assignment rule for classifying future observations. For a general review of the subject, see Everitt and Hand [1981].

6.8 CASES OF DOUBT

There may be occasions when a researcher wishes to reserve judgment on the classification of a member of the population $\mathscr{P}$. In this case the sample space R

is partitioned into three mutually exclusive regions R_1, R_2, and R_{12}, where $\mathbf{x}$ is assigned to G_i if $\mathbf{x} \in R_i$, and $\mathbf{x}$ is left in doubt if $\mathbf{x} \in R_{12}$. A simple procedure would be to decide what misclassification probabilities are to be tolerated, use these to define R_1 and R_2, and designate the remainder of R as R_{12}. J. A. Anderson [1969] (see also Lachenbruch [1975a: pp. 86–88]) described a procedure that looks for a rule which minimizes the total probability of misclassification subject to upper bounds on the individual misclassification probabilities $P(i|j)$. We can either classify every observation using this conditional rule, which may not be readily computed, or else simply use the usual unconditional rule and bear in mind that there will be some observations not classified, as the restrictions imposed by the bounds will not be satisfied. For another approach to this problem, see McLachlan [1977b].

Habbema et al. [1974b] argued that it is only sensible to have a region R_{12} if the cost of reserving judgment when the observation actually comes from $G_1(G_2)$ is less than the cost of wrongly assigning to $G_2(G_1)$. The authors gave a Bayesian analysis of this problem.

A related problem, which also involves uncertainty, arises when each sample observation is not classified but has an associated probability p $(0 < p < 1)$ of belonging to G_1, where p varies over the population $\mathscr{P}$. A Bayesian solution of this problem is given by Aitchison and Begg [1976].

In the case of normal data, McDonald et al. [1976] gave a test of the hypothesis that $\mathbf{x}$ comes from $G_1 \cup G_2$ versus the alternative that it does not.

6.9 MORE THAN TWO GROUPS

Suppose that we have a population $\mathscr{P}$ consisting of g mutually exclusive groups and let π_i be the proportion of $\mathscr{P}$ in group G_i $(i = 1, 2, \ldots, g; \Sigma_i \pi_i = 1)$. The theory developed so far for $g = 2$ carries over naturally to the case of more than two groups. We define $f_i(\mathbf{x})$ to be the probability or probability density function of $\mathbf{x}$ if $\mathbf{x} \in G_i$, and we wish to find a suitable partition $\{R_1, R_2, \ldots, R_g\}$ of the sample space R such that we assign a member of $\mathscr{P}$ to G_i if $\mathbf{x} \in R_i$. The probability of assigning a member to G_j when it actually comes from G_i is then

$$P(j|i) = \int_{R_j} f_i(\mathbf{x})\, d\mathbf{x},$$

and the probability of misclassifying a member of G_i is

$$P(i) = \sum_{j=1,\, j \neq i}^{g} P(j|i) = 1 - P(i|i).$$

The total probability of misclassification is then [see (6.2)]

$$P(\mathbf{R},\mathbf{f}) = \sum_{i=1}^{g} \pi_i P(i)$$

$$= 1 - \sum_{i=1}^{g} \pi_i P(i|i). \tag{6.94}$$

When $g = 2$ we saw from (6.17) that the rule that minimizes $P(\mathbf{R},\mathbf{f})$ is to assign $\mathbf{x}$ to the group with the larger posterior probability $q_i(\mathbf{x})$. Noting from (6.16) that

$$q_i(\mathbf{x}) = \pi_i f_i(\mathbf{x}) / \sum_{j=1}^{g} \pi_j f_j(\mathbf{x}) = \frac{\pi_i f_i(\mathbf{x})}{f(\mathbf{x})}, \tag{6.95}$$

say, we find that the same result generalizes, as seen by the following theorem.

THEOREM 6.3 Given the rule Assign to G_i if $\mathbf{x} \in R_i$, then $P(\mathbf{R},\mathbf{f})$ is minimized when

$$R_i = R_{0i} = \{\mathbf{x}\colon q_i(\mathbf{x}) \geq q_j(\mathbf{x}),\ j = 1,2,\ldots,g\}.$$

Proof Define the indicator function $\chi_A(\mathbf{x}) = 1$ if $\mathbf{x} \in A$ and 0 otherwise. For any partition $\{R_1, R_2,\ldots,R_g\}$.

$$\sum_{i=1}^{g} \pi_i P(i|i) = \sum_{i=1}^{g} \int_{R_i} \pi_i f_i(\mathbf{x})\, d\mathbf{x}$$

$$= \sum_{i=1}^{g} \int_R \chi_{R_i}(\mathbf{x}) q_i(\mathbf{x}) f(\mathbf{x})\, d\mathbf{x} \qquad \text{[from (6.95)]}$$

$$= \int_R \sum_{i=1}^{g} \chi_{R_i}(\mathbf{x}) q_i(\mathbf{x}) f(\mathbf{x})\, d\mathbf{x}$$

$$= \int_R g(\mathbf{x})\, d\mathbf{x}, \tag{6.96}$$

say. Let $g_0(\mathbf{x})$ be the corresponding function for the partition $\{R_{01}, R_{02},\ldots, R_{0g}\}$. Suppose $\mathbf{x} \in R_j$, then $\mathbf{x} \in R_{0m}$ for some m as $R = \cup_i R_{0i}$. Therefore

$q_m(\mathbf{x}) \geq q_i(\mathbf{x})$ for all i, and

$$\begin{aligned} g_0(\mathbf{x}) &= \sum_{i=1}^{g} \chi_{R_{0i}}(\mathbf{x}) q_i(\mathbf{x}) f(\mathbf{x}) \\ &= q_m(\mathbf{x}) f(\mathbf{x}) \\ &= \sum_{i=1}^{g} \chi_{R_i}(\mathbf{x}) q_m(\mathbf{x}) f(\mathbf{x}) \\ &\geq \sum_{i=1}^{g} \chi_{R_i}(\mathbf{x}) q_i(\mathbf{x}) f(\mathbf{x}) \\ &= g(\mathbf{x}). \end{aligned}$$

Thus $g(\mathbf{x})$ is maximized for all $\mathbf{x}$. From (6.94) and (6.96),

$$P(\mathbf{R}, \mathbf{f}) = 1 - \int_R g(\mathbf{x})\, d\mathbf{x}$$

is minimized when $R_i = R_{0i}$ $(i = 1, 2, \ldots, g)$. □

From the above theorem, our optimal assignment rule is Assign to G_i if $\pi_i f_i(\mathbf{x}) \geq \pi_j f_j(\mathbf{x}) (j = 1, 2, \ldots, g)$, that is, if $\max_j \pi_j f_j(\mathbf{x}) = \pi_i f_i(\mathbf{x})$, with the assignment on the boundary of R_{0i} being arbitrary. Clearly all the methods for estimating $f_i(\mathbf{x})$ for the case $g = 2$ apply to general g. For example, suppose we have n_i observations $\mathbf{x}_{ij}$ $(i = 1, 2, \ldots, g;\ j = 1, 2, \ldots, n_i)$ from $N_d(\boldsymbol{\mu}_i, \boldsymbol{\Sigma})$. Let $\mathbf{S}_p = \sum_{i=1}^{g}(n_i - 1)S_i/(n - g)$, where $n = \sum_{i=1}^{g} n_i$, be the pooled estimate of $\boldsymbol{\Sigma}$. Then

$$\log[\pi_i \hat{f}_i(\mathbf{x})] = \log \pi_i + c - \tfrac{1}{2}(\mathbf{x} - \bar{\mathbf{x}}_{i\cdot})' \mathbf{S}_p^{-1} (\mathbf{x} - \bar{\mathbf{x}}_{i\cdot}), \tag{6.97}$$

and, subtracting off the common part $c - \frac{1}{2}\mathbf{x}'\mathbf{S}_p^{-1}\mathbf{x}$, we obtain the linear function

$$L_i(\mathbf{x}) = \log \pi_i + \bar{\mathbf{x}}_{i\cdot}' \mathbf{S}_p^{-1} (\mathbf{x} - \tfrac{1}{2}\bar{\mathbf{x}}_{i\cdot}). \tag{6.98}$$

We assign $\mathbf{x}$ to the group with the largest value of $L_i(\mathbf{x})$. With regard to this particular model, it should be noted that there are differences in terminology, particularly in some computer software packages. For example, the SPSS system, under the title "discriminant analysis," refers to the linear functions derived in discriminant coordinate analysis (Section 5.8) as discriminant functions (some authors call them canonical variates), and refers to the discriminant functions (6.98) as classification functions. To confuse the issue further, there is a third set of linear functions, sometimes quoted in the literature, given by the differences $L_i(\mathbf{x}) - L_j(\mathbf{x}) = 0$. This difference defines a hyperplane for separating group i from group j; for two groups this leads to the LDF. Jennrich [1977] describes any linear function of the variables as a discriminant function and refers to the three types as canonical, group classification, and group separation functions, respectively.

When the dispersion matrices are unequal, (6.98) is replaced by a quadratic. However, Wilf [1977] suggests a "coalition" method in which dispersion matrices may be equal within coalitions of groups. This method is a compromise between the LDF and QDF techniques and attempts to reduce the number of unknown parameters. A test for multivariate normality and equal dispersion matrices is given by Hawkins [1981].

EXAMPLE 6.7 In Table 6.3 we have bivariate observations given by Lubischew [1962: x_{14} and x_{18} from Tables 4–6] on specimens of three species of male flea beetles, namely, 21 specimens of *Chaetocnema concinna*, 31 specimens of *Chaetocnema heikertingeri*, and 22 specimens of *Chaetocnema heptapotamica*. The variables measured were

x_1 = the maximal width of the aedeagus in the forepart (in microns),

x_2 = the front angle of the aedeagus (1 unit $\equiv 7.5°$).

Rounded to two decimal places for ease of exposition, the three sample means for the species are, respectively,

$$\bar{\mathbf{x}}_{1\cdot} = \begin{pmatrix} 146.19 \\ 14.10 \end{pmatrix}, \qquad \bar{\mathbf{x}}_{2\cdot} = \begin{pmatrix} 124.65 \\ 14.29 \end{pmatrix}, \qquad \bar{\mathbf{x}}_{3\cdot} = \begin{pmatrix} 138.27 \\ 10.09 \end{pmatrix},$$

and the corresponding sample dispersion matrices $\mathbf{S}_i$ $(i = 1, 2, 3)$ are

$$\begin{pmatrix} 31.66 & -0.97 \\ -0.97 & 0.79 \end{pmatrix}, \qquad \begin{pmatrix} 21.37 & -0.33 \\ -0.33 & 1.21 \end{pmatrix}, \qquad \begin{pmatrix} 17.16 & -0.50 \\ -0.50 & 0.94 \end{pmatrix}.$$

Since these are similar, and bearing in mind the robustness of linear discriminant functions to unequal group dispersion matrices, we can use a pooled estimate

$$\mathbf{S}_p = \sum_{i=1}^{3} (n_i - 1)\mathbf{S}_i \Big/ \sum_{i=1}^{3} (n_i - 1) = \begin{pmatrix} 23.02 & -0.56 \\ -0.56 & 1.01 \end{pmatrix}.$$

In the face of ignorance about species proportions, we assume $\pi_1 = \pi_2 = \pi_3 = \frac{1}{3}$ for the prior probabilities and calculate three discriminant functions [see (6.98)]

$$L_i(\mathbf{x}) = \log \pi_i - \tfrac{1}{2}\bar{\mathbf{x}}_{i\cdot}'\mathbf{S}_p^{-1}\bar{\mathbf{x}}_{i\cdot} + \bar{\mathbf{x}}_{i\cdot}'\mathbf{S}_p^{-1}\mathbf{x} \qquad (i = 1, 2, 3).$$

TABLE 6.3 Two Measurements on Samples of Three Species of Flea-Beetles[a]

C. concinna		C. heikertingeri		C. heptapotamica	
x_1	x_2	x_1	x_2	x_1	x_2
150	15	120	14	145	8
147	13	123	16	140	11
144	14	130	14	140	11
144	16	131	16	131	10
153	13	116	16	139	11
140	15	122	15	139	10
151	14	127	15	136	12
143	14	132	16	129	11
144	14	125	14	140	10
142	15	119	13	137	9
141	13	122	13	141	11
150	15	120	15	138	9
148	13	119	14	143	9
154	15	123	15	142	11
147	14	125	15	144	10
137	14	125	14	138	10
134	15	129	14	140	10
157	14	130	13	130	9
149	13	129	13	137	11
147	13	122	12	137	10
148	14	129	15	136	9
		124	15	140	10
		120	13		
		119	16		
		119	14		
		133	13		
		121	15		
		128	14		
		129	14		
		124	13		
		129	14		

[a]Reproduced from A. A. Lubischew (1962). On the use of discriminant functions in taxonomy. *Biometrics*, **18**, 455–477. With permission from The Biometric Society.

These are

$$L_1(\mathbf{x}) = \log\tfrac{1}{3} - 619.746 + 6.778x_1 + 17.636x_2,$$

$$L_2(\mathbf{x}) = \log\tfrac{1}{3} - 487.284 + 5.834x_1 + 17.308x_2,$$

$$L_3(\mathbf{x}) = \log\tfrac{1}{3} - 505.619 + 6.332x_1 + 13.442x_2.$$

We assign a new observation $\mathbf{x}$ to the species with the largest value of $L_i(\mathbf{x}) - \log\frac{1}{3}$.

Alternatively, we can use the functions

$$D_{ij}(\mathbf{x}) = L_i(\mathbf{x}) - L_j(\mathbf{x}),$$

namely,

$$D_{12}(\mathbf{x}) = -132.46 + 0.94x_1 + 0.33x_2,$$

$$D_{23}(\mathbf{x}) = 18.34 - 0.50x_1 + 3.87x_2,$$

$$D_{13}(\mathbf{x}) = -114.13 + 0.45x_1 + 4.19x_2.$$

The assignment rule is now as follows: If $D_{12}(\mathbf{x}) > 0$ and $D_{13}(\mathbf{x}) > 0$, assign to G_1; if $D_{12}(\mathbf{x}) < 0$ and $D_{23}(\mathbf{x}) > 0$, assign to G_2; otherwise assign to G_3. In Fig. 6.4 we have drawn in the three lines $D_{ij}(\mathbf{x}) = 0$ showing the three assignment regions. The three lines meet at a point as

$$D_{13}(\mathbf{x}) = D_{12}(\mathbf{x}) + D_{23}(\mathbf{x}),$$

and $a_1 D_{12}(\mathbf{x}) + a_2 D_{23}(\mathbf{x}) = 0$ is the equation of any line through the intersection of $D_{12}(\mathbf{x}) = 0$ and $D_{23}(\mathbf{x}) = 0$. When the rule is applied to the data in Table 6.3, we find that only one observation, in group 1, is incorrectly assigned (to group 2). □

The logistic method also extends naturally to more than two groups using the assumption

$$\log[f_i(\mathbf{x})/f_g(\mathbf{x})] = \alpha_i + \boldsymbol{\beta}_i'x \qquad (i = 1, 2, \ldots, g-1).$$

This implies that

$$q_i(\mathbf{x}) = \pi_i f_i(\mathbf{x})/f(\mathbf{x})$$

$$= e^{z_i} \Big/ \sum_{j=1}^{g} e^{z_j} \qquad (i = 1, 2, \ldots, g),$$

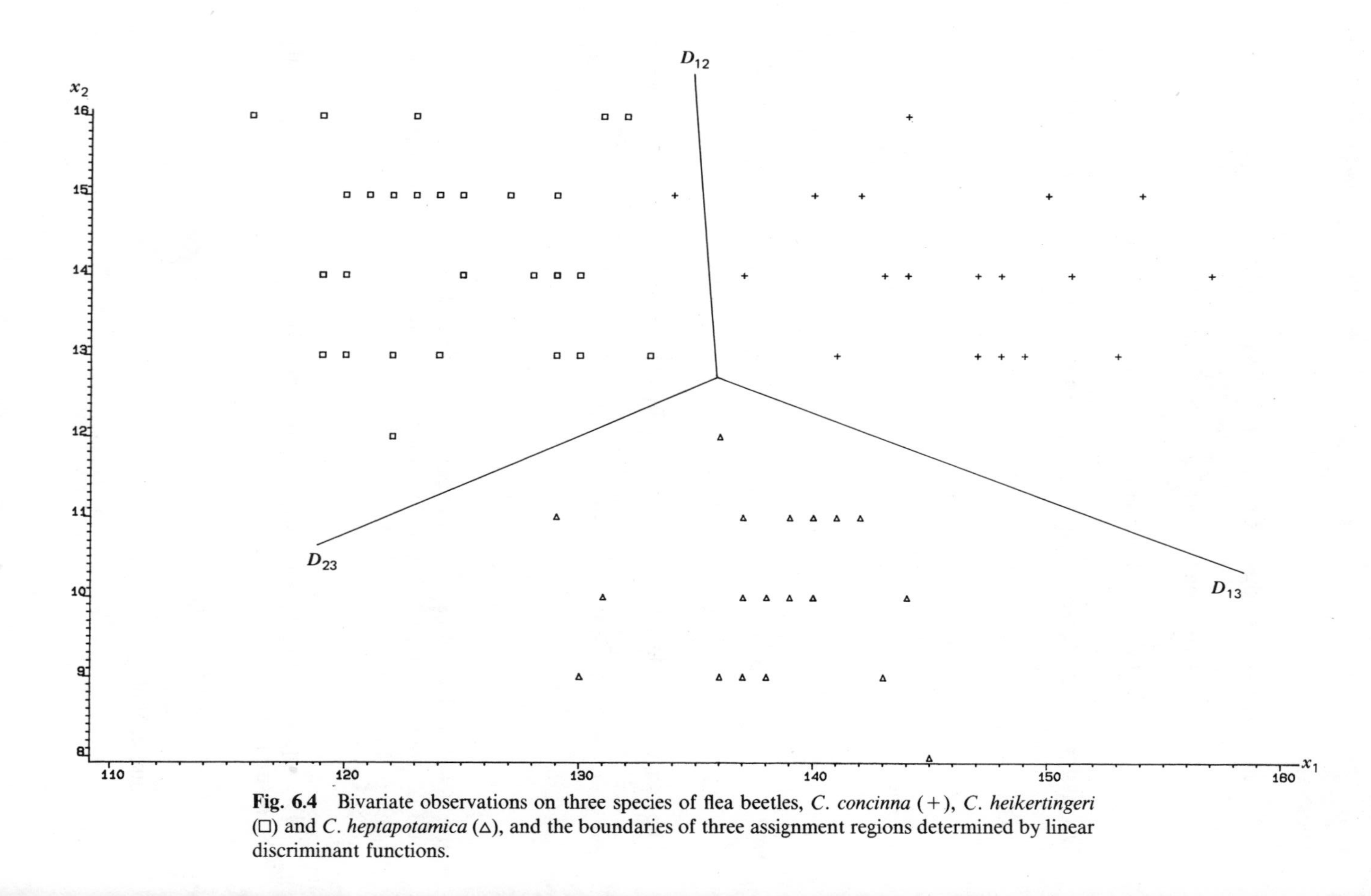

Fig. 6.4 Bivariate observations on three species of flea beetles, *C. concinna* (+), *C. heikertingeri* (□) and *C. heptapotamica* (△), and the boundaries of three assignment regions determined by linear discriminant functions.

where

$$z_i = \alpha_i + \log(\pi_i/\pi_g) + \boldsymbol{\beta}_i'\mathbf{x} \qquad (i = 1, 2, \ldots, g-1)$$

$$= \beta_{0i} + \boldsymbol{\beta}_i'\mathbf{x}, \qquad \text{say, and}$$

$$z_g = 0.$$

We assign $\mathbf{x}$ to G_i if, for $j = 1, 2, \ldots, g$,

$$\begin{aligned} 0 &\le \log\left[q_i(\mathbf{x})/q_j(\mathbf{x})\right] \\ &= z_i - z_j \\ &= (\beta_{0i} - \beta_{0j}) + (\boldsymbol{\beta}_i - \boldsymbol{\beta}_j)'\mathbf{x}. \end{aligned} \tag{6.99}$$

For conditional and mixture sampling we can find estimates of $\{\beta_{0i}\}$ and $\{\boldsymbol{\beta}_i\}$ by maximizing the conditional likelihood function (see Section 6.4.3)

$$L_C = \prod_{i=1}^{g} \prod_{j=1}^{m} \{q_i(\mathbf{x}_j)\}^{n_i(\mathbf{x}_j)}$$

using quasi-Newton methods. This likelihood can also be used for separate sampling, except that it leads to estimates of $\alpha_i + \log(n_i/n_g)$ rather than of $\beta_{0i} = \alpha_i + \log(\pi_i/\pi_g)$.

We note that Theorem 6.3 above can be generalized to the case of unequal costs of misclassification (see Exercise 6.13). Also the various estimates of error rates for the two group case carry over naturally to more than two groups (Exercise 6.18). Further models can be developed using a latent variable approach (Lauder [1981]).

6.10 SELECTION OF VARIABLES

6.10.1 *Two Groups*

Although misclassification probabilities tend to decrease as d increases (Urbakh [1971], Kokolakis [1981], Exercise 6.15), the cost of a vector observation goes up with d. Also, both the precision of estimation and the robustness of various discriminant functions such as the LDF and QDF fall off with increasing d. One might therefore ask whether $\mathbf{x}^{(1)}$, a subset of k of the d variables in $\mathbf{x}'$ $[= (\mathbf{x}^{(1)\prime}, \mathbf{x}^{(2)\prime})]$, will discriminate just as well. For MVN data with common dispersion matrix $\boldsymbol{\Sigma}$ we can use a number of tests due to Rao (see Rao [1970] for a review). Let $\boldsymbol{\delta} = (\boldsymbol{\mu}_1 - \boldsymbol{\mu}_2)$ and $\boldsymbol{\delta}_1 = (\boldsymbol{\mu}_1^{(1)} - \boldsymbol{\mu}_2^{(1)})$. If $\Delta_d^2 = \boldsymbol{\delta}'\boldsymbol{\Sigma}^{-1}\boldsymbol{\delta}$ and $\Delta_k^2 = \boldsymbol{\delta}_1'\boldsymbol{\Sigma}_{11}^{-1}\boldsymbol{\delta}_1$ are the Mahalanobis squared distances for d and k $(< d)$ variables, respectively, then a test of $H_0: \Delta_k^2 = \Delta_d^2$ is given by (Exercise 6.16)

$$F = \frac{n_1 + n_2 - d - 1}{d - k}\left\{\frac{D_d^2 - D_k^2}{c + D_k^2}\right\}, \tag{6.100}$$

where $c = (n_1 + n_2)(n_1 + n_2 - 2)/n_1 n_2$, and D_d^2 and D_k^2 are the sample squared distances, for example, $D_d^2 = (\bar{\mathbf{x}}_{1\cdot} - \bar{\mathbf{x}}_{2\cdot})'\mathbf{S}_p^{-1}(\bar{\mathbf{x}}_{1\cdot} - \bar{\mathbf{x}}_{2\cdot})$. When H_0 is true, $F \sim F_{d-k,\, n_1+n_2-d-1}$. This result follows from Theorem 2.11 [Equation (2.91)] applied to the difference $\bar{\mathbf{x}}_{1\cdot} - \bar{\mathbf{x}}_{2\cdot}$. Also, from (2.88), H_0 is true if and only if

$$\left(\boldsymbol{\mu}_1^{(2)} - \boldsymbol{\mu}_2^{(2)}\right) - \boldsymbol{\Sigma}_{21}\boldsymbol{\Sigma}_{11}^{-1}\left(\boldsymbol{\mu}_1^{(1)} - \boldsymbol{\mu}_2^{(1)}\right) = \mathbf{0},$$

that is,

$$\boldsymbol{\mu}_1^{(2)} - \boldsymbol{\Sigma}_{21}\boldsymbol{\Sigma}_{11}^{-1}\boldsymbol{\mu}_1^{(1)} = \boldsymbol{\mu}_2^{(2)} - \boldsymbol{\Sigma}_{21}\boldsymbol{\Sigma}_{11}^{-1}\boldsymbol{\mu}_2^{(1)}. \tag{6.101}$$

Therefore testing H_0 is equivalent to testing that the conditional MVN distribution of $\mathbf{x}^{(2)}$, given that $\mathbf{x}^{(1)}$ is the same for both groups [Theorem 2.1(viii) in Section 2.2]; that is, $\mathbf{x}^{(1)}$ is sufficient for discrimination, as $\mathbf{x}^{(2)}$ provides no further information about differences between the two groups.

There is also a direct relationship between testing H_0 and testing whether certain regression coefficients in a multiple linear regression are zero. From (6.69) the LDF takes the form $\alpha + \boldsymbol{\beta}'\mathbf{x}$, where $\boldsymbol{\beta} = \boldsymbol{\lambda} = \boldsymbol{\Sigma}^{-1}\boldsymbol{\delta}$. Writing

$$\boldsymbol{\beta}'\mathbf{x} = \boldsymbol{\beta}_1'\mathbf{x}^{(1)} + \boldsymbol{\beta}_2'\mathbf{x}^{(2)}, \tag{6.102}$$

we see, from Exercise 6.17, that testing H_0 is equivalent to testing $\boldsymbol{\beta}_2 = \mathbf{0}$. Because of this close link with regression, it is clear that many of the techniques for selecting the "best" regression subset (Seber [1977: Chapter 12]) can be applied to discrimination, for example, stepwise and forward selection procedures (Lachenbruch [1975a: Chapter 6], Jennrich [1977], Costanza and Afifi [1979]), "adequate" subsets based on simultaneous testing (McKay [1976, 1977, 1978: p. 261], McKay and Campbell [1982a]), Spjøtvoll's [1977] P_k plot, where P_k is the probability that F exceeds its observed value in (6.100) (see McKay [1978] for details), and generating all subsets (McCabe [1975]). Several procedures based on canonical variates (discriminant coordinates) have been proposed, but these have been criticized by McKay and Campbell [1982a].

In the stepwise procedure the F-statistic (6.100) for testing $\Delta_k^2 = \Delta_{k+1}^2$ serves as an F-to-enter or F-to-remove. At any given stage the variable with the largest F-to-enter is added to the current subset if its F-value is larger than F_{IN}, a specified threshold. After a variable has been entered, all the variables in the subset are reexamined and the one with the smallest F to remove is deleted if its F-value is less than F_{OUT}, a second threshold. At each step we do not try to eliminate the variable that was just brought in or add the variable just eliminated. This can be automatically taken care of by choosing $F_{\mathrm{IN}} \geq F_{\mathrm{OUT}}$.

Instead of a full stepwise procedure, we can use a forward selection method in which variables are brought in one at a time and are not tested for removal once they are in the subset. As with all stepwise and forward selection procedures, the choice of significance levels for testing is a problem, as at each stage we are dealing with the maximum (or minimum) of several correlated F-variables (see Draper et al. [1979] for references). For this reason Costanza and Afifi [1979] suggested using a significance level of $0.10 \leq \alpha \leq 0.25$ (preferably the larger values) for testing $\Delta_k^2 = \Delta_{k+1}^2$ in the forward selection procedure. The conventional levels ($\alpha \leq 0.10$) lead to premature stopping, unless it is certain that there are only a few important variables.

Unfortunately, stepwise procedures based on (6.100), testing for adequate subsets, and P_k plots depend very much on the assumptions of normality and equal dispersion matrices. These assumptions are not easy to test and may not be tenable. An alternative method for assessing the usefulness of a given subset is to use an estimate, $\tilde{e}_{\mathrm{act}}$, say, of the actual error rate (e.g., e_c, e_{app}, or e_{boot} of Section 6.3.1). For normal data (6.100) and $\tilde{e}_{\mathrm{act}}$ are similar criteria, as larger values of F correspond to smaller values of e_{act} (see McLachlan [1980b]). In general, we would like to consider subsets that lead to an allocation rule with low $\tilde{e}_{\mathrm{act}}$ values. For example, Habbema et al. [1974a] used a forward selection procedure in conjunction with a kernel method of estimating the discriminating function. Variables are brought in one at a time until the addition of a further variable leads to a negligible decrease in e_c. Beginning with $d = 1$, the variable x_j is selected that minimizes e_c when $\mathbf{x} = x_j$. Suppose this is x_{j_1}. For $d = 2$ we now consider all pairs $(x_{j_1}, x_j), j \neq j_1$, and choose the pair, (x_{j_1}, x_{j_2}), say, that minimizes e_c. For $d = 3$ we consider all triples (x_{j_1}, x_{j_2}, x_j), and so on. This procedure gives an ordering of the x_j and the process stops when the decrease in e_c through introducing a further variable is less than some prescribed small number.

If d is not too large (say, $d < 20$), it would be preferable to look at all subsets of k variables for $k = 1, 2, \ldots, d$ and use some criterion for choosing the "best" overall subsets, or the best k-subset for each k. McCabe [1975] proposed using Furnival's [1971] algorithm and the U-statistic (Wilks' Λ of Section 9.2.1) for selecting subsets, with good subsets corresponding to small U-values. For the case of two groups, small values of U correspond to large values of D_k^2 [see (2.58)] and the best k-subset is the one with the largest value of D_k^2. Alternatively, we can use $\tilde{e}_{\mathrm{act}}$ as a criterion for selecting best subsets. However, as noted by Murray [1977b], there is a serious problem underlying all statistical subset procedures, irrespective of the measure used for comparing k-subsets. For a given k the number of values of $\tilde{e}_{\mathrm{act}}$ to be compared, $\binom{d}{k}$, increases rapidly as k approaches $\frac{1}{2}d$. Even if $\tilde{e}_{\mathrm{act}}$ is an unbiased estimate of e_{act}, we are essentially dealing with ordered values of $\tilde{e}_{\mathrm{act}}$ when looking for the best k-subset, so that the smallest $\tilde{e}_{act}$ will tend to be an underestimate of the e_{act} corresponding to that subset. The situation is even worse if we use an underestimate like e_{app} as our estimator of e_{act}, though it should be emphasized that we use $\tilde{e}_{act}$ for *comparative* purposes only in subset selection. Murray [1977b] gave the following results for the "best" k-subsets in a simulated experiment with normal populations, $d = 10$, and $n_1 = n_2 = 25$:

k:	1	2	3	4	5	6	7	8	9	10
e_{act}:	40.1	36.2	33.3	30.8	28.8	27.0	25.4	24.0	22.7	21.5
e_{app}:	25.2	18.0	13.9	11.5	10.6	10.0	10.8	12.2	15.3	21.4

We note that e_{act} decreases as k increases, while e_{app} dips as k approaches 5. For this reason Murray suggests that once the best subset is found, then e_{act} should be estimated from an independent sample.

Anderson [1982] has suggested a forward selection procedure based on logistic discrimination. Testing whether we can ignore $d - k$ of the x-variables is equivalent to testing the hypothesis H_0 that the corresponding β's in (6.102) are zero. This can be done using a likelihood ratio test based on

$$y = -2\log\left[\hat{L}_{(k)}/\hat{L}_{(d)}\right], \tag{6.103}$$

where $\hat{L}_{(k)}$ is the maximum value of the likelihood function L_C of (6.70). When H_0 is true, y is approximately χ^2_{d-k} and y plays the same role for logistic discrimination as F of (6.100) plays for normal data. In the forward selection procedure variables are brought in one at a time. Given k variables in the subset, the next variable for consideration is the one that maximizes $\hat{L}_{(k+1)}$, and it is brought in if $-2\log[\hat{L}_{(k)}/\hat{L}_{(k+1)}]$ is significantly large. As with all stepwise procedures, the test statistic is the largest of several chi-square statistics so that its distribution is not strictly asymptotically chi square. Since we are using maximum likelihood estimation, it transpires that any nonlinear nonnormal model can generally be approximated by a linear normal regression (see Seber [1980: Chapter 11]). Lawless and Singhal [1978] exploited this relationship further and adapted the efficient algorithms of Furnival and Wilson [1974] for generating all the subsets, or just the m best k-subsets ($k = 1, 2, \ldots, d$), to nonnormal models including the logistic. They chose subsets corresponding to the smallest values of y in (6.103).

A number of special procedures and associated computer programs for handling discrete data are described by Goldstein and Dillon [1978] and Sturt [1981]. Instead of comparing Δ^2_k with Δ^2_d, Wolde-Tsadik and Yu [1979] suggested a different measure for comparing a set with a subset that they called the probability of concordance.

6.10.2 More Than Two Groups

For the case of n_i observations ($i = 1, 2, \ldots, g$) from each of g normal populations with a common dispersion matrix, a generalization of (6.100) is available. If Λ_k is Wilks' ratio (see Section 9.2.1) for testing the equality of the g means, based on k variables, and Λ_{k+1} is the corresponding ratio when another variable is added, then a test for a significant change in Λ_k (the so-called test for additional information of Section 9.5.3) is given by [see (2.73) and following (2.75)]

$$F = \frac{n-g-k}{g-1}\left(\frac{\Lambda_k}{\Lambda_{k+1}} - 1\right), \tag{6.104}$$

where $n = n_1 + n_2 + \cdots + n_g$. We can select the variable that maximizes F or, equivalently, minimizes Λ_{k+1} (commonly called the U-statistic; McCabe [1975]). When $g = 2$, (6.104) reduces to (6.100) with $d = k + 1$. The above F-statistic is used in a number of popular computer programs such as SPSS and BMDP (see also Jennrich [1977]). However, in contrast to the two-group case, a larger value of F does not necessarily imply a smaller e_{act}. In fact, Habbema and Hermans [1977] provided an example where the smallest rather than the largest F-value gave the smallest e_{act}, and demonstrated the problem with simulated data. In general, the F-criterion tends to separate well-separated groups further instead of trying to separate badly separated groups. For these reasons, and the dependence on normality and equal dispersion matrices, the above F-test is not recommended, as it does not necessarily yield a subset with minimum e_{act}.

All the methods of the previous section apply to more than two groups. We noted there that the use of $\tilde{e}_{\text{act}}$ as a criterion for the selection of a best k-subset leads to an underestimate of e_{act}. The same problem was demonstrated by Rencher and Larson [1980] for the criterion Λ_k, which is biased downward for the best or near-best k-subsets. This bias is particularly pronounced if the number of variables d exceeds the degrees of freedom $n - g$. In this case a value of Λ_d cannot be obtained for the entire set of variables. We can expect Λ to share the same shortcomings as (6.100), in that it may lead to "best" subsets that do not have minimum e_{act}.

Habbema and Hermans [1977] compared five selection of variables programs: DISCRIM (McCabe [1975]), BMD07M, BMDP7M, SPSS, and ALLOC-I (Habbema et al. [1974a]). They recommend the use of ALLOC-I, based on kernel estimation and the use of e_c, for selecting the best variables, as the other methods require assumptions of normality and equal dispersion matrices and use F-statistics or Λ for subset selection. However, their program is much slower and operates as a forward selection procedure, since "an investigation of all possible subsets with the kernel method will take in general too much computer time." Even with the forward selection procedure the computer time "is prohibitive for straightforward analysis of large scale problems." An extension of their program is ALLOC 80 (Hermans et al. [1982]).

A helpful review of selection procedures is given by McKay and Campbell [1982b].

6.11 SOME CONCLUSIONS

The choice of method for estimating the allocation rule will depend very much on the nature of the data. If the data are continuous and transformed to multivariate normality, then the LDF methods of Section 6.3.2 can be used, provided that the dispersion matrices for the transformed variables are not too different. However, it is not clear what differences in the sample dispersion matrices can be tolerated. If the dispersion matrices are widely different, the QDF can only be used if the training samples are big enough. Generally the QDF is rather sensitive to moderate departures from normality, particularly skewness, and in most cases is probably not appropriate. However, the ranked versions RLDF and RQDF of Section 6.5.2d look promising. For nonnormal data the LDF and QDF no longer provide optimal allocation rules, and this is generally reflected in widely different individual error rates. The total error rate is rather insensitive to departures from the optimal allocation rule (e.g., Beauchamp et al. [1980]) and perhaps should not be used for comparing different models. Although the above comments refer to continuous data, most data, particularly in medicine, are a mixture of discrete and continuous, and the LDF can then have poor properties for certain underlying correlation structures [e.g., when "reversals" can occur; see (6.45) and (6.46)]. The QDF is generally unsatisfactory for binary data.

The logistic approach is an attractive one, as it can be applied to a wide spectrum of distributions and is straightforward to compute. It is much more robust to nonnormality than the LDF, it requires the estimation of fewer parameters, and it can be used for discrete, continuous, or mixed data. However, as the form of the density function f_i is unknown, error rates must be estimated nonparametrically, using, for example, e_c or e_{boot} (which is always a good idea anyway), and monitoring devices such as atypicality assessments [see (6.37)] cannot be used. However, the posterior probabilities $q_i(\mathbf{x})$ can be estimated, though they may not be too accurate.

Nonparametric procedures are also attractive, as no distributional assumptions are required. In particular, the kernel methods show considerable promise, though they are at present slow computationally and further research is needed on selecting the appropriate kernel and estimating the smoothing parameters. Using simulation, Van Ness and Simpson [1976] compared the LDF, QDF, and two kernel methods with normal and Cauchy kernels, respectively, for normal data, equal dispersion matrices, and different values of d. There was essentially no difference between the two kernel methods, thus supporting what appears to be a general consensus that the detailed shape of the kernel is not so important as the more general properties of unimodality, smoothness, symmetry, and so on. They also showed that the kernel methods performed better than the LDF and QDF and were less sensitive to increasing d. Van Ness [1979] also provided evidence that a similar result holds for the case of unequal dispersion matrices.

EXERCISES 6

6.1 Show that (6.12) is equivalent to "$\boldsymbol{\lambda}'\mathbf{x}$ is closer to $\boldsymbol{\lambda}'\boldsymbol{\mu}_1$ than $\boldsymbol{\lambda}'\boldsymbol{\mu}_2$."

6.2 (a) A diagnostic test for gout is based on the serum uric acid level. Using appropriate units, the level is approximately distributed as $N_1(5,1)$ among healthy adults and $N_1(8.5,1)$ among people with gout. If 1% of the population under investigation have gout, how should a patient with level x be classified? Calculate the total probability of misclassification e_{opt} and the posterior probabilities $q_i(x)$, $i = 1,2$, for $x = 7$.

(b) Calculate the total probability of misclassification for the minimax rule.

(c) What assignment rule would you use if the probability of wrongly asserting that a person is free of gout is to be no greater than 0.1? Calculate the total probability of misclassification for your rule.

(d) If the level in people with gout is approximately $N(8.5,2)$, how should a patient with level x be classified? Find e_{opt} for your rule.

6.3 In the two-group discrimination problem suppose that

$$f_i(x) = \binom{n}{x}\theta_i^x(1-\theta_i)^{n-x}, \qquad 0 < \theta_i < 1, i = 1,2,$$

where θ_1 and θ_2 are known. If π_1 and π_2 are the usual prior probabilities, show that the optimal classification rule (6.4) leads to a linear discriminant function. Give an expression for the probability of misclassification $P(1|2)$. Assume $\theta_1 > \theta_2$.

6.4 An observation x comes from one of the two populations with prior probabilities $\pi_1 = \pi_2 = \frac{1}{2}$ and probability density functions

$$f_i(x) = \lambda_i e^{-\lambda_i x}, \qquad x \geq 0,$$

where $\lambda_1 = 2$ and $\lambda_2 = 1$. Find the optimal rule of (6.4) and the minimax rule. Calculate the total probability of misclassification in each case.

6.5 Omitting $\mathbf{x}_{1j}$ from the observations $\mathbf{x}_{ij}$ $(i = 1,2; j = 1,2,\ldots,n_i)$, let $\overline{\mathbf{x}}_{1(j)}$ be the mean of the $n_1 - 1$ observations from G_1, and $\mathbf{S}_{p(j)}$ the pooled unbiased estimate of the dispersion matrix. If $\mathbf{u}_j = \mathbf{x}_{1j} - \overline{\mathbf{x}}_{1\cdot}$, show the following:

(a) $\bar{\mathbf{x}}_{1(j)} = \bar{\mathbf{x}}_{1\cdot} - \mathbf{u}_j/(n_1 - 1)$.

(b) $(n_1 + n_2 - 3)\mathbf{S}_{p(j)} = (n_1 + n_2 - 2)\mathbf{S}_p - n_1\mathbf{u}_j\mathbf{u}_j'/(n_1 - 1)$.

Lachenbruch [1975a]

6.6 Show that the linear discriminant function $D_s(\mathbf{x})$ of (6.39) takes the form $D_s(\mathbf{x}) = \frac{1}{2}[Q_{22}(\mathbf{x}) - Q_{11}(\mathbf{x})]$, where

$$Q_{rs}(\mathbf{x}) = (\mathbf{x} - \bar{\mathbf{x}}_{r\cdot})'\mathbf{S}_p^{-1}(\mathbf{x} - \bar{\mathbf{x}}_{s\cdot}) \qquad (r, s = 1, 2).$$

If one of the $n_1 + n_2$ sample observations, say, $\mathbf{z}$, is left out, show that the resulting LDF, $D_L(\mathbf{z})$, say, takes the form

$$D_L(\mathbf{z}) = \frac{1}{2}\left\{\frac{\nu - 1}{\nu}Q_{22}(\mathbf{z}) + \frac{C_1(\nu - 1)}{\nu^2}\frac{Q_{12}^2(\mathbf{z})}{1 - (C_1/\nu)Q_{11}(\mathbf{z})}\right.$$

$$\left. - C_1^2\left[\frac{\nu - 1}{\nu}Q_{11}(\mathbf{z}) + \frac{C_1(\nu - 1)}{\nu^2}\frac{Q_{11}^2(\mathbf{z})}{1 - (C_1/\nu)Q_{11}(\mathbf{z})}\right]\right\}$$

when $\mathbf{z}$ comes from G_1. Here $\nu = n_1 + n_2 - 2$ and $C_i = n_i/(n_i - 1)$. Write down the corresponding expression for when $\mathbf{z}$ comes from G_2. Derive similar results for the quadratic discriminant function

$$Q_s(\mathbf{x}) = \tfrac{1}{2}\left[Q_{22}(\mathbf{x}) - Q_{11}(\mathbf{x}) + \log(|\mathbf{S}_2|/|\mathbf{S}_1|)\right].$$

[Hint: Use Exercise 6.5 and A3.3.]

Lachenbruch [1975a]

6.7 Derive Equations (6.41) and (6.42).

6.8 Derive Equation (6.50).

6.9 Show that (6.49) tends to the density function for the MVN distribution as $\nu \to \infty$. [Hint: Stirling's approximation is

$$\Gamma(k + 1) \sim (2\pi k)^{1/2}k^k e^{-k} \qquad (k > 0).]$$

6.10 If $Q_{rr}(\mathbf{x})$ is defined as in Exercise 6.6, the data are normal, and $\boldsymbol{\Sigma}_1 = \boldsymbol{\Sigma}_2 = \boldsymbol{\Sigma}$, show that

$$\mathrm{E}\left[\frac{1}{2}\left(\frac{n_1 + n_2 - d - 3}{n_1 + n_2 - 2}\right)\{Q_{22}(\mathbf{x}) - Q_{11}(\mathbf{x})\} + \tfrac{1}{2}d\left(\frac{1}{n_1} - \frac{1}{n_2}\right)\right]$$

$$= (\boldsymbol{\mu}_1 - \boldsymbol{\mu}_2)'\boldsymbol{\Sigma}^{-1}\left[\mathbf{x} - \tfrac{1}{2}(\boldsymbol{\mu}_1 + \boldsymbol{\mu}_2)\right].$$

[Hint: Use Exercise 2.23.]

Moran and Murphy [1979]

6.11 Using (6.56), find the maximum likelihood estimates of p_{ij} and π_i.

6.12 Using Lagrange multipliers, show that maximizing (6.74) subject to (6.75) and (6.76) is equivalent to maximizing (6.77).

6.13 Let $c(j|i)$ and $P(j|i)$ be the respective cost and probability of assigning an observation from G_i to G_j, and let $c(i|i) = 0$. The expected cost of assigning an observation from G_i is

$$C(i) = \sum_{j=1}^{g} c(j|i)P(j|i)$$

and the total expected cost is

$$C(\mathbf{R}, \mathbf{f}) = \sum_{i=1}^{g} \pi_i C(i).$$

The conditional expected cost of saying that an observation comes from G_i, given its observed value is $\mathbf{x}$, is

$$c(i|\mathbf{x}) = \sum_{j=1}^{g} c(i|j)q_j(\mathbf{x}).$$

Given the assignment rule Assign to G_i if $\mathbf{x} \in R_i$, show that $C(\mathbf{R}, \mathbf{f})$ is minimized when

$$R_i = R_{0i} = \{\mathbf{x}: c(i|\mathbf{x}) \le c(j|\mathbf{x}), j = 1, 2, \ldots, g\}.$$

If $c(j|i) = c$ for all i, j, $i \neq j$, show that the above result reduces to that given in Theorem 6.3 of Section 6.9.

6.14 Given the model in Exercise 13, suppose that the π_i are unknown. Let $\{R_{0i}\}$ be the partition defined above for some prior distribution, $\{\pi_i^*\}$, say, such that $C^*(1) = C^*(2) = \cdots = C^*(g)$. Show that the assignment rule based on this partition minimizes $\max_{1 \le i \le g} C(i)$ with respect to all partitions.

6.15 Show that reducing the dimension d reduces $P(1|2)$ of (6.11) when $\pi_1 = \pi_2$. [Hint: Use Exercise 2.20.]

6.16 Using Theorem 2.11, derive the test given by (6.100).

6.17 Show that testing $H_0: \Delta_k^2 = \Delta_d^2$ is equivalent to testing $\boldsymbol{\beta}_2 = (\beta_{k+1}, \ldots, \beta_d)' = \mathbf{0}$ in (6.102).

6.18 How would you estimate $P(i|j)$, $P(i|i)$ and the total probability of misclassification $P(\mathbf{R}, \mathbf{f})$ for more than two groups?

6.19 Let $\mathbf{x} = (x_1, x_2)'$, where x_1 is normally distributed and x_2 is a binary random variable. Given that $\mathbf{x}$ comes from $G_i (i = 1, 2)$, x_2 takes values 1 and 0 with respective probabilities p_i and $1 - p_i$, while x_1 given $x_2 = s$ is $N_1(\mu_i^{(s)}, \sigma^2)$. Determine the optimal allocation rule and the corresponding probabilities of misclassification $P(i|j)$.

6.20 Suppose we wish to assign $\mathbf{x}$ to one of two bivariate normal populations $N_2(\mathbf{0}, \boldsymbol{\Sigma})$ and $N_2(\boldsymbol{\mu}, \boldsymbol{\Sigma})$, where $\boldsymbol{\mu} = (\mu_1, \mu_2)'$ and

$$\boldsymbol{\Sigma} = \begin{pmatrix} 1 & \rho \\ \rho & 1 \end{pmatrix}.$$

Find $\Delta^2 = \boldsymbol{\mu}'\boldsymbol{\Sigma}^{-1}\boldsymbol{\mu}$. If $\pi_1 = \pi_2$, for what values of ρ is the total probability of misclassification greater than that when $\rho = 0$? What happens when $\mu_1 = \mu_2$?

Cochran [1962]

CHAPTER 7

Cluster Analysis

7.1 INTRODUCTION

Clustering is the grouping of similar objects using data from the objects. It is part of the general scientific process of searching for patterns in data and then trying to construct laws that explain the pattern. However, the goals of cluster analysis are varied and include widely different activities such as data "snooping," looking for "natural" groups of like objects to form the first stage of a stratified sampling scheme, hypothesis generation, and searching for a suitable classification scheme, two notable examples being the classification of plants and animals (taxonomy) and the classification of diseases. Clearly, different goals lead to different techniques, and to a casual observer there seems to be as many different cluster techniques as there are applications. For example, Good [1977] gives a classification of cluster methods based on 45 characteristics, each mainly of an either/or nature, so that the number of conceivable categories of methods is of order 2^{45}! The extent of the literature is seen in the reviews and books by Cormack [1971: 229 references], Jardine and Sibson [1971b], Anderberg [1973], Duran and Odell [1974: 409 references], Everitt [1974], Gregson [1974], Hartigan [1975], Sneath and Sokal [1973: approximately 1600 references], Clifford and Stephenson [1975], Orlóci [1978: Chapter 4, ecological applications], and Gordon [1981]. In 1971 Cormack estimated the publication rate of articles on clustering and classification at over 1000 publications per year! Beginning with taxonomy (see Clifford and Stephenson [1975] for a historical perspective), applications have been made to many subjects among which might be mentioned archaeology, anthropology, agriculture, economics, education, geography, geology, linguistics, market research, genetics, medicine, political science, psychology, psychiatry, and sociology. The book by Hartigan [1975] provides a happy hunting ground for numerous applications, including over 40 data sets related to such diverse subjects as the contents of mammals' milk, jigsaw puzzles, civil wars, car repairs, the appearance times of British butterflies, and, one is tempted to add, "cabbages and kings." Three detailed and interesting case studies are given by Gordon [1981: Chapter 7].

With such diversity it is perhaps not surprising that the subject of cluster analysis lacks coherence, though there have been some attempts at providing a suitable structure using, for example, graph theory (Matula [1977], Hansen and Delattre [1978]); and in the related subject of pattern recognition, fuzzy sets (Zadeh [1977], Gordon [1981: Section 4.2] and even category theory (Gordesch and Sint [1974]). Much of the popularity of the subject is due to the advent of electronic computers, though the explosion of methods in the 1960s is now being followed by a more critical evaluation. Fortunately some effort is going into providing efficient algorithms for the more useful methods instead of into generating still more techniques (e.g., Andersberg [1973: Appendix], Hartigan [1975], Enslein et al. [1977: Part IV], Wishart [1978a], Everitt [1979b], Späth [1980]). We recall the words of Dr. R. Sibson in his comments on Cormack's [1971] paper: "I think the time has come for putting more effort into the *efficient* implementation of methods which are known to be useful and whose theoretical background is well understood, rather than the development of yet more *ad hoc* techniques." Sokal [1977] also commented, "I believe that the major effort in classificatory work in the next few years should be devoted to comparisons of different approaches and to tests of significance of classifications. Work in these fields has so far been quite unsatisfactory," We shall see later that much recent research has been along the lines suggested by Sibson and Sokal.

It should be noted that cluster analysis goes under a number of names, including classification, pattern recognition (with "unsupervised learning"), numerical taxonomy, and morphometrics. The reader is referred to Cormack's [1971] paper for a condensed summary of the many problems and differences of opinion in cluster analysis, and to Everitt [1974, 1979b] for a readable account of the more popular techniques. Because of lack of unanimity in the subject, I propose discussing the subject in fairly broad outlines, with emphasis on the strengths and weaknesses of the various approaches.

There are three main types of data used in clustering (Kruskal [1977a]). The first is d-dimensional vector data $\mathbf{x}_1, \mathbf{x}_2, \ldots, \mathbf{x}_n$ arising from measuring or observing d characteristics on each of n objects or individuals. We saw in Section 1.2 that the characteristics or variables may be quantitative (discrete or continuous) or qualitative (ordinal or nominal). It is usual to treat present–absent (dichotomous) qualitative variables separately. Although such variables are simply two-state qualitative variables, the two states are not generally regarded as being comparable, as the presence of a given character can be of much greater significance than its absence. A further complication arises if the "present" category is itself subdivided further on a quantitative or qualitative basis, a so-called conditionally present variable. A detailed discussion on methods of changing from one type of variable to another is given by Anderberg [1973: Section 3.2]: such conversions are useful in handling mixed data. Whatever the method used for coding the qualitative variables, the data can be expressed as a matrix $\mathbf{X} = [(x_{ij})]$, where

$$\mathbf{X} = \begin{pmatrix} \mathbf{x}_1' \\ \mathbf{x}_2' \\ \vdots \\ \mathbf{x}_n' \end{pmatrix} = \left(\mathbf{x}^{(1)}, \mathbf{x}^{(2)}, \ldots, \mathbf{x}^{(d)}\right),$$

and the aim of cluster analysis is to devise a classification scheme for grouping the $\mathbf{x}_i$ into g clusters (groups, types, classes, etc.). In contrast to discriminant analysis (see Chapter 6), the characteristics of the clusters and, in most cases, the number of clusters have to be determined from the data itself. We may also wish to cluster the variables, that is, group the vectors $\mathbf{x}^{(j)}$. For example, suppose n students fill out a teacher assessment questionnaire and give a rank value of 1 to 7 on each of d teaching attributes such as the organization of the course, legibility of writing, clarity of voice, approachability, and apparent knowledge of the subject. We may wish to see if the students form "natural" groups on the basis of their scores $\mathbf{x}_i$, or if the various attributes $\mathbf{x}^{(j)}$ group naturally together.

A second type of data encountered for clustering consists of an $n \times n$ proximity matrix $[(c_{rs})]$ or $[(d_{rs})]$, where c_{rs} (d_{rs}) is a measure of the similarity (dissimilarity) between the rth and sth objects. A c_{rs} or d_{rs} is called a proximity and the data is referred to as proximity data.

A third type of data that is already in a cluster format is what might be called sorting data. For example, each of several subjects may be asked to sort n items or stimuli into a number of similar, possibly overlapping groups.

All three types of data can be converted into proximity data and Cormack [1971] lists 10 proximity measures. For example, the dissimilarity d_{rs} between observations $\mathbf{x}_r$ and $\mathbf{x}_s$ could simply be the Euclidean distance $\|\mathbf{x}_r - \mathbf{x}_s\|$, and the use of such dissimilarity measures is described below. For sorting data we can base c_{rs} on the number of subjects who place items r and s in the same group. Once we have a proximity matrix, we can then proceed to form clusters of objects that are similar or close to one another. An alternative approach, which avoids the use of proximity data, starts with vector data and produces clustering directly. The data are assumed to come from a mixture of g distributions, for example, MVN distributions, with unknown parameters and unknown mixing proportions π_i to be estimated from the data. However, the distribution theory is difficult, and some of the problems are highlighted by Hartigan [1977, 1978], who provides some conjectures and a few solutions (see also Section 7.5.4).

There are basically three types of clustering available:

(1) *Hierarchical clustering*. The clusters are themselves grouped into bigger clusters, the process being repeated at different levels to form what is technically known as a tree of clusters. Such a tree can be constructed

from the bottom up (see Fig. 7.5) using an agglomerative method that proceeds by a series of successive fusions of the n objects into clusters, or from the top down using a divisive method which partitions the total set of n objects into finer and finer partitions. The former method begins with n clusters, each of one object, and ultimately ends up with a single cluster of n objects, whereas the latter method consists of the reverse process. The graphical representation of hierarchical clustering given by Fig. 7.5 is called a *dendrogram*. Clearly the investigator must decide at what stage he wants to stop the process of fusion or division.

(2) *Partitioning*. Here the objects are partitioned into nonoverlapping clusters.

(3) *Overlapping clusters*.

In order to carry out a cluster analysis a number of algorithms are needed. These may be described as (Hartigan [1975]) *sorting* (e.g., objects are sorted or partitioned into groups on the basis of a single variable), *switching* or reallocation (after a partition an object may be switched from one cluster to another), *joining* or fusion (near clusters are fused to form a new cluster), *splitting* or division (the reverse of joining), *addition* (a clustering structure or a set of cluster centers is already available and each object is added to it in turn in some optimal fashion), and *searching* through the eligible cluster structures for the one that satisfies some optimality condition.

The biggest problem with trying to develop clustering methods is that the notion of a cluster is not easy to define. Williams et al. [1971b] commented that "there are now so many variant uses of the word that it would probably be better if we all avoided it completely." Clusters come in all shapes and sizes and some of the problems are indicated in Fig. 7.1. For spherical clusters, like those in Fig. 7.1*a*, most methods would arrive at an adequate description; however, for long thin clusters like those in Fig. 7.1*b*, a method that uses interpoint distances may give misleading results, as points B and C in the same cluster are further apart than A and B, which are in different clusters. Also, if the algorithm tends to produce spherical clusters, it may produce four clusters (as shown with dotted lines) instead of two: This effect is demonstrated by Everitt [1974: p. 74]. Some algorithms use a nearest neighbor concept of linking up adjacent points in a cluster, and this can lead to a chaining effect. For example, in Fig. 7.1*c*, A is close to B, B is close to C, and so on, and if we work with just pairs of points, it is likely that we end up with one cluster instead of two.

In the light of the above examples, clusters could be defined as simply regions of high density. However, we are then faced with deciding between the compactness of the cluster as a primary criterion, or the clarity of the separation between the clusters. Figure 7.2 gives a picture of univariate density of points along a line. If compactness is important, then we might decide on three or more clusters; if clarity of separation is important, then we would probably decide on two clusters.

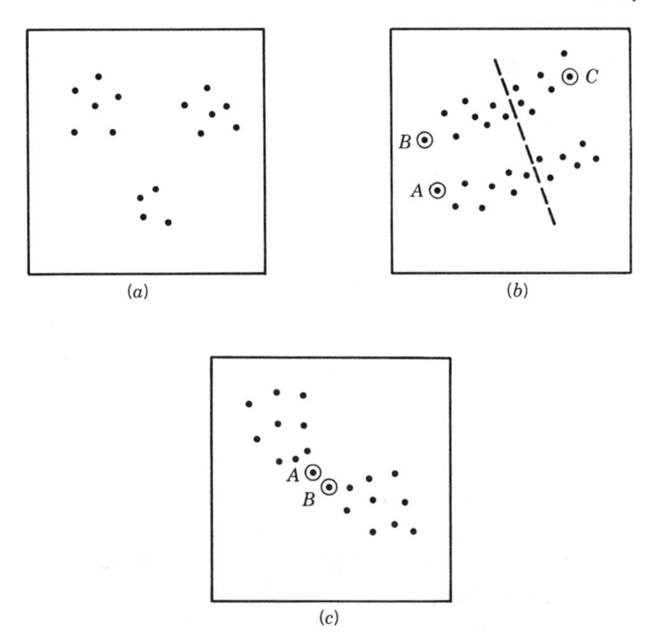

Fig. 7.1 Three types of clustering. (*a*) Spherical clusters. (*b*) Two or four clusters. (*c*) Chaining.

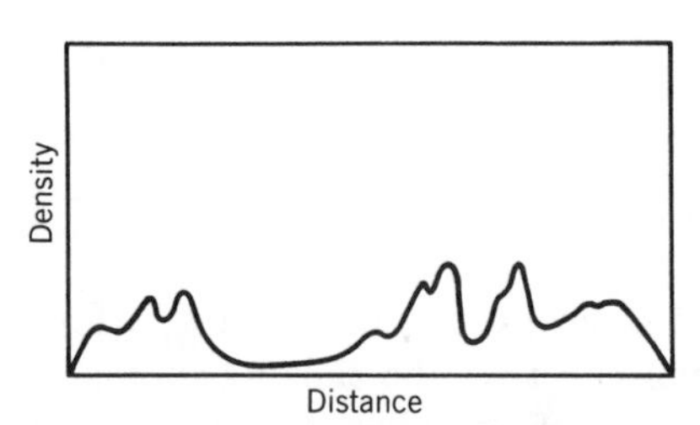

Fig. 7.2 Density of points along a line.

7.2 PROXIMITY DATA

7.2.1 Dissimilarities

Perhaps the most common dissimilarity measure for measuring the nearness of two points is a metric Δ that maps $R^d \times R^d$ onto R^1 and satisfies the following axioms:

(1) $\Delta(\mathbf{x}, \mathbf{y}) \geq 0$, all $\mathbf{x}, \mathbf{y}$ in R^d.
(2) $\Delta(\mathbf{x}, \mathbf{y}) = 0$ if and only if $\mathbf{x} = \mathbf{y}$.
(3) $\Delta(\mathbf{x}, \mathbf{y}) = \Delta(\mathbf{y}, \mathbf{x})$, all $\mathbf{x}, \mathbf{y}$ in R^d.
(4) $\Delta(\mathbf{x}, \mathbf{y}) \leq \Delta(\mathbf{x}, \mathbf{z}) + \Delta(\mathbf{y}, \mathbf{z})$, all $\mathbf{x}, \mathbf{y}, \mathbf{z}$ in R^d.

Assumptions (1) and (2) imply that Δ is positive definite, assumption (3) implies symmetry of Δ, and (4) is the well-known triangle inequality. Using the so-called "L_p-norm," we have the family of Minkowski metrics

$$\Delta_p(\mathbf{x},\mathbf{y}) = \left\{ \sum_{j=1}^{d} |x_j - y_j|^p \right\}^{1/p}, \tag{7.1}$$

the most common being the L_1 or "city block" metric, the L_2 or Euclidean metric $\|\mathbf{x} - \mathbf{y}\|$, and the sup-norm ($p = \infty$) metric

$$\Delta_\infty(\mathbf{x},\mathbf{y}) = \sup_{1 \le j \le d} |x_j - y_j|.$$

We note that $\Delta_p(\mathbf{x},\mathbf{y}) \le \Delta_q(\mathbf{x},\mathbf{y})$ for $p \le q$ and all $\mathbf{x},\mathbf{y}$.

The L_1 metric is easy to evaluate and is used by Carmichael and Sneath [1969] in their TAXMAP procedure. The sup-norm is also simple computationally, but it involves a ranking procedure that could be very sensitive to a change in scale of one of the variables. The Euclidean distance is very popular, and, given observations $\mathbf{x}_1, \mathbf{x}_2, \ldots, \mathbf{x}_n$, we can define

$$d_{rs} = \Delta_2(\mathbf{x}_r, \mathbf{x}_s)$$

$$= \left\{ \sum_{j=1}^{d} |x_{rj} - x_{sj}|^2 \right\}^{1/2} \tag{7.2}$$

as the dissimilarity (distance) between $\mathbf{x}_r$ and $\mathbf{x}_s$. However, a change in scale can have a substantial effect on the ranking of the distances. For example, suppose we have bivariate measurements, height and weight, on three objects, namely,

Object	Weight (gm)	Height (cm)
1 (A)	10	7
2 (B)	20	2
3 (C)	30	10

Then $d_{12} = 11.2$, $d_{13} = 20.2$, and $d_{23} = 12.8$, so that object A is closer to B than C. However, if the height is measured in millimeters, the respective distances become 51.0, 36.1, and 80.6, so that object A is now closer to C. In an attempt to overcome this problem each variable could be scaled by dividing by its range or sample standard deviation $s_j = \{\Sigma_i (x_{ij} - \bar{x}_{\cdot j})^2/(n-1)\}^{1/2}$, that is, x_{ij} is replaced by x_{ij}/s_j. This method will remove the dependence on the units of measurement, but it has other problems. For example, it can dilute the differences between clusters with respect to the variables that are the best discriminators. The interpoint distances within clusters become larger relative to between-cluster distances, and clusters become less clear-cut (see Fleiss and Zubin [1969], Hartigan [1975: pp. 59–60]). Looking at it another way, suppose we initially scale the variables so that within-cluster sample variances of the variables are unity. If we now rescale all the variables to have unit variances with respect to the *whole* data set, then those variables with relatively large between-cluster variances will be reduced in importance, as their scaling factors are larger. This means that the overall between-cluster variance will be reduced relative to the within-cluster variance and the clusters will become less distinct. Ideally we would like to scale the data so that within-cluster variances are approximately equal. For example, suppose Fig. 7.3 gives probability contour maps C_1 and C_2 for two bivariate normal distributions, that is, the probability that a point from the distribution lies within the outside ellipse of C_i is 0.9, and so on. Clearly point P should be grouped with C_1, as, in terms of probability, it is "closer" to the center (mean) of C_1; the unweighted Euclidean distance is not appropriate here, as it would group P with C_2. However, scaling so that within-cluster variances are approximately equal would help to even up the two contour maps. Unfortunately this method of scaling begs the question, as the clusters are not known a priori.

The effect of scaling can depend very much on the skewness of the data. An extreme case is a highly skewed binary variable taking values 1 and 0 with relative frequencies p and q $(= 1 - p)$ in the n objects, where p is very small. The sample variance pq $(\approx p)$ will be small and division by $\sqrt{pq}$ will inflate the importance of this variable.

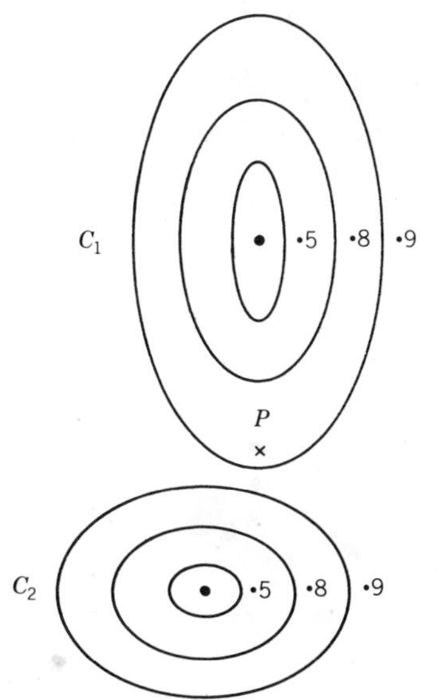

Fig. 7.3 Probability contour ellipses for two bivariate normal distributions.

To overcome not only the scaling problem, but also correlation effects among the variables, the Mahalanobis distance

$$\Delta(\mathbf{x}_r, \mathbf{x}_s) = \left\{(\mathbf{x}_r - \mathbf{x}_s)'\mathbf{S}^{-1}(\mathbf{x}_r - \mathbf{x}_s)\right\}^{1/2}, \tag{7.3}$$

where $\mathbf{S} = \Sigma_i(\mathbf{x}_i - \overline{\mathbf{x}})(\mathbf{x}_i - \overline{\mathbf{x}})'/(n-1)$, has been proposed as a distance measure (see Exercise 7.1). This measure is invariant to transformations of the form $\mathbf{y}_i = \mathbf{A}\mathbf{x}_i + \mathbf{b}$, where $\mathbf{A}$ is nonsingular (see Exercise 7.2). However, as Hartigan [1975: p. 63] comments, invariance under general linear transformations seems less compelling than invariance under simply a change of measuring units for each variable. Hartigan then goes on to argue that use of (7.3) reduces the clarity of the clusters even more than scaling to unit sample variances. Suppose, for example, that the data consists of two similar, well-separated spherical clusters, each with n_k points and centroids $\overline{\mathbf{x}}_k$ ($k = 1, 2$), as indicated in Fig. 7.4. The grand mean $\overline{\mathbf{x}} = (n_1\overline{\mathbf{x}}_1 + n_2\overline{\mathbf{x}}_2)/(n_1 + n_2)$ will lie on the line L joining the two centroids. Now the first principal component is $\mathbf{a}'\mathbf{x}$, where $\mathbf{a}$ is the direction of the line through $\overline{\mathbf{x}}$ that minimizes the sum of the squares of the distances of all the points from the line [see after (5.17)]. Clearly this line will be close to L and the second component will correspond approximately to the line M through $\overline{\mathbf{x}}$ perpendicular to L. Since the distances from M are much greater than from L, the first principal component (obtained by projecting orthogonally onto L) will have a large sample variance compared with the second component: These two variances are the eigenvalues of $\mathbf{S}$. If we use the Mahalanobis distance instead of the Euclidean distance, we are effectively replacing the $\mathbf{x}_i$ by $\mathbf{S}^{-1/2}\mathbf{x}_i$, which have a sample covariance matrix $\mathbf{S}^{-1/2}\mathbf{S}\mathbf{S}^{-1/2} = \mathbf{I}_d$. This means that the two principal components are standardized to have equal sample variances. With this new scaling, the within-cluster distances increase relative to the between-cluster distances and the clusters become less distinct.

It would seem preferable to replace $\mathbf{S}$ in (7.3) by the pooled cluster estimate $\mathbf{S}_p$ [$= \mathbf{W}/(n - g)$ of Section 5.8]. Although this involves the same circular problem as with scaling when the clusters are unknown, the estimate $\mathbf{S}_p$ could be recalculated, as the cluster structure changes. However, from their simula-

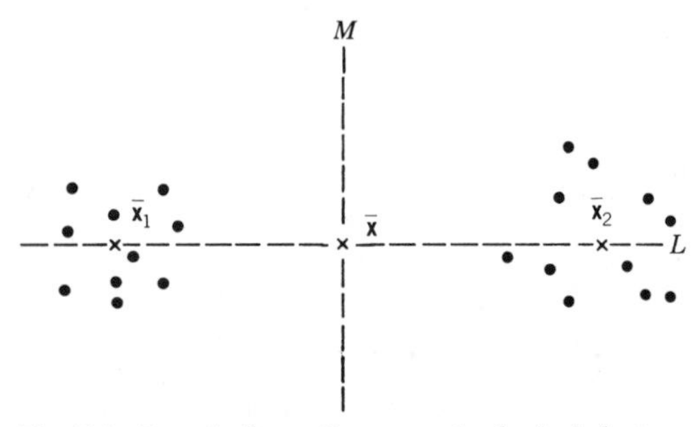

Fig. 7.4 Two similar well-separated spherical clusters.

tion study, Maronna and Jacovkis [1974] felt that neither version was worth using.

Various scaled versions of the L_1 metric have also been used. For example, Gower [1971a] scales each variable by its range, which leads to the metric

$$d_{rs} = \frac{1}{d} \sum_{j=1}^{d} \frac{|x_{rj} - x_{sj}|}{R_j}, \tag{7.4}$$

where R_j is the range of the variable j. The Bray–Curtis [1957] measure, which is still used occasionally in ecology, takes the form

$$\begin{aligned} d_{rs} &= \left(\sum_j |x_{rj} - x_{sj}|\right) \Big/ \left(\sum_j x_{rj} + \sum_j x_{sj}\right) \\ &= 1 - 2C/(A + B), \end{aligned} \tag{7.5}$$

where $A = \Sigma_j x_{rj}$, $B = \Sigma_j x_{sj}$, and $C = \Sigma_j \min(x_{rj}, x_{sj})$. This measure is not necessarily a metric (see Exercise 7.3). Another measure, which is a metric for positive variables (see Exercise 7.4), is the so-called Canberra metric,

$$d_{rs} = \frac{1}{d} \sum_j \left\{ \frac{|x_{rj} - x_{sj}|}{x_{rj} + x_{sj}} \right\}, \tag{7.6}$$

introduced by Lance and Williams [1966, 1967b]. It is generally insensitive to skewness and outlying values.

The question of scaling is clearly a difficult one and there is a general lack of guidance in the literature. Applying some form of scaling is equivalent to weighting the variables and, unfortunately, no satisfactory scheme for weighting variables has been proposed (Sokal [1977]). In some experimental situations it may be possible to use certain reference objects as a model for standardization (Stoddard [1979]). Clearly, programs with automatic scaling built in should be avoided; the experimenter should have the option of whether or not to scale.

A measure of dissimilarity d_{rs} between objects r and s need not be a metric. In Section 5.5.1 we defined it as a function satisfying $d_{rs} \geq 0$, $d_{rr} = 0$, and $d_{rs} = d_{sr}$. It is then possible to have $d_{rs} = 0$, but object r different from object s. As Sibson [1972] argues convincingly, order relationships are more important than numerical values, and a monotonic increasing function of a dissimilarity coefficient might do just as well. For this reason the metric triangle inequality is not an important requirement, as a monotonic transformation of a metric need not satisfy the triangle inequality, for example, the square of the Euclidean metric (see Exercise 7.6).

7.2.2 Similarities

Similarity coefficients have a long history and in the early literature were usually known as association coefficients. For example, suppose the d variables or characteristics are dichotomous, indicating the presence (+) or absence (−) of each characteristic. Then, for each pair of the n objects, we can form the usual two-way contingency table given by Table 7.1, where, for example, β is the number of characteristics present in object r but absent in object s. Numerous measures of association c_{rs}, satisfying $0 \leq c_{rs} \leq 1$, have been proposed and Clifford and Stephenson [1975: pp. 54–55] list 11 (see also Anderberg [1973: Table 4.53]). The most popular measures are Jaccard's [1908] coefficient $\alpha/(\alpha + \beta + \gamma)$, Czekanowski's [1913] coefficient $2\alpha/(2\alpha + \beta + \gamma)$, and the simple matching proportion $(\alpha + \delta)/d$. The choice of a coefficient depends very much on the relative weights given to positive matches (α) and negative matches (δ). Clearly there are situations, in numerical taxonomy, for example, where the joint absence of a characteristic would carry little or no weight in comparison with the joint presence, and the matching coefficient, with equal emphasis on both categories, would not be appropriate. However, the matching coefficient would be appropriate if the variables were all nominal with two states, the states simply being alternatives with equal weight. The above measures of association can be extended to nominal variables with more than two states (Anderberg [1973: Section 5.4]). For example, we can calculate the matching coefficient as the proportion of the nominal variables that match for two objects.

With quantitative variables, one measure of similarity between $\mathbf{x}_r$ and $\mathbf{x}_s$, the observations on objects r and s, is the correlation of the pairs (x_{rj}, x_{sj}), $j = 1, 2, \ldots, d$, namely,

$$c_{rs} = \frac{\sum_j (x_{rj} - \bar{x}_{r\cdot})(x_{sj} - \bar{x}_{s\cdot})}{\left\{\sum_j (x_{rj} - \bar{x}_{r\cdot})^2 \sum_j (x_{sj} - \bar{x}_{s\cdot})^2\right\}^{1/2}}, \tag{7.7}$$

TABLE 7.1 Number of Characteristics Occurring in, or Absent from, Two Objects: α Common to Both Objects; β and γ Occurring in Only One Object; δ Absent from Both

		Object s		
		Present (+)	Absent (−)	Sum
Object r	Present (+)	α	β	$\alpha + \beta$
	Absent (−)	γ	δ	$\gamma + \delta$
	Sum	$\alpha + \gamma$	$\beta + \delta$	$\alpha + \beta + \gamma + \delta = d$

and $-1 \leq c_{rs} \leq 1$. Apart from not satisfying axiom (1) below, this measure, however, has certain disadvantages. For example, if $c_{rs} = 1$, it does not follow that $\mathbf{x}_r = \mathbf{x}_s$, only that the elements of $\mathbf{x}_r$ are linearly related to those of $\mathbf{x}_s$ (see Exercise 7.9). Also, what meaning can we give $\bar{x}_{r\cdot}$, the mean over the different variables for object r? For these and other reasons, the correlation coefficient has been criticized by a number of authors (e.g., Fleiss and Zubin [1969]). Although there is some difference of opinion, the evidence would suggest that dissimilarities based on metrics are better proximity measures than correlations. Cormack [1971] states that the "use of the correlation coefficient must be restricted to situations in which variables are uncoded, comparable measurements or counts; it is not invariant under scaling of variables, or even under alterations in the direction of coding of some variables (Minkoff [1965])."

A large number of similarity measures have been proposed in the literature and they can be categorized mathematically in several ways (e.g., using trees in Hartigan [1967]; see also Duran and Odell [1974: Chapter 4]). If $\mathscr{P}$ is the population of objects, then we can define a similarity as a function that maps $\mathscr{P} \times \mathscr{P}$ into R^1 and satisfies the following axioms:

(1) $0 \leq C(r, s) \leq 1$ for all r, s in $\mathscr{P}$.
(2a) $C(r, r) = 1$.
(2b) $C(r, s) = 1$ only if $r = s$.
(3) $C(r, s) = C(s, r)$.

We shall write $c_{rs} = C(r, s)$ and use the notation $C(\mathbf{x}_r, \mathbf{x}_s)$ for vector data. The Jaccard and Czekanowski coefficients satisfy the above axioms.

Gower [1971a]) has proposed an all-purpose measure of similarity

$$c_{rs} = \left(\sum_{j=1}^{d} c_{rsj} \right) \Big/ \sum_{j=1}^{d} w_{rsj}, \tag{7.8}$$

where c_{rsj} is a measure of similarity between objects r and s for variable j. Here w_{rsj} is unity except when a comparison is not possible, as with missing observations or negative matches of dichotomous variables, in which case we set $c_{rsj} = w_{rsj} = 0$. In Table 7.2 we have the appropriate coefficients for a dichotomous variable or a two-state qualitative variable. With a multistate variable (ordinal or nominal) of more than two states, we set $c_{rsj} = 1$ if objects r and s agree in variable j, and $c_{rsj} = 0$ otherwise: In both cases $w_{rsj} = 1$. For a quantitative variable, $w_{rsj} = 1$ and

$$\begin{aligned} c_{rsj} &= 1 - |x_{rj} - x_{sj}|/R_j \\ &= 1 - |x'_{rj} - x'_{sj}|, \end{aligned}$$

where R_j is the range of variable j and $x'_{rj} = x_{rj}/R_j$. Thus if we have d_1 quantitative variables, d_2 dichotomous variables, and d_3 multistate variables,

then

$$c_{rs} = \frac{\sum_{j=1}^{d_1} \left(1 - |x'_{rj} - x'_{sj}|\right) + \alpha_2 + m_3}{\left[d_1 + (d_2 - \delta_2) + d_3\right]},$$

where α_2 and δ_2 are the number of positive and negative matches, respectively, for the dichotomous variables, and m_3 are the number of matches for the multistate variables. If all the variables are dichotomous, then c_{rs} reduces to Jaccard's coefficient, whereas if all the variables are two-state, then c_{rs} reduces to the matching coefficient. Williams and Lance [1977] do not recommend its use if the continuous data are highly skewed, as the range is very sensitive to skewness.

Gower [1971a] showed that $[(c_{rs})]$ is positive semidefinite for his coefficient. From (5.73), $d_{rs} = (2 - 2c_{rs})^{1/2}$ satisfies the triangle inequality, being the Euclidean distance measure for some configuration of points (the factor 2 may be omitted).

A dissimilarity coefficient can always be obtained from a similarity by setting $d_{rs} = 1 - c_{rs}$, though d_{rs} will not be a metric unless c_{rs} satisfies Axiom (2b) above and d_{rs} satisfies the triangle inequality. For example, if we apply the scaled Euclidean metric $\|\mathbf{x}_r - \mathbf{x}_s\|/d$ or the Canberra metric (7.6) to binary 0–1 data, we get $(\beta + \gamma)/d$, the "one-complement" of the matching coefficient c_{rs}, so that $1 - c_{rs}$ is a metric. Similarly, we find that the one-complements of the Jaccard and Czekanowski coefficients satisfy the triangle inequality (Ihm [1965]) so that they are also metrics.

TABLE 7.2 Gower's Similarity Coefficients for (a) Dichotomous and (b) Two State Qualitative Variables [see Equation (7.8)]

(a) Presence/absence of dichotomous variable j

Object r	+	+	−	−
Object s	+	−	+	−
c_{rsj}	1	0	0	0
w_{rsj}	1	1	1	0

(b) Two state qualitative variable j

Object r	1	1	2	2
Object s	1	2	1	2
c_{rsj}	1	0	0	1
w_{rsj}	1	1	1	1

If we are interested in clustering *variables* rather than objects, the correlation coefficient for variables j and k is

$$r_{jk} = \frac{\sum_i (x_{ij} - \bar{x}_{\cdot j})(x_{ik} - \bar{x}_{\cdot k})}{\left\{\sum_i (x_{ij} - \bar{x}_{\cdot j})^2 \sum_i (x_{ik} - \bar{x}_{\cdot k})^2\right\}^{1/2}} \tag{7.9}$$

$$= \sum_{i=1}^{n} \tilde{x}_{ij}\tilde{x}_{ik}, \tag{7.10}$$

where $\tilde{x}_{ij}$ is x_{ij} suitably "standardized." If we define the Euclidean distance d_{jk} between standardized variables j and k by

$$\begin{aligned} d_{jk}^2 &= \sum_i (\tilde{x}_{ij} - \tilde{x}_{ik})^2 \\ &= \sum_i \tilde{x}_{ij}^2 + \sum_i \tilde{x}_{ik}^2 - 2\sum_i \tilde{x}_{ij}\tilde{x}_{ik} \\ &= 2(1 - r_{jk}), \end{aligned}$$

then we can use $d_{jk} = [2(1 - r_{jk})]^{1/2}$ to transform the similarity measure r_{jk} into a distance; the factor 2 can be dropped. However, there are problems with using a correlation coefficient if one or both of the variables j and k are nominal (disordered multistate) variables. One solution has been proposed by Lance and Williams [1968] (see also Anderberg [1973: pp. 96–97]). Dichotomous variables can be handled using the values 0 and 1 (see Exercise 7.12). Some measures of association between nominal and ordinal variables are described by Agresti [1981]. Methods for estimating missing values are discussed by Wishart [1978b] and Gordon [1981: Section 2.4.3].

7.3 HIERARCHICAL CLUSTERING: AGGLOMERATIVE TECHNIQUES

The agglomerative methods all begin with n clusters each containing just one object, a proximity matrix for the n objects (we assume, for the moment, that this is an $n \times n$ matrix $\mathbf{D} = [(d_{rs})]$ of dissimilarities), and a measure of distance between two clusters, where each cluster contains one or more objects. The first step is to fuse the two nearest objects into a single cluster so that we now have $n - 2$ clusters containing one object each and a single cluster of two objects. The second step is to fuse the two nearest of the $n - 1$ clusters to form

$n - 2$ clusters. We continue in this manner until at the $(n - 1)$th step we fuse the two clusters left into a single cluster of n objects. A number of different distance measures for clusters have been proposed and the ones that are or have been widely used are described below.

7.3.1 *Some Commonly Used Methods*

a SINGLE LINKAGE (NEAREST NEIGHBOR) METHOD

If C_1 and C_2 are two clusters, then the distance between them is defined to be the smallest dissimilarity between a member of C_1 and a member of C_2 (Sneath [1957], Sokal and Sneath [1963], Johnson [1967]), namely,

$$d_{(C_1)(C_2)} = \min\{d_{rs} : r \in C_1, s \in C_2\}, \tag{7.11}$$

where r denotes "object r." We demonstrate the fusion process with the following simple example. Let

$$\mathbf{D} = \begin{matrix} \\ 1 \\ 2 \\ 3 \\ 4 \\ 5 \end{matrix} \begin{matrix} \begin{matrix} 1 & 2 & 3 & 4 & 5 \end{matrix} \\ \begin{pmatrix} 0 & 7.0 & 1.0 & 9.0 & 8.0 \\ 7.0 & 0 & 6.0 & 3.0 & 5.0 \\ 1.0 & 6.0 & 0 & 8.0 & 7.0 \\ 9.0 & 3.0 & 8.0 & 0 & 4.0 \\ 8.0 & 5.0 & 7.0 & 4.0 & 0 \end{pmatrix} \end{matrix}. \tag{7.12}$$

The minimum d_{rs} is $a_1 = d_{13} = 1.0$ so that objects 1 and 3 are joined and our clusters are (1, 3), (2), (4), and (5). Now

$$d_{(2)(1,3)} = \min\{d_{21}, d_{23}\} = d_{23} = 6.0,$$

$$d_{(4)(1,3)} = \min\{d_{41}, d_{43}\} = d_{43} = 8.0,$$

$$d_{(5)(1,3)} = \min\{d_{51}, d_{53}\} = d_{53} = 7.0,$$

and the distance matrix for the clusters is

$$\mathbf{D}_1 = \begin{matrix} \\ (1,3) \\ 2 \\ 4 \\ 5 \end{matrix} \begin{matrix} \begin{matrix} (1,3) & 2 & 4 & 5 \end{matrix} \\ \begin{pmatrix} 0 & 6.0 & 8.0 & 7.0 \\ 6.0 & 0 & 3.0 & 5.0 \\ 8.0 & 3.0 & 0 & 4.0 \\ 7.0 & 5.0 & 4.0 & 0 \end{pmatrix} \end{matrix}.$$

The smallest entry is $a_2 = d_{24} = 3.0$, so that objects 2 and 4 are joined and our clusters become (1, 3), (2, 4), and (5), with

$$d_{(1,3)(2,4)} = \min\{d_{(2)(1,3)}, d_{(4)(1,3)}\} = 6.0,$$

$$d_{(5)(2,4)} = \min\{d_{52}, d_{54}\} = d_{54} = 4.0$$

and

$$\mathbf{D}_2 = \begin{array}{c} \\ (1,3) \\ (2,4) \\ 5 \end{array} \begin{array}{c} \begin{array}{ccc} (1,3) & (2,4) & 5 \end{array} \\ \begin{pmatrix} 0 & 6.0 & 7.0 \\ 6.0 & 0 & 4.0 \\ 7.0 & 4.0 & 0 \end{pmatrix} \end{array}.$$

The smallest entry is $a_3 = d_{(5)(2,4)} = 4.0$, so that object 5 is joined to cluster (2, 4) and the clusters are now (1, 3) and (2, 4, 5). Finally, these two clusters are fused to give the single cluster (1, 2, 3, 4, 5). We note that

$$\begin{aligned} d_{(1,3)(2,4,5)} &= \min\{d_{(1,3)(2,4)}, d_{(1,3)(5)}\} \\ &= d_{32} = 6.0 \qquad (= a_4, \text{ say}). \end{aligned}$$

The above process can be described diagrammatically in the form of a dendrogram as shown in Fig. 7.5. The vertical scale gives a measure of the size of a cluster; tight clusters tend to have lower values. In constructing dendrograms, some relabeling is generally needed so that each cluster is a contiguous sequence of objects, for example, the interchange of 2 and 3 as in Fig. 7.5 (for algorithms, see Exercises 7.13 and 7.19). Although the above technique of spelling out $\mathbf{D}, \mathbf{D}_1, \mathbf{D}_2$, and so on, is a general one and can be applied to other agglomerative methods, it can be simplified for single linkage using the ordered d_{rs} as in Table 7.3.

b COMPLETE LINKAGE (FARTHEST NEIGHBOR) METHOD

This method is the opposite of the single linkage method in that the distance between two clusters is defined in terms of the largest dissimilarity between a

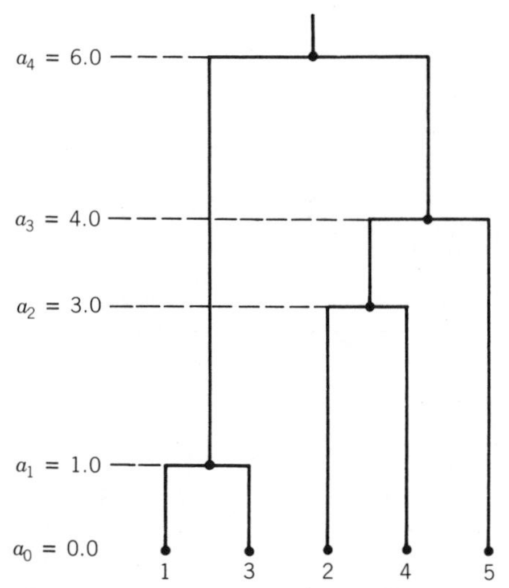

Fig. 7.5 Single linkage dendrogram for dissimilarity matrix (7.12).

member of C_1 and a member of C_2 (Sokal and Sneath [1963], McQuitty [1964]), namely,

$$d_{(C_1)(C_2)} = \max\{d_{rs} : r \in C_1, s \in C_2\}.$$

At each step we fuse the two clusters that are closest, that is, have minimum $d_{(C_1)(C_2)}$. A dendrogram can be constructed as for the single linkage method using these minimum values (see Exercise 7.14).

c CENTROID METHOD

The distance between two clusters is defined to be the "distance" between the cluster centroids (Sokal and Michener [1958], King [1966, 1967]). If

$$\bar{\mathbf{x}}_1 = \sum_{i \in C_1} \frac{\mathbf{x}_i}{n_1} \tag{7.13}$$

is the centroid of the n_1 members of C_1 and $\bar{\mathbf{x}}_2$ is similarly defined for C_2, then

$$d_{(C_1)(C_2)} = P(\bar{\mathbf{x}}_1, \bar{\mathbf{x}}_2), \tag{7.14}$$

where P is a proximity measure such as the correlation of (7.7), the squared Euclidean distance $\|\bar{\mathbf{x}}_1 - \bar{\mathbf{x}}_2\|^2$, or some other dissimilarity such as $\Delta_1(\bar{\mathbf{x}}_1, \bar{\mathbf{x}}_2)$ [see (7.1)]. We would start with a proximity matrix with elements $P(\mathbf{x}_r, \mathbf{x}_s)$ and, at each stage, the two nearest clusters are fused and replaced by the centroid of the new cluster. The centroid of $C_1 \cup C_2$, the fusion of C_1 and C_2, is given by the weighted average

$$\bar{\mathbf{x}} = \frac{n_1\bar{\mathbf{x}}_1 + n_2\bar{\mathbf{x}}_2}{n_1 + n_2}. \tag{7.15}$$

TABLE 7.3 Single-Linkage Clustering for Dissimilarity Matrix (7.12)

Ordered Distances	Clusters
$d_{13} = 1.0$	(1, 3), (2), (4), (5)
$d_{24} = 3.0$	(1, 3), (2, 4), (5)
$d_{45} = 4.0$	(1, 3), (2, 4, 5)
$d_{25} = 5.0$	(1, 3), (2, 4, 5)
$d_{23} = 6.0$	(1, 2, 3, 4, 5)
$d_{35} = 7.0$	(1, 2, 3, 4, 5)
$d_{15} = 8.0$	(1, 2, 3, 4, 5)
$d_{14} = 9.0$	(1, 2, 3, 4, 5)

d INCREMENTAL SUM OF SQUARES METHOD

Using an idea suggested by Ward [1963] for the case of univariate data, Wishart [1969a] suggested fusing the two clusters that minimize $I_{(C_1)(C_2)}$, the increase in the total within-cluster sum of squares of the distances from the respective centroids on fusing. If $\bar{\mathbf{x}}$ is defined by (7.15), then (see Exercise 7.15)

$$I_{(C_1)(C_2)} = \sum_{i \in C_1 \cup C_2} \|\mathbf{x}_i - \bar{\mathbf{x}}\|^2 - \left\{ \sum_{i \in C_1} \|\mathbf{x}_i - \bar{\mathbf{x}}_1\|^2 + \sum_{i \in C_2} \|\mathbf{x}_i - \bar{\mathbf{x}}_2\|^2 \right\}$$

$$= \sum_{\alpha=1}^{2} n_\alpha \|\bar{\mathbf{x}}_\alpha - \bar{\mathbf{x}}\|^2 \tag{7.16}$$

$$= \frac{n_1 n_2}{n_1 + n_2} \|\bar{\mathbf{x}}_1 - \bar{\mathbf{x}}_2\|^2. \tag{7.17}$$

In particular, for objects r and s,

$$I_{(r)(s)} = \tfrac{1}{2}\|\mathbf{x}_r - \mathbf{x}_s\|^2 = \tfrac{1}{2} d_{rs}^2.$$

Beginning with $\mathbf{D} = [(d_{rs}^2)]$, we can define the "distance" between two clusters by

$$d_{(C_1)(C_2)} = 2 I_{(C_1)(C_2)}. \tag{7.18}$$

This method was proposed independently by several other authors under the names of minimum variance clustering, sum of squares method [Orlóci [1967]), incremental sum of squares (Burr [1968, 1970]), and Ward's method (Wishart [1969a]). A generalization of the method that allows for missing observations has been given by Wishart [1978b].

e MEDIAN METHOD

This is the same as the centroid method, except that a new cluster is replaced by the *unweighted* average, $\bar{\mathbf{x}} = \frac{1}{2}(\bar{\mathbf{x}}_1 + \bar{\mathbf{x}}_2)$ (Gower [1967a]). The method was introduced to overcome a shortcoming of the centroid method, namely, that if a small group fuses with a large one, it loses its identity and the new centroid may lie in the large group.

f GROUP AVERAGE METHOD

The distance between C_1 and C_2 is defined to be the average of the $n_1 n_2$ dissimilarities between all pairs (Sokal and Michener [1958], McQuitty [1964],

Lance and Williams [1966]), namely,

$$d_{(C_1)(C_2)} = \frac{1}{n_1 n_2} \sum_{r \in C_1} \sum_{s \in C_2} d_{rs}.$$

g LANCE AND WILLIAMS FLEXIBLE METHOD

Lance and Williams [1967a] showed that the preceding methods labeled a–c, e, and f (with $P(\bar{\mathbf{x}}_1, \bar{\mathbf{x}}_2) = \|\bar{\mathbf{x}}_1 - \bar{\mathbf{x}}_2\|^2$ in the centroid method) are special cases of the following formula for the distance between clusters C_3 and $C_1 \cup C_2$:

$$d_{(C_3)(C_1 \cup C_2)} = \alpha_1 d_{(C_3)(C_1)} + \alpha_2 d_{(C_3)(C_2)} + \beta d_{(C_1)(C_2)} + \gamma |d_{(C_3)(C_1)} - d_{(C_3)(C_2)}|. \tag{7.19}$$

Wishart [1969a] then showed that the incremental sum of squares method also satisfies the above formula (see Exercise 7.17), and the values of the parameters are given in Table 7.4; n_i is the number of objects in cluster C_i $(i = 1, 2, 3)$. Lance and Williams [1967a] suggested using a flexible scheme satisfying the constraints $\alpha_1 + \alpha_2 + \beta = 1$, $\alpha_1 = \alpha_2$, $\beta < 1$, and $\gamma = 0$, and recommended a small negative value of β such as $\beta = -0.25$. Sibson [1971] noted that (7.19) is symmetric with regard to C_1 and C_2 for all the methods in Table 7.4 so that these methods are independent of labeling.

Using the recurrence relationship (7.19), it is easy to devise a general program that incorporates all the above methods (e.g., CLUSTAN, produced by Wishart [1978a], and CLASS, described by Lance and Williams [1967a] and

TABLE 7.4 Parameters for Lance and Williams [1967] Recurrence Formula (7.19)

	Parameter		
	α_i	β	γ
1. Nearest neighbor	$\frac{1}{2}$	0	$-\frac{1}{2}$
2. Farthest neighbor	$\frac{1}{2}$	0	$\frac{1}{2}$
3. Centroid	$\dfrac{n_i}{n_1 + n_2}$	$\dfrac{-n_1 n_2}{(n_1 + n_2)^2}$	0
4. Incremental	$\dfrac{n_i + n_3}{n_1 + n_2 + n_3}$	$\dfrac{-n_3}{n_1 + n_2 + n_3}$	0
5. Median	$\frac{1}{2}$	$-\frac{1}{4}$	0
6. Group average	$\dfrac{n_i}{n_1 + n_2}$	0	0
7. Flexible	$\frac{1}{2}(1 - \beta)$	$\beta(< 1)$	0

Williams and Lance [1977]). It is appropriate to start with $\mathbf{D} = [(d_{(r)(s)})]$, so that the definition of $d_{(C_1)(C_2)}$ holds for clusters of single objects. This means that in some methods (c, d, and e) we use $d_{(r)(s)} = d_{rs}^2$, whereas in others (a, b, and f) $d_{(r)(s)} = d_{rs}$, though in the latter cases (7.19) still holds if we work with d_{rs}^2 instead of d_{rs}.

Most of the methods can also be applied to similarities instead of dissimilarities, provided that the measure makes sense. For example, an average correlation in the group average method is generally unacceptable, as correlations can be negative. Also, the intuitive geometrical properties of the centroid and median methods are lost if correlations are used. In general, we can simply interpret nearness in terms of a small dissimilarity or a large similarity, and all that is required is to interchange the words *maximum* and *minimum* in the appropriate places. In the single linkage method, for example, we can define similarity as

$$c_{(C_1)(C_2)} = \max\{c_{rs} : r \in C_1, s \in C_2\},$$

and we fuse the two clusters with maximum similarity.

h INFORMATION MEASURES

For noncontinuous data, another method, based on Shannon's information measure (Shannon and Weaver [1963]), has been proposed. If the d variables are all dichotomous, and a_j objects in a cluster C of m objects possess characteristic j ($j = 1, 2, \ldots, d$), then the Shannon information content of C is

$$I_C = dm \log m - \sum_{j=1}^{d} \left[a_j \log a_j + (m - a_j)\log(m - a_j)\right]. \qquad (7.20)$$

We fuse the two clusters that have the smallest gain

$$\Delta I = I_{C_1 \cup C_2} - \left(I_{C_1} + I_{C_2}\right). \qquad (7.21)$$

To start the process we note that I_C is zero for a cluster of one object. The method is insensitive to "skewed" variables, for example, rare attributes missing from most of the objects. It can also be adapted to categorical and continuous variables, though there are difficulties (Williams and Lance [1977]). For example, the information gain from a multistate variable can be rather large and tend to dominate the clustering process; the number of states should be kept small (no more than, say, 5) where possible. Continuous variables have to be converted into discrete variables with a consequent loss of information. For this reason the method is best avoided in most situations where the variables are largely quantitative.

Another information measure that has been used is that of Brillouin [1962], which uses the function $\log n!$ rather than $n \log n$ in (7.20). However, there is not a great deal of difference between the two measures, as one is monotonic with respect to the other (see Clifford and Stephenson [1975: Appendix A] for an elementary summary). In general, the Shannon measure applies to a sample, whereas Brillouin's measure is appropriate for a population. The latter is discussed in some detail by Pielou [1966, 1969].

A new entropy-based technique for binary data has been proposed by Buser and Baroni-Urbani [1982].

i OTHER METHODS

Two other methods that provide a compromise between the extremes of single and complete linkage are k-clustering (see Ling [1972, 1973a], Matula [1977: p. 105]) and r-diameter clustering (Hubert and Baker [1977]). These methods have not yet received much attention in the literature and their main use at present seems to be for comparing clustered data with "random" data. D'Andrade [1978] has proposed a third method, called U-statistic hierarchical clustering, that apparently performs better than either single or complete linkage on the basis of the γ-criterion [see (7.25)].

EXAMPLE 7.1 In a study of English dialects, Morgan [1981] compared 25 East Midland villages to see if they use the same words for 60 items. The measure of similarity between two villages is the percentage of items for which

TABLE 7.5 Similarity Matrix (Lower Triangle) of Dialect Similarities Between 25 East Midland Villages[a]

	Nt1	Nt3	Nt4	L1	L3	L4	L6	L8	L11	L12	L13	L15	Lei1	Lei4	Lei8	Lei10	R1	Nth1	Nth3	Nth4	Nth5	Hu1	C1	Bk1
Nt3	71																							
Nt4	58	57																						
L1	49	45	48																					
L3	63	63	47	59																				
L4	64	66	50	53	71																			
L6	71	75	52	53	71	68																		
L8	52	56	36	34	60	58	69																	
L11	46	50	57	33	42	43	43	44																
L12	61	49	52	40	58	61	56	61	52															
L13	57	60	56	35	53	48	55	48	63	59														
L15	39	46	45	30	42	40	47	44	50	53	60													
Lei1	42	50	53	28	41	36	43	39	48	47	58	48												
Lei4	32	34	47	20	27	29	31	23	44	39	43	39	63											
Lei8	32	39	50	19	25	25	36	37	43	41	48	49	64	63										
Lei10	23	27	42	14	20	22	24	28	36	27	38	35	54	62	68									
R1	41	47	56	25	38	42	46	38	48	54	54	48	72	57	61	59								
Nth1	39	42	48	24	37	34	36	42	43	49	60	56	59	51	56	47	54							
Nth3	32	36	43	22	22	24	34	29	47	34	45	47	38	46	51	42	44	53						
Nth4	27	36	38	19	22	20	25	25	40	25	40	40	45	49	54	49	42	44	63					
Nth5	28	37	37	20	25	25	31	33	42	29	41	37	46	48	49	47	43	44	58	59				
Hu1	26	26	30	20	21	28	28	28	41	33	39	55	34	33	40	33	38	40	58	54	47			
C1	30	33	32	16	25	26	33	32	41	37	37	46	47	46	49	39	46	58	42	44	42	50		
Bk1	36	49	45	26	29	31	41	32	47	32	52	46	57	49	56	49	54	53	63	68	73	51	51	
Bd1	31	44	40	23	29	32	32	31	47	33	43	45	45	47	53	43	46	53	60	61	62	55	54	72

[a]From Morgan [1981: Table 1].

the same word is used. The similarity matrix is given in Table 7.5 and the coding is the following: Nt (Nottinghamshire), Lei (Leicestershire), Nth (Northamptonshire), L (Lincolnshire), R (Rutland), Hu (Huntingdonshire), C (Cambridgeshire), Bk (Buckinghamshire), and Bd (Bedfordshire). The data are part of a larger survey of over 300 villages. Using single linkage, with similarities instead of dissimilarities, the dendrogram in Fig. 7.6 was constructed, and from this Morgan obtained two sets of clusters corresponding to "threshold" similarities of 57.5% and 60.5%, as shown in Fig. 7.7. This separation of clusters suggest a north–south dichotomy of the villages that in fact corresponds to a series of east–west pronounciation boundaries, or isoglosses, thus implying a relationship between what we speak and how we speak.

In Section 5.9 the minimum spanning tree (MST) is mentioned as a useful graphical tool to assess the distortion that may exist in a two-dimensional representation of proximity data, in this case a map. Morgan [1981] produced the MST for the dialect data (Fig. 7.8) and it highlights villages Bd1, Lei1 and L13, which have four or more links with other villages. In this example we need to watch out for ties. Morgan noted that Fig. 7.7 gives an acceptable picture of geographically coherent groups of villages. Also, outlying villages are less well attached to their clusters, for example, C1 is probably more similar to East Anglian dialects.

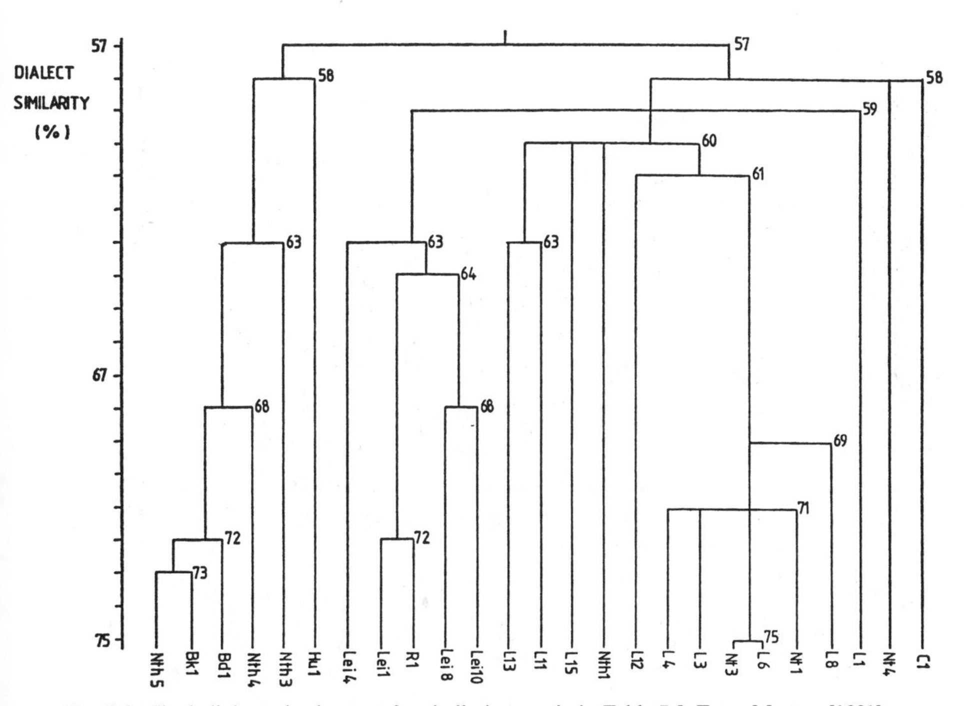

Fig. 7.6 Single linkage dendrogram for similarity matrix in Table 7.5. From Morgan [1981].

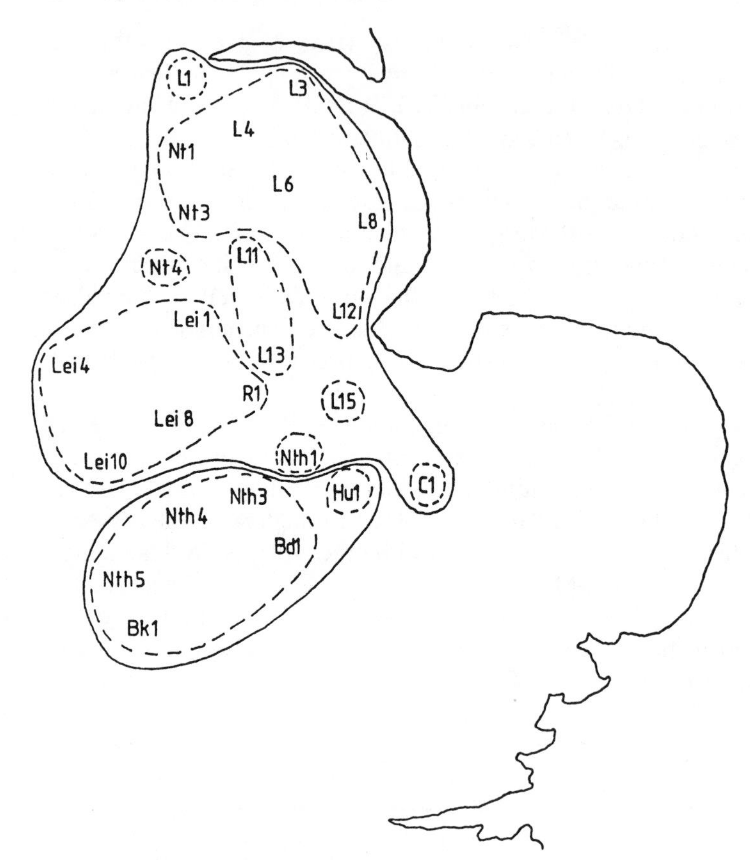

Fig. 7.7 Two sets of single linkage clusters resulting from sections of the dendrogram of Fig. 7.6: ——, 51.5% similarity; ---, 60.5% similarity. From Morgan [1981].

7.3.2 *Comparison of Methods*

a MONOTONICITY

If the fusion process is to be visually represented by a dendrogram with its reasonable property of increasing ordinate values (see Fig. 7.5), then we require $a_k = \min d_{(C_1)(C_2)}$ at step k to be monotonically increasing with k. Now what happens at steps k and $k+1$ can be described in terms of four clusters C_i $(i = 1, 2, 3, 4)$. Suppose at step k we form $C_1 \cup C_2$, that is,

$$a_k = d_{(C_1)(C_2)} < \min\left\{ d_{(C_1)(C_3)}, d_{(C_2)(C_3)}, d_{(C_3)(C_4)} \right\}, \tag{7.22}$$

and at step $k+1$ we form either (1) $C_3 \cup C_4$ or (2) $C_1 \cup C_2 \cup C_3$. In case (1) (7.22) implies that $a_{k+1} = d_{(C_3)(C_4)} > a_k$. In case (2) we find that for the single

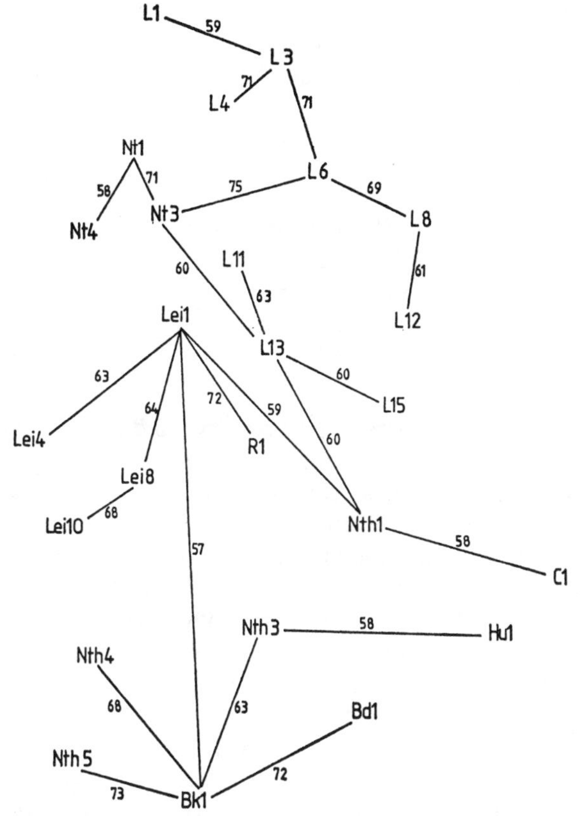

Fig. 7.8 Minimum spanning tree for data in Table 7.5 showing similarities between linked pairs of villages. From Morgan [1981].

linkage and complete linkage methods [with $\gamma \neq 0$ in (7.19)], $d_{(C_3)(C_1 \cup C_2)} = a_{k+1}$ is the minimum (respectively maximum) of $d_{(C_3)(C_1)}$ and $d_{(C_3)(C_2)}$. As both the latter are greater than $d_{(C_1)(C_2)}$, $a_{k+1} > a_k$. For the other methods (Table 7.4; $\gamma = 0$, $\alpha_i > 0$ for $i = 1, 2$), we find from (7.19) and (7.22) that

$$d_{(C_3)(C_1 \cup C_2)} > (\alpha_1 + \alpha_2 + \beta) d_{(C_1)(C_2)},$$

so that we require $\alpha_1 + \alpha_2 + \beta \geq 1$ for monotonicity (Lance and Williams [1967a]). Hence all the methods in Table 7.4 are monotonic except the centroid and median methods with which frequent "reversals" are possible at any particular stage, that is, $a_{k+1} < a_k$. These latter methods are therefore not recommended and are regarded as obsolete by several authors. Although Burr [1970] has shown that information methods using ΔI as ordinate of the dendrogram are not necessarily monotonic, Williams and Lance [1977] state that "monotonicity failure in these cases is so rare that it can be disregarded." The method is monotonic if we use I as the ordinate rather than ΔI.

Referring to Table 7.3, we have the following steps in the single linkage method:

	Group of Clusters	a_k
Step 0:	$\mathscr{C}_0 = \{(1),(2),(3),(4),(5)\}$	$a_0 = 0.0$
Step 1:	$\mathscr{C}_1 = \{(1,3),(2),(4),(5)\}$	$a_1 = 1.0$
Step 2:	$\mathscr{C}_2 = \{(1,3),(2,4),(5)\}$	$a_2 = 3.0$
Step 3:	$\mathscr{C}_3 = \{(1,3),(2,4,5)\}$	$a_3 = 4.0$
Step 4:	$\mathscr{C}_4 = \{(1,2,3,4,5)\}$	$a_4 = 6.0$

where $\mathscr{C}_k$ is the group of clusters formed at step k and a_k is the current value of minimum $d_{(C_1)(C_2)}$ ($a_0 = 0$, by definition). In general, any monotonic method that gives rise to a hierarchical clustering scheme with $\mathscr{C}_0, \mathscr{C}_1, \ldots, \mathscr{C}_m$ and $0 = a_0 < a_1 < \cdots < a_m$ can also be described in terms of a certain function called an ultrametric. Following Johnson [1967], let k be the lowest level at which objects r and s occur in the same cluster, that is, k is the smallest number in $(0, 1, \ldots, m)$ such that objects r and s are in the same cluster in $\mathscr{C}_k$ (called the *partition rank* by Hubert [1974]). Then we define the function δ from $\mathscr{P} \times \mathscr{P}$, $\mathscr{P}$ being the population of objects, to the real line by

$$\delta(r,s) = a_k \qquad (k = 0, 1, \ldots, m).$$

Now r and r are in the same cluster for all $\mathscr{C}_r$ so that $\delta(r,r) = 0$ as $k = 0$ is the smallest number. Conversely, if $\delta(r,s) = 0$ for some r and s, then r and s are in the same cluster in $\mathscr{C}_0$ and $r = s$. Clearly, $\delta(r,s) = \delta(s,r)$ and δ will be a metric if it satisfies the triangle inequality. Let r, s, t be three objects with $\delta(r,s) = a_i$ and $\delta(s,t) = a_j$. Define $k = \max[i, j]$. Then objects r and s are in the same cluster in $\mathscr{C}_i$, and s and t are in the same cluster in $\mathscr{C}_j$. Because the groups of clusters are hierarchical, one of these clusters equals or includes the other, namely, the one belonging to $\mathscr{C}_k$. Thus r, s, t all belong to the same cluster in $\mathscr{C}_k$ so that

$$\delta(r,t) \leq a_k = \max[a_i, a_j]$$

or

$$\delta(r,t) \leq \max[\delta(r,s), \delta(s,t)]. \tag{7.23}$$

This is called the ultrametric inequality and is stronger than the triangle inequality in that

$$\max[\delta(r,s), \delta(s,t)] \leq \delta(r,s) + \delta(s,t). \tag{7.24}$$

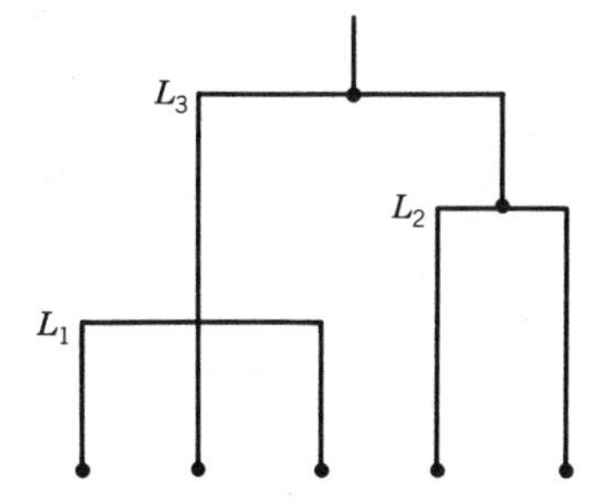

Fig. 7.9 Dendrogram with a tie at L_1.

Thus δ is a metric satisfying the ultrametric inequality, and constructing a (monotonic) dendrogram from proximity data is equivalent to imposing an ultrametric on the proximity matrix. This equivalence was also demonstrated by Hartigan [1967] and Jardine et al. [1967] (see also Jardine and Sibson [1971b]). We note that, as in Johnson's proof, the above argument still holds for $a_0 \le a_1 \le \cdots \le a_m$, where the equality of two a_k's represents a tie (see Fig. 7.9).

The above approach can also be used as a basis for comparing a dendrogram with the original proximity matrix by comparing the pairs d_{rs} and $\delta(r, s)$. Various measures of "distortion" have been proposed (see Cormack [1971: Table 3] and Orlóci [1978: pp. 264–276]) such as the correlation r_0, the so-called cophenetic correlation coefficient (see Farris [1969], Sneath and Sokal [1973], Rohlf [1974a]). It is suggested (Everitt [1978: p. 46]) that if r_0 is very high, say, about 0.95, then the hierarchical clustering model represented by the dendrogram is a reasonable fit to the data. However, if r_0 is much lower, say, 0.6 or 0.7, then the model is questionable. In this case one could also carry out one of the dimension-reducing or ordination techniques such as multidimensional scaling and compare the clusters suggested by the hierarchical cluster analysis with a two-dimensional reduction, as in Fig. 7.10.

In the case of single and complete linkage, ranks are more appropriate for comparison, as both methods are invariant under monotonic transformations of the proximities. We can compare the ranks of the d_{rs} with the corresponding partition ranks, which are the same as the ranks given by the $\delta(r, s)$. A popular measure based on ranks is the γ-statistic proposed by Goodman and Kruskal [1954], which, although somewhat arbitrary, has several desirable properties (Hubert [1974]). The γ-index can be characterized as the difference between two conditional probabilities, namely,

$$\gamma = \text{pr}[\text{consistent ranking}] - \text{pr}[\text{inconsistent ranking}], \qquad (7.25)$$

where a consistent ranking for two pairs of objects implies that both orderings give the higher rank to the same pair, that is, if $d_{ij} < d_{rs}$, then $\delta(i, j) < \delta(r, s)$. As several objects may link up in a cluster for the first time (as in Fig. 7.9), there may be some ties in the partition rank. However, omitting any ties in the

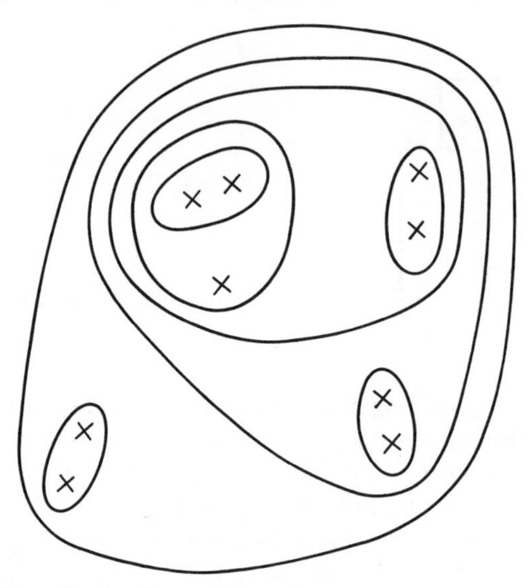

Fig. 7.10 Multidimensional scaling and hierarchical clustering shown on same diagram.

original proximities or partition ranks and using "untied" data only,

$$\gamma = \frac{C - D}{C + D}, \tag{7.26}$$

where C is the number of concordances and D the number of discordances. Baker and Hubert [1975] also suggest the measure $\alpha = D/(C + D)$ $(= \frac{1}{2}[1 - \gamma])$.

In general, a dendrogram representation of a proximity matrix is likely to be useful only if the data are strongly clustered and have a hierarchical type of structure; otherwise, as demonstrated by Everitt [1974: Chapter 4], the dendrogram can be very misleading. For instance, in a dendrogram displayed on a printed page the objects have to be linearly ordered (see Exercise 7.19). Such an order may be misleading, as two adjacent objects on the page may be far apart and may stay in different clusters right up until the final fusion of all the data into a single cluster. At the same time objects in widely separated stems of the dendrogram may be more similar to each other than individuals in their own stems. For example, in Fig. 7.11, where we have two clusters, A is closer to C than D is to C. If the cluster started growing from point E, then C and D would join up at about the same time and could even end up next to each other in the linear ordering on the page.

Once a cluster analysis has been carried out, the sequence of objects can be reordered so that in the corresponding permuted proximity matrix, members of the same cluster always lie in consecutive rows and columns. The sizes of the individual proximities can then be represented by dots or symbols of increas-

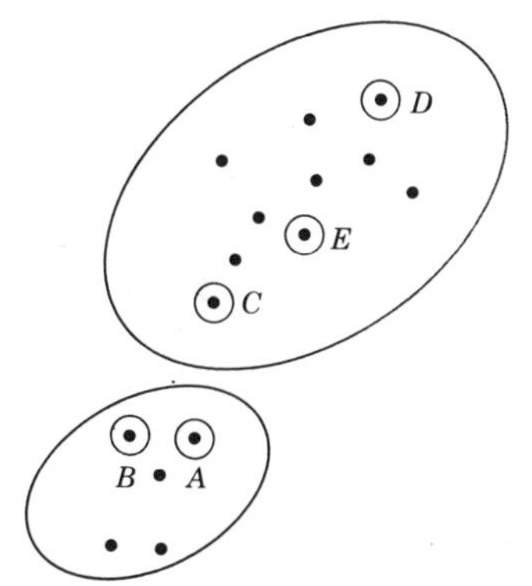

Fig. 7.11 Two clusters with points in different clusters being closer than points in the same cluster.

ing size and darkness so that, in the case of dissimilarities, for example, clusters will show up as dark triangles near the diagonal of the dissimilarity matrix. This graphical representation of a proximity matrix forms the basis of a procedure, suggested by Ling [1973b], called SHADE (see Everitt [1978] for details and examples).

A measure of stability or replicability has been proposed by Mezzich [1975] (see Solomon [1977]). The data are split randomly in two and the clustering method is applied to both sets. The two sets of $\delta(r, s)$ can be compared using a measure of distortion.

b SPATIAL PROPERTIES

The initial proximities may be regarded as defining a space with known properties. However, as clusters form, it does not follow that intercluster distances define a space with the original properties. If the properties are retained, Lance and Williams [1967a] (see also Williams et al. [1971a], and Williams and Lance [1977]) describe the method as space conserving; otherwise they call it space distorting with *contraction* or *dilation*. Although these terms are not formally defined, the authors describe a space-contracting method as one in which a cluster will appear, on formation, to move closer to some or all of the remaining objects so that an individual will tend to add to a preexisting cluster rather than act as the nucleus of a new cluster. A more common description of the phenomenon is *chaining*, whereby objects are linked together by a chain of intermediaries, giving straggly clusters.

In a space-dilating system, clusters appear to "recede" on formation and growth, so that individuals not yet in clusters are more likely to form the nuclei of new clusters. We now have the reverse of chaining, with clusters appearing to be more distinct than they really are. In particular, "nonconformist" clusters (e.g., outliers) can develop consisting largely of rejects from other clusters (see Watson et al. [1966] for an example).

The single linkage (nearest neighbor) method is intensely space contracting and its chaining tendencies are well known (e.g., Williams et al. [1966]). For example, in Fig. 7.12 the cluster of three points on the left would grow first and spread to the second cluster through the link in the middle so that a single cluster would result. We thus find that a single nonrepresentative proximity value can force the union of two otherwise dissimilar clusters at a premature level in the hierarchy. For these and other reasons there has been considerable controversy over the usefulness of the single linkage method (see the papers by Sibson [1971], Jardine and Sibson [1971a], and Williams et al. [1971b] in Vol. 14 of *The Computer Journal*). We saw, in the previous section, that a dendrogram with monotonically increasing levels is characterized by an ultrametric transformation. Jardine and Sibson [1968] argue that this transformation should satisfy a number of criteria and show that only the single linkage method satisfied all the criteria, the implication being that it is the only acceptable method. Sibson [1971] considers the problem of ties (equal proximities or intercluster distances) and, assuming random choice, demonstrates that of all the methods in Table 7.4, only single linkage is independent of choice. Although ties may be rare in practice, they are inadequately discussed in the literature. Several solutions are suggested by Williams et al. [1971b], who question the need for cluster methods to satisfy all of Jardine and Sibson's proposed criteria and suggest a more pragmatic approach to clustering on the grounds that they do not find single linkage useful. They note that the ultrametric approach excludes the information statistic method: Here the ordinate values of the dendrogram must be calculated from the original data, and not from a proximity matrix.

With regard to single linkage, Williams and Lance [1977] go as far as to say that "we never advocate its use unless there are special reasons for doing so." Their stance is supported by the fact that single linkage is not very popular. Also, several studies comparing derived clusters with known structures (see Cunningham and Ogilvie [1972], Kuiper and Fisher [1975], Blashfield [1976], Edelbrock and McLaughlin [1980], Golden and Meehl [1980]) suggest that single linkage is generally the least successful, with the group average and incremental sum of squares methods doing reasonably well overall. The power studies of Baker and Hubert [1975] also indicate that single linkage is of limited usefulness. Baker [1974] shows that single linkage is more sensitive to certain types of data errors than complete linkage, while Hubert [1974] shows

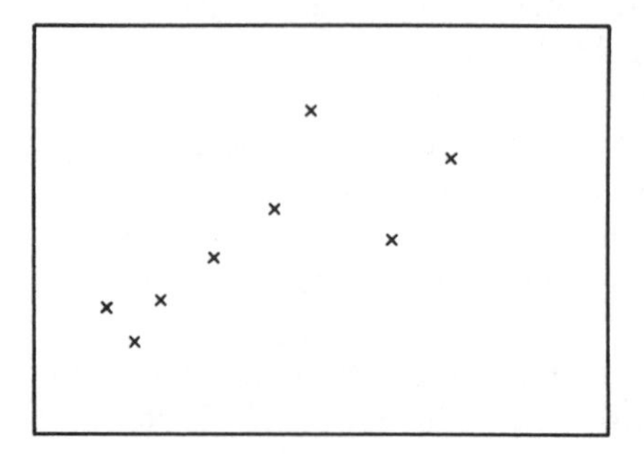

Fig. 7.12 Effect of chaining on single linkage.

that the latter tends to produce a partition rank closer to the original ranking when measured by γ of (7.25). The superiority of complete linkage over single linkage is also demonstrated by Solomon [1977]. However, the relative performances of the various methods are very dependent on the type of input data and the distortion measure used for the comparison (Hartigan [1977: p. 61]). Also, Hartigan [1981] notes that single linkage has certain consistency properties for high-density clusters.

While single linkage is intensely space contracting, complete linkage is intensely space dilating, with the tendency to artificially sharpen cluster boundaries. A major weakness of complete linkage is its overall sensitivity to a single change in the rank order of the proximities. Matula [1977: p. 115] gives an example of this and states that "this extreme sensitivity of the overall clustering provided by complete-linkage to a single transposition in the proximity data order suggests that a practitioner is well advised to avoid the complete-linkage method."

The spatial properties of the other agglomerative methods tend to be between the two extremes of single and complete linkage. The centroid and group average methods are space conserving, the incremental sum of squares is space dilating, and the information statistic approach is strongly space dilating and must be used with caution (Williams and Lance [1977: p. 280]). Because of its space-dilating properties, the information statistic can also produce nonconformist groups. For instance, Edye et al. [1970] gave an example where a set of 51 objects produced two such groups. Sneath [1969] has pointed out that although clustering tends to produce low entropy (a quantity closely related to information), a regular distribution also has low entropy.

The properties of the flexible method depend on the value of β. If $\beta = 0$, the method is space conserving; as β becomes positive, the method becomes increasingly space contracting; and as β becomes negative, the method becomes increasingly space dilating (Williams et al. [1971a]). It appears that some dilation is appropriate to sharpen up cluster boundaries, but too much can lead to an artificial proliferation of clusters. A value of $\beta = -0.25$ is used by Williams and his co-workers in the "Canberra school."

It is clear that the above methods will be affected by outliers (Kuiper and Fisher [1975]) and, from a practical point of view, hierarchical clustering can provide a useful visual method for pointing out outliers.

c COMPUTATIONAL EFFORT

For the iterative schemes in Table 7.4, based on (7.19), the distance between clusters can be calculated directly from the distances for the previous step so that the original proximity matrix need not be stored. A hierarchical method with this property is called *combinatorial* by Lance and Williams [1967a]. The information method, and the centroid method with a different dissimilarity such as the L_1 metric, require all the data at each step and are therefore not combinatorial. Given the $n \times n$ proximity matrix, combinatorial methods require about $\frac{1}{2}n(n-1)$ operations, whereas noncombinatorial methods require about $d(n-1)^2$ operations. If the proximities have to be calculated from the raw data, then a further $\frac{1}{2}dn(n-1)$ operations involving $\binom{n}{2}$ pairs and d variables are required.

Sibson [1973] has produced a very efficient algorithm for single linkage that can handle very large data sets of, say, 10^3–10^4 observations, while Defay [1977] and Hansen and Delattre [1978] have given efficient algorithms for complete linkage. Williams and Lance [1977] describe a suite of programs for both agglomerative methods and the divisive methods (described below). Dixon and Brown [1977] use the group average method in their algorithm BMDP-2M. Some general algorithms for handling very large data sets are given by Bruynooghe [1978]. For a useful computer package see CLUSTAN (available from St Andrews University, Scotland).

7.4 HIERARCHICAL CLUSTERING: DIVISIVE TECHNIQUES

We mentioned in Section 7.1 that hierarchical methods of clustering are of two types, agglomerative and divisive. Williams and Lance [1977] note the following advantages of the divisive over agglomerative procedures: (1) The process begins at maximum information content; (2) the division need not be continued until we have n clusters of one object; and (3) if there are fewer variables or characteristics than objects, the computation needed is less, the time depending on d^2 rather than approximately $(n-1)^2$, as with the agglomerative methods.

With divisive methods, the first step is to divide the n objects into two groups. This can be done in $2^{n-1}-1$ ways, so that even with a large computer it is only possible to examine every such division if n is small. With moderate values of n, only a restricted search is possible. For this reason most of the early techniques are *monothetic*, that is, the split is based on a single variable only. Unfortunately, such a method is sensitive to errors in recording or coding the variable used for the division so that outliers (e.g., a nontypical measurement, the presence of a rare trait, the absence of a common trait) can lead to progression down a wrong branch of the hierarchy. By contrast, agglomerative methods are *polythetic* by nature, as the fusion process is based on all the variables. Although a monothetic system can be made polythetic using iterative relocation of all objects at each division (Section 7.5.3), there are some problems in doing this with a mixture of quantitative and qualitative data, particularly if nominal variables are present. Williams and Lance [1977] conclude that monothetic divisive methods are still the only comparatively realistic approach to cluster analysis when n is very large and d is moderate. However, monothetic divisive programs are not favored in taxonomic studies, as they frequently lead to misclassifications (Clifford and Stephenson [1975: p. 103]).

7.4.1 Monothetic Methods

If all the variables are dichotomous, we could divide the objects into two groups on the basis of the presence or absence of a single attribute A. We could choose A to maximize some given measure of distance between the two groups. For example, for each pair of attributes, j and k, say, we can construct a 2×2 table of presences and absences like Table 7.1, but based on two variables rather than two objects, and calculate the usual chi-square statistic for a 2×2 contingency table (see Exercise 7.12), namely,

$$y_{jk} = \frac{n(\alpha\delta - \beta\gamma)^2}{(\alpha + \beta)(\alpha + \gamma)(\beta + \delta)(\gamma + \delta)}. \tag{7.27}$$

One criterion is to choose A to maximize

$$\sum_{j:\, j \neq A} y_{jA}. \tag{7.28}$$

Once the first division has taken place, the same method is applied separately to each group. However, the method is not monotonic, is affected by skewed data, and is apparently being supplanted by the following information method based on the information content, I, say [see (7.20)], of a cluster. For each attribute the objects are divided into two groups, depending on whether j is present or absent. If the corresponding information contents of each of the two groups are I_{+j} and I_{-j}, respectively, then the reduction in information due to the division is $I - (I_{+j} + I_{-j})$; A is chosen to minimize this reduction. The technique is widely used for handling large sets of dichotomous or binary data (Williams and Lance [1977: p. 282]).

Multistate qualitative variables with k states can be treated as a string of $k - 1$ binary variables, with the division made on the presence or absence of a specified state, though a more satisfactory method is needed for ordered states.

With observations on a quantitative variable, x_i, say $(i = 1, 2, \ldots, n)$, a natural split would be that which minimized the within-group sum of squares or, equivalently, maximized the between-group sum of squares [see (7.33) with $d = 1$]

$$\begin{aligned} B &= n_1(\bar{x}_1 - \bar{x})^2 + n_2(\bar{x}_2 - \bar{x})^2 \\ &= n_1\bar{x}_1^2 + n_2\bar{x}_2^2 - n\bar{x}^2, \end{aligned}$$

where $\bar{x}_k$ is the mean of the n_k values in group k $(k = 1, 2)$, and $\bar{x}$ is the overall mean. However, instead of considering all possible splits, the following restricted procedure of Dale [1964] can be followed (Williams and Lance [1977]).

We rank the data $x_{(1)} < x_{(2)} < \cdots < x_{(n)}$, and consider

$$R = r\bar{x}_1^2 + (n-r)\bar{x}_2^2$$

$$= \left(\sum_{i=1}^{r} x_{(i)}\right)^2 \Big/ r + \left(\sum_{i=r+1}^{n} x_{(i)}\right)^2 \Big/ (n-r) \tag{7.29}$$

for each $r = 0, 1, \ldots, n-1$, and make the split at the value of r that maximizes R.

The above procedure can be applied to the values of the first principal component of vector data, as in the polythetic POLDIV system of Williams and Lance [1977]. It can also be applied to each of the variables in turn so that we obtain a maximum value of R or, equivalently, a maximum value of B, $B_{\max}$. If $T = \Sigma_i(x_i - \bar{x})^2$, then the variable with the highest $B_{\max}/T$ ratio can be used to split the parent group in two. This approach is similar to that used in the automatic interaction detection method (AID) developed by Sonquist and Morgan [1963]. In AID we have a single response variable y_i and predictor (independent) variables $x_{i1}, x_{i2}, \ldots, x_{id}$ $(i = 1, 2, \ldots, n)$. The first split is determined by the predictor that gives the best split for the response, that is, which minimizes

$$B = n_1(\bar{y}_1 - \bar{y})^2 + n_2(\bar{y}_2 - \bar{y})^2.$$

A generalization of this approach that uses several predictors at once, called MAID-M (multivariate version of monitored AID), is given by Gillo and Shelly [1974]. A similar technique to AID has been used to partition sample means in analysis of variance models (see Worsley [1977] for references).

7.4.2 *Polythetic Methods*

MacNaughton-Smith et al. [1964] have proposed the following technique. At each step of the procedure we have a splinter group and the remainder. For each object in the remainder we calculate its average dissimilarity with the other objects in the remainder and subtract its average dissimilarity with objects in the splinter group. If these numbers are all negative, the process stops; otherwise the object with the largest positive value is shifted to the splinter group and the process repeated. The splinter group is started by separating out the object with the largest average dissimilarity with respect to the other $n - 1$ objects. When the process steps, we have the first binary split into two clusters and we repeat the process on each cluster (see Everitt [1974: pp. 19–20] for a numerical example).

7.5 PARTITIONING METHODS

7.5.1 Number of Partitions

Suppose that the number of clusters g is determined in advance. A natural procedure would be to partition the n objects into g clusters so as to optimize some criterion. The number of possible partitions is (Duran and Odell [1974: Chapter 2])

$$P_{g,n} = \frac{1}{g!} \sum_{i=1}^{g} \binom{g}{i} (-1)^{g-i} i^n$$

$$\sim \frac{g^n}{g!} \qquad \text{as } n \to \infty, \tag{7.30}$$

where $P_{g,n}$ is a Stirling number of the second kind. As with the divisive methods, this approach is impractical, as $P_{g,n}$ is astronomically large, even for small values of g. Fortier and Solomon [1966] discuss this problem and from their Table 1 we have, for example, $P_{3,19} \approx 1.9 \times 10^8$ and $P_{6,19} \approx 6.9 \times 10^{11}$. If g is not specified, the situation is even worse, as the total number of possibilities is

$$\sum_{g=1}^{n} P_{g,n}.$$

When $n = 25$, this number exceeds 4×10^{18}! Clearly a more restricted, less than optimal approach to the problem is required. For example, the hierarchical techniques can be used to produce a partition with a given number g of clusters, but a major weakness of these methods is that an improper fusion or split at an early stage cannot be corrected later. However, we now consider partitioning methods that allow objects to be relocated. Such relocation techniques can, for example, be applied to the divisive but not the agglomerative methods, as the latter use a different criterion for the formation of a cluster from that used for relocation.

7.5.2 Starting the Process

The first step is to choose g points in d-dimensional space as nuclei for initiating the formation of clusters. A variety of techniques for choosing the cluster centers have been proposed, namely, choose the first g points (MacQueen [1967]); select g points at random (Ball and Hall [1965], MacQueen [1967]), regularly spaced (Beale [1969a, b]), mutually farthest apart (Thorndike [1953]; see also Kennard and Stone [1969]), or at suitably spaced positions of maximum density (Astrahan [1970], Anderberg [1973: pp. 157–158]); or choose g points "at random" in R^d, assign the objects to form g clusters, and use the g centroids as nuclei for a further partition (Jancey [1966], Forgy [1965]). Ball and Hall [1967] use the overall mean as the first nucleus, then select subsequent nuclei by examining the data points in their input sequence and accepting any data point that is at least some specified distance, δ, say, from all previous nuclei, and continue until they have g nuclei or have run out of data. Several values of δ can be tried if the first value gives too few nuclei or examines too little of the data set. Their algorithm is called the sphere-factor starting algorithm and is part of a cluster analysis package ISODATA developed by the authors in 1965 and described in detail by Hall and Khanna [1977]. A randomized version of this algorithm was suggested by Bonner [1964]. Several other procedures are described by Anderberg [1973: Chapter 7].

Once g cluster nuclei are chosen, the remaining objects are assigned to their nearest nucleus using an appropriate proximity measure, generally the Euclidean squared distance. The cluster center, usually the centroid, can be updated either after each addition to the cluster (e.g., MacQueen's [1967] "k-means" method) or only after all the points have been allocated (Forgy [1965], Jancey [1966]). Nagy [1969] gives a modification that permits overlapping clusters. An initial partition can also be obtained by stopping a divisive hierarchical procedure at an appropriate level. Hartigan [1975: Chapter 3] gives a number of "quick partition" algorithms.

7.5.3 *Reassigning the Objects*

The next step is to search for poor fits, that is, objects that should be reallocated to another cluster. For example, in Fig. 7.13 point P is closer to nucleus B than nucleus A; but once the clusters are formed, P is closer to the new nucleus A' than the new nucleus B'. Each individual is scrutinized in turn and reassignment takes place if it causes an increase, or decrease in the case of minimization, in the value of a given clustering criterion. Repeated passes are made through all the objects until a local optimum of the criterion is reached, that is, no further improvement can be obtained by moving a single object. Unfortunately, we do not know if a global optimum has been achieved and it may be appropriate to try and improve the solution using different starting nuclei. For example, Hodson [1971] found the global maximum only 3 times in 24 random starts.

A simple procedure, suggested by Forgy [1965], is to consider each object in turn and reassign it if it is closer to the centroid of another cluster. After a pass

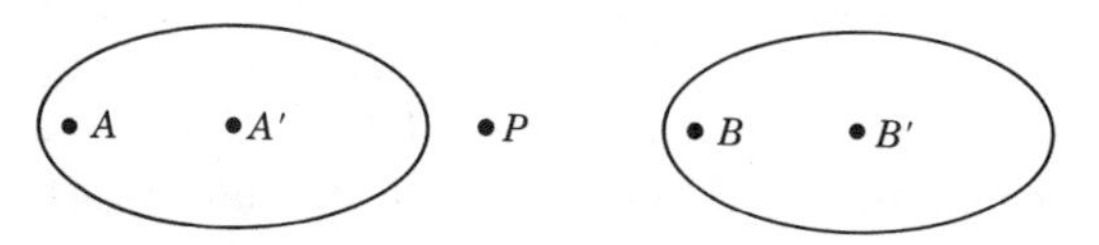

Fig. 7.13 Reassignment of P after the clusters are formed.

through the data, the centroids are updated and the process repeated until convergence, that is, until no objects change their cluster membership. This procedure also forms the basis of ISODATA (Hall and Khanna [1977]). Modifications of the procedure were proposed by Jancey [1966], MacQueen [1967], and Hartigan [1975: p. 102]. Anderberg [1973: Chapter 7] gives a helpful discussion on some of these modifications. An efficient version of Hartigan's algorithm is given by Hartigan and Wong [1979]. The above collection of methods and their variants are commonly described as "k-means," or, in our notation, "g-means" algorithms, and tend to produce clusters of similar overall size (Everitt [1979a]). Some methods for handling missing values are given by Wishart [1978b].

Several other clustering criteria have been proposed and three of them are based on the fundamental matrix equation

$$\mathbf{T} = \mathbf{W} + \mathbf{B}, \tag{7.31}$$

where

$$\mathbf{T} = \sum_{i=1}^{n} (\mathbf{x}_i - \overline{\mathbf{x}})(\mathbf{x}_i - \overline{\mathbf{x}})', \tag{7.32}$$

and $\mathbf{W}$ and $\mathbf{B}$ are the matrices of within-cluster and between-cluster variation. Using a different notation, let $\mathbf{y}_{ij}$ be the observation for the jth object in the ith cluster ($i = 1, 2, \ldots, g$; $j = 1, 2, \ldots, n_i$ and $\sum_i n_i = n$). Then

$$\overline{\mathbf{x}} = \left(\sum_{i=1}^{g} \sum_{j=1}^{n_i} \mathbf{y}_{ij} \right) \bigg/ \sum_{i=1}^{g} n_i \qquad (= \overline{\mathbf{y}}_{\cdot\cdot}, \text{ say})$$

and (7.31) now takes the form

$$\sum_i \sum_j (\mathbf{y}_{ij} - \overline{\mathbf{y}}_{\cdot\cdot})(\mathbf{y}_{ij} - \overline{\mathbf{y}}_{\cdot\cdot})' = \sum_i \sum_j (\mathbf{y}_{ij} - \overline{\mathbf{y}}_{i\cdot})(\mathbf{y}_{ij} - \overline{\mathbf{y}}_{i\cdot})' + \sum_i n_i (\overline{\mathbf{y}}_{i\cdot} - \overline{\mathbf{y}}_{\cdot\cdot})(\overline{\mathbf{y}}_{i\cdot} - \overline{\mathbf{y}}_{\cdot\cdot})'. \tag{7.33}$$

(In the multivariate analysis of variance context of Chapter 9 we use the notation $\mathbf{E}_H = \mathbf{E} + \mathbf{H}$.) For univariate data ($d = 1$) the above expression reduces to the usual sum of squares decomposition for the one-way analysis of variance. In this case, for sharp clusters, we would like W to be small and B large, so that distances within the clusters are small compared with distances between cluster centers. For d dimensions a similar criterion would be appropriate so that an intuitive procedure for choosing clusters is to minimize the "size" of $\mathbf{W}$ or, as $\mathbf{T}$ is constant, maximize $\mathbf{B}$. Several functions of $\mathbf{W}$ have been proposed and three are discussed below. Suitable algorithms for these functions are given, for example, by Friedman and Rubin [1967], McRae [1971], and Gordon and Henderson [1977]. An object is reallocated to the cluster that gives an optimum value of the criterion.

a MINIMIZE TRACE W

Edwards and Cavalli-Sforza [1965] and Singleton (see Friedman and Rubin [1967]) proposed minimizing

$$\begin{aligned}\operatorname{tr}\mathbf{W} &= \sum_i \sum_j (\mathbf{y}_{ij} - \bar{\mathbf{y}}_{i\cdot})'(\mathbf{y}_{ij} - \bar{\mathbf{y}}_{i\cdot}) \qquad \text{(by A1.1)}\\ &= \sum_i \sum_j \|\mathbf{y}_{ij} - \bar{\mathbf{y}}_{i\cdot}\|^2\\ &= \sum_i \operatorname{tr}\mathbf{W}_i,\end{aligned}$$

say. This method is equivalent to minimizing the sum of squared interpoint Euclidean distances within the clusters because of the identity given in Exercise 5.14. Jensen [1968] gives a dynamic programming algorithm for finding the global maximum, though, as noted by Everitt [1979b], it does not seem to offer realistic practical solutions. The algorithm is described by Duran and Odell [1974: Chapter 3], who also discuss some integer programming methods. The nearest centroid methods of Forgy and others mentioned above may be regarded as attempts at minimizing $\operatorname{tr}\mathbf{W}$ (see also Maronna and Jacovkis [1974]). Although these approximations are loosely called "k-means clustering," the term might be more appropriately reserved for describing the minimum $\operatorname{tr}\mathbf{W}$ criterion. Some asymptotic properties of the method are given by Hartigan [1978] and Pollard [1981].

b MINIMIZE $|W|$

This can be motivated in two ways. Suppose that the g clusters represent samples from g MVN distributions $N_d(\boldsymbol{\mu}_i, \boldsymbol{\Sigma})$ $(i = 1, 2, \ldots, g)$ with a common dispersion matrix. From Section 9.2.1 the usual likelihood ratio test for H_0: $\boldsymbol{\mu}_1 = \boldsymbol{\mu}_2 = \cdots = \boldsymbol{\mu}_g$ is related to $\Lambda = |\mathbf{W}|/|\mathbf{T}|$. The hypothesis H_0 is rejected if Λ is too small so that a possible clustering criterion is to minimize Λ or, equivalently, minimize $|\mathbf{W}|$. Alternatively, suppose we wish to assign the n observations to the g normal distributions with the same dispersion matrix in such a way that the likelihood function is maximized (Scott and Symons [1971]). For a *given* allocation, the maximum of the log likelihood is (see Exercise 7.23) a constant minus $\frac{1}{2}n\log|\mathbf{W}|$ so that we choose the allocation to minimize $\log|\mathbf{W}|$, that is, minimize $|\mathbf{W}|$.

Suppose that a further point $\mathbf{x}$ is added to a cluster with centroid $\mathbf{m}$ and let $\mathbf{d} = \mathbf{x} - \mathbf{m}$. Then if $\mathbf{m}$ is not immediately updated with the addition of $\mathbf{x}$, the matrix $\mathbf{W}$ becomes $\mathbf{W} + \mathbf{dd}'$ and the increase in the criterion is

$$|\mathbf{W} + \mathbf{dd}'| - |\mathbf{W}| = |\mathbf{W}|\mathbf{d}'\mathbf{W}^{-1}\mathbf{d} \qquad \text{[by (2.57)]} \qquad (7.34)$$

The smallest increase in $|\mathbf{W}|$ therefore occurs when $\mathbf{x}$ is added to the cluster for which the Mahalanobis distance $\mathbf{d}'\mathbf{W}^{-1}\mathbf{d}$ of $\mathbf{x}$ from the centroid is a minimum. This contrasts with the $\operatorname{tr}\mathbf{W}$ criterion that uses $\mathbf{d}'\mathbf{d}$.

We note that $|\mathbf{W}|$ and $\operatorname{tr}\mathbf{W}$ can also be justified by maximizing certain approximated posterior probabilities (Binder [1978]).

C MAXIMIZE TRACE BW^{-1}

Using another test statistic for testing $H_0: \boldsymbol{\mu}_1 = \boldsymbol{\mu}_2 = \cdots = \boldsymbol{\mu}_g$ above, Friedman and Rubin [1967] proposed maximizing $\operatorname{tr}[\mathbf{BW}^{-1}]$.

Which of the three criteria should be used? The trace criterion is very popular, no doubt because of its simplicity and ease of computation. It is invariant under orthogonal transformations, but not under any nonsingular linear transformation $\mathbf{y}_{ij} = \mathbf{Rz}_{ij}$, so that minimizing $\operatorname{tr}\mathbf{W}$ may give different solutions on the raw and standardized data. As only diagonal elements are involved, the trace does not take into account correlations among the variables. For this reason the method tends to produce spherical clusters.

The criterion of minimizing $|\mathbf{W}|$ is invariant under nonsingular transformations, since $|\mathbf{W}_y| = |\mathbf{R}||\mathbf{W}_z||\mathbf{R}'| = |\mathbf{R}|^2|\mathbf{W}_z| = c|\mathbf{W}_z|$ $(c > 0)$. Friedman and Rubin [1967] conclude from their experiments that this criterion has "a greater sensitivity to the local structure of the data," that is, it performs the best. Marriott [1971] also prefers $|\mathbf{W}|$ to $\operatorname{tr}\mathbf{W}$ because of correlation effects, though he notes that the former may be strongly influenced by a single well-clustered variable. The method searches for any natural grouping, even one based on a single variable rather than on all the variables. The $|\mathbf{W}|$ criterion does not aim to produce spherical clusters and will identify strongly elliptical clusters; however, as might be expected from its motivation, it does tend to produce clusters of the same shape. Scott and Symons [1971] also confirm the usefulness of $|\mathbf{W}|$ when the clusters are clearly separated or approximately of the same geometrical size. They give theoretical and empirical evidence that the method tends to divide the data into approximately equal-sized clusters when the separation is not great. This tendency is also demonstrated by Everitt [1974: Chapter 4], so that some caution is necessary if such an analysis produces clusters of about the same size. One suggestion for getting around the problem is to use $\prod_i |\mathbf{W}_i|^{n_i}$ (Scott and Symons [1971]), and Everitt [1974: pp. 76–77] gives an example that demonstrates its superiority over $|\mathbf{W}|$. Further modifications are proposed by Symons [1981] and Marriott [1982].

The criterion $\operatorname{tr}[\mathbf{BW}^{-1}]$ can be increased most easily by increasing the largest eigenvalue of $\mathbf{BW}^{-1}$. This means that if the groups formed in the initial or subsequent partition are very elongated in the wrong direction, the error will not be corrected but will be increased in further iterations. This criterion can therefore be very unreliable (Maronna and Jacovkis [1974]).

In addition to the criteria based on **W**, Rubin [1967] proposed optimizing the "average entity stability." Wallace and Boulton [1968] and Boulton and Wallace [1970] suggested minimizing an information measure based on multinomial distributions for multistate data and normal distributions for continuous data.

EXAMPLE 7.2 Pilowsky and Spence [1975, 1976] in Adelaide used the behavior profiles from an illness behavior questionnaire (IBQ) to identify six groups of patients, and described the characteristics of their groups. A similar experiment was carried out by Large and Mullins [1981] with 200 patients with chronic pain at the Auckland Hospital Pain Clinic, New Zealand. The same IBQ was used to obtain 10 clusters, and the analysis is described below.

The IBQ has 30 items that are used to calculate patient scores on seven subscales labeled as follows: general hypochondriasis, x_1; disease conviction, x_2; psychological versus somatic perception of illness, x_3; affective inhibition, x_4; affective disturbance, x_5; denial, x_6; and irritability, x_7. The cluster method was essentially a so-called "k-means" (or, in our notation, g-means) technique using a suitable initial partition and the minimum $\operatorname{tr} \mathbf{W}$ criterion. The number of clusters g was allowed to vary from 2 to 12 and, on the basis of $\operatorname{tr} \mathbf{W}$, the authors chose $g = 10$. Because the algorithm is only locally optimal, different initial partitions were used and the best results, in terms of $\operatorname{tr} \mathbf{W}$, were obtained using the single linkage algorithm to produce clusters. The 10 clusters were then compared with the six groups obtained by Pilowsky and Spence by calculating the Euclidean distance between the centroids for each group (G) and cluster (C). These distances are given in Table 7.6 and we see that row and column minima coincide for five of the six groups, yielding initial pairs of

TABLE 7.6 Euclidean Distances Between the Centroids of the Clusters and Groups[a]

	Groups (G)					
Clusters (C)	1	2	3	4	5	6
1	2.54	2.31	1.50	2.67	3.44	6.29
2	0.82	2.06	3.36	3.01	4.55	7.43
3	1.26	1.22	2.87	2.92	4.19	6.64
4	4.33	3.68	3.05	4.21	4.33	4.33
5	4.51	3.42	3.13	2.35	2.80	4.54
6	2.48	1.74	3.30	3.50	3.51	6.80
7	3.14	2.45	4.24	1.37	3.53	5.74
8	3.00	3.10	3.72	1.41	2.54	5.41
9	4.92	4.19	4.84	3.59	4.29	2.66
10	7.80	7.27	6.91	6.04	6.04	1.23

[a]From Large and Mullins [1981: Table 3].

(G1–C2), (G2–C3), (G3–C1), (G4–C7) and (G6–C10). The pairs (G2–C6) and (G4–C8) are also suggested by minimal distances.

Using Euclidean distances between centroids for comparing the groups and clusters seems appropriate, as both the k-means algorithm and the method of Pilowsky and Spence are based on a Euclidean view of the space spanned by the observations. However, Large and Mullins point out that these distances may not take into account any similarity of profile across subscales that may be clinically relevant. The pairs (G3–C4), (G4–C5), (G5–C8), and (G6–C9), while representing minimum distances between clusters and groups, are quite dissimilar in profile, on inspection of the pattern of mean subscale scores. Therefore, taking into account both minimal distances and similarity of profiles, the authors obtained Table 7.7. They concluded that "this independent replication of the Pilowsky and Spence study shows a close similarity

TABLE 7.7 Matching of Clusters and Groups[a]

	%	Subscale mean scores						
		1	2	3	4	5	6	7
Adelaide group 1	29	0.52	1.97	0.31	0.52	0.35	2.46	0.31
Auckland cluster 2	15	0.17	2.17	0.73	0.27	0.17	2.90	0.53
Adelaide group 2	18	0.50	2.44	0.61	0.94	2.06	2.83	0.67
Auckland cluster 3	22.5	0.93	1.96	1.02	0.82	1.13	2.89	0.51
Auckland cluster 6	6.5	0.69	1.70	0.92	1.15	2.38	1.38	0.38
Adelaide group 3	9	0.78	1.44	0.67	1.00	1.78	2.89	3.22
Auckland cluster 1	10	0.90	2.20	1.25	0.90	0.90	2.80	2.50
Adelaide group 4	25	1.00	4.38	0.07	1.00	1.10	2.31	1.68
Auckland cluster 7	11	1.09	4.73	0.55	1.18	1.50	2.68	0.59
Auckland cluster 8	10	1.35	4.10	0.35	0.40	0.65	1.25	1.80
Adelaide group 5	8	0.63	4.00	1.25	0.63	2.50	0.38	2.75
Adelaide group 6	11	6.36	4.72	0.55	1.18	2.18	1.73	2.55
Auckland cluster 9	8.5	4.18	4.06	0.82	1.41	2.41	2.06	1.29
Auckland cluster 10	2.5	6.40	5.00	0.60	1.20	2.40	1.20	3.60
Auckland cluster 4	5.5	3.55	1.55	1.09	1.00	2.45	2.36	2.36
Auckland cluster 5	8.5	2.00	4.18	0.41	1.06	2.59	2.59	3.12
Total								
Adelaide sample	100	1.35	3.28	0.44	0.88	1.43	2.38	1.38
Auckland sample	100	1.51	2.94	0.80	0.86	1.38	2.43	1.31

[a]From Large and Mullins [1981: Table 4].

between the populations studied at both centers and adds validity to the groups described by them. Further experience with these profiles may well produce therapeutic and prognostic guidelines of considerable clinical utility."

7.5.4 Other Techniques

A completely different approach to partitioning is the density search method in which one attempts to separate regions with a high density of points $\mathbf{x}_i$ from those of lower density. Usually each mode is taken to signify the presence of a cluster. Everitt [1974: Section 23] describes a number of such methods that are mainly derivatives of single linkage with a criterion for deciding when to stop adding points to clusters. These include the TAXMAP method (Carmichael et al. [1968], Carmichael and Sneath [1969]), an algorithm for detecting unimodal fuzzy sets (Gitman and Levine [1970]), the Cartet count method of Cattell and Coulter [1966], and the mode analysis of Wishart [1969b]. Unfortunately, these methods suffer from the problem of having a number of controlling parameters whose values have to be arbitrarily chosen. Also, Wishart's mode analysis is scale dependent and assumes spherical modes. However, these methods essentially ignore "noise" points between clusters so that the effect of chaining is reduced.

Another density-type procedure based on a modal analysis of the histogram for each variable was proposed by Chhikara and Register [1979]. Through a process of splitting and merging, clusters are eventually obtained so that for the observations in any one cluster, all one-dimensional histograms are unimodal. Alternatively, one could fit a probability density function to the n points using a multivariate kernel function (Section 6.5.1) and then use the modes of the distributions to determine clusters (Bryan [1971]; see also Duran and Odell [1974: Section 5.2] for a description of the method).

Finally we mention one other technique called the method of mixtures. It is assumed that the n observations come from g different probability distributions, and the problem is to develop an assignment rule for assigning each of the n observations to the most likely distribution. The mixture problem has already been mentioned in discriminant analysis (Section 6.7), where the emphasis is on finding an assignment rule for assigning a future observation. In the multivariate case research has largely been confined to mixtures of g multivariate normals $N_d(\boldsymbol{\mu}_i, \boldsymbol{\Sigma}_i)$ [with density function $f_i(\mathbf{x}|\boldsymbol{\mu}_i, \boldsymbol{\Sigma}_i)$], and one approach is to assign each of the n observations to one of the g groups using the rule Assign to the group with the largest value of [see (6.97)]

$$\log \hat{\pi}_i - \tfrac{1}{2}\log \hat{\boldsymbol{\Sigma}}_i - \tfrac{1}{2}(\mathbf{x} - \hat{\boldsymbol{\mu}}_i)'\hat{\boldsymbol{\Sigma}}_i^{-1}(\mathbf{x} - \hat{\boldsymbol{\mu}}_i).$$

Here $\hat{\pi}_i$, $\hat{\boldsymbol{\mu}}_i$, and $\hat{\boldsymbol{\Sigma}}_i$ are obtained by maximizing the likelihood $\prod_r f(\mathbf{x}_r)$, where

$$f(\mathbf{x}) = \sum_{i=1}^{g} \pi_i f_i(\mathbf{x}|\boldsymbol{\mu}_i, \boldsymbol{\Sigma}_i). \tag{7.35}$$

However, in contrast to this mixture maximum likelihood method with its emphasis on estimating an assignment rule, there is another technique that may be called the classification method. Observations are now assigned to the g populations and the maximization is over all possible assignments as well as over all values of the unknown parameters. The likelihood to be maximized is therefore

$$\prod_{r=1}^{n} f_{h(r)}(\mathbf{x}_r | \boldsymbol{\mu}_{h(r)}, \boldsymbol{\Sigma}_{h(r)}), \tag{7.36}$$

where $h(r) = i$ when $\mathbf{x}_r$ is assigned to group G_i. Scott and Symons [1971] showed that for equal dispersion matrices this leads to the criterion of minimizing $|\mathbf{W}|$ (see also Symons [1981]). However, for this case, Marriott [1975] noted that the resulting maximum likelihood estimates of the parameters are inconsistent; the same is true for nonnormal populations (Bryant and Williamson [1978]). We conclude, therefore, that when the emphasis is on the estimation of parameters or an assignment rule (as in discrimination and cluster analysis), the mixture method is preferred to the classification method. If the classification method is used, then the following comments of Marriott [1975] relating to the use of $|\mathbf{W}|$ should be borne in mind: "The assumption of underlying normal distributions with equal dispersion matrices is seldom strictly true in practice, and in many practical situations, when the proportions in the underlying distributions are approximately equal, minimizing $|\mathbf{W}|$ gives a sensible and reasonably robust clustering procedure. It is better regarded, however, as a heuristic approach, rather than an estimation process applied to a particular model." For a further discussion on the two approaches, see Sclove [1977] and Ganesalingam and McLachlan [1980].

Some of the problems associated with testing for the existence of a mixture are discussed by Hartigan [1975: Section 4.8; 1977, 1978]. Bayesian approaches to the mixture model are given by Binder [1978, 1981] and Symons [1981]. A sequential Bayes-type procedure in which observations are assigned one at a time in the order they are taken is proposed by Smith and Makov [1978].

7.6 OVERLAPPING CLUSTERS (CLUMPING)

Uptil now we have considered methods that produce nonoverlapping (disjoint) clusters. However, there are situations where it is more meaningful to allow "clumping," that is, some overlap between clusters. For example, in linguistics, words have several meanings and may belong to several groups, whereas in human populations mixed marriages lead to many racial overlaps (Rao [1977]). Some methods of clumping are given by Needham [1967], Parker-Rhodes and Jackson [1969], Jones and Jackson [1967], and Rao [1977] (see also Clifford and Stephenson [1975: pp. 118–119]). Jardine and Sibson [1968, 1971b] introduced their family of overlapping agglomerative methods (B_k clustering) that allow clusters to overlap by as many as $k - 1$ objects, and showed that the single link method could be generalized to a k link method. Their algorithm was considerably improved by Cole and Wishart [1970] and it is incorporated in the CLUSTAN suite of programs (see also Rohlf [1974b]). However, although the method has a number of desirable mathematical properties that it shares with the single link method (Sibson [1970]), it has received little or no support, as it is computationally complex and the resulting dendrogram is difficult to draw and interpret (Sneath and Sokal [1973: p. 208], Kruskal [1977a]). An example of B_2 clustering is given by Morgan [1981].

7.7 CHOOSING THE NUMBER OF CLUSTERS

A major problem of cluster analysis is the choice of g, the number of clusters. It is helpful if a rough idea of the likely range of g is known, but such information may not be available. In agglomerative methods it is suggested that a large change in the dendrogram level is indicative of the correct number of clusters. However, Everitt [1974: Section 4.3] demonstrated that such an approach can be misleading, as a large change in the fusion level of a dendrogram is a necessary but not a sufficient condition for clear-cut clusters. Two "stopping rules" have been proposed by Mojena [1977], and one of them seems worthy of further consideration. In the same vein it has been suggested that for partitioning methods, a plot of the criterion value against g will show a sharp increase or decrease (depending on whether maximizing or minimizing) at the correct number of clusters; however, this procedure is very subjective and can be misleading.

If the n observations $\mathbf{x}_i$ are distributed uniformly on a suitable d-dimensional subset, then the number of observations, N, say, within Euclidean distance δ of a given observation will be proportional to δ^d. Hence if δ_N is the distance of the Nth nearest point to the given observation, then $\delta_N \simeq \delta$ and (Hartigan [1975: p. 65])

$$\log N \simeq a + d \log \delta_N. \tag{7.37}$$

If N_0 observations form a cluster near the given observation, then the distances δ_N for $N > N_0$ will be greater than expected from a plot of (7.37) of objects within the cluster, and there will be a break in the plot at the boundary of the cluster. Such plots may be useful for cluster detection.

Once an initial g-partition is obtained, the value of g can be changed iteratively by allowing small close clusters to be merged with other clusters, and larger loose clusters to be split. Procedures for doing this are given, for example, by MacQueen [1967] and ISODATA (see Hall and Khanna [1977]).

Marriott [1971] investigated the criterion $|\mathbf{W}|$ and suggests choosing g to minimize $g^2|\mathbf{W}|$. He provides a crude but useful graph for the decision maker and recommends that the data be regarded as a single cluster if $g^2|\mathbf{W}|/|\mathbf{T}| > 1$

for all values of g. However, Everitt [1974: Section 4.2] gives simple examples where the criterion $g^2|\mathbf{W}|$ fails to give the correct value of g. As already noted, there is a tendency for this method to produce clusters of the same shape.

The question of a single cluster is important, as spurious clusters can be found even in random data. Ling and Killough [1976] give some exact and approximate tables for assessing the randomness of the data using the notion of random graphs. Their tables allow us to compare the global structure of the data with the expected global structure of a random graph. They emphasize that their tables should be used as an informal screening device for the spurious clustering of random data rather than for precise statistical inference or formal hypothesis testing. Their article should be consulted for further references and background. A possible probabilistic approach for assessing the reality of an individual cluster in a classification is given by Ling [1973a]. Another approach, based on computing the probability distribution of all dendrograms under randomness, is given by Frank and Svensson [1981].

A number of methods have been developed on the assumption of multivariate normality. For example, K. L. Lee [1979] gives a test of the hypothesis H_0 that n observations come from one MVN distribution, versus the alternative H_1 that they come from two normal distributions with common dispersion matrix but different means. He gives some approximate percentage points for $d = 1, 2, 3$ and suggests a possible formula for $d > 3$. This approach could be used for deciding whether or not to split a given cluster in two. Using normal theory, Beale [1969a, b] proposed an F-statistic

$$F = \left(R_{g_1} - R_{g_2}\right) \Big/ R_{g_2}\left[\frac{n - g_1}{n - g_2}\left(\frac{g_2}{g_1}\right)^{2/d} - 1\right] \qquad (g_2 > g_1)$$

based on $d(g_2 - g_1)$ and $d(n - g_2)$ degrees of freedom. Here R_{g_i} is the value of $\operatorname{tr}\mathbf{W}$ for g_i clusters and a significant result indicates that a subdivision into g_2 clusters is significantly better than a subdivision into a smaller number of clusters g_1. Beale starts with a value of g larger than thought necessary and endeavors to reduce it. The "derivation" of the statistic is highly heuristic so that the test should be used as a rough guide only (Kendall [1975: p. 41]).

Another criterion

$$C = \frac{\operatorname{tr}\mathbf{B}}{g - 1} \Big/ \frac{\operatorname{tr}\mathbf{W}}{n - g}$$

has been proposed by Caliński and Harabasz [1974]. A value of C increasing monotonically with g suggests no cluster structure, whereas C decreasing monotonically with g suggests a hierarchical structure. However, C rising to a maximum at g suggests the presence of g clusters. The authors, in fact, choose for each g that partition of the minimum spanning tree into g sections that maximizes C. The above techniques based on the normal distribution are only likely to be useful for spherical-type clusters.

If the technique of fitting mixtures of normal distributions described in Section 7.5.4 is used, then a likelihood ratio statistic $-2\log \ell$ is available based on fitting g_1 and g_2 distributions, respectively. Wolfe [1970] has suggested referring this statistic to a chi-squared distribution, but this is not correct (Binder [1978]). Hartigan [1977] gives a helpful discussion on inference problems for mixtures.

7.8 GENERAL COMMENTS

Unfortunately, there is a general lack of guidelines as to which clustering method should be used in any given situation, though some attempts have been made in formulating principles (Fisher and Van Ness [1971], Williams et al. [1971a], Jardine and Sibson [1971a], Rand [1971]). Agglomerative methods are clearly preferred in taxonomic problems where a treelike structure can have, for example, an evolutionary interpretation. However, as demonstrated by Everitt [1974: Section 4.3], different agglomerative methods can give rise to very different dendrograms, and one is forced to consider each cluster problem on its own merits. For example, if one suspects the presence of certain "natural" clusters, then a space-dilating algorithm for sharpening cluster boundaries could be appropriate. However, if one has no idea what to look for in the data, then pushing the data through some cluster program in the hope that something will turn up may simply be forcing the data into fictitious clusters.

Graphical techniques such as those described in Section 4.1 may give some indication about the presence of possible clusters, and the dimension-reducing techniques described in Chapter 5 can be used in conjunction with cluster methods, though some care is needed. Clifford and Stephenson [1975: p. 177] give the example in Fig. 7.14 of bivariate data in three distinct clusters. The projections of the observations onto the first principal components would produce a continuous frequency distribution, suggesting that the observations belong to a single group. However, the three clusters would show up if the projections onto the second principal component were considered. A one-dimensional reduction using principal components would therefore be completely misleading.

Multidimensional scaling and clustering are sometimes regarded as competitive methods in that if one method fits perfectly, the other fits poorly, and vice versa (Holman [1972]). However, Kruskal [1977a] argues that this conclusion refers to a "boundary" situation. With real data, in which fits are not perfect, the two methods are complementary: If one model fits well, then the other model also fits well. What happens is that cluster methods essentially extract meaning from the small dissimilarities, whereas multidimensional scaling extracts meaning from the large dissimilarities. For example, in hierarchical clustering it is often found that small clusters fit well and are meaningful, while larger clusters farther up the tree do not seem to be meaningful, except, say, in

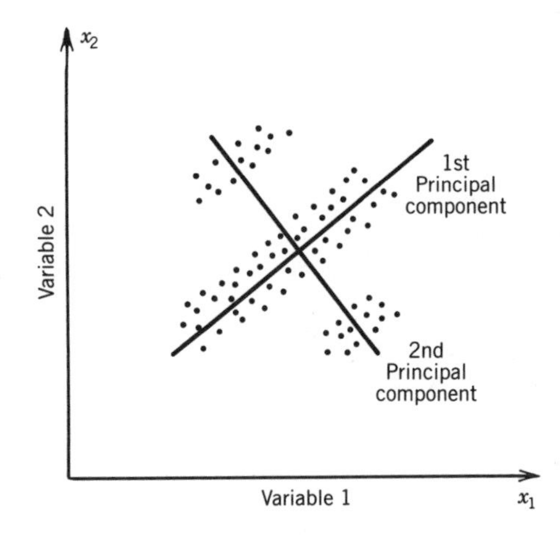

Fig. 7.14 Three clusters of bivariate data showing the two principal components.

taxonomy, where the whole tree may provide an evolutionary model. The small clusters are based on small dissimilarities, and the large clusters on large dissimilarities. Simulation experiments by Graef and Spence [1979] demonstrate that small and medium distances play a lesser role than large distances in multidimensional scaling. Also, small changes in proximity data can cause drastic local changes in the resulting ordination from multidimensional scaling without changing the general overall configuration. Since both methods are sensitive to complementary aspects of the same data, it would seem appropriate to use them both on the same data in many cases. When scaling in two dimensions is appropriate, both results can be combined on a single diagram, as in Fig. 7.10. The positions of the points are obtained by multidimensional scaling and the loops indicate the successive clusters. Even if a two-dimensional reduction is not very appropriate, the method will generally still give a useful picture. Kruskal [1977a] opposes the practice of using *just* multidimensional scaling, generally in two dimensions, to pick out clusters visually because of reasons given above: The scaling configuration reflects large dissimilarities and does not faithfully indicate which points are nearest to which other points.

In conclusion, we mention a few miscellaneous topics. Methods and references for space–time clustering (Are points that are close geographically also close in time?) are given by Klauber [1971, 1975] and Abe [1973]. Some interesting graphical methods analyzing how clustering effects change with time are given by Dunn and Landwehr [1980]. Conover et al. [1979] introduced a window-scanning technique for detecting clusters on a two-dimensional map. They proposed moving a rectangular window in a systematic way over the map and recording when the number of points in the window exceeded K, say. The method was applied to detecting clusters of possible uranium deposits. A novel method for determining bird territories was given by North [1977]. Sometimes objects have a natural ordering that imposes certain constraints on the allowable clusters. This topic of constrained clustering is reviewed by Gordon [1981: Section 4.3].

EXERCISES 7

7.1 Let $\Delta(\mathbf{x}, \mathbf{y}) = \{(\mathbf{x} - \mathbf{y})'\mathbf{B}(\mathbf{x} - \mathbf{y})\}^{1/2}$, where $\mathbf{B}$ is positive definite. Verify that Δ is a metric.

7.2 Verify that the Mahalanobis distance defined by (7.3) is invariant to linear transformations of the form $\mathbf{y}_i = \mathbf{A}\mathbf{x}_i + \mathbf{b}$, where $\mathbf{A}$ is nonsingular.

7.3 Suppose that

$$\mathbf{x}_1 = \begin{pmatrix} 2 \\ 5 \\ 2 \\ 5 \\ 3 \end{pmatrix}, \quad \mathbf{x}_2 = \begin{pmatrix} 3 \\ 5 \\ 2 \\ 4 \\ 3 \end{pmatrix}, \quad \mathbf{x}_3 = \begin{pmatrix} 9 \\ 1 \\ 1 \\ 1 \\ 1 \end{pmatrix}.$$

If d_{rs} is the Bray–Curtis measure defined by Equation (7.5), show that $d_{13} > d_{12} + d_{23}$ (i.e., d_{rs} is not a metric).

7.4 Let

$$\Delta(x, y) = \frac{|x - y|}{x + y} \qquad (x, y \geq 0).$$

Given that $x < y < z$, verify that Δ satisfies

$$\Delta(x, y) \leq \Delta(x, z) + \Delta(y, z).$$

7.5 If $\Delta(\mathbf{x}, \mathbf{y})$ is a metric, show that

$$\Delta_1(\mathbf{x}, \mathbf{y}) = \frac{\Delta(\mathbf{x}, \mathbf{y})}{[w + \Delta(\mathbf{x}, \mathbf{y})]} \qquad (w > 0)$$

is also a metric.

7.6 Show that $D(\mathbf{x}, \mathbf{y}) = \|\mathbf{x} - \mathbf{y}\|^2$ is not a metric.

7.7 Suppose that we have $n = 3$ objects, each with 10 binary attributes as follows:

		Attribute									
		1	2	3	4	5	6	7	8	9	10
Object	1	1	0	0	0	1	1	0	0	1	0
	2	0	0	0	0	1	0	0	1	1	0
	3	0	0	0	0	0	0	0	1	0	0

Calculate c_{12}, c_{13} and c_{23} for the Jaccard coefficient $\alpha/(\alpha + \beta + \gamma)$ and the matching coefficient $(\alpha + \delta)/d$ (see Table 7.1).

7.8 Show that the similarity measures $\alpha/(\alpha + \beta + \gamma)$ and $2\alpha/(2\alpha + \beta + \gamma)$ are monotonic with respect to each other.

7.9 Given $\mathbf{x}' = (1, 4, 7)$ and $\mathbf{y}' = (3, 7, 11)$, show that the correlation coefficient for the pairs (x_j, y_j), $j = 1, 2, 3$, is unity.

7.10 Show that the Bray–Curtis dissimilarity measure for binary (1, 0) data reduces to the one-complement of Czekanowski's coefficient $2\alpha/(2\alpha + \beta + \gamma)$ [see (7.5)].

7.11 Show that Gower's similarity measure (7.8) reduces to $\alpha/(\alpha + \beta + \gamma)$ when all the variables are dichotomous, and to $(\alpha + \delta)/d$ when all the variables are two-state qualitative variables.

7.12 Show that the sample correlation r_{jk} for two binary variables j and k, scored 1 for presence and 0 for absence, is given by [see (7.9)]

$$r_{jk} = \frac{(\alpha\delta - \beta\gamma)}{\{(\alpha + \beta)(\gamma + \delta)(\alpha + \gamma)(\beta + \delta)\}^{1/2}}.$$

7.13 For single linkage, the following algorithm relabels the objects so that each cluster is a contiguous sequence of objects: Label any object as O_1, and label the nearest object to O_1 as O_2. For $i = 3, 4, \ldots, n$, label the object that is nearest to one of $O_1, O_2, \ldots, O_{i-1}$ as O_i. Apply this algorithm to (7.12), beginning with $O_1 = 4$, and verify that the clusters have the contiguous properties.

Hartigan [1975: p. 194]

7.14 Apply the complete linkage and group average methods to the dissimilarity matrix (7.12) and draw the corresponding dendrograms.

7.15 Verify Equations (7.16) and (7.17).

7.16 Given three clusters with centroids and numbers of individuals $\bar{\mathbf{x}}_i$ and n_i, respectively ($i = 1, 2, 3$), verify that

$$\left\| \bar{\mathbf{x}}_3 - \left(\frac{n_1\bar{\mathbf{x}}_1 + n_2\bar{\mathbf{x}}_2}{n_1 + n_2} \right) \right\|^2 = \frac{n_1}{n_1 + n_2}\|\bar{\mathbf{x}}_3 - \bar{\mathbf{x}}_1\|^2 + \frac{n_2}{n_1 + n_2}\|\bar{\mathbf{x}}_3 - \bar{\mathbf{x}}_2\|^2 - \frac{n_1 n_2}{(n_1 + n_2)^2}\|\bar{\mathbf{x}}_1 - \bar{\mathbf{x}}_2\|^2$$

Lance and Williams [1967a]

7.17 Verify Table 7.4.

7.18 Find the ultrametric $\delta(r, s)$ for Fig. 7.5.

7.19 After hierarchical clustering is carried out, it is always possible to order the objects so that every cluster consists of a set of objects contiguous in the order. Let O_i ($i = 1, 2, \ldots, n$) denote the name of the ith object in the order. Such a procedure is as follows:

(a) Select the first object O_1 arbitrarily.

(b) Select the second object O_2 from the smallest cluster containing O_1 but not included in the set $\{O_1\}$. If no such cluster exists, select the second object to be any object other than O_1.

(c) For $k = 3, 4, \ldots, n$, select the kth object O_k from the smallest cluster containing O_{k-1} but not contained in the set $\{O_1, O_2, \ldots, O_{k-1}\}$. If no such cluster exists, select the kth object to be any object other than one of $\{O_1, O_2, \ldots, O_{k-1}\}$.

(i) Beginning with $O_1 = 4$, verify the above procedure for the clusters arising from (7.12), namely,

$$(1), \ldots, (5), (1,3), (2,4), (2,4,5), (1,2,3,4,5).$$

(ii) Apply the above procedure to the following clusters ($O_1 = 2$):

$$(1), \ldots, (9), (3,7), (8,9), (1,5), (2,6), (3,7,10), (8,9,2,6),$$
$$(3,7,10,4), (1,5,8,9,2,6), (1,2,3,4,5,6,7,8,9),$$

and draw the corresponding dendrogram.

Hartigan [1975: p. 156]

7.20 Given observations $\mathbf{x}_1, \mathbf{x}_2, \ldots, \mathbf{x}_n$, prove that

$$\sum_{i=1}^{n} \|\mathbf{x}_i - \mathbf{a}\|^2 = \sum_{i=1}^{n} \|\mathbf{x}_i - \overline{\mathbf{x}}\|^2 + n\|\overline{\mathbf{x}} - \mathbf{a}\|^2.$$

7.21 Suppose that the observations $\mathbf{x}_1, \mathbf{x}_2, \ldots, \mathbf{x}_n$ are in two clusters C_i with n_i observations per cluster and cluster means $\overline{\mathbf{x}}_i$. Prove that

$$\frac{1}{n_1 n_2} \sum_{r \in C_1} \sum_{s \in C_2} \|\mathbf{x}_r - \mathbf{x}_s\|^2 = \frac{1}{n_1} \sum_{r \in C_1} \|\mathbf{x}_r - \overline{\mathbf{x}}_1\|^2$$
$$+ \frac{1}{n_2} \sum_{s \in C_2} \|\mathbf{x}_s - \overline{\mathbf{x}}_2\|^2 + \|\overline{\mathbf{x}}_1 - \overline{\mathbf{x}}_2\|^2.$$

7.22 Three methods for splitting n d-dimensional observations into g groups are based on the following:
(a) Minimizing $|\mathbf{W}|$.
(b) Maximizing $\operatorname{tr}[\mathbf{B}\mathbf{W}^{-1}]$.
(c) Minimizing $\operatorname{tr}[\mathbf{W}\mathbf{T}^{-1}]$.
Here $\mathbf{T}$, $\mathbf{W}$, and $\mathbf{B}$ are given by (7.31) and (7.33). Show that the methods all lead to the same partition when $g = 2$. [Hint: Use A2.4 and assume $\mathbf{W}$ is positive definite with probability 1.]

7.23 Suppose that the $\mathbf{y}_{ij}$ are independently distributed as $N_d(\boldsymbol{\mu}_i, \boldsymbol{\Sigma})$, $i = 1, 2, \ldots, g$; $j = 1, 2, \ldots, n_i$. Show that the maximum of the log likelihood function with respect to the $\boldsymbol{\mu}_i$ and $\boldsymbol{\Sigma}$ is a constant minus $\frac{1}{2} n \log|\mathbf{W}|$, where

$$\mathbf{W} = \sum_{i=1}^{g} \sum_{j=1}^{n_i} (\mathbf{y}_{ij} - \overline{\mathbf{y}}_{i\cdot})(\mathbf{y}_{ij} - \overline{\mathbf{y}}_{i\cdot})'.$$

CHAPTER 8

Multivariate Linear Models

8.1 LEAST SQUARES ESTIMATION

In this chapter we consider several multivariate extensions of the (univariate) linear model

$$\mathbf{y} = \boldsymbol{\theta} + \mathbf{u}, \tag{8.1}$$

where $\mathbf{y} = [(y_i)]$ represents n independent measurements on a single response variable y, $\mathscr{E}[\mathbf{y}] = \boldsymbol{\theta} = \mathbf{X}\boldsymbol{\beta}$, and $\mathbf{X}$ is an $n \times p$ matrix of rank r $(r \leq p)$. If the elements of $\mathbf{X}$ are quantitative, for example, measurements on controlled regressor or predictor variables, then (8.1) is the usual regression model with

$$\theta_i = \beta_0 + \beta_1 x_{i1} + \cdots + \beta_{p-1} x_{i,p-1}, \tag{8.2}$$

and, generally, $r = p$. However, if the elements of $\mathbf{X}$ are 1 or 0 so that $\mathbf{X}$ represents underlying qualitative factors, as, for example, in the randomized block design

$$\theta_{ij} = \mu + \alpha_i + \tau_j \qquad (i = 1, 2, \ldots, I; j = 1, 2, \ldots, J), \tag{8.3}$$

where $\boldsymbol{\theta}' = (\theta_{11}, \theta_{12}, \ldots, \theta_{1J}, \ldots, \theta_{I1}, \theta_{I2}, \ldots, \theta_{IJ})$, then we have an analysis of variance model. In this case $\mathbf{X}$ is sometimes called the design matrix, as it expresses the structure imposed by the underlying experimental design, usually $r < p$. When $\mathbf{X}$ is a mixture of quantitative and qualitative elements as in

$$\theta_{ij} = \mu + \alpha_i + \tau_j + \gamma z_{ij}, \tag{8.4}$$

a randomized block design with one concomitant variable z, we refer to the model as an analysis of covariance model. The general theory for (8.1) is considered in detail in, for example, Seber [1977], and the main features of the theory there can be summarized as follows.

The least squares estimate of $\boldsymbol{\theta}$ is obtained by minimizing $\|\mathbf{u}\|^2 = \|\mathbf{y} - \boldsymbol{\theta}\|^2$ subject to $\boldsymbol{\theta} \in \Omega = \mathscr{R}[\mathbf{X}]$, the range space of $\mathbf{X}$. The minimum occurs at $\hat{\boldsymbol{\theta}} = \mathbf{P}_\Omega \mathbf{y}$, where $\mathbf{P}_\Omega$ is the unique symmetric idempotent matrix representing the orthogonal projection of n-dimensional Euclidean space R^n onto Ω. Since $(\mathbf{y} - \hat{\boldsymbol{\theta}}) \perp \Omega$, $\mathbf{X}'(\mathbf{y} - \hat{\boldsymbol{\theta}}) = \mathbf{0}$. Hence, if $\hat{\boldsymbol{\beta}}$ satisfies $\hat{\boldsymbol{\theta}} = \mathbf{X}\hat{\boldsymbol{\beta}}$, then $\hat{\boldsymbol{\beta}}$ is a solution of the normal equations $\mathbf{X}'(\mathbf{y} - \mathbf{X}\hat{\boldsymbol{\beta}}) = \mathbf{0}$ or

$$\mathbf{X}'\mathbf{X}\hat{\boldsymbol{\beta}} = \mathbf{X}'\mathbf{y}. \tag{8.5}$$

Conversely, if $\hat{\boldsymbol{\beta}}$ satisfies (8.5), $\hat{\boldsymbol{\beta}}'\mathbf{X}'(\mathbf{y} - \mathbf{X}\hat{\boldsymbol{\beta}}) = 0$ and

$$\mathbf{y} = \mathbf{X}\hat{\boldsymbol{\beta}} + (\mathbf{y} - \mathbf{X}\hat{\boldsymbol{\beta}}) = \mathbf{u} + \mathbf{v},$$

say, is an orthogonal decomposition with $\mathbf{u} \in \Omega$ and $\mathbf{v} \perp \Omega$. As this decomposition is unique (see B1.1), $\hat{\boldsymbol{\theta}} = \mathbf{u} = \mathbf{X}\hat{\boldsymbol{\beta}}$.

When $\mathbf{X}$ has less than full rank $(r < p)$, (8.5) has a solution (see B1.7) $\hat{\boldsymbol{\beta}} = (\mathbf{X}'\mathbf{X})^-\mathbf{X}'\mathbf{y}$, where $(\mathbf{X}'\mathbf{X})^-$ is a generalized inverse of $\mathbf{X}'\mathbf{X}$. Although $\hat{\boldsymbol{\beta}}$ is not unique, $\mathbf{X}\hat{\boldsymbol{\beta}}$ $(= \hat{\boldsymbol{\theta}})$ is unique. The uniqueness of $\mathbf{P}_\Omega$ in $\mathbf{X}\hat{\boldsymbol{\beta}} = \mathbf{P}_\Omega\mathbf{y}$ implies that $\mathbf{P}_\Omega = \mathbf{X}(\mathbf{X}'\mathbf{X})^-\mathbf{X}'$. When $\mathbf{X}$ has full rank $(r = p)$, $(\mathbf{X}'\mathbf{X})^- = (\mathbf{X}'\mathbf{X})^{-1}$ and $\hat{\boldsymbol{\beta}}$ is unique.

In model (8.1) we assumed that $\mathscr{E}[\mathbf{u}] = \mathbf{0}$. If we further assume that $\mathscr{D}[\mathbf{u}] = \mathscr{D}[\mathbf{y}] = \sigma^2\mathbf{I}_n$, then the least squares estimates have certain optimal properties. For example, suppose that $\mathbf{X}$ has less than full rank and $\mathbf{a}'\boldsymbol{\beta}$ is an estimable function, that is, $\mathbf{a}$ is linearly dependent on the rows of $\mathbf{X}$. Then $\mathbf{a}'\hat{\boldsymbol{\beta}}$ is unique and is the best linear unbiased estimate (BLUE) of $\mathbf{a}'\boldsymbol{\beta}$, that is, the linear unbiased estimate with the smallest variance. Also, $s^2 = \|\mathbf{y} - \hat{\boldsymbol{\theta}}\|^2/(n - r)$ is an unbiased estimate of σ^2 and, under certain conditions, has optimal properties. If we now add the assumption of normality so that the elements u_i of $\mathbf{u}$ are i.i.d. $N_1(0, \sigma^2)$, then $\mathbf{u} \sim N_d(\mathbf{0}, \sigma^2\mathbf{I}_n)$ and we can obtain the distributions of $\hat{\boldsymbol{\beta}}$ and s^2. An F-statistic for testing the hypothesis $H_0: \mathbf{A}\boldsymbol{\beta} = \mathbf{c}$ is also available.

In developing a multivariate analogue for (8.1), we would hope to follow a similar path to that outlined above. An obvious generalization would be to assume that we are now interested in measuring d response variables instead of one variable on each of n sampling units. Then y_i is replaced by a $1 \times d$ row vector $\mathbf{y}_i'$, and $\mathbf{y} = [(y_i)]$ is replaced by the data matrix

$$\mathbf{Y} = \begin{pmatrix} \mathbf{y}_1' \\ \mathbf{y}_2' \\ \vdots \\ \mathbf{y}_n' \end{pmatrix} = \left(\mathbf{y}^{(1)}, \mathbf{y}^{(2)}, \ldots, \mathbf{y}^{(d)}\right),$$

say. Here $\mathbf{y}^{(j)}$ $(j = 1, 2, \ldots, d)$ represents n independent observations on the jth variable. Writing $\mathbf{y}^{(j)} = \boldsymbol{\theta}^{(j)} + \mathbf{u}^{(j)}$ with $\mathscr{E}[\mathbf{u}^{(j)}] = \mathbf{0}$, as in (8.1), we have the multivariate linear model

$$\mathbf{Y} = \boldsymbol{\Theta} + \mathbf{U}, \tag{8.6}$$

where $\boldsymbol{\Theta} = (\boldsymbol{\theta}^{(1)}, \boldsymbol{\theta}^{(2)}, \ldots, \boldsymbol{\theta}^{(d)})$, $\mathbf{U} = (\mathbf{u}^{(1)}, \mathbf{u}^{(2)}, \ldots, \mathbf{u}^{(d)})$, and $\mathscr{E}[\mathbf{U}] = \mathbf{O}$. If we are in an experimental design situation in which some sort of structure is imposed on the n sampling units, then the design will be the same for each of the d variables. This means that $\boldsymbol{\theta}^{(j)} = \mathbf{X}\boldsymbol{\beta}^{(j)}$, $\boldsymbol{\Theta} = \mathbf{X}(\boldsymbol{\beta}^{(1)}, \boldsymbol{\beta}^{(2)}, \ldots, \boldsymbol{\beta}^{(d)}) = \mathbf{XB}$, say, and

$$\underset{(n\times d)}{\mathbf{Y}} = \underset{(n\times p)}{\mathbf{X}} \; \underset{(p\times d)}{\mathbf{B}} + \underset{(n\times d)}{\mathbf{U}}. \tag{8.7}$$

With this specification of a common design matrix $\mathbf{X}$, we now find that the univariate least squares theory for (8.1) generalizes naturally if we use the usual partial ordering for symmetric matrices, namely, $\mathbf{C} \geq \mathbf{D}$ when $\mathbf{C} - \mathbf{D}$ is positive semidefinite, that is, $\mathbf{C} - \mathbf{D} \geq \mathbf{O}$ (Siotani [1967]). Thus if $\mathbf{C}(\boldsymbol{\Theta})$ is a symmetric matrix valued function, we say that $\mathbf{C}$ is minimized at $\boldsymbol{\Theta} = \hat{\boldsymbol{\Theta}}$ if $\mathbf{C}(\boldsymbol{\Theta}) \geq \mathbf{C}(\hat{\boldsymbol{\Theta}})$ for all $\boldsymbol{\Theta}$.

By analogy with univariate least squares, we can estimate $\boldsymbol{\Theta} = \mathbf{XB}$ by minimizing $\mathbf{U}'\mathbf{U} = (\mathbf{Y} - \boldsymbol{\Theta})'(\mathbf{Y} - \boldsymbol{\Theta})$ subject to the columns $\boldsymbol{\theta}^{(j)}$ of $\boldsymbol{\Theta}$ belonging to $\Omega = \mathscr{R}[\mathbf{X}]$. Now if $\mathbf{P}_\Omega$ once again represents the orthogonal projection of R^n onto Ω, then $\mathbf{P}_\Omega\boldsymbol{\theta}^{(j)} = \boldsymbol{\theta}^{(j)}$ $(j = 1, 2, \ldots, d)$ and

$$\mathbf{P}_\Omega\boldsymbol{\Theta} = \boldsymbol{\Theta}. \tag{8.8}$$

Setting $\hat{\boldsymbol{\Theta}} = \mathbf{P}_\Omega\mathbf{Y}$ and using $\mathbf{P}_\Omega(\mathbf{I}_n - \mathbf{P}_\Omega) = \mathbf{O}$ (see B1.6), we see that

$$\begin{aligned}(\mathbf{Y} - \hat{\boldsymbol{\Theta}})'(\hat{\boldsymbol{\Theta}} - \boldsymbol{\Theta}) &= \mathbf{Y}'(\mathbf{I}_n - \mathbf{P}_\Omega)\mathbf{P}_\Omega(\mathbf{Y} - \boldsymbol{\Theta}) \\ &= \mathbf{O}.\end{aligned} \tag{8.9}$$

Hence for all $\boldsymbol{\Theta}$ with columns in Ω,

$$\begin{aligned}\mathbf{C}(\boldsymbol{\Theta}) &= (\mathbf{Y} - \boldsymbol{\Theta})'(\mathbf{Y} - \boldsymbol{\Theta}) \\ &= (\mathbf{Y} - \hat{\boldsymbol{\Theta}} + \hat{\boldsymbol{\Theta}} - \boldsymbol{\Theta})'(\mathbf{Y} - \hat{\boldsymbol{\Theta}} + \hat{\boldsymbol{\Theta}} - \boldsymbol{\Theta}) \\ &= (\mathbf{Y} - \hat{\boldsymbol{\Theta}})'(\mathbf{Y} - \hat{\boldsymbol{\Theta}}) + (\hat{\boldsymbol{\Theta}} - \boldsymbol{\Theta})'(\hat{\boldsymbol{\Theta}} - \boldsymbol{\Theta}) \qquad (8.10) \\ &\geq (\mathbf{Y} - \hat{\boldsymbol{\Theta}})'(\mathbf{Y} - \hat{\boldsymbol{\Theta}}) \\ &= \mathbf{C}(\hat{\boldsymbol{\Theta}}),\end{aligned}$$

since $(\hat{\mathbf{\Theta}} - \mathbf{\Theta})'(\hat{\mathbf{\Theta}} - \mathbf{\Theta}) \geq \mathbf{O}$, and $\hat{\mathbf{\Theta}}$ gives the required minimum. Equality occurs in (8.10) only when $(\hat{\mathbf{\Theta}} - \mathbf{\Theta})'(\hat{\mathbf{\Theta}} - \mathbf{\Theta}) = \mathbf{O}$ or, by A4.5, when $\mathbf{\Theta} = \hat{\mathbf{\Theta}}$. Since $\mathbf{P}_\Omega$ is unique, $\hat{\mathbf{\Theta}}$ is unique and it is called the least squares estimator of $\mathbf{\Theta}$. We also note that the minimum value of $\mathbf{U}'\mathbf{U}$ is

$$\mathbf{E} = (\mathbf{Y} - \hat{\mathbf{\Theta}})'(\mathbf{Y} - \hat{\mathbf{\Theta}}) \tag{8.11}$$

$$= \mathbf{Y}'(\mathbf{I}_n - \mathbf{P}_\Omega)'(\mathbf{I}_n - \mathbf{P}_\Omega)\mathbf{Y}$$

$$= \mathbf{Y}'(\mathbf{I}_n - \mathbf{P}_\Omega)\mathbf{Y}, \tag{8.12}$$

since $\mathbf{I}_n - \mathbf{P}_\Omega$ is symmetric and idempotent (see B1.4). We have called this minimum value $\mathbf{E}$, as it is the multivariate analogue of the univariate residual or "error" sum of squares.

Now $\mathbf{P}_\Omega\mathbf{X} = \mathbf{X}$ so that $\mathbf{X}'(\mathbf{Y} - \hat{\mathbf{\Theta}}) = \mathbf{X}'(\mathbf{I}_n - \mathbf{P}_\Omega)\mathbf{Y} = \mathbf{O}$. Therefore, if $\hat{\mathbf{B}}$ satisfies $\hat{\mathbf{\Theta}} = \mathbf{X}\hat{\mathbf{B}}$, it satisfies the so-called normal equations

$$\mathbf{X}'\mathbf{X}\hat{\mathbf{B}} = \mathbf{X}'\mathbf{Y}, \tag{8.13}$$

the multivariate analogue of (8.5). Conversely, if $\hat{\mathbf{B}}$ satisfies (8.13), we can argue as for the univariate case and show that $\mathbf{X}\hat{\mathbf{B}} = \hat{\mathbf{\Theta}}$. Extracting the jth column from each side of (8.13), we have

$$\mathbf{X}'\mathbf{X}\hat{\boldsymbol{\beta}}^{(j)} = \mathbf{X}'\mathbf{y}^{(j)}, \tag{8.14}$$

so that, as far as least squares estimation is concerned, we can treat each of the d response variables separately, even though the $\mathbf{y}^{(j)}$ are correlated. This means that any technique for finding $\hat{\boldsymbol{\beta}}$ in the corresponding univariate model can be used to find each $\hat{\boldsymbol{\beta}}^{(j)}$. For the same reason, univariate computational techniques extend naturally to the multivariate case. For example, replacing $\mathbf{y}$ by $\mathbf{Y}$, the efficient algorithms described in Seber [1977: Chapter 11] can be used for calculating $\hat{\mathbf{B}}$, and $\mathbf{E}$ of (8.11). A brief survey of these methods is given in Section 10.1.1. Also, some univariate methods for finding the "best" subset of x-variables (Seber [1977: Chapter 12], Berk [1978], Thompson [1978a, b]) extend to the multivariate case (see Section 10.1.6). When $\mathbf{X}$ is ill conditioned so that $\mathbf{X}'\mathbf{X}$ is near singularity, the ridge regression biased estimates of $\boldsymbol{\beta}$ can be generalized to the multivariate case (Brown and Zidek [1980]).

If $\mathbf{X}$ has less than full rank, the univariate methods for handling questions relating to identifiability and estimability (see Seber 1977: Chapter 3) carry over to the multivariate case. For example, we can introduce identifiability restrictions $\mathbf{F}\boldsymbol{\beta}^{(j)} = \mathbf{0}$ $(j = 1, 2, \ldots, d)$, or $\mathbf{F}\mathbf{B} = \mathbf{O}$, where the rank of $\mathbf{G}' = (\mathbf{X}', \mathbf{F}')$ is p and the rows of $\mathbf{F}$ are linearly independent of the rows of $\mathbf{X}$. We then have a solution $\hat{\mathbf{B}} = (\mathbf{G}'\mathbf{G})^{-1}\mathbf{X}'\mathbf{Y}$. Also, estimable linear functions of the elements of **B** are readily handled. We say that

$$\phi = \sum_{j=1}^{d} \mathbf{a}_j' \boldsymbol{\beta}^{(j)}$$

is estimable if it has a linear unbiased estimate, that is, an unbiased estimate of the form $\sum_{j=1}^{d} \mathbf{b}_j' \mathbf{y}^{(j)}$. Then

$$\sum_{j=1}^{d} \mathbf{a}_j' \boldsymbol{\beta}^{(j)} = \mathscr{E}\left[\sum_{j=1}^{d} \mathbf{b}_j' \mathbf{y}^{(j)}\right] = \sum_{j=1}^{d} \mathbf{b}_j' \mathbf{X} \boldsymbol{\beta}^{(j)}$$

identically in the $\boldsymbol{\beta}^{(j)}$. Setting $\boldsymbol{\beta}^{(r)} = 0$ for all r, $r \neq j$, implies that

$$\mathbf{a}_j = \mathbf{X}'\mathbf{b}_j \qquad (j = 1, 2, \ldots, d), \tag{8.15}$$

that is, $\mathbf{a}_j$ is linearly dependent on the rows of **X**. Thus ϕ is estimable if and only if all the "univariate" functions $\mathbf{a}_j' \boldsymbol{\beta}^{(j)}$ are estimable. In practice, however, one's interest is usually confined to linear functions of individual columns of **B**, like $\mathbf{a}_j' \boldsymbol{\beta}^{(j)}$, rather than linear functions ϕ of all the elements of **B**. We note that various characterizations of univariate estimability are collected together by Alalouf and Styan [1979].

In conclusion, we note from the ith row of (8.7) that

$$\mathbf{y}_i = \mathbf{B}'\mathbf{x}_i + \mathbf{u}_i \qquad (i = 1, 2, \ldots, n), \tag{8.16}$$

where $\mathbf{x}_i$ is the ith row of **X**. This row representation of (8.7) is sometimes more convenient to use than the column representation

$$\mathbf{y}^{(j)} = \mathbf{X}\boldsymbol{\beta}^{(j)} + \mathbf{u}^{(j)} \qquad (j = 1, 2, \ldots, d), \tag{8.17}$$

8.2 PROPERTIES OF LEAST SQUARES ESTIMATES

Uptil now the only assumption we have made about the "error" matrix $\mathbf{U} = (\mathbf{u}_1, \mathbf{u}_2, \ldots, \mathbf{u}_n)'$ is that $\mathscr{E}[\mathbf{U}] = \mathbf{O}$. With this assumption it follows from (8.8) that

$$\mathscr{E}[\hat{\boldsymbol{\Theta}}] = \mathbf{P}_\Omega \mathscr{E}[\mathbf{Y}] = \mathbf{P}_\Omega \boldsymbol{\Theta} = \boldsymbol{\Theta}, \tag{8.18}$$

and $\hat{\boldsymbol{\Theta}}$ is unbiased. When **X** has full rank, we can solve (8.13) and obtain

$\hat{\mathbf{B}} = (\mathbf{X}'\mathbf{X})^{-1}\mathbf{X}'\mathbf{Y}$. Then

$$\begin{aligned}\mathscr{E}[\hat{\mathbf{B}}] &= (\mathbf{X}'\mathbf{X})^{-1}\mathbf{X}'\mathscr{E}[\mathbf{Y}] \\ &= (\mathbf{X}'\mathbf{X})^{-1}\mathbf{X}'\mathbf{X}\mathbf{B} \\ &= \mathbf{B}, \end{aligned} \tag{8.19}$$

and $\hat{\mathbf{B}}$ is unbiased.

To consider variance properties we now generalize the univariate assumption $E[u_h u_i] = \delta_{hi}\sigma^2$ and assume that the $\mathbf{u}_i$ are uncorrelated with a common dispersion matrix $\mathbf{\Sigma} = [(\sigma_{jk})]$, namely,

$$\begin{aligned}\mathscr{C}[\mathbf{y}_h, \mathbf{y}_i] &= \mathscr{C}[\mathbf{u}_h, \mathbf{u}_i] \\ &= \mathscr{E}[\mathbf{u}_h \mathbf{u}_i'] \\ &= \delta_{hi}\mathbf{\Sigma} \qquad (h, i = 1, 2, \ldots, n), \end{aligned} \tag{8.20}$$

where $\delta_{hi} = 1$ when $h = i$ and 0 otherwise. Referring to (8.17), we have

$$\mathscr{C}[\mathbf{y}^{(j)}, \mathbf{y}^{(k)}] = \mathscr{C}[\mathbf{u}^{(j)}, \mathbf{u}^{(k)}] = \sigma_{jk}\mathbf{I}_d, \tag{8.21}$$

and, since $\hat{\boldsymbol{\beta}}^{(j)} = \mathbf{X}'\mathbf{X})^{-1}\mathbf{X}'\mathbf{y}^{(j)}$,

$$\begin{aligned}\mathscr{C}[\hat{\boldsymbol{\beta}}^{(j)}, \hat{\boldsymbol{\beta}}^{(k)}] &= (\mathbf{X}'\mathbf{X})^{-1}\mathbf{X}'\mathscr{C}[\mathbf{y}^{(j)}, \mathbf{y}^{(k)}]\mathbf{X}(\mathbf{X}'\mathbf{X})^{-1} \\ &= \sigma_{jk}(\mathbf{X}'\mathbf{X})^{-1}. \end{aligned} \tag{8.22}$$

We now give a multivariate generalization of the so-called Gauss–Markov theorem.

THEOREM 8.1 Let $\mathbf{Y} = \mathbf{\Theta} + \mathbf{U}$, where the rows $\mathbf{u}_i$ of $\mathbf{U}$ are uncorrelated with mean $\mathbf{0}$ and common dispersion matrix $\mathbf{\Sigma}$, and $\mathbf{\Theta} = (\boldsymbol{\theta}^{(1)}, \boldsymbol{\theta}^{(2)}, \ldots, \boldsymbol{\theta}^{(d)})$. Let $\phi = \Sigma_{j=1}^{d}\mathbf{b}_j'\boldsymbol{\theta}^{(j)}$ and let $\hat{\mathbf{\Theta}}$ be the least squares estimate of $\mathbf{\Theta}$ subject to the columns of $\mathbf{\Theta}$ belonging to Ω. Then $\hat{\phi} = \Sigma_{j=1}^{d}\mathbf{b}_j'\hat{\boldsymbol{\theta}}^{(j)}$ is the BLUE of ϕ, that is, the linear unbiased estimate with minimum variance.

Proof From (8.18), $\hat{\phi}$ is an unbiased estimator of ϕ. Since $\hat{\mathbf{\Theta}} = \mathbf{P}_\Omega\mathbf{Y}$, $\hat{\boldsymbol{\theta}}^{(j)} = \mathbf{P}_\Omega\mathbf{y}^{(j)}$, and $\hat{\phi} = \Sigma_j\mathbf{b}_j'\mathbf{P}_\Omega\mathbf{y}^{(j)} = \Sigma_j(\mathbf{P}_\Omega\mathbf{b}_j)'\mathbf{y}^{(j)}$ is linear in the elements of $\mathbf{Y}$. Let $\phi^* = \Sigma_j\mathbf{c}_j'\mathbf{y}^{(j)}$ be any other linear unbiased estimator of ϕ. Then, if we take expected values, $\Sigma_j\mathbf{c}_j'\boldsymbol{\theta}^{(j)} = \phi = \Sigma_j\mathbf{b}_j'\boldsymbol{\theta}^{(j)}$ for all $\boldsymbol{\theta}^{(j)} \in \Omega$, that is, $(\mathbf{b}_j - \mathbf{c}_j)'\boldsymbol{\theta}^{(j)} = 0$ for all $\boldsymbol{\theta}^{(j)} \in \Omega$ $(j = 1, 2, \ldots, d)$. Hence $(\mathbf{b}_j - \mathbf{c}_j)$ is perpendicular to Ω, and its projection onto Ω is zero; that is, $\mathbf{P}_\Omega(\mathbf{b}_j - \mathbf{c}_j) = \mathbf{0}$, or $\mathbf{P}_\Omega\mathbf{b}_j = \mathbf{P}_\Omega\mathbf{c}_j$

$(j = 1, 2, \ldots, d)$. We now compare the variances of the two estimators $\hat{\phi}$ and ϕ^*:

$$\begin{aligned}
\operatorname{var} \hat{\phi} &= \operatorname{var}\left[\sum_j \mathbf{b}_j' \mathbf{P}_\Omega \mathbf{y}^{(j)}\right] \\
&= \operatorname{cov}\left[\sum_j \mathbf{c}_j' \mathbf{P}_\Omega \mathbf{y}^{(j)}, \sum_k \mathbf{c}_k' \mathbf{P}_\Omega \mathbf{y}^{(k)}\right] \\
&= \sum_j \sum_k \mathbf{c}_j' \mathbf{P}_\Omega \mathscr{C}\left[\mathbf{y}^{(j)}, \mathbf{y}^{(k)}\right] \mathbf{P}_\Omega \mathbf{c}_k \\
&= \sum_j \sum_k \mathbf{c}_j' \mathbf{P}_\Omega \mathbf{c}_k \sigma_{jk} \qquad [\text{by } (8.21)]
\end{aligned}$$

and, similarly,

$$\operatorname{var} \phi^* = \sum_j \sum_k \mathbf{c}_j' \mathbf{c}_k \sigma_{jk}.$$

Setting $\mathbf{C} = (\mathbf{c}_1, \mathbf{c}_2, \ldots, \mathbf{c}_n)$ and $\boldsymbol{\Sigma} = \mathbf{R}\mathbf{R}'$, where $\mathbf{R}$ is nonsingular (see A5.3), we have

$$\begin{aligned}
\operatorname{var} \phi^* - \operatorname{var} \hat{\phi} &= \sum_j \sum_k \mathbf{c}_j' (\mathbf{I}_n - \mathbf{P}_\Omega) \mathbf{c}_k \sigma_{jk} \\
&= \operatorname{tr}\left[\mathbf{C}'(\mathbf{I}_n - \mathbf{P}_\Omega)\mathbf{C}\boldsymbol{\Sigma}\right] \\
&= \operatorname{tr}\left[\mathbf{R}'\mathbf{C}'(\mathbf{I}_n - \mathbf{P}_\Omega)'(\mathbf{I}_n - \mathbf{P}_\Omega)\mathbf{C}\mathbf{R}\right] \qquad (\text{by A1.1}) \\
&= \operatorname{tr}[\mathbf{D}'\mathbf{D}], \qquad \text{say}, \\
&\geq 0,
\end{aligned}$$

since $\mathbf{D}'\mathbf{D}$ is positive semidefinite and its trace is the sum of its eigenvalues (see A4.2). Equality occurs only if $\mathbf{D}'\mathbf{D} = \mathbf{O}$ or $\mathbf{D} = \mathbf{O}$ (see A4.5), that is, if $(\mathbf{I}_n - \mathbf{P}_\Omega)\mathbf{C} = \mathbf{O}$ or $\mathbf{c}_j = \mathbf{P}_\Omega \mathbf{c}_j = \mathbf{P}_\Omega \mathbf{b}_j$. Thus $\operatorname{var} \phi^* \geq \operatorname{var} \hat{\phi}$, with equality if and only if $\phi^* = \hat{\phi}$, and $\hat{\phi}$ is the unique estimate with minimum variance.

COROLLARY If $\hat{\boldsymbol{\Theta}} = [(\hat{\theta}_{ij})]$, then $\hat{\theta}_{ij}$ is the BLUE of $\hat{\theta}_{ij}$. This follows by appropriately choosing the $\mathbf{b}_j$ in ϕ. □

The above theorem is quite general, as Ω is not specified; for example, Ω could be expressed as the null space of a matrix, say, $\{\mathbf{x} : \mathbf{K}\mathbf{x} = \mathbf{0}\}$. However, suppose $\Omega = \mathscr{R}[\mathbf{X}]$, as in Section 8.1, where $\mathbf{X}$ has less than full rank. If

$\phi = \Sigma_j \mathbf{a}_j' \boldsymbol{\beta}^{(j)}$ is an estimable function, then $\mathbf{a}_j = \mathbf{X}'\mathbf{b}_j$ for some $\mathbf{b}_j$, by (8.15), and $\phi = \Sigma_j \mathbf{b}_j' \mathbf{X} \boldsymbol{\beta}^{(j)} = \Sigma_j \mathbf{b}_j' \boldsymbol{\theta}^{(j)}$. It follows, from the above theorem, that $\Sigma_j \mathbf{a}_j' \hat{\boldsymbol{\beta}}^{(j)}$ $(= \Sigma_j \mathbf{b}_j' \hat{\boldsymbol{\theta}}^{(j)} = \hat{\phi})$ is the BLUE of ϕ. When $\mathbf{X}$ has full rank, all linear functions like ϕ are estimable. A coordinate-free approach to the above theory is given by Eaton [1970, 1983]. For a general reference on multivariate linear models see Arnold [1981].

We now focus our attention on the estimation of $\boldsymbol{\Sigma}$ and we begin with the following theorem.

THEOREM 8.2 Consider the model $\mathbf{Y} = \boldsymbol{\Theta} + \mathbf{U}$ of Theorem 8.1 with r the dimension of Ω. Then $\mathbf{E}/(n - r)$, where $\mathbf{E}$ is given by (8.11), is an unbiased estimate of $\boldsymbol{\Sigma}$.

Proof As $\mathbf{P}_\Omega$ is symmetric and idempotent,

$$\begin{aligned} \operatorname{tr}[\mathbf{I}_n - \mathbf{P}_\Omega] &= n - \operatorname{tr} \mathbf{P}_\Omega \\ &= n - \operatorname{rank} \mathbf{P}_\Omega \qquad \text{(by A6.3)} \\ &= n - r. \end{aligned} \tag{8.23}$$

Also, $\mathbf{P}_\Omega \boldsymbol{\Theta} = \boldsymbol{\Theta}$ so that

$$\begin{aligned} \mathbf{E} &= \mathbf{Y}'(\mathbf{I}_n - \mathbf{P}_\Omega)\mathbf{Y} \\ &= (\mathbf{Y} - \boldsymbol{\Theta})'(\mathbf{I}_n - \mathbf{P}_\Omega)(\mathbf{Y} - \boldsymbol{\Theta}) \\ &= \mathbf{U}'(\mathbf{I}_n - \mathbf{P}_\Omega)\mathbf{U} \qquad (8.24) \\ &= \sum_h \sum_i (\mathbf{I}_n - \mathbf{P}_\Omega)_{hi} \mathbf{u}_h \mathbf{u}_i'. \end{aligned}$$

Hence, by (8.20),

$$\begin{aligned} \mathscr{E}[\mathbf{E}] &= \sum_h \sum_i (\mathbf{I}_n - \mathbf{P}_\Omega)_{hi} \delta_{hi} \boldsymbol{\Sigma} \\ &= \{\operatorname{tr}[(\mathbf{I}_n - \mathbf{P}_\Omega)\mathbf{I}_n]\} \boldsymbol{\Sigma} \\ &= (n - r)\boldsymbol{\Sigma}. \end{aligned} \tag{8.25}$$

□

We note that if $\boldsymbol{\Sigma} > \mathbf{O}$ (which we have assumed throughout) and $n - r \geq d$, then, under fairly general conditions, $\mathbf{E}$ is positive definite with probability 1 (see A5.13).

8.3 LEAST SQUARES WITH LINEAR CONSTRAINTS

Given the model $\mathbf{Y} = \mathbf{\Theta} + \mathbf{U} = \mathbf{XB} + \mathbf{U}$, where $\mathbf{X}$ is $n \times p$ of rank p, suppose we wish to find the (restricted) least squares estimate of $\mathbf{B}$ subject to H_0: $\mathbf{AB} = \mathbf{O}$, where $\mathbf{A}$ is a known $q \times p$ matrix of rank q. Our problem, then, is to minimize $(\mathbf{Y} - \mathbf{XB})'(\mathbf{Y} - \mathbf{XB})$ subject to H_0. Now $\mathbf{B} = (\mathbf{X}'\mathbf{X})^{-1}\mathbf{X}'\mathbf{\Theta}$, and H_0 becomes $\mathbf{O} = \mathbf{AB} = \mathbf{A}(\mathbf{X}'\mathbf{X})^{-1}\mathbf{X}'\mathbf{\Theta} = \mathbf{A}_1\mathbf{\Theta}$. We therefore wish to minimize $(\mathbf{Y} - \mathbf{\Theta})'(\mathbf{Y} - \mathbf{\Theta})$ subject to the columns of $\mathbf{\Theta}$ lying in $\omega = \Omega \cap \mathscr{N}[\mathbf{A}_1]$, where $\Omega = \mathscr{R}[\mathbf{X}]$. Since ω is a vector subspace, we are still in the same situation as the unconstrained model, except that Ω is replaced by ω. Hence the unique least squares estimator of $\mathbf{\Theta}$ subject to H_0 is $\hat{\mathbf{\Theta}}_H = \mathbf{P}_\omega \mathbf{Y}$, where $\mathbf{P}_\omega$ represents the orthogonal projection onto ω. Since $\mathbf{X}$ has full rank, we have the unique representation $\hat{\mathbf{\Theta}}_H = \mathbf{X}\hat{\mathbf{B}}_H$, and $\hat{\mathbf{B}}_H$ is the required restricted least squares estimate of $\mathbf{B}$. Now

$$\mathbf{P}_\Omega \mathbf{A}_1' = \mathbf{X}(\mathbf{X}'\mathbf{X})^{-1}\mathbf{X}'\mathbf{X}(\mathbf{X}'\mathbf{X})^{-1}\mathbf{A}' = \mathbf{X}(\mathbf{X}'\mathbf{X})^{-1}\mathbf{A}',$$

and

$$\begin{aligned}
\mathbf{X}\hat{\mathbf{B}}_H &= \hat{\mathbf{\Theta}}_H \\
&= \mathbf{P}_\omega \mathbf{Y} \\
&= \mathbf{P}_\Omega \mathbf{Y} + (\mathbf{P}_\omega - \mathbf{P}_\Omega)\mathbf{Y} \\
&= \hat{\mathbf{\Theta}} - \mathbf{P}_{\omega^\perp \cap \Omega}\mathbf{Y} \qquad \text{(by B3.2).}
\end{aligned} \tag{8.26}$$

Since $\mathbf{P}_\Omega \mathbf{A}_1'$ is $n \times q$ of rank q (by B3.4 and $\mathscr{R}[\mathbf{A}_1'] \subset \mathscr{R}[\mathbf{X}] = \Omega$),

$$\begin{aligned}
\mathbf{P}_{\omega^\perp \cap \Omega} &= \mathbf{P}_\Omega \mathbf{A}_1'\left[\mathbf{A}_1 \mathbf{P}_\Omega^2 \mathbf{A}_1'\right]^{-1}(\mathbf{P}_\Omega \mathbf{A}_1')' \qquad \text{(B3.3 and B1.8)} \\
&= \mathbf{X}(\mathbf{X}'\mathbf{X})^{-1}\mathbf{A}'\left[\mathbf{A}(\mathbf{X}'\mathbf{X})^{-1}\mathbf{A}'\right]^{-1}\mathbf{A}(\mathbf{X}'\mathbf{X})^{-1}\mathbf{X}',
\end{aligned} \tag{8.27}$$

and, premultiplying (8.27) by $\mathbf{X}'$, (8.26) becomes

$$\mathbf{X}'\mathbf{X}\hat{\mathbf{B}}_H = \mathbf{X}'\mathbf{X}\hat{\mathbf{B}} - \mathbf{A}'\left[\mathbf{A}(\mathbf{X}'\mathbf{X})^{-1}\mathbf{A}'\right]^{-1}\mathbf{A}\hat{\mathbf{B}}. \tag{8.28}$$

If $\mathbf{X}$ is not of full rank, then the constraints $\mathbf{a}_i'\boldsymbol{\beta}^{(j)}$ must be estimable, that is, the rows $\mathbf{a}_i'$ of $\mathbf{A}$ are linear combinations of the rows of $\mathbf{X}$, or $\mathbf{A} = \mathbf{MX}$. Otherwise, if $\mathbf{a}_i'$ is linearly independent of the rows of $\mathbf{X}$, then $\mathbf{a}_i'\boldsymbol{\beta}^{(j)} = 0$ is not a bona fide constraint, but, instead, acts as an identifiability restriction. Now the $q \times n$ matrix $\mathbf{M}$ is of rank q, as $q = \text{rank } \mathbf{A} \le \text{rank } \mathbf{M} \le q$; also $\mathbf{M\Theta} = \mathbf{MXB} = \mathbf{AB} = \mathbf{O}$. Thus $\omega = \Omega \cap \mathscr{N}[\mathbf{M}]$ and

$$\mathbf{P}_\Omega \mathbf{M}' = \mathbf{X}(\mathbf{X}'\mathbf{X})^{-}\mathbf{X}'\mathbf{M}' = \mathbf{X}(\mathbf{X}'\mathbf{X})^{-}\mathbf{A}'. \tag{8.29}$$

Suppose that this $n \times q$ matrix has linearly dependent columns so that, for some $\mathbf{c}$ ($\neq \mathbf{0}$), $\mathbf{P}_\Omega \mathbf{M}'\mathbf{c} = \mathbf{0}$. Then $\mathbf{M}'\mathbf{c} \perp \Omega = \mathscr{R}[\mathbf{X}]$ and $\mathbf{A}'\mathbf{c} = \mathbf{X}'\mathbf{M}'\mathbf{c} = \mathbf{0}$, which is a contradiction, as the rows of $\mathbf{A}$ are linearly independent. Hence (8.29) has full rank q and

$$\mathbf{P}_{\omega^\perp \cap \Omega} = \mathbf{P}_\Omega \mathbf{M}'[\mathbf{M}\mathbf{P}_\Omega \mathbf{M}']^{-1}(\mathbf{P}_\Omega \mathbf{M}')' \tag{8.30}$$

$$= \mathbf{X}(\mathbf{X}'\mathbf{X})^- \mathbf{A}'[\mathbf{A}(\mathbf{X}'\mathbf{X})^- \mathbf{A}']^{-1}\mathbf{A}(\mathbf{X}'\mathbf{X})^- \mathbf{X}', \tag{8.31}$$

a symmetric indempotent matrix of rank (trace) q. Since $\mathbf{X}'\mathbf{P}_\Omega = \mathbf{X}'$, it follows from (8.26) (which does not depend on the rank of $\mathbf{X}$) that Equation (8.28) still holds with $(\mathbf{X}'\mathbf{X})^{-1}$ replaced by $(\mathbf{X}'\mathbf{X})^-$, and $\hat{\mathbf{B}}$ any solution of the normal Equations (8.13).

If we now consider constraints of the form $\mathbf{AB} = \mathbf{C}$, we can still use the above theory if we let $\mathbf{B}_0$ be any solution satisfying the constraints and set $\tilde{\mathbf{Y}} = \mathbf{Y} - \mathbf{X}\mathbf{B}_0$. Then

$$\begin{aligned}\tilde{\mathbf{Y}} &= \mathbf{X}(\mathbf{B} - \mathbf{B}_0) + \mathbf{U}\\ &= \mathbf{X}\boldsymbol{\Lambda} + \mathbf{U}\\ &= \boldsymbol{\Phi} + \mathbf{U},\end{aligned} \tag{8.32}$$

say, where $\mathbf{A}\boldsymbol{\Lambda} = \mathbf{AB} - \mathbf{AB}_0 = \mathbf{C} - \mathbf{C} = \mathbf{O}$, and our "translated" model (8.32) is of the same form as (8.7) above, but with $\mathbf{Y}$ replaced by $\tilde{\mathbf{Y}}$. Hence, from (8.26) and $\mathbf{P}_\Omega \mathbf{X} = \mathbf{X}$,

$$\begin{aligned}\hat{\boldsymbol{\Phi}}_H &= \mathbf{P}_\Omega \tilde{\mathbf{Y}} + (\mathbf{P}_\omega - \mathbf{P}_\Omega)\tilde{\mathbf{Y}}\\ &= \mathbf{P}_\Omega(\mathbf{Y} - \mathbf{X}\mathbf{B}_0) - \mathbf{P}_{\omega^\perp \cap \Omega}\tilde{\mathbf{Y}}\\ &= \hat{\boldsymbol{\Theta}} - \mathbf{X}\mathbf{B}_0 - \mathbf{P}_{\omega^\perp \cap \Omega}\tilde{\mathbf{Y}}.\end{aligned} \tag{8.33}$$

Since

$$\boldsymbol{\Theta} = \boldsymbol{\Phi} + \mathbf{X}\mathbf{B}_0, \tag{8.34}$$

we have

$$\hat{\boldsymbol{\Theta}}_H = \mathbf{X}\hat{\mathbf{B}}_H$$

$$= \hat{\boldsymbol{\Phi}}_H + \mathbf{X}\mathbf{B}_0 \tag{8.35}$$

$$= \hat{\boldsymbol{\Theta}} - \mathbf{P}_{\omega^\perp \cap \Omega}\tilde{\mathbf{Y}} \qquad [\text{by } (8.33)], \tag{8.36}$$

which is similar to (8.26). Premultiplying by $\mathbf{X}'$ and using (8.30), we have

$$\begin{aligned}\mathbf{X}'\mathbf{X}\hat{\mathbf{B}}_H &= \mathbf{X}'\mathbf{P}_\Omega\mathbf{Y} - \mathbf{X}'\mathbf{P}_{\omega^\perp\cap\Omega}(\mathbf{Y} - \mathbf{X}\mathbf{B}_0)\\ &= \mathbf{X}'\mathbf{Y} - \mathbf{X}'\mathbf{P}_\Omega\mathbf{M}'[\mathbf{M}\mathbf{P}_\Omega\mathbf{M}']^{-1}\mathbf{M}\mathbf{P}_\Omega(\mathbf{Y} - \mathbf{X}\mathbf{B}_0)\\ &= \mathbf{X}'\mathbf{Y} - \mathbf{X}'\mathbf{M}'[\mathbf{M}\mathbf{P}_\Omega\mathbf{M}']^{-1}(\mathbf{M}\mathbf{P}_\Omega\mathbf{Y} - \mathbf{M}\mathbf{X}\mathbf{B}_0)\\ &= \mathbf{X}'\mathbf{Y} - \mathbf{A}'[\mathbf{A}(\mathbf{X}'\mathbf{X})^-\mathbf{A}']^{-1}[\mathbf{A}(\mathbf{X}'\mathbf{X})^-\mathbf{X}'\mathbf{Y} - \mathbf{C}], \qquad (8.37)\end{aligned}$$

which can be compared with (8.28). A solution for $\hat{\mathbf{B}}_H$ takes the form

$$\hat{\mathbf{B}}_H = \hat{\mathbf{B}} - (\mathbf{X}'\mathbf{X})^-\mathbf{A}'[\mathbf{A}(\mathbf{X}'\mathbf{X})^-\mathbf{A}']^{-1}(\mathbf{A}\hat{\mathbf{B}} - \mathbf{C}),$$

where $\hat{\mathbf{B}} = (\mathbf{X}'\mathbf{X})^-\mathbf{X}'\mathbf{Y}$. Although $\hat{\mathbf{B}}$ and $\hat{\mathbf{B}}_H$ are not unique, $\hat{\mathbf{\Theta}}_H = \mathbf{X}\hat{\mathbf{B}}_H$ and $\hat{\mathbf{\Theta}} = \mathbf{X}\hat{\mathbf{B}}$ are unique (given $\mathbf{B}_0$). For full rank $\mathbf{X}$, $(\mathbf{X}'\mathbf{X})^-$ reduces to $(\mathbf{X}'\mathbf{X})^{-1}$, and $\hat{\mathbf{B}}_H$ and $\hat{\mathbf{B}}$ are unique. If the $q \times p$ matrix $\mathbf{A}$ has rank less than q, then we use a generalized inverse $[\mathbf{A}(\mathbf{X}'\mathbf{X})^-\mathbf{A}']^-$.

If H_0 holds, we have, from (8.34), (8.35) and $\mathbf{P}_\omega\mathbf{\Phi} = \mathbf{\Phi}$,

$$\begin{aligned}(\mathbf{Y} - \hat{\mathbf{\Theta}}_H)'(\hat{\mathbf{\Theta}}_H - \mathbf{\Theta}) &= (\tilde{\mathbf{Y}} - \hat{\mathbf{\Phi}}_H)'(\hat{\mathbf{\Phi}}_H - \mathbf{\Phi})\\ &= (\tilde{\mathbf{Y}} - \mathbf{P}_\omega\tilde{\mathbf{Y}})'(\mathbf{P}_\omega\tilde{\mathbf{Y}} - \mathbf{P}_\omega\mathbf{\Phi})\\ &= \tilde{\mathbf{Y}}'(\mathbf{I}_n - \mathbf{P}_\omega)\mathbf{P}_\omega(\tilde{\mathbf{Y}} - \mathbf{\Phi})\\ &= \mathbf{O},\end{aligned}$$

and we have the identity [see (8.9) and (8.10)]

$$(\mathbf{Y} - \mathbf{\Theta})'(\mathbf{Y} - \mathbf{\Theta}) = (\mathbf{Y} - \hat{\mathbf{\Theta}}_H)'(\mathbf{Y} - \hat{\mathbf{\Theta}}_H) + (\hat{\mathbf{\Theta}}_H - \mathbf{\Theta})'(\hat{\mathbf{\Theta}}_H - \mathbf{\Theta}). \qquad (8.38)$$

Using a similar argument, we have

$$\begin{aligned}\mathbf{E}_H &= (\mathbf{Y} - \hat{\mathbf{\Theta}}_H)'(\mathbf{Y} - \hat{\mathbf{\Theta}}_H)\\ &= \tilde{\mathbf{Y}}'(\mathbf{I}_n - \mathbf{P}_\omega)\tilde{\mathbf{Y}}. \qquad (8.39)\end{aligned}$$

8.4 DISTRIBUTION THEORY

So far the only assumptions we have made about the rows $\mathbf{u}_i$ of $\mathbf{U}$ are $\mathscr{E}[\mathbf{u}_i] = \mathbf{0}$ and $\mathscr{E}[\mathbf{u}_h\mathbf{u}_i'] = \delta_{hi}\mathbf{\Sigma}$. We now also assume that the $\mathbf{u}_i$ are normally distributed and obtain the following theorem.

THEOREM 8.3 Let $\mathbf{Y} = \mathbf{\Theta} + \mathbf{U}$, where the columns of $\mathbf{\Theta}$ belong to an r-dimensional subspace Ω, and the rows of $\mathbf{U}$ are i.i.d. $N_d(\mathbf{0}, \mathbf{\Sigma})$. Then $\hat{\mathbf{\Theta}}$, the least squares estimate of $\mathbf{\Theta}$, is statistically independent of $\mathbf{E} = (\mathbf{Y} - \hat{\mathbf{\Theta}})'(\mathbf{Y} - \hat{\mathbf{\Theta}})$, and $\mathbf{E} \sim W_d(n - r, \mathbf{\Sigma})$.

Proof We note that $\hat{\mathbf{\Theta}} = \mathbf{P}_\Omega \mathbf{Y}$ and $\mathbf{E} = [(\mathbf{I}_n - \mathbf{P}_\Omega)\mathbf{Y}]'[(\mathbf{I}_n - \mathbf{P}_\Omega)\mathbf{Y}]$. Since $\mathbf{P}_\Omega(\mathbf{I}_n - \mathbf{P}_\Omega) = \mathbf{O}$, we see that $\mathbf{P}_\Omega \mathbf{Y}$ and $(\mathbf{I}_n - \mathbf{P}_\Omega)\mathbf{Y}$ are statistically independent (by Exercise 2.18 with $\mathbf{B} = \mathbf{D} = \mathbf{I}_d$); hence $\hat{\mathbf{\Theta}}$ and $\mathbf{E}$ are statistically independent. Also, from (8.24), $\mathbf{E} = \mathbf{U}'(\mathbf{I}_n - \mathbf{P}_\Omega)\mathbf{U}$, where $\mathbf{I}_n - \mathbf{P}_\Omega$ is idempotent with rank $n - r$. We then apply Theorem 2.4 Corollary 1, of Section 2.3.2.

COROLLARY If $\hat{\mathbf{\Theta}} = \mathbf{X}\hat{\mathbf{B}}$, where $\mathbf{X}$ has full rank, then $\hat{\mathbf{B}}$ and $\mathbf{E}$ are statistically independent. This follows from the above theorem by noting that $\hat{\mathbf{B}} = (\mathbf{X}'\mathbf{X})^{-1}\mathbf{X}'\hat{\mathbf{\Theta}}$ is a function of $\hat{\mathbf{\Theta}}$. □

In the univariate model (8.1) we find that under normality assumptions the least squares estimate $\hat{\boldsymbol{\theta}}$ of $\boldsymbol{\theta}$ is also the maximum likelihood estimate, and $(\mathbf{y} - \hat{\boldsymbol{\theta}})'(\mathbf{y} - \hat{\boldsymbol{\theta}})/n$ is the maximum likelihood estimate of σ^2. It is therefore not surprising that we have the following theorem and corollaries.

THEOREM 8.4 Given the model of Theorem 8.3 with $n - r \geq d$, then $\hat{\mathbf{\Theta}} = \mathbf{X}\hat{\mathbf{B}}$ and

$$\hat{\mathbf{\Sigma}} = \frac{1}{n}(\mathbf{Y} - \hat{\mathbf{\Theta}})'(\mathbf{Y} - \hat{\mathbf{\Theta}}) = \frac{1}{n}\mathbf{E} \tag{8.40}$$

are the maximum likelihood estimates of $\mathbf{\Theta}$ and $\mathbf{\Sigma}$, respectively.

Proof We use the same approach as that given in Section 3.2.1. The likelihood function is the joint density function of the rows of $\mathbf{Y}$, namely,

$$L(\mathbf{\Theta}, \mathbf{\Sigma}) = (2\pi)^{-nd/2}|\mathbf{\Sigma}|^{-n/2}\exp\left(-\frac{1}{2}\sum_{i=1}^{n}(\mathbf{y}_i - \boldsymbol{\theta}_i)'\mathbf{\Sigma}^{-1}(\mathbf{y}_i - \boldsymbol{\theta}_i)\right), \tag{8.41}$$

where $\boldsymbol{\theta}_i'$ is the ith row of $\mathbf{\Theta}$, and we wish to maximize this subject to the columns of $\mathbf{\Theta}$ belonging to $\Omega = \mathscr{R}[\mathbf{X}]$. The last term of (8.41) is

$$\begin{aligned}\operatorname{tr}\left[\mathbf{\Sigma}^{-1}\sum_{i=1}^{n}(\mathbf{y}_i - \boldsymbol{\theta}_i)(\mathbf{y}_i - \boldsymbol{\theta}_i)'\right] &= \operatorname{tr}\left[\mathbf{\Sigma}^{-1}(\mathbf{Y} - \mathbf{\Theta})'(\mathbf{Y} - \mathbf{\Theta})\right] \\ &= \operatorname{tr}\left[\mathbf{\Sigma}^{-1}(\mathbf{Y} - \hat{\mathbf{\Theta}})'(\mathbf{Y} - \hat{\mathbf{\Theta}}) + \mathbf{\Sigma}^{-1}(\hat{\mathbf{\Theta}} - \mathbf{\Theta})'(\hat{\mathbf{\Theta}} - \mathbf{\Theta})\right] \qquad [\text{by } (8.10)].\end{aligned}$$

Hence

$$\log L(\mathbf{\Theta}, \mathbf{\Sigma}) = c - \tfrac{1}{2}n\log|\mathbf{\Sigma}| - \tfrac{1}{2}\operatorname{tr}[\mathbf{\Sigma}^{-1}\mathbf{E}] - b, \tag{8.42}$$

where

$$b = \tfrac{1}{2}\mathrm{tr}\left[\boldsymbol{\Sigma}^{-1}(\hat{\boldsymbol{\Theta}} - \boldsymbol{\Theta})'(\hat{\boldsymbol{\Theta}} - \boldsymbol{\Theta})\right]$$

$$= \tfrac{1}{2}\mathrm{tr}\left[(\hat{\boldsymbol{\Theta}} - \boldsymbol{\Theta})\boldsymbol{\Sigma}^{-1}(\hat{\boldsymbol{\Theta}} - \boldsymbol{\Theta})'\right]$$

$$\geq 0 \qquad \text{(by A4.2)},$$

since $\boldsymbol{\Sigma}^{-1} > \mathbf{O}$ implies that $(\hat{\boldsymbol{\Theta}} - \boldsymbol{\Theta})\boldsymbol{\Sigma}^{-1}(\hat{\boldsymbol{\Theta}} - \boldsymbol{\Theta})' \geq \mathbf{O}$ (see A4.4). Therefore $\log L$ is maximized for any $\boldsymbol{\Sigma} > \mathbf{O}$ when $b = 0$, that is, when $\boldsymbol{\Theta} = \hat{\boldsymbol{\Theta}}$. Thus

$$\log L(\hat{\boldsymbol{\Theta}}, \boldsymbol{\Sigma}) \geq \log L(\boldsymbol{\Theta}, \boldsymbol{\Sigma}) \qquad \text{for all } \boldsymbol{\Sigma} > \mathbf{O}.$$

The next step is to maximize

$$\log L(\hat{\boldsymbol{\Theta}}, \boldsymbol{\Sigma}) = c - \tfrac{1}{2}n\left(\log|\boldsymbol{\Sigma}| + \mathrm{tr}\left[\boldsymbol{\Sigma}^{-1}\mathbf{E}/n\right]\right)$$

subject to $\boldsymbol{\Sigma} > \mathbf{O}$. Since $n - r \geq d$, $\mathbf{E} > \mathbf{O}$ with probability 1 (see Theorem 8.3 and A5.13), and, by A7.1, the maximum occurs at $\boldsymbol{\Sigma} = \hat{\boldsymbol{\Sigma}} = \mathbf{E}/n$. Thus

$$\log L(\hat{\boldsymbol{\Theta}}, \hat{\boldsymbol{\Sigma}}) \geq \log L(\hat{\boldsymbol{\Theta}}, \boldsymbol{\Sigma}) \geq \log L(\boldsymbol{\Theta}, \boldsymbol{\Sigma}), \tag{8.43}$$

and $\hat{\boldsymbol{\Theta}}$ and $\hat{\boldsymbol{\Sigma}}$ are the maximum likelihood estimates.

COROLLARY 1 The maximum value of L is

$$L(\hat{\boldsymbol{\Theta}}, \hat{\boldsymbol{\Sigma}}) = (2\pi)^{nd/2}|\hat{\boldsymbol{\Sigma}}|^{-n/2}e^{-nd/2}. \tag{8.44}$$

COROLLARY 2 If $\mathbf{X}$ is $n \times p$ of rank p, then $\hat{\boldsymbol{\Theta}}$ has the unique representation $\hat{\boldsymbol{\Theta}} = \mathbf{X}\hat{\mathbf{B}}$ and $\hat{\mathbf{B}}$ is the maximum likelihood estimate of $\mathbf{B}$ (see A7.2). □

The key steps in the derivation of (8.43) are the decomposition (8.10) and the positive definiteness of $\mathbf{E}$. Working with $\tilde{\mathbf{Y}}$ instead of $\mathbf{Y}$ [see (8.32)], we find that the same steps apply if we maximize $L(\boldsymbol{\Theta}, \boldsymbol{\Sigma})$ subject to the constraints $H_0\colon \mathbf{AB} = \mathbf{C}$; ω now replaces Ω. Firstly, (8.38) leads to

$$(\tilde{\mathbf{Y}} - \boldsymbol{\Phi})'(\tilde{\mathbf{Y}} - \boldsymbol{\Phi}) = (\tilde{\mathbf{Y}} - \hat{\boldsymbol{\Phi}}_H)'(\tilde{\mathbf{Y}} - \hat{\boldsymbol{\Phi}}_H) + (\hat{\boldsymbol{\Phi}}_H - \boldsymbol{\Phi})'(\hat{\boldsymbol{\Phi}}_H - \boldsymbol{\Phi}).$$

Secondly, from (8.39), we see that $\mathbf{E}_H > \mathbf{O}$ with probability 1, as $n - r + q \geq d$, $r - q$ being the dimension of ω (see Exercise 8.8). Therefore $\hat{\boldsymbol{\Phi}}_H$, $\hat{\boldsymbol{\Theta}}_H$, and $\boldsymbol{E}_H$ are the respective maximum likelihood estimates of $\boldsymbol{\Phi}$, $\boldsymbol{\Theta}$, and $\boldsymbol{\Sigma}$, and (8.44) holds for the restricted case if we add the subscript H.

For a Bayesian approach to inference see Press [1980].

8.5 ANALYSIS OF RESIDUALS

In a multiple linear regression model, the residuals obtained after fitting the model have been found useful for investigating the validity of the model and its associated distributional assumptions. By the same token, we can expect the matrix of residuals

$$\hat{\mathbf{U}} = \begin{pmatrix} \hat{\mathbf{u}}_1' \\ \hat{\mathbf{u}}_2' \\ \vdots \\ \hat{\mathbf{u}}_n' \end{pmatrix} = \left(\hat{\mathbf{u}}^{(1)}, \hat{\mathbf{u}}^{(2)}, \dots, \hat{\mathbf{u}}^{(d)}\right),$$

where $\hat{\mathbf{U}} = \mathbf{Y} - \mathbf{X}\hat{\mathbf{B}} = (\mathbf{I}_n - \mathbf{P}_\Omega)(\mathbf{y}^{(1)}, \mathbf{y}^{(2)}, \dots, \mathbf{y}^{(d)})$, to provide a similar insight. As a first step we could apply the univariate plotting methods (see Seber [1977: Section 6.6]) to each column $\hat{\mathbf{u}}^{(j)}$. For example, we could (1) carry out normal probability plots on the ranked elements of $\hat{\mathbf{u}}^{(j)}$, (2) plot the elements of $\hat{\mathbf{u}}^{(j)}$ against the elements of $\mathbf{y}^{(j)}$, and (3) plot the elements of $\hat{\mathbf{u}}^{(j)}$ against the corresponding values of any regressor variables or an omitted factor (e.g., time).

We note, from (8.21), that

$$\begin{aligned} \mathscr{C}\left[\hat{\mathbf{u}}^{(j)}, \hat{\mathbf{u}}^{(k)}\right] &= (\mathbf{I}_n - \mathbf{P}_\Omega)\mathscr{C}\left[\mathbf{y}^{(j)}, \mathbf{y}^{(k)}\right](\mathbf{I}_n - \mathbf{P}_\Omega)' \\ &= \sigma_{jk}(\mathbf{I}_n - \mathbf{P}_\Omega). \end{aligned} \tag{8.45}$$

Since, as far as univariate graphical methods are concerned, $\mathbf{P}_\Omega$ can generally be ignored in (8.45) when $j = k$, it can probably be ignored for $j \neq k$. Therefore, under the normality assumptions of the model, the rows $\hat{\mathbf{u}}_i$ of $\hat{\mathbf{U}}$ can be regarded as being approximately i.i.d. $N_d(\mathbf{0}, \boldsymbol{\Sigma})$ and the graphical methods of Chapter 4 for detecting outliers and checking for multivariate normality can be used. For example, a gamma plot of the ordered $\hat{\mathbf{u}}_i'\mathbf{S}^{-1}\hat{\mathbf{u}}_i$, where $\mathbf{S}$ is an estimate of $\boldsymbol{\Sigma}$, would be useful. As outliers can inflate $\mathbf{S}$, a robust version of it with or without an "unbiasing" constant may be more appropriate.

There are some problems in analyzing residuals for an experimental design, whether they come from a univariate or multivariate model. First, there are constraints on subsets of the residuals, namely, $\mathbf{P}_\Omega\hat{\mathbf{U}} = \mathbf{P}_\Omega(\mathbf{I}_n - \mathbf{P}_\Omega)\mathbf{Y} = \mathbf{O}$. For example, in a univariate one-way classification with rows representing factor levels, the row sums of residuals are all zero. This can cause problems when the number of columns is small; one aberrant residual in a row will affect the other residuals in the row. Second, the presence of outliers may seriously bias the usual main effects and interactions that are subtracted from an observation and thus mask the local effect of an outlier in the corresponding residual. To avoid this masking effect of outliers, $\hat{\mathbf{B}}$ could be replaced by a robust estimate $\mathbf{B}^*$ and the residuals $\mathbf{U}^* = \mathbf{Y} - \mathbf{X}\mathbf{B}^*$ used instead. These modified residuals will not satisfy the constraints $\mathbf{P}_\Omega\mathbf{U}^* = \mathbf{O}$, but this is not important, as their main use is to accentuate the presence of outliers. Although there is a rapidly growing literature on robust estimation for univariate linear models (see Huber [1981]), the multivariate case has received little attention. One possibility is to apply a univariate technique to each column $\boldsymbol{\beta}^{(j)}$ of $\mathbf{B}$.

8.6 HYPOTHESIS TESTING

8.6.1 *Extension of Univariate Theory*

Consider the linear model $\mathbf{Y} = \mathbf{XB} + \mathbf{U}$, where $\mathbf{X}$ is $n \times p$ of rank r, and the rows of $\mathbf{U}$ are i.i.d. $N_d(\mathbf{0}, \boldsymbol{\Sigma})$. We saw, in Section 8.1, that this model arises naturally in the experimental design situation where the same design applies to each of d variables. This common design implies that main effects and interactions have the same form for each of the d variables so that the constraint matrix, $\mathbf{A}$, say, for any corresponding test will also be the same for each variable. Thus we might wish to test $H_{0j}: \mathbf{A}\boldsymbol{\beta}^{(j)} = \mathbf{c}^{(j)}$ $(j = 1, 2, \ldots, d)$ simultaneously, that is, test

$$H_0 = \bigcap_{j=1}^{d} H_{0j}: \mathbf{AB} = \mathbf{C}, \tag{8.46}$$

where $\mathbf{A}$ is $q \times p$ of rank q. If $r < p$, then H_0 must be testable; that is, the constraints must be estimable so that $\mathbf{A} = \mathbf{MX}$ for some matrix $\mathbf{M}$ of rank q (see Section 8.3).

In the univariate model it was found that a test statistic for H_{0j} based on two residual or error sums of squares could be derived either intuitively from the distribution of $\mathbf{A}\hat{\boldsymbol{\beta}}^{(j)} - \mathbf{c}^{(j)}$, or using the likelihood ratio principle (Seber [1977: Section 4.1.2]). It is therefore not surprising that we can base a multivariate test of H_0 on the matrix generalizations of these sum of squares, namely,

$$\mathbf{E} = (\mathbf{Y} - \mathbf{X}\hat{\mathbf{B}})'(\mathbf{Y} - \mathbf{X}\hat{\mathbf{B}})$$

and

$$\mathbf{E}_H = (\mathbf{Y} - \mathbf{X}\hat{\mathbf{B}}_H)'(\mathbf{Y} - \mathbf{X}\hat{\mathbf{B}}_H).$$

As the Wishart distribution takes over the role of the chi-square distribution, we have the following multivariate generalization.

THEOREM 8.5 When $H_0: \mathbf{AB} = \mathbf{C}$ is true, $\mathbf{E}$ and $\mathbf{H} = \mathbf{E}_H - \mathbf{E}$ are independently distributed as $W_d(n - r, \boldsymbol{\Sigma})$ and $W_d(q, \boldsymbol{\Sigma})$, respectively.

Proof Using the notation of Section 8.3, let $\tilde{\mathbf{Y}} = \mathbf{Y} - \mathbf{XB}_0$ and let $\omega = \Omega \cap \mathcal{N}[\mathbf{M}]$. Then, from (8.24),

$$\mathbf{E} = \mathbf{U}'(\mathbf{I}_n - \mathbf{P}_\Omega)\mathbf{U},$$

where $\mathbf{I}_n - \mathbf{P}_\Omega$ has rank $n - r$. Also, from (8.39) and (8.32),

$$\begin{aligned}\mathbf{E}_H &= \tilde{\mathbf{Y}}'(\mathbf{I}_n - \mathbf{P}_\omega)\tilde{\mathbf{Y}} \\ &= (\boldsymbol{\Phi} + \mathbf{U})'(\mathbf{I}_n - \mathbf{P}_\omega)(\boldsymbol{\Phi} + \mathbf{U}) \\ &= \mathbf{U}'(\mathbf{I}_n - \mathbf{P}_\omega)\mathbf{U},\end{aligned}$$

since $\mathbf{P}_\omega\boldsymbol{\Phi} = \boldsymbol{\Phi}$ when H_0 is true. Thus, when H_0 is true,

$$\begin{aligned}\mathbf{E}_H - \mathbf{E} &= \mathbf{U}'(\mathbf{P}_\Omega - \mathbf{P}_\omega)\mathbf{U} \\ &= \mathbf{U}'\mathbf{P}_{\omega^\perp \cap \Omega}\mathbf{U},\end{aligned}$$

where $\mathbf{P}_{\omega^\perp \cap \Omega}$ is a symmetric idempotent matrix of rank q [see (8.31)]. Thus $\mathbf{I}_n - \mathbf{P}_\Omega$ and $\mathbf{P}_\Omega - \mathbf{P}_\omega$ are symmetric and idempotent with respective ranks $n - r$ and q, and (by B3.1)

$$(\mathbf{I}_n - \mathbf{P}_\Omega)(\mathbf{P}_\omega - \mathbf{P}_\Omega) = \mathbf{P}_\omega - \mathbf{P}_\Omega\mathbf{P}_\omega = \mathbf{O}.$$

The proof is now completed by evoking Theorem 2.4, Corollaries 1 and 2, from Section 2.3.2.

COROLLARY 1

$$\mathbf{H} = (\mathbf{A}\hat{\mathbf{B}} - \mathbf{C})'[\mathbf{A}(\mathbf{X}'\mathbf{X})^-\mathbf{A}']^{-1}(\mathbf{A}\hat{\mathbf{B}} - \mathbf{C}), \tag{8.47}$$

Proof Since $\mathbf{P}_\Omega\mathbf{X} = \mathbf{X}$,

$$\mathbf{E} = (\mathbf{Y} - \mathbf{X}\mathbf{B}_0)'(\mathbf{I}_n - \mathbf{P}_\Omega)(\mathbf{Y} - \mathbf{X}\mathbf{B}_0) = \tilde{\mathbf{Y}}'(\mathbf{I}_n - \mathbf{P}_\Omega)\tilde{\mathbf{Y}}. \tag{8.48}$$

Hence, by (8.39),

$$\begin{aligned}\mathbf{E}_H - \mathbf{E} &= \tilde{\mathbf{Y}}'(\mathbf{P}_\Omega - \mathbf{P}_\omega)\tilde{\mathbf{Y}} \\ &= (\mathbf{Y} - \mathbf{X}\mathbf{B}_0)'\mathbf{P}_\Omega\mathbf{M}'[\mathbf{M}\mathbf{P}_\Omega\mathbf{M}']^{-1}\mathbf{M}\mathbf{P}_\Omega(\mathbf{Y} - \mathbf{X}\mathbf{B}_0) \qquad \text{[by (8.30)]} \\ &= (\mathbf{M}\mathbf{P}_\Omega\mathbf{Y} - \mathbf{M}\mathbf{X}\mathbf{B}_0)'[\mathbf{A}(\mathbf{X}'\mathbf{X})^-\mathbf{A}']^{-1}(\mathbf{M}\mathbf{P}_\Omega\mathbf{Y} - \mathbf{M}\mathbf{X}\mathbf{B}_0) \qquad \text{[by (8.31)]} \\ &= (\mathbf{A}(\mathbf{X}'\mathbf{X})^-\mathbf{X}'\mathbf{Y} - \mathbf{C})'[\mathbf{A}(\mathbf{X}'\mathbf{X})^-\mathbf{A}']^{-1}(\mathbf{A}(\mathbf{X}'\mathbf{X})^-\mathbf{X}'\mathbf{Y} - \mathbf{C}) \\ &= (\mathbf{A}\hat{\mathbf{B}} - \mathbf{C})'[\mathbf{A}(\mathbf{X}'\mathbf{X})^-\mathbf{A}']^{-1}(\mathbf{A}\hat{\mathbf{B}} - \mathbf{C}),\end{aligned}$$

since $\mathbf{M}\mathbf{X}\mathbf{B}_0 = \mathbf{A}\mathbf{B}_0 = \mathbf{C}$.

COROLLARY 2 Given that H_0 is false,

$$\begin{aligned}\mathscr{E}[\mathbf{H}] &= q\boldsymbol{\Sigma} + (\mathbf{A}\mathbf{B} - \mathbf{C})'[\mathbf{A}(\mathbf{X}'\mathbf{X})^-\mathbf{A}']^{-1}(\mathbf{A}\mathbf{B} - \mathbf{C}) \\ &= q\boldsymbol{\Sigma} + \mathbf{D},\end{aligned} \tag{8.49}$$

say, and $\mathbf{H}$ has a noncentral Wishart distribution $W_d(q, \boldsymbol{\Sigma}; \boldsymbol{\Delta})$ with noncentrality matrix $\boldsymbol{\Delta} = \boldsymbol{\Sigma}^{-1/2}\mathbf{D}\boldsymbol{\Sigma}^{-1/2}$.

Proof The rows of $\mathbf{Y} - \mathbf{XB}_0$ are independently distributed with common dispersion matrix $\boldsymbol{\Sigma}$; and $\text{tr}[\mathbf{P}_\Omega - \mathbf{P}_\omega] = q$. Hence, by (1.9),

$$\begin{aligned}\mathscr{E}[\mathbf{H}] &= \mathscr{E}\left[(\mathbf{Y} - \mathbf{XB}_0)'(\mathbf{P}_\Omega - \mathbf{P}_\omega)(\mathbf{Y} - \mathbf{XB}_0)\right] \\ &= q\boldsymbol{\Sigma} + (\mathbf{XB} - \mathbf{XB}_0)'(\mathbf{P}_\Omega - \mathbf{P}_\omega)(\mathbf{XB} - \mathbf{XB}_0) \\ &= q\boldsymbol{\Sigma} + \mathbf{D},\end{aligned}$$

the last step following from the same algebra used in proving Corollary 1, but with $\mathbf{Y}$ replaced by $\mathbf{XB}$. We now invoke Section 2.3.3 with $\boldsymbol{\Delta} = \boldsymbol{\Sigma}^{-1/2}\mathbf{D}\boldsymbol{\Sigma}^{-1/2}$.

8.6.2 Test Procedures

In looking for a test statistic to test H_0, we note that $\mathscr{E}[\mathbf{E}] = (n - r)\boldsymbol{\Sigma}$ and $\mathscr{E}[\mathbf{H}] = q\boldsymbol{\Sigma} + \mathbf{D}$, where $\mathbf{D} \geq \mathbf{O}$ with equality if and only if H_0 is true. Clearly, we would reject H_0 if $\mathbf{H}$ or $\mathbf{E}_H$ is too "large" compared with $\mathbf{E}$. Two test statistics based on this general principle are derived below. We shall assume that $n - r \geq d$ so that $\mathbf{E}$ is positive definite with probability 1.

a UNION–INTERSECTION TEST

Suppose we reduce the multivariate model to a univariate one by the transformation $\mathbf{y} = \mathbf{Y}\boldsymbol{\ell}$, $\boldsymbol{\beta} = \mathbf{B}\boldsymbol{\ell}$ and $\mathbf{u} = \mathbf{U}\boldsymbol{\ell}$, where $\boldsymbol{\ell} \neq \mathbf{0}$ (see Theorem 2.4 of Section 2.3.2). Then $\mathbf{y} = \mathbf{X}\boldsymbol{\beta} + \mathbf{u}$, where $\mathbf{u} \sim N_n(\mathbf{0}, \sigma_\ell^2\mathbf{I}_n)$ and $\sigma_\ell^2 = \boldsymbol{\ell}'\boldsymbol{\Sigma}\boldsymbol{\ell}$, and H_0 reduces to $H_{0\ell}: \mathbf{A}\boldsymbol{\beta} = \mathbf{C}\boldsymbol{\ell} = \mathbf{c}$, say. Let

$$Q = (\mathbf{y} - \mathbf{X}\hat{\boldsymbol{\beta}})'(\mathbf{y} - \mathbf{X}\hat{\boldsymbol{\beta}}) = \mathbf{y}'(\mathbf{I}_n - \mathbf{P}_\Omega)\mathbf{y} = \boldsymbol{\ell}'\mathbf{E}\boldsymbol{\ell},$$

and

$$Q_H = (\mathbf{y} - \mathbf{X}\hat{\boldsymbol{\beta}}_H)'(\mathbf{y} - \mathbf{X}\hat{\boldsymbol{\beta}}_H) = \tilde{\mathbf{y}}'(\mathbf{I}_n - \mathbf{P}_\Omega)\tilde{\mathbf{y}} = \boldsymbol{\ell}'\mathbf{E}_H\boldsymbol{\ell},$$

where $\tilde{\mathbf{y}} = \tilde{\mathbf{Y}}\boldsymbol{\ell}$. Then we can test $H_{0\ell}$ using the F-ratio

$$F_\ell = \frac{(Q_H - Q)/q}{Q/(n - r)} = \frac{\boldsymbol{\ell}'\mathbf{H}\boldsymbol{\ell}/q}{\boldsymbol{\ell}'\mathbf{E}\boldsymbol{\ell}/(n - r)},$$

and we reject H_ℓ if F_ℓ is too large. Now $H_0 = \cap_\ell H_{0\ell}$, so that, using the union–intersection principle, a test of H_0 has acceptance region

$$\begin{aligned}\bigcap_\ell \{\mathbf{Y} : F_\ell \leq k\} &= \left\{\mathbf{Y} : \sup_\ell F_\ell \leq k\right\} \\ &= \left\{\mathbf{Y} : \sup_\ell \frac{\boldsymbol{\ell}'\mathbf{H}\boldsymbol{\ell}}{\boldsymbol{\ell}'\mathbf{E}\boldsymbol{\ell}} \leq \frac{kq}{n - r} = k_1\right\} \\ &= \{\mathbf{Y} : \phi_{\max} \leq k_1\},\end{aligned}$$

where $\phi_{\max}$ is the largest eigenvalue of $\mathbf{HE}^{-1}$ (see A7.5 and A1.4), that is, the largest root of $|\mathbf{H} - \phi\mathbf{E}| = 0$. This test, based on the largest eigenvalue, is due to Roy [1953], and we reject H_0 if $\phi_{\max}$ is too large. The tables of critical values that we use are based on [see (2.34)] $\theta_{\max} = \phi_{\max}/(1 + \phi_{\max})$, the maximum eigenvalue of $\mathbf{H}(\mathbf{E} + \mathbf{H})^{-1}$. When H_0 is true we have, from Theorem 8.5, that $\mathbf{H}$ and $\mathbf{E}$ are independent Wishart matrices with degrees of freedom $m_H = q$ and $m_E = n - r$ $(\geq d)$. Referring to Section 2.5.5, we reject H_0 if $\theta_{\max} \geq \theta_\alpha$, where θ_α is obtained from Appendix D14 with $s = \min(q, d)$, $\nu_1 = \frac{1}{2}(|q - d| - 1)$, and $\nu_2 = \frac{1}{2}(n - r - d - 1)$. When $s = 1$, we can use tables of the F-distribution (Section 2.5.5e). An alternative derivation of $\phi_{\max}$, also based on the union–intersection principle, is given by Mudholkar et al. [1974a].

b LIKELIHOOD RATIO TEST

Since $n - r \geq d$, both $\mathbf{E}$ and $\mathbf{E}_H$ are positive definite with probability 1. Therefore if $L(\boldsymbol{\Theta}, \boldsymbol{\Sigma})$ is the likelihood function for the rows of $\mathbf{Y}$, the likelihood ratio test statistic foı H_0 is (see Theorem 8.4 and the following discussion in Section 8.4)

$$\ell = \frac{L(\hat{\boldsymbol{\Theta}}_H, \hat{\boldsymbol{\Sigma}}_H)}{L(\hat{\boldsymbol{\Theta}}, \hat{\boldsymbol{\Sigma}})}$$

$$= \frac{|\hat{\boldsymbol{\Sigma}}_H|^{-n/2}}{|\hat{\boldsymbol{\Sigma}}|^{-n/2}} \qquad [\text{by } (8.44)],$$

from which we obtain

$$\Lambda = \ell^{2/n}$$

$$= \frac{|\hat{\boldsymbol{\Sigma}}|}{|\hat{\boldsymbol{\Sigma}}_H|}$$

$$= \frac{|\mathbf{E}|}{|\mathbf{E}_H|}$$

$$= \frac{|\mathbf{E}|}{|\mathbf{E} + \mathbf{H}|} \tag{8.50}$$

$$= |\mathbf{I}_d - \mathbf{V}| \qquad [\text{by } (2.25)]$$

$$= \prod_{j=1}^{d} (1 - \theta_j) \tag{8.51}$$

$$= \prod_{j=1}^{d} (1 + \phi_j)^{-1}, \tag{8.52}$$

where $\mathbf{V} = \mathbf{E}_H^{-1/2}\mathbf{H}\mathbf{E}_H^{-1/2}$, as introduced in Section 2.5, and the θ_j are the ordered eigenvalues of $\mathbf{V}$. The test statistic Λ, a simple function of the likelihood ratio statistic ℓ, was proposed by Wilks [1932]. We see from (2.48) that when H_0 is true, $\Lambda \sim U_{d,q,n-r}$, and, by the likelihood ratio principle, we reject H_0 if Λ is too small, that is, if $|\mathbf{E}_H|$ is much greater than $|\mathbf{E}|$. Properties of the U-distribution are discussed in detail in Chapter 2 and summarized in Section 2.5.5b, while methods of computing Λ are described in Section 10.1.4a. If we use the tables in Appendix D13, we enter α, d, $m_H = q$, and $M = n - r - d + 1$ and obtain $C_\alpha = C_\alpha(d, q, M)$. We reject H_0 at the α level of significance if $-f\log\Lambda \geq C_\alpha \chi^2_{dq}(\alpha)$, where $f = n - r - \frac{1}{2}(d - q + 1)$ and $\text{pr}[\chi^2_{dq} \geq \chi^2_{dq}(\alpha)] = \alpha$. Since $C_\alpha > 1$, we do not need to look up C_α if $-f\log\Lambda < \chi^2_{dq}(\alpha)$.

We note, from (8.52), that if $\phi_1 \geq \phi_2 \geq \cdots \geq \phi_d$ are the ordered eigenvalues of $\mathbf{H}\mathbf{E}^{-1}$, then

$$-f\log\Lambda = f\sum_{j=1}^{d}\log(1 + \phi_j).$$

Since $-f\log\Lambda$ is approximately chi square when H_0 is true, it is tempting to assert that each $f\log(1 + \phi_j)$ is approximately chi square; however, this is not the case. Although the individual population eigenvalues generate distributions of $f\log(1 + \phi_j)$, there is no way of knowing which of the d population values is generating the jth largest sample value ϕ_j. There is interest in the literature in testing the sequence of hypotheses $H_{0k}: \phi_{k+1} = \phi_{k+2} = \cdots = \phi_d = 0$ using

$$f\sum_{j=k+1}^{d}\log(1 + \phi_j),$$

which is stated to be asymptoticallyally $\chi^2_{(d-k)(q-k)}$ when H_{0k} is true (see Muirhead [1982: Section 10.7.4] for a good survey of the theory). However, Harris [1975: p. 112] claims that the appropriate distribution is not chi square and that such tests should be abandoned. Clearly there are some problems associated with such tests when f is not large.

c OTHER TEST STATISTICS

We note that $\Lambda = \prod_j(1 - \theta_j)$ and $\phi_{\max} = \theta_{\max}/(1 - \theta_{\max})$ are functions of the eigenvalues θ_j. Since these eigenvalues have certain invariant properties (see Anderson [1958: pp. 222–223]), other functions of the θ_j have been proposed as test statistics for H_0, or special cases of H_0. If $s = \min(d, m_H)$, these statistics include the following (see Section 2.5.5):

(1) Wilks' ratio (Wilks [1932], Hsu [1940])

$$\frac{|\mathbf{H}|}{|\mathbf{E} + \mathbf{H}|} = \prod_{j=1}^{d}\theta_j \qquad (s = d).$$

(2) The Lawley–Hotelling trace

$$T_g^2 = (n-r)\,\mathrm{tr}[\mathbf{HE}^{-1}]$$
$$= (n-r)\sum_{j=1}^{s}\left[\frac{\theta_j}{1-\theta_j}\right].$$

(3) Pillai's trace

$$V^{(s)} = \mathrm{tr}\left[\mathbf{H}(\mathbf{E}+\mathbf{H})^{-1}\right] = \sum_{j=1}^{s}\theta_j.$$

Several other statistics have been proposed by Pillai [1955]. In the literature the upper limit s of the above sums is sometimes replaced by d so that some of the θ_j may then be zero. A promising plotting method for assessing H_0, based on a Q–Q plot (see Appendix C4) of the θ_j versus the appropriate quantile values of the limiting null distribution, is given by Wachter [1980]. However, limiting forms under alternative hypotheses are not yet available.

d COMPARISON OF FOUR TEST STATISTICS

As Wilks' ratio has been shown to be generally inferior to the likelihood ratio statistic Λ (e.g., Hart and Money [1976]), we shall drop it from our list and compare just the four test statistics Roy's $\phi_{\max}$, Wilks' Λ, Hotelling's T_g^2, and Pillai's $V^{(s)}$. A brief review of some of the optimality properties of these tests is given by Giri [1977: Section 8.4.4].

When H_0 is false, $\mathbf{E} \sim W_d(n-r, \mathbf{\Sigma})$, $\mathbf{E}$ is independent of $\mathbf{H}$, but now $\mathbf{H} \sim W_d(q, \mathbf{\Sigma}; \mathbf{\Delta})$, a noncentral Wishart distribution with noncentrality matrix [see (8.49)]

$$\mathbf{\Delta} = \mathbf{\Sigma}^{-1/2}(\mathbf{AB}-\mathbf{C})'\left[\mathbf{A}(\mathbf{X}'\mathbf{X})^{-}\mathbf{A}'\right]^{-1}(\mathbf{AB}-\mathbf{C})\mathbf{\Sigma}^{-1/2}. \qquad (8.53)$$

Unfortunately the corresponding noncentral distributions of the four test statistics are very complicated, making power comparisons difficult. When rank $\mathbf{\Delta} = 1$, the so-called linear case (Section 2.3.3), the noncentral distribution function of Λ can be expressed as a mixture of incomplete beta functions (Tretter and Walster [1975]) and can therefore be readily computed. This mixture approach has been extended to general $\mathbf{\Delta}$ by Walster and Tretter [1980]. An asymptotic expansion up to order n^{-2} has been given by Sugiura and Fujikoshi [1969]. Further asymptotic expansions are given by Y. S. Lee [1971] for $(n-r)V^{(s)}$, and by Siotani [1956] and Ito [1960] for T_g^2. For a helpful survey and further references see Johnson and Kotz [1972: pp. 199–201] (in their notation $\mathbf{S}_1 = \mathbf{H}$, $\mathbf{S}_2 = \mathbf{E}$, $m = d$, $\nu_1 = q$, $\nu_2 = n-r$, and $p = s$).

A very general study was carried out by Schatzoff [1966b] using the concept of expected significance level (ESL), proposed by Dempster and Schatzoff

[1965], to compare the powers of the four test statistics (and two others) over a wide range of alternatives. Working with the eigenvalues ϕ_j, Schatzoff concluded that no single statistic was uniformly better than any of the others, though $\phi_{\max}$ was slightly worse than the others, except in the one case where $\phi_{\max}$ dominated the other ϕ_j. This latter result is not surprising when the following theory, based on Rao [1973], is considered.

Replacing the matrices **E** and **H** by their expected values [see (8.25) and (8.49)], we have a population version of $|\mathbf{H} - \phi\mathbf{E}| = 0$, namely,

$$\begin{aligned} 0 &= |\mathscr{E}[\mathbf{H}] - \phi^*\mathscr{E}[\mathbf{E}]| \\ &= |q\boldsymbol{\Sigma} + \mathbf{D} - \phi^*(n-r)\boldsymbol{\Sigma}| \\ &= |\mathbf{D} - \{\phi^*(n-r) - q\}\boldsymbol{\Sigma}| \\ &= |\mathbf{D} - \gamma\boldsymbol{\Sigma}|, \end{aligned}$$

say, where $\gamma = \phi^*(n-r) - q$ is an eigenvalue of $\mathbf{D}\boldsymbol{\Sigma}^{-1}$, that is, of $\boldsymbol{\Delta} = \boldsymbol{\Sigma}^{-1/2}\mathbf{D}\boldsymbol{\Sigma}^{-1/2}$ (see A1.4). Here large values of ϕ^* correspond to large values of γ. Also, H_0 is the hypothesis that $\boldsymbol{\Delta} = \mathbf{O}$ or, equivalently, $\gamma_1 = \gamma_2 = \cdots = \gamma_d = 0$. Usually any departures from H_0 will be reflected in all the γ_j, which, in turn, will be reflected in the ϕ_j^* and their corresponding sample estimates ϕ_j. Hence for general departures from H_0 ($\boldsymbol{\Delta}$ has full rank), we would expect a better performance from test statistics that use all the sample roots. However, if rank $\boldsymbol{\Delta} = 1$, the entire deviation from H_0 is concentrated in the one nonzero root $\gamma_{\max}$. This root corresponds to $\phi_{\max}^*$, the other ϕ_j^* being all equal to $q/(n-r)$. In this situation we would expect the maximum root test to perform the best.

The asymptotic equivalence of the other three tests is not surprising, as when $n \to \infty$, $-f\log\Lambda$, T_g^2, and $(n-r)V^{(s)}$ are all asymptotically distributed as noncentral chi square with dq degrees of freedom and noncentrality parameter tr $\boldsymbol{\Delta}$. For example, Ito [1962] compared the powers of T_g^2 and Λ for the linear alternative rank $\boldsymbol{\Delta} = 1$ and showed that for $m_E = n - r$ equal to 100 or 200, the powers differed at most in the second decimal place. Pillai and Jayachandran [1967: Table 10] compared the powers of Λ, T_g^2, and $V^{(2)}$ for $d = 2$, $\alpha = 0.05$, and small samples [$\nu_2 = \frac{1}{2}(m_E - d - 1)$ in the range 5–40]. They showed that the powers differed only in the third decimal place for small departures from H_0, and differed by no more than 0.02 with larger deviations. The maximum root test was included in a subsequent study (Pillai and Jayachandran [1968: Table 4]) and their power comparisons supported the general conclusions of the previous paragraph. However, in spite of the small differences, the authors were able to come to the following conclusions ($d = 2$):

(1) For small departures from H_0, or large deviations but with $\gamma_1 \approx \gamma_2$, the tests may be ranked $V^{(2)} \geq \Lambda \geq T_g^2$.

(2) For large deviations from H_0, and γ_1 and γ_2 widely different, the order is reversed, that is, $T_g^2 \geq \Lambda \geq V^{(2)}$.

These conclusions are also supported by Y. S. Lee [1971]. Roy et al. [1971] carried out simulations and included smaller values of ν_2. Although power differences were greater, their conclusions generally agreed with the order in (1).

For general d, the ranking for small departures from H_0 would appear to be $V^{(s)} \geq \Lambda \geq T_g^2 \geq \phi_{\max}$, with the reverse order for the linear case of rank $\Delta = 1$.

e MARDIA'S PERMUTATION TEST

Consider the model

$$\mathbf{XB} = (\mathbf{1}_n, \mathbf{X}_2)\begin{pmatrix} \boldsymbol{\beta}_1' \\ \mathbf{B}_2 \end{pmatrix} = \mathbf{1}_n\boldsymbol{\beta}_1' + \mathbf{X}_2\mathbf{B}_2, \tag{8.54}$$

where the $n \times (p-1)$ matrix $\mathbf{X}_2$ has full rank $p-1$ and $\mathbf{X}_2'\mathbf{1}_n = \mathbf{0}$. Mardia [1971] investigated the permutation distribution of $V^* = V^{(s)}/s$ under the null hypothesis $H_0: \mathbf{B}_2 = \mathbf{O}$. Instead of the usual normality assumptions, which imply that the $\mathbf{y}_i$ are i.i.d. $N_d(\boldsymbol{\beta}_1, \boldsymbol{\Sigma})$ under H_0, we now make the weaker assumption that the $\mathbf{y}_i$ are simply i.i.d. under H_0. Each permutation of the set $\{\mathbf{y}_1, \mathbf{y}_2, \ldots, \mathbf{y}_n\}$ will then have the same probability $1/n!$. Using the approach of Box and Watson [1962], Mardia [1971] showed that the permutation distribution of V^* is approximately beta with aP and aQ degrees of freedom, respectively, where

$$\begin{aligned} a^{-1} &= 1 + \frac{c[s(n-1)+2]}{s(n-1)-2c}, \\ c &= \tfrac{1}{2}(n-3)C_X\frac{\mathrm{E}[C_Y]}{n(n-1)}, \\ P &= \tfrac{1}{2}dq, \\ Q &= \tfrac{1}{2}[s(n-1)-dq], \\ q &= p-1, \\ s &= \min(d, q). \end{aligned}$$

Here C_X and C_Y are measures of the nonnormality of the rows of $\mathbf{X}_1$ and of the $\mathbf{y}_i$, respectively. In particular,

$$C_Y = \frac{(n-1)(n+1)}{d(n-3)(n-d-1)}\left\{b_{2,d} - \frac{(n-1)d(d+2)}{n+1}\right\},$$

where $b_{2,d}$ is the sample multivariate skewness of the $\mathbf{y}_i$ (see Section 2.7). When the $\mathbf{y}_i$ are MVN, the expected value of the above term in braces is zero (Mardia [1974: Equation (5.2)]), so that $\mathrm{E}[C_Y] = 0$ and $a = 1$. In this case the beta approximation is the one proposed by Pillai [1955], and it is exact when $s = 1$.

It transpires that

$$-2 \leq \frac{C_X(n-3)}{n-1} \leq n-1, \tag{8.55}$$

so that c is of order n^{-1} near the lower bound of C_X, and of order 1 near the upper bound. Therefore, when the $\mathbf{y}_i$ are not normally distributed, C_X determines the robustness of $V^{(s)}$, a result that Box and Watson [1962] demonstrated for the case $d = 1$. Mardia [1971] showed that

$$C_X = \frac{(n-1)}{q(n-3)(n-q-1)}\left\{n(n+1)\sum_{r=1}^{n} m_{rr}^2 - (n-1)q(q+2)\right\}, \tag{8.56}$$

where $[(m_{rs})] = \mathbf{M} = \mathbf{X}_2(\mathbf{X}_2'\mathbf{X}_2)^{-1}\mathbf{X}_2'$. If the diagonal elements of $\mathbf{M}$ are all equal, then, from $\operatorname{tr}\mathbf{M} = \operatorname{tr}[\mathbf{X}_2'\mathbf{X}_2(\mathbf{X}_2'\mathbf{X}_2)^{-1}] = \operatorname{tr}\mathbf{I}_q = q$, we have $m_{rr} = q/n$, $\Sigma_r m_{rr}^2 = q^2/n$, and $C_X = -2(n-1)/(n-3)$. Thus C_X attains its lower bound in (8.55), $c = -\mathrm{E}[C_Y]/n$,

$$a^{-1} = 1 - \frac{\mathrm{E}[C_Y]}{n} + O(n^{-2}) \approx 1,$$

and, for large n, the permutation distribution of V^* is insensitive to nonnormality. From symmetry, we see that any cross-classification with equal cell frequencies (e.g., Section 9.3) and any hierarchical classification with equal cell frequencies at each stage of the hierarchy are designs with equal elements m_{rr}. Applications of the method are given in Chapter 9. From Exercise 8.7, we have

$$\begin{aligned}\mathbf{H} &= (\mathbf{Y} - \mathbf{1}_n\bar{\mathbf{y}}')'\mathbf{M}(\mathbf{Y} - \mathbf{1}_n\bar{\mathbf{y}}') \\ &= \mathbf{Y}'\mathbf{M}\mathbf{Y} \\ &= \sum_r \sum_s m_{rs}\mathbf{y}_r\mathbf{y}_s',\end{aligned} \tag{8.57}$$

since $\mathbf{X}_2'\mathbf{1}_n = \mathbf{0}$ implies that $\mathbf{M}\mathbf{1}_n = \mathbf{0}$. If $\mathbf{H}$ is known, then the diagonal elements of $\mathbf{M}$ can be found directly. Replacing $(\mathbf{X}_2'\mathbf{X}_2)^{-1}$ by $(\mathbf{X}_2'\mathbf{X}_2)^{-}$ we find that the above theory still applies when $\mathbf{X}_2$ has less than full rank.

We note that the above theory assumes that the $\mathbf{y}_i$ have the same dispersion matrix. It is not clear what effect the violation of this assumption will have on the permutation test.

8.6.3 Simultaneous Confidence Intervals

Associated with the largest root test of $H_0: \mathbf{AB} = \mathbf{O}$ we have a set of simultaneous confidence intervals for all linear combinations of the form $\mathbf{a}'\mathbf{AB}\mathbf{b}$ that we can derive as follows on the assumption that $\mathbf{X}$ has full rank.

Now H_0 is true if and only if $H_{0ab}: \mathbf{a'ABb} = 0$ is true for all $\mathbf{a}$, $\mathbf{b}$; that is, $H_0 = \cap_a \cap_b H_{0ab}$. Setting $\mathbf{y} = \mathbf{Yb}$, $\boldsymbol{\beta} = \mathbf{Bb}$, $\mathbf{a}'_0 = \mathbf{a'A}$, and $\hat{\boldsymbol{\beta}} = (\mathbf{X'X})^{-1}\mathbf{X'y}$, we can test $H_{0ab}: \mathbf{a}'_0\boldsymbol{\beta} = 0$ using the F-ratio (with $q = 1$)

$$F_{(a,b)} = \frac{Q_H - Q}{Q/(n-p)},$$

where

$$\begin{aligned} Q_H - Q &= (\mathbf{a}'_0\hat{\boldsymbol{\beta}})'\left[\mathbf{a}'_0(\mathbf{X'X})^{-1}\mathbf{a}_0\right]^{-1}\mathbf{a}'_0\hat{\boldsymbol{\beta}} \\ &= \mathbf{y'X}(\mathbf{X'X})^{-1}\mathbf{A'a}\left[\mathbf{a'A}(\mathbf{X'X})^{-1}\mathbf{A'a}\right]^{-1}\mathbf{a'A}(\mathbf{X'X})^{-1}\mathbf{X'y} \\ &= \left\{\mathbf{a'A}(\mathbf{X'X})^{-1}\mathbf{X'Yb}\right\}^2/\left\{\mathbf{a'A}(\mathbf{X'X})^{-1}\mathbf{A'a}\right\} \\ &= \frac{(\mathbf{a'Lb})^2}{\mathbf{a'Ma}}, \end{aligned} \tag{8.58}$$

say, and

$$\begin{aligned} Q &= \mathbf{y}'(\mathbf{I}_n - \mathbf{P}_\Omega)\mathbf{y} \\ &= \mathbf{b'Y}'(\mathbf{I}_n - \mathbf{P}_\Omega)\mathbf{Yb} \\ &= \mathbf{b'Eb}. \end{aligned} \tag{8.59}$$

Using the union–intersection principle, a test of H_0 has acceptance region

$$\begin{aligned} \bigcap_a \bigcap_b \{\mathbf{Y}: F_{(a,b)} \le k\} &= \left\{\mathbf{Y}: \sup_{\mathbf{a},\mathbf{b}\,(\neq \mathbf{0})} F_{(a,b)} \le k\right\} \\ &= \left\{\mathbf{Y}: \sup_{\mathbf{a},\mathbf{b}} \frac{(\mathbf{a'Lb})^2}{(\mathbf{a'Ma})(\mathbf{b'Eb})} \le \frac{k}{n-p} = k_1\right\} \qquad (8.60) \\ &= \{\mathbf{Y}: \phi_{\max} \le k_1\}, \end{aligned}$$

where $\phi_{\max}$ is the maximum eigenvalue of $\mathbf{M}^{-1}\mathbf{L}\mathbf{E}^{-1}\mathbf{L}'$ (by A7.7), that is, of (see A1.4)

$$\begin{aligned} \mathbf{L'M}^{-1}\mathbf{LE}^{-1} &= \mathbf{Y'X}(\mathbf{X'X})^{-1}\mathbf{A}'\left[\mathbf{A}(\mathbf{X'X})^{-1}\mathbf{A}'\right]^{-1}\mathbf{A}(\mathbf{X'X})^{-1}\mathbf{X'YE}^{-1} \\ &= (\mathbf{A}\hat{\mathbf{B}})'\left[\mathbf{A}(\mathbf{X'X})^{-1}\mathbf{A}'\right]^{-1}\mathbf{A}\hat{\mathbf{B}}\mathbf{E}^{-1} \\ &= \mathbf{HE}^{-1} \qquad [\text{by } (8.47)]. \end{aligned}$$

We have therefore arrived at Roy's maximum root test once again; however, we can use (8.60) with $\hat{\mathbf{B}}$ replaced by $\hat{\mathbf{B}} - \mathbf{B}$ in $\mathbf{L}$ to obtain the following:

$$\begin{aligned} 1 - \alpha &= \text{pr}[\phi_{\max} \le \phi_\alpha] \\ &= \text{pr}\left[|\mathbf{a}'\mathbf{A}(\hat{\mathbf{B}} - \mathbf{B})\mathbf{b}| \le \left\{\phi_\alpha \mathbf{a}'\mathbf{A}(\mathbf{X}'\mathbf{X})^{-1}\mathbf{A}'\mathbf{a} \cdot \mathbf{b}'\mathbf{E}\mathbf{b}\right\}^{1/2} \quad \text{for all } \mathbf{a}, \mathbf{b}\right]. \end{aligned}$$

Thus a set of multiple confidence intervals for all linear combinations is given by

$$\mathbf{a}'\mathbf{A}\hat{\mathbf{B}}\mathbf{b} \pm \left\{\phi_\alpha \mathbf{a}'\mathbf{A}(\mathbf{X}'\mathbf{X})^{-1}\mathbf{A}'\mathbf{a} \cdot \mathbf{b}'\mathbf{E}\mathbf{b}\right\}^{1/2}, \tag{8.61}$$

and the set has an overall confidence of $100(1-\alpha)\%$. The largest root test of H_0 will be significant at the α level of significance if and only if at least one of the intervals (8.61) does not contain zero. Although the maximum root test is not as powerful as the likelihood ratio test, it does have the advantage of being linked directly to a set of simultaneous confidence intervals.

Since $\text{var}[\mathbf{a}'\mathbf{A}\hat{\mathbf{B}}\mathbf{b}] = \mathbf{a}'\mathbf{A}(\mathbf{X}'\mathbf{X})^{-1}\mathbf{A}'\mathbf{a}\sigma_b^2$, where $\sigma_b^2 = \mathbf{b}'\boldsymbol{\Sigma}\mathbf{b}$, the intervals (8.61) may be written in the form

$$\mathbf{a}'\mathbf{A}\hat{\mathbf{B}}\mathbf{b} \pm \left\{\phi_\alpha \widetilde{\text{var}}[\mathbf{a}'\mathbf{A}\hat{\mathbf{B}}\mathbf{b}]\right\}^{1/2}, \tag{8.62}$$

where $\widetilde{\text{var}}[\mathbf{a}'\mathbf{A}\hat{\mathbf{B}}\mathbf{b}]$ is $\text{var}[\mathbf{a}'\mathbf{A}\hat{\mathbf{B}}\mathbf{b}]$ with $\boldsymbol{\Sigma}$ replaced by $\mathbf{E}$. The term $\mathbf{a}'\mathbf{A}(\mathbf{X}'\mathbf{X})^{-1}\mathbf{A}'\mathbf{a}$ can usually be found directly from the corresponding "reduced" univariate model by considering $\text{var}[\mathbf{a}'\mathbf{A}\hat{\boldsymbol{\beta}}]$.

By setting $\mathbf{A} = \mathbf{I}_p$ $(q = p)$, we obtain the set of intervals

$$\mathbf{a}'\hat{\mathbf{B}}\mathbf{b} \pm \left\{\phi_\alpha \mathbf{a}'(\mathbf{X}'\mathbf{X})^{-1}\mathbf{a} \cdot \mathbf{b}'\mathbf{E}\mathbf{b}\right\}^{1/2}. \tag{8.63}$$

Included in this set is each element β_{ij} of $\mathbf{B}$, obtained by setting $\mathbf{a}$ and $\mathbf{b}$ equal to vectors with 1 in the ith and jth positions, respectively, and zeros elsewhere.

If we are interested in just a certain number, say, m, of the elements of $\mathbf{B}$, we can use the Bonferroni method of constructing m conservative confidence intervals with an overall confidence of at least $100(1-\alpha)\%$ (see Exercise 8.9). When m is small, these intervals may be shorter than those given by (8.63).

Finally, we note that if $\mathbf{X}$ does not have full rank, we simply replace $(\mathbf{X}'\mathbf{X})^{-1}$ by $(\mathbf{X}'\mathbf{X})^{-}$; Equation (8.62) still holds.

8.6.4 *Comparing Two Populations*

Let $\mathbf{v}_1, \mathbf{v}_2, \ldots, \mathbf{v}_{n_1}$ and $\mathbf{w}_1, \mathbf{w}_2, \ldots, \mathbf{w}_{n_2}$ be independent random samples from $N_d(\boldsymbol{\mu}_1, \boldsymbol{\Sigma})$ and $N_d(\boldsymbol{\mu}_2, \boldsymbol{\Sigma})$, respectively. Setting $\mathbf{y}_i = \mathbf{v}_i$ $(i = 1, 2, \ldots, n_1)$, $\mathbf{y}_{n_1+j} =$

$\mathbf{w}_j$ ($j = 1, 2, \ldots, n_2$), and $n = n_1 + n_2$, we have the linear model

$$\begin{pmatrix} \mathbf{y}_1' \\ \mathbf{y}_2' \\ \vdots \\ \mathbf{y}_n' \end{pmatrix} = \begin{pmatrix} \mathbf{1}_{n_1} & \mathbf{0} \\ \mathbf{0} & \mathbf{1}_{n_2} \end{pmatrix} \begin{pmatrix} \boldsymbol{\mu}_1' \\ \boldsymbol{\mu}_2' \end{pmatrix} + \begin{pmatrix} \mathbf{u}_1' \\ \mathbf{u}_2' \\ \vdots \\ \mathbf{u}_n' \end{pmatrix} \tag{8.64}$$

or

$$\mathbf{Y} = \mathbf{X}\mathbf{B} + \mathbf{U}.$$

The hypothesis $\boldsymbol{\mu}_1 = \boldsymbol{\mu}_2$ is of the form $(1, -1)\mathbf{B} = \boldsymbol{\mu}_1' - \boldsymbol{\mu}_2' = \mathbf{0}'$ or $\mathbf{AB} = \mathbf{O}$, so that this example is a special case of the general theory. Using this theory, we have

$$\begin{pmatrix} \hat{\boldsymbol{\mu}}_1' \\ \hat{\boldsymbol{\mu}}_2' \end{pmatrix} = \hat{\mathbf{B}} = (\mathbf{X}'\mathbf{X})^{-1}\mathbf{X}'\mathbf{Y} = \begin{pmatrix} n_1^{-1} & 0 \\ 0 & n_2^{-1} \end{pmatrix} \begin{pmatrix} \sum_i \mathbf{v}_i' \\ \sum_j \mathbf{w}_j' \end{pmatrix} = \begin{pmatrix} \bar{\mathbf{v}}' \\ \bar{\mathbf{w}}' \end{pmatrix}$$

and

$$\begin{aligned} \mathbf{E} &= (\mathbf{Y} - \mathbf{X}\hat{\mathbf{B}})'(\mathbf{Y} - \mathbf{X}\hat{\mathbf{B}}) \\ &= \sum_{i=1}^{n_1} (\mathbf{v}_i - \bar{\mathbf{v}})(\mathbf{v}_i - \bar{\mathbf{v}})' + \sum_{j=1}^{n_2} (\mathbf{w}_j - \bar{\mathbf{w}})(\mathbf{w}_j - \bar{\mathbf{w}})' \\ &= (n_1 + n_2 - 2)\mathbf{S}_p, \end{aligned}$$

say. Also,

$$\mathbf{A}\hat{\mathbf{B}} = (1, -1)\begin{pmatrix} \bar{\mathbf{v}}' \\ \bar{\mathbf{w}}' \end{pmatrix} = \bar{\mathbf{v}}' - \bar{\mathbf{w}}',$$

so that from (8.47),

$$\begin{aligned} \mathbf{H} &= (\mathbf{A}\hat{\mathbf{B}})'\left[\mathbf{A}(\mathbf{X}'\mathbf{X})^{-1}\mathbf{A}'\right]^{-1}\mathbf{A}\hat{\mathbf{B}} \\ &= (\bar{\mathbf{v}} - \bar{\mathbf{w}})\left[(1, -1)\begin{pmatrix} n_1^{-1} & 0 \\ 0 & n_2^{-1} \end{pmatrix}(1, -1)'\right]^{-1}(\bar{\mathbf{v}} - \bar{\mathbf{w}})' \\ &= \frac{n_1 n_2}{n_1 + n_2}(\bar{\mathbf{v}} - \bar{\mathbf{w}})(\bar{\mathbf{v}} - \bar{\mathbf{w}})', \end{aligned}$$

and $\Lambda = |\mathbf{E}|/|\mathbf{E}+\mathbf{H}|$. Since $\mathbf{H}$ has rank 1, we can use the test statistic [see (2.56) with $m_E = n_1 + n_2 - 2$]

$$\begin{aligned} T_0^2 &= m_E \operatorname{tr}[\mathbf{E}^{-1}\mathbf{H}] \\ &= m_E \operatorname{tr}\left[\frac{1}{m_E}\mathbf{S}_p^{-1}(\bar{\mathbf{v}}-\bar{\mathbf{w}})(\bar{\mathbf{v}}-\bar{\mathbf{w}})'\frac{n_1 n_2}{n_1+n_2}\right] \\ &= \frac{n_1 n_2}{n_1+n_2}(\bar{\mathbf{v}}-\bar{\mathbf{w}})'\mathbf{S}_p^{-1}(\bar{\mathbf{v}}-\bar{\mathbf{w}}) \qquad (8.65) \\ &\sim T^2_{d,\, n_1+n_2-2}, \end{aligned}$$

which is the same as the test statistic (3.86). Thus we find that Hotelling's T_0^2 test of $H_0: \boldsymbol{\theta} = \boldsymbol{\mu}_1 - \boldsymbol{\mu}_2 = \mathbf{0}$ is equivalent to the likelihood ratio test.

8.6.5 Canonical Form

Consider the general linear model $\mathbf{Y} = \mathbf{XB} + \mathbf{U}$ and the testable hypothesis $H_0: \mathbf{AB} = \mathbf{C}$, where $\mathbf{A}$ is $q \times p$ of rank q and $\mathbf{X}$ is $n \times p$ of rank r. It is possible to shift the origin (to "remove" $\mathbf{C}$) and make an orthogonal transformation $\mathbf{Z} = \mathbf{T}'\mathbf{X}$ of the data (see Seber [1980: p. 80]) so that the model and hypothesis become

$$\begin{pmatrix} \mathbf{Z}_1 \\ \mathbf{Z}_2 \\ \mathbf{Z}_3 \end{pmatrix} = \begin{pmatrix} \boldsymbol{\Phi}_1 \\ \boldsymbol{\Phi}_2 \\ \mathbf{O} \end{pmatrix} + \mathbf{U} \quad \text{and} \quad H_0: \boldsymbol{\Phi}_1 = \mathbf{O}, \qquad (8.66)$$

where $\boldsymbol{\Phi}_1$ is $q \times d$, $\boldsymbol{\Phi}_2$ is $(r-q) \times d$, and the rows of $\mathbf{U}$ are i.i.d. $N_d(\mathbf{0}, \boldsymbol{\Sigma}_z)$. Such canonical forms provide a simpler framework for theoretical derivations (such as powers of tests) and simulation (as in robustness studies). For example, using (8.66) we have, from Exercise 8.6, that

$$\mathbf{H} = \mathbf{Z}_1'\mathbf{Z}_1 = \sum_{i=1}^{q} \mathbf{z}_i \mathbf{z}_i' \qquad (8.67)$$

and

$$\mathbf{E} = \mathbf{Z}_3'\mathbf{Z}_3 = \sum_{i=r+1}^{n} \mathbf{z}_i \mathbf{z}_i'. \qquad (8.68)$$

As $\mathbf{Z}_2$ makes no contribution to $\mathbf{H}$ or $\mathbf{E}$, (8.66) can be reduced further by simply omitting $\mathbf{Z}_2$ and $\boldsymbol{\Phi}_2$.

8.6.6 Missing Observations

If a sampling unit is missing in a design, for example, a patient withdraws from the treatment, then all d measurements on that unit are missing and a row of $\mathbf{Y}$ is missing. Suppose then that $\mathscr{E}[\mathbf{Y}] = \mathbf{XB}$, where

$$\mathbf{Y} = \begin{pmatrix} \mathbf{Y}_{(1)} \\ \mathbf{Y}_{(2)} \end{pmatrix} \begin{matrix} \}n-m \\ \}m \end{matrix}, \qquad \mathbf{X} = \begin{pmatrix} \mathbf{X}_1 \\ \mathbf{X}_2 \end{pmatrix} \begin{matrix} \}n-m \\ \}m \end{matrix},$$

and m rows of $\mathbf{Y}$, $\mathbf{Y}_{(2)}$, say, are missing. The least squares estimate $\hat{\mathbf{B}}$ for the *available* data is a solution of $\mathbf{X}_1'\mathbf{X}_1\mathbf{B} = \mathbf{X}_1'\mathbf{Y}_{(1)}$ [by (8.13)], and we have $\mathbf{E} = (\mathbf{Y}_{(1)} - \mathbf{X}_1\hat{\mathbf{B}})'(\mathbf{Y}_{(1)} - \mathbf{X}_1\hat{\mathbf{B}})$. However, using

$$(\mathbf{Y} - \mathbf{XB})'(\mathbf{Y} - \mathbf{XB}) = (\mathbf{Y}_{(1)} - \mathbf{X}_1\mathbf{B})'(\mathbf{Y}_{(1)} - \mathbf{X}_1\mathbf{B}) + (\mathbf{Y}_{(2)} - \mathbf{X}_2\mathbf{B})'(\mathbf{Y}_{(2)} - \mathbf{X}_2\mathbf{B}), \tag{8.69}$$

an alternative procedure is available.

Let $\hat{\mathbf{Y}}_{(2)} = \mathbf{X}_2\hat{\mathbf{B}}$ and $\hat{\mathbf{Y}} = (\mathbf{Y}_{(1)}', \hat{\mathbf{Y}}_{(2)}')'$, then

$$\begin{aligned}(\hat{\mathbf{Y}} - \mathbf{X}\hat{\mathbf{B}})'(\hat{\mathbf{Y}} - \mathbf{X}\hat{\mathbf{B}}) &= (\mathbf{Y}_{(1)} - \mathbf{X}_1\hat{\mathbf{B}})'(\mathbf{Y}_{(1)} - \mathbf{X}_1\hat{\mathbf{B}}) \\ &\quad + (\hat{\mathbf{Y}}_{(2)} - \mathbf{X}_2\hat{\mathbf{B}})'(\hat{\mathbf{Y}}_{(2)} - \mathbf{X}_2\hat{\mathbf{B}}) \\ &= \mathbf{E}.\end{aligned}$$

Also, adding $\mathbf{X}_1'\mathbf{X}_1\hat{\mathbf{B}} = \mathbf{X}_1'\mathbf{Y}_{(1)}$ and $\mathbf{X}_2'\mathbf{X}_2\hat{\mathbf{B}} = \mathbf{X}_2'\hat{\mathbf{Y}}_{(2)}$, we have

$$\mathbf{X}'\mathbf{X}\hat{\mathbf{B}} = \mathbf{X}'\hat{\mathbf{Y}},$$

and $\hat{\mathbf{B}}$ is obtained by minimizing $(\hat{\mathbf{Y}} - \mathbf{XB})'(\hat{\mathbf{Y}} - \mathbf{XB})$. Our procedure, therefore, consists of finding the appropriate estimate $\hat{\mathbf{Y}}_{(2)}$ of $\mathbf{Y}_{(2)}$, completing the data set to $\hat{\mathbf{Y}}$, and then finding $\hat{\mathbf{B}}$ and $\mathbf{E}$ for this completed model. This approach allows one to make use of the symmetry of the complete design matrix $\mathbf{X}$ and the simplicity of the corresponding analysis.

Replacing $\mathbf{Y}$ by $\mathbf{y}$, we see that the method is the same as the usual univariate missing observation procedure (see Seber [1977: Section 10.2]). We can therefore find $\hat{\mathbf{Y}}_{(2)}$ by applying a standard univariate procedure to each column of $\mathbf{Y}$. Since both constrained and unconstrained least squares estimation for the multivariate model are equivalent to separate least squares estimation for each of the d univariate models, the same procedure can be used for finding $\mathbf{E}_H$ when there are missing rows of $\mathbf{Y}$. We then have $\mathbf{H} = \mathbf{E}_H - \mathbf{E}$ and we can test H_0 using $\mathbf{H}$ and $\mathbf{E}$ with q and $m_E - m$ $(= n - r - m)$ degrees of freedom respectively, $\mathbf{E}$ losing one degree of freedom for every missing vector observation.

Although the above case, where whole rows of $\mathbf{Y}$ are missing, is the natural one to consider, it may happen that single elements of $\mathbf{Y}$ are missing. This situation is more difficult to handle and some general procedures are described in Section 10.3.

8.6.7 *Other Topics*

Rogers and Young [1978] give methods for testing a linear hypothesis when $\boldsymbol{\Sigma}$ satisfies

$$\boldsymbol{\Sigma} = \sum_{r=1}^{k} \phi_r \mathbf{G}_r \quad \text{and} \quad \boldsymbol{\Sigma}^{-1} = \sum_{r=1}^{k} \psi_r \mathbf{G}_r,$$

for $\mathbf{G}_1, \mathbf{G}_2, \ldots, \mathbf{G}_k$ known linearly independent symmetric $d \times d$ matrices, and $k \leq \frac{1}{2}d(d+1)$. When $k = \frac{1}{2}d(d+1)$, the ϕ_r can be taken as the elements in the upper triangle of $\boldsymbol{\Sigma}$, and $\boldsymbol{\Sigma}$ is unstructured or "unpatterned." Many of the special covariance patterns are included in the case $k < \frac{1}{2}d(d+1)$.

A number of authors (see Tso [1981] for references) have considered the problem of estimating the $p \times d$ matrix of parameters $\mathbf{B}$ subject to rank $\mathbf{B} \leq s < \min(p, d)$, and testing rank $\mathbf{B} = s$ versus the alternative that $\mathbf{B}$ has full rank (or has rank $s + 1$). This topic is usually referred to as the reduced-rank regression model, and it is closely related to discriminant coordinates (canonical variate analysis). A further problem is testing rank $\mathbf{AB} = t$ versus rank $\mathbf{AB} > t$. When $t = 0$, this reduces to testing $\mathbf{AB} = \mathbf{O}$ (Fujikoshi [1974]).

8.7 A GENERALIZED LINEAR HYPOTHESIS

8.7.1 *Theory*

Consider the model $\mathbf{Y} = \mathbf{XB} + \mathbf{U}$, where $\mathbf{X}$ is $n \times p$ of rank p, and the rows of $\mathbf{U}$ are i.i.d. $N_d(\mathbf{0}, \boldsymbol{\Sigma})$. Suppose we wish to test

$$H_0: \mathbf{ABD} = \mathbf{O}, \tag{8.70}$$

where $\mathbf{A}$ is $q \times p$ of rank q ($q \leq p$), $\mathbf{B}$ is $p \times d$ of rank d, and $\mathbf{D}$ is $d \times v$ of rank v ($v \leq d$). As H_0 reduces to the hypothesis $\mathbf{AB} = \mathbf{O}$ when $\mathbf{D} = \mathbf{I}_d$, a natural procedure for handling H_0 is to try and carry out the same reduction using a suitable transformation. This can be achieved by setting $\mathbf{Y}_D = \mathbf{YD}$ so that

$$\begin{aligned}\mathbf{Y}_D &= \mathbf{XBD} + \mathbf{UD} \\ &= \mathbf{X\Lambda} + \mathbf{U}_0,\end{aligned}$$

say, where the rows of

$$\mathbf{U}_0 = \begin{pmatrix} \mathbf{u}_1' \\ \mathbf{u}_2' \\ \vdots \\ \mathbf{u}_n' \end{pmatrix} \mathbf{D} = \begin{pmatrix} (\mathbf{D}'\mathbf{u}_1)' \\ (\mathbf{D}'\mathbf{u}_2)' \\ \vdots \\ (\mathbf{D}'\mathbf{u}_n)' \end{pmatrix}$$

are i.i.d. $N_v(\mathbf{0}, \mathbf{D}'\mathbf{\Sigma}\mathbf{D})$ by Theorem 2.1(i) of Section 2.2. Since H_0 is now $\mathbf{A}\mathbf{\Lambda} = \mathbf{O}$, we can apply the general theory of this chapter with

$$\begin{aligned} \mathbf{H} &= (\mathbf{A}\hat{\mathbf{\Lambda}})'\left[\mathbf{A}(\mathbf{X}'\mathbf{X})^{-1}\mathbf{A}'\right]^{-1}\mathbf{A}\hat{\mathbf{\Lambda}} \qquad [\text{by } (8.47)] \\ &= \mathbf{Y}_D'\mathbf{X}(\mathbf{X}'\mathbf{X})^{-1}\mathbf{A}'\left[\mathbf{A}(\mathbf{X}'\mathbf{X})^{-1}\mathbf{A}'\right]^{-1}\mathbf{A}(\mathbf{X}'\mathbf{X})^{-1}\mathbf{X}'\mathbf{Y}_D \\ &= (\mathbf{A}\hat{\mathbf{B}}\mathbf{D})'\left[\mathbf{A}(\mathbf{X}'\mathbf{X})^{-1}\mathbf{A}'\right]\mathbf{A}\hat{\mathbf{B}}\mathbf{D} \end{aligned} \tag{8.71}$$

and

$$\mathbf{E} = \mathbf{Y}_D'(\mathbf{I}_n - \mathbf{P}_\Omega)\mathbf{Y}_D = \mathbf{D}'\mathbf{Y}'(\mathbf{I}_n - \mathbf{P}_\Omega)\mathbf{Y}\mathbf{D}. \tag{8.72}$$

The only change is that $\mathbf{Y}$ is replaced by $\mathbf{Y}_D$, and d by v. Thus $\mathbf{E} \sim W_v(n - p, \mathbf{D}'\mathbf{\Sigma}\mathbf{D})$, and, when H_0 is true, $\mathbf{H} \sim W_v(q, \mathbf{D}'\mathbf{\Sigma}\mathbf{D})$. If $\mathbf{X}$ has less than full rank, namely rank r, then the above theory still holds, with $(\mathbf{X}'\mathbf{X})^{-1}$ replaced by $(\mathbf{X}'\mathbf{X})^{-}$, and p by r.

Simultaneous test procedures for testing several hypotheses like H_0 are given by Gabriel [1968, 1969] and Mudholkar et al. [1974b]. Two examples of H_0, which we have already met in another context, are given below.

8.7.2 *Profile Analysis for* K *Populations*

In Section 3.6.4 we considered a profile analysis of $K = 2$ populations with means $\boldsymbol{\eta}$ and $\boldsymbol{\nu}$. As a step toward generalizing to $K > 2$, we now change the notation and refer to these mean vectors as $\boldsymbol{\mu}_1$ and $\boldsymbol{\mu}_2$, respectively. Then, with $\mathbf{C}_1$ defined by (3.18), the hypothesis H_{01} of parallelism for $K = 2$ is $\mathbf{C}_1(\boldsymbol{\mu}_1 - \boldsymbol{\mu}_2) = \mathbf{0}$, that is,

$$\begin{aligned} \mathbf{0}' &= (\mathbf{C}_1\boldsymbol{\mu}_1)' - (\mathbf{C}_1\boldsymbol{\mu}_2)' \\ &= (1, -1)\begin{pmatrix} \boldsymbol{\mu}_1' \\ \boldsymbol{\mu}_2' \end{pmatrix}\mathbf{C}_1' \\ &= (1, -1)\mathbf{B}\mathbf{C}_1', \end{aligned} \tag{8.73}$$

say. For the case of K populations $N_d(\boldsymbol{\mu}_k, \boldsymbol{\Sigma})$, $k = 1, 2, \ldots, K$, the corresponding hypothesis of parallelism is $H_{01}: \mathbf{C}_1\boldsymbol{\mu}_1 = \mathbf{C}_1\boldsymbol{\mu}_2 = \cdots = \mathbf{C}_1\boldsymbol{\mu}_K$, or

$$\mathbf{O} = \begin{pmatrix} 1 & -1 & 0 & \cdots & 0 \\ 0 & 1 & -1 & \cdots & 0 \\ \vdots & \vdots & \vdots & & \vdots \\ 0 & 0 & \cdots & \cdots & -1 \end{pmatrix} \begin{pmatrix} \boldsymbol{\mu}_1' \\ \boldsymbol{\mu}_2' \\ \vdots \\ \boldsymbol{\mu}_K' \end{pmatrix} \mathbf{C}_1'$$

$$= \mathbf{A}_1 \mathbf{B} \mathbf{C}_1', \tag{8.74}$$

where the $(K-1) \times K$ matrix $\mathbf{A}_1$ is of the same form as the $(d-1) \times d$ matrix $\mathbf{C}_1$. We find, then, that (8.73) and (8.74) are special cases of (8.70). We note in passing that some nonparametric tests of parallelism are given by Bhapkar and Patterson [1977, 1978].

In a similar fashion H_{02}, the hypothesis that the profiles are all at the same level, generalizes from $\mathbf{1}_d'(\boldsymbol{\mu}_1 - \boldsymbol{\mu}_2) = 0$ for the case of two populations to $\mathbf{A}_1\mathbf{B}\mathbf{1}_d = \mathbf{0}$ for K populations. However, this test is equivalent to carrying out a one-way analysis of the differences between K means for the data consisting of the sum of the elements of each observation vector (e.g., $\mathbf{1}_d'\mathbf{v}_i$)

The hypothesis H_{03} of no test main effects, namely,

$$\mathbf{0}' = [\mathbf{C}_1(\boldsymbol{\mu}_1 + \boldsymbol{\mu}_2)] = (1, 1)\begin{pmatrix} \boldsymbol{\mu}_1' \\ \boldsymbol{\mu}_2' \end{pmatrix}\mathbf{C}_1'$$

generalizes to

$$\mathbf{0}' = [\mathbf{C}_1(\boldsymbol{\mu}_1 + \boldsymbol{\mu}_2 + \cdots + \boldsymbol{\mu}_K)] = \mathbf{1}_K'\mathbf{B}\mathbf{C}_1'.$$

In the same way the test statistic, given by (3.103), for the case $K = 2$ generalizes to

$$T_0^2 = n(\mathbf{C}_1\overline{\mathbf{x}})'\left[\mathbf{C}_1\mathbf{S}_p\mathbf{C}_1'\right]^{-1}\mathbf{C}_1\overline{\mathbf{x}}, \tag{8.75}$$

when $n = n_1 + n_2 + \cdots + n_K$, $\overline{\mathbf{x}}$ is the mean of all n observations, and $\mathbf{S}_p$ is the usual pooled estimate of $\boldsymbol{\Sigma}$. When $H_{03} \cap H_{01}$ is true, $T_0^2 \sim T_{d-1, n-K}^2$

8.7.3 Tests for Mean of Multivariate Normal

We shall now show that the hypothesis tests in Section 3.4 for the mean of a MVN distribution are a special case of the generalized linear hypothesis described above. Let $\mathbf{v}_1, \mathbf{v}_2, \ldots, \mathbf{v}_n$ be i.i.d. $N_d(\boldsymbol{\mu}, \boldsymbol{\Sigma})$, and suppose we wish to test $H_0: \mathbf{D}'\boldsymbol{\mu} = \mathbf{0}$, where $\mathbf{D}'$ is a known $q \times d$ matrix of rank q. Setting $\mathbf{Y}' = (\mathbf{v}_1, \mathbf{v}_2, \ldots, \mathbf{v}_n)$ and $\mathbf{X}\mathbf{B} = \mathbf{1}_n\boldsymbol{\mu}'$, we have the linear model $\mathbf{Y} = \mathbf{X}\mathbf{B} + \mathbf{U}$, where the rows of $\mathbf{U}$ are i.i.d. $N_d(\mathbf{0}, \boldsymbol{\Sigma})$. Then H_0 becomes $\mathbf{0}' = \boldsymbol{\mu}'\mathbf{D} = \mathbf{B}\mathbf{D}$, which is a special case of $\mathbf{A}\mathbf{B}\mathbf{D} = \mathbf{O}$ with $\mathbf{A} = 1$. Hence, from (8.71) and (8.72), the test statistic for H_0 is $\Lambda = |\mathbf{E}|/|\mathbf{E}+\mathbf{H}|$, where

$$\begin{aligned}\mathbf{H} &= \mathbf{D}'\mathbf{Y}'\mathbf{X}(\mathbf{X}'\mathbf{X})^{-1}\mathbf{A}'\left[\mathbf{A}(\mathbf{X}'\mathbf{X})^{-1}\mathbf{A}'\right]^{-1}\mathbf{A}(\mathbf{X}'\mathbf{X})^{-1}\mathbf{X}'\mathbf{Y}\mathbf{D}\\ &= \mathbf{D}'\mathbf{Y}'\mathbf{1}_n(\mathbf{1}_n'\mathbf{1}_n)^{-1}(\mathbf{1}_n'\mathbf{1}_n)(\mathbf{1}_n'\mathbf{1}_n)^{-1}\mathbf{1}_n'\mathbf{Y}\mathbf{D}\\ &= n\mathbf{D}'\bar{\mathbf{v}}\bar{\mathbf{v}}'\mathbf{D}\end{aligned}$$

and

$$\begin{aligned}\mathbf{E} &= \mathbf{D}'\mathbf{Y}'\left\{\mathbf{I}_n - \mathbf{1}_n(\mathbf{1}_n'\mathbf{1}_n)^{-1}\mathbf{1}_n'\right\}\mathbf{Y}\mathbf{D}\\ &= \mathbf{D}'\sum_i(\mathbf{v}_i-\bar{\mathbf{v}})(\mathbf{v}_i-\bar{\mathbf{v}})'\mathbf{D}\\ &= \mathbf{D}'\mathbf{Q}\mathbf{D},\end{aligned}$$

say. However, as rank $\mathbf{H} = 1$ and $\mathbf{E} \sim W_q(n-1, \mathbf{\Sigma})$, we can test H_0 using (see Section 2.5.5e)

$$\begin{aligned}T_0^2 &= (n-1)\,\mathrm{tr}[\mathbf{E}^{-1}\mathbf{H}]\\ &= n(\mathbf{D}'\bar{\mathbf{v}})'[\mathbf{D}'\mathbf{S}\mathbf{D}]^{-1}\mathbf{D}'\bar{\mathbf{v}} \qquad [\text{see } (8.65)],\end{aligned}$$

where $\mathbf{S} = \mathbf{Q}/(n-1)$. We have arrived once again at (3.17) with $\mathbf{A} = \mathbf{D}'$ and $\mathbf{b} = \mathbf{0}$.

In conclusion, we see that the general hypothesis $\mathbf{ABD} = \mathbf{0}$ reduces to the usual multivariate linear hypothesis when $\mathbf{D} = \mathbf{I}_d$, but reduces to a linear hypothesis about the mean of a MVN distribution when $\mathbf{A} = \mathbf{I}_p$. In Chapter 9 we consider a further generalization where we replace the model $\mathscr{E}[\mathbf{Y}] = \mathbf{XB}$ by $\mathscr{E}[\mathbf{Y}] = \mathbf{X}\mathbf{\Delta}\mathbf{K}'$.

8.8 STEP-DOWN PROCEDURES

For completeness we mention a step-down procedure, generalizing that of Section 3.3.4, for testing $H_0: \mathbf{AB} = \mathbf{O}$ for the model $\mathbf{Y} = \mathbf{XB} + \mathbf{U}$, where the rows of $\mathbf{U}$ are i.i.d. $N_d(\mathbf{0}, \mathbf{\Sigma})$ and $\mathbf{\Sigma} = [(\sigma_{jk})]$. To set the scene, let $\mathbf{Y}_k = (\mathbf{y}^{(1)}, \mathbf{y}^{(2)}, \ldots, \mathbf{y}^{(k)})$ be the first k columns of $\mathbf{Y}$ $(k \le d)$; let $\mathbf{B}_k = (\boldsymbol{\beta}^{(1)}, \boldsymbol{\beta}^{(2)}, \ldots, \boldsymbol{\beta}^{(k)})$, $\boldsymbol{\sigma}_{k-1,k} = (\sigma_{1k}, \sigma_{2k}, \ldots, \sigma_{k-1,k})'$, and let $\mathbf{\Sigma}_k$ be the leading $k \times k$ submatrix of $\mathbf{\Sigma}$. From $\mathbf{Y} = [(y_{ij})]$ and $\mathscr{E}[\mathbf{Y}] = \mathbf{XB}$, it follows that $E[y_{ij}] = \mathbf{x}_i'\boldsymbol{\beta}^{(j)}$ and the distribution of y_{ik} given $\mathbf{y}_i^* = (y_{i1}, y_{i2}, \ldots, y_{i,k-1})'$ is [by Theorem 2.1(viii) of Section 2.2 and (2.17)]

$$N_1\left(\mathbf{x}_i'\boldsymbol{\beta}^{(k)} + \boldsymbol{\sigma}_{k-1,k}'\mathbf{\Sigma}_{k-1}^{-1}\left\{\mathbf{y}_i^* - \mathbf{B}_{k-1}'\mathbf{x}_i\right\}, \sigma_k^2\right), \tag{8.76}$$

where $\sigma_k^2 = |\mathbf{\Sigma}_k|/|\mathbf{\Sigma}_{k-1}|$. The elements $y_{1k}, y_{2k}, \ldots, y_{nk}$ of $\mathbf{y}^{(k)}$ are mutually independent, and conditioning y_{ik} on $\mathbf{y}_i^*$ for each i does not alter this independence, as the rows of $\mathbf{Y}_k$ are mutually independent. Hence, if we transpose the second term in the mean of (8.76), the conditional distribution of $\mathbf{y}^{(k)}$, given $\mathbf{Y}_{k-1}$, is

$$N_n\left(\mathbf{X}\boldsymbol{\beta}^{(k)} + \{\mathbf{Y}_{k-1} - \mathbf{X}\mathbf{B}_{k-1}\}\mathbf{\Sigma}_{k-1}^{-1}\boldsymbol{\sigma}_{k-1,k}, \sigma_k^2\mathbf{I}_n\right),$$

or

$$N_n\left(\mathbf{X}\boldsymbol{\eta}^{(k)} + \mathbf{Y}_{k-1}\boldsymbol{\gamma}^{(k-1)}, \sigma_k^2\mathbf{I}_n\right), \tag{8.77}$$

where

$$\boldsymbol{\eta}^{(k)} = \boldsymbol{\beta}^{(k)} - \mathbf{B}_{k-1}\boldsymbol{\gamma}^{(k-1)} \quad \text{and} \quad \boldsymbol{\gamma}^{(k-1)} = \mathbf{\Sigma}_{k-1}^{-1}\boldsymbol{\sigma}_{k-1,k}.$$

To this set of conditional distributions ($k = 2, 3, \ldots, d$) we add $\mathbf{y}^{(1)} \sim N_n(\mathbf{X}\boldsymbol{\beta}^{(1)}, \sigma_{11}\mathbf{I}_n)$.

Consider now the following decomposition of the testable hypothesis H_0: $\mathbf{AB} = \mathbf{O}$, where $\mathbf{A}$ is $q \times p$ of rank q:

$$\begin{aligned}
\mathbf{AB} = \mathbf{O} &\Leftrightarrow \bigcap_{k=1}^{d}\left[\mathbf{A}\boldsymbol{\beta}^{(k)} = \mathbf{0}\right] \\
&\Leftrightarrow \bigcap_{k=1}^{d}\left[\mathbf{A}\boldsymbol{\eta}^{(k)} = \mathbf{0}\right] \\
&\Leftrightarrow \bigcap_{k=1}^{d}\left[(\mathbf{A}|\mathbf{O})\begin{pmatrix}\boldsymbol{\eta}^{(k)} \\ \boldsymbol{\gamma}^{(k-1)}\end{pmatrix} = \mathbf{0}\right],
\end{aligned} \tag{8.78}$$

that is,

$$H_0 = \bigcap_{k=1}^{d} H_{0k},$$

say. Equation (8.78) follows from the fact that if $H_{01}, H_{02}, \ldots, H_{0,k-1}$ are true, then $\mathbf{AB}_{k-1} = \mathbf{O}$ and $H_{0k}: \mathbf{A}\boldsymbol{\eta}^{(k)} = \mathbf{0}$ reduces to $\mathbf{A}\boldsymbol{\beta}^{(k)} = \mathbf{0}$. For fixed $\mathbf{Y}_{k-1}$ we have model (8.77), and, as this is a standard analysis of covariance model (see Section 9.5), we can test H_{0k} *conditionally* using a standard F-statistic, F_k, say. If $r = \text{rank } \mathbf{X}$, then, conditional on fixed $\mathbf{Y}_{k-1}$, $F_k \sim F_{q, n-r-(k-1)}$ when H_{0k} is true. Since this null distribution does not depend on $\mathbf{Y}_{k-1}$, F_k is independent of $\mathbf{Y}_{k-1}$ (and therefore of F_{k-1}, F_{k-2}, etc.) and is *unconditionally* distributed as $F_{q, n-r-k+1}$ when H_{0k} is true. Thus when H_0 is true, each H_{0k} is true and the joint distribution of the F_k satisfies

$$\begin{aligned}
g(F_1, F_2, \ldots, F_d) &= g_1(F_1)g_2(F_2|F_1) \cdots g_d(F_d|F_1, F_2, \ldots, F_{d-1}) \\
&= g_1(F_1)g_2(F_2) \cdots g_d(F_d),
\end{aligned} \tag{8.79}$$

that is, the F_k are mutually independent. We can therefore test H_0 at the α level of significance by testing the sequence $H_{01}, H_{02}, H_{03}, \dots,$ using significance levels α_k, where

$$\text{pr}\left[\bigcap_{k=1}^{d}\left(F_k \leq F_{q,n-r-k+1}^{\alpha_k} | H_{0k}\right)\right] = \prod_{k=1}^{d}(1-\alpha_k) = 1-\alpha. \qquad (8.80)$$

This so-called step-down procedure can be used where there is a physically meaningful basis for considering the response variables in a certain order; different orders may lead to different conclusions. The method has been used to construct simultaneous confidence intervals (e.g., Roy et al. [1971]). A method for computing the successive F-statistics is given in Chapter 10 [see (10.12) and (10.22)].

8.9 MULTIPLE DESIGN MODELS

A natural extension of the linear model described above is to allow the matrix $\mathbf{X}$ to depend on the response variable, namely, $\mathbf{y}^{(j)} = \boldsymbol{\theta}^{(j)} + \mathbf{u}^{(j)}$, $j = 1, 2, \dots, d$, where $\boldsymbol{\theta}^{(j)} = \mathbf{X}_j\boldsymbol{\beta}^{(j)}$ and $\mathbf{X}_j$ is $n \times p_j$ of rank p_j. The null hypothesis H_0 also can be generalized: We now wish to test $H_0 = \cap_j H_{0j}$, where $H_{0j}: \mathbf{A}_j\boldsymbol{\beta}^{(j)} = \mathbf{c}^{(j)}$ and $\mathbf{A}_j$ is $q_j \times p_j$ of rank q_j.

EXAMPLE 8.1 Suppose we have two correlated regression models

$$\text{E}[y_{i1}] = \theta_{i1} = \beta_{10} + \beta_{11}x_i$$

and

$$\text{E}[y_{i2}] = \theta_{i2} = \beta_{20} + \beta_{21}x_i + \beta_{22}x_i^2 \qquad (i = 1, 2, \dots, n).$$

Here $\mathbf{Y} = [(y_{ij})] = (\mathbf{y}^{(1)}, \mathbf{y}^{(2)})$,

$$\boldsymbol{\theta}^{(1)} = \begin{pmatrix} \theta_{11} \\ \theta_{21} \\ \vdots \\ \theta_{n1} \end{pmatrix} = \begin{pmatrix} 1 & x_1 \\ 1 & x_2 \\ \vdots & \vdots \\ 1 & x_n \end{pmatrix} \begin{pmatrix} \beta_{10} \\ \beta_{11} \end{pmatrix} = \mathbf{X}_1\boldsymbol{\beta}^{(1)}$$

and

$$\boldsymbol{\theta}^{(2)} = \begin{pmatrix} \theta_{12} \\ \theta_{22} \\ \vdots \\ \theta_{n2} \end{pmatrix} = \begin{pmatrix} 1 & x_1 & x_1^2 \\ 1 & x_2 & x_2^2 \\ \vdots & \vdots & \vdots \\ 1 & x_n & x_n^2 \end{pmatrix} \begin{pmatrix} \beta_{20} \\ \beta_{21} \\ \beta_{22} \end{pmatrix} = \mathbf{X}_2\boldsymbol{\beta}^{(2)}.$$

The hypothesis of no regression on x then takes the form

$$H_0: \beta_{11} = 0, \qquad \begin{pmatrix} \beta_{21} \\ \beta_{22} \end{pmatrix} = \begin{pmatrix} 0 \\ 0 \end{pmatrix}, \tag{8.81}$$

so that

$$\mathbf{A}_1 = (0, 1) \quad \text{and} \quad \mathbf{A}_2 = \begin{pmatrix} 0 & 1 & 0 \\ 0 & 0 & 1 \end{pmatrix}. \qquad \square$$

We now return to the general formulation given at the beginning of this section and derive a maximum root test due to McDonald [1975]. Writing $\mathbf{Y} = \mathbf{\Theta} + \mathbf{U}$ as before, we assume that the rows of $\mathbf{U}$ are i.i.d. $N_d(\mathbf{0}, \mathbf{\Sigma})$. In order to use the union–intersection principle we reduce the multivariate model to a univariate one by using the transformation $\mathbf{y} = \mathbf{Y}\boldsymbol{\ell}$, where $\boldsymbol{\ell} = (\ell_1, \ell_2, \ldots, \ell_d)'$. Then

$$\begin{aligned} \mathbf{\Theta}\boldsymbol{\ell} &= \sum_{j=1}^{d} \boldsymbol{\theta}^{(j)} \ell_j \\ &= \sum_{j=1}^{d} \mathbf{X}_j \boldsymbol{\beta}^{(j)} \ell_j \\ &= \mathbf{X}\boldsymbol{\beta}, \end{aligned}$$

say, where $\mathbf{X} = (\mathbf{X}_1, \mathbf{X}_2, \ldots, \mathbf{X}_d)$ is $n \times p$ $(p = \Sigma_j p_j)$ and

$$\boldsymbol{\beta}' = \left(\ell_1 \boldsymbol{\beta}^{(1)\prime}, \ell_2 \boldsymbol{\beta}^{(2)\prime}, \ldots, \ell_d \boldsymbol{\beta}^{(d)\prime}\right).$$

We see that H_0 is true if and only if $H_{0\ell}: \cap_j (\mathbf{A}_j \boldsymbol{\beta}^{(j)} \ell_j = \mathbf{c}^{(j)} \ell_j)$ for all $\boldsymbol{\ell}$, that is, $H_0 = \cap_\ell H_{0\ell}$. If $q = \max(q_1, q_2, \ldots, q_d)$, then $H_{0\ell}$ can be written in the form $\Sigma_j \mathbf{A}_j^* \boldsymbol{\beta}^{(j)} \ell_j = \Sigma_j \mathbf{c}^{*(j)} \ell_j$ $(= \mathbf{c}$, say), where

$$\mathbf{A}_j^* = \begin{pmatrix} \mathbf{A}_j \\ \mathbf{O}_j \end{pmatrix} \begin{matrix} \}\, q_j \\ \}\, q - q_j \end{matrix} \quad \text{and} \quad \mathbf{c}^{*(j)} = \begin{pmatrix} \mathbf{c}^{(j)} \\ \mathbf{0}_j \end{pmatrix}.$$

Here $\mathbf{O}_j$ is a $(q - q_j) \times p_j$ block of zeros and $\mathbf{0}_j$ is a vector of $q - q_j$ zeros. Therefore setting $\mathbf{A} = (\mathbf{A}_1^*, \mathbf{A}_2^*, \ldots, \mathbf{A}_d^*)$, we see that $H_{0\ell}$ takes the form $\mathbf{A}\boldsymbol{\beta} = \mathbf{c}$ and we assume that this hypothesis is testable, that is, the rows of $\mathbf{A}$ are linearly dependent on the rows of $\mathbf{X}$. With the univariate theory, the appropriate sums of squares are

$$Q = \mathbf{y}'(\mathbf{I}_n - \mathbf{P}_\Omega)\mathbf{y} = \boldsymbol{\ell}'\mathbf{E}\boldsymbol{\ell}$$

and

$$Q_H - Q = \ell' \mathbf{H} \ell.$$

It then follows from Section 8.6.2a, that H_0 can be tested using $\phi_{\max}$, the largest root of $|\mathbf{H} - \phi \mathbf{E}| = 0$. If $\mathbf{X}$ has rank r, $\mathbf{A}$ has rank q, and $\sigma^2 = \ell' \boldsymbol{\Sigma} \ell$, then Q/σ^2 and $(Q_H - Q)/\sigma^2$ are independently distributed as χ_q^2 and χ_{n-r}^2, respectively, for all ℓ when H_0 is true. Hence, from Theorem 2.4(ii) in Section 2.3.2, $\mathbf{E}$ and $\mathbf{H}$ are independently distributed as $W_d(n - r, \boldsymbol{\Sigma})$ and $W_d(q, \boldsymbol{\Sigma})$, respectively when H_0 is true. Here

$$\mathbf{E} = \mathbf{Y}'\left(\mathbf{I}_n - \mathbf{X}(\mathbf{X}'\mathbf{X})^{-}\mathbf{X}'\right)\mathbf{Y}$$

and, since $\mathbf{c} = \mathbf{C}\ell$, we have from (8.47),

$$\mathbf{H} = \left\{\mathbf{A}(\mathbf{X}'\mathbf{X})^{-}\mathbf{X}'\mathbf{Y} - \mathbf{C}\right\}'\left[\mathbf{A}(\mathbf{X}'\mathbf{X})^{-}\mathbf{A}'\right]^{-1}\left\{\mathbf{A}(\mathbf{X}'\mathbf{X})^{-}\mathbf{X}'\mathbf{Y} - \mathbf{C}\right\}, \quad (8.82)$$

where $\mathbf{C} = (\mathbf{c}^{*(1)}, \mathbf{c}^{*(2)}, \ldots, \mathbf{c}^{*(d)})$. If rank $\mathbf{A} = q_0$, then, when H_0 is true, $\mathbf{H} \sim \mathbf{W}_d(q_0, \boldsymbol{\Sigma})$ and we use $[\mathbf{A}(\mathbf{X}'\mathbf{X})^{-}\mathbf{A}']^{-}$ in (8.82). Since $\mathbf{E}$ and $\mathbf{H}$ are independent d-dimensional Wishart distributions when H_0 is true, we can use all four test statistics given in Section 8.6.2 to test H_0.

Further insight into the above approach is gained by noting that

$$\begin{aligned}
\boldsymbol{\Theta} &= \left(\mathbf{X}_1\boldsymbol{\beta}^{(1)}, \mathbf{X}_2\boldsymbol{\beta}^{(2)}, \ldots, \mathbf{X}_d\boldsymbol{\beta}^{(d)}\right) \\
&= (\mathbf{X}_1, \mathbf{X}_2, \ldots, \mathbf{X}_d)\begin{pmatrix} \boldsymbol{\beta}^{(1)} & \mathbf{0} & \cdots & \mathbf{0} \\ \mathbf{0} & \boldsymbol{\beta}^{(2)} & \cdots & \mathbf{0} \\ \vdots & \vdots & & \vdots \\ \mathbf{0} & \mathbf{0} & \cdots & \boldsymbol{\beta}^{(d)} \end{pmatrix} \\
&= \mathbf{X}\mathbf{B}_d, \qquad (8.83)
\end{aligned}$$

say. Then H_0 is $\mathbf{A}\mathbf{B}_d = \mathbf{C}$ and the matrices $\mathbf{E}$ and $\mathbf{H}$ of (8.82) arise naturally by applying the general theory of this chapter to (8.83). However, when $\mathbf{X}$ has full rank, $\mathbf{H}$ of (8.82) is based on the estimator $\hat{\mathbf{B}}_d = (\mathbf{X}'\mathbf{X})^{-1}\mathbf{X}'\mathbf{Y}$. This estimator will not be efficient, as it estimates some elements of $\mathbf{B}_d$ that are known to be zero. We can therefore expect the above test procedure to be less powerful than what might be suggested by the general theory.

More efficient estimators of the $\boldsymbol{\beta}^{(j)}$ can be found as follows (Zellner [1962], Press [1972a: Section 8.5]. Let

$$\begin{pmatrix} \mathbf{y}^{(1)} \\ \mathbf{y}^{(2)} \\ \vdots \\ \mathbf{y}^{(d)} \end{pmatrix} = \begin{pmatrix} \mathbf{X}_1 & \mathbf{O} & \cdots & \mathbf{O} \\ \mathbf{O} & \mathbf{X}_2 & \cdots & \mathbf{O} \\ \vdots & \vdots & & \vdots \\ \mathbf{O} & \mathbf{O} & \cdots & \mathbf{X}_d \end{pmatrix}\begin{pmatrix} \boldsymbol{\beta}^{(1)} \\ \boldsymbol{\beta}^{(2)} \\ \vdots \\ \boldsymbol{\beta}^{(d)} \end{pmatrix} + \begin{pmatrix} \mathbf{u}^{(1)} \\ \mathbf{u}^{(2)} \\ \vdots \\ \mathbf{u}^{(d)} \end{pmatrix}$$

or

$$\mathbf{y}_t = \mathbf{X}_t\boldsymbol{\beta}_t + \mathbf{u}_t \qquad (t = \Sigma_i q_i) \tag{8.84}$$

say. Then $\mathscr{D}[\mathbf{u}_t] = \mathbf{V}_t$, where $\mathbf{V}_t$ is a function of the elements of $\boldsymbol{\Sigma}$ (see Exercise 8.13), and the minimum variance linear unbiased estimate of $\boldsymbol{\beta}_t$ is $\hat{\boldsymbol{\beta}}_t = (\mathbf{X}_t'\mathbf{V}_t^{-1}\mathbf{X}_t)^{-1}\mathbf{X}_t'\mathbf{V}_t^{-1}\mathbf{y}_t$ (see Seber [1977: Section 3.6]). Estimating $\boldsymbol{\Sigma}$ by $(\mathbf{Y} - \mathbf{X}\hat{\mathbf{B}}_d)'(\mathbf{Y} - \mathbf{X}\hat{\mathbf{B}}_d)/n$ $(= \mathbf{E}/n)$, we can substitute this into $\mathbf{V}_t$ to get $\hat{\mathbf{V}}_t$ and an approximation, $\tilde{\boldsymbol{\beta}}_t$, say, for $\hat{\boldsymbol{\beta}}_t$. Although $\tilde{\boldsymbol{\beta}}_t$ will be a biased estimate of $\boldsymbol{\beta}_t$, the bias will be small (of order n^{-1}) when n is sufficiently large. Plugging $\tilde{\boldsymbol{\beta}}_t$ back into $\mathbf{E}$ will lead to a more efficient estimate of $\boldsymbol{\Sigma}$. To test H_0 we set $\mathbf{A}_t = \text{diag}(\mathbf{A}_1, \mathbf{A}_2, \dots, \mathbf{A}_d)$, as in the definition of $\mathbf{X}_t$, and $\mathbf{c}_t' = (\mathbf{c}^{(1)\prime}, \mathbf{c}^{(2)\prime}, \dots, \mathbf{c}^{(d)\prime})$. Then $\mathbf{A}_t\tilde{\boldsymbol{\beta}}_t$ is approximately $N_t(\mathbf{A}_t\boldsymbol{\beta}_t, \mathbf{A}_t(\mathbf{X}_t'\hat{\mathbf{V}}^{-1}\mathbf{X}_t)^{-1}\mathbf{A}_t')$ and, when H_0 is true,

$$(\mathbf{A}_t\tilde{\boldsymbol{\beta}}_t - \mathbf{c}_t)'\left[\mathbf{A}_t(\mathbf{X}_t'\hat{\mathbf{V}}_t^{-1}\mathbf{X}_t)^{-1}\mathbf{A}_t'\right]^{-1}(\mathbf{A}_t\tilde{\boldsymbol{\beta}}_t - \mathbf{c}_t) \quad \text{is approximately} \quad \chi_t^2.$$

The multiple design multivariate (MDM) model described above was considered by Roy and Srivastava [1964] and Srivastava [1966, 1967]. Some examples relating to multivariate regression systems, like Example 8.1 above, and their estimation problems are given by Zellner [1962, 1963]. McDonald [1975] mentions that the advantages of the above tests over the step-down procedure proposed by Roy and Srivastava [1964] are that the testability conditions are relatively simple, and the standard computational techniques and tables based on $\mathbf{E}$ and $\mathbf{H}$ are available. McDonald also gives a simple necessary and sufficient condition for the testability of a special case of the above theory that he calls the monotone MDM model.

EXERCISES 8

8.1 Let $\mathbf{Y} = \mathbf{XB} + \mathbf{U}$, where $\mathbf{X}$ is $n \times p$ of rank p and $\mathscr{E}[\mathbf{U}] = \mathbf{O}$. Show that the least squares estimate $\hat{\mathbf{B}} = (\mathbf{X}'\mathbf{X})^{-1}\mathbf{X}'\mathbf{Y}$ uniquely minimizes $\text{tr}[(\mathbf{Y} - \mathbf{XB})'(\mathbf{Y} - \mathbf{XB})]$. [Hint: Use A7.9.]

8.2 Let $\mathbf{y}_1$, $\mathbf{y}_2$, and $\mathbf{y}_3$ be mutually independent, d-dimensional MVN random vectors with means $\boldsymbol{\theta}_1$, $\boldsymbol{\theta}_1 - \boldsymbol{\theta}_2$, and $\boldsymbol{\theta}_1 + \boldsymbol{\theta}_2$, respectively, and common dispersion matrix $\boldsymbol{\Sigma}$. Find the least squares estimates of $\boldsymbol{\theta}_1$ and $\boldsymbol{\theta}_2$.

8.3 Suppose $\mathbf{y}_i$ $(i = 1, 2, \dots, n)$ are independently distributed as $N_d(\boldsymbol{\mu}_i, \boldsymbol{\Sigma})$, where $\boldsymbol{\mu}_i = \mathbf{B}'\mathbf{x}_i$, $\mathbf{B}$ is a $p \times d$ matrix of unknown parameters and $\mathbf{X} = (\mathbf{x}_1, \mathbf{x}_2, \dots, \mathbf{x}_n)'$ is a given $n \times p$ matrix of rank p. If $\hat{\mathbf{B}} = (\mathbf{X}'\mathbf{X})^{-1}\mathbf{XY}$, find the dispersion matrix of $\hat{\mathbf{B}}'\mathbf{x}_0$, the predicted value of $\mathbf{y}_0$ corresponding to $\mathbf{x} = \mathbf{x}_0$. How would you estimate this matrix? Construct a confidence interval for $\mathbf{a}'\mathbf{y}_0$.

8.4 Let $\mathbf{y}_1, \mathbf{y}_2, \dots, \mathbf{y}_n$ be a random sample from $N_d(\boldsymbol{\mu}, \boldsymbol{\Sigma})$. Show that the sample mean $\bar{\mathbf{y}}$ is the least squares estimate of $\boldsymbol{\mu}$.

8.5 Use Exercise 2.23 and (8.49) to find an unbiased estimate of $\mathbf{\Sigma}^{-1}\mathbf{D}$.

8.6 Verify (8.67) and (8.68).

8.7 Let $\mathbf{Y} = \mathbf{XB} + \mathbf{U}$, where $\mathbf{XB} = \mathbf{1}_n\boldsymbol{\beta}_1' + \mathbf{X}_2\mathbf{B}_2, \mathbf{X} = (\mathbf{1}_n, \mathbf{X}_2)$ is $n \times p$ of rank p, $\mathbf{X}_2'\mathbf{1}_n = \mathbf{0}$, and the rows of $\mathbf{U}$ are i.i.d. $N_d(\mathbf{0}, \mathbf{\Sigma})$.

(a) Show that the least squares estimates of $\boldsymbol{\beta}_1$ and $\mathbf{B}_2$ are $\bar{\mathbf{y}}$ and $(\mathbf{X}_2'\mathbf{X}_2)^{-1}\mathbf{X}_2'\tilde{\mathbf{Y}}$, respectively, where $\tilde{\mathbf{Y}} = (\mathbf{y}_1 - \bar{\mathbf{y}}, \mathbf{y}_2 - \bar{\mathbf{y}}, \dots, \mathbf{y}_n - \bar{\mathbf{y}})'$.

(b) For testing $H_0: \mathbf{B}_2 = \mathbf{O}$, show that

$$\mathbf{E} = \tilde{\mathbf{Y}}'(\mathbf{I}_n - \mathbf{M})\tilde{\mathbf{Y}} \quad \text{and} \quad \mathbf{H} = \tilde{\mathbf{Y}}'\mathbf{M}\tilde{\mathbf{Y}},$$

where $\mathbf{M} = \mathbf{X}_2(\mathbf{X}_2'\mathbf{X}_2)^{-1}\mathbf{X}_2'$.

Note: A special case of this result is as follows. Let

$$\mathbf{XB} = (\mathbf{1}_n, \mathbf{X}_0)\begin{pmatrix} \boldsymbol{\alpha}' \\ \mathbf{B}_2 \end{pmatrix} = (\mathbf{1}_n, \tilde{\mathbf{X}}_0)\begin{pmatrix} \boldsymbol{\beta}_1' \\ \mathbf{B}_2 \end{pmatrix},$$

where $\mathbf{X}_0 = (\mathbf{x}_1, \mathbf{x}_2, \dots, \mathbf{x}_n)'$ and $\tilde{\mathbf{X}}_0 = (\mathbf{x}_1 - \bar{\mathbf{x}}, \mathbf{x}_2 - \bar{\mathbf{x}}, \dots, \mathbf{x}_n - \bar{\mathbf{x}})'$. Then $\tilde{\mathbf{X}}_0'\mathbf{1}_n = \mathbf{0}$.

8.8 Using the notation of Section 8.3, let $\omega = \Omega \cap \mathcal{N}[\mathbf{M}]$, where Ω has dimension r and $\mathbf{M}$ is $q \times n$ of rank q. Show that $\mathbf{I}_n - \mathbf{P}_\omega$ has rank $n - r + q$.

8.9 Using the notation of (8.63), construct a t-confidence interval for a given linear combination $\mathbf{a}'\mathbf{Bb}$. Apply the Bonferroni method to obtain confidence intervals for m such linear combinations that have an overall confidence of at least $100(1 - \alpha)\%$.

8.10 Let $\mathbf{Y} = \mathbf{XB} + \mathbf{U}$, where $\mathbf{X}$ is $n \times p$ of rank p, and the rows of $\mathbf{U}$ are i.i.d. $N_d(\mathbf{0}, \mathbf{\Sigma})$. Let $\mathbf{XB} = \mathbf{X}_1\mathbf{B}_1 + \mathbf{X}_2\mathbf{B}_2$, where $\mathbf{X}_2$ is $n \times p_2$. Derive the likelihood ratio test for testing $H_0: \mathbf{B}_2 = \mathbf{O}$.

8.11 Verify that (8.75) has a Hotelling's T^2 distribution under the null hypothesis $H_{03} \cap H_{01}$.

8.12 For example 8.1 of Section 8.9 prove that the first three hypotheses below are each testable, and that the last two are not.

(a) $\beta_{11} = 0, \beta_{21} = \beta_{22} = 0$.

(b) $\beta_{22} = 0$.

(c) $\beta_{11} = \beta_{21} = 0$.

(d) $\beta_{11} = 0$.

(e) $\beta_{11} = \beta_{22} = 0$.

8.13 Find the dispersion matrix $\mathbf{V}_t$ of $\mathbf{u}_t$, where $\mathbf{u}_t$ is given by (8.84).

CHAPTER 9

Multivariate Analysis of Variance and Covariance

9.1 INTRODUCTION

In Chapter 8 we developed a general theory for multivariate linear models. We saw that the usual univariate linear model generalizes naturally to a multivariate model in which a vector $\mathbf{y}$ of independent observations y_i is replaced by a matrix of observations $\mathbf{Y}$ with statistically independent rows $\mathbf{y}_i'$. In particular, a quadratic form $\mathbf{y}'\mathbf{G}\mathbf{y} = \sum_i \sum_j g_{ij} y_i y_j$, for example, $(\mathbf{y} - \mathbf{X}\hat{\boldsymbol{\beta}})'(\mathbf{y} - \mathbf{X}\hat{\boldsymbol{\beta}})$ $[= \mathbf{y}'(\mathbf{I}_n - \mathbf{P}_\Omega)\mathbf{y}]$, simply becomes $\mathbf{Y}'\mathbf{G}\mathbf{Y} = \sum_i \sum_j g_{ij} \mathbf{y}_i \mathbf{y}_j'$. As fixed effects univariate analysis of variance (ANOVA) models are special cases of the linear model (e.g., Seber [1977: Chapter 9]), it is not surprising that ANOVA models generalize naturally to MANOVA (multivariate analysis of variance) models. We now demonstrate this extension with several simple examples.

9.2 ONE-WAY CLASSIFICATION

9.2.1 Hypothesis Testing

In Section 8.6.4 we considered the problem of comparing the means of two MVN populations, assuming a common dispersion matrix $\boldsymbol{\Sigma}$, using the general theory of multivariate linear models. We now extend this theory to the case of comparing I normal populations when there are n_i independent d-dimensional observations from the ith population.

Let $\mathbf{y}_{ij}$ be the jth sample observation ($j = 1, 2, \ldots, n_i$) from the ith MVN distribution $N_d(\boldsymbol{\mu}_i, \boldsymbol{\Sigma})$ ($i = 1, 2, \ldots, I$) so that we have the following array of data:

		Sample mean
Population 1:	$\mathbf{y}_{11}, \mathbf{y}_{12}, \ldots, \mathbf{y}_{1n_1}$	$\bar{\mathbf{y}}_{1\cdot}$
Population 2:	$\mathbf{y}_{21}, \mathbf{y}_{22}, \ldots, \mathbf{y}_{2n_2}$	$\bar{\mathbf{y}}_{2\cdot}$
$\vdots$	$\vdots$	$\vdots$
Population I:	$\mathbf{y}_{I1}, \mathbf{y}_{I2}, \ldots, \mathbf{y}_{In_I}$	$\bar{\mathbf{y}}_{I\cdot}$

We also require the grand mean

$$\bar{\mathbf{y}}_{\cdot\cdot} = \sum_i \sum_j \mathbf{y}_{ij} \Big/ \sum_i n_i = \frac{1}{n} \sum_i n_i \bar{\mathbf{y}}_{i\cdot},$$

where $n = \sum_i n_i$. In order to apply the general theory of Chapter 8, we combine the above array of vectors into the single model

$$\mathbf{y}_{ij} = \boldsymbol{\mu}_i + \mathbf{u}_{ij} \qquad (i = 1, 2, \ldots, I : j = 1, 2, \ldots, n_i),$$

where the $\mathbf{u}_{ij}$ are i.i.d. $N_d(\mathbf{0}, \boldsymbol{\Sigma})$. Then

$$\begin{pmatrix} \mathbf{y}'_{11} \\ \mathbf{y}'_{12} \\ \vdots \\ \mathbf{y}'_{1n_1} \\ \hline \vdots \\ \hline \mathbf{y}'_{I1} \\ \mathbf{y}'_{I2} \\ \vdots \\ \mathbf{y}'_{In_I} \end{pmatrix} = \begin{pmatrix} \mathbf{1}_{n_1} & \mathbf{0} & \cdots & \mathbf{0} \\ \mathbf{0} & \mathbf{1}_{n_2} & \cdots & \mathbf{0} \\ \vdots & \vdots & & \vdots \\ \mathbf{0} & \mathbf{0} & \cdots & \mathbf{1}_{n_I} \end{pmatrix} \begin{pmatrix} \boldsymbol{\mu}'_1 \\ \boldsymbol{\mu}'_2 \\ \vdots \\ \boldsymbol{\mu}'_I \end{pmatrix} + \begin{pmatrix} \mathbf{u}'_{11} \\ \mathbf{u}'_{12} \\ \vdots \\ \mathbf{u}'_{1n_1} \\ \hline \vdots \\ \hline \mathbf{u}'_{I1} \\ \mathbf{u}'_{I2} \\ \vdots \\ \mathbf{u}'_{In_I} \end{pmatrix}$$

or

$$\mathbf{Y} = \mathbf{XB} + \mathbf{U}. \tag{9.1}$$

The hypothesis of equal population means, $H_0 : \boldsymbol{\mu}_1 = \boldsymbol{\mu}_2 = \cdots = \boldsymbol{\mu}_I$, can be written in the form

$$\mathbf{O} = \begin{pmatrix} 1 & 0 & \cdots & 0 & -1 \\ 0 & 1 & \cdots & 0 & -1 \\ \vdots & \vdots & & \vdots & \vdots \\ 0 & 0 & \cdots & 1 & -1 \end{pmatrix} \begin{pmatrix} \boldsymbol{\mu}'_1 \\ \boldsymbol{\mu}'_2 \\ \vdots \\ \boldsymbol{\mu}'_I \end{pmatrix} = \mathbf{AB}. \tag{9.2}$$

TABLE 9.1 One-Way Univariate Analysis of Variance Table

Source	Sum of Squares (SS)	Degrees of Freedom (df)	SS/df
Between populations	$\Sigma n_i(\bar{y}_{i\cdot} - \bar{y}_{\cdot\cdot})^2$	$q = I - 1$	s_H^2
Error	$\Sigma\Sigma(y_{ij} - \bar{y}_{i\cdot})^2$	$n - p = \Sigma n_i - I$	s_E^2
Corrected total	$\Sigma\Sigma(y_{ij} - \bar{y}_{\cdot\cdot})$	$\Sigma n_i - 1$	

To test H_0 we consider the quadratic forms for the univariate analysis of the model $y_{ij} = \mu_i + u_{ij}$. These are given in the ANOVA table, Table 9.1. It is now easy to write down the corresponding matrices for the MANOVA model, and these are given in Table 9.2. Instead of using the F-ratio s_H^2/s_E^2, we use one of the various test statistics based on the roots of $|\mathbf{H} - \theta(\mathbf{E} + \mathbf{H})| = 0$. For example, Wilks' likelihood ratio statistic is

$$\Lambda = \frac{|\mathbf{E}|}{|\mathbf{E} + \mathbf{H}|} = \frac{|\mathbf{E}|}{|\mathbf{E}_H|},$$

and $\Lambda \sim U_{d, m_H, m_E}$ when H_0 is true. Procedures for computing the various test statistics are described in Section 10.1.4. When $I = 2$, we have $m_H = 1$ and, from Section 2.5.5e, all four test statistics of Section 8.6.2 are equivalent to Hotelling's T_0^2 test for comparing two means (see Exercise 9.1). For general I, the four tests are compared in Section 9.2.3.

Methods for testing H_0 against the alternative $\boldsymbol{\mu}_1 = \cdots = \boldsymbol{\mu}_r \neq \boldsymbol{\mu}_{r+1} = \cdots = \boldsymbol{\mu}_I$ (r unknown) are given by Sen and Srivastava [1973].

EXAMPLE 9.1 In Table 9.3 we have a selection of transformed measurements from Reeve [1941] on anteater skulls for the subspecies *chapadensis* from three different localities and now deposited in the British Museum. Three measurements were taken on each skull (see Example 5.8 in Section 5.8), and Table 9.3 shows the common logarithms of the original measurements in millimeters, along with their means. To test $H_0: \boldsymbol{\mu}_1 = \boldsymbol{\mu}_2 = \boldsymbol{\mu}_3$ for the three

TABLE 9.2 One-Way Multivariate Analysis of Variance Table

Source	Matrix	df
Between populations	$\mathbf{H} = \Sigma n_i(\bar{\mathbf{y}}_{i\cdot} - \bar{\mathbf{y}}_{\cdot\cdot})(\bar{\mathbf{y}}_{i\cdot} - \bar{\mathbf{y}}_{\cdot\cdot})'$	$m_H = I - 1$
Error	$\mathbf{E} = \Sigma\Sigma(\mathbf{y}_{ij} - \bar{\mathbf{y}}_{i\cdot})(\mathbf{y}_{ij} - \bar{\mathbf{y}}_{i\cdot})'$	$m_E = \Sigma n_i - I$
Corrected total	$\mathbf{E}_H = \Sigma\Sigma(\mathbf{y}_{ij} - \bar{\mathbf{y}}_{\cdot\cdot})(\mathbf{y}_{ij} - \bar{\mathbf{y}}_{\cdot\cdot})'$	

TABLE 9.3 Logarithms of Multiple Measurements on Anteater Skulls at Three Localities[a]

	Minas Graes, Brazil			Matto Grosso, Brazil			Santa Cruz, Bolivia		
	x_1	x_2	x_3	x_1	x_2	x_3	x_1	x_2	x_3
	2·068	2·070	1·580	2·045	2·054	1·580	2·093	2·098	1·653
	2·068	2·074	1·602	2·076	2·088	1·602	2·100	2·106	1·623
	2·090	2·090	1·613	2·090	2·093	1·643	2·104	2·101	1·653
	2·097	2·093	1·613	2·111	2·114	1·643	—	—	—
	2·117	2·125	1·663	—	—	—	—	—	—
	2·140	2·146	1·681	—	—	—	—	—	—
Means	2·097	2·100	1·625	2·080	2·087	1·617	2·099	2·102	1·643

[a]From Reeve [1941], courtesy of The Zoological Society of London.

locations, we compute

$$\mathbf{H} = \begin{pmatrix} 0.80597 \times 10^{-3} & 0.62324 \times 10^{-3} & 0.74979 \times 10^{-3} \\ - & 0.48202 \times 10^{-3} & 0.58595 \times 10^{-3} \\ - & - & 1.18436 \times 10^{-3} \end{pmatrix}$$

and

$$\mathbf{E} = \begin{pmatrix} 0.634233 \times 10^{-2} & 0.624183 \times 10^{-2} & 0.761567 \times 10^{-2} \\ - & 0.635275 \times 10^{-2} & 0.761267 \times 10^{-2} \\ - & - & 1.096733 \times 10^{-2} \end{pmatrix}.$$

Here $d = 3$, $I = 3$, $n_1 = 6$, $n_2 = 4$, $n_3 = 3$, $n = \sum_i n_i = 13$, $s = \min(I - 1, d) = 2$, $m_H = I - 1 = 2$, and $m_E = n - I = 13 - 3 = 10$. The various test statistics for testing H_0 are

$$\Lambda = |\mathbf{E}|/|\mathbf{E} + \mathbf{H}| = 0.6014,$$

$$\operatorname{tr}[\mathbf{HE}^{-1}] = 0.5821,$$

$$V^{(2)} = \operatorname{tr}\left[\mathbf{H}(\mathbf{E} + \mathbf{H})^{-1}\right] = 0.4470,$$

$$\phi_{\max} = 0.35531,$$

where $\phi_{\max}$ is the maximum eigenvalue of $\mathbf{HE}^{-1}$. Since $m_H = 2$, we can transform Λ into an exact F-statistic using (2.45), namely,

$$F_0 = \frac{1 - \Lambda^{1/t}}{\Lambda^{1/t}} \frac{ft - g}{2d} \sim F_{2d, ft-g},$$

where $f = m_E - \frac{1}{2}(d - m_H + 1) = 9$, $t = [(4d^2 - 4)/(d^2 - 1)]^{1/2} = 2$, and

$g = \frac{1}{2}(2d - 2) = 2$. Thus

$$F_0 = \frac{1 - \Lambda^{1/2}}{\Lambda^{1/2}} \frac{16}{6} = 0.77,$$

which is not significant, as $F_{6,16}^{0.05} = 2.74$.

For Hotelling's T_g^2 test we use Appendix D15. Since $m_H < d$, we make the transformation

$$(d, m_H, m_E) \to (m_H, d, m_E + m_H - d) = (2, 3, 9)$$

and calculate

$$\frac{m_E}{m_H} \operatorname{tr}[\mathbf{HE}^{-1}] = \frac{9}{3}(0.58210) = 1.75.$$

Interpolating in Appendix D15 with the above transformed triple, we get a 5% critical value of about 7.8, which is well above 1.75. Hence T_g^2 is not significant.

Now $s = 2$, $\nu_1 = \frac{1}{2}(|d - m_H| - 1) = 0$, and $\nu_2 = \frac{1}{2}(m_E - d - 1) = 3$ and, from Appendix D16, the 5% critical value for Pillai's test is 0.890, which is much greater than $V^{(2)} = 0.4470$. Also, the maximum eigenvalue of $\mathbf{H}(\mathbf{E} + \mathbf{H})^{-1}$ is

$$\theta_{\max} = \frac{\phi_{\max}}{1 + \phi_{\max}} = 0.26,$$

and, from Appendix D14, the 5% critical value is $\theta_{0.05} = 0.70174$. In both cases the 5% critical value is not exceeded.

9.2.2 Multiple Comparisons

For a general linear model we have the result [see (8.61)]

$$1 - \alpha = \operatorname{pr}\left[\mathbf{a}'\mathbf{ABb} \in \mathbf{a}'\mathbf{A}\hat{\mathbf{B}}\mathbf{b} \pm \left\{\phi_\alpha \mathbf{a}'\mathbf{A}(\mathbf{X}'\mathbf{X})^{-1}\mathbf{A}'\mathbf{a} \cdot \mathbf{b}'\mathbf{E}\mathbf{b}\right\}^{1/2} \text{ for all } \mathbf{a}, \mathbf{b}\right]. \tag{9.3}$$

Applying this to our one-way model [see (9.2)], we have

$$\mathbf{AB} = \begin{pmatrix} (\boldsymbol{\mu}_1 - \boldsymbol{\mu}_I)' \\ (\boldsymbol{\mu}_2 - \boldsymbol{\mu}_I)' \\ \vdots \\ (\boldsymbol{\mu}_{I-1} - \boldsymbol{\mu}_I)' \end{pmatrix} \tag{9.4}$$

so that

$$\begin{aligned}\mathbf{a}'\mathbf{AB} &= a_1(\boldsymbol{\mu}_1 - \boldsymbol{\mu}_I)' + a_2(\boldsymbol{\mu}_2 - \boldsymbol{\mu}_I)' + \cdots + a_{I-1}(\boldsymbol{\mu}_{I-1} - \boldsymbol{\mu}_I)' \\ &= \sum_{i=1}^{I} c_i \boldsymbol{\mu}_i',\end{aligned}$$

where $c_i = a_i$ $(i = 1, 2, \ldots, I-1)$, $c_I = -\sum_{i=1}^{I-1} a_i$ and $\sum_{i=1}^{I} c_i = 0$. We therefore find that the class of all vectors $\mathbf{a}'\mathbf{AB}$ is the same as the class of all contrasts in the $\boldsymbol{\mu}_i$. Furthermore, from Exercise 9.2,

$$\mathbf{a}'\mathbf{A}\hat{\mathbf{B}} = \sum_{i=1}^{I} c_i \bar{\mathbf{y}}_{i\cdot}',$$

and, when $d = 1$,

$$\text{var}\left[\sum_i c_i \bar{y}_{i\cdot}\right] = \sigma^2 \sum_i \frac{c_i^2}{n_i}.$$

Hence, from (8.62), the probability is $1 - \alpha$ that [with $\phi_\alpha = \theta_\alpha/(1 - \theta_\alpha)$]

$$\sum_i c_i \boldsymbol{\mu}_i' \mathbf{b} \in \sum_i c_i \bar{\mathbf{y}}_{i\cdot}' \mathbf{b} \pm \left\{ \phi_\alpha \left(\sum_i \frac{c_i^2}{n_i} \right) \mathbf{b}'\mathbf{E}\mathbf{b} \right\}^{1/2} \tag{9.5}$$

simultaneously for all contrasts and all $\mathbf{b}$. The condition $\mathbf{b}'\mathbf{b} = 1$ can be imposed, as (9.5) is unaffected by a scale change in $\mathbf{b}$. If $\boldsymbol{\mu}_i = [(\mu_{ij})]$, the most interesting members of this set of intervals would generally be $\sum_i c_i \mu_{ij}$ $(j = 1, 2, \ldots, d)$, a contrast in the jth elements of the $\boldsymbol{\mu}_i$ obtained by setting $\mathbf{b}' = (0, \ldots, 0, 1, 0, \ldots, 0)$ with a 1 in the jth position. In particular, we would be interested in pairwise contrasts $\boldsymbol{\mu}_r - \boldsymbol{\mu}_s$ and the corresponding subset of (9.5), namely,

$$(\boldsymbol{\mu}_r - \boldsymbol{\mu}_s)'\mathbf{b} \in (\bar{\mathbf{y}}_{r\cdot} - \bar{\mathbf{y}}_{s\cdot})'\mathbf{b} \pm \left\{ \phi_\alpha \left(\frac{1}{n_r} + \frac{1}{n_s} \right) \mathbf{b}'\mathbf{E}\mathbf{b} \right\}^{1/2}. \tag{9.6}$$

If the maximum root test of $H_0: \mathbf{AB} = \mathbf{O}$ is significant, then at least one of the intervals given by (9.5) does not contain zero and we can use the simultaneous intervals to look at any contrast suggested by the data. We have already noted above that the set (9.5) includes contrasts for each of d univariate models with respective means $\mu_{1j}, \mu_{2j}, \ldots, \mu_{Ij}$ $(j = 1, 2, \ldots, d)$. However, it is well known that for a univariate model the so-called Scheffé confidence intervals for all contrasts are very wide, so that the set (9.5) will be wider still and may not be very useful.

A set of simultaneous intervals for all linear combinations $(\boldsymbol{\mu}_r - \boldsymbol{\mu}_s)'\mathbf{b}$, based on a multivariate analogue of Tukey's Studentized range, was proposed by Krishnaiah [1965, 1969, 1979]. Writing $H_0: \boldsymbol{\mu}_1 = \boldsymbol{\mu}_2 = \cdots = \boldsymbol{\mu}_I$ as

$$H_0 = \bigcap_{\substack{r \\ r<s}} \bigcap_{s} H_{0rs},$$

where $H_{0rs}: \boldsymbol{\mu}_r - \boldsymbol{\mu}_s = \mathbf{0}$, we can test each of the $K = \binom{I}{2} = \frac{1}{2}I(I-1)$ hypotheses H_{0rs} using a Hotelling's T^2 statistic, T_k^2, say, based on a pooled estimate $\mathbf{S}_E = \mathbf{E}/\nu$ of $\boldsymbol{\Sigma}$, where $\nu = \sum_i (n_i - 1) = n - I$ (see Exercise 9.3). Then, arguing as in Section 1.6.1, we can test H_0 using

$$T_{\max}^2 = \max_{1 \le k \le K} T_k^2.$$

If c_α satisfies $\text{pr}[T_{\max}^2 \le c_\alpha | H_0] = 1 - \alpha$, then

$$\text{pr}\left[T_k^2 \le c_\alpha, k = 1, 2, \ldots, K | H_0\right] = 1 - \alpha,$$

and, if we argue as in (3.90), the probability is $1 - \alpha$ that

$$(\boldsymbol{\mu}_r - \boldsymbol{\mu}_s)'\mathbf{b} \in (\bar{\mathbf{y}}_{r\cdot} - \bar{\mathbf{y}}_{s\cdot})'\mathbf{b} \pm \left\{ c_\alpha \left(\frac{1}{n_r} + \frac{1}{n_s} \right) \mathbf{b}'\mathbf{S}_E\mathbf{b} \right\}^{1/2}$$

simultaneously for all r, s $(r \ne s)$ and for all $\mathbf{b}$. These intervals are the same as (9.6), except that ϕ_α is replaced by c_α. Since the intervals of (9.6) are a subset of (9.5), their overall probability exceeds $1 - \alpha$ and $\phi_\alpha > c_\alpha/\nu$. Unfortunately, extensive tables of c_α are not available, though a Bonferroni approach can be used. Since $T_k^2 \sim T_{d, n-I}^2$ when H_0 is true, an approximation for c_α is then the α/k quantile of $T_{d, n-I}^2$. For further details and another set of simultaneous intervals based on $T_{\max}^2$, see Subbaiah and Mudholkar [1981: there is a misprint in their Equation (2.8)].

If we are interested in a single linear combination $\theta = \mathbf{h}'\mathbf{B}\mathbf{b} = \sum_i h_i(\boldsymbol{\mu}_i'\mathbf{b})$, chosen before seeing the data, we can construct a confidence interval as follows. If we set $\hat{\theta} = \sum_i h_i(\bar{\mathbf{y}}_{i\cdot}'\mathbf{b})$, then $\hat{\theta} \sim N_1(\theta, \sum_i h_i^2 \mathbf{b}'\boldsymbol{\Sigma}\mathbf{b}/n_i)$, $\mathbf{b}'\mathbf{E}\mathbf{b} \sim \mathbf{b}'\boldsymbol{\Sigma}\mathbf{b}\chi_\nu^2$, and $\hat{\theta}$ is independent of $\mathbf{b}'\mathbf{E}\mathbf{b}$, as $\hat{\mathbf{B}}$ is independent of $\mathbf{E}$ (Theorem 8.3, Corollary, in Section 8.4). Hence

$$(\hat{\theta} - \theta) \bigg/ \left(\mathbf{b}'\mathbf{S}_E\mathbf{b} \sum_i \frac{h_i^2}{n_i} \right)^{1/2} \sim t_\nu,$$

so that a $100(1 - \alpha)\%$ confidence interval for θ is then

$$\hat{\theta} \pm t_\nu^{\alpha/2} \left(\mathbf{b}'\mathbf{S}_E\mathbf{b} \sum_i \frac{h_i^2}{n_i} \right)^{1/2}. \tag{9.7}$$

If we are interested in r such confidence intervals, then we can use the Bonferroni method with t-value $t_\nu^{\alpha/2r}$.

EXAMPLE 9.2 To demonstrate the above method of calculating simultaneous confidence intervals, we refer, once again, to Example 9.1, where we consider testing $H_0: \boldsymbol{\mu}_1 = \boldsymbol{\mu}_2 = \boldsymbol{\mu}_3$ with $\boldsymbol{\mu}_i = (\mu_{i1}, \mu_{i2}, \mu_{i3})'$. Looking at the three sample means

$$\bar{\mathbf{y}}_{1\cdot} = \begin{pmatrix} 2.097 \\ 2.100 \\ 1.625 \end{pmatrix}, \qquad \bar{\mathbf{y}}_{2\cdot} = \begin{pmatrix} 2.080 \\ 2.087 \\ 1.617 \end{pmatrix}, \qquad \bar{\mathbf{y}}_{3\cdot} = \begin{pmatrix} 2.099 \\ 2.102 \\ 1.643 \end{pmatrix},$$

we might be interested in the largest pairwise difference $1.643 - 1.617 = 0.026$. Using $\alpha = 0.05$, the confidence interval for $\mu_{33} - \mu_{23}$ is given by (9.6) with $\mathbf{b}' = (0, 0, 1)$, and, from Example 9.1, $\mathbf{b}'\mathbf{E}\mathbf{b} = e_{33} = 1.096733 \times 10^{-2}$ and

$$\phi_\alpha = \frac{\theta_\alpha}{1 - \theta_\alpha} = \frac{0.70174}{1 - 0.70174} = 2.3528.$$

The interval is therefore

$$0.026 \pm \left\{2.3528\left(\tfrac{1}{4} + \tfrac{1}{3}\right)(0.0197633)\right\}^{1/2} \quad \text{or} \quad 0.026 \pm 0.16.$$

This interval contains zero, which is to be expected, as the test of H_0 using $\theta_{\max}$ is not significant at the 5% level.

9.2.3 Comparison of Test Statistics

Assuming that the $\mathbf{u}_{ij}$ are independently distributed as $N_d(\mathbf{0}, \boldsymbol{\Sigma}_i)$, Ito and Schull [1964] investigated the robustness of the null distribution of Hotelling's generalized statistic $T_g^2 = (n - I)\mathrm{tr}[\mathbf{H}\mathbf{E}^{-1}]$ with respect to the inequality of the $\boldsymbol{\Sigma}_i$ for very large n_i, that is, large enough for $\mathbf{S}_i \approx \boldsymbol{\Sigma}_i$, where $\mathbf{S}_i = \sum_j (\mathbf{y}_{ij} - \bar{\mathbf{y}}_{i\cdot})(\mathbf{y}_{ij} - \bar{\mathbf{y}}_{i\cdot})'/(n_i - 1)$. They demonstrated that when the n_i are equal, moderate inequality of the $\boldsymbol{\Sigma}_i$ does not affect T_g^2 seriously, but when some or all of the n_i are unequal, both the significance level and power of T_g^2 can be seriously affected. Ito [1969, 1980] showed that the asymptotic properties of T_g^2 are little affected by any nonnormality of the $\mathbf{u}_{ij}$.

Since the null distribution of $-f\log\Lambda$, where $f = m_E - \frac{1}{2}(d - m_H + 1)$ and $m_E = n - I$, is asymptotically $\chi^2_{dm_E}$, it is convenient to compare $-f\log\Lambda$ rather than Λ with T_g^2. Letting $n \to \infty$ and $n_i/n \to r_i$, we have

$$\frac{\mathbf{E}}{n - I} = \sum_{i=1}^{I} \frac{(n_i - 1)\mathbf{S}_i}{n - I} \to \sum_{i=1}^{I} r_i\boldsymbol{\Sigma}_i = \dot{\boldsymbol{\Sigma}},$$

say. Hence by A2.4,

$$
\begin{aligned}
-f\log\Lambda &\approx -(n-I)\log\Lambda \\
&= (n-I)\log|\mathbf{I}_d + \mathbf{E}^{-1}\mathbf{H}| \\
&\approx (n-I)\log\left|\mathbf{I}_d + \frac{\dot{\boldsymbol{\Sigma}}^{-1}\mathbf{H}}{n-I}\right| \\
&= (n-I)\log\left\{1 + \operatorname{tr}\left[\frac{\dot{\boldsymbol{\Sigma}}^{-1}\mathbf{H}}{n-I}\right]\right\} \\
&= \operatorname{tr}[\dot{\boldsymbol{\Sigma}}^{-1}\mathbf{H}] - \frac{\operatorname{tr}\left[(\dot{\boldsymbol{\Sigma}}^{-1}\mathbf{H})^2\right]}{2(n-I)} + O(n^{-2})
\end{aligned}
$$

and

$$T_g^2 = (n-I)\operatorname{tr}[\mathbf{E}^{-1}\mathbf{H}] \to \operatorname{tr}[\dot{\boldsymbol{\Sigma}}^{-1}\mathbf{H}].$$

The above argument follows Ito [1969: p. 111] and demonstrates that $-f\log\Lambda$ and T_g^2 are asymptotically equivalent tests, even in the presence of nonnormality and inequality of dispersion matrices.

When permutation theory is valid, Mardia [1971] showed that the permutation distribution of $V^{(s)}$ is insensitive to nonnormality when the n_i are equal. This test is discussed in the next section. Unfortunately, the permutation theory does not allow for unequal $\boldsymbol{\Sigma}_i$, and we would expect $V^{(s)}$ to show a similar sensitivity to this departure as the other tests. Clearly, in designing a one-way classification, we must make every effort to achieve equal n_i.

Which of the four statistics Λ, T_g^2, $V^{(s)}$, and $\phi_{\max}$ do we choose? When the underlying normality assumptions are valid, we know, from the general recommendations in Section 8.6.2d and special studies like that of Pillai and Sudjana [1975], that they have similar powers. In general, the ranking is $V^{(s)} \geq \Lambda \geq T_g^2 \geq \phi_{\max}$ for small departures from H_0, with the order being reversed in the special case when the noncentrality matrix $\boldsymbol{\Delta}$ has rank 1. However, the differences are small, and, following Olson [1974], it seems more appropriate to base our choice of test on a comparison of their robustness.

Davis [1980b] has investigated the robustness of Λ to departures from normality. He showed that correction terms involving the measures of multivariate skewness and kurtosis are needed for the calculation of the true significance level. However, these correction terms are generally small for large m_E, thus indicating a general robustness of Λ to nonnormality. He demonstrated that for sufficiently large skewness and kurtosis, the effects on the true significance level may be quite serious for moderate m_E, say, $m_E \approx 30$. Increasing the skewness tends to raise the significance level, whereas increasing the kurtosis tends to lower it. The effect of skewness is more pronounced for smaller m_E, and that of kurtosis more pronounced for larger m_E. Increasing the inequality of sample sizes generally seems to influence the true significance level in the opposite directions to those of the basic skewness and kurtosis effects. Similar conclusions apply to $\phi_{\max}$ (Davis [1982]).

Using simulation, Olson [1974] compared the four statistics under *equal* n_i, *symmetric* (contaminated normal) distributions, and unequal Σ_i. He concluded that kurtotic departures from normality have relatively mild effects on the type I error rates (significance levels), although the effects are more serious for $\phi_{\max}$ under certain conditions. Positive kurtosis usually leads to H_0 being accepted too frequently, that is, to a reduced type I error rate, a fact demonstrated theoretically by Davis [1980b]. The powers of all the tests suffer under kurtosis, but the effect is greatest for $\phi_{\max}$ in the general case of rank $\boldsymbol{\Delta} \neq 1$. As expected from the above asymptotic results, inequality of the dispersion matrices is a more serious violation. Type I error rates can become excessively high for T_g^2, Λ, and $\phi_{\max}$, while $V^{(s)}$ is much less disturbed. Thus positive kurtosis and inequality of the dispersion matrices tend to have opposite effects on the significance level.

Since departures from the underlying assumptions affect all four statistics, the choice of a "best" statistic depends to some extend on balancing type I against type II errors. Olson's view is that "very high type I errors rates make a test dangerous; low power merely makes it less useful." On this basis the $\phi_{\max}$ test, which produces excessive rejections of H_0 under both kurtosis and covariance heterogeneity, is ruled out. For protection against kurtosis, any of the remaining three would be acceptable. The $V^{(s)}$ test is generally better than the other two in terms of type I error rates, but it is sometimes less powerful. It also stands up best to covariance heterogeneity, though its type I error rate is somewhat high. In this respect the other two statistics Λ and T_g^2 behave like $\phi_{\max}$ and perhaps should be avoided if covariance heterogeneity is moderate.

Olson also made some helpful comments about the parameters d, I, and the common group size n/I. Generally robustness improves as d is reduced, so that variables should not be thoughtlessly included in an analysis simply because the data are available. Methods for choosing variables are described in Section 10.1.6. Similarly, if I is under the experimenter's control, then, from a robustness viewpoint, I should generally be as small as possible. Surprisingly, robustness properties do not always improve with increasing group size: small groups are preferred when there is covariance heterogeneity and $V^{(s)}$ is used. However, any robustness advantages for smaller groups must be offset against loss of power, as the power increases with group size.

Finally, Olson demonstrated that the multivariate measures of skewness and kurtosis (Section 2.7) are not very informative, and both are sensitive to kurtotic nonnormality and covariance heterogeneity. They may indicate that there are departures from the underlying assumptions, but not their nature, so that we cannot tell when the departures are damaging.

9.2.4 Robust Tests for Equal Means

a PERMUTATION TEST

Since the $V^{(s)}$ test scored so well with regard to robustness, it is appropriate to consider a test due to Mardia [1971] based on the permutation distribution of $V^* = V^{(s)}/s$, where (see Table 9.2 in Section 9.2.1)

$$V^{(s)} = \mathrm{tr}\left[\mathbf{H}(\mathbf{E}+\mathbf{H})^{-1}\right]$$

$$= \mathrm{tr}\left[\sum_i n_i(\bar{\mathbf{y}}_{i\cdot} - \bar{\mathbf{y}}_{\cdot\cdot})(\bar{\mathbf{y}}_{i\cdot} - \bar{\mathbf{y}}_{\cdot\cdot})'\mathbf{E}_H^{-1}\right]$$

$$= \sum_i n_i(\bar{\mathbf{y}}_{i\cdot} - \bar{\mathbf{y}}_{\cdot\cdot})'\mathbf{E}_H^{-1}(\bar{\mathbf{y}}_{i\cdot} - \bar{\mathbf{y}}_{\cdot\cdot}), \tag{9.8}$$

and $s = \text{minimum } (I-1, d)$. Setting $n = \sum_i n_i$, $\bar{\boldsymbol{\mu}} = \sum_i n_i \boldsymbol{\mu}_i / n$ $(= \boldsymbol{\beta}_1$, say), and

$$\boldsymbol{\mu}_i = \bar{\boldsymbol{\mu}} + (\boldsymbol{\mu}_i - \boldsymbol{\mu}_I) - \sum_{i=1}^{I} \frac{n_i(\boldsymbol{\mu}_i - \boldsymbol{\mu}_I)}{n}$$

$$= \bar{\boldsymbol{\mu}} + \boldsymbol{\phi}_i - \sum_{i=1}^{I-1} \frac{n_i \boldsymbol{\phi}_i}{n} \qquad (\boldsymbol{\phi}_I = \mathbf{0}),$$

our model

$$\mathbf{y}_{ij} = \boldsymbol{\mu}_i + \mathbf{u}_{ij}$$

can be written in the form

$$\mathbf{Y} = \mathbf{1}_n \boldsymbol{\beta}_1' + \mathbf{X}_2 \mathbf{B}_2 + \mathbf{u}, \tag{9.9}$$

where $\mathbf{B}_2 = (\boldsymbol{\phi}_1, \boldsymbol{\phi}_2, \ldots, \boldsymbol{\phi}_{I-1})'$. Since H_0 is now $\boldsymbol{\phi}_1 = \boldsymbol{\phi}_2 = \cdots = \boldsymbol{\phi}_{I-1} = \mathbf{0}$, or $\mathbf{B}_2 = \mathbf{O}$, we can now apply the general theory of Section 8.6.2e to the one-way MANOVA (see Exercise 9.4). Reparameterizing the model does not change $\mathbf{E}$ and $\mathbf{H}$, so that

$$\mathbf{H} = \sum_i n_i \bar{\mathbf{y}}_{i\cdot}\bar{\mathbf{y}}_{i\cdot}' - n\bar{\mathbf{y}}_{\cdot\cdot}\bar{\mathbf{y}}_{\cdot\cdot}' \qquad (= \mathbf{Y}'\mathbf{M}\mathbf{Y}, \text{say}),$$

and the coefficient of $\mathbf{y}_{ij}\mathbf{y}_{ij}'$ is $(n_i^{-1} - n^{-1})$. Thus, from (8.56) and (8.57),

$$\sum_r m_{rr}^2 = \sum_i \sum_j (n_i^{-1} - n^{-1})^2 = \sum_i n_i^{-1} - (2I-1)n^{-1}$$

and

$$C_X = \frac{n-1}{(I-1)(n-3)(n-I)}\left[n(n+1)\sum_r m_{rr}^2 - (n-1)(I^2-1)\right]$$

$$= \frac{n-1}{(I-1)(n-3)(n-I)}\left[n(n+1)\left(\sum_i n_i^{-1} - I^2 n^{-1}\right)\right.$$

$$\left. -2(I-1)(n-I)\right].$$

The permutation distribution of V^* is approximately beta with degrees of freedom aP and aQ, respectively, where

$$a = \frac{[s(n-1)-2c]}{[s(n-1)(c+1)]},$$

$$c = \tfrac{1}{2}(n-3)\frac{C_X C_Y}{[n(n-1)]}$$

$$P = \tfrac{1}{2}d(I-1), \qquad Q = \tfrac{1}{2}[s(n-1)-d(I-1)]$$

$$C_Y = \frac{(n-1)(n+1)}{d(n-3)(n-d-1)}\left[b_{2,d} - \frac{(n-1)d(d+2)}{n+1}\right],$$

$$b_{2,d} = \frac{1}{n}\sum_{i=1}^{I}\sum_{j=1}^{n_i}\left\{(\mathbf{y}_{ij}-\bar{\mathbf{y}}_{\cdot\cdot})'(\mathbf{E}_H/n)^{-1}(\mathbf{y}_{ij}-\bar{\mathbf{y}}_{\cdot\cdot})\right\}^2,$$

and C_X is given above. To carry out the test we can use the statistic $F^* = QV^*/P(1-V^*)$, which has an approximate F-distribution with $2aP$ and $2aQ$ degrees of freedom, respectively, under the null hypothesis that the $\mathbf{y}_{ij}$ all come from the same distribution. Noninteger degrees of freedom can be handled using, for example, the tables of Mardia and Zemroch [1975b] (see also Pearson and Hartley [1972: Table 4]). We reject the null hypothesis if F^* is significantly large.

The above test applies in two situations: Either the randomization occurs at the design stage when a completely randomized design is used to allocate the I treatments to n sampling units, or the randomization occurs after the experiment when the n observations are randomly assigned to the I populations. In either case, the permutation distribution associated with the $\mathbf{y}_{ij}$ is *symmetric.* When the n_i are all equal, we note that the m_{rr} are all equal and the test is robust to kurtosis (see Section 8.6.2e). When the n_i are unequal, we would expect the test to be sensitive to differences in the $\boldsymbol{\Sigma}_i$. An approximate test that allows for unequal $\boldsymbol{\Sigma}_i$ is described below.

EXAMPLE 9.3 Mardia [1971] demonstrated his test on the anteater data given in Table 9.3. Here $I = 3$, $d = 3$, $s = \min(I-1, d) = 2$, $n_1 = 6$, $n_2 = 4$, $n_3 = 3$, and $n = 13$. For Example 9.1 we have

$$\bar{\mathbf{y}}_{..} = \begin{pmatrix} 2.092 \\ 2.096 \\ 1.627 \end{pmatrix}, \qquad \mathbf{E}_H = \begin{pmatrix} 0.007148 & 0.006865 & 0.008365 \\ - & 0.006835 & 0.008199 \\ - & - & 0.012152 \end{pmatrix},$$

$$b_{1,3} = 2.0440, \qquad b_{2,3} = 11.0134,$$

$$C_x = \frac{12}{2(10)(10)}\left[13(14)\left(\tfrac{1}{6} + \tfrac{1}{4} + \tfrac{1}{3} - \tfrac{9}{13}\right) - 40\right] = -1.7700,$$

$$C_y = \frac{12(14)}{3(10)(9)}\left[11.0134 - \frac{12(3)(5)}{14}\right] = -1.1472,$$

$$c = \tfrac{1}{2}(10)(1.7700)(1.1472)/(13)(12) = 0.065082,$$

$$P = 3, \qquad Q = 9, \qquad a = 0.9338.$$

Now $V^* = \frac{1}{2}V^{(2)} = 0.2235$ and $F^* = 3V^*/(1 - V^*) = 0.8635$, which is clearly not significant. We note that $F^* \sim F_{5.6,16.8}$, approximately, when the null hypothesis is true.

b JAMES' TEST

James [1954] has given an approximate test for $H_0: \boldsymbol{\mu}_1 = \boldsymbol{\mu}_2 = \cdots = \boldsymbol{\mu}_I$ when the $\boldsymbol{\Sigma}_i$ may be unequal. Let $\mathbf{z}_i = \bar{\mathbf{y}}_{i\cdot} - \bar{\mathbf{y}}_{I\cdot}$ $(i = 1, 2, \ldots, I-1)$; then, for $h \neq i$,

$$\mathscr{C}[\mathbf{z}_h, \mathbf{z}_i] = \mathscr{D}[\bar{\mathbf{y}}_{I\cdot}] = \frac{\boldsymbol{\Sigma}_I}{n_I} = \mathbf{A}_I,$$

say, and

$$\mathscr{D}[\mathbf{z}_i] = \mathbf{A}_i + \mathbf{A}_I.$$

Writing $\mathbf{z}' = (\mathbf{z}_1', \mathbf{z}_2', \ldots, \mathbf{z}_{I-1}')$, we then have

$$\mathscr{D}[\mathbf{z}] = \begin{pmatrix} \mathbf{A}_1 + \mathbf{A}_I, & \mathbf{A}_I, & \ldots, & \mathbf{A}_I \\ \mathbf{A}_I, & \mathbf{A}_2 + \mathbf{A}_I, & \ldots, & \mathbf{A}_I \\ \vdots & \vdots & \vdots & \vdots \\ \mathbf{A}_I, & \mathbf{A}_I, & \ldots, & \mathbf{A}_{I-1} + \mathbf{A}_I \end{pmatrix} = \mathbf{V},$$

say. We assume normality, so that when H_0 is true, $\mathbf{z} \sim N_{d(I-1)}(\mathbf{0}, \mathbf{V})$ and $\mathbf{z}'\mathbf{V}^{-1}\mathbf{z} \sim \chi^2_{d(I-1)}$. If $\mathbf{V}$ is now estimated by $\hat{\mathbf{V}}$, where $\hat{\mathbf{V}}$ is obtained from $\mathbf{V}$ by replacing $\boldsymbol{\Sigma}_i$ by $\mathbf{S}_i$, then we have a test statistic that is approximately distributed as chi square. Here

$$\hat{\mathbf{V}}^{-1} = \begin{pmatrix} \mathbf{W}_1 - \mathbf{W}_1\mathbf{W}^{-1}\mathbf{W}_1, & -\mathbf{W}_1\mathbf{W}^{-1}\mathbf{W}_2, & \ldots, & -\mathbf{W}_1\mathbf{W}^{-1}\mathbf{W}_{I-1} \\ -\mathbf{W}_2\mathbf{W}^{-1}\mathbf{W}_1, & \mathbf{W}_2 - \mathbf{W}_2\mathbf{W}^{-1}\mathbf{W}_2, & \ldots, & -\mathbf{W}_2\mathbf{W}^{-1}\mathbf{W}_{I-1} \\ \vdots & \vdots & \vdots & \vdots \\ -\mathbf{W}_{I-1}\mathbf{W}^{-1}\mathbf{W}_1, & -\mathbf{W}_{I-1}\mathbf{W}^{-1}\mathbf{W}_2, & \ldots, & \mathbf{W}_{I-1} - \mathbf{W}_{I-1}\mathbf{W}^{-1}\mathbf{W}_{I-1} \end{pmatrix},$$

where $\mathbf{W}_i = \hat{\mathbf{A}}_i^{-1} = (\mathbf{S}_i/n_i)^{-1}$ and $\mathbf{W} = \Sigma_{i=1}^{I}\mathbf{W}_i$. Multiplying out, we have

$$\begin{aligned} \mathbf{z}'\hat{\mathbf{V}}^{-1}\mathbf{z} &= \sum_{h=1}^{I}\sum_{i=1}^{I}(\bar{\mathbf{y}}_{h\cdot} - \bar{\mathbf{y}}_{I\cdot})'(\delta_{hi}\mathbf{W}_h - \mathbf{W}_h\mathbf{W}^{-1}\mathbf{W}_i)(\bar{\mathbf{y}}_{i\cdot} - \bar{\mathbf{y}}_{I\cdot}) \\ &= \sum_{i=1}^{I}\bar{\mathbf{y}}_{i\cdot}'\mathbf{W}_i\bar{\mathbf{y}}_{i\cdot} - \sum_{h=1}^{I}\sum_{i=1}^{I}\bar{\mathbf{y}}_{h\cdot}'\mathbf{W}_h\mathbf{W}^{-1}\mathbf{W}_i\bar{\mathbf{y}}_{i\cdot} \qquad (9.10) \\ &= \sum_{i=1}^{I}(\bar{\mathbf{y}}_{i\cdot} - \bar{\mathbf{y}}_w)'\mathbf{W}_i(\bar{\mathbf{y}}_{i\cdot} - \bar{\mathbf{y}}_w) \qquad \text{(see Exercise 9.5)}, \qquad (9.11) \end{aligned}$$

where

$$\bar{\mathbf{y}}_w = \mathbf{W}^{-1}\sum_{i=1}^{I}\mathbf{W}_i\bar{\mathbf{y}}_{i\cdot}. \qquad (9.12)$$

James [1954] showed that the upper α quantile of the test statistic $\mathbf{z}'\hat{\mathbf{V}}^{-1}\mathbf{z}$ is, to order n_i^{-1}, $\chi_r^2(\alpha)[A + B\chi_r^2(\alpha)]$, where $r = d(I-1)$,

$$A = 1 + \frac{1}{2r}\sum_{i=1}^{I}\frac{\left(\operatorname{tr}[\mathbf{I}_d - \mathbf{W}^{-1}\mathbf{W}_i]\right)^2}{n_i - 1},$$

$$B = \frac{1}{r(r+2)}\left\{\sum_{i=1}^{I}\frac{\operatorname{tr}\left[(\mathbf{I}_d - \mathbf{W}^{-1}\mathbf{W}_i)^2\right]}{n_i - 1} + \frac{1}{2}\sum_{i=1}^{I}\frac{\left(\operatorname{tr}[\mathbf{I}_d - \mathbf{W}^{-1}\mathbf{W}_i]\right)^2}{n_i - 1}\right\},$$

and $\chi_r^2(\alpha)$ is the upper α quantile value of χ_r^2. For very large n_i, $A \approx 1$, $B \approx 0$, and our test statistic is approximately distributed as χ_r^2 when H_0 is true. The same result is also true even when the underlying distribution is not normal, as the central limit theorem applies to the sample means $\bar{\mathbf{y}}_{i\cdot}$, and hence to $\mathbf{z}$. We can therefore expect some measure of robustness against nonnormality for moderate sample sizes.

EXAMPLE 9.4 James [1954] considered an example in which there were samples of sizes 16, 11, and 11, respectively from three bivariate normal populations. The sample means are

$$\bar{\mathbf{y}}_{1\cdot} = \begin{pmatrix} 9.82 \\ 15.06 \end{pmatrix}, \quad \bar{\mathbf{y}}_{2\cdot} = \begin{pmatrix} 13.05 \\ 22.57 \end{pmatrix}, \quad \bar{\mathbf{y}}_{3\cdot} = \begin{pmatrix} 14.67 \\ 25.17 \end{pmatrix};$$

the sample dispersion matrices $\mathbf{S}_i$ $(i = 1, 2, 3)$ are

$$\begin{pmatrix} 120.0 & -16.3 \\ -16.3 & 17.8 \end{pmatrix}, \quad \begin{pmatrix} 81.8 & 32.1 \\ 32.1 & 53.8 \end{pmatrix}, \quad \begin{pmatrix} 100.3 & 23.2 \\ 23.2 & 97.1 \end{pmatrix};$$

and the $\mathbf{W}_i$ $(i = 1, 2, 3)$ are

$$\begin{pmatrix} 0.1523 & 0.1396 \\ 0.1396 & 1.0272 \end{pmatrix}, \quad \begin{pmatrix} 0.1756 & -0.1048 \\ -0.1048 & 0.2670 \end{pmatrix}, \quad \begin{pmatrix} 0.1161 & -0.0277 \\ -0.0277 & 0.1199 \end{pmatrix}.$$

Now $\mathbf{W} = \Sigma_i \mathbf{W}_i$,

$$\bar{\mathbf{y}}_w = \mathbf{W}^{-1} \sum_i \mathbf{W}_i \bar{\mathbf{y}}_{i\cdot} = \begin{pmatrix} 9.9314 \\ 16.9998 \end{pmatrix},$$

and substituting in (9.11) gives 18.75 for the value of our test statistic. Furthermore, $r = d(I - 1) = 4$,

$$A = 1 + \tfrac{1}{8}(0.5311) = 1.0664,$$

and

$$B = \tfrac{1}{24}\left[0.2820 + \tfrac{1}{2}(0.5311)\right] = 0.02281.$$

Using a significance level of $\alpha = 0.01$, $\chi_4^2(0.01) = 13.277$ and

$$\chi_4^2(0.01)\left[A + B\chi_4^2(0.01)\right] = 18.18,$$

so that our value of 18.75 appears to be significant at the 1% level. However, James cautions that "such results should not of course be interpreted too literally, particularly when working far out in the tail of the distribution."

9.2.5 *Reparameterization*

In Section 9.2.4a we considered a reparameterization of the one-way model, namely,

$$\boldsymbol{\mu}_i = \bar{\boldsymbol{\mu}} + \boldsymbol{\phi}_i - \sum_{i=1}^{I-1} n_i \boldsymbol{\phi}_i / n,$$

where $\boldsymbol{\phi}_i = \boldsymbol{\mu}_i - \boldsymbol{\mu}_I$ $(i = 1, 2, \ldots, I)$ and $\bar{\boldsymbol{\mu}} = \Sigma_i n_i \boldsymbol{\mu}_i / n$. This led to the model

$$\mathbf{Y} = \mathbf{X}\mathbf{B} + \mathbf{U}$$

$$= (\mathbf{1}_n, \mathbf{X}_2)\begin{pmatrix} \boldsymbol{\beta}_1' \\ \mathbf{B}_2 \end{pmatrix} + \mathbf{U},$$

where $\boldsymbol{\beta}_1 = \bar{\boldsymbol{\mu}}$ and $\mathbf{B}_2 = (\boldsymbol{\phi}_1, \boldsymbol{\phi}_2, \ldots, \boldsymbol{\phi}_{I-1})'$. With this particular formulation, $\mathbf{X}_2'\mathbf{1}_n = \mathbf{0}$ and $\mathbf{X}$ has full rank. However, other reparameterizations are possible; for example, if $\bar{\boldsymbol{\mu}} = \Sigma_i \boldsymbol{\mu}_i / I$, we have

$$\boldsymbol{\mu}_i = \bar{\boldsymbol{\mu}} + (\boldsymbol{\mu}_i - \bar{\boldsymbol{\mu}}) = \boldsymbol{\mu} + \boldsymbol{\alpha}_i,$$

say, with identifiability restriction $\Sigma_i \boldsymbol{\alpha}_i = \mathbf{0}$. In this case $\mathbf{B}_2 = (\boldsymbol{\alpha}_1, \boldsymbol{\alpha}_2, \ldots, \boldsymbol{\alpha}_I)'$, and $\mathbf{X}$ no longer has full rank.

Another full rank formulation is to use $\boldsymbol{\mu} + \boldsymbol{\alpha}_i$ with the constraint $\boldsymbol{\alpha}_I = \mathbf{0}$ and $\mathbf{B}_2 = (\boldsymbol{\alpha}_1, \boldsymbol{\alpha}_2, \ldots, \boldsymbol{\alpha}_{I-1})'$. In this case $\boldsymbol{\mu} = \boldsymbol{\mu}_I$ and $\boldsymbol{\alpha}_i = \boldsymbol{\phi}_i$. This type of model, in which $\mathbf{X}$ has full rank and can be fitted in parts (in this case the column $\mathbf{1}_n$ first, corresponding to H_0, followed by $\mathbf{X}_2$), can be used for more complex designs. Multivariate computing procedures using this type of model are given in Section 10.1.4a. The corresponding univariate model forms the basis of the statistical package GLIM (Baker and Nelder [1978]).

9.2.6 *Comparing Dispersion Matrices*

a TEST FOR EQUAL DISPERSION MATRICES

As the assumption of equal dispersion matrices is so critical with regard to tests for the means, it is natural to consider a likelihood ratio test for H_0: $\boldsymbol{\Sigma}_1 = \boldsymbol{\Sigma}_2 = \cdots = \boldsymbol{\Sigma}_I$, given that the $\mathbf{y}_{ij}$ are i.i.d. $N_d(\boldsymbol{\mu}_i, \boldsymbol{\Sigma}_i)$. Arguing as in Section 3.6.1a for the case $I = 2$, we obtain a similar equation to (3.78) for the likelihood ratio, namely,

$$\ell = \frac{|\hat{\boldsymbol{\Sigma}}|^{-n/2}}{|\hat{\boldsymbol{\Sigma}}_1|^{-n_1/2}|\hat{\boldsymbol{\Sigma}}_2|^{-n_2/2} \cdots |\hat{\boldsymbol{\Sigma}}_I|^{-n_I/2}}$$

$$= \left(\prod_{i=1}^{I} |\mathbf{Q}_i|^{n_i/2} n^{dn/2}\right) \Big/ \left|\sum_{i=1}^{I} \mathbf{Q}_i\right|^{n/2} \prod_{i=1}^{I} n_i^{dn_i/2}, \tag{9.13}$$

where

$$n_i\hat{\boldsymbol{\Sigma}}_i = \mathbf{Q}_i = \sum_{j=1}^{n_i} (\mathbf{y}_{ij} - \bar{\mathbf{y}}_{i\cdot})(\mathbf{y}_{ij} - \bar{\mathbf{y}}_{i\cdot})', \tag{9.14}$$

and $n = \Sigma_i n_i$. When H_0 is true and n is large, $-2\log \ell$ is approximately distributed as $\chi^2_{\nu_1}$, where $\nu_1 = \frac{1}{2}d(d+1)(I-1)$ (the difference in the number of parameters to be estimated under the unrestricted and restricted models, respectively). However, a better chi-square approximation can be obtained by using a slight modification of ℓ. This modification, in which f_i $(= n_i - 1)$, the degrees of freedom associated with $\mathbf{Q}_i$, replaces n_i, and which leads to an unbiased test (Perlman [1980]), is

$$M = \left(\prod_{i=1}^{I} |\mathbf{S}_i|^{f_i/2}\right) \Big/ |\mathbf{S}|^{f/2}$$

$$= \left(\prod_{i=1}^{I} |\mathbf{Q}_i|^{f_i/2} f^{df/2}\right) \Big/ \left|\sum_{i=1}^{I} \mathbf{Q}_i\right|^{f/2} \prod_{i=1}^{I} f_i^{df_i/2}, \tag{9.15}$$

where $f = \Sigma_i f_i = n - I$, $\mathbf{S}_i = \mathbf{Q}_i/f_i$, and $\mathbf{S} = \Sigma_i \mathbf{Q}_i/f = \mathbf{E}/f$. For the case $d = 2$ and $I = 2$ (Section 3.6.1c), Layard [1974] showed that the M-test is so sensitive to nonnormality that its usefulness is suspect. Ito [1969: p. 117] demonstrated the nonrobustness of a statistic Y that is asymptotically equal to $n(1 - \ell^{2/n})$, and Mardia [1974: pp. 119–120] showed that under the null hypothesis of a common underlying distribution for the I populations,

$$\mathrm{E}[Y] = \tfrac{1}{2}(I-1)d(d+1)\left\{1 + \frac{\beta_{2,d} - d(d+2)}{d(d+1)}\right\}.$$

Here $\beta_{2,d}$ is the multivariate kurtosis of the underlying distribution and is equal to $d(d+2)$ under normality [see (2.95)].

Olson [1974] has also studied the robustness of M for equal n_i and demonstrated that M is very sensitive to kurtosis. A significant M could equally be due to H_0 being false or to kurtosis, or to both. Unfortunately, M is overly sensitive to departures from H_0 in that it detects differences among the dispersion matrices that have little effect on tests for the means. For the same reason it is much too sensitive to kurtosis. All that can be said is that it gives a somewhat inflated measure of departures from normality and covariance homogeneity, and seems to be as useful as the multivariate measures of skewness and kurtosis, the latter having similar shortcomings.

For completeness we mention two approximations due to Box [1949] for the distribution of M (see Pearson [1969]), Krishnaiah and Lee [1980: pp. 526–528]).

(1) The chi-square approximation

$$-2(1 - c_1)\log M \qquad \text{is approximately } X^2_{\nu_1}, \tag{9.16}$$

where $\nu_1 = \frac{1}{2}d(d+1)(I-1)$ and

$$c_1 = \frac{2d^2+3d-1}{6(d+1)(I-1)}\left\{\sum_i f_i^{-1} - f^{-1}\right\}$$

$$= \frac{(2d^2+3d-1)(I+1)}{6(d+1)If_0} \qquad \text{if } f_i = f_0,\ i = 1,2,\ldots,I.$$

This is the multivariate analogue of Bartlett's test for homogeneity (see Seber [1977: p. 147]), and it's significance level is closer to the nominal significance level than either $-2\log \ell$ or $-2\log M$ (Greenstreet and Connor [1974]). If more accuracy is needed, further terms in the expression leading to the above approximation can be included (Anderson [1958: p. 255]); however, the following approximation, which is more accurate, can also be used.

(2) The F-approximation

$$-2b\log M \qquad \text{is approximately } F_{\nu_1,\nu_2}, \tag{9.17}$$

where $\nu_2 = (\nu_1+2)/|c_2 - c_1^2|$, $b = [1 - c_1 - (\nu_1/\nu_2)]/\nu_1$, and

$$c_2 = \frac{(d-1)(d+2)}{6(I-1)}\left\{\sum_i f_i^{-2} - f^{-2}\right\}$$

$$= \frac{(d-1)(d+2)(I^2+I+1)}{6I^2 f_0^2} \qquad \text{if } f_i = f_0,\ i = 1,2,\ldots,I.$$

The above holds for $c_2 - c_1^2 > 0$. If $c_2 - c_1^2 < 0$, then

$$-\frac{2b_1\nu_2\log M}{\nu_1 + 2b_1\nu_1\log M} \qquad \text{is approximately } F_{\nu_1,\nu_2},$$

where

$$b_1 = \frac{1 - c_1 - (2/\nu_2)}{\nu_2}.$$

We note that ν_1 is always an integer but it may be necessary to interpolate for ν_2, particularly when ν_1 is large. The tables of Mardia and Zemroch [1975b], with fractional degrees of freedom, may be useful.

How accurate are the above approximations? When the f_i are unequal, little information is available, though Box has made a few comparisons. His tentative conclusion was that the F-approximation could be satisfactory if $d \le 5$, $I \le 5$, and $f_i \ge 9$ $(i = 1,2,\ldots,I)$. For small values of f_i, further terms are needed in Box's series expansion. An accurate Pearson-type 1 approximation, along with some tables, is also available for $-2\log M$ (J. C. Lee et al. [1977]).

The most common situation, which usually occurs in planned experiments, is that of equal f_i. In this case exact upper 5% significant points for the distribution of $-2\log M$ are available from J. C. Lee et al. [1977: $d = 2(1)5$]; these are reproduced in Appendix D18. Pearson [1969] compared the above two approximations with a similar table by Korin [1969] and concluded that the chi-square approximation seems adequate for most practical purposes, and the F-approximation is remarkably accurate. It appears, from Box's comparisons, that the errors in approximating the 1% points are, proportionately, only slightly greater than those for the 5% points. Thus if the f_i are equal and a 5% point is required, use Appendix D18; otherwise use the F-approximation.

We note that asymptotic expansions of the nonnull distribution of $-2\log M$ under various alternatives were given by Sugiura [1969] and Nagao [1970] (see also Muirhead [1982]). Also, Nagao [1973a] gave another test statistic based on the asymptotic variance of $-2\log M$. In conclusion, we reference a promising test (Hawkins [1981]) for simultaneously testing multivariate normality and equal dispersion matrices.

b GRAPHICAL COMPARISONS

Under the assumptions of normality and equal dispersion matrices, the $\mathbf{Q}_i$ of (9.14) are independently distributed as $W_d(n_i - 1, \boldsymbol{\Sigma})$. For the case of equal n_i ($= n_0$, say), Gnanadesikan and Lee [1970] gave two methods for comparing the $\mathbf{Q}_i$ graphically, based on summary measures of the eigenvalues of $\mathbf{Q}_i$. If $c_1 \geq c_2 \geq \cdots \geq c_t > 0$ are the positive eigenvalues of the $d \times d$ matrix $\mathbf{Q}_i$, they propose using the sum and geometric mean, namely,

$$a_i = \sum_{j=1}^{t} c_j = \operatorname{tr} \mathbf{Q}_i \tag{9.18}$$

and

$$g_i = \left(\prod_{j=1}^{t} c_j \right)^{1/t}. \tag{9.19}$$

Usually $n_0 - 1 \geq d$, so that $t = d$ and $\mathbf{Q}_i$ is positive definite. However, if $n_0 - 1 < d$, then $t = n_0 - 1$ and $\mathbf{Q}_i$ is positive semidefinite. Now the trace of a Wishart matrix is distributed as a linear combination of independent chi-square variables [see Exercise 2.12(c)], which can be approximated by a scaled chi square, that is, a gamma variable (see Appendix C2). We can therefore plot the ordered values $a_{(1)} < a_{(2)} < \cdots < a_{(I)}$ against quantiles of the gamma distribution (see Appendix C4) and, if the underlying assumptions are satisfied, the plot will be approximately linear. Gnanadesikan and Lee [1970] suggested that a similar plot could be made for the ordered g_i.

Another graphical technique, based on the transformation $\mathbf{z} = \boldsymbol{\phi}(\mathbf{s})$ [see (3.67)] of the r [$= \frac{1}{2}d(d+1)$] distinct elements of the sample dispersion matrix, was given by Campbell [1981]. A matrix, $\mathbf{Z} = [(z_{hi})]$, say, is formed whose ith column is the value of $\mathbf{z}$ for the ith sample covariance matrix. Each of these column vectors can be regarded as a "profile" (see Section 3.6.4), and similar dispersion matrices will lead to similar profiles. Campbell suggested treating $\mathbf{Z}$ like a data matrix from a two-way analysis of variance model with one observation per cell, namely,

$$E[z_{hi}] = \mu + \alpha_h + \beta_i + \gamma_{hi},$$

with the usual constraints $\Sigma_h \alpha_h = 0$, and so on. If the dispersion matrices are the same, then we would expect, approximately,

$$E[z_{hi}] = \mu + \alpha_h = E[\bar{z}_{h\cdot}].$$

Campbell suggested plotting the pairs $(z_{hi}, \bar{z}_{h\cdot})$ $(h = 1, 2, \dots, r)$ for each i and seeing if the graphs approximate to straight lines of unit slope through the origin. He calls the plots I-A or individual-average plots and suggests treating separately the two parts of $\mathbf{z}$, corresponding to the sample variances and sample correlations, respectively. He also presented a number of test procedures and other plotting methods.

9.2.7 *Exact Procedures for Means Assuming Unequal Dispersion Matrices*

a HYPOTHESIS TESTING

Suppose that $\mathbf{y}_{ij} \sim N_d(\boldsymbol{\mu}_i, \boldsymbol{\Sigma}_i)$ and we wish to test $H_0: \boldsymbol{\mu}_1 = \boldsymbol{\mu}_2 = \cdots = \boldsymbol{\mu}_I$ without assuming equality of the $\boldsymbol{\Sigma}_i$. Scheffé [1943] proposed a solution for the univariate case ($d = 1$) and $I = 2$, and this was generalized to the multivariate case by Bennett [1951] (see Section 3.6.3) and Anderson [1963a: $I > 2$]. A more general method is given by Eaton [1969], which includes all previous work as special cases, and we follow his approach.

Let $n_0 = \text{minimum}(n_1, n_2, \dots, n_I)$ and suppose, for each i, that we randomly select n_0 of the n_i observations. For ease of exposition we assume that they are the first n_0, namely, $\mathbf{y}_{ij}$ $(j = 1, 2, \dots, n_0)$. If $\bar{\mathbf{y}}_{i\cdot}$ is the mean of the n_i observations, let

$$\mathbf{z} = \begin{pmatrix} \bar{\mathbf{y}}_{1\cdot} \\ \bar{\mathbf{y}}_{2\cdot} \\ \vdots \\ \bar{\mathbf{y}}_{I\cdot} \end{pmatrix} \quad \text{and} \quad \boldsymbol{\theta} = \begin{pmatrix} \boldsymbol{\mu}_1 \\ \boldsymbol{\mu}_2 \\ \vdots \\ \boldsymbol{\mu}_I \end{pmatrix}.$$

Then $\mathbf{z} \sim N_{dI}(\boldsymbol{\theta}, \boldsymbol{\Sigma})$, $\boldsymbol{\Sigma} = \text{diag}(\boldsymbol{\Sigma}_1/n_1, \boldsymbol{\Sigma}_2/n_2, \ldots, \boldsymbol{\Sigma}_I/n_I)$. For $j = 1, 2, \ldots, n_0$ define

$$\mathbf{v}_j = \begin{pmatrix} n_1^{-1/2}\mathbf{y}_{1j} \\ n_2^{-1/2}\mathbf{y}_{2j} \\ \vdots \\ n_I^{-1/2}\mathbf{y}_{Ij} \end{pmatrix},$$

and let

$$\bar{\mathbf{v}} = \sum_{j=1}^{n_0} \frac{\mathbf{v}_j}{n_0} \quad \text{and} \quad \mathbf{Q}_0 = \sum_{j=1}^{n_0} (\mathbf{v}_j - \bar{\mathbf{v}})(\mathbf{v}_j - \bar{\mathbf{v}})'.$$

Then $\mathbf{Q}_0 \sim W_{dI}(n_0 - 1, \boldsymbol{\Sigma})$ and, since $\mathbf{Q}_0$ is statistically independent of $\bar{\mathbf{v}}$ (Theorem 3.1 of Section 3.2.2) and of the observations left after removing n_0 from each group, $\mathbf{Q}_0$ is independent of $\mathbf{z}$. To test H_0, we note that H_0 is true if and only if $\boldsymbol{\mu}_1 - \boldsymbol{\mu}_2 = \cdots = \boldsymbol{\mu}_{I-1} - \boldsymbol{\mu}_I = \mathbf{0}$ or $\mathbf{C}\boldsymbol{\theta} = \mathbf{0}$, where $\mathbf{C}$ is the $r \times dI$ matrix $[r = d(I-1)]$

$$\mathbf{C} = \begin{pmatrix} \mathbf{I}_d, & -\mathbf{I}_d, & \mathbf{O}, & \ldots, & \mathbf{O} \\ \mathbf{O}, & \mathbf{I}_d, & -\mathbf{I}_d, & \ldots, & \mathbf{O} \\ \vdots & \vdots & \vdots & \vdots & \vdots \\ \mathbf{O}, & \mathbf{O}, & \mathbf{O}, & \ldots, & -\mathbf{I}_d \end{pmatrix}. \tag{9.20}$$

Ignoring the information that we have about the structure of $\boldsymbol{\Sigma}$, we can now construct a Hotelling's T^2 test using

$$T_0^2 = (n_0 - 1)(\mathbf{Cz})'[\mathbf{CQ}_0\mathbf{C}']^{-1}\mathbf{Cz}.$$

When H_0 is true, $T_0^2 \sim T^2_{r, n_0 - 1}$. The above procedure holds for any matrix $\mathbf{C}$ of rank r, provided that $r \leq \text{minimum}(dI, n_0 - 1)$. Eaton [1969] also allows for the possibility that d may vary with i. However, the above test has the following disadvantages: (1) when the n_i are unequal, some of the data are discarded in calculating $\mathbf{Q}_0$; (2) the test statistic involves estimates of known elements of $\boldsymbol{\Sigma}$, the off-diagonal blocks of zeros, so that the test will be inefficient; and (3) randomization is required in selecting n_0 observations from each group. In the light of the comments made in Section 3.6.3 on the case $I = 2$, it would appear, therefore, that this test should not be recommended and that the approximate test of James [see (9.11)] is preferred. However, the above theory has been included, as the method used is informative.

b MULTIPLE CONFIDENCE INTERVALS

Using a useful general inequality, Dalal [1978] proved that

$$\text{pr}\left[\sum_{i=1}^{I} a_i \boldsymbol{\mu}_i' \mathbf{b}_i \in \sum_{i=1}^{I} a_i \bar{\mathbf{y}}_{i\cdot}' \mathbf{b}_i \pm T_\alpha \sum_{i=1}^{I} |a_i| \left(\mathbf{b}_i' \mathbf{S}_i \mathbf{b}_i / n_i\right)^{1/2} \quad \text{for all } a_i, \mathbf{b}_i\right] = 1 - \alpha,$$

where T_α is the solution of

$$\prod_{i=1}^{I} \left\{ G_{d, n_i - d}\left(\frac{d[n_i - 1]}{n_i - d} T_\alpha^2 \right) \right\} = 1 - \alpha,$$

and $G_{d, n_i - d}(\cdot)$ denotes the distribution function of the $F_{d, n_i - d}$ distribution. Unfortunately, these intervals will be even wider than those based on equal dispersion matrices [given by (9.5)] and for this reason may not be too useful, particularly if $n_i - d$ is not large.

9.3 RANDOMIZED BLOCK DESIGN

9.3.1 *Hypothesis Testing and Confidence Intervals*

If y_{ij} is the observation corresponding to the ith treatment in the jth block of a randomized block design, then the linear model for this design is

$$y_{ij} = \mu + \alpha_i + \beta_j + u_{ij} \qquad (i = 1, 2, \ldots, I; j = 1, 2, \ldots, J),$$

where $\Sigma_i \alpha_i = \Sigma_j \beta_j = 0$, and the u_{ij} are i.i.d. $N_1(0, \sigma^2)$. For this model we have the analysis of variable table, Table 9.4, for testing $H_{01}: \alpha_i = 0$ (all i) and $H_{02}: \beta_j = 0$ (all j). Replacing each y_{ij} by a vector observation $\mathbf{y}_{ij}$, the corresponding multivariate model is

$$\mathbf{y}_{ij} = \boldsymbol{\mu} + \boldsymbol{\alpha}_i + \boldsymbol{\beta}_j + \mathbf{u}_{ij}, \tag{9.21}$$

TABLE 9.4 Univariate Randomized Block Design

Source	Sum of Squares (SS)	Degrees of Freedom (df)	$\frac{\text{SS}}{\text{df}}$
Treatments	$J\Sigma(\bar{y}_{i\cdot} - \bar{y}_{\cdot\cdot})^2$	$I - 1$	s_1^2
Blocks	$I\Sigma(\bar{y}_{\cdot j} - \bar{y}_{\cdot\cdot})^2$	$J - 1$	s_2^2
Error	$\Sigma\Sigma(y_{ij} - \bar{y}_{i\cdot} - \bar{y}_{\cdot j} + \bar{y}_{\cdot\cdot})^2$	$(I - 1)(J - 1)$	s^2
Corrected total	$\Sigma\Sigma(y_{ij} - \bar{y}_{\cdot\cdot})^2$	$IJ - 1$	

TABLE 9.5 Multivariate Randomized Block Design

Source	Matrix	df
Treatments	$\mathbf{H}_1 = J\Sigma(\bar{\mathbf{y}}_{i\cdot} - \bar{\mathbf{y}}_{\cdot\cdot})(\bar{\mathbf{y}}_{i\cdot} - \bar{\mathbf{y}}_{\cdot\cdot})'$	$m_1 = I - 1$
Blocks	$\mathbf{H}_2 = I\Sigma(\bar{\mathbf{y}}_{\cdot j} - \bar{\mathbf{y}}_{\cdot\cdot})(\bar{\mathbf{y}}_{\cdot j} - \bar{\mathbf{y}}_{\cdot\cdot})'$	$m_2 = J - 1$
Error	$\mathbf{E} = \Sigma\Sigma(\mathbf{y}_{ij} - \bar{\mathbf{y}}_{i\cdot} - \bar{\mathbf{y}}_{\cdot j} + \bar{\mathbf{y}}_{\cdot\cdot})(\mathbf{y}_{ij} - \bar{\mathbf{y}}_{i\cdot} - \bar{\mathbf{y}}_{\cdot j} + \bar{\mathbf{y}}_{\cdot\cdot})'$	$m_E = (I-1)(J-1)$

where $\Sigma_i \boldsymbol{\alpha}_i = \Sigma_j \boldsymbol{\beta}_j = \mathbf{0}$, and the $\mathbf{u}_{ij}$ are i.i.d. $N_d(\mathbf{0}, \boldsymbol{\Sigma})$. We can readily write down the matrices corresponding to the quadratic forms in Table 9.4; these are given in Table 9.5. Corresponding to the univariate F-ratios s_i^2/s^2 $(i = 1, 2)$, we use one of the various test statistics based on the roots of $|\mathbf{H}_i - \theta(\mathbf{E} + \mathbf{H}_i)| = 0$. For example, the likelihood ratio test statistic for H_{0i} is

$$\Lambda_i = \frac{|\mathbf{E}|}{|\mathbf{E} + \mathbf{H}_i|},$$

where $\Lambda_i \sim U_{d, m_i, m_E}$ when H_{0i} is true.

The procedures for multiple comparisons are essentially the same as for the one-way classification, as (9.21) leads to

$$\begin{aligned}\bar{\mathbf{y}}_{i\cdot} &= \boldsymbol{\mu} + \boldsymbol{\alpha}_i + \bar{\mathbf{u}}_{i\cdot} \\ &= \bar{\boldsymbol{\mu}}_{i\cdot} + \bar{\mathbf{u}}_{i\cdot},\end{aligned}$$

where $\boldsymbol{\mu}_{ij} = \mathscr{E}[\mathbf{y}_{ij}]$ and testing $H_{01}: \boldsymbol{\alpha}_i = \mathbf{0}$ (all i) is equivalent to testing $\bar{\boldsymbol{\mu}}_{1\cdot} = \bar{\boldsymbol{\mu}}_{2\cdot} = \cdots = \bar{\boldsymbol{\mu}}_{I\cdot}$. Since $\Sigma_i c_i \boldsymbol{\alpha}_i = \Sigma_i c_i \bar{\boldsymbol{\mu}}_{i\cdot}$, it follows [see (9.5)] that, with probability $1 - \alpha$,

$$\sum_i c_i \boldsymbol{\alpha}_i' \mathbf{b} \in \sum_i c_i \bar{\mathbf{y}}_{i\cdot}' \mathbf{b} \pm \left\{ \frac{\theta_\alpha}{1 - \theta_\alpha} \frac{\sum_i c_i^2}{J} \mathbf{b}'\mathbf{E}\mathbf{b} \right\}^{1/2} \tag{9.22}$$

simultaneously for all contrasts and all $\mathbf{b}$. Corresponding to (9.6), we have the subset

$$(\boldsymbol{\alpha}_r - \boldsymbol{\alpha}_s)'\mathbf{b} \in (\bar{\mathbf{y}}_{r\cdot} - \bar{\mathbf{y}}_{s\cdot})'\mathbf{b} \pm \left\{ \frac{\theta_\alpha}{1 - \theta_\alpha} \frac{2}{J} \mathbf{b}'\mathbf{E}\mathbf{b} \right\}^{1/2} \tag{9.23}$$

If we use a similar argument, the probability is $1 - \alpha$ that

$$\sum_j c_j \boldsymbol{\beta}_j' \mathbf{b} \in \sum_j c_j \bar{\mathbf{y}}_{\cdot j}' \mathbf{b} \pm \left\{ \phi_\alpha^* \left(\sum_j \frac{c_j^2}{I} \right) \mathbf{b}'\mathbf{E}\mathbf{b} \right\}^{1/2} \tag{9.24}$$

TABLE 9.6 Observations on Bean Plants Infested with Leaf Miners.[a]

Treatment	Blocks 1			2			3			4		
	y_1	y_2	y_3	y_1	y_2	y_3	y_1	y_2	y_3	y_1	y_2	y_3
1	1.7	0.4	0.20	1.2	1.4	0.20	1.3	0.6	0.36	1.7	1.1	0.39
2	1.7	1.0	0.40	1.2	0.6	0.25	1.7	0.1	0.32	1.1	0.0	0.29[b]
3	1.4	0.8	0.28	1.5	0.8	0.83	1.1	0.7	0.58	1.1	0.9	0.50
4	0.1	0.8	0.10	0.2	1.2	0.08	0.3	1.2	0.00	0.0	0.4	0.00
5	1.3	1.0	0.12	1.4	1.2	0.20	1.3	0.8	0.30	1.2	0.6	0.36
6	1.7	0.5	0.74	2.1	1.0	0.59	2.3	0.4	0.50	1.3	0.9	0.28

[a] Data courtesy of Dr. R. Fullerton.
[b] Estimate of missing value.

simultaneously for all contrasts and all **b**. We note that $\phi_\alpha = \theta_\alpha/(1 - \theta_\alpha)$ and ϕ_α^* are quantile values for the maximum eigenvalues of $\mathbf{H}_1\mathbf{E}^{-1}$ and $\mathbf{H}_2\mathbf{E}^{-1}$, respectively. Bonferroni confidence intervals are also available for r prespecified comparisons: We simply replace $\mathbf{E}$ by $\mathbf{E}/\nu$ and $\phi_\alpha^{1/2}$ by $t_\nu^{\alpha/(2r)}$, with $\nu = (I-1)(J-1)$, in (9.22).

EXAMPLE 9.5 In Table 9.6 we have the results of a randomized block experiment in the Cook Islands involving the effects of six different treatments on plots of bean plants infested by the serpentine leaf miner insect. Three variables were measured on each plot, namely,

y_1 = the number of miners per leaf,

y_2 = the weight of beans per plot (in kilograms),

$y_3 = \sin^{-1}(\sqrt{p})$, where p is the proportion of leaves infested with borer.

In Table 9.7 we have the matrices $\mathbf{H}_1$, $\mathbf{H}_2$, and $\mathbf{E}$ with their elements given to three decimal places only for convenience. Here $d = 3$, $I = 6$, and $J = 4$.

The likelihood ratio statistic for H_{01}, the hypothesis of no treatment effects, is

$$\Lambda_1 = \frac{|\mathbf{E}|}{|\mathbf{E} + \mathbf{H}_1|} = 0.05038 \sim U_{3,5,15}.$$

Now $f_1 = m_E + m_1 - \frac{1}{2}(m_1 + d + 1) = 15.5$, $M = m_E - d + 1 = 13$, and $-f_1 \log_e \Lambda_1 = 46.32 \sim C_\alpha \chi_{15}^2$. Entering Appendix D13 with $d = 3$, $m_H = 5$, $M = 13$, and $\alpha = 0.005$, we obtain

$$C_\alpha \chi_{15}^2(\alpha) = 1.015(32.801) = 33.29.$$

We therefore reject H_{01} at the 0.5% level of significance.

TABLE 9.7 MANOVA Table for Bean Infestation Data Showing Hypothesis and Error Matrices

Source	Matrix Elements						df
	(1,1)	(1,2)	(1,3)	(2,2)	(2,3)	(3,3)	
Treatments ($\mathbf{H}_1$)	6.617	−0.858	1.684	0.678	−0.219	0.707	$m_1 = 5$
Blocks ($\mathbf{H}_2$)	0.271	0.082	0.029	0.617	0.056	0.013	$m_2 = 3$
Error ($\mathbf{E}$)	1.031	−0.007	0.156	1.578	−0.135	0.338	$m_E = 15$

For demonstration purposes we note that $\phi_{\max} = 6.69842$ and

$$\theta_{\max} = \frac{\phi_{\max}}{1 + \phi_{\max}} = 0.87.$$

Entering Appendix D14 with $s = 3$, $\nu_1 = \frac{1}{2}(|d - m_1| - 1) = \frac{1}{2}$, and $\nu_2 = \frac{1}{2}(m_E - d - 1) = 5.5$, we find that $\theta_{0.01} < 0.80$ (from $\nu_1 = 1$, $\nu_2 = 5$). Since $\theta_{\max} > \theta_{0.01}$, we reject H_{01} at the 1% level of significance.

The likelihood ratio statistic for H_{02}, the hypothesis of no block effects, is

$$\Lambda_2 = \frac{|\mathbf{E}|}{|\mathbf{E} + \mathbf{H}_2|} = 0.53795 \sim U_{3,3,15},$$

so that

$$-f_2 \log \Lambda_2 = 8.99 < \chi_9^2(0.10) = 14.68$$

and we do not reject H_{02}. We note that C_α is not required when the test statistic is not significant, as $C_\alpha > 1$.

Since H_{01} is rejected, we now focus our attention on how the different treatments affected the number of leaf miners (y_1). Simultaneous intervals for all treatment pairs are given by (9.23) with $\mathbf{b}' = (1, 0, 0)$. Here $J = 4$, $\mathbf{b}'\mathbf{E}\mathbf{b} = e_{11} = 1.03125$, and, interpolating linearly in Appendix D14 with $s = 3$, $\nu_1 = \frac{1}{2}$, and $\nu_2 = 5.5$, $\theta_{0.05} \approx 0.67$. Hence

$$\delta = \left\{ \frac{\theta_\alpha}{1 - \theta_\alpha} \frac{2}{J} \mathbf{b}'\mathbf{E}\mathbf{b} \right\}^{1/2} = 1.02.$$

The six ranked treatment means for the first variable, the first elements of the $\bar{\mathbf{y}}_{i\cdot}$, are

T_4	T_3	T_5	T_2	T_1	T_6
0.15	1.28	1.30	1.43	1.48	1.85

Here $T_3 - T_4$ exceeds 1.02, so that T_4 is clearly different from the other treatments with respect to y_1.

If we now carry out the same procedure for variable y_2, the weight of beans per plot, the only change is to set $\mathbf{b}' = (0, 1, 0)$ and $\mathbf{b}'\mathbf{E}\mathbf{b} = e_{22} = 1.57833$. We now have $\delta = 1.27$, and the ranked means are

T_2	T_6	T_3	T_1	T_4	T_5
0.43	0.70	0.80	0.88	0.90	0.90

All the intervals $T_r - T_s \pm \delta$ contain zero, so that y_2 does not appear to be a contributing factor to the rejection of H_{01}.

Finally, for y_3, $\mathbf{b}' = (0, 0, 1)$, $\mathbf{b}'\mathbf{E}\mathbf{b} = e_{33} = 0.33781$, and $\delta = 0.59$. The ranked means are

T_4	T_5	T_1	T_2	T_6	T_3
0.05	0.25	0.29	0.32	0.53	0.55

Although all the intervals $T_r - T_s \pm \delta$ contain zero, T_4 is much lower than the rest. We therefore conclude that treatment four is more effective than the other treatments in causing a reduction in both the leaf miner and borer without affecting bean yield.

9.3.2 *Underlying Assumptions*

Since treatments are assigned at random within blocks, the associated permutation distribution can be used to study the effects of nonnormality on the test statistics. Arnold [1964] has done this for the case of $I = 2$, when all four test statistics of Section 8.6.2 are equivalent to a Hotelling's T^2 (i.e., $m_H = 1$ in Section 2.5.5e). By simulating bivariate data from the rectangular, normal, and double exponential distributions, he demonstrated that for $d = 2$, the significance level of T^2 for H_{01} is not likely to be biased by more than 1 or 2% at the 5% level of significance when there is nonnormality. He also gives a correction term, based on estimating certain cumulants, that leads to a type I error closer to its nominal level α when there is nonnormality.

9.4 TWO-WAY CLASSIFICATION WITH EQUAL OBSERVATIONS PER MEAN

Suppose we have a (univariate) two-factor experiment, and let y_{ijk} be the kth observation on the combination of the ith level of factor A and the jth level of factor B. Then the usual univariate model is (see Seber [1977: Section 9.2])

$$\begin{aligned} y_{ijk} &= \mu_{ij} + u_{ijk} \qquad (i = 1, 2, \ldots, I;\, j = 1, 2, \ldots, J;\, k = 1, 2, \ldots, K) \\ &= \mu + \alpha_i + \beta_j + \gamma_{ij} + u_{ijk}, \end{aligned}$$

TABLE 9.8 Univariate Two-Way Analysis of Variance Table

Source	Sum of Squares (SS)	Degrees of Freedom (df)	$\frac{\text{SS}}{\text{df}}$
Factor A	$JK\Sigma(\bar{y}_{i\cdot\cdot} - \bar{y}_{\cdot\cdot\cdot})^2$	$I - 1$	s_1^2
Factor B	$IK\Sigma(\bar{y}_{\cdot j\cdot} - \bar{y}_{\cdot\cdot\cdot})^2$	$J - 1$	s_2^2
$A \times B$	$K\Sigma\Sigma(\bar{y}_{ij\cdot} - \bar{y}_{i\cdot\cdot} - \bar{y}_{\cdot j\cdot} + \bar{y}_{\cdot\cdot\cdot})^2$	$(I-1)(J-1)$	s_3^2
Error	$\Sigma\Sigma\Sigma(y_{ijk} - \bar{y}_{ij\cdot})^2$	$IJ(K-1)$	s^2
Corrected total	$\Sigma\Sigma\Sigma(y_{ijk} - \bar{y}_{\cdot\cdot\cdot})^2$	$IJK - 1$	

where $\Sigma_i \alpha_i = \Sigma_j \beta_j = \Sigma_i \gamma_{ij} = \Sigma_j \gamma_{ij} = 0$, and the u_{ijk} are i.i.d. $N_1(0, \sigma^2)$. Here α_i and β_j are main effects, and γ_{ij} is the interaction between the ith level of A and the jth level of B. For the above model we have the ANOVA table, Table 9.8, for testing $H_{03}: \gamma_{ij} = 0$ (all i, j), $H_{01}: \alpha_i = 0$ (all i), and $H_{02}: \beta_j = 0$ (all j). Replacing y_{ijk} by $\mathbf{y}_{ijk}$, we have the multivariate model

$$\mathbf{y}_{ijk} = \boldsymbol{\mu}_{ij} + \mathbf{u}_{ijk}$$

$$= \boldsymbol{\mu} + \boldsymbol{\alpha}_i + \boldsymbol{\beta}_j + \boldsymbol{\gamma}_{ij} + \mathbf{u}_{ijk},$$

where $\Sigma_i \boldsymbol{\alpha}_i = \Sigma_j \boldsymbol{\beta}_j = \Sigma_i \boldsymbol{\gamma}_{ij} = \Sigma_j \boldsymbol{\gamma}_{ij} = \mathbf{0}$, and the $\mathbf{u}_{ijk}$ are i.i.d. $N_d(\mathbf{0}, \boldsymbol{\Sigma})$. The corresponding MANOVA table is given by Table 9.9 and we test H_{0i} $(i = 1, 2, 3)$ using, for example, $\Lambda_i = |\mathbf{E}|/|\mathbf{E} + \mathbf{H}_i|$, where $\Lambda_i \sim U_{d, m_i, m_E}$ when H_{0i} is true. We test the interaction hypothesis H_{03} first, as H_{01} and H_{02} are difficult to interpret if H_{03} is false.

The above procedure can be expected to be reasonably robust to small departures from the assumption of equal dispersion matrices, as we have equal numbers of observations per group. However, if $\mathbf{u}_{ijk} \sim N_d(\mathbf{0}, \boldsymbol{\Sigma}_{ij})$, then an alternative test for H_{01} (or H_{02}) is available from Garcia Ben and Yohai [1980]. Unfortunately, these authors do not give a test of H_{03}, though, in a randomized block design with one of the factors representing the block effects, we can

TABLE 9.9 Multivariate Two-Way Analysis of Variance Table

Source	Matrix	df
A	$\mathbf{H}_1 = JK\Sigma(\bar{\mathbf{y}}_{i\cdot\cdot} - \bar{\mathbf{y}}_{\cdot\cdot\cdot})(\bar{\mathbf{y}}_{i\cdot\cdot} - \bar{\mathbf{y}}_{\cdot\cdot\cdot})'$	$m_1 = I - 1$
B	$\mathbf{H}_2 = IK\Sigma(\bar{\mathbf{y}}_{\cdot j\cdot} - \bar{\mathbf{y}}_{\cdot\cdot\cdot})(\bar{\mathbf{y}}_{\cdot j\cdot} - \bar{\mathbf{y}}_{\cdot\cdot\cdot})'$	$m_2 = J - 1$
$A \times B$	$\mathbf{H}_3 = K\Sigma\Sigma(\bar{\mathbf{y}}_{ij\cdot} - \bar{\mathbf{y}}_{i\cdot\cdot} - \bar{\mathbf{y}}_{\cdot j\cdot} + \bar{\mathbf{y}}_{\cdot\cdot\cdot})$ $\times(\bar{\mathbf{y}}_{ij\cdot} - \bar{\mathbf{y}}_{i\cdot\cdot} - \bar{\mathbf{y}}_{\cdot j\cdot} + \bar{\mathbf{y}}_{\cdot\cdot\cdot})'$	$m_3 = (I-1)(J-1)$
Error	$\mathbf{E} = \Sigma\Sigma\Sigma(\mathbf{y}_{ijk} - \bar{\mathbf{y}}_{ij\cdot})(\mathbf{y}_{ijk} - \bar{\mathbf{y}}_{ij\cdot})'$	$m_E = IJ(K-1)$

expect the interactions to be small. We could also test H_{0i} using the general procedure of Eaton [1969] described in Section 9.2.7 by writing

$$\boldsymbol{\theta} = (\boldsymbol{\mu}'_{11}, \boldsymbol{\mu}'_{12}, \ldots, \boldsymbol{\mu}'_{1J}, \boldsymbol{\mu}'_{21}, \ldots, \boldsymbol{\mu}'_{2J}, \ldots, \boldsymbol{\mu}'_{I1}, \ldots, \boldsymbol{\mu}'_{IJ})'$$

and choosing an appropriate matrix like (9.20). However, his test for H_{02} is not as powerful as the one given by Garcia Ben and Yohai [1980].

Since $\boldsymbol{\alpha}_i = \bar{\boldsymbol{\mu}}_{i\cdot} - \bar{\boldsymbol{\mu}}_{\cdot\cdot}$, we can express H_{01} in the form $\bar{\boldsymbol{\mu}}_{1\cdot} = \bar{\boldsymbol{\mu}}_{2\cdot} = \cdots = \bar{\boldsymbol{\mu}}_{I\cdot}$, or $\bar{\boldsymbol{\mu}}_{1\cdot} - \bar{\boldsymbol{\mu}}_{I\cdot} = \bar{\boldsymbol{\mu}}_{2\cdot} - \bar{\boldsymbol{\mu}}_{I\cdot} = \cdots = \bar{\boldsymbol{\mu}}_{I-1\cdot} - \bar{\boldsymbol{\mu}}_{I\cdot}$. This implies $\mathbf{AB} = \mathbf{O}$, where $\mathbf{B} = (\boldsymbol{\mu}_{11}, \boldsymbol{\mu}_{12}, \ldots, \boldsymbol{\mu}_{IJ})'$ and

$$\mathbf{AB} = \begin{pmatrix} \bar{\boldsymbol{\mu}}_{1\cdot} - \bar{\boldsymbol{\mu}}_{I\cdot} \\ \bar{\boldsymbol{\mu}}_{2\cdot} - \bar{\boldsymbol{\mu}}_{I\cdot} \\ \vdots \\ \bar{\boldsymbol{\mu}}_{I-1\cdot} - \bar{\boldsymbol{\mu}}_{I\cdot} \end{pmatrix}.$$

Arguing as in Section 9.2.2, we can write

$$\mathbf{a}'\mathbf{AB} = \sum_{i=1}^{I} c_i \bar{\boldsymbol{\mu}}'_{i\cdot} \qquad \left(\sum_{i=1}^{I} c_i = 0\right)$$

and

$$\mathbf{a}'\mathbf{A}\hat{\mathbf{B}} = \sum_{i=1}^{I} c_i \bar{\mathbf{y}}'_{i\cdot\cdot},$$

(by Exercise 9.6). Therefore the probability is $1 - \alpha$ that for all contrasts and all $\mathbf{b}$,

$$\sum_{i=1}^{I} c_i \bar{\boldsymbol{\mu}}'_{i\cdot}\mathbf{b} \in \sum_{i=1}^{I} c_i \bar{\mathbf{y}}'_{i\cdot\cdot}\mathbf{b} \pm \left\{ \phi_\alpha \left(\sum_i \frac{c_i^2}{JK} \right) \mathbf{b}'\mathbf{E}\mathbf{b} \right\}^{1/2} \tag{9.25}$$

or, equivalently,

$$\sum_{i=1}^{I} c_i \boldsymbol{\alpha}'_i \mathbf{b} \in \sum_{i=1}^{I} c_i \hat{\boldsymbol{\alpha}}'_i \mathbf{b} \pm \left\{ \phi_\alpha \left(\sum_i \frac{c_i^2}{JK} \right) \mathbf{b}'\mathbf{E}\mathbf{b} \right\}^{1/2},$$

where $\hat{\boldsymbol{\alpha}}_i = \bar{\mathbf{y}}_{i\cdot\cdot} - \bar{\mathbf{y}}_{\cdot\cdot\cdot}$. In a similar fashion we can construct simultaneous confidence intervals for $\sum_j c_j \boldsymbol{\beta}'_j \mathbf{b}$ (see Exercise 9.6).

EXAMPLE 9.6 A laboratory experiment was set up to investigate the effect on growth of inoculating paspalum grass with a fungal infection applied at four different temperatures (14, 18, 22, 26°C). For each pot of paspalum, measurements were made on three variables:

$$y_1 = \text{the fresh weight of roots (gm)},$$

$$y_2 = \text{the maximum root length (mm)},$$

$$y_3 = \text{the fresh weight of tops (gm)}.$$

The inoculated group was compared with a control group and six three-dimensional observations were made on each treatment–temperature combination: These are given in Table 9.10. All observations were transformed by taking logarithms, and the matrices for testing the hypotheses H_{01} : no treatment main effects, H_{02} : no temperature main effects, and H_{03} : no treatment × temperature interactions are given in Table 9.11 (rounded to three decimal places).

TABLE 9.10 Observations on Paspalum Plants[a]

	Control (Treat. 1)			Inoculated (Treat. 2)		
	y_1	y_2	y_3	y_1	y_2	y_3
	2.2	23.5	1.7	2.3	23.5	2.0
	3.0	27.0	2.3	3.0	21.0	2.7
	3.3	24.5	3.2	2.3	22.0	1.8
Temp. 1	2.2	20.5	1.5	2.5	22.5	2.4
	2.0	19.0	2.0	2.4	21.5	1.1
	3.5	23.5	2.9	2.7	25.0	2.6
	21.8	41.5	23.0	10.1	43.5	14.2
	11.0	32.5	15.4	7.6	27.0	14.7
	16.4	46.5	22.8	19.7	32.5	21.4
Temp. 2	13.1	31.0	21.5	4.3	28.5	9.7
	15.4	41.5	20.8	5.2	33.5	12.2
	14.5	46.0	20.3	3.9	24.5	8.2
	13.6	29.5	30.8	10.0	21.0	23.6
	6.2	23.5	14.6	12.3	49.0	28.1
	16.7	58.5	36.0	4.9	28.5	13.3
Temp. 3	12.2	40.5	23.9	9.6	27.0	24.6
	8.7	37.0	20.3	6.5	29.0	19.3
	12.3	41.5	27.7	13.6	30.5	31.5
	3.0	24.0	10.2	4.2	25.5	13.3
	5.3	26.5	15.6	2.2	23.5	8.5
	3.1	24.5	14.7	2.8	19.5	11.8
Temp. 4	4.8	34.0	20.5	1.3	21.5	7.8
	3.4	22.5	14.3	4.2	28.5	15.1
	7.4	32.0	23.2	3.0	25.0	11.8

[a]Data courtesy of Peter Buchanan.

TABLE 9.11 MANOVA Table for Paspalum Data Showing Hypothesis and Error Matrices (from Log Observations in Table 9.10)

	Matrix elements						
Source	(1,1)	(1,2)	(1,3)	(2,2)	(2,3)	(3,3)	df
Treatments	0.294	0.121	0.203	0.049	0.083	0.140	$m_1 = 1$
Temperatures	3.514	1.014	4.209	0.296	1.243	7.876	$m_2 = 3$
Treat. × Temp.	0.156	0.037	0.097	0.017	0.017	0.066	$m_3 = 3$
Error	1.035	0.291	0.693	0.294	0.217	0.646	$m_E = 40$

The likelihood ratio test statistic for H_{01} is

$$\Lambda_1 = \frac{|\mathbf{E}|}{|\mathbf{E} + \mathbf{H}_1|} = 0.76499 \sim U_{3,1,40}.$$

Since $m_1 = 1$,

$$F_0 = \frac{1 - \Lambda_1}{\Lambda_1} \frac{m_E + 1 - d}{d} \sim F_{d,\, m_E + 1 - d},$$

i.e., $F_0 = 3.89 \sim F_{3,38}$, which is not quite significant at the 1% level as $F_{3,38}^{0.01} = 4.3$. The corresponding test statistic for H_{02} is

$$\Lambda_2 = \frac{|\mathbf{E}|}{|\mathbf{E} + \mathbf{H}_2|} = 0.01257 \sim U_{3,3,40}.$$

Here $f_2 = m_E + m_2 - \frac{1}{2}(m_2 + d + 1) = 39.5$, $M = m_E - d + 1 = 38$, and $-f_2 \log \Lambda_2 = 172.9 \sim C_\alpha \chi_9^2$. Entering Appendix D13 with $d = 3$, $m_H = 3$, $M = 38$, and $\alpha = 0.005$ we obtain

$$C_\alpha \chi_9^2(\alpha) = 1.001(23.59) = 23.6 < 172.9.$$

We therefore reject H_{02} at the 0.5% level of significance.

The test statistic for H_{03} is

$$\Lambda_3 = \frac{|\mathbf{E}|}{|\mathbf{E} + \mathbf{H}_3|} = 0.80170 \sim U_{3,3,40},$$

and, as $f_3 = f_2$, $-f_3 \log \Lambda_3 = 8.73$, which is below $\chi_9^2(0.1) = 14.7$, that is, Λ_3 is not significant. In practice, we would look at H_{03} first, since the presence of interactions makes the interpretation of H_{01} and H_{02} difficult.

In conclusion, it is appropriate to plot the sample mean of the transformed data for each variable against temperature, and to compare the temperature

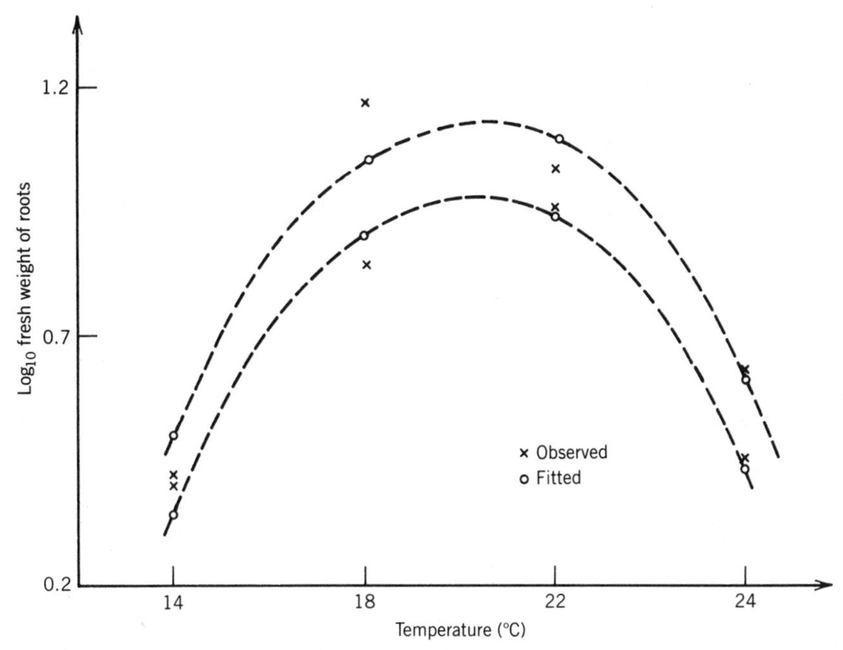

Fig. 9.1 Plot of $\log_{10}$(fresh weight of roots) versus temperature for the control and inoculated groups. Fitted quadratics are drawn in for each group.

curves of the control group and inoculated group. Such a plot for the sample mean of $\log y_1$ is given in Fig. 9.1. We find that quadratic curves fit the data well and the two curves are approximately parallel, as might be expected from negligible interactions. The curves for the other variables y_2 and y_3 are similar.

9.5 ANALYSIS OF COVARIANCE

9.5.1 *Univariate Theory*

It is assumed that the reader is familiar with the usual univariate analysis of covariance model

$$\begin{aligned} G: \mathbf{y} &= \mathbf{X}\boldsymbol{\beta} + \mathbf{Z}\boldsymbol{\gamma} + \mathbf{u} \\ &= (\mathbf{X}, \mathbf{Z})\begin{pmatrix} \boldsymbol{\beta} \\ \boldsymbol{\gamma} \end{pmatrix} + \mathbf{u} \\ &= \mathbf{W}\boldsymbol{\phi} + \mathbf{u}, \end{aligned} \tag{9.26}$$

where $\mathbf{X}$ is $n \times p$ of rank p, $\mathbf{Z}$ is $n \times t$ of rank t consisting of values taken by the so-called concomitant variables, the columns of $\mathbf{Z}$ are linearly independent of the columns of $\mathbf{X}$, and $\mathbf{u} \sim N_n(\mathbf{0}, \sigma^2\mathbf{I}_n)$. The general theory is given, for example, in Seber [1977: Sections 3.7, 3.8.3, and Chapter 10], and some of the basic results are summarized below.

The least squares estimates for model G are

$$\hat{\boldsymbol{\gamma}}_G = (\mathbf{Z}'\mathbf{R}\mathbf{Z})^{-1}\mathbf{Z}'\mathbf{R}\mathbf{y} \quad \text{and} \quad \hat{\boldsymbol{\beta}}_G = (\mathbf{X}'\mathbf{X})^{-1}\mathbf{X}'(\mathbf{y} - \mathbf{Z}\hat{\boldsymbol{\gamma}}),$$

where $\mathbf{R} = \mathbf{I}_n - \mathbf{X}(\mathbf{X}'\mathbf{X})^{-1}\mathbf{X}'$. The residual sum of squares is

$$\begin{aligned} \text{RSS}_G &= (\mathbf{y} - \mathbf{X}\hat{\boldsymbol{\beta}}_G - \mathbf{Z}\hat{\boldsymbol{\gamma}}_G)'(\mathbf{y} - \mathbf{X}\hat{\boldsymbol{\beta}}_G - \mathbf{Z}\hat{\boldsymbol{\gamma}}_G) \\ &= (\mathbf{y} - \mathbf{Z}\hat{\boldsymbol{\gamma}}_G)'\mathbf{R}(\mathbf{y} - \mathbf{Z}\hat{\boldsymbol{\gamma}}_G) \\ &= \mathbf{y}'\mathbf{R}\mathbf{y} - \hat{\boldsymbol{\gamma}}_G'\mathbf{Z}'\mathbf{R}\mathbf{y}. \end{aligned} \tag{9.27}$$

If we wish to test $H_{0\gamma}: \boldsymbol{\gamma} = \mathbf{0}$, that is, the concomitant variables have no effect, we require $\text{RSS} = \mathbf{y}'\mathbf{R}\mathbf{y}$, the residual sum of squares for the model $\mathscr{E}[\mathbf{y}] = \mathbf{X}\boldsymbol{\beta}$, and

$$\text{RSS} - \text{RSS}_G = \hat{\boldsymbol{\gamma}}_G'\mathbf{Z}'\mathbf{R}\mathbf{y}. \tag{9.28}$$

The appropriate F-statistic is then

$$F_\gamma = \frac{(\text{RSS} - \text{RSS}_G)/t}{\text{RSS}_G/(n-p-t)} = \frac{\hat{\boldsymbol{\gamma}}_G'\mathbf{Z}'\mathbf{R}\mathbf{y}/t}{s^2},$$

say, where $F_\gamma \sim F_{t,n-p-t}$ when $H_{0\gamma}$ is true. To test $H_0: \mathbf{A}\boldsymbol{\beta} = \mathbf{0}$, where $\mathbf{A}$ is $q \times p$ of rank q, we can use the general linear hypothesis theory and write

$$\mathbf{A}\boldsymbol{\beta} = (\mathbf{A}, \mathbf{O})\begin{pmatrix} \boldsymbol{\beta} \\ \boldsymbol{\gamma} \end{pmatrix} = \mathbf{A}_1\boldsymbol{\phi},$$

say. If $\hat{\boldsymbol{\phi}}_G = (\mathbf{W}'\mathbf{W})^{-1}\mathbf{W}'\mathbf{y}$ is the least squares estimate of $\boldsymbol{\phi}$, then it follows from A3.1 that

$$\begin{aligned} \mathscr{D}\begin{bmatrix} \hat{\boldsymbol{\beta}}_G \\ \hat{\boldsymbol{\gamma}}_G \end{bmatrix} &= \mathscr{D}[\hat{\boldsymbol{\phi}}_G] \\ &= \sigma^2(\mathbf{W}'\mathbf{W})^{-1} \\ &= \sigma^2\begin{pmatrix} (\mathbf{X}'\mathbf{X})^{-1} + \mathbf{L}\mathbf{M}\mathbf{L}', & -\mathbf{L}\mathbf{M} \\ -\mathbf{M}\mathbf{L}', & \mathbf{M} \end{pmatrix}, \end{aligned}$$

where $\mathbf{L} = (\mathbf{X}'\mathbf{X})^{-1}\mathbf{X}'\mathbf{Z}$ and $\mathbf{M} = (\mathbf{Z}'\mathbf{R}\mathbf{Z})^{-1}$. Here the F-statistic for testing H_0 is

$$\begin{aligned} F &= (\mathbf{A}_1\hat{\boldsymbol{\phi}}_G)'\left[\mathbf{A}_1(\mathbf{W}'\mathbf{W})^{-1}\mathbf{A}_1'\right]^{-1}\mathbf{A}_1\hat{\boldsymbol{\phi}}_G/qs^2 \\ &= (\mathbf{A}\hat{\boldsymbol{\beta}}_G)'\left[\mathbf{A}(\mathbf{X}'\mathbf{X})^{-1}\mathbf{A}' + \mathbf{A}\mathbf{L}\mathbf{M}\mathbf{L}'\mathbf{A}'\right]^{-1}\mathbf{A}\hat{\boldsymbol{\beta}}_G/qs^2, \end{aligned}$$

where $F \sim F_{q,n-p-t}$ when H_0 is true. If $\mathbf{X}$ has less than full rank, we simply replace p by the rank of $\mathbf{X}$, and $(\mathbf{X}'\mathbf{X})^{-1}$ by a generalized inverse $(\mathbf{X}'\mathbf{X})^{-}$. In concluding this summary, we note that (see Exercise 9.7)

$$\mathscr{C}[\hat{\boldsymbol{\beta}}, \hat{\boldsymbol{\gamma}}_G] = \mathbf{O}. \tag{9.29}$$

9.5.2 Multivariate Theory

The above theory readily generalizes to deal with the multivariate analysis of covariance model [see (9.26)]

$$\begin{aligned} G: \mathbf{Y} &= \mathbf{XB} + \mathbf{Z\Gamma} + \mathbf{U} \\ &= \mathbf{W\Phi} + \mathbf{U}, \end{aligned}$$

where the rows of $\mathbf{U}$ are i.i.d. $N_d(\mathbf{0}, \boldsymbol{\Sigma})$. From the general theory of linear models developed in Chapter 8 we simply replace $\mathbf{y}$ by $\mathbf{Y}$. To test $H_{0\Gamma}: \boldsymbol{\Gamma} = \mathbf{O}$ we use the matrices [see (9.28) and (9.27)]

$$\mathbf{H} = \hat{\boldsymbol{\Gamma}}_G'\mathbf{Z}'\mathbf{RY} \quad \text{and} \quad \mathbf{E} = \mathbf{Y}'\mathbf{RY} - \hat{\boldsymbol{\Gamma}}_G'\mathbf{Z}'\mathbf{RY},$$

where $\hat{\boldsymbol{\Gamma}}_G = (\mathbf{Z}'\mathbf{RZ})^{-1}\mathbf{Z}'\mathbf{RY}$. When $H_{0\Gamma}$ is true, $\mathbf{H}$ and $\mathbf{E}$ are independently distributed as $W_d(t, \boldsymbol{\Sigma})$ and $W_d(n-p-t, \boldsymbol{\Sigma})$, respectively.

To test $H_0: \mathbf{AB} = \mathbf{O}$, $\mathbf{E}$ remains the same and $\mathbf{H}$ now becomes

$$\mathbf{H} = (\mathbf{A}\hat{\mathbf{B}}_G)'\left[\mathbf{A}(\mathbf{X}'\mathbf{X})^{-1}\mathbf{A}' + \mathbf{ALML}'\mathbf{A}'\right]^{-1}\mathbf{A}\hat{\mathbf{B}}_G, \tag{9.30}$$

where

$$\hat{\mathbf{B}}_G = (\mathbf{X}'\mathbf{X})^{-1}\mathbf{X}'(\mathbf{Y} - \mathbf{Z}\hat{\boldsymbol{\Gamma}}_G) = \hat{\mathbf{B}} - (\mathbf{X}'\mathbf{X})^{-1}\mathbf{X}'\mathbf{Z}\hat{\boldsymbol{\Gamma}}_G.$$

When H_0 is true, $\mathbf{H} \sim W_d(q, \boldsymbol{\Sigma})$. Finally, to test $\mathbf{ABD} = \mathbf{O}$, where $\mathbf{D}$ is $d \times v$ of rank v, we simply replace $\mathbf{H}$, $\mathbf{E}$, and d by $\mathbf{D}'\mathbf{HD}$, $\mathbf{D}'\mathbf{ED}$, and v (see Section 8.7). Simultaneous confidence intervals for all linear combinations $\mathbf{a}'\mathbf{ABb}$ can be constructed using the general theory of Section 8.6.3 applied to $\mathbf{Y} = \mathbf{W\Phi} + \mathbf{U}$ and $\mathbf{A}_1\boldsymbol{\Phi} = (\mathbf{A}, \mathbf{O})\boldsymbol{\Phi} = \mathbf{AB}$. However a direct approach is considered in Exercise 9.15.

Univariate analysis of covariance models can be readily handled using a two-stage minimization technique that can be described as the two-step least squares procedure (Seber [1977]). The same procedure can be generalized to multivariate models and the relevant steps are as follows:

(1) Find $\hat{\mathbf{B}} = (\mathbf{X}'\mathbf{X})^{-1}\mathbf{X}'\mathbf{Y}$ and $\mathbf{Y}'\mathbf{RY}$ ($= \mathbf{E}_{YY}$, say) for the MANOVA model $\mathscr{E}[\mathbf{Y}] = \mathbf{XB}$.

(2)

$$\begin{aligned} \hat{\boldsymbol{\Gamma}}_G &= (\mathbf{Z}'\mathbf{RZ})^{-1}\mathbf{Z}'\mathbf{RY} \\ &= \mathbf{E}_{ZZ}^{-1}\mathbf{E}_{ZY}, \quad \text{say}. \end{aligned} \tag{9.31}$$

(3) We then have

$$\mathbf{E} = (\mathbf{Y} - \mathbf{Z}\hat{\boldsymbol{\Gamma}}_G)'\mathbf{R}(\mathbf{Y} - \mathbf{Z}\hat{\boldsymbol{\Gamma}}_G)$$

$$= \mathbf{Y}'\mathbf{R}\mathbf{Y} - \hat{\boldsymbol{\Gamma}}_G'\mathbf{Z}'\mathbf{R}\mathbf{Y}$$

$$= \mathbf{E}_{YY} - \mathbf{E}_{YZ}\mathbf{E}_{ZZ}^{-1}\mathbf{E}_{ZY}. \tag{9.32}$$

(4) Replace $\mathbf{Y}$ by $\mathbf{Y} - \mathbf{Z}\hat{\boldsymbol{\Gamma}}_G$ in $\hat{\mathbf{B}}$ to get

$$\hat{\mathbf{B}}_G = (\mathbf{X}'\mathbf{X})^{-1}\mathbf{X}'(\mathbf{Y} - \mathbf{Z}\hat{\boldsymbol{\Gamma}}_G). \tag{9.33}$$

(5) To obtain $\mathbf{E}_H$ for $H_0: \mathbf{AB} = \mathbf{O}$ we first obtain $\mathbf{Y}'\mathbf{R}_H\mathbf{Y}$ ($= \mathbf{E}_{HYY}$, say) for the model $\mathscr{E}[\mathbf{Y}] = \mathbf{XB}$ and $\mathbf{AB} = \mathbf{O}$ and then repeat steps (2) and (3). Thus, from (9.32),

$$\mathbf{E}_H = \mathbf{Y}'\mathbf{R}_H\mathbf{Y} - \mathbf{Y}'\mathbf{R}_H\mathbf{Z}(\mathbf{Z}'\mathbf{R}_H\mathbf{Z})^{-1}\mathbf{Z}'\mathbf{R}_H\mathbf{Y}$$

$$= \mathbf{E}_{HYY} - \mathbf{E}_{HYZ}\mathbf{E}_{HZZ}^{-1}\mathbf{E}_{HZY},$$

and Wilks' likelihood ratio statistic for H_0 is

$$\Lambda = \frac{|\mathbf{E}|}{|\mathbf{E}_H|} = \frac{|\mathbf{E}_{YY} - \mathbf{E}_{YZ}\mathbf{E}_{ZZ}^{-1}\mathbf{E}_{ZY}|}{|\mathbf{E}_{HYY} - \mathbf{E}_{HYZ}\mathbf{E}_{HZZ}^{-1}\mathbf{E}_{HZY}|}. \tag{9.34}$$

This is distributed as $U_{d,q,n-p-t}$ when H_0 is true, and it may be compared with the corresponding test statistic for the MANOVA model ($\boldsymbol{\Gamma} = \mathbf{O}$), namely,

$$\Lambda = \frac{|\mathbf{E}_{YY}|}{|\mathbf{E}_{HYY}|} = \frac{|\mathbf{E}_{YY}|}{|\mathbf{E}_{YY} + \mathbf{H}_{YY}|}.$$

If $\mathbf{H}_{ab} = \mathbf{E}_{Hab} - \mathbf{E}_{ab}$ $(a, b = Y, Z)$, then, for the maximum root test, we require

$$\mathbf{H} = \mathbf{E}_H - \mathbf{E}$$

$$= \mathbf{H}_{YY} - (\mathbf{E}_{YZ} + \mathbf{H}_{YZ})(\mathbf{E}_{ZZ} + \mathbf{H}_{ZZ})^{-1}(\mathbf{E}_{ZY} + \mathbf{H}_{ZY}) + \mathbf{E}_{YZ}\mathbf{E}_{ZZ}^{-1}\mathbf{E}_{ZY}.$$

EXAMPLE 9.7 Consider the one-way analysis of covariance model with one concomitant variable z, namely,

$$y_{ij} = \mu_i + \gamma z_{ij} + u_{ij} \qquad (i = 1,2,\ldots,I;\ j = 1,2,\ldots,J),$$

or, writing

$$\mathbf{y}' = (y_{11}, y_{12}, \ldots, y_{IJ}),\ \mathbf{z}' = (z_{11}, z_{12}, \ldots, z_{IJ}),$$

and $\boldsymbol{\beta} = (\mu_1, \mu_2, \ldots, \mu_I)'$, we have

$$\mathbf{y} = \mathbf{X}\boldsymbol{\beta} + \mathbf{z}\gamma + \mathbf{u}.$$

In the multivariate generalization we replace each y_{ij} by a row vector $\mathbf{y}_{ij}' = (y_{ij1}, y_{ij2}, \ldots, y_{ijd})$, where

$$y_{ijr} = \mu_{ir} + \gamma_r z_{ij} + u_{ijr} \qquad (r = 1, 2, \ldots, d).$$

Thus

$$\mathbf{y}_{ij}' = \boldsymbol{\mu}_i' + z_{ij}\boldsymbol{\gamma}' + \mathbf{u}_{ij}'$$

and

$$G: \mathbf{Y} = \mathbf{XB} + \mathbf{Z}\boldsymbol{\Gamma} + \mathbf{U},$$

where $\mathbf{B} = (\boldsymbol{\mu}_1, \boldsymbol{\mu}_2, \ldots, \boldsymbol{\mu}_I)'$, $\mathbf{Z} = \mathbf{z}$, and $\boldsymbol{\Gamma} = \boldsymbol{\gamma}'$.

Suppose that we wish to test $H_0: \boldsymbol{\mu}_1 = \boldsymbol{\mu}_2 = \cdots = \boldsymbol{\mu}_I$ for the model G. Then, from Table 9.2 in Section 9.2.1,

$$\mathbf{E}_{YY} = \sum_i \sum_j (\mathbf{y}_{ij} - \bar{\mathbf{y}}_{i\cdot})(\mathbf{y}_{ij} - \bar{\mathbf{y}}_{i\cdot})',$$

$$\mathbf{H}_{YY} = \sum_i \sum_j (\bar{\mathbf{y}}_{i\cdot} - \bar{\mathbf{y}}_{\cdot\cdot})(\bar{\mathbf{y}}_{i\cdot} - \bar{\mathbf{y}}_{\cdot\cdot})',$$

and

$$\mathbf{E}_{HYY} = \sum_i \sum_j (\mathbf{y}_{ij} - \bar{\mathbf{y}}_{\cdot\cdot})(\mathbf{y}_{ij} - \bar{\mathbf{y}}_{\cdot\cdot})'.$$

Hence

$$\mathbf{E}_{YZ} = \sum_i \sum_j (\mathbf{y}_{ij} - \bar{\mathbf{y}}_{i\cdot})(z_{ij} - \bar{z}_{i\cdot}),$$

$$\mathbf{E}_{ZZ} = \sum_i \sum_j (z_{ij} - \bar{z}_{i\cdot})^2,$$

and

$$\begin{aligned} \hat{\boldsymbol{\gamma}}_G &= (\mathbf{E}_{ZZ}^{-1}\mathbf{E}_{ZY})' \qquad [\text{from } (9.31)] \\ &= \mathbf{E}_{YZ}\mathbf{E}_{ZZ}^{-1} \\ &= \sum_i \sum_j (\mathbf{y}_{ij} - \bar{\mathbf{y}}_{i\cdot})(z_{ij} - \bar{z}_{i\cdot}) \Big/ \sum_i \sum_j (z_{ij} - \bar{z}_{i\cdot})^2, \end{aligned}$$

so that

$$\hat{\gamma}_{Gr} = \sum_i \sum_j (y_{ijr} - \bar{y}_{i\cdot r})(z_{ij} - \bar{z}_{i\cdot}) \Big/ \sum_i \sum_j (z_{ij} - \bar{z}_{i\cdot})^2,$$

which is a pooled estimate of slope for variable r. The statistic Λ is readily calculated and the degrees of freedom of $\mathbf{H}$ and $\mathbf{E}$ are $q = I - 1$ and $n - p - t = IJ - I - 1$, respectively. We note, from (9.33), that the least squares estimator $\hat{\boldsymbol{\mu}}_{i,G}$ of $\boldsymbol{\mu}_i$ for model G is obtained by replacing $\mathbf{Y}$ by $\mathbf{Y} - \mathbf{Z}\hat{\boldsymbol{\Gamma}}$ in $\hat{\boldsymbol{\mu}}_i = \bar{\mathbf{y}}_{i\cdot}$; thus

$$\hat{\boldsymbol{\mu}}_{i,G} = \bar{\mathbf{y}}_{i\cdot} - \hat{\boldsymbol{\gamma}}_G \bar{z}_{i\cdot}. \qquad \square$$

The case when the I populations have different dispersion matrices is considered by Chakravorti [1980].

From a computational viewpoint it is usually more convenient to test a hypothesis for an analysis of covariance model using the representation $\mathbf{Y} = \mathbf{W}\boldsymbol{\Phi} + \mathbf{U}$, where any identifiability restrictions are used to eliminate parameters and reduce $\mathbf{W}$ to full rank. All the hypotheses considered above can be expressed in the form $H_0: \mathbf{A}_1\boldsymbol{\Phi} = \mathbf{O}$ so that from the general theory of Chapter 8

$$\mathbf{H} = (\mathbf{A}_1\hat{\boldsymbol{\Phi}})'\left[\mathbf{A}_1(\mathbf{W}'\mathbf{W})^{-1}\mathbf{A}_1'\right]^{-1}\mathbf{A}_1\hat{\boldsymbol{\Phi}}$$

and

$$\mathbf{E} = \mathbf{Y}'\left(\mathbf{I}_n - \mathbf{W}(\mathbf{W}'\mathbf{W})^{-1}\mathbf{W}'\right)\mathbf{Y},$$

where $\hat{\boldsymbol{\Phi}} = (\mathbf{W}'\mathbf{W})^{-1}\mathbf{W}'\mathbf{Y}$. These matrices can be calculated using the general methods of Sections 10.1.1 and 10.1.2, but applied to $\mathbf{W}$ instead of $\mathbf{X}$. However, in most cases, H_0 can be expressed in the form $\boldsymbol{\Phi}_2 = \mathbf{O}$, where $\boldsymbol{\Phi}' = (\boldsymbol{\Phi}_1', \boldsymbol{\Phi}_2')$ so that, under H_0, $\mathscr{E}[\mathbf{Y}] = \mathbf{W}_1\boldsymbol{\Phi}_1$. A method for deriving the various test statistics for this particular H_0 is described in Section 10.1.4a. Alternatively, since

$$\mathbf{E}_H = \mathbf{Y}'\left(\mathbf{I}_n - \mathbf{W}_1(\mathbf{W}_1'\mathbf{W}_1)^{-1}\mathbf{W}_1'\right)\mathbf{Y},$$

we only need a general program for computing $\mathbf{E}$; $\mathbf{E}_H = \mathbf{E} + \mathbf{H}$ follows by simply omitting columns of $\mathbf{W}$ and applying the same program. An example of this follows.

EXAMPLE 9.8 In Table 9.12 we have observations on men in four weight categories with $x_2 = 1$ referring to group 1, $x_3 = 1$ to group 2, $x_4 = 1$ to group 3, and $x_2 = x_3 = x_4 = 0$ to group 4. Urine samples were taken from each man and the following 13 variables, including two concomitant variables z_1 and z_2,

TABLE 9.12 Forty-five 11-Dimensional Observations on Men in Four Weight Groups (Coded by the x-Variables) with Two Concomitant Variables z_1 and z_2[a]

y_1	y_2	y_3	y_4	y_5	y_6	y_7	y_8	y_9	y_{10}	y_{11}	x_1	x_2	x_3	x_4	z_1	z_2
5.7	4.67	17.6	1.50	.104	1.50	1.88	5.15	8.40	7.5	.14	1	1	0	0	205	24
5.5	4.67	13.4	1.65	.245	1.32	2.24	5.75	4.50	7.1	.11	1	1	0	0	160	32
6.6	2.70	20.3	.90	.097	.89	1.28	4.35	1.20	2.3	.10	1	1	0	0	480	17
5.7	3.49	22.3	1.75	.174	1.50	2.24	7.55	2.75	4.0	.12	1	1	0	0	230	30
5.6	3.49	20.5	1.40	.210	1.19	2.00	8.50	3.30	2.0	.12	1	1	0	0	235	30
6.0	3.49	18.5	1.20	.275	1.03	1.84	10.25	2.00	2.0	.12	1	1	0	0	215	27
5.3	4.84	12.1	1.90	.170	1.87	2.40	5.95	2.60	16.8	.14	1	1	0	0	215	25
5.4	4.84	12.0	1.65	.164	1.68	3.00	6.30	2.72	14.5	.14	1	1	0	0	190	30
5.4	4.84	10.1	2.30	.275	2.08	2.68	5.45	2.40	.9	.20	1	1	0	0	190	28
5.6	4.48	14.7	2.35	.210	2.55	3.00	3.75	7.00	2.0	.21	1	1	0	0	175	24
5.6	4.48	14.8	2.35	.050	1.32	2.84	5.10	4.00	.4	.12	1	1	0	0	145	26
5.6	4.48	14.4	2.50	.143	2.38	2.84	4.05	8.00	3.8	.18	1	1	0	0	155	27
5.2	3.48	18.1	1.50	.153	1.20	2.60	9.00	2.35	14.5	.13	1	0	1	0	220	31
5.2	3.48	19.7	1.65	.203	1.73	1.88	5.30	2.52	12.5	.20	1	0	1	0	300	23
5.6	3.48	16.9	1.40	.074	1.15	1.72	9.85	2.45	8.0	.07	1	0	1	0	305	32
5.8	2.63	23.7	1.65	.155	1.58	1.60	3.60	3.75	4.9	.10	1	0	1	0	275	20
6.0	2.63	19.2	.90	.155	.96	1.20	4.05	3.30	.2	.10	1	0	1	0	405	18
5.3	2.63	18.0	1.60	.129	1.68	2.00	4.40	3.00	3.6	.18	1	0	1	0	210	23
5.4	4.46	14.8	2.45	.245	2.15	3.12	7.15	1.81	12.0	.13	1	0	1	0	170	31
5.6	4.46	15.6	1.65	.422	1.42	2.56	7.25	1.92	5.2	.15	1	0	1	0	235	28
5.3	2.80	16.2	1.65	.063	1.62	2.04	5.30	3.90	10.2	.12	1	0	1	0	185	21
5.4	2.80	14.1	1.25	.042	1.62	1.84	3.10	4.10	8.5	.30	1	0	1	0	255	20
5.5	2.80	17.5	1.05	.030	1.56	1.48	2.40	2.10	9.6	.20	1	0	1	0	265	15
5.4	2.57	14.1	2.70	.194	2.77	2.56	4.25	2.60	6.9	.17	1	0	1	0	305	26
5.4	2.57	19.1	1.60	.139	1.59	1.88	5.80	2.30	4.7	.16	1	0	1	0	440	24
5.2	2.57	22.5	.85	.046	1.65	1.20	1.55	1.50	3.5	.21	1	0	1	0	430	16
5.5	1.26	17.0	.70	.094	.97	1.24	4.55	2.90	1.9	.12	1	0	0	1	350	18
5.9	1.26	12.5	.80	.039	.80	.64	2.65	0.72	.7	.13	1	0	0	1	475	10
5.6	2.52	21.5	1.80	.142	1.77	2.60	6.50	2.48	8.3	.17	1	0	0	1	195	33
5.6	2.52	22.2	1.05	.080	1.17	1.48	4.85	2.20	9.3	.14	1	0	0	1	375	25
5.3	2.52	13.0	2.20	.215	1.85	3.84	8.75	2.40	13.0	.11	1	0	0	1	160	35
5.6	3.24	13.0	3.55	.166	3.18	3.48	5.20	3.50	18.3	.22	1	0	0	1	240	33
5.5	3.24	10.9	3.30	.111	2.79	3.04	4.75	2.52	10.5	.21	1	0	0	1	205	31
5.6	3.24	12.0	3.65	.180	2.40	3.00	5.85	3.00	14.5	.21	1	0	0	1	270	34
5.4	1.56	22.8	.55	.069	1.00	1.14	2.85	2.90	3.3	.15	1	0	0	1	475	16
5.3	1.56	16.5	2.05	.222	1.49	2.40	6.55	3.90	6.3	.11	1	0	0	0	430	31
5.2	1.56	18.4	1.05	.267	1.17	1.36	6.60	2.00	4.9	.11	1	0	0	0	490	28
5.8	4.12	12.5	5.90	.093	3.80	3.84	2.90	3.00	22.5	.24	1	0	0	0	105	32
5.7	4.12	8.7	4.25	.147	3.62	5.32	3.00	3.55	19.5	.20	1	0	0	0	115	25
5.5	4.12	9.4	3.85	.217	3.36	5.52	3.40	5.20	1.3	.31	1	0	0	0	97	28
5.4	2.14	15.0	2.45	.418	2.38	2.40	5.40	1.81	20.0	.17	1	0	0	0	325	27
5.4	2.14	12.9	1.70	.323	1.74	2.48	4.45	1.88	1.0	.15	1	0	0	0	310	23
4.9	2.03	12.1	1.80	.205	2.00	2.24	4.30	3.70	5.0	.19	1	0	0	0	245	25
5.0	2.03	13.2	3.65	.348	1.95	2.12	5.00	1.80	3.0	.15	1	0	0	0	170	26
4.9	2.03	11.5	2.25	.320	2.25	3.12	3.40	2.50	5.1	.18	1	0	0	0	220	34

[a]Reproduced from H. Smith, R. Gnanadesikan, and J. B. Hughes (1962) Multivariate analysis of variance (MANOVA). *Biometrics*, **18**, 22–41, Tables 2, 3. With permission from the Biometric Society.

were measured:

$$\begin{aligned}
y_1 &= \text{pH},\\
y_2 &= \text{modified creatinine coefficient},\\
y_3 &= \text{pigment creatinine},\\
y_4 &= \text{phosphate (mg/ml)},\\
y_5 &= \text{calcium (mg/ml)},\\
y_6 &= \text{phosphorus (mg/ml)},\\
y_7 &= \text{creatinine (mg/ml)},\\
y_8 &= \text{chloride (mg/ml)},\\
y_9 &= \text{boron } (\mu\text{g/ml}),\\
y_{10} &= \text{choline } (\mu\text{g/ml}),\\
y_{11} &= \text{copper } (\mu\text{g/ml}),\\
z_1 &= \text{volume (ml)},\\
z_2 &= (\text{specific gravity } -1) \times 10^3.
\end{aligned}$$

The model proposed by Smith et al. [1962] for this data is the one-way analysis of covariance with two concomitant variables, namely,

$$\mathbf{y}'_{ij} = \boldsymbol{\mu}' + \boldsymbol{\alpha}'_i + z_{ij1}\boldsymbol{\gamma}'_1 + z_{ij2}\boldsymbol{\gamma}'_2 + \mathbf{u}'_{ij}$$

or

$$\mathbf{Y} = \mathbf{XB} + \mathbf{Z\Gamma} + \mathbf{U} = \mathbf{W\Phi} + \mathbf{U},$$

say, where

$$\boldsymbol{\Phi}' = (\boldsymbol{\mu}, \boldsymbol{\alpha}_1, \boldsymbol{\alpha}_2, \boldsymbol{\alpha}_3, \boldsymbol{\gamma}_1, \boldsymbol{\gamma}_2).$$

I have used the identifiability constraint $\boldsymbol{\alpha}_4 = \mathbf{0}$ (see Section 9.2.5) so that the matrix $\mathbf{X}$ takes the form given in Table 9.12. The hypotheses considered by the authors were

$$H_{01}: \boldsymbol{\alpha}_1 = \boldsymbol{\alpha}_2 = \boldsymbol{\alpha}_3 = \mathbf{0},$$

$$H_{02}: \boldsymbol{\gamma}_1 = \mathbf{0},$$

$$H_{03}: \boldsymbol{\gamma}_2 = \mathbf{0}.$$

These are all special cases of $H_0: \boldsymbol{\Phi}_2 = \mathbf{O}$ and can be expressed in the form $\mathbf{A}_1\boldsymbol{\Phi} = \mathbf{O}$, where $\mathbf{A}_1$ is given by

$$\begin{pmatrix} 0 & 1 & 0 & 0 & 0 & 0 \\ 0 & 0 & 1 & 0 & 0 & 0 \\ 0 & 0 & 0 & 1 & 0 & 0 \end{pmatrix}, \quad (0\ \ 0\ \ 0\ \ 0\quad 1\ \ 0), \quad (0\ \ 0\ \ 0\ \ 0\quad 0\ \ 1),$$

respectively, with ranks 3, 1, and 1. The appropriate degrees of freedom for hypothesis testing are $m_1 = 3$, $m_2 = m_3 = 1$, and $m_E = n - p - t = 45 - 4 - 2 = 39$. For H_{01} (see Appendix D13),

$$\Lambda_1 = \frac{|\mathbf{E}|}{|\mathbf{E} + \mathbf{H}_1|} = 0.0734,$$

$$f_1 = m_E + m_1 - \tfrac{1}{2}(m_1 + d + 1) = 39 + 3 - \tfrac{1}{2}(3 + 11 + 1) = 34.5,$$

and $-f_1 \log \Lambda_1 = 90.11$. Since $C_\alpha \approx 1$ and $\chi^2_{33}(0.01) = 54.78$, we strongly reject H_{01}. For H_{02}, $\Lambda_2 = 0.4718$, and, since $m_2 = 1$,

$$\begin{aligned} F_0 &= \frac{1 - \Lambda_2}{\Lambda_2} \frac{m_E + 1 - d}{d} \\ &= \frac{0.5282}{0.4718} \frac{29}{11} = 2.95 \sim F_{11,29}. \end{aligned}$$

Since $F^{0.05}_{11,29} = 2.14$, we reject H_{02} at the 5% level of significance. Similarly, $\Lambda_3 = 0.2657$, so that F_0 is larger and we also reject H_{03} at the 5% level.

9.5.3 *Test for Additional Information*

Although this topic arises in several places in this book, it is appropriate to highlight it here, as it is basically an analysis of covariance technique. We begin with the usual linear model $\mathbf{Y} = \mathbf{XB} + \mathbf{U}$ and hypothesis $H_0: \mathbf{AB} = \mathbf{O}$, where $\mathbf{X}$ is $n \times p$ of rank p, $\mathbf{A}$ is $q \times p$ of rank q, and the rows of $\mathbf{U}$ are i.i.d. $N_d(\mathbf{0}, \mathbf{\Sigma})$. Suppose we partition the data matrix $\mathbf{Y} = (\mathbf{Y}_1, \mathbf{Y}_2)$, where $\mathbf{Y}_1$ represents n observations on d_1 y-variables and $\mathbf{Y}_2$ represents n observations on a further d_2 $(= d - d_1)$ variables. Partitioning $\mathbf{B} = (\mathbf{B}_1, \mathbf{B}_2)$ in the same way, we consider the model and hypothesis

$$\mathbf{Y}_1 = \mathbf{XB}_1 + \mathbf{U}_1, \qquad H_{01}: \mathbf{AB}_1 = \mathbf{O},$$

where the rows of $\mathbf{U}_1$ are i.i.d. $N_{d_1}(\mathbf{0}, \mathbf{\Sigma}_{11})$, and $\mathbf{\Sigma}_{11}$ is given by

$$\mathbf{\Sigma} = \begin{pmatrix} \mathbf{\Sigma}_{11} & \mathbf{\Sigma}_{12} \\ \mathbf{\Sigma}_{21} & \mathbf{\Sigma}_{22} \end{pmatrix} \begin{matrix} \}d_1 \\ \}d_2 \end{matrix}.$$

Furthermore, conditional on $\mathbf{Y}_1$, it follows from Theorem 2.1(viii) of Section 2.2 that the rows of $\mathbf{Y}_2$ are independently normally distributed with common dispersion matrix $\mathbf{\Sigma}_{22\cdot1} = \mathbf{\Sigma}_{22} - \mathbf{\Sigma}_{21}\mathbf{\Sigma}_{11}^{-1}\mathbf{\Sigma}_{12}$, and that

$$\begin{aligned} \mathscr{E}[\mathbf{Y}_2|\mathbf{Y}_1] &= \mathbf{XB}_2 + (\mathbf{Y}_1 - \mathbf{XB}_1)\mathbf{\Sigma}_{11}^{-1}\mathbf{\Sigma}_{12} \\ &= \mathbf{X}(\mathbf{B}_2 - \mathbf{B}_1\mathbf{\Sigma}_{11}^{-1}\mathbf{\Sigma}_{12}) + \mathbf{Y}_1\mathbf{\Sigma}_{11}^{-1}\mathbf{\Sigma}_{12} \\ &= \mathbf{X\Delta} + \mathbf{Y}_1\mathbf{\Gamma}, \end{aligned} \tag{9.35}$$

say. The corresponding hypothesis for this conditional model is $H_{02\cdot 1}: \mathbf{A}\boldsymbol{\Delta} = \mathbf{A}(\mathbf{B}_2 - \mathbf{B}_1\boldsymbol{\Gamma}) = \mathbf{O}$, and we now relate the likelihood ratio tests for all three hypotheses.

In the first instance we note that $\mathbf{A}(\mathbf{B}_1, \boldsymbol{\Delta}) = \mathbf{O}$ if and only if $\mathbf{AB} = \mathbf{O}$ so that $H_0 = H_{01} \cap H_{02\cdot 1}$. Then, if we partition $\mathbf{E}$ and $\mathbf{E}_H$ in the same manner as $\boldsymbol{\Sigma}$,

$$\Lambda = \frac{|\mathbf{E}|}{|\mathbf{E}_H|} \quad \text{and} \quad \Lambda_1 = \frac{|\mathbf{E}_{11}|}{|\mathbf{E}_{H11}|}$$

are the likelihood ratio test statistics for testing H_0 and H_{01}, respectively. As the conditional model is an analysis of covariance model (with $\mathbf{Y} \to \mathbf{Y}_2$ and $\mathbf{Z} \to \mathbf{Y}_1$), the likelihood ratio test statistic for $H_{02\cdot 1}$ is, by (9.34),

$$\Lambda_{2\cdot 1} = \frac{|\mathbf{E}_{22} - \mathbf{E}_{21}\mathbf{E}_{11}^{-1}\mathbf{E}_{12}|}{|\mathbf{E}_{H22} - \mathbf{E}_{H21}\mathbf{E}_{H11}^{-1}\mathbf{E}_{H12}|},$$

which is distributed as $U_{d_2, q, n-p-d_1}$ when $H_{02\cdot 1}$ is true. By (2.76) we have the factorization $\Lambda = \Lambda_1\Lambda_{2.1}$ and, when H_0 is true, we can write symbolically [as in (2.78)],

$$U_{d, q, n-p} \equiv U_{d_1, q, n-p} U_{d_2, q, n-p-d_1}.$$

In particular, if $\mathbf{AB}_1 = \mathbf{O}$, then testing $H_{02\cdot 1}$ can be regarded as testing that the additional observations $\mathbf{Y}_2$ make no further contribution, that is, $\mathbf{AB} = \mathbf{O}$. For this reason $H_{02\cdot 1}$ is commonly called a test for "additional information" (Rao [1973]). The power of this test is considered by Fujikoshi [1981]. A "stepwise" testing of H_0 using H_{01} followed by $H_{02\cdot 1}$ forms the basis of the so-called step-down procedure (see Section 8.8 and the special case in Section 3.3.4).

Perhaps a more illuminating example of the procedure is the special case of $\mathbf{A} = \mathbf{I}_p$, that is, $H_{02\cdot 1}: \boldsymbol{\Delta} = \mathbf{O}$. If this hypothesis is true, then the conditional distribution of $\mathbf{Y}_2$, given $\mathbf{Y}_1$, does not contain the parameter $\mathbf{B}_1$, and $\mathbf{Y}_2$ can give no additional information on $\mathbf{B}_1$. The power of the test for $\boldsymbol{\Delta} = \mathbf{O}$ is considered by Kabe [1973]. When $\mathbf{X} = \mathbf{1}_n$ and $\mathbf{B} = \boldsymbol{\mu}'$, the test statistic $\boldsymbol{\Delta} = \mathbf{O}$ reduces to the F-statistic of Theorem 2.11 in Section 2.6. This theorem forms the basis of testing whether a subset of variables gives adequate discrimination (see Section 6.10.1).

9.6 MULTIVARIATE COCHRAN'S THEOREM ON QUADRATICS

We now briefly introduce a multivariate version of a theorem due to Cochran on the decomposition of quadratic forms. The theorem depends on the following lemma. A proof, using projection matrices, is given in Seber [1966, 1980: Appendix 1, Lemma 4].

LEMMA 9.1 Suppose $\mathbf{A}_0, \mathbf{A}_1, \ldots, \mathbf{A}_t$ is a sequence of symmetric $n \times n$ matrices such that $\Sigma_{i=0}^t \mathbf{A}_i = \mathbf{I}_n$, then the following conditions are equivalent:

(i) $\Sigma_{i=0}^t \text{rank } \mathbf{A}_i = n$.
(ii) $\mathbf{A}_i \mathbf{A}_j = \mathbf{O}$ (all $i, j, i \neq j$).
(iii) $\mathbf{A}_i^2 = \mathbf{A}_i$ ($i = 0, 1, \ldots, t$). □

An experimental design is described by the model $\mathbf{Y} = \mathbf{\Theta} + \mathbf{U}$, where the columns of $\mathbf{\Theta}$ belong to Ω, the range space of some matrix $\mathbf{X}$. Typically we wish to test a sequence of hypotheses H_{0i} ($i = 1, 2, \ldots, k$) that the columns of $\mathbf{\Theta}$ belong to ω_i, a subspace of Ω. We can test each H_{0i} by comparing $\mathbf{H}_i = \mathbf{Y}'(\mathbf{P}_\Omega - \mathbf{P}_{\omega_i})\mathbf{Y}$ with $\mathbf{E} = \mathbf{Y}'(\mathbf{I}_n - \mathbf{P}_\Omega)\mathbf{Y}$, using, for example, $\Lambda = |\mathbf{E}|/|\mathbf{E} + \mathbf{H}_i|$. In the case of the one-way classification of Section 9.2.1, $k = 1$ and

$$\begin{aligned}\mathbf{I}_n &= (\mathbf{I}_n - \mathbf{P}_\Omega) + (\mathbf{P}_\Omega - \mathbf{P}_{\omega_1}) + \mathbf{P}_{\omega_1} \\ &= \mathbf{A}_0 + \mathbf{A}_1 + \mathbf{A}_2,\end{aligned}$$

which clearly satisfies condition (iii) of the above lemma. In the so-called *balanced* design corresponding to two-way and higher-way classifications with *equal* observations per cell, we find that the hypotheses are "orthogonal" (see Seber [1980: Chapter 6]), that is,

$$\mathbf{P}_{\omega_1 \cap \omega_2 \cap \cdots \cap \omega_{i-1}} - \mathbf{P}_{\omega_1 \cap \omega_2 \cap \cdots \cap \omega_i} = \mathbf{P}_\Omega - \mathbf{P}_{\omega_i},$$

so that

$$\mathbf{I}_n = (\mathbf{I}_n - \mathbf{P}_\Omega) + (\mathbf{P}_\Omega - \mathbf{P}_{\omega_1}) + (\mathbf{P}_{\omega_1} - \mathbf{P}_{\omega_1 \cap \omega_2}) + \cdots + \mathbf{P}_{\omega_1 \cap \omega_2 \cap \cdots \cap \omega_k}$$

becomes

$$\mathbf{I}_n = (\mathbf{I}_n - \mathbf{P}_\Omega) + (\mathbf{P}_\Omega - \mathbf{P}_{\omega_1}) + \cdots + (\mathbf{P}_\Omega - \mathbf{P}_{\omega_k}) + \mathbf{P}_{\omega_1 \cap \omega_2 \cap \cdots \cap \omega_k}. \quad (9.36)$$

Again, this decomposition satisfies condition (iii) and it provides a motivation for the following generalization of Cochran's theorem on quadratic forms.

THEOREM 9.2 Suppose that the rows $\mathbf{y}_\alpha$ ($\alpha = 1, 2, \ldots, n$) of $\mathbf{Y}$ are independently distributed as $N_d(\boldsymbol{\theta}_\alpha, \mathbf{\Sigma})$, and let $\mathbf{A}_0, \mathbf{A}_1, \ldots, \mathbf{A}_t$ be a sequence of $n \times n$ matrices with ranks $r_0, r_1, \ldots, r_t$ such that $\Sigma_{i=0}^t \mathbf{A}_i = \mathbf{I}_n$. If one (and therefore all) of conditions of Lemma 9.1 hold, then the $\mathbf{Y}'\mathbf{A}_i\mathbf{Y}$ ($i = 0, 1, \ldots, t$) are independently distributed as the noncentral Wishart, $W_d(r_i, \mathbf{\Sigma}; \mathbf{\Delta}_i)$, where $\mathbf{\Delta}_i = \mathbf{\Sigma}^{-1/2}\mathbf{\Theta}'\mathbf{A}_i\mathbf{\Theta}\mathbf{\Sigma}^{-1/2}$ is the noncentrality matrix and $\mathbf{\Theta} = (\boldsymbol{\theta}_1, \boldsymbol{\theta}_2, \ldots, \boldsymbol{\theta}_n)'$.

Proof The special case $\mathbf{\Theta} = \mathbf{O}$ is proved using, for example, Exercise 2.25. The noncentral case follows in a similar fashion using Section 2.3.3. □

For testing k orthogonal hypotheses, the decomposition corresponding to (9.36) is

$$\mathbf{Y}'\mathbf{Y} = \mathbf{E} + \mathbf{H}_1 + \mathbf{H}_2 + \cdots + \mathbf{H}_k + \mathbf{Y}'\mathbf{P}_{\omega_1 \cap \cdots \cap \omega_k}\mathbf{Y},$$

so that by Theorem 9.2, with $t = k + 1$, each matrix on the right-hand side has a noncentral Wishart distribution. Since $\mathbf{P}_\Omega \boldsymbol{\Theta} = \boldsymbol{\Theta}$, $\mathbf{A}_0 \boldsymbol{\Theta} = (\mathbf{I}_n - \mathbf{P}_\Omega)\boldsymbol{\Theta} = \mathbf{O}$ and $\boldsymbol{\Delta}_0 = \mathbf{O}$; thus $\mathbf{E}$ has a central Wishart distribution, as already proved in Theorem 8.3 of Section 8.4. When H_{0i} is true, $\mathbf{P}_{\omega_i}\boldsymbol{\Theta} = \boldsymbol{\Theta}$ so that $\mathbf{A}_i\boldsymbol{\Theta} = (\mathbf{P}_\Omega - \mathbf{P}_{\omega_i})\boldsymbol{\Theta} = \mathbf{O}$ and $\boldsymbol{\Delta}_i = \mathbf{O}$; then $\mathbf{H}_i$ also has a central distribution. When $H_{01}, H_{02}, \ldots, H_{0k}$ are simultaneously true, $\mathbf{E}, \mathbf{H}_1, \mathbf{H}_2, \ldots, \mathbf{H}_k$ all have independent central distributions. If α is the overall significance level for testing all the H_{0i}, then, from Dykstra [1979],

$$\alpha < 1 - \prod_{i=1}^{k} (1 - \alpha_i),$$

where α_i is the significance level for H_{0i}. Some generalizations of Theorem 9.2 are given by Khatri [1977].

If the $\mathbf{H}_i$ all have the same degrees of freedom (e.g., $m_H = 1$ for all main effects and interactions in a multivariate 2^r design), then the graphical methods of Section 9.2.6b can be used for comparing the $\mathbf{H}_i$. For example, if $a_i = \operatorname{tr} \mathbf{H}_i$, the k ordered values $a_{(1)} < a_{(2)} < \cdots < a_{(k)}$ could be plotted against a gamma distribution estimated from the first few ordered values (see Appendix C4). Under null conditions, that is, all the H_{0i} true, the points should lie approximately on a straight line; however, any significant $\mathbf{H}_i$ will give rise to a point above the linear trend [by (8.49)]. Such points will occur with the largest ordered values at the right-hand end of the plot.

9.7 GROWTH CURVE ANALYSIS

9.7.1 Examples

a SINGLE GROWTH CURVE

In Section 3.4.2 we considered the example of sampling from a polynomial growth curve (3.23). Using a change in notation, let $\mathbf{v}_1, \mathbf{v}_2, \ldots, \mathbf{v}_n$ be i.i.d. $N_d(\boldsymbol{\eta}, \boldsymbol{\Sigma})$, where

$$\eta_r = \gamma_0 + \gamma_1 t_r + \cdots + \gamma_{k-1} t_r^{k-1} \qquad (r = 1, 2, \ldots, d)$$

is the expected "size" of an organism at time t_r. In vector notation

$$\boldsymbol{\eta} = \begin{pmatrix} 1 & t_1 & \cdots & t_1^{k-1} \\ 1 & t_2 & \cdots & t_2^{k-1} \\ \vdots & \vdots & & \vdots \\ 1 & t_d & \cdots & t_d^{k-1} \end{pmatrix} \begin{pmatrix} \gamma_0 \\ \gamma_1 \\ \vdots \\ \gamma_{k-1} \end{pmatrix} = \mathbf{K}\boldsymbol{\gamma}, \qquad (9.37)$$

say. The elements of $\mathbf{v}_i$, plotted against time, give the observed growth curve for the ith individual. In Section 3.4.2 we were interested in testing the adequacy of the above model, that is, in testing $\boldsymbol{\eta} = \mathbf{K}\boldsymbol{\gamma}$. However, given that this specification is acceptable, we may now wish to test that a lower-degree polynomial is acceptable, for example, $H_0: \gamma_{k-1} = \gamma_{k-2} = \cdots = \gamma_{k-v} = 0$. To do this, we can reformulate the model as follows.

Writing $\mathbf{Y} = (\mathbf{v}_1, \mathbf{v}_2, \ldots, \mathbf{v}_n)'$, we have $\mathscr{E}[\mathbf{Y}] = \mathbf{1}_n\boldsymbol{\eta}'$, which, combined with (9.37), gives us $\mathscr{E}[\mathbf{Y}] = \mathbf{1}_n\boldsymbol{\gamma}'\mathbf{K}'$. Then H_0 takes the form

$$\mathbf{0}' = \boldsymbol{\gamma}'\begin{pmatrix} \mathbf{O} \\ \mathbf{I}_v \end{pmatrix} = \boldsymbol{\gamma}'\mathbf{D},$$

say. This is a special case of testing $\mathbf{A\Delta D} = \mathbf{O}$ for the model $\mathbf{Y} = \mathbf{X\Delta K}' + \mathbf{U}$ with $\mathbf{A} = 1$, $\boldsymbol{\Delta} = \boldsymbol{\gamma}'$, and $\mathbf{X} = \mathbf{1}_n$. A general theory for handling this type of model and hypothesis is derived below, and a test statistic for the above example is given in Section 9.7.3.

b TWO GROWTH CURVES

Suppose that we wish to compare the polynomial growth trends for two populations. Let $\mathbf{v}_1, \mathbf{v}_2, \ldots, \mathbf{v}_{n_1}$ be i.i.d. $N_d(\boldsymbol{\eta}, \boldsymbol{\Sigma})$ and let $\mathbf{w}_1, \mathbf{w}_2, \ldots, \mathbf{w}_{n_2}$ be i.i.d. $N_d(\boldsymbol{\nu}, \boldsymbol{\Sigma})$. For example, $\mathbf{v}_i$ could represent the measurements at d points in time of the height of the ith animal in the first group. Writing $\boldsymbol{\eta} = [(\eta_r)]$ and $\boldsymbol{\nu} = [(\nu_r)]$, we assume, for ease of exposition, that the growth curves are quadratic, namely,

$$\eta_r = \gamma_0 + \gamma_1 t_r + \gamma_2 t_r^2$$

and

$$\nu_r = \delta_0 + \delta_1 t_r + \delta_2 t_r^2 \qquad (r = 1, 2, \ldots, d).$$

Suppose that we wish to test H_0 that the two quadratics in time t are "parallel", that is, $H_0: \gamma_1 = \delta_1$, $\gamma_2 = \delta_2$. One method is to use the linear model representation

$$\mathbf{Y} = \boldsymbol{\Theta} + \mathbf{U} = \mathbf{XB} + \mathbf{U}, \tag{9.38}$$

where

$$\mathbf{Y} = \begin{pmatrix} \mathbf{v}_1' \\ \vdots \\ \mathbf{v}_{n_1}' \\ \mathbf{w}_1' \\ \vdots \\ \mathbf{w}_{n_2}' \end{pmatrix}, \qquad \mathbf{X} = \begin{pmatrix} \mathbf{1}_{n_1} & \mathbf{0} \\ \mathbf{0} & \mathbf{1}_{n_2} \end{pmatrix}, \qquad \mathbf{B} = \begin{pmatrix} \boldsymbol{\eta}' \\ \boldsymbol{\nu}' \end{pmatrix} \tag{9.39}$$

and the rows of $\mathbf{U}$ are i.i.d $N_d(\mathbf{0}, \mathbf{\Sigma})$. Noting, from (9.37), that $\boldsymbol{\eta} = \mathbf{K}\boldsymbol{\gamma}$ and $\boldsymbol{\nu} = \mathbf{K}\boldsymbol{\delta}$, where

$$\mathbf{K} = \begin{pmatrix} 1 & t_1 & t_1^2 \\ 1 & t_2 & t_2^2 \\ \vdots & \vdots & \vdots \\ 1 & t_d & t_d^2 \end{pmatrix}, \qquad \boldsymbol{\gamma} = \begin{pmatrix} \gamma_0 \\ \gamma_1 \\ \gamma_2 \end{pmatrix}, \qquad \boldsymbol{\delta} = \begin{pmatrix} \delta_0 \\ \delta_1 \\ \delta_2 \end{pmatrix}, \tag{9.40}$$

we can write

$$\mathbf{B} = \begin{pmatrix} \boldsymbol{\gamma}'\mathbf{K}' \\ \boldsymbol{\delta}'\mathbf{K}' \end{pmatrix} = \begin{pmatrix} \boldsymbol{\gamma}' \\ \boldsymbol{\delta}' \end{pmatrix}\mathbf{K}' = \mathbf{\Delta}\mathbf{K}', \tag{9.41}$$

say. Here $\mathbf{B}$ is $2 \times d$ and $\mathbf{K}' = 3 \times d$ $(d \geq 3)$. The hypothesis H_0 can also be written in matrix form, namely,

$$\begin{aligned} \mathbf{0}' &= (\gamma_1 - \delta_1, \gamma_2 - \delta_2) \\ &= (1, -1)\begin{pmatrix} \gamma_0 & \gamma_1 & \gamma_2 \\ \delta_0 & \delta_1 & \delta_2 \end{pmatrix}\begin{pmatrix} 0 & 0 \\ 1 & 0 \\ 0 & 1 \end{pmatrix} \\ &= (1, -1)\,\mathbf{\Delta}\mathbf{D}, \end{aligned}$$

say. Combining (9.41) with (9.38), we have the model

$$\mathbf{Y} = \mathbf{X}\mathbf{\Delta}\mathbf{K}' + \mathbf{U} \tag{9.42}$$

and we wish to test a hypothesis of the form $\mathbf{A}\mathbf{\Delta}\mathbf{D} = \mathbf{O}$, where $\mathbf{A} = (1, -1)$. In much of the earlier work on growth curves, t^r was replaced by an orthogonal polynomial of degree r in the matrix $\mathbf{K}$.

Suppose we now change the notation and formulate the model (9.38) as a one-way classification for two populations. To do this we write

$$\mathbf{Y}' = (\mathbf{y}_{11}, \mathbf{y}_{12}, \ldots, \mathbf{y}_{1n_1}, \mathbf{y}_{21}, \mathbf{y}_{22}, \ldots, \mathbf{y}_{2n_2})$$

and $\mathscr{E}[\mathbf{y}_{ij}] = \boldsymbol{\theta}_{ij}$. We can now describe this model as having (1) an *external* design structure, namely, a one-way classification with $\boldsymbol{\theta}_{1j} = \boldsymbol{\eta}$ and $\boldsymbol{\theta}_{2j} = \boldsymbol{\nu}$, imposed by the design matrix $\mathbf{X}$ of (9.39) on the $n_1 + n_2$ vectors $\boldsymbol{\theta}_{ij}$, and (2) an *internal* structure of the form $\boldsymbol{\theta}_{ij} = \mathbf{K}\boldsymbol{\lambda}_{ij}$, say, where $\boldsymbol{\lambda}_{1j} = \boldsymbol{\gamma}$ and $\boldsymbol{\lambda}_{2j} = \boldsymbol{\delta}$. Clearly, we can generalize this model by simply changing the design matrix $\mathbf{X}$, as in the following example.

C SINGLE GROWTH CURVE FOR A RANDOMIZED BLOCK DESIGN

The linear model for a randomized block design,

$$\mathbf{y}_{ij} = \boldsymbol{\theta}_{ij} + \mathbf{u}_{ij} \qquad (i = 1, 2, \ldots, I;\ j = 1, 2, \ldots, J),$$

has an external structure $\boldsymbol{\theta}_{ij} = \boldsymbol{\mu} + \boldsymbol{\alpha}_i + \boldsymbol{\beta}_j$ [see (9.21)]. Assuming, for simplicity, a quadratic growth curve, we can write the internal structure as

$$\theta_{ijr} = \lambda_{0ij} + \lambda_{1ij}t_r + \lambda_{2ij}t_r^2 \qquad (r = 1, 2, \ldots, d),$$

or $\boldsymbol{\theta}_{ij} = \mathbf{K}\boldsymbol{\lambda}_{ij}$, where $\boldsymbol{\lambda}'_{ij} = (\lambda_{0ij}, \lambda_{1ij}, \lambda_{2ij})$. Substituting for $\boldsymbol{\mu}$ $(= \bar{\boldsymbol{\theta}}_{..})$, $\boldsymbol{\alpha}_i$ $(= \bar{\boldsymbol{\theta}}_{i.} - \bar{\boldsymbol{\theta}}_{..})$, and so on, and using $\boldsymbol{\theta}_{ij} = \mathbf{K}\boldsymbol{\lambda}_{ij}$, we find that $\boldsymbol{\lambda}_{ij}$ has the same "additive" structure as $\boldsymbol{\theta}_{ij}$. Therefore combining the external and internal structures gives us the model

$$\boldsymbol{\theta}_{ij} = \mathbf{K}(\boldsymbol{\phi} + \boldsymbol{\psi}_i + \boldsymbol{\xi}_j), \tag{9.43}$$

where $\Sigma_i \boldsymbol{\psi}_i = \Sigma_j \boldsymbol{\xi}_j = \mathbf{0}$. Writing $\boldsymbol{\Theta}' = (\boldsymbol{\theta}_{11}, \boldsymbol{\theta}_{12}, \ldots, \boldsymbol{\theta}_{IJ})$, we have $\boldsymbol{\Theta} = \mathbf{X}\boldsymbol{\Delta}\mathbf{K}'$, where $\mathbf{X}$ is the usual design matrix for a randomized block design and

$$\boldsymbol{\Delta}' = (\boldsymbol{\phi}, \boldsymbol{\psi}_1, \ldots, \boldsymbol{\psi}_I, \boldsymbol{\xi}_1, \ldots, \boldsymbol{\xi}_J).$$

The usual identifiability constraints $\Sigma_i \boldsymbol{\alpha}_i = \Sigma_j \boldsymbol{\beta}_j = \mathbf{0}$, or $\mathbf{FB} = \mathbf{O}$, where

$$\mathbf{B} = (\boldsymbol{\mu}, \boldsymbol{\alpha}_1, \ldots, \boldsymbol{\alpha}_I, \boldsymbol{\beta}_1, \ldots, \boldsymbol{\beta}_J)' = \boldsymbol{\Delta}\mathbf{K}',$$

also apply to $\boldsymbol{\Delta}$. This is true for *any* design matrix $\mathbf{X}$, as $\mathbf{B}'\mathbf{F}' = \mathbf{O}$ implies that $\mathbf{K}\boldsymbol{\Delta}'\mathbf{F}' = \mathbf{O}$, or $\boldsymbol{\Delta}'\mathbf{F}' = \mathbf{O}$, as the columns of $\mathbf{K}$ are linearly independent.

d TWO-DIMENSIONAL GROWTH CURVE

We now consider a different generalization of the example in Section 9.7.1b above. Suppose we measure the height and weight of each organism at d times t_r $(r = 1, 2, \ldots, d)$, and we assume that the corresponding growth curves are linear and quadratic respectively. Then, for the sample from the first population, $\mathbf{v}_i$ has $2d$ elements, the first d being height, and the second d weight. Since height and weight will be correlated, we assume that $\mathbf{v}_i \sim N_{2d}(\boldsymbol{\eta}, \boldsymbol{\Sigma}_0)$, where

$$\boldsymbol{\eta}' = \left(\eta_1^{(1)}, \eta_2^{(1)}, \ldots, \eta_d^{(1)}, \eta_1^{(2)}, \eta_2^{(2)}, \ldots, \eta_d^{(2)}\right) = \left(\boldsymbol{\eta}^{(1)\prime}, \boldsymbol{\eta}^{(2)\prime}\right),$$

$$\eta_r^{(1)} = \gamma_0^{(1)} + \gamma_1^{(1)}t_r,$$

$$\eta_r^{(2)} = \gamma_0^{(2)} + \gamma_1^{(2)}t_r + \gamma_2^{(2)}t_r^2,$$

with superscripts (1) and (2) referring to height and weight, respectively. Then, writing $\boldsymbol{\eta}^{(i)} = \mathbf{K}_i\boldsymbol{\gamma}^{(i)}$, we have

$$\boldsymbol{\eta} = \begin{pmatrix} \boldsymbol{\eta}^{(1)} \\ \boldsymbol{\eta}^{(2)} \end{pmatrix} = \begin{pmatrix} \mathbf{K}_1 & \mathbf{O} \\ \mathbf{O} & \mathbf{K}_2 \end{pmatrix} \begin{pmatrix} \boldsymbol{\gamma}^{(1)} \\ \boldsymbol{\gamma}^{(2)} \end{pmatrix} = \mathbf{K}\boldsymbol{\gamma},$$

say. For the second population we assume that $\mathbf{w}_i \sim N_{2d}(\boldsymbol{\nu}, \boldsymbol{\Sigma}_0)$, where $\boldsymbol{\nu} = \mathbf{K}\boldsymbol{\delta}$. Putting the two samples together leads to the model (9.42) once again, with $\mathbf{X}$

given by (9.39), but

$$\boldsymbol{\Delta} = \begin{pmatrix} \boldsymbol{\gamma}^{(1)\prime}, \boldsymbol{\gamma}^{(2)\prime}, \\ \boldsymbol{\delta}^{(1)\prime}, \boldsymbol{\delta}^{(2)\prime} \end{pmatrix} \quad \text{and} \quad \mathbf{K} = \begin{pmatrix} \mathbf{K}_1 & \mathbf{O} \\ \mathbf{O} & \mathbf{K}_2 \end{pmatrix}. \tag{9.44}$$

Clearly, further generalizations are possible that still lead to the same general model (9.42) above (Potthoff and Roy [1964: pp. 315–316]). For example, the heights and weights could be measured at different times, or more than two growth measurements could be made on each individual. However, before giving a general theory, some comments from Sandland and McGilchrist [1979] are particularly appropriate. They note that in many cases of practical interest, for example, in a very short series, no sensible alternative model is available and the polynomial model will be adequate. However, such models are often biologically unsatisfactory, as the parameters may not have a satisfactory biological interpretation. Richards [1969] puts it more strongly when he says, "the polynomial curves commonly adopted in statistical work are usually quite inappropriate for growth studies." Sandland and McGilchrist give four reasons why polynomials are not appropriate for modeling long structured growth series; (1) Growth processes undergo changes of phase, which are not appropriately fitted by a global polynomial; (2) an ill-fitting polynomial will lead to a distorted estimate of the correlation structure; (3) polynomials tend to be poor at fitting asymptotic approaches to limiting values; and (4) growth data usually exhibit nonstationary second-order properties (e.g., Guttman and Guttman [1965]), which are not readily analyzed by standard procedures. However, some authors, for example, Grizzle and Allen [1969] and Jöreskog [1970: pp. 245–248; 1973: pp. 278–280], have considered models that allow for $\boldsymbol{\Sigma}$ to be internally structured, as, for example, when the errors of measurement on a growth curve are stationary and follow an autoregressive moving average (ARMA) process.

With the above warnings in mind, we now develop a general mathematical theory for the so-called growth curve or repeated-measurement model.

9.7.2 *General Theory*

Consider the linear model $\mathbf{Y} = \mathbf{X}\boldsymbol{\Delta}\mathbf{K}' + \mathbf{U}$, where $\mathbf{X}$ is $n \times p$ of rank p, $\boldsymbol{\Delta}$ is $p \times k$, $\mathbf{K}'$ is $k \times d$ of rank k $(k < d)$, and the rows of $\mathbf{U}$ are i.i.d. $N_d(\mathbf{0}, \boldsymbol{\Sigma})$. We wish to develop a test statistic for testing $H_0: \mathbf{A}\boldsymbol{\Delta}\mathbf{D} = \mathbf{O}$, where $\mathbf{A}$ is $q \times d$ of rank q and $\mathbf{D}$ is $k \times v$ of rank v. It should be noted that d is the total number of growth measurements made on each individual and is, in fact, "$2d$" in the example in Section 9.7.1d above. Also, the above linear model and hypothesis can be reduced to a canonical form (Gleser and Olkin [1970], Kariya [1978]), and a likelihood ratio test constructed (Kabe [1975]). Unfortunately the test statistic has an intractable distribution and the following methods have been proposed instead.

a POTTHOFF AND ROY'S METHOD

The first method, suggested by Potthoff and Roy [1964], is to choose a nonsingular $d \times d$ matrix $\mathbf{G}$ (usually positive definite) such that the $k \times k$ matrix $\mathbf{K}'\mathbf{G}^{-1}\mathbf{K}$ is nonsingular, and transform $\mathbf{y}_i$ to $\mathbf{C}_1'\mathbf{y}_i$, where

$$\mathbf{C}_1 = \mathbf{G}^{-1}\mathbf{K}(\mathbf{K}'\mathbf{G}^{-1}\mathbf{K})^{-1} \tag{9.45}$$

is $d \times k$ of rank k. Then

$$\mathbf{K}'\mathbf{C}_1 = \mathbf{I}_k, \tag{9.46}$$

and the $k \times 1$ transformed observation vectors are independently normally distributed with dispersion matrix $\mathbf{\Sigma}_1 = \mathbf{C}_1'\mathbf{\Sigma}\mathbf{C}_1$. Writing the transformed observations as rows of $\mathbf{Y}_1$, we have

$$\begin{aligned} \mathscr{E}[\mathbf{Y}_1] &= \mathscr{E}[\mathbf{Y}\mathbf{C}_1] \\ &= \mathbf{X}\mathbf{\Delta}\mathbf{K}'\mathbf{C}_1 \\ &= \mathbf{X}\mathbf{\Delta} \qquad [\text{by } (9.46)]. \end{aligned} \tag{9.47}$$

We have therefore reduced our original model to $\mathbf{Y}_1 = \mathbf{X}\mathbf{\Delta} + \mathbf{U}_1$, where the rows of $\mathbf{U}_1$ are i.i.d. $N_k(\mathbf{0}, \mathbf{\Sigma}_1)$, and H_0 can be tested using the general theory of Section 8.7.1 with $\mathbf{Y} \to \mathbf{Y}_1$, $\mathbf{B} \to \mathbf{\Delta}$, $\mathbf{Y}_D \to \mathbf{D}\mathbf{Y}_1$, and $d \to k$. The basic matrices are

$$\mathbf{H}_1 = (\mathbf{A}\hat{\mathbf{\Delta}}_1\mathbf{D})'\left[\mathbf{A}(\mathbf{X}'\mathbf{X})^{-1}\mathbf{A}'\right]^{-1}\mathbf{A}\hat{\mathbf{\Delta}}_1\mathbf{D}$$

and

$$\mathbf{E}_1 = \mathbf{D}'\mathbf{Y}_1'\left(\mathbf{I}_n - \mathbf{X}(\mathbf{X}'\mathbf{X})^{-1}\mathbf{X}'\right)\mathbf{Y}_1\mathbf{D},$$

where

$$\begin{aligned} \hat{\mathbf{\Delta}}_1 &= (\mathbf{X}'\mathbf{X})^{-1}\mathbf{X}'\mathbf{Y}_1 \\ &= (\mathbf{X}'\mathbf{X})^{-1}\mathbf{X}'\mathbf{Y}\mathbf{G}^{-1}\mathbf{K}(\mathbf{K}'\mathbf{G}^{-1}\mathbf{K})^{-1}, \end{aligned} \tag{9.48}$$

by (9.45). From the general theory, $\mathbf{H}_1$ and $\mathbf{E}_1$ are independently distributed as $W_v(q, \mathbf{D}'\mathbf{\Sigma}_1\mathbf{D})$ and $W_v(n-p, \mathbf{D}'\mathbf{\Sigma}_1\mathbf{D})$, respectively, when H_0 is true.

If $k = d$, then $\mathbf{K}$ is nonsingular and we simply make the transformation $\mathbf{Y}_1 = \mathbf{Y}\mathbf{K}^{-1}$. For $k < d$, $\mathbf{G}$ can be arbitrary and we face the problem of choosing a suitable $\mathbf{G}$ that can either be a matrix of constants or even a random matrix distributed independently of $\mathbf{Y}$. The simplest choice is $\mathbf{G} = \mathbf{I}_d$, so that

$$\mathbf{Y}_1 = \mathbf{Y}\mathbf{K}(\mathbf{K}'\mathbf{K})^{-1}. \tag{9.49}$$

Working with the transformed data $(\mathbf{K}'\mathbf{K})^{-1}\mathbf{K}'\mathbf{y}_i$ is equivalent to using the estimated regression coefficients of the associated polynomials instead of the original data $\mathbf{y}_i$. The calculations are somewhat simplified if normalized orthogonal polynomials are used in $\mathbf{K}$. We then have $\mathbf{K}'\mathbf{K} = \mathbf{I}_k$ and $\mathbf{Y}_1 = \mathbf{YK}$. This is essentially the method adopted in the earlier papers of Wishart [1938] and Leech and Healy [1959].

Using a minimum variance criterion, Potthoff and Roy [1964] showed that the optimal choice of $\mathbf{G}$ is $\mathbf{G} = \boldsymbol{\Sigma}$, so that variances increase as $\mathbf{G}$ moves away from $\boldsymbol{\Sigma}$. However, $\boldsymbol{\Sigma}$ is unknown and most estimates of $\boldsymbol{\Sigma}$ are statistically dependent on $\mathbf{Y}$. A natural choice is $\mathbf{G} = \mathbf{S}$, where

$$\begin{aligned}\mathbf{S} &= \frac{\mathbf{Y}'\mathbf{R}\mathbf{Y}}{n-p} \\ &= \frac{\mathbf{Y}'\left(\mathbf{I}_n - \mathbf{X}(\mathbf{X}'\mathbf{X})^{-1}\mathbf{X}'\right)\mathbf{Y}}{n-p}. \end{aligned} \tag{9.50}$$

As shown below, $\hat{\boldsymbol{\Delta}}_1$ of (9.48) is then the maximum likelihood estimate of $\boldsymbol{\Delta}$ for the original model. However, $\mathbf{S}$ depends on $\mathbf{Y}$ so that the usual distribution theory for the matrices $\mathbf{H}_1$ and $\mathbf{E}_1$ no longer applies.

b RAO–KHATRI ANALYSIS OF COVARIANCE METHOD

Rao [1965, 1966, 1967] and Khatri [1966] independently proposed an alternative reduction that gets around the problem of an arbitrary $\mathbf{G}$. A helpful exposition is given by Grizzle and Allen [1969]. The first step is to choose a $d \times k$ matrix $\mathbf{C}_1$ of rank k such that $\mathbf{K}'\mathbf{C}_1 = \mathbf{I}_k$ [as in (9.46)], and a $d \times (d-k)$ matrix $\mathbf{C}_2$ of rank $d-k$ such that $\mathbf{K}'\mathbf{C}_2 = \mathbf{O}$ and $\mathbf{C} = (\mathbf{C}_1, \mathbf{C}_2)$ is a $d \times d$ nonsingular matrix. We can now transform the observations $\mathbf{y}_i$ to $\mathbf{C}'\mathbf{y}_i$ and write the transformed observations as rows of $\mathbf{Y}_C = \mathbf{YC} = (\mathbf{Y}_1, \mathbf{Y}_2)$, where $\mathbf{Y}_r = \mathbf{YC}_r$ $(r = 1, 2)$. Then $\mathscr{E}[\mathbf{Y}_1] = \mathbf{X}\boldsymbol{\Delta}\mathbf{K}'\mathbf{C}_1 = \mathbf{X}\boldsymbol{\Delta}$, $\mathscr{E}[\mathbf{Y}_2] = \mathbf{X}\boldsymbol{\Delta}\mathbf{K}'\mathbf{C}_2 = \mathbf{O}$, and we can write down the analysis of covariance model [see (9.35) with $\mathbf{XB}_1 = \mathbf{O}$ and subscripts 1 and 2 interchanged]

$$\mathscr{E}[\mathbf{Y}_1|\mathbf{Y}_2] = \mathbf{X}\boldsymbol{\Delta} + \mathbf{Y}_2\boldsymbol{\Gamma} = (\mathbf{X}, \mathbf{Y}_2)\begin{pmatrix}\boldsymbol{\Delta} \\ \boldsymbol{\Gamma}\end{pmatrix}. \tag{9.51}$$

The matrix $(\mathbf{X}, \mathbf{Y}_2)$ will have rank $p + d - k$ with probability 1, as the p columns of $\mathbf{X}$ are linearly independent, and $\mathbf{Y}_2$ is an $n \times (d-k)$ unrestricted random matrix with linearly independent columns that are also linearly independent of the columns of $\mathbf{X}$ (otherwise $\mathbf{Y}_2$ is restricted). Since the rows of $\mathbf{Y}_1$ are conditionally distributed as independent multivariate normals with a common covariance matrix, we can apply the analysis of covariance theory of Section 9.5 (with $\mathbf{Z} = \mathbf{Y}_2$ and $\mathbf{B} = \boldsymbol{\Delta}$) as follows.

The least squares (and therefore maximum likelihood) estimates of Γ and Δ for the conditional model (9.51) are [see (9.31) and (9.33)]

$$\begin{aligned}\hat{\Gamma} &= (\mathbf{Z}'\mathbf{R}\mathbf{Z})^{-1}\mathbf{Z}'\mathbf{R}\mathbf{Y}_1 \\ &= \left(\mathbf{C}_2'\mathbf{Y}'\mathbf{R}\mathbf{Y}\mathbf{C}_2\right)^{-1}\mathbf{C}_2'\mathbf{Y}'\mathbf{R}\mathbf{Y}\mathbf{C}_1 \\ &= \left(\mathbf{C}_2'\mathbf{S}\mathbf{C}_2\right)^{-1}\mathbf{C}_2'\mathbf{S}\mathbf{C}_1 \qquad [\text{by } (9.50)] \end{aligned} \tag{9.52}$$

and

$$\begin{aligned}\hat{\Delta} &= (\mathbf{X}'\mathbf{X}^{-1})\mathbf{X}'(\mathbf{Y}_1 - \mathbf{Z}\hat{\Gamma}) \\ &= (\mathbf{X}'\mathbf{X})^{-1}\mathbf{X}'(\mathbf{Y}\mathbf{C}_1 - \mathbf{Y}\mathbf{C}_2\hat{\Gamma}) \\ &= (\mathbf{X}'\mathbf{X})^{-1}\mathbf{X}'\mathbf{Y}\left(\mathbf{C}_1 - \mathbf{C}_2\left[\mathbf{C}_2'\mathbf{S}\mathbf{C}_2\right]^{-1}\mathbf{C}_2'\mathbf{S}\mathbf{C}_1\right).\end{aligned}$$

From B3.5 we have (with $\mathbf{K}'\mathbf{C}_2 = \mathbf{O}$)

$$\mathbf{C}_2\left(\mathbf{C}_2'\mathbf{S}\mathbf{C}_2\right)^{-1}\mathbf{C}_2' = \mathbf{S}^{-1} - \mathbf{S}^{-1}\mathbf{K}(\mathbf{K}'\mathbf{S}^{-1}\mathbf{K})^{-1}\mathbf{K}'\mathbf{S}^{-1}, \tag{9.53}$$

so that

$$\begin{aligned}\hat{\Delta} &= (\mathbf{X}'\mathbf{X})^{-1}\mathbf{X}'\mathbf{Y}\mathbf{S}^{-1}\mathbf{K}(\mathbf{K}'\mathbf{S}^{-1}\mathbf{K})^{-1}\mathbf{K}'\mathbf{C}_1 \\ &= (\mathbf{X}'\mathbf{X})^{-1}\mathbf{X}'\mathbf{Y}\mathbf{S}^{-1}\mathbf{K}(\mathbf{K}'\mathbf{S}^{-1}\mathbf{K})^{-1},\end{aligned} \tag{9.54}$$

since $\mathbf{K}'\mathbf{C}_1 = \mathbf{I}_k$. If f and g are the joint density functions of the normally distributed rows of $\mathbf{Y}$ and $\mathbf{Y}_C$, respectively, then (by A9.2)

$$\begin{aligned}f(\mathbf{Y}) &= g(\mathbf{Y}_C)|\mathbf{C}|^n \\ &= g(\mathbf{Y}_1|\mathbf{Y}_2)g_2(\mathbf{Y}_2)|\mathbf{C}|^n,\end{aligned}$$

where $g_2(\mathbf{Y}_2)$ does not contain Δ, as $\mathscr{E}[\mathbf{Y}_2] = \mathbf{O}$. Thus maximizing $f(\mathbf{Y})$ with respect to Δ is equivalent to maximizing $g(\mathbf{Y}_1|\mathbf{Y}_2)$ with respect to Δ, so that $\hat{\Delta}$ of (9.54) is the maximum likelihood estimate of Δ with respect to the original model, as well as the conditional model.

Having cast the general growth model into an analysis of covariance framework, we now consider the problem of testing $H_0: \mathbf{A}\Delta\mathbf{D} = \mathbf{O}$. To do this we first of all require $\mathbf{E}$ and $\mathbf{H}$ matrices for testing $\mathbf{A}\Delta = \mathbf{O}$. From Section 9.5,

Equation (9.32),

$$\begin{aligned}
\mathbf{E} &= \mathbf{Y}_1'\mathbf{R}\mathbf{Y}_1 - \hat{\mathbf{\Gamma}}'\mathbf{Y}_2'\mathbf{R}\mathbf{Y}_1 \\
&= \mathbf{C}_1'\mathbf{Y}'\mathbf{R}\mathbf{Y}\mathbf{C}_1 - \hat{\mathbf{\Gamma}}'\mathbf{C}_2'\mathbf{Y}'\mathbf{R}\mathbf{Y}\mathbf{C}_1 \\
&= (n-p)\left\{\mathbf{C}_1'\mathbf{S}\mathbf{C}_1' - \mathbf{C}_1'\mathbf{S}\mathbf{C}_2\left(\mathbf{C}_2'\mathbf{S}\mathbf{C}_2\right)^{-1}\mathbf{C}_2'\mathbf{S}\mathbf{C}_1\right\} \qquad [\text{by } (9.52)] \\
&= (n-p)\left\{\mathbf{C}_1'\mathbf{K}(\mathbf{K}'\mathbf{S}^{-1}\mathbf{K})^{-1}\mathbf{K}'\mathbf{C}_1\right\} \qquad [\text{by } (9.53)] \\
&= (n-p)(\mathbf{K}'\mathbf{S}^{-1}\mathbf{K})^{-1} \qquad (\text{since } \mathbf{K}'\mathbf{C}_1 = \mathbf{I}_k).
\end{aligned}$$

Using (9.30), we have

$$\mathbf{H} = (\mathbf{A}\hat{\mathbf{\Delta}})'\left[\mathbf{A}(\mathbf{X}'\mathbf{X})^{-1}\mathbf{A}' + \mathbf{A}\mathbf{L}\mathbf{M}\mathbf{L}'\mathbf{A}'\right]^{-1}\mathbf{A}\hat{\mathbf{\Delta}},$$

where, by (9.53),

$$\begin{aligned}
\mathbf{L}\mathbf{M}\mathbf{L}' &= (\mathbf{X}'\mathbf{X})^{-1}\mathbf{X}'\mathbf{Y}_2\left(\mathbf{Y}_2'\mathbf{R}\mathbf{Y}_2\right)^{-1}\mathbf{Y}_2'\mathbf{X}(\mathbf{X}'\mathbf{X})^{-1} \\
&= \frac{1}{n-p}(\mathbf{X}'\mathbf{X})^{-1}\mathbf{X}'\mathbf{Y}\mathbf{C}_2\left(\mathbf{C}_2'\mathbf{S}\mathbf{C}_2\right)^{-1}\mathbf{C}_2'\mathbf{Y}'\mathbf{X}(\mathbf{X}'\mathbf{X})^{-1} \\
&= \frac{1}{n-p}(\mathbf{X}'\mathbf{X})^{-1}\mathbf{X}'\mathbf{Y}\left[\mathbf{S}^{-1} - \mathbf{S}^{-1}\mathbf{K}(\mathbf{K}'\mathbf{S}^{-1}\mathbf{K})^{-1}\mathbf{K}'\mathbf{S}^{-1}\right]\mathbf{Y}'\mathbf{X}(\mathbf{X}'\mathbf{X})^{-1}.
\end{aligned} \tag{9.55}$$

The extension to test $H_0: \mathbf{A\Delta D} = \mathbf{O}$ is straightforward. From Section 8.7, we simply replace the matrices $\mathbf{H}$ and $\mathbf{E}$ by $\mathbf{H}_D = \mathbf{D}'\mathbf{H}\mathbf{D}$ and $\mathbf{E}_D = \mathbf{D}'\mathbf{E}\mathbf{D}$. When H_0 is true, these latter matrices are (conditionally) independently distributed as v-dimensional Wishart distributions with a common dispersion matrix and degrees of freedom $m_H = q$ and $m_E = n - (p + d - k)$, respectively. Since these Wishart distributions do not depend on the condition that $\mathbf{Y}_2$ is fixed, they are also the unconditional distributions of $\mathbf{D}'\mathbf{H}\mathbf{D}$ and $\mathbf{D}'\mathbf{E}\mathbf{D}$ when H_0 is true. We note that $n - p$ can be omitted from the above expressions if it is omitted from the definition of $\mathbf{S}$. If $\mathbf{X}$ has less than full rank, then we simply use $(\mathbf{X}'\mathbf{X})^-$ and replace p by rank $\mathbf{X}$.

We note that $\hat{\mathbf{\Delta}}$, $\mathbf{E}$, $\mathbf{L}\mathbf{M}\mathbf{L}'$, and $\mathbf{H}$ do not depend on $\mathbf{C}$, so that the above test of H_0 does not depend on $\mathbf{C}$. Therefore if we use Potthoff and Roy's $\mathbf{C}_1 = \mathbf{G}^{-1}\mathbf{K}(\mathbf{K}'\mathbf{G}^{-1}\mathbf{K})^{-1}$, the above test is independent of $\mathbf{G}$ and we can set $\mathbf{G} = \mathbf{I}_d$. In this case $\mathbf{C}_1 = \mathbf{K}(\mathbf{K}'\mathbf{K})^{-1}$, and choosing $\mathbf{C}_2$ such that $\mathbf{K}'\mathbf{C}_2 = \mathbf{O}$ implies that $\mathbf{C}_1'\mathbf{C}_2 = \mathbf{O}$. We can therefore choose $\mathbf{C}_2'$ to be the linearly independent rows of the projection matrix $\mathbf{I}_d - \mathbf{K}(\mathbf{K}'\mathbf{K})^{-1}\mathbf{K}$ (see B2.3). If normalized orthogonal polynomials are used, then $\mathbf{K}'\mathbf{K} = \mathbf{I}_k$, $\mathbf{C}_1 = \mathbf{K}$, and we can choose $\mathbf{C}_2$ such that $(\mathbf{C}_1, \mathbf{C}_2)$ is orthogonal. For instance, in the growth curve example given by (9.37) we can use $\phi_a(t_r)$ instead of t_r^a, where $\phi_a(t)$ is a polynomial of degree a and

$$\sum_{r=1}^{d} \phi_a(t_r)\phi_b(t_r) = \delta_{ab}.$$

With such a $\mathbf{K}$ we can then choose $\mathbf{C}_1$ to be the matrix of normalized orthogonal polynomials of degrees 0 to $k-1$, and $\mathbf{C}_2$ to be a similar matrix for degrees k to $d-1$ (see Rao [1966: pp. 100–101]; Grizzle and Allen [1969]).

The above method uses all the $d-k$ concomitant variables in $\mathbf{Y}_2$ and, in the context of the single growth curve example, Rao [1965] suggests that a better procedure might be to select only a subset of the concomitant variables, particularly if the correlations between any of the columns of $\mathbf{Y}_1$ and $\mathbf{Y}_2$ are small. If the covariance of each column of $\mathbf{Y}_2$ with each of $\mathbf{Y}_1$ is zero, then $\mathbf{Y}_2$ provides no information about $\mathbf{Y}_1$, and $\mathbf{Y}_2$ should be discarded. In this case Roy and Potthoff's method with $\mathbf{G} = \mathbf{I}_d$ is appropriate, the so-called "unweighted" estimation procedure.

When $\mathbf{G} = \mathbf{S}$, Roy and Potthoff's estimate $\hat{\boldsymbol{\Delta}}_1$ of (9.48) is the same as the maximum likelihood estimate $\hat{\boldsymbol{\Delta}}$ of (9.54). This equivalence between the two classes of estimates when all the concomitant variables are used was noted independently by Grizzle and Allen [1969] and Y. H. Lee [1974], and extended by Baksalary et al. [1978] to the case when not all the concomitant variables are used.

We can readily extend the theory of Section 8.6.3 and obtain the following set of simultaneous confidence intervals corresponding to the maximum root test of $H_0: \mathbf{A}\boldsymbol{\Delta}\mathbf{D} = \mathbf{O}$. With probability $1-\alpha$, $\mathbf{a}'\mathbf{A}\boldsymbol{\Delta}\mathbf{D}\mathbf{b}$ is contained in the interval

$$\mathbf{a}'\mathbf{A}\hat{\boldsymbol{\Delta}}\mathbf{D}\mathbf{b} \pm \left\{\frac{\theta_\alpha}{1-\theta_\alpha}\cdot \mathbf{a}'\mathbf{F}\mathbf{a}\cdot \mathbf{b}'\mathbf{E}_D\mathbf{b}\right\}^{1/2} \tag{9.56}$$

for all $\mathbf{a}$ and $\mathbf{b}$, where

$$\mathbf{F} = \mathbf{A}(\mathbf{X}'\mathbf{X})^{-1}\mathbf{A}' + \mathbf{A}\mathbf{L}\mathbf{M}\mathbf{L}'\mathbf{A}',$$

and θ_α is obtained from Appendix D14 with

$$s = \min(v, q), \qquad \nu_1 = \tfrac{1}{2}(|v-q|-1), \qquad \nu_2 = \tfrac{1}{2}[n-p-(d-k)-v-1].$$

If we are interested in r predetermined linear combinations, we can use the Bonferroni intervals

$$\mathbf{a}'\mathbf{A}\hat{\boldsymbol{\Delta}}\mathbf{D}\mathbf{b} \pm t_\nu^{\alpha/(2r)}\{\mathbf{a}'\mathbf{F}\mathbf{a}\cdot \mathbf{b}'\mathbf{E}_D\mathbf{b}/\nu\}^{1/2} \tag{9.57}$$

where $\nu = \eta - p - (d-k)$

c CHOICE OF METHOD

Which procedure do we use, the Roy–Potthoff transformation method or the Rao–Khatri reduction method? We have already mentioned above that both methods lead to the maximum likelihood estimate $\hat{\boldsymbol{\Delta}}$ when $\mathbf{G} = \mathbf{S}$ and all $d-k$

covariates (concomitant variables) are used. With $\mathbf{G} = \mathbf{I}_d$, the Roy–Potthoff method is equivalent to not using any covariates in the Rao–Khatri reduction. However, in hypothesis testing, $\mathbf{G} = \mathbf{S}$ is not appropriate in the Roy–Potthoff approach, as $\mathbf{S}$ is not independent of $\mathbf{Y}$ and the two methods have different test statistics for testing $H_0: \mathbf{A\Delta D} = \mathbf{O}$ (see Timm [1980] for a good discussion). Although the Rao–Khatri method avoids the choice of an arbitrary matrix, both methods require choosing the matrix $\mathbf{K}$. This is also equivalent to choosing the covariate matrix $\mathbf{Y}_2$ in the partition $\mathbf{YC} = (\mathbf{YC}_1, \mathbf{YC}_2) = (\mathbf{Y}_1, \mathbf{Y}_2)$. A test for the adequacy of $\mathbf{K}$ is described in Section 9.7.4 below. Also, by evaluating the correlations between the columns of $\mathbf{Y}_1$ and $\mathbf{Y}_2$, a selection of the $d - k$ covariates can be made to increase the efficiency of estimation. Those columns of $\mathbf{Y}_2$ that have small correlations with all the columns of $\mathbf{Y}_1$ are omitted (see Grizzle and Allen [1969] for a helpful discussion and some numerical examples).

In conclusion, we mention three queries raised by Timm [1975] with respect to the Rao–Khatri method. First, when considering more than one growth curve (see Section 9.7.5), the analysis of covariance method is only appropriate if the covariates do not affect the treatment conditions. This may not be the case here, as the covariates are not prechosen but are selected after the experiment. Second, different covariates may be appropriate for different groups, and, third, the selected set of covariates may vary from sample to sample. However, a major advantage of the Rao–Khatri method is that one can apply a standard multivariate analysis of covariance program to the data.

9.7.3. Single Growth Curve

We now apply the analysis of covariance method to the example in Section 9.7.1a. Here we have a random sample $\mathbf{v}_1, \mathbf{v}_2, \ldots, \mathbf{v}_n$ from $N_d(\mathbf{K}\boldsymbol{\gamma}, \boldsymbol{\Sigma})$ and we wish to test $H_0: \mathbf{D}'\boldsymbol{\gamma} = \mathbf{0}$. Setting $\mathbf{A} = 1$, $\boldsymbol{\Delta} = \boldsymbol{\gamma}'$, $\mathbf{X} = \mathbf{1}_n$, and $p = 1$ in the general theory,

$$(\mathbf{X}'\mathbf{X})^{-1}\mathbf{X}'\mathbf{Y} = \frac{1}{n}\mathbf{1}_n' \begin{pmatrix} \mathbf{v}_1' \\ \vdots \\ \mathbf{v}_n' \end{pmatrix} = \bar{\mathbf{v}}',$$

$$\mathbf{S} = \sum_{i=1}^{n} \frac{(\mathbf{v}_i - \bar{\mathbf{v}})(\mathbf{v}_i - \bar{\mathbf{v}})'}{n-1},$$

$$\hat{\boldsymbol{\gamma}} = \hat{\boldsymbol{\Delta}}' = (\mathbf{K}'\mathbf{S}^{-1}\mathbf{K})^{-1}\mathbf{K}'\mathbf{S}^{-1}\bar{\mathbf{v}} \qquad [\text{by } (9.54)]$$

$$\mathbf{H}_D = \mathbf{D}'\mathbf{H}\mathbf{D}$$

$$= \mathbf{D}'\hat{\boldsymbol{\gamma}}\left[\frac{1}{n} + \frac{1}{n-1}\bar{\mathbf{v}}'\{\mathbf{S}^{-1} - \mathbf{S}^{-1}\mathbf{K}(\mathbf{K}'\mathbf{S}^{-1}\mathbf{K})^{-1}\mathbf{K}'\mathbf{S}^{-1}\}\bar{\mathbf{v}}\right]^{-1}\hat{\boldsymbol{\gamma}}'\mathbf{D} \qquad [\text{by } (9.55)]$$

$$= \frac{\mathbf{D}'\hat{\boldsymbol{\gamma}}\hat{\boldsymbol{\gamma}}'\mathbf{D}}{w},$$

where w is the expression in square brackets, and

$$\mathbf{E}_D = (n-1)\mathbf{D}'(\mathbf{K}'\mathbf{S}^{-1}\mathbf{K})^{-1}\mathbf{D}.$$

When H_0 is true, $\mathbf{H}_D$ and $\mathbf{E}_D$ have v-dimensional Wishart distributions with $m_H = q = 1$ and $m_E = n - (1 + d - k)$ degrees of freedom, respectively, and one of the different test statistics described in Section 8.6.2 can be used for testing H_0. However, as $\mathbf{H}_D$ has rank 1, the test statistics are all equivalent to Hotelling's T^2 statistic $m_E \text{tr}[\mathbf{H}_D\mathbf{E}_D^{-1}]$ of Section 2.5.5e where

$$\begin{aligned}
\text{tr}\left[\mathbf{H}_D\mathbf{E}_D^{-1}\right] &= \frac{\text{tr}\left[\mathbf{D}'\hat{\boldsymbol{\gamma}}\hat{\boldsymbol{\gamma}}'\mathbf{D}\mathbf{E}_D^{-1}\right]}{w} \\
&= \frac{\text{tr}\left[(\mathbf{D}'\hat{\boldsymbol{\gamma}})'\mathbf{E}_D^{-1}\mathbf{D}'\hat{\boldsymbol{\gamma}}\right]}{w} \qquad \text{(by A1.1)} \\
&= \frac{[n/(n-1)](\mathbf{D}'\hat{\boldsymbol{\gamma}})'\left\{\mathbf{D}'[\mathbf{K}'\mathbf{S}^{-1}\mathbf{K}]^{-1}\mathbf{D}\right\}^{-1}\mathbf{D}'\hat{\boldsymbol{\gamma}}}{1 + [n/(n-1)]\bar{\mathbf{v}}'\left\{\mathbf{S}^{-1} - \mathbf{S}^{-1}\mathbf{K}(\mathbf{K}'\mathbf{S}^{-1}\mathbf{K})^{-1}\mathbf{K}\mathbf{S}^{-1}\right\}\bar{\mathbf{v}}}.
\end{aligned} \tag{9.58}$$

When H_0 is true,

$$\left[\frac{(m_E - v + 1)}{v}\right]\text{tr}\left[\mathbf{H}_D\mathbf{E}_D^{-1}\right] \sim F_{v,\, m_E - v + 1}.$$

This statistic was derived by Rao [1959] using a different approach.

The maximum likelihood estimator $\hat{\boldsymbol{\gamma}}$ for the analysis of covariance method, the so-called weighted least squares estimate, may be compared with the minimum variance unbiased estimator (MVUE),

$$\tilde{\boldsymbol{\gamma}} = (\mathbf{K}'\boldsymbol{\Sigma}^{-1}\mathbf{K})^{-1}\mathbf{K}'\boldsymbol{\Sigma}^{-1}\bar{\mathbf{v}}, \tag{9.59}$$

and Roy and Potthoff's estimator (9.48), namely,

$$\hat{\boldsymbol{\gamma}}_1 = (\mathbf{K}'\mathbf{G}^{-1}\mathbf{K})^{-1}\mathbf{K}'\mathbf{G}^{-1}\bar{\mathbf{v}}.$$

An alternative approach to modeling growth curves that has been used in the past and which leads to the unweighted estimator $\hat{\boldsymbol{\gamma}}_1$ with $\mathbf{G} = \mathbf{I}_d$ is as follows. Suppose that each of the n individuals has a separate growth curve with independent normal observations, that is $\mathbf{v}_i \sim N_d(\mathbf{K}\boldsymbol{\beta}_i, \sigma^2\mathbf{I}_d)$, and suppose further that the regression coefficients are a random sample from another normal distribution so that $\boldsymbol{\beta}_i \sim N_k(\boldsymbol{\gamma}, \boldsymbol{\Phi})$. Combining these two models, we have $\mathbf{v}_i \sim N_d(\mathbf{K}\boldsymbol{\gamma}, \mathbf{K}\boldsymbol{\Phi}\mathbf{K}' + \sigma^2\mathbf{I}_d)$ or

$$\bar{\mathbf{v}} \sim N_d\left(\mathbf{K}\boldsymbol{\gamma}, \frac{1}{n}\boldsymbol{\Sigma}\right) \quad \text{and} \quad \boldsymbol{\Sigma} = \mathbf{K}\boldsymbol{\Phi}\mathbf{K}' + \sigma^2\mathbf{I}_d. \tag{9.60}$$

In this case the MVUE of $\boldsymbol{\gamma}$ is $(\mathbf{K}'\mathbf{K})^{-1}\mathbf{K}'\bar{\mathbf{v}}$ (Rao [1967: Lemma 5a]; see also Exercise 9.14). Since the analysis of covariance method does not depend on the choice of $\mathbf{C}$, we can choose $\mathbf{C}_1 = \mathbf{K}(\mathbf{K}'\mathbf{K})^{-1}$. Then

$$\begin{aligned}\mathscr{D}\left[\mathbf{C}_1'\mathbf{y}_i, \mathbf{C}_2'\mathbf{y}_i\right] &= \mathbf{C}_1'\boldsymbol{\Sigma}\mathbf{C}_2 \\ &= \mathbf{C}_1'\left(\mathbf{K}\boldsymbol{\Phi}\mathbf{K}' + \sigma^2\mathbf{I}_d\right)\mathbf{C}_2 \\ &= \mathbf{O},\end{aligned}$$

as $\mathbf{C}_1'\mathbf{C}_2 = (\mathbf{K}'\mathbf{K})^{-1}\mathbf{K}'\mathbf{C}_2$ and $\mathbf{K}'\mathbf{C}_2 = \mathbf{O}$. We therefore have a special case of the situation where the covariances between corresponding rows of $\mathbf{Y}_1$ and $\mathbf{Y}_2$ are zero. As argued in the previous section, $\mathbf{Y}_2$ should then be discarded, and the reason is now clear: The introduction of $\mathbf{Y}_2$ leads to a less efficient estimate of $\boldsymbol{\gamma}$.

Fearn [1975, 1977] gives a Bayesian analysis of the growth model using the above two-stage approach. In his 1975 paper he also considers the prediction problem and compares several Bayesian predictors with those given by Lee and Geisser [1972]. The two-stage model can also be written in the form

$$\bar{\mathbf{v}} = \mathbf{K}\boldsymbol{\beta} + \boldsymbol{\varepsilon},$$

where $\boldsymbol{\varepsilon} \sim N_d(\mathbf{0}, \{\sigma^2/n\}\mathbf{I}_d)$, $\boldsymbol{\beta} \sim N_k(\boldsymbol{\gamma}, n^{-1}\boldsymbol{\Phi})$, and $\mathscr{C}[\boldsymbol{\varepsilon}, \boldsymbol{\beta}] = \mathbf{O}$. The above regression model with stochastic coefficients $\boldsymbol{\beta}$ arises naturally in random-effects analysis of variance models. Bowden and Steinhorst [1973] use model (9.60) to construct a tolerance band so that a given proportion of individuals have their (conditional) expected growth curves $(1, t, t^2, \ldots)\boldsymbol{\beta}$ lying in the band for all t, with an overall probability of approximately $1 - \alpha$. They also give an interesting numerical example relating to the problem of deciding when to move fish from one hatchery pond to another for larger fish. For further references, see Geisser [1980].

If the appropriate assumptions are satisfied, it may be possible to impose a simpler covariance structure on the general growth model, as for example in the random effects model of Reinsel [1982]. Another approach is to use randomization tests (Zerbe [1979a, b]).

9.7.4 *Test for Internal Adequacy*

The linear model we have been considering is $\mathscr{E}[\mathbf{Y}] = \mathbf{XB}$, where $\mathbf{B} = \boldsymbol{\Delta}\mathbf{K}'$ and, since $\mathbf{K}'\mathbf{C}_2 = \mathbf{O}$,

$$\mathscr{E}[\mathbf{Y}_2] = \mathbf{XBC}_2 = \mathbf{X}\boldsymbol{\Delta}\mathbf{K}'\mathbf{C}_2 = \mathbf{O}.$$

Since $\mathbf{X}$ is $n \times p$ of rank p, $\mathbf{XBC}_2 = \mathbf{O}$ if and only if $\mathbf{BC}_2 = \mathbf{O}$. We can therefore test the adequacy of the model $\mathbf{B} = \boldsymbol{\Delta}\mathbf{K}'$ by testing $\mathscr{E}[\mathbf{Y}_2] = \mathbf{O}$, that is, by testing $H_0: \mathbf{I}_p\mathbf{BC}_2 = \mathbf{O}$ for the model $\mathscr{E}[\mathbf{Y}] = \mathbf{XB}$. This test is a special case of the linear hypothesis test described in Section 8.7 with $\mathbf{A} = \mathbf{I}_p$.

For the single growth curve of Section 9.7.3, $p = 1$, $\mathbf{B} = \boldsymbol{\eta}'$, and, from (8.71) with $\mathbf{D} = \mathbf{C}_2$, $\mathbf{x} = \mathbf{1}_n$ and $\hat{\mathbf{B}}$ now equal to $(\mathbf{X}'\mathbf{X})^{-1}\mathbf{X}'\mathbf{Y} = \bar{\mathbf{v}}'$,

$$\begin{aligned}\mathbf{H} &= (\hat{\mathbf{B}}\mathbf{C}_2)'(\mathbf{X}'\mathbf{X})\hat{\mathbf{B}}\mathbf{C}_2\\ &= n(\bar{\mathbf{v}}'\mathbf{C}_2)'(\bar{\mathbf{v}}'\mathbf{C}_2)\end{aligned}$$

and

$$\mathbf{E} = \mathbf{C}_2' \sum_{i=1}^{n} (\mathbf{v}_i - \bar{\mathbf{v}})(\mathbf{v}_i - \bar{\mathbf{v}})'\mathbf{C}_2.$$

Since $m_H = 1$, the usual multivariate tests all reduce to Hotelling's T^2 statistic (with $m_E = n - 1$), namely,

$$\begin{aligned}m_E \operatorname{tr}[\mathbf{H}\mathbf{E}^{-1}] &= (n-1)n(\mathbf{C}_2'\bar{\mathbf{v}})'\mathbf{E}^{-1}(\mathbf{C}_2'\bar{\mathbf{v}})\\ &= n(\mathbf{C}_2'\bar{\mathbf{v}})'[\mathbf{C}_2'\mathbf{S}\mathbf{C}_2]^{-1}\mathbf{C}_2'\bar{\mathbf{v}}.\end{aligned}$$

This is, of course, the same test statistic as that given in Section 3.4.2d (with $\mathbf{A} = \mathbf{C}_2'$).

9.7.5 A Case Study

Grizzle and Allen [1969] applied the Rao–Khatri method to the data in Table 9.13. Here we have observations on four groups of dogs showing the response of each dog at times 1, 3, 5, 7, 9, 11, and 13 min after an occlusion. This model, an extension of the second example in Section 9.7.1 from two to four populations, may be described as a one-way growth curve model with four groups (one control and three treatment groups) and 9, 10, 8, and 9 seven-dimensional observations, respectively, in the groups. A plot of the sample means for each group against time suggests that a polynomial curve of at least third degree is required. Since the times are equally spaced, orthogonal polynomials can be used. For a third-degree polynomial (see Pearson and Hartley [1972: Table 47],

$$\mathbf{K} = \begin{pmatrix} 1 & -3 & 5 & -1\\ 1 & -2 & 0 & 1\\ 1 & -1 & -3 & 1\\ 1 & 0 & -4 & 0\\ 1 & 1 & -3 & -1\\ 1 & 2 & 0 & -1\\ 1 & 3 & 5 & 1 \end{pmatrix}.$$

TABLE 9.13 Observations on Four Groups of Dogs at Given Times after Occlusion[a]

Control								Treatment 1							
	Time after occlusion (min)								Time after occlusion (min)						
Dog	1	3	5	7	9	11	13	Dog	1	3	5	7	9	11	13
1	4.0	4.0	4.1	3.6	3.6	3.8	3.1	10	3.4	3.4	3.5	3.1	3.1	3.7	3.3
2	4.2	4.3	3.7	3.7	4.8	5.0	5.2	11	3.0	3.2	3.0	3.0	3.1	3.2	3.1
3	4.3	4.2	4.3	4.3	4.5	5.8	5.4	12	3.0	3.1	3.2	3.0	3.3	3.0	3.0
4	4.2	4.4	4.6	4.9	5.3	5.6	4.9	13	3.1	3.2	3.2	3.2	3.3	3.1	3.1
5	4.6	4.4	5.3	5.6	5.9	5.9	5.3	14	3.8	3.9	4.0	2.9	3.5	3.5	3.4
6	3.1	3.6	4.9	5.2	5.3	4.2	4.1	15	3.0	3.6	3.2	3.1	3.0	3.0	3.0
7	3.7	3.9	3.9	4.8	5.2	5.4	4.2	16	3.3	3.3	3.3	3.4	3.6	3.1	3.1
8	4.3	4.2	4.4	5.2	5.6	5.4	4.7	17	4.2	4.0	4.2	4.1	4.2	4.0	4.0
9	4.6	4.6	4.4	4.6	5.4	5.9	5.6	18	4.1	4.2	4.3	4.3	4.2	4.0	4.2
								19	4.5	4.4	4.3	4.5	5.3	4.4	4.4
Average	4.11	4.18	4.51	4.77	5.07	5.22	4.72	Average	3.54	3.63	3.62	3.56	3.56	3.50	3.46

Treatment 2								Treatment 3							
	Time after occlusion (min)								Time after occlusion (min)						
Dog	1	3	5	7	9	11	13	Dog	1	3	5	7	9	11	13
20	3.2	3.3	3.8	3.8	4.4	4.2	3.7	28	3.1	3.5	3.5	3.2	3.0	3.0	3.2
21	3.3	3.4	3.4	3.7	3.7	3.6	3.7	29	3.3	3.2	3.6	3.7	3.7	4.2	4.4
22	3.1	3.3	3.2	3.1	3.2	3.1	3.1	30	3.5	3.9	4.7	4.3	3.9	3.4	3.5
23	3.6	3.4	3.5	4.6	4.9	5.2	4.4	31	3.4	3.4	3.5	3.3	3.4	3.2	3.4
24	4.5	4.5	5.4	5.7	4.9	4.0	4.0	32	3.7	3.8	4.2	4.3	3.6	3.8	3.7
25	3.7	4.0	4.4	4.2	4.6	4.8	5.4	33	4.0	4.6	4.8	4.9	5.4	5.6	4.8
26	3.5	3.9	5.8	5.4	4.9	5.3	5.6	34	4.2	3.9	4.5	4.7	3.9	3.8	3.7
27	3.9	4.0	4.1	5.0	5.4	4.4	3.9	35	4.1	4.1	3.7	4.0	4.1	4.6	4.7
								36	3.5	3.6	3.6	4.2	4.8	4.9	5.0
Average	3.60	3.73	4.20	4.44	4.50	4.53	4.26	Average	3.64	3.78	4.01	4.07	3.98	4.07	4.04

[a]Reproduced from J. E. Grizzle and D. M. Allen (1969). Analysis of growth and dose response curves. *Biometrics*, **25**, 357–381, Table 1. With permission from The Biometric Society.

If $\mathbf{y}_{ij} = \boldsymbol{\mu}_i + \mathbf{u}_{ij}$ represents the jth observation in group i ($i = 1,2,3,4$; $j = 1,2,\ldots,n_i$), then $\boldsymbol{\mu}_i = \mathbf{K}\boldsymbol{\delta}_i$, say, and

$$\mathbf{B} = \begin{pmatrix} \boldsymbol{\mu}_1' \\ \boldsymbol{\mu}_2' \\ \vdots \\ \boldsymbol{\mu}_4' \end{pmatrix} = \begin{pmatrix} \boldsymbol{\delta}_1' \\ \boldsymbol{\delta}_2' \\ \vdots \\ \boldsymbol{\delta}_4' \end{pmatrix} \mathbf{K}' = \boldsymbol{\Delta}\mathbf{K}'.$$

To test the adequacy of the third-degree polynomial model we can choose

$$\mathbf{C}_2 = \begin{pmatrix} 3 & -1 & 1 \\ -7 & 4 & -6 \\ 1 & -5 & 15 \\ 6 & 0 & -20 \\ 1 & 5 & 15 \\ -7 & -4 & 6 \\ 3 & 1 & 1 \end{pmatrix},$$

the coefficients corresponding to the orthogonal polynomials of degrees 4, 5, and 6. Since $\mathbf{K}'\mathbf{C}_2 = \mathbf{O}$, our test for adequacy given in Section 9.7.4 is equivalent to testing $H_0\colon \mathbf{B}\mathbf{C}_2 = \mathbf{O}$ for the model $\mathscr{E}[\mathbf{Y}] = \mathbf{X}\mathbf{B}$. Setting $\mathbf{A} = \mathbf{I}_p = \mathbf{I}_4$, $\mathbf{D} = \mathbf{C}_2$, $v = 3$, $q = 4$, and $n - p = 20$ in Section 8.7.1, we have

$$\mathbf{H} = (\hat{\mathbf{B}}\mathbf{C}_2)'(\mathbf{X}'\mathbf{X})\hat{\mathbf{B}}\mathbf{C}_2,$$

$$\mathbf{E} = \mathbf{C}_2'\mathbf{Y}'\left(\mathbf{I}_n - \mathbf{X}(\mathbf{X}'\mathbf{X})^{-1}\mathbf{X}'\right)\mathbf{Y}\mathbf{C}_2$$

$$= \mathbf{C}_2'\sum_i \sum_j (\mathbf{y}_{ij} - \bar{\mathbf{y}}_{i\cdot})(\mathbf{y}_{ij} - \bar{\mathbf{y}}_{i\cdot})'\mathbf{C}_2,$$

and, when H_0 is true,

$$\Lambda = \frac{|\mathbf{E}|}{|\mathbf{E} + \mathbf{H}|} \sim U_{3,4,20}.$$

Since $-f \log \Lambda = 8.29$ is less than $\chi_{12}^2(0.05) = 21.03$ and $C_\alpha > 1$ (see Appendix D13), we do not reject H_0.

In estimating $\boldsymbol{\Delta}$ we first consider which of the three covariates corresponding to the fourth-, fifth-, and sixth-degree orthogonal polynomials we shall use. From Table 9.14 we see that none of the columns in the (1, 2) part of the table has all its correlations small, so that all three covariates are appropriate.

TABLE 9.14 Correlation Matrix for the Columns of the Transformed Data $(\mathbf{YC}_1, \mathbf{YC}_2)^a$

	$\mathbf{YC}_1(1)$				$\mathbf{YC}_2(2)$		
	1.00	0.42	−0.17	−0.16	0.22	−0.19	−0.11
	—	1.00	0.25	−0.37	−0.30	0.02	0.02
$\mathbf{YC}_1$	—	—	1.00	−0.05	−0.70	−0.04	−0.20
(1)	—	—	—	1.00	−0.31	−0.57	0.43
	—	—	—	—	1.00	−0.12	0.18
$\mathbf{YC}_2$	—	—	—	—	—	1.00	−0.08
(2)	—	—	—	—	—	—	1.00

[a] Reproduced from J. E. Grizzle and D. M. Allen (1969). Analysis of growth and dose response curves. *Biometrics*, **25**, 357–381, Table 3. With permission from The Biometric Society.

Therefore, given

$$\mathbf{S} = \sum_i \sum_j \frac{(\mathbf{y}_{ij} - \bar{\mathbf{y}}_{i\cdot})(\mathbf{y}_{ij} - \bar{\mathbf{y}}_{i\cdot})'}{n - p}$$

$$= \begin{pmatrix} 0.2261 & 0.1721 & 0.1724 & 0.2054 & 0.1705 & 0.1958 & 0.1817 \\ - & 0.1696 & 0.1840 & 0.1919 & 0.1628 & 0.1700 & 0.1644 \\ - & - & 0.3917 & 0.3473 & 0.2370 & 0.1876 & 0.2194 \\ - & - & - & 0.4407 & 0.3689 & 0.2870 & 0.2582 \\ - & - & - & - & 0.4337 & 0.3733 & 0.3178 \\ - & - & - & - & - & 0.5235 & 0.4606 \\ - & - & - & - & - & - & 0.5131 \end{pmatrix}$$

we have, from (9.54),

$$\hat{\boldsymbol{\Delta}} = (\mathbf{X}'\mathbf{X})^{-1}\mathbf{X}'\mathbf{Y}\mathbf{S}^{-1}\mathbf{K}(\mathbf{K}'\mathbf{S}^{-1}\mathbf{K})^{-1}$$

$$= \begin{pmatrix} \hat{\boldsymbol{\delta}}_1' \\ \vdots \\ \hat{\boldsymbol{\delta}}_4' \end{pmatrix}$$

$$= \begin{array}{c} \begin{array}{cccc} \text{Intercept} & \text{Linear} & \text{Quadratic} & \text{Cubic} \end{array} \\ \begin{pmatrix} 32.70 & 4.14 & -4.68 & -1.14 \\ 25.07 & -0.69 & -1.00 & 0.11 \\ 28.59 & 4.10 & -2.90 & -0.48 \\ 27.12 & 1.91 & -1.55 & 0.04 \end{pmatrix} \end{array} \begin{array}{l} \text{group 1} \\ \text{group 2} \\ \text{group 3} \\ \text{group 4} \end{array}$$

Thus, for the control group, the fitted curve is

$$y = 32.70 + 4.14x - 4.68(x^2 - 4) - 1.14(x^3 - 7x)/6,$$

where $x = \frac{1}{2}(t - 7)$.

To test whether the four growth curves are the same, we test $\mathbf{A\Delta D} = \mathbf{O}$, where $\mathbf{D} = \mathbf{I}_4$ and

$$\mathbf{A} = \begin{pmatrix} 1 & -1 & 0 & 0 \\ 0 & 1 & -1 & 0 \\ 0 & 0 & 1 & -1 \end{pmatrix}.$$

This hypothesis is rejected, as might be expected from $\hat{\mathbf{\Delta}}$, and we can compare the coefficients of the four curves using the simultaneous intervals (9.56). For example, we can compare the coefficients of the quadratic terms of the first two curves by setting $\mathbf{a}' = (1,0,0)$ and $\mathbf{b}' = (0,0,1,0)$, that is,

$$\begin{aligned} \mathbf{a}'\mathbf{A\Delta Db} &= (1, -1, 0, 0)\mathbf{\Delta b} \\ &= (\boldsymbol{\delta}_1 - \boldsymbol{\delta}_2)'\mathbf{b} \\ &= \delta_{13} - \delta_{23}, \end{aligned}$$

say. In a similar fashion, we can compare the first and third curves by setting $\mathbf{a}' = (1,1,0)$ as $\mathbf{a}'\mathbf{A} = (1, 0, -1, 0)$. The vector $\mathbf{a}$ determines the pair of curves, and the vector $\mathbf{b}$ the corresponding coefficients. However, if we decided before the experiment that we would be interested in the $\binom{4}{2} \times 4 = 24$ pairs of coefficients, then we could use the Bonferroni intervals (9.57) with $r = 24$. To compare the two sets of intervals we note that $n = 36$ observations, $d = 7$ dimensions, $p = 4$ groups, $k = 4$ (third-degree polynomial), $q = 3$ (rank of $\mathbf{A}$), and $v = 4$ (columns in $\mathbf{D}$). Hence $s = 3$, $\nu_1 = 0$, and $\nu_2 = 12$, and, from Appendix D14, $\theta_{0.05} = 0.42$ and $[\theta_{0.05}/(1 - \theta_{0.05}]^{1/2} = 0.85$. This may be compared with (by linear interpolation in Appendix D1 with $r = 24$ and $\nu = m_E = 29$)

$$t_{29}^{0.05/48} m_E^{-1/2} = \frac{3.38}{\sqrt{29}} = 0.63.$$

The Bonferroni intervals are slightly shorter. Grizzle and Allen [1969] gave both sets of intervals in the form *difference* $\pm$ *LSD*, where LSD, the half-width, is referred to as the least significant difference. For a given $\mathbf{b}$, LSD varies only slightly with $\mathbf{a}$, and the authors, for convenience, used the average LSD; their values are reproduced in Table 9.15. Clearly, the intercept difference for groups 1 and 2 is a strong contributor to the significance of the test for the identity of the four curves.

TABLE 9.15 Tests of Differences Among Groups[a]

	Estimated Differences			
Group Comparison	Intercept	Linear	Quadratic	Cubic
1 vs. 4	5.59	2.23	−3.13	−1.15
2 vs. 4	−2.04	−2.59	0.55	−0.11
3 vs. 4	1.47	2.19	−0.85	−0.48
1 vs. 3	4.12	0.03	−2.28	−0.66
2 vs. 3	−3.51	−4.79	1.41	0.59
1 vs. 2	7.63	4.82	−3.68	−1.25
LSD based on:				
θ_α	8.10	7.13	7.13	1.46
t	5.94	5.23	5.23	1.07

[a]Reproduced from J. E. Grizzle and D. M. Allen (1969). Analysis of growth and dose response curves. *Biometrics*, **25**, 357–381, Table 4. With permission from The Biometric Society.

9.7.6 Further Topics

Kleinbaum [1973] (see also Woolson and Leeper [1980] and Schwertman et al. [1981]) has given a generalization of the growth curve model that allows for data which are missing, either by accident or design. Machin [1975] has considered a related problem in which certain observations are omitted by design but are compensated for by the introduction of new subjects so that the total number of observations remains constant. Chakravorti [1974, 1980] has studied the problem of testing the equality of several growth curves when the underlying dispersion matrices are unequal. Referring to (9.39), if we ignore the polynomial structure, then testing $\boldsymbol{\eta} = \boldsymbol{\nu}$ for the case of unequal dispersion matrices is the classical Behrens–Fisher problem discussed in Section 3.6.3. Unfortunately, the polynomial structure adds further complications. Following the approach of Ito and Schull [1964], Chakravorti showed that their results also applied to the growth model. For example, the T_g^2 statistic for testing H_0: $\boldsymbol{\gamma} = \boldsymbol{\delta}$ in (9.41) (or, more generally, $\boldsymbol{\gamma}_1 = \boldsymbol{\gamma}_2 = \cdots = \boldsymbol{\gamma}_m$), using the Rao–Khatri analysis of covariance model, is robust to differences in the dispersion matrices if equal numbers of observations are taken on each growth curve. Chakravorti also considered two exact test procedures for testing H_0 under unequal dispersion matrices.

We note that Jöreskog [1970, 1973, 1978] has also investigated a model like (9.60), but with a more general covariance structure of the form

$$\boldsymbol{\Sigma} = \mathbf{F}(\boldsymbol{\Lambda}\boldsymbol{\Phi}\boldsymbol{\Lambda}' + \boldsymbol{\Psi}^2)\mathbf{F}' + \boldsymbol{\Theta}^2.$$

He considered estimation and hypothesis testing for this general structure and gave a number of examples, including special models in the behavioral sciences, factor analysis, variance components, path analysis models and linear structural relationships, and growth curves with serially correlated errors. Khatri [1973] also considered tests on $\mathbf{\Sigma}$ for the growth curve, but focused his attention on the type of problems considered in Section 3.5, such as independence and sphericity. A test for $\mathbf{\Sigma} = \mathbf{\Sigma}_0$ is given by Khatri and Srivastava [1975].

An alternative approach to growth models, which fits polynomials to group differences rather than the groups, is given by Madsen [1977]. In (9.39) and (9.40) it is assumed that the growth curves for the two populations have the same degree and are sampled at the same points (t_r) in time. A method for relaxing both these assumptions is given by Zerbe and Walker [1977]. A useful class of models that uses gains (e.g., in weight) rather than the actual measurements was proposed by Snee and Acuff [1979]. Gafarian [1978] gave confidence bands for a multivariate polynomial regression, while Zerbe and Jones [1980] applied growth curve techniques to time series data. The problem of prediction is considered by Lee and Geisser [1972, 1975]. For further references and reviews of growth curve analysis, see Woolson and Leeper [1980] and Timm [1975, 1980].

A closely related topic to growth curve analysis is allometry, which may be described as the study of differences in shape associated with size. For example, instead of relating two size measurements x and y (e.g., lengths of two bones) to time, we may be interested in relating them to each other so that both absolute (i.e., size) and relative (i.e., shape) measurements are important. This bivariate case has been studied for over 50 years, with attention devoted largely to the model $y = \alpha x^{\beta}$, or

$$\log y = \log \alpha + \beta \log x. \tag{9.61}$$

Growth is often proportional, so that a simple model would assume that an increase dx in x during a small time interval of duration dt is proportional to x and dt, that is, $dx = k_x x\, dt$ or

$$\frac{1}{x}\frac{dx}{dt} = k_x. \tag{9.62}$$

Assuming the same model for y, and defining $\beta = k_y/k_x$, then

$$\frac{1}{y}\frac{dy}{dt} = \beta \frac{1}{x}\frac{dx}{dt}. \tag{9.63}$$

Canceling out dt and integrating leads to (9.61). The ratios on the left-hand side of (9.62) and (9.63) are called the specific growth rates. The assumption that these rates are constant (independent of x and y) is sufficient, but not necessary, for (9.63) and (9.61). The extension of this approach to more than two variables is not straightforward, and some authors have proposed using principal components and factor analysis. However, one of the first attempts to give clear definitions of what are meant by size and shape is the article by Mosimann [1970]. For further references on the subject, see Sprent [1972], Mosimann et al. [1978], and Siegel and Benson [1982].

EXERCISES 9

9.1 Show that when $I = 2$, the likelihood ratio statistic for testing $\boldsymbol{\mu}_1 = \boldsymbol{\mu}_2$ in the one-way classification is equivalent to T_0^2 of (3.86).

9.2 In the one-way classification model show that $\bar{\mathbf{y}}_{i\cdot}$ is the least squares estimate of $\boldsymbol{\mu}_i$.

9.3 Derive a Hotelling's T^2 statistic for testing $\boldsymbol{\mu}_r = \boldsymbol{\mu}_s$ in the one-way classification model based on the pooled estimate of $\boldsymbol{\Sigma}$ [i.e., based on $\mathbf{E}/(n - I)$].

9.4 For the linear model of Equation (9.9), verify that $\mathbf{X}_2'\mathbf{1}_n = \mathbf{0}$.

9.5 Verify Equation (9.11).

9.6 For the two-way classification model of Section 9.4, show that $\bar{\mathbf{y}}_{ij\cdot}$ is the least squares estimate of $\boldsymbol{\mu}_{ij}$. Obtain a set of simultaneous confidence intervals for $\Sigma_j c_j \boldsymbol{\beta}_j'\mathbf{b}$ for all contrasts $(\Sigma_j c_j = 0)$ and all $\mathbf{b}$.

9.7 Verify Equation (9.29).

9.8 In Section 9.5.2 prove that any column of $\hat{\mathbf{B}}$ and any column of $\hat{\boldsymbol{\Gamma}}_G$ have zero covariance.

9.9 Referring to Example 9.7 in Section 9.5.2, show that

$$\hat{\boldsymbol{\gamma}}_G = \sum_i \sum_j (\mathbf{y}_{ij} - \bar{\mathbf{y}}_{i\cdot})(z_{ij} - \bar{z}_{i\cdot}) \Big/ \sum_i \sum_j (z_{ij} - \bar{z}_{i\cdot})^2$$

is an unbiased estimate of $\boldsymbol{\gamma}$. Find the dispersion matrix of $\hat{\boldsymbol{\gamma}}_G$. Derive a test statistic for testing $\gamma_1 = \gamma_2 = \cdots = \gamma_d$.

9.10 In multivariate bioassay we have two models (for the control and treatment), namely,

$$\mathbf{y}_{1jk} = \boldsymbol{\mu}_1 + \boldsymbol{\gamma}_1 z_j + \boldsymbol{\varepsilon}_{1jk},$$

$$\mathbf{y}_{2jk} = \boldsymbol{\mu}_2 + \boldsymbol{\gamma}_2 z_j + \boldsymbol{\varepsilon}_{2jk} \qquad (j = 1, 2, \ldots, J;\ k = 1, 2, \ldots, n),$$

where the $\boldsymbol{\varepsilon}_{ijk}$ are i.i.d. $N_d(\mathbf{0}, \boldsymbol{\Sigma})$. Derive a Hotelling's T^2 test for H_0: $\boldsymbol{\gamma}_1 = \boldsymbol{\gamma}_2$.

9.11 Let $\mathbf{Y} = \mathbf{XB} + \mathbf{U}$, where $\mathbf{X}$ is $n \times p$ of rank p and the rows of $\mathbf{U}$ are i.i.d. $N_d(\mathbf{0}, \boldsymbol{\Sigma})$. Let $\mathbf{a}_j'$ be the jth row of $(\mathbf{X}'\mathbf{X})^{-1}$, and $\boldsymbol{\beta}_j'$ the jth row of $\mathbf{B}$. If $\hat{\boldsymbol{\beta}}_j$ is the least squares estimate of $\boldsymbol{\beta}_j$, show that $\hat{\boldsymbol{\beta}}_j \sim N_d(\boldsymbol{\beta}_j, \|\mathbf{X}\mathbf{a}_j\|^2\boldsymbol{\Sigma})$. Derive a Hotelling's T^2 statistic for testing $H_0\!:\boldsymbol{\beta}_j = \mathbf{0}$. Use the general theory of testing the linear hypothesis $\mathbf{AB} = \mathbf{O}$ for the model $\mathbf{Y} = \mathbf{XB} + \mathbf{U}$ to obtain the same test. Derive the same test for testing $\boldsymbol{\beta}_p = \mathbf{0}$ using the analysis of covariance theory with $\boldsymbol{\Gamma} = \boldsymbol{\beta}_p'$.

9.12 Let $\mathbf{v}_1, \mathbf{v}_2, \ldots, \mathbf{v}_n$ be i.i.d. $N_d(\mathbf{K}\boldsymbol{\gamma}, \boldsymbol{\Sigma})$, where $\mathbf{K}$ is $d \times k$ of rank k. If $\mathbf{S} = \Sigma_i(\mathbf{v}_i - \overline{\mathbf{v}})(\mathbf{v}_i - \overline{\mathbf{v}})'/(n-1)$, show that $\hat{\boldsymbol{\gamma}} = (\mathbf{K}'\mathbf{S}^{-1}\mathbf{K})^{-1}\mathbf{K}'\mathbf{S}^{-1}\overline{\mathbf{v}}$ is the maximum likelihood estimate of $\boldsymbol{\gamma}$.

9.13 Use the Rao–Khatri analysis of covariance method to derive a Hotelling's T^2 test for testing $\gamma_2 = \delta_2$ in (9.40). Express the hypothesis that the two quadratic growth curves in (9.40) are identical in the form $\mathbf{A}\boldsymbol{\Delta}\mathbf{D} = \mathbf{O}$.

9.14 Prove that (9.59) reduces to $(\mathbf{K}'\mathbf{K})^{-1}\mathbf{K}'\overline{\mathbf{v}}$ when $\boldsymbol{\Sigma} = \mathbf{K}\boldsymbol{\Phi}\mathbf{K}' + \sigma^2\mathbf{I}_d$, where $\boldsymbol{\Sigma}$ and $\boldsymbol{\Phi}$ are positive definite and the columns of $\mathbf{K}$ are linearly independent. [Hint: Use B3.5]

9.15 For the general analysis of covariance model of Section 9.5.2, show that

$$\operatorname{var}[\mathbf{a}'\mathbf{A}\hat{\mathbf{B}}_G\mathbf{b}] = \mathbf{a}'\left[\mathbf{A}(\mathbf{X}'\mathbf{X})^{-1}\mathbf{A}' + \mathbf{ALML'A'}\right]\mathbf{a} \cdot \mathbf{b}'\boldsymbol{\Sigma}\mathbf{b},$$

where $\mathbf{L}$ and $\mathbf{M}$ are defined in Section 9.5.1. Find a set of simultaneous confidence intervals for all linear combinations $\mathbf{a}'\mathbf{A}\mathbf{B}\mathbf{b}$.

9.16 Consider the one-way classification model of Section 9.2.5 with typical row vector

$$\mathscr{E}[\mathbf{y}'|\mathbf{x}'] = \boldsymbol{\beta}_1' + \mathbf{x}'\mathbf{B}_2,$$

where $\boldsymbol{\beta}_1 = \overline{\boldsymbol{\mu}}_{\cdot}$, $\mathbf{B}_2 = (\boldsymbol{\alpha}_1, \boldsymbol{\alpha}_2, \ldots, \boldsymbol{\alpha}_{I-1})'$, and $\mathbf{x}'$ is a row of $\mathbf{X}_2$ indicating which of the I populations $\mathbf{y}$ comes from, that is, $\mathbf{x}'$ takes values $(1,0,\ldots,0)$, $(0,1,\ldots,0), \ldots, (0,0,\ldots,0,1)$, and $(0,0,\ldots,0)$, respectively. Here $\mathbf{x}$ is usually described as the vector of dummy variables. The likelihood ratio test for equal means is [by (2.48)]

$$\Lambda = \frac{|\mathbf{E}|}{|\mathbf{E} + \mathbf{H}|} = \prod_j (1 - \theta_j).$$

Show that $\theta_j = r_j^2$, the square of the jth sample canonical correlation between the $\mathbf{y}$ and $\mathbf{x}$ variables. [Hint: Use (10.25) and (10.26).]

CHAPTER 10

Special Topics

10.1 COMPUTATIONAL TECHNIQUES

10.1.1 Solving the Normal Equations

In least squares estimation for linear models, we have to solve the so-called normal equations

$$\mathbf{F}\hat{\mathbf{B}} = \mathbf{X}'\mathbf{Y}, \tag{10.1}$$

where $\mathbf{F} = \mathbf{X}'\mathbf{X}$, $\mathbf{X}$ is $n \times p$ and $\hat{\mathbf{B}}$ is $p \times d$. If $\mathbf{X}$ has rank p, $\mathbf{F}$ is nonsingular (in fact, positive definite), and (10.1) has a unique solution $\hat{\mathbf{B}} = \mathbf{F}^{-1}\mathbf{X}'\mathbf{Y}$. Two basic methods for finding this solution are described briefly below. They both consist of reducing the problem to that of solving a triangular system of equations, as such systems can be solved very accurately (Wilkinson [1965, 1967]). A third method is given in Section 10.1.5b, and all three methods can be adapted to handle the case when $\mathbf{X}$ has rank less than p. A number of general references are available, including Golub [1969], Lawson and Hanson [1974], Seber [1977: Chapter 11], Chambers [1977: Chapter 5], Kennedy and Gentle [1980], and, in particular, Maindonald [1984]. Useful algorithms for solving various matrix problems are given by Wilkinson and Reinsch [1971]. A helpful general reference on matrix computations is given by Stewart [1973].

a CHOLESKY DECOMPOSITION

Since $\mathbf{F}$ is positive definite, it can be expressed *uniquely* in the form (see A5.11) $\mathbf{F} = \mathbf{U}'\mathbf{U}$, where $\mathbf{U}$ is a real upper triangular $p \times p$ matrix with positive diagonal elements. Once $\mathbf{U}$ is found, we solve $\mathbf{U}'\mathbf{U}\hat{\mathbf{B}} = \mathbf{X}'\mathbf{Y}$ by solving two triangular systems of equations,

$$\mathbf{U}'\mathbf{Z} = \mathbf{X}'\mathbf{Y} \text{ for } \mathbf{Z} \quad \text{and} \quad \mathbf{U}\hat{\mathbf{B}} = \mathbf{Z} \text{ for } \hat{\mathbf{B}}. \tag{10.2}$$

We can then calculate the matrix of residuals $\mathbf{Y} - \mathbf{X}\hat{\mathbf{B}}$, and $\mathbf{E} = (\mathbf{Y} - \mathbf{X}\hat{\mathbf{B}})'(\mathbf{Y} - \mathbf{X}\hat{\mathbf{B}})$. For later reference we note that

$$\begin{aligned}\mathbf{E} &= \mathbf{Y}'\mathbf{Y} - \hat{\mathbf{B}}'\mathbf{X}'\mathbf{X}\hat{\mathbf{B}} \\ &= \mathbf{Y}'\mathbf{Y} - \hat{\mathbf{B}}'\mathbf{U}'\mathbf{U}\hat{\mathbf{B}} \\ &= \mathbf{Y}'\mathbf{Y} - \mathbf{Z}'\mathbf{Z}. \end{aligned} \tag{10.3}$$

To find $\mathbf{F}^{-1}$, we note that

$$\mathbf{F}^{-1} = (\mathbf{U}'\mathbf{U})^{-1} = \mathbf{U}^{-1}(\mathbf{U}^{-1})' = \mathbf{V}\mathbf{V}',$$

say, where $\mathbf{V}$ is also upper triangular, since the inverse of an upper triangular matrix is upper triangular. Details of algorithms for finding $\mathbf{U}$, $\mathbf{V}$, and $\mathbf{F}^{-1}$ are given, for example, by Seber [1977: Section 11.2.2] and Maindonald [1977, 1984].

A helpful modification is to use the augmented matrix

$$\begin{aligned}\mathbf{F}_{\text{AUG}} &= (\mathbf{X}, \mathbf{Y})'(\mathbf{X}, \mathbf{Y}) \\ &= \begin{pmatrix} \mathbf{X}'\mathbf{X}, & \mathbf{X}'\mathbf{Y} \\ \mathbf{Y}'\mathbf{X}, & \mathbf{Y}'\mathbf{Y} \end{pmatrix}, \end{aligned} \tag{10.4}$$

and find the Cholesky decomposition $\mathbf{F}_{\text{AUG}} = \mathbf{U}'_{\text{AUG}}\mathbf{U}_{\text{AUG}}$. If $\mathbf{E} = \mathbf{T}'\mathbf{T}$ is the Cholesky decomposition of $\mathbf{E}$, then, by (10.3), $\mathbf{Z}'\mathbf{Z} = \mathbf{T}'\mathbf{T} + \mathbf{Y}'\mathbf{Y}$ and

$$\mathbf{U}_{\text{AUG}} = \begin{pmatrix} \mathbf{U}, & \mathbf{Z} \\ \mathbf{O}, & \mathbf{T} \end{pmatrix}, \tag{10.5}$$

as

$$\begin{aligned}\mathbf{U}'_{\text{AUG}}\mathbf{U}_{\text{AUG}} &= \begin{pmatrix} \mathbf{U}'\mathbf{U}, & \mathbf{U}'\mathbf{Z} \\ \mathbf{Z}'\mathbf{U}, & \mathbf{Z}'\mathbf{Z} + \mathbf{T}'\mathbf{T} \end{pmatrix} \\ &= \begin{pmatrix} \mathbf{X}'\mathbf{X}, & \mathbf{X}'\mathbf{Y} \\ \mathbf{Y}'\mathbf{X}, & \mathbf{Y}'\mathbf{Y} \end{pmatrix} \qquad [\text{by (10.2)}]. \end{aligned}$$

Thus if we find the Cholesky decomposition of $\mathbf{F}_{\text{AUG}}$ instead of $\mathbf{F}$, we obtain $\mathbf{U}$, $\mathbf{Z}$, $\mathbf{T}$, and $\mathbf{E} = \mathbf{T}'\mathbf{T}$. For further statistical properties, see Hawkins and Eplett [1982].

b *QR*-ALGORITHM

A more accurate method, which avoids the errors in computing $\mathbf{X}'\mathbf{X}$ (and squaring the condition number of $\mathbf{X}$), goes for $\mathbf{U}$ directly. Using the

Gram–Schmidt orthogonalization process, we can find an orthogonal matrix $\mathbf{Q} = (\mathbf{Q}_p, \mathbf{Q}_{n-p})$, where the p columns of $\mathbf{Q}_p$ form an orthonormal basis for $\mathbf{X}$ (i.e., $\mathbf{Q}'_{n-p}\mathbf{X} = \mathbf{O}$), such that

$$\mathbf{X} = \mathbf{Q}_p\mathbf{U} \tag{10.6}$$

$$= \mathbf{Q}\begin{pmatrix} \mathbf{U} \\ \mathbf{O} \end{pmatrix} \qquad (= \mathbf{QR}, \text{ say}). \tag{10.7}$$

From (10.2), $\mathbf{Z} = (\mathbf{U}^{-1})'\mathbf{X}'\mathbf{Y} = \mathbf{Q}'_p\mathbf{Y}$. Also, $\mathbf{Q}'_p\mathbf{X} = \mathbf{U}$, so that

$$\mathbf{Q}'(\mathbf{X}, \mathbf{Y}) = \begin{pmatrix} \mathbf{Q}'_p \\ \mathbf{Q}'_{n-p} \end{pmatrix}(\mathbf{X}, \mathbf{Y})$$

$$= \begin{pmatrix} \mathbf{U}, & \mathbf{Z} \\ \mathbf{O}, & \mathbf{K} \end{pmatrix},$$

say. From $\mathbf{F}_{\text{AUG}} = (\mathbf{X}'\mathbf{Y})'\mathbf{QQ}'(\mathbf{X}, \mathbf{Y})$ of (10.4), we find that $\mathbf{K}'\mathbf{K} = \mathbf{E}$. If $\mathbf{Q}$ is chosen so that $\mathbf{K}$ is upper triangular with positive diagonal elements, then $\mathbf{K} = \mathbf{T}$, by the uniqueness of the Cholesky decomposition.

Several methods for finding the orthogonal matrix $\mathbf{Q}$ are available, namely, the modified Gram–Schmidt algorithm based on constructing an orthonormal basis for the space spanned by the columns of $\mathbf{X}$; the Householder method, which uses column transformations (reflections) to change $\mathbf{X}$ into $\mathbf{R}$; and the Givens method, which rotates the rows of $\mathbf{X}$ two at a time. These methods are preferable to the Cholesky method when $\mathbf{X}'\mathbf{X}$ is close to singularity (i.e., ill conditioned): it is then better to work directly with $\mathbf{X}$ rather than form $\mathbf{X}'\mathbf{X}$.

Both the Cholesky and QR-methods can be adapted to deal with the case when $\mathbf{X}$ does not have rull rank (Golub and Styan [1973]; Seber [1977: Section 11.5]). One approach is to obtain a solution of the normal equations in the form $\hat{\mathbf{B}} = \mathbf{F}^{-}(\mathbf{X}'\mathbf{Y})$, where $\mathbf{F}^{-}$ is a generalized inverse of $\mathbf{F}$, or in the form $\mathbf{X}^*\mathbf{Y}$, where $\mathbf{X}^*$ is a generalized inverse of $\mathbf{X}$ with certain properties, for example, $\mathbf{X}^* = \mathbf{X}^+$, the Moore–Penrose inverse (see Noble [1976] for a helpful discussion). These methods can also be modified to avoid square roots and to incorporate weighted least squares. When the rank of $\mathbf{X}$ is unknown, the singular value decomposition method of Section 10.1.5b can be used.

10.1.2 Hypothesis Matrix

Consider the linear model $\mathscr{E}[\mathbf{Y}] = \mathbf{XB}$, where $\mathbf{X}$ is $n \times p$ of rank p, and hypothesis $H_0: \mathbf{AB} = \mathbf{C}$, where $\mathbf{A}$ is $q \times p$ of rank q. We have already discussed the computation of $\mathbf{E} = (\mathbf{Y} - \mathbf{X}\hat{\mathbf{B}})'(\mathbf{Y} - \mathbf{X}\hat{\mathbf{B}})$ in the previous section and we now turn our attention to [from (8.47)]

$$\mathbf{H} = (\mathbf{A}\hat{\mathbf{B}} - \mathbf{C})'\left[\mathbf{A}(\mathbf{X}'\mathbf{X})^{-1}\mathbf{A}'\right]^{-1}(\mathbf{AB} - \mathbf{C}).$$

The following method for computing $\mathbf{H}$ was essentially proposed by Golub and Styan [1973]:

(1) Find the Cholesky decomposition $\mathbf{X}'\mathbf{X} = \mathbf{U}'\mathbf{U}$.
(2) Solve the triangular system of equations $\mathbf{U}'\mathbf{K}_1 = \mathbf{A}'$ for $\mathbf{K}_1$.
(3) Find the Cholesky decompositon $\mathbf{K}_1'\mathbf{K}_1 = \mathbf{U}_1'\mathbf{U}_1$.
(4) Solve the triangular system $\mathbf{U}_1'\mathbf{K}_2 = \mathbf{A}\hat{\mathbf{B}} - \mathbf{C}$ for $\mathbf{K}_2$.
(5) Calculate $\mathbf{K}_2'\mathbf{K}_2$.

We note that

$$\mathbf{A}(\mathbf{X}'\mathbf{X})^{-1}\mathbf{A}' = \mathbf{A}\mathbf{U}^{-1}(\mathbf{U}^{-1})'\mathbf{A}' = \mathbf{K}_1'\mathbf{K}_1 = \mathbf{U}_1'\mathbf{U}_1$$

and

$$\begin{aligned}\mathbf{K}_2'\mathbf{K}_2 &= (\mathbf{A}\hat{\mathbf{B}} - \mathbf{C})'\mathbf{U}_1^{-1}(\mathbf{U}_1^{-1})'(\mathbf{A}\hat{\mathbf{B}} - \mathbf{C}) \\ &= (\mathbf{A}\hat{\mathbf{B}} - \mathbf{C})'(\mathbf{U}_1'\mathbf{U}_1)^{-1}(\mathbf{A}\hat{\mathbf{B}} - \mathbf{C}) \\ &= \mathbf{H}.\end{aligned}$$

The matrices $\mathbf{U}$ and $\mathbf{U}_1$ can also be found using a *QR*-algorithm, thus avoiding the formation of $\mathbf{X}'\mathbf{X}$ and $\mathbf{K}_1'\mathbf{K}_1$.

10.1.3 *Calculating Hotelling's* T^2

Hotelling's T^2 test takes the form $T_0^2 = k\mathbf{y}'\mathbf{W}^{-1}\mathbf{y}$, and this can be computed as follows. First find the Cholesky decomposition $\mathbf{W} = \mathbf{U}'\mathbf{U}$, where $\mathbf{U}$ is upper triangular, and then solve the triangular set $\mathbf{U}'\mathbf{z} = \mathbf{y}$ for $\mathbf{z}$. The answer we require is $T_0^2 = k\mathbf{z}'\mathbf{z}$, as

$$\mathbf{y}'\mathbf{W}^{-1}\mathbf{y} = \mathbf{y}'(\mathbf{U}'\mathbf{U})^{-1}\mathbf{y} = \left(\mathbf{U}'^{-1}\mathbf{y}\right)'\left(\mathbf{U}'^{-1}\mathbf{y}\right) = \mathbf{z}'\mathbf{z}.$$

For example, suppose we are given a random sample $\mathbf{x}_1, \mathbf{x}_2, \ldots, \mathbf{x}_n$ from $N_d(\boldsymbol{\mu}, \boldsymbol{\Sigma})$. Then to test $H_0: \boldsymbol{\mu} = \boldsymbol{\mu}_0$ we use (see Section 3.3.1)

$$T_0^2 = n(n-1)(\bar{\mathbf{x}} - \boldsymbol{\mu}_0)'\mathbf{W}^{-1}(\bar{\mathbf{x}} - \boldsymbol{\mu}_0),$$

where

$$\mathbf{W} = \sum_{i=1}^{n} (\mathbf{x}_i - \bar{\mathbf{x}})(\mathbf{x}_i - \bar{\mathbf{x}})' = \tilde{\mathbf{X}}'\tilde{\mathbf{X}} \tag{10.8}$$

and $\tilde{\mathbf{X}}' = (\mathbf{x}_1 - \bar{\mathbf{x}}, \mathbf{x}_2 - \bar{\mathbf{x}}, \ldots, \mathbf{x}_n - \bar{\mathbf{x}})$. We can find $\mathbf{U}$ either from the Cholesky decomposition of $\mathbf{W}$, or else by applying the *QR*-algorithm to $\tilde{\mathbf{X}}$. The separate computation of $\mathbf{W}$ or $\tilde{\mathbf{X}}$ can be avoided if we begin with the augmented matrix $(\mathbf{1}_n, \mathbf{X})$. Instead of $\mathbf{U}$, we now get

$$\begin{pmatrix} \sqrt{n} & \sqrt{n}\,\bar{\mathbf{x}} \\ \mathbf{0} & \mathbf{U} \end{pmatrix}.$$

Alternatively, as kindly pointed out by Professor Gene Golub, we can obtain T_0^2 directly by applying just d steps of the Cholesky algorithm as follows:

$$\begin{pmatrix} \mathbf{W} & \mathbf{y} \\ \mathbf{y}' & 0 \end{pmatrix} \to \begin{pmatrix} \mathbf{U} & (\mathbf{U}')^{-1}\mathbf{y} \\ * & -\mathbf{y}'\mathbf{W}^{-1}\mathbf{y} \end{pmatrix}.$$

Another application of Hotelling's T^2 is in testing for equality of two normal means, given samples $\mathbf{v}_1, \mathbf{v}_2, \ldots, \mathbf{v}_{n_1}$ and $\mathbf{w}_1, \mathbf{w}_2, \ldots, \mathbf{w}_{n_2}$. The test statistic is

$$T_0^2 = \frac{n_1 n_2 (n_1 + n_2 - 2)}{n_1 + n_2} (\bar{\mathbf{v}} - \bar{\mathbf{w}})' \mathbf{W}^{-1} (\bar{\mathbf{v}} - \bar{\mathbf{w}}),$$

where

$$\mathbf{W} = \sum_{i=1}^{n_1} (\mathbf{v}_i - \bar{\mathbf{v}})(\mathbf{v}_i - \bar{\mathbf{v}})' + \sum_{i=1}^{n_2} (\mathbf{w}_i - \bar{\mathbf{w}})(\mathbf{w}_i - \bar{\mathbf{w}})'.$$

The steps in the triangular reduction are now

$$\begin{pmatrix} 1 & 0 & \mathbf{v}_1' \\ \vdots & \vdots & \vdots \\ 1 & 0 & \mathbf{v}_{n_1}' \\ 0 & 1 & \mathbf{w}_1' \\ \vdots & \vdots & \vdots \\ 0 & 1 & \mathbf{w}_{n_2}' \end{pmatrix} \to \begin{pmatrix} \sqrt{n_1} & 0 & \sqrt{n_1}\,\bar{\mathbf{v}}' \\ 0 & \sqrt{n_2} & \sqrt{n_2}\,\bar{\mathbf{w}}' \\ \mathbf{0} & \mathbf{0} & \mathbf{U} \end{pmatrix}, \tag{10.9}$$

where $\mathbf{W} = \mathbf{U}'\mathbf{U}$.

10.1.4 Generalized Symmetric Eigenproblem

In Section 10.1.5c we mention procedures for finding eigenvalues and eigenvectors in relation to principal components. However, in multivariate analysis, a common problem is the calculation of the eigenvalues ϕ_i of $\mathbf{D}^{-1}\mathbf{C}$ (or, equivalently, of $\mathbf{C}\mathbf{D}^{-1}$ by A1.4), where $\mathbf{C}$ is symmetric and $\mathbf{D}$ is positive definite. A simple procedure is as follows: (1) Find the Cholesky decomposition $\mathbf{D} = \mathbf{T}'\mathbf{T}$, where $\mathbf{T}$ is upper triangular; (2) form the matrix $\mathbf{A}_1 = (\mathbf{T}')^{-1}\mathbf{C}\mathbf{T}^{-1}$, which has the same eigenvalues as $\mathbf{D}^{-1}\mathbf{C}$ (by A1.4); and (3) find the eigenvalues ϕ_i (and the eigenvectors $\mathbf{a}_i$ if necessary) of the symmetric matrix $\mathbf{A}$ using a standard package.

The matrix $\mathbf{A}_1$ is found by forward substituting for the columns of $\mathbf{Z}$ in $\mathbf{T}'\mathbf{Z} = \mathbf{C}$ and the columns of $\mathbf{A}_1$ in $\mathbf{T}'\mathbf{A}_1 = \mathbf{Z}'$. If $\mathbf{c}_i$ is an eigenvector of $\mathbf{D}^{-1}\mathbf{C}$ corresponding to ϕ_i, then

$$\mathbf{D}^{-1}\mathbf{C}\mathbf{c}_i = \phi_i\mathbf{c}_i,$$

or

$$\mathbf{T}^{-1}(\mathbf{T}')^{-1}\mathbf{C}\mathbf{T}^{-1}\mathbf{T}\mathbf{c}_i = \phi_i\mathbf{c}_i,$$

that is,

$$\mathbf{A}_1\mathbf{T}\mathbf{c}_i = \phi_i\mathbf{T}\mathbf{c}_i.$$

We therefore find $\mathbf{c}_i$ by solving $\mathbf{T}\mathbf{c}_i = \mathbf{a}_i$ for each i.

Algorithms for the above reduction to $\mathbf{A}_1$ are given, for example, by Martin and Wilkinson [1971]. The eigenproblem is also discussed briefly by C. Cohen [1977]. We now consider some applications of the above procedure.

a MULTIVARIATE LINEAR HYPOTHESIS

When testing a linear hypothesis H_0 for a multivariate linear model, we require the eigenvalues ϕ_i of $\mathbf{H}\mathbf{E}^{-1}$, which are the same as those of $\mathbf{E}^{-1}\mathbf{H}$, where $\mathbf{H}$ is the "hypothesis" matrix and $\mathbf{E}$ is the "error" matrix. The four test statistics that are used, namely, Wilks' likelihood ratio statistic Λ, Hotelling's T_g^2, Pillai's $V^{(s)}$, and Roy's $\phi_{\max}$, are all functions of the ϕ_i (see Sections 8.6.2 and 2.5.5). Thus

$$\Lambda = \frac{|\mathbf{E}|}{|\mathbf{E}+\mathbf{H}|} \qquad \left(= \frac{|\mathbf{E}|}{|\mathbf{E}_H|}\right)$$

$$= \frac{1}{|\mathbf{I}_d + \mathbf{H}\mathbf{E}^{-1}|}$$

$$= \prod_{i=1}^{d}(1+\phi_i)^{-1},$$

$$T_g^2 = m_E \operatorname{tr}[\mathbf{H}\mathbf{E}^{-1}] = m_E \sum_i \phi_i,$$

$$V^{(s)} = \operatorname{tr}\left[\mathbf{H}(\mathbf{E}+\mathbf{H})^{-1}\right] = \sum_i \frac{\phi_i}{1+\phi_i}.$$

If the linear model is $\mathscr{E}[\mathbf{Y}] = \mathbf{X}\mathbf{B}$, then the reduction of $(\mathbf{X}, \mathbf{Y})$ to upper triangular form (10.5) using the Cholesky or QR-methods automatically gives us $\mathbf{T}$, where $\mathbf{E} = \mathbf{T}'\mathbf{T}$ is the Cholesky decomposition of $\mathbf{E}$. We then find $\mathbf{A}_1 = (\mathbf{T}')^{-1}\mathbf{H}\mathbf{T}^{-1}$, as described above; $\mathbf{H}$ can be found using Section 10.1.2.

Alternative methods are available for the most common form of hypothesis testing in linear models, namely, $H_0: \mathbf{B}_2 = \mathbf{O}$, where

$$\mathbf{XB} = (\mathbf{X}_1, \mathbf{X}_2)\begin{pmatrix} \mathbf{B}_1 \\ \mathbf{B}_2 \end{pmatrix}$$

(see Section 9.2.5). Partitioning $\mathbf{Z}$ in the same way as $\mathbf{B}$, that is,

$$\mathbf{Z} = \begin{pmatrix} \mathbf{Z}_1 \\ \mathbf{Z}_2 \end{pmatrix},$$

we have, from (10.3),

$$\mathbf{E} = \mathbf{Y}'\mathbf{Y} - \mathbf{Z}'\mathbf{Z} = \mathbf{Y}'\mathbf{Y} - \mathbf{Z}_1'\mathbf{Z}_1' - \mathbf{Z}_2'\mathbf{Z}_2. \tag{10.10}$$

The effect of deleting $\mathbf{X}_2$ and $\mathbf{B}_2$ from $\mathbf{XB}$ on the triangular reduction (10.5) of $(\mathbf{X}, \mathbf{Y})$ is to delete $\mathbf{Z}_2$ from $\mathbf{Z}$. Hence $\mathbf{E}_H = \mathbf{Y}'\mathbf{Y} - \mathbf{Z}_1'\mathbf{Z}_1$, $\mathbf{H} = \mathbf{E}_H - \mathbf{E} = \mathbf{Z}_2'\mathbf{Z}_2$, and

$$\begin{aligned} T_g^2 &= m_E \operatorname{tr}[\mathbf{H}\mathbf{E}^{-1}] \\ &= m_E \operatorname{tr}\left[\mathbf{Z}_2'\mathbf{Z}_2\mathbf{T}^{-1}(\mathbf{T}')^{-1}\right] \\ &= m_E \operatorname{tr}\left[(\mathbf{T}')^{-1}\mathbf{Z}_2'\mathbf{Z}_2\mathbf{T}^{-1}\right] \\ &= m_E \operatorname{tr}[\mathbf{G}'\mathbf{G}], \end{aligned} \tag{10.11}$$

where $\mathbf{G}$ is the solution of the triangular set of equations $\mathbf{T}'\mathbf{G}' = \mathbf{Z}_2'$. If the matrix

$$\begin{pmatrix} \mathbf{Z}_2 \\ \mathbf{T} \end{pmatrix}$$

is also reducted to the upper triangular matrix, $\mathbf{V}$, say, then

$$\mathbf{T}'\mathbf{T} + \mathbf{Z}_2'\mathbf{Z}_2 = \mathbf{E} + \mathbf{H} = \mathbf{V}'\mathbf{V}. \tag{10.12}$$

Since the determinant of a triangular matrix is the product of its diagonal elements,

$$\begin{aligned}\Lambda &= \frac{|\mathbf{E}|}{|\mathbf{E}+\mathbf{H}|}\\ &= \frac{|\mathbf{T}'\mathbf{T}|}{|\mathbf{V}'\mathbf{V}|}\\ &= \prod_{j=1}^{d}\left(\frac{t_{jj}^2}{v_{jj}^2}\right) \qquad (10.13)\\ &= b_1 b_2 \cdots b_d \qquad [\text{see } (2.70)].\end{aligned}$$

Also, if we repeat the argument that leads to (10.11),

$$\begin{aligned}V^{(s)} &= \operatorname{tr}\left[\mathbf{H}(\mathbf{E}+\mathbf{H})^{-1}\right]\\ &= \operatorname{tr}\left[\mathbf{G}_1'\mathbf{G}_1\right], \qquad (10.14)\end{aligned}$$

where $\mathbf{G}_1$ is the solution of $\mathbf{V}'\mathbf{G}_1' = \mathbf{Z}_2'$.

In the above application we do not need to find the eigenvalues of $\mathbf{A}_1 = (\mathbf{T}')^{-1}\mathbf{Z}_2'\mathbf{Z}_2\mathbf{T}^{-1}$, as prescribed by the general eigenvalue method.

b ONE-WAY CLASSIFICATION

For the one-way classification described in Section 9.2.1 we have the model $\mathbf{y}_{ij} \sim N_d(\boldsymbol{\mu}_i, \boldsymbol{\Sigma})$ $(i = 1,2,\ldots,I;\ j = 1,2,\ldots,n_i)$ and hypothesis $H_0: \boldsymbol{\mu}_1 = \boldsymbol{\mu}_2 = \cdots = \boldsymbol{\mu}_I$. The required matrices are

$$\mathbf{H} = \sum_i n_i(\bar{\mathbf{y}}_{i\cdot} - \bar{\mathbf{y}}_{\cdot\cdot})(\bar{\mathbf{y}}_{i\cdot} - \bar{\mathbf{y}}_{\cdot\cdot})' \qquad (10.15)$$

and

$$\mathbf{E} = \sum_i \sum_j (\mathbf{y}_{ij} - \bar{\mathbf{y}}_{i\cdot})(\mathbf{y}_{ij} - \bar{\mathbf{y}}_{i\cdot})'. \qquad (10.16)$$

In this case we have

$$(\mathbf{X},\mathbf{Y}) = \begin{pmatrix} \mathbf{1}_{n_1} & \mathbf{0} & \cdots & \mathbf{0} & \mathbf{y}_{11}' \\ \mathbf{0} & \mathbf{1}_{n_2} & \cdots & \mathbf{0} & \mathbf{y}_{12}' \\ \vdots & \vdots & & \vdots & \vdots \\ \mathbf{0} & \mathbf{0} & \cdots & \mathbf{1}_{n_I} & \mathbf{y}_{In_I}' \end{pmatrix},$$

and (10.5) becomes

$$\mathbf{U}_{\text{AUG}} = \begin{pmatrix} \sqrt{n_1} & 0 & \cdots & 0 & \sqrt{n_1}\,\bar{\mathbf{y}}_{1\cdot}' \\ 0 & \sqrt{n_2} & \cdots & 0 & \sqrt{n_2}\,\bar{\mathbf{y}}_{2\cdot}' \\ \vdots & \vdots & & \vdots & \vdots \\ 0 & 0 & \cdots & \sqrt{n_I} & \sqrt{n_I}\,\bar{\mathbf{y}}_{I\cdot}' \\ & \mathbf{O} & & & \mathbf{T} \end{pmatrix},$$

where $\mathbf{T}'\mathbf{T} = \mathbf{E}$. If we now set

$$\mathbf{Z}_2' = \left(\sqrt{n_1}\,[\bar{\mathbf{y}}_{1\cdot} - \bar{\mathbf{y}}_{\cdot\cdot}], \ldots, \sqrt{n_I}\,[\bar{\mathbf{y}}_{I\cdot} - \bar{\mathbf{y}}_{\cdot\cdot}]\right),$$

we have $\mathbf{H} = \mathbf{Z}_2'\mathbf{Z}_2$, and we can use the methods of (10.11), (10.13), and (10.14).

c DISCRIMINANT COORDINATES

From Section 5.8, the matrices under consideration are $\mathbf{B} = \mathbf{H}$ and $\mathbf{W} = \mathbf{E}$, where $\mathbf{H}$ and $\mathbf{E}$ are given by (10.15) and (10.16). Finding the discriminant coordinates therefore reduces to finding the eigenvalues and eigenvectors of $\mathbf{W}^{-1}\mathbf{B}$ $(= \mathbf{E}^{-1}\mathbf{H})$.

10.1.5 Singular Value Decomposition

a DEFINITION

Any $n \times m$ matrix $\mathbf{A}$ of rank r $(r \le m \le n)$ can be expressed in the form (Seber [1977: Appendix A10])

$$\underset{(n\times m)}{\mathbf{A}} = \underset{(n\times m)(m\times m)(m\times m)}{\mathbf{L}_m \boldsymbol{\Delta} \mathbf{M}'}, \tag{10.17}$$

where $\mathbf{L}_m$ is an $n \times m$ matrix with columns consisting of m orthogonal unit eigenvectors associated with the m largest eigenvalues of $\mathbf{AA}'$ (i.e., $\mathbf{L}_m'\mathbf{L}_m = \mathbf{I}_m$), $\mathbf{M}$ is an $m \times m$ orthogonal matrix consisting of the orthogonal unit eigenvectors of the $m \times m$ matrix $\mathbf{A}'\mathbf{A}$, and $\boldsymbol{\Delta} = \text{diag}(\delta_1, \delta_2, \ldots, \delta_m)$ is an $m \times m$ diagonal matrix. Here

$$\delta_1 \ge \delta_2 \ge \cdots \ge \delta_r > \delta_{r+1} = \delta_{r+2} = \cdots = \delta_m = 0,$$

and these diagonal elements of $\boldsymbol{\Delta}$, called the singular values of $\mathbf{A}$, are the positive square roots of the eigenvalues of $\mathbf{A}'\mathbf{A}$. We note that $\mathbf{AA}'$ and $\mathbf{A}'\mathbf{A}$ are positive semidefinite with m common eigenvalues, including some zeros when $r < m$ (see A1.4). The remaining $n - m$ eigenvalues of $\mathbf{AA}'$ are equal to zero. The decomposition (10.17) is called the singular value decomposition (SVD) of $\mathbf{A}$.

An alternative version of (10.17) is

$$\mathbf{A} = \mathbf{L}\begin{pmatrix} \boldsymbol{\Delta} \\ \mathbf{O} \end{pmatrix}\mathbf{M}, \tag{10.18}$$

where $\mathbf{L}$ is an orthogonal $n \times n$ matrix with columns that are eigenvectors of $\mathbf{AA}'$; the first m columns of $\mathbf{L}$ form $\mathbf{L}_m$. If $m \geq n$, then

$$\mathbf{A} = \mathbf{L}(\boldsymbol{\Delta}, \mathbf{O})\mathbf{M}' = \mathbf{L}\boldsymbol{\Delta}\mathbf{M}_n', \tag{10.19}$$

where $\mathbf{M}_n$ consists of the first n columns of $\mathbf{M}$. A more general form, which includes all the above versions, is

$$\mathbf{A} = \mathbf{LDM}', \tag{10.20}$$

where $\mathbf{D}$ is an $n \times m$ matrix with r positive leading diagonal elements $\delta_1, \delta_2, \ldots, \delta_r$, and the remaining elements zero. All three forms (10.17), (10.19), and (10.20) are used in the literature, though, in practice, only the first s ($=$ minimum $[m, n]$) columns of $\mathbf{L}$ and $\mathbf{M}$ are required in (10.20). We can also write

$$a_{ij} = \sum_{k=1}^{r} \delta_k l_{ik} m_{jk}. \tag{10.21}$$

Is the decomposition unique? If we assume that $m \leq n$, $\boldsymbol{\Delta}$ will be unique, as the eigenvalues of $\mathbf{A}'\mathbf{A}$ are unique. However, the eigenvectors making up $\mathbf{L}_m$ and $\mathbf{M}$ in (10.17) will not be unique unless the eigenvalues are distinct and an appropriate sign convention is adopted for eigenvectors (see A1.3).

b SOLUTION OF NORMAL EQUATIONS

Let $\mathbf{X}$ be an $n \times p$ matrix of rank r and consider the SVD $\mathbf{X} = \mathbf{L}_p\boldsymbol{\Delta}\mathbf{M}'$. Then the Moore–Penrose generalized inverse of $\mathbf{X}$ is $\mathbf{X}^+ = \mathbf{M}\boldsymbol{\Delta}^+\mathbf{L}_p'$, where

$$\boldsymbol{\Delta}^+ = \operatorname{diag}\left(\delta_1^{-1}, \ldots, \delta_r^{-1}, 0, \ldots, 0\right).$$

A solution of the normal equations $\mathbf{X}'\mathbf{X}\hat{\mathbf{B}} = \mathbf{X}'\mathbf{Y}$ is now given by $\hat{\mathbf{B}} = \mathbf{X}^+\mathbf{Y}$, since $\mathbf{X}'\mathbf{X}\hat{\mathbf{B}} = \mathbf{X}'\mathbf{X}\mathbf{X}^+\mathbf{Y} = \mathbf{X}'\mathbf{Y}$. This method is particularly useful when $r = p$ but $\mathbf{X}'\mathbf{X}$ is highly ill conditioned, and when $r < p$ and r is unknown; a δ_i is deemed to be zero if its value is less than a certain tolerance. The most common algorithm consists of the following steps: (1) reduce $\mathbf{X}$ to upper bidiagonal form [only the (i, i) and $(i, i+1)$ elements are nonzero] using Householder reflections from the left and the right; (2) compute the SVD of the bidiagonal form; and (3) transform back using reverse reflections (Lawson and Hanson [1974: Chapter 18]). Algorithms are given in FORTRAN (e.g., Businger and Golub [1965], Lawson and Hanson [1974]) and in ALGOL60 (e.g., Golub and Reinsch [1971]). Another algorithm, suitable for a small computer, is given by Nash and Lefkovitch [1976].

Noble [1976] proved a theorem that states that if a matrix $\mathbf{X} = \mathbf{A} + \mathbf{K}$ is near $\mathbf{A}$ but has rank greater than $\mathbf{A}$, then $\mathbf{X}^+$ can be "larger" and completely different from $\mathbf{A}^+$: The smaller $\mathbf{K}$, the worse the trouble can be. It is clear, therefore, that r needs to be known at some stage of the calculation. If r is known from the beginning, then a *QR*-algorithm using, for example, Householder transformations is preferred, as the SVD method is two to four times as expensive and is less adaptable with regard to other aspects of regression such as updating.

c PRINCIPAL COMPONENTS

Given $\mathbf{W}$ of (10.8), we can find the eigenvalues and eigenvectors of $\mathbf{W}$, $\mathbf{S} = \mathbf{W}/(n-1)$, or $\hat{\mathbf{\Sigma}} = \mathbf{W}/n$ directly using a standard eigenvalue package such as the ALGOL procedures of Wilkinson and Reinsch [1971: Part II], and the FORTRAN procedures of Sparks and Todd [1973]. An alternative method, however, is to calculate the SVD of the $n \times d$ matrix $\tilde{\mathbf{X}}$. Writing $\tilde{\mathbf{X}} = \mathbf{L}_d \mathbf{\Delta} \mathbf{M}'$, the diagonal elements of $\mathbf{\Delta}^2$ are the eigenvalues of $\tilde{\mathbf{X}}'\tilde{\mathbf{X}} = \mathbf{W}$, and the columns of $\mathbf{M}$ are the corresponding eigenvectors.

d CANONICAL CORRELATIONS

Let $\mathbf{z}_1, \mathbf{z}_2, \ldots, \mathbf{z}_n$ be a random sample and suppose $\mathbf{z}_i' = (\mathbf{x}_i', \mathbf{y}_i')$. Then

$$\begin{aligned}
\tilde{\mathbf{Z}}' &= (\mathbf{z}_1 - \bar{\mathbf{z}}, \mathbf{z}_2 - \bar{\mathbf{z}}, \ldots, \mathbf{z}_n - \bar{\mathbf{z}}) \\
&= \begin{pmatrix} \tilde{\mathbf{X}}' \\ \tilde{\mathbf{Y}}' \end{pmatrix} \begin{matrix} \}\, d_1 \\ \}\, d_2 \end{matrix},
\end{aligned}$$

say. Using the *QR*-algorithm (10.6) twice, we have $\tilde{\mathbf{X}} = \mathbf{Q}_{d_1}\mathbf{T}_{11}$, where $\mathbf{Q}_{d_1}$ is an $n \times d_1$ matrix with orthogonal columns and $\mathbf{T}_{11}$ is a $d_1 \times d_1$ upper triangular matrix; similarly, we have $\tilde{\mathbf{Y}} = \mathbf{Q}_{d_2}\mathbf{T}_{22}$, where $\mathbf{Q}_{d_2}$ is $n \times d_2$ and $\mathbf{T}_{22}$ is $d_2 \times d_2$. Referring to Section 5.7.2, we have the Cholesky factorizations $\mathbf{Q}_{11} = \tilde{\mathbf{X}}'\tilde{\mathbf{X}} = \mathbf{T}_{11}'\mathbf{T}_{11}$ and $\mathbf{Q}_{22} = \tilde{\mathbf{Y}}'\tilde{\mathbf{Y}} = \mathbf{T}_{22}'\mathbf{T}_{22}$. Then

$$\begin{aligned}
\mathbf{Q}_{d_1}'\mathbf{Q}_{d_2} &= (\mathbf{T}_{11}^{-1})'\tilde{\mathbf{X}}'\tilde{\mathbf{Y}}\mathbf{T}_{22}^{-1} \\
&= (\mathbf{T}_{11}^{-1})'\mathbf{Q}_{12}\mathbf{T}_{22}^{-1} \\
&= \hat{\mathbf{C}},
\end{aligned}$$

say, where $\hat{\mathbf{C}}$ is the $d_1 \times d_2$ matrix of (5.97). Assuming that $d_1 \leq d_2$, consider the SVD of $\hat{\mathbf{C}}$, namely,

$$\hat{\mathbf{C}} = \mathbf{L}\mathbf{\Delta}\mathbf{M}_{d_1}'.$$

Then, referring to Section 5.7.2, we can identify

$$\mathbf{L} = (\hat{\mathbf{a}}_{01}, \hat{\mathbf{a}}_{02}, \ldots, \hat{\mathbf{a}}_{0d_1}) = \hat{\mathbf{A}}_0' = \mathbf{T}_{11}\hat{\mathbf{A}}',$$

and $\mathbf{M}_{d_1} = \hat{\mathbf{B}}_0' = \mathbf{T}_{22}\hat{\mathbf{B}}'$. We therefore have the following computational procedure (Golub [1969: pp. 393–1395]): Compute $\mathbf{Q}_{d_1}$ and $\mathbf{Q}_{d_2}$ using a *QR*-algorithm, form $\hat{\mathbf{C}}$ using an accurate (e.g., double precision) inner product procedure, compute the SVD of $\hat{\mathbf{C}}$, and use back-substitution to solve the triangular sets of equations

$$\mathbf{T}_{11}\hat{\mathbf{A}}' = \mathbf{L} \quad \text{and} \quad \mathbf{T}_{22}\hat{\mathbf{B}}' = \mathbf{M}_{d_1}$$

for $\hat{\mathbf{A}}'$ and $\hat{\mathbf{B}}'$. From (5.98), the canonical variates for $\mathbf{x}_1, \mathbf{x}_2, \ldots, \mathbf{x}_n$ are given by the rows of $\mathbf{U} = \tilde{\mathbf{X}}\hat{\mathbf{A}}'$, and those for the $\mathbf{y}_i$ by the rows of $\mathbf{V} = \tilde{\mathbf{Y}}\hat{\mathbf{B}}'$. The diagonal elements of $\boldsymbol{\Delta}$ (the square roots of the eigenvalues of $\hat{\mathbf{C}}'\hat{\mathbf{C}}$ and of $\mathbf{Q}_{11}^{-1}\mathbf{Q}_{12}\mathbf{Q}_{22}^{-1}\mathbf{Q}_{21}$) are the sample canonical correlations $\sqrt{r_j^2}$.

10.1.6 Selecting the Best Subset

In designing a multivariate experiment, one is faced with the problem of choosing appropriate variables for measurement. Usually experimenters quite reasonably tend to measure as many variables as possible to avoid missing any that might prove relevant. Unfortunately this approach commonly leads to excessive values of the dimension d of the data vectors, and very large samples are then usually required for an adequate analysis. Furthermore, the inclusion of irrelevant variables not only increases the cost of the experiment, but it can also cause a marked reduction in the effectiveness of a multivariate technique. For example, in multiple linear regression, an additional x-variable leads to an increase in the variance of prediction; in discriminant analysis (see Chapter 6) the precision of estimation and the robustness of various discriminant functions falls off with increasing d. Test procedures, like Hotelling's T^2 test for comparing two population means, can also be adversely affected by increasing d (Das Gupta and Perlman [1974]).

Graphical methods for reducing dimensionality are given in Chapter 5 and we shall now consider more formal procedures for dimension reduction. The problem is one of two types, depending on whether we are considering response (y) variables or regressor (x) variables.

a RESPONSE VARIABLES

Suppose we have a multivariate linear model $\mathbf{Y} = \mathbf{XB} + \mathbf{U}$, where $\mathbf{Y}$ is $n \times d$, and a linear hypothesis $H_0: \mathbf{AB} = \mathbf{C}$. From Chapter 8, H_0 can be tested using

$$U = \frac{|\mathbf{E}|}{|\mathbf{E} + \mathbf{H}|} = \frac{|\mathbf{E}|}{|\mathbf{E}_H|},$$

where $U \sim U_{d, m_H, m_E}$ when H_0 is true. If we wished to reduce the dimension d, we would try and eliminate the y-variables that make an insignificant contribution to the test of H_0. From Section 2.5.6 we have the factorization of (2.62),

$$U = b_1 b_2 \cdots b_d,$$

where $b_k = \tilde{e}_{kk}/\tilde{e}_{Hkk}$, and, from (2.60),

$$\tilde{e}_{kk} = e_{kk} - \mathbf{e}'_{k-1}\mathbf{E}_{k-1}^{-1}\mathbf{e}_{k-1} = e_{kk \cdot 1,2,\ldots,k-1}, \tag{10.22}$$

say. The statistic b_k provides a likelihood ratio test for additional information from the kth variable. Under the null hypothesis of no additional information (see also Section 9.5.3), we have, from (2.72),

$$F_k = \frac{m_E - (k-1)}{m_H}\left(\frac{1}{b_k} - 1\right) \sim F_{m_H, m_E - k + 1}. \tag{10.23}$$

As $\Lambda_r = b_1 b_2 \cdots b_r$ is the test statistic for H_0 based on r variables, a common criterion for choosing the best subset of r variables is to find the subset that minimizes Λ_r. To find the best overall subsets then requires, in principle, the calculation of $2^d - 1$ values of Λ_r. However, this is not feasible for large d (say, $d > 20$), and an alternative procedure is to use a stepwise method. Hawkins [1976] described the following algorithm for doing this using the so-called sweep and reverse sweep operations, widely used in stepwise regression analysis (Seber [1977: Section 12.4]), for removing and adding variables.

Step 1. Search the diagonal elements of $\mathbf{E}$ and $\mathbf{E}_H$ to find k for which e_{kk}/e_{Hkk} is a minimum. If this ratio is sufficiently small, then include variable k. Renumber it variable 1 and sweep it out of $\mathbf{E}$ and $\mathbf{E}_H$.

Step 2. Suppose that repeated applications of step 3 have introduced variables $1, 2, \ldots, r$. For $k = 1$ to r, test whether variable k has retained its significance by examining the "b-ratio" [see (10.22)]

$$\frac{e_{kk \cdot 1,2,\ldots,k-1,k+1,\ldots,r}}{e_{Hkk \cdot 1,2,\ldots,k-1,k+1,\ldots,r}}. \tag{10.24}$$

If the largest such ratio is sufficiently large [or the corresponding F-ratio of (10.23) is sufficiently small, say, less than F_{OUT}), then eliminate the corresponding variable. Reverse sweep it into $\mathbf{E}$ and $\mathbf{E}_H$ and renumber the variables $1, 2, \ldots, r - 1$.

Step 3. Suppose that the variables $1, 2, \ldots, r$ have been introduced. For $k = r + 1$ to d, test whether the variable k is significant by examining the b-ratio

$$\frac{e_{kk \cdot 1,2,\ldots,r}}{e_{Hkk \cdot 1,2,\ldots,r}}. \tag{10.25}$$

If the smallest such ratio is sufficiently small (i.e., the F-ratio greater than F_{IN}), then introduce the corresponding variable k. Renumber it $r + 1$ and sweep it out of $\mathbf{E}$ and $\mathbf{E}_H$.

Steps 2 and 3 are repeated until a situation is reached in which all the variables that have been introduced retain their significance and none of those excluded is significant. The procedure will terminate in a finite number of steps if $F_{\text{OUT}} \le F_{\text{IN}}$. Of course, the renumbering of the variables is not necessary, having been introduced for ease of exposition. If $\mathbf{E}$ and $\mathbf{E}_H$ are swept "in place," and e_{kk} and e_{Hkk} are the *current* kth diagonal elements, then it transpires that the b-ratio for inclusion is e_{kk}/e_{Hkk}, and that for elimination is e_{Hkk}/e_{kk}. Using a Bonferroni approach, Hawkins [1976] suggests the bounds

$$F_{\text{IN}} = F_{m_H, m_E - r}^{\alpha/(d-r)} \quad \text{and} \quad F_{\text{out}} = F_{m_H, m_E - r + 1}^{\alpha'/(d-r+1)},$$

where $\alpha < \alpha'$. These values are conservative in that they tend to lead to a subset with fewer variables.

If d is large, the major source of computation is the sweeping operation at each step. According to Hawkins, this involves $6d^2 + O(d)$ computer operations. Assuming that a stepwise MANOVA typically involves $\frac{1}{2}d$ sweeps, the total number of operations is roughly $3d^3 + O(d^2)$. This is much less than that required for computing all $2^d - 1$ subsets. However, by careful coding, it is possible to obtain Λ_r for any subset by means of a single sweep of a matrix computed earlier, so that the number of operations can be reduced to about $3d^3 \times 2^d$. Such a method was described, for example, by Furnival [1971] for regression, and adapted by McCabe [1975] to MANOVA. McCabe focused his discussion on the one-way MANOVA of Section 9.2; this is sometimes referred to as the discrimination problem. Using the significance of Λ_r as a measure of discrimination, the aim is to find the subset that gives the maximum discrimination (minimum Λ_r) among the I samples (see Section 6.10). McCabe compared an "all subset" program called DISCRIM with the stepwise procedure in BMD07M using simulated data and demonstrated that the stepwise procedure frequently did not produce the best r-subsets.

Because of the close parallel in the computational procedures for the above stepwise algorithm and stepwise regression, the so-called branch-and-bound algorithms of stepwise regression (Furnival and Wilson [1974], Seber [1977: Section 12.3]) can be readily adapted so that one can obtain the best r-subset without having to look at all r-subsets. In this case about $175(1.39)^d$ operations are required.

Another procedure, which is a compromise between the stepwise and all subset procedures, has been proposed by McHenry [1978]. The basic algorithm utilizes the close association of the factorization of U with the Cholesky decomposition [see (2.70)]. Given a subset S_r of r variables, we now consider it as a candidate for the best r-subset. This is done by examining the remaining

$d - r$ variables, denoted by S_{d-r}. For each variable k $(1 \leq k \leq r)$ in S_r, we compute the ratio (10.24) and compare it with the $d - r$ ratios obtained by replacing variable k by a member of S_{d-r}. If variable k gives the smallest ratio, then it stays in; otherwise it is replaced by the member of S_{d-r} that gives the smallest ratio. The process is continued until none of the variables in the final r-subset can be replaced. This usually, but not invariably, produces the best r-subset in terms of minimizing Λ_r. A further check to see if a better r-subset is available can also be incorporated: We delete one variable at a time, bringing in the best of the remainder and repeating the above procedure (the deleted variable is not allowed to reenter). Once we have finalized our r-subset, we obtain a candidate for the best $(r + 1)$-subset by bringing in the variable that minimizes (10.25). Although the above procedure does not guarantee the best subsets for each r, experiments by McHenry indicates that there is a very high probability of obtaining the best subsets.

b REGRESSOR VARIABLES

We now focus our attention on the x-variables in the d-dimensional multivariate linear regression model $\mathbf{Y} = \mathbf{XB} + \mathbf{U}$, where the first column of $\mathbf{X}$ is $\mathbf{1}_n$, that is,

$$\mathbf{X} = (\mathbf{1}_n, \mathbf{X}_1, \mathbf{X}_2) = (\mathbf{1}_n, \mathbf{X}_0),$$

say, and $\mathbf{XB} = \mathbf{1}_n\boldsymbol{\beta}_0' + \mathbf{X}_1\mathbf{B}_1 + \mathbf{X}_2\mathbf{B}_2$, where $\mathbf{X}_1$ represents n measurements on the first r x-variables. It is assumed that $\mathbf{X}$ is $n \times p$ of rank p. A test for no regression [i.e., $(\mathbf{B}_1, \mathbf{B}_2) = \mathbf{O}$] is given by $U_0 = |\mathbf{E}|/|\mathbf{E}_H|$, where

$$\begin{aligned} \mathbf{E} &= \mathbf{Y}'\left(\mathbf{I}_n - \mathbf{X}(\mathbf{X}'\mathbf{X})^{-1}\mathbf{X}'\right)\mathbf{Y}, \\ \mathbf{E}_H &= \mathbf{Y}'\left(\mathbf{I}_n - \mathbf{1}_n\left(\mathbf{1}_n'\mathbf{1}_n\right)^{-1}\mathbf{1}_n'\right)\mathbf{Y} \\ &= \sum_{i=1}^{n}(\mathbf{y}_i - \bar{\mathbf{y}})(\mathbf{y}_i - \bar{\mathbf{y}})' \\ &= \tilde{\mathbf{Y}}'\tilde{\mathbf{Y}} = \mathbf{Q}_{yy}, \end{aligned} \tag{10.26}$$

say, and $\tilde{\mathbf{Y}} = (\mathbf{y}_1 - \bar{\mathbf{y}}, \mathbf{y}_2 - \bar{\mathbf{y}}, \ldots, \mathbf{y}_n - \bar{\mathbf{y}})'$. Defining $\tilde{\mathbf{X}}_0$ the same way, we can also write (see Exercise 8.7)

$$\begin{aligned} \mathbf{E} &= \tilde{\mathbf{Y}}'\tilde{\mathbf{Y}} - \tilde{\mathbf{Y}}'\tilde{\mathbf{X}}_0\left(\tilde{\mathbf{X}}_0'\tilde{\mathbf{X}}_0\right)^{-1}\tilde{\mathbf{X}}_0'\tilde{\mathbf{Y}} \\ &= \mathbf{Q}_{yy} - \mathbf{Q}_{yx}\mathbf{Q}_{xx}^{-1}\mathbf{Q}_{xy}, \end{aligned} \tag{10.27}$$

say. We "accept" the hypothesis of no regression at the α level of significance if $U_0 > U^{\alpha}_{d,\,p-1,\,n-p} = U_0^{\alpha}$.

To test whether the subset S of the first r x-variables is "adequate" (i.e., H_0: $\mathbf{B}_2 = \mathbf{O}$ is "true"), we require

$$\mathbf{E}_H = \tilde{\mathbf{Y}}'\tilde{\mathbf{Y}} - \tilde{\mathbf{Y}}'\tilde{\mathbf{X}}_1(\tilde{\mathbf{X}}_1'\tilde{\mathbf{X}}_1)^{-1}\tilde{\mathbf{X}}_1'\tilde{\mathbf{Y}}$$

$$= \mathbf{Q}_{yy} - \mathbf{Q}_{yS}\mathbf{Q}_{SS}^{-1}\mathbf{Q}_{Sy},$$

say, and a test statistic for H_0 is

$$U_S = \frac{|\mathbf{Q}_{yy} - \mathbf{Q}_{yx}\mathbf{Q}_{xx}^{-1}\mathbf{Q}_{xy}|}{|\mathbf{Q}_{yy} - \mathbf{Q}_{yS}\mathbf{Q}_{SS}^{-1}\mathbf{Q}_{Sy}|}.$$

We accept H_0 if $U_S > U^{\alpha}_{d,p-1-r,n-p}$. McKay [1979] has extended Aitkin's [1974] ideas about adequate subsets in regression analysis (see also Seber [1977: Section 12.2.3]) to the above multivariate situation. McKay suggested *simultaneously* testing all subsets for adequacy by comparing U_S with U_0^{α}. A subset is said to be *adequate* if $U_S > U_0^{\alpha}$, and we can use such a subset as a candidate for the subset stakes. Unfortunately, the actual significance level α_S of an individual test in the simultaneous procedure (i.e., α_S satisfying $U^{\alpha_S}_{d,p-1-r,n-p} = U_0^{\alpha}$) may be too small, so that too many subsets may be adequate. McKay [1979] used an approach of Spjøtvoll [1977] to reduce the number of subsets, and provided a plotting technique for assessing the relative merits of subsets. He noted that the above procedure can also be based on one of the other test statistics such as Roy's maximum root test instead of the likelihood ratio statistic U.

Another univariate procedure, due to Newton and Spurrell [1967a, b], called "elemental analysis" has been extended to the multivariate case by Hintze [1980]. His procedure is also based on testing for no regression using [see (10.27) and (10.26)]

$$U_0 = \frac{|\mathbf{Q}_{yy} - \mathbf{Q}_{yx}\mathbf{Q}_{xx}^{-1}\mathbf{Q}_{xy}|}{|\mathbf{Q}_{yy}|}$$

$$= \begin{vmatrix} \mathbf{Q}_{yy}\mathbf{Q}_{yx} \\ \mathbf{Q}_{xy}\mathbf{Q}_{xx} \end{vmatrix} \Big/ |\mathbf{Q}_{xx}|\,|\mathbf{Q}_{yy}| \qquad \text{(by A3.2)}$$

$$= |\mathbf{Q}_{xx} - \mathbf{Q}_{xy}\mathbf{Q}_{yy}^{-1}\mathbf{Q}_{yx}|/|\mathbf{Q}_{xx}|. \tag{10.28}$$

This statistic is also the test statistic for testing the independence of the $\mathbf{Y}$ and $\mathbf{X}$ data sets [see (3.42)], so that it is not surprising that x and y can be interchanged in U_0. Hintze uses the latter formulation (10.28), as it is more convenient computationally. What he is in fact doing is simply changing the problem into one of selecting y-variables by interchanging the roles of x and y.

The gain in U_0 when an x-variable is dropped is computed and the difference is used to measure the "elements." These elements can then be used as a basis for selecting good subsets.

10.2 LOG-LINEAR MODELS FOR BINARY DATA

In this section it is convenient to distinguish between random variables and their values. Let $\mathbf{X} = (X_1, X_2)'$, where X_1 and X_2 are binary random variables with joint probability function

$$f(x_1, x_2) = \begin{cases} \pi_{00}, & x_1 = 0, \quad x_2 = 0, \\ \pi_{01}, & x_1 = 0, \quad x_2 = 1, \\ \pi_{10}, & x_1 = 1, \quad x_2 = 0, \\ \pi_{11}, & x_1 = 1, \quad x_2 = 1, \end{cases}$$

where $\pi_{00} + \pi_{01} + \pi_{10} + \pi_{11} = 1$. Then, the so-called log-linear model for binary data uses the representation

$$\begin{aligned} \log \pi_{00} &= u + u_1 + u_2 + u_{12}, \\ \log \pi_{01} &= u + u_1 - u_2 - u_{12}, \\ \log \pi_{10} &= u - u_1 + u_2 - u_{12}, \\ \log \pi_{11} &= u - u_1 - u_2 + u_{12}. \end{aligned}$$

If there is no structure on the π_{ab}, then we simply have a reparameterization with four π-parameters replaced by four u-parameters. Any hypothesis about the π_{ab} generates a corresponding hypothesis about the u's. For example, the hypothesis that X_1 and X_2 are independent, namely, $\pi_{00}\pi_{11} = \pi_{01}\pi_{10}$, corresponds to $u_{12} = 0$ so that u_{12} can be interpreted as an "interaction" between X_1 and X_2. We also note that

$$\begin{aligned} \log \pi_{ab} &= \log \mathrm{pr}[X_1 = a, X_2 = b] \\ &= u + u_1(1 - 2a) + u_2(1 - 2b) + u_{12}(1 - 2a)(1 - 2b) \end{aligned} \tag{10.29}$$

$$= u + u_1(-1)^a + u_2(-1)^b + u_{12}(-1)^{a+b}. \tag{10.30}$$

If $Z_i = 1 - 2X_i$, then Z_i takes values ± 1 and

$$\log \mathrm{pr}[Z_1 = z_1, Z_2 = z_2] = u + u_1 z_1 + u_2 z_2 + u_{12} z_1 z_2. \tag{10.31}$$

A further linear representation, which may be more familiar to the readers, is

$$\log \pi_{ab} = \mu + \alpha_a^{(1)} + \alpha_b^{(2)} + \alpha_{ab}^{(12)}, \tag{10.32}$$

where, from (10.29), $\mu = u$, $\alpha_a^{(1)} = u_1(1 - 2a)$, and so on. It is readily verified that $\Sigma_a \alpha_a^{(1)} = \Sigma_b \alpha_b^{(2)} = \Sigma_a \alpha_{ab}^{(12)} = \Sigma_b \alpha_{ab}^{(12)} = 0$, and we recognize the usual main effects and interactions associated with a two-way classification.

If $\mathbf{X} = (X_1, X_2, \dots, X_d)'$ is now a d-dimensional vector of binary variables, then $\mathbf{X}$ can take 2^d different values with probabilities $\pi_{ab\cdots r}$ $(a, b, \dots, r = 0$ or $1)$. The log-linear model now reparameterizes $\log \pi_{ab\cdots r}$ as a sum and difference of an overall mean u, main effects u_i for X_i, two-factor interactions u_{ij} for X_i and X_j, three-factor interactions u_{ijk} for X_i, X_j, and X_k, and higher-order interactions, ending with $u_{12\cdots d}$, the d-factor interaction. Since

$$\binom{d}{1} + \binom{d}{2} + \cdots + \binom{d}{d} = 2^d,$$

the above u's can be listed in a 2^d-dimensional vector, $\mathbf{u}$, say. If $\mathbf{L} = \log \boldsymbol{\pi}$ is the vector of $\log \pi_{ab\cdots r}$, then $\mathbf{L} = \mathbf{Ku}$, where $\mathbf{K}$ is a $2^d \times 2^d$ orthogonal matrix with elements ± 1. As the notation, and to some extent the interpretation, is reminiscent of a 2^d factorial design, we can use the "standard order" in writing out $\mathbf{u}$. For example, if $d = 3$,

$$\begin{pmatrix} L_{000} \\ L_{100} \\ L_{010} \\ L_{110} \\ L_{001} \\ L_{101} \\ L_{011} \\ L_{111} \end{pmatrix} = \begin{pmatrix} 1 & 1 & 1 & 1 & 1 & 1 & 1 & 1 \\ 1 & -1 & 1 & -1 & 1 & -1 & 1 & -1 \\ 1 & 1 & -1 & -1 & 1 & 1 & -1 & -1 \\ 1 & -1 & -1 & 1 & 1 & -1 & -1 & 1 \\ 1 & 1 & 1 & 1 & -1 & -1 & -1 & -1 \\ 1 & -1 & 1 & -1 & -1 & 1 & -1 & 1 \\ 1 & 1 & -1 & -1 & -1 & -1 & 1 & 1 \\ 1 & -1 & -1 & 1 & -1 & 1 & 1 & -1 \end{pmatrix} \begin{pmatrix} u \\ u_1 \\ u_2 \\ u_{12} \\ u_3 \\ u_{13} \\ u_{23} \\ u_{123} \end{pmatrix},$$

or $\mathbf{L} = \mathbf{Ku}$. The corresponding model for $Z_j = 1 - 2X_j$ is

$$\log \mathrm{pr}[\mathbf{Z} = \mathbf{z}] = u + u_1 z_1 + \cdots + u_{12} z_1 z_2 + \cdots + u_{12\cdots d} z_1 z_2 \cdots z_d. \tag{10.33}$$

When there are no constraints on the $\pi_{ab\cdots r}$, we have the same number of u's as π's and the model is said to be (fully) saturated. However, if the elements of $\mathbf{X}$ are mutually independent, then all the interactions are zero and we only have $d + 1$ parameters $u, u_1, u_2, \dots, u_d$ to estimate: The model (10.33) is then linear in $\mathbf{z}$. These two models, saturation and complete independence, may be regarded as extremes, and we would hope that an intermediate model with not too many parameters is applicable, for example,

$$\log \text{pr}[\mathbf{Z} = \mathbf{z}] = u + u_1 z_1 + \cdots + u_d z_d + u_{12} z_1 z_2 + \cdots + u_{d-1,d} z_{d-1} z_d,$$

where the three-factor and higher interactions are assumed to be zero.

If we have n observations on $\mathbf{X}$ with $n_{ab\cdots r}$ falling in the category with probability $\pi_{ab\cdots r}$, then we can construct a 2^d contingency table and estimate $\pi_{ab\cdots r}$ by $\hat{\pi}_{ab\cdots r} = n_{ab\cdots r}/n$. This leads to an estimate $\hat{\mathbf{L}}$ of $\mathbf{L}$, and an estimate of $\mathbf{u}$. To avoid dividing by n, we can add $\log n$ to u and model $\log(\text{E}[n_{ab\cdots r}]) = \log(n\pi_{ab\cdots r})$ as above. The general theory and analysis of such models is described in detail by A. H. Anderson [1974], Haberman [1974], and Bishop et al. [1975]. An interesting application of this method to capture–recapture data is given by Cormack [1981]. For a general review, see Imrey et al. [1981, 1982].

A number of other models have been developed for handling multivariate binary data, and these are surveyed briefly by Cox [1972]. Log-linear models are also used in discriminant analysis (Section 6.3.3). A related topic is logistic regression (see Kleinbaum et al. [1982]).

10.3 INCOMPLETE DATA

Dempster et al. [1977] give a general method of iteratively computing maximum likelihood estimates where there is incomplete data due, for example, to missing observations. Since each iteration of the algorithm consists of an expectation step followed by a maximization step, the authors call it the EM algorithm. Let $\mathbf{x} = (\mathbf{x}_1', \mathbf{x}_2', \ldots, \mathbf{x}_n')'$ denote the vector of "complete" data, and let $\overset{*}{\mathbf{x}} = (\overset{*}{\mathbf{x}}_1', \overset{*}{\mathbf{x}}_2', \ldots, \overset{*}{\mathbf{x}}_n')$ denote the data actually available; the elements of $\overset{*}{\mathbf{x}}_i$ are a subset of the elements of $\mathbf{x}_i$. It is assumed that sampling is from an exponential family, of which the MVN is a member. Let $\boldsymbol{\Phi}$ be the matrix (or vector) of unknown parameters and let $\mathbf{T}(\mathbf{x})$ be a sufficient statistic for $\boldsymbol{\Phi}$ based on the complete data. Since $\mathbf{x}$ is only partly known, the EM procedure amounts to simultaneously estimating $\mathbf{T}$ and $\boldsymbol{\Phi}$. If $\boldsymbol{\Phi}^{(k)}$ is the current value of $\boldsymbol{\Phi}$ after k cycles of the algorithm, then the next cycle has two steps, the E-step, which estimates $\mathbf{T}(\mathbf{x})$ by

$$\mathbf{T}^{(k)} = \text{E}\left[\mathbf{T} | \overset{*}{\mathbf{x}}, \boldsymbol{\Phi}^{(k)}\right], \tag{10.34}$$

and the M-step, which determines $\boldsymbol{\Phi}^{(k+1)}$ as the solution of

$$\text{E}[\mathbf{T} | \boldsymbol{\Phi}] = \mathbf{T}^{(k)}. \tag{10.35}$$

Eventually $\boldsymbol{\Phi}^{(k)}$ converges to the maximum likelihood estimate of $\boldsymbol{\Phi}$ based on $\overset{*}{\mathbf{x}}$.

For the special case of the MVN distribution, $\boldsymbol{\Phi} = (\boldsymbol{\mu}, \boldsymbol{\Sigma})$, $\mathbf{T} = (\overline{\mathbf{x}}, \mathbf{S})$, and (10.35) reduces to

$$\boldsymbol{\Phi}^{(k+1)} = \mathbf{T}^{(k)}. \tag{10.36}$$

The main computational burden then consists of finding for each i the parameters of the conditional normal distribution of the missing values of $\mathbf{x}_i$ conditional on $\overset{*}{\mathbf{x}}_i$, evaluated at the current value of $\mathbf{\Phi}$. Equation (10.34) amounts to replacing the missing parts of $\mathbf{T}$ by their conditional expectations, and Equation (10.36) tells us to use this estimated $\mathbf{T}$ as the next estimate of $\mathbf{\Phi}$. A detailed example is given below.

EXAMPLE 10.1 Suppose that we have a random sample $\mathbf{x}_1, \mathbf{x}_2, \dots, \mathbf{x}_n$ from $N_d(\boldsymbol{\mu}, \boldsymbol{\Sigma})$. Using a slight change in notation, let $\mathbf{x}_{i(1)}$ denote the available part of $\mathbf{x}_i$ (instead of $\overset{*}{\mathbf{x}}_i$ above) and $\mathbf{x}_{i(2)}$ the missing part. Let $\mathbf{y}_i = (\mathbf{x}'_{i(1)}, \mathbf{x}'_{i(2)})'$ with mean $\boldsymbol{\mu}_i$ and dispersion matrix $\boldsymbol{\Sigma}_i$, where $\mathbf{y}_i$, $\boldsymbol{\mu}_i$, and $\boldsymbol{\Sigma}_i$ are appropriate permutations of $\mathbf{x}_i$, $\boldsymbol{\mu}$, and $\boldsymbol{\Sigma}$ (the permutation depending on i). Instead of choosing $\bar{\mathbf{x}}$ and $\mathbf{S}$ as our sufficient statistics, we can also use

$$\mathbf{T} = \left(\sum_{i=1}^{n} \mathbf{x}_i, \sum_{i=1}^{n} \mathbf{x}_i \mathbf{x}'_i \right) = (\mathbf{t}_1, \mathbf{T}_2),$$

say. Now, from Theorem 2.1(viii) of Section 2.2, the kth iteration estimate of $\mathbf{x}_{i(2)}$, given estimates $\boldsymbol{\mu}^{(k)}$ and $\boldsymbol{\Sigma}^{(k)}$ of $\boldsymbol{\mu}$ and $\boldsymbol{\Sigma}$, respectively, is

$$\begin{aligned}
\mathbf{x}_{i(2)}^{(k)} &= \mathscr{E}\left[\mathbf{x}_{i(2)} | \mathbf{x}_{i(1)}, \boldsymbol{\mu}_i^{(k)}, \boldsymbol{\Sigma}_i^{(k)}\right] \\
&= \mathscr{E}\left[\mathbf{x}_{i(2)} | \mathbf{A}^{(k)}\right], \qquad \text{say}, \\
&= \boldsymbol{\mu}_{i(2)}^{(k)} + \boldsymbol{\Sigma}_{i21}^{(k)} \left\{ \boldsymbol{\Sigma}_{i11}^{(k)} \right\}^{-1} \left(\mathbf{x}_{i(1)} - \boldsymbol{\mu}_{i(1)}^{(k)} \right).
\end{aligned}$$

Similarly,

$$\begin{aligned}
\left(\mathbf{x}_{i(2)} \mathbf{x}'_{i(2)} \right)^{(k)} &= \mathscr{E}\left[\mathbf{x}_{i(2)} \mathbf{x}'_{i(2)} | \mathbf{A}^{(k)}\right] \\
&= \mathscr{D}\left[\mathbf{x}_{i(2)} | \mathbf{A}^{(k)}\right] + \left\{ \mathscr{E}\left[\mathbf{x}_{i(2)} | \mathbf{A}^{(k)}\right] \right\} \left\{ \mathscr{E}\left[\mathbf{x}_{i(2)} | \mathbf{A}^{(k)}\right] \right\}' \\
&= \boldsymbol{\Sigma}_{i22}^{(k)} - \boldsymbol{\Sigma}_{i21}^{(k)} \left\{ \boldsymbol{\Sigma}_{i11}^{(k)} \right\}^{-1} \boldsymbol{\Sigma}_{i12}^{(k)} + \mathbf{x}_{i(2)}^{(k)} \mathbf{x}_{i(2)}^{(k)\prime},
\end{aligned}$$

and

$$\left(\mathbf{x}_{i(1)} \mathbf{x}'_{i(2)} \right)^{(k)} = \mathscr{E}\left[\mathbf{x}_{i(1)} \mathbf{x}'_{i(2)} | \mathbf{A}^{(k)}\right] = \mathbf{x}_{i(1)} \mathbf{x}_{i(2)}^{(k)\prime}.$$

Setting $\mathbf{x}_{i(1)}^{(k)} = \mathbf{x}_{i(1)}$ for the available observations, then, from (10.34),

$$\mathbf{T}^{(k)} = \left(\sum_i \mathbf{x}_i^{(k)}, \sum_i \mathbf{x}_i^{(k)} \mathbf{x}_i^{(k)\prime} \right) = \left(\mathbf{t}_1^{(k)}, \mathbf{T}_2^{(k)} \right)$$

can be found by summing over the above equations. Since the maximum likelihood estimates for the complete data are

$$\hat{\boldsymbol{\mu}} = \frac{\mathbf{t}_1}{n} \quad \text{and} \quad \hat{\boldsymbol{\Sigma}} = \frac{T_2}{n} - \hat{\boldsymbol{\mu}}\hat{\boldsymbol{\mu}}',$$

Equation (10.35) leads to the estimates

$$\boldsymbol{\mu}^{(k+1)} = \frac{\mathbf{t}_1^{(k)}}{n} \quad \text{and} \quad \boldsymbol{\Sigma}^{(k+1)} = \frac{\mathbf{T}_2^{(k)}}{n} - \boldsymbol{\mu}^{(k)}\boldsymbol{\mu}^{(k)\prime}.$$

The procedure is then repeated. □

Dempster et al. [1977] apply the EM algorithm to more general families of distributions and review several applications, including (1) missing data for the MVN and multinomial distributions; (2) grouped, censored, and truncated data; (3) variance components; (4) iteratively reweighted least squares; and (5) finite mixtures. In the mixture problem (see Sections 6.7 and 7.5.4), we have n observations $\mathbf{x}_1, \mathbf{x}_2, \ldots, \mathbf{x}_n$ from g MVN distributions, and the missing information is a corresponding set of g-dimensional vectors $\mathbf{y}_1, \mathbf{y}_2, \ldots, \mathbf{y}_n$, where $\mathbf{y}_i$ is an indicator vector with all its elements zero except for one element equal to unity in the position indicating from which population $\mathbf{x}_i$ was taken. The article by Dempster et al. [1977], along with the discussion, gives a useful survey of the whole area of missing observations.

In the special case of sampling from $N_d(\boldsymbol{\mu}, \boldsymbol{\Sigma})$, the same algorithm for finding the maximum likelihood estimates of $\boldsymbol{\mu}$ and $\boldsymbol{\Sigma}$, when there are missing data, is given by Beale and Little [1975]: A method for pooled data is given by Huseby et al. [1980], and questions of inference are discussed by Little [1976]. The bivariate case is considered by Naik [1975], Little [1976], A. Cohen [1977], and Dahiya and Korwar [1980: equal variances] for the case of missing observations on one variable only, and by Bhoj [1978, 1979] for the case of missing observations on both variables. For general d, likelihood ratio tests have been given by Bhargava [1962] for several one- and two-population problems for the special case when the pattern of missing observations in the data matrix $\mathbf{X}$ is triangular (called "monotonic"), that is, there are $n_1, n_2, \ldots, n_d$ observations on each of the respective x-variables $x_1, x_2, \ldots, x_d$, with $n_1 \geq n_2 \geq \cdots \geq n_d \geq d$. Approximations for these tests are given by Bhoj [1973a, b].

APPENDIX A

Some Matrix Algebra

Below we list a number of matrix results used in this book. Most of the omitted proofs are given, for example, in Seber [1977: Appendix A].

A1 TRACE AND EIGENVALUES

1 Provided that the matrices are conformable, we have the following:
  (a) $\text{tr}[\mathbf{A} + \mathbf{B}] = \text{tr}\,\mathbf{A} + \text{tr}\,\mathbf{B}$.
  (b) $\text{tr}[\mathbf{AC}] = \text{tr}[\mathbf{CA}]$.

2 For any $n \times n$ matrix $\mathbf{A}$ with eigenvalues $\lambda_1, \lambda_2, \ldots, \lambda_n$, we have the following:
  (a) $\text{tr}[\mathbf{A}] = \sum_{i=1}^{n} \lambda_i$
  (b) $|\mathbf{A}| = \prod_{i=1}^{n} \lambda_i$.
  (c) $|\mathbf{I}_n \pm \mathbf{A}| = \prod_{i=1}^{n} (1 \pm \lambda_i)$.

Proof $|\lambda \mathbf{I}_n - \mathbf{A}| = \Pi_i (\lambda - \lambda_i) = \lambda^n - \lambda^{n-1}(\lambda_1 + \lambda_2 + \cdots + \lambda_n) + \cdots + (-1)^n \lambda_1 \lambda_2 \cdots \lambda_n$. Expanding $|\lambda \mathbf{I}_n - \mathbf{A}|$, we see that the coefficient of λ^{n-1} is $-(a_{11} + a_{22} + \cdots + a_{nn})$, and the constant term is $|-\mathbf{A}| = (-1)^n |\mathbf{A}|$. Hence the sum of the roots is $\text{tr}\,\mathbf{A}$, and the product $|\mathbf{A}|$. Furthermore, $|\mathbf{I}_n + \mathbf{A}| = \Pi_i \theta_i$, where θ_i is the ith root of $0 = |\theta \mathbf{I}_n - (\mathbf{I}_n + \mathbf{A})| = |(\theta - 1)\mathbf{I}_n - \mathbf{A}|$. Hence $\theta_i - 1 = \lambda_i$, and so on.

3 *Spectral decomposition.* Let $\mathbf{A}$ be an $n \times n$ symmetric matrix. There exists an orthogonal matrix $\mathbf{T} = (\mathbf{t}_1, \mathbf{t}_2, \ldots, \mathbf{t}_n)$ such that

$$\mathbf{T}'\mathbf{A}\mathbf{T} = \text{diag}(\lambda_1, \lambda_2, \ldots, \lambda_n) = \mathbf{\Lambda},$$

where $\lambda_1 \geq \lambda_2 \geq \cdots \geq \lambda_n$ are the ordered eigenvalues of $\mathbf{A}$. With this ordering, $\mathbf{\Lambda}$ is unique and $\mathbf{T}$ is unique up to a postfactor

$$\mathbf{S} = \begin{pmatrix} \mathbf{S}_1 & \mathbf{O} & \cdots & \mathbf{O} \\ \mathbf{O} & \mathbf{S}_2 & \cdots & \mathbf{O} \\ \vdots & \vdots & & \vdots \\ \mathbf{O} & \mathbf{O} & \cdots & \mathbf{S}_k \end{pmatrix}, \qquad \mathbf{S}_i \in T(m_i),$$

where k is the number of different eigenvalues of $\mathbf{A}$; $m_1, m_2, \ldots, m_k$ are the multiplicities, that is, $\lambda_1 = \lambda_2 = \cdots = \lambda_{m_1} > \lambda_{m_1+1} = \cdots = \lambda_{m_1+m_2}$, and so on; and $T(m_i)$ stands for the set of all $m_i \times m_i$ orthogonal matrices. If all the eigenvalues are distinct, each m_i is unity and $\mathbf{S}$ reduces to a diagonal matrix with diagonal elements equal to ± 1. In this case the columns $\mathbf{t}_i$ of $\mathbf{T}$ are unique except for their signs. If we stipulate, for example, that the element of $\mathbf{t}_i$ with the largest magnitude is positive, then $\mathbf{S} = \mathbf{I}_n$ and $\mathbf{T}$ is unique. We note that $\mathbf{A} = \mathbf{T\Lambda T}'$, that is,

$$\mathbf{A} = \sum_{i=1}^{n} \lambda_i \mathbf{t}_i \mathbf{t}_i'.$$

4 For conformable matrices, the nonzero eigenvalues of $\mathbf{AB}$ are the same as those of $\mathbf{BA}$. The eigenvalues are identical for square matrices.

Proof Let λ be a nonzero eigenvalue of $\mathbf{AB}$. Then there exists $\mathbf{u}$ ($\neq \mathbf{0}$) such that $\mathbf{ABu} = \lambda\mathbf{u}$, that is, $\mathbf{BABu} = \lambda\mathbf{Bu}$. Hence $\mathbf{BAv} = \lambda\mathbf{v}$, where $\mathbf{v} = \mathbf{Bu} \neq \mathbf{0}$ (as $\mathbf{ABu} \neq \mathbf{0}$), and λ is an eigenvalue of $\mathbf{BA}$. The argument reverses by interchanging the roles of $\mathbf{A}$ and $\mathbf{B}$.

A2 RANK

1 If $\mathbf{A}$ and $\mathbf{B}$ are conformable matrices, then

$$\text{rank}[\mathbf{AB}] \leq \text{minimum}(\text{rank}[\mathbf{A}], \text{rank}[\mathbf{B}]).$$

2 If $\mathbf{A}$ is any matrix and $\mathbf{P}$ and $\mathbf{Q}$ are conformable nonsingular matrices, then $\text{rank}[\mathbf{PAQ}] = \text{rank}\,\mathbf{A}$.

3 $\text{rank}\,\mathbf{A} = \text{rank}\,\mathbf{A}' = \text{rank}[\mathbf{A}'\mathbf{A}] = \text{rank}[\mathbf{AA}']$.

4 If $\mathbf{A}$ is symmetric and $\text{rank}\,\mathbf{A} = 1$, then

$$|\mathbf{I} + \mathbf{A}| = 1 + \text{tr}\,\mathbf{A}.$$

Proof From A1.3, there exists orthogonal $\mathbf{T}$ such that $\mathbf{T}'\mathbf{AT} = \mathbf{\Lambda}$. Then $1 = \text{rank}\,\mathbf{A} = \text{rank}\,\mathbf{\Lambda}$ (by A2.2), so that there is only one nonzero eigenvalue, λ_j, say. Hence, by A1.2, $|\mathbf{I} + \mathbf{A}| = 1 + \lambda_j = 1 + \text{tr}\,\mathbf{A}$.

5 If $\mathbf{A}$ is $m \times n$, then $\text{rank}\,\mathbf{A} + \text{nullity}\,\mathbf{A} = n$, where the nullity of $\mathbf{A}$ is the dimension of the null space or kernel of $\mathbf{A}$.

6 If $\mathbf{A}$ is $n \times m$ of rank m and $\mathbf{B}$ is $m \times p$ of rank p, then $\mathbf{AB}$ has rank p.

7 The rank of a symmetric matrix is equal to the number of nonzero eigenvalues.

8 Let $\mathbf{X}' = (\mathbf{x}_1, \mathbf{x}_2, \ldots, \mathbf{x}_n)$ be a $d \times n$ matrix of random variables and let $\mathbf{A}$ be a symmetric $n \times n$ matrix of rank r. If the joint distribution of the elements of $\mathbf{X}$ is absolutely continuous with respect to the nd-dimensional Lebesgue measure, then the following statement holds with probability 1: $\text{rank}[\mathbf{X}'\mathbf{A}\mathbf{X}] = \min(d, r)$ and the nonzero eigenvalues of $\mathbf{X}'\mathbf{A}\mathbf{X}$ are distinct. For a proof see Okamoto [1973].

The conditions in 8 are satisfied if the $\mathbf{x}_i$ have independent nonsingular multivariate normal distributions. Therefore, if we set $\mathbf{A} = \mathbf{I}_n$, the eigenvalues of a Wishart matrix (represented by $\mathbf{X}'\mathbf{X}$) are positive and distinct with probability 1 if $d \leq n$. Furthermore, from (1.13), the same applies to $\mathbf{Q} = \sum_i (\mathbf{x}_i - \bar{\mathbf{x}})(\mathbf{x}_i - \bar{\mathbf{x}})'$ if $(n-1) \geq d$ (since $\mathbf{A} = \mathbf{I}_n - \mathbf{P}_1$ has rank $n - 1$).

A3 PATTERNED MATRICES

1 If $\mathbf{A}$ and $\mathbf{D}$ are symmetric and all inverses exist,

$$\begin{pmatrix} \mathbf{A} & \mathbf{B} \\ \mathbf{B}' & \mathbf{D} \end{pmatrix}^{-1} = \begin{pmatrix} \mathbf{A}^{-1} + \mathbf{F}\mathbf{E}^{-1}\mathbf{F}' & -\mathbf{F}\mathbf{E}^{-1} \\ -\mathbf{E}^{-1}\mathbf{F}' & \mathbf{E}^{-1} \end{pmatrix},$$

where $\mathbf{E} = \mathbf{D} - \mathbf{B}'\mathbf{A}^{-1}\mathbf{B}$ and $\mathbf{F} = \mathbf{A}^{-1}\mathbf{B}$.

2

$$\begin{vmatrix} \mathbf{A} & \mathbf{B} \\ \mathbf{C} & \mathbf{D} \end{vmatrix} = \begin{cases} |\mathbf{D}|\,|\mathbf{A} - \mathbf{B}\mathbf{D}^{-1}\mathbf{C}| & \text{if } \mathbf{D}^{-1} \text{ exists,} \\ |\mathbf{A}|\,|\mathbf{D} - \mathbf{C}\mathbf{A}^{-1}\mathbf{B}|, & \text{if } \mathbf{A}^{-1} \text{ exists.} \end{cases}$$

Proof If $\mathbf{D}^{-1}$ exists, then

$$\begin{pmatrix} \mathbf{A} & \mathbf{B} \\ \mathbf{C} & \mathbf{D} \end{pmatrix} \begin{pmatrix} \mathbf{I}_r & \mathbf{O} \\ -\mathbf{D}^{-1}\mathbf{C} & \mathbf{I}_s \end{pmatrix} = \begin{pmatrix} \mathbf{A} - \mathbf{B}\mathbf{D}^{-1}\mathbf{C} & \mathbf{B} \\ \mathbf{O} & \mathbf{D} \end{pmatrix}.$$

The first result follows immediately by taking determinants and noting that the determinant of the lower triangular matrix on the left-hand side is the product of the diagonal elements, that is, unity. The second result follows in a similar fashion.

3

$$(\mathbf{A} + \mathbf{U}\mathbf{B}\mathbf{V})^{-1} = \mathbf{A}^{-1} - \mathbf{A}^{-1}\mathbf{U}\mathbf{B}(\mathbf{B} + \mathbf{B}\mathbf{V}\mathbf{A}^{-1}\mathbf{U}\mathbf{B})^{-1}\mathbf{B}\mathbf{V}\mathbf{A}^{-1}$$

(provided that inverses exist).

The above result and A3.1 are proved by verifying that the given matrix multiplied by its inverse gives the identity matrix. For the particular case $\mathbf{B} = 1$, $\mathbf{U} = \mathbf{u}$, and $\mathbf{V} = \mathbf{v}'$, we have

$$(\mathbf{A} + \mathbf{uv}')^{-1} = \mathbf{A}^{-1} - \mathbf{A}^{-1}\mathbf{uv}'\mathbf{A}^{-1}(1 + \mathbf{v}'\mathbf{A}^{-1}\mathbf{u})^{-1}.$$

Furthermore,

$$\mathbf{x}'(\mathbf{A} + \mathbf{xx}')^{-1}\mathbf{x} = \frac{\mathbf{x}'\mathbf{A}^{-1}\mathbf{x}}{1 + \mathbf{x}'\mathbf{A}^{-1}\mathbf{x}}.$$

4 Let $\boldsymbol{\Sigma} = \boldsymbol{\Gamma}\boldsymbol{\Gamma}' + \boldsymbol{\Psi}$, where $\boldsymbol{\Sigma}$ and $\boldsymbol{\Psi}$ are nonsingular. Then, from A3.3 above,

$$\boldsymbol{\Sigma}^{-1} = \boldsymbol{\Psi}^{-1} - \boldsymbol{\Psi}^{-1}\boldsymbol{\Gamma}(\mathbf{I} + \boldsymbol{\Gamma}'\boldsymbol{\Psi}^{-1}\boldsymbol{\Gamma})^{-1}\boldsymbol{\Gamma}'\boldsymbol{\Psi}^{-1}.$$

If $\mathbf{J} = \boldsymbol{\Gamma}'\boldsymbol{\Psi}^{-1}\boldsymbol{\Gamma}$, then, from $\mathbf{J} = \mathbf{I} + \mathbf{J} - \mathbf{I}$, we have $\mathbf{J}(\mathbf{I} + \mathbf{J})^{-1} = \mathbf{I} - (\mathbf{I} + \mathbf{J})^{-1}$.

5 Consider the $n \times n$ patterned matrix

$$\mathbf{A} = \begin{pmatrix} a & b & \cdots & b & b \\ b & a & \cdots & b & b \\ \vdots & \vdots & & \vdots & \vdots \\ b & b & \cdots & b & a \end{pmatrix} = (a - b)\mathbf{I}_n + b\mathbf{J}_n \qquad (a \neq b),$$

where $\mathbf{J}_n$ is an $n \times n$ matrix of 1's. Then we have the following:

(a)
$$|\mathbf{A}| = (a-b)^n \left| \mathbf{I}_n + \frac{\mathbf{1}_n\mathbf{1}_n' b}{a-b} \right|$$

$$= (a-b)^n \left\{ 1 + \frac{\operatorname{tr}[\mathbf{1}_n\mathbf{1}_n']b}{a-b} \right\} \qquad \text{(by A2.4)}$$

$$= (a-b)^n + nb(a-b)^{n-1}.$$

(b) If $|\mathbf{A}| \neq 0$, then $\mathbf{A}^{-1}$ has the same form as $\mathbf{A}$, namely,

$$\mathbf{A}^{-1} = (c - d)\mathbf{I}_n + d\mathbf{J}_n,$$

where

$$c = \frac{-[b(n-2) + a]}{(b-a)[b(n-1) + a]}$$

and

$$d = \frac{b}{(b-a)[b(n-1) + a]}.$$

A4 POSITIVE SEMIDEFINITE MATRICES

A symmetric matrix $\mathbf{A}$ is said to be positive semidefinite† (p.s.d) if and only if $\mathbf{x}'\mathbf{A}\mathbf{x} \geq 0$ for all $\mathbf{x}$. We shall write $\mathbf{A} \geq \mathbf{O}$, where $\mathbf{O}$ is a matrix of zeros.

1 A symmetric matrix is p.s.d. if and only if its eigenvalues are nonnegative.
2 If $\mathbf{A} \geq \mathbf{O}$, then, by A1.2, $\operatorname{tr} \mathbf{A} \geq 0$.
3 $\mathbf{A}$ is p.s.d. of rank r if and only if there exists a square matrix $\mathbf{R}$ of rank r such that $\mathbf{A} = \mathbf{R}'\mathbf{R}$.
4 Given $\mathbf{A} \geq \mathbf{O}$, then $\mathbf{C}'\mathbf{A}\mathbf{C} \geq \mathbf{O}$; in particular, $\mathbf{C}'\mathbf{C} \geq \mathbf{O}$.
5 If $\mathbf{A}$ is p.s.d. and $\mathbf{C}'\mathbf{A}\mathbf{C} = \mathbf{O}$, then $\mathbf{A}\mathbf{C} = \mathbf{O}$; in particular, $\mathbf{C}'\mathbf{C} = \mathbf{O}$ implies $\mathbf{C} = \mathbf{O}$.
6 Given $\mathbf{A} \geq \mathbf{O}$, there exists a p.s.d. matrix $\mathbf{A}^{1/2}$ such that $\mathbf{A} = (\mathbf{A}^{1/2})^2$. The eigenvalues of $\mathbf{A}^{1/2}$ are the square roots of those of $\mathbf{A}$.

   Proof From A1.3, $\mathbf{A} = \mathbf{T}\mathbf{\Lambda}\mathbf{T}' = \mathbf{T}\mathbf{\Lambda}^{1/2}\mathbf{T}'\mathbf{T}\mathbf{\Lambda}^{1/2}\mathbf{T}' = (\mathbf{A}^{1/2})^2$, where $\mathbf{\Lambda}^{1/2} = \operatorname{diag}(\lambda_1^{1/2}, \lambda_2^{1/2}, \ldots, \lambda_n^{1/2}) \geq \mathbf{O}$ and $\mathbf{A}^{1/2} = \mathbf{T}\mathbf{\Lambda}^{1/2}\mathbf{T}' \geq \mathbf{O}$ (by A4.4). Also, $\mathbf{T}'\mathbf{A}^{1/2}\mathbf{T} = \mathbf{\Lambda}^{1/2}$.
7 Given any symmetric matrix $\mathbf{C}$, $\operatorname{tr}[(\mathbf{C}'\mathbf{C})^{1/2}] = \operatorname{tr}[(\mathbf{C}\mathbf{C}')^{1/2}]$.

   Proof By A4.1 and A1.4, $\mathbf{C}'\mathbf{C}$ and $\mathbf{C}\mathbf{C}'$ have the same positive eigenvalues, their remaining eigenvalues being equal to zero. The result follows from A4.6 and A1.2.

A5 POSITIVE DEFINITE MATRICES

A symmetric matrix $\mathbf{A}$ is said to be positive definite (p.d.) if $\mathbf{x}'\mathbf{A}\mathbf{x} > 0$ for all $\mathbf{x}$, $\mathbf{x} \neq \mathbf{0}$. We shall write $\mathbf{A} > \mathbf{O}$. We note that a p.d. matrix is also p.s.d. A symmetric matrix $\mathbf{A}$ is said to be negative definite if $-\mathbf{A}$ is p.d.

1 A symmetric matrix is p.d. if and only if its eigenvalues are all positive.
2 If $\mathbf{A} \geq \mathbf{O}$ and $|\mathbf{A}| \neq 0$, then $\mathbf{A} > \mathbf{O}$ [by A4.1 and A1.2(a)].
3 $\mathbf{A}$ is p.d. if and only if there exists a nonsingular matrix $\mathbf{R}$ such that $\mathbf{A} = \mathbf{R}'\mathbf{R}$.
4 Given that $\mathbf{A}$ is p.d., there exists a p.d. matrix $\mathbf{A}^{1/2}$ such that $\mathbf{A} = (\mathbf{A}^{1/2})^2$ (see A4.6).
5 If $\mathbf{A}$ is p.d., so is $\mathbf{A}^{-1}$.
6 If $\mathbf{A}$ is p.d., then $\operatorname{rank}[\mathbf{C}\mathbf{A}\mathbf{C}'] = \operatorname{rank}[\mathbf{C}]$.
7 Given that $\mathbf{A}$ is an $n \times n$ p.d. matrix and $\mathbf{C}$ is $p \times n$ of rank p, then $\mathbf{C}\mathbf{A}\mathbf{C}'$ is p.d.
8 If $\mathbf{A}$ is p.d. and $\mathbf{B} = [(b_{ij})]$ is symmetric, then $\mathbf{A} - \mathbf{B}$ is p.d., provided that all the $|b_{ij}|$ are sufficiently small. In particular, $\mathbf{A} - t\mathbf{B}$ is p.d. for $|t|$ sufficiently small.

†Some authors use the term *nonnegative definite*.

9 **A** is p.d. if and only if all the leading minor determinants of **A**, including $|\mathbf{A}|$ itself, are positive.

10 If $\mathbf{A} > \mathbf{O}$, $\mathbf{B} \geq \mathbf{O}$, $c > 0$, and $d > 0$, then $c\mathbf{A} + d\mathbf{B} > \mathbf{O}$.

11 *Cholesky decomposition.* If **A** is p.d., there exists a unique upper triangular matrix

$$\mathbf{U} = \begin{pmatrix} u_{11} & u_{12} & \cdots & u_{1n} \\ 0 & u_{22} & \cdots & u_{2n} \\ \vdots & \vdots & & \vdots \\ 0 & 0 & \cdots & u_{nn} \end{pmatrix}$$

with positive diagonal elements ($u_{ii} > 0$), such that $\mathbf{A} = \mathbf{U}'\mathbf{U}$. We note that $|\mathbf{U}| = u_{11}u_{22} \cdots u_{nn}$.

There also exists a unique lower triangular matrix **L** with positive diagonal elements such that $\mathbf{A} = \mathbf{L}'\mathbf{L}$.

12 Let **C** and **D** be $n \times n$ symmetric matrices and suppose that **C** is positive definite. Then there exists a nonsingular matrix **R** such that $\mathbf{R}'\mathbf{C}\mathbf{R} = \mathbf{\Gamma}$ and $\mathbf{R}'\mathbf{D}\mathbf{R} = \mathbf{I}_n$, where $\mathbf{\Gamma} = \text{diag}(\gamma_1, \gamma_2, \ldots, \gamma_n)$ and the γ_i are the eigenvalues of $\mathbf{D}^{-1}\mathbf{C}$ (or $\mathbf{C}\mathbf{D}^{-1}$, by A1.4).

13 Let $\mathbf{X}' = (\mathbf{x}_1, \mathbf{x}_2, \ldots, \mathbf{x}_n)$, where the $\mathbf{x}_i$ are n independent d-dimensional vectors of random variables, and let **A** be a positive semidefinite $n \times n$ matrix of rank r ($r \geq d$). Suppose that for each $\mathbf{x}_i$ and all **b** and c ($\mathbf{b} \neq \mathbf{0}$), $\text{pr}[\mathbf{b}'\mathbf{x}_i = c] = 0$. Then

$$\text{pr}[\mathbf{X}'\mathbf{A}\mathbf{X} > \mathbf{O}] = 1;$$

that is, $\mathbf{X}'\mathbf{A}\mathbf{X}$ is positive definite with probability 1.

Proof This is given independently by Das Gupta [1971: Theorem 5] and Eaton and Perlman [1973: Theorem 2.3].

If $\mathbf{x}_i \sim N_d(\boldsymbol{\theta}_i, \mathbf{\Sigma})$ and $\mathbf{\Sigma} > \mathbf{O}$, $\mathbf{b}'\mathbf{x}_i \sim N_d(\mathbf{b}'\boldsymbol{\theta}_i, \mathbf{b}'\mathbf{\Sigma}\mathbf{b})$. This distribution is nondegenerate as $\mathbf{b}'\mathbf{\Sigma}\mathbf{b} > 0$; hence $\text{pr}[\mathbf{b}'\mathbf{x}_i = c] = 0$. Hence the above result 13 holds for the multivariate normal. For this case, an elementary proof that $\mathbf{Q} = \Sigma_i(\mathbf{x}_i - \overline{\mathbf{x}})(\mathbf{x}_i - \overline{\mathbf{x}})' > \mathbf{O}$ with probability 1 when $d \leq n - 1$ ($= \text{rank}[\mathbf{A}]$) is given by Dykstra [1970].

If $\mathbf{W} \sim W_d(m, \mathbf{\Sigma})$, where $\mathbf{\Sigma} > \mathbf{O}$ and $m \geq d$, then we can express **W** in the form $\mathbf{W} = \mathbf{X}'\mathbf{X}$, where the rows of **X** are i.i.d. $N_d(\mathbf{0}, \mathbf{\Sigma})$. Hence $\mathbf{W} > \mathbf{O}$ with probability 1 (see also A2.8).

A6 IDEMPOTENT MATRICES

A matrix **P** is idempotent if $\mathbf{P}^2 = \mathbf{P}$. A symmetric idempotent matrix is called a projection matrix.

1 If **P** is an $n \times n$ symmetric matrix, then **P** is idempotent of rank r if and only if it has r eigenvalues equal to 1 and $n - r$ equal to 0.

2 If $\mathbf{P}$ is a projection matrix of rank r, then it can be expressed in the form

$$\mathbf{P} = \sum_{i=1}^{r} \mathbf{t}_i \mathbf{t}_i',$$

where $\mathbf{t}_1, \mathbf{t}_2, \ldots, \mathbf{t}_r$ form an orthonormal set.

Proof From the end of A1.3 we have $\mathbf{A} = \Sigma_{i=1}^{n} \lambda_i \mathbf{t}_i \mathbf{t}_i'$, and the result follows from A6.1.

3 If $\mathbf{P}$ is a projection matrix, then $\operatorname{rank} \mathbf{P} = \operatorname{tr} \mathbf{P}$ (by A1.2 and A6.1).

4 If $\mathbf{P}$ is idempotent, then so is $\mathbf{I} - \mathbf{P}$.

5 Given $\mathbf{y} \sim N_d(\mathbf{0}, \sigma^2 \mathbf{I}_d)$ and $\mathbf{P}$ a $d \times d$ symmetric matrix, then $\mathbf{y}'\mathbf{P}\mathbf{y}/\sigma^2 \sim \chi_r^2$ if and only if $\mathbf{P}$ is idempotent of rank r.

6 Suppose $\mathbf{y} \sim N_d(\mathbf{0}, \sigma^2 \mathbf{I}_d)$ and $Q_i = \mathbf{y}_i'\mathbf{P}_i\mathbf{y}_i/\sigma^2$ is distributed as $\chi_{r_i}^2$ $(i = 1, 2)$. Then Q_1 and Q_2 are statistically independent if and only if $\mathbf{P}_1\mathbf{P}_2 = \mathbf{O}$.

Proof Given that Q_1 and Q_2 are independent, then, since the sum of independent chi-square variables is chi square, $\mathbf{y}'(\mathbf{P}_1 + \mathbf{P}_2)\mathbf{y}/\sigma^2$ is $\chi_{r_1+r_2}^2$. Hence, by A6.5, $\mathbf{P}_1$, $\mathbf{P}_2$ and $\mathbf{P}_1 + \mathbf{P}_2$ are idempotent, so that $\mathbf{P}_1 + \mathbf{P}_2 + \mathbf{P}_1\mathbf{P}_2 + \mathbf{P}_2\mathbf{P}_1 = (\mathbf{P}_1 + \mathbf{P}_2)^2 = \mathbf{P}_1 + \mathbf{P}_2$, or $\mathbf{P}_1\mathbf{P}_2 + \mathbf{P}_2\mathbf{P}_1 = \mathbf{O}$. Multiplying this last equation on the left, respectively right, by $\mathbf{P}_1$ gives us two equations that lead to $\mathbf{P}_1\mathbf{P}_2 = \mathbf{P}_2\mathbf{P}_1$; that is, $\mathbf{P}_1\mathbf{P}_2 = \mathbf{O}$.

Conversely, if $\mathbf{P}_1\mathbf{P}_2 = \mathbf{O}$, then $\mathscr{C}[\mathbf{P}_1\mathbf{y}, \mathbf{P}_2\mathbf{y}] = \mathbf{P}_1\mathscr{D}[\mathbf{y}]\mathbf{P}_2' = \sigma^2\mathbf{P}_1\mathbf{P}_2 = \mathbf{O}$, and $\mathbf{P}_1\mathbf{y}$ and $\mathbf{P}_2\mathbf{y}$ are statistically independent [Theorem 2.1(v)]. Hence the Q_i are independent at $\|\mathbf{P}_i\mathbf{y}\|^2 = \mathbf{y}'\mathbf{P}_i^2\mathbf{y} = \mathbf{y}'\mathbf{P}_i\mathbf{y} = \sigma^2 Q_i$. (The assumption that Q_i is chi square is not actually needed; however, the proof is then much longer and not as instructive; and the interested reader is referred to Lancaster [1969]).

7 Given that $\mathbf{X}$ is $n \times p$ of rank p, then $\mathbf{P} = \mathbf{X}(\mathbf{X}'\mathbf{X})^{-1}\mathbf{X}'$ is a projection matrix, $\mathbf{P}\mathbf{X} = \mathbf{X}$, and (by A1.1)

$$\operatorname{tr} \mathbf{P} = \operatorname{tr}\left[(\mathbf{X}'\mathbf{X})^{-1}\mathbf{X}'\mathbf{X}\right] = \operatorname{tr}\left[\mathbf{I}_p\right] = p.$$

A7 OPTIMIZATION AND INEQUALITIES

1 Consider the matrix function f, where

$$f(\mathbf{\Sigma}) = \log|\mathbf{\Sigma}| + \operatorname{tr}[\mathbf{\Sigma}^{-1}\mathbf{A}].$$

If $\mathbf{A} > \mathbf{O}$, then, subject to $\mathbf{\Sigma} > \mathbf{O}$, $f(\mathbf{\Sigma})$ is minimized uniquely at $\mathbf{\Sigma} = \mathbf{A}$.

Proof Let $\lambda_1, \lambda_2, \ldots, \lambda_d$ be the eigenvalues of $\mathbf{\Sigma}^{-1}\mathbf{A}$, that is, of $\mathbf{\Sigma}^{-1/2}\mathbf{A}\mathbf{\Sigma}^{-1/2}$ (by A5.4 and A1.4). Since the latter matrix is positive definite (A5.7), the λ_i are positive. Hence

$$\begin{aligned} f(\boldsymbol{\Sigma})-f(\mathbf{A}) &= \log|\boldsymbol{\Sigma}\mathbf{A}^{-1}| + \operatorname{tr}[\boldsymbol{\Sigma}^{-1}\mathbf{A}] - \operatorname{tr}\mathbf{I}_d \\ &= -\log|\boldsymbol{\Sigma}^{-1}\mathbf{A}| + \operatorname{tr}[\boldsymbol{\Sigma}^{-1}\mathbf{A}] - d \\ &= -\log\Big(\prod_i \lambda_i\Big) + \sum_i \lambda_i - d \qquad \text{(by A1.2)} \\ &= \sum_{i=1}^{d} (-\log\lambda_i + \lambda_i - 1) \geq 0, \end{aligned}$$

as $\log x \leq x - 1$ for $x > 0$. Equality occurs when each λ_i is unity, that is, when $\boldsymbol{\Sigma} = \mathbf{A}$. (This proof is due to Watson [1964].)

2 Let $f: \boldsymbol{\theta} \to f(\boldsymbol{\theta})$ be a real-valued function with domain Θ, and let $\mathbf{g}: \boldsymbol{\theta} \to \mathbf{g}(\boldsymbol{\theta}) = \boldsymbol{\phi}$ be a bijective (one-to-one) function from Θ onto Φ. Since $\mathbf{g}$ is bijective, it has an inverse, $\mathbf{g}^{-1}$, say, and we can define $h(\boldsymbol{\phi}) = f(\mathbf{g}^{-1}(\boldsymbol{\phi}))$ for $\boldsymbol{\phi} \in \Phi$.
   (a) If $f(\boldsymbol{\theta})$ attains a maximum at $\boldsymbol{\theta} = \hat{\boldsymbol{\theta}}$, $h(\boldsymbol{\phi})$ attains its maximum at $\hat{\boldsymbol{\phi}} = \mathbf{g}(\hat{\boldsymbol{\theta}})$.
   (b) If the maximum of $f(\boldsymbol{\theta})$ occurs uniquely at $\hat{\boldsymbol{\theta}}$, then the maximum of $h(\boldsymbol{\phi})$ occurs uniquely at $\hat{\boldsymbol{\phi}}$.

Proof (a) $f(\hat{\boldsymbol{\theta}}) \geq f(\boldsymbol{\theta})$ for all $\boldsymbol{\theta} \in \Theta$, and

$$h(\hat{\boldsymbol{\phi}}) = f(\mathbf{g}^{-1}(\hat{\boldsymbol{\phi}})) = f(\hat{\boldsymbol{\theta}}) \geq f(\boldsymbol{\theta}) = f(\mathbf{g}^{-1}(\boldsymbol{\phi})) = h(\boldsymbol{\phi}).$$

Hence
$h(\boldsymbol{\phi})$ attains a maximum at $\hat{\boldsymbol{\phi}}$.
(b) If the maximum occurs uniquely at $\hat{\boldsymbol{\theta}}$, then the above inequalities are strict for $\boldsymbol{\theta} \neq \hat{\boldsymbol{\theta}}$, that is, $h(\hat{\boldsymbol{\phi}}) > h(\boldsymbol{\phi})$ for $\hat{\boldsymbol{\phi}} \neq \boldsymbol{\phi}$.

3 *Frobenius norm approximation*. Let $\mathbf{B}$ be a $p \times q$ matrix of rank r with singular value decomposition [see (10.20)] $\Sigma_{i=1}^{r} \delta_i \boldsymbol{l}_i \mathbf{m}'_i$, and let $\mathbf{C}$ be a $p \times q$ matrix of rank s ($s < r$). Then

$$\|\mathbf{B} - \mathbf{C}\|^2 = \sum_{i=1}^{p} \sum_{j=1}^{q} (b_{ij} - c_{ij})^2$$

is minimized when

$$\mathbf{C} = \mathbf{B}_{(s)} = \sum_{i=1}^{s} \delta_i \boldsymbol{l}_i \mathbf{m}'_i .$$

The minimum value is $\delta_{s+1}^2 + \delta_{s+2}^2 + \cdots + \delta_r^2$.

This result is due to Eckart and Young [1936] (see also Householder and Young [1938]). When $\mathbf{B}$ is symmetric, $\mathbf{m}_i = \boldsymbol{l}_i$. For a general reference on matrix approximation see Rao [1980].

4 Let $\mathbf{A}$ be an $n \times n$ symmetric matrix with eigenvalues $\lambda_1 \geq \lambda_2 \geq \cdots \geq \lambda_n$, and a corresponding set of orthonormal eigenvectors $\mathbf{t}_1, \mathbf{t}_2, \ldots, \mathbf{t}_n$ (see A1.3). Define $\mathbf{T}_k = (\mathbf{t}_1, \mathbf{t}_2, \ldots, \mathbf{t}_k)$ $(k = 1, 2, \ldots, n-1)$, and $\mathbf{T} = (\mathbf{t}_1, \mathbf{t}_2, \ldots, \mathbf{t}_n)$. Then, if we assume that $\mathbf{x} \neq \mathbf{0}$, we have the following:

(a)

$$\underset{\mathbf{x}}{\text{supremum}} \left\{ \frac{\mathbf{x}'\mathbf{A}\mathbf{x}}{\mathbf{x}'\mathbf{x}} \right\} = \lambda_1,$$

and the supremum is attained if $\mathbf{x} = \mathbf{t}_1$.

(b)

$$\underset{\mathbf{T}_k'\mathbf{x}=\mathbf{0}}{\text{supremum}} \left\{ \frac{\mathbf{x}'\mathbf{A}\mathbf{x}}{\mathbf{x}'\mathbf{x}} \right\} = \lambda_{k+1},$$

and the supremum is attained if $\mathbf{x} = \mathbf{t}_{k+1}$.

(c)

$$\underset{\mathbf{x}}{\text{infimum}} \left\{ \frac{\mathbf{x}'\mathbf{A}\mathbf{x}}{\mathbf{x}'\mathbf{x}} \right\} = \lambda_n,$$

and the infimum is attained if $\mathbf{x} = \mathbf{t}_n$.

(d) If $\mathbf{T}_{n-k} = (\mathbf{t}_{n-k+1}, \mathbf{t}_{n-k+2}, \ldots, \mathbf{t}_n)$, then

$$\underset{\mathbf{T}_{n-k}'\mathbf{x}=\mathbf{0}}{\text{infimum}} \left\{ \frac{\mathbf{x}'\mathbf{A}\mathbf{x}}{\mathbf{x}'\mathbf{x}} \right\} = \lambda_{n-k},$$

and the infimum is attained if $\mathbf{x} = \mathbf{t}_{n-k}$.

(e) *Courant–Fischer min–max theorem.*

$$\inf_{\mathbf{L}} \sup_{\mathbf{L}'\mathbf{x}=\mathbf{0}} \left\{ \frac{\mathbf{x}\mathbf{A}\mathbf{x}}{\mathbf{x}'\mathbf{x}} \right\} = \lambda_{k+1},$$

where inf denotes the infimum with respect to an $n \times k$ matrix $\mathbf{L}$, while sup denotes the supremum with respect to an $n \times 1$ vector $\mathbf{x}$ satisfying $\mathbf{L}'\mathbf{x} = \mathbf{0}$. The above result is attained when $\mathbf{L} = \mathbf{T}_k$ and $\mathbf{x} = \mathbf{t}_{k+1}$.

(f)

$$\sup_{\mathbf{L}} \inf_{\mathbf{L}'\mathbf{x}=\mathbf{0}} \left\{ \frac{\mathbf{x}'\mathbf{A}\mathbf{x}}{\mathbf{x}'\mathbf{x}} \right\} = \lambda_{n-k},$$

and the right-hand side is attained when

$$\mathbf{L} = (\mathbf{t}_{n-k+1}, \mathbf{t}_{n-k+2}, \ldots, \mathbf{t}_n) \quad \text{and} \quad \mathbf{x} = \mathbf{t}_{n-k}.$$

Proof (a) Let $\mathbf{x} = \mathbf{Ty} = y_1\mathbf{t}_1 + y_2\mathbf{t}_2 + \cdots + y_n\mathbf{t}_n$. Then

$$\frac{\mathbf{x}'\mathbf{Ax}}{\mathbf{x}'\mathbf{x}} = \frac{\mathbf{y}'\mathbf{T}'\mathbf{ATy}}{\mathbf{y}'\mathbf{y}} = \frac{\left(\sum_i \lambda_i y_i^2\right)}{\mathbf{y}'\mathbf{y}} \leq \frac{\lambda_1 \mathbf{y}'\mathbf{y}}{\mathbf{y}'\mathbf{y}} = \lambda_1,$$

with equality when $y_1 = 1, y_2 = y_3 = \cdots = y_n = 0$, that is, when $\mathbf{x} = \mathbf{t}_1$.
(b) If $\mathbf{x} \perp \{\mathbf{t}_1, \mathbf{t}_2, \ldots, \mathbf{t}_k\}$, then $y_1 = y_2 = \cdots = y_k = 0$. The result then follows with the argument used in proving (a).
(c) and (d). These follow in a similar fashion to (a) and (b), but with the inequality reversed.
(e) Let $\mathbf{x} = \mathbf{Ty}$; then $\mathbf{L}'\mathbf{x} = \mathbf{0}$ if and only if $\mathbf{L}'\mathbf{Ty} = \mathbf{0}$. If $\mathbf{M} = \mathbf{T}'\mathbf{L}$, then $\mathbf{M}$ is $n \times k$ of rank less than or equal to k. The problem now reduces to finding

$$\inf_{\mathbf{M}} \sup_{\mathbf{M}'\mathbf{y}=\mathbf{0}} \left\{ \left(\sum_{i=1}^{n} \lambda_i y_i^2 \right) \Big/ \mathbf{y}'\mathbf{y} \right\}.$$

We first note that

$$\sup_{\mathbf{M}'\mathbf{y}=\mathbf{0}} \left\{ \left(\sum_{i=1}^{n} \lambda_i y_i^2 \right) \Big/ \sum_{i=1}^{n} y_i^2 \right\} \geq \sup_{\mathbf{M}'\mathbf{y}=\mathbf{0}} \left\{ \left(\sum_{i=1}^{k+1} \lambda_i y_i^2 \right) \Big/ \sum_{i=1}^{k+1} y_i^2 \right\}$$

$$\geq \sup_{\mathbf{M}'\mathbf{y}=\mathbf{0}} \{\lambda_{k+1}\} = \lambda_{k+1},$$

provided that the set of $\mathbf{y}$ satisfying $\mathbf{M}'\mathbf{y} = \mathbf{0}$ and $y_{k+2} = y_{k+3} = \cdots = y_n = 0$ is not empty. If we write $\mathbf{M}' = (\mathbf{M}_1', \mathbf{M}_2')$, where $\mathbf{M}_1'$ is $k \times (k+1)$, then rank $\mathbf{M}_1' \leq k$, and (by A2.5) $\mathbf{M}_1'$ has nullity at least 1. The required set $\{\mathbf{y}_1 : \mathbf{M}_1'\mathbf{y}_1 = \mathbf{0}\}$ is therefore not empty. We have found a lower bound λ_{k+1} that is independent of $\mathbf{M}$ so that it is a lower bound if we take the infimum with respect to $\mathbf{M}$. Using (b), we see that the lower bound is achieved with $\mathbf{L} = \mathbf{T}_k$.
(f) This follows in a similar fashion to (e).

5 Let $\mathbf{A}$ be an $n \times n$ symmetric matrix and let $\mathbf{D}$ be any $n \times n$ positive definite matrix. Let $\gamma_1 \geq \gamma_2 \geq \cdots \geq \gamma_n$ be the eigenvalues of $\mathbf{D}^{-1}\mathbf{A}$ with corresponding eigenvectors $\mathbf{v}_1, \mathbf{v}_2, \ldots, \mathbf{v}_n$. Then

$$\sup_{\mathbf{x}} \left\{ \frac{\mathbf{x}'\mathbf{Ax}}{\mathbf{x}'\mathbf{Dx}} \right\} = \gamma_1 \quad \text{and} \quad \inf_{\mathbf{x}} \left\{ \frac{\mathbf{x}'\mathbf{Ax}}{\mathbf{x}'\mathbf{Dx}} \right\} = \gamma_n,$$

with the bounds being attained when $\mathbf{x} = \mathbf{v}_1$ and $\mathbf{x} = \mathbf{v}_n$, respectively.

Proof By A5.3 there exists a nonsingular matrix $\mathbf{R}$ such that $\mathbf{D} = \mathbf{R}'\mathbf{R}$. Let $\mathbf{y} = \mathbf{R}\mathbf{x}$, then

$$\sup_{\mathbf{x}} \left\{ \frac{\mathbf{x}'\mathbf{A}\mathbf{x}}{\mathbf{x}'\mathbf{D}\mathbf{x}} \right\} = \sup_{\mathbf{y}} \left\{ \frac{\mathbf{y}'(\mathbf{R}^{-1})'\mathbf{A}\mathbf{R}^{-1}\mathbf{y}}{\mathbf{y}'\mathbf{y}} \right\} = \gamma_1,$$

where γ_1 is the maximum eigenvalue of $(\mathbf{R}^{-1})'\mathbf{A}\mathbf{R}^{-1}$, that is, of $(\mathbf{R}'\mathbf{R})^{-1}\mathbf{A} = \mathbf{D}^{-1}\mathbf{A}$ (by A1.4). The supremum occurs when $\mathbf{y}$ satisfies $(\mathbf{R}^{-1})'\mathbf{A}\mathbf{R}^{-1}\mathbf{y} = \gamma_1\mathbf{y}$, that is, $\mathbf{D}^{-1}\mathbf{A}\mathbf{x} = \gamma_1\mathbf{x}$. The second result follows in a similar fashion.

6 If $\mathbf{D}$ is positive definite, then for any $\mathbf{a}$

$$\sup_{\mathbf{x}} \left\{ \frac{(\mathbf{a}'\mathbf{x})^2}{\mathbf{x}'\mathbf{D}\mathbf{x}} \right\} = \mathbf{a}'\mathbf{D}^{-1}\mathbf{a}.$$

The supremum occurs when $\mathbf{x}$ is proportional to $\mathbf{D}^{-1}\mathbf{a}$. Although this result is usually proved using the Cauchy–Schwartz inequality ($\mathbf{D} = \mathbf{I}$), it can be deduced directly from A7.5 above.

7 If $\mathbf{M}$ and $\mathbf{N}$ are positive definite, then

$$\sup_{\mathbf{x},\mathbf{y}} \left\{ \frac{(\mathbf{x}'\mathbf{L}\mathbf{y})^2}{\mathbf{x}'\mathbf{M}\mathbf{x} \cdot \mathbf{y}'\mathbf{N}\mathbf{y}} \right\} = \theta_{max}$$

where θ_{max} is the largest eigenvalue of $\mathbf{M}^{-1}\mathbf{L}\mathbf{N}^{-1}\mathbf{L}'$, and of $\mathbf{N}^{-1}\mathbf{L}'\mathbf{M}^{-1}\mathbf{L}$. The supremum occurs when $\mathbf{x}$ is an eigenvector of $\mathbf{M}^{-1}\mathbf{L}\mathbf{N}^{-1}\mathbf{L}'$ corresponding to θ_{max}, and $\mathbf{y}$ is an eigenvector of $\mathbf{N}^{-1}\mathbf{L}'\mathbf{M}^{-1}\mathbf{L}$ corresponding to θ_{max}.

Proof Let $\mathbf{z} = \mathbf{L}\mathbf{y}$. Then, if we assume nonzero vectors $\mathbf{x}$ and $\mathbf{y}$, we have

$$\begin{aligned} \sup_{\mathbf{x},\mathbf{y}} \left\{ \frac{(\mathbf{x}'\mathbf{L}\mathbf{y})^2}{\mathbf{x}'\mathbf{M}\mathbf{x} \cdot \mathbf{y}'\mathbf{N}\mathbf{y}} \right\} &= \sup_{\mathbf{y}} \left\{ \frac{1}{\mathbf{y}'\mathbf{N}\mathbf{y}} \sup_{\mathbf{x}} \left[\frac{(\mathbf{x}'\mathbf{z})^2}{\mathbf{x}'\mathbf{M}\mathbf{x}} \right] \right\} \\ &= \sup_{\mathbf{y}} \left\{ \frac{\mathbf{y}'\mathbf{L}'\mathbf{M}^{-1}\mathbf{L}\mathbf{y}}{\mathbf{y}'\mathbf{N}\mathbf{y}} \right\} \qquad \text{(by A7.6)} \\ &= \theta_{max}, \end{aligned}$$

where, by A7.5, θ_{max} is the largest eigenvalue of $\mathbf{N}^{-1}\mathbf{L}'\mathbf{M}^{-1}\mathbf{L}$, and therefore of $\mathbf{M}^{-1}\mathbf{L}\mathbf{N}^{-1}\mathbf{L}'$ (by A1.4). The first supremum is attained when $\mathbf{x} = \mathbf{M}^{-1}\mathbf{z} = \mathbf{M}^{-1}\mathbf{L}\mathbf{y}$, and the second when $\mathbf{y}$ is an eigenvector of $\mathbf{N}^{-1}\mathbf{L}'\mathbf{M}^{-1}\mathbf{L}$ corresponding to θ_{max}, that is, $\mathbf{N}^{-1}\mathbf{L}'\mathbf{M}^{-1}\mathbf{L}\mathbf{y} = \theta_{max}\mathbf{y}$. Multiplying this last equation on the left by $\mathbf{M}^{-1}\mathbf{L}$, we have $\mathbf{M}^{-1}\mathbf{L}\mathbf{N}^{-1}\mathbf{L}'\mathbf{x} = \theta_{max}\mathbf{x}$, and $\mathbf{x}$ is the required eigenvector that gives the supremum.

8 Let $\mathbf{C}$ be a $p \times q$ matrix of rank m and let $\rho_1^2 \geq \rho_2^2 \geq \cdots \geq \rho_m^2 > 0$ be the nonzero eigenvalues of $\mathbf{CC}'$ (and of $\mathbf{C}'\mathbf{C}$). Let $\mathbf{t}_1, \mathbf{t}_2, \ldots, \mathbf{t}_m$ be the corresponding eigenvectors of $\mathbf{CC}'$ and let $\mathbf{w}_1, \mathbf{w}_2, \ldots, \mathbf{w}_m$ be the corresponding eigenvectors of $\mathbf{C}'\mathbf{C}$. If $\mathbf{T}_k = (\mathbf{t}_1, \mathbf{t}_2, \ldots, \mathbf{t}_k)$ and $\mathbf{W}_k = (\mathbf{w}_1, \mathbf{w}_2, \ldots, \mathbf{w}_k)$ $(k < m)$, then

$$\sup_{\mathbf{T}_k'\mathbf{x}=\mathbf{0},\, \mathbf{W}_k'\mathbf{y}=\mathbf{0}} \left\{ \frac{(\mathbf{x}'\mathbf{C}\mathbf{y})^2}{\mathbf{x}'\mathbf{x} \cdot \mathbf{y}'\mathbf{y}} \right\} = \rho_{k+1}^2,$$

and the supremum occurs when $\mathbf{x} = \mathbf{t}_{k+1}$ and $\mathbf{y} = \mathbf{w}_{k+1}$.

Proof Adding in the eigenvectors corresponding to zero eigenvalues, let the columns of $\mathbf{T} = (\mathbf{t}_1, \mathbf{t}_2, \ldots, \mathbf{t}_p)$ be a set of orthonormal eigenvectors of $\mathbf{CC}'$, and let the columns of $\mathbf{W} = (\mathbf{w}_1, \mathbf{w}_2, \ldots, \mathbf{w}_q)$ be a set for $\mathbf{C}'\mathbf{C}$. Then we have the singular value decomposition [see (10.20)] $\mathbf{C} = \mathbf{TDW}'$, where $\mathbf{D} = \mathrm{diag}(\rho_1, \rho_2, \ldots, \rho_m, 0, \ldots, 0)$. Setting $\mathbf{x} = \mathbf{Tu}$ and $\mathbf{y} = \mathbf{Wv}$, we have

$$\mathbf{x}'\mathbf{C}\mathbf{y} = \mathbf{u}'\mathbf{T}'\mathbf{C}\mathbf{W}\mathbf{v} = \mathbf{u}'\mathbf{D}\mathbf{v} = \sum_{i=1}^{m} \rho_i u_i v_i.$$

If we argue as in the proof of A7.4(b) above, the constraint $\mathbf{x} \perp \{\mathbf{t}_1, \mathbf{t}_2, \ldots, \mathbf{t}_k\}$ implies that $u_1 = u_2 = \cdots = u_k = 0$. Similarly, $v_1 = v_2 = \cdots = v_k = 0$. Hence, subject to $\mathbf{T}_k'\mathbf{x} = \mathbf{0}$ and $\mathbf{W}_k'\mathbf{y} = \mathbf{0}$,

$$\begin{aligned}\frac{(\mathbf{x}'\mathbf{C}\mathbf{y})^2}{\mathbf{x}'\mathbf{x} \cdot \mathbf{y}'\mathbf{y}} &= \left(\sum_{i=k+1}^{m} \rho_i u_i v_i \right)^2 \Big/ \left(\sum_{i=k+1}^{p} u_i^2 \right) \left(\sum_{i=k+1}^{q} v_i^2 \right) \\ &\leq \left(\sum_{i=k+1}^{m} \rho_i u_i^2 \right) \left(\sum_{i=k+1}^{m} \rho_i v_i^2 \right) \Big/ \left(\sum_{i=k+1}^{p} u_i^2 \right) \left(\sum_{i=k+1}^{q} v_i^2 \right) \\ &\qquad \left(\text{since } [\mathbf{a}'\mathbf{b}]^2 \leq \mathbf{a}'\mathbf{a} \cdot \mathbf{b}'\mathbf{b} \right) \\ &\leq \rho_{k+1}^2.\end{aligned}$$

Equality occurs when $u_{k+1} = 1$, $u_{k+2} = u_{k+3} = \cdots = u_p = 0$, $v_{k+1} = 1$, and $v_{k+2} = v_{k+3} = \cdots = v_q = 0$, that is, when $\mathbf{x} = \mathbf{t}_{k+1}$ and $\mathbf{y} = \mathbf{w}_{k+1}$.

9 Let $\mathbf{A}$ and $\mathbf{B}$ be $n \times n$ symmetric matrices with eigenvalues $\mu_1(\mathbf{A}) \geq \mu_2(\mathbf{A}) \geq \cdots \geq \mu_n(\mathbf{A})$ and $\mu_1(\mathbf{B}) \geq \mu_2(\mathbf{B}) \geq \cdots \geq \mu_n(\mathbf{B})$, respectively. If $\mathbf{A} - \mathbf{B} \geq \mathbf{O}$, then we have the following:

(a) $\mu_i(\mathbf{A}) \geq \mu_i(\mathbf{B})$ $(i = 1, 2, \ldots, n)$.

(b) $\mathrm{tr}\,\mathbf{A} \geq \mathrm{tr}\,\mathbf{B}$.

(c) $|\mathbf{A}| \geq |\mathbf{B}|$.

(d) $\|\mathbf{A}\| \geq \|\mathbf{B}\|$, where $\|\mathbf{A}\| = \{\mathrm{tr}[\mathbf{A}\mathbf{A}']\}^{1/2}$.

Proof

(a) Since $\mathbf{x}'(\mathbf{A}-\mathbf{B})\mathbf{x} \geq 0$, $\mathbf{x}'\mathbf{A}\mathbf{x} \geq \mathbf{x}'\mathbf{B}\mathbf{x}$. By A7.4 there exists an $n \times (i-1)$ matrix $\mathbf{K}$ such that

$$\begin{aligned}\mu_i(\mathbf{A}) &= \sup_{\mathbf{K}'\mathbf{x}=\mathbf{0}} \left\{\frac{\mathbf{x}'\mathbf{A}\mathbf{x}}{\mathbf{x}'\mathbf{x}}\right\} \\ &\geq \sup_{\mathbf{K}'\mathbf{x}=\mathbf{0}} \left\{\frac{\mathbf{x}'\mathbf{B}\mathbf{x}}{\mathbf{x}'\mathbf{x}}\right\} \\ &\geq \inf_{\mathbf{L}} \sup_{\mathbf{L}'\mathbf{x}=\mathbf{0}} \left\{\frac{\mathbf{x}'\mathbf{B}\mathbf{x}}{\mathbf{x}'\mathbf{x}}\right\} \\ &= \mu_i(\mathbf{B}).\end{aligned}$$

(b) and (c) follow from A1.2, while (d) follows by noting that

$$\operatorname{tr}[\mathbf{A}\mathbf{A}'] = \operatorname{tr}\mathbf{A}^2 = \sum_{i=1}^{n} \mu_i^2(\mathbf{A}).$$

We note that we must have at least one strict inequality in (a) if $\mathbf{A} \neq \mathbf{B}$; otherwise $\operatorname{tr}[\mathbf{A}-\mathbf{B}] = \operatorname{tr}\mathbf{A} - \operatorname{tr}\mathbf{B} = 0$, and $\mathbf{A}-\mathbf{B} = \mathbf{O}$, as the eigenvalues of $\mathbf{A}-\mathbf{B}$ are nonnegative. Property (a) is frequently expressed in the form $\mu_i(\mathbf{A}-\mathbf{B}+\mathbf{B}) \geq \mu_i(\mathbf{B})$.

10 Let $\mathbf{A}$, $\mathbf{B}$, and $\mathbf{A}-\mathbf{B}$ be $n \times n$ positive semidefinite matrices, with $\operatorname{rank}\mathbf{B} \leq r$, and let $\mu_i(\cdot)$ represent the ith largest eigenvalue. Then

$$\mu_i(\mathbf{A}-\mathbf{B}) \geq \begin{cases} \mu_{r+i}(\mathbf{A}), & i = 1,2,\ldots,n-r \\ 0, & i = n-r+1, n-r+2,\ldots,n. \end{cases}$$

Equality occurs if

$$\mathbf{B} = \mathbf{B}_0 = \sum_{i=1}^{r} \mu_i(\mathbf{A})\mathbf{t}_i\mathbf{t}_i',$$

where $\mathbf{t}_1, \mathbf{t}_2, \ldots, \mathbf{t}_n$ are orthonormal eigenvectors corresponding to $\mu_1(\mathbf{A}), \mu_2(\mathbf{A}), \ldots, \mu_n(\mathbf{A})$.

Proof By A7.4(b) there exists, for each i, an $n \times (i-1)$ matrix $\mathbf{K}$ such that

$$\begin{aligned}
\mu_i(\mathbf{A}-\mathbf{B}) &= \sup_{\mathbf{K}'\mathbf{x}=\mathbf{0}} \left\{ \frac{\mathbf{x}'(\mathbf{A}-\mathbf{B})\mathbf{x}}{\mathbf{x}'\mathbf{x}} \right\} \\
&\geq \sup_{\mathbf{K}'\mathbf{x}=\mathbf{0},\, \mathbf{B}'\mathbf{x}=\mathbf{0}} \left\{ \frac{\mathbf{x}'(\mathbf{A}-\mathbf{B})\mathbf{x}}{\mathbf{x}'\mathbf{x}} \right\} \\
&= \sup_{(\mathbf{K},\mathbf{B})'\mathbf{x}=\mathbf{0}} \left\{ \frac{\mathbf{x}'\mathbf{A}\mathbf{x}}{\mathbf{x}'\mathbf{x}} \right\} \\
&\geq \mu_{r+i}(\mathbf{A}), \qquad \text{by A7.4(e),}
\end{aligned}$$

since the matrix $(\mathbf{K},\mathbf{B})$ is, at most , of rank $r+i-1$.

Finally, by A1.3,

$$\mathbf{A} = \sum_{i=1}^{n} \mu_i(\mathbf{A})\mathbf{t}_i\mathbf{t}_i' \quad \text{and} \quad \mathbf{A}-\mathbf{B}_0 = \sum_{i=r+1}^{n} \mu_i(\mathbf{A})\mathbf{t}_i\mathbf{t}_i'.$$

Postmultiplying $\mathbf{A}-\mathbf{B}_0$ by $\mathbf{t}_k$, we see that the eigenvalues of $\mathbf{A}-\mathbf{B}_0$ are $\mu_{r+1}(\mathbf{A}),\ldots,\mu_n(\mathbf{A}),0,\ldots,0$, and we have shown that $\mathbf{B}=\mathbf{B}_0$ is a sufficient condition for equality; it is also a necessary condition (Okamoto [1969]).

Note: Our inequality can be expressed in the form $\mu_i(\mathbf{A}-\mathbf{B}) \geq \mu_{r+i}(\mathbf{A} - \mathbf{B} + \mathbf{B})$. However, a more general inequality due to Weyl (Bellman [1960: p. 119]) exists for symmetric matrices. If $\mathbf{C}$ and $\mathbf{D}$ are symmetric $n \times n$ matrices, then

$$\mu_i(\mathbf{C}) + \mu_j(\mathbf{D}) \geq \mu_{i+j-1}(\mathbf{C}+\mathbf{D}) \qquad (i+j-1 \leq n).$$

Our particular result follows by setting $\mathbf{C}=\mathbf{A}-\mathbf{B}$, $\mathbf{D}=\mathbf{B}$, and $j=r+1$. Similar inequalities relating to products of eigenvalues are given by Anderson and Das Gupta [1963] and Thompson and Therianos [1973].

A8 VECTOR AND MATRIX DIFFERENTIATION

1 If $d/d\boldsymbol{\beta} = [(\partial/\partial\beta_i)]$, then we have the following:

(a) $$\frac{d(\boldsymbol{\beta}'\mathbf{a})}{d\boldsymbol{\beta}} = \mathbf{a}.$$

(b) $$\frac{d(\boldsymbol{\beta}'\mathbf{A}\boldsymbol{\beta})}{d\boldsymbol{\beta}} = 2\mathbf{A}\boldsymbol{\beta} \qquad (\mathbf{A} \text{ symmetric}).$$

2 If $\mathbf{M} = [(m_{ij})]$ is a matrix with distinct elements, and $d/d\mathbf{M} = [(\partial/\partial m_{ij})]$, then we have the following:

(a) $$\frac{d\,\mathrm{tr}[\mathbf{LMN}]}{d\mathbf{M}} = \mathbf{L}'\mathbf{N}'.$$

(b) $$\frac{d\,\mathrm{tr}[\mathbf{M}'\mathbf{L}\mathbf{M}\mathbf{N}]}{d\mathbf{M}} = 2\mathbf{L}\mathbf{M}\mathbf{N} \qquad (\mathbf{L} \text{ and } \mathbf{N} \text{ symmetric}).$$

(c) $$\frac{d\log|\mathbf{M}|}{d\mathbf{M}} = (\mathbf{M}')^{-1} \qquad (\mathbf{M} \text{ nonsingular}).$$

Since $\mathrm{tr}[\mathbf{LMN}] = \mathrm{tr}[\mathbf{N}'\mathbf{M}'\mathbf{L}']$ and $\mathrm{tr}[(\mathbf{M}'\mathbf{L})(\mathbf{MN})] = \mathrm{tr}[(\mathbf{MN})(\mathbf{M}'\mathbf{L})]$, we can deduce the following:

(d) $$\frac{d\,\mathrm{tr}[\mathbf{L}\mathbf{M}'\mathbf{N}]}{d\mathbf{M}} = \mathbf{NL}.$$

(e) $$\frac{d\,\mathrm{tr}[\mathbf{M}\mathbf{N}\mathbf{M}'\mathbf{L}]}{d\mathbf{M}} = 2\mathbf{L}\mathbf{M}\mathbf{N} \qquad (\mathbf{L} \text{ and } \mathbf{N} \text{ symmetric}).$$

A9 JACOBIANS AND TRANSFORMATIONS

1 If the distinct elements of a symmetric $d \times d$ matrix $\mathbf{W}$ have a joint density function of the form $g(\lambda_1, \lambda_2, \ldots, \lambda_d)$, where $\lambda_1 > \lambda_2 > \cdots > \lambda_d$ are the eigenvalues of $\mathbf{W}$, then the joint density function of the eigenvalues is

$$\pi^{d^2/2} g(\lambda_1, \lambda_2, \ldots, \lambda_d) \Big\{ \prod_{j<k} (\lambda_j - \lambda_k) \Big\} \Big/ \Gamma_d(\tfrac{1}{2}d),$$

where

$$\Gamma_d(\tfrac{1}{2}d) = \pi^{d(d-1)/4} \prod_{j=1}^{d} \Gamma(\tfrac{1}{2}[d+1-j]).$$

For a proof see Anderson [1958: Theorem 13.3.1].

2 Let $\mathbf{X}$ be an $m \times n$ matrix of distinct random variables and let $\mathbf{Y} = \mathbf{a}(\mathbf{X})$, where $\mathbf{Y}$ is $m \times n$ and $\mathbf{a}$ is a bijective (one-to-one) function. Then there exists an inverse function $\mathbf{b} = \mathbf{a}^{-1}$, so that $\mathbf{X} = \mathbf{b}(\mathbf{Y})$. If $\mathbf{X}$ has density function f and $\mathbf{Y}$ has density function g, then

$$g(\mathbf{Y}) = f(\mathbf{b}[\mathbf{Y}]) \left| \frac{d\mathbf{X}}{d\mathbf{Y}} \right|,$$

where the symbol $d\mathbf{X}/d\mathbf{Y}$ represents the Jacobian of the transformation from $\mathbf{X}$ to $\mathbf{Y}$. Specifically, if x_i is the ith element of vec $\mathbf{X}$ [see (1.10) and the following discussion there], and y_j is the jth element of vec $\mathbf{Y}$, then $d\mathbf{X}/d\mathbf{Y}$ is the determinant of the matrix with (i, j)th element dx_i/dy_j. Some useful Jacobians are as follows:

(a) If $\mathbf{X} = \mathbf{AYB}$, where $\mathbf{A}$ and $\mathbf{B}$ are $m \times m$ and $n \times n$ nonsingular matrices, respectively, then

$$\frac{d\mathbf{X}}{d\mathbf{Y}} = |\mathbf{A}|^n |\mathbf{B}|^m.$$

Special cases are obtained by setting $\mathbf{A} = \mathbf{I}_m$ or $\mathbf{B} = \mathbf{I}_n$.

(b) If $\mathbf{X}$ and $\mathbf{Y}$ are $n \times n$ symmetric matrices, $\mathbf{A}$ is nonsingular, and $\mathbf{X} = \mathbf{AYA}'$, then

$$\frac{d\mathbf{X}}{d\mathbf{Y}} = |\mathbf{A}|^{n+1}.$$

(c) Let $\mathbf{E}$ and $\mathbf{H}$ be $d \times d$ positive definite matrices, and let $\mathbf{Z} = \mathbf{E} + \mathbf{H}$ and $\mathbf{V} = (\mathbf{E} + \mathbf{H})^{-1/2}\mathbf{H}(\mathbf{E} + \mathbf{H})^{-1/2}$. Then

$$\frac{d(\mathbf{H}, \mathbf{E})}{d(\mathbf{V}, \mathbf{Z})} = |\mathbf{Z}|^{(d+1)/2}.$$

Proof For (a) and (b) see, for example, Deemer and Olkin [1951] or Muirhead [1982: Chapter 2]. To prove (c) we note that Jacobians are multiplicative, so that, symbolically,

$$\frac{d(\mathbf{V}, \mathbf{Z})}{d(\mathbf{H}, \mathbf{E})} = \frac{d(\mathbf{V}, \mathbf{Z})}{d(\mathbf{H}, \mathbf{Z})}\frac{d(\mathbf{H}, \mathbf{Z})}{d(\mathbf{H}, \mathbf{E})}.$$

Setting $\mathbf{V} = \mathbf{Z}^{-1/2}\mathbf{H}\mathbf{Z}^{-1/2}$ and $\mathbf{W} = \mathbf{Z}$, we see that $\partial w_i/\partial h_j = 0$ and $\partial w_i/\partial z_j = \delta_{ij}$. Hence

$$\frac{d(\mathbf{V}, \mathbf{Z})}{d(\mathbf{H}, \mathbf{Z})} = \begin{vmatrix} \dfrac{\partial \mathbf{V}}{\partial \mathbf{H}} & \dfrac{\partial \mathbf{W}}{\partial \mathbf{H}} \\ \dfrac{\partial \mathbf{V}}{\partial \mathbf{Z}} & \dfrac{\partial \mathbf{W}}{\partial \mathbf{Z}} \end{vmatrix} = \begin{vmatrix} \dfrac{\partial \mathbf{V}}{\partial \mathbf{H}} & \mathbf{O} \\ \dfrac{\partial \mathbf{V}}{\partial \mathbf{Z}} & \mathbf{I} \end{vmatrix} = \left|\frac{\partial \mathbf{V}}{\partial \mathbf{H}}\right|,$$

or $d\mathbf{V}/d\mathbf{H}$ with $\mathbf{Z}$ fixed, that is, $|\mathbf{Z}|^{-(d+1)/2}$ [by (b) above]. Similarly, $d(\mathbf{H}, \mathbf{Z})/d(\mathbf{H}, \mathbf{E}) = 1$, and the result follows from

$$\frac{d(\mathbf{H}, \mathbf{E})}{d(\mathbf{V}, \mathbf{Z})} = \left\{\frac{d(\mathbf{V}, \mathbf{Z})}{d(\mathbf{H}, \mathbf{E})}\right\}^{-1}.$$

A10 ASYMPTOTIC NORMALITY

1 *Central limit theorem.* Let $\mathbf{x}_1, \mathbf{x}_2, \ldots, \mathbf{x}_n$ be a random sample from a d-dimensional distribution with mean $\boldsymbol{\mu}$ and dispersion matrix $\boldsymbol{\Sigma}$. Then, as $n \to \infty$, $\sqrt{n}(\overline{\mathbf{x}} - \boldsymbol{\mu})$ is asymptotically $N_d(\mathbf{0}, \boldsymbol{\Sigma})$.

2 Suppose $\sqrt{n}(\mathbf{y} - \boldsymbol{\theta})$ is asymptotically $N_d(\mathbf{0}, \boldsymbol{\Sigma})$, and let $\mathbf{f} = (f_1, f_2, \ldots, f_q)'$ be a q-dimensional real-valued function differentiable at $\boldsymbol{\theta}$. If $\mathbf{F} = [(f_{ij})]$, where $f_{ij} = \partial f_j(\mathbf{y})/\partial y_i$ evaluated at $\mathbf{y} = \boldsymbol{\theta}$, then $\sqrt{n}[\mathbf{f}(\mathbf{y}) - \mathbf{f}(\boldsymbol{\theta})]$ is asymptotically $N_q(\mathbf{0}, \mathbf{F}'\boldsymbol{\Sigma}\mathbf{F})$.

APPENDIX B

Orthogonal Projections

B1 ORTHOGONAL DECOMPOSITION OF VECTORS

1 Given Ω, a vector subspace of R^n (n-dimensional Euclidean space), every $n \times 1$ vector $\mathbf{y}$ can be expressed uniquely in the form $\mathbf{y} = \mathbf{u} + \mathbf{v}$, where $\mathbf{u} \in \Omega$ and $\mathbf{v} \in \Omega^{\perp}$.

Proof Suppose there are two such decompositions $\mathbf{y} = \mathbf{u}_i + \mathbf{v}_i$ $(i = 1, 2)$; then $\mathbf{u}_1 - \mathbf{u}_2 + \mathbf{v}_1 - \mathbf{v}_2 = \mathbf{0}$. Because $\mathbf{u}_1 - \mathbf{u}_2 \in \Omega$ and $\mathbf{v}_1 - \mathbf{v}_2 \in \Omega^{\perp}$, we must have $\mathbf{u}_1 = \mathbf{u}_2$ and $\mathbf{v}_1 = \mathbf{v}_2$.

2 $\mathbf{u} = \mathbf{P}_{\Omega}\mathbf{y}$ and $\mathbf{P}_{\Omega}$ is unique.

Proof Given two such matrices $\mathbf{P}_i$ $(i = 1, 2)$, then, since $\mathbf{u}$ is unique for *every* $\mathbf{y}$, $(\mathbf{P}_1 - \mathbf{P}_2)\mathbf{y} = \mathbf{0}$ for all $\mathbf{y}$; hence $\mathbf{P}_1 - \mathbf{P}_2 = \mathbf{O}$ [set $\mathbf{y} = (1, 0, \ldots, 0)'$, etc.] and $\mathbf{P}_{\Omega}$ is unique. (The existence of $\mathbf{P}_{\Omega}$ is proved in B1.7.)

3 $\mathbf{I}_n - \mathbf{P}_{\Omega}$ represents the orthogonal projection on $\Omega^{\perp}$.

Proof Using the identity $\mathbf{y} = \mathbf{P}_{\Omega}\mathbf{y} + (\mathbf{I}_n - \mathbf{P}_{\Omega})\mathbf{y}$, it follows, from B1.1 above, that $\mathbf{v} = (\mathbf{I}_n - \mathbf{P}_{\Omega})\mathbf{y}$.

4 $\mathbf{P}_{\Omega}$ and $\mathbf{I}_n - \mathbf{P}_{\Omega}$ are symmetric and idempotent.

Proof $\mathbf{P}_{\Omega}\mathbf{a} \in \Omega$ and $(\mathbf{I}_n - \mathbf{P}_{\Omega})\mathbf{b} \in \Omega^{\perp}$ so that $0 = \mathbf{a}'\mathbf{P}_{\Omega}'(\mathbf{I}_n - \mathbf{P}_{\Omega})\mathbf{b}$, that is, $\mathbf{P}_{\Omega}'(\mathbf{I}_n - \mathbf{P}_{\Omega}) = \mathbf{O}$. Hence $\mathbf{P}_{\Omega}' = \mathbf{P}_{\Omega}'\mathbf{P}_{\Omega}$ so that $\mathbf{P}_{\Omega}$ is symmetric, and $\mathbf{P}_{\Omega}^2 = \mathbf{P}_{\Omega}$. Also, $(\mathbf{I}_n - \mathbf{P}_{\Omega})^2 = \mathbf{I}_n - 2\mathbf{P}_{\Omega} + \mathbf{P}_{\Omega}^2 = \mathbf{I}_n - \mathbf{P}_{\Omega}$.
Note: We could have also written $\mathbf{P}_{\Omega}\mathbf{y} = \mathbf{u} = \mathbf{P}_{\Omega}\mathbf{u} = \mathbf{P}_{\Omega}^2\mathbf{y}$ for all $\mathbf{y}$, so that $\mathbf{P}_{\Omega}^2 = \mathbf{P}_{\Omega}$. Thus idempotency follows from the fact that $\mathbf{P}_{\Omega}$ represents a projection; it is also symmetric, as the projection is orthogonal. Oblique projections are also used in statistics (e.g., Rao [1974]).

5 $\mathscr{R}[\mathbf{P}_{\Omega}] = \Omega$, and the dimension of Ω is $\operatorname{tr} \mathbf{P}_{\Omega}$.

Proof $\mathbf{P}_{\Omega}\mathbf{y} = \mathbf{u} \in \Omega$ so that $\mathscr{R}[\mathbf{P}_{\Omega}] \subset \Omega$. Conversely, if $\mathbf{x} \in \Omega$, it follows from B1.1 that the unique orthogonal decomposition of $\mathbf{x}$ is $\mathbf{x} = \mathbf{x} + \mathbf{0}$, so that $\mathbf{x} = \mathbf{P}_\Omega \mathbf{x} \in \mathscr{R}[\mathbf{P}_\Omega]$. Hence the two spaces are the same. Finally $\dim \Omega = \operatorname{rank} \mathbf{P}_\Omega = \operatorname{tr} \mathbf{P}_\Omega$. By A6.3.

6 If $\mathbf{P}$ is a symmetric idempotent $n \times n$ matrix, then $\mathbf{P}$ represents an orthogonal projection onto $\mathscr{R}[\mathbf{P}]$.

Proof Let $\mathbf{y} = \mathbf{P}\mathbf{y} + (\mathbf{I}_n - \mathbf{P})\mathbf{y}$. Then $(\mathbf{P}\mathbf{y})'(\mathbf{I}_n - \mathbf{P})\mathbf{y} = \mathbf{y}'(\mathbf{P} - \mathbf{P}^2)\mathbf{y} = 0$, so that this decomposition gives orthogonal components of $\mathbf{y}$. The result now follows from B1.5.

7 If $\Omega = \mathscr{R}[\mathbf{X}]$, then $\mathbf{P}_\Omega = \mathbf{X}(\mathbf{X}'\mathbf{X})^- \mathbf{X}'$, where $(\mathbf{X}'\mathbf{X})^-$ is any generalized inverse of $\mathbf{X}'\mathbf{X}$ (i.e., if $\mathbf{B} = \mathbf{X}'\mathbf{X}$, then $\mathbf{B}\mathbf{B}^-\mathbf{B} = \mathbf{B}$).

Proof Let $\mathbf{c} = \mathbf{X}'\mathbf{y}$. Then $\hat{\boldsymbol{\beta}} = \mathbf{B}^-\mathbf{c}$ is a solution of $\mathbf{B}\boldsymbol{\beta} = \mathbf{c}$, that is, of $\mathbf{X}'\mathbf{X}\boldsymbol{\beta} = \mathbf{X}'\mathbf{y}$, since $\mathbf{B}(\mathbf{B}^-\mathbf{c}) = \mathbf{B}\mathbf{B}^-\mathbf{B}\boldsymbol{\beta} = \mathbf{B}\boldsymbol{\beta}$. Hence, writing $\hat{\boldsymbol{\theta}} = \mathbf{X}\hat{\boldsymbol{\beta}}$, we have $\mathbf{y} = \hat{\boldsymbol{\theta}} + (\mathbf{y} - \hat{\boldsymbol{\theta}})$, where

$$\mathbf{X}'(\mathbf{y} - \hat{\boldsymbol{\theta}}) = \mathbf{X}'\mathbf{y} - \mathbf{X}'\mathbf{X}\hat{\boldsymbol{\beta}} = \mathbf{0}.$$

Thus we have an orthogonal decomposition of $\mathbf{y}$ such that $\hat{\boldsymbol{\theta}} \in \mathscr{R}[\mathbf{X}]$ and $(\mathbf{y} - \hat{\boldsymbol{\theta}}) \perp \mathscr{R}[\mathbf{X}]$. Hence $\mathbf{u} = \hat{\boldsymbol{\theta}} = \mathbf{X}\hat{\boldsymbol{\beta}} = \mathbf{X}(\mathbf{X}'\mathbf{X})^-\mathbf{X}'\mathbf{y} = \mathbf{P}_\Omega \mathbf{y}$ and $\mathbf{P}_\Omega = \mathbf{X}(\mathbf{X}'\mathbf{X})^-\mathbf{X}'$, by B1.2.

8 When the columns of $\mathbf{X}$ are linearly independent in B1.7, $\mathbf{P}_\Omega = \mathbf{X}(\mathbf{X}'\mathbf{X})^{-1}\mathbf{X}$.

9 If the r columns of $\mathbf{T}$ form an orthonormal basis for Ω, then, setting $\mathbf{X} = \mathbf{T}$, we have $\mathbf{P}_\Omega = \mathbf{T}\mathbf{T}'$. Conversely, if $\mathbf{P}_\Omega = \mathbf{T}\mathbf{T}'$ with $\mathbf{T}'\mathbf{T} = \mathbf{I}_r$, then $\Omega = \mathscr{R}[\mathbf{T}]$.

B2 ORTHOGONAL COMPLEMENTS

1 For any matrix $\mathbf{A}$, the null space (kernel) of $\mathbf{A}$ is the orthogonal complement of the range space of $\mathbf{A}'$, that is, $\mathscr{N}[\mathbf{A}] = \{\mathscr{R}[\mathbf{A}']\}^\perp$.

2 If the columns of $\mathbf{A}'$ are linearly independent, then the orthogonal projector onto $\mathscr{N}[\mathbf{A}]$ is $\mathbf{I} - \mathbf{A}'(\mathbf{A}\mathbf{A}')^{-1}\mathbf{A}$.

Proof If $\Omega = \mathscr{N}[\mathbf{A}]$, then $\Omega^\perp = \mathscr{R}[\mathbf{A}']$ (by B2.1) and $\mathbf{P}_{\Omega^\perp} = \mathbf{A}'(\mathbf{A}\mathbf{A}')^{-1}\mathbf{A}$ (by B1.8). Finally, $\mathbf{P}_\Omega = \mathbf{I} - \mathbf{P}_{\Omega^\perp}$.

3 If $V = \mathscr{R}[\mathbf{C}]$, where $\mathbf{C}$ is $p \times q$ of rank q, then there exists a $(p - q) \times p$ matrix $\mathbf{B}$ satisfying $\mathbf{B}\mathbf{C} = \mathbf{O}$ such that $\mathbf{V} = \mathscr{N}[\mathbf{B}]$.

Proof By B1.8 and B1.5, $\mathbf{P}_V = \mathbf{C}(\mathbf{C}'\mathbf{C})^{-1}\mathbf{C}'$ has rank q and $\mathbf{I}_p - \mathbf{P}_V$ is a $p \times p$ matrix of rank $p - q$. Let $\mathbf{B}$ be any $p - q$ linearly independent rows of $\mathbf{I}_p - \mathbf{P}_V$. Then $(\mathbf{I}_p - \mathbf{P}_V)\mathbf{P}_V = \mathbf{O}$ implies that $\mathbf{B}\mathbf{C} = \mathbf{O}$. Also, $\mathscr{N}[\mathbf{B}] = \{\mathscr{R}[\mathbf{B}']\}^\perp = \{\mathscr{R}[\mathbf{I}_p - \mathbf{P}_V]\}^\perp = V$.

4 $(\Omega_1 \cap \Omega_2)^\perp = \Omega_1^\perp + \Omega_2^\perp$.

Proof Let $\mathbf{C}_i$ be such that $\Omega_i = \mathcal{N}[\mathbf{C}_i]$ $(i = 1, 2)$. Then

$$\begin{aligned}(\Omega_1 \cap \Omega_2)^{\perp} &= \left\{\mathcal{N}\begin{bmatrix}\mathbf{C}_1\\ \mathbf{C}_2\end{bmatrix}\right\}^{\perp}\\ &= \mathcal{R}\left[\mathbf{C}_1', \mathbf{C}_2'\right] \qquad \text{(by B2.1)}\\ &= \mathcal{R}\left[\mathbf{C}_1'\right] + \mathcal{R}[\mathbf{C}_2]\\ &= \Omega_1^{\perp} + \Omega_2^{\perp}.\end{aligned}$$

B3 PROJECTIONS ON SUBSPACES

1 Given $\omega \subset \Omega$, then $\mathbf{P}_{\Omega}\mathbf{P}_{\omega} = \mathbf{P}_{\omega}\mathbf{P}_{\Omega} = \mathbf{P}_{\omega}$.

Proof Since $\omega \subset \Omega$ and $\omega = \mathcal{R}[\mathbf{P}_{\omega}]$ (by B1.5), we have $\mathbf{P}_{\Omega}\mathbf{P}_{\omega} = \mathbf{P}_{\omega}$. The result then follows by the symmetry of $\mathbf{P}_{\omega}$ and $\mathbf{P}_{\Omega}$.

2 $\mathbf{P}_{\Omega} - \mathbf{P}_{\omega} = \mathbf{P}_{\omega^{\perp} \cap \Omega}$.

Proof Consider $\mathbf{P}_{\Omega}\mathbf{y} = \mathbf{P}_{\omega}\mathbf{y} + (\mathbf{P}_{\Omega} - \mathbf{P}_{\omega})\mathbf{y}$. Now $\mathbf{P}_{\Omega}\mathbf{y}$ and $\mathbf{P}_{\omega}\mathbf{y}$ belong to Ω so that $(\mathbf{P}_{\Omega} - \mathbf{P}_{\omega})\mathbf{y} \in \Omega$. Hence the equation for $\mathbf{P}_{\Omega}\mathbf{y}$ represents an orthogonal decomposition of Ω into ω and $\omega^{\perp} \cap \Omega$, since $\mathbf{P}_{\omega}(\mathbf{P}_{\Omega} - \mathbf{P}_{\omega}) = \mathbf{O}$ (by B3.1).

3 If $\mathbf{A}_1$ is any matrix such that $\omega = \mathcal{N}[\mathbf{A}_1] \cap \Omega$, then $\omega^{\perp} \cap \Omega = \mathcal{R}[\mathbf{P}_{\Omega}\mathbf{A}_1']$.

Proof

$$\begin{aligned}\omega^{\perp} \cap \Omega &= \{\Omega \cap \mathcal{N}[\mathbf{A}_1]\}^{\perp} \cap \Omega\\ &= \left\{\Omega^{\perp} + \mathcal{R}\left[\mathbf{A}_1'\right]\right\} \cap \Omega \qquad \text{(by B.1 and B2.4).}\end{aligned}$$

If $\mathbf{x}$ belongs to the right side, then

$$\mathbf{x} = \mathbf{P}_{\Omega}\mathbf{x} = \mathbf{P}_{\Omega}\{(\mathbf{I}_n - \mathbf{P}_{\Omega})\boldsymbol{\alpha} + \mathbf{A}_1'\boldsymbol{\beta}\} = \mathbf{P}_{\Omega}\mathbf{A}_1'\boldsymbol{\beta} \in \mathcal{R}\left[\mathbf{P}_{\Omega}\mathbf{A}_1'\right].$$

Conversely, if $\mathbf{x} \in \mathcal{R}[\mathbf{P}_{\Omega}\mathbf{A}_1']$, then $\mathbf{x} \in \mathcal{R}[\mathbf{P}_{\Omega}] = \Omega$. Also, if $\mathbf{z} \in \omega$, $\mathbf{A}_1\mathbf{z} = \mathbf{0}$ and $\mathbf{x}'\mathbf{z} = \boldsymbol{\beta}'\mathbf{A}_1\mathbf{P}_{\Omega}\mathbf{z} = \boldsymbol{\beta}'\mathbf{A}_1\mathbf{z} = 0$, that is, $\mathbf{x} \in \omega^{\perp}$. Thus $\mathbf{x} \in \omega^{\perp} \cap \Omega$.

4 If $\mathbf{A}_1$ is a $q \times n$ matrix of rank q, then $\operatorname{rank}[\mathbf{P}_{\Omega}\mathbf{A}_1'] = q$ if and only if $\mathcal{R}[\mathbf{A}_1'] \cap \Omega^{\perp} = \mathbf{0}$.

Proof $\operatorname{rank}[\mathbf{P}_{\Omega}\mathbf{A}_1'] \leq \operatorname{rank}\mathbf{A}_1$ (by A2.1). Let the rows of $\mathbf{A}_1$ be $\mathbf{a}_i'$ $(i = 1, 2, \ldots, q)$ and suppose that $\operatorname{rank}[\mathbf{P}_{\Omega}\mathbf{A}_1'] < q$. Then the columns of $\mathbf{P}_{\Omega}\mathbf{A}_1'$ are linearly dependent so that $\sum_{i=1}^{q} c_i\mathbf{P}_{\Omega}\mathbf{a}_i = \mathbf{0}$, that is, there exists a vector $\sum_i c_i\mathbf{a}_i \in \mathcal{R}[\mathbf{A}_1']$ that is perpendicular to Ω. Hence $\mathcal{R}[\mathbf{A}_1'] \cap \Omega^{\perp} \neq \mathbf{0}$, which is a contradiction. [By selecting the linearly independent rows of $\mathbf{A}_1$, we find that the above result is true if $\mathbf{A}_1$ is $k \times n$ $(k \geq q)$.]

5 Let $\mathbf{V}$ be a $d \times d$ positive definite matrix, $\mathbf{G}$ a $d \times g$ matrix of rank g ($g \leq d$), and $\mathbf{F}$ a $d \times f$ matrix ($f = d - g$) of rank f such that $\mathbf{G}'\mathbf{F} = \mathbf{O}$. Then

$$\mathbf{F}(\mathbf{F}'\mathbf{V}\mathbf{F})^{-1}\mathbf{F}' = \mathbf{V}^{-1} - \mathbf{V}^{-1}\mathbf{G}(\mathbf{G}'\mathbf{V}^{-1}\mathbf{G})^{-1}\mathbf{G}'\mathbf{V}^{-1}.$$

Proof Using A5.4, let $\mathbf{F}_1 = \mathbf{V}^{1/2}\mathbf{F}$ and $\mathbf{G}_1 = \mathbf{V}^{-1/2}\mathbf{G}$. Then we have to prove that

$$\mathbf{F}_1(\mathbf{F}_1'\mathbf{F}_1)^{-1}\mathbf{F}_1' + \mathbf{G}_1(\mathbf{G}_1'\mathbf{G}_1')^{-1}\mathbf{G}_1 = \mathbf{I}_d.$$

Both matrices are projection matrices (B1.8), so that the above result will hold if they project onto orthogonal complements, that is, if $\mathscr{R}[\mathbf{F}_1] = \{\mathscr{R}[\mathbf{G}_1]\}^\perp$ (by B1.2). Now $\mathbf{G}_1'\mathbf{F}_1 = \mathbf{G}'\mathbf{V}^{-1/2}\mathbf{V}^{1/2}\mathbf{F} = \mathbf{G}'\mathbf{F} = \mathbf{O}$ and, by B2.1, $\mathscr{R}[\mathbf{F}_1] \subset \mathscr{N}[\mathbf{G}_1'] = \{\mathscr{R}[\mathbf{G}_1]\}^\perp$. However, by A2.2,

$$\dim\{\mathscr{R}[\mathbf{G}_1]\}^\perp = d - \operatorname{rank}\mathbf{G}_1 = d - g = \operatorname{rank}\mathbf{F} = \operatorname{rank}\mathbf{F}_1.$$

If a vector space is contained in another yet both spaces have the same dimension, then they are identical and the required result follows.

6 Let $\mathbf{y} = \mathbf{K}\boldsymbol{\beta} + \mathbf{v}$, where $\mathbf{v} \sim N_n(\mathbf{0}, \sigma^2\mathbf{I}_n)$ and $\mathbf{K}$ is $n \times p$ of rank p. Let $\mathbf{y} = (\mathbf{y}_1', \mathbf{y}_2')'$ and $\mathbf{K} = (\mathbf{K}_1', \mathbf{K}_2')'$, where $\mathbf{y}_1$ is $n_1 \times 1$ and $\mathbf{K}_1$ is $n_1 \times p$ of rank p. Then

$$b = \min_{\boldsymbol{\beta}} \|\mathbf{y}_1 - \mathbf{K}_1\boldsymbol{\beta}\|^2 \Big/ \min_{\boldsymbol{\beta}} \|\mathbf{y} - \mathbf{K}\boldsymbol{\beta}\|^2 = \frac{E}{E+H}$$

has a beta distribution with $\frac{1}{2}(n_1 - p)$ and $\frac{1}{2}(n - n_1)$ degrees of freedom, respectively.

Proof

$$\begin{aligned} E &= \mathbf{y}_1'\left[\mathbf{I}_{n_1} - \mathbf{K}_1(\mathbf{K}_1'\mathbf{K}_1)^{-1}\mathbf{K}_1'\right]\mathbf{y}_1 \\ &= \mathbf{y}_1'(\mathbf{I}_{n_1} - \mathbf{P}_{11})\mathbf{y}_1, \qquad \text{say}, \\ &= (\mathbf{y}_1 - \mathbf{K}_1\boldsymbol{\beta})'(\mathbf{I}_{n_1} - \mathbf{P}_{11})(\mathbf{y}_1 - \mathbf{K}_1\boldsymbol{\beta}) \\ &= \mathbf{v}_1'(\mathbf{I}_{n_1} - \mathbf{P}_{11})\mathbf{v}_1, \end{aligned}$$

since $(\mathbf{I}_{n_1} - \mathbf{P}_{11})\mathbf{K}_1 = \mathbf{O}$. As $\mathbf{P}_{11}$ is symmetric and idempotent of rank p, it follows from A6.5 that $E/\sigma^2 \sim \chi^2_{n_1 - p}$. Setting

$$\mathbf{P}_1 = \begin{pmatrix} \mathbf{P}_{11} & \mathbf{O} \\ \mathbf{O} & \mathbf{I}_{n-n_1} \end{pmatrix},$$

we have $E = \mathbf{v}'(\mathbf{I}_n - \mathbf{P}_1)\mathbf{v}$. With a similar argument to that above,

$$E + H = \mathbf{v}'\left[\mathbf{I}_n - \mathbf{K}(\mathbf{K}'\mathbf{K})^{-1}\mathbf{K}'\right]\mathbf{v}$$

$$= \mathbf{v}'(\mathbf{I}_n - \mathbf{P})\mathbf{v},$$

say. Since $\mathbf{P}_{11}\mathbf{K}_1 = \mathbf{K}_1$,

$$\mathbf{P}_1\mathbf{P} = \begin{pmatrix} \mathbf{P}_{11} & \mathbf{O} \\ \mathbf{O} & \mathbf{I}_{n-n_1} \end{pmatrix} \begin{pmatrix} \mathbf{K}_1(\mathbf{K}'\mathbf{K})^{-1}\mathbf{K}_1', \mathbf{K}_1(\mathbf{K}'\mathbf{K})^{-1}\mathbf{K}_2' \\ \mathbf{K}_2(\mathbf{K}'\mathbf{K})^{-1}\mathbf{K}_1', \mathbf{K}_2(\mathbf{K}'\mathbf{K})^{-1}\mathbf{K}_2' \end{pmatrix}$$

$$= \mathbf{P}.$$

Transposing, we also get $\mathbf{P}\mathbf{P}_1 = \mathbf{P}$, so that

$$(\mathbf{P}_1 - \mathbf{P})^2 = \mathbf{P}_1^2 - \mathbf{P}_1\mathbf{P} - \mathbf{P}\mathbf{P}_1 + \mathbf{P}^2 = \mathbf{P}_1 - \mathbf{P}.$$

Now $H = E + H - E = \mathbf{v}'(\mathbf{P}_1 - \mathbf{P})\mathbf{v}$ and, since $\mathbf{P}_1 - \mathbf{P}$ is symmetric and idempotent, we have (A6.3)

$$\operatorname{rank}[\mathbf{P}_1 - \mathbf{P}] = \operatorname{tr}[\mathbf{P}_1 - \mathbf{P}]$$

$$= \operatorname{tr}\mathbf{P}_1 - \operatorname{tr}\mathbf{P}$$

$$= (p + n - n_1) - p = n - n_1.$$

Hence, by A6.5, $H/\sigma^2 \sim \chi^2_{n-n_1}$. Also, $(\mathbf{P}_1 - \mathbf{P})(\mathbf{I}_n - \mathbf{P}_1) = \mathbf{O}$, so that E and H are independent (A6.6). Finally,

$$F = \frac{H/(n - n_1)}{E/(n_1 - p)} \sim F_{n-n_1, n_1-p},$$

and, from (2.21),

$$b = \frac{1}{1 + HE^{-1}}$$

has a beta distribution with $\frac{1}{2}(n_1 - p)$ and $\frac{1}{2}(n - n_1)$ degrees of freedom.

Note: Using generalized inverses, the above result can be generalized to the case when $\operatorname{rank}\mathbf{K} = r$ and $\operatorname{rank}\mathbf{K}_1 = r_1$. Then b has a beta distribution with $\frac{1}{2}(n_1 - r_1)$ and $\frac{1}{2}(n - n_1 + r_1 - r)$ degrees of freedom, respectively.

7 Let $\mathbf{Y} = \mathbf{KB} + \mathbf{V}$, where the rows of $\mathbf{V}$ are i.i.d. $N_d(\mathbf{0}, \boldsymbol{\Sigma})$. Using the notation of B3.6 above, let

$$\mathbf{E} = \mathbf{V}'(\mathbf{I}_n - \mathbf{P}_1)\mathbf{V} \quad \text{and} \quad \mathbf{E} + \mathbf{H} = \mathbf{V}'(\mathbf{I}_n - \mathbf{P})\mathbf{V}.$$

If $b = |\mathbf{E}|/|\mathbf{E} + \mathbf{H}|$, then $b \sim U_{d,\, n-n_1,\, n_1-p}$.

Proof Using the same arguments in B3.6, but with $\mathbf{v}$ replaced by $\mathbf{V}$, we can apply Corollaries 1 and 2 of Theorem 2.4. Thus $\mathbf{E} \sim W_d(n_1 - p, \boldsymbol{\Sigma})$, $\mathbf{H} \sim W_d(n - n_1, \boldsymbol{\Sigma})$, and $\mathbf{E}$ is independent of $\mathbf{H}$. The result follows from (2.48).

APPENDIX C

Order Statistics and Probability Plotting

C1 SAMPLE DISTRIBUTION FUNCTIONS

Suppose we have a random sample of ordered observations $x_{(1)} \le x_{(2)} \le \cdots \le x_{(n)}$ from a distribution with density function f and distribution function F. If $u = F(x)$, then $x = F^{-1}(u)$ $(= x[(u]$, say) and u has density function

$$g(u) = f(x[u])\left|\frac{dx}{du}\right|$$

$$= f(x[u])\Big/\left|\frac{du}{dx}\right|$$

$$= \frac{f(x[u])}{F'(x[u])}$$

$$= 1.$$

Setting $u_{(i)} = F(x_{(i)})$, then $u_{(1)} \le u_{(2)} \le \cdots \le u_{(n)}$ is an ordered random sample from the uniform distribution on $[0, 1]$. From symmetry, we would expect the $u_{(i)}$ to, on the average, divide the interval $[0, 1]$ into $n + 1$ equal parts. Thus (see David [1970])

$$\mathrm{E}\left[F(x_{(i)})\right] = \frac{i}{n+1}. \tag{1}$$

For future reference we also note that

$$\mathrm{E}[x_{(i)}] = F^{-1}(p_i), \tag{2}$$

for some probability p_i that will depend on F. Various approximations for p_i have been used such as $i/(n+1)$ or, more commonly, $(i-\frac{1}{2})/n$, which works well when F is normal (David [1970: pp. 64–67]). These are both special cases of $p_i \approx (i-\alpha)/(n+1-2\alpha)$, which can be used for symmetric distributions.

C2 GAMMA DISTRIBUTION

The gamma distribution, which includes the chi-square and exponential distributions as special cases, has numerous applications in statistics. In particular, it is a useful approximation to the distribution of a positive semidefinite (p.s.d.) quadratic form in normal variables (see Box [1954]). For example, consider the quadratic form $\mathbf{x}'\mathbf{A}\mathbf{x}$, where $\mathbf{x} \sim N_d(\boldsymbol{\mu}, \boldsymbol{\Sigma})$ and $\mathbf{A}$ is a $d \times d$ p.s.d. matrix of rank r. From A5.4 we can define $\boldsymbol{\Sigma}^{1/2} > \mathbf{O}$, and, by A4.4 and A2.2, $\boldsymbol{\Sigma}^{1/2}\mathbf{A}\boldsymbol{\Sigma}^{1/2}$ is p.s.d. of rank r. Hence there exists orthogonal $\mathbf{T}$ such that

$$\mathbf{T}'\boldsymbol{\Sigma}^{1/2}\mathbf{A}\boldsymbol{\Sigma}^{1/2}\mathbf{T} = \operatorname{diag}(\lambda_1, \lambda_2, \ldots, \lambda_r, 0, \ldots, 0) = \boldsymbol{\Lambda},$$

where each $\lambda_i > 0$. Let $\mathbf{w} = (w_1, w_2, \ldots, w_d)' = \mathbf{T}'\boldsymbol{\Sigma}^{-1/2}\mathbf{x}$; then

$$\begin{aligned}\mathbf{x}'\mathbf{A}\mathbf{x} &= \mathbf{w}'\mathbf{T}'\boldsymbol{\Sigma}^{1/2}\mathbf{A}\boldsymbol{\Sigma}^{1/2}\mathbf{T}\mathbf{w} \\ &= \mathbf{w}'\boldsymbol{\Lambda}\mathbf{w} \\ &= \lambda_1 w_1^2 + \lambda_2 w_2^2 + \cdots + \lambda_r w_r^2,\end{aligned}$$

where $\mathscr{E}[\mathbf{w}] = \mathbf{T}'\boldsymbol{\Sigma}^{-1/2}\boldsymbol{\mu} = \boldsymbol{\theta}$, say, and

$$\mathscr{D}[\mathbf{w}] = \mathbf{T}'\boldsymbol{\Sigma}^{-1/2}\mathscr{D}[\mathbf{x}]\boldsymbol{\Sigma}^{-1/2}\mathbf{T} = \mathbf{T}'\boldsymbol{\Sigma}^{-1/2}\boldsymbol{\Sigma}\boldsymbol{\Sigma}^{-1/2}\mathbf{T} = \mathbf{I}_d.$$

Hence $\mathbf{w} \sim N_d(\boldsymbol{\theta}, \mathbf{I}_d)$ and the w_i are independently distributed as $N_1(\theta_i, 1)$. Since w_i^2 has a noncentral chi-square distribution with one degree of freedom, $\mathbf{x}'\mathbf{A}\mathbf{x}$ is a linear combination of independent noncentral chi-square variables. By, for example, equating the first two moments (Patnaik [1949]), one can approximate this linear combination by a scaled chi-square variable, that is, by a gamma variable. In some applications $\boldsymbol{\theta} = \mathbf{0}$ so that $\mathbf{x}'\mathbf{A}\mathbf{x}$ is then a linear combination of independent (central) chi squares.

A random variable Y with a gamma distribution has density function

$$f(y; \lambda, \eta) = \frac{\lambda^\eta}{\Gamma(\eta)} y^{\eta-1} e^{-\lambda y}, \qquad y \geq 0, \tag{3}$$

where $\lambda > 0$ and $\eta > 0$ are the "scale" and "shape" parameters, respectively. As order statistics are frequently used for assessing underlying distributions, Wilk et al. [1962b] developed the following method for finding the maximum likelihood estimates $\hat{\lambda}$ and $\hat{\eta}$ based on the m smallest order statistics $y_{(1)} \le y_{(2)} \le \cdots \le y_{(m)}$ for a random sample of size n from (3). The estimates are functions of

$$P = \left\{ \prod_{i=1}^{m} y_{(i)}^{1/m} \right\} \Big/ y_{(m)} \quad \text{and} \quad S = \left(\sum_{i=1}^{m} y_{(i)} \right) \Big/ m y_{(m)},$$

so that given P, S, and n/m, $\hat{\eta}$ and $\hat{\mu} = \hat{\eta}/(\hat{\lambda} y_{(m)})$ can be read from tables given by Wilk et al. [1962b: $n/m \to K/M$]. Details of bilinear interpolation are also given. How do we choose n/m? Although clear guidelines are not given, Roy et al. [1971: pp. 103–104] stated that estimates are rather insensitive to n for a fixed value of m, provided that n is not too close to m (say, $n/m > \frac{3}{2}$). The loss of efficiency in choosing m/n small appears to have little effect on conclusions drawn from gamma plots (see Appendix C4 below).

When the complete sample is used, that is, $m = n$, $\hat{\eta}$ satisfies

$$\frac{\Gamma'(\hat{\eta})}{\Gamma(\hat{\eta})} - \log \hat{\eta} = \log R, \tag{4}$$

where $R = P/S \ (\le 1)$, the ratio of the geometric mean to the arithmetic mean. $\hat{\lambda}$ is then given by

$$\hat{\lambda} = \frac{\hat{\eta}}{\bar{y}}.$$

Wilk et al. [1962b] noted that the root of (4) is nearly linear in $(1 - R)^{-1} = S/(S - P)$, and that linear interpolation in their table (reproduced here as Appendix D2) will give results accurate to four decimal places everywhere in the table, except between the first two values. Various approximate solutions are available as functions of $-\log R$ (Shenton and Bowman [1972]).

C3 BETA DISTRIBUTION

Suppose we have order statistics $y_{(1)} \le y_{(2)} \le \cdots \le y_{(m)}$ for a random sample of n observations from the beta distribution

$$f(y; \alpha, \beta) = \frac{1}{B(\alpha, \beta)} y^{\alpha - 1}(1 - y)^{\beta - 1}, \qquad 0 \le y \le 1, \alpha > 0, \beta > 0.$$

Then the maximum likelihood estimates $\hat{\alpha}$ and $\hat{\beta}$ of the shape parameters α and β are complex functions of m/n, $y_{(m)}$, and the two geometric means

$$G_1 = \left\{ \prod_{i=1}^{m} y_{(i)} \right\}^{1/m} \quad \text{and} \quad G_2 = \left\{ \prod_{i=1}^{m} (1 - y_{(i)}) \right\}^{1/m}.$$

When $m = n$, the estimates satisfy

$$\log G_1 = \Psi(\hat{\alpha}) - \Psi(\hat{\alpha} + \hat{\beta})$$

and

$$\log G_2 = \Psi(\hat{\beta}) - \Psi(\hat{\alpha} + \hat{\beta}),$$

where $\Psi(x) = d \log \Gamma(x)/dx = \Gamma'(x)/\Gamma(x)$, the so-called digamma function. Gnanadesikan et al. [1967] give a small table of $\hat{\alpha}$ and $\hat{\beta}$ for a few values of G_1 and G_2 and provide a useful appendix on the numerical methods used.

C4 PROBABILITY PLOTTING

Let $y = F(x) = \text{pr}[X \leq x]$ be the graph of the distribution function for a random variable X. Wilk and Gnanadesikan [1968] described two plotting techniques for comparing the "closeness" of two such graphs. We can either compare their y (probability) values for a set of common x-values (a P–P plot), or else compare their x (quantile) values for a set of common y values (a quantile–quantile or Q–Q plot) using the inverse relationship $x = F^{-1}(y)$. If the two random variables to be compared are X_1 and $X_2 = (X_1 - \mu)/\sigma$, then

$$y = \text{pr}[X_1 \leq x_1] = \text{pr}\left[X_2 \leq \frac{x_1 - \mu}{\sigma}\right] \qquad (= \text{pr}[X_2 \leq x_2], \text{say}),$$

so that for a common y, the x-values satisfy the linear relation $x_1 = \sigma x_2 + \mu$. Therefore if the distributions are identical, $x_1 = x_2$ and the Q–Q plot is a straight line through the origin with unit slope. However, if the distributions differ only in location and/or scale, then the Q–Q plot is still linear, with a nonzero intercept if there is a location difference, and a slope different from unity if there is a scale difference. This linear invariance property of a Q–Q plot is not shared by the P–P plot which is linear only if the two distributions are identical.

The most common type of Q–Q plot consists of comparing an empirical distribution function $\hat{F}(x_{(i)}) = i/n$ based on an ordered random sample $x_{(1)} \leq x_{(2)} \leq \cdots \leq x_{(n)}$, with a theoretical distribution function $G(x; \boldsymbol{\theta})$, where $\boldsymbol{\theta}$ is specified or, more frequently, estimated from the sample. The Q–Q plot then consists of plotting the pairs $(G^{-1}\{i/n\}, x_{(i)})$ and looking for departures from linearity. However, the point corresponding to $i = n$ cannot be plotted and i/n has commonly been replaced by $(i + 1)/n$ [see (1)] or $(i - \frac{1}{2})/n$. We can also regard the plot as a comparison of $x_{(i)}$ with its expected value under G, namely [see (2)], $\text{E}[x_{(i)}] = G^{-1}(p_i)$, where p_i is estimated by, for example, $(i - \frac{1}{2})/n$, $(i + 1)/n$, or $(i - \frac{3}{8})/(n + \frac{1}{4})$ (see Mage [1982]).

To reduce the number of parameters $\boldsymbol{\theta}$ to be estimated, we can standardize G so that the origin or location parameter is zero and the scale parameter is unity. This will have no effect on the linearity of the Q–Q plot, as we noted above. For example, if we wish to see if X comes from a normal distribution, we can set $G = \Phi$, the distribution function for the standard $N_1(0, 1)$ variable, and there are no parameters to be estimated. In the normal case, special graph paper, called normal probability paper or arithmetic probability paper, is available that has a special nonlinear scale that converts $(i - \frac{1}{2})/n$ into $\Phi^{-1}(i - \frac{1}{2})/n$. We then simply plot the pairs $(\{i - \frac{1}{2}\}/n, x_{(i)})$ and check for linearity (see Barnett [1975] and Gerson [1975]). If the plot is linear, it can be used to provide estimates of location and scale (Barnett [1976b]). These days we do not need special graph paper, as most statistical packages (e.g., SAS, MINITAB) provide automatic plots of the pairs $(\Phi^{-1}\{p_i\}, x_{(i)})$. If the plot is not linear, its shape will suggest the likely departure from normality. For example, plots for short-tailed (e.g., uniform) and long-tailed (e.g., Cauchy) distributions will look like Figs. C1*a* and *b*. Since the exponential distribution is short tailed on the left and long tailed on the right, it will be a combination of Figs. C1*a* and *b*, namely, Fig. C1*c*. A number of tests for normality based on the nonlinearity of the plot are available (see LaBrecque [1977]). Several quick graphical techniques for Lillefors' test (Lillefors [1967]) can also be used

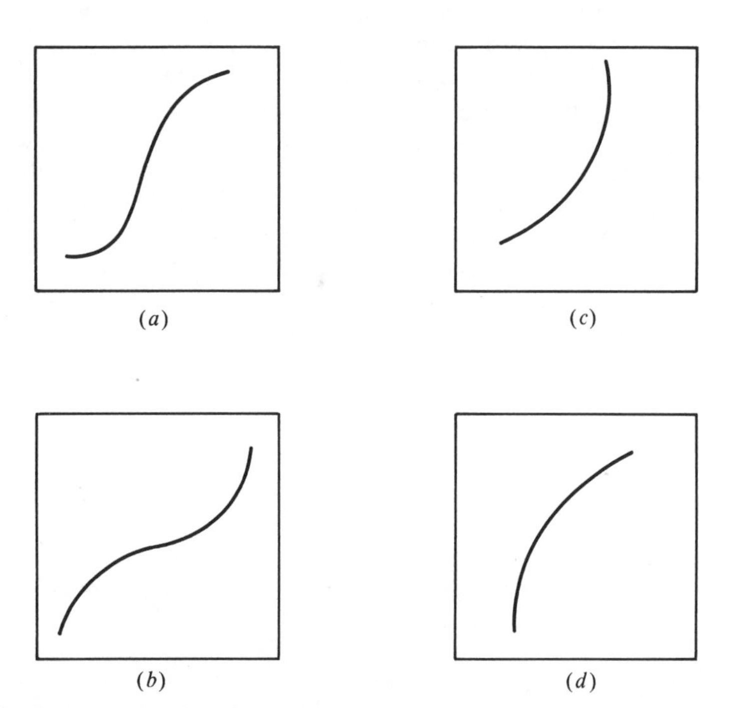

Fig. C1 Q–Q plots for four distributions compared with the normal. (*a*) Short-tailed. (*b*) Long-tailed. (*c*) Skewed to the right. (*d*) Skewed to the left.

(Iman [1982, Mage [1982]). A related technique, called half-normal plotting, has been developed for investigating a series of hypotheses on contrasts from a 2^k design (Daniel [1959], Zahn [1975a, b]). An application to correlation coefficients is mentioned briefly at the end of Section 3.5.7. Experiments with normal distribution would suggest that we must have $n > 20$ (preferably $n > 50$) for useful plots.

To see if G is a gamma distribution, we use $x_{(1)}, x_{(2)}, \ldots, x_{(m)}$ $(m \leq n)$ to obtain $\hat{\eta}$ from, say, Wilk et al. [1962b: $m < n$] or Appendix D2 $(m = n)$. Since we can change the scale without affecting the linearity of a Q–Q plot, we use the values $\hat{\eta}$ and $\lambda = 1$ and solve [see (3)]

$$p_i = \int_0^{x_i^*} \frac{1}{\Gamma(\hat{\eta})} y^{\hat{\eta}-1} e^{-y}\, dy$$

for x_i^* using a suitable algorithm (Wilk et al. [1962a]). For example, the statistical package SAS has a procedure GAMIN(P, ETA) that uses the bisection method for finding x_i^*. If only a crude estimate of η is available, then plots of the pairs $(x_i^*, x_{(i)})$ for several values of $\hat{\eta}$ could be carried out. In most practical situations we would use all the data (i.e., $m = n$), as $\hat{\eta}$ is apparently not greatly affected by the large ordered observations (Rohlf [1975]); it is reasonably insensitive to any outliers we may be trying to identify using the plot. The use of truncated data arises, for example, in analysis of variance models such as 2^k designs where there are a number of squared contrasts, each with one degree of freedom, representing main effects and interactions. The smallest of these will generally be nonsignificant, with the same null gamma distribution, and can be used to estimate η and λ. However, the larger significant squares will not belong to this gamma distribution and will deviate from the linear plot determined by the first few squares. As we have noted in Appendix C2 above, $\hat{\eta}$ is insensitive to n/m, provided that m is not too close to n. Gnanadesikan [1977: p. 236] endorses the recommendation $n/m > \frac{3}{2}$ and gives graphs to support it. Any loss of efficiency in estimating $\hat{\eta}$ for large n/m seems to have little effect on the interpretation of the plot.

For a helpful review of plotting methods, see Gnanadesikan [1980].

APPENDIX D

Statistical Tables

D1 BONFERRONI t-PERCENTAGE POINTS

Table of $t_\nu^{\alpha/(2r)}$, where $\text{pr}[t_\nu > t_\nu^{\alpha/(2r)}] = \alpha/(2r)$ and t_ν is the t-distribution with ν degrees of freedom (from Bailey [1977: Table 1]: values for $\alpha = 0.01$ are also available there).

	$\alpha = 0.05$									
ν/r	1	2	$3 = \binom{3}{2}$	4	5	$6 = \binom{4}{2}$	7	8	9	$10 = \binom{5}{2}$
$100\alpha/r$	5.0000	2.5000	1.6667	1.2500	1.0000	0.8333	0.7143	0.6250	0.5556	0.5000
2	4.3027	6.2053	7.6488	8.8602	9.9248	10.8859	11.7687	12.5897	13.3604	14.0890
3	3.1824	4.1765	4.8567	5.3919	5.8409	6.2315	6.5797	6.8952	7.1849	7.4533
4	2.7764	3.4954	3.9608	4.3147	4.6041	4.8510	5.0675	5.2611	5.4366	5.5976
5	2.5706	3.1634	3.5341	3.8100	4.0321	4.2193	4.3818	4.5257	4.6553	4.7733
6	2.4469	2.9687	3.2875	3.5212	3.7074	3.8630	3.9971	4.1152	4.2209	4.3168
7	2.3646	2.8412	3.1276	3.3353	3.4995	3.6358	3.7527	3.8552	3.9467	4.0293
8	2.3060	2.7515	3.0158	3.2060	3.3554	3.4789	3.5844	3.6766	3.7586	3.8325
9	2.2622	2.6850	2.9333	3.1109	3.2498	3.3642	3.4616	3.5465	3.6219	3.6897
10	2.2281	2.6338	2.8701	3.0382	3.1693	3.2768	3.3682	3.4477	3.5182	3.5814
11	2.2010	2.5931	2.8200	2.9809	3.1058	3.2081	3.2949	3.3702	3.4368	3.4966
12	2.1788	2.5600	2.7795	2.9345	3.0545	3.1527	3.2357	3.3078	3.3714	3.4284
13	2.1604	2.5326	2.7459	2.8961	3.0123	3.1070	3.1871	3.2565	3.3177	3.3725
14	2.1448	2.5096	2.7178	2.8640	2.9768	3.0688	3.1464	3.2135	3.2727	3.3257
15	2.1314	2.4899	2.6937	2.8366	2.9467	3.0363	3.1118	3.1771	3.2346	3.2860
16	2.1199	2.4729	2.6730	2.8131	2.9208	3.0083	3.0821	3.1458	3.2019	3.2520
17	2.1098	2.4581	2.6550	2.7925	2.8982	2.9840	3.0563	3.1186	3.1735	3.2224
18	2.1009	2.4450	2.6391	2.7745	2.8784	2.9627	3.0336	3.0948	3.1486	3.1966
19	2.0930	2.4334	2.6251	2.7586	2.8609	2.9439	3.0136	3.0738	3.1266	3.1737
20	2.0860	2.4231	2.6126	2.7444	2.8453	2.9271	2.9958	3.0550	3.1070	3.1534
21	2.0796	2.4138	2.6013	2.7316	2.8314	2.9121	2.9799	3.0382	3.0895	3.1352
22	2.0739	2.4055	2.5912	2.7201	2.8188	2.8985	2.9655	3.0231	3.0737	3.1188
23	2.0687	2.3979	2.5820	2.7097	2.8073	2.8863	2.9525	3.0095	3.0595	3.1040
24	2.0639	2.3909	2.5736	2.7002	2.7969	2.8751	2.9406	2.9970	3.0465	3.0905
25	2.0595	2.3846	2.5660	2.6916	2.7874	2.8649	2.9298	2.9856	3.0346	3.0782
26	2.0555	2.3788	2.5589	2.6836	2.7787	2.8555	2.9199	2.9752	3.0237	3.0669
27	2.0518	2.3734	2.5525	2.6763	2.7707	2.8469	2.9107	2.9656	3.0137	3.0565
28	2.0484	2.3685	2.5465	2.6695	2.7633	2.8389	2.9023	2.9567	3.0045	3.0469
29	2.0452	2.3638	2.5409	2.6632	2.7564	2.8316	2.8945	2.9485	2.9959	3.0380
30	2.0423	2.3596	2.5357	2.6574	2.7500	2.8247	2.8872	2.9409	2.9880	3.0298
35	2.0301	2.3420	2.5145	2.6334	2.7238	2.7966	2.8575	2.9097	2.9554	2.9960
40	2.0211	2.3289	2.4989	2.6157	2.7045	2.7759	2.8355	2.8867	2.9314	2.9712
45	2.0141	2.3189	2.4868	2.6021	2.6896	2.7599	2.8187	2.8690	2.9130	2.9521
50	2.0086	2.3109	2.4772	2.5913	2.6778	2.7473	2.8053	2.8550	2.8984	2.9370
55	2.0040	2.3044	2.4694	2.5825	2.6682	2.7370	2.7944	2.8436	2.8866	2.9247
60	2.0003	2.2990	2.4630	2.5752	2.6603	2.7286	2.7855	2.8342	2.8768	2.9146
70	1.9944	2.2906	2.4529	2.5639	2.6479	2.7153	2.7715	2.8195	2.8615	2.8987
80	1.9901	2.2844	2.4454	2.5554	2.6387	2.7054	2.7610	2.8086	2.8502	2.8870
90	1.9867	2.2795	2.4395	2.5489	2.6316	2.6978	2.7530	2.8002	2.8414	2.8779
100	1.9840	2.2757	2.4349	2.5437	2.6259	2.6918	2.7466	2.7935	2.8344	2.8707
110	1.9818	2.2725	2.4311	2.5394	2.6213	2.6868	2.7414	2.7880	2.8287	2.8648
120	1.9799	2.2699	2.4280	2.5359	2.6174	2.6827	2.7370	2.7835	2.8240	2.8599
250	1.9695	2.2550	2.4102	2.5159	2.5956	2.6594	2.7124	2.7577	2.7972	2.8322
500	1.9647	2.2482	2.4021	2.5068	2.5857	2.6488	2.7012	2.7460	2.7850	2.8195
1000	1.9623	2.2448	2.3980	2.5022	2.5808	2.6435	2.6957	2.7402	2.7790	2.8133
∞	1.9600	2.2414	2.3940	2.4977	2.5758	2.6383	2.6901	2.7344	2.7729	2.8070

D1 (Continued)

	α = 0.05								
	11	12	13	14	$15 = \binom{6}{2}$	16	17	18	19
ν / r									
$100\alpha / r$	0.4545	0.4167	0.3846	0.3571	0.3333	0.3125	0.2941	0.2778	0.2632
2	14.7818	15.4435	16.0780	16.6883	17.2772	17.8466	18.3984	18.9341	19.4551
3	7.7041	7.9398	8.1625	8.3738	8.5752	8.7676	8.9521	9.1294	9.3001
4	5.7465	5.8853	6.0154	6.1380	6.2541	6.3643	6.4693	6.5697	6.6659
5	4.8819	4.9825	5.0764	5.1644	5.2474	5.3259	5.4005	5.4715	5.5393
6	4.4047	4.4858	4.5612	4.6317	4.6979	4.7604	4.8196	4.8759	4.9295
7	4.1048	4.1743	4.2388	4.2989	4.3553	4.4084	4.4586	4.5062	4.5514
8	3.8999	3.9618	4.0191	4.0724	4.1224	4.1693	4.2137	4.2556	4.2955
9	3.7513	3.8079	3.8602	3.9088	3.9542	3.9969	4.0371	4.0752	4.1114
10	3.6388	3.6915	3.7401	3.7852	3.8273	3.8669	3.9041	3.9394	3.9728
11	3.5508	3.6004	3.6462	3.6887	3.7283	3.7654	3.8004	3.8335	3.8648
12	3.4801	3.5274	3.5709	3.6112	3.6489	3.6842	3.7173	3.7487	3.7783
13	3.4221	3.4674	3.5091	3.5478	3.5838	3.6176	3.6493	3.6793	3.7076
14	3.3736	3.4173	3.4576	3.4949	3.5296	3.5621	3.5926	3.6214	3.6487
15	3.3325	3.3749	3.4139	3.4501	3.4837	3.5151	3.5447	3.5725	3.5989
16	3.2973	3.3386	3.3765	3.4116	3.4443	3.4749	3.5036	3.5306	3.5562
17	3.2667	3.3070	3.3440	3.3783	3.4102	3.4400	3.4680	3.4944	3.5193
18	3.2399	3.2794	3.3156	3.3492	3.3804	3.4095	3.4369	3.4626	3.4870
19	3.2163	3.2550	3.2906	3.3235	3.3540	3.3826	3.4094	3.4347	3.4585
20	3.1952	3.2333	3.2683	3.3006	3.3306	3.3587	3.3850	3.4098	3.4332
21	3.1764	3.2139	3.2483	3.2802	3.3097	3.3373	3.3632	3.3876	3.4106
22	3.1595	3.1965	3.2304	3.2618	3.2909	3.3181	3.3436	3.3676	3.3903
23	3.1441	3.1807	3.2142	3.2451	3.2739	3.3007	3.3259	3.3495	3.3719
24	3.1302	3.1663	3.1994	3.2300	3.2584	3.2849	3.3097	3.3331	3.3552
25	3.1175	3.1532	3.1859	3.2162	3.2443	3.2705	3.2950	3.3181	3.3400
26	3.1058	3.1412	3.1736	3.2035	3.2313	3.2572	3.2815	3.3044	3.3260
27	3.0951	3.1301	3.1622	3.1919	3.2194	3.2451	3.2691	3.2918	3.3132
28	3.0852	3.1199	3.1517	3.1811	3.2084	3.2339	3.2577	3.2801	3.3013
29	3.0760	3.1105	3.1420	3.1712	3.1982	3.2235	3.2471	3.2694	3.2904
30	3.0675	3.1017	3.1330	3.1620	3.1888	3.2138	3.2373	3.2594	3.2802
35	3.0326	3.0658	3.0962	3.1242	3.1502	3.1744	3.1971	3.2185	3.2386
40	3.0069	3.0393	3.0690	3.0964	3.1218	3.1455	3.1676	3.1884	3.2081
45	2.9872	3.0191	3.0482	3.0751	3.1000	3.1232	3.1450	3.1654	3.1846
50	2.9716	3.0030	3.0318	3.0582	3.0828	3.1057	3.1271	3.1472	3.1661
55	2.9589	2.9900	3.0184	3.0446	3.0688	3.0914	3.1125	3.1324	3.1511
60	2.9485	2.9792	3.0074	3.0333	3.0573	3.0796	3.1005	3.1202	3.1387
70	2.9321	2.9624	2.9901	3.0156	3.0393	3.0613	3.0818	3.1012	3.1194
80	2.9200	2.9500	2.9773	3.0026	3.0259	3.0476	3.0679	3.0870	3.1050
90	2.9106	2.9403	2.9675	2.9924	3.0156	3.0371	3.0572	3.0761	3.0939
100	2.9032	2.9327	2.9596	2.9844	3.0073	3.0287	3.0487	3.0674	3.0851
110	2.8971	2.9264	2.9532	2.9778	3.0007	3.0219	3.0417	3.0604	3.0779
120	2.8921	2.9212	2.9479	2.9724	2.9951	3.0162	3.0360	3.0545	3.0720
250	2.8635	2.8919	2.9178	2.9416	2.9637	2.9842	3.0034	3.0213	3.0383
500	2.8505	2.8785	2.9041	2.9276	2.9494	2.9696	2.9885	3.0063	3.0230
1000	2.8440	2.8719	2.8973	2.9207	2.9423	2.9624	2.9812	2.9988	3.0154
∞	2.8376	2.8653	2.8905	2.9137	2.9352	2.9552	2.9738	2.9913	3.0078

D1 (Continued)

	$\alpha = 0.05$							
ν / r	20	$21 = \binom{7}{2}$	$28 = \binom{8}{2}$	$36 = \binom{9}{2}$	$45 = \binom{10}{2}$	$55 = \binom{11}{2}$	$66 = \binom{12}{2}$	$78 = \binom{13}{2}$
$100\alpha / r$	0.2500	0.2381	0.1786	0.1389	0.1111	0.0909	0.0758	0.0641
2	19.9625	20.4573	23.6326	26.8049	29.9750	33.1436	36.3112	39.4778
3	9.4649	9.6242	10.6166	11.5632	12.4715	13.3471	14.1943	15.0165
4	6.7583	6.8471	7.3924	7.8998	8.3763	8.8271	9.2558	9.6655
5	5.6042	5.6665	6.0447	6.3914	6.7126	7.0128	7.2952	7.5625
6	4.9807	5.0297	5.3255	5.5937	5.8399	6.0680	6.2810	6.4813
7	4.5946	4.6359	4.8839	5.1068	5.3101	5.4973	5.6712	5.8339
8	4.3335	4.3699	4.5869	4.7810	4.9570	5.1183	5.2675	5.4065
9	4.1458	4.1786	4.3744	4.5485	4.7058	4.8494	4.9818	5.1048
10	4.0045	4.0348	4.2150	4.3747	4.5184	4.6492	4.7695	4.8810
11	3.8945	3.9229	4.0913	4.2400	4.3735	4.4947	4.6059	4.7087
12	3.8065	3.8334	3.9925	4.1327	4.2582	4.3719	4.4761	4.5722
13	3.7345	3.7602	3.9118	4.0452	4.1643	4.2721	4.3706	4.4614
14	3.6746	3.6992	3.8448	3.9725	4.0865	4.1894	4.2833	4.3698
15	3.6239	3.6477	3.7882	3.9113	4.0209	4.1198	4.2099	4.2928
16	3.5805	3.6036	3.7398	3.8589	3.9649	4.0604	4.1473	4.2272
17	3.5429	3.5654	3.6980	3.8137	3.9165	4.0091	4.0933	4.1706
18	3.5101	3.5321	3.6614	3.7742	3.8744	3.9644	4.0463	4.1214
19	3.4812	3.5027	3.6292	3.7395	3.8373	3.9251	4.0050	4.0781
20	3.4554	3.4765	3.6006	3.7087	3.8044	3.8903	3.9683	4.0398
21	3.4325	3.4532	3.5751	3.6812	3.7750	3.8593	3.9357	4.0056
22	3.4118	3.4322	3.5522	3.6564	3.7487	3.8314	3.9064	3.9750
23	3.3931	3.4132	3.5314	3.6341	3.7249	3.8062	3.8800	3.9474
24	3.3761	3.3960	3.5126	3.6139	3.7033	3.7834	3.8560	3.9223
25	3.3606	3.3803	3.4955	3.5954	3.6836	3.7626	3.8342	3.8995
26	3.3464	3.3659	3.4797	3.5785	3.6656	3.7436	3.8142	3.8787
27	3.3334	3.3526	3.4653	3.5629	3.6491	3.7261	3.7959	3.8595
28	3.3214	3.3404	3.4520	3.5486	3.6338	3.7101	3.7790	3.8419
29	3.3102	3.3291	3.4397	3.5354	3.6198	3.6952	3.7634	3.8256
30	3.2999	3.3186	3.4282	3.5231	3.6067	3.6814	3.7489	3.8105
35	3.2577	3.2758	3.3816	3.4730	3.5534	3.6252	3.6900	3.7490
40	3.2266	3.2443	3.3473	3.4362	3.5143	3.5840	3.6468	3.7040
45	3.2028	3.2201	3.3211	3.4081	3.4845	3.5525	3.6138	3.6696
50	3.1840	3.2010	3.3003	3.3858	3.4609	3.5277	3.5878	3.6425
55	3.1688	3.1856	3.2836	3.3679	3.4418	3.5076	3.5668	3.6206
60	3.1562	3.1728	3.2697	3.3530	3.4260	3.4910	3.5494	3.6025
70	3.1366	3.1529	3.2481	3.3299	3.4015	3.4652	3.5224	3.5744
80	3.1220	3.1381	3.2321	3.3127	3.3833	3.4460	3.5024	3.5536
90	3.1108	3.1267	3.2197	3.2995	3.3693	3.4313	3.4870	3.5375
100	3.1018	3.1176	3.2099	3.2890	3.3582	3.4196	3.4747	3.5248
110	3.0945	3.1102	3.2018	3.2804	3.3491	3.4100	3.4648	3.5144
120	3.0885	3.1041	3.1952	3.2733	3.3416	3.4021	3.4565	3.5058
250	3.0543	3.0694	3.1577	3.2332	3.2991	3.3575	3.4099	3.4573
500	3.0387	3.0537	3.1406	3.2150	3.2798	3.3373	3.3887	3.4354
1000	3.0310	3.0459	3.1322	3.2059	3.2703	3.3272	3.3783	3.4245
∞	3.0233	3.0381	3.1237	3.1970	3.2608	3.3172	3.3678	3.4136

D1 (Continued)

	$\alpha = 0.05$						
ν / r	$91 = \binom{14}{2}$	$105 = \binom{15}{2}$	$120 = \binom{16}{2}$	$136 = \binom{17}{2}$	$153 = \binom{18}{2}$	$171 = \binom{19}{2}$	$190 = \binom{20}{2}$
$100\alpha / r$	0.0549	0.0476	0.0417	0.0368	0.0327	0.0292	0.0263
2	42.6439	45.8094	48.9745	52.1392	55.3037	58.4679	61.6320
3	15.8165	16.5964	17.3582	18.1035	18.8336	19.5497	20.2528
4	10.0585	10.4367	10.8016	11.1545	11.4966	11.8288	12.1519
5	7.8166	8.0591	8.2913	8.5143	8.7290	8.9362	9.1365
6	6.6705	6.8500	7.0210	7.1844	7.3410	7.4914	7.6363
7	5.9868	6.1313	6.2684	6.3990	6.5236	6.6430	6.7577
8	5.5368	5.6594	5.7755	5.8857	5.9906	6.0909	6.1869
9	5.2197	5.3276	5.4295	5.5260	5.6177	5.7051	5.7888
10	4.9849	5.0823	5.1740	5.2608	5.3431	5.4215	5.4963
11	4.8044	4.8939	4.9781	5.0576	5.1330	5.2046	5.2729
12	4.6615	4.7450	4.8233	4.8972	4.9672	5.0336	5.0969
13	4.5457	4.6243	4.6981	4.7675	4.8332	4.8956	4.9549
14	4.4500	4.5247	4.5947	4.6606	4.7228	4.7818	4.8379
15	4.3695	4.4410	4.5079	4.5708	4.6302	4.6865	4.7400
16	4.3011	4.3698	4.4341	4.4946	4.5516	4.6056	4.6568
17	4.2421	4.3085	4.3706	4.4289	4.4839	4.5360	4.5853
18	4.1907	4.2551	4.3154	4.3719	4.4251	4.4755	4.5232
19	4.1456	4.2083	4.2669	4.3218	4.3736	4.4225	4.4688
20	4.1057	4.1669	4.2240	4.2776	4.3280	4.3756	4.4208
21	4.0701	4.1300	4.1858	4.2381	4.2874	4.3339	4.3780
22	4.0382	4.0969	4.1516	4.2028	4.2510	4.2966	4.3397
23	4.0095	4.0671	4.1207	4.1710	4.2183	4.2629	4.3052
24	3.9834	4.0400	4.0928	4.1422	4.1886	4.2325	4.2739
25	3.9597	4.0154	4.0674	4.1160	4.1616	4.2047	4.2455
26	3.9380	3.9929	4.0441	4.0920	4.1370	4.1794	4.2196
27	3.9181	3.9723	4.0228	4.0700	4.1144	4.1562	4.1958
28	3.8997	3.9533	4.0032	4.0498	4.0936	4.1349	4.1739
29	3.8828	3.9357	3.9850	4.0311	4.0744	4.1151	4.1537
30	3.8671	3.9195	3.9682	4.0138	4.0566	4.0969	4.1350
35	3.8032	3.8533	3.8999	3.9434	3.9842	4.0226	4.0590
40	3.7564	3.8049	3.8499	3.8919	3.9314	3.9684	4.0035
45	3.7208	3.7680	3.8118	3.8527	3.8911	3.9271	3.9612
50	3.6926	3.7389	3.7818	3.8218	3.8594	3.8946	3.9279
55	3.6699	3.7154	3.7576	3.7969	3.8337	3.8684	3.9010
60	3.6511	3.6960	3.7376	3.7763	3.8126	3.8467	3.8789
70	3.6220	3.6658	3.7065	3.7444	3.7798	3.8131	3.8445
80	3.6004	3.6435	3.6835	3.7207	3.7555	3.7883	3.8191
90	3.5837	3.6263	3.6658	3.7025	3.7369	3.7691	3.7995
100	3.5705	3.6127	3.6517	3.6880	3.7220	3.7539	3.7840
110	3.5598	3.6016	3.6403	3.6763	3.7100	3.7416	3.7714
120	3.5509	3.5924	3.6308	3.6665	3.7000	3.7313	3.7609
250	3.5007	3.5405	3.5774	3.6117	3.6437	3.6737	3.7020
500	3.4779	3.5170	3.5532	3.5868	3.6182	3.6477	3.6754
1000	3.4666	3.5054	3.5412	3.5745	3.6056	3.6348	3.6622
∞	3.4554	3.4938	3.5293	3.5623	3.5931	3.6219	3.6491

D2 MAXIMUM LIKELIHOOD ESTIMATES FOR THE GAMMA DISTRIBUTION

A table (reproduced from Wilk et al. [1962b: Table 1], by permission of the Biometrika Trustees) is given for finding the maximum likelihood estimates $\hat{\eta}$ and $\hat{\lambda}$, based on a random sample of size n from the gamma distribution

$$f(y; \lambda, \eta) = \frac{\lambda^\eta}{\Gamma(\eta)} y^{\eta-1} e^{-\lambda y}, \qquad y \geq 0.$$

We enter the table with $1/(1 - R)$, where

$$R = \text{geometric mean of } y_i / \text{arithmetic mean of } y_i,$$

and read off $\hat{\eta}$. Linear interpolation will give four decimal accuracy everywhere in the table, except between the first two values. Although $\hat{\lambda}$ is not needed for probability plotting (we use $\lambda = 1$), it is given by $\hat{\lambda} = \hat{\eta}/\bar{y}$.

$1/(1-R)$	$\hat{\eta}$	$1/(1-R)$	$\hat{\eta}$	$1/(1-R)$	$\hat{\eta}$
1·000	0·00000	1·11	0·29854	2·4	1·06335
1·001	0·11539	1·12	0·30749	2·5	1·11596
1·002	0·12663	1·13	0·31616	2·6	1·16833
1·003	0·13434	1·14	0·32461	2·7	1·22050
1·004	0·14043	1·15	0·33285	2·8	1·27248
1·005	0·14556	1·16	0·34090	2·9	1·32430
1·006	0·15005	1·18	0·35654	3·0	1·37599
1·007	0·15408	1·20	0·37165	3·2	1·47899
1·008	0·15775	1·22	0·38631	3·4	1·58158
1·009	0·16115	1·24	0·40060	3·6	1·68385
1·010	0·16432	1·26	0·41457	3·8	1·78585
1·012	0·17011	1·28	0·42825	4·0	1·88763
1·014	0·17535	1·30	0·44170	4·2	1·98921
1·016	0·18017	1·32	0·45492	4·4	2·09063
1·018	0·18464	1·34	0·46795	4·6	2·19191
1·020	0·18884	1·36	0·48081	4·8	2·29308
1·022	0·19282	1·38	0·49351	5·0	2·39414
1·024	0·19659	1·40	0·50608	5·5	2·64643
1·026	0·20020	1·45	0·53694	6·0	2·89830
1·028	0·20366	1·50	0·56715	6·5	3·14984
1·030	0·20700	1·55	0·59682	7·0	3·40115
1·035	0·21486	1·60	0·62604	8·0	3·90325
1·040	0·22217	1·65	0·65488	9·0	4·40482
1·045	0·22904	1·70	0·68340	10·0	4·90608
1·050	0·23554	1·75	0·71163	12·0	5·90790
1·055	0·24175	1·80	0·73961	14·0	6·90914
1·060	0·24771	1·85	0·76737	16·0	7·91005
1·065	0·25345	1·90	0·79494	18·0	8·91073
1·070	0·25899	1·95	0·82233	20·0	9·91125
1·075	0·26437	2·00	0·84957	30·0	14·91301
1·080	0·26959	2·10	0·90364	40·0	19·91401
1·090	0·27966	2·20	0·95724	50·0	24·91457
1·100	0·28928	2·30	1·01046		

n \ %	10	5	2·5	1	0·5	0·1
4	0.831	0.987	1.070	1.120	1.137	1.151
5	0.821	1.049	1.207	1.337	1.396	1.464
6	0.795	1.042	1.239	1.429	1.531	1.671
7	0.782	1.018	1.230	1.457	1.589	1.797
8	0.765	0.998	1.208	1.452	1.605	1.866
9	0.746	0.977	1.184	1.433	1.598	1.898
10	0.728	0.954	1.159	1.407	1.578	1.906
11	0.710	0.931	1.134	1.381	1.553	1.899
12	0.693	0.910	1.109	1.353	1.526	1.882
13	0.677	0.890	1.085	1.325	1.497	1.859
14	0.662	0.870	1.061	1.298	1.468	1.832
15	0.648	0.851	1.039	1.272	1.440	1.803
16	0.635	0.834	1.018	1.247	1.412	1.773
17	0.622	0.817	0.997	1.222	1.385	1.744
18	0.610	0.801	0.978	1.199	1.359	1.714
19	0.599	0.786	0.960	1.176	1.334	1.685
20	0.588	0.772	0.942	1.155	1.310	1.657
21	0.578	0.758	0.925	1.134	1.287	1.628
22	0.568	0.746	0.909	1.114	1.265	1.602
23	0.559	0.733	0.894	1.096	1.243	1.575
24	0.550	0.722	0.880	1.078	1.223	1.550
25	0.542	0.710	0.866	1.060	1.203	1.526

D3 UPPER TAIL PERCENTAGE POINTS FOR $\sqrt{b_1}$

The above table (reproduced from Mulholland [1977] by permission of the Biometrika Trustees) gives the upper tail percentage points of

$$\sqrt{b_1} = \sqrt{n} \sum_{i=1}^{n} (x_i - \bar{x})^3 \bigg/ \left\{ \sum_{i=1}^{n} (x_i - \bar{x})^2 \right\}^{3/2},$$

the sample coefficient of skewness, for a random sample of size n from a normal distribution. $\sqrt{b_1}$ is independent of the mean and variance of x and its distribution is symmetric about the origin. For other values of n see Appendix D4.

D4 COEFFICIENTS IN A NORMALIZING TRANSFORMATION OF $\sqrt{b_1}$

On the following page are given values of δ and $1/\lambda$ (reproduced from D'Agostino and Pearson [1973: Table 4] by permission of the Biometrika Trustees) such that

$$X\left(\sqrt{b_1}\right) = \delta \sinh^{-1}\left(\frac{\sqrt{b_1}}{\lambda}\right)$$

is approximately $N(0,1)$. b_1 is defined in Appendix D3.

n	δ	$1/\lambda$
8	5·563	0·3030
9	4·260	0·4080
10	3·734	0·4794
11	3·447	0·5339
12	3·270	0·5781
13	3·151	0·6153
14	3·069	0·6473
15	3·010	0·6753
16	2·968	0·7001
17	2·937	0·7224
18	2·915	0·7426
19	2·900	0·7610
20	2·890	0·7779
21	2·884	0·7934
22	2·882	0·8078
23	2·882	0·8211
24	2·884	0·8336
25	2·889	0·8452
26	2·895	0·8561
27	2·902	0·8664
28	2·910	0·8760
29	2·920	0·8851
30	2·930	0·8938
31	2·941	0·9020
32	2·952	0·9097
33	2·964	0·9171
34	2·977	0·9241
35	2·990	0·9308
36	3·003	0·9372
37	3·016	0·9433
38	3·030	0·9492
39	3·044	0·9548
40	3·058	0·9601
41	3·073	0·9653
42	3·087	0·9702
43	3·102	0·9750
44	3·117	0·9795
45	3·131	0·9840
46	3·146	0·9882
47	3·161	0·9923
48	3·176	0·9963
49	3·192	1·0001
50	3·207	1·0038
52	3·237	1·0108
54	3·268	1·0174
56	3·298	1·0235
58	3·329	1.0293
60	3·359	1·0348

n	δ	$1/\lambda$
62	3·389	1·0400
64	3·420	1·0449
66	3·450	1·0495
68	3·480	1·0540
70	3·510	1·0581
72	3·540	1·0621
74	3·569	1·0659
76	3·599	1·0695
78	3·628	1·0730
80	3·657	1·0763
82	3·686	1·0795
84	3·715	1·0825
86	3·744	1·0854
88	3·772	1·0882
90	3·801	1·0909
92	3·829	1·0934
94	3·857	1·0959
96	3·885	1·0983
98	3·913	1·1006
100	3·940	1·1028
105	4·009	1·1080
110	4·076	1·1128
115	4·142	1·1172
120	4·207	1·1212
125	4·272	1·1250
130	4·336	1·1285
135	4·398	1·1318
140	4·460	1·1348
145	4·521	1·1377
150	4·582	1·1403
155	4·641	1·1428
160	4·700	1·1452
165	4·758	1·1474
170	4·816	1·1496
175	4·873	1·1516
180	4·929	1·1535
185	4·985	1·1553
190	5·040	1·1570
195	5·094	1·1586
200	5·148	1·1602
205	5·202	1·1616
210	5·255	1·1631
215	5·307	1·1644
220	5·359	1·1657
225	5·410	1·1669
230	5·461	1·1681
235	5·511	1·1693
240	5·561	1·1704
245	5·611	1·1714
250	5·660	1·1724

n	δ	$1/\lambda$
260	5·757	1·1744
270	5·853	1·1761
280	5·946	1·1779
290	6·039	1·1793
300	6·130	1·1808
310	6·220	1·1821
320	6·308	1·1834
330	6·396	1·1846
340	6·482	1·1858
350	6·567	1·1868
360	6·651	1·1879
370	6·733	1·1888
380	6·815	1·1897
390	6·896	1·1906
400	6·976	1·1914
410	7·056	1·1922
420	7·134	1·1929
430	7·211	1·1937
440	7·288	1·1943
450	7·363	1·1950
460	7·438	1·1956
470	7·513	1·1962
480	7·586	1·1968
490	7·659	1·1974
500	7·731	1·1979
520	7·873	1·1989
540	8·013	1·1998
560	8·151	1·2007
580	8·286	1·2015
600	8·419	1·2023
620	8·550	1·2030
640	8·679	1·2036
660	8·806	1·2043
680	8·931	1·2049
700	9·054	1·2054
720	9·176	1·2060
740	9·297	1·2065
760	9·415	1·2069
780	9·533	1·2073
800	9·649	1·2078
820	9·763	1·2082
840	9·876	1·2086
860	9·988	1·2089
880	10·098	1·2093
900	10·208	1·2096
920	10·316	1·2100
940	10·423	1·2103
960	10·529	1·2106
980	10·634	1·2109
1000	10·738	1·2111

D5 SIMULATION PERCENTILES FOR b_2

Below is a table of percentiles (reproduced from D'Agostino and Tietjen [1971: Table 1] by permission of the Biometrika Trustees) for

$$b_2 = n \sum_{i=1}^{n} (x_i - \bar{x})^4 \bigg/ \left\{ \sum_{i=1}^{n} (x_i - \bar{x})^2 \right\}^2,$$

the sample coefficient of kurtosis, from a random sample of size n from a normal distribution. b_2 is independent of the mean and variance of x. For other values of n see Appendix D6.

Sample size	Percentiles											
	1	2	2·5	5	10	20	80	90	95	97·5	98	99
7	1·25	1·30	1·34	1·41	1·53	1·70	2·78	3·20	3·55	3·85	3·93	4·23
8	1·31	1·37	1·40	1·46	1·58	1·75	2·84	3·31	3·70	4·09	4·20	4·53
9	1·35	1·42	1·45	1·53	1·63	1·80	2·98	3·43	3·86	4·28	4·41	4·82
10	1·39	1·45	1·49	1·56	1·68	1·85	3·01	3·53	3·95	4·40	4·55	5·00
12	1·46	1·52	1·56	1·64	1·76	1·93	3·06	3·55	4·05	4·56	4·73	5·20
15	1·55	1·61	1·64	1·72	1·84	2·01	3·13	3·62	4·13	4·66	4·85	5·30
20	1·65	1·71	1·74	1·82	1·95	2·13	3·21	3·68	4·17	4·68	4·87	5·36
25	1·72	1·79	1·83	1·91	2·03	2·20	3·23	3·68	4·16	4·65	4·82	5·30
30	1·79	1·86	1·90	1·98	2·10	2·26	3·25	3·68	4·11	4·59	4·75	5·21
35	1·84	1·91	1·95	2·03	2·14	2·31	3·27	3·68	4·10	4·53	4·68	5·13
40	1·89	1·96	1·98	2·07	2·19	2·34	3·28	3·67	4·06	4·46	4·61	5·04
45	1·93	2·00	2·03	2·11	2·22	2·37	3·28	3·65	4·00	4·39	4·52	4·94
50	1·95	2·03	2·06	2·15	2·25	2·41	3·28	3·62	3·99	4·33	4·45	4·88

D6 CHARTS FOR THE PERCENTILES OF b_2

Charts for the percentiles of b_2 (defined in Appendix D5), based on a sample of size n from a normal distribution, are given below. We enter P and read off $b_2(P)$, where

$$\text{pr}[b_2 \leq b_2(P)] = P.$$

(Reproduced from D'Agostino and Pearson [1973: Figs. 1 and 2] by permission of the Biometrika Trustees.)

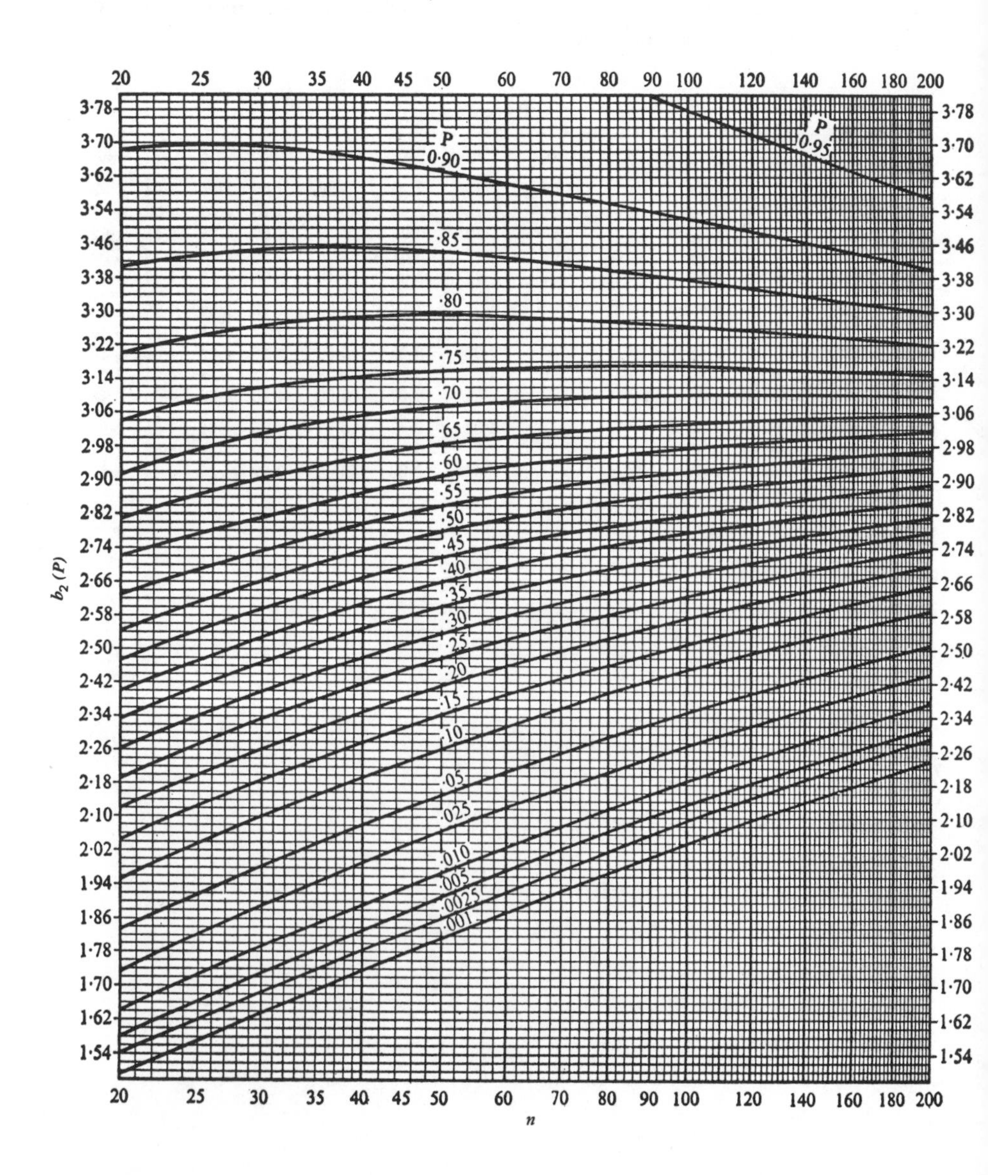

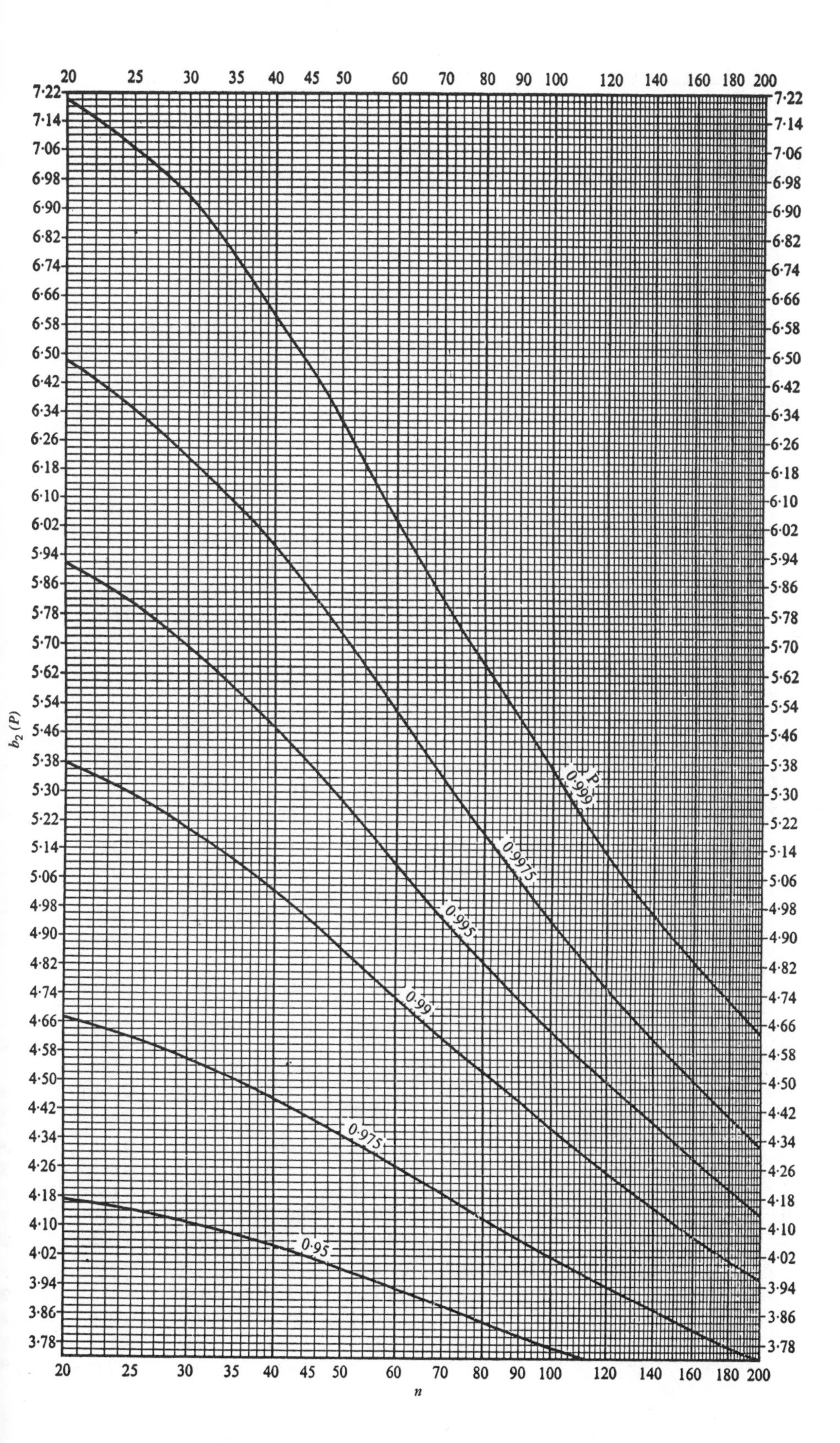

20 25 30 35 40 45 50 60 70 80 90 100 120 140 160 180 200
7·22 7·14 7·06 6·98 6·90 6·82 6·74 6·66 6·58 6·50 6·42 6·34 6·26 6·18 6·10 6·02 5·94 5·86 5·78 5·70 5·62 5·54 5·46 5·38 5·30 5·22 5·14 5·06 4·98 4·90 4·82 4·74 4·66 4·58 4·50 4·42 4·34 4·26 4·18 4·10 4·02 3·94 3·86 3·78
$b_2(P)$
P
0·999
0·9975
0·995
0·99
0·975
0·95
n

D7 COEFFICIENTS FOR THE WILK–SHAPIRO (W) TEST

Coefficients $\{a_{n-i+1}^{(n)}\}$ are given for the Wilk–Shapiro statistic

$$W = \left\{ \sum_{i=1}^{k} a_{n-i+1}^{(n)} \left(x_{(n-i+1)} - x_{(i)}\right) \right\}^2 \Big/ \sum_{i=1}^{n} \left(x_{(i)} - \bar{x}\right)^2,$$

where $x_{(1)} \leq x_{(2)} \leq \cdots \leq x_{(n)}$ is an ordered random sample of size n from the normal distribution, and $n = 2k$ or $2k + 1$ (reproduced from Shapiro and Wilks [1965: Table 5] by permission of the Biometrika Trustees). For the percentiles of W see Appendix D8.

Coefficients $a_{n-i+1}^{(n)}$

$i \backslash n$	2	3	4	5	6	7	8	9	10
1	0·7071	0·7071	0·6872	0·6646	0·6431	0·6233	0·6052	0·5888	0·5739
2	—	·0000	·1677	·2413	·2806	·3031	·3164	·3244	·3291
3	—	—	—	·0000	·0875	·1401	·1743	·1976	·2141
4	—	—	—	—	—	·0000	·0561	·0947	·1224
5	—	—	—	—	—	—	—	·0000	·0399

$i \backslash n$	11	12	13	14	15	16	17	18	19	20
1	0·5601	0·5475	0·5359	0·5251	0·5150	0·5056	0·4968	0·4886	0·4808	0·4734
2	·3315	·3325	·3325	·3318	·3306	·3290	·3273	·3253	·3232	·3211
3	·2260	·2347	·2412	·2460	·2495	·2521	·2540	·2553	·2561	·2565
4	·1429	·1586	·1707	·1802	·1878	·1939	·1988	·2027	·2059	·2085
5	·0695	·0922	·1099	·1240	·1353	·1447	·1524	·1587	·1641	·1686
6	0·0000	0·0303	0·0539	0·0727	0·0880	0·1005	0·1109	0·1197	0·1271	0·1334
7	—	—	·0000	·0240	·0433	·0593	·0725	·0837	·0932	·1013
8	—	—	—	—	·0000	·0196	·0359	·0496	·0612	·0711
9	—	—	—	—	—	—	·0000	·0163	·0303	·0422
10	—	—	—	—	—	—	—	—	·0000	·0140

$i \backslash n$	21	22	23	24	25	26	27	28	29	30
1	0·4643	0·4590	0·4542	0·4493	0·4450	0·4407	0·4366	0·4328	0·4291	0·4254
2	·3185	·3156	·3126	·3098	·3069	·3043	·3018	·2992	·2968	·2944
3	·2578	·2571	·2563	·2554	·2543	·2533	·2522	·2510	·2499	·2487
4	·2119	·2131	·2139	·2145	·2148	·2151	·2152	·2151	·2150	·2148
5	·1736	·1764	·1787	·1807	·1822	·1836	·1848	·1857	·1864	·1870
6	0·1399	0·1443	0·1480	0·1512	0·1539	0·1563	0·1584	0·1601	0·1616	0·1630
7	·1092	·1150	·1201	·1245	·1283	·1316	·1346	·1372	·1395	·1415
8	·0804	·0878	·0941	·0997	·1046	·1089	·1128	·1162	·1192	·1219
9	·0530	·0618	·0696	·0764	·0823	·0876	·0923	·0965	·1002	·1036
10	·0263	·0368	·0459	·0539	·0610	·0672	·0728	·0778	·0822	·0862
11	0·0000	0·0122	0·0228	0·0321	0·0403	0·0476	0·0540	0·0598	0·0650	0·0697
12	—	—	·0000	·0107	·0200	·0284	·0358	·0424	·0483	·0537
13	—	—	—	—	·0000	·0094	·0178	·0253	·0320	·0381
14	—	—	—	—	—	—	·0000	·0084	·0159	·0227
15	—	—	—	—	—	—	—	—	·0000	·0076

D7 (Continued)

i \ n	31	32	33	34	35	36	37	38	39	40
1	0·4220	0·4188	0·4156	0·4127	0·4096	0·4068	0·4040	0·4015	0·3989	0·3964
2	·2921	·2898	·2876	·2854	·2834	·2813	·2794	·2774	·2755	·2737
3	·2475	·2463	·2451	·2439	·2427	·2415	·2403	·2391	·2380	·2368
4	·2145	·2141	·2137	·2132	·2127	·2121	·2116	·2110	·2104	·2098
5	·1874	·1878	·1880	·1882	·1883	·1883	·1883	·1881	·1880	·1878
6	0·1641	0·1651	0·1660	0·1667	0·1673	0·1678	0·1683	0·1686	0·1689	0·1691
7	·1433	·1449	·1463	·1475	·1487	·1496	·1505	·1513	·1520	·1526
8	·1243	·1265	·1284	·1301	·1317	·1331	·1344	·1356	·1366	·1376
9	·1066	·1093	·1118	·1140	·1160	·1179	·1196	·1211	·1225	·1237
10	·0899	·0931	·0961	·0988	·1013	·1036	·1056	·1075	·1092	·1108
11	0·0739	0·0777	0·0812	0·0844	0·0873	0·0900	0·0924	0·0947	0·0967	0·0986
12	·0585	·0629	·0669	·0706	·0739	·0770	·0798	·0824	·0848	·0870
13	·0435	·0485	·0530	·0572	·0610	·0645	·0677	·0706	·0733	·0759
14	·0289	·0344	·0395	·0441	·0484	·0523	·0559	·0592	·0622	·0651
15	·0144	·0206	·0262	·0314	·0361	·0404	·0444	·0481	·0515	·0546
16	0·0000	0·0068	0·0131	0·0187	0·0239	0·0287	0·0331	0·0372	0·0409	0·0444
17	—	—	·0000	·0062	·0119	·0172	·0220	·0264	·0305	·0343
18	—	—	—	—	·0000	·0057	·0110	·0158	·0203	·0244
19	—	—	—	—	—	—	·0000	·0053	·0101	·0146
20	—	—	—	—	—	—	—	—	·0000	·0049

i \ n	41	42	43	44	45	46	47	48	49	50
1	0·3940	0·3917	0·3894	0·3872	0·3850	0·3830	0·3808	0·3789	0·3770	0·3751
2	·2719	·2701	·2684	·2667	·2651	·2635	·2620	·2604	·2589	·2574
3	·2357	·2345	·2334	·2323	·2313	·2302	·2291	·2281	·2271	·2260
4	·2091	·2085	·2078	·2072	·2065	·2058	·2052	·2045	·2038	·2032
5	·1876	·1874	·1871	·1868	·1865	·1862	·1859	·1855	·1851	·1847
6	0·1693	0·1694	0·1695	0·1695	0·1695	0·1695	0·1695	0·1693	0·1692	0·1691
7	·1531	·1535	·1539	·1542	·1545	·1548	·1550	·1551	·1553	·1554
8	·1384	·1392	·1398	·1405	·1410	·1415	·1420	·1423	·1427	·1430
9	·1249	·1259	·1269	·1278	·1286	·1293	·1300	·1306	·1312	·1317
10	·1123	·1136	·1149	·1160	·1170	·1180	·1189	·1197	·1205	·1212
11	0·1004	0·1020	0·1035	0·1049	0·1062	0·1073	0·1085	0·1095	0·1105	0·1113
12	·0891	·0909	·0927	·0943	·0959	·0972	·0986	·0998	·1010	·1020
13	·0782	·0804	·0824	·0842	·0860	·0876	·0892	·0906	·0919	·0932
14	·0677	·0701	·0724	·0745	·0765	·0783	·0801	·0817	·0832	·0846
15	·0575	·0602	·0628	·0651	·0673	·0694	·0713	·0731	·0748	·0764
16	0·0476	0·0506	0·0534	0·0560	0·0584	0·0607	0·0628	0·0648	0·0667	0·0685
17	·0379	·0411	·0442	·0471	·0497	·0522	·0546	·0568	·0588	·0608
18	·0283	·0318	·0352	·0383	·0412	·0439	·0465	·0489	·0511	·0532
19	·0188	·0227	·0263	·0296	·0328	·0357	·0385	·0411	·0436	·0459
20	·0094	·0136	·0175	·0211	·0245	·0277	·0307	·0335	·0361	·0386
21	0·0000	0·0045	0·0087	0·0126	0·0163	0·0197	0·0229	0·0259	0·0288	0·0314
22		—	·0000	·0042	·0081	·0118	·0153	·0185	·0215	·0244
23	—	—	—	—	·0000	·0039	·0076	·0111	·0143	·0174
24	—	—	—	—	—	—	·0000	·0037	·0071	·0104
25	—	—	—	—	—	—	—	—	·0000	·0035

n	5%	1%	n	5%	1%
3	0·767	0·753	**26**	0·920	0·891
4	·748	·687	**27**	·923	·894
5	·762	·686	**28**	·924	·896
			29	·926	·898
			30	·927	·900
6	0·788	0·713	**31**	0·929	0·902
7	·803	·730	**32**	·930	·904
8	·818	·749	**33**	·931	·906
9	·829	·764	**34**	·933	·908
10	·842	·781	**35**	·934	·910
11	0·850	0·792	**36**	0·935	0·912
12	·859	·805	**37**	·936	·914
13	·866	·814	**38**	·938	·916
14	·874	·825	**39**	·939	·917
15	·881	·835	**40**	·940	·919
16	0·887	0·844	**41**	0·941	0·920
17	·892	·851	**42**	·942	·922
18	·897	·858	**43**	·943	·923
19	·901	·863	**44**	·944	·924
20	·905	·868	**45**	·945	·926
21	0·908	0·873	**46**	0·945	0·927
22	·911	·878	**47**	·946	·928
23	·914	·881	**48**	·947	·929
24	·916	·884	**49**	·947	·929
25	·918	·888	**50**	·947	·930

D8 PERCENTILES FOR THE WILK–SHAPIRO (W) TEST

Percentage points for the W-test of normality, defined in Appendix D7, are given above (reproduced from Shapiro and Wilk [1965: Table 6] by permission of the Biometrika Trustees). Any departure from normality is detected by a small value of W; n is the number of observations.

D9 D'AGOSTINO'S TEST FOR NORMALITY

Percentage points (reproduced from D'Agostino [1971, 1972] by permission of the Biometrika Trustees) are given for the statistic

$$Y = \frac{\sqrt{n}\left[D - (2\sqrt{\pi})^{-1}\right]}{0.02998598},$$

where
$$D = \sum_{i=1}^{n} \left[i - \tfrac{1}{2}(n+1)\right] x_{(i)} \Big/ n^{3/2}\left[\sum_{i=1}^{n} \left(x_{(i)} - \bar{x}\right)^2\right]^{1/2},$$

and $x_{(1)} \le x_{(2)} \le \cdots \le x_{(n)}$ is an ordered random sample of n observations from a normal distribution. The hypothesis of normality is rejected if $|Y|$ is too large. As the range of Y is small, D should be calculated to five decimal places.

D9 D'AGOSTINO'S TEST FOR NORMALITY

	Percentiles of Y									
n	0·5	1·0	2·5	5	10	90	95	97·5	99	99·5
10	−4·66	−4·06	−3·25	−2·62	−1·99	0·149	0·235	0·299	0·356	0·385
12	−4·63	−4·02	−3·20	−2·58	−1·94	0·237	0·329	0·381	0·440	0·479
14	−4·57	−3·97	−3·16	−2·53	−1·90	0·308	0·399	0·460	0·515	0·555
16	−4·52	−3·92	−3·12	−2·50	−1·87	0·367	0·459	0·526	0·587	0·613
18	−4·47	−3·87	−3·08	−2·47	−1·85	0·417	0·515	0·574	0·636	0·667
20	−4·41	−3·83	−3·04	−2·44	−1·82	0·460	0·565	0·628	0·690	0·720
22	−4·36	−3·78	−3·01	−2·41	−1·81	0·497	0·609	0·677	0·744	0·775
24	−4·32	−3·75	−2·98	−2·39	−1·79	0·530	0·648	0·720	0·783	0·822
26	−4·27	−3·71	−2·96	−2·37	−1·77	0·559	0·682	0·760	0·827	0·867
28	−4·23	−3·68	−2·93	−2·35	−1·76	0·586	0·714	0·797	0·868	0·910
30	−4·19	−3·64	−2·91	−2·33	−1·75	0·610	0·743	0·830	0·906	0·941
32	−4·16	−3·61	−2·88	−2·32	−1·73	0·631	0·770	0·862	0·942	0·983
34	−4·12	−3·59	−2·86	−2·30	−1·72	0·651	0·794	0·891	0·975	1·02
36	−4·09	−3·56	−2·85	−2·29	−1·71	0·669	0·816	0·917	1·00	1·05
38	−4·06	−3·54	−2·83	−2·28	−1·70	0·686	0·837	0·941	1·03	1·08
40	−4·03	−3·51	−2·81	−2·26	−1·70	0·702	0·857	0·964	1·06	1·11
42	−4·00	−3·49	−2·80	−2·25	−1·69	0·716	0·875	0·986	1·09	1·14
44	−3·98	−3·47	−2·78	−2·24	−1·68	0·730	0·892	1·01	1·11	1·17
46	−3·95	−3·45	−2·77	−2·23	−1·67	0·742	0·908	1·02	1·13	1·19
48	−3·93	−3·43	−2·75	−2·22	−1·67	0·754	0·923	1·04	1·15	1·22
50	−3·91	−3·41	−2·74	−2·21	−1·66	0·765	0·937	1·06	1·18	1·24
60	−3·81	−3·34	−2·68	−2·17	−1·64	0·812	0·997	1·13	1·26	1·34
70	−3·73	−3·27	−2·64	−2·14	−1·61	0·849	1·05	1·19	1·33	1·42
80	−3·67	−3·22	−2·60	−2·11	−1·59	0·878	1·08	1·24	1·39	1·48
90	−3·61	−3·17	−2·57	−2·09	−1·58	0·902	1·12	1·28	1·44	1·54
100	−3·57	−3·14	−2·54	−2·07	−1·57	0·923	1·14	1·31	1·48	1·59
150	−3·409	−3·009	−2·452	−2·004	−1·520	0·990	1·233	1·423	1·623	1·746
200	−3·302	−2·922	−2·391	−1·960	−1·491	1·032	1·290	1·496	1·715	1·853
250	−3·227	−2·861	−2·348	−1·926	−1·471	1·060	1·328	1·545	1·779	1·927
300	−3·172	−2·816	−2·316	−1·906	−1·456	1·080	1·357	1·528	1·826	1·983
350	−3·129	−2·781	−2·291	−1·888	−1·444	1·096	1·379	1·610	1·863	2·026
400	−3·094	−2·753	−2·270	−1·873	−1·434	1·108	1·396	1·633	1·893	2·061
450	−3·064	−2·729	−2·253	−1·861	−1·426	1·119	1·411	1·652	1·918	2·090
500	−3·040	−2·709	−2·239	−1·850	−1·419	1·127	1·423	1·668	1·938	2·114
550	−3·019	−2·691	−2·226	−1·841	−1·413	1·135	1·434	1·682	1·957	2·136
600	−3·000	−2·676	−2·215	−1·833	−1·408	1·141	1·443	1·694	1·972	2·154
650	−2·984	−2·663	−2·206	−1·826	−1·403	1·147	1·451	1·704	1·986	2·171
700	−2·969	−2·651	−2·197	−1·820	−1·399	1·152	1·458	1·714	1·999	2·185
750	−2·956	−2·640	−2·189	−1·814	−1·395	1·157	1·465	1·722	2·010	2·199
800	−2·944	−2·630	−2·182	−1·809	−1·392	1·161	1·471	1·730	2·020	2·211
850	−2·933	−2·621	−2·176	−1·804	−1·389	1·165	1·476	1·737	2·029	2·221
900	−2·923	−2·613	−2·170	−1·800	−1·386	1·168	1·481	1·743	2·037	2·231
950	−2·914	−2·605	−2·164	−1·796	−1·383	1·171	1·485	1·749	2·045	2·241
1000	−2·906	−2·599	−2·159	−1·792	−1·381	1·174	1·489	1·754	2·052	2·249

D10 ANDERSON–DARLING (A_n^2) TEST FOR NORMALITY

Let $x_{(1)} \le x_{(2)} \le \cdots \le x_{(n)}$ be an ordered random sample of n observations from a normal distribution. Percentage points for the statistic

$$A_n^2 = -\frac{1}{n}\left\{\sum_{i=1}^{n}(2i-1)\left[\log z_i + \log(1 - z_{n+1-i})\right]\right\} - n,$$

where $z_i = \Phi([x_{(i)} - \bar{x}]/s)$, $s^2 = \Sigma_i(x_{(i)} - \bar{x})^2/(n-1)$, and Φ is the distribution function for $N(0,1)$, can be obtained from the table below (reproduced from Pettitt [1977: Table 1]). Given $p = \text{pr}[A_n^2 \le a_n]$, a_n is calculated from the expression

$$a_n \approx a_\infty\left(1 + c_1 n^{-1} + c_2 n^{-2}\right).$$

The hypothesis of normality is rejected if A_n^2 is too large.

p	c_1	c_2	a_∞
·05	−·512	2·10	·1674
·10	−·552	1·25	·1938
·15	−·608	1·07	·2147
·20	−·643	·93	·2333
·25	−·707	1·03	·2509
·30	−·735	1·02	·2681
·35	−·772	1·04	·2853
·40	−·770	·90	·3030
·45	−·778	·80	·3213
·50	−·779	·67	·3405
·55	−·803	·70	·3612
·60	−·818	·58	·3836
·65	−·818	·42	·4085
·70	−·801	·12	·4367
·75	−·800	−·09	·4695
·80	−·756	−·39	·5091
·85	−·749	−·59	·5597
·90	−·750	−·80	·6305
·95	−·795	−·89	·7514
·975	−·881	−·94	·8728
·99	−1·013	−·93	1·0348
·995	−1·063	−1·34	1·1578

D11 DISCORDANCY TEST FOR SINGLE GAMMA OUTLIER

Below are critical values from Kimber [1979: Table 1, $n = 5(1)20$] and Barnett and Lewis [1978: Table I, $r = 0.5$, $n > 20$] for 5% and 1% discordancy tests of a single outlier in a sample $x_1, x_2, \ldots, x_n$ of size n from a gamma distribution (defined in Appendix D2). The test statistic is

$$Z = \max_{1 \leq i \leq n} \left(\frac{v_i}{n\bar{v}} \right),$$

where $v_i = -\log u_i - (n-1)\log\{(n - u_i)/(n-1)\}$ and $u_i = x_i/\bar{x}$. Z is significant if it exceeds the tabulated value. For further details see Section 4.5.

n	5% level	1% level
5	0.756	0.834
6	0.721	0.810
7	0.684	0.785
8	0.649	0.756
9	0.615	0.726
10	0.584	0.696
11	0.555	0.667
12	0.529	0.640
13	0.505	0.614
14	0.484	0.590
15	0.464	0.568
16	0.446	0.547
17	0.430	0.528
18	0.414	0.510
19	0.440	0.494
20	0.387	0.478
24	0.343	0.425
30	0.293	0.363
40	0.237	0.294
60	0.174	0.215
120	0.100	0.123
∞	0	0

n	$d=2$		$d=3$		$d=4$		$d=5$	
	5%	1%	5%	1%	5%	1%	5%	1%
5	3.17	3.19						
6	4.00	4.11	4.14	4.16				
7	4.71	4.95	5.01	5.10	5.12	5.14		
8	5.32	5.70	5.77	5.97	6.01	6.09	6.11	6.12
9	5.85	6.37	6.43	6.76	6.80	6.97	7.01	7.08
10	6.32	6.97	7.01	7.47	7.50	7.79	7.82	7.98
12	7.10	8.00	7.99	8.70	8.67	9.20	9.19	9.57
14	7.74	8.84	8.78	9.71	9.61	10.37	10.29	10.90
16	8.27	9.54	9.44	10.56	10.39	11.36	11.20	12.02
18	8.73	10.15	10.00	11.28	11.06	12.20	11.96	12.98
20	9.13	10.67	10.49	11.91	11.63	12.93	12.62	13.81
25	9.94	11.73	11.48	13.18	12.78	14.40	13.94	15.47
30	10.58	12.54	12.24	14.14	13.67	15.51	14.95	16.73
35	11.10	13.20	12.85	14.92	14.37	16.40	15.75	17.73
40	11.53	13.74	13.36	15.56	14.96	17.13	16.41	18.55
45	11.90	14.20	13.80	16.10	15.46	17.74	16.97	19.24
50	12.23	14.60	14.18	16.56	15.89	18.27	17.45	19.83
100	14.22	16.95	16.45	19.26	18.43	21.30	20.26	23.17
200	15.99	18.94	18.42	21.47	20.59	23.72	22.59	25.82
500	18.12	21.22	20.75	23.95	23.06	26.37	25.21	28.62

D12 DISCORDANCY TEST FOR SINGLE MULTIVARIATE NORMAL OUTLIER

Above are critical values from Barnett and Lewis [1978: Table XXVIII] for 5% and 1% discordancy tests of a single outlier in a sample of size n from $N_d(\boldsymbol{\mu}, \boldsymbol{\Sigma})$; $\boldsymbol{\mu}$ and $\boldsymbol{\Sigma}$ are unknown. The test statistic is

$$D_{(n)}^2 = \max_{1 \leq i \leq n} (\mathbf{x}_i - \overline{\mathbf{x}})' \mathbf{S}^{-1} (\mathbf{x}_i - \overline{\mathbf{x}}),$$

where $\mathbf{S} = \sum_i (\mathbf{x}_i - \overline{\mathbf{x}})(\mathbf{x}_i - \overline{\mathbf{x}})'/(n-1)$. The table gives upper bounds only for the true critical values, so that a test is significant if $D_{(n)}^2$ exceeds the tabulated value.

D13 WILKS' LIKELIHOOD RATIO TEST

Let $U = |\mathbf{E}|/|\mathbf{E} + \mathbf{H}|$, where $\mathbf{H}$ and $\mathbf{E}$ have independent Wishart distributions $W_d(m_H, \boldsymbol{\Sigma})$ and $W_d(m_E, \boldsymbol{\Sigma})$, respectively. Then $U \sim U_{d, m_H, m_E}$ and

$$-\left(\frac{f}{C_\alpha}\right) \log_e U \sim \chi^2_{d m_H},$$

where

$$f = m_E - \tfrac{1}{2}(d - m_H + 1) = m_E + m_H - \tfrac{1}{2}(m_H + d + 1).$$

Given α and $\chi^2_{dm_H}(\alpha)$, where $\text{pr}[\chi^2_{dm_H} > \chi^2_{dm_H}(\alpha)] = \alpha$, the upper α quantile value for $-f \log U$ is $C_\alpha \chi^2_{dm_H}(\alpha)$. Values of C_α and $\chi^2_{dm_H}(\alpha)$ are given in the following tables for the parameters d, m_H ($d \leq m_H$, m_E), and

$$M = m_E - d + 1 = (m_E + m_H) - m_H - d + 1.$$

For $d > m_H$ we interchange d and m_H. Both f and M are unchanged, as the value of $n_0 = m_E + m_H$ is unchanged with regard to the distribution of U (now $U_{m_H, d, m_E + m_H - d}$). If the parameter values lie outside the tables, we can use the F-approximation (2.45). When $s = \min(d, m_H)$ is 1 or 2, we use the upper tail of an appropriate F-distribution [see (2.49) and (2.50)]. The following tables, pages 565 to 592, are taken from Schatzoff [1966a], Pillai and Gupta [1969], Lee [1972], and Davis [1979] by permission of the Biometrika Trustees.

D14 ROY'S MAXIMUM ROOT STATISTIC

Let $\theta_{\max}$ be the largest eigenvalue of $\mathbf{H}(\mathbf{E} + \mathbf{H})^{-1}$, where $\mathbf{H}$ and $\mathbf{E}$ have independent Wishart distributions $W_d(m_H, \boldsymbol{\Sigma})$ and $W_d(m_E, \boldsymbol{\Sigma})$, respectively. Define $s = \min(d, m_H)$, $\nu_1 = \tfrac{1}{2}(|d - m_H| - 1)$ and $\nu_2 = \tfrac{1}{2}(m_E - d - 1)$. Upper percentage points for $\theta_{\max}$ are given in the tables on pages 593 to 598 (from Davis, personal communication, and Pillai [1960]) for $s = 2(1)7$. For $\nu_2 > 50$, $240/\nu_2$ may be used as an argument in harmonic interpolation. For example, $240/\nu_2 = 5, 4, 3, 2, 1, 0$ for $\nu_2 = 48, 60, 80, 120, 240, \infty$. Entries for $s = 6(1)10$ and selected values of ν_2 are given in Pearson and Hartley [1972: Table 48] (see also Pillai [1964, 1965]); $s = 11, 12$ are in Pillai [1970]; and $s = 14, 16, 18, 20$ are in Pillai [1967]. Percentage points for odd values of s are obtained by interpolation. Charts are also available for $s = 2(1)5$ from Heck [1960], and more extensive percentage points for $s = 2, 3, 4$ are given by Pearson and Hartley [1972: Table 49].

D15 LAWLEY–HOTELLING TRACE STATISTIC

Let $T_g^2 = m_E \text{tr}[\mathbf{H}\mathbf{E}^{-1}]$, where $\mathbf{H}$ and $\mathbf{E}$ have independent Wishart distributions $W_d(m_H, \boldsymbol{\Sigma})$ and $W_d(m_E, \boldsymbol{\Sigma})$, respectively. Upper percentage points for

$$\frac{T_g^2}{m_H} = \left(\frac{m_E}{m_H}\right) \text{tr}[\mathbf{H}\mathbf{E}^{-1}]$$

are given in the tables on pages 599 to 607 (reprinted from Davis [1970a by permission of the Biometrika Trustees, 1970b, 1980a by courtesy of Marcel Dekker, Inc., and personal communication: $d = 2$]) for values of d, m_H, and m_E, with $d \leq m_H, m_E$. If $m_H < d$, we make the transformation $d \to m_H$, $m_H \to d$, $m_E \to m_E + m_H - d$, and enter these tables using the transformed values of the triple (d, m_H, m_E).

For other values of the parameters we have, approximately,

$$\frac{\operatorname{tr}[\mathbf{HE}^{-1}]}{c} \sim F_{a,b},$$

where a, b, and c are defined in Section 2.5.3a. We use the upper tail values of this F-distribution.

The statistic T_g^2 is also known as Hotelling's generalized T_0^2 statistic.

D16 PILLAI'S TRACE STATISTIC

Let $V^{(s)} = \operatorname{tr}[\mathbf{H}(\mathbf{E} + \mathbf{H})^{-1}]$, where $\mathbf{H}$ and $\mathbf{E}$ have independent Wishart distributions $W_d(m_H, \mathbf{\Sigma})$ and $W_d(m_E, \mathbf{\Sigma})$, respectively. Define $s = \min(d, m_H)$, $\nu_1 = \frac{1}{2}(|d - m_H| - 1)$, and $\nu_2 = \frac{1}{2}(m_E - d - 1)$. Upper percentage points are given in the tables on pages 608 to 611 (from Schuurmann et al. [1975]) for values of s, ν_1, and ν_2.

For other parameter values we have, approximately,

$$\frac{2\nu_2 + s + 1}{2\nu_1 + s + 1} \frac{V^{(s)}}{s - V^{(s)}} \sim F_{s(2\nu_1 + s + 1),\, s(2\nu_2 + s + 1)}.$$

D17 TEST FOR MUTUAL INDEPENDENCE

Upper 5% and 1% points for $-m \log \Lambda$, the likelihood ratio test statistic for testing that the dispersion matrix of a d-dimensional multivariate normal distribution is diagonal, are given on page 612 (reproduced from Mathai and Katiyar [1979a] by permission of the Biometrika Trustees). Here n is the sample size, $m = n - (2d + 11)/6$, and Λ is given by (3.51). As $n \to \infty$, $-m \log \Lambda \to \chi_\nu^2$, where $\nu = \frac{1}{2}d(d-1)$. Percentage points for this limiting chi-square distribution are also tabulated for comparison.

D18 TEST FOR EQUAL DISPERSION MATRICES WITH EQUAL SAMPLE SIZES

A modified likelihood ratio test statistic for testing the equality of I dispersion matrices is $-2 \log M$, where M is given by (9.15). Assuming multivariate normality and equal sample sizes ($n_1 = n_2 = \cdots = n_I = n_0$), upper 5% points for the test statistic are given on pages 613 and 614 (from J. C. Lee et al. [1977]). Here d is the number of variables and $f_0 = n_0 - 1$. Chi-square and F-approximations are given by (9.16) and (9.17).

$d = 3$

	$m_H = 3$					$m_H = 4$				
$M \backslash \alpha$	0·100	0·050	0·025	0·010	0·005	0·100	0·050	0·025	0·010	0·005
1	1·322	1·359	1·394	1·437	1·468	1·379	1·422	1·463	1·514	1·550
2	1·127	1·140	1·153	1·168	1·179	1·159	1·174	1·188	1·207	1·220
3	1·071	1·077	1·084	1·092	1·098	1·091	1·099	1·107	1·116	1·123
4	1·045	1·049	1·053	1·058	1·062	1·060	1·065	1·070	1·076	1·080
5	1·032	1·035	1·037	1·041	1·043	1·043	1·046	1·050	1·054	1·057
6	1·023	1·026	1·028	1·030	1·032	1·032	1·035	1·037	1·040	1·042
7	1·018	1·020	1·021	1·023	1·025	1·025	1·027	1·029	1·031	1·033
8	1·014	1·016	1·017	1·018	1·019	1·020	1·022	1·023	1·025	1·026
9	1·012	1·013	1·014	1·015	1·016	1·017	1·018	1·019	1·021	1·022
10	1·010	1·011	1·011	1·012	1·013	1·014	1·015	1·016	1·017	1·018
12	1·007	1·008	1·008	1·009	1·009	1·010	1·011	1·012	1·012	1·013
14	1·005	1·006	1·006	1·007	1·007	1·008	1·008	1·009	1·010	1·010
16	1·004	1·005	1·005	1·005	1·006	1·006	1·007	1·007	1·007	1·008
18	1·003	1·004	1·004	1·004	1·005	1·005	1·005	1·006	1·006	1·006
20	1·003	1·003	1·003	1·004	1·004	1·004	1·004	1·005	1·005	1·005
24	1·002	1·002	1·002	1·002	1·003	1·003	1·003	1·003	1·004	1·004
30	1·001	1·001	1·001	1·002	1·002	1·002	1·002	1·002	1·002	1·002
40	1·001	1·001	1·001	1·001	1·001	1·001	1·001	1·001	1·001	1·001
60	1·000	1·000	1·000	1·000	1·000	1·000	1·001	1·001	1·001	1·001
120	1·000	1·000	1·000	1·000	1·000	1·000	1·000	1·000	1·000	1·000
∞	1·000	1·000	1·000	1·000	1·000	1·000	1·000	1·000	1·000	1·000
$\chi^2_{dm_H}$	14·6837	16·9190	19·0228	21·6660	23·5894	18·5494	21·0261	23·3367	26·2170	28·2995

	$m_H = 5$					$m_H = 6$				
$M \backslash \alpha$	0·100	0·050	0·025	0·010	0·005	0·100	0·050	0·025	0·010	0·005
1	1·433	1·481	1·527	1·584	1·025	1·482	1·535	1·586	1·649	1·694
2	1·191	1·208	1·224	1·245	1·260	1·222	1·241	1·259	1·282	1·298
3	1·113	1·122	1·131	1·142	1·150	1·135	1·145	1·155	1·167	1·176
4	1·076	1·082	1·087	1·094	1·099	1·092	1·099	1·105	1·113	1·119
5	1·055	1·059	1·063	1·068	1·071	1·068	1·072	1·077	1·082	1·086
6	1·042	1·045	1·048	1·051	1·054	1·052	1·056	1·059	1·063	1·066
7	1·033	1·035	1·038	1·040	1·042	1·041	1·044	1·047	1·050	1·052
8	1·027	1·029	1·030	1·033	1·034	1·034	1·036	1·038	1·041	1·042
9	1·022	1·024	1·025	1·027	1·028	1·028	1·030	1·032	1·034	1·035
10	1·019	1·020	1·021	1·023	1·024	1·024	1·025	1·027	1·028	1·030
12	1·014	1·015	1·015	1·017	1·017	1·018	1·019	1·020	1·021	1·022
14	1·010	1·011	1·012	1·013	1·013	1·014	1·014	1·015	1·016	1·017
16	1·008	1·009	1·009	1·010	1·011	1·011	1·012	1·012	1·013	1·014
18	1·007	1·007	1·008	1·008	1·009	1·009	1·009	1·010	1·011	1·011
20	1 006	1·006	1·006	1·007	1·007	1·007	1·008	1·008	1·009	1·009
24	1·004	1·004	1·005	1·005	1·005	1·005	1·006	1·006	1·006	1·007
30	1·003	1·003	1·003	1·003	1·003	1·004	1·004	1·004	1·004	1·004
40	1·002	1·002	1·002	1·002	1·002	1·002	1·002	1·002	1·002	1·003
60	1·001	1·001	1·001	1·001	1·001	1·001	1·001	1·001	1·001	1·001
120	1·000	1·000	1·000	1·000	1·000	1·000	1·000	1·000	1·000	1·000
∞	1·000	1·000	1·000	1·000	1·000	1·000	1·000	1·000	1·000	1·000
$\chi^2_{dm_H}$	22·3071	24·9958	27·4884	30·5779	32·8013	25·9894	28·8693	31·5264	34·8053	37·1564

D13 (Continued)

$d = 3$

$M \backslash \alpha$	$m_H = 7$ 0·100	0·050	0·025	0·010	0·005
1	1·529	1·585	1·640	1·708	1·758
2	1·251	1·272	1·292	1·317	1·335
3	1·156	1·168	1·178	1·192	1·202
4	1·109	1·116	1·123	1·132	1·138
5	1·081	1·086	1·091	1·097	1·102
6	1·063	1·067	1·070	1·075	1·078
7	1·050	1·053	1·056	1·060	1·062
8	1·041	1·044	1·046	1·049	1·051
9	1·034	1·037	1·038	1·041	1·043
10	1·029	1·031	1·033	1·035	1·036
12	1·022	1·023	1·024	1·026	1·027
14	1·017	1·018	1·019	1·020	1·021
16	1·014	1·014	1·015	1·016	1·017
18	1·011	1·012	1·012	1·013	1·014
20	1·009	1·010	1·010	1·011	1·011
24	1·007	1·007	1·008	1·008	1·008
30	1·005	1·005	1·005	1·005	1·006
40	1·003	1·003	1·003	1·003	1·003
60	1·001	1·001	1·001	1·001	1·002
120	1·000	1·000	1·000	1·000	1·000
∞	1·000	1·000	1·000	1·000	1·000
$\chi^2_{dm_H}$	29·6151	32·6706	35·4789	38·9322	41·4011

$M \backslash \alpha$	$m_H = 8$ 0·100	0·050	0·025	0·010	0·005
1	1·572	1·632	1·690	1·763	1·816
2	1·280	1·302	1·324	1·350	1·370
3	1·177	1·190	1·201	1·216	1·227
4	1·125	1·133	1·141	1·150	1·157
5	1·094	1·100	1·105	1·112	1·117
6	1·073	1·078	1·082	1·087	1·091
7	1·059	1·063	1·066	1·070	1·073
8	1·049	1·052	1·054	1·058	1·060
9	1·041	1·043	1·046	1·048	1·050
10	1·035	1·037	1·039	1·041	1·043
12	1·026	1·028	1·029	1·031	1·032
14	1·021	1·022	1·023	1·024	1·025
16	1·017	1·018	1·018	1·019	1·020
18	1·014	1·014	1·015	1·016	1·017
20	1·011	1·012	1·013	1·013	1·014
24	1·008	1·009	1·009	1·010	1·010
30	1·006	1·006	1·006	1·007	1·007
40	1·003	1·004	1·004	1·004	1·004
60	1·002	1·002	1·002	1·002	1·002
120	1·000	1·000	1·000	1·000	1·001
∞	1·000	1·000	1·000	1·000	1·000
$\chi^2_{dm_H}$	33·1963	36·4151	39·3641	42·9798	45·5585

$M \backslash \alpha$	$m_H = 9$ 0·100	0·050	0·025	0·010	0·005
1	1·612	1·676	1·737	1·814	1·871
2	1·307	1·331	1·354	1·382	1·403
3	1·198	1·211	1·224	1·240	1·251
4	1·141	1·150	1·158	1·169	1·176
5	1·107	1·113	1·119	1·127	1·132
6	1·084	1·089	1·094	1·099	1·103
7	1·068	1·072	1·076	1·080	1·084
8	1·057	1·060	1·063	1·067	1·069
9	1·048	1·051	1·053	1·056	1·058
10	1·041	1·043	1·045	1·048	1·050
12	1·031	1·033	1·034	1·036	1·038
14	1·025	1·026	1·027	1·028	1·030
16	1·020	1·021	1·022	1·023	1·024
18	1·016	1·017	1·018	1·019	1·020
20	1·014	1·014	1·015	1·016	1·016
24	1·010	1·011	1·011	1·012	1·012
30	1·007	1·007	1·008	1·008	1·008
40	1·004	1·004	1·004	1·005	1·005
60	1·002	1·002	1·002	1·002	1·002
120	1·001	1·001	1·001	1·001	1·001
∞	1·000	1·000	1·000	1·000	1·000
$\chi^2_{dm_H}$	36·7412	40·1133	43·1945	46·9629	49·6449

$M \backslash \alpha$	$m_H = 10$ 0·100	0·050	0·025	0·010	0·005
1	1·650	1·716	1·781	1·862	1·921
2	1·333	1·359	1·383	1·413	1·435
3	1·218	1·232	1·245	1·262	1·274
4	1·157	1·167	1·175	1·187	1·195
5	1·120	1·127	1·133	1·141	1·147
6	1·095	1·101	1·106	1·112	1·116
7	1·078	1·082	1·086	1·091	1·094
8	1·065	1·068	1·072	1·075	1·078
9	1·055	1·058	1·061	1·064	1·066
10	1·047	1·050	1·052	1·055	1·057
12	1·036	1·038	1·040	1·042	1·043
14	1·029	1·030	1·031	1·033	1·034
16	1·023	1·024	1·025	1·027	1·028
18	1·019	1·020	1·021	1·022	1·023
20	1·016	1·017	1·018	1·019	1·019
24	1·012	1·012	1·013	1·014	1·014
30	1·008	1·009	1·009	1·009	1·010
40	1·005	1·005	1·005	1·006	1·006
60	1·002	1·002	1·003	1·003	1·003
120	1·001	1·001	1·001	1·001	1·001
∞	1·000	1·000	1·000	1·000	1·000
$\chi^2_{dm_H}$	40·2560	43·7730	46·9792	50·8922	53·6720

$d = 3$

$m_H = 11$

M \ α	0·100	0·050	0·025	0·010	0·005
1	1·685	1·754	1·821	1·907	1·969
2	1·358	1·385	1·410	1·442	1·466
3	1·237	1·252	1·266	1·284	1·297
4	1·173	1·183	1·192	1·204	1·213
5	1·133	1·140	1·147	1·156	1·162
6	1·106	1·112	1·117	1·124	1·128
7	1·087	1·092	1·096	1·101	1·105
8	1·073	1·077	1·080	1·084	1·087
9	1·062	1·065	1·068	1·072	1·074
10	1·054	1·056	1·059	1·062	1·064
12	1·041	1·043	1·045	1·047	1·049
14	1·033	1·034	1·036	1·037	1·039
16	1·027	1·028	1·029	1·030	1·031
18	1·022	1·023	1·024	1·025	1·026
20	1·019	1·020	1·020	1·021	1·022
24	1·014	1·014	1·015	1·016	1·016
30	1·009	1·010	1·010	1·011	1·011
40	1·006	1·006	1·006	1·007	1·007
60	1·003	1·003	1·003	1·003	1·003
120	1·001	1·001	1·001	1·001	1·001
∞	1·000	1·000	1·000	1·000	1·000
$\chi^2_{dm_H}$	43·745	47·400	50·725	54·776	57·648

$m_H = 12$

M \ α	0·100	0·050	0·025	0·010	0·005
1	1·718	1·791	1·860	1·949	2·013
2	1·382	1·410	1·437	1·470	1·495
3	1·256	1·272	1·287	1·306	1·319
4	1·188	1·199	1·209	1·221	1·230
5	1·146	1·154	1·161	1·170	1·176
6	1·117	1·123	1·129	1·136	1·141
7	1·097	1·101	1·106	1·111	1·115
8	1·081	1·085	1·089	1·093	1·097
9	1·069	1·073	1·076	1·080	1·082
10	1·060	1·063	1·066	1·069	1·071
12	1·046	1·048	1·050	1·053	1·054
14	1·037	1·039	1·040	1·042	1·043
16	1·030	1·032	1·033	1·034	1·035
18	1·025	1·026	1·027	1·029	1·029
20	1·021	1·022	1·023	1·024	1·025
24	1·016	1·017	1·017	1·018	1·019
30	1·011	1·011	1·012	1·012	1·013
40	1·007	1·007	1·007	1·008	1·008
60	1·003	1·003	1·004	1·004	1·004
120	1·001	1·001	1·001	1·001	1·001
∞	1·000	1·000	1·000	1·000	1·000
$\chi^2_{dm_H}$	47·2122	50·9985	54·4373	58·6192	61·5812

$m_H = 13$

M \ α	0·100	0·050	0·025	0·010	0·005
1	1·750	1·824	1·896	1·988	2·055
2	1·405	1·434	1·462	1·497	1·522
3	1·274	1·291	1·306	1·326	1·340
4	1·203	1·214	1·225	1·238	1·247
5	1·158	1·167	1·174	1·184	1·191
6	1·128	1·134	1·140	1·148	1·153
7	1·106	1·111	1·116	1·122	1·126
8	1·089	1·094	1·098	1·102	1·106
9	1·076	1·080	1·083	1·088	1·090
10	1·066	1·069	1·072	1·076	1·078
12	1·052	1·054	1·056	1·059	1·061
14	1·041	1·043	1·045	1·047	1·048
16	1·034	1·035	1·037	1·038	1·040
18	1·028	1·029	1·031	1·032	1·033
20	1·024	1·025	1·026	1·027	1·028
24	1·018	1·019	1·019	1·020	1·021
30	1·012	1·013	1·013	1·014	1·014
40	1·008	1·008	1·008	1·009	1·009
60	1·004	1·004	1·004	1·004	1·004
120	1·001	1·001	1·001	1·001	1·001
∞	1·000	1·000	1·000	1·000	1·000
$\chi^2_{dm_H}$	50·660	54·572	58·120	62·428	65·476

$m_H = 14$

M \ α	0·100	0·050	0·025	0·010	0·005
1	1·780	1·857	1·931	2·026	2·095
2	1·427	1·458	1·486	1·523	1·549
3	1·292	1·309	1·326	1·346	1·361
4	1·217	1·229	1·240	1·254	1·264
5	1·171	1·179	1·188	1·198	1·205
6	1·138	1·145	1·152	1·159	1·165
7	1·115	1·121	1·126	1·132	1·136
8	1·097	1·102	1·106	1·111	1·115
9	1·084	1·088	1·091	1·095	1·099
10	1·073	1·076	1·079	1·082	1·085
12	1·057	1·059	1·061	1·064	1·066
14	1·046	1·048	1·049	1·052	1·053
16	1·037	1·039	1·041	1·042	1·044
18	1·031	1·033	1·034	1·035	1·036
20	1·027	1·028	1·029	1·030	1·031
24	1·020	1·021	1·022	1·023	1·023
30	1·014	1·015	1·015	1·016	1·016
40	1·009	1·009	1·009	1·010	1·010
60	1·004	1·004	1·005	1·005	1·005
120	1·001	1·001	1·001	1·001	1·001
∞	1·000	1·000	1·000	1·000	1·000
$\chi^2_{dm_H}$	54·0902	58·1240	61·7768	66·2062	69·3360

$d = 3$

$m_H = 15$

M \ α	0·100	0·050	0·050	0·010	0·005
1	1·808	1·887	1·964	2·061	2·133
2	1·449	1·480	1·510	1·547	1·575
3	1·309	1·327	1·344	1·365	1·381
4	1·232	1·244	1·256	1·270	1·280
5	1·183	1·192	1·200	1·211	1·218
6	1·149	1·156	1·163	1·171	1·177
7	1·124	1·130	1·135	1·142	1·147
8	1·105	1·110	1·115	1·120	1·124
9	1·091	1·095	1·099	1·103	1·107
10	1·079	1·083	1·086	1·090	1·093
12	1·062	1·065	1·067	1·070	1·072
14	1·050	1·052	1·054	1·056	1·058
16	1·041	1·043	1·045	1·047	1·048
18	1·035	1·036	1·037	1·039	1·040
20	1·030	1·031	1·032	1·033	1·034
24	1·022	1·023	1·024	1·025	1·026
30	1·016	1·016	1·017	1·017	0·018
40	1·010	1·010	1·010	1·011	1·011
60	1·005	1·005	1·005	1·005	1·005
120	1·001	1·001	1·001	1·001	1·002
∞	1·000	1·000	1·000	1·000	1·000
$\chi^2_{dm_H}$	57·505	61·656	65·410	69·957	73·166

$m_H = 16$

M \ α	0·100	0·050	0·025	0·010	0·005
1	1·835	1·916	1·995	2·095	2·169
2	1·469	1·501	1·532	1·571	1·599
3	1·325	1·344	1·362	1·384	1·400
4	1·245	1·258	1·271	1·285	1·296
5	1·195	1·204	1·213	1·224	1·232
6	1·159	1·167	1·174	1·182	1·188
7	1·133	1·139	1·145	1·152	1·157
8	1·114	1·119	1·123	1·129	1·133
9	1·098	1·102	1·106	1·111	1·115
10	1·085	1·089	1·092	1·097	1·099
12	1·067	1·070	1·073	1·076	1·078
14	1·054	1·057	1·059	1·061	1·063
16	1·045	1·047	1·049	1·051	1·052
18	1·038	1·039	1·041	1·043	1·044
20	1·032	1·034	1·035	1·036	1·037
24	1·025	1·026	1·026	1·027	1·028
30	1·017	1·018	1·018	1·019	1·020
40	1·011	1·011	1·011	1·012	1·012
60	1·005	1·006	1·006	1·006	1·006
120	1·002	1·002	1·002	1·002	1·002
∞	1·000	1·000	1·000	1·000	1·000
$\chi^2_{dm_H}$	60·9066	65·1708	69·0226	73·6826	76·9688

$m_H = 17$

M \ α	0·100	0·050	0·025	0·010	0·005
1	1·861	1·944	2·025	2·127	2·203
2	1·489	1·522	1·554	1·594	1·623
3	1·341	1·361	1·379	1·402	1·419
4	1·259	1·273	1·285	1·300	1·312
5	1·206	1·216	1·225	1·237	1·245
6	1·169	1·177	1·184	1·193	1·200
7	1·142	1·149	1·154	1·162	1·167
8	1·122	1·127	1·132	1·138	1·142
9	1·105	1·110	1·114	1·119	1·123
10	1·092	1·096	1·100	1·104	1·107
12	1·073	1·076	1·079	1·082	1·084
14	1·059	1·061	1·064	1·066	1·068
16	1·049	1·051	1·053	1·055	1·056
18	1·041	1·043	1·044	1·046	1·047
20	1·035	1·037	1·038	1·040	1·041
24	1·027	1·028	1·029	1·030	1·031
30	1·019	1·020	1·020	1·021	1·022
40	1·012	1·012	1·013	1·013	1·013
60	1·006	1·006	1·006	1·006	1·007
120	1·002	1·002	1·002	1·002	1·002
∞	1·000	1·000	1·000	1·000	1·000
$\chi^2_{dm_H}$	64·295	68·669	72·616	77·386	80·747

$m_H = 18$

M \ α	0·100	0·050	0·025	0·010	0·005
1	1·886	1·971	2·053	2·158	2·235
2	1·508	1·542	1·575	1·616	1·646
3	1·357	1·377	1·396	1·420	1·437
4	1·272	1·286	1·299	1·315	1·327
5	1·218	1·228	1·238	1·249	1·258
6	1·179	1·188	1·195	1·204	1·211
7	1·151	1·158	1·164	1·171	1·177
8	1·129	1·135	1·140	1·146	1·151
9	1·112	1·117	1·121	1·127	1·130
10	1·099	1·103	1·107	1·111	1·114
12	1·078	1·081	1·084	1·087	1·090
14	1·064	1·066	1·068	1·071	1·073
16	1·053	1·055	1·057	1·059	1·061
18	1·045	1·046	1·048	1·050	1·051
20	1·038	1·040	1·041	1·043	1·044
24	1·029	1·030	1·031	1·032	1·033
30	1·021	1·021	1·022	1·023	1·023
40	1·013	1·013	1·014	1·014	1·015
60	1·006	1·007	1·007	1·007	1·007
120	1·002	1·002	1·002	1·002	1·002
∞	1·000	1·000	1·000	1·000	1·000
$\chi^2_{dm_H}$	67·6728	72·1532	76·1920	81·0688	84·5019

$d = 3$

$m_H = 19$

$M \backslash \alpha$	0·100	0·050	0·025	0·010	0·005
1	1·909	1·996	2·080	2·188	2·267
2	1·526	1·561	1·595	1·637	1·668
3	1·372	1·393	1·412	1·437	1·454
4	1·285	1·300	1·313	1·330	1·341
5	1·229	1·240	1·250	1·262	1·271
6	1·189	1·198	1·205	1·215	1·222
7	1·160	1·167	1·173	1·181	1·186
8	1·137	1·143	1·148	1·155	1·159
9	1·119	1·124	1·129	1·134	1·138
10	1·105	1·109	1·113	1·118	1·121
12	1·084	1·087	1·090	1·093	1·096
14	1·068	1·071	1·073	1·076	1·078
16	1·057	1·059	1·061	1·063	1·065
18	1·048	1·050	1·052	1·054	1·055
20	1·041	1·043	1·044	1·046	1·047
24	1·032	1·033	1·034	1·035	1·036
30	1·022	1·023	1·024	1·025	1·025
40	1·014	1·015	1·015	1·016	1·016
60	1·007	1·007	1·008	1·008	1·008
120	1·002	1·002	1·002	1·002	1·002
∞	1·000	1·000	1·000	1·000	1·000
$\chi^2_{dm_H}$	71·040	75·624	79·752	84·733	88·236

$m_H = 20$

$M \backslash \alpha$	0·100	0·050	0·025	0·010	0·005
1	1·932	2·021	2·106	2·216	2·297
2	1·544	1·580	1·614	1·657	1·689
3	1·387	1·408	1·428	1·453	1·472
4	1·298	1·313	1·327	1·344	1·356
5	1·240	1·251	1·261	1·274	1·283
6	1·199	1·208	1·216	1·226	1·233
7	1·168	1·176	1·182	1·190	1·196
8	1·145	1·151	1·157	1·163	1·168
9	1·127	1·132	1·136	1·142	1·146
10	1·112	1·116	1·120	1·125	1·128
12	1·089	1·092	1·095	1·099	1·102
14	1·073	1·075	1·078	1·081	1·083
16	1·061	1·063	1·065	1·067	1·069
18	1·052	1·053	1·055	1·057	1·059
20	1·044	1·046	1·048	1·049	1·050
24	1·034	1·035	1·036	1·038	1·039
30	1·024	1·025	1·026	1·027	1·027
40	1·015	1·016	1·016	1·017	1·017
60	1·008	1·008	1·008	1·009	1·009
120	1·002	1·002	1·002	1·002	1·003
∞	1·000	1·000	1·000	1·000	1·000
$\chi^2_{dm_H}$	74·3970	79·0819	83·2976	88·3794	91·9517

$m_H = 21$

$M \backslash \alpha$	0·100	0·050	0·025	0·010	0·005
1	1·954	2·044	2·131	2·243	2·325
2	1·561	1·598	1·633	1·677	1·709
3	1·401	1·423	1·444	1·470	1·488
4	1·310	1·325	1·340	1·357	1·370
5	1·250	1·262	1·273	1·286	1·295
6	1·208	1·217	1·226	1·236	1·243
7	1·177	1·184	1·191	1·200	1·205
8	1·153	1·159	1·165	1·172	1·176
9	1·133	1·139	1·144	1·150	1·154
10	1·118	1·122	1·127	1·132	1·135
12	1·094	1·098	1·101	1·105	1·108
14	1·077	1·080	1·083	1·086	1·088
16	1·065	1·067	1·069	1·072	1·074
18	1·055	1·057	1·059	1·061	1·063
20	1·048	1·049	1·051	1·053	1·054
24	1·036	1·038	1·039	1·040	1·041
30	1·026	1·027	1·028	1·029	1·029
40	1·016	1·017	1·018	1·018	1·019
60	1·008	1·009	1·009	1·009	1·009
120	1·002	1·002	1·003	1·003	1·003
∞	1·000	1·000	1·000	1·000	1·000
$\chi^2_{dm_H}$	77·745	82·529	86·830	92·010	95·649

$m_H = 22$

$M \backslash \alpha$	0·100	0·050	0·025	0·010	0·005
1	1·975	2·067	2·156	2·269	2·353
2	1·578	1·616	1·651	1·696	1·729
3	1·415	1·438	1·459	1·485	1·504
4	1·322	1·338	1·353	1·371	1·384
5	1·261	1·273	1·284	1·297	1·307
6	1·218	1·227	1·236	1·246	1·254
7	1·185	1·193	1·200	1·209	1·215
8	1·160	1·167	1·173	1·180	1·185
9	1·142	1·147	1·151	1·157	1·161
10	1·124	1·129	1·133	1·139	1·141
12	1·099	1·103	1·106	1·110	1·113
14	1·082	1·085	1·087	1·091	1·093
16	1·069	1·071	1·073	1·076	1·078
18	1·059	1·061	1·063	1·065	1·066
20	1·051	1·052	1·054	1·056	1·057
24	1·039	1·040	1·041	1·043	1·044
30	1·028	1·029	1·030	1·031	1·031
40	1·018	1·018	1·019	1·020	1·020
60	1·009	1·009	1·010	1·010	1·010
120	1·003	1·003	1·003	1·003	1·003
∞	1·000	1·000	1·000	1·000	1·000
$\chi^2_{dm_H}$	81·0855	85·9649	90·3489	95·6257	99·3304

$d = 4$

M \ α	$m_H = 4$: 0·100	0·050	0·025	0·010	0·005	$m_H = 5$: 0·100	0·050	0·025	0·010	0·005
1	1·405	1·451	1·494	1·550	1·589	1·435	1·483	1·530	1·589	1·632
2	1·178	1·194	1·209	1·229	1·243	1·199	1·216	1·233	1·253	1·269
3	1·105	1·114	1·122	1·132	1·139	1·121	1·130	1·139	1·150	1·158
4	1·071	1·076	1·081	1·088	1·092	1·083	1·089	1·094	1·101	1·106
5	1·051	1·055	1·058	1·063	1·066	1·061	1·065	1·069	1·074	1·077
6	1·039	1·042	1·044	1·048	1·050	1·047	1·050	1·053	1·056	1·059
7	1·031	1·033	1·035	1·037	1·039	1·037	1·040	1·042	1·044	1·046
8	1·025	1·027	1·028	1·030	1·032	1·030	1·032	1·034	1·036	1·038
9	1·020	1·022	1·023	1·025	1·026	1·025	1·027	1·028	1·030	1·031
10	1·017	1·018	1·019	1·021	1·022	1·021	1·023	1·024	1·025	1·026
12	1·013	1·014	1·014	1·015	1·016	1·016	1·017	1·018	1·019	1·020
14	1·010	1·010	1·011	1·012	1·012	1·012	1·013	1·014	1·014	1·015
16	1·008	1·008	1·009	1·009	1·010	1·010	1·010	1·011	1·012	1·012
18	1·006	1·007	1·007	1·008	1·008	1·008	1·008	1·009	1·009	1·010
20	1·005	1·006	1·006	1·006	1·007	1·007	1·007	1·007	1·008	1·008
24	1·004	1·004	1·004	1·004	1·005	1·005	1·005	1·005	1·006	1·006
30	1·002	1·003	1·003	1·003	1·003	1·003	1·003	1·004	1·004	1·004
40	1·001	1·002	1·002	1·002	1·002	1·002	1·002	1·002	1·002	1·002
60	1·001	1·001	1·001	1·001	1·001	1·001	1·001	1·001	1·001	1·001
120	1·000	1·000	1·000	1·000	1·000	1·000	1·000	1·000	1·000	1·000
∞	1·000	1·000	1·000	1·000	1·000	1·000	1·000	1·000	1·000	1·000
$\chi^2_{dm_H}$	23·5418	26·2962	28·8454	31·9999	34·2672	28·4120	31·4104	34·1696	37·5662	39·9968

M \ α	$m_H = 6$: 0·100	0·050	0·025	0·010	0·005	$m_H = 7$: 0·100	0·050	0·025	0·010	0·005
1	1·466	1·517	1·566	1·628	1·674	1·497	1·550	1·601	1·667	1·715
2	1·222	1·240	1·257	1·279	1·295	1·244	1·263	1·281	1·305	1·322
3	1·138	1·148	1·157	1·168	1·177	1·155	1·165	1·175	1·188	1·197
4	1·096	1·102	1·108	1·115	1·121	1·109	1·116	1·122	1·130	1·136
5	1·071	1·076	1·080	1·085	1·089	1·082	1·087	1·092	1·097	1·101
6	1·055	1·059	1·062	1·066	1·068	1·064	1·068	1·071	1·076	1·079
7	1·044	1·047	1·049	1·052	1·055	1·052	1·055	1·057	1·061	1·063
8	1·036	1·038	1·040	1·043	1·045	1·043	1·045	1·047	1·050	1·052
9	1·030	1·032	1·034	1·036	1·037	1·036	1·038	1·040	1·042	1·044
10	1·026	1·027	1·029	1·030	1·032	1·031	1·032	1·034	1·036	1·037
12	1·019	1·020	1·021	1·023	1·024	1·023	1·024	1·026	1·027	1·028
14	1·015	1·016	1·017	1·018	1·018	1·018	1·019	1·020	1·021	1·022
16	1·012	1·013	1·013	1·014	1·015	1·015	1·015	1·016	1·017	1·017
18	1·010	1·010	1·011	1·012	1·012	1·012	1·013	1·013	1·014	1·014
20	1·008	1·009	1·009	1·010	1·010	1·010	1·011	1·011	1·012	1·012
24	1·006	1·006	1·007	1·007	1·007	1·007	1·008	1·008	1·008	1·009
30	1·004	1·004	1·004	1·005	1·005	1·005	1·005	1·005	1·006	1·006
40	1·002	1·002	1·003	1·003	1·003	1·003	1·003	1·003	1·003	1·004
60	1·001	1·001	1·001	1·001	1·001	1·001	1·001	1·002	1·002	1·002
120	1·000	1·000	1·000	1·000	1·000	1·000	1·000	1·000	1·000	1·000
∞	1·000	1·000	1·000	1·000	1·000	1·000	1·000	1·000	1·000	1·000
$\chi^2_{dm_H}$	33·1963	36·4151	39·3641	42·9798	45·5585	37·9159	41·3372	44·4607	48·2782	50·9933

$d = 4$

$M \backslash \alpha$	$m_H = 8$ 0·100	0·050	0·025	0·010	0·005	$m_H = 9$ 0·100	0·050	0·025	0·010	0·005
1	1·528	1·583	1·636	1·704	1·754	1·557	1·614	1·669	1·740	1·792
2	1·266	1·286	1·305	1·330	1·348	1·288	1·309	1·329	1·355	1·373
3	1·172	1·183	1·193	1·207	1·216	1·189	1·201	1·212	1·226	1·236
4	1·123	1·130	1·137	1·146	1·152	1·137	1·144	1·152	1·161	1·167
5	1·093	1·099	1·103	1·109	1·114	1·105	1·110	1·115	1·122	1·127
6	1·074	1·078	1·081	1·086	1·089	1·083	1·088	1·091	1·096	1·100
7	1·060	1·063	1·066	1·070	1·072	1·068	1·071	1·075	1·078	1·081
8	1·050	1·052	1·055	1·058	1·060	1·057	1·060	1·062	1·065	1·068
9	1·042	1·044	1·046	1·048	1·050	1·048	1·050	1·053	1·055	1·057
10	1·036	1·038	1·039	1·041	1·043	1·041	1·043	1·045	1·047	1·049
12	1·027	1·029	1·030	1·031	1·033	1·032	1·033	1·034	1·036	1·037
14	1·021	1·023	1·023	1·025	1·026	1·025	1·026	1·027	1·029	1·029
16	1·017	1·018	1·019	1·020	1·021	1·020	1·021	1·022	1·023	1·024
18	1·014	1·015	1·016	1·016	1·017	1·017	1·018	1·018	1·019	1·020
20	1·012	1 013	1·013	1·014	1·014	1·014	1·015	1·015	1·016	1·017
24	1·009	1·009	1·010	1·010	1·010	1·010	1·011	1·011	1·012	1·012
30	1·006	1·006	1·007	1·007	1·007	1·007	1·007	1·008	1·008	1·008
40	1·003	1·004	1·004	1·004	1·004	1·004	1·004	1·005	1·005	1·005
60	1·002	1·002	1·002	1·002	1·002	1·002	1·002	1·002	1·002	1·002
120	1·000	1·000	1·000	1·001	1·001	1·001	1·001	1·001	1·001	1·001
∞	1·000	1·000	1·000	1·000	1·000	1·000	1·000	1·000	1·000	1·000
$\chi^2_{dm_H}$	42·5847	46·1943	49·4804	53·4858	56·3281	47·2122	50·9985	54·4373	58·6192	61·5812

$M \backslash \alpha$	$m_H = 10$ 0·100	0·050	0·025	0·010	0·005	$m_H = 11$ 0·100	0·050	0·025	0·010	0·005
1	1·585	1·644	1·701	1·774	1·828	—	—	—	—	—
2	1·309	1·331	1·352	1·379	1·398	1·330	1·352	1·374	1·402	1·422
3	1·206	1·218	1·230	1·244	1·255	1·222	1·235	1·247	1·262	1·274
4	1·150	1·159	1·166	1·176	1·183	1·164	1·173	1·181	1·191	1·198
5	1·116	1·122	1·128	1·134	1·139	1·127	1·134	1·140	1·147	1·152
6	1·093	1·097	1·102	1·107	1·111	1·103	1·107	1·112	1·118	1·122
7	1·076	1·080	1·083	1·088	1·090	1·085	1·089	1·092	1·097	1·100
8	1·064	1·067	1·070	1·073	1·076	1·071	1·075	1·077	1·081	1·084
9	1·054	1·057	1·059	1·062	1·064	1·061	1·064	1·066	1·069	1·071
10	1·047	1·049	1·051	1·054	1·055	1·053	1·055	1·057	1·060	1·062
12	1·036	1·038	1·039	1·041	1·042	1·041	1·043	1·044	1·046	1·047
14	1·029	1·030	1·031	1·033	1·034	1·033	1·034	1·035	1·037	1·038
16	1·023	1·024	1·025	1·026	1·027	1·027	1·028	1·029	1·030	1·031
18	1·019	1·020	1·021	1·022	1·023	1·022	1·023	1·024	1·025	1·026
20	1·016	1·017	1·018	1·019	1·019	1·019	1·020	1·020	1·021	1·022
24	1·012	1·013	1·013	1·014	1·014	1·014	1·015	1·015	1·016	1·016
30	1·008	1·009	1·009	1·009	1·010	1·010	1·010	1·010	1·011	1·011
40	1·005	1·005	1·005	1·006	1·006	1·006	1·006	1·006	1·007	1·007
60	1·002	1·003	1·003	1·003	1·003	1·003	1·003	1·003	1·003	1·003
120	1·001	1·001	1·001	1·001	1·001	1·001	1·001	1·001	1·001	1·001
∞	1·000	1·000	1·000	1·000	1·000	1·000	1·000	1·000	1·000	1·000
$\chi^2_{dm_H}$	51·8050	55·7585	59·3417	63·6907	66·7659	56·369	60·481	64·201	68·710	71·893

$d = 4$

M \ α	$m_H = 12$ 0·100	0·050	0·025	0·010	0·005	$m_H = 13$ 0·100	0·050	0·025	0·010	0·005
1	1·638	1·700	1·760	1·838	1·895	1·369	1·393	1·417	1·446	1·468
2	1·350	1·373	1·396	1·424	1·446	1·254	1·268	1·281	1·298	1·310
3	1·238	1·252	1·264	1·280	1·292	1·190	1·200	1·209	1·220	1·228
4	1·177	1·186	1·195	1·205	1·213	1·150	1·157	1·163	1·171	1·177
5	1·139	1·145	1·152	1·159	1·165	1·122	1·127	1·132	1·139	1·143
6	1·112	1·118	1·122	1·128	1·132	1·102	1·106	1·110	1·115	1·118
7	1·093	1·097	1·101	1·106	1·109	1·086	1·090	1·093	1·097	1·100
8	1·079	1·082	1·085	1·089	1·092	1·074	1·077	1·080	1·083	1·086
9	1·068	1·070	1·073	1·076	1·079	1·065	1·067	1·070	1·073	1·075
10	1·059	1·061	1·063	1·066	1·068	1·050	1·052	1·054	1·056	1·058
12	1·046	1·047	1·049	1·051	1·053	1·041	1·042	1·044	1·045	1·047
14	1·037	1·038	1·039	1·041	1·042	1·033	1·035	1·036	1·037	1·038
16	1·030	1·031	1·032	1·033	1·034	1·028	1·029	1·030	1·031	1·032
18	1·025	1·026	1·027	1·028	1·029	1·024	1·025	1·026	1·027	1·027
20	1·021	1·022	1·023	1·024	1·024	1·018	1·019	1·019	1·020	1·020
24	1·016	1·017	1·017	1·018	1·018	1·012	1·013	1·013	1·014	1·014
30	1·011	1·011	1·012	1·012	1·013	1·008	1·008	1·008	1·008	1·009
40	1·007	1·007	1·007	1·008	1·008	1·004	1·004	1·004	1·004	1·004
60	1·003	1·003	1·004	1·004	1·004	1·001	1·001	1·001	1·001	1·001
120	1·001	1·001	1·001	1·001	1·001	1·000	1·000	1·000	1·000	1·000
∞	1·000	1·000	1·000	1·000	1·000	—	—	—	—	—
$\chi^2_{dm_H}$	60·9066	65·1708	69·0226	73·6826	76·9688	65·422	69·832	73·810	78·616	82·001

M \ α	$m_H = 14$ 0·100	0·050	0·025	0·010	0·005	$m_H = 15$ 0·100	0·050	0·025	0·010	0·005
1	1·686	1·751	1·814	1·896	1·956	1·406	1·432	1·456	1·488	1·511
2	1·388	1·413	1·436	1·467	1·489	1·284	1·299	1·313	1·331	1·344
3	1·269	1·284	1·297	1·314	1·327	1·216	1·226	1·236	1·248	1·256
4	1·203	1·213	1·222	1·234	1·242	1·172	1·179	1·187	1·195	1·202
5	1·161	1·168	1·175	1·183	1·189	1·141	1·147	1·153	1·159	1·164
6	1·131	1·137	1·142	1·149	1·154	1·118	1·123	1·128	1·133	1·137
7	1·110	1·115	1·119	1·124	1·128	1·101	1·105	1·109	1·113	1·116
8	1·094	1·097	1·101	1·105	1·109	1·087	1·091	1·094	1·098	1·101
9	1·081	1·084	1·087	1·091	1·093	1·077	1·080	1·082	1·085	1·088
10	1·071	1·073	1·076	1·079	1·081	1·060	1·063	1·065	1·067	1·069
12	1·055	1·058	1·059	1·062	1·064	1·049	1·051	1·052	1·054	1·056
14	1·045	1·046	1·048	1·050	1·051	1·040	1·042	1·043	1·045	1·046
16	1·037	1·038	1·039	1·041	1·042	1·034	1·035	1·036	1·038	1·039
18	1·031	1·032	1·033	1·034	1·035	1·029	1·030	1·031	1·032	1·033
20	1·026	1·027	1·028	1·029	1·030	1·022	1·023	1·023	1·024	1·025
24	1·020	1·021	1·021	1·022	1·023	1·015	1·016	1·016	1·017	1·017
30	1·014	1·014	1·015	1·015	1·016	1·010	1·010	1·010	1·011	1·011
40	1·009	1·009	1·009	1·009	1·010	1·005	1·005	1·005	1·005	1·005
60	1·004	1·004	1·005	1·005	1·005	1·001	1·001	1·001	1·001	1·001
120	1·001	1·001	1·001	1·001	1·001	1·000	1·000	1·000	1·000	1·000
∞	1·000	1·000	1·000	1·000	1·000	—	—	—	—	—
$\chi^2_{dm_H}$	69·9185	74·4683	78·5671	83·5134	86·9937	74·397	79·082	83·298	88·379	91·952

$d = 4$

M \ α	$m_H = 16$					$m_H = 17$				
	0·100	0·050	0·025	0·010	0·005	0·100	0·050	0·025	0·010	0·005
1	1·731	1·799	1·864	1·949	2·012	—	—	—	—	—
2	1·423	1·450	1·475	1·507	1·531	1·440	1·468	1·494	1·527	1·551
3	1·299	1·314	1·329	1·347	1·360	1·313	1·329	1·344	1·363	1·377
4	1·228	1·239	1·249	1·261	1·270	1·240	1·252	1·262	1·275	1·284
5	1·182	1·190	1·198	1·207	1·213	1·193	1·201	1·209	1·218	1·225
6	1·150	1·157	1·163	1·169	1·174	1·160	1·166	1·172	1·180	1·185
7	1·127	1·132	1·136	1·142	1·146	1·135	1·140	1·145	1·151	1·155
8	1·108	1·113	1·117	1·121	1·125	1·116	1·120	1·124	1·129	1·133
9	1·094	1·098	1·101	1·105	1·108	1·101	1·105	1·108	1·112	1·115
10	1·083	1·086	1·089	1·092	1·094	1·089	1·092	1·095	1·098	1·101
12	1·065	1·068	1·070	1·073	1·074	1·070	1·073	1·075	1·078	1·080
14	1·053	1·055	1·056	1·058	1·060	1·057	1·059	1·061	1·063	1·065
16	1·044	1·045	1·047	1·049	1·050	1·048	1·049	1·051	1·053	1·054
18	1·037	1·038	1·040	1·041	1·042	1·040	1·042	1·043	1·045	1·046
20	1·032	1·033	1·034	1·035	1·036	1·035	1·036	1·037	1·038	1·039
24	1·024	1·025	1·026	1·027	1·027	1·026	1·027	1·028	1·029	1·030
30	1·017	1·018	1·018	1·019	1·019	1·019	1·019	1·020	1·020	1·021
40	1·011	1·011	1·011	1·012	1·012	1·012	1·012	1·012	1·013	1·013
60	1·005	1·005	1·006	1·006	1·006	1·006	1·006	1·006	1·006	1·007
120	1·001	1·002	1·002	1·002	1·002	1·002	1·002	1·002	1·002	1·002
∞	1·000	1·000	1·000	1·000	1·000	1·000	1·000	1·000	1·000	1·000
$\chi^2_{dm_H}$	78·8597	83·6753	88·0040	93·2168	96·8781	83·308	88·250	92·689	98·028	101·776

M \ α	$m_H = 18$					$m_H = 19$				
	0·100	0·050	0·025	0·010	0·005	0·100	0·050	0·025	0·010	0·005
1	1·773	1·843	1·911	1·999	2·065	—	—	—	—	—
2	1·457	1·485	1·511	1·545	1·570	1·473	1·502	1·529	1·563	1·588
3	1·327	1·343	1·359	1·378	1·392	1·340	1·357	1·373	1·393	1·408
4	1·252	1·264	1·274	1·287	1·297	1·264	1·276	1·287	1·300	1·310
5	1·203	1·212	1·220	1·230	1·237	1·214	1·223	1·231	1·241	1·248
6	1·169	1·176	1·182	1·189	1·195	1·178	1·185	1·191	1·199	1·205
7	1·143	1·149	1·154	1·160	1·164	1·151	1·157	1·162	1·169	1·173
8	1·123	1·128	1·132	1·137	1·141	1·130	1·135	1·140	1·145	1·149
9	1·107	1·111	1·115	1·119	1·122	1·114	1·118	1·122	1·126	1·130
10	1·095	1·098	1·101	1·105	1·108	1·101	1·104	1·107	1·111	1·114
12	1·075	1·078	1·080	1·083	1·085	1·080	1·083	1·086	1·089	1·091
14	1·061	1·063	1·065	1·068	1·069	1·066	1·068	1·070	1·073	1·074
16	1·051	1·053	1·054	1·056	1·058	1·055	1·057	1·059	1·061	1·062
18	1·044	1·045	1·046	1·048	1·049	1·047	1·048	1·050	1·051	1·053
20	1·037	1·039	1·040	1·041	1·042	1·040	1·042	1·043	1·044	1·045
24	1·029	1·030	1·030	1·031	1·032	1·031	1·032	1·033	1·034	1·035
30	1·020	1·021	1·022	1·022	1·023	1·022	1·023	1·023	1·024	1·025
40	1·013	1·013	1·014	1·014	1·014	1·014	1·014	1·015	1·015	1·015
60	1·006	1·007	1·007	1·007	1·007	1·007	1·007	1·007	1·008	1·008
120	1·002	1·002	1·002	1·002	1·002	1·002	1·002	1·002	1·002	1·002
∞	1·000	1·000	1·000	1·000	1·000	1·000	1·000	1·000	1·000	1·000
$\chi^2_{dm_H}$	87·7431	92·8083	97·3531	102·816	106·648	92·166	97·351	101·999	107·583	111·495

$d = 4$

$M \backslash \alpha$	$m_H = 20$ 0·100	0·050	0·025	0·010	0·005	$m_H = 21$ 0·100	0·050	0·025	0·010	0·005
1	1·812	1·884	1·954	2·045	2·113	—	—	—	—	—
2	1·488	1·518	1·545	1·580	1·606	1·504	1·533	1·562	1·598	1·624
3	1·353	1·371	1·387	1·408	1·422	1·367	1·384	1·401	1·422	1·437
4	1·275	1·288	1·299	1·313	1·323	1·287	1·299	1·311	1·325	1·335
5	1·224	1·233	1·241	1·252	1·259	1·234	1·243	1·252	1·262	1·270
6	1·187	1·194	1·201	1·208	1·215	1·196	1·203	1·210	1·218	1·224
7	1·159	1·165	1·170	1·177	1·182	1·167	1·173	1·179	1·186	1·190
8	1·138	1·143	1·147	1·153	1·157	1·145	1·150	1·155	1·160	1·164
9	1·121	1·125	1·129	1·133	1·137	1·127	1·132	1·136	1·140	1·144
10	1·107	1·110	1·114	1·118	1·121	1·113	1·116	1·120	1·124	1·127
12	1·086	1·088	1·091	1·094	1·096	1·091	1·094	1·096	1·099	1·102
14	1·070	1·072	1·074	1·077	1·078	1·075	1·077	1·079	1·082	1·084
16	1·059	1·061	1·062	1·064	1·066	1·063	1·065	1·066	1·069	1·070
18	1·050	1·052	1·053	1·055	1·056	1·054	1·055	1·057	1·059	1·060
20	1·043	1·045	1·046	1·047	1·048	1·046	1·048	1·049	1·051	1·052
24	1·033	1·034	1·035	1·036	1·037	1·036	1·037	1·038	1·039	1·040
30	1·024	1·024	1·025	1·026	1·026	1·025	1·026	1·027	1·028	1·028
40	1·015	1·016	1·016	1·016	1·017	1·016	1·017	1·017	1·018	1·018
60	1·008	1·008	1·008	1·008	1·008	1·008	1·008	1·009	1·009	1·009
120	1·002	1·002	1·002	1·002	1·002	1·002	1·002	1·003	1·003	1·003
∞	1·000	1·000	1·000	1·000	1·000	1·000	1·000	1·000	1·000	1·000
$\chi^2_{dm_H}$	96·5782	101·879	106·629	112·329	116·321	100·980	106·395	111·242	117·057	121·126

$M \backslash \alpha$	$m_H = 22$ 0·100	0·050	0·025	0·010	0·005
1	1·848	1·922	1·994	2·088	2·158
2	1·518	1·549	1·577	1·614	1·641
3	1·379	1·397	1·414	1·436	1·451
4	1·298	1·310	1·322	1·337	1·347
5	1·243	1·253	1·262	1·273	1·281
6	1·204	1·212	1·219	1·228	1·234
7	1·175	1·181	1·187	1·194	1·199
8	1·152	1·157	1·162	1·168	1·172
9	1·134	1·138	1·142	1·147	1·151
10	1·119	1·123	1·126	1·130	1·134
12	1·095	1·098	1·101	1·104	1·107
14	1·079	1·081	1·083	1·086	1·088
16	1·066	1·068	1·070	1·072	1·074
18	1·057	1·058	1·060	1·062	1·063
20	1·049	1·051	1·052	1·053	1·055
24	1·038	1·039	1·040	1·041	1·042
30	1·027	1·028	1·029	1·030	1·030
40	1·017	1·018	1·018	1·019	1·019
60	1·009	1·009	1·009	1·010	1·010
120	1·003	1·003	1·003	1·003	1·003
∞	1·000	1·000	1·000	1·000	1·000
$\chi^2_{dm_H}$	105·372	110·898	115·841	121·767	125·913

$d = 5$

$M \backslash \alpha$	$m_H = 5$ 0·100	0·050	0·025	0·010	0·005	$m_H = 6$ 0·100	0·050	0·025	0·010	0·005
1	1·448	1·496	1·544	1·604	1·649	1·465	1·514	1·563	1·625	1·671
2	1·212	1·230	1·246	1·267	1·283	1·228	1·245	1·262	1·284	1·300
3	1·132	1·141	1·150	1·161	1·169	1·144	1·154	1·163	1·175	1·183
4	1·092	1·098	1·103	1·110	1·116	1·102	1·108	1·114	1·121	1·127
5	1·068	1·072	1·076	1·081	1·085	1·077	1·081	1·085	1·090	1·094
6	1·053	1·056	1·059	1·063	1·065	1·060	1·063	1·066	1·070	1·073
7	1·042	1·045	1·047	1·050	1·052	1·048	1·051	1·053	1·056	1·059
8	1·035	1·037	1·039	1·041	1·043	1·040	1·042	1·044	1·046	1·048
9	1·029	1·031	1·032	1·034	1·035	1·034	1·035	1·037	1·039	1·040
10	1·025	1·026	1·027	1·029	1·030	1·029	1·030	1·031	1·033	1·034
12	1·018	1·020	1·020	1·022	1·022	1·022	1·023	1·024	1·025	1·026
14	1·014	1·015	1·016	1·017	1·017	1·017	1·018	1·019	1·019	1·020
16	1·011	1·012	1·013	1·013	1·014	1·014	1·014	1·015	1·016	1·016
18	1·009	1·010	1·010	1·011	1·011	1·011	1·012	1·012	1·013	1·013
20	1·008	1·008	1·009	1·009	1·009	1·009	1·010	1·010	1·011	1·011
24	1·006	1·006	1·006	1·007	1·007	1·007	1·007	1·007	1·008	1·008
30	1·004	1·004	1·004	1·004	1·005	1·005	1·005	1·005	1·005	1·005
40	1·002	1·002	1·002	1·003	1·003	1·003	1·003	1·003	1·003	1·003
60	1·001	1·001	1·001	1·001	1·001	1·001	1·001	1·001	1·001	1·002
120	1·000	1·000	1·000	1·000	1·000	1·000	1·000	1·000	1·000	1·000
∞	1·000	1·000	1·000	1·000	1·000	1·000	1·000	1·000	1·000	1·000
$\chi^2_{dm_H}$	34·3816	37·6525	40·6465	44·3141	46·9279	40·2560	43·7730	46·9792	50·8922	53·6720

$M \backslash \alpha$	$m_H = 7$ 0·100	0·050	0·025	0·010	0·005	$m_H = 8$ 0·100	0·050	0·025	0·010	0·005
1	1·484	1·535	1·584	1·648	1·695	1·505	1·556	1·607	1·672	1·721
2	1·244	1·262	1·280	1·302	1·319	1·261	1·280	1·298	1·321	1·338
3	1·158	1·168	1·177	1·189	1·198	1·171	1·182	1·192	1·204	1·213
4	1·113	1·119	1·125	1·133	1·139	1·124	1·131	1·137	1·145	1·151
5	1·086	1·090	1·095	1·100	1·104	1·095	1·100	1·105	1·110	1·114
6	1·068	1·071	1·074	1·078	1·081	1·076	1·079	1·083	1·087	1·090
7	1·055	1·058	1·060	1·063	1·066	1·062	1·065	1·068	1·071	1·073
8	1·046	1·048	1·050	1·052	1·054	1·052	1·054	1·056	1·059	1·061
9	1·038	1·040	1·042	1·044	1·046	1·044	1·046	1·048	1·050	1·052
10	1·033	1·035	1·036	1·038	1·039	1·038	1·039	1·041	1·043	1·044
12	1·025	1·026	1·027	1·029	1·030	1·029	1·030	1·031	1·033	1·034
14	1·020	1·021	1·021	1·022	1·023	1·023	1·024	1·025	1·026	1·027
16	1·016	1·017	1·017	1·018	1·019	1·018	1·019	1·020	1·021	1·022
18	1·013	1·014	1·014	1·015	1·015	1·015	1·016	1·017	1·017	1·018
20	1·011	1·012	1·012	1·013	1·013	1·013	1·013	1·014	1·015	1·015
24	1·008	1·008	1·009	1·009	1·010	1·009	1·010	1·010	1·011	1·011
30	1·005	1·006	1·006	1·006	1·006	1·006	1·007	1·007	1·007	1·008
40	1·003	1·003	1·004	1·004	1·004	1·004	1·004	1·004	1·004	1·005
60	1·002	1·002	1·002	1·002	1·002	1·002	1·002	1·002	1·002	1·002
120	1·000	1·000	1·000	1·000	1·000	1·001	1·001	1·001	1·001	1·001
∞	1·000	1·000	1·000	1·000	1·000	1·000	1·000	1·000	1·000	1·000
$\chi^2_{dm_H}$	46·0588	49·8019	53·2033	57·3421	60·2748	51·8050	55·7585	59·3417	63·6907	66·7659

D13 (Continued)

$d = 5$

	$m_H = 9$					$m_H = 10$				
M \ α	0·100	0·050	0·025	0·010	0·005	0·100	0·050	0·025	0·010	0·005
1	1·526	1·578	1·630	1·697	1·746	1·547	1·600	1·653	1·721	1·772
2	1·278	1·298	1·316	1·340	1·358	1·295	1·315	1·334	1·359	1·377
3	1·185	1·196	1·206	1·219	1·229	1·199	1·211	1·221	1·235	1·244
4	1·136	1·143	1·150	1·158	1·164	1·147	1·155	1·162	1·171	1·177
5	1·105	1·110	1·115	1·121	1·125	1·115	1·120	1·125	1·131	1·136
6	1·084	1·088	1·092	1·096	1·099	1·092	1·097	1·101	1·105	1·109
7	1·069	1·072	1·075	1·079	1·081	1·076	1·080	1·083	1·087	1·089
8	1·058	1·060	1·063	1·066	1·068	1·064	1·067	1·070	1·073	1·075
9	1·049	1·051	1·053	1·056	1·058	1·055	1·057	1·059	1·062	1·064
10	1·043	1·044	1·046	1·048	1·050	1·048	1·050	1·051	1·054	1·055
12	1·033	1·034	1·035	1·037	1·038	1·037	1·038	1·040	1·041	1·043
14	1·026	1·027	1·028	1·029	1·030	1·029	1·031	1·032	1·033	1·034
16	1·021	1·022	1·023	1·024	1·024	1·024	1·025	1·026	1·027	1·028
18	1·018	1·018	1·019	1·020	1·020	1·020	1·021	1·022	1·022	1·023
20	1·015	1·015	1·016	1·017	1·017	1·017	1·018	1·018	1·019	1·019
24	1·011	1·011	1·012	1·012	1·013	1·013	1·013	1·014	1·014	1·014
30	1·008	1·008	1·008	1·008	1·009	1·009	1·009	1·009	1·010	1·010
40	1·005	1·005	1·005	1·005	1·005	1·005	1·005	1·006	1·006	1·006
60	1·002	1·002	1·002	1·002	1·003	1·003	1·003	1·003	1·003	1·003
120	1·001	1·001	1·001	1·001	1·001	1·001	1·001	1·001	1·002	1·002
∞	1·000	1·000	1·000	1·000	1·000	1·000	1·000	1·000	1·000	1·000
$\chi^2_{dm_H}$	57·5053	61·6562	65·4102	69·9568	73·1661	63·1671	67·5048	71·4202	76·1539	79·4900

	$m_H = 11$					$m_H = 12$				
M \ α	0·100	0·050	0·025	0·010	0·005	0·100	0·050	0·025	0·010	0·005
1	—	—	—	—	—	1·587	1·643	1·697	1·768	1·821
2	1·312	1·333	1·352	1·378	1·396	1·329	1·350	1·370	1·396	1·415
3	1·213	1·225	1·236	1·250	1·260	1·227	1·239	1·251	1·265	1·275
4	1·159	1·167	1·174	1·183	1·190	1·171	1·179	1·186	1·196	1·203
5	1·125	1·130	1·136	1·142	1·147	1·135	1·141	1·146	1·153	1·158
6	1·101	1·105	1·110	1·115	1·118	1·110	1·114	1·119	1·124	1·128
7	1·084	1·087	1·091	1·095	1·098	1·092	1·095	1·099	1·103	1·106
8	1·071	1·074	1·077	1·080	1·082	1·078	1·081	1·084	1·087	1·089
9	1·061	1·063	1·066	1·068	1·070	1·067	1·070	1·072	1·075	1·077
10	1·053	1·055	1·057	1·059	1·061	1·058	1·061	1·063	1·065	1·067
12	1·041	1·043	1·044	1·046	1·047	1·046	1·047	1·049	1·051	1·052
14	1·033	1·034	1·035	1·037	1·038	1·037	1·038	1·039	1·041	1·042
16	1·027	1·028	1·029	1·030	1·031	1·030	1·031	1·032	1·033	1·034
18	1·023	1·023	1·024	1·025	1·026	1·025	1·026	1·027	1·028	1·029
20	1·019	1·020	1·021	1·021	1·022	1·022	1·022	1·023	1·024	1·024
24	1·014	1·015	1·015	1·016	1·016	1·016	1·017	1·017	1·018	1·018
30	1·010	1·010	1·011	1·011	1·011	1·011	1·012	1·012	1·012	1·013
40	1·006	1·006	1·006	1·007	1·007	1·007	1·007	1·007	1·008	1·008
60	1·003	1·003	1·003	1·003	1·003	1·003	1·003	1·004	1·004	1·004
120	1·001	1·001	1·001	1·001	1·001	1·001	1·001	1·001	1·001	1·001
∞	1·000	1·000	1·000	1·000	1·000	1·000	1·000	1·000	1·000	1·000
$\chi^2_{dm_H}$	68·796	73·311	77·380	82·292	85·749	74·3970	79·0819	83·2977	88·3794	91·9517

$d = 5$

M \ α	$m_H = 13$: 0·100	0·050	0·025	0·010	0·005	$m_H = 14$: 0·100	0·050	0·025	0·010	0·005
1	—	—	—	—	—	1·626	1·683	1·740	1·813	1·867
2	1·345	1·367	1·387	1·414	1·433	1·361	1·383	1·404	1·431	1·451
3	1·241	1·253	1·265	1·280	1·290	1·254	1·267	1·279	1·294	1·305
4	1·182	1·191	1·199	1·208	1·215	1·194	1·203	1·211	1·221	1·228
5	1·145	1·151	1·157	1·164	1·169	1·155	1·161	1·167	1·174	1·180
6	1·118	1·123	1·128	1·133	1·137	1·127	1·132	1·137	1·143	1·147
7	1·099	1·103	1·107	1·111	1·114	1·107	1·111	1·115	1·119	1·123
8	1·084	1·088	1·091	1·094	1·097	1·091	1·095	1·098	1·102	1·104
9	1·073	1·076	1·078	1·081	1·083	1·079	1·082	1·085	1·088	1·090
10	1·064	1·066	1·068	1·071	1·073	1·069	1·072	1·074	1·077	1·079
12	1·050	1·052	1·054	1·055	1·057	1·055	1·057	1·058	1·060	1·062
14	1·040	1·042	1·043	1·045	1·046	1·044	1·046	1·047	1·049	1·050
16	1·033	1·035	1·036	1·037	1·038	1·037	1·038	1·039	1·040	1·041
18	1·028	1·029	1·030	1·031	1·032	1·031	1·032	1·033	1·034	1·035
20	1·024	1·025	1·026	1·026	1·027	1·026	1·027	1·028	1·029	1·030
24	1·018	1·019	1·019	1·020	1·020	1·020	1·021	1·021	1·022	1·022
30	1·013	1·013	1·013	1·014	1·014	1·014	1·014	1·015	1·015	1·016
40	1·008	1·008	1·008	1·009	1·009	1·009	1·009	1·009	1·010	1·010
60	1·004	1·004	1·004	1·004	1·004	1·004	1·004	1·005	1·005	1·005
120	1·001	1·001	1·001	1·001	1·001	1·001	1·001	1·001	1·001	1·001
∞	1·000	1·000	1·000	1·000	1·000	1·000	1·000	1·000	1·000	1·000
$\chi^2_{dm_H}$	79·973	84·821	89·177	94·422	98·105	85·5270	90·5312	95·0232	100·4252	104·2149

M \ α	$m_H = 15$: 0·100	0·050	0·025	0·010	0·005	$m_H = 16$: 0·0100	0·050	0·025	0·010	0·005
1	—	—	—	—	—	1·663	1·722	1·780	1·855	1·911
2	1·377	1·399	1·421	1·449	1·469	1·392	1·415	1·437	1·465	1·486
3	1·267	1·281	1·293	1·309	1·320	1·280	1·294	1·307	1·323	1·334
4	1·205	1·214	1·223	1·233	1·240	1·216	1·226	1·234	1·245	1·253
5	1·164	1·171	1·177	1·185	1·190	1·174	1·181	1·188	1·195	1·201
6	1·136	1·141	1·146	1·152	1·156	1·144	1·150	1·155	1·161	1·165
7	1·115	1·119	1·123	1·127	1·131	1·122	1·127	1·131	1·136	1·139
8	1·098	1·102	1·105	1·109	1·112	1·105	1·109	1·112	1·116	1·119
9	1·085	1·088	1·091	1·094	1·097	1·091	1·095	1·098	1·101	1·104
10	1·075	1·078	1·080	1·083	1·085	1·081	1·083	1·086	1·089	1·091
12	1·059	1·061	1·063	1·065	1·067	1·064	1·066	1·068	1·070	1·072
14	1·048	1·050	1·051	1·053	1·054	1·052	1·054	1·055	1·057	1·059
16	1·040	1·041	1·043	1·044	1·045	1·043	1·045	1·046	1·048	1·049
18	1·034	1·035	1·036	1·037	1·038	1·037	1·038	1·039	1·040	1·041
20	1·029	1·030	1·031	1·032	1·033	1·032	1·033	1·033	1·035	1·035
24	1·022	1·023	1·023	1·024	1·025	1·024	1·025	1·025	1·026	1·027
30	1·015	1·016	1·016	1·017	1·017	1·017	1·018	1·018	1·019	1·019
40	1·010	1·010	1·010	1·011	1·011	1·011	1·011	1·011	1·012	1·012
60	1·005	1·005	1·005	1·005	1·005	1·005	1·005	1·006	1·006	1·006
120	1·001	1·001	1·001	1·001	1·002	1·002	1·002	1·002	1·002	1·002
∞	1·000	1·000	1·000	1·000	1·000	1·000	1·000	1·000	1·000	1·000
$\chi^2_{dm_H}$	91·061	96·217	100·839	106·393	110·286	96·5782	101·8795	106·6286	112·3288	116·3211

$d = 5$

$M \backslash \alpha$	$m_H = 17$ 0·100	0·050	0·025	0·010	0·005	$m_H = 18$ 0·100	0·050	0·025	0·010	0·005
1	—	—	—	—	—	1·698	1·758	1·818	1·895	1·953
2	1·407	1·431	1·453	1·482	1·503	1·421	1·445	1·468	1·498	1·519
3	1·293	1·307	1·320	1·336	1·348	1·305	1·320	1·333	1·350	1·362
4	1·227	1·237	1·246	1·257	1·265	1·238	1·248	1·257	1·268	1·277
5	1·184	1·191	1·198	1·206	1·212	1·193	1·201	1·208	1·216	1·222
6	1·153	1·159	1·164	1·170	1·175	1·161	1·167	1·173	1·179	1·184
7	1·130	1·134	1·139	1·144	1·147	1·137	1·142	1·147	1·152	1·156
8	1·112	1·116	1·119	1·124	1·127	1·119	1·123	1·126	1·131	1·134
9	1·098	1·101	1·104	1·108	1·110	1·104	1·107	1·110	1·114	1·117
10	1·086	1·089	1·092	1·095	1·097	1·092	1·095	1·098	1·101	1·103
12	1·069	1·071	1·073	1·075	1·077	1·073	1·076	1·078	1·080	1·082
14	1·056	1·058	1·060	1·062	1·063	1·060	1·062	1·064	1·066	1·067
16	1·047	1·048	1·050	1·051	1·052	1·050	1·052	1·053	1·055	1·056
18	1·040	1·041	1·042	1·044	1·044	1·043	1·044	1·045	1·047	1·048
20	1·034	1·035	1·036	1·037	1·038	1·037	1·038	1·039	1·040	1·041
24	1·026	1·027	1·028	1·029	1·029	1·028	1·029	1·030	1·031	1·031
30	1·019	1·019	1·020	1·020	1·021	1·020	1·021	1·021	1·022	1·022
40	1·012	1·012	1·012	1·013	1·013	1·013	1·013	1·013	1·014	1·014
60	1·006	1·006	1·006	1·006	1·006	1·006	1·007	1·007	1·007	1·007
120	1·002	1·002	1·002	1·002	1·002	1·002	1·002	1·002	1·002	1·002
∞	1·000	1·000	1·000	1·000	1·000	1·000	1·000	1·000	1·000	1·000
$\chi^2_{dm_H}$	102·079	107·522	112·393	118·236	122·325	107·5650	113·1453	118·1359	124·1163	128·2989

$M \backslash \alpha$	$m_H = 19$ 0·100	0·050	0·025	0·010	0·005	$m_H = 20$ 0·100	0·050	0·025	0·010	0·005
1	—	—	—	—	—	1·731	1·793	1·853	1·933	1·992
2	1·436	1·460	1·483	1·513	1·535	1·449	1·474	1·498	1·528	1·551
3	1·318	1·332	1·346	1·363	1·375	1·330	1·345	1·358	1·376	1·388
4	1·249	1·259	1·268	1·280	1·288	1·259	1·270	1·279	1·291	1·300
5	1·203	1·210	1·217	1·226	1·232	1·212	1·220	1·227	1·236	1·242
6	1·170	1·176	1·181	1·188	1·193	1·178	1·184	1·190	1·197	1·202
7	1·145	1·150	1·154	1·160	1·164	1·152	1·157	1·162	1·168	1·172
8	1·126	1·130	1·134	1·138	1·141	1·132	1·137	1·141	1·145	1·149
9	1·110	1·114	1·117	1·121	1·124	1·116	1·120	1·123	1·127	1·130
10	1·097	1·101	1·103	1·107	1·109	1·103	1·106	1·109	1·113	1·115
12	1·078	1·081	1·083	1·085	1·087	1·083	1·086	1·088	1·091	1·092
14	1·064	1·066	1·068	1·070	1·072	1·069	1·071	1·072	1·075	1·076
16	1·054	1·056	1·057	1·059	1·060	1·058	1·059	1·061	1·063	1·064
18	1·046	1·047	1·049	1·050	1·051	1·049	1·051	1·052	1·053	1·054
20	1·040	1·041	1·042	1·043	1·044	1·043	1·044	1·045	1·046	1·047
24	1·031	1·031	1·032	1·033	1·034	1·033	1·034	1·035	1·036	1·036
30	1·022	1·022	1·023	1·024	1·024	1·024	1·024	1·025	1·025	1·026
40	1·014	1·014	1·015	1·015	1·015	1·015	1·015	1·016	1·016	1·016
60	1·007	1·007	1·007	1·008	1·008	1·008	1·008	1·008	1·008	1·008
120	1·002	1·002	1·002	1·002	1·002	1·002	1·002	1·002	1·002	1·002
∞	1·000	1·000	1·000	1·000	1·000	1·000	1·000	1·000	1·000	1·000
$\chi^2_{dm_H}$	113·038	118·752	123·858	129·973	134·247	118·4980	124·3421	129·5612	135·8067	140·1695

D13 (Continued)

$d = 6$

$m_H = 6$

$M \backslash \alpha$	0·100	0·050	0·025	0·010	0·005
1	1·471	1·520	1·568	1·631	1·677
2	1·237	1·255	1·272	1·294	1·310
3	1·153	1·163	1·172	1·183	1·192
4	1·109	1·116	1·122	1·129	1·134
5	1·083	1·088	1·092	1·097	1·101
6	1·066	1·069	1·072	1·076	1·079
7	1·053	1·056	1·058	1·061	1·064
8	1·044	1·046	1·048	1·051	1·053
9	1·037	1·039	1·041	1·043	1·044
10	1·032	1·034	1·035	1·037	1·038
12	1·024	1·025	1·026	1·028	1·029
14	1·019	1·020	1·021	1·022	1·022
16	1·015	1·016	1·017	1·018	1·018
18	1·013	1·013	1·014	1·014	1·015
20	1·011	1·011	1·012	1·012	1·013
24	1·008	1·008	1·009	1·009	1·009
30	1·005	1·006	1·006	1·006	1·006
40	1·003	1·003	1·003	1·004	1·004
60	1·002	1·002	1·002	1·002	1·002
120	1·000	1·000	1·000	1·000	1·000
∞	1·000	1·000	1·000	1·000	1·000
$\chi^2_{dm_H}$	47·2122	50·9985	54·4373	58·6192	61·5812

$m_H = 7$

$M \backslash \alpha$	0·100	0·050	0·025	0·010	0·005
1	1·481	1·530	1·579	1·642	1·688
2	1·249	1·266	1·284	1·306	1·322
3	1·163	1·173	1·182	1·194	1·203
4	1·118	1·124	1·131	1·138	1·144
5	1·090	1·095	1·099	1·105	1·109
6	1·072	1·075	1·079	1·083	1·086
7	1·059	1·062	1·064	1·067	1·070
8	1·049	1·051	1·053	1·056	1·058
9	1·042	1·043	1·045	1·047	1·049
10	1·036	1·037	1·039	1·041	1·042
12	1·027	1·029	1·030	1·031	1·032
14	1·022	1·023	1·023	1·024	1·025
16	1·018	1·018	1·019	1·020	1·020
18	1·014	1·015	1·016	1·016	1 017
20	1·012	1·013	1·013	1·014	1·014
24	1·009	1·009	1·010	1·010	1·010
30	1·006	1·006	1·007	1·007	1·007
40	1·004	1·004	1·004	1·004	1·004
60	1·002	1·002	1·002	1·002	1·002
120	1·000	1·000	1·001	1·001	1·001
∞	1·000	1·000	1·000	1·000	1·000
$\chi^2_{dm_H}$	54·0902	58·1240	61·7768	66·2062	69·3360

$m_H = 8$

$M \backslash \alpha$	0·100	0·050	0·025	0·010	0·005
1	1·494	1·543	1·592	1·656	1·703
2	1·261	1·279	1·297	1·319	1·336
3	1·174	1·184	1·194	1·205	1·214
4	1·127	1·134	1·140	1·148	1·153
5	1·098	1·103	1·108	1·113	1·117
6	1·079	1·082	1·086	1·090	1·093
7	1·065	1·068	1·070	1·074	1·076
8	1·054	1·057	1·059	1·062	1·063
9	1·046	1·048	1·050	1·052	1·054
10	1·040	1·042	1·043	1·045	1·046
12	1·031	1·032	1·033	1·035	1·036
14	1·024	1·025	1·026	1·027	1·028
16	1·020	1·021	1·021	1·022	1·022
18	1·016	1·017	1·018	1·018	1·019
20	1·014	1·014	1·015	1·015	1·016
24	1·010	1·011	1·011	1·011	1·012
30	1·007	1·007	1·008	1·008	1·008
40	1·004	1·004	1·005	1·005	1·005
60	1·002	1·002	1·002	1·002	1·002
120	1·001	1·001	1·001	1·001	1·001
∞	1·000	1·000	1·000	1·000	1·000
$\chi^2_{dm_H}$	60·9066	65·1708	69·0226	73·6826	76·9688

$m_H = 9$

$M \backslash \alpha$	0·100	0·050	0·025	0·010	0·005
1	1·508	1·558	1·607	1·671	1·719
2	1·275	1·293	1·311	1·333	1·350
3	1·185	1·196	1·205	1·218	1·227
4	1·137	1·144	1·150	1·158	1·164
5	1·107	1·112	1·116	1·122	1·126
6	1·086	1·090	1·093	1·098	1·101
7	1·071	1·074	1·077	1·080	1·083
8	1·060	1·062	1·065	1·067	1·069
9	1·051	1·053	1·055	1·058	1·059
10	1·044	1·046	1·048	1·050	1·051
12	1·034	1·035	1·036	1·037	1·037
14	1·027	1·028	1·029	1·031	1·031
16	1·022	1·023	1·024	1·024	1·025
18	1·019	1·019	1·020	1·020	1·021
20	1·016	1·016	1·017	1·017	1·018
24	1·012	1·012	1·013	1·013	1·013
30	1·008	1·008	1·009	1·009	1·009
40	1·005	1·005	1·005	1·005	1·006
60	1·002	1·002	1·003	1·003	1·003
120	1·001	1·001	1·001	1·001	1·002
∞	1·000	1·000	1·000	1·000	1·000
$\chi^2_{dm_H}$	67·6728	72·1532	76·1921	81·0688	84·5016

D13 (Continued)

$d = 6$

α / M	$m_H = 10$ 0.100	0.050	0.025	0.010	0.005	$m_H = 11$ 0.100	0.050	0.025	0.010	0.005
1	1.523	1.573	1.623	1.687	1.736	1.538	1.589	1.639	1.704	—
2	1.288	1.307	1.325	1.348	1.365	1.302	1.321	1.339	1.363	1.380
3	1.197	1.208	1.218	1.230	1.239	1.209	1.220	1.230	1.243	1.252
4	1.147	1.154	1.161	1.169	1.175	1.157	1.164	1.171	1.180	1.186
5	1.115	1.120	1.125	1.131	1.135	1.124	1.129	1.134	1.140	1.145
6	1.093	1.097	1.101	1.106	1.109	1.101	1.105	1.109	1.114	1.117
7	1.078	1.081	1.084	1.087	1.090	1.084	1.088	1.091	1.094	1.097
8	1.066	1.068	1.071	1.074	1.076	1.072	1.074	1.077	1.080	1.082
9	1.056	1.059	1.061	1.063	1.065	1.062	1.064	1.066	1.069	1.070
10	1.049	1.051	1.053	1.055	1.056	1.054	1.055	1.058	1.060	1.061
12	1.038	1.040	1.041	1.042	1.043	1.042	1.044	1.045	1.047	1.048
14	1.031	1.032	1.033	1.034	1.034	1.034	1.035	1.036	1.037	1.038
16	1.025	1.026	1.026	1.027	1.028	1.028	1.029	1.030	1.030	1.031
18	1.021	1.021	1.022	1.023	1.023	1.023	1.024	1.025	1.026	1.026
20	1.018	1.018	1.019	1.019	1.020	1.020	1.020	1.021	1.022	1.022
24	1.013	1.014	1.014	1.015	1.015	1.015	1.015	1.016	1.016	1.017
30	1.009	1.010	1.010	1.010	1.010	1.010	1.011	1.011	1.011	1.012
40	1.006	1.006	1.006	1.006	1.006	1.006	1.007	1.007	1.007	1.007
60	1.003	1.003	1.003	1.003	1.003	1.003	1.003	1.003	1.003	1.003
120	1.001	1.001	1.001	1.001	1.001	1.001	1.001	1.001	1.001	1.001
∞	1.000	1.000	1.000	1.000	1.000	1.000	1.000	1.000	1.000	1.000
$\chi^2_{dm_H}$	74.3970	79.0819	83.2976	88.3794	91.9517	81.085	85.9649	90.349	95.6257	99.330

α / M	$m_H = 12$ 0.100	0.050	0.025	0.010	0.005	$m_H = 13$ 0.100	0.050	0.025	0.010	0.005
1	1.554	1.605	1.655	1.722	1.771	1.569	1.621	1.672	1.739	—
2	1.316	1.335	1.354	1.378	1.395	1.330	1.349	1.368	1.393	1.410
3	1.221	1.232	1.242	1.255	1.265	1.233	1.244	1.255	1.268	1.278
4	1.167	1.175	1.182	1.190	1.197	1.178	1.185	1.192	1.201	1.208
5	1.133	1.138	1.144	1.150	1.154	1.142	1.148	1.153	1.159	1.164
6	1.109	1.113	1.117	1.122	1.125	1.117	1.121	1.125	1.130	1.134
7	1.091	1.095	1.098	1.102	1.104	1.098	1.102	1.105	1.109	1.112
8	1.078	1.081	1.083	1.086	1.089	1.084	1.088	1.090	1.093	1.095
9	1.067	1.070	1.072	1.074	1.076	1.073	1.075	1.078	1.080	1.082
10	1.059	1.061	1.063	1.065	1.067	1.064	1.066	1.068	1.070	1.072
12	1.046	1.048	1.049	1.051	1.052	1.050	1.052	1.053	1.055	1.057
14	1.037	1.039	1.040	1.041	1.042	1.041	1.042	1.043	1.044	1.046
16	1.031	1.032	1.033	1.034	1.035	1.034	1.035	1.036	1.037	1.038
18	1.026	1.027	1.027	1.028	1.029	1.028	1.029	1.030	1.031	1.032
20	1.022	1.023	1.023	1.024	1.025	1.024	1.025	1.026	1.026	1.027
24	1.017	1.017	1.018	1.018	1.019	1.018	1.019	1.020	1.020	1.021
30	1.012	1.012	1.012	1.013	1.013	1.013	1.013	1.014	1.014	1.014
40	1.007	1.007	1.008	1.008	1.008	1.008	1.008	1.008	1.009	1.009
60	1.004	1.004	1.004	1.004	1.004	1.004	1.004	1.004	1.004	1.004
120	1.001	1.001	1.001	1.001	1.001	1.001	1.001	1.001	1.001	1.001
∞	1.000	1.000	1.000	1.000	1.000	1.001	1.000	1.000	1.000	1.000
$\chi^2_{dm_h}$	87.7430	92.8083	97.3531	102.8163	106.6476	94.374	99.6169	104.316	109.9580	113.911

D13 (Continued)

$d = 6$

$m_H = 14$

M \ α	0·100	0·050	0·025	0·010	0·005
1	1·585	1·637	1·688	1·756	1·806
2	1·343	1·363	1·383	1·407	1·425
3	1·244	1·256	1·267	1·281	1·291
4	1·188	1·196	1·203	1·212	1·219
5	1·151	1·157	1·162	1·169	1·174
6	1·125	1·129	1·133	1·139	1·142
7	1·105	1·109	1·112	1·117	1·119
8	1·090	1·093	1·096	1·100	1·102
9	1·078	1·081	1·083	1·086	1·088
10	1·069	1·071	1·073	1·076	1·078
12	1·055	1·056	1·058	1·060	1·061
14	1·044	1·046	1·047	1·049	1·050
16	1·037	1·038	1·039	1·040	1·041
18	1·031	1·032	1·033	1·034	1·035
20	1·027	1·028	1·028	1·029	1·030
24	1·020	1·021	1·021	1·022	1·023
30	1·014	1·015	1·015	1·016	1·016
40	1·009	1·009	1·009	1·010	1·010
60	1·004	1·005	1·005	1·005	1·005
120	1·001	1·001	1·001	1·001	1·001
∞	1·000	1·000	1·000	1·000	1·000
$\chi^2_{dm_H}$	100·9800	106·3948	111·2423	117·0565	121·1263

$m_H = 15$

M \ α	0·100	0·050	0·025	0·010	0·005
1	—	—	—	—	—
2	1·357	1·377	1·397	1·422	1·440
3	1·256	1·268	1·279	1·293	1·303
4	1·198	1·206	1·214	1·223	1·230
5	1·160	1·166	1·171	1·178	1·183
6	1·132	1·137	1·142	1·147	1·151
7	1·112	1·116	1·120	1·124	1·127
8	1·097	1·100	1·103	1·106	1·109
9	1·084	1·087	1·089	1·092	1·095
10	1·074	1·076	1·079	1·081	1·083
12	1·059	1·061	1·062	1·064	1·066
14	1·048	1·050	1·051	1·053	1·054
16	1·040	1·041	1·042	1·044	1·045
18	1·034	1·035	10·36	1·037	1·038
20	1·029	1·030	1·031	1·032	1·032
24	1·022	1·023	1·023	1·024	1·025
30	1·016	1·016	1·017	1·017	1·017
40	1·010	1·010	1·010	1·011	1·011
60	1·005	1·005	1·005	1·005	1·005
120	1·001	1·001	1·001	1·002	1·002
∞	1·000	1·000	1·000	1·000	1·000
$\chi^2_{dm_H}$	107·565	113·145	118·136	124·116	128·299

$m_H = 16$

M \ α	0·100	0·050	0·025	0·010	0·005
1	1·615	1·668	1·721	1·789	1·841
2	1·370	1·391	1·411	1·436	1·455
3	1·267	1·280	1·291	1·305	1·316
4	1·208	1·216	1·224	1·234	1·241
5	1·168	1·175	1·181	1·188	1·193
6	1·140	1·145	1·150	1·155	1·159
7	1·119	1·123	1·127	1·131	1·135
8	1·103	1·106	1·109	1·113	1·116
9	1·090	1·093	1·095	1·099	1·101
10	1·079	1·082	1·084	1·087	1·089
12	1·063	1·065	1·067	1·069	1·071
14	1·052	1·053	1·055	1·056	1·058
16	1·043	1·045	1·046	1·047	1·048
18	1·037	1·038	1·039	1·040	1·041
20	1·032	1·033	1·033	1·034	1·035
24	1·024	1·025	1·025	1·026	1·027
30	1·017	1·018	1·018	1·019	1·019
40	1·011	1·011	1·011	1·012	1·012
60	1·005	1·006	1·006	1·006	1·006
120	1·002	1·002	1·002	1·002	1·002
∞	1·000	1·000	1·000	1·000	1·000
$\chi^2_{dm_H}$	114·1307	119·8709	125·0001	131·1412	135·4330

$m_H = 17$

M \ α	0·100	0·050	0·025	0·010	0·005
1	—	—	—	—	—
2	1·383	1·404	1·424	1·450	1·469
3	1·279	1·291	1·303	1·317	1·328
4	1·218	1·226	1·234	1·244	1·251
5	1·177	1·184	1·190	1·197	1·202
6	1·148	1·153	1·158	1·164	1·168
7	1·126	1·130	1·134	1·139	1·142
8	1·109	1·113	1·116	1·120	1·123
9	1·096	1·099	1·101	1·105	1·107
10	1·085	1·087	1·090	1·092	1·094
12	1·068	1·070	1·072	1·074	1·075
14	1·056	1·057	1·059	1·061	1·062
16	1·047	1·048	1·049	1·051	1·052
18	1·040	1·041	1·042	1·043	1·044
20	1·034	1·035	1·036	1·037	1·038
24	1·026	1·027	1·028	1·028	1·029
30	1·019	1·019	1·020	1·020	1·021
40	1·012	1·012	1·012	1·013	1·013
60	1·006	1·006	1·006	1·006	1·007
120	1·002	1·002	1·002	1·002	1·002
∞	1·000	1·000	1·000	1·000	1·000
$\chi^2_{dm_H}$	120·679	126·574	131·838	138·134	142·532

$d = 6$

$M \backslash \alpha$	$m_H = 18$: 0·100	0·050	0·025	0·010	0·005	$m_H = 19$: 0·100	0·050	0·025	0·010	0·005
1	1·644	1·698	1·752	1·822	1·874	—	—	—	—	—
2	1·396	1·417	1·438	1·464	1·483	1·408	1·430	1·451	1·477	1·497
3	1·290	1·303	1·315	1·329	1·340	1·301	1·314	1·326	1·341	1·352
4	1·228	1·237	1·245	1·255	1·262	1·237	1·246	1·255	1·265	1·273
5	1·186	1·193	1·199	1·206	1·212	1·195	1·201	1·208	1·215	1·221
6	1·156	1·161	1·166	1·172	1·176	1·164	1·169	1·174	1·180	1·184
7	1·133	1·138	1·142	1·146	1·150	1·140	1·145	1·149	1·154	1·157
8	1·116	1·119	1·123	1·127	1·129	1·122	1·126	1·129	1·133	1·136
9	1·101	1·105	1·107	1·111	1·113	1·107	1·110	1·113	1·117	1·119
10	1·090	1·093	1·095	1·098	1·100	1·095	1·098	1·101	1·104	1·106
12	1·072	1·074	1·076	1·079	1·080	1·077	1·079	1·081	1·083	1·085
14	1·060	1·061	1·063	1·065	1·066	1·063	1·065	1·067	1·069	1·070
16	1·050	1·051	1·053	1·054	1·055	1·053	1·055	1·056	1·058	1·059
18	1·043	1·044	1·045	1·046	1·047	1·046	1·047	1·048	1·049	1·050
20	1·037	1·038	1·039	1·040	1·041	1·039	1·041	1·041	1·043	1·043
24	1·028	1·029	1·030	1·031	1·031	1·030	1·031	1·032	1·033	1·033
30	1·020	1·021	1·021	1·022	1·022	1·022	1·022	1·023	1·023	1·024
40	1·013	1·013	1·013	1·014	1·014	1·014	1·014	1·015	1·015	1·015
60	1·006	1·007	1·007	1·007	1·007	1·007	1·007	1·007	1·008	1·008
120	1·002	1·002	1·002	1·002	1·002	1·002	1·002	1·002	1·002	1·002
∞	1·000	1·000	1·000	1·000	1·000	1·000	1·000	1·000	1·000	1·000
$\chi^2_{dm_H}$	127·2111	133·2569	138·6506	145·0988	149·5994	133·729	139·921	145·441	152·037	156·637

$m_H = 20$

$M \backslash \alpha$	0·100	0·050	0·025	0·010	0·005
1	1·672	1·727	1·781	1·853	1·906
2	1·420	1·443	1·464	1·490	1·510
3	1·312	1·325	1·337	1·353	1·364
4	1·247	1·256	1·265	1·275	1·283
5	1·203	1·210	1·217	1·224	1·230
6	1·171	1·177	1·182	1·188	1·193
7	1·147	1·152	1·156	1·161	1·165
8	1·128	1·132	1·136	1·140	1·143
9	1·113	1·116	1·119	1·123	1·126
10	1·101	1·103	1·106	1·109	1·111
12	1·081	1·084	1·086	1·088	1·090
14	1·067	1·069	1·071	1·073	1·074
16	1·057	1·058	1·060	1·061	1·062
18	1·049	1·050	1·051	1·052	1·053
20	1·042	1·043	1·044	1·045	1·046
24	1·033	1·033	1·034	1·035	1·036
30	1·023	1·024	1·025	1·025	1·026
40	1·015	1·015	1·016	1·016	1·016
60	1·008	1·008	1·008	1·008	1·008
120	1·002	1·002	1·002	1·002	1·002
∞	1·000	1·000	1·000	1·000	1·000
$\chi^2_{dm_H}$	140·2326	146·5674	152·2114	158·9502	163·6482

$d = 7$

$m_H = 7$

M \ α	0·100	0·050	0·025	0·010	0·005
1	1·484	1·532	1·580	1·643	1·689
2	1·256	1·273	1·290	1·312	1·329
3	1·170	1·180	1·189	1·201	1·210
4	1·125	1·131	1·137	1·145	1·150
5	1·096	1·101	1·105	1·111	1·114
6	1·077	1·081	1·084	1·088	1·091
7	1·063	1·066	1·069	1·072	1·074
8	1·053	1·055	1·058	1·060	1·062
9	1·045	1·047	1·049	1·051	1·053
10	1·039	1·041	1·042	1·044	1·045
12	1·030	1·031	1·032	1·034	1·035
14	1·024	1·025	1·026	1·027	1·028
16	1·019	1·020	1·021	1·022	1·022
18	1·016	1·017	1·017	1·018	1·019
20	1·014	1·014	1·015	1·015	1·016
24	1·010	1·010	1·011	1·011	1·012
30	1·007	1·007	1·007	1·008	1·008
40	1·004	1·004	1·004	1·005	1·005
60	1·002	1·002	1·002	1·002	1·002
120	1·001	1·001	1·001	1·001	1·001
∞	1·000	1·000	1·000	1·000	1·000
$\chi^2_{dm_H}$	62·0375	66·3386	70·2224	74·9195	78·2307

$m_H = 8$

M \ α	0·100	0·050	0·025	0·010	0·005
1	1·490	1·538	1·586	1·648	1·694
2	1·265	1·282	1·299	1·321	1·337
3	1·179	1·189	1·198	1·210	1·218
4	1·132	1·139	1·145	1·152	1·158
5	1·103	1·108	1·112	1·117	1·121
6	1·083	1·086	1·090	1·094	1·097
7	1·068	1·071	1·074	1·077	1·080
8	1·058	1·060	1·062	1·065	1·067
9	1·049	1·051	1·053	1·055	1·057
10	1·043	1·044	1·046	1·048	1·049
12	1·033	1·034	1·035	1·037	1·038
14	1·026	1·027	1·028	1·028	1·029
16	1·021	1·022	1·023	1·023	1·024
18	1·018	1·018	1·019	1·020	1·020
20	1·015	1·016	1·016	1·017	1·017
24	1·011	1·012	1·012	1·012	1·013
30	1·008	1·008	1·008	1·009	1·009
40	1·005	1·005	1·005	1·005	1·005
60	1·002	1·002	1·002	1·003	1·003
120	1·001	1·001	1·001	1·001	1·001
∞	1·000	1·000	1·000	1·000	1·000
$\chi^2_{dm_H}$	69·9185	74·4683	78·5672	83·5134	86·9938

$m_H = 9$

M \ α	0·100	0·050	0·025	0·010	0·005
1	1·499	1·547	1·594	1·656	1·703
2	1·275	1·292	1·309	1·331	1·348
3	1·188	1·198	1·207	1·219	1·228
4	1·140	1·147	1·153	1·161	1·166
5	1·110	1·115	1·119	1·125	1·129
6	1·089	1·093	1·096	1·100	1·103
7	1·074	1·077	1·080	1·083	1·085
8	1·063	1·065	1·067	1·070	1·072
9	1·054	1·056	1·058	1·060	1·062
10	1·047	1·048	1·050	1·052	1·053
12	1·036	1·038	1·039	1·040	1·041
14	1·029	1·030	1·031	1·032	1·033
16	1·024	1·025	1·025	1·026	1·027
18	1·020	1·021	1·021	1·022	1·023
20	1·017	1·018	1·018	1·019	1·019
24	1·013	1·013	1·013	1·014	1·014
30	1·009	1·009	1·009	1·010	1·010
40	1·005	1·006	1·006	1·006	1·006
60	1·003	1·003	1·003	1·003	1·003
120	1·001	1·001	1·001	1·001	1·001
∞	1·000	1·000	1·000	1·000	1·000
$\chi^2_{dm_H}$	77·7454	82·5287	86·8296	92·0100	95·6493

$m_H = 10$

M \ α	0·100	0·050	0·025	0·010	0·005
1	1·509	1·557	1·604	1·666	1·713
2	1·285	1·303	1·320	1·342	1·359
3	1·197	1·208	1·217	1·229	1·238
4	1·148	1·155	1·162	1·169	1·175
5	1·117	1·122	1·127	1·132	1·136
6	1·096	1·099	1·103	1·107	1·110
7	1·080	1·083	1·086	1·089	1·091
8	1·068	1·070	1·073	1·075	1·077
9	1·058	1·060	1·062	1·065	1·066
10	1·051	1·053	1·054	1·056	1·058
12	1·040	1·042	1·042	1·044	1·045
14	1·032	1·034	1·034	1·036	1·036
16	1·026	1·028	1·028	1·029	1·029
18	1·022	1·023	1·023	1·024	1·024
20	1·019	1·019	1·020	1·020	1·021
24	1·014	1·014	1·015	1·015	1·016
30	1·010	1·010	1·010	1·011	1·011
40	1·006	1·006	1·006	1·007	1·007
60	1·003	1·003	1·003	1·003	1·003
120	1·001	1·001	1·001	1·001	1·001
∞	1·000	1·000	1·000	1·000	1·000
$\chi^2_{dm_H}$	85·5271	90·5312	95·0231	100·4250	104·2150

D13 (Continued)

$d = 7$

$m_H = 11$

M \ α	0·100	0·050	0·025	0·010	0·005
1	—	—	—	—	—
2	1·297	1·315	1·332	1·354	1·371
3	1·207	1·218	1·227	1·239	1·248
4	1·157	1·164	1·171	1·179	1·184
5	1·125	1·130	1·135	1·140	1·145
6	1·102	1·106	1·110	1·114	1·118
7	1·086	1·089	1·092	1·095	1·098
8	1·073	1·076	1·078	1·081	1·083
9	1·063	1·065	1·067	1·070	1·072
10	1·055	1·057	1·059	1·061	1·062
12	1·043	1·045	1·046	1·048	1·049
14	1·035	1·036	1·037	1·038	1·039
16	1·029	1·030	1·031	1·032	1·032
18	1·024	1·025	1·026	1·027	1·027
20	1·021	1·021	1·022	1·023	1·023
24	1·016	1·016	1·017	1·017	1·017
30	1·011	1·011	1·012	1·012	1·012
40	1·007	1·007	1·007	1·007	1·008
60	1·003	1·003	1·004	1·004	1·004
120	1·001	1·001	1·001	1·001	1·001
∞	1·000	1·000	1·000	1·000	1·000
$\chi^2_{dm_H}$	93·270	98·484	103·158	108·771	112·704

$m_H = 12$

M \ α	0·100	0·050	0·025	0·100	0·005
1	1·531	1·579	1·627	1·690	1·737
2	1·308	1·326	1·344	1·366	1·383
3	1·217	1·228	1·238	1·250	1·259
4	1·166	1·173	1·180	1·188	1·194
5	1·133	1·138	1·143	1·149	1·153
6	1·109	1·113	1·117	1·122	1·125
7	1·092	1·095	1·098	1·102	1·104
8	1·079	1·081	1·084	1·087	1·089
9	1·068	1·070	1·073	1·075	1·077
10	1·060	1·062	1·064	1·066	1·067
12	1·047	1·049	1·050	1·052	1·053
14	1·038	1·039	1·041	1·042	1·043
16	1·032	1·033	1·034	1·035	1·035
18	1·027	1·028	1·028	1·029	1·030
20	1·023	1·024	1·024	1·025	1·025
24	1·017	1·018	1·018	1·019	1·019
30	1·012	1·012	1·013	1·013	1·013
40	1·008	1·008	1·008	1·008	1·008
60	1·004	1·004	1·004	1·004	1·004
120	1·001	1·001	1·001	1·001	1·001
∞	1·000	1·000	1·000	1·000	1·000
$\chi^2_{dm_H}$	100·9800	106·3948	111·2423	117·0565	121·1263

$m_H = 13$

M \ α	0·100	0·050	0·025	0·010	0·005
1	—	—	—	—	—
2	1·320	1·338	1·356	1·378	1·395
3	1·228	1·238	1·248	1·261	1·270
4	1·175	1·182	1·189	1·197	1·203
5	1·141	1·146	1·151	1·157	1·161
6	1·116	1·120	1·124	1·129	1·132
7	1·098	1·102	1·105	1·108	1·111
8	1·084	1·087	1·090	1·093	1·095
9	1·073	1·076	1·078	1·080	1·082
10	1·064	1·066	1·068	1·071	1·072
12	1·051	1·053	1·054	1·056	1·057
14	1·042	1·043	1·044	1·045	1·046
16	1·035	1·036	1·036	1·038	1·038
18	1·029	1·030	1·031	1·032	1·032
20	1·025	1·026	1·026	1·027	1·028
24	1·019	1·020	1·020	1·021	1·021
30	1·013	1·014	1·014	1·014	1·015
40	1·008	1·009	1·009	1·009	1·009
60	1·004	1·004	1·004	1·004	1·005
120	1·001	1·001	1·001	1·001	1·001
∞	1·000	1·000	1·000	1·000	1·000
$\chi^2_{dm_H}$	108·661	114·268	119·282	125·289	129·491

$m_H = 14$

M \ α	0·100	0·050	0·025	0·010	0·005
1	1·556	1·604	1·652	1·715	1·763
2	1·331	1·350	1·368	1·391	1·408
3	1·238	1·249	1·259	1·272	1·281
4	1·184	1·192	1·198	1·207	1·213
5	1·149	1·154	1·159	1·165	1·170
6	1·123	1·128	1·132	1·137	1·140
7	1·105	1·108	1·111	1·115	1·118
8	1·090	1·093	1·096	1·099	1·101
9	1·078	1·081	1·083	1·086	1·088
10	1·069	1·071	1·073	1·076	1·077
12	1·055	1·057	1·058	1·060	1·061
14	1·045	1·046	1·047	1·049	1·050
16	1·037	1·039	1·039	1·041	1·041
18	1·032	1·033	1·033	1·034	1·035
20	1·027	1·028	1·029	1·030	1·030
24	1·021	1·021	1·022	1·022	1·023
30	1·015	1·015	1·015	1·016	1·016
40	1·009	1·009	1·010	1·010	1·010
60	1·005	1·005	1·005	1·005	1·005
120	1·001	1·001	1·001	1·001	1·001
∞	1·000	1·000	1·000	1·000	1·000
$\chi^2_{dm_H}$	116·3153	122·1077	127·2821	133·4757	137·8032

D13 (Continued)

$d = 7$

M \ α	$m_H = 15$: 0·100	0·050	0·025	0·010	0·005	$m_H = 16$: 0·100	0·050	0·025	0·010	0·005
1	—	—	—	—	—	1·580	1·629	1·678	1·742	1·790
2	1·343	1·362	1·380	1·403	1·420	1·354	1·373	1·392	1·415	1·432
3	1·248	1·259	1·270	1·283	1·292	1·258	1·270	1·280	1·293	1·303
4	1·193	1·201	1·208	1·216	1·223	1·202	1·210	1·217	1·226	1·232
5	1·157	1·162	1·168	1·174	1·179	1·165	1·171	1·176	1·182	1·187
6	1·131	1·135	1·139	1·144	1·148	1·138	1·142	1·147	1·152	1·155
7	1·111	1·115	1·118	1·122	1·125	1·118	1·121	1·125	1·129	1·132
8	1·096	1·099	1·102	1·105	1·107	1·102	1·105	1·108	1·111	1·114
9	1·084	1·086	1·089	1·092	1·094	1·089	1·092	1·094	1·097	1·099
10	1·074	1·076	1·078	1·081	1·082	1·079	1·081	1·083	1·086	1·088
12	1·059	1·061	1·062	1·064	1·066	1·063	1·065	1·067	1·069	1·070
14	1·048	1·050	1·051	1·053	1·054	1·052	1·053	1·055	1·056	1·057
16	1·040	1·042	1·043	1·044	1·045	1·044	1·045	1·046	1·047	1·048
18	1·034	1·035	1·036	1·037	1·038	1·037	1·038	1·039	1·040	1·041
20	1·030	1·030	1·031	1·032	1·033	1·032	1·033	1·034	1·034	1·035
24	1·023	1·023	1·024	1·024	1·025	1·024	1·025	1·026	1·026	1·027
30	1·016	1·016	1·017	1·017	1·018	1·017	1·018	1·018	1·019	1·019
40	1·010	1·010	1·011	1·011	1·011	1·011	1·011	1·012	1·012	1·012
60	1·005	1·005	1·005	1·005	1·006	1·006	1·006	1·006	1·006	1·006
120	1·001	1·001	1·002	1·002	1·002	1·002	1·002	1·002	1·002	1·002
∞	1·000	1·000	1·000	1·000	1·000	1·000	1·000	1·000	1·000	1·000
$\chi^2_{dm_H}$	123·947	129·918	135·247	141·620	146·070	131·5576	137·7015	143·1801	149·7269	154·2944

M \ α	$m_H = 17$: 0·100	0·050	0·025	0·010	0·005	$m_H = 18$: 0·100	0·050	0·025	0·010	0·005
1	—	—	—	—	—	1·605	1·654	1·703	1·768	1·816
2	1·365	1·385	1·403	1·427	1·445	1·377	1·396	1·415	1·439	1·457
3	1·268	1·280	1·291	1·304	1·314	1·278	1·290	1·301	1·315	1·324
4	1·211	1·219	1·226	1·235	1·242	1·220	1·228	1·236	1·245	1·252
5	1·173	1·179	1·184	1·191	1·196	1·181	1·187	1·192	1·199	1·204
6	1·145	1·150	1·154	1·159	1·163	1·152	1·157	1·161	1·167	1·171
7	1·124	1·128	1·131	1·136	1·139	1·131	1·134	1·138	1·142	1·146
8	1·108	1·111	1·114	1·117	1·120	1·114	1·117	1·120	1·124	1·126
9	1·095	1·097	1·100	1·103	1·105	1·100	1·103	1·105	1·108	1·111
10	1·084	1·086	1·088	1·091	1·093	1·089	1·091	1·093	1·096	1·098
12	1·067	1·069	1·071	1·073	1·074	1·072	1·074	1·075	1·077	1·079
14	1·056	1·057	1·058	1·060	1·061	1·059	1·061	1·062	1·064	1·065
16	1·047	1·048	1·049	1·050	1·051	1·050	1·051	1·052	1·054	1·055
18	1·040	1·041	1·042	1·043	1·044	1·043	1·044	1·045	1·046	1·047
20	1·034	1·035	1·036	1·037	1·038	1·037	1·038	1·039	1·040	1·040
24	1·026	1·027	1·028	1·028	1·029	1·028	1·029	1·030	1·031	1·031
30	1·019	1·019	1·020	1·020	1·021	1·020	1·021	1·021	1·022	1·022
40	1·012	1·012	1·013	1·013	1·013	1·013	1·013	1·014	1·014	1·014
60	1·006	1·006	1·006	1·007	1·007	1·007	1·007	1·007	1·007	1·007
120	1·002	1·002	1·002	1·002	1·002	1·002	1·002	1·002	1·002	1·002
∞	1·000	1·000	1·000	1·000	1·000	1·000	1·000	1·000	1·000	1·000
$\chi^2_{dm_H}$	139·149	145·461	151·084	157·800	162·481	146·7241	153·1979	158·9624	165·8410	170·6341

	$d = 7$									
	$m_H = 19$					$m_H = 20$				
M \ α	0·100	0·050	0·025	0·010	0·005	0·100	0·050	0·025	0·010	0·005
1	—	—	—	—	—	1·629	1·679	1·728	1·793	1·843
2	1·388	1·408	1·427	1·451	1·469	1·399	1·419	1·438	1·462	1·480
3	1·288	1·300	1·311	1·325	1·335	1·298	1·310	1·321	1·335	1·346
4	1·229	1·237	1·245	1·254	1·261	1·238	1·246	1·254	1·263	1·270
5	1·189	1·195	1·201	1·208	1·213	1·196	1·203	1·209	1·216	1·221
6	1·159	1·164	1·169	1·174	1·178	1·166	1·172	1·176	1·182	1·186
7	1·137	1·141	1·145	1·149	1·152	1·144	1·148	1·151	1·156	1·159
8	1·119	1·123	1·126	1·130	1·132	1·125	1·129	1·132	1·136	1·139
9	1·105	1·108	1·111	1·114	1·116	1·111	1·114	1·117	1·120	1·122
10	1·094	1·096	1·099	1·101	1·103	1·099	1·101	1·104	1·107	1·109
12	1·076	1·078	1·080	1·082	1·083	1·080	1·082	1·084	1·086	1·088
14	1·063	1·065	1·066	1·068	1·069	1·067	1·068	1·070	1·072	1·073
16	1·053	1·054	1·056	1·057	1·058	1·056	1·058	1·059	1·060	1·062
18	1·045	1·047	1·048	1·049	1·050	1·048	1·050	1·051	1·052	1·053
20	1·039	1·040	1·041	1·042	1·043	1·042	1·043	1·044	1·045	1·046
24	1·030	1·031	1·032	1·033	1·033	1·033	1·033	1·034	1·035	1·035
30	1·022	1·022	1·023	1·023	1·024	1·024	1·024	1·025	1·025	1·026
40	1·014	1·014	1·015	1·015	1·015	1·015	1·015	1·016	1·016	1·016
60	1·007	1·007	1·007	1·008	1·008	1·008	1·008	1·008	1·008	1·008
120	1·002	1·002	1·002	1·002	1·002	1·002	1·002	1·002	1·002	1·002
∞	1·000	1·000	1·000	1·000	1·000	1·000	1·000	1·000	1·000	1·000
$\chi^2_{dm_H}$	154·283	160·915	166·816	173·854	178·755	161·8270	168·6130	174·6478	181·8403	186·8468

$m_H = 8$

$M \backslash \alpha$	0·100	0·050	0·025	0·010	0·005
1	1·491	1·538	1·585	1·646	1·692
2	1·270	1·288	1·305	1·326	1·342
3	1·185	1·195	1·204	1·215	1·224
4	1·138	1·144	1·150	1·158	1·163
5	1·108	1·113	1·117	1·123	1·126
6	1·088	1·091	1·095	1·099	1·102
7	1·073	1·076	1·078	1·082	1·084
8	1·061	1·064	1·066	1·069	1·071
9	1·053	1·055	1·057	1·059	1·060
10	1·046	1·048	1·049	1·051	1·052
12	1·036	1·038	1·039	1·040	1·041
14	1·028	1·030	1·031	1·031	1·032
16	1·023	1·025	1·026	1·026	1·027
18	1·020	1·021	1·022	1·022	1·023
20	1·017	1·017	1·018	1·018	1·018
24	1·012	1·013	1·013	1·014	1·014
30	1·009	1·009	1·009	1·009	1·010
40	1·005	1·005	1·006	1·006	1·006
60	1·003	1·003	1·003	1·003	1·003
120	1·001	1·001	1·001	1·001	1·001
∞	1·000	1·000	1·000	1·000	1·000
$\chi^2_{dm_H}$	78·8596	83·6753	88·0041	93·2169	96·8781

$m_H = 9$

$M \backslash \alpha$	0·100	0·050	0·025	0·010	0·005
1	1·495	1·541	1·587	1·648	1·694
2	1·277	1·295	1·311	1·333	1·349
3	1·192	1·202	1·211	1·222	1·231
4	1·144	1·151	1·157	1·165	1·170
5	1·114	1·119	1·123	1·129	1·132
6	1·093	1·097	1·100	1·104	1·107
7	1·077	1·080	1·083	1·086	1·089
8	1·066	1·068	1·070	1·073	1·075
9	1·057	1·059	1·061	1·063	1·064
10	1·049	1·051	1·053	1·055	1·056
12	1·039	1·040	1·041	1·043	1·044
14	1·031	1·032	1·033	1·034	1·035
16	1·026	1·026	1·027	1·028	1·029
18	1·021	1·022	1·023	1·024	1·024
20	1·018	1·019	1·019	1·020	1·020
24	1·014	1·014	1·015	1·015	1·015
30	1·010	1·010	1·010	1·010	1·011
40	1·006	1·006	1·006	1·006	1·007
60	1·003	1·003	1·003	1·003	1·003
120	1·001	1·001	1·001	1·001	1·001
∞	1·000	1·000	1·000	1·000	1·000
$\chi^2_{dm_H}$	87·7430	92·8083	97·3531	102·8163	106·6476

$m_H = 10$

$M \backslash \alpha$	0·100	0·050	0·025	0·010	0·005
1	1·501	1·547	1·593	1·653	1·698
2	1·286	1·303	1·319	1·341	1·357
3	1·200	1·209	1·219	1·230	1·239
4	1·151	1·158	1·164	1·172	1·177
5	1·120	1·125	1·130	1·135	1·139
6	1·098	1·102	1·106	1·110	1·113
7	1·082	1·086	1·088	1·092	1·094
8	1·070	1·073	1·075	1·078	1·080
9	1·061	1·063	1·065	1·067	1·069
10	1·053	1·055	1·057	1·059	1·060
12	1·042	1·043	1·044	1·046	1·047
14	1·034	1·035	1·036	1·037	1·038
16	1·028	1·029	1·030	1·030	1·031
18	1·023	1·024	1·025	1·026	1·026
20	1·020	1·021	1·021	1·022	1·022
24	1·015	1·106	1·106	1·016	1·017
30	1·011	1·011	1·011	1·011	1·012
40	1·006	1·007	1·007	1·007	1·007
60	1·003	1·003	1·003	1·003	1·004
120	1·001	1·001	1·001	1·001	1·001
∞	1·000	1·000	1·000	1·000	1·000
$\chi^2_{dm_H}$	96·5782	101·8795	106·6286	112·3288	116·3211

$m_H = 11$

$M \backslash \alpha$	0·100	0·050	0·025	0·010	0·005
1	—	—	—	—	—
2	1·294	1·312	1·328	1·350	1·366
3	1·208	1·218	1·227	1·239	1·247
4	1·159	1·165	1·172	1·179	1·185
5	1·127	1·132	1·136	1·142	1·146
6	1·104	1·108	1·112	1·116	1·119
7	1·088	1·091	1·094	1·097	1·100
8	1·075	1·078	1·080	1·083	1·085
9	1·065	1·067	1·069	1·072	1·073
10	1·057	1·059	1·061	1·063	1·064
12	1·045	1·046	1·048	1·049	1·050
14	1·037	1·038	1·039	1·040	1·041
16	1·030	1·031	1·032	1·033	1·034
18	1·026	1·026	1·027	1·028	1·028
20	1·022	1·022	1·023	1·024	1·024
24	1·017	1·017	1·017	1·018	1·018
30	1·012	1·012	1·012	1·013	1·013
40	1·007	1·007	1·008	1·008	1·008
60	1·004	1·004	1·004	1·004	1·004
120	1·001	1·001	1·001	1·001	1·001
∞	1·000	1·000	1·000	1·000	1·000
$\chi^2_{dm_H}$	105·372	110·898	115·841	121·767	125·913

$d = 8$

	$m_H = 12$					$m_H = 13$				
$M \backslash \alpha$	0·100	0·050	0·025	0·010	0·005	0·100	0·050	0·025	0·010	0·005
1	1·516	1·562	1·608	1·667	1·713	—	—	—	—	—
2	1·304	1·321	1·338	1·359	1·375	1·313	1·331	1·347	1·369	1·385
3	1·216	1·226	1·236	1·248	1·256	1·225	1·235	1·245	1·257	1·266
4	1·166	1·173	1·180	1·187	1·193	1·174	1·181	1·188	1·196	1·201
5	1·134	1·139	1·143	1·149	1·153	1·141	1·146	1·151	1·156	1·161
6	1·111	1·114	1·118	1·122	1·126	1·117	1·121	1·125	1·129	1·132
7	1·093	1·097	1·099	1·103	1·105	1·099	1·102	1·105	1·109	1·111
8	1·080	1·083	1·085	1·088	1·090	1·085	1·088	1·090	1·093	1·096
9	1·070	1·072	1·074	1·076	1·078	1·074	1·077	1·079	1·081	1·083
10	1·061	1·063	1·065	1·067	1·068	1·066	1·067	1·069	1·071	1·073
12	1·049	1·050	1·051	1·053	1·054	1·052	1·054	1·055	1·057	1·058
14	1·039	1·041	1·042	1·043	1·044	1·043	1·044	1·045	1·046	1·047
16	1·033	1·034	1·035	1·036	1·036	1·035	1·036	1·037	1·038	1·039
18	1·028	1·029	1·029	1·030	1·031	1·030	1·031	1·032	1·033	1·033
20	1·024	1·024	1·025	1·026	1·026	1·026	1·027	1·027	1·028	1·028
24	1·018	1·019	1·019	1·020	1·020	1·020	1·020	1·021	1·021	1·022
30	1·013	1·013	1·013	1·014	1·014	1·014	1·014	1·015	1·015	1·015
40	1·008	1·008	1·008	1·009	1·009	1·009	1·009	1·009	1·009	1·010
60	1·004	1·004	1·004	1·004	1·004	1·004	1·004	1·005	1·005	1·005
120	1·001	1·001	1·001	1·001	1·001	1·001	1·001	1·001	1·001	1·001
∞	1·000	1·000	1·000	1·000	1·000	1·000	1·000	1·000	1·000	1·000
$\chi^2_{dm_H}$	114·1307	119·8709	125·0001	131·1412	135·4330	122·858	128·804	134·111	140·459	144·891

	$m_H = 14$					$m_H = 15$				
$M \backslash \alpha$	0·100	0·050	0·025	0·010	0·005	0·100	0·050	0·025	0·010	0·005
1	1·535	1·581	1·626	1·686	1·731	—	—	—	—	—
2	1·323	1·341	1·357	1·379	1·395	1·333	1·351	1·368	1·389	1·406
3	1·234	1·244	1·254	1·266	1·275	1·243	1·253	1·263	1·275	1·284
4	1·182	1·189	1·196	1·204	1·210	1·190	1·198	1·204	1·212	1·218
5	1·148	1·153	1·158	1·164	1·168	1·155	1·160	1·165	1·171	1·176
6	1·123	1·127	1·131	1·136	1·139	1·130	1·134	1·138	1·143	1·146
7	1·105	1·108	1·111	1·115	1·118	1·111	1·114	1·117	1·121	1·124
8	1·091	1·093	1·096	1·099	1·101	1·096	1·099	1·101	1·105	1·107
9	1·079	1·082	1·084	1·086	1·088	1·084	1·087	1·089	1·091	1·093
10	1·070	1·072	1·074	1·076	1·078	1·074	1·076	1·078	1·081	1·082
12	1·056	1·057	1·059	1·061	1·062	1·060	1·061	1·063	1·065	1·066
14	1·046	1·047	1·048	1·050	1·051	1·049	1·050	1·052	1·053	1·054
16	1·038	1·039	1·040	1·041	1·042	1·041	1·042	1·043	1·044	1·045
18	1·032	1·033	1·034	1·035	1·036	1·035	1·036	1·037	1·038	1·038
20	1·028	1·029	1·029	1·030	1·031	1·030	1·031	1·032	1·032	1·033
24	1·021	1·022	1·022	1·023	1·023	1·023	1·024	1·024	1·025	1·025
30	1·015	1·016	1·016	1·016	1·017	1·016	1·017	1·017	1·018	1·018
40	1·010	1·010	1·010	1·010	1·010	1·010	1·011	1·011	1·011	1·011
60	1·005	1·005	1·005	1·005	1·005	1·005	1·005	1·005	1·006	1·006
120	1·001	1·001	1·001	1·001	1·001	1·002	1·002	1·002	1·002	1·002
∞	1·000	1·000	1·000	1·000	1·000	1·000	1·000	1·000	1·000	1·000
$\chi^2_{dm_H}$	131·5576	137·7015	143·1801	149·7269	154·2944	140·233	146·567	152·211	158·950	163·648

$d = 8$

	$m_H = 16$					$m_H = 17$				
$M \backslash \alpha$	0·100	0·050	0·025	0·010	0·005	0·100	0·050	0·025	0·010	0·005
1	1·555	1·601	1·646	1·706	1·751	—	—	—	—	—
2	1·343	1·361	1·378	1·400	1·416	1·353	1·371	1·388	1·410	—
3	1·252	1·263	1·272	1·285	1·294	1·261	1·272	1·282	1·294	1·303
4	1·198	1·206	1·212	1·221	1·227	1·207	1·214	1·221	1·229	1·235
5	1·162	1·168	1·173	1·179	1·183	1·170	1·175	1·180	1·187	1·191
6	1·136	1·141	1·145	1·149	1·153	1·143	1·147	1·151	1·156	1·160
7	1·117	1·120	1·123	1·127	1·130	1·123	1·126	1·130	1·134	1·136
8	1·101	1·104	1·107	1·110	1·113	1·107	1·110	1·113	1·116	1·118
9	1·089	1·092	1·094	1·097	1·099	1·094	1·097	1·099	1·102	1·104
10	1·079	1·081	1·083	1·085	1·087	1·084	1·086	1·088	1·090	1·092
12	1·064	1·065	1·067	1·069	1·070	1·067	1·069	1·071	1·073	1·074
14	1·052	1·054	1·055	1·056	1·057	1·056	1·057	1·058	1·060	1·061
16	1·044	1·045	1·046	1·047	1·048	1·047	1·048	1·049	1·050	1·051
18	1·038	1·038	1·039	1·040	1·041	1·040	1·041	1·042	1·043	1·044
20	1·032	1·033	1·034	1·035	1·035	1·035	1·036	1·036	1·037	1·038
24	1·025	1·026	1·026	1·027	1·027	1·027	1·027	1·028	1·029	1·029
30	1·018	1·018	1·019	1·019	1·019	1·019	1·020	1·020	1·021	1·021
40	1·011	1·012	1·012	1·012	1·012	1·012	1·013	1·013	1·013	1·013
60	1·006	1·006	1·006	1·006	1·006	1·006	1·006	1·007	1·007	1·007
120	1·002	1·002	1·002	1·002	1·002	1·002	1·002	1·002	1·002	1·002
∞	1·000	1·000	1·000	1·000	1·000	1·000	1·000	1·000	1·000	1·000
$\chi^2_{dm_H}$	148·8853	155·4047	161·2087	168·1332	172·9575	157·518	164·216	170·175	177·280	182·226

	$m_H = 18$				
$M \backslash \alpha$	0·100	0·050	0·025	0·010	0·005
1	1·575	1·621	1·667	1·727	1·773
2	1·363	1·381	1·398	1·420	1·437
3	1·270	1·281	1·291	1·304	1·313
4	1·215	1·222	1·229	1·238	1·244
5	1·177	1·183	1·188	1·194	1·199
6	1·150	1·154	1·158	1·163	1·167
7	1·129	1·133	1·136	1·140	1·143
8	1·112	1·115	1·118	1·122	1·124
9	1·099	1·102	1·104	1·107	1·109
10	1·088	1·091	1·093	1·095	1·097
12	1·071	1·073	1·075	1·077	1·078
14	1·059	1·061	1·062	1·064	1·065
16	1·050	1·051	1·052	1·054	1·055
18	1·043	1·044	1·045	1·046	1·047
20	1·037	1·038	1·039	1·040	1·040
24	1·029	1·029	1·030	1·031	1·031
30	1·021	1·021	1·022	1·022	1·022
40	1·013	1·014	1·014	1·014	1·014
60	1·007	1·007	1·007	1·007	1·007
120	1·002	1·002	1·002	1·002	1·002
∞	1·000	1·000	1·000	1·000	1·000
$\chi^2_{dm_H}$	166·1318	173·0041	179·1137	186·3930	191·4585

$d = 9$

$M \backslash \alpha$	$m_H = 9$ 0·100	0·050	0·025	0·010	0·005	$m_H = 10$ 0·100	0·050	0·025	0·010	0·005
1	1·495	1·540	1·585	1·645	1·690	1·497	1·542	1·586	1·645	1·690
2	1·282	1·299	1·315	1·337	1·353	1·288	1·305	1·321	1·342	1·357
3	1·197	1·207	1·216	1·227	1·236	1·203	1·213	1·222	1·233	1·242
4	1·149	1·156	1·162	1·169	1·175	1·155	1·162	1·168	1·175	1·181
5	1·119	1·123	1·128	1·133	1·137	1·124	1·129	1·133	1·139	1·143
6	1·097	1·101	1·104	1·108	1·111	1·102	1·106	1·109	1·113	1·116
7	1·081	1·084	1·087	1·090	1·093	1·086	1·089	1·092	1·095	1·097
8	1·069	1·072	1·074	1·077	1·079	1·073	1·076	1·078	1·081	1·083
9	1·060	1·062	1·064	1·066	1·068	1·064	1·066	1·068	1·070	1·072
10	1·052	1·054	1·056	1·058	1·059	1·056	1·058	1·059	1·061	1·063
12	1·041	1·043	1·044	1·045	1·046	1·044	1·045	1·047	1·048	1·049
14	1·033	1·034	1·035	1·036	1·037	1·036	1·037	1·038	1·039	1·040
16	1·027	1·028	1·029	1·030	1·031	1·030	1·030	1·031	1·032	1·033
18	1·023	1·024	1·024	1·025	1·026	1·025	1·026	1·026	1·027	1·028
20	1·020	1·020	1·021	1·022	1·022	1·021	1·022	1·023	1·023	1·024
24	1·015	1·015	1·016	1·016	1·017	1·016	1·017	1·017	1·018	1·018
30	1·010	1·011	1·011	1·011	1·012	1·011	1·012	1·012	1·012	1·013
40	1·006	1·007	1·007	1·007	1·007	1·007	1·007	1·007	1·008	1·008
60	1·003	1·003	1·003	1·003	1·003	1·003	1·004	1·004	1·004	1·004
120	1·001	1·001	1·001	1·001	1·001	1·001	1·001	1·001	1·001	1·001
∞	1·000	1·000	1·000	1·000	1·000	1·000	1·000	1·000	1·000	1·000
$\chi^2_{dm_H}$	97·6796	103·0095	107·7834	113·5124	117·5242	107·5650	113·1453	118·1359	124·1163	128·2989

$M \backslash \alpha$	$m_H = 11$ 0·100	0·050	0·025	0·010	0·005	$m_H = 12$ 0·100	0·050	0·025	0·010	0·005
1	—	—	—	—	—	1·506	1·550	1·594	1·652	1·696
2	1·294	1·311	1·327	1·348	1·364	1·302	1·319	1·335	1·355	1·371
3	1·210	1·219	1·229	1·240	1·248	1·217	1·227	1·236	1·247	1·256
4	1·161	1·168	1·174	1·182	1·187	1·168	1·175	1·181	1·188	1·194
5	1·130	1·134	1·139	1·144	1·148	1·136	1·141	1·145	1·151	1·155
6	1·107	1·111	1·114	1·119	1·122	1·113	1·116	1·120	1·124	1·127
7	1·091	1·094	1·096	1·100	1·102	1·095	1·099	1·101	1·105	1·107
8	1·078	1·080	1·083	1·085	1·087	1·082	1·085	1·087	1·090	1·092
9	1·068	1·070	1·072	1·074	1·076	1·072	1·074	1·076	1·078	1·080
10	1·059	1·061	1·063	1·065	1·066	1·063	1·065	1·067	1·069	1·070
12	1·047	1·048	1·050	1·051	1·052	1·050	1·052	1·053	1·055	1·056
14	1·038	1·039	1·040	1·042	1·043	1·041	1·042	1·043	1·044	1·045
16	1·032	1·033	1·034	1·035	1·035	1·034	1·035	1·036	1·037	1·038
18	1·027	1·028	1·028	1·029	1·030	1·029	1·030	1·030	1·031	1·032
20	1·023	1·024	1·024	1·025	1·025	1·025	1·026	1·026	1·027	1·027
24	1·018	1·018	1·018	1·019	1·019	1·019	1·019	1·020	1·020	1·021
30	1·012	1·013	1·013	1·013	1·014	1·013	1·014	1·014	1·014	1·015
40	1·008	1·008	1·008	1·008	1·008	1·008	1·009	1·009	1·009	1·009
60	1·004	1·004	1·004	1·004	1·004	1·004	1·004	1·004	1·005	1·005
120	1·001	1·001	1·001	1·001	1·001	1·001	1·001	1·001	1·001	1·001
∞	1·000	1·000	1·000	1·000	1·000	1·000	1·000	1·000	1·000	1·000
$\chi^2_{dm_H}$	117·407	123·225	128·422	134·642	138·987	127·2111	133·2569	138·6506	145·0988	149·5994

$d = 9$

$m_H = 13$

M \ α	0·100	0·050	0·025	0·010	0·005
1	—	—	—	—	—
2	1·310	1·326	1·343	1·363	1·379
3	1·224	1·234	1·243	1·255	1·263
4	1·175	1·181	1·188	1·195	1·201
5	1·142	1·147	1·151	1·157	1·161
6	1·118	1·122	1·126	1·130	1·133
7	1·101	1·104	1·107	1·110	1·113
8	1·087	1·089	1·092	1·095	1·097
9	1·076	1·078	1·080	1·083	1·084
10	1·067	1·069	1·071	1·073	1·074
12	1·054	1·055	1·056	1·058	1·059
14	1·044	1·045	1·046	1·047	1·048
16	1·037	1·038	1·039	1·040	1·040
18	1·031	1·032	1·033	1·034	1·034
20	1·027	1·028	1·028	1·029	1·029
24	1·020	1·021	1·021	1·022	1·022
30	1·015	1·015	1·015	1·016	1·016
40	1·009	1·009	1·010	1·010	1·010
60	1·005	1·005	1·005	1·005	1·005
120	1·001	1·001	1·001	1·001	1·001
∞	1·000	1·000	1·000	1·000	1·000
$\chi^2_{dm_H}$	136·982	143·246	148·829	155·496	160·146

$m_H = 14$

M \ α	0·100	0·050	0·025	0·010	0·005
1	1·520	1·563	1·607	1·664	1·708
2	1·318	1·335	1·351	1·371	1·387
3	1·232	1·242	1·251	1·263	1·271
4	1·182	1·189	1·195	1·203	1·208
5	1·148	1·153	1·158	1·164	1·168
6	1·124	1·128	1·132	1·136	1·139
7	1·106	1·109	1·112	1·116	1·118
8	1·092	1·094	1·097	1·100	1·102
9	1·080	1·083	1·085	1·087	1·089
10	1·071	1·073	1·075	1·077	1·079
12	1·057	1·059	1·060	1·062	1·063
14	1·047	1·048	1·049	1·051	1·051
16	1·039	1·040	1·041	1·042	1·043
18	1·033	1·034	1·035	1·036	1·037
20	1·029	1·030	1·030	1·031	1·032
24	1·022	1·023	1·023	1·024	1·024
30	1·016	1·016	1·016	1·017	1·017
40	1·010	1·010	1·010	1·011	1·011
60	1·005	1·005	1·005	1·005	1·005
120	1·001	1·001	1·002	1·002	1·002
∞	1·000	1·000	1·000	1·000	1·000
$\chi^2_{dm_H}$	146·7241	153·1979	158·9624	165·8410	170·6341

$m_H = 15$

M \ α	0·100	0·050	0·025	0·010	0·005
1	—	—	—	—	—
2	1·326	1·343	1·359	—	—
3	1·240	1·250	1·259	1·271	1·279
4	1·189	1·196	1·202	1·210	1·216
5	1·155	1·160	1·165	1·170	1·174
6	1·130	1·134	1·138	1·142	1·145
7	1·111	1·115	1·118	1·121	1·124
8	1·097	1·099	1·102	1·105	1·107
9	1·085	1·087	1·089	1·092	1·094
10	1·075	1·077	1·079	1·081	1·083
12	1·061	1·062	1·064	1·065	1·066
14	1·050	1·051	1·052	1·054	1·055
16	1·042	1·043	1·044	1·045	1·046
18	1·036	1·037	1·037	1·038	1·039
20	1·031	1·032	1·032	1·033	1·034
24	1·024	1·024	1·025	1·025	1·026
30	1·017	1·017	1·018	1·018	1·018
40	1·011	1·011	1·011	1·012	1·012
60	1·005	1·006	1·006	1·006	1·006
120	1·002	1·002	1·002	1·002	1·002
∞	1·000	1·000	1·000	1·000	1·000
$\chi^2_{dm_H}$	156·440	163·116	169·056	176·138	181·070

$m_H = 16$

M \ α	0·100	0·050	0·025	0·010	0·005
1	1·536	1·579	1·622	1·679	1·722
2	1·335	1·352	1·368	1·389	1·404
3	1·248	1·258	1·267	1·279	1·288
4	1·196	1·203	1·210	1·218	1·223
5	1·161	1·166	1·171	1·177	1·181
6	1·136	1·140	1·144	1·148	1·152
7	1·117	1·120	1·123	1·127	1·130
8	1·102	1·104	1·107	1·110	1·112
9	1·089	1·092	1·094	1·097	1·099
10	1·079	1·082	1·083	1·086	1·087
12	1·064	1·066	1·067	1·069	1·070
14	1·053	1·054	1·056	1·057	1·058
16	1·045	1·046	1·047	1·048	1·049
18	1·038	1·039	1·040	1·041	1·042
20	1·033	1·034	1·035	1·035	1·036
24	1·026	1·026	1·027	1·027	1·028
30	1·018	1·019	1·019	1·020	1·020
40	1·012	1·012	1·012	1·012	1·013
60	1·006	1·006	1·006	1·006	1·006
120	1·002	1·002	1·002	1·002	1·002
∞	1·000	1·000	1·000	1·000	1·000
$\chi^2_{dm_H}$	166·1318	173·0041	179·1137	186·3930	191·4585

D13 (Continued)

$d = 10$

M \ α	$m_H = 10$: 0·100	0·050	0·025	0·010	0·005	$m_H = 11$: 0·100	0·050	0·025	0·010	0·005
1	1·496	1·540	1·584	1·641	1·685	—	—	—	—	—
2	1·291	1·308	1·324	1·345	1·360	1·296	1·313	1·329	1·349	—
3	1·208	1·217	1·226	1·238	1·246	1·213	1·222	1·231	1·243	1·251
4	1·160	1·166	1·172	1·180	1·185	1·165	1·171	1·177	1·185	1·190
5	1·128	1·133	1·137	1·143	1·147	1·133	1·138	1·142	1·148	1·152
6	1·106	1·110	1·113	1·117	1·120	1·111	1·114	1·118	1·122	1·125
7	1·090	1·093	1·095	1·099	1·101	1·094	1·097	1·099	1·103	1·105
8	1·077	1·079	1·082	1·084	1·086	1·081	1·083	1·085	1·088	1·090
9	1·067	1·069	1·071	1·073	1·075	1·070	1·072	1·074	1·077	1·078
10	1·059	1·061	1·062	1·064	1·066	1·062	1·064	1·065	1·067	1·069
12	1·047	1·048	1·049	1·051	1·052	1·049	1·051	1·052	1·054	1·055
14	1·038	1·039	1·040	1·041	1·042	1·040	1·041	1·042	1·044	1·044
16	1·031	1·032	1·033	1·034	1·035	1·034	1·034	1·035	1·036	1·037
18	1·027	1·027	1·028	1·029	1·029	1·028	1·029	1·030	1·031	1·031
20	1·023	1·023	1·024	1·025	1·025	1·024	1·025	1·026	1·026	1·027
24	1·017	1·018	1·018	1·019	1·019	1·019	1·019	1·020	1·020	1·020
30	1·012	1·013	1·013	1·013	1·013	1·013	1·013	1·014	1·014	1·014
40	1·008	1·008	1·008	1·008	1·008	1·008	1·008	1·009	1·009	1·009
60	1·004	1·004	1·004	1·004	1·004	1·004	1·004	1·004	1·004	1·005
120	1·001	1·001	1·001	1·001	1·001	1·001	1·001	1·001	1·001	1·001
∞	1·000	1·000	1·000	1·000	1·000	1·000	1·000	1·000	1·000	1·000
$\chi^2_{dm_H}$	118·4980	124·3421	129·5612	135·8067	140·1695	129·385	135·480	140·917	147·414	151·948

M \ α	$m_H = 12$: 0·100	0·050	0·025	0·010	0·005	$m_H = 13$: 0·100	0·050	0·025	0·010	0·005
1	1·500	1·543	1·585	1·641	1·684	1·509	1·551	1·593	1·648	1·690
2	1·302	1·318	1·334	1·354	1·369	1·315	1·331	1·347	1·367	1·382
3	1·219	1·228	1·237	1·248	1·257	1·232	1·241	1·250	1·261	1·269
4	1·170	1·177	1·183	1·190	1·196	1·182	1·189	1·195	1·203	1·208
5	1·138	1·143	1·148	1·153	1·157	1·149	1·154	1·159	1·164	1·168
6	1·115	1·119	1·123	1·127	1·130	1·126	1·129	1·133	1·137	1·141
7	1·098	1·101	1·104	1·107	1·110	1·107	1·111	1·113	1·117	1·119
8	1·085	1·087	1·090	1·092	1·094	1·093	1·096	1·098	1·101	1·103
9	1·074	1·076	1·078	1·081	1·082	1·082	1·084	1·086	1·089	1·090
10	1·065	1·067	1·069	1·071	1·072	1·073	1·075	1·076	1·078	1·080
12	1·052	1·054	1·055	1·057	1·058	1·058	1·060	1·061	1·063	1·064
14	1·043	1·044	1·045	1·046	1·047	1·048	1·049	1·051	1·052	1·053
16	1·036	1·037	1·038	1·039	1·039	1·040	1·042	1·042	1·043	1·044
18	1·030	1·031	1·032	1·033	1·033	1·035	1·035	1·036	1·037	1·038
20	1·026	1·027	1·027	1·028	1·029	1·030	1·031	1·031	1·032	1·033
24	1·020	1·020	1·021	1·021	1·022	1·023	1·024	1·024	1·025	1·025
30	1·014	1·015	1·015	1·015	1·015	1·016	1·017	1·017	1·018	1·018
40	1·009	1·009	1·009	1·010	1·010	1·010	1·011	1·011	1·011	1·011
60	1·004	1·005	1·005	1·005	1·005	1·005	1·005	1·006	1·006	1·006
120	1·001	1·001	1·001	1·001	1·001	1·002	1·002	1·002	1·002	1·002
∞	1·000	1·000	1·000	1·000	1·000	1·000	1·000	1·000	1·000	1·000
$\chi^2_{dm_H}$	140·2326	146·5674	152·2114	158·9502	163·6482	161·8270	168·6130	174·6478	181·8403	186·8468

$\alpha = 0.05$ ν_2 \ ν_1	s = 2								
	0	1	2	3	4	5	7	10	15
2	.7919	.3514	.8839	.9045	.9189	.9295	.9441	.9573	.9693
3	.7017	.7761	.8197	.8487	.8696	.8853	.9075	.9283	.9478
4	.6267	.7090	.7600	.7953	.8213	.8413	.8702	.8980	.9247
5	.5646	.6507	.7063	.7459	.7758	.7992	.8337	.8676	.9011
6	.5130	.6002	.6585	.7010	.7337	.7598	.7988	.8380	.8775
7	.4696	.5564	.6160	.6604	.6952	.7232	.7658	.8095	.8544
8	.4328	.5182	.5782	.6238	.6599	.6893	.7348	.7822	.8318
9	.4011	.4846	.5445	.5906	.6276	.6581	.7058	.7562	.8100
10	.3737	.4550	.5143	.5606	.5980	.6293	.6786	.7316	.7889
15	.2781	.3477	.4015	.4455	.4826	.5145	.5670	.6266	.6955
20	.2211	.2810	.3287	.3688	.4034	.4339	.4855	.5463	.6198
25	.1835	.2355	.2780	.3143	.3463	.3478	.4239	.4835	.5580
30	.1568	.2027	.2408	.2738	.3031	.3296	.3760	.4333	.5071
35	.1369	.1780	.2124	.2425	.2696	.2924	.3377	.3924	.4644
40	.1242	.1585	.1898	.2175	.2425	.2655	.3064	.3585	.4282
48	.1031	.1352	.1626	.1870	.2093	.2299	.2670	.3150	.3807
60	.0836	.1103	.1333	.1540	.1731	.1909	.2233	.2661	.3260
80	.0638	.0846	.1027	.1192	.1346	.1409	.1756	.2114	.2630
120	.0433	.0577	.0704	.0821	.0931	.1035	.1230	.1498	.1896
240	.0220	.0295	.0362	.0424	.0483	.0540	.0647	.0798	.1030
∞	.0000	.0000	.0000	.0000	.0000	.0000	.0000	.0000	.0000

$\alpha = 0.01$ ν_2 \ ν_1	s = 2								
	0	1	2	3	4	5	7	10	15
2	.8826	.9173	.9358	.9475	.9556	.9615	.9695	.9768	.9834
3	.8074	.8575	.8863	.9051	.9185	.9286	.9427	.9557	.9679
4	.7381	.7989	.8357	.8607	.8789	.8929	.9129	.9318	.9499
5	.6770	.7446	.7873	.8171	.8394	.8568	.8820	.9066	.9306
6	.6237	.6954	.7422	.7758	.8013	.8215	.8514	.8810	.9106
7	.5773	.6512	.7008	.7371	.7652	.7877	.8215	.8556	.8903
8	.5369	.6116	.6630	.7013	.7313	.7556	.7927	.8308	.8702
9	.5014	.5762	.6285	.6683	.6997	.7255	.7652	.8067	.8503
10	.4701	.5443	.5971	.6378	.6703	.6972	.7391	.7834	.8309
15	.3573	.4247	.4757	.5168	.5511	.5803	.6279	.6812	.7418
20	.2876	.3473	.3941	.4329	.4661	.4951	.5435	.5998	.6670
25	.2404	.2935	.3360	.3719	.4032	.4309	.4782	.5347	.6045
30	.2065	.2540	.2926	.3258	.3550	.3812	.4265	.4819	.5521
35	.1811	.2239	.2592	.2898	.3171	.3417	.3847	.4383	.5077
40	.1610	.2000	.2325	.2608	.2863	.3094	.3503	.4017	.4697
48	.1372	.1712	.1999	.2251	.2480	.2689	.3064	.3544	.4193
60	.1117	.1403	.1646	.1863	.2061	.2244	.2576	.3008	.3607
80	.0855	.1080	.1273	.1448	.1609	.1759	.2035	.2402	.2925
120	.0582	.0740	.0877	.1002	.1118	.1228	.1433	.1711	.2120
240	.0297	.0380	.0453	.0520	.0583	.0644	.0758	.0917	.1160
∞	.0000	.0000	.0000	.0000	.0000	.0000	.0000	.0000	.0000

D14 (Continued)

$\alpha = 0.05$ ν_2 \ ν_1	$s = 3$ 0	1	2	3	4	5	7	10	15
2	.8646	.8986	.9188	.9322	.9417	.9489	.9590	.9684	.9771
3	.7922	.8386	.8676	.8876	.9022	.9134	.9296	.9450	.9596
4	.7266	.7815	.8172	.8426	.8617	.8766	.8983	.9195	.9402
5	.6689	.7292	.7698	.7994	.8221	.8400	.8668	.8934	.9199
6	.6185	.6820	.7261	.7589	.7844	.8040	.8358	.8672	.8992
7	.5745	.6398	.6861	.7212	.7489	.7715	.8060	.8416	.8785
8	.5359	.6019	.6497	.6864	.7158	.7400	.7774	.8167	.8580
9	.5019	.5679	.6165	.6544	.6850	.7104	.7503	.7926	.8380
10	.4718	.5373	.5862	.6249	.6564	.6828	.7246	.7696	.8185
15	.3620	.4219	.4690	.5079	.5407	.5691	.6158	.6687	.7299
20	.2931	.3465	.3898	.4265	.4582	.4862	.5334	.5889	.6559
25	.2461	.2937	.3332	.3671	.3970	.4237	.4697	.5252	.5944
30	.2120	.2548	.2907	.3221	.3500	.3752	.4192	.4734	.5429
35	.1863	.2250	.2579	.2869	.3129	.3366	.3784	.4308	.4993
40	.1660	.2013	.2316	.2584	.2828	.3050	.3447	.3905	.4620
48	.1417	.1726	.1994	.2234	.2452	.2654	.3018	.3486	.4125
60	.1157	.1417	.1644	.1850	.2042	.2217	.2538	.2961	.3550
80	.0888	.1093	.1274	.1440	.1592	.1740	.2008	.2366	.2880
120	.0606	.0750	.0879	.0999	.1111	.1217	.1415	.1687	.2089
240	.0310	.0386	.0455	.0519	.0580	.0639	.0750	.0905	.1143
∞	.0000	.0000	.0000	.0000	.0000	.0000	.0000	.0000	.0000

$\alpha = 0.01$ ν_2 \ ν_1	$s = 3$ 0	1	2	3	4	5	7	10	15
2	.9248	.9441	.9554	.9629	.9682	.9721	.9777	.9828	.9876
3	.8680	.8983	.9170	.9298	.9391	.9462	.9564	.9660	.9751
4	.8113	.8505	.8757	.8934	.9066	.9169	.9318	.9462	.9602
5	.7562	.8040	.8344	.8564	.8731	.8862	.9056	.9247	.9437
6	.7096	.7601	.7947	.8201	.8397	.8554	.8789	.9025	.9263
7	.6657	.7195	.7571	.7853	.8074	.8252	.8523	.8799	.9084
8	.6262	.6821	.7220	.7524	.7764	.7960	.8262	.8576	.8903
9	.5906	.6478	.6893	.7214	.7470	.7681	.8010	.8356	.8723
10	.5586	.6164	.6590	.6923	.7192	.7416	.7767	.8141	.8544
15	.4375	.4937	.5374	.5730	.6029	.6285	.6703	.7172	.7708
20	.3586	.4104	.4519	.4867	.5167	.5428	.5866	.6376	.6985
25	.3034	.3506	.3893	.4223	.4511	.4767	.5203	.5726	.6370
30	.2629	.3058	.3416	.3726	.3999	.4245	.4670	.5189	.5846
35	.2319	.2712	.3043	.3332	.3591	.3824	.4233	.4741	.5397
40	.2073	.2434	.2742	.3012	.3256	.3477	.3869	.4361	.5010
48	.1776	.2095	.2369	.2613	.2835	.3038	.3401	.3865	.4491
60	.1456	.1727	.1963	.2175	.2369	.2549	.2874	.3298	.3883
80	.1121	.1338	.1528	.1701	.1861	.2010	.2284	.2648	.3166
120	.0769	.0922	.1059	.1185	.1302	.1413	.1619	.1899	.2309
240	.0395	.0477	.0551	.0619	.0684	.0746	.0863	.1025	.1272
∞	.0000	.0000	.0000	.0000	.0000	.0000	.0000	.0000	.0000

D14 (Continued)

$\alpha = 0.05$ ν_2 \ ν_1	0	1	2	3	4	5	7	10	15
					$s = 4$				
2	.9045	.9259	.9393	.9486	.9554	.9606	.9680	.9751	.9818
3	.8463	.8773	.8976	.9121	.9229	.9313	.9436	.9555	.9671
4	.7904	.8287	.8548	.8738	.8884	.8998	.9169	.9337	.9504
5	.7388	.7825	.8132	.8360	.8537	.8679	.8892	.9108	.9326
6	.6920	.7396	.7737	.8000	.8199	.8364	.8616	.8875	.9141
7	.6499	.7000	.7367	.7650	.7875	.8060	.8345	.8643	.8954
8	.6120	.6638	.7024	.7325	.7568	.7769	.8083	.8414	.8767
9	.5779	.6307	.6706	.7021	.7277	.7492	.7830	.8192	.8583
10	.5472	.6004	.6412	.6737	.7004	.7229	.7588	.7976	.8401
15	.4307	.4822	.5235	.5578	.5869	.6121	.6538	.7012	.7561
20	.3543	.4017	.4409	.4742	.5031	.5286	.5719	.6228	.6843
25	.3006	.3439	.3802	.4117	.4395	.4644	.5072	.5590	.6235
30	.2609	.3004	.3341	.3636	.3899	.4137	.4552	.5064	.5720
35	.2306	.2667	.2978	.3254	.3502	.3728	.4127	.4626	.5279
40	.2063	.2396	.2685	.2943	.3177	.3391	.3773	.4256	.4899
48	.1770	.2065	.2323	.2555	.2768	.2964	.3317	.3772	.4391
60	.1454	.1704	.1927	.2129	.2315	.2488	.2805	.3219	.3796
80	.1122	.1322	.1501	.1666	.1820	.1964	.2230	.2586	.3094
120	.0770	.0913	.1042	.1162	.1274	.1381	.1581	.1854	.2257
240	.0397	.0473	.0542	.0608	.0670	.0730	.0843	.1002	.1243
∞	.0000	.0000	.0000	.0000	.0000	.0000	.0000	.0000	.0000

$\alpha = 0.01$ ν_2 \ ν_1	0	1	2	3	4	5	7	10	15
					$s = 4$				
2	.9473	.9593	.9668	.9719	.9757	.9785	.9826	.9865	.9901
3	.9032	.9231	.9361	.9453	.9521	.9574	.9651	.9726	.9797
4	.8567	.8836	.9017	.9149	.9248	.9327	.9443	.9557	.9670
5	.8110	.8436	.8663	.8830	.8959	.9062	.9216	.9371	.9526
6	.7677	.8049	.8312	.8510	.8665	.8790	.8980	.9174	.9371
7	.7275	.7680	.7973	.8197	.8375	.8519	.8742	.8972	.9211
8	.6903	.7333	.7650	.7895	.8092	.8254	.8505	.8769	.9047
9	.6561	.7010	.7345	.7607	.7820	.7996	.8273	.8567	.8882
10	.6247	.6708	.7057	.7334	.7560	.7748	.8047	.8369	.8717
15	.5016	.5490	.5867	.6177	.6439	.6664	.7034	.7452	.7930
20	.4175	.4627	.4997	.5309	.5579	.5815	.6213	.6678	.7234
25	.3570	.3992	.4343	.4645	.4910	.5146	.5550	.6033	.6631
30	.3117	.3502	.3837	.4125	.4380	.4609	.5007	.5494	.6111
35	.2765	.3126	.3435	.3707	.3951	.4171	.4558	.5039	.5661
40	.2483	.2819	.3108	.3365	.3596	.3807	.4181	.4651	.5270
48	.2138	.2438	.2699	.2933	.3145	.3341	.3691	.4139	.4742
60	.1763	.2021	.2249	.2454	.2643	.2818	.3135	.3548	.4118
80	.1367	.1575	.1760	.1930	.2087	.2234	.2505	.2864	.3373
120	.0943	.1092	.1227	.1350	.1469	.1579	.1785	.2065	.2474
240	.0488	.0568	.0642	.0711	.0777	.0839	.0957	.1122	.1371
∞	.0000	.0000	.0000	.0000	.0000	.0000	.0000	.0000	.0000

D14 (Continued)

$\alpha = 0.05$ $\nu_2 \backslash \nu_1$	$s = 5$								
	0	1	2	3	4	5	7	10	15
2	.9289	.9432	.9527	.9594	.9645	.9684	.9741	.9796	.9850
3	.8815	.9032	.9181	.9289	.9372	.9437	.9534	.9629	.9724
4	.8338	.8617	.8814	.8960	.9074	.9165	.9302	.9439	.9577
5	.7882	.8210	.8447	.8627	.8768	.8883	.9058	.9236	.9419
6	.7456	.7822	.8091	.8299	.8465	.8600	.8809	.9026	.9252
7	.7063	.7457	.7752	.7983	.8169	.8323	.8563	.8815	.9082
8	.6702	.7117	.7432	.7681	.7884	.8054	.8321	.8605	.8911
9	.6372	.6801	.7131	.7395	.7612	.7795	.8085	.8399	.8740
10	.6069	.6507	.6849	.7125	.7354	.7547	.7858	.8197	.8571
15	.4883	.5328	.5690	.5993	.6252	.6477	.6850	.7277	.7773
20	.4072	.4495	.4847	.5150	.5414	.5647	.6043	.6511	.7077
25	.3488	.3881	.4215	.4507	.4764	.4995	.5394	.5877	.6480
30	.3049	.3413	.3726	.4003	.4250	.4474	.4865	.5349	.5967
35	.2708	.3045	.3338	.3599	.3834	.4049	.4428	.4904	.5525
40	.2434	.2746	.3021	.3267	.3490	.3696	.4061	.4525	.5141
48	.2097	.2377	.2625	.2849	.3054	.3244	.3585	.4026	.4624
60	.1732	.1973	.2188	.2385	.2567	.2736	.3045	.3450	.4013
80	.1344	.1539	.1714	.1877	.2028	.2171	.2433	.2785	.3286
120	.0928	.1068	.1196	.1316	.1428	.1535	.1735	.2008	.2409
240	.0481	.0557	.0627	.0693	.0756	.0816	.0931	.1091	.1335
∞	.0000	.0000	.0000	.0000	.0000	.0000	.0000	.0000	.0000

$\alpha = 0.01$ $\nu_2 \backslash \nu_1$	$s = 5$								
	0	1	2	3	4	5	7	10	15
2	.9610	.9689	.9742	.9779	.9806	.9828	.9859	.9889	.9918
3	.9258	.9396	.9490	.9559	.9611	.9651	.9712	.9771	.9830
4	.8871	.9064	.9200	.9300	.9378	.9440	.9533	.9625	.9718
5	.8478	.8719	.8892	.9023	.9125	.9208	.9334	.9461	.9591
6	.8094	.8376	.8582	.8739	.8865	.8967	.9124	.9285	.9453
7	.7729	.8043	.8278	.8457	.8603	.8723	.8909	.9103	.9308
8	.7385	.7724	.7980	.8181	.8345	.8480	.8693	.8918	.9158
9	.7062	.7422	.7696	.7914	.8093	.8242	.8479	.8732	.9000
10	.6762	.7136	.7426	.7658	.7850	.8011	.8269	.8548	.8853
15	.5544	.5948	.6274	.6546	.6777	.6977	.7306	.7680	.8111
20	.4677	.5074	.5404	.5685	.5928	.6143	.6505	.6930	.7440
25	.4038	.4415	.4735	.5011	.5255	.5473	.5846	.6295	.6581
30	.3549	.3904	.4208	.4475	.4713	.4927	.5301	.5757	.6337
35	.3165	.3498	.3786	.4041	.4270	.4478	.4844	.5299	.5889
40	.2854	.3166	.3438	.3681	.3900	.4101	.4457	.4906	.5497
48	.2469	.2751	.2999	.3222	.3426	.3614	.3950	.4382	.4962
60	.2048	.2293	.2512	.2710	.2893	.3063	.3371	.3772	.4326
80	.1596	.1796	.1977	.2142	.2296	.2441	.2706	.3059	.3559
120	.1108	.1253	.1386	.1509	.1625	.1735	.1940	.2217	.2624
240	.0577	.0656	.0730	.0799	.0865	.0927	.1047	.1212	.1464
∞	.0000	.0000	.0000	.0000	.0000	.0000	.0000	.0000	.0000

D14 (Continued)

$\alpha = 0.05$ ν_2 \ ν_1	s = 6								
	0	1	2	3	4	5	7	10	15
2	.9450	.9551	.9620	.9670	.9709	.9740	.9785	.9829	.9873
3	.9058	.9216	.9328	.9411	.9476	.9528	.9606	.9684	.9763
4	.8649	.8858	.9010	.9126	.9217	.9290	.9402	.9517	.9633
5	.8246	.8499	.8686	.8830	.8945	.9039	.9185	.9335	.9491
6	.7861	.8149	.8365	.8535	.8671	.8784	.8960	.9145	.9340
7	.7499	.7814	.8054	.8245	.8401	.8531	.8735	.8952	.9184
8	.7160	.7496	.7757	.7966	.8138	.8282	.8511	.8758	.9026
9	.6845	.7198	.7474	.7698	.7884	.8041	.8292	.8566	.8867
10	.6552	.6917	.7206	.7442	.7640	.7808	.8078	.8377	.8708
15	.5372	.5759	.6077	.6346	.6577	.6779	.7115	.7500	.7951
20	.4535	.4913	.5231	.5506	.5746	.5960	.6324	.6754	.7278
25	.3919	.4276	.4583	.4852	.5091	.5306	.5677	.6128	.6692
30	.3447	.3782	.4074	.4333	.4565	.4775	.5144	.5600	.6184
35	.3076	.3390	.3665	.3912	.4135	.4338	.4699	.5151	.5743
40	.2775	.3069	.3329	.3563	.3777	.3973	.4322	.4767	.5357
48	.2403	.2668	.2905	.3120	.3317	.3500	.3830	.4265	.4833
60	.1995	.2226	.2434	.2624	.2801	.2966	.3267	.3662	.4210
80	.1556	.1745	.1916	.2075	.2224	.2364	.2623	.2969	.3462
120	.1081	.1218	.1344	.1463	.1574	.1681	.1880	.2151	.2551
240	.0563	.0638	.0708	.0775	.0838	.0899	.1014	.1176	.1422
∞	.0000	.0000	.0000	.0000	.0000	.0000	.0000	.0000	.0000

$\alpha = 0.01$ ν_2 \ ν_1	s = 6								
	0	1	2	3	4	5	7	10	15
2	.9699	.9755	.9793	.9820	.9842	.9858	.9883	.9907	.9931
3	.9412	.9512	.9583	.9635	.9676	.9708	.9757	.9805	.9854
4	.9086	.9230	.9334	.9413	.9475	.9525	.9600	.9677	.9756
5	.8745	.8930	.9065	.9169	.9252	.9320	.9424	.9531	.9642
6	.8406	.8625	.8789	.8917	.9019	.9104	.9236	.9373	.9517
7	.8075	.8323	.8512	.8661	.8782	.8883	.9040	.9207	.9384
8	.7758	.8031	.8241	.8409	.8546	.8661	.8842	.9037	.9247
9	.7458	.7750	.7978	.8161	.8313	.8441	.8645	.8865	.9106
10	.7173	.7482	.7724	.7922	.8086	.8225	.8449	.8694	.8964
15	.5986	.6334	.6619	.6858	.7063	.7241	.7535	.7872	.8262
20	.5111	.5462	.5757	.6010	.6231	.6426	.6757	.7147	.7616
25	.4450	.4792	.5081	.5335	.5559	.5760	.6106	.6524	.7042
30	.3936	.4261	.4542	.4789	.5011	.5211	.5561	.5990	.6536
35	.3527	.3835	.4103	.4342	.4557	.4754	.5100	.5531	.6090
40	.3194	.3484	.3740	.3969	.4177	.4367	.4706	.5134	.5698
48	.2775	.3040	.3276	.3488	.3683	.3863	.4187	.4602	.5160
60	.2315	.2548	.2757	.2948	.3125	.3289	.3588	.3977	.4515
80	.1814	.2006	.2181	.2342	.2493	.2634	.2894	.3240	.3730
120	.1266	.1407	.1538	.1659	.1774	.1882	.2085	.2360	.2763
240	.0663	.0741	.0841	.0883	.0949	.1012	.1132	.1298	.1550
∞	.0000	.0000	.0000	.0000	.0000	.0000	.0000	.0000	.0000

$\alpha = 0.05$ $\nu_2 \backslash \nu_1$	$s = 7$								
	0	1	2	3	4	5	7	10	15
2	.9561	.9635	.9687	.9726	.9757	.9781	.9817	.9854	.9890
3	.9232	.9351	.9438	.9504	.9556	.9598	.9662	.9727	.9793
4	.8879	.9040	.9160	.9253	.9327	.9388	.9481	.9577	.9677
5	.8523	.8722	.8872	.8990	.9085	.9163	.9285	.9413	.9548
6	.8176	.8406	.8582	.8722	.8837	.8932	.9082	.9241	.9410
7	.7843	.8099	.8298	.8457	.8589	.8700	.8875	.9063	.9267
8	.7527	.7804	.8022	.8199	.8345	.8469	.8668	.8884	.9120
9	.7229	.7523	.7757	.7948	.8108	.8244	.8463	.8705	.8972
10	.6949	.7257	.7503	.7707	.7878	.8025	.8263	.8527	.8823
15	.5792	.6130	.6411	.6651	.6858	.7039	.7342	.7692	.8103
20	.4944	.5282	.5570	.5820	.6040	.6236	.6570	.6968	.7453
25	.4305	.4631	.4914	.5162	.5384	.5583	.5930	.6351	.6879
30	.3809	.4119	.4390	.4632	.4850	.5048	.5395	.5825	.6378
35	.3415	.3707	.3965	.4198	.4409	.4603	.4944	.5374	.5939
40	.3093	.3369	.3615	.3837	.4040	.4227	.4561	.4986	.5552
48	.2690	.2941	.3167	.3373	.3562	.3738	.4056	.4466	.5024
60	.2244	.2465	.2665	.2850	.3021	.3181	.3474	.3858	.4392
80	.1759	.1942	.2109	.2264	.2401	.2547	.2801	.3141	.3626
120	.1229	.1363	.1487	.1605	.1715	.1820	.2018	.2287	.2684
240	.0644	.0718	.0788	.0854	.0918	.0979	.1095	.1257	.1504
∞	.0000	.0000	.0000	.0000	.0000	.0000	.0000	.0000	.0000

$\alpha = 0.01$ $\nu_2 \backslash \nu_1$	$s = 7$								
	0	1	2	3	4	5	7	10	15
2	.9761	.9801	.9830	.9851	.9868	.9881	.9901	.9921	.9940
3	.9522	.9597	.9651	.9693	.9725	.9751	.9791	.9831	.9872
4	.9244	.9354	.9436	.9499	.9549	.9590	.9653	.9718	.9785
5	.8947	.9091	.9199	.9284	.9352	.9408	.9496	.9587	.9682
6	.8645	.8819	.8952	.9057	.9143	.9214	.9325	.9443	.9568
7	.8346	.8546	.8701	.8825	.8927	.9013	.9147	.9292	.9447
8	.8055	.8278	.8452	.8594	.8710	.8809	.8965	.9135	.9320
9	.7776	.8017	.8209	.8365	.8495	.8605	.8782	.8976	.9189
10	.7508	.7766	.7971	.8141	.8282	.8403	.8600	.8815	.9056
15	.6363	.6665	.6915	.7126	.7308	.7467	.7732	.8037	.8392
20	.5491	.5803	.6068	.6297	.6497	.6675	.6978	.7337	.7770
25	.4817	.5125	.5391	.5624	.5831	.6016	.6338	.6726	.7210
30	.4286	.4583	.4843	.5073	.5280	.5467	.5795	.6198	.6713
35	.3858	.4142	.4393	.4617	.4820	.5005	.5332	.5740	.6272
40	.3506	.3777	.4017	.4233	.4431	.4612	.4934	.5342	.5881
48	.3060	.3309	.3533	.3736	.3922	.4094	.4405	.4803	.5341
60	.2565	.2787	.2987	.3171	.3341	.3500	.3789	.4167	.4689
80	.2020	.2205	.2375	.2532	.2678	.2816	.3070	.3409	.3889
120	.1418	.1556	.1683	.1803	.1916	.2023	.2223	.2496	.2894
240	.0747	.0824	.0897	.0966	.1031	.1094	.1214	.1380	.1633
∞	.0000	.0000	.0000	.0000	.0000	.0000	.0000	.0000	.0000

	m_E \ m_H	2	3	4	5	6	8	$d = 2$ 10	12	15	20	25	40	60
5%	2	9.8591†	10.659†	11.098†	11.373†	11.562†	11.804†	11.952†	12.052†	12.153†	12.254†	12.316†	12.409†	12.461†
	3	58.428	58.915	59.161	59.308	59.407	59.531	59.606	59.655	59.705	59.755	59.785	59.830	59.855
	4	23.999	23.312	22.918	22.663	22.484	22.250	22.104	22.003	21.901	21.797	21.733	21.636	21.582
	5	15.639	14.864	14.422	14.135	13.934	13.670	13.504	13.391	13.275	13.156	13.083	12.972	12.909
	6	12.175	11.411	10.975	10.691	10.491	10.228	10.063	9.9489	9.8320	9.7118	9.6381	9.5251	9.4610
	7	10.334	9.5937	9.1694	8.8927	8.6975	8.4396	8.2765	8.1639	8.0480	7.9285	7.8549	7.7417	7.6773
	8	9.2069	8.4881	8.0752	7.8054	7.6145	7.3614	7.2008	7.0896	6.9748	6.8560	6.7826	6.6694	6.6048
	10	7.9095	7.2243	6.8294	6.5702	6.3860	6.1405	5.9837	5.8745	5.7612	5.6433	5.5701	5.4564	5.3910
	12	7.1902	6.5284	6.1461	5.8942	5.7147	5.4744	5.3200	5.2122	5.0997	4.9820	4.9085	4.7938	4.7274
	14	6.7350	6.0902	4.7168	5.4703	5.2941	5.0574	4.9048	4.7977	4.6856	4.5678	4.4939	4.3780	4.3105
	16	6.4217	5.7895	5.4230	5.1804	5.0067	4.7727	4.6213	4.5147	4.4028	4.2846	4.2102	4.0930	4.0243
	18	6.1932	5.5708	5.2095	4.9700	4.7982	4.5663	4.4157	4.3094	4.1976	4.0791	4.0042	3.8855	3.8158
	20	6.0192	5.4046	5.0475	4.8105	4.6402	4.4099	4.2600	4.1539	4.0420	3.9231	3.8477	3.7278	3.6569
	25	5.7244	5.1237	4.7741	4.5415	4.3740	4.1465	3.9977	3.8919	3.7798	3.6598	3.5832	3.4605	3.3868
	30	5.5401	4.9487	4.6040	4.3743	4.2086	3.9829	3.8347	3.7291	3.6166	3.4957	3.4181	3.2926	3.2168
	35	5.4140	4.8291	4.4880	4.2604	4.0959	3.8715	3.7237	3.6181	3.5054	3.3836	3.3051	3.1774	3.1000
	40	5.3224	4.7424	4.4039	4.1778	4.0143	3.7908	3.6433	3.5377	3.4247	3.3022	3.2230	3.0933	3.0140
	50	5.1981	4.6249	4.2900	4.0661	3.9039	3.6817	3.5346	3.4289	3.3154	3.1919	3.1115	2.9787	2.8965
	60	5.1178	4.5490	4.2166	3.9941	3.8328	3.6114	3.4646	3.3588	3.2450	3.1206	3.0392	2.9041	2.8196
	70	5.0616	4.4960	4.1653	3.9439	3.7831	3.5624	3.4157	3.3099	3.1957	3.0706	2.9886	2.8516	2.7652
	80	5.0200	4.4569	4.1275	3.9068	3.7465	3.5262	3.3796	3.2737	3.1594	3.0338	2.9512	2.8126	2.7247
	100	4.9628	4.4030	4.0754	3.8557	3.6961	3.4764	3.3300	3.2240	3.1093	2.9829	2.8994	2.7586	2.6683
	200	4.8514	4.2982	3.9742	3.7567	3.5983	3.3798	3.2336	3.1275	3.0120	2.8838	2.7984	2.6520	2.5559
	∞	4.7442	4.1973	3.8769	3.6614	3.5044	3.2870	3.1410	3.0346	2.9182	2.7879	2.7002	2.5470	2.4428
1%	2	2.4673*	2.6666*	2.7758*	2.8444*	2.8914*	2.9517*	2.9886*	3.0135*	3.0387*	3.0641*	3.0796*	3.1025*	3.1154*
	3	2.9849†	2.9898†	2.9923†	2.9938†	2.9948†	2.9961†	2.9968†	2.9973†	2.9978†	2.9983†	2.9985†	2.9991†	2.9993†
	4	74.275	71.026	69.244	68.116	67.337	66.332	65.712	65.290	64.862	64.427	64.163	63.763	63.538
	5	38.295	35.567	34.070	33.121	32.465	31.615	31.088	30.729	30.364	29.993	29.766	29.422	29.228
	6	26.118	23.794	22.517	21.706	21.143	20.413	19.958	19.648	19.332	19.009	18.812	18.511	18.341
	7	20.388	18.326	17.191	16.469	15.967	15.313	14.905	14.626	14.341	14.049	13.870	13.596	13.442
	8	17.152	15.268	14.229	13.567	13.106	12.504	12.127	11.868	11.603	11.331	11.165	10.909	10.764
	10	13.701	12.038	11.120	10.531	10.121	9.5819	9.2431	9.0096	8.7694	8.5215	8.3688	8.1334	7.9990
	12	11.920	10.388	9.5405	8.9961	8.6148	8.1132	7.7962	7.5770	7.3505	7.1159	6.9707	6.7457	6.6166
	14	10.844	9.3990	8.5974	8.0816	7.7196	7.2419	6.9389	6.7287	6.5109	6.2843	6.1435	5.9244	5.7980
	16	10.128	8.7432	7.9743	7.4786	7.1301	6.6691	6.3758	6.1718	5.9598	5.7385	5.6002	5.3847	5.2596
	18	9.6174	8.2781	7.5334	7.0527	6.7142	6.2655	5.9793	5.7797	5.5718	5.3540	5.2178	5.0040	4.8794
	20	9.2360	7.9316	7.2056	6.7365	6.4057	5.9664	5.6855	5.4893	5.2844	5.0692	4.9342	4.7214	4.5970
	25	8.6044	7.3601	6.6663	6.2169	5.8993	5.4760	5.2042	5.0134	4.8135	4.6021	4.4686	4.2577	4.1314
	30	8.2188	7.0127	6.3394	5.9026	5.5933	5.1801	4.9138	4.7264	4.5293	4.3199	4.1870	3.9745	3.8479
	35	7.9592	6.7796	6.1205	5.6923	5.3888	4.9825	4.7200	4.5348	4.3395	4.1312	3.9985	3.7850	3.6581
	40	7.7727	6.6125	5.9638	5.5420	5.2426	4.8413	4.5816	4.3980	4.2038	3.9962	3.8635	3.6488	3.5191
	50	7.5228	6.3891	5.7545	5.3414	5.0477	4.6533	4.3972	4.2156	4.0230	3.8161	3.6830	3.4661	3.3335
	60	7.3630	6.2465	5.6212	5.2138	4.9238	4.5338	4.2800	4.0998	3.9081	3.7014	3.5676	3.3490	3.2139
	70	7.2520	6.1478	5.5290	5.1255	4.8382	4.4512	4.1991	4.0197	3.8285	3.6219	3.4880	3.2675	3.1303
	80	7.1705	6.0753	5.4613	5.0608	4.7754	4.3908	4.1398	3.9610	3.7703	3.5636	3.4294	3.2074	3.0685
	100	7.0588	5.9761	5.3688	4.9723	4.6896	4.3081	4.0587	3.8808	3.6906	3.4839	3.3491	3.1247	2.9831
	200	6.8435	5.7852	5.1910	4.8025	4.5251	4.1497	3.9034	3.7271	3.5377	3.3304	3.1941	2.9640	2.8152
	∞	6.6385	5.6040	5.0226	4.6419	4.3695	4.0000	3.7566	3.5817	3.3928	3.1845	3.0462	2.8082	2.6492

†Multiply entry by 100. *Multiply entry by 10^4.

D15 (Continued)

	m_E \ m_H	3	4	5	6	8	10	12	15	20	25	40	60
						$d = 3$							
5%	3	25·930*	26·996*	27·665*	28·125*	28·712*	29·073*	29·316*	29·561*	29·809*	29·959*	30·19*	30·31*
	4	1·1880*	1·1929*	1·1959*	1·1978*	1·2003*	1·2018*	1·2028*	1·2038*	1·2048*	1·2054*	1·2063*	1·2068*
	5	42·474	41·764	41·305	40·983	40·562	40·300	40·120	39·937	39·750	39·635	39·462	39·366
	6	25·456	24·715	24·235	23·899	23·458	23·182	22·992	22·799	22·600	22·479	22·294	22·190
	7	18·752	18·056	17·605	17·288	16·870	16·608	16·427	16·241	16·051	15·934	15·755	15·653
	8	15·308	14·657	14·233	13·934	13·540	13·290	13·118	12·941	12·758	12·646	12·473	12·375
	10	11·893	11·306	10·921	10·649	10·287	10·057	9·8974	9·7320	9·5603	9·4541	9·2897	9·1955
	12	10·229	9·6825	9·3234	9·0680	8·7271	8·5088	8·3566	8·1982	8·0330	7·9301	7·7700	7·6777
	14	9·2550	8·7356	8·3935	8·1495	7·8225	7·6122	7·4649	7·3110	7·1497	7·0488	6·8908	6·7991
	16	8·6180	8·1183	7·7884	7·5526	7·2355	7·0307	6·8868	6·7360	6·5772	6·4774	6·3204	6·2287
	18	8·1701	7·6851	7·3644	7·1347	6·8251	6·6244	6·4830	6·3343	6·1771	6·0780	5·9212	5·8292
	20	7·8384	7·3649	7·0513	6·8263	6·5224	6·3249	6·1853	6·0383	5·8822	5·7834	5·6266	5·5341
	25	7·2943	6·8407	6·5394	6·3227	6·0287	5·8365	5·7001	5·5555	5·4010	5·3025	5·1446	5·0503
	30	6·9654	6·5245	6·2311	6·0196	5·7319	5·5431	5·4085	5·2654	5·1116	5·0129	4·8535	4·7575
	35	6·7453	6·3132	6·0253	5·8175	5·5341	5·3476	5·2143	5·0720	4·9185	4·8195	4·6586	4·5608
	40	6·5877	6·1621	5·8783	5·6732	5·3929	5·2081	5·0757	4·9340	4·7806	4·6813	4·5189	4·4195
	50	6·3773	5·9606	5·6823	5·4809	5·2050	5·0224	4·8911	4·7502	4·5967	4·4968	4·3319	4·2297
	60	6·2433	5·8324	5·5577	5·3587	5·0856	4·9044	4·7739	4·6334	4·4798	4·3793	4·2123	4·1078
	70	6·1504	5·7436	5·4715	5·2742	5·0031	4·8229	4·6929	4·5526	4·3988	4·2979	4·1292	4·0227
	80	6·0823	5·6786	5·4084	5·2122	4·9426	4·7632	4·6336	4·4935	4·3395	4·2381	4·0680	3·9600
	100	5·9891	5·5896	5·3220	5·1276	4·8601	4·6817	4·5525	4·4126	4·2583	4·1563	3·9840	3·8734
	200	5·8099	5·4186	5·1562	4·9653	4·7017	4·5252	4·3970	4·2574	4·1023	3·9988	3·8212	3·7042
	∞	5·6397	5·2565	4·9992	4·8116	4·5519	4·3773	4·2499	4·1104	3·9541	3·8487	3·6642	3·5384
1%	3	6·4845†	6·7500†	6·9169†	7·0313†	7·1778†	7·2675†	7·3281†	7·3891†	7·4511†	7·4883†	—	—
	4	5·9896*	5·9946*	5·9976*	5·9996*	6·0021*	6·0035*	6·0046*	6·0056*	6·0067*	6·0071*	6·008*	6·008*
	5	1·2738*	1·2420*	1·2219*	1·2080*	1·1901*	1·1790*	1·1715*	1·1638*	1·1561*	1·1514*	1·144*	1·141*
	6	59·507	57·032	55·462	54·377	52·973	52·102	51·509	50·906	50·292	49·918	49·349	49·04
	7	37·994	35·993	34·721	33·840	32·695	31·984	31·498	31·002	30·496	30·188	29·718	29·452
	8	28·308	26·599	25·511	24·755	23·771	23·157	22·737	22·308	21·868	21·599	21·188	20·955
	10	19·737	18·355	17·471	16·855	16·050	15·544	15·197	14·840	14·472	14·246	13·899	13·702
	12	15·973	14·765	13·990	13·448	12·737	12·288	11·978	11·659	11·328	11·124	10·809	10·628
	14	13·905	12·803	12·096	11·599	10·945	10·530	10·243	9·9462	9·6377	9·4463	9·1490	8·9780
	16	12·610	11·581	10·918	10·452	9·8359	9·4444	9·1724	8·8900	8·5955	8·4121	8·1260	7·9605
	18	11·729	10·751	10·120	9·6756	9·0870	8·7117	8·4503	8·1782	7·8934	7·7154	7·4365	7·2743
	20	11·091	10·152	9·5452	9·1173	8·5492	8·1861	7·9325	7·6679	7·3901	7·2159	6·9419	6·7818
	25	10·075	9·2005	8·6339	8·2333	7·6992	7·3560	7·1152	6·8627	6·5958	6·4273	6·1598	6·0019
	30	9·4785	8·6441	8·1022	7·7183	7·2050	6·8739	6·6407	6·3953	6·1346	5·9690	5·7042	5·5464
	35	9·0874	8·2798	7·7548	7·3822	6·8829	6·5598	6·3317	6·0909	5·8339	5·6700	5·4063	5·2478
	40	8·8113	8·0233	7·5108	7·1460	6·6564	6·3392	6·1147	5·8771	5·6227	5·4598	5·1962	5·0367
	50	8·4479	7·6861	7·1894	6·8358	6·3599	6·0503	5·8305	5·5970	5·3457	5·1838	4·9196	4·7578
	60	8·2195	7·4745	6·9882	6·6416	6·1744	5·8696	5·6528	5·4218	5·1722	5·0108	4·7455	4·5815
	70	8·0627	7·3295	6·8504	6·5087	6·0474	5·7460	5·5312	5·3019	5·0535	4·8922	4·6258	4·4598
	80	7·9485	7·2239	6·7502	6·4120	5·9551	5·6562	5·4428	5·2147	4·9670	4·8058	4·5383	4·3706
	100	7·7932	7·0805	6·6141	6·2809	5·8300	5·5344	5·3230	5·0965	4·8497	4·6883	4·4190	4·2484
	200	7·4980	6·8083	6·3561	6·0323	5·5930	5·3037	5·0961	4·8725	4·6270	4·4650	4·1906	4·0124
	∞	7·2220	6·5542	6·1156	5·8009	5·3725	5·0892	4·8849	4·6638	4·4190	4·2557	3·9738	3·7843

* Multiply entry by 100. † Multiply entry by 10^4.

$d = 4$

	$m_E \backslash m_H$	4	5	6	8	10	12	15	20	25	40	60
5%	4	49·964*	51·204*	52·054*	53·142*	53·808*	54·258*	54·71*	55·17*	55·46*	—	—
	5	1·9964*	2·0013*	2·0046*	2·0087*	2·0112*	2·0128*	2·0145*	2·0161*	2·0171*	2·019*	—
	6	65·715	64·999	64·497	63·841	63·432	63·151	62·866	62·573	62·396	62·13	—
	7	37·343	36·629	36·129	35·474	35·064	34·782	34·495	34·200	34·019	33·75	—
	8	26·516	25·868	25·413	24·814	24·437	24·178	23·912	23·639	23·471	23·214	23·072
	10	17·875	17·326	16·938	16·424	16·098	15·872	15·640	15·399	15·250	15·021	14·891
	12	14·338	13·848	13·500	13·037	12·741	12·535	12·321	12·099	11·961	11·747	11·624
	14	12·455	12·002	11·680	11·248	10·972	10·778	10·577	10·366	10·234	10·029	9·9103
	16	11·295	10·868	10·563	10·154	9·8904	9·7054	9·5119	9·3085	9·1810	8·9808	8·8644
	18	10·512	10·104	9·8121	9·4190	9·1647	8·9857	8·7978	8·5996	8·4748	8·2778	8·1626
	20	9·9500	9·5560	9·2736	8·8926	8·6453	8·4708	8·2871	8·0926	7·9696	7·7748	7·6601
	25	9·0585	8·6884	8·4223	8·0616	7·8261	7·6590	7·4821	7·2933	7·1730	6·9805	6·8659
	30	8·5377	8·1825	7·9265	7·5784	7·3502	7·1876	7·0147	6·8291	6·7101	6·5181	6·4026
	35	8·1968	7·8517	7·6026	7·2631	7·0397	6·8801	6·7099	6·5262	6·4079	6·2156	6·0989
	40	7·9566	7·6188	7·3746	7·0413	6·8214	6·6640	6·4955	6·3131	6·1952	6·0023	5·8844
	50	7·6404	7·3125	7·0751	6·7501	6·5350	6·3804	6·2143	6·0334	5·9157	5·7214	5·6011
	60	7·4417	7·1202	6·8872	6·5676	6·3555	6·2027	6·0381	5·8581	5·7403	5·5446	5·4222
	70	7·3054	6·9884	6·7584	6·4426	6·2325	6·0809	5·9173	5·7378	5·6200	5·4230	5·2987
	80	7·2061	6·8924	6·6646	6·3515	6·1430	5·9924	5·8294	5·6503	5·5323	5·3343	5·2084
	100	7·0711	6·7619	6·5372	6·2279	6·0215	5·8721	5·7101	5·5313	5·4131	5·2133	5·0849
	200	6·8143	6·5139	6·2952	5·9933	5·7910	5·6439	5·4836	5·3053	5·1863	4·9819	4·8471
	∞	6·5741	6·2821	6·0692	5·7743	5·5758	5·4309	5·2721	5·0940	4·9737	4·7629	4·6190
1%	4	12·491†	12·800†	13·012†	13·283†	13·449†	13·561†	13·67†	13·79†	13·87†	—	—
	5	9·9992*	10·004*	10·008*	10·012*	10·014*	10·016*	10·018*	10·02*	10·02*	—	—
	6	1·9377*	1·9064*	1·8848*	1·8570*	1·8398*	1·8281*	1·8162*	1·8041*	1·7969*	—	—
	7	85·053	82·731	81·125	79·047	77·759	76·882	75·989	75·082	74·522	—	—
	8	51·991	50·178	48·921	47·290	46·276	45·583	44·877	44·156	43·715	43·04	—
	10	29·789	28·478	27·566	26·376	25·632	25·121	24·597	24·060	23·731	23·224	22·95
	12	21·965	20·889	20·138	19·154	18·534	18·108	17·668	17·215	16·936	16·505	16·261
	14	18·142	17·199	16·539	15·670	15·121	14·742	14·349	13·943	13·691	13·301	13·077
	16	15·916	15·059	14·457	13·662	13·157	12·807	12·444	12·066	11·831	11·466	11·255
	18	14·473	13·674	13·112	12·368	11·894	11·564	11·221	10·863	10·639	10·289	10·086
	20	13·466	12·710	12·177	11·470	11·018	10·703	10·374	10·030	9·8138	9·4748	9·2771
	25	11·924	11·237	10·751	10·103	9·6871	9·3951	9·0890	8·7658	8·5618	8·2383	8·0476
	30	11·055	10·409	9·9509	9·3382	8·9430	8·6646	8·3715	8·0602	7·8626	7·5468	7·3586
	35	10·499	9·8801	9·4405	8·8511	8·4695	8·2000	7·9153	7·6115	7·4177	7·1059	6·9186
	40	10·114	9·5138	9·0872	8·5142	8·1424	7·8791	7·6002	7·3015	7·1102	6·8006	6·6132
	50	9·6141	9·0400	8·6308	8·0795	7·7204	7·4652	7·1938	6·9015	6·7131	6·4054	6·2168
	60	9·3053	8·7472	8·3490	7·8114	7·4603	7·2101	6·9434	6·6548	6·4679	6·1606	5·9704
	70	9·0954	8·5485	8·1578	7·6297	7·2840	7·0373	6·7736	6·4875	6·3015	5·9940	5·8022
	80	8·9437	8·4048	8·0197	7·4984	7·1567	6·9124	6·6510	6·3666	6·1812	5·8733	5·6799
	100	8·7388	8·2111	7·8334	7·3214	6·9851	6·7443	6·4858	6·2036	6·0189	5·7100	5·5139
	200	8·3542	7·8476	7·4842	6·9900	6·6639	6·4293	6·1763	5·8979	5·7138	5·4012	5·1976
	∞	8·0000	7·5132	7·1633	6·6857	6·3691	6·1402	5·8920	5·6164	5·4323	5·1133	4·8981

* Multiply entry by 100. † Multiply entry by 10^4.

	$m_E \backslash m_H$	5	6	8	10	12	15	20	25	40	60
						$d = 5$					
5%	5	81.991 +	83.352 +	85.093 +	86.160 +	86.88 +	—	—	—	—	—
	6	3.0093+	3.0142+	3.0204+	3.0241+	3.0266+	3.0291+	3.032 +	—	—	—
	7	93.762	93.042	92.102	91.515	91.113	90.705	90.29	90.04	—	—
	8	51.339	50.646	49.739	49.170	48.780	48.382	47.973	47.723	47.35	—
	10	27.667	27.115	26.387	25.927	25.610	25.284	24.947	24.740	24.422	—
	12	20.169	19.701	19.079	18.683	18.409	18.124	17.830	17.647	17.365	17.20
	14	16.643	16.224	15.666	15.309	15.059	14.800	14.530	14.361	14.100	13.95
	16	14.624	14.239	13.722	13.389	13.157	12.914	12.659	12.499	12.250	12.105
	18	13.326	12.963	12.476	12.161	11.939	11.708	11.463	11.310	11.068	10.928
	20	12.424	12.078	11.612	11.310	11.097	10.874	10.637	10.488	10.252	10.113
	25	11.046	10.728	10.297	10.016	9.8168	9.6061	9.3814	9.2386	9.0102	8.8745
	30	10.270	9.9689	9.5592	9.2907	9.0995	8.8964	8.6785	8.5389	8.3141	8.1790
	35	9.7739	9.4836	9.0879	8.8277	8.6419	8.4437	8.2301	8.0926	7.8693	7.7339
	40	9.4292	9.1469	8.7613	8.5070	8.3250	8.1303	7.9195	7.7833	7.5607	7.4247
	50	8.9825	8.7107	8.3385	8.0921	7.9150	7.7248	7.5177	7.3829	7.1605	7.0229
	60	8.7057	8.4406	8.0769	7.8355	7.6615	7.4741	7.2692	7.1351	6.9124	6.7730
	70	8.5174	8.2570	7.8991	7.6612	7.4894	7.3039	7.1004	6.9667	6.7434	6.6024
	80	8.3811	8.1241	7.7705	7.5351	7.3648	7.1807	6.9782	6.8448	6.6208	6.4785
	100	8.1969	7.9446	7.5969	7.3649	7.1968	7.0145	6.8133	6.6801	6.4550	6.3103
	200	7.8505	7.6070	7.2706	7.0451	6.8811	6.7023	6.5032	6.3702	6.1416	5.9908
	∞	7.5305	7.2955	6.9698	6.7505	6.5902	6.4144	6.2171	6.0838	5.8499	5.6899
1%	5	20.495 *	20.834 *	21.267 *	21.53 *	—	—	—	—	—	—
	6	15.014 +	15.019 +	15.025 +	15.029 +	15.033 +	15.03 +	15.06 +	—	—	—
	7	2.7354+	2.7045+	2.6646+	2.6400+	2.6232+	2.6064+	2.590 +	2.579 +	—	—
	8	1.1498+	1.1276+	1.0989+	1.0811+	1.0689+	1.0567+	1.0440+	1.0364+	—	—
	10	48.048	46.670	44.877	43.758	42.992	42.210	41.408	40.921	—	—
	12	31.108	30.065	28.701	27.846	27.257	26.653	26.031	25.648	25.06	24.71
	14	24.016	23.145	22.001	21.279	20.781	20.268	19.736	19.408	18.90	18.61
	16	20.240	19.472	18.459	17.817	17.373	16.913	16.435	16.138	15.678	15.412
	18	17.929	17.228	16.302	15.713	15.304	14.878	14.435	14.159	13.727	13.478
	20	16.380	15.727	14.862	14.310	13.925	13.525	13.105	12.843	12.431	12.192
	25	14.107	13.529	12.759	12.265	11.918	11.555	11.172	10.930	10.547	10.322
	30	12.880	12.345	11.629	11.167	10.842	10.500	10.136	9.9059	9.5378	9.3188
	35	12.115	11.607	10.926	10.486	10.174	9.8453	9.4944	9.2706	8.9106	8.6946
	40	11.593	11.105	10.448	10.022	9.7204	9.4006	9.0581	8.8387	8.4838	8.2691
	50	10.928	10.465	9.8408	9.4336	9.1441	8.8361	8.5041	8.2901	7.9404	7.7261
	60	10.523	10.076	9.4712	9.0758	8.7938	8.4930	8.1674	7.9563	7.6090	7.3940
	70	10.251	9.8142	9.2229	8.8354	8.5586	8.2626	7.9411	7.7319	7.3858	7.1697
	80	10.055	9.6261	9.0446	8.6629	8.3899	8.0973	7.7787	7.5708	7.2251	7.0078
	100	9.7929	9.3742	8.8058	8.4319	8.1638	7.8758	7.5611	7.3547	7.0093	6.7897
	200	9.3055	8.9065	8.3629	8.0036	7.7448	7.4652	7.1572	6.9532	6.6062	6.3798
	∞	8.8628	8.4820	7.9613	7.6154	7.3650	7.0929	6.7903	6.5878	6.2361	5.9984

+ **Multiply entry by 100.** * Multiply entry by 10^4.

D15 (Continued)

$d = 6$

	m_E \ m_H	6	8	10	12	15	20	25	30	35
5%	10	45.722	44.677	44.019	43.567	43.103	42.626	42.334	42.136	41.993
	12	28.959	28.121	27.590	27.223	26.843	26.451	26.209	26.044	25.925
	14	22.321	21.600	21.141	20.821	20.489	20.144	19.929	19.783	19.677
	16	18.858	18.210	17.795	17.505	17.202	16.886	16.688	16.553	16.455
	18	16.755	16.157	15.772	15.501	15.218	14.921	14.735	14.607	14.513
	20	15.351	14.788	14.424	14.168	13.899	13.615	13.436	13.313	13.223
	25	13.293	12.786	12.456	12.222	11.975	11.711	11.544	11.428	11.343
	30	12.180	11.705	11.395	11.173	10.939	10.687	10.526	10.414	10.331
	35	11.484	11.031	10.733	10.520	10.293	10.049	9.8921	9.7820	9.7003
	40	11.009	10.571	10.282	10.075	9.8535	9.6142	9.4596	9.3508	9.2699
	50	10.402	9.9832	9.7060	9.5067	9.2927	9.0598	8.9082	8.8009	8.7207
	60	10.031	9.6246	9.3547	9.1602	8.9507	8.7215	8.5717	8.4651	8.3851
	70	9.7813	9.3830	9.1182	8.9269	8.7204	8.4938	8.3450	8.2388	8.1589
	80	9.6014	9.2093	8.9480	8.7591	8.5548	8.3300	8.1819	8.0759	7.9959
	100	9.3598	8.9760	8.7197	8.5340	8.3326	8.1102	7.9629	7.8572	7.7771
	200	8.9099	8.5419	8.2950	8.1153	7.9193	7.7011	7.5552	7.4494	7.3685
	500	8.6594	8.3002	8.0587	7.8823	7.6894	7.4734	7.3280	7.2219	7.1403
	1000	8.5788	8.2226	7.9827	7.8075	7.6155	7.4002	7.2550	7.1487	7.0668
	∞	8.4997	8.1463	7.9082	7.7340	7.5430	7.3284	7.1832	7.0768	6.9945
1%	10	86.397	83.565	81.804	80.602	79.376	78.124	77.360	76.845	76.474
	12	46.027	44.103	42.899	42.073	41.227	40.359	39.826	39.466	39.206
	14	32.433	30.918	29.966	29.309	28.634	27.936	27.507	27.215	27.004
	16	25.977	24.689	23.875	23.311	22.729	22.126	21.753	21.498	21.314
	18	22.292	21.146	20.418	19.913	19.389	18.844	18.505	18.273	18.105
	20	19.935	18.886	18.217	17.752	17.267	16.761	16.445	16.229	16.071
	25	16.642	15.737	15.156	14.749	14.324	13.875	13.592	13.397	13.254
	30	14.944	14.118	13.586	13.211	12.816	12.398	12.133	11.949	11.814
	35	13.913	13.138	12.635	12.281	11.906	11.506	11.252	11.074	10.943
	40	13.223	12.482	12.000	11.659	11.298	10.911	10.663	10.490	10.361
	50	12.358	11.661	11.206	10.882	10.538	10.167	9.9271	9.7587	9.6333
	60	11.839	11.169	10.730	10.417	10.083	9.7206	9.4860	9.3202	9.1963
	70	11.493	10.841	10.413	10.107	9.7795	9.4238	9.1922	9.0281	8.9050
	80	11.246	10.607	10.187	9.8859	9.5634	9.2121	8.9826	8.8195	8.6968
	100	10.917	10.295	9.8857	9.5918	9.2758	8.9301	8.7033	8.5414	8.4193
	200	10.312	9.7231	9.3330	9.0517	8.7476	8.4121	8.1897	8.0295	7.9075
	500	9.9799	9.4089	9.0296	8.7553	8.4576	8.1275	7.9072	7.7473	7.6249
	1000	9.8738	9.3085	8.9328	8.6607	8.3651	8.0366	7.8168	7.6570	7.5344
	∞	9.7699	9.2103	8.8379	8.5680	8.2744	7.9475	7.7283	7.5685	7.4456

D15 (Continued)

$d = 7$

	m_E \ m_H	8	10	12	15	20	25	30	35
5%	10	85.040	84.082	83.426	82.755	82.068	81.648	81.364	81.159
	12	42.850	42.126	41.627	41.113	40.583	40.257	40.037	39.877
	14	29.968	29.373	28.961	28.534	28.091	27.817	27.631	27.495
	16	24.038	23.519	23.158	22.781	22.389	22.145	21.978	21.857
	18	20.692	20.222	19.893	19.549	19.189	18.964	18.809	18.696
	20	18.561	18.125	17.819	17.498	17.159	16.947	16.800	16.694
	25	15.587	15.202	14.930	14.642	14.337	14.143	14.009	13.911
	30	14.049	13.693	13.440	13.172	12.884	12.701	12.573	12.478
	35	13.113	12.776	12.535	12.278	12.002	11.825	11.700	11.608
	40	12.485	12.160	11.927	11.679	11.411	11.237	11.115	11.025
	50	11.695	11.386	11.165	10.927	10.668	10.500	10.381	10.292
	60	11.219	10.921	10.706	10.475	10.221	10.056	9.9383	9.8500
	70	10.901	10.610	10.400	10.173	9.9233	9.7596	9.6429	9.5550
	80	10.674	10.388	10.181	9.9567	9.7102	9.5478	9.4317	9.3440
	100	10.371	10.091	9.8886	9.6688	9.4259	9.2652	9.1498	9.0622
	200	9.8118	9.5448	9.3504	9.1384	8.9021	8.7441	8.6295	8.5419
	500	9.5037	9.2438	9.0539	8.8462	8.6134	8.4568	8.3424	8.2543
	1000	9.4051	9.1475	8.9591	8.7527	8.5211	8.3648	8.2504	8.1621
	∞	9.3085	9.0531	8.8662	8.6612	8.4306	8.2747	8.1603	8.0718
1%	10	185.93	182.94	180.90	178.83	176.73	175.44	174.57	173.92
	12	71.731	69.978	68.779	67.552	66.296	65.528	65.010	64.636
	14	44.255	42.978	42.099	41.197	40.269	39.698	39.311	39.032
	16	33.097	32.057	31.339	30.599	29.834	29.361	29.039	28.806
	18	27.273	26.374	25.750	25.105	24.435	24.019	23.735	23.529
	20	23.757	22.949	22.388	21.804	21.195	20.816	20.556	20.367
	25	19.117	18.440	17.965	17.469	16.947	16.619	16.392	16.227
	30	16.848	16.239	15.810	15.360	14.882	14.580	14.370	14.216
	35	15.512	14.945	14.544	14.121	13.670	13.383	13.183	13.036
	40	14.634	14.095	13.713	13.309	12.876	12.599	12.405	12.262
	50	13.553	13.049	12.691	12.310	11.899	11.634	11.448	11.309
	60	12.914	12.432	12.088	11.720	11.323	11.065	10.882	10.746
	70	12.492	12.024	11.690	11.332	10.942	10.689	10.509	10.374
	80	12.193	11.736	11.408	11.056	10.673	10.422	10.244	10.110
	100	11.797	11.353	11.034	10.691	10.316	10.070	9.8935	9.7607
	200	11.077	10.658	10.356	10.028	9.6670	9.4273	9.2545	9.1228
	500	10.685	10.280	9.9866	9.6679	9.3140	9.0776	8.9060	8.7744
	1000	10.561	10.160	9.8693	9.5533	9.2017	8.9663	8.7950	8.6634
	∞	10.439	10.043	9.7547	9.4413	9.0920	8.8575	8.6865	8.5548

D15 (Continued)

$d = 8$

m_E \ m_H		8	10	12	15	20	25	30	35
5%	14	42.516	41.737	41.198	40.641	40.066	39.711	39.470	39.296
	16	31.894	31.242	30.788	30.318	29.829	29.525	29.318	29.167
	18	26.421	25.847	25.446	25.028	24.591	24.319	24.132	23.996
	20	23.127	22.605	22.239	21.856	21.454	21.201	21.028	20.902
	25	18.770	18.324	18.009	17.677	17.325	17.102	16.947	16.834
	30	16.626	16.221	15.934	15.629	15.303	15.095	14.950	14.843
	35	15.356	14.977	14.707	14.418	14.109	13.910	13.771	13.668
	40	14.518	14.156	13.898	13.621	13.322	13.129	12.994	12.893
	50	13.482	13.142	12.898	12.636	12.351	12.165	12.034	11.936
	60	12.866	12.540	12.305	12.051	11.774	11.593	11.465	11.368
	70	12.459	12.142	11.912	11.665	11.393	11.215	11.088	10.992
	80	12.169	11.858	11.634	11.390	11.122	10.946	10.820	10.725
	100	11.785	11.483	11.264	11.026	10.763	10.590	10.465	10.370
	200	11.084	10.798	10.589	10.362	10.108	9.9389	9.8159	9.7218
	500	10.701	10.423	10.221	9.9993	9.7509	9.5836	9.4614	9.3673
	1000	10.579	10.304	10.104	9.8840	9.6371	9.4704	9.3484	9.2543
	∞	10.459	10.188	9.9892	9.7712	9.5258	9.3598	9.2379	9.1437
1%	14	65.793	64.035	62.828	61.592	60.323	59.545	59.019	58.639
	16	44.977	43.633	42.707	41.754	40.771	40.164	39.753	39.456
	18	35.265	34.146	33.373	32.573	31.745	31.232	30.882	30.629
	20	29.786	28.808	28.129	27.425	26.691	26.235	25.924	25.697
	25	23.001	22.212	21.661	21.085	20.480	20.100	19.838	19.647
	30	19.867	19.173	18.686	18.173	17.631	17.288	17.051	16.876
	35	18.077	17.440	16.991	16.516	16.011	15.690	15.466	15.301
	40	16.924	16.324	15.900	15.451	14.970	14.662	14.447	14.288
	50	15.528	14.975	14.582	14.163	13.711	13.420	13.216	13.063
	60	14.715	14.190	13.815	13.414	12.980	12.698	12.499	12.351
	70	14.184	13.677	13.313	12.925	12.502	12.226	12.031	11.885
	80	13.810	13.315	12.960	12.580	12.165	11.894	11.701	11.556
	100	13.317	12.839	12.496	12.127	11.722	11.457	11.267	11.124
	200	12.429	11.983	11.660	11.311	10.925	10.669	10.484	10.343
	500	11.951	11.521	11.210	10.871	10.495	10.244	10.061	9.9208
	1000	11.800	11.375	11.067	10.732	10.359	10.109	9.9270	9.7870
	∞	11.652	11.233	10.928	10.597	10.227	9.9778	9.7963	9.6564

D15 (Continued)

$d = 9$

m_E \ m_H		10	12	15	20	25	30	35
5%	14	61.915	61.196	60.456	59.694	59.224	58.907	58.68
	16	42.157	41.583	40.988	40.372	39.990	39.730	39.542
	18	33.171	32.680	32.169	31.637	31.305	31.079	30.914
	20	28.140	27.702	27.245	26.766	26.466	26.260	26.110
	25	21.911	21.548	21.165	20.759	20.503	20.326	20.196
	30	19.020	18.695	18.350	17.982	17.747	17.584	17.464
	35	17.361	17.059	16.737	16.391	16.170	16.015	15.900
	40	16.287	16.000	15.694	15.363	15.150	15.000	14.889
	50	14.982	14.714	14.427	14.115	13.912	13.768	13.661
	60	14.218	13.962	13.687	13.385	13.188	13.049	12.944
	70	13.717	13.469	13.201	12.907	12.714	12.577	12.473
	80	13.364	13.121	12.859	12.570	12.380	12.243	12.141
	100	12.898	12.663	12.407	12.125	11.938	11.804	11.703
	200	12.055	11.833	11.591	11.321	11.141	11.010	10.910
	500	11.600	11.385	11.150	10.887	10.710	10.580	10.480
	1000	11.455	11.243	11.011	10.749	10.573	10.444	10.344
	∞	11.315	11.105	10.874	10.615	10.440	10.311	10.211
1%	14	101.87	100.15	98.387	96.583	95.478	94.74	94.2
	16	60.990	59.770	58.518	57.229	56.437	55.90	55.51
	18	44.668	43.697	42.697	41.662	41.022	40.587	40.272
	20	36.244	35.419	34.564	33.676	33.126	32.750	32.477
	25	26.601	25.960	25.292	24.591	24.152	23.850	23.629
	30	22.443	21.890	21.310	20.696	20.308	20.040	19.843
	35	20.153	19.650	19.120	18.556	18.198	17.949	17.765
	40	18.708	18.238	17.741	17.210	16.870	16.632	16.457
	50	16.993	16.563	16.105	15.612	15.294	15.071	14.905
	60	16.011	15.604	15.169	14.698	14.393	14.177	14.016
	70	15.375	14.983	14.563	14.107	13.810	13.599	13.441
	80	14.930	14.548	14.139	13.693	13.401	13.194	13.038
	100	14.348	13.981	13.585	13.152	12.868	12.664	12.511
	200	13.310	12.968	12.597	12.188	11.915	11.719	11.569
	500	12.756	12.427	12.070	11.673	11.407	11.214	11.066
	1000	12.581	12.257	11.904	11.511	11.247	11.054	10.907
	∞	12.412	12.092	11.743	11.353	11.091	10.899	10.752

D15 (Continued)

$d = 10$

m_E \ m_H	10	12	15	20	25	30	35
5% 14	98.999	98.013	97.002	95.963	95.326	94.9	94.6
16	58.554	57.814	57.050	56.260	55.772	55.44	55.20
18	43.061	42.454	41.824	41.169	40.762	40.485	40.284
20	35.146	34.620	34.071	33.497	33.140	32.895	32.716
25	26.080	25.660	25.219	24.753	24.458	24.255	24.107
30	22.140	21.773	21.384	20.970	20.706	20.523	20.388
35	19.955	19.618	19.260	18.876	18.630	18.458	18.331
40	18.569	18.252	17.914	17.550	17.316	17.151	17.029
50	16.913	16.622	16.309	15.969	15.748	15.592	15.476
60	15.960	15.684	15.385	15.059	14.847	14.695	14.582
70	15.341	15.074	14.786	14.469	14.261	14.113	14.002
80	14.907	14.647	14.365	14.055	13.851	13.705	13.595
100	14.338	14.087	13.814	13.513	13.313	13.170	13.061
200	13.319	13.085	12.828	12.542	12.351	12.212	12.106
500	12.774	12.548	12.301	12.023	11.836	11.699	11.594
1000	12.602	12.379	12.134	11.859	11.674	11.538	11.432
∞	12.434	12.214	11.972	11.700	11.515	11.380	11.275
1% 14	180.90	178.28	175.62	172.91	171.24	170	--
16	89.068	87.414	85.270	83.980	82.91	82.2	81.7
18	59.564	58.328	57.055	55.742	54.933	54.384	53.990
20	45.963	44.951	43.905	42.821	42.150	41.693	41.362
25	31.774	31.029	30.253	29.440	28.932	28.583	28.328
30	26.115	25.489	24.832	24.139	23.701	23.399	23.177
35	23.116	22.556	21.966	21.338	20.939	20.663	20.459
40	21.267	20.749	20.201	19.615	19.241	18.980	18.787
50	19.114	18.646	18.148	17.611	17.266	17.023	16.842
60	17.901	17.462	16.992	16.484	16.154	15.922	15.748
70	17.124	16.703	16.252	15.762	15.443	15.216	15.046
80	16.583	16.175	15.738	15.260	14.948	14.726	14.559
100	15.881	15.490	15.069	14.608	14.305	14.088	13.925
200	14.641	14.280	13.889	13.457	13.169	12.962	12.803
500	13.986	13.641	13.266	12.848	12.569	12.366	12.210
1000	13.780	13.441	13.070	12.658	12.381	12.179	12.023
∞	13.581	13.246	12.881	12.472	12.198	11.997	11.842

D16 PILLAI'S TRACE STATISTIC

	ν_1 \ ν_2	s = 2													
		0	1	2	3	4	5	6	7	8	9	10	15	20	25
5%	0	1.536	1.232	1.031	0.890	0.782	0.698	0.629	0.573	0.526	0.485	0.451	0.333	0.263	0.218
	1	1.706	1.452	1.258	1.109	0.991	0.896	0.817	0.751	0.694	0.646	0.604	0.455	0.364	0.304
	2	1.784	1.573	1.397	1.254	1.137	1.039	0.956	0.886	0.825	0.772	0.725	0.556	0.451	0.379
	3	1.829	1.649	1.492	1.358	1.245	1.149	1.065	0.993	0.930	0.875	0.825	0.643	0.526	0.445
	4	1.859	1.703	1.560	1.436	1.329	1.235	1.153	1.081	1.018	0.961	0.910	0.719	1.594	0.506
	5	1.880	1.742	1.613	1.497	1.395	1.305	1.226	1.155	1.091	1.034	0.983	0.786	0.655	0.561
	6	1.895	1.772	1.654	1.546	1.450	1.364	1.286	1.217	1.154	1.098	1.046	0.846	0.710	0.612
	7	1.907	1.796	1.687	1.586	1.495	1.413	1.338	1.270	1.209	1.153	1.102	0.901	0.761	0.658
	8	1.917	1.815	1.714	1.620	1.534	1.455	1.383	1.317	1.257	1.202	1.151	0.950	0.808	0.702
	9	1.924	1.831	1.737	1.649	1.567	1.491	1.422	1.358	1.299	1.245	1.195	0.995	0.851	0.743
	10	1.931	1.844	1.757	1.673	1.595	1.523	1.456	1.394	1.337	1.284	1.235	1.036	0.891	0.781
	15	1.951	1.888	1.822	1.758	1.695	1.636	1.580	1.527	1.477	1.430	1.386			
	20	1.963	1.913	1.860	1.807	1.756	1.706	1.658	1.612	1.568	1.527	1.487			
	25	1.969	1.929	1.885	1.840	1.796	1.753	1.711	1.671	1.632	1.595	1.559			
1%	0	1.729	1.444	1.230	1.072	0.950	0.854	0.776	0.711	0.656	0.609	0.569	0.426	0.340	0.283
	1	1.834	1.618	1.431	1.278	1.153	1.050	0.964	0.891	0.828	0.773	0.725	0.553	0.447	0.375
	2	1.880	1.707	1.545	1.405	1.286	1.185	1.097	1.022	0.956	0.898	0.846	0.657	0.536	0.453
	3	1.906	1.762	1.621	1.494	1.382	1.284	1.199	1.123	1.057	0.997	0.944	0.744	0.614	0.522
	4	1.923	1.800	1.674	1.558	1.454	1.361	1.279	1.205	1.139	1.079	1.025	0.819	0.682	0.584
	5	1.934	1.827	1.715	1.608	1.511	1.423	1.344	1.272	1.207	1.148	1.094	0.885	0.743	0.639
	6	1.943	1.848	1.746	1.648	1.557	1.474	1.398	1.328	1.265	1.207	1.154	0.944	0.798	0.690
	7	1.950	1.864	1.771	1.680	1.595	1.516	1.443	1.376	1.315	1.258	1.206	0.996	0.848	0.737
	8	1.955	1.877	1.792	1.707	1.627	1.552	1.482	1.418	1.358	1.303	1.251	1.043	0.893	0.780
	9	1.959	1.888	1.809	1.730	1.654	1.583	1.516	1.454	1.396	1.342	1.292	1.086	0.935	0.821
	10	1.963	1.897	1.823	1.749	1.678	1.610	1.546	1.485	1.430	1.377	1.328	1.125	0.974	0.858
	15	1.974	1.926	1.872	1.816	1.759	1.705	1.652	1.602	1.554	1.508	1.465			
	20	1.980	1.943	1.900	1.854	1.808	1.762	1.718	1.675	1.633	1.593	1.555			
	25	1.984	1.953	1.917	1.879	1.840	1.801	1.763	1.725	1.689	1.653	1.619			

	ν_1 \ ν_2	0	1	2	3	4	5	6	7	8	9	10	15	20	25
								$s = 3$							
5%	0	2.037	1.710	1.473	1.294	1.153	1.040	0.947	0.869	0.803	0.746	0.697	0.524	0.420	0.350
	1	2.297	1.988	1.751	1.564	1.412	1.287	1.183	1.094	1.017	0.950	0.892	0.682	0.552	0.463
	2	2.447	2.168	1.943	1.759	1.606	1.477	1.367	1.273	1.190	1.117	1.053	0.818	0.668	0.565
	3	2.544	2.294	2.084	1.907	1.757	1.628	1.517	1.420	1.334	1.258	1.190	0.937	0.772	0.656
	4	2.612	2.386	2.191	2.023	1.878	1.752	1.641	1.543	1.456	1.378	1.308	1.042	0.866	0.740
	5	2.662	2.457	2.276	2.117	1.978	1.854	1.745	1.648	1.561	1.482	1.411	1.137	0.952	0.818
	6	2.701	2.514	2.345	2.194	2.061	1.941	1.835	1.739	1.652	1.573	1.502	1.222	1.030	0.890
	7	2.732	2.559	2.402	2.259	2.131	2.016	1.912	1.818	1.732	1.654	1.582	1.300	1.103	0.957
	8	2.757	2.597	2.449	2.314	2.192	2.081	1.979	1.887	1.803	1.726	1.655	1.371	1.170	1.020
	9	2.777	2.629	2.490	2.362	2.244	2.137	2.039	1.949	1.866	1.790	1.720	1.436	1.23	
	10	2.795	2.656	2.525	2.403	2.291	2.187	2.092	2.004	1.923	1.848	1.779	1.496	1.3	
	15	2.853	2.748	2.646	2.549	2.457	2.370	2.288	2.211	2.139	2.071	2.007			
	20	2.885	2.802	2.718	2.637	2.560	2.485	2.414	2.347	2.283	2.222	2.163			
	25	2.906	2.836	2.766	2.697	2.630	2.565	2.503	2.443	2.385					
1%	0	2.263	1.925	1.678	1.487	1.336	1.212	1.110	1.023	0.949	0.885	0.829	0.630	0.508	0.426
	1	2.483	2.183	1.943	1.750	1.591	1.459	1.346	1.250	1.166	1.093	1.029	0.795	0.647	0.546
	2	2.660	2.340	2.119	1.933	1.776	1.642	1.527	1.426	1.338	1.260	1.191	0.933	0.767	0.651
	3	2.674	2.445	2.243	2.069	1.917	1.786	1.671	1.569	1.479	1.398	1.326	1.053	0.873	0.746
	4	2.725	2.522	2.337	2.173	2.028	1.901	1.788	1.687	1.596	1.515	1.441	1.159	0.968	1.831
	5	2.761	2.579	2.410	2.256	2.119	1.996	1.885	1.786	1.697	1.615	1.541	1.253	1.054	0.910
	6	2.790	2.624	2.468	2.324	2.194	2.075	1.968	1.871	1.783	1.702	1.629	1.337	1.133	0.983
	7	2.812	2.661	2.516	2.380	2.256	2.143	2.040	1.945	1.858	1.779	1.706	1.413	1.205	1.050
	8	2.830	2.691	2.555	2.428	2.310	2.202	2.102	2.010	1.925	1.847	1.774	1.482	1.3	1.11
	9	2.845	2.716	2.589	2.469	2.357	2.252	2.156	2.067	1.984	1.907	1.836	1.545		
	10	2.857	2.737	2.618	2.504	2.397	2.297	2.204	2.117	2.037	1.962	1.892	1.603		
	15	2.898	2.809	2.718	2.628	2.541	2.459	2.380	2.306	2.235	2.168	2.105			
	20	2.921	2.850	2.776	2.702	2.630	2.560	2.492	2.427	2.365	2.306	2.249			
	25	2.935	2.876	2.814	2.752	2.690	2.629	2.570	2.513	2.457					

D16 (Continued)

	$\nu_1 \backslash \nu_2$	$s = 4$												
		0	1	2	3	4	5	6	7	8	9	10	15	20
5%	0	2.549	2.194	1.926	1.717	1.548	1.410	1.294	1.196	1.112	1.038	0.974	0.744	0.602
	1	2.852	2.510	2.241	2.023	1.844	1.693	1.566	1.456	1.360	1.277	1.203	0.932	0.761
	2	3.052	2.733	2.472	2.256	2.074	1.919	1.786	1.670	1.567	1.477	1.396	1.097	0.903
	3	3.193	2.898	2.650	2.440	2.260	2.104	1.969	1.849	1.743	1.649	1.564	1.243	1.032
	4	3.298	3.025	2.791	2.589	2.413	2.259	2.123	2.002	1.895	1.798	1.710	1.375	1.149
	5	3.378	3.126	2.905	2.711	2.541	2.390	2.255	2.135	2.027	1.929	1.840	1.494	
	6	3.442	3.208	2.999	2.814	2.649	2.502	2.370	2.251	2.143	2.044	1.955	1.602	
	7	3.494	3.276	3.079	2.902	2.743	2.600	2.470	2.353	2.246	2.148	2.058	1.70	
	8	3.537	3.333	3.146	2.977	2.824	2.685	2.559	2.444	2.338	2.241	2.151	1.8	
	9	3.574	3.382	3.205	3.043	2.896	2.761	2.638	2.525	2.421	2.325	2.236		
	10	3.605	3.424	3.256	3.101	2.959	2.829	2.708	2.598	2.496	2.401	2.313		
	15	3.710	3.570	3.436	3.310	3.191	3.079	2.974	2.876	2.783	2.696	2.615		
	20	3.771	3.657	3.546	3.440	3.338	3.241	3.149						
1%	0	2.773	2.415	2.139	1.919	1.741	1.593	1.468	1.362	1.270	1.189	1.118	0.862	0.701
	1	3.057	2.718	2.445	2.221	2.034	1.877	1.741	1.624	1.522	1.432	1.352	1.056	0.867
	2	3.234	2.923	2.664	2.445	2.258	2.098	1.959	1.837	1.729	1.633	1.547	1.224	1.013
	3	3.355	3.072	2.828	2.618	2.436	2.276	2.136	2.012	1.902	1.803	1.713	1.372	1.144
	4	3.442	3.184	2.956	2.756	2.580	2.424	2.285	2.161	2.050	1.949	1.858	1.504	1.26
	5	3.509	3.272	3.059	2.869	2.699	2.547	2.411	2.288	2.177	2.076	1.984	1.62	
	6	3.561	3.343	3.143	2.963	2.800	2.653	2.520	2.399	2.289	2.188	2.096	1.7	
	7	3.603	3.401	3.214	3.042	2.886	2.744	2.614	2.496	2.387	2.288	2.196		
	8	3.638	3.450	3.273	3.110	2.961	2.823	2.697	2.582	2.475	2.377	2.285		
	9	3.667	3.491	3.325	3.169	3.026	2.893	2.771	2.658	2.554	2.457	2.366		
	10	3.692	3.527	3.369	3.221	3.083	2.995	2.837	2.727	2.624	2.529	2.440		
	15	3.776	3.649	3.525	3.405	3.292	3.184	3.082	2.985	2.894	2.808	2.726		
	20	3.824	3.721	3.619	3.519	3.423	3.330	3.241						

	ν_1 \ ν_2	$s = 5$										
		0	1	2	3	4	5	6	7	8	9	10
5%	0	3.055	2.681	2.389	2.155	1.962	1.801	1.664	1.547	1.445	1.356	1.277
	1	3.390	3.025	2.731	2.488	2.285	2.122	1.964	1.835	1.722	1.622	1.533
	2	3.628	3.281	2.993	2.751	2.545	2.367	2.213	2.077	1.957	1.850	1.754
	3	3.805	3.478	3.201	2.964	2.759	2.580	2.423	2.284	2.160	2.048	1.948
	4	3.941	3.635	3.370	3.140	2.938	2.761	2.604	2.463	2.337	2.222	2.119
	5	4.050	3.762	3.510	3.288	3.091	2.916	2.760	2.619	2.492	2.377	2.271
	6	4.138	3.868	3.627	3.414	3.223	3.052	2.897	2.758	2.630	2.514	2.408
	7	4.212	3.957	3.728	3.522	3.337	3.170	3.018	2.880			
	8	4.274	4.033	3.815	3.617	3.438	3.275	3.126				
	9	4.327	4.099	3.890	3.700	3.527	3.369					
	10	4.372	4.156	3.957	3.774	3.607	3.45					
1%	0	3.283	2.906	2.607	2.364	2.162	1.993	1.848	1.722	1.613	1.517	1.431
	1	3.605	3.240	2.942	2.694	2.484	2.304	2.149	2.013	1.893	1.787	1.692
	2	3.825	3.483	3.194	2.949	2.738	2.555	2.395	2.254	2.128	2.016	1.915
	3	3.985	3.666	3.391	3.153	2.945	2.763	2.602	2.458	2.329	2.213	2.108
	4	4.107	3.810	3.549	3.320	3.117	2.937	2.777	2.633	2.502	2.384	2.277
	5	4.203	3.926	3.679	3.459	3.262	3.086	2.928	2.784	2.654	2.535	2.427
	6	4.280	4.021	3.787	3.576	3.387	3.215	3.059	2.918	2.788	2.670	2.560
	7	4.343	4.100	3.879	3.677	3.494	3.327	3.175	3.036			
	8	4.396	4.168	3.958	3.765	3.588	3.426	3.278				
	9	4.441	4.226	4.026	3.841	3.671	3.514					
	10	4.480	4.277	4.086	3.909	3.745						

	n	$d=3$	$d=4$	$d=5$	$d=6$	$d=7$	$d=8$	$d=9$	$d=10$
5%	4	8.020							
	5	7.834	15.22						
	6	7.814	13.47	24.01					
	7	7.811	13.03	20.44	34.30				
	8	7.811	12.85	19.45	28.75	46.05			
	9	7.811	12.76	19.02	27.11	38.41	59.25		
	10	7.812	12.71	18.80	26.37	36.03	49.42	73.79	
	11	7.812	12.68	18.67	25.96	34.91	46.22	61.76	89.92
	12	7.813	12.66	18.58	25.71	34.28	44.67	57.68	75.45
	13	7.813	12.65	18.52	25.55	33.89	43.78	55.65	70.43
	14	7.813	12.64	18.48	25.44	33.63	43.21	54.46	67.87
	15	7.813	12.63	18.45	25.36	33.44	42.82	53.69	66.34
	16	7.814	12.62	18.43	25.30	33.31	42.55	53.15	65.33
	17	7.814	12.62	18.41	25.25	33.20	42.34	52.77	64.63
	18	7.814	12.62	18.40	25.21	33.12	42.19	52.48	64.12
	19	7.814	12.61	18.38	25.19	22.06	42.06	52.26	63.73
	20	7.814	12.61	18.37	25.16	33.01	41.97	52.08	63.43
χ^2_ν		7.815	12.59	18.31	25.00	32.67	41.34	51.00	61.66
1%	4	11.79							
	5	11.41	21.18						
	6	11.36	18.27	32.16					
	7	11.34	17.54	26.50	44.65				
	8	11.34	17.24	24.95	36.09	58.61			
	9	11.34	17.10	24.29	33.63	47.05	74.01		
	10	11.34	17.01	23.95	32.54	43.59	59.36	90.87	
	11	11.34	16.96	23.75	31.95	42.00	54.83	73.03	109.53
	12	11.34	16.93	23.62	31.60	41.13	52.70	67.37	88.05
	13	11.34	16.90	23.53	31.36	40.59	51.49	64.64	81.20
	14	11.34	16.89	23.47	31.20	40.23	50.73	63.06	77.83
	15	11.34	16.87	23.42	31.09	39.97	50.22	62.05	75.84
	16	11.34	16.86	23.39	31.00	39.79	49.85	61.36	74.56
	17	11.34	16.86	23.36	30.94	39.65	49.59	60.86	73.66
	18	11.34	16.85	23.34	30.88	39.54	49.38	60.49	73.01
	19	11.34	16.85	23.32	30.84	39.46	49.22	60.21	72.52
	20	11.34	16.84	23.31	30.81	39.39	49.09	59.99	72.15
χ^2_ν		11.34	16.81	23.21	30.58	38.93	48.28	58.57	69.92

D18 TEST FOR EQUAL DISPERSION MATRICES WITH EQUAL SAMPLE SIZES

$d = 2$

f_0/I	2	3	4	5	6	7	8	9	10
3	12.18	18.70	24.55	30.09	35.45	40.68	45.81	50.87	55.86
4	10.70	16.65	22.00	27.07	31.97	36.75	41.45	46.07	50.64
5	9.97	15.63	20.73	25.57	30.23	34.79	39.26	43.67	48.02
6	9.53	15.02	19.97	24.66	29.19	33.61	37.95	42.22	46.45
7	9.24	14.62	19.46	24.05	28.49	32.83	37.08	41.26	45.40
8	9.04	14.33	19.10	23.62	27.99	32.26	36.44	40.57	44.64
9	8.88	14.11	18.83	23.30	27.62	31.84	35.98	40.05	44.08
10	8.76	13.94	18.61	23.05	27.33	31.51	35.61	39.65	43.64
11	8 67	13.81	18.44	22.85	27.10	31.25	35.32	39.33	43.29
12	8.59	13.70	18.30	22.68	26.90	31.03	35.08	39.07	43.00
13	8.52	13.60	18.19	22.54	26.75	30.85	34.87	38.84	42.76
14	8.47	13.53	18.10	22.42	26.61	30.70	34.71	38.66	42.56
15	8.42	13.46	18.01	22.33	26.50	30.57	34.57	38.50	42.38
16	8.38	13.40	17.94	22.24	26.40	30.45	34.43	38.36	42.23
17	8.35	13.35	17.87	22.17	26.31	30.35	34.32	38.24	42.10
18	8.32	13.30	17.82	22.10	26.23	30.27	34.23	38.13	41.99
19	8.28	13.26	17.77	22.04	26.16	30.19	34.14	38.04	41.88
20	8.26	13.23	17.72	21.98	26.10	30.12	34.07	37.95	41.79
25	8.17	13.10	17.55	21.79	25.87	29.86	33.78	37.63	41.44
30	8.11	13.01	17.44	21.65	25.72	29.69	33.59	37.42	41.21

$d = 3$

f_0/I	2	3	4	5	6	7	8	9	10
4	22.41	35.00	46.58	57.68	68.50	79.11	89.60	99.94	110.21
5	19.19	30.52	40.95	50.95	60.69	70.26	79.69	89.03	98.27
6	17.57	28.24	38.06	47.49	56.67	65.69	74.58	83.39	92.09
7	16.59	26.84	36.29	45.37	54.20	62.89	71.44	79.90	88.30
8	15.93	25.90	35.10	43.93	52.54	60.99	69.32	77.57	85.73
9	15.46	25.22	34.24	42.90	51.33	59.62	67.78	75.86	83.87
10	15.11	24.71	33.59	42.11	50.42	58.57	66.62	74.58	82.46
11	14.83	24.31	33.08	41.50	49.71	57.76	65.71	73.57	81.36
12	14.61	23.99	32.67	41.00	49.13	57.11	64.97	72.75	80.45
13	14.43	23.73	32.33	40.60	48.65	56.56	64.36	72.09	79.72
14	14.28	23.50	32.05	40.26	48.26	56.11	63.86	71.53	79.11
15	14.15	23.32	31.81	39.97	47.92	55.73	63.43	71.05	78.60
16	14.04	23.16	31.60	39.72	47.63	55.40	63.06	70.64	78.14
17	13.94	23.02	31.43	39.50	47.38	55.11	62.73	70.27	77.76
18	13.86	22.89	31.26	39.31	47.16	54.86	62.45	69.97	77.41
19	13.79	22.78	31.13	39.15	46.96	54.64	62.21	69.69	77.11
20	13.72	22.69	31.01	39.00	46.79	54.44	61.98	69.45	76.84
25	13.48	22.33	30.55	38.44	46.15	53.70	61.16	68.54	75.84
30	13.32	22.10	30.25	38.09	45.73	53.22	60.62	67.94	75.18

D18 (Continued)

$d = 4$

f_0/I	2	3	4	5	6	7	8	9	10
5	35.39	56.10	75.36	93.97	112.17	130.11	147.81	165.39	182.80
6	30.06	48.62	65.90	82.60	98.93	115.03	130.94	146.69	162.34
7	27.31	44.69	60.89	76.56	91.88	106.98	121.90	136.71	151.39
8	25.61	42.24	57.77	72.77	87.46	101.94	116.23	130.43	144.50
9	24.45	40.57	55.62	70.17	84.42	98.46	112.32	126.08	139.74
10	23.62	39.34	54.04	68.26	82.19	95.90	109.46	122.91	136.24
11	22.98	38.41	52.84	66.81	80.48	93.95	107.27	120.46	133.57
12	22.48	37.67	51.90	65.66	79.14	92.41	105.54	118.55	131.45
13	22.08	37.08	51.13	64.73	78.04	91.15	104.12	116.98	129.74
14	21.75	36.59	50.50	63.95	77.13	90.12	102.97	115.69	128.32
15	21.47	36.17	49.97	63.30	76.37	89.26	101.99	114.59	127.14
16	21.24	35.82	49.51	62.76	75.73	88.51	101.14	113.67	126.10
17	21.03	35.52	49.12	62.28	75.16	87.87	100.42	112.87	125.22
18	20.86	35.26	48.78	61.86	74.68	87.31	99.80	112.17	124.46
19	20.70	35.02	48.47	61.50	74.25	86.82	99.25	111.56	123.79
20	20.56	34.82	48.21	61.17	73.87	86.38	98.75	111.02	123.18
25	20.06	34.06	47.23	59.98	72.47	84.78	96.95	109.01	120.99
30	19.74	33.59	46.61	59.21	71.58	83.74	95.79	107.71	119.57

$d = 5$

f_0/I	2	3	4	5	6	7	8	9	10
6	51.11	81.99	110.92	138.98	166.54	193.71	220.66	247.37	273.88
7	43.40	71.06	97.03	122.22	146.95	171.34	195.49	219.47	243.30
8	39.29	65.15	89.45	113.03	136.18	159.04	181.65	204.14	226.48
9	36.71	61.39	84.62	107.17	129.30	151.17	172.80	194.27	215.64
10	34.93	58.78	81.25	103.06	124.48	145.64	166.56	187.37	208.02
11	33.62	56.85	78.75	100.02	120.92	141.54	161.98	182.24	202.37
12	32.62	55.37	76.83	97.68	118.15	138.38	158.38	178.23	198.03
13	31.83	54.19	75.30	95.82	115.96	135.86	155.54	175.10	194.51
14	31.19	53.23	74.05	94.29	114.16	133.80	153.21	172.49	191.68
15	30.66	52.44	73.01	93.02	112.66	132.07	151.29	170.36	189.38
16	30.22	51.76	72.14	91.94	111.41	130.61	149.66	166.53	187.32
17	29.83	51.19	71.39	91.03	110.34	129.38	148.25	166.99	185.61
18	29.51	50.69	70.74	90.23	109.39	128.29	147.03	165.65	184.10
19	29.22	50.26	70.17	89.54	108.57	127.36	145.97	164.45	182.81
20	28.97	49.88	69.67	88.93	107.85	126.52	145.02	163.38	181.65
25	28.05	48.48	67.86	86.70	105.21	123.51	141.62	159.60	177.49
30	27.48	47.61	66.71	85.29	103.56	121.60	139.47	157.22	174.87

Outline Solutions to Exercises

EXERCISES 1

1 (a) $\mathscr{C}[\mathbf{Ax}, \mathbf{By}] = \mathscr{E}[(\mathbf{Ax} - \mathbf{A}\boldsymbol{\mu}_1)(\mathbf{By} - \mathbf{B}\boldsymbol{\mu}_2)'] = \mathbf{A}\mathscr{E}[(\mathbf{x} - \boldsymbol{\mu}_1)(\mathbf{y} - \boldsymbol{\mu}_2)']\mathbf{B}' = \mathbf{A}\mathscr{C}[\mathbf{x}, \mathbf{y}]\mathbf{B}'$.
(b) Expand $(\mathbf{x} - \boldsymbol{\mu}_1)((\mathbf{y} - \boldsymbol{\mu}_2)'$ and take expected values.
(c) If $\mathbf{y} = \mathbf{x} - \mathbf{a}$, then $\mathbf{y} - \boldsymbol{\mu}_y = \mathbf{x} - \boldsymbol{\mu}_x$.

3 (a) $\mathscr{C}[\Sigma_i a_i \mathbf{x}_i, \Sigma_j a_j \mathbf{x}_j] = \Sigma_i \Sigma_j a_i a_j \delta_{ij} \boldsymbol{\Sigma}$.
(b) $\mathscr{C}[\Sigma_i a_i \mathbf{x}_i, \Sigma_j b_j \mathbf{x}_j] = \Sigma_i \Sigma_j a_i b_j \delta_{ij} \boldsymbol{\Sigma} = (\Sigma_i a_i b_i)\boldsymbol{\Sigma}$.

4 $\text{E}[\mathbf{x}'\mathbf{A}\mathbf{x}] = \text{E}[(\mathbf{x} - \boldsymbol{\theta})'\mathbf{A}(\mathbf{x} - \boldsymbol{\theta})] + \boldsymbol{\theta}'\mathbf{A}\boldsymbol{\theta} = c + \boldsymbol{\theta}'\mathbf{A}\boldsymbol{\theta}$,
$c = \text{E}[\Sigma_i \Sigma_j a_{ij}(x_i - \theta_i)(x_j - \theta_j)] = \Sigma_i \Sigma_j a_{ij} \sigma_{ji} = \text{tr}[\mathbf{A}\boldsymbol{\Sigma}]$.

5 (a) $(\mathbf{I}_n - \mathbf{P})^2 = \mathbf{I}_n - 2\mathbf{P} + \mathbf{P}^2 = \mathbf{I}_n - \mathbf{P}$.
(c) Use A6.3 and A6.7.
(d) $(\mathbf{I}_n - \mathbf{P})(\mathbf{y} - \mathbf{K}\boldsymbol{\beta}) = (\mathbf{I}_n - \mathbf{P})\mathbf{y}$, by (b).

6 (a) $\Sigma_r g_{rr} = \Sigma_r \text{tr}[g_{rr}] = \text{tr}[\hat{\boldsymbol{\Sigma}}^{-1}\Sigma_r(\mathbf{x}_r - \overline{\mathbf{x}})(\mathbf{x}_r - \overline{\mathbf{x}})'] = n\, \text{tr}[\hat{\boldsymbol{\Sigma}}^{-1}\hat{\boldsymbol{\Sigma}}] = nd$.
Take expected values.
(b) $0 = \Sigma_s g_{rs} = g_{rr} + \Sigma_{s \neq r} g_{rs}$ implies $\text{E}[g_{rr}] + (n - 1)\text{E}[g_{rs}] = 0$.

7 $(\delta_{rs} - n^{-1})\boldsymbol{\Sigma}$.

9 $\boldsymbol{\Sigma} = \mathbf{R}'\mathbf{R}$ (A5.3). Let $\mathbf{z} = (\mathbf{R}')^{-1}\mathbf{y}$; then $\mathbf{z} \sim N_d(\mathbf{0}, \mathbf{I}_d)$. From A6.5, $\mathbf{z}'\mathbf{RAR}'\mathbf{z} \sim \chi_r^2 \Leftrightarrow \mathbf{RAR}'\mathbf{RAR}' = \mathbf{RAR}'$.

EXERCISES 2

1 $\mathbf{x} = \boldsymbol{\Sigma}^{-1/2}\mathbf{y} \sim N_n(\boldsymbol{\mu}, \mathbf{I}_n)$, where $\boldsymbol{\mu} = \boldsymbol{\Sigma}^{-1/2}\boldsymbol{\theta}$ (see A5.4). Hence $\mathbf{y}'\boldsymbol{\Sigma}^{-1}\mathbf{y} = \mathbf{x}'\mathbf{x}$ is noncentral chi square, with $\delta = \boldsymbol{\mu}'\boldsymbol{\mu} = \boldsymbol{\theta}'\boldsymbol{\Sigma}^{-1}\boldsymbol{\theta}$.

2 Suppose $\mathbf{y} \sim N_d(\boldsymbol{\theta}, \boldsymbol{\Sigma})$ by Definition 1b. Then, for all $\boldsymbol{\ell}$, $\boldsymbol{\ell}'\mathbf{y} \sim N_1(\mu, \sigma^2)$. with moment generating function (m.g.f.) $\exp(t\mu + \frac{1}{2}\sigma^2 t^2)$, where $\mu = \boldsymbol{\ell}'\boldsymbol{\theta}$, $\sigma^2 = \boldsymbol{\ell}'\boldsymbol{\Sigma}\boldsymbol{\ell}$. With $t = 1$, $\text{E}[\exp(\boldsymbol{\ell}'\mathbf{y})]$ gives the correct m.g.f, which, by the uniqueness of the m.g.f, implies that $\mathbf{y}$ has density (2.1).

3 (i) $\boldsymbol{\ell}'\mathbf{C}\mathbf{y} = \mathbf{a}'\mathbf{y}$ is univariate normal for all $\boldsymbol{\ell}$, $\mathbf{C}\boldsymbol{\Sigma}\mathbf{C}' > \mathbf{O}$ (A5.7)
(ii) Set $\boldsymbol{\ell}' = (\boldsymbol{\ell}_1', \mathbf{0}')$.
(iii) Since $\mathbf{t}'\mathbf{y}$ is univariate normal, $\text{E}[\exp(s\mathbf{t}'\mathbf{y})] = \exp(s\mathbf{t}'\boldsymbol{\theta} + \frac{1}{2}s^2\mathbf{t}'\boldsymbol{\Sigma}\mathbf{t})$. Set $s = 1$.
(iv) Use (iii) and $\boldsymbol{\Sigma}_{12} = \mathbf{O}$ to factorize (2.2).

4 If $\mathbf{a}'\mathbf{y} = b$, $0 = \text{var}[\mathbf{a}'\mathbf{y}] = \mathbf{a}'\boldsymbol{\Sigma}\mathbf{a}$ and $\boldsymbol{\Sigma} > \mathbf{O}$ implies $\mathbf{a} = \mathbf{0}$??

5 (a) $\mathscr{D}[\mathbf{e}] = \sigma^2(\mathbf{I}_n - \mathbf{P})$, $\text{rank}(\mathbf{I}_n - \mathbf{P}) = n - p < n$ (by Exercise 1.5).
(b) $\boldsymbol{\ell}'(\mathbf{I}_n - \mathbf{P})\mathbf{y} = \mathbf{a}'\mathbf{y}$ is univariate normal for all $\boldsymbol{\ell}$.
(c) $\mathbf{X}$ has column $\mathbf{1}_n$. Then $\mathbf{1}_n'(\mathbf{I}_n - \mathbf{P})\mathbf{y} = 0$, as $\mathbf{X}'(\mathbf{I}_n - \mathbf{P}) = \mathbf{O}$.

6 (a) $\text{E}[\exp(\mathbf{t}'\Sigma_i a_i \mathbf{y}_i)] = \exp[\mathbf{t}'(\Sigma_i a_i \boldsymbol{\theta}_i) + \frac{1}{2}\mathbf{t}'(\Sigma_i a_i^2 \boldsymbol{\Sigma}_i)\mathbf{t}]$.
(b) $\Sigma_i a_i \boldsymbol{\ell}'\mathbf{y}_i = \Sigma_i a_i x_i$ is univariate normal for all $\boldsymbol{\ell}$.

7 Use Definition 1b with $\ell' = (\ell_1', \ell_2', \ldots, \ell_n')$. Then $\ell'\mathbf{y} = \Sigma_i \ell_i' \mathbf{y}_i = \Sigma_i z_i$ is univariate normal. Hence $\mathbf{y}$ is multivariate normal with parameters $\mathscr{E}[\mathbf{y}]$ and $\mathscr{D}[\mathbf{y}]$.

8 (a) Use Theorem 2.1(v) with $\mathbf{y} \sim N_n(\theta \mathbf{1}_n, \sigma^2 \mathbf{I}_n)$ to prove that $\bar{y}$ is independent of $\mathbf{y} - \mathbf{1}_n \bar{y}$, that is, of $Q = \|\mathbf{y} - \mathbf{1}_n \bar{y}\|^2$.
(b)

$$\mathbf{A} = \begin{pmatrix} 1 - 1/n & -1/n & \cdots & -1/n \\ \vdots & \vdots & & \vdots \\ -1/n & -1/n & \cdots & 1 - 1/n \end{pmatrix}, \qquad \mathbf{A}^2 = \mathbf{A},\ \operatorname{tr} \mathbf{A} = n - 1.$$

9 The integral of (2.3) is unity; that is, $\mathrm{E}[\operatorname{etr}(\mathbf{UW})]$ equals $c^{-1} \int |\mathbf{W}|^{m-d-1} \operatorname{etr}[-\frac{1}{2}(\boldsymbol{\Sigma}^{-1} - 2\mathbf{U})\mathbf{W}]\, d\mathbf{x} = |\boldsymbol{\Sigma}^{-1} - 2\mathbf{U}|^{-m/2} |\boldsymbol{\Sigma}|^{-m/2}$.

10 $\mathscr{E}[\mathbf{W}] = \Sigma_i \mathscr{E}[\mathbf{x}_i \mathbf{x}_i'] = m\boldsymbol{\Sigma}$, where the $\mathbf{x}_i$ are i.i.d. $N_d(\mathbf{0}, \boldsymbol{\Sigma})$.

11 Using Definition 2a, set $d = 1$ in (2.4) to get $f(w_{11})$. Using Definition 2b, $W_1 = \Sigma_i x_i^2$, where the x_i are i.i.d. $N_1(0, \sigma^2)$.

12 (a) $\mathrm{E}[\operatorname{etr}(t\ell'\mathbf{W}\ell)] = \mathrm{E}[\operatorname{etr}(t\ell\ell'\mathbf{W}] = |\mathbf{I}_d - 2t\ell\ell'\boldsymbol{\Sigma}|^{-m/2} = [1 - 2t \operatorname{tr}(\ell\ell'\boldsymbol{\Sigma})]^{-m/2} = (1 - 2t\ell'\boldsymbol{\Sigma}\ell)^{-m/2}$, using A2.4. The m.g.f. is $(1 - 2t)^{-m/2}$.
(b)

$$\mathrm{E}\left[\operatorname{etr}(\mathbf{U}_{11}\mathbf{W}_{11} + \mathbf{U}_{22}\mathbf{W}_{22})\right] = \left|\mathbf{I}_d - 2\begin{pmatrix} \mathbf{U}_{11} & \mathbf{O} \\ \mathbf{O} & \mathbf{U}_{22} \end{pmatrix}\begin{pmatrix} \boldsymbol{\Sigma}_{11} & \mathbf{O} \\ \mathbf{O} & \boldsymbol{\Sigma}_{22} \end{pmatrix}\right|^{-m/2}$$

$$= |\mathbf{I}_r - 2\mathbf{U}_{11}\boldsymbol{\Sigma}_{11}|^{-m/2} |\mathbf{I}_{d-r} - 2\mathbf{U}_{22}\boldsymbol{\Sigma}_{22}|^{-m/2}.$$

(c) $\mathrm{E}[\exp(t\Sigma_i w_{ii})] = \mathrm{E}[\operatorname{etr}(\mathbf{UW})] = |\mathbf{I}_d - 2\mathbf{U}\boldsymbol{\Sigma}|^{-m/2} = |\mathbf{I}_d - 2t\boldsymbol{\Sigma}|^{-m/2} = \Pi_i(1 - 2t\lambda_i)^{-m/2} = \mathrm{E}[\exp(t\Sigma_i \lambda_i x_i)]$, where the λ_i are the eigenvalues of $\boldsymbol{\Sigma}$, the x_i are i.i.d. χ_m^2 and $u_{ij} = \delta_{ij} t$.
(d) Multiply the m.g.f.'s of $\mathbf{W}_1$ and $\mathbf{W}_2$.

14 $\mathbf{W}_{11} = \Sigma_i \mathbf{x}_i^{(1)} \mathbf{x}_i^{(1)\prime}$, where the $\mathbf{x}_i^{(1)}$ are i.i.d. $N_r(\mathbf{0}, \boldsymbol{\Sigma}_{11})$.

15 $\boldsymbol{\Sigma}^{-1} > \mathbf{O}$ and, by A5.8, $\boldsymbol{\Sigma}^{-1} - 2t\mathbf{A} > \mathbf{O}$ for $|t| < \delta$.

16 $\mathrm{E}[\operatorname{etr}(\mathbf{WU})] = \Pi_i \mathrm{E}[\operatorname{etr}(\mathbf{x}_i'\mathbf{U}\mathbf{x}_i)] = \Pi_i |\mathbf{I}_d - 2\mathbf{U}\boldsymbol{\Sigma}|^{-1/2} = |\mathbf{I}_d - 2\mathbf{U}\boldsymbol{\Sigma}|^{-m/2}$, which uniquely determines the distribution.

17 Show that the covariance is zero and invoke Theorem 2.1(v).

18 $\operatorname{cov}[y_{ij}, z_{rs}] = \Sigma_m \Sigma_n \Sigma_p \Sigma_q a_{im} b_{nj} c_{rp} d_{qs} \operatorname{cov}[x_{mn}, x_{pq}] = \Sigma_m \Sigma_n \Sigma_p \Sigma_q a_{im} b_{nj} c_{rp} d_{qs} \delta_{mp} \delta_{nq} \sigma_{nq} = (\Sigma_m a_{im} c_{mr}')(\Sigma_n \Sigma_q b_{jn}' \sigma_{nq} d_{qs}) = (\mathbf{AC}')_{ir} (\mathbf{B\Sigma D})_{js}$.

19 $f_1(x) = \int f(x, y)\, dy = \int f(x|y) f_2(y)\, dy = f(x|y)$; thus $f(x, y) = f_1(x) f_2(y)$.

20 Partition $\boldsymbol{\Sigma}$ and use A3.1 with $\mathbf{E} = \boldsymbol{\Sigma}_{22 \cdot 1}$ and $\mathbf{F} = \boldsymbol{\Sigma}_{11}^{-1}\boldsymbol{\Sigma}_{12}$. Multiply out the quadratic and show that the difference is $(\boldsymbol{\mu}_2 - \mathbf{F}'\boldsymbol{\mu}_1)'\mathbf{E}^{-1}(\boldsymbol{\mu}_2 - \mathbf{F}'\boldsymbol{\mu}_1)$. Since $\boldsymbol{\Sigma}^{-1} > \mathbf{O}$, we use modified A5.9 to show $\mathbf{E}^{-1} > \mathbf{O}$.

21 $|\mathbf{\Sigma}^{-1/2}\mathbf{W}_1\mathbf{\Sigma}^{-1/2}|/|\mathbf{\Sigma}^{-1/2}(\mathbf{W}_1 + \mathbf{W}_2)\mathbf{\Sigma}^{-1/2}| = |\mathbf{Z}_1|/|\mathbf{Z}_1 + \mathbf{Z}_2|$, where, by Theorem 2.2, $\mathbf{Z}_i \sim W_d(m_i, \mathbf{I}_d)$.

22 Use (2.60) and Lemma 2.10 successively, with Theorem 2.2, Corollary 2.

23 $\mathrm{E}[\boldsymbol{l}'\mathbf{W}^{-1}\boldsymbol{l}] = \boldsymbol{l}'\mathbf{\Sigma}^{-1}\boldsymbol{l}/k$ $(k = m - d - 1)$. If $\boldsymbol{l}' = (1, 0, \ldots, 0)$, then $\mathrm{E}[w^{11}] = \sigma^{11}/k$. If $\boldsymbol{l}' = (1, 1, 0, \ldots, 0)$, then $\mathrm{E}[w^{11} + w^{22} + 2w^{12}] = (\sigma^{11} + \sigma^{22} + 2\sigma^{12})/k$. Hence $\mathrm{E}[w^{12}] = \sigma^{12}/k$, and so on. Then $\mathrm{E}[\mathbf{y}'\mathbf{W}^{-1}\mathbf{y}] = \mathrm{tr}\{\mathscr{E}[\mathbf{W}^{-1}\mathbf{y}\mathbf{y}']\} = \mathrm{tr}\{\mathscr{E}[\mathbf{W}^{-1}]\mathscr{E}[\mathbf{y}\mathbf{y}']\} = \mathrm{tr}\{\mathbf{\Sigma}^{-1}\mathbf{\Sigma}/k] = d/k$.

24 Use Theorem 2.4 Corollary 1, with $a = \frac{1}{2}$ and $b = \pm\frac{1}{2}$.

25 Expand the argument following (2.12) to r independent sets of i.i.d $N_d(\mathbf{0}, \mathbf{\Sigma})$.

26 $\{\Sigma_i\Sigma_j(x_i - \mu_i)(y_j - \mu_j)\sigma^{ij}\}^3$ has zero expectation, as $\mathrm{E}[(x_i - \mu_i)^{2m+1}] = 0$ for univariate normal. Furthermore, $z = (\mathbf{x} - \boldsymbol{\mu})'\mathbf{\Sigma}^{-1}(\mathbf{x} - \boldsymbol{\mu}) \sim \chi_d^2$ and $\mathrm{E}[z^2] = d(d + 2)$.

EXERCISES 3

1 Use (3.17) with $\mathbf{A} = (1, -1)$.

2 Use B3.5 with $\mathbf{F} = \mathbf{A}'$, $\mathbf{V} = \mathbf{S}^{-1}$, and $\mathbf{G} = \mathbf{1}_d$.

3 Steps are as follows:

(a) $\mathrm{cov}[\bar{x}_{\cdot k} - \bar{x}_{\cdot\cdot}, x_{ij} - \bar{x}_{i\cdot} - \bar{x}_{\cdot j} + \bar{x}_{\cdot\cdot}] = 0$ implies that Q_H and Q_E are independent by Theorem 2.1(v).

(b) $Q_H/(1 - \rho) = \bar{\mathbf{z}}'\mathbf{A}\bar{\mathbf{z}}$, where $[n/(1 - \rho)]^{1/2}(\bar{\mathbf{x}} - \mu\mathbf{1}_d) = \bar{\mathbf{z}}$ is $N_d(\mathbf{0}, \mathbf{\Sigma}/[1 - \rho])$ when H_0 is true, and $\bar{\mathbf{x}}' = (\bar{\mathbf{x}}_{\cdot 1}, \bar{\mathbf{x}}_{\cdot 2}, \ldots, \bar{\mathbf{x}}_{\cdot d})$. Since $\mathbf{A}\mathbf{\Sigma}\mathbf{A}/(1 - \rho) = \mathbf{A}$, $Q_H/(1 - \rho) \sim \chi_{d-1}^2$ (by Exercise 1.9);

(c) $\mathbf{z} = (1 - \rho)^{-1/2}(\mathbf{x} - \mu\mathbf{1}_{nd}) \sim N_{nd}(\mathbf{0}, \mathbf{\Sigma}_0)$ when H_0 is true, where $\mathbf{x}$ is the vector of x_{ij} and $\mathbf{\Sigma}_0 = \mathrm{diag}(\mathbf{\Sigma}, \mathbf{\Sigma}, \ldots, \mathbf{\Sigma})/(1 - \rho)$. Finally, $Q_H = \mathbf{z}'\mathbf{A}_0\mathbf{z} \sim \chi_{(n-1)(d-1)}^2$, as $\mathbf{A}_0\mathbf{\Sigma}_0\mathbf{A}_0 = \mathbf{A}_0$.

4 Under $H_0: \mu_1 = \mu_2 = \cdots = \mu_d\ (= \mu)$ and $\hat{\mu} = \Sigma_i\Sigma_j s^{ij}(\bar{x}_i + \bar{x}_j)/2\Sigma_i\Sigma_j s^{ij}$. Substitute in $T_0^2 = n(\bar{\mathbf{x}}'\mathbf{S}^{-1}\bar{\mathbf{x}} - \bar{\mathbf{x}}'\mathbf{S}^{-1}\mathbf{1}_n\hat{\mu})$.

5 $\Sigma_{i=1}^{d-1}h_i\phi_i = \Sigma_{i=1}^{d-1}h_i\mu_i - (\Sigma_{i=1}^{d-1}h_i)\mu_d$. Set $c_i = h_i$ $(i = 1, 2, \ldots, d - 1)$ and $c_d = -\Sigma_{i=1}^{d-1}h_i$. Hence show that a member of one set is a member of the other, and vice versa.

6

$$\left(\mathbf{Q}\mathbf{Q}_{(2)}^{-1} - \mathbf{I}_d\right)^2 = \begin{pmatrix} \mathbf{Q}_{12}\mathbf{Q}_{22}^{-1}\mathbf{Q}_{21}\mathbf{Q}_{11}^{-1} & \mathbf{O} \\ \mathbf{O} & \mathbf{Q}_{21}\mathbf{Q}_{11}^{-1}\mathbf{Q}_{12}\mathbf{Q}_{22}^{-1} \end{pmatrix} = \begin{pmatrix} \mathbf{A}_1 & \mathbf{O} \\ \mathbf{O} & \mathbf{A}_2 \end{pmatrix}$$

and $\mathrm{tr}\,\mathbf{A}_1 = \mathrm{tr}\,\mathbf{A}_2$ (by A1.1).

7 When H_0 is true, $t = \hat{\beta}_1\{\Sigma_i(x_{i1} - \bar{x}_{\cdot 1})^2/s^2\}^{1/2} = r(n - 2)^{1/2}/(1 - r^2)^{1/2}$. Obtain $f(r)$ from the t-density function.

8 Under H_0 the maximum likelihood estimates are $\hat{\boldsymbol{\mu}} = \bar{\mathbf{x}}$ (arguing as in Theorem 3.2) and, by differentiation, $\hat{\sigma}^2 = \Sigma_i(\mathbf{x}_i - \bar{\mathbf{x}})'(\mathbf{x}_i - \bar{\mathbf{x}})/nd = \mathrm{tr}[\hat{\mathbf{\Sigma}}/d]$. Then $\ell_2 = L(\bar{\mathbf{x}}, \hat{\sigma}^2\mathbf{I}_d)/L(\bar{\mathbf{x}}, \hat{\mathbf{\Sigma}})$.

9 $\ell_3 = L(\bar{\mathbf{x}}, \mathbf{I}_d)/L(\bar{\mathbf{x}}, \hat{\boldsymbol{\Sigma}})$ and apply (3.6) to $L(\bar{\mathbf{x}}, \hat{\boldsymbol{\Sigma}})$.

10 $\ell_4 = L(\boldsymbol{\mu}_0, \boldsymbol{\Sigma}_0)/L(\bar{\mathbf{x}}, \hat{\boldsymbol{\Sigma}})$ and apply (3.4) to $L(\boldsymbol{\mu}_0, \boldsymbol{\Sigma}_0)$.

11

$$\mathbf{C} = \begin{pmatrix} b^{-1}\mathbf{I}_{d_0} & b^{-1}\mathbf{I}_{d_0} & \cdots & b^{-1}\mathbf{I}_{d_0} \\ -b^{-1}\mathbf{I}_{d_0} & (1-b^{-1})\mathbf{I}_{d_0} & \cdots & -b^{-1}\mathbf{I}_{d_0} \\ \vdots & \vdots & & \vdots \\ -b^{-1}\mathbf{I}_{d_0} & -b^{-1}\mathbf{I}_{d_0} & \cdots & (1-b^{-1})\mathbf{I}_{d_0} \end{pmatrix}.$$

By adding the first row to other rows, we see that the following is nonsingular:

$$\begin{pmatrix} b^{-1} & b^{-1} & \cdots & b^{-1} \\ -b^{-1} & (1-b^{-1}) & \cdots & -b^{-1} \\ \vdots & \vdots & & \vdots \\ -b^{-1} & -b^{-1} & \cdots & (1-b^{-1}) \end{pmatrix}.$$

The columns of $\mathbf{C}$ are therefore orthogonal sets of independent vectors.

12 Use $a = \sigma^2$ and $b = \sigma^2\rho$ so that $\boldsymbol{\Sigma} = \boldsymbol{\Sigma}(a, b)$. Differentiate $\log L[\bar{\mathbf{x}}, \boldsymbol{\Sigma}(a, b)]$ with respect to a and b using A3.5.

13 Use Exercise 3.12 and $\ell_6 = L[\bar{\mathbf{x}}, \boldsymbol{\Sigma}(\hat{a}, \hat{b})]/L(\bar{\mathbf{x}}, \hat{\boldsymbol{\Sigma}})$ [see (3.6)].

15 Use the method of Section 3.2.1 to show that $\ell = L_1(\tilde{\boldsymbol{\mu}}, \tilde{\boldsymbol{\Sigma}})L_2(\tilde{\boldsymbol{\mu}}, \tilde{\boldsymbol{\Sigma}})/L_1(\bar{\mathbf{v}}, \hat{\boldsymbol{\Sigma}})L_2(\bar{\mathbf{w}}, \hat{\boldsymbol{\Sigma}})$, where $\tilde{\boldsymbol{\mu}} = (n_1\bar{\mathbf{v}} + n_2\bar{\mathbf{w}})/n$, $n = n_1 + n_2$, $\tilde{\boldsymbol{\Sigma}} = [\Sigma_i(\mathbf{v}_i - \tilde{\boldsymbol{\mu}})(\mathbf{v}_i - \tilde{\boldsymbol{\mu}})' + \Sigma_j(\mathbf{w}_j - \tilde{\boldsymbol{\mu}})(\mathbf{w}_j - \tilde{\boldsymbol{\mu}})']/n = \mathbf{Q}_0/n$, and $\hat{\boldsymbol{\Sigma}} = [\Sigma_i(\mathbf{v}_i - \bar{\mathbf{v}})(\mathbf{v}_i - \bar{\mathbf{v}})' + \Sigma_j(\mathbf{w}_j - \bar{\mathbf{w}})(\mathbf{w}_j - \bar{\mathbf{w}})']/n = \mathbf{Q}/n$. By writing $\mathbf{v}_i - \tilde{\boldsymbol{\mu}} = \mathbf{v}_i - \bar{\mathbf{v}} + \bar{\mathbf{v}} - \tilde{\boldsymbol{\mu}}$, and so on, we have $\ell^{2/n} = |\mathbf{Q}|/|\mathbf{Q}_0| = |\mathbf{Q}|/|\mathbf{Q} + n_1n_2(n_1 + n_2)^{-1}(\bar{\mathbf{v}} - \bar{\mathbf{w}})(\bar{\mathbf{v}} - \bar{\mathbf{w}})'| = [1 + (n_1 + n_2 - 2)^{-1}T_0^2]^{-1}$, as in Theorem 3.2.

16 $T_0^2 \sim T_{d, n_1 - 1}$.

17 $\bar{\mathbf{v}} - \bar{\mathbf{w}} \sim N_d(\boldsymbol{\mu}_1 - \boldsymbol{\mu}_2, [n_1^{-1} + (kn_2)^{-1}]\boldsymbol{\Sigma}_1)$, $\mathbf{Q}_1 = \Sigma_i(\mathbf{v}_i - \bar{\mathbf{v}})(\mathbf{v}_i - \bar{\mathbf{v}})' \sim W_d(n_1 - 1, \boldsymbol{\Sigma}_1)$, $k\mathbf{Q}_2 \sim W_d(n_2 - 1, \boldsymbol{\Sigma}_1)$, $(\mathbf{Q}_1 + k\mathbf{Q}_2) \sim W_d(\nu, \boldsymbol{\Sigma}_1)$, where $\nu = n_1 + n_2 - 2$. Use (2.19) to show that

$$T_0^2 = \frac{kn_1n_2\nu}{n_1 + kn_2}(\bar{\mathbf{v}} - \bar{\mathbf{w}})'(\mathbf{Q}_1 + k\mathbf{Q}_2)^{-1}(\bar{\mathbf{v}} - \bar{\mathbf{w}}) \sim T_{d,\nu}^2.$$

18 $r = \boldsymbol{l}'\mathbf{Q}\boldsymbol{l}/\boldsymbol{l}'\boldsymbol{\Sigma}_0\boldsymbol{l} \sim \chi_{n-1}^2$ when H_0 is true. If $b = \chi_{n-1}^2(\alpha/2)$ and $a = \chi_{n-1}^2(1 - \alpha/2)$, the acceptance region is $\cap_{\boldsymbol{l}}\{\mathbf{x}_1, \mathbf{x}_2, \ldots, \mathbf{x}_n : a \le r \le b\} = \{\mathbf{x}_1, \mathbf{x}_2, \ldots, \mathbf{x}_n : a \le \inf r \le \sup r \le b\} = \{\mathbf{x}_1, \mathbf{x}_2, \ldots, \mathbf{x}_n : a \le \phi_{\min} \le \phi_{\max} \le b\}$, where $\phi_{\min}$ and $\phi_{\max}$ are eigenvalues of $\boldsymbol{\Sigma}_0^{-1}\mathbf{Q}$, by A7.5.

EXERCISES 4

1 $\int_{-\pi}^{\pi}\cos at \cos bt\, dt = \int_{-\pi}^{\pi}\sin at \sin bt\, dt = \delta_{ab}\pi$, $\int_{-\pi}^{\pi}\sin at \cos bt\, dt = 0$.

2 $z_i^{(1)} = x_i + \xi - 1$ so that $z_i^{(1)} - \bar{z}^{(1)} = x_i - \bar{x}$.

3. See Section 3.3.1.

4 $2[L_{\max}(\hat{\boldsymbol{\lambda}}) - L_{\max}(\hat{\theta}\mathbf{1}_d)]$ is approximately χ^2_{d-1}, where $\hat{\theta}$ minimizes $-\frac{1}{2}n\log|\hat{\boldsymbol{\Sigma}}| + \Sigma_i\Sigma_j(\theta - 1)x_{ij}$.

5 From Exercise 1.6, $\mathrm{E}[D_i^2] = (n-1)d/n$.

6 See Rao [1973: p. 143] for $\beta_2 - \beta_1 \geq 1$. Since $[\Sigma_i(x_i - \bar{x})^2]^2 \geq \Sigma_i(x_i - \bar{x})^4$, $b_2 \leq n$. Furthermore, $\mathrm{E}[U] = \bar{x}$ $(= \mu)$ and $\mathrm{E}[(U-\mu)^3] = \Sigma_i(x_i - \bar{x})^3/n$, and so on, so that $b_2 - b_1 \geq 1$, that is, $b_1 \leq b_2 - 1 \leq n - 1$.

7 Substitute for $\bar{\mathbf{x}}$ in $\mathbf{Q}$ using $\bar{\mathbf{x}}_{-i} = \bar{\mathbf{x}} - \mathbf{d}_i/(n-1)$, where $\mathbf{d}_i = \mathbf{x}_i - \bar{\mathbf{x}}$. Thus $\mathbf{Q} = \mathbf{Q}_{-i} + \mathbf{d}_i\mathbf{d}_i'/(n-1) + \mathbf{d}_i\mathbf{d}_i'$, and we take traces.

8 Let $y_i = (x_i - \mu)/\sigma$; then $0 = \Sigma_i\psi[(x_i - \tilde{\mu})/\sigma] \approx \Sigma_i\psi(y_i) - \Sigma_i\psi'(y_i)\sigma^{-1} \times (\tilde{\mu} - \mu)$, $\sqrt{n}(\tilde{\mu} - \mu) \approx \sigma n^{-1/2}\Sigma_i\psi(y_i)/n^{-1}\Sigma_i\psi'(y_i) \approx \sigma n^{-1/2}\Sigma_i\psi(y_i)/\mathrm{E}[\psi'(y)]$, and $\mathrm{var}[\sqrt{n}(\tilde{\mu} - \mu)] \approx \sigma^2\mathrm{E}[\psi^2(y)]/\{\mathrm{E}[\psi'(y)]\}^2$. Replace expected values by sample means, σ by s, and μ by $\tilde{\mu}$.

9 Using the notation of Exercise 4.7 and A2.4, we have $|\mathbf{Q}_{-i}\mathbf{Q}^{-1}| = |\mathbf{I}_d - [n/(n-1)]\mathbf{Q}^{-1}\mathbf{d}_i\mathbf{d}_i'| = 1 - [n/(n-1)^2]\mathbf{d}_i'\mathbf{S}^{-1}\mathbf{d}_i$.

EXERCISES 5

1 $\mathrm{tr}[\mathbf{T}'\mathbf{C}\mathbf{T}] = \mathrm{tr}[\mathbf{T}\mathbf{T}'\mathbf{C}]$; $|\mathbf{T}'\mathbf{C}\mathbf{T}| = |\mathbf{T}'||\mathbf{C}||\mathbf{T}| = |\mathbf{T}'\mathbf{T}||\mathbf{C}|$.

2 With A5.12, $\mathbf{O} \leq \mathbf{R}'(\mathbf{A} - \mathbf{B})\mathbf{R} = \boldsymbol{\Gamma} - \mathbf{I}$ or $\gamma_i \geq 1$. Then $|\mathbf{A}|/|\mathbf{B}| = |\mathbf{R}'\mathbf{A}\mathbf{R}|/|\mathbf{R}'\mathbf{B}\mathbf{R}| = |\boldsymbol{\Gamma}| = \Pi_i\gamma_i \geq 1$.

3 Equality occurs in (5.3) as

$$f(\mathbf{A} - \mathbf{B}\mathbf{B}') = f\left[\mathbf{T}'(\mathbf{A} - \mathbf{T}_1\boldsymbol{\Lambda}_1\mathbf{T}_1')\mathbf{T}\right] = f\left[\boldsymbol{\Lambda} - \begin{pmatrix} \boldsymbol{\Lambda}_1 & \mathbf{O} \\ \mathbf{O} & \mathbf{O} \end{pmatrix}\right]$$

$$= g(\lambda_{k+1},\ldots,\lambda_d,0,\ldots,0).$$

4 $y_1 = (x_1 + x_2)/\sqrt{2}$, $y_2 = (x_1 - x_2)/\sqrt{2}$.

5 $\mathrm{cov}[x_i, y_j] = \mathscr{C}[\mathbf{e}_i'\mathbf{x}, \mathbf{t}_j'\mathbf{x}] = \mathbf{e}_i'\boldsymbol{\Sigma}\mathbf{t}_j = \lambda_j\mathbf{e}_i'\mathbf{t}_j = \lambda_j t_{ij}$.

6 $\boldsymbol{\Sigma}\mathbf{1}_n = (n+1)\sigma^2\mathbf{1}_n$ and $\mathbf{1}_n/\sqrt{n}$ is a unit eigenvector. Furthermore, $\mathrm{var}[\sqrt{n}\,\bar{x}] = (n+1)\sigma^2$, which is more than half of $\mathrm{tr}\boldsymbol{\Sigma}$ $(= 2n\sigma^2)$, so that, $y_1 = \sqrt{n}\,\bar{x}$.

7 $3 + \alpha, \alpha, \alpha$ and $y_1 = \sqrt{3}\,\bar{x}$.

8 Since $\mathbf{T}_1'\mathbf{T}_1 = \mathbf{I}_k$, $Q_1 = (\mathbf{x} - \boldsymbol{\mu})'\mathbf{T}'\mathbf{T}(\mathbf{x} - \boldsymbol{\mu}) - \mathbf{y}_k'\mathbf{y}_k = \mathbf{y}'\mathbf{y} - \mathbf{y}_k'\mathbf{y}_k$.

9 $\sqrt{n}(\hat{\boldsymbol{\lambda}} - \boldsymbol{\lambda}) \sim N_d(\mathbf{0}, 2\boldsymbol{\Lambda}^2)$, approximately, where $\boldsymbol{\Lambda} = \mathrm{diag}(\lambda_1, \lambda_2,\ldots,\lambda_d)$.

Also, $\rho_q = f(\boldsymbol{\lambda})$ and

$$f_j = \frac{\partial f}{\partial \lambda_j} = \begin{cases} \dfrac{1-\rho_q}{\operatorname{tr} \boldsymbol{\Sigma}}, & i \leq q, \\ \dfrac{-\rho_q}{\operatorname{tr} \boldsymbol{\Sigma}}, & i > q. \end{cases}$$

If $\mathbf{f} = [(f_j)]$, then $v = 2\mathbf{f}'\boldsymbol{\Lambda}^2\mathbf{f}$.

11 Let $\mathbf{A} = \mathbf{M}\boldsymbol{\Delta}^{-2}\mathbf{M}'$. Since $\mathbf{B}'\mathbf{B} = \mathbf{M}\boldsymbol{\Delta}^2\mathbf{M}'$ [by (5.34)] and $\mathbf{M}'\mathbf{M} = \mathbf{I}_r$, $\mathbf{B}'\mathbf{B}\mathbf{A}\mathbf{B}'\mathbf{B} = \mathbf{B}'\mathbf{B}$.

12 $\mathbf{H}_i$ is $m \times r$ of rank r so that $\mathbf{H}_i'(\mathbf{H}_i\mathbf{H}_i')^-\mathbf{H}_i$ is an $r \times r$ projection matrix projecting onto an r-dimensional space, that is, it must equal $\mathbf{I}_r$. By B1.9, $\mathbf{B}(\mathbf{B}'\mathbf{B})^-\mathbf{B}'$ projects onto Ω, say, where the columns of $\mathbf{G}_i$ form an orthonormal basis for $\boldsymbol{\Omega}$. Thus $\mathbf{G}_1 = \mathbf{G}_2\mathbf{R}$ ($\mathbf{R}$ nonsingular) and $\mathbf{G}_i'\mathbf{G}_i = \mathbf{I}_r$ imply $\mathbf{R}'\mathbf{R} = \mathbf{I}_r$.

13 $\|\mathbf{A}\|^2 = \operatorname{tr}[\mathbf{A}'\mathbf{A}]$, $\|\mathbf{B} - \mathbf{B}_{(s)}\|^2 = \|\Sigma_{i=s+1}^r \delta_i \boldsymbol{l}_i \mathbf{m}_i'\|^2 = \Sigma_i\Sigma_j\delta_i\delta_j\boldsymbol{l}_i'\boldsymbol{l}_j\mathbf{m}_j'\mathbf{m}_i = \Sigma_{i=s+1}^r\delta_i^2$.

14 Use $\mathbf{x}_r - \mathbf{x}_s = \mathbf{x}_r - \overline{\mathbf{x}} - (\mathbf{x}_s - \overline{\mathbf{x}})$.

15 Suppose $\gamma_{r1} \neq 0$; then, in $\boldsymbol{\Gamma}\mathbf{f}$, γ_{r1} and f_1 are confounded, as they only occur in the product $\gamma_{r1}f_1$.

17 $g(\mathbf{f}|\mathbf{x}) = h_1(\mathbf{x}|\mathbf{f})h_2(\mathbf{f})/h(\mathbf{x})$. Furthermore, $|\mathbf{I}_m + \boldsymbol{\Psi}^{-1/2}\boldsymbol{\Gamma}\boldsymbol{\Gamma}'\boldsymbol{\Psi}^{-1/2}| = |\mathbf{I}_m + \boldsymbol{\Gamma}'\boldsymbol{\Psi}^{-1}\boldsymbol{\Gamma}|$, as both matrices have the same eigenvalues.

18 Show that $\hat{\boldsymbol{\Gamma}}'\hat{\boldsymbol{\Psi}}^{-1}\hat{\boldsymbol{\Gamma}} = \boldsymbol{\Theta} - \mathbf{I}_m$; then use $\hat{\boldsymbol{\Psi}}^{-1/2}\mathbf{S}\hat{\boldsymbol{\Psi}}^{-1/2}\boldsymbol{\Omega} = \boldsymbol{\Omega}\boldsymbol{\Theta}$ to prove (5.53).

19 Use A3.4 and (5.53) to show that $\operatorname{tr}[\mathbf{S}\hat{\boldsymbol{\Sigma}}^{-1}] = \operatorname{tr}[\mathbf{S}\hat{\boldsymbol{\Psi}}^{-1}] - \operatorname{tr}[\hat{\boldsymbol{\Gamma}}\hat{\boldsymbol{\Gamma}}'\hat{\boldsymbol{\Psi}}^{-1}] = \operatorname{tr}[\operatorname{diag}(\mathbf{S} - \hat{\boldsymbol{\Gamma}}\hat{\boldsymbol{\Gamma}}')\hat{\boldsymbol{\Psi}}^{-1}] = \operatorname{tr}[\hat{\boldsymbol{\Psi}}\hat{\boldsymbol{\Psi}}^{-1}]$ by (5.54).

20 $\mathscr{E}[\hat{\mathbf{f}}] = \mathscr{E}[\mathbf{f}] = \mathbf{0}$ and, conditioning first on $\mathbf{f}$, $\mathscr{E}[\hat{\mathbf{f}}\mathbf{f}'] = (\mathbf{I}_m + \mathbf{J})^{-1}\mathbf{J}$ $(= \mathscr{D}[\hat{\mathbf{f}}] = \mathscr{E}[\hat{\mathbf{f}}\hat{\mathbf{f}}'])$. Then $\mathscr{E}[(\hat{\mathbf{f}} - \mathbf{f})(\hat{\mathbf{f}} - \mathbf{f})'] = \mathscr{D}[\mathbf{f}] - \mathscr{E}[\hat{\mathbf{f}}\hat{\mathbf{f}}']$ [use A3.4 (second part)].

21 $\beta = \rho$, the correlation of f_1 and f_2. $\gamma_1 = \boldsymbol{\theta}_1 - \rho\boldsymbol{\theta}_2$, $\gamma_2 = \boldsymbol{\theta}_2$. No, not without knowing ρ.

22 In general, $\mathbf{y} - \bar{y}\mathbf{1}_n = (\mathbf{I}_n - n^{-1}\mathbf{1}_n\mathbf{1}_n')\mathbf{y}$. Apply this to each column of $[(a_{rs} - \bar{a}_{\cdot s})]$. Now consider the rows of $\mathbf{B}$.

23 The points are $(-1,1)$, $(1,1)$, $(-1,-1)$, $(1,-1)$.

24 The maximum value of $\mathbf{TA}$ is $(\mathbf{A}'\mathbf{A})^{-1}$, which is p.s.d. Conversely, if $\mathbf{TA}$ is p.s.d., there exists an orthogonal matrix $\mathbf{M}$ such that $\mathbf{TA} = \mathbf{MDM}'$, where $\mathbf{D}$ is the diagonal matrix of nonnegative eigenvalues listed in decreasing order of magnitude. Then $\mathbf{A} = \mathbf{T}'\mathbf{MDM}' = \mathbf{LDM}'$ is a singular value decomposition of $\mathbf{A}$, that is, $\mathbf{D} = \boldsymbol{\Delta}$ and $\operatorname{tr}[\mathbf{TA}] = \operatorname{tr}\boldsymbol{\Delta}$

25 $\rho^2(z_1, \boldsymbol{\beta}'\mathbf{y}) = (\boldsymbol{\sigma}_{21}'\boldsymbol{\beta})^2/\sigma_{11}\boldsymbol{\beta}'\boldsymbol{\Sigma}_{22}\boldsymbol{\beta}$; then use A7.6.

26 From A3.2 $|\boldsymbol{\Sigma}| = |\boldsymbol{\Sigma}_{22}||\boldsymbol{\Sigma}_{11} - \boldsymbol{\Sigma}_{12}\boldsymbol{\Sigma}_{22}^{-1}\boldsymbol{\Sigma}_{21}|$. If $\rho^2 = 1$, then $|\boldsymbol{\Sigma}_{11}^{-1}\boldsymbol{\Sigma}_{12}\boldsymbol{\Sigma}_{22}^{-1}\boldsymbol{\Sigma}_{21} - \mathbf{I}| = 0$ and $|\boldsymbol{\Sigma}| = 0$, a contradiction.

27 $\mathbf{R} = \mathbf{D}^{-1/2}\mathbf{\Sigma}\mathbf{D}^{-1/2}$, where $\mathbf{D} = \text{diag}(\sigma_{11}, \sigma_{22}, \ldots, \sigma_{dd})$. $0 = |\mathbf{\Sigma}_{11}^{-1}\mathbf{\Sigma}_{12}\mathbf{\Sigma}_{22}^{-1}\mathbf{\Sigma}_{21} - \rho^2\mathbf{I}_d|$ implies $0 = |\mathbf{D}^{1/2}||\mathbf{\Sigma}_{11}^{-1}\mathbf{\Sigma}_{12}\mathbf{\Sigma}_{22}^{-1}\mathbf{\Sigma}_{21} - \rho^2\mathbf{I}_d||\mathbf{D}^{-1/2}| = |\mathbf{R}_{11}^{-1}\mathbf{R}_{12}\mathbf{R}_{22}^{-1}\mathbf{R}_{21} - \rho^2\mathbf{I}_d|$.

28 $\text{cov}[u_i, v_j] = \mathbf{a}_i'\mathbf{\Sigma}_{12}\mathbf{b}_j$. Show that $\mathbf{a}_j = \mathbf{\Sigma}_{11}^{-1}\mathbf{\Sigma}_{12}\mathbf{b}_j$ satisfies $\mathbf{\Sigma}_{11}^{-1}\mathbf{\Sigma}_{12}\mathbf{\Sigma}_{22}^{-1}\mathbf{\Sigma}_{21}\mathbf{a}_j = \rho_j^2\mathbf{a}_j$. Then $\mathbf{a}_i'\mathbf{\Sigma}_{12}\mathbf{b}_j = \mathbf{a}_i'\mathbf{\Sigma}_{11}\mathbf{a}_j = 0$ [Theorem 5.9(ii)].

29 Let $L = (\mathbf{a}'\mathbf{\Sigma}_{12}\boldsymbol{\beta})^2 + \lambda_1(\mathbf{a}'\mathbf{\Sigma}_{11}\mathbf{a} - 1) + \lambda_2(\boldsymbol{\beta}'\mathbf{\Sigma}_{22}\boldsymbol{\beta})$. Differentiate partially with respect to $\boldsymbol{\alpha}$ and $\boldsymbol{\beta}$ using A8.1, and show that $\lambda_1 = \lambda_2 = (\boldsymbol{\alpha}'\mathbf{\Sigma}_{12}\boldsymbol{\beta})^2 = \lambda^2$, say, and $\mathbf{\Sigma}_{11}^{-1}\mathbf{\Sigma}_{12}\mathbf{\Sigma}_{22}^{-1}\mathbf{\Sigma}_{21}\boldsymbol{\alpha} = \lambda^2\boldsymbol{\alpha}$. Thus $\lambda^2 = \rho_1^2$.

30 $\rho^2 = \mathbf{c}'\mathbf{\Sigma}_{12}\mathbf{\Sigma}_{22}^{-1}\mathbf{\Sigma}_{21}\mathbf{c}/\mathbf{c}'\mathbf{\Sigma}_{11}\mathbf{c}$; then use A7.5.

31 Show that $\mathbf{\Sigma}_{11}^{-1}\mathbf{\Sigma}_{12}\mathbf{\Sigma}_{22}^{-1}\mathbf{\Sigma}_{21} = [2b^2/(1 + a)(1 + c)]\mathbf{J}_2 = d\mathbf{J}_2$, where $\mathbf{J}_2$ is a 2×2 matrix of 1's. As $\mathbf{J}_2$ has eigenvalues 2 and 0, $\rho_1^2 = 2d$. Solve $\mathbf{J}_2\boldsymbol{\alpha} = 2\boldsymbol{\alpha}$ to get $u_1 = (x_1 + x_2)/\sqrt{2}$ and, by symmetry, $v_1 = (y_1 + y_2)/\sqrt{2}$.

EXERCISES 6

1 $\boldsymbol{\lambda}'\boldsymbol{\mu}_1 \geq \boldsymbol{\lambda}'\boldsymbol{\mu}_2$ as $\mathbf{\Sigma} \geq \mathbf{O}$ implies $\boldsymbol{\lambda}'(\boldsymbol{\mu}_1 - \boldsymbol{\mu}_2) \geq 0$.

2 Choose $G_1 : x \sim N_1(8.5, 1)$.
 (a) Assign to G_1 if $x > 8.0628$. Using (6.10) and (6.11), $P(2|1) = 0.3311$, $P(1|2) = 0.0011$, and $e_{\text{opt}} = 0.0044$; $q_1(7) = 0.024$, and $q_2(7) = 0.976$.
 (b) Assign to G_1 if $x > 6.75$, $\tilde{e} = P(2|1) = 0.0401$.
 (c) If $P(2|1) = 0.1$, then assign to G_1 if $x > 7.2184$; $P(1|2) = 0.0133$ and $\tilde{e} = 0.0142$.
 (d) $c_0 = \frac{1}{2}\log(\sigma_2^2/\sigma_1^2) - \frac{1}{2}\mu_1^2/\sigma_1^2 + \frac{1}{2}\mu_2^2/\sigma_2^2$. Assign to G_1 if $x^2 - 3x - 42.017 > 0$ or $x > 8.1533$; $P(2|1) = 0.4033$, $P(1|2) = 0.008$, and $e_{\text{opt}} = 0.0048$.

3 Assign to G_1 if $x\log[\theta_1(1 - \theta_2)/\theta_2(1 - \theta_1)] + n\log[(1 - \theta_1)/(1 - \theta_2)] > \log(\pi_2/\pi_1)$, or $x > c$ (as $\theta_1 > \theta_2$). $P(1|2) = \Sigma_{[c]+1}^{n} f_2(x)$.

4 Optimal rule: Assign to G_1 if $x < \log 2$. $P(1|2) = \frac{1}{2}$, $P(2|1) = \frac{1}{4}$, $e_{\text{opt}} = \frac{3}{8}$. Minimax rule: Assign to G_1 if $x < c$, where $P(1|2) = 1 - e^{-c} = e^{-2c} = P(2|1)$, that is, $c = 0.481$ and $\tilde{e} = 0.382$ $(= e^{-2c})$.

5 See Exercise 4.7.

6 When $\mathbf{z}$ comes from G_2, interchange suffixes 1 and 2 and change the sign in $D_L(\mathbf{z})$. If $\mathbf{z}$ comes from G_1, then

$$Q_L(\mathbf{z}) = Q_s(\mathbf{z}) - \frac{1}{2}\left\{\frac{Q_{11}(\mathbf{z}) + Q_{11}^2(\mathbf{z})}{n_1 - 1 - Q_{11}(\mathbf{z})} + \log\left(1 - \frac{1}{n_1 - 1}Q_{11}(\mathbf{z})\right)\right.$$
$$\left. + d\log\left(\frac{n_1}{n_1 - 1}\right)\right\}.$$

If $\mathbf{z}$ comes from G_2, replace 1 by 2 and change the sign of the term in braces.

7 Conditional on the data and $\mathbf{x} \in G_1$, $\text{E}[D_s(\mathbf{x})] = D_s(\boldsymbol{\mu}_i)$, $\text{var}[D_s(\mathbf{x})] = \hat{\boldsymbol{\lambda}}'\mathbf{\Sigma}\hat{\boldsymbol{\lambda}} = \sigma^2$. Then argue as in (6.10) and (6.11).

9 Use $(1 + x/n)^n \to \exp(x)$ as $n \to \infty$.

10 Since $\bar{\mathbf{x}}_{1\cdot}$, $\bar{\mathbf{x}}_{2\cdot}$, and $\mathbf{S}_p$ are mutually independent, $E[Q_{rr}(\mathbf{x})] = (\mathbf{x} - \boldsymbol{\mu}_r)'\mathscr{E}[\mathbf{S}_p^{-1}](\mathbf{x} - \boldsymbol{\mu}_r) + \text{tr}\{\mathscr{E}[(\bar{\mathbf{x}}_{r\cdot} - \boldsymbol{\mu}_r)(\bar{\mathbf{x}}_{r\cdot} - \boldsymbol{\mu}_r)']\mathscr{E}[\mathbf{S}_p^{-1}]\}$. Use Exercise 2.23 with $\nu\mathbf{S}_p \sim W_d(\nu, \boldsymbol{\Sigma})$.

12 Differentiate $L = \log L_S + \lambda_1(\Sigma_{\mathbf{x}} p_{\mathbf{x}} - 1) + \lambda_2[\Sigma_{\mathbf{x}} p_{\mathbf{x}} \exp(\alpha + \boldsymbol{\beta}'\mathbf{x}) - 1]$ with respect to α, $p_{\mathbf{x}}$, and $\boldsymbol{\beta}$. Solve for λ_1 and λ_2 and substitute in L_S.

13 $\Sigma_i \pi_i C(i) = \Sigma_i \Sigma_j \int_{R_j} c(j|i)\pi_i f_i(\mathbf{x})\, d\mathbf{x} = \Sigma_j \int_{R_j} c(j|\mathbf{x}) f(\mathbf{x})\, d\mathbf{x} = \int_R \Sigma_j \chi_{R_j}(\mathbf{x}) c(j|\mathbf{x}) f(\mathbf{x})\, d\mathbf{x} = \int_R g(\mathbf{x})\, d\mathbf{x}$. Proceed as in Theorem 6.3 in Section 6.9 with $c(i|\mathbf{x})$ instead of $q_i(\mathbf{x})$.

14 Use Exercise 6.13 with $\{\pi_i^*\}$: $\max C^*(i) = \Sigma_i \pi_i^* C^*(i) \le \Sigma_i \pi_i^* C(i) \le (\Sigma_i \pi_i^*) \max C(i)$.

15 $\Phi(-\frac{1}{2}\Delta)$ decreases as Δ increases. From Exercise 2.20, $\Delta^2 - \Delta_{-1}^2 = \boldsymbol{\mu}_{2\cdot 1}' \boldsymbol{\Sigma}_{22\cdot 1}^{-1} \boldsymbol{\mu}_{2\cdot 1} \ge 0$.

16 $\mathbf{W} = \nu\mathbf{S}_p \sim W_d(\nu, \boldsymbol{\Sigma})$, $\nu = n_1 + n_2 - 2$, and $\mathbf{x} = c_0(\bar{\mathbf{x}}_1 - \bar{\mathbf{x}}_2) \sim N_d(\boldsymbol{\mu}, \boldsymbol{\Sigma})$, where $\boldsymbol{\mu} = c_0(\boldsymbol{\mu}_1 - \boldsymbol{\mu}_2)$, $c_0 = [n_1 n_2/(n_1 + n_2)]^{1/2}$. Then

$$\frac{T_d^2 - T_k^2}{\nu + T_k^2} \frac{\nu - d + 1}{d - k} = \frac{D_d^2 - D_k^2}{\nu/c_0^2 + D_k^2} \frac{n_1 + n_2 - d - 1}{d - k} = F.$$

17 Use A3.1 to show that $\boldsymbol{\beta}_2 = \mathbf{E}^{-1}\boldsymbol{\delta}_2 - \mathbf{E}^{-1}\mathbf{F}'\boldsymbol{\delta}_1$.

18 Suppose that m_{ij} of the n_i observations from G_i are classified in G_j by the assignment rule. Let m_{ii} be correctly classified in G_i. Then $P(j|i)$ can be estimated using the apparent error rate m_{ij}/n_i $(i \ne j)$, and $P(i|i)$ can be estimated by m_{ii}/n_i. Use a similar approach for the leaving-one-out and bootstrap methods. Plug-in estimates are more difficult, as they require integration of the multivariate normal over unbounded convex regions $\hat{R}_{0j}$. Use (6.94) to estimate $P(\mathbf{R}, \mathbf{f})$.

20 $\Delta^2 = (\mu_1^2 + \mu_2^2 - 2\rho\mu_1\mu_2)/(1 - \rho^2)$. When $\rho = 0$, $\Delta^2 = \mu_1^2 + \mu_2^2 = \Delta_0^2$. Following (6.11), the total probability of misclassification is $\Phi(-\frac{1}{2}\Delta)$. $\Delta^2 < \Delta_0^2$ if $0 < \rho < 2\mu_1\mu_2\Delta_0^{-2}$. If $\mu_1 = \mu_2$, *any* positive correlation will increase $\Phi(-\frac{1}{2}\Delta)$.

EXERCISES 7

1 Let $\mathbf{x}_0 = \mathbf{B}^{1/2}\mathbf{x}$ (A5.4); then $\Delta(\mathbf{x}, \mathbf{y}) = \|\mathbf{x}_0 - \mathbf{y}_0\|$.

2 See Section 3.3.1.

3 $d_{13} = \frac{18}{30}$, $d_{12} = \frac{2}{34}$, $d_{23} = \frac{16}{30}$.

4 Let $\Delta_{12} = w^{-1}\Delta(\mathbf{x}, \mathbf{y})$, and so on. Then

$$\frac{\Delta_{13}}{1 + \Delta_{13}} = 1 - \frac{1}{1 + \Delta_{13}} < 1 - \frac{1}{(1 + \Delta_{12})(1 + \Delta_{23})}$$

$$< \frac{\Delta_{12}}{1 + \Delta_{12}} + \frac{\Delta_{23}}{1 + \Delta_{23}}.$$

6 Choose $\mathbf{x}, \mathbf{y}, \mathbf{z}$ such that $a = (\mathbf{x} - \mathbf{y})'(\mathbf{y} - \mathbf{z}) > 0$. Then $D(\mathbf{x}, \mathbf{y}) + D(\mathbf{y}, \mathbf{z}) = \|\mathbf{x} - \mathbf{y} + \mathbf{y} - \mathbf{z}\|^2 - 2a < D(\mathbf{x}, \mathbf{z})$.

7 Jaccard: $\frac{2}{5}, 0, \frac{1}{3}$. Matching: $\frac{7}{10}, \frac{5}{10}, \frac{8}{10}$.

8 $s_1 = 1 + (\beta + \gamma)/\alpha$ and $s_2 = 1 + (\beta + \gamma)/2\alpha$ increase together.

9 $y_i = \frac{4}{3}x_i + \frac{5}{3}$.

10 $d_{rs} = (\beta + \gamma)/[(\alpha + \beta) + (\alpha + \gamma)]$.

12 Here $n = \alpha + \beta + \gamma + \delta$ and

$$r_{jk} = \frac{\alpha - (\alpha + \beta)(\alpha + \gamma)/n}{\left\{\left[(\alpha + \beta) - (\alpha + \beta)^2/n\right]\left[(\alpha + \gamma) - (\alpha + \gamma)^2/n\right]\right\}^{1/2}}.$$

13 $O_1 = 4, O_2 = 2, O_3 = 5, O_4 = 3, O_5 = 1$.

14 Same clusters as for single linkage.

15 Prove (7.33) using $\mathbf{y}_{ij} - \bar{\mathbf{y}}_{..} = \mathbf{y}_{ij} - \bar{\mathbf{y}}_{i.} + \bar{\mathbf{y}}_{i.} - \bar{\mathbf{y}}_{..}$.

17 Use (7.17) and Exercise 7.16 for the median and incremental methods.

18 $\delta(1,3) = 1$, $\delta(2,4) = 2$, $\delta(2,5) = \delta(4,5) = 3$, $\delta(1,2) = \delta(1,4) = \delta(1,5) = \delta(3,2) = \delta(3,4) = \delta(3,5) = 4$.

19 (a) $O_1 = 4, O_2 = 2, O_3 = 5, O_4 = 3, O_5 = 1$.
(b) $O_1 = 2, O_2 = 6, O_3 = 8, O_4 = 9, O_5 = 1, O_6 = 5, O_7 = 3, O_8 = 7, O_9 = 10, O_{10} = 4$ (this order is not unique.)

20 $\mathbf{x}_i - \mathbf{a} = \mathbf{x}_i - \bar{\mathbf{x}} + \bar{\mathbf{x}} - \mathbf{a}$.

21 $\|\mathbf{x}_r - \mathbf{x}_s\|^2 = \|\mathbf{x}_r - \bar{\mathbf{x}}_1 - (\mathbf{x}_s - \bar{\mathbf{x}}_1)\|^2$ and $\|\mathbf{x}_s - \bar{\mathbf{x}}_1\|^2 = \|\mathbf{x}_s - \bar{\mathbf{x}}_2 - (\bar{\mathbf{x}}_1 - \bar{\mathbf{x}}_2)\|^2$.

22 $\mathbf{B} = (\bar{\mathbf{y}}_{1.} - \bar{\mathbf{y}}_{2.})(\bar{\mathbf{y}}_{1.} - \bar{\mathbf{y}}_{2.})'n_1 n_2/(n_1 + n_2)$ has rank 1.

23 (a) $\Leftrightarrow$ (c) as $|\mathbf{T}^{-1/2}\mathbf{W}\mathbf{T}^{-1/2}| = |\mathbf{I}_d - \mathbf{T}^{-1/2}\mathbf{B}\mathbf{T}^{-1/2}| = 1 - \text{tr}[\mathbf{B}\mathbf{T}^{-1}] = 1 - d + \text{tr}[\mathbf{W}\mathbf{T}^{-1}]$.
(b) $\Leftrightarrow$ (c) as $\text{tr}[\mathbf{W}\mathbf{T}^{-1}] = \{\text{tr}[\mathbf{T}\mathbf{W}^{-1}]\}^{-1} = \{d + \text{tr}[\mathbf{B}\mathbf{W}^{-1}]\}^{-1}$.

EXERCISES 8

1 From A7.9, $(\mathbf{Y} - \mathbf{X}\mathbf{B})'(\mathbf{Y} - \mathbf{X}\mathbf{B}) \geq (\mathbf{Y} - \mathbf{X}\hat{\mathbf{B}})'(\mathbf{Y} - \mathbf{X}\hat{\mathbf{B}})$ if and only if the inequality holds for traces.

2 $\hat{\boldsymbol{\theta}}_1 = \frac{1}{3}(\mathbf{y}_1 + \mathbf{y}_2 + \mathbf{y}_3)$, $\hat{\boldsymbol{\theta}}_2 = \frac{1}{2}(\mathbf{y}_3 - \mathbf{y}_2)/2$.

3 $\mathbf{x}_0'(\mathbf{X}'\mathbf{X})^{-1}\mathbf{x}_0\boldsymbol{\Sigma}$ [use $\mathscr{C}[\hat{\boldsymbol{\beta}}_i, \hat{\boldsymbol{\beta}}_j] = \sigma_{ij}(\mathbf{X}'\mathbf{X})^{-1}$]. Estimate $\boldsymbol{\Sigma}$ by $\mathbf{S} = (\mathbf{Y} - \mathbf{X}\hat{\mathbf{B}})'(\mathbf{Y} - \mathbf{X}\hat{\mathbf{B}})/n - p)$. $u = \mathbf{a}'\mathbf{y}_0 - \mathbf{a}'\hat{\mathbf{B}}'\mathbf{x}_0 \sim N_1(0, \mathbf{a}'\boldsymbol{\Sigma}\mathbf{a}[1 + \mathbf{x}_0'(\mathbf{X}'\mathbf{X})^{-1}\mathbf{x}_0])$, $(n - p)\mathbf{a}'\mathbf{S}\mathbf{a}/\mathbf{a}'\boldsymbol{\Sigma}\mathbf{a} \sim \chi^2_{n-p}$ and independent of $\hat{\mathbf{B}}$. A 100 $(1 - \alpha)$% confidence interval is $\mathbf{a}'\hat{\mathbf{B}}'\mathbf{x}_0 \pm t_{n-p}^{\alpha/2}\{\mathbf{a}'\mathbf{S}\mathbf{a}[1 + \mathbf{x}_0'(\mathbf{X}'\mathbf{X})^{-1}\mathbf{x}_0]\}^{1/2}$.

4 $\mathbf{Y} = \mathbf{1}_n\boldsymbol{\mu}' + \mathbf{U}$, $\hat{\boldsymbol{\mu}}' = (\mathbf{1}_n'\mathbf{1}_n)^{-1}\mathbf{1}_n'\mathbf{Y} = \bar{\mathbf{y}}'$.

5 $\mathbf{H}$ is independent of $\mathbf{E}$. $\mathscr{E}[\mathbf{E}^{-1}\mathbf{H}] = \mathscr{E}[\mathbf{E}^{-1}]\mathscr{E}[\mathbf{H}] = \boldsymbol{\Sigma}^{-1}(q\boldsymbol{\Sigma} + \mathbf{D})/(n - p - d - 1)$.

6 $\hat{\mathbf{\Phi}}_1 = \mathbf{Z}_1, \hat{\mathbf{\Phi}}_2 = \mathbf{Z}_2, \hat{\mathbf{\Phi}}_{2H} = \mathbf{Z}_2, \mathbf{E}_H = \mathbf{Z}_1'\mathbf{Z}_1 + \mathbf{Z}_3'\mathbf{Z}_3.$

7 (a) Use

$$\mathbf{X}'\mathbf{X} = \begin{pmatrix} n & \mathbf{0}' \\ \mathbf{0} & \mathbf{X}_2'\mathbf{X}_2 \end{pmatrix}, \qquad \mathbf{X}_2'(\mathbf{Y} - \mathbf{1}_n\bar{\mathbf{y}}') = \mathbf{X}_2'\mathbf{Y}.$$

(b) $\mathbf{E}_H = \tilde{\mathbf{Y}}'\tilde{\mathbf{Y}}$ and $\mathbf{H} = \mathbf{E}_H - \mathbf{E}$.

8 $\operatorname{tr}\mathbf{P}_\omega = \operatorname{tr}[\mathbf{P}_\Omega - \mathbf{P}_{\omega^\perp \cap \Omega}] = r - q.$

9 From Exercise 8.3, $\mathbf{a}'\hat{\mathbf{B}}\mathbf{b} = \mathbf{b}'\hat{\mathbf{B}}'\mathbf{a} \sim N_1(\mathbf{a}'\mathbf{B}\mathbf{b}, \mathbf{b}'\mathbf{\Sigma}\mathbf{b} \cdot \mathbf{a}'(\mathbf{X}'\mathbf{X})^{-1}\mathbf{a})$ is independent of $\mathbf{b}'\mathbf{E}\mathbf{b} = (n-p)\mathbf{b}'\mathbf{S}\mathbf{b} \sim \mathbf{b}'\mathbf{\Sigma}\mathbf{b}\chi^2_{n-p}$. The confidence interval is $\mathbf{a}'\hat{\mathbf{B}}\mathbf{b} \pm t^{\alpha/2}_{n-p}\{\mathbf{b}'\mathbf{S}\mathbf{b} \cdot \mathbf{a}'(\mathbf{X}'\mathbf{X})^{-1}\mathbf{a}\}^{1/2}$. Use $t^{\alpha/2m}_{n-p}$.

10 $\mathbf{E} = \mathbf{Y}'(\mathbf{I}_n - \mathbf{X}(\mathbf{X}'\mathbf{X})^{-1}\mathbf{X}')\mathbf{Y}$, $\mathbf{E}_H = \mathbf{Y}'(\mathbf{I}_n - \mathbf{X}_1(\mathbf{X}_1'\mathbf{X}_1)^{-1}\mathbf{X}_1')\mathbf{Y}$, and $\Lambda = |\mathbf{E}|/|\mathbf{E}_H \sim U_{d, p_2, n-p}$ when H_0 is true.

11 Since $\mathbf{C}_1$ is $(d-1)\times d$ of rank $d-1$, $(n-K)\mathbf{C}_1\mathbf{S}_p\mathbf{C}_1' \sim W_{d-1}(n-K, \mathbf{C}_1\mathbf{\Sigma}\mathbf{C}_1')$. When H_0 is true, $\mathbf{C}_1\bar{\mathbf{x}} \sim N_{d-1}(\mathbf{0}, \mathbf{C}_1\mathbf{\Sigma}\mathbf{C}_1')$ independently of $\mathbf{S}_p$. Use (2.19).

12 (a)

$$\begin{pmatrix} \mathbf{X} \\ \mathbf{A} \end{pmatrix} = \begin{pmatrix} 1 & x_1 & 1 & x_1 & x_1^2 \\ \vdots & \vdots & \vdots & \vdots & \vdots \\ 1 & x_n & 1 & x_n & x_n^2 \\ 0 & 1 & 0 & 1 & 0 \\ 0 & 0 & 0 & 0 & 1 \end{pmatrix}.$$

Since $\mathbf{X}_2$ has rank 3, three rows of $\mathbf{X}_2$ are linearly independent and form a basis of R^3. Then $(0,1,0)$ and $(0,0,1)$ are linearly dependent on this basis. Similarly for (b) and (c). For (d) and (e) to be testable we must have $\Sigma_i x_i = 1$ and $\Sigma_i x_i = 0$, a contradiction.

13 $\mathscr{C}[\mathbf{u}^{(j)}, \mathbf{u}^{(k)}] = \sigma_{jk}\mathbf{I}_d$.

EXERCISES 9

1 $\mathbf{H} = \Sigma_{i=1}^2 n_i(\bar{\mathbf{y}}_{i\cdot} - \bar{\mathbf{y}}_{\cdot\cdot})(\bar{\mathbf{y}}_{i\cdot} - \bar{\mathbf{y}}_{\cdot\cdot})' = n_1 n_2(n_1+n_2)^{-1}(\bar{\mathbf{y}}_{1\cdot} - \bar{\mathbf{y}}_{2\cdot})(\bar{\mathbf{y}}_{1\cdot} - \bar{\mathbf{y}}_{2\cdot})'$. Arguing as in Theorem 3.2, $|\mathbf{E}|/|\mathbf{E}+\mathbf{H}| = [1 + n_1 n_2(n_1+n_2)^{-1}(\bar{\mathbf{y}}_{1\cdot} - \bar{\mathbf{y}}_{2\cdot})'\mathbf{E}^{-1}(\bar{\mathbf{y}}_{1\cdot} - \bar{\mathbf{y}}_{2\cdot})]^{-1} = [1 + T_0^2/(n_1+n_2-2)]^{-1}$.

2 From (9.1), $\hat{\mathbf{B}} = (\mathbf{X}'\mathbf{X})^{-1}\mathbf{X}'\mathbf{Y} = (\bar{\mathbf{y}}_{1\cdot}, \bar{\mathbf{y}}_{2\cdot}, \ldots, \bar{\mathbf{y}}_{I\cdot})'$.

3 $(\bar{\mathbf{y}}_{r\cdot} - \bar{\mathbf{y}}_{s\cdot}) \sim N_d(\boldsymbol{\mu}_r - \boldsymbol{\mu}_s, [n_r^{-1} + n_s^{-1}]\mathbf{\Sigma})$, $\mathbf{E} \sim W_d(n-I, \mathbf{\Sigma})$. $T_0^2 = n_r n_s(n_r + n_s)^{-1}(\bar{\mathbf{y}}_{r\cdot} - \bar{\mathbf{y}}_{s\cdot})'[\mathbf{E}/(n-I)]^{-1}(\bar{\mathbf{y}}_{r\cdot} - \bar{\mathbf{y}}_{s\cdot}) \sim T^2_{d, n-I}$.

4

$$\mathbf{X}_2 = \begin{pmatrix} \mathbf{1}_{n_1} & \mathbf{0} & \cdots & \mathbf{0} \\ \vdots & \vdots & & \vdots \\ \mathbf{0} & \mathbf{0} & \cdots & \mathbf{1}_{n_{I-1}} \\ \mathbf{0} & \mathbf{0} & \cdots & \mathbf{0} \end{pmatrix} - \frac{1}{n}\mathbf{1}_n(n_1, n_2, \ldots, n_{I-1}).$$

5 (9.11) $= \Sigma_i \bar{\mathbf{y}}_{i\cdot}'\mathbf{W}_i\bar{\mathbf{y}}_{i\cdot} - \bar{\mathbf{y}}_w'\mathbf{W}\bar{\mathbf{y}}_w =$ (9.10).

6 $(\mathbf{X}'\mathbf{X})^{-1}\mathbf{X}'\mathbf{Y} = (\bar{\mathbf{y}}_{11\cdot}', \ldots, \bar{\mathbf{y}}_{IJ\cdot}')$. $\Sigma_j c_j \hat{\boldsymbol{\beta}}_j'\mathbf{b} \pm \{\phi_\alpha^*\{\Sigma_j c_j^2/IK)\mathbf{b}'\mathbf{E}\mathbf{b}\}^{1/2}$, where ϕ_α^* is the α significance level of the maximum root test of H_{02}.

7 As $\mathscr{D}[\mathbf{y}] = \sigma^2\mathbf{I}_n$ and $\mathbf{X}'\mathbf{R} = \mathbf{O}$, $\mathscr{C}[\mathbf{X}'\mathbf{X})^{-1}\mathbf{X}'\mathbf{y}, (\mathbf{Z}'\mathbf{R}\mathbf{Z})^{-1}\mathbf{Z}'\mathbf{R}\mathbf{y}] = \mathbf{O}$ by Exercise 1.1.

8 $\mathscr{C}[\mathbf{y}^{(j)}, \mathbf{y}^{(k)}] = \sigma^{jk}\mathbf{I}_n$ and $\mathscr{C}[\mathbf{X}'\mathbf{X})^{-1}\mathbf{X}'\mathbf{y}^{(j)}, (\mathbf{Z}'\mathbf{R}\mathbf{Z})^{-1}\mathbf{Z}'\mathbf{R}\mathbf{y}^{(k)}] = \mathbf{O}$.

9 $\hat{\boldsymbol{\gamma}}_G = \Sigma_i\Sigma_j \mathbf{y}_{ij}(z_{ij} - \bar{z}_{i\cdot})/R$ and $\mathscr{D}[\hat{\boldsymbol{\gamma}}_G] = R\boldsymbol{\Sigma}/R^2 = \boldsymbol{\Sigma}/R$. $H_0: \mathbf{C}_1\boldsymbol{\gamma} = \mathbf{0}$ [see (3.18)]. There are two approaches.

(a) Use the general theory with $\mathbf{A}_1\boldsymbol{\Phi} = (\mathbf{O}, \mathbf{C}_1)\binom{\mathbf{B}}{\boldsymbol{\gamma}'} = \mathbf{O}$.

(b) Consider a T^2 test based on $\mathbf{C}_1\hat{\boldsymbol{\gamma}}_G \sim N_{d-1}(\mathbf{C}_1\boldsymbol{\gamma}, \mathbf{C}_1\boldsymbol{\Sigma}\mathbf{C}_1'/R)$ that is independent of $\nu\mathbf{C}_1\mathbf{S}\mathbf{C}_1' = \mathbf{C}_1\mathbf{E}\mathbf{C}_1' \sim W_{d-1}(\nu, \mathbf{C}_1\boldsymbol{\Sigma}\mathbf{C}_1')$, $\nu = IJ - I - 1$, namely, $T_0^2 = (\mathbf{C}_1\hat{\boldsymbol{\gamma}}_G)'[\mathbf{C}_1\mathbf{S}\mathbf{C}_1']^{-1}\mathbf{C}_1\hat{\boldsymbol{\gamma}}_G R \sim T_{d-1,\nu}$ when H_0 is true [by (2.19)].

10 Use $\bar{\mathbf{y}}_{ij\cdot}$; then $\hat{\boldsymbol{\gamma}}_i = \Sigma_j(\bar{\mathbf{y}}_{ij\cdot} - \bar{\mathbf{y}}_{i\cdot\cdot})(\bar{z}_j - \bar{z}_\cdot)/\Sigma_j(z_j - \bar{z}_\cdot)^2 = \mathbf{E}_{iyz}/E_{zz} \sim N_d(\boldsymbol{\gamma}_i, \boldsymbol{\Sigma}/[nE_{zz}])$, by Exercise 9.9. Furthermore, $\mathbf{E}_i = \mathbf{E}_{iyy} - \mathbf{E}_{iyz}\mathbf{E}_{iyz}'/E_{zz} \sim W_d(Jn - 2, \boldsymbol{\Sigma}/n)$. $T_0^2 = (\hat{\boldsymbol{\gamma}}_1 - \hat{\boldsymbol{\gamma}}_2)'[(\mathbf{E}_1 + \mathbf{E}_2)/(2Jn - 4)]^{-1}(\hat{\boldsymbol{\gamma}}_1 - \hat{\boldsymbol{\gamma}}_2)E_{zz} \sim T_{d,2(Jn-2)}^2$ when $H_0: \boldsymbol{\gamma}_1 = \boldsymbol{\gamma}_2$ is true, by (2.19).

11 Use $\hat{\boldsymbol{\beta}}_j = \mathbf{Y}'(\mathbf{X}\mathbf{a}_j)$ and Lemma 2.3 of Section 2.3.2. $T_0^2 = \hat{\boldsymbol{\beta}}_j'\mathbf{S}^{-1}\hat{\boldsymbol{\beta}}_j/\|\mathbf{X}\mathbf{a}_j\|^2$, where $\mathbf{S} = (\mathbf{Y} - \mathbf{X}\hat{\mathbf{B}})'(\mathbf{Y} - \mathbf{X}\hat{\mathbf{B}})$. Then $T_0^2 \sim T_{d,n-p}^2$ when H_0 is true [see (2.19)]. Finally, use the technique leading to (3.11).

12 Let $L(\boldsymbol{\gamma}, \boldsymbol{\Sigma})$ be the log likelihood. By differentiation, $L(\tilde{\boldsymbol{\gamma}}, \boldsymbol{\Sigma}) \geq L(\boldsymbol{\gamma}, \boldsymbol{\Sigma})$ for all $\boldsymbol{\Sigma} > \mathbf{O}$, where $\tilde{\boldsymbol{\gamma}} = (\mathbf{K}'\boldsymbol{\Sigma}^{-1}\mathbf{K})^{-1}\mathbf{K}'\boldsymbol{\Sigma}^{-1}\bar{\mathbf{v}}$. Show that $L(\tilde{\boldsymbol{\gamma}}, \boldsymbol{\Sigma}) = c - \frac{1}{2}n\log\boldsymbol{\Sigma} - \frac{1}{2}\mathrm{tr}[\boldsymbol{\Sigma}^{-1}\Sigma_i(\mathbf{v}_i - \bar{\mathbf{v}})'(\mathbf{v}_i - \bar{\mathbf{v}})]$. Use A7.1 to show that $L(\hat{\boldsymbol{\gamma}}, \hat{\boldsymbol{\Sigma}}) \geq L(\tilde{\boldsymbol{\gamma}}, \boldsymbol{\Sigma})$, where $\hat{\boldsymbol{\Sigma}} = \Sigma_i(\mathbf{v}_i - \bar{\mathbf{v}})(\mathbf{v}_i - \bar{\mathbf{v}})'/n$.

13 $\mathbf{D}' = (0, 0, 1)$, $(\mathbf{X}'\mathbf{X})^{-1}\mathbf{X}'\mathbf{Y} = (\bar{\mathbf{v}}, \bar{\mathbf{w}})'$, $\mathbf{E}_D = \nu\mathbf{D}'(\mathbf{K}'\mathbf{S}^{-1}\mathbf{K})^{-1}\mathbf{D} = \nu b$, where $\nu = n_1 + n_2 - 2$ and $\nu\mathbf{S} = [\Sigma_i(\mathbf{u}_i - \bar{\mathbf{u}})(\mathbf{u}_i - \bar{\mathbf{u}})' + \Sigma_j(\mathbf{v}_j - \bar{\mathbf{v}})(\mathbf{v}_j - \bar{\mathbf{v}})']$. Also $\mathbf{H}_D = (\hat{\gamma}_2 - \hat{\delta}_2)^2/c$ where $\hat{\gamma}_2 - \hat{\delta}_2 = (\bar{\mathbf{v}} - \bar{\mathbf{w}})'\mathbf{S}^{-1}\mathbf{K}(\mathbf{K}'\mathbf{S}^{-1}\mathbf{K})^{-1}\mathbf{D}$ and $c = n_1^{-1} + n_2^{-1} + (\bar{\mathbf{v}} - \bar{\mathbf{w}})'[\mathbf{S}^{-1} - \mathbf{S}^{-1}\mathbf{K}(\mathbf{K}'\mathbf{S}^{-1}\mathbf{K}')^{-1}\mathbf{K}'\mathbf{S}^{-1}](\bar{\mathbf{v}} - \bar{\mathbf{w}})/\nu$. As $\mathbf{H}_D$ has rank 1 (see Section 9.7.3), $T_0^2 = m_E\mathrm{tr}[\mathbf{H}_D\mathbf{E}_D^{-1}] = m_E(\hat{\gamma}_2 - \hat{\delta}_2)^2/bc\nu \sim T_{1,m_E}^2$, $m_E = n_1 + n_2 - d + 1$, when H_0 is true. $(1, -1)\boldsymbol{\Delta}\mathbf{I}_3 = \mathbf{0}'$.

14 Let $\mathbf{Z}$ be $d \times (d - k)$ of rank $d - k$ such that $\mathbf{K}'\mathbf{Z} = \mathbf{O}$. From B3.5, $\mathbf{K}(\mathbf{K}'\boldsymbol{\Sigma}^{-1}\mathbf{K})^{-1}\mathbf{K}'\boldsymbol{\Sigma}^{-1} = \mathbf{I}_d - \boldsymbol{\Sigma}\mathbf{Z}(\mathbf{Z}'\boldsymbol{\Sigma}\mathbf{Z})^{-1}\mathbf{Z}' = \mathbf{I}_d - \mathbf{Z}(\mathbf{Z}'\mathbf{Z})^{-1}\mathbf{Z}' = \mathbf{K}(\mathbf{K}'\mathbf{K})^{-1}\mathbf{K}'$. Multiply on the left by $(\mathbf{K}'\mathbf{K})^{-1}\mathbf{K}'$.

15 $\hat{\mathbf{B}}_G = (\mathbf{X}'\mathbf{X})^{-1}\mathbf{X}'[\mathbf{I}_n - \mathbf{Z}(\mathbf{Z}'\mathbf{R}\mathbf{Z})^{-1}\mathbf{Z}'\mathbf{R}]\mathbf{Y} = \mathbf{C}\mathbf{Y}$. Then $\mathbf{b}'\hat{\mathbf{B}}_G'\mathbf{A}'\mathbf{a} = \mathbf{b}'\mathbf{Y}'\mathbf{C}'\mathbf{A}'\mathbf{a} = \mathbf{b}'\mathbf{Y}'\mathbf{c}$. By Lemma 2.3(ii), $\mathscr{D}[\mathbf{Y}'\mathbf{c}] = \boldsymbol{\Sigma}\|\mathbf{c}\|^2$ and the variance is $\mathbf{b}'\boldsymbol{\Sigma}\mathbf{b}\|\mathbf{c}\|^2 = \mathbf{b}'\boldsymbol{\Sigma}\mathbf{b} \cdot \mathbf{a}'\mathbf{A}\mathbf{C}\mathbf{C}'\mathbf{A}'\mathbf{a}$. Use $\mathbf{R}\mathbf{X} = \mathbf{O}$ to simplify $\mathbf{C}\mathbf{C}'$. From Section 8.6.3 the intervals are $\mathbf{a}'\mathbf{A}\hat{\mathbf{B}}_G\mathbf{b} \pm \{\phi_\alpha^*\mathbf{a}'\mathbf{A}\mathbf{C}\mathbf{C}'\mathbf{A}'\mathbf{a} \cdot \mathbf{b}'\mathbf{E}\mathbf{b}\}^{1/2}$, where ϕ_α^* is the α level of the max root test of $\mathbf{A}\mathbf{B} = \mathbf{A}_1\boldsymbol{\Phi} = \mathbf{O}$.

16 $|\mathbf{E}|/|\mathbf{E}_H| = |\mathbf{I}_d - \mathbf{Q}_{yx}\mathbf{Q}_{xx}^{-1}\mathbf{Q}_{xy}\mathbf{Q}_{yy}^{-1}|$.

References

Abe, O. (1973). A note on the methodology of Knox's tests of "Time and space interaction." *Biometrics*, **29**, 67–77.

Agresti, A. (1981). Measures of nominal–ordinal association. *J. Am. Stat. Assoc.*, **76**, 524–529.

Ahmed, S. W., and Lachenbruch, P. A. (1975). Discriminant analysis when one or both of the initial samples is contaminated: large sample results. *EDV in Medizin und Biologie*, **6**, 35–42.

Ahmed, S. W., and Lachenbruch, P. A. (1977). Discriminant analysis when scale contamination is present in the initial sample. *In*. J. Van Ryzin (Ed.), *Classification and Clustering*, pp. 331–353. Academic Press: New York.

Aitchison, J., and Aitken, C. G. G. (1976). Multivariate binary discrimination by the kernel method. *Biometrika*, **63**, 413–420.

Aitchison, J., and Begg, C. B. (1976). Statistical diagnosis when basic cases are not classified with certainty. *Biometrika*, **63**, 1–12.

Aitchison, J., and Shen, S. M. (1980). Logistic-normal distributions: some properties and uses. *Biometrika*, **67**, 261–272.

Aitchison, J., Habbema, J. D. F., and Kay, J. W. (1977). A critical comparison of two methods of statistical discrimination. *Appl. Stat.*, **26**, 15–25.

Aitken, C. G. G. (1978). Methods of discrimination in multivariate binary data. *In* L. C. A. Corsten and J. Hermans (Eds.), *Compstat 1978*, pp. 155–161. Physica-Verlag: Vienna.

Aitkin, M. A. (1969). Some tests for correlation matrices. *Biometrika*, **56**, 443–446.

Aitkin, M. A. (1971). Correction to "Some tests for correlation matrices." *Biometrika*, **58**, 245.

Aitkin, M. A. (1974). Simultaneous inference and the choice of variable subsets. *Technometrics*, **16**, 221–227.

Akaike, H. (1973). Information theory and an extension of the maximum likelihood principle. *In* B. N. Petrov and F. Czâki (Eds.), *Second International Symposium on Information Theory*, pp. 267–281. Akademiai Kiadó: Budapest.

Alalouf, I. S., and Styan, G. P. H (1979). Characterizations of estimability in the general linear model. *Ann. Stat.*, **7**, 194–200.

Alam, K. (1975). Minimax and admissible minimax estimators of the mean of a multivariate normal distribution for unknown covariance matrix. *J. Multivar. Anal.*, **5**, 83–95.

Albert, A. (1978). *Quelques apports nouveaux à l'analyse discriminante*. Ph.D. thesis, Faculté des Sciences, Université de Liège, Liège, France.

Albert, A., and Anderson, J. A. (1981). Probit and logistic discriminant functions. *Commun. Stat. Theor. Methods A*, **10**, 641–657.

Anderberg, M. R. (1973). *Cluster Analysis for Applications*. Academic Press: New York.

Andersen, E. B. (1982). Latent structure analysis. *Scand. J. Stat.*, **9**, 1–12.

Anderson, A. H. (1974). Multidimensional contingency tables. *Scand. J. Stat.* **1**, 115–127.

Anderson, E. (1957). A semi-graphical method for the analysis of complex problems. *Proc. Nat. Acad. Sci. U.S.A.*, **43**, 923–927. [Reprinted, with an appended note, in *Technometrics* (1960), **2**, 387–391.]

Anderson, J. A. (1969). Discrimination between k populations with constraints on the probabilities of misclassification. *J. R. Stat. Soc. B*, **31**, 123–139.

Anderson, J. A. (1972). Separate sample logistic discrimination. *Biometrika*, **59**, 19–35.

Anderson, J. A. (1974). Diagnosis by logistic discriminant function: further practical problems and results. *Appl. Stat.*, **23**, 397–404.

Anderson, J. A. (1975). Quadratic logistic discrimination. *Biometrika*, **62**, 149–154.

Anderson, J. A. (1979). Multivariate logistic compounds. *Biometrika*, **66**, 17–26.

Anderson, J. A. (1982). Logistic discrimination. *In* P. R. Krishnaiah and L. Kanal (Eds.), *Handbook of Statistics*. Vol. II. *Classification, Pattern Recognition, and Reduction of Dimension*, North-Holland: Amsterdam.

Anderson, J. A., and Blair, V. (1982). Penalized maximum likelihood estimation in logistic regression and discrimination. *Biometrika*, **69**, 123–136.

Anderson, J. A., and Richardson, S. C. (1979). Logistic discrimination and bias correction in maximum likelihood estimation. *Technometrics*, **21**, 71–78.

Anderson, T. W. (1951). Classification by multivariate analysis. *Psychometrika*, **16**, 31–50.

Anderson, T. W. (1958). *Introduction to Multivariate Statistical Analysis*. Wiley: New York.

Anderson, T. W. (1963a). A test for equality of means when covariance matrices are unequal. *Ann. Math. Stat.*, **34**, 671–672.

Anderson, T. W. (1963b). Asymptotic theory for principal component analysis. *Ann. Stat. Math.*, **34**, 122–148.

Anderson, T. W. (1965). Some optimum confidence bounds for roots of determinantal equations. *Ann. Stat.*, **36**, 468–488.

Anderson, T. W. (1966). Some nonparametric multivariate procedures based on statistically equivalent blocks. *In* P. R. Krishnaiah (Ed.), *Multivariate Analysis*, pp. 5–27. Academic Press: New York.

Anderson, T. W. (1969). Statistical inference for covariance matrices with linear structure. *In* P. R. Krishnaiah (Ed.), *Multivariate Analysis*, Vol. II, pp. 55–66. Academic Press: New York.

Anderson, T. W. (1970). Estimation of covariance matrices which are linear combinations or whose inverses are linear combinations of given matrices. *In* R. C. Bose et al. (Eds.), *Essays in Probability and Statistics*, pp. 1–24. University of North Carolina Press: Chapel Hill, N.C.

Anderson, T. W. (1973). Asymptotically efficient estimation of covariance matrices with linear structure. *Ann. Stat.*, **1**, 135–141.

Anderson, T. W., and Das Gupta, S. (1963). Some inequalities on characteristic roots of matrices. *Biometrika*, **50**, 522–524.

Andrews, D. F. (1971). A note on the selection of data transformations. *Biometrika*, **58**, 249–254.

Andrews, D. F. (1972). Plots of high-dimensional data. *Biometrics*, **28**, 125–136.

Andrews, D. F. (1973). Graphical techniques for high dimensional data. *In* T. Cacoullos (Ed.), *Discriminant Analysis and Applications*, pp. 37–59. Academic Press: New York.

Andrews, D. F., Gnanadesikan, R., and Warner, J. L. (1971). Transformations of multivariate data. *Biometrics*, **27**, 825–840.

Andrews, D. F., Bickel, P. J., Hampel, F. R., Huber, P. J., Rogers, W. H., and Tukey, J. W. (1972a). *Robust Estimates of Location—Survey and Advances*. Princeton University Press: Princeton, N.J.

Andrews, D. F., Gnanadesikan, R., and Warner, J. L. (1972b). Methods for assessing multivariate normality. Bell Laboratories Memorandum. Murray Hill, N.J.

Andrews, D. F., Gnanadesikan, R., and Warner, J. L. (1973). Methods for assessing multivariate normality. *In* P. R. Krishnaiah (Ed.), *Multivariate Analysis*, Vol. III, pp. 95–116. Academic Press: New York.

Arnold, H. J. (1964). Permutation support for multivariate techniques. *Biometrika*, **51**, 65–70.

Arnold, S. F. (1979). A coördinate-free approach to finding optimal procedures for repeated measures designs. *Ann. Stat.*, **7**, 812–822.

Arnold, S. F. (1981). *The Theory of Linear Models and Multivariate Analysis*. Wiley: New York.

Ashikaga, T., and Chang, P. C. (1981). Robustness of Fisher's linear discriminant function under two-component mixed normal models. *J. Am. Stat. Assoc.*, **76**, 676–680.

Astrahan, M. M. (1970). *Speech Analysis by Clustering, or the Hyperphoneme Method*. Stanford Artificial Intelligence Project Memorandum AIM-124, AD 709067. Stanford University, Palo Alto, Calif.

Atkinson, A. C. (1973). Testing transformations to normality. *J. R. Stat. Soc. B*, **35**, 473–479.

Bahadur, R. R. (1961). A representation of the joint distribution of responses to *n* dichotomous items. *In* H. Solomon (Ed.), *Studies in Item Analysis and Prediction*, pp. 158–168. Stanford University Press: Palo Alto, Calif.

Bailey, B. J. R. (1977). Tables of the Bonferroni *t* statistic. *J. Am. Stat. Assoc.*, **72**, 469–478.

Baker, F. B. (1974). Stability of two hierarchical grouping techniques. Case I: sensitivity to data errors. *J. Am. Stat. Assoc.*, **69**, 440–445.

Baker, F. B., and Hubert, L. J. (1975). Measuring the power of hierarchical cluster analysis. *J. Am. Stat. Assoc.*, **70**, 31–38.

Baker, R. J., and Nelder, J. A. (1978). *The GLIM System, Release 3*. Numerical Algorithms Group, Oxford.

Baksalary, J. K., Corsten, L. C. A., and Kala, R. (1978). Reconciliation of two different views on estimation of growth curve parameters. *Biometrika*, **65**, 662–665.

Balasooriya, U., and Chan, L. K. (1981). Robust estimation of Stigler's data sets by cross validation. *Appl. Stat.*, **30**, 170–177.

Ball, G. H., and Hall, D. J. (1965). *ISODATA, a Novel Method of Data Analysis and Pattern Classification*. Technical Report, Stanford Research Institute: Menlo Park, Calif.

Ball, G. H., and Hall, D.J. (1967). A clustering technique for summarizing multivariate data. *Behav. Sci.*, **12**, 153–155.

Banfield, C. F., and Gower, J. C. (1980). A note on the graphical representation of multivariate binary data. *Appl. Stat.*, **29**, 238–245.

Barnett, V. (1975). Probability plotting methods and order statistics. *Appl. Stat.*, **24**, 95–108.

Barnett, V. (1976a). The ordering of multivariate data (with discussion). *J. R. Stat. Soc. A*, **139**, 318–354.

Barnett, V. (1976b). Convenient probability plotting positions for the normal distribution. *Appl. Stat.*, **25**, 47–50.

Barnett, V. (Ed.) (1981). *Interpreting Multivariate Data*. Wiley: Chichester.

Barnett, V., and Lewis, T. (1978). *Outliers in Statistical Data*. Wiley: Chicester, England.

Bartholomew, D. J. (1980). Factor analysis for categorical data. *J. R. Stat. Soc. B*, **42**, 293–321.

Bartholomew, D. J. (1981). Posterior analysis of the factor model. *Br. J. Math. Stat. Psychol.*, **34**, 93–99.

Bartlett, M. S. (1938a). Further aspects of the theory of multiple regression. *Proc. Cambridge Philos. Soc.*, **34**, 33–40.

Bartlett, M. S. (1938b). Methods of estimating mental factors. *Br. J. Psychol.*, **28**, 97–104.

Bartlett, M. S. (1947). Multivariate analysis. *J. R. Stat. Soc. Suppl.*, **9**, 176–190.

Bartlett, M. S. (1951). The effect of standardisation on an approximation in factor analysis. *Biometrika*, **38**, 337–344.

Bartlett, M. S., and Please, N. W. (1963). Discrimination in the case of zero mean differences. *Biometrika*, **50**, 17–21.

Beale, E. M. L. (1969a). *Cluster Analysis*. Scientific Control Systems: London.

Beale, E. M. L. (1969b). Euclidean cluster analysis. *Bull. Int. Stat. Inst.*, **43**, 92–94.

Beale, E. M. L., and Little, R. J. A. (1975). Missing values in multivariate analysis. *J. R. Stat. Soc. B*, **37**, 129–145.

Beauchamp, J. J., Folkert, J. E., and Robson, D. S. (1980). A note on the effect of logarithmic transformation on the probability of misclassfication. *Commun. Stat. Theor. Methods A*, **9**, 777–794.

Bebbington, A. C. (1978). A method of bivariate trimming for robust estimation of the correlation coefficient. *Appl. Stat.*, **27**, 221–226.

Beckman, R. J., and Johnson, M. E. (1981). A ranking procedure for partial discriminant analysis. *J. Am. Stat. Assoc.*, **76**, 671–675.

Bellmann, R. (1960). *Introduction to Matrix Analysis*. McGraw-Hill: New York.

Bendel, R. B. and Mickey, M. R. (1978). Population correlation matrices for sampling experiments. *Commun. Stat. Simul. Comput. B*, **7**, 163–182.

Bennett, B. M. (1951). Note on a solution of the generalized Behrens–Fisher problem. *Ann. Inst. Stat. Math.*, **2**, 87–90.

Berger, J. O. (1976a). Admissible minimax estimation of a multivariate normal mean with arbitrary quadratic loss. *Ann. Stat.*, **4**, 223–226.

Berger, J. O. (1976b). Minimax estimation of a multivariate normal mean under arbitrary quadratic loss. *J. Multivar. Anal.*, **6**, 256–264.

Berger, J. O. (1978). Minimax estimation of a multivariate normal mean under polynomial loss. *J. Multivar. Anal.*, **8**, 173–180.

Berger, J. O. (1980a). A robust generalized Bayes estimator and confidence region for a multivariate normal mean. *Ann. Stat.*, **8**, 716–761.

Berger, J. O. (1980b). Improving on inadmissable estimators in continuous exponential families with applications to simultaneous estimation of gamma scale parameters. *Ann. Stat.*, **8**, 545–571.

Berger, J. O. (1982). Selecting a minimax estimator of a multivariate normal mean. *Ann. Stat.*, **10**, 81–92.

Berger, J. O., and Bock, M. E. (1976a). Eliminating singularities of Stein-type estimators of location vectors. *J. R. Stat. Soc. B*, **38**, 166–170.

Berger, J. O., and Bock, M. E. (1976b). Combining independent normal mean estimation problems with unknown variances. *Ann. Stat.*, **4**, 642–648.

Berk, K. N. (1978). Comparing subset regression procedures. *Technometrics*, **20**, 1–6.

Bertin, T. (1967). *Semiologie Graphique*. Gauthier-Villars: Paris.

Bhapkar, V. P., and Patterson, K. W. (1977). On some nonparametric tests for profile analysis of several multivariate samples. *J. Multivar. Anal.*, **7**, 265–277.

Bhapkar, V. P., and Patterson, K. W. (1978). A Monte Carlo study of some multivariate nonparametric statistics for profile analysis of several samples. *J. Stat. Comput. Simul.*, **6**, 223–237.

Bhargava, R. P. (1962). *Multivariate Tests of Hypotheses with Incomplete Data*. Technical Rep. No. 3. Appl. Math. and Labs., Stanford University, Palo Alto, Calif.

Bhoj, D. S. (1973a). Percentage points of the statistics for testing hypotheses on mean vectors of multivariate normal distributions with missing observations. *J. Stat. Comput. Simul.*, **2**, 211–224.

Bhoj, D. S. (1973b). On the distribution of a statistic used for testing a multinormal mean vector with incomplete data. *J. Stat. Comput. Simul.*, **2**, 309–316.

Bhoj, D. S. (1978). Testing equality of means of correlated variates with missing observations on both responses. *Biometrika*, **65**, 225–228.

Bhoj, D. S. (1979). Testing equality of variances of correlated variates with incomplete data on both responses. *Biometrika*, **66**, 681–683.

Bickel, P. J. (1964). On some alternative estimates for shift in the *p*-variate one-sample problem. *Ann. Math. Stat.*, **35**, 1079–1090.

Bickel, P. J. (1976). Another look at robustness: a review of reviews and some new developments. *Scand. J. Stat.*, **3**, 145–168.

Bickel, P. J., and Doksum, K. A. (1981). An analysis of transformations revisited. *J. Am. Stat. Assoc.*, **76**, 296–311.

Binder, D. A. (1978). Bayesian cluster analysis. *Biometrika*, **65**, 31–38.

Binder, D. A. (1981). Approximations to Bayesian cluster analysis. *Biometrika*, **68**, 275–285.

Bishop, Y., Fienberg, S., and Holland, P. (1975). *Discrete Multivariate Analysis*. MIT Press: Cambridge, Mass.

Blackith, R. E., and Reyment, R. A. (1971). *Multivariate Morphometrics*. Academic Press: London.

Blashfield, R. K. (1976). Mixture model tests of cluster analysis: accuracy of four agglomerative hierarchical methods. *Psychol. Bull.*, **83**, 377–388.

Block, H. W. (1977). Multivariate reliability classes. *In* P. R. Krishnaiah, (Ed.), *Applications of Statistics*, pp. 79–88. North-Holland: Amsterdam.

Bonner, R. E. (1964). On some clustering techniques. *IBM J.*, **22**, 22–32.

Boos, D. D. (1980). A new method for constructing approximate confidence intervals from *M* estimates. *J. Am. Stat. Assoc.*, **75**, 142–145.

Boulton, D. M., and Wallace, C. S. (1970). A program for numerical classification. *Comput. J.*, **13**, 63–69.

Bowden, D. C., and Steinhorst, R. K. (1973). Tolerance bands for growth curves. *Biometrics*, **29**, 361–371.

Bowman, K. O., and Shenton, L. R. (1973a). Notes on the distribution of $\sqrt{b_1}$ in sampling from Pearson distributions. *Biometrika*, **60**, 155–167.

Bowman, K. O., and Shenton, L. R. (1973b). Remarks on the distribution of $\sqrt{b_1}$ in sampling from a normal mixture and normal Type A distribution. *J. Am. Stat. Assoc.*, **68**, 998–1003.

Bowman, K. O., and Shenton, L. R. (1975). Omnibus test contours for departures from normality based on $\sqrt{b_1}$ and b_2. *Biometrika*, **62**, 243–249.

Box, G. E. P. (1949). A general distribution theory for a class of likelihood criteria. *Biometrika*, **36**, 317–346.

Box, G. E. P. (1953). Non-normality and tests on variances. *Biometrika*, **40**, 318–335.

Box, G. E. P. (1979). Robustness in the strategy of scientific model building. *In* R. L. Launer and G. N. Wilkinson (Eds.), *Robustness in Statistics*, pp. 201–236. Academic Press: New York.

Box, G. E. P., and Cox, D. R. (1964). An analysis of transformations. *J. R. Stat. Soc. B*, **26**, 211–252.

Box, G. E. P., and Cox, D. R. (1982). An analysis of transformations revisited, rebutted. *J. Am. Stat. Assoc.*, **77**, 209–210.

Box, G. E. P., and Watson, G. S. (1962). Robustness to non-normality of regression tests. *Biometrika*, **49**, 93–106.

Box, G. E. P., Hunter, W. G., MacGregor, J. F., and Erjavec, J. (1973). Some problems associated with the analysis of multiresponse data. *Technometrics*, **15**, 33–51.

Bradu, D., and Gabriel, K. R. (1978). The biplot as a diagnostic tool for models of two-way tables. *Technometrics*, **20**, 47–68.

Brandwein, A. R. C., and Strawderman, W. E. (1980). Minimax estimation of location parameters for spherically symmetric distributions with concave loss. *Ann. Stat.*, **8**, 279–284.

Bray, J. R., and Curtis, J. T. (1957). An ordination of the upland forest communities of southern Wisconsin. *Ecol. Monogr.*, **27**, 325–349.

Breiman, L., Meisel, W., and Purcell, E. (1977). Variable kernel estimates of multivariate densities. *Technometrics*, **19**, 135–144.

Brillinger, D. R. (1975). *Time Series: Data Analysis and Theory*. Holt, Rinehart, and Winston: New York.

Brillouin, L. (1962). *Science and Information Theory*, 2nd ed. Academic Press: New York.

Broffitt, B., Clarke, W. R., and Lachenbruch, P. A. (1980). The effect of Huberizing and trimming on the quadratic discriminant function. *Commun. Stat. Theor. Methods A*, **9**, 13–25.

Broffitt, B., Clarke, W. R., and Lachenbruch, P. A. (1981). Measurement errors—a location contamination problem in discriminant analysis. *Commun. Stat. Simul. Comput. B*, **10**, 129–141.

Broffitt, J. D., Randles, R. H., and Hogg, R. V. (1976). Distribution-free partial discriminant analysis. *J. Am. Stat. Assoc.*, **71**, 934–939.

Brown, L. (1966). On the admissibility of estimators of one or more location parameters. *Ann. Math. Stat.*, **37**, 1087–1136.

Brown, P. J., and Zidek, J. V. (1980). Adaptive multivariate ridge regression. *Ann. Stat.*, **8**, 64–74.

Brunk, H. D., and Pierce, D. A. (1974). Estimation of discrete multivariate densities for computer-aided differential diagnosis of disease. *Biometrika*, **61**, 493–499.

Bruntz, S. M., Cleveland, W. S., Kleiner, B., and Warner, J. L. (1974). The dependence of ambient ozone on solar radiation, wind, temperature, and mixing height. *Proc. Symp. Atmos. Diffus. Air Pollution Am. Meterol. Soc.*, 125–128.

Bruynooghe, M. (1978). Large data set clustering methods using the concept of space contraction. *In* L. C. A. Corsten and J. Hermans (Eds.), *Compstat 1978*, pp. 239–245. Physica-Verlag: Vienna.

Bryan, J. K. (1971). *Classification and Clustering Using Density Estimation*. Ph.D. thesis, University of Missouri, Columbia, Mo.

Bryant, P., and Williamson, J. A. (1978). Asymptotic behaviour of classification maximum likelihood estimates. *Biometrika*, **65**, 273–281.

Burr, E. J. (1968). Cluster sorting with mixed character types. I. Standardization of character values. *Aust. Comput. J.*, **1**, 97–99.

Burr, E. J. (1970). Cluster sorting with mixed character types. II. Fusion strategies. *Aust. Comput. J.*, **2**, 98–103.

Buser, M. W., and Baroni-Urbani, C. (1982). A direct nondimensional clustering method for binary data. *Biometrics*, **38**, 351–360.

Businger, P., and Golub, G. H. (1965). Linear least squares solutions by Householder transformations. *Numer. Math.*, **7**, 269–276.

Byth, K., and McLachlan, G. J. (1980). Logistic regression compared to normal discrimination for non-normal populations. *Aust. J. Stat.*, **22**, 188–196.

Cacoullos, T. (1966). Estimation of a multivariate density. *Ann. Inst. Stat. Math.*, **18**, 179–189.

Cacoullos, T. (Ed.) (1973). *Discriminant Analysis and Applications*. Academic Press: New York.

Caliński, T., and Harabasz, J. (1974). A dendrite method for cluster analysis. *Commun. Stat.*, **3**, 1–27.

Campbell, N. A. (1978). The influence function as an aid in outlier detection in discriminant analysis. *Appl. Stat.*, **27**, 251–258.

Campbell, N. A. (1980a). Robust procedures in multivariate analysis. I. Robust covariance estimation. *Appl. Stat.*, **29**, 231–237.

Campbell, N. A. (1980b). Shrunken estimators in discriminant and canonical variate analysis. *Appl. Stat.*, **29**, 5–14.

Campbell, N. A. (1981). Graphical comparison of covariance matrices. *Aust. J. Stat.*, **23**, 21–37.

Campbell, N. A. (1982). Robust procedures in multivariate analysis. II. Robust canonical variate analysis. *Appl. Stat.*, **31**, 1–8.

Campbell, N. A., and Atchley, W. R. (1981). The geometry of canonical variate analysis. *Syst. Zool.*, **30**, 268–280.

Carmichael, J. W., and Sneath, P. H. A. (1969). Taxometric maps. *Syst. Zool.*, **18**, 402–415.

Carmichael, J. W., George, J. A., and Juluis, R. S. (1968). Finding natural clusters. *Syst. Zool.*, **17**, 144–150.

Carroll, J. D., and Arabie, P. (1980). Multidimensional scaling. *Annu. Rev. Psychol.*, **31**, 607–649.

Carroll, J. D., and Chang, J. J. (1970). Analysis of individual differences in multidimensional scaling via an N-way generalization of "Eckart–Young" decomposition. *Psychometrika*, **35**, 283–319.

Carroll, R. J. (1978). On the asymptotic distribution of multivariate M-estimates. *J. Multivar. Anal.*, **8**, 361–371.

Carroll, R. J. (1979). On estimating variances of robust estimators when the errors are asymmetric. *J. Am. Stat. Assoc.*, **74**, 674–679.

Carroll, R. J. (1980). A robust method for testing transformations to achieve approximate normality. *J. R. Stat. Soc. B*, **42**, 71–78.

Cattell, R. B., and Coulter, M. A. (1966). Principles of behavioural taxonomy and the mathematical basis of the taxonome computer program. *Br. J. Math. Stat. Psychol.*, **19**, 237–269.

Cattell, R. B., and Khanna, D. K. (1977). Principles and procedures for unique rotation in factor analysis. *In* K. Enslein, A. Ralston, and H. S. Wilf (Eds.), *Statistical Methods for Digital Computers*, Vol. 3, pp. 166–202. Wiley: New York.

Chakravorti, S. R. (1974). On some tests of growth curve model under Behrens–Fisher situation. *J. Multivar. Anal.*, **4**, 31–51.

Chakravorti, S. R. (1980). On some tests in MANCOVA model with different dispersion matrices. *Commun. Stat. Theor. Methods A*, **9**, 291–308.

Chalmers, C. P. (1975). Generation of correlation matrices with a given eigen-structure. *J. Stat. Comp. Simul.*, **4**, 133–139.

Chambers, J. M. (1977). *Computational Methods for Data Analysis*. Wiley: New York.

Chan, L. S., and Dunn, O. J. (1974). A note on the asymptotic aspect of the treatment of missing values in discriminant analysis. *J. Am. Stat. Assoc.*, **69**, 672–673.

Chan, L. S., Gilman, J. A., and Dunn, O. J. (1976). Alternative approaches to missing values in discriminant analysis. *J. Am. Stat. Assoc.*, **71**, 842–844.

Chang, C. L., and Lee, R. C. T. (1973). A heuristic relaxation method for nonlinear mapping in cluster analysis. *IEEE Trans. Syst. Man Cybern.*, **SMC-2**, 197–200.

Chang, T. C. (1974). Upper percentage points of the extreme roots of the MANOVA matrix. *Ann. Inst. Stat. Math. Suppl.* **8**, 59–66.

Chase, G. R., and Bulgren, W. G. (1971). A Monte Carlo investigation of the robustness of T^2. *J. Am. Stat. Assoc.*, **66**, 499–502.

Chatfield, C., and Collins, A. J. (1980). *Introduction to Multivariate Analysis*. Chapman and Hall: London.

Chatterjee, S., and Price, B. (1977). *Regression Analysis by Example*. Wiley: New York.

Chen, C. F. (1979). Baysian inference for a normal dispersion matrix and its application to stochastic multiple regression analysis. *J. R. Stat. Soc. B*, **41**, 235–248.

Chen, C. W. (1974). An optimal property of principal components. *Commun. Stat.*, **3**, 979–983.

Chernoff, H. (1973). Using faces to represent points in k-dimensional space graphically. *J. Am. Stat. Assoc.*, **68**, 361–368.

Chernoff, H., and Rizvi, H. M. (1975). Effect on classification error of random permutations of features in representing multivariate data by faces. *J. Am. Stat. Assoc.*, **70**, 548–554.

Chhikara, R. S., and Register, D. T. (1979). A numerical classification method for partitioning of a large multidimensional mixed data set. *Technometrics*, **21**, 531–537.

Chi, P. Y., and Van Ryzin, J. (1977). A simple histogram method for nonparametric classification. *In* J. Van Ryzin (Ed.), *Classification and Clustering*, pp. 395–421. Academic Press: New York.

Chinganda, E. F., and Subrahmaniam, K. (1979). Robustness of the linear discriminant function to nonnormality: Johnson's system. *J. Stat. Plan. Infer.*, **3**, 69–77.

Chmielewski, M. A. (1981). Elliptically symmetric distributions: a review and bibliography. *Int. Stat. Rev.*, **49**, 67–74.

Choi, S. C. (1977). Tests of equality of dependent correlation coefficients. *Biometrika*, **64**, 645–647.

Choi, S. C., and Wette, R. (1972). A test for the homogeneity of variances among correlated variables. *Biometrics*, **28**, 589–591.

Chu, S. S., and Pillai, K. C. S. (1979). Power comparisons of two-sided tests of equality of two covariance matrices based on six criteria. *Ann. Inst. Stat. Math.*, **31**, 185–205.

Clarke, M. R. B. (1971). Algorithm AS 41: Updating the sample mean and dispersion matrix. *Appl. Stat.*, **20**, 206–209.

Clarke, W. R., Lachenbruch, P. A., and Broffitt, B. (1979). How non-normality affects the quadratic discriminant function. *Commun. Stat. Theor. Methods A*, **8**, 1285–1301.

Clemm, D. S., Krishnaiah, P. R., and Waikar, V. B. (1973). Tables for the extreme roots of the Wishart matrix. *J. Stat. Comput. Simul.*, **2**, 65–92.

Clifford, H. T., and Stephenson, W. (1975). *An Introduction to Numerical Classification*. Academic Press: New York.

Cochran, W. G. (1962). On the performance of the linear discriminant function. *Bull. Inst. Int. Stat.*, **39**, 435–447.

Cohen, A. (1977). A result on hypothesis testing for a multivariate normal distribution when some observations are missing. *J. Multivar. Anal.*, **7**, 454–460.

Cohen, C. (1977). The generalized symmetric eigenproblem in multivariate statistical models. *Commun. Stat. Theor. Methods A*, **6**, 277–288.

Cole, A. J., and Wishart, D. (1970). An improved algorithm for the Jardine–Sibson method of generating overlapping clusters. *Comput. J.*, **13**, 156–163.

Collins, J. R. (1976). Robust estimation of a location parameter in the presence of asymmetry. *Ann. Stat.*, **4**, 68–85.

Comrey, A. L. (1973). *A First Course in Factor Analysis*. Academic Press: New York.

Conover, W. J., and Iman, R. L. (1980). The rank transformation as a method of discrimination with some examples. *Commun. Stat. Theor. Methods A*, **9**, 465–487.

Conover, W. J., and Iman, R. L. (1981). Rank transformations as a bridge between parametric and nonparametric statistics. *Am. Stat.*, **35**, 124–129.

Conover, W. J., Bement, T. R., and Iman, R. L. (1979). On a method of detecting clusters of possible uranium deposits. *Technometrics*, **21**, 277–282.

Constantine, A. G. (1963). Some noncentral distributions in multivariate analysis. *Ann. Math. Stat.*, **34**, 1270–1285.

Constantine, A. G. (1966). The distribution of Hotelling's generalized measure of multivariate dispersion. *Ann. Math. Stat.*, **37**, 215–225.

Constantine, A. G., and Gower, J. C. (1978). Graphical representation of asymmetric matrices. *Appl. Stat.*, **27**, 297–304.

Constantine, A. G., and Muirhead, R. J. (1976). Asymptotic expansions for distributions of latent roots in multivariate analysis. *J. Multivar. Anal.*, **6**, 369–391.

Cook, M. B. (1951). Bivariate k-statistics and cumulants of their joint sampling distribution. *Biometrika*, **38**, 179–195.

Cook, R. D., and Johnson, M. E. (1981). A family of distributions for modelling non-elliptically symmetric multivariate data. *J. R. Stat. Soc. B*, **43**, 210–218.

Cooper, P. W. (1963). Statistical classification with quadratic forms. *Biometrika*, **50**, 439–448.

Cormack, R. M. (1971). A review of classification. *J. R. Stat. Soc. A*, **134**, 321–367.

Cormack, R. M. (1981). Loglinear models for capture—recapture experiments on open populations. *In* R. W. Hiorns and D. Cooke (Eds.), *The Mathematical Theory of the Dynamics of Biological Populations*. Academic Press: London.

Cornelius, P. L. (1980). Functions approximating Mandel's tables for the means and standard deviations of the first three roots of a Wishart matrix. *Technometrics*, **22**, 613–616.

Corsten, L. C. A., and Gabriel, K. R. (1976). Graphical exploration in comparing variance matrices. *Biometrics*, **32**, 851–863.

Costanza, M. C., and Afifi, A. A. (1979). Comparison of stopping rules in forward stepwise discriminant analysis. *J. Am. Stat. Assoc.*, **74**, 777–785.

Cox, D. R. (1966). Some procedures associated with the logistic qualitative response curve. *In* F. N. David (Ed.), *Research Papers in Statistics*: *Festschrift for J. Neyman*, pp. 55–71. Wiley: New York.

Cox, D. R. (1968). Notes on some aspects of regression analysis. *J. R. Stat. Soc. A*, **131**, 265–279.

Cox, D. R. (1972). The analysis of multivariate binary data. *Appl. Stat.*, **21**, 113–120.

Cox, D. R., and Small, N. J. H. (1978). Testing multivariate normality. *Biometrika*, **65**, 263–272.

Cramer, E. M. (1973). A simple derivation of the canonical correlation equations. *Biometrics*, **29**, 379–380.

Cran, G. W., Martin, K. J., and Thomas, G. E. (1977). Remark AS R19 and Algorithm AS 109. A remark on Algorithms AS 63: the incomplete beta integral; AS 64: inverse of the incomplete beta function ratio. *Appl. Stat.*, **26**, 111–114.

Crawley, D. R. (1979). Logistic discrimination as an alternative to Fisher's linear discrimination function. *N. Z. Stat.*, **14**(*2*), 21–25.

Cunningham, K. M., and Ogilvie, J. C. (1972). Evaluation of hierarchical grouping techniques: a preliminary study. *Comput. J.*, **15**, 209–213.

Czekanowski, J. (1913). *Zarys Metod Statystycznck*. E. Wendego: Warsaw.

D'Agostino, R. B. (1971). An omnibus test of normality for moderate and large size samples. *Biometrika*, **58**, 341–348.

D'Agostino, R. B. (1972). Small sample probability points for the D test of normality. *Biometrika*, **59**, 219–221.

D'Agostino, R. B., and Lee, A. F. S. (1977). Robustness of location estimators under changes of population kurtosis. *J. Am. Stat. Assoc.*, **72**, 393–396.

D'Agostino, R. B., and Pearson, E. S. (1973). Tests for departure from normality. Empirical results for the distributions of b_2 and $\sqrt{b_1}$. *Biometrika*, **60**, 613–622. [Correction: *Biometrika*, **61**, 647].

D'Agostino, R. B., and Tietjen, G. L. (1971). Simulated probability points of b_2 for small samples. *Biometrika*, **58**, 669–672.

Dahiya, R. C., and Korwar, R. M. (1980). Maximum likelihood estimates for a bivariate normal distribution with missing data. *Ann. Stat.*, **8**, 687–692.

Dalal, S. R. (1978). Simultaneous confidence procedures for univariate and multivariate Behrens–Fisher type problems. *Biometrika*, **65**, 221–225.

Dale, M. B. (1964). *The Application of Multivariate Methods to Heterogeneous Data*. Ph.D. thesis, University of Southampton, Southampton, England.

D'Andrade, R. G. (1978). U-Statistic hierarchical clustering. *Psychometrika*, **43**, 59–67.

Daniel, C. (1959). Use of half-normal plots in interpreting factorial two-level experiments. *Technometrics*, **1**, 311–341.

Daniel, D., and Wood, F. S. (1981). *Fitting Equations to Data*, 2nd ed. Wiley: New York.

Darroch, J. N. (1965). An optimal property of principal components. *Ann. Math. Stat.*, **36**, 1579–1582.

Das Gupta, S. (1965). Optimum classification rules for classification into two multivariate normal populations. *Ann. Math. Stat.*, **36**, 1174–1184.

Das Gupta, S. (1971). Nonsingularity of the sample covariance matrix. *Sankhya A*, **33**, 475–478.

Das Gupta, S., and Perlman, M. D. (1974). Power of the noncentral *F*-test: effect of additional variates on Hotelling's T^2-test. *J. Am. Stat. Assoc.*, **69**, 174–180.

David, H. A. (1970). *Order Statistics*. Wiley: New York.

Davis, A. W. (1970a). Exact distributions of Hotelling's generalized T_0^2. *Biometrika*, **57**, 187–191.

Davis, A. W. (1970b). Further applications of a differential equation for Hotelling's generalized T_0^2. *Ann. Inst. Stat. Math.* **22**, 77–87.

Davis, A. W. (1977). Asymptotic theory for principal components analysis: non-normal case. *Aust. J. Stat.*, **19**, 206–212.

Davis, A. W. (1979). On the differential equation for Meijer's $G_{p,p}^{p,0}$ function, and further tables of Wilks's likelihood ratio criterion. *Biometrika*, **66**, 519–531.

Davis, A. W. (1980a). Further tabulation of Hotelling's generalized T_0^2. *Comm. Stat. Simul. Comput. B*, **9**, 321–336.

Davis, A. W. (1980b). On the effects of moderate multivariate nonnormality on Wilks's likelihood ratio criterion. *Biometrika*, **67**, 419–427.

Davis, A. W. (1982). On the effects of moderate multivariate nonnormality on Roy's largest root test. *J. Am. Stat. Assoc.*, **77**, 896–900.

Davis, A. W., and Field, J. B. F. (1971). Tables of some multivariate test criteria. Division of Mathematical Statistics Technical Paper No. 32, CSIRO, Melbourne, Australia.

Dawid, A. P. (1976) Properties of diagnostic data distributions. *Biometrics*, **32**, 647–658.

Dawid, A. P. (1981). Some matrix-variate distribution theory: notational considerations and a Bayesian application. *Biometrika*, **68**, 265–274.

Day, N. E. (1969). Estimating the components of a mixture of normal distributions. *Biometrika*, **56**, 463–474.

Day, N. E., and Kerridge, D. F. (1967). A general maximum likelihood discriminant. *Biometrics*, **23**, 313–323.

Deemer, W. L., and Olkin, I. (1951). The Jacobians of certain matrix transformations useful in multivariate analysis. *Biometrika*, **38**, 345–367.

Defay, D. (1977). An efficient algorithm for a complete link method. *Comput. J.*, **20**, 364–366.

DeLeeuw, J., and Heiser, W. (1980). Multidimensional scaling with restrictions on the configuration. *In* P. R. Krishnaiah (Ed.), *Multivariate Analysis*, Vol. V, pp. 501–522. North-Holland: Amsterdam.

Dempster, A. P., and Schatzoff, M. (1965). Expected significance level as a sensitivity index for test statistics. *J. Am. Stat. Assoc.*, **60**, 420–436.

Dempster, A. P., Laird, N. M., and Rubin, D. B. (1977). Maximum likelihood from incomplete data via the *EM* algorithm (with discussion). *J. R. Stat. Soc. B*, **39**, 1–38.

Desu, M. M., and Geisser, S. (1973). Methods and applications of equal-mean discrimination. *In* T. Cacoullos (Ed.), *Discriminant Analysis and Applications*, pp. 139–159. Academic Press: New York.

Devlin, S. J., Gnanadesikan, R., and Kettenring, J. R. (1975). Robust estimation and outlier detection with correlation coefficients. *Biometrika*, **62**, 531–545.

Devlin, S. J., Gnanadesikan, R., and Kettenring, J. R. (1976). Some multivariate applications of elliptical distributions. *In* S. Ikeda et al. (Eds.), *Essays in Probability and Statistics* (dedicated to J. Ogawa), pp. 365–395. Shinko Tsusho: Tokyo.

Devlin, S. J., Gnanadesikan, R., and Kettenring, J. R. (1981). Robust estimation of dispersion matrices and principal components. *J. Am. Stat. Assoc.*, **76**, 354–362.

De Wet, T., and van Wyk, J. W. J. (1979). Efficiency and robustness of Hogg's adaptive trimmed means. *Commun. Stat. Theor. Methods A*, **8**, 117–128.

Dillon, W. R., and Goldstein, M. (1978). On the performance of some multinomial classification rules. *J. Am. Stat. Assoc.*, **73**, 305–313.

Di Pillo, P. J. (1976). The application of bias to discriminant analysis. *Commun. Stat. Theor. Methods A*, **5**, 834–844.

Di Pillo, P. J. (1977). Further applications of bias to discriminant analysis. *Commun. Stat. Theor. Methods A*, **6**, 933–943.

Di Pillo, P. J. (1979). Biased discriminant analysis: evaluation of the optimum probability of misclassification. *Commun. Stat. Theor. Methods A*, **8**, 1447–1457.

Dixon, W. J., and Brown, M. B. (1977). *BMDP-77: Biomedical Computer Programs, P Series*. University of California Press: Berkeley, Calif.

Dolby, G. R., and Freeman, T. G. (1975). Functional relationships having many independent variates and errors with multivariate normal distribution. *J. Multivar. Anal.*, **5**, 466–479.

Dolby, J. L. (1970). Some statistical aspects of character recognition. *Technometrics*, **12**, 231–245.

Downton, F. (1966). Linear estimates with polynomial coefficients. *Biometrika*, **53**, 129–141.

Draper, N. R., and Smith, H. (1980). *Applied Regression Analysis*, 2nd ed. Wiley: New York.

Draper, N. R., Guttman, I., and Lapczak, L. (1979). Actual rejection levels in a certain stepwise test. *Commun. Stat. Theor. Methods A*, **8**, 99–105.

Dudziński, M. L., and Arnold, G. W. (1973). Comparisons of diets of sheep and cattle grazing together on sown pastures on the southern tablelands of New South Wales by principal components analysis. *Aust. J. Agric. Res.*, **24**, 899–912.

Dunn, D. M., and Landwehr, J. M. (1980). Analyzing clustering effects across time. *J. Am. Stat. Assoc.*, **75**, 8–15.

Duran, B. S., and Odell, P. L. (1974). *Cluster Analysis: A Survey*. Springer-Verlag: Berlin.

Dyer, A. R. (1974). Comparison of tests for normality with a cautionary note. *Biometrika*, **61**, 185–189.

Dykstra, R. L. (1970). Establishing the positive definiteness of the sample covariance matrix. *Ann. Math. Stat.*, **41**, 2153–2154.

Dykstra, R. L. (1979). On dependent tests of significance in the multivariate analysis of variance. *Ann. Stat.*, **7**, 459–461.

Eastment, H. T., and Krzanowski, W. J. (1982). Cross-validatory choice of the number of components from a principal component analysis. *Technometrics*, **24**, 73–77.

Eaton, M. L. (1969). Some remarks on Scheffé's solution to the Behrens–Fisher problem. *J. Am. Stat. Assoc.*, **64**, 1318–1322.

Eaton, M. L. (1970). Gauss–Markov estimation for multivariate linear models: a coordinate free approach. *Ann. Math. Stat.*, **41**, 528–538.

Eaton, M. L. (1983). *Multivariate Statistics: A Vector Space Approach*. Wiley: New York.

Eaton, M. L., and Perlman, M. D. (1973). The non-singularity of generalized sample covariance matrices. *Ann. Stat.*, **1**, 710–717.

Eckart, C., and Young, G. (1936). The approximation of one matrix by another of lower rank. *Psychometrika*, **1**, 211–218.

Edelbrock, C., and McLaughlin, B. (1980). Hierarchical cluster analysis using intraclass correlations: a mixture model study. *Multivar. Behav. Res.*, **15**, 299–318.

Edwards, A. W. F., and Cavalli-Sforza, L. (1965). A method for cluster analysis. *Biometrics*, **21**, 362–375.

Edye, L. A., Williams, W. T., and Pritchard, A. J. (1970). A numerical analysis of variation pattern in Australian introductions of *Glycine wightii* (*G. javanica*). *Aust. J. Agric. Res.*, **21**, 57–69.

Efron, B. (1975). The efficiency of logistic regression compared to normal discriminant analysis. *J. Am. Stat. Assoc.*, **70**, 892–898.

Efron, B. (1979). Bootstrap methods: another look at the jackknife. *Ann. Stat.*, **7**, 1–26.

Efron, B. (1981). Nonparametric standard errors and confidence intervals. *Can. J. Stat.*, **9**, 139–172.

Efron, B., and Morris, C. (1975). Data analysis using Stein's estimator and its generalizations. *J. Am. Stat. Assoc.*, **70**, 311–319.

Efron, B., and Morris, C. (1976a). Families of minimax estimators of the mean of a multivariate normal distribution. *Ann. Stat.*, **4**, 11–21.

Efron, B., and Morris, C. (1976b). Multivariate empirical Bayes and estimation of covariance matrices. *Ann. Stat.*, **4**, 22–32.

Ehrenberg, A. S. C. (1975). *Data Reduction*. Wiley: London.

Ehrenberg, A. S. C. (1977). Rudiments of numeracy. *J. R. Stat. Soc. A*, **140**, 277–297.

Ehrenberg, A. S. C. (1981). The problem of numeracy. *Am. Stat.*, **35**, 67–71.

Elston, R. C., and Grizzle, J. E. (1962). Estimation of time-response curves and their confidence bands. *Biometrics*, **18**, 148–159.

Enslein, K., Ralston, A., and Wilf, H. S. (Eds.) (1977). *Statistical Methods for Digital Computers*, Vol. 3. Wiley: New York.

Escoufier, Y., and Robert, P. (1980). Choosing variables and metrics by optimizing the RV-coefficient. *In* J. S. Rustagi (Ed.), *Optimizing Methods in Statistics*, pp. 205–219. Academic Press: New York.

Everitt, B. (1974). *Cluster Analysis*. Heinemann: London.

Everitt, B. (1978). *Graphical Techniques for Multivariate Data*. Heinemann: London.

Everitt, B. S. (1979a). A Monte Carlo investigation of the robustness of Hotelling's one and two-sample T^2 statistic. *J. Am. Stat. Assoc.*, **74**, 48–51.

Everitt, B. (1979b). Unresolved problems in cluster analysis. *Biometrics*, **35**, 169–181.

Everitt, B. S., and Hand, D. J. (1981). *Finite Mixture Distributions*. Chapman and Hall: London.

Faith, R. E. (1978). Minimax Bayes estimators of a multivariate normal mean. *J. Multivar. Anal.*, **8**, 372–379.

Farris, J. S. (1969). On the cophenetic correlation coefficient. *Systematic Zool.*, **18**, 279–285.

Fay, R. E., III, and Herriott, R. A. (1979). Estimates of income for small places: an application of James–Stein procedures to census data. *J. Am. Stat. Assoc.*, **74**, 269–277.

Fearn, T. (1975). A Bayesian approach to growth curves. *Biometrika*, **62**, 89–100.

Fearn, T. (1977). A two-stage model for growth curves which leads to Rao's covariance adjusted estimators. *Biometrika*, **64**, 141–143.

Ferguson, T. S. (1961). On the rejection of outliers. *Proc. Fourth Berkeley Symp. Math. Stat. Prob.*, **1**, 253–287.

Fienberg, S. E. (1979). Graphical methods in statistics. *Am. Stat.*, **33**, 165–178.

Fisher, L., and Van Ness, J. W. (1971). Admissable clustering procedures. *Biometrika*, **58**, 91–104.

Fisher, N. I., and Hudson, H. M. (1980). A note on the multivariate linear model with constraints on the dependent vector. *Aust. J. Stat.*, **22**, 75–78.

Fisher, R. A. (1936). The use of multiple measurement in taxonomic problems. *Ann. Eugen.*, **7**, 179–188.

Fisher, R. A. (1939). The sampling distribution of some statistics obtained from non-linear equations. *Ann. Eugen.*, **9**, 238–249.

Fix, E., and Hodges, J. L. (1951). Discriminatory analysis, nonparametric discrimination: consistency properties. Report No. 4, Project No. 21-49-004, USAF School of Aviation Medicine, Brooks Air Force Base, Randolph Field, Texas.

Fleiss, J. L., and Zubin, J. (1969). On the methods and theory of clustering. *Multivar. Behav. Res.*, **4**, 235–250.

Flury, B., and Riedwyl, H. (1981). Graphical representation of multivariate data by means of asymmetrical faces. *J. Am. Stat. Assoc.*, **76**, 757–765.

Fomby, T. B., Hill, R. C., and Johnson, S. R. (1978). An optimal property of principal components in the context of restricted least squares. *J. Am. Stat. Assoc.*, **73**, 191–193.

Forgy, E. W. (1965). Cluster analysis of multivariate data: efficiency versus interpretability of classifications. *Biometrics*, **21**, 768–769.

Forst, F. R., and Ali, M. M. (1981). Monte Carlo studies of some adaptive robust procedures for location. *Can. J. Stat.*, **9**, 229–235.

Fortier, J. J., and Solomon, H. (1966). Clustering procedures. *In* P. R. Krishnaiah (Ed.), *Multivariate Analysis*, pp. 493–506. Academic Press: New York.

Foster, F. G. (1957). Upper percentage points of the generalized beta distribution. II. *Biometrika*, **44**, 441–453.

Foster, F. G. (1958). Upper percentage points of the generalized beta distribution. III. *Biometrika*, **45**, 492–502.

Foster, F. G., and Rees, D. H. (1957). Upper percentage points of the generalized beta distribution. *Biometrika*, **44**, 237–247.

Francis, I. (1973). Factor analysis: its purpose, practice, and packaged programs. Invited paper, American Statistical Association, New York, December 1973.

Francis, I. (1974). Factor analysis: fact or fabrication. *Math. Chron.*, **3**, 9–44.

Frank, O., and Svensson, K. (1981). On probability distributions of single-linkage dendrograms. *J. Stat. Comput. Simul.*, **12**, 121–131.

Frets, G. P. (1921). Heredity of head form in man. *Genetica*, **3**, 193–384.

Friedman, H. P., and Rubin, J. (1967). On some invariant criteria for grouping data. *J. Am. Stat. Assoc.*, **62**, 1159–1178.

Friedman, J. H., and Rafsky, L. C. (1979). Multivariate generalisations of the Wald–Wolfowitz and Smirnov two-sample tests. *Ann. Stat.*, **7**, 697–717.

Friedman, J. H., and Rafsky, L. C. (1981). Graphics for the multivariate two-sample problem (with discussion). *J. Am. Stat. Assoc.*, **76**, 277–295.

Fryer, M. J. (1977). A review of some non-parametric methods of density estimation. *J. Inst. Math. Appl.*, **20**, 335–354.

Fu, K. S. (1977). Linguistic approach to pattern recognition. *In* J. Van Ryzin (Ed.), *Classification and Clustering*, pp. 199–250. Academic Press: New York.

Fujikoshi, Y. (1974). The likelihood ratio tests for the dimensionality of regression coefficients. *J. Multivar. Anal.*, **4**, 327–340.

Fujikoshi, Y. (1977a). Asymptotic expansions of the distributions of the latent roots in MANOVA and the canonical correlations. *J. Multivar. Anal.*, **7**, 386–396.

Fujikoshi, Y. (1977b). Asymptotic expansions for the distributions of some multivariate tests. *In* P. R. Krishnaiah (Ed.), *Multivariate Analysis*, Vol. IV, pp. 55–71. North-Holland: Amsterdam.

Fujikoshi, Y. (1978). Asymptotic expansions for the distributions of some functions of the latent roots of matrices in three situations. *J. Multivar. Anal.*, **8**, 63–72.

Fujikoshi, Y. (1980). Asymptotic expansions for the distributions of the sample roots under nonnormality. *Biometrika*, **67**, 45–51.

Fujikoshi, Y. (1981). The power of the likelihood ratio test for additional information in a multivariate linear model. *Ann. Inst. Stat. Math.*, **33**, 279–285.

Fujikoshi, Y., and Veitch, L. G. (1979). Estimation of dimensionality in canonical correlation analysis. *Biometrika*, **66**, 345–351.

Fukunaga, K., and Kessel, D. (1971). Estimation of classification error. *IEEE Trans. Comput.*, **C-20**, 1521–1527.

Furnival, G. M. (1971). All possible regressions with less computation. *Technometrics*, **13**, 403–408.

Furnival, G. M., and Wilson, R. W. (1974). Regression by leaps and bounds. *Technometrics*, **16**, 499–511.

Gabriel, K. R. (1968). Simultaneous test procedures in multivariate analysis of variance. *Biometrika*, **55**, 489–504.

Gabriel, K. R. (1969). Simultaneous test procedures—some theory of multiple comparisons. *Ann. Math. Stat.*, **40**, 224–250.

Gabriel, K. R. (1971). The biplot graphic display of matrices with application to principal component analysis. *Biometrika*, **58**, 453–467.

Gabriel, K. R. (1978). Least squares approximation of matrices by additive and multiplicative models. *J. R. Stat. Soc. B*, **40**, 186–196.

Gabriel, K. R., and Zamir, S. (1979). Lower rank approximation of matrices by least squares with any choice of weights. *Technometrics*, **21**, 489–498.

Gafarian, A. V. (1978). Confidence bands in multivariate polynomial regression. *Technometrics*, **20**, 141–149.

Ganesalingam, S., and McLachlan, G. J. (1978). The efficiency of a linear discriminant function based on unclassified initial samples. *Biometrika*, **65**, 658–662.

Ganesalingam, S., and McLachlan, G. J. (1979). Small sample results for a linear discriminant function estimated from a mixture of normal populations. *J. Stat. Comput. Simul.*, **9**, 151–158.

Ganesalingam, S., and McLachlan, G. J. (1980). A comparison of the mixture and classification approaches to cluster analysis. *Commun. Stat. Theor. Methods A*, **9**, 923–933.

Ganesalingam, S., and McLachlan, G. J. (1981). Some efficiency results for the estimation of the mixing proportion in a mixture of two normal distributions. *Biometrics*, **37**, 23–33.

Garcia Ben, M. S., and Yohai, V. J. (1980). Multivariate analysis of variance in a randomized blocks design when covariance matrices are unequal. *Biometrics*, **36**, 127–133.

Gardner, M. J., and Barker, D. J. B. (1975). A case study in techniques of allocation. *Biometrics*, **31**, 931–942.

Geisser, S. (1973). Multiple birth discrimination. *In* D. G. Kabe and R. P. Gupta (Eds.), *Multivariate Statistical Inference*, pp. 49–55. North-Holland: Amsterdam.

Geisser, S. (1977). Discrimination, allocatory and separatory, linear aspects. *In* J. Van Ryzin (Ed.), *Classification and Clustering*, pp. 301–330. Academic Press: New York.

Geisser, S. (1980). Growth curve analysis. *In* P. R. Krishnaiah (Ed.), *Handbook in Statistics*, Vol. 1, pp. 89–115. North-Holland: Amsterdam.

Gerson, M. (1975). The techniques and uses of probability plotting. *Statistician*, **4**, 235–257.

Gessaman, M. P. (1970). A consistent nonparametric multivariate density estimator based on statistically equivalent blocks. *Ann. Math. Stat.*, **41**, 1344–1346.

Gessaman, M. P., and Gessaman, P. H. (1972). A comparison of some multivariate discrimination procedures. *J. Am. Stat. Assoc.*, **67**, 468–472.

Geweke, J. F., and Singleton, K. J. (1980). Interpreting the likelihood ratio statistic in factor models when sample size is small. *J. Am. Stat. Assoc.*, **75**, 133–137.

Gilbert, E. S. (1969). The effect of unequal variance–covariance matrices on Fisher's linear discriminant function. *Biometrics*, **25**, 505–515.

Gill, P. E., and Murray, W. (1972). Quasi-Newton methods for unconstrained optimisation. *J. Inst. Math. Appl.*, **9**, 91–108.

Gillo, M. W., and Shelly, M. W. (1974). Predictive modeling of multivariable and multivariate data. *J. Am. Stat. Assoc.*, **69**, 646–653.

Giri, N. C. (1977). *Multivariate Statistical Inference*. Academic Press: New York.

Girshick, M. A. (1939). On the sampling theory of roots of determinantal equations. *Ann. Math. Stat.*, **10**, 203–224.

Gitman, I., and Levine, M. D. (1970). An algorithm for detecting unimodal fuzzy sets and its application as a clustering technique. *IEEE Trans. Comput.*, **C19**, 583–593.

Gleser, L. J. (1976). A canonical representation for the noncentral Wishart distribution useful for simulation. *J. Am. Stat. Assoc.*, **71**, 690–695.

Gleser, L. J. (1979). Minimax estimation of a normal mean vector when the covariance matrix is unknown. *Ann. Stat.*, **7**, 838–846.

Gleser, L. J. (1981). Estimation in a multivariate "errors in variables" regression model: large sample results. *Ann. Stat.*, **9**, 24–44.

Gleser, L. J., and Olkin, I. (1970). Linear models in multivariate analysis. *In* R. C. Bose, et al. (Eds.), *Essays in Probability and Statistics* (dedicated to S. N. Roy), pp. 267–292. University of North Carolina Press: Chapel Hill, N.C.

Glick, N. (1972). Sample-based classification procedures derived from density estimators. *J. Am. Stat. Assoc.*, **67**, 116–122.

Glick, N. (1973). Sample-based multinomial classification. *Biometrics*, **29**, 241–256.

Glick, N. (1978). Additive estimators for probabilities of correct classification. *Pattern Recognition*, **10**, 211–222.

Glynn, W. J., and Muirhead, R. J. (1978). Inference in canonical correlation analysis. *J. Multivar. Anal.*, **8**, 468–478.

Gnanadesikan, R. (1977). *Methods for Statistical Data Analysis of Multivariate Observations*. Wiley: New York.

Gnanadesikan, R. (1980). Graphical methods for internal comparisons in ANOVA and MANOVA. *In* P. R. Krishnaiah (Ed.), *Handbook of Statistics*, Vol. 1, pp. 133–177. North-Holland: Amsterdam.

Gnanadesikan, R., and Kettenring, J. R. (1972). Robust estimates, residuals, and outlier detection with multiresponse data. *Biometrics*, **28**, 81–124.

Gnanadesikan, R., and Lee, E. T. (1970). Graphical techniques for internal comparisons amongst equal degree of freedom groupings in multiresponse experiments. *Biometrika*, **57**, 229–337.

Gnanadesikan, R., and Wilk, M. B. (1966) Data analytic methods in multivariate statistical analysis. General Methodology Lecture on Multivariate Analysis, 126th Annual Meeting of the American Statistical Association, Los Angeles, Calif.

Gnanadesikan, R., and Wilk, M. B. (1969). Data analytic methods in multivariate statistical analysis. *In* P. R. Krishnaiah (Ed.), *Multivariate Analysis*, Vol. II, pp. 593–638. Academic Press: New York.

Gnanadesikan, R., Pinkham, R. S., and Hughes, L. P. (1967). Maximum likelihood estimation of the parameters of the beta distribution from smallest order statistics. *Technometrics*, **9**, 607–620.

Golden, R. R., and Meehl, P. E. (1980). Detection of biological sex: an empirical test of cluster methods. *Multivar. Behav. Res.*, **15**, 475–496.

Goldstein, M., and Dillon, W. R. (1978). *Discrete Discriminant Analysis*. Wiley: New York.

Goldstein, M., and Wolf, E. (1977). On the problem of bias in multinomial classification.

Biometrics, **33**, 325–331.

Golub, G. H. (1969). Matrix decompositions and statistical calculation. *In* R. C. Milton and J. A. Nelder (Eds.), *Statistical Computation*, pp. 365–397. Academic Press: New York.

Golub, G. H., and Reinsch, C. (1971). Singular value decomposition and least squares solutions. *In* J. H. Wilkinson and C. H. Reinsch (Eds.), *Handbook for Automatic Computation, Linear Algebra*, Vol. II, pp. 134–151. Springer-Verlag: Berlin.

Golub, G. H., and Styan, G. P. (1973). Numerical computations for univariate linear models. *J. Stat. Comput. Simul.*, **2**, 253–274.

Good, I. J. (1977). The botryology of botryology. *In* J. Van Ryzin (Ed.), *Classification and Clustering*, pp. 73–94. Academic Press: New York.

Goodchild, N. A., and Vijayan, K. (1974). Significance tests in plots of multi-dimensional data in two dimensions. *Biometrics*, **30**, 209–210.

Goodman, L. A., and Kruskal, W. H. (1954). Measures of association for cross-classifications. *J. Am. Stat. Assoc.*, **49**, 732–764.

Gordesch, J., and Sint, P. P. (1974). Clustering structures. *In* G. Bruckmann, F. Ferschl, and L. Schmetterer (Eds.), *Compstat 1974*, pp. 82–92. Physica-Verlag: Vienna.

Gordon, A. D. (1981). *Classification*. Chapman and Hall: London.

Gordon, A. D., and Henderson, J. T. (1977). An algorithm for Euclidean sum of squares classification. *Biometrics*, **33**, 355–362.

Gordon, L., and Olshen, R. A. (1978). Asymptotically efficient solutions to the classification problem. *Ann. Stat.*, **6**, 515–533.

Gower, J. C. (1966). Some distance properties of latent root and vector methods used in multivariate analysis. *Biometrika*, **53**, 325–338.

Gower, J. C. (1967a). A comparison of some methods of cluster analysis. *Biometrics*, **23**, 623–628.

Gower, J. C. (1967b). Multivariate analysis and multidimensional geometry. *Statistician*, **17**, 13–25.

Gower, J. C. (1971a). A general coefficient of similarity and some of its properties. *Biometrics*, **27**, 857–874.

Gower, J. C. (1971b). Statistical methods of comparing different multivariate analyses of the same data. *In* F. R. Hodson, D. G. Kendall, and P. Tǎutu (Eds.), *Mathematics in the Archaelogical and Historical Sciences*, pp. 138–149. Edinburgh University Press: Edinburgh.

Gower, J. C., and Ross, G. J. S. (1969). Minimum spanning trees and single linkage cluster analysis. *Appl. Stat.*, **18**, 54–64.

Graef, J., and Spence, I. (1979). Using distance information in the design of large multidimensional scaling experiments. *Psychol. Bull.*, **36**, 60–66.

Green, J. R., and Hegazy, Y. A. S. (1976). Powerful modified-EDF goodness-of-fit tests. *J. Am. Stat. Assoc.*, **71**, 204–209.

Greenstreet, R. L., and Connor, R. J. (1974). Power of tests for equality of covariance matrices. *Technometrics*, **16**, 27–30.

Gregson, R. A. M. (1974). *Psychometrics of Similarity*. Academic Press: New York.

Grizzle, J. E., and Allen, D. M. (1969). Analysis of growth and dose response curves. *Biometrics*, **25**, 357–381.

Gross, A. M. (1976). Confidence interval robustness with long-tailed symmetric distributions. *J. Am. Stat. Assoc.*, **71**, 409–416.

Gupta, A. K. (1977). On the distribution of sphericity test criterion in the multivariate Gaussian distribution. *Aust. J. Stat.*, **19**, 202–205.

Guttman, L. (1954). A new approach to factor analysis: the radex. *In* P. F. Lazarsteld (Ed.), *Mathematical Thinking in the Social Sciences*, pp. 258–348, 430–433. The Free Press: Glencoe, Ill.

Guttman, R., and Guttman, L. (1965). A new approach to the analysis of growth patterns: the simplex structure of intercorrelations of measurement. *Growth*, **29**, 145–152.

Habbema, J. D. F. (1976). A discriminant analysis approach to the identification of human chromosomes. *Biometrics*, **32**, 919–928.

Habbema, J. D. F. (1979). Statistical methods for classification of human chromosomes. *Biometrics*, **35**, 103–118.

Habbema, J. D. F., and Hermans, J. (1977). Selection of variables in discriminant analysis by *F*-statistic and error rate. *Technometrics*, **19**, 487–493.

Habbema, J. D. F., Hermans, J., and van den Broek, K. (1974a). A stepwise discriminant analysis program using density estimation. *In* G. Bruckmann, F. Ferschl, and L. Schmetterer (Eds.), *Compstat 1974*, pp. 101–110. Physica-Verlag: Vienna.

Habbema, J. D. F., Hermans, J., and van der Burgt, A. T. (1974b). Cases of doubt in allocation problems. *Biometrika*, **61**, 313–324.

Habbema, J. D. F., Hermans, J., and Remme, J. (1978). Variable kernel density estimation in discriminant analysis. *In* L. C. A. Corsten and J. Hermans (Eds.), *Compstat 1978*, pp. 178–185. Physica-Verlag: Vienna.

Haberman, S. J. (1974). *The Analysis of Frequency Data*. University of Chicago Press: Chicago.

Haff, L. R. (1979). Estimation of the inverse covariance matrix: random mixtures of the inverse Wishart matrix and the identity. *Ann. Stat.*, **7**, 1264–1276.

Haff, L. R. (1980). Empirical Bayes estimation of the multivariate normal covariance matrix. *Ann. Stat.*, **8**, 586–597.

Hall, D. J., and Khanna, D. K. (1977). The ISODATA method: computation for the relative perception of similarities and differences in complex and real data. *In* K. Enslein, A. Ralston, and H. S. Wilf (Eds.), *Statistical Methods for Digital Computers*, Vol. 3, pp. 340–373, Wiley: New York.

Hall, P. (1981). On nonparametric multivariate binary discrimination. *Biometrika*, **68**, 287–294.

Halperin, M., Blackwelder, W. C., and Verter, J. I. (1971). Estimation of the multivariate logistic risk function: a comparison of the discriminant function and maximum likelihood approaches. *J. Chronic Diseases*, **24**, 125–128.

Hampel, F. R. (1971). A general qualitative definition of robustness. *Ann. Math. Stat.*, **4**, 1887–1896.

Hampel, F. R. (1973). Robust estimation: a condensed partial survey. *Z. Wahrsch. Verw. Gebiete*, **27**, 87–104.

Hampel, F. R. (1974). The influence curve and its role in robust estimation. *J. Am. Stat. Assoc.*, **69**, 383–393.

Hampel, F. R. (1978). Modern trends in the theory of robustness. *Math. Operationsforsch. Ser. Stat.*, **9**, 425–442.

Hampel, F. R., Rousseew, P. J., and Ronchetti, E. (1981). The change-of-variance curve and optimal redescending *M*-estimators. *J. Am. Stat. Assoc.*, **76**, 643–648.

Han, C.-P. (1979). Alternative methods of estimating the likelihood ratio in classification of multivariate normal observations. *Am. Stat.*, **33**, 204–206.

Hannan, E. J. (1967). Canonical correlation and multiple equation systems in economics. *Econometrica*, **35**, 123–138.

Hansen, P., and Delattre, M. (1978). Complete-link cluster analysis by graph coloring. *J. Am. Stat. Assoc.*, **73**, 397–403.

Hanumara, R. C., and Strain, W. F. (1980). A note on the application of the percentage points of the extreme roots of a Wishart matrix. *Biometrika*, **67**, 501–502.

Hanumara, R. C., and Thompson, W. A., Jr. (1968). Percentage points of the extreme roots of a Wishart matrix. *Biometrika*, **55**, 505–512.

Harman, H. H. (1977). Minres method of factor analysis. *In* K. Enslein, A. Ralston, and H. S. Wilf (Eds.), *Statistical Methods for Digital Computers*, Vol. 3, pp. 154–165. Wiley: New York.

Harris, C. W. (1978). Note on the squared multiple correlation as a lower bound to communality. *Psychometrika*, **43**, 283–284.

Harris, R. J. (1975). *A Primer of Multivariate Statistics*. Academic Press: New York.

Hart, M. L., and Money, A. H. (1976). On Wilks's multivariate generalization of the correlation ratio. *Biometrika*, **63**, 59–67.

Hartigan, J. A. (1967). Representation of similarity matrices by trees. *J. Am. Stat. Assoc.*, **62**, 1140–1158.

Hartigan, J. A. (1975). *Clustering Algorithms*. Wiley: New York.

Hartigan, J. A. (1977). Distribution problems in clustering. *In* J. Van Ryzin (Ed.), *Classification and Clustering*, pp. 45–71. Academic Press: New York.

Hartigan, J. A. (1978). Asymptotic distributions for clustering criteria. *Ann. Stat.*, **6**, 117–131.

Hartigan, J. A. (1981). Consistency of single linkage for high-density clusters. *J. Am. Stat. Assoc.*, **76**, 388–394.

Hartigan, J. A., and Wong, M. A. (1979). Algorithm AS 136: A *K*-means clustering algorithm. *Appl. Stat.*, **28**, 100–108.

Hawkins, D. M. (1974). The detection of errors in multivariate data using principal components. *J. Am. Stat. Assoc.*, **69**, 340–344.

Hawkins, D. M. (1976). The subset problem in multivariate analysis of variance. *J. R. Stat. Soc. B*, **38**, 132–139.

Hawkins, D. M. (1980). *Identification of Outliers*. Chapman and Hall: New York.

Hawkins, D. M. (1981). A new test for multivariate normality and homoscedasticity. *Technometrics*, **23**, 105–110.

Hawkins, D. M., and Eplett, W. J. R. (1982). The Cholesky factorization of the inverse correlation or covariance matrix in multiple regression. *Technometrics*, **24**, 191–198.

Healy, M. J. R. (1968). Multivariate normal plotting. *Appl. Stat.*, **17**, 157–161.

Healy, M. J. R. (1978). A mean difference estimator of standard deviation in symmetrically censored normal samples. *Biometrika*, **65**, 643–646.

Heck, D. L. (1960). Charts of some upper percentage points of the distribution of the largest characteristic root. *Ann. Math. Stat.*, **31**, 625–642.

Hedayat, A., and Afsarinejad, K. (1975). Repeated measurement designs, I. *In* J. N. Srivastava (Ed.), *A Survey of Statistical Design and Linear Models*, pp. 229–242. North-Holland: Amsterdam.

Henderson, H., and Searle, S. R. (1979). Vec and vech operators for matrices, with some uses in Jacobians and multivariate statistics. *Can. J. Stat.*, **7**, 65–81.

Hensler, G. L., Mehrotra, K. G., and Michalek, J. E. (1977). A goodness of fit test for multivariate normality. *Commun. Stat. Theor. Methods A*, **6**, 33–41.

Hermans, J., and Habbema, J. D. F. (1975). Comparison of five methods to estimate posterior probabilities. *EDV in Medizin und Biologie*, **6**, 14–19.

Hermans, J., Habbema, J. D. F., and Schäfer, J. R. (1982). The ALLOC80 package for discriminant analysis. *Stat. Software Newsl.*, **8**(*1*), 15–20.

Hernandez, F., and Johnson, R. A. (1980). The large-sample behaviour of transformations to normality. *J. Am. Stat. Assoc.*, **75**, 855–861.

Hertsgaard, D. (1979). Distribution of asymmetric trimmed means. *Commun. Stat. Simul. Comput. B*, **8**, 359–367.

Hill, G. W., and Davis, A. W. (1968). Generalized asymptotic expansions of Cornish–Fisher type. *Ann. Math. Stat.*, **39**, 1264–1273.

Hill, M. O. (1974). Correspondence analysis: a neglected multivariate method. *Appl. Stat.*, **23**, 340–354.

Hill, R. C., Fomby, T. B., and Johnson, S. R. (1977). Component selection norms for principal components regression. *Commun. Stat. Theor. Methods A*, **6**, 309–334.

Hills, M. (1966). Allocation rules and their error rates. *J. R. Stat. Soc. B*, **28**, 1–31.

Hills, M. (1967). Discrimination and allocation with discrete data. *Appl. Stat.*, **16**, 237–250.

Hills, M. (1969). On looking at large correlation matrices. *Biometrika*, **56**, 249–253.

Hinkley, D. (1975). On power transformations to normality. *Biometrika*, **62**, 101–111.

Hinkley, D. (1977). On quick choice of power transformation. *Appl. Stat.*, **26**, 67–69.

Hintze, J. L. (1980). On the use of "elemental analysis" in multivariate variable selection. *Technometrics*, **22**, 609–612.

Hodges, J. L., Jr., and Lehmann, E. L. (1963). Estimates of location based on rank tests. *Ann. Math. Stat.*, **34**, 598–611.

Hodson, F. R. (1971). Numerical typology and prehistoric archaeology. *In* F. R. Hodson, D. G. Kendall, and P. A. Tăutu (Eds.), *Mathematics in the Archaeological and Historical Sciences*, pp. 30–45. Edinburgh University Press: Edinburgh.

Hodson, F. R., Kendall, D. G., and Tăutu, P. A. (Eds.) (1971). *Mathematics in the Archaelogical and Historical Sciences*. Edinburgh University Press: Edinburgh.

Hogg, R. V. (1967). Some observations on robust estimation. *J. Am. Stat. Assoc.*, **62**, 1179–1186.

Hogg, R. V. (1972). More light on the kurtosis and related statistics. *J. Am. Stat. Assoc.*, **67**, 422–424.

Hogg, R. V. (1974). Adaptive robust procedures: a partial review and some suggestions for future applications and theory (with comments). *J. Am. Stat. Assoc.*, **69**, 909–927.

Hogg, R. V. (1979a). Statistical robustness: one view of its use in applications today. *Am. Stat.*, **33**, 108–115.

Hogg, R. V. (1979b). An introduction to robust estimation. *In* R. L. Launer and G. N. Wilkinson (Eds.), *Robustness in Statistics*, pp. 1–17. Academic Press: New York.

Holman, E. W. (1972). The relation between hierarchical and Euclidean models for psychological distances. *Psychometrika*, **37**, 417–423.

Hooper, J. W. (1959). Simultaneous equations and canonical correlation theory. *Econometrica*, **27**, 245–256.

Hora, S. C. (1980). Sequential discrimination. *Commun. Stat. Theor. Methods A*, **9**, 905–916.

Hosmer, D. W., Jr. (1973a). A comparison of iterative maximum likelihood estimates of the parameters of a mixture of two normal distributions under three different types of sample. *Biometrics*, **29**, 761–770.

Hosmer, D. W., Jr. (1973b). On MLE of the parameters of a mixture of two normal distributions when the sample size is small. *Commun. Stat.*, **1**, 217–227.

Hosmer, D. W., Jr., and Dick, N. P. (1977). Information and mixtures of two normal distributions. *J. Stat. Comput. Simul.*, **6**, 137–148.

Hotelling, H. (1931). The generalization of Student's ratio. *Ann. Math. Stat.*, **2**, 360–378.

Hotelling, H. (1933). Analysis of a complex of statistical variables into principal components. *J. Educ. Psychol.*, **24**, 417–441.

Hotelling, H. (1936). Relations between two sets of variates. *Biometrika*, **28**, 321–327.

Hotelling, H. (1951). A generalised *T*-test and measure of multivariate dispersion. *Proc. Second Berkeley Symp. Math. Stat. Prob.*, **1**, 23–41.

Householder, A. S., and Young, G. (1938). Matrix approximation and latent roots. *Am. Math. Mon.*, **45**, 165–171.

Howarth, R. J. (1973). Preliminary assessment of a non-linear mapping algorithm in a geological context. *Math. Geol.*, **5**, 39–57.

Howe, W. G. (1955). *Some Contributions to Factor Analysis*. USAEC Report ORNL-1919.

Hsu, P. L. (1938). Notes on Hotelling's generalized *T*. *Ann. Math. Stat.*, **9**, 231–243.

Hsu, P. L. (1939). On the distribution of the roots of certain determinantal equations. *Ann. Eugen.*, **9**, 250–258.

Hsu, P. L. (1940). On generalised analysis of variance. *Biometrika*, **31**, 221–237.

Hsu, P. L. (1941). On the limiting distribution of canonical correlations. *Biometrika*, **33**, 38–45.

Huang, C. J., and Bolch, B. W. (1974). On testing of regression disturbances for normality. *J. Am. Stat. Assoc.*, **69**, 330–335.

Huber, P. J. (1964). Robust estimation of a location parameter. *Ann. Math. Stat.*, **35**, 73–101.

Huber, P. J. (1970). Studentizing robust estimates. *In* M. L. Puri (Ed.), *Nonparametric Techniques in Statistical Inference*, pp. 453–463. Cambridge University Press: Cambridge.

Huber, P. J. (1972). Robust statistics: a review. *Ann. Math. Stat.*, **43**, 1041–1067.

Huber, P. J. (1977a). *Robust Statistical Procedures*. CBMS-NSF Regional Conference Series in Applied Mathematics No. 27. Published by SIAM., Philadelphia, Penn.

Huber, P. J. (1977b). Robust covariances. *In* S. S. Gupta and D. S. Moore (Eds.), *Statistical Decision Theory and Related Topics*, Vol. II, pp. 165–191. Academic Press: New York.

Huber, P. J. (1981). *Robust Statistics*. Wiley: New York.

Hubert, L. (1974). Approximate evaluation techniques for the single-link and complete-link hierarchical clustering procedures. *J. Am. Stat. Assoc.*, **69**, 698–704.

Hubert, L. J., and Baker, F. B. (1977). An empirical comparison of baseline models for goodness-of-fit in *r*-diameter hierarchical clustering. *In* J. Van Ryzin (Ed.), *Classification and Clustering*, pp. 131–153. Academic Press: New York.

Huberty, C. J. (1975). Discriminant analysis. *Rev. Educ. Res.*, **45**, 543–593.

Hughes, D. T., and Saw, J. G. (1972). Approximating the percentage points of Hotelling's generalized T_0^2 statistic. *Biometrika*, **59**, 224–226.

Hurley, J. R., and Cattell, R. B. (1962). The procrustes program: producing direct rotation to test a hypothesised factor structure. *Behav. Sci.*, **7**, 258–262.

Huseby, J. R., Schwertman, N. C., and Allen, D. M. (1980). Computation of the mean vector and dispersion matrix for incomplete multivariate data. *Commun. Stat. Simul. Comput. B*, **8**, 301–309.

Ihm, P. (1965). Automatic classification in anthropology. *In* D. Hymes (Ed.), *The Use of Computers in Anthropology*, pp. 357–376. Mouton: The Netherlands.

Iman, R. L. (1982). Graphs for use with the Lilliefors test for normal and exponential distributions. *Am. Stat.*, **36**, 109–112.

Imrey, P. B., Koch, G. G., and Stokes, M. E. (1981). Categorical data analysis: some reflections on the log linear model and logistic regression. Part I: Historical and methodological overview. *Int. Stat. Rev.*, **49**, 265–283.

Imrey, P. B., G. G. Koch, and Stokes, M. E. (1982). Categorical data analysis: some reflections on the log linear model and logistic regression. Part II: Data analysis. *Int. Stat. Rev.*, **50**, 35–63.

Isogawa, Y., and Okamoto, M. (1980). Linear prediction in the factor analysis model. *Biometrika*, **67**, 482–484.

Ito, K. (1956). Asymptotic formulae for the distribution of Hotelling's generalized T_0^2 statistic. *Ann. Math. Stat.*, **27**, 1091–1105.

Ito, K. (1960). Asymptotic formulae for the distribution of Hotelling's generalized T_0^2 statistic, II. *Ann. Math. Stat.*, **31**, 1148–1153.

Ito, K. (1962). A comparison of the powers of two multivariate analysis of variance tests. *Biometrika*, **49**, 455–462.

Ito, K. (1969). On the effect of heteroscedasticity and nonnormality upon some multivariate test procedures. *In* P. R. Krishnaiah (Ed.), *Multivariate Analysis*, Vol. II, pp. 87–120. Academic Press: New York.

Ito, K., and Schull, W. J. (1964). On the robustness of the T_0^2 test in multivariate analysis of variance when variance-covariance matrices are not equal. *Biometrika*, **51**, 71–82.

Ito, P. K. (1980). Robustness of ANOVA and MANOVA test procedures. *In* P. R. Krishnaiah (Ed.), *Handbook of Statistics*, Vol. 1, pp. 199–236. North-Holland: Amsterdam.

Izenman, A. J. (1975). Reduced-rank regression for the multivariate linear model. *J. Multivar. Anal.*, **5**, 248–264.

Jaccard, P. (1908). Nouvelles recherches sur la distribution florale. *Bull. Soc. Vaudoise Sci. Nat.*, **44**, 223–270.

Jackson, J. E., and Hearne, F. T. (1979). Hotelling's T_M^2 for principal components—what about absolute values? *Technometrics*, **21**, 253–255.

Jackson, J. E., and Mudholkar, G. S. (1979). Control procedures for residuals associated with principal component analysis. *Technometrics*, **21**, 341–349.

Jacob, R. J. K. (1981). Comment on a paper by Kleiner and Hartigan. *J. Am. Stat. Assoc.*, **76**, 270–272.

Jacob, R. J. K., Egeth, H. E., and Bevan, W. (1976). The face as a data display. *Hum. Factors*, **18**, 189–200.

James, A. T. (1960). The distribution of roots of the covariance matrix. *Ann. Math. Stat.*, **31**, 151–158.

James, A. T. (1964). Distributions of matrix variates and latent roots derived from normal samples. *Ann. Math. Stat.*, **35**, 475–501.

James, A. T. (1969). Tests of equality of latent roots of the covariance matrix. *In* P. R. Krishnaiah (Ed.), *Multivariate Analysis*, Vol. II, pp. 205–218. Academic Press: New York.

James, G. S. (1954). Tests of linear hypotheses in univariate and multivariate analysis when the ratios of the population variances are unknown. *Biometrika*, **41**, 19–43.

James, I. R. (1978). Estimation of the mixing proportion in a mixture of two normal distributions from simple rapid measurements. *Biometrics*, **34**, 265–275.

James, W., and Stein, C. (1961). Estimation with quadratic loss. *Proc. Fourth Berkeley Symp. Math. Stat. Prob.*, **1**, 362–379.

Jancey, R. C. (1966). Multidimensional group analysis. *Aust. J. Bot.*, **14**, 127–130.

Jardine, C. J., Jardine, N., and Sibson, R. (1967). The structure and construction of taxonomic hierarchies. *Math. Biosci.*, **1**, 173–179.

Jardine, N., and Sibson, R. (1968). The construction of hierarchic and nonhierarchic classifications. *Comput. J.*, **11**, 177–184.

Jardine, N., and Sibson, R. (1971a). Choice of methods for automatic classification. *Comput. J.*, **14**, 404–406.

Jardine, N., and Sibson, R. (1971b). *Mathematical Taxonomy*. Wiley: London.

Jeffers, J. N. R. (1967). Two case studies in the application of principal component analysis. *Appl. Stat.*, **16**, 225–236.

Jennrich, R. I. (1977). Stepwise discriminant analysis. *In* K. Enslein, A. Ralston, and H. S. Wilf (Eds.), *Statistical Methods for Digital Computers*, Vol. 3, pp. 76–95. Wiley: New York.

Jensen, D. R. (1972). Some simultaneous multivariate procedures using Hotelling's T^2 statistics. *Biometrics*, **28**, 39–53.

Jensen, R. E. (1968). A dynamic programming algorithm for cluster analysis. *Oper. Res.*, **12**, 1034–1057.

Johansen, S. (1980). The Welch–James approximation to the distribution of the residual sum of squares in a weighted linear regression. *Biometrika*, **67**, 85–92.

John, J. A., and Draper, N. R. (1980). An alternative family of transformations. *Appl. Stat.*, **29**, 190–197.

John, S. (1960). On some classification problems. *Sankhya*, **22**, 301–308.

John, S. (1963). On classification by the statistics R and Z. *Ann. Inst. Stat. Math.*, **14**, 237–246.

John, S. (1976). Fitting sampling distribution agreeing in support and moments and tables of critical values of sphericity criterion. *J. Multivar. Anal.*, **6**, 601–607.

John, S. (1977). Unbiased and upper critical values of mean trace of multivariate beta for testing difference of two covariance matrices or several mean vectors. *Commun. Stat. Simul. Comput. B*, **6**, 89–96.

Johns, M. V. (1979). Robust Pitman-like estimators. *In* R. L. Launer and G. N. Wilkinson (Eds.), *Robustness in Statistics*, pp. 49–59. Academic Press: New York.

Johnson, M. E., and Lowe, V. W., Jr. (1979). Bounds on the sample skewness and kurtosis. *Technometrics*, **21**, 377–378.

Johnson, N. L. (1965). Tables to facilitate fitting S_u frequency curves. *Biometrika*, **52**, 547–558.

Johnson, N. L., and Kotz, S. (1970). *Continuous Univariate Distributions*, Vol. 2. Wiley: New York.

Johnson, N. L., and Kotz, S. (1972). *Distributions in Statistics*, Vol. 4. *Continuous Multivariate Distributions*, Wiley: New York.

Johnson, N. L., and Leone, F. C. (1964). *Statistics and Experimental Design in Engineering and the Physical Sciences*, Vol. 1. Wiley: New York.

Johnson, S. C. (1967). Hierarchical clustering schemes. *Psychometrika*, **32**, 241–254.

Jolliffe, I. T. (1972). Discarding variables in a principal component analysis. I. Artificial data. *Appl. Stat.*, **21**, 160–173.

Jolliffe, I. T. (1973). Discarding variables in a principal component analysis. II. Real data. *Appl. Stat.*, **22**, 21–31.

Jolliffe, I. T. (1982). A note on the use of principal components in regression. *Appl. Stat.*, **31**, 300–303.

Jolliffe, I. T., Jones, B., and Morgan, B. J. T. (1982). Utilising clusters: a case-study involving the elderly. *J. R. Stat. Soc. A*, **145**, 224–236.

Jones, K. S., and Jackson, D. M. (1967). Current approaches to classification and clump finding at the Cambridge Language Research Unit. *Comput. J.*, **10**, 29–37.

Jones, R. H. (1975). Probability estimation using a multinomial logistic function. *J. Stat. Comput. Simul.*, **3**, 315–329.

Jöreskog, K. G. (1967). Some contributions to maximum likelihood factor analysis. *Psychometrika*, **32**, 443–482.

Jöreskog, K. G. (1970). A general method for analysis of covariance structure. *Biometrika*, **57**, 239–251.

Jöreskog, K. G. (1973). Analysis of covariance structures. *In* P. R. Krishnaiah (Ed.), *Multivariate Analysis*, Vol. III, pp. 263–285. Academic Press: New York.

Jöreskog, K. G. (1977). Factor analysis by least-squares and maximum-likelihood methods. *In* K. Enslein, A. Ralston, and H. S. Wilf (Eds.), *Statistical Methods for Digital Computers*, Vol. 3, pp. 125–153. Wiley: New York.

Jöreskog, K. G. (1978). Structural analysis of covariance and correlation matrices. *Psychometrika*, **43**, 443–477.

Jöreskog, K. G. (1981). Analysis of covariance structures. *Scand. J. Stat.*, **8**, 65–92.

Jöreskog, K. G., and Van Thillo, M. (1971). New rapid algorithms for factor analysis by unweighted least squares, generalized least squares and maximum likelihood. Unpublished Research Memorandum, Educational Testing Service, Princeton, N. J.

Kabe, D. G. (1973). On Rao's generalized U statistic. *In* D. G. Kabe and R P. Gupta (Eds.), *Multivariate Statistical Inference*, pp. 129–135. North-Holland: Amsterdam.

Kabe, D. G. (1975). Some results for the reduced form model. *In* R. P. Gupta (Ed.), *Applied Statistics*, pp. 169–175. North-Holland: Amsterdam.

Kagan, A. M., Linnik, Y. V., and Rao, C. R. (1973). *Characterization Problems in Mathematical Statistics*. Wiley: New York.

Kaiser, H. F. (1958). The varimax criterion for analytic rotation in factor analysis. *Psychometrika*, **23**, 187–200.

Kariya, T. (1978). The general MANOVA problem. *Ann. Stat.*, **6**, 200–214.

Kariya, T., and Kanazawa, M. (1978). A locally most powerful invariant test for the equality of means associated with covariate discriminant analysis. *J. Multivar. Anal.*, **8**, 134–140.

Kelker, D. (1970). Distribution theory of spherical distributions and a location-scale parameter generalization. *Sankhya A*, **32**, 419–430.

Kendall, D. G. (1971). Seriation from abundance matrices. *In* F. R. Hodson, D. G. Kendall and P. A. P. Tăutu (Eds.), *Mathematics in the Archaelogical and Historical Sciences*, pp. 215–252. Edinburgh University Press: Edinburgh.

Kendall, M. G. (1975). *Multivariate Analysis*. Griffin: London.

Kendall, M. G., and Babington Smith, B. (1950). Factor analysis. *J. R. Stat. Soc. B*, **12**, 60–94.

Kendall, M. G., and Lawley, D. N. (1956). The principles of factor analysis. *J. R. Stat. Soc. A*, **119**, 83–84.

Kendall, M. G., and Stuart, A. (1966). *The Advanced Theory of Statistics*, Vol. III. Griffin: London.

Kennard, R. W., and Stone, L. A. (1969). Computer aided design of experiments. *Technometrics*, **11**, 137–148.

Kennedy, W. J., Jr., and Gentle, J. E. (1980). *Statistical Computing*. Marcel Dekker: New York.

Kettenring, J. R. (1971). Canonical analysis of several sets of variables. *Biometrika*, **58**, 433–451.

Khatri, C. G. (1966). A note on a MANOVA model applied to problems in growth curves. *Ann. Inst. Stat. Math.*, **18**, 75–86.

Khatri, C. G. (1973). Testing some covariance structures under a growth curve model. *J. Multivar. Anal.*, **3**, 102–116.

Khatri, C. G. (1977). Quadratic forms and extension of Cochran's theorem to normal vector variables. *In* P. R. Krishnaiah (Ed.), *Multivariate Analysis*, Vol. IV, pp. 79–94. North-Holland: Amsterdam.

Khatri, C. G. (1978). A remark on the necessary and sufficient conditions for quadratic form to be distributed as chi-squared. *Biometrika*, **65**, 239–240.

Khatri, C. G., and Srivastava, M. S. (1975). On the likelihood ratio test for covariance matrix in growth curve model. *In* R. P. Gupta (Ed.), *Applied Statistics*, pp. 187–198. North-Holland: Amsterdam.

Kimber, A. C. (1979). Tests for a single outlier in a gamma sample with unknown shape and scale parameters. *Appl. Stat.*, **28**, 243–250.

King, B. F. (1966). Market and industry factors in stock price behaviour. *J. Business*, **39**, 139–190.

King, B. F. (1967). Step-wise clustering procedures. *J. Am. Stat. Assoc.*, **62**, 86–101.

Klauber, M. R. (1971). Two-sample randomization tests for space-time clustering. *Biometrics*, **27**, 129–142.

Klauber, M. R. (1975). Space-time clustering for more than two samples. *Biometrics*, **31**, 719–726.

Kleinbaum, D. G. (1973). A generalization of the growth curve model which allows missing data. *J. Multivar. Anal.*, **3**, 117–124.

Kleinbaum, D. G., Kupper, L. L., and Chambless, L. E. (1982). Logistic regression analysis of epidemiologic data: theory and practice. *Commun. Stat. Theor. Methods*, **11**, 485–547.

Kleiner, B., and Hartigan, J. A. (1981). Representing points in many dimensions by trees and castles (with comments). *J. Am. Stat. Assoc.*, **76**, 260–276.

Kokolakis, G. E. (1981). On the expected probability of correct classification. *Biometrika*, **68**, 477–483.

Konishi, S. (1978). An approximation to the distribution of the sample correlation coefficient. *Biometrika*, **65**, 654–656.

Konishi, S., and Sugiyama, T. (1981). Improved approximations to distributions of the largest and smallest latent roots of a Wishart matrix. *Ann. Inst. Stat. Math.*, **33**, 27–33.

Korin, B. P. (1969). On testing the equality of k covariance matrices. *Biometrika*, **56**, 216–218.

Korin, B. P., and Stevens, E. H. (1973). Some approximations for the distribution of a multivariate likelihood ratio criterion. *J. R. Stat. Soc. B*, **35**, 24–27.

Kowal, R. R. (1971). Disadvantages of the generalised variance as a measure of variability. *Biometrics*, **27**, 213–216.

Kramer, H. P., and Mathews, M. V. (1956). A linear coding for transmitting a set of correlated signals. *IRE Trans. Inf. Theory*, **IT-2**, 41–46.

Krishnaiah, P. R. (1965). Multiple comparison tests in multiresponse experiments. *Sankhya A*, **27**, 65–72.

Krishnaiah, P. R. (1969). Simultaneous test procedures under general MANOVA models. In P. R. Krishnaiah (Ed.), *Multivariate Analysis*, Vol. II, pp. 121–143. Academic Press: New York.

Krishnaiah, P. R. (1975). Tests for the equality of the covariance matrices of correlated multivariate normal populations. *In* J. N. Srivastava (Ed.), *A Survey of Statistical Design and Linear Models*, pp. 355–366. North-Holland: Amsterdam.

Krishnaiah, P. R. (1976). Some recent developments on complex multivariate distribution. *J. Multivar. Anal.*, **6**, 1–30.

Krishnaiah, P. R. (1978). Some recent developments on real multivariate distributions. *In* P. R. Krishnaiah (Ed.), *Developments in Statistics*, Vol. 1, pp. 135–169. Academic Press: New York.

Krishnaiah, P. R. (1979). Some developments on simultaneous test procedures. *In* P. R. Krishnaiah (Ed.), *Developments in Statistics*, Vol. 2, pp. 157–201. Academic Press: New York.

Krishnaiah, P. R. (1980). Computations of some multivariate distributions. *In* P. R. Krishnaiah (Ed.), *Handbook of Statistics*, Vol. 1, pp. 745–971. North-Holland: Amsterdam.

Krishnaiah, P. R., and Chang, T. C. (1971). On the exact distribution of the extreme roots of the Wishart and MANOVA matrices. *J. Multivar. Anal.*, **1**, 108–117.

Krishnaiah, P. R., and Chang, T. C. (1972). On the exact distributions of the traces of $S_1(S_1 + S_2)^{-1}$ and $S_1S_2^{-1}$. *Sankhya A*, **34**, 153–160.

Krishnaiah, P. R., and Chattopadhyay, A. K. (1975). On some noncentral distributions in multivariate analysis. *S. Afr. Stat. J.*, **9**, 37–46.

Krishnaiah, P. R., and Lee, J. C. (1976). On covariance structures. *Sankhya A*, **38**, 357–371.

Krishnaiah, P. R., and Lee, J. C. (1980). Likelihood ratio tests for mean vectors and covariance matrices. *In* P. R. Krishnaiah (Ed.), *Handbook of Statistics*, Vol. 1, pp. 513–570. North-Holland: Amsterdam.

Krishnaiah, P. R., and Schuurmann, F. J. (1974). On the evaluation of some distributions that arise in simultaneous tests for the equality of the latent roots of the covariance matrix. *J. Multivar. Anal.*, **4**, 265–282.

Krishnaiah, P. R., and Waikar, V. P. (1971). Exact joint distributions of any few ordered roots of a class of random matrices. *J. Multivar. Anal.*, **1**, 308–315.

Krishnaiah, P. R., and Yochmowitz, M. G. (1980). Inference on the structure of interaction in two-way classification model. *In* P. R. Krishnaiah (Ed.), *Handbook of Statistics*, Vol. 1, 973–994. North-Holland: Amsterdam.

Krishnaiah, P. R., Lee, J. C., and Chang, T. C. (1976). The distributions of the likelihood ratio statistics for tests of certain covariance structures of complex multivariate normal populations. *Biometrika*, **63**, 543–549.

Krishnaiah, P. R., Mudholkar, G. S., and Subbaiah, P. (1980). *In* P. R. Krishnaiah (Ed.), *Handbook of Statistics*, Vol. 1, pp. 631–671. North-Holland: Amsterdam.

Kruskal, J. B. (1964a). Multidimensional scaling by optimizing goodness of fit to a nonmetric hypothesis. *Psychometrika*, **29**, 1–27.

Kruskal, J. B. (1964b). Nonmetric multidimensional scaling: a numerical method. *Psychometrika*, **29**, 115–129.

Kruskal, J. B. (1977a). The relationship between multidimensional scaling and clustering. *In* J. Van Ryzin (Ed.), *Classification and Clustering*, pp. 17–44. Academic Press: New York.

Kruskal, J. B. (1977b). Multidimensional scaling and other methods for discovering structure. *In* K. Enslein, A. Ralston, and H. S. Wilf (Eds.), *Statistical Methods for Digital Computers*, Vol. 3, pp. 296–339. Wiley: New York.

Kruskal, J. B., and Wish, M. (1978). *Multidimensional Scaling*. Sage, Beverly Hills, Calif.

Krzanowski, W. J. (1975). Discrimination and classification using both binary and continuous variables. *J. Am. Stat. Assoc.*, **70**, 782–790.

Krzanowski, W. J. (1976). Canonical representation of the location model for discrimination or classification. *J. Am. Stat. Assoc.*, **71**, 845–848.

Krzanowski, W. J. (1977). The performance of Fisher's linear discriminant function under non-optimal conditions. *Technometrics*, **19**, 191–200.

Krzanowski, W. J. (1979a). Between-groups comparison of principal components. *J. Am. Stat. Assoc.*, **74**, 703–707.

Krzanowski, W. J. (1979b). Some exact percentage points of a statistic useful in analysis of variance and principal component analysis. *Technometrics*, **21**, 261–263.

Krzanowski, W. J. (1979c). Some linear transformations for mixtures of binary and continuous variables, with particular reference to linear discriminant analysis. *Biometrika*, **66**, 33–39.

Krzanowski, W. J. (1980). Mixtures of continuous and categorical variables in discriminant analysis. *Biometrics*, **36**, 493–499.

Kshirsagar, A. M. (1972). *Multivariate Analysis*. Marcel Dekker: New York.

Kuiper, F. K., and Fisher, L. (1975). A Monte Carlo comparison of six clustering procedures. *Biometrics*, **31**, 777–783.

Kullback, S., and Liebler, R. A. (1951). On information and sufficiency. *Ann. Math. Stat.*, **22**, 79–86.

LaBrecque, J. (1977). Goodness-of-fit tests based on nonlinearity in probability plots. *Technometrics*, **19**, 293–306.

Lachenbruch, P. A. (1966). Discriminant analysis when the initial samples are misclassified. *Technometrics*, **8**, 657–662.

Lachenbruch, P. A. (1974). Discriminant analysis when the initial samples are misclassified, II: non-random misclassification models. *Technometrics*, **16**, 419–424.

Lachenbruch, P. A. (1975a). *Discriminant Analysis*. Hafner: New York.

Lachenbruch, P. A. (1975b). Zero-mean difference discrinimation and the absoulute linear discriminant function. *Biometrika*, **62**, 397–401.

Lachenbruch, P. A. (1979). Note on initial misclassification effects on the quadratic discrimination function. *Technometrics*, **21**, 129–132.

Lachenbruch, P. A., and Goldstein, M. (1979). Discriminant analysis. *Biometrics*, **35**, 69–85.

Lachenbruch, P. A., and Mickey, M. R. (1968). Estimation of error rates in discriminant analysis. *Technometrics*, **10**, 1–11.

Lachenbruch, P. A., Sneeringer, C., and Revo, L. T. (1973). Robustness of the linear and quadratic discriminant function to certain types of non-normality. *Commun. Stat.*, **1**, 39–57.

Lancaster, H. O. (1963). Canonical correlations and partitions of χ^2. *Q. J. Math.*, **14**, 220–224.

Lancaster, H. O. (1969). *The Chi-Squared Distribution*. Wiley: New York.

Lance, G. N., and Williams, W. T. (1966). Computer programs for hierarchical polythetic classification ("similarity analyses"). *Comput. J.*, **9**, 60–64.

Lance, G. N., and Williams, W. T. (1967a). A general theory of classificatory sorting strategies. I. Hierarchical systems. *Comput. J.*, **9**, 373–380.

Lance, G. N., and Williams, W. T. (1967b). Mixed-data classificatory programs. I. Agglomerative systems. *Aust. Comput. J.*, **1**, 15–20.

Lance, G. N., and Williams, W. T. (1968). Mixed-data classificatory programs. II. Divisive systems. *Aust. Comput. J.*, **1**, 82–85.

Large, R. G., and Mullins, P. R. (1981). Illness behaviour profiles in chronic pain: the Auckland experience. *Pain*, **10**, 231–239.

Lauder, I. J. (1978). Computational problems in predictive diagnosis. *In* L. C. A. Corsten and J. Hermans (Eds.), *Compstat 1978*, pp. 186–192. Physica-Verlag: Vienna.

Lauder, I. J. (1981). Latent variable models for statistical diagnosis. *Biometrika*, **68**, 365–372.

Lawless, J. F., and Singhal, K. (1978). Efficient screening of nonormal regression models. *Biometrics*, **34**, 318–327.

Lawley, D. N. (1938). A generalisation of Fisher's Z-test. *Biometrika*, **30**, 180–187.

Lawley, D. N. (1956). Tests of significance of the latent roots of covariance and correlation matrices. *Biometrika*, **43**, 128–136.

Lawley, D. N. (1967). Some new results in maximum likelihood factor analysis. *Proc. R. Soc. Edinburgh A*, **67**, 256–264.

Lawley, D. N., and Maxwell, A. E. (1971). *Factor Analysis as a Statistical Method*, 2nd ed. Butterworths: London.

Lawson, C. L., and Hanson, R. J. (1974). *Linear Least-Squares Solutions*. Prentice-Hall, Englewood Cliffs, N.J.

Layard, M. W. J. (1972). Large sample tests for the equality of two covariance matrices. *Ann Math. Stat.*, **43**, 123–141.

Layard, M. W. J. (1973). Robust large-sample tests for homogeneity of variances. *J. Am. Stat. Assoc.*, **68**, 195–198.

Layard, M. W. J. (1974). A Monte Carlo comparison tests for equality of covariance matrices. *Biometrika*, **16**, 461–465.

Lee, J. C., and Geisser, S. (1972). Growth curve prediction. *Sankhya A*, **34**, 393–412.

Lee, J. C., and Geisser, S. (1975). Applications of growth curve prediction. *Sankhya A*, **37**, 239–256.

Lee, J. C., Krishnaiah, P. R., and Chang, T. C. (1976). On the distribution of the likelihood ratio test statistic for compound symmetry. *S. Afr. Stat. J.*, **10**, 49–62.

Lee, J. C., Chang, T. C., and Krishnaiah, P. R. (1977). Approximations to the distributions of the likelihood ratio statistics for testing certain structures on the covariance matrices of real multivariate normal populations. *In* P. R. Krishnaiah (Ed.), *Multivariate Analysis*, Vol. IV, pp. 105–118. North-Holland: Amsterdam.

Lee, K. L. (1979). Multivariate tests for clusters. *J. Am. Stat. Assoc.*, **74**, 708–714.

Lee, S. Y. (1979). Constrained estimation in covariance structure analysis. *Biometrika*, **66**, 539–545.

Lee, S. Y. (1981). The multiplier method in constrained estimation of covariance structure models. *J. Stat. Comput. Simul.*, **12**, 247–257.

Lee, S. Y., and Bentler, P. M. (1980). Some asymptotic properties of constrained generalized least squares estimation in covariance structure models. *S. Afr. Stat. J.*, **14**, 121–136.

Lee, Y.-H. K. (1974). A note on Rao's reduction of Potthoff and Roy's generalized linear model. *Biometrika*, **61**, 349–351.

Lee, Y.-S. (1971). Asymptotic formulae for the distribution of a multivariate test statistic: power comparisons of certain multivariate tests. *Biometrika*, **58**, 647–651.

Lee, Y.-S. (1972). Some results on the distribution of Wilks's likelihood-ratio criterion. *Biometrika*, **95**, 649–663.

Leech, F. B., and Healy, M. J. R. (1959). The analysis of experiments on growth rate. *Biometrics*, **15**, 98–106.

Levine, D. M. (1978). A Monte Carlo study of Kruskal's variance based measure on stress. *Psychometrika*, **43**, 307–315.

Lewis, T. (1970). The Statistician as a Member of Society. Inaugural lecture, University of Hull: Hull, England.

Lilliefors, H. W. (1967). On the Kolmogorov–Smirnov test for normality with mean and variance unknown. *J. Am. Stat. Assoc.*, **62**, 399–402.

Lin, C.-C., and Mudholkar, G. S. (1980). A simple test for normality against asymmetric alternatives. *Biometrika*, **67**, 455–461.

Ling, R. F. (1972). On the theory and construction of *k*-clusters. *Comput. J.*, **15**, 326–333.

Ling, R. F. (1973a). A probability theory of cluster analysis. *J. Am. Stat. Assoc.*, **68**, 159–164.

Ling, R. F. (1973b). A computer generated aid for cluster analysis. *Commun. Assoc. Comput. Mach.*, **16**, 355–361.

Ling, R. F., and Killough, G. G. (1976). Probability tables for cluster analysis based on a theory of random graphs. *J. Am. Stat. Assoc.*, **71**, 293–300.

Little, R. J. A. (1976). Inference about means from incomplete multivariate data. *Biometrika*, **63**, 593–604.

Little, R. J. A. (1978). Consistent regression methods for discriminant analysis with incomplete data. *J. Am. Stat. Assoc.*, **73**, 319–322.

Lubischew, A. A. (1962). On the use of discriminant functions in taxonomy. *Biometrics*, **18**, 455–477

MacDonald, P. D. M. (1975). Estimation of finite distribution mixtures. *In* R. P. Gupta (Ed.), *Applied Statistics*, pp. 231–245. North-Holland; Amsterdam.

MacNaughton-Smith, P., Williams, W. T., Dale, N. B., and Mockett, L. G. (1964). Dissimilarity analysis: a new technique of hierarchical subdivision. *Nature* (*London*), **202**, 1034–1035.

MacQueen, J. B. (1967). Some methods for classification and analysis of multivariate observations. *Proc. Fifth Berkely Symp. Math. Stat. Prob.*, **1**, 281–297.

McCabe, G. P., Jr. (1975). Computations for variable selection in discriminant analysis. *Technometrics*, **17**, 103–109..

McDonald, L. (1975). Tests for the general linear hypothesis under the multiple design multivariate linear model. *Ann. Stat.*, **3**, 461–466.

McDonald, L. L., Lowe, V. W., Smidt, R. K., and Meister, K. A. (1976). A preliminary test for discriminant analysis based on small samples. *Biometrics*, **32**, 417–422.

McDonald, R. P., and Burr, E. J. (1967). A comparison of four methods of constructing factor scores. *Psychometrika*, **32**, 381–401.

McHenry, C. E. (1978). Computation of a best subset in multivariate analysis. *Appl. Stat.*, **27**, 291–296.

McKay, R. J. (1976). Simultaneous procedures in discriminant analysis involving two groups. *Technometrics*, **18**, 47–53.

McKay, R. J. (1977). Simultaneous procedures for variable selection in multiple discriminant analysis. *Biometrika*, **64**, 283–290.

McKay, R. J. (1978). A graphical aid to selection of variables in two-group discriminant analysis. *Appl. Stat.*, **27**, 259–263.

McKay, R. J. (1979). The adequacy of variable subsets in multivariate regression. *Technometrics*, **21**, 475–479.

McKay, R. J., and Campbell, N. A. (1982a). Variable selection techniques in discriminant analysis. I. Description. *Br. J. Math. Stat. Psychol.*, **35**, 1–29.

McKay, R. J., and Campbell, N. A. (1982b). Variable selection techniques in discriminant analysis. II. Allocation. *Br. J. Math. Stat. Psychol.*, **35**, 30–41.

McKeon, J. J. (1974). F approximations to the distribution of Hotelling's T_0^2. *Biometrika*, **61**, 381–383.

McLachlan, G. J. (1972). Asymptotic results for discriminant analysis when the initial samples are misclassified. *Technometrics*, **14**, 414–422.

McLachlan, G. J. (1975). An interactive reclassification procedure for constructing an asymptotically optimal rule of allocation in discriminant analysis. *J. Am. Stat. Assoc.*, **70**, 365–369.

McLachlan, G. J. (1976). The bias of the apparent error rate in discriminant analysis. *Biometrika*, **63**, 239–244.

McLachlan, G. J. (1977a). Estimating the linear discriminant function from initial samples containing a small number of unclassified observations. *J. Am. Stat. Assoc.*, **72**, 403–406.

McLachlan, G. J. (1977b). Constrained sample discrimination with the studentized classification statistic W. *Commun. Stat. Theor. Methods A*, **6**, 575–583.

McLachlan, G. J. (1979). A comparison of the estimative and predictive methods of estimating posterior probabilities. *Commun. Stat. Theor. Methods A*, **8**, 919–929.

McLachlan, G. J. (1980a). The efficiency of Efron's "bootstrap" approach applied to error rate estimation in discriminant analysis. *J. Stat. Comput. Simul.*, **11**, 273–279.

McLachlan, G. J. (1980b). On the relationship between the F-test and the overall error rate for variable selection in two-group discriminant analysis. *Biometrics*, **36**, 501–510.

McQuitty, L. L. (1964). Capabilities and improvements of linkage analysis as a clustering method. *Educ. Psychol. Meas.*, **24**, 441–456.

McRae, D. J. (1971). MIKCA: A FORTRAN IV iterative k-means cluster analysis program. *Behav. Sci.*, **16**, 423–424.

Mabbett, A., Stone, M., and Washbrook, J. (1980). Cross-validatory selection of binary variables in differential diagnosis. *Appl. Stat.*, **29**, 198–204.

Machin, D. (1975). On a design problem in growth studies. *Biometrics*, **31**, 749–753.

Madsen, K. S. (1977). A growth curve model for studies in morphometrics. *Biometrics*, **33**, 659–669.

Mage, D. T. (1982). An objective graphical method for testing normal distributional assumptions using probability plots. *Am. Stat.*, **36**, 116–120.

Maindonald, J. H. (1977). Least squares Computations based on the Cholesky decomposition of the correlation matrix. *J. Stat. Comput. Simul.*, **5**, 247–258.

Maindonald, J. H. (1984). *Statistical Computation*. Wiley: New York.

Malkovich, J. F., and Afifi, A. A. (1973). On tests for multivariate normality. *J. Am. Stat. Assoc.*, **68**, 176–179.

Mallows, C. L. (1973). Some comments on C_p. *Technometrics*, **15**, 661–675.

Mandel, J. (1971). A new analysis of variance model for non-additive data. *Technometrics*, **13**, 1–18.

Mandel, J. (1972). Principal components, analysis of variance and data structure. *Stat. Neerl.* **26**, 119–129.

Mandel, J. (1982). Use of the singular value decomposition in regression analysis. *Am. Stat.*, **36**, 15–24.

Mansfield, E. R., Webster, J. T., and Gunst, R. F. (1977). An analytic variable selection technique for principal component regression. *Appl. Stat.*, **26**, 34–40.

Marasinghe, M. G., and Johnson, D. E. (1981). Testing subhypotheses in the multiplicative interaction model. *Technometrics*, **23**, 385–393.

Mardia, K. V. (1970). Measures of multivariate skewness and kurtosis with applications. *Biometrika*, **57**, 519–520.

Mardia, K, V. (1971). The effect of nonnormality on some multivariate tests and robustness to nonormality in the linear model. *Biometrika*, **58**, 105–121.

Mardia, K. V. (1974). Applications of some measures of multivariate skewness and kurtosis for testing normality and robustness studies. *Sankhya B*, **36**, 115–128.

Mardia, K. V. (1975). Assessment of multinormality and the robustness of Hotelling's T^2 test.*Appl. Stat.* **24**, 163–171.

Mardia, K. V. (1977). Mahalanobis distances and angles. *In* P. R. Krishnaiah (Ed.), *Multivariate Analysis*, Vol. IV., pp. 495–511. North-Holland: Amsterdam.

Mardia, K. V. (1978). Some properties of classical multi-dimensional scaling. *Commun. Stat. Theor. Methods A*, **7**, 1233–1241.

Mardia, K. V. (1980). Tests of univariate and multivariate normality. *In* P. R. Krishnaiah (Ed.), *Handbook in Statistics*, Vol. 1, pp. 279–320. North-Holland: Amsterdam.

Mardia, K. V., and Zemroch, P. J. (1975a). Algorithm AS 84: Measures of multivariate skewness and kurtosis. *Appl. Stat.*, **24**, 262–265.

Mardia, K. V., and Zemroch, P. J. (1975b). *Tables of the F-distribution*. Academic Press: London.

Mardia, K. V., Kent, J. T., and Bibby, J. M. (1979). *Multivariate Analysis*. Academic Press: London.

Marks, S., and Dunn, O. J. (1974). Discriminant functions when covariance matrices are unequal. *J. Am. Stat. Assoc.*, **69**, 555–559.

Maronna, R. A. (1976). Robust M-estimators of multivariate location and scatter. *Ann. Stat.*, **4**, 51–67.

Maronna, R. A., and Jacovkis, P. M. (1974). Multivariate clustering procedures with variable metrics. *Biometrics*, **30**, 499–505.

Marriott, F. H. C. (1971). Practical problems in a method of cluster analysis. *Biometrics*, **27**, 501–514.

Marriott, F. H. C. (1975). Separating mixtures of normal distributions. *Biometrics*, **31**, 767–769.

Marriott, F. H. C. (1982). Optimization methods of cluster analysis. *Biometrika*, **69**, 417–421.

Martin, D. C. and Bradley, R. A. (1972). Probability models, estimation, and classification for multivariate dichotomous populations. *Biometrics*, **28**, 203–222.

Martin, R. D. (1979). Robust estimation for time series autoregressions. *In* R. L. Launer and G. N. Wilkinson (Eds.), *Robustness in Statistics*, pp. 147–176. Academic Press: New York.

Martin, R. S., and Wilkinson, J. H. (1971). Reduction of the symmetric eigenproblem $Ax = \lambda Bx$ and related problems to standard form. *In* J. H. Wilkinson and C. H. Reinsch (Eds.), *Handbook for Automatic Computation*, Vol. II, *Linear Algebra*, pp. 303–314. Springer-Verlag: Berlin.

Mathai, A. M. (1971). On the distribution of the likelihood-ratio criterion for testing linear hypotheses on regression coefficients. *Ann. Inst. Stat. Math.*, **23**, 181–197.

Mathai, A. M. (1980). Moments of the trace of a noncentral Wishart matrix. *Commun. Stat. Theor. Methods A*, **9**, 795–801.

Mathai, A. M., and Katiyar, R. S. (1979a). Exact percentage points for testing independence. *Biometrika*, **66**, 353–356.

Mathai, A. M., and Katiyar, R. S. (1979b). The distribution and the exact percentage points for Wilks' L_{mvc} criterior. *Ann. Inst. Stat. Math.*, **31**, 215–224.

Matula, D. W. (1977). Graph theoretic techniques for cluster analysis algorithms. *In* J. Van Ryzin (Ed.), *Classification and Clustering*, pp. 95–129. Academic Press: New York.

Matusita, K. (1964). Distance and decision rules. *Ann. Inst. Stat. Math.*, **16**, 305–315.

Memon, A. Z., and Okamoto, M. (1971). Asymptotic expansion of the distribution of the Z statistic in discriminant analysis. *J. Multivar. Anal.*, **1**, 249–307.

Mezzich, J. E. (1975). *An Evaluation of Quantitative Taxonomic Methods*. Ph.D. thesis, Ohio State University, Ames.

Mijares, T. A. (1964). *Percentage Points of the Sum V_1^s of s Roots* ($s = 1 - 50$). Statistical Center, University of the Phillipines, Manila.

Miller, R. G., Jr. (1977). Developments in multiple comparisons 1966–1976. *J. Am. Stat. Assoc.*, **72**, 779–788.

Miller, R. G., Jr. (1981). *Simultaneous Statistical Inference*, 2nd ed. McGraw-Hill: New York.

Minkoff, E. C. (1965). The effects on classification of slight alterations in numerical technique. *Syst. Zool.*, **14**, 196–213.

Mitra, S. K. (1969). Some characteristic and non-characteristic properties of the Wishart distributions. *Sankhya A*, **31**, 19–22.

Moberg, T. F., Ramberg, J. S. and Randles, R. H. (1978). An adaptive M-estimator and its application to a selection problem. *Technometrics*, **20**, 255–263.

Mojena, R. (1977). Hierarchical grouping methods and stopping rules: an evaluation. *Comput. J.*, **20**, 359–363.

Mood, A. M. (1941). On the joint distribution of the median in samples from a multivariate population. *Ann. Math. Stat.*, **12**, 268–278.

Mood, A. M. (1951). On the distribution of the characteristic roots of normal second-moment matrices. *Ann. Math. Stat.*, **22**, 266–273.

Moore, D. H., II (1973). Evaluation of five discrimination procedures for binary variables. *J. Am. Stat. Assoc.*, **68**, 399–404.

Moran, M. A. (1975). On the expectation of errors of allocation associated with a linear discriminant function. *Biometrika*, **62**, 141–148.

Moran, M. A., and Murphy, B. J. (1979). A closer look at two alternative methods of statistical discrimination. *Appl. Stat.*, **28**, 223–232.

Morgan, B. J. T. (1981). Three applications of methods of cluster-analysis. *Statistician*, **30**, 205–223.

Morrison, D. F. (1972). The analysis of a single sample of repeated measurements. *Biometrics*, **28**, 55–71.

Morrison, D. F. (1976). *Multivariate Statistical Methods*, 2nd ed. McGraw-Hill: New York.

Mosier, C. I. (1939). Determining a simple structure when loadings for certain tests are known. *Psychometrika*, **4**, 149–162.

Mosimann, J. E. (1970). Size allometry; size and shape variables with characterizations of the lognormal and gamma distributions. *J. Am. Stat. Assoc.*, **65**, 930–945.

Mosimann, J. E., Malley, J. D., Cheever, A. W., and Clark, C. B. (1978). Size and shape analysis of schistosome egg-counts in Egyptian autopsy data. *Biometrics*, **34**, 341–356.

Mosteller, F., and Tukey, J. W. (1977). *Data Analysis and Regression*. Addison-Wesley: Reading, Mass.

Mudholkar, G. S., and Trivedi, M. C. (1980). A normal approximation for the distribution of the likelihood ratio statistic in multivariate analysis of variance. *Biometrika*, **67**, 485–488.

Mudholkar, G. S., Davidson, M. L., and Subbaiah, P. (1974a). A note on the union–intersection character of some MANOVA procedures. *J. Multivar. Anal.*, **4**, 486–493.

Mudholkar, G. S., Davidson, M. L., and Subbaiah, P. (1974b). Extended linear hypotheses and simultaneous tests in multivariate analysis of variance. *Biometrika*, **61**, 467–477.

Mudholkar, G. S., Trivedi, M. C., and Lin, C. T. (1982). An approximation to the distribution of the likelihood ratio statistic for testing complete independence. *Technometrics*, **24**, 139–143.

Muirhead, R. J. (1978). Latent roots and matrix variates: a review of some asymptotic results. *Ann. Stat.*, **6**, 5–33.

Muirhead, R. J. (1982). *Aspects of Multivariate Statistical Theory*. Wiley: New York.

Muirhead, R. J., and Waternaux, C. M. (1980). Asymptotic distributions in canonical correlation analysis and other multivariate procedures for nonnormal populations. *Biometrika*, **67**, 31–43.

Mulholland, H. P. (1977). On the null distribution of $\sqrt{b_1}$ for samples of size at most 25, with tables. *Biometrika*, **64**, 401–409.

Murray, G. D. (1977b). A cautionary note on selection of variables in discriminant analysis. *Appl. Stat.*, **26**, 246–250.

Murray, G. D., and Titterington, D. M. (1978). Estimation problems with data from a mixture. *Appl. Stat.*, **27**, 325–334.

Murthy, V. K. (1966). Nonparametric estimation of multivariate densities with applications. *In* P. R. Krishnaiah (Ed.), *Multivariate Analysis*, pp. 43–56. Academic Press: New York.

Nagao, H. (1970). Asymptotic expansions of some test criteria for homogeneity of variances and covariance matrices. *J. Sci. Hiroshima Univ. Ser. A-I*, **34**, 153–247.

Nagao, H. (1973a). On some test criteria for covariance matrix. *Ann. Stat.*, **1**, 700–709.

Nagao, H. (1973b). Nonnull distributions of two test criteria for independence under local alternatives. *J. Multivar. Anal.*, **3**, 435–444.

Nagao, H. (1974). Asymptotic nonnull distributions of certain test criteria for a covariance matrix. *J. Multivar. Anal.*, **4**, 409–418.

Nagarsenker, B. N. (1975). Percentage points of Wilks' L_{vc} criterion. *Commun. Stat.*, **4**, 629–641.

Nagarsenker, B. N. (1978). Nonnull distributions of some statistics associated with testing for the equality of two covariance matrices. *J. Multivar. Anal.*, **8**, 396–404.

Nagarsenker, B. N., and Pillai, K. C. S. (1973a). The distribution of the sphericity test criterion. *J. Multivar. Anal.*, **3**, 226–235.

Nagarsenker, B. N., and Pillai, K. C. S. (1973b). Distribution of the likelihood ratio criterion for testing a hypothesis specifying a covariance matrix. *Biometrika*, **60**, 359–364.

Nagarsenker, B. N., and Pillai, K. C. S. (1974). Distribution of the likelihood ratio criterion for testing $\Sigma = \Sigma_0$ and $\mu = \mu_0$. *J. Multivar. Anal.*, **4**, 114–122.

Nagy, G. (1969). Feature extraction on binary patterns. *IEEE Trans. Syst. Sci.*, **SSC5**, 273–278.

Naik, U. D. (1975). On testing equality of means of correlated variables with incomplete data. *Biometrika*, **62**, 615–622.

Nanda, D. N. (1950). Distribution of the sum of roots of a determinantal equation under a condition. *Ann. Math. Stat.*, **21**, 432–439.

Nash, J. C., and Lefkovitch, L. P. (1976). Principal components and regression by singular value decomposition on a small computer. *Appl. Stat.*, **25**, 210–216.

Needham, R. M. (1967). Automatic classification in linguistics. *Statistician*, **17**, 45–54.

Newton, R. G., and Spurrell, D. J. (1967a). A development of multiple regression for the analysis of routine data. *Appl. Stat.*, **16**, 51–64.

Newton, R. G., and Spurrell, D. J. (1967b). Examples of the use of elements for clarifying regression analysis. *Appl. Stat.*, **16**, 165–172.

Neyman, J. (1937). "Smooth" test for goodness-of-fit. *Skand. Aktuarie Tidskr.*, **20**, 149–199.

Noble, B. (1976). Methods for computing the Moore–Penrose generalised inverse, and related matters. *In* M. Z. Nashed (Ed) *Generalized Inverses and Applications*, pp. 245–301. Academic Press: New York.

North, P. M. (1977). A novel method for estimating numbers of bird territories. *Appl. Stat.*, **26**, 149–155.

Obenchain, R. L. (1972). Regression optimality of principal components. *Am. Math. Stat.*, **43**, 1317–1319.

Odell, P. L., and Basu, J. P. (1976). Concerning several methods for estimating crop acreages using remote sensing data. *Comm. Stat. Theor. Methods A*, **5**, 1091–1114.

O'Hara, T. F., Hosmer, D. W., Jr., Lemeshow, S., and Hartz, S. C. (1982). A comparison of discriminant function and maximum likelihood estimates of logistic coefficients for categorical-scaled data. *J. Stat. Comput. Simul.* **14**, 169–178.

Okamoto, M. (1963). An asymptotic expansion for the distribution of the linear discriminant function. *Ann. Math. Stat.* **34**, 1286–1301; correction, **39** (1968), 1358–1359.

Okamoto, M. (1969). Optimality of principal components. *In* P. R. Krishnaiah (Ed.), *Multivariate Analysis*, Vol. II, pp. 673–686. Academic Press: New York.

Okamoto, M. (1973). Distinctness of the eigenvalues of a quadratic form in a multivariate sample. *Ann. Stat.*, **1**, 763–765.

Okamoto, M. (1976). Random model and fixed model of principal component analysis. *In* S. Ikeda, et al. (Eds.), *Essays in Probability and Statistics* (dedicated to J. Ogawa), pp. 339–351. Shinko Tsusho: Tokyo.

Okamoto, M., and Kanazawa, M. (1968). Minimization of eigenvalues of a matrix and optimality of principal components. *Ann. Math. Stat.*, **39**, 859–863.

Olkin, I. (1973). Testing and estimation for structures which are circularly symmetric in blocks. *In* D. G. Kabe and R. P. Gupta (Eds.), *Multivariate Statistical Inference*, pp. 183–195. North-Holland: Amsterdam.

Olkin, I., and Press, S. J. (1969). Testing and estimation for a circular stationary model. *Ann. Math. Stat.*, **40**, 1358–1373.

Olkin, I., and Rubin, H. (1964). Multivariate beta distributions and independence properties of the Wishart distribution. *Ann. Math. Stat.*, **35**, 261–269.

Olkin, I., and Siotani, M. (1976). Asymptotic distribution of functions of a correlation matrix. *In* S. Ikeda, et al. (Eds.), *Essays in Probability and Statistics* (dedicated to J. Ogawa), pp. 235–251. Shinko Tsusho: Tokyo.

Olkin, I., and Tomsky, J. L. (1981). A new class of multivariate tests based on the union–intersection principle. *Ann. Stat.*, **9**, 792–802.

Olshen, R. A. (1973). The conditional level of the F-test. *J. Am. Stat. Assoc.*, **68**, 692–698.

Olson, C. L. (1974). Comparative robustness of six tests in multivariate analysis of variance. *J. Am. Stat. Assoc.*, **69**, 894–908.

O'Neill, T. J. (1978). Normal discrimination with unclassified observations. *J. Am. Stat. Assoc.*, **73**, 821–826.

O'Neill, T. J. (1980). The general distribution of the error rate of a classification procedure with application to logistic regression discrimination. *J. Am. Stat. Assoc.*, **75**, 154–160.

Orlóci, L. (1967). Data centering: a review and evaluation with reference to component analysis. *Syst. Zool.* **16**, 208–212.

Orlóci, L. (1978). *Multivariate Analysis in Vegetation Research*, 2nd ed. Junk: The Hague.

Ott, J., and Kronmal, R. A. (1976). Some classification procedures for binary data using orthogonal functions. *J. Am. Stat. Assoc.*, **71**, 391–399.

Parker-Rhodes, A. F., and Jackson, D. M. (1969). Automatic classification in the ecology of the higher fungi. *In* A. J. Cole (Ed.), *Numerical Taxonomy*, pp. 181–215. Academic Press: New York.

Patefield, W. M. (1981). Multivariate linear relationships: maximum likelihood estimation and regression bounds. *J. R. Stat. Soc. B*, **43**, 342–352.

Patnaik, P. B. (1949). The non-central χ^2 and F-distributions and their approximations. *Biometrika*, **36**, 202–232.

Payne, R., and Preece, D. A. (1980). Identification keys and diagnostic tables; a review (with discussion). *J. R. Stat. Soc. A*, **143**, 253–292.

Pearson, E. S. (1969). Some comments on the accuracy of Box's approximations to the distribution of *M*. *Biometrika*, **56**, 219–220.

Pearson, E. S., and Hartley, H. O. (1972). *Biometrika Tables for Statisticians*, Vol. 2. Cambridge University Press: Cambridge

Pearson, E. S., D'Agostino, R. B., and Bowman, K. O. (1977). Tests for departure from normality: comparison of powers. *Biometrika*, **64**, 321–246.

Pearson, K. (1901). On lines and planes of closest fit to systems of points in space. *Philos. Mag.*, **2**, 559–572.

Perlman, M. D. (1980). Unbiasedness of the likelihood ratio tests for equality of several covariance matrices and equality of several multivariate normal populations. *Ann. Stat.*, **8**, 247–263.

Pettitt, A. N. (1977). Testing the normality of several independent samples using the Anderson–Darling statistic. *Appl. Stat.*, **26**, 156–161.

Pickett, R., and White, B. W. (1966). Constructing data pictures. *Proc. 7th Nat. Symp. Inform. Display*, 75–81.

Pielou, E. C. (1966). Species-diversity and pattern-diversity in the study of ecological succession. *J. Theor. Biol.*, **10**, 370–383.

Pielou, E. C. (1969). *An Introduction to Mathematical Ecology*. Wiley: New York.

Pillai, K. C. S. (1955). Some new test criterial in multivariate analysis. *Ann. Math. Stat.*, **26**, 117–121.

Pillai, K. C. S. (1960). *Statistical Tables for Tests of Multivariate Hypotheses*. Statistical Center, University of the Philipines, Manila.

Pillai, K. C. S. (1964). On the distribution of the largest of seven roots of a matrix in multivariate analysis. *Biometrika*, **51**, 270–275.

Pillai, K. C. S. (1965). On the distribution of the largest characteristic root of a matrix in multivariate analysis. *Biometrika*, **52**, 405–414.

Pillai, K. C. S. (1967). Upper percentage points of the largest root of a matrix in multivariate analysis. *Biometrika*, **54**, 189–194.

Pillai, K. C. S. (1970). On the noncentral distributions of the largest root of two matrices in multivariate analysis. *In* R. C. Bose et al. (Eds.), *Essays in Probability and Statistics*, pp. 557–586. University of North Carolina Press: Chapel Hill, N. C.

Pillai, K. C. S. (1976). Distributions of characteristic roots in multivariate analysis. Part I: Null distributions. *Can. J. Stat. A & B*, **4**, 157–184.

Pillai, K. C. S. (1977). Distributions of characteristic roots in multivariate analysis. Part II: Non-null distributions. *Can. J. Stat. A & B*, **5**, 1–62.

Pillai, K. C. S., and Gupta, A. K. (1969). On the exact distribution of Wilks's criterion. *Biometrika*, **56**, 109–118.

Pillai, K. C. S., and Hsu, Y. S. (1979). Exact robustness studies of the test of independence based on four multivariate criteria and their distribution problems under violations. *Ann. Inst. Stat. Math.*, **31**, 85–101.

Pillai, K. C. S., and Jayachandran, K. (1967). Power comparisons of tests of two multivariate hypotheses based on four criteria. *Biometrika*, **54**, 195–210.

Pillai, K. C. S., and Jayachandran, K. (1968). Power comparisons of tests of equality of two covariance matrices based on four criteria. *Biometrika*, **55**, 335–342.

Pillai, K. C. S., and Jayachandran, K. (1970). On the exact distribution of Pillai's V^s criterion. *J. Am. Stat. Assoc.*, **65**, 447–454.

Pillai, K. C. S., and Nagarsenker, B. N. (1972). On the distributions of a class of statistics in multivariate analysis. *J. Multivar. Anal.*, **2**, 96–114.

Pillai, K. C. S., and Samson, P. (1959). On Hotelling's generalization of T^2. *Biometrika*, **46**, 160–168.

Pillai, K. C. S., and Sudjana (1975). Exact robustness studies of tests of two multivariate hypotheses based on four criteria and their distribution problems under violations. *Ann. Stat.*, **3**, 617–636.

Pillai, K. C. S., and Young, D. C. (1971). On the exact distribution of Hotelling's generalized T_0^2. *J. Multivar. Analysis*, **1**, 90–107.

Pilowsky, I., and Spence, N. D. (1975). Patterns of illness behaviour in patients with intractable pain. *J. Psychosom. Res.*, **19**, 279–287.

Pilowsky, I., and Spence, N. D. (1976). Illness behaviour syndromes associated with intractable pain. *Pain*, **2**, 61–71.

Pollard, D. (1981). Strong consistency of k-means clustering. *Ann. Stat.*, **9**, 135–140.

Potthoff, R. F., and Roy, S. N. (1964). A generalized multivariate analysis of variance model useful especially for growth curve problems. *Biometrika*, **51**, 313–326.

Prentice, R. L., and Pyke, R. (1979). Logistic disease incidence models and case-control studies. *Biometrika*, **66**, 403–411.

Prescott, P. (1975). A simple alternative to Student's *t*. *Appl. Stat.*, **24**, 210–217.

Prescott, P. (1978). Selection of trimming proportions for robust adaptive trimmed means. *J. Am. Stat. Assoc.* **73**, 133–140.

Prescott, P. (1979). A mean difference estimator of standard deviation in asymmetrically censored normal samples. *Biometrika*, **66**, 684–686.

Press, S. J. (1972a). *Applied Multivariate Analysis*. Holt, Rinehart, and Winston: New York.

Press, S. J. (1972b). Multivariate stable distributions. *J. Multivar. Anal.*, **2**, 444–462.

Press, S. J. (1980). Bayesian inference in MANOVA. *in* P. R. Krishnaiah (Ed.), *Handbook of Statistics*, Vol. 1, pp. 117–132. North-Holland: Amsterdam.

Press, S. J. and Wilson, S. (1978). Choosing between logistic regression and discriminant analysis. *J. Am. Stat. Assoc.*, **73**, 699–705.

Priestly, M. B., Rao, T. S., and Tong, H. (1973). Identification of the structure of multivariable systems. *In* P. R. Krishnaiah (Ed.), *Multivariate Analysis*, Vol. III, pp. 351–368. Academic Press: New York.

Quesenberry, C. P., Whitaker, T. B., and Dickens, J. W. (1976). On testing normality using several samples: an analysis of peanut aflatoxin data. *Biometrics*, **32**, 753–759.

Radhakrishnan, R., and Kshirsagar, A. M. (1981). Influence functions for certain parameters in multivariate analysis. *Commum. Stat. Theor. Methods A*, **10**, 515–529.

Ramsay, J. O. (1978). Confidence regions for multidimensional scaling. *Psychometrika*, **43**, 145–160.

Rand, W. M. (1971). Objective criteria for evaluating clustering methods. *J. Am. Stat. Assoc.*, **66**, 846–850.

Randles, R. H., Broffitt, J. D., Ramberg, J. S., and Hogg, R. V. (1978a). Discriminant analysis based on ranks. *J. Am. Stat. Assoc.*, **73**, 379–384.

Randles, R. H., Broffitt, J. D., Ramberg, J. S., and Hogg, R. V. (1978b). Generalized linear and quadratic discriminant functions using robust estimates. *J. Am. Stat. Assoc.*, **73**, 564–568.

Rao, C. R. (1948). Tests of significance in multivariate analysis. *Biometrika*, **35**, 58–79.

Rao, C. R. (1951). An asymptotic expansion of the distribution of Wilks' criterion. *Bull. Inst. Int. Stat.*, **33**, 177–180.

Rao, C. R. (1952). *Advanced Statistical Methods in Biometric Research*. Wiley: New York.

Rao, C. R. (1959). Some problems involving linear hypotheses in multivariate analysis. *Biometrika*, 49–58.

Rao, C. R. (1964a). The use and interpretation of principal components in applied research. *Sankhya A*, **26**, 329–358.

Rao, C. R. (1965). The theory of least squares when the parameters are stochastic and its application to the analysis of growth curves. *Biometrika*, **52**, 447–458.

Rao, C. R. (1966). Covariance adjustment and related problems in multivariate analysis. *In* P. R. Krishnaiah (Ed.), *Multivariate Analysis*, Vol. II, pp. 87–103. Academic Press: New York.

Rao, C. R. (1967). Least squares theory using an estimated dispersion matrix and its application to measurement of signals. *Proc. Fifth Berkeley Symp. Math. Stat. Prob.*, **1**, 355–372.

Rao, C. R. (1970). Inference on discriminant function coefficients. *In* R. C. Bose et al. (Eds.), *Essays in Probability and Statistics*, pp. 587–602. University of North Carolina Press: Chapel Hill, N.C.

Rao, C. R. (1973). *Linear Statistical Inference and Its Applications*, 2nd ed. Wiley: New York.

Rao, C. R. (1974). Projectors, generalized inverses and the BLUE's. *J. R. Stat. Soc. B*, **36**, 442–448.

Rao, C. R. (1977). Cluster analysis applied to a study of race mixture in human populations. *In* J. Van Ryzin (Ed.), *Classification and Clustering*, pp. 175–197. Academic Press: New York.

Rao, C. R. (1980). Matrix approximations and reduction of dimensionality in multivariate statistical analysis. *In* P. R. Krishnaiah (Ed.), *Multivariate Analysis*, Vol. V, pp. 3–22. North-Holland: Amsterdam.

Rao, C. R., and Shinozaki, N. (1978). Precision of individual estimators in simultaneous estimation of parameters. *Biometrika*, **65**, 23–30.

Reeve, E. C. R. (1941). A statistical analysis of taxonomic differences within the genus *Tamandua* Gray (Xenarthra). *Proc. Zool. Soc. London A*, **111**, 279–302.

Reinsel, G. (1982). Multivariate repeated-measurement or growth curve models with multivariate random-effects covariance structure. *J. Am. Stat. Assoc.*, **77**, 190–195.

Relles, D. A., and Rogers, W. H. (1977). Statisticians are fairly robust estimators of location. *J. Am. Stat. Assoc.*, **72**, 107–111.

Remme, J., Habbema, J. D. F., and Hermans, J. (1980). A simulative comparison of linear quadratic and kernel discrimination. *J. Stat. Comput. Simul.*, **11**, 87–106.

Rencher, A. C., and Larson, S. F. (1980). Bias in Wilks' Λ in stepwise discriminant analysis. *Technometrics*, **22**, 349–356.

Reyment, R. A. (1961). A note on the geographical variation in European *Rana*. *Growth*, **25**, 219–227.

Richards, F. J. (1969). The quantitative analysis of plant growth. *In* F. C. Steward (Ed.), *Plant Physiology*, Vol. VA. Academic Press: New York.

Richards, L. E. (1972). Refinement and extension of distribution free discriminant analysis. *Appl. Stat.*, **21**, 174–176.

Richardson, M. W. (1938). Multidimensional psychophysics. *Psychol. Bull.*, **35**, 659–660.

Rincón-Gallardo, S., and Quesenberry, C. P. (1982). Testing multivariate normality using several samples: applications techniques. *Commun. Stat. Theor. Methods A*, **11**, 343–358.

Rincón-Gallardo, S., Quesenberry, C. P., and O'Reilly, F. J. (1979). Conditional probability integral transformations and goodness-of-fit tests for multivariate normal distributions. *Ann. Stat.*, **7**, 1052–1057.

Robert, P., and Escoufier, Y. (1976). A unifying tool for linear multivariate statistical methods: the *RV*-coefficient. *Appl. Stat.*, **25**, 257–265.

Robinson, P. M. (1977). The estimation of a multivariate linear relation. *J. Multivar. Anal.*, **7**, 409–423.

Robinson, W. S. (1951). A method for chronologically ordering archeological deposits. *Am. Antiq.*, **16**, 293–301.

Rocke, D. M., and Downs, G. W. (1981). Estimating the variances of robust estimators of location: influence curve, jackknife and bootstrap. *Commun. Stat. Simul. Comput. B*, **10**, 221–248.

Rogers, G. S., and Young, D. L. (1978). On testing a multivariate linear hypothesis when the covariance matrix and its inverse have the same pattern. *J. Am. Stat. Assoc.*, **73**, 203–207.

Rohlf, F. J. (1974a). Methods of comparing classifications. *Annu. Rev. Ecol. Syst.*, **5**, 101–113.

Rohlf, F. J. (1974b). Graphs implied by the Jardine–Sibson overlapping clustering methods, B_k. *J. Am. Stat. Assoc.*, **69**, 705–710.

Rohlf, F. J. (1975). Generalization of the gap test for the detection of multivariate outliers. *Biometrics*, **31**, 93–101.

Ronning, G. (1977). A simple scheme for generating multivariate gamma distributions with non-negative covariance matrix. *Technometrics*, **19**, 179–183.

Rothkopf, E. J. (1957). A measure of stimulus similarity and errors in some paired-associate learning tasks. *J. Exp. Psychol.*, **53**, 94–101.

Roy, J. (1951). The distribution of certain likelihood criteria useful in multivariate analysis. *Bull. Inst. Int. Stat.*, **33**, 219–230.

Roy, J. (1958). Step-down procedure in multivariate analysis. *Ann. Math. Stat.*, **29**, 1177–1187.

Roy, S. N. (1939). *p*-Statistics, or some generalizations in analysis of variance appropriate to multivariate problems. *Sankhya*, **4**, 381–396.

Roy, S. N. (1953). On a heuristic method of test construction and its use in multivariate analysis. *Ann. Math. Stat.*, **24**, 220–238.

Roy, S. N. (1957). *Some Aspects of Multivariate Analysis*. Wiley: New York.

Roy, S. N., and Srivastrava, J. N. (1964). Hierarchical and *p*-block multiresponse designs and their analysis. *In* C. R. Rao et al. (Eds.), *Contributions to Statistics*, Mahalanobis Dedicatory Volume, pp. 419–428. Pergamon Press: New York.

Roy, S. N., Gnanadesikan, R., and Srivastava, J. N. (1971). *Analysis and Design of Certain Quantitative Multiresponse Experiments*. Pergamon Press: New York.

Rubin, J. (1967). Optimal classification into groups: an approach for solving the taxonomy problem. *J. Theor. Biol.*, **15**, 103–144.

Ryan, T. A., Jr., Joiner, B. L. and Ryan, B. F. (1976). *Minitab: Student Handbook*. Duxbury Press: North Scituate, Mass.

Sager, T. W. (1979). An iterative method for estimating a multivariate mode and isopleth. *J. Am. Stat. Assoc.*, **74**, 329–339.

Saito, T. (1978). The problem of the additive constant and eigenvalues in metric multidimensional scaling. *Psychometrika*, **43**, 193–201.

Sammon, J. W. (1969). A non-linear mapping for data structure analysis. *IEEE Trans. Comput.*, **C18**, 401–409.

Sandland, R. L., and McGilchrist, C. A. (1979). Stochastic growth curve analysis. *Biometrics*, **35**, 255–271.

Sasabuchi, S. (1980). A test of a multivariate normal mean with composite hypotheses determined by linear inequalities. *Biometrika*, **67**, 429–439.

Saxena, A. K. (1978). Complex multivariate statistical analysis: an annotated bibliography. *Int. Stat. Rev.*, **46**, 209–214.

Schatzoff, M. (1966a). Exact distributions of Wilks's likelihood ratio criterion. *Biometrika*, **53**, 347–358.

Schatzoff, M. (1966b). Sensitivity comparisons among tests of the general linear hypothesis. *J. Am. Stat. Assoc.*, **61**, 415–435.

Scheffé, H. (1943). On solutions of the Behrens–Fisher problem based on the t-distribution. *Ann. Math. Stat.*, **14**, 35–44.

Scheffé, H. (1959). *The Analysis of Variance.* Wiley: New York.

Scheffé, H. (1970). Practical solutions of the Behrens–Fisher problem. *J. Am. Stat. Assoc.*, **65**, 1501–1508.

Scheffé, H. (1977). A note on the reformulation of the S-method of multiple comparison. *J. Am. Stat. Assoc.*, **72**, 143–144.

Schervish, M. J. (1981). Asymptotic expansions for correct classification rates in discriminant analysis. *Ann. Stat.*, **9**, 1002–1009.

Schuenemeyer, J. H., and Bargmann, R. E. (1978). Maximum eccentricity as a union–intersection test statistic in multivariate analysis. *J. Multivar. Anal.*, **8**, 268–273.

Schuurmann, F. J., and Waikar, V. B. (1974). Upper percentage points of the smallest root of the MANOVA matrix. *Ann. Inst. Stat. Math. Suppl.*, **8**, 79–94.

Schuurmann, F. J., Waikar, V. B., and Krishnaiah, P. R. (1973a). Percentage points of the joint distribution of the extreme roots of the random matrix $(S_1 + S_2)^{-1}$. *J. Stat. Comput. Simul.*, **2**, 17–38.

Schuurmann, F. J., Krishnaiah, P. R., and Chattopadhyay, A. K. (1973b). On the distributions of the ratios of the extreme roots to the trace of the Wishart matrix. *J. Multivar. Anal.*, **3**, 445–453.

Schuurmann, F. J., Krishnaiah, P. R., and Chattopadhyay, A. K. (1975). Exact percentage points of the distribution of the trace of a multivariate beta matrix. *J. Stat. Comput. Simul.*, **3**, 331–343.

Schwager, S. J, and Margolin, B. H. (1982). Detection of multivariate normal outliers. *Ann. Stat.*, **10**, 943–954.

Schwemer, G. T., and Mickey, M. R. (1980). A note on the linear discriminant function when group means are equal. *Commun. Stat. Simul. Computat. B*, **9**, 633–638.

Schwertman, N. C., Fridshal, D., and Magrey, J. M. (1981). On the analysis of incomplete growth curve data, a Monte Carlo study of the two nonparametric procedures. *Commun. Stat. Simul. Computat. B*, **10**, 51–66.

Sclove, S. L. (1977). Population mixture models and clustering algorithms. *Commun. Stat. Theor. Methods A.*, **6**, 417–434.

Scott, A. J., and Symons, M. J. (1971). Clustering methods based on maximum likelihood. *Biometrics*, **27**, 387–398.

Seal, H. L. (1964). *Multivariate Statistical Analysis for Biologists.* Methuen: London.

Seber, G. A. F. (1966). *The Linear Hypothesis: A General Theory*, 1st ed. Griffin: London.

Seber, G. A. F. (1977). *Linear Regression Analysis.* Wiley: New York.

Seber, G. A. F. (1980). *The Linear Hypothesis: A General Theory*, 2nd ed. Griffin: London.

Sen, A. K., and Srivastava, M. S. (1973). On multivariate tests for detecting change in mean. *Sankhya A*, **35**, 173–186.

Shannon, C. E., and Weaver, W. (1963). *The Mathematical Theory of Communication.* University of Illinois Press: Urbana, Ill.

Shapiro, S. S., and Francia, R. S. (1972). An approximate analysis of variance test for normality. *J. Am. Stat. Assoc.*, **67**, 215–216.

Shapiro, S. S., and Wilk, M. B. (1965). An analysis-of-variance test for normality (complete samples). *Biometrika*, **52**, 591–611.

Shapiro, S. S, Wilk, M. B., and Chen, H. J. (1968). A comparative study of various tests for normality. *J. Am. Stat. Assoc.*, **63**, 1343–1372.

Shenton, L. R., and Bowman, K. O. (1972). Further remarks on maximum likelihood estimates for the gamma distribution. *Technometrics*, **14**, 725–733.

Shenton, L. R., and Bowman, K. O. (1977). A bivariate model for the distribution of $\sqrt{b_i}$ and b_2. *J. Am. Stat. Assoc.*, **72**, 206–211.

Shepard, R. N. (1962a). The analysis of proximities: multidimensional scaling with an unknown distance function—I. *Psychometrika*, **27**, 125–140.

Shepard, R. N. (1962b). The analysis of proximities: multidimensional scaling with an unknown distance function—II. *Psychometrika*, **27**, 219–246.

Shepard, R. N. (1963). Analysis of proximities as a technique for the study of information processing in man. *Hum. Factors*, **5**, 33–48.

Shepard, R. N. (1980). Multidimensional scaling, tree-fitting and clustering. *Science*, **210**, 390–398.

Shinozaki, N. (1980). Estimation of a multivariate normal mean with a class of quadratic loss functions. *J. Am. Stat. Assoc.*, **75**, 973–976.

Shorack, G. R. (1974). Random means. *Ann. Stat.*, **2**, 661–675.

Shorack, G. R. (1976). Robust Studentisation of location estimates. *Stat. Neerl.*, **30**, 119–142.

Shorrock, R. W., and Zidek, J. V. (1976). An improved estimator of the generalized variance. *Ann. Stat.*, **4**, 629–638.

Sibson, R. (1970). A model for taxonomy II. *Math. Biosci.*, **6**, 405–430.

Sibson, R. (1971). Some observations on a paper by Lance and Williams. *Comput. J.*, **14**, 156–157.

Sibson, R. (1972). Order invariance methods for data analysis (with discussion). *J. R. Stat. Soc. B*, **34**, 311–349.

Sibson, R. (1973). SLINK: An optimally efficient algorithm for the single-link cluster method. *Comput. J.*, **16**, 30–34.

Sibson, R. (1978). Studies in the robustness of multidimensional scaling: procrustes statistics. *J. R. Stat. Soc. B*, **40**, 234–238.

Sibson, R. (1979). Studies in the robustness of multidimensional scaling; perturbation analysis of classical scaling. *J. R. Stat. Soc. B*, **41**, 217–229.

Siegel, A. F., and Benson, R. H. (1982). A robust comparison of biological shapes. *Biometrics*, **28**, 341–350.

Siegel, J. H., Goldwyn, R. M., and Friedman, H. P. (1971). Pattern and process of the evolution of human septic shock. *Surgery*, **70**, 232–245.

Singh, A. (1980). On the exact distribution of the likelihood ratio criterion for testing $H:\mu = \mu_0$; $\Sigma = \sigma^2 I$. *Common. Stat. Simul. Comput. B*, **9**, 611–619.

Sinha, B. K. (1976). On improved estimators of the generalized variance. *J. Multivar. Anal.*, **6**, 617–625.

Sinha, B. K., and Wieand, H. S. (1979). Union–intersection test for the mean vector when the covariance matrix is totally reducible. *J. Am. Stat. Assoc.*, **74**, 340–343.

Siotani, M. (1956). On the distributions of the Hotelling's T^2-statistics. *Ann. Inst. Stat. Math.*, **8**, 1–14.

Siotani, M. (1967). Some applications of Loewner's ordering of symmetric matrices. *Ann. Inst. Stat. Math.*, **19**, 245–259.

Siotani, M. (1968). Some methods for asymptotic distributions in the multivariate analysis. Mimeo Series No. 595, Institute of Statistics, University of North Carolina, Chapel Hill, N.C.

Siotani, M., and Wang, R.-H. (1977). Asymptotic expansions for error rates and comparison of the W-procedure and the Z-procedure in discriminant analysis. *In* P. R. Krishnaiah (Ed.), *Multivariate Analysis*, Vol. IV, pp. 523–545. North-Holland: Amsterdam.

Siskind, V. (1972). Second moments of inverse Wishart-matrix elements. *Biometrika*, **59**, 691–692.

Small, N. J. H. (1978). Plotting squared radii. *Biometrika*, **65**, 657–658.

Small, N. J. H. (1980). Marginal skewness and kurtosis in testing multivariate normality. *Appl. Stat.*, **29**, 85–87.

Smith, A. F. M., and Makov, U. E. (1978). A quasi-Bayes sequential procedure for mixtures. *J. R. Stat. Soc. B*, **40**, 106–112.

Smith, H., Gnanadesikan, R., and Hughes, J. B. (1962). Multivariate analysis of variance (MANOVA). *Biometrics*, **18**, 22–41.

Sneath, P. H. A., (1957). The application of computers to taxonomy. *J. Gen. Microbiol.*, **17**, 201–226.

Sneath, P. H. A. (1969). Evaluation of clustering methods. *In* A. J. Cole (Ed.), *Numerical Taxonomy*, pp. 257–271. Academic Press: New York.

Sneath, P. H. A., and Sokal, R. R. (1973). *Numerical Taxonomy. The Principles and Practice of Numerical Classification*. Freeman: San Francisco.

Snee, R. D., and Acuff, S. J. (1979). A useful method for the analysis of growth studies. *Biometrics*, **35**, 835–848.

Sokal, R. R. (1977). Clustering and classification: background and current directions. *In* J. Van Ryzin (Ed.), *Classification and Clustering*, pp. 1-15. Academic Press: New York.

Sokal, R. R., and Michener, C. D. (1958). A statistical method for evaluating systematic relationships. *Univ. Kansas Sci. Bull.*, **38**, 1409–1438.

Sokal, R. R., and Sneath, P. H. A. (1963). *Principles of Numerical Taxonomy*. Freeman: San Francisco.

Solomon, H. (1977). Data dependent clustering techniques. *In* J. Van Ryzin (Ed.), *Classification and Clustering*, pp. 155–173. Academic Press: New York.

Sonquist, J. A., and Morgan, J. N. (1963). Problems in the analysis of survey data and a proposal. *J. Am. Stat. Assoc.*, **58**, 415–435.

Sparks, D. N., and Todd, A. D. (1973). Algorithms AS 60: Latent roots and vectors of a symmetric matrix. *Appl. Stat.*, **22**, 260–262.

Späth, H. (1980). *Cluster Analysis Algorithms for Data Reduction and Classification of Objects*. Ellis Harwood: Chichester, England.

Spaulding, A. C. (1971). Some elements of quantitative archaeology. *In* F. R. Hodson, D. G. Kendall, and P. A. P. Tăutu (eds), *Mathematics in the Archaelogical and Historical Sciences*, pp. 3–16. Edinburgh University Press: Edinburgh.

Spearman, C. (1904). The proof and measurement of association between two things. *Am. J. Psychol.*, **15**, 72 and 202.

Spence, I. (1972). *An Aid to the Estimation of Dimensionality in Nonmetric Multidimensional Scaling*. University of Western Ontario Research Bulletin No. 229.

Spence, I. (1979). A simple approximation for random rankings stress values. *Multivar. Behav. Res.*, **14**, 355–365.

Spence, I. (1982). Incomplete experimental designs for multidimensional scaling. *In* R. G. Golledge and J. N. Rayner (Eds.), *Proximity and Preference: Problems in the Multidimensional Analysis of Large Data Sets*. University of Minnesota Press: Minneapolis.

Spence, I., and Domoney, D. W. (1974). Single subject incomplete designs for nonmetric multidimensional scaling. *Psychometrika*, **39**, 469–490.

Spence, I., and Graef, J. (1974). The determination of the underlying dimensionality of an empirically obtained matrix of proximities. *Multivar. Behav. Res.*, **9**, 331–342.

Spiegelhalter, D. J. (1977). A test for normality against symmetric alternatives. *Biometrika*, **64**, 415–418.

Spiegelhalter, D. J. (1980). An omnibus test for normality for small samples. *Biometrika*, **67**, 493–496.

Spjøtvoll, E. (1977). Alternatives to plotting C_p in multiple regression. *Biometrika*, **64**, 1–8.

Sprent, P. (1972). The mathematics of size and shape. *Biometrics*, **28**, 23–37.

Srivastava, J. N. (1966). Some generalizations of multivariate analysis of variance. *In* P. R. Krishnaiah (Ed.), *Multivariate Analysis*, pp. 129–145. Academic Press: New York.

Srivastava, J. N. (1967). On the extension of Gauss–Markov theorem to complex multivariate linear models. *Ann. Inst. Stat. Math.*, **19**, 417–437.

Srivastava, M. S. (1973). A sequential approach to classification: cost of not knowing the covariance matrix. *J. Multivar. Anal.*, **3**, 173–183.

Srivastava, M. S., and Awan, H. M. (1982). On the robustness of Hotelling's T^2-test and distribution of linear and quadratic forms in sampling from a mixture of two multivariate normal populations. *Commun. Stat. Theor. Methods A.*, **11**, 81–107.

Steiger, J. H. (1979). Factor indeterminacy in the 1930's and the 1970's: some interesting parallels. *Psychometrika*, **44**, 157–167.

Stein, C. (1956). Inadmissibility of the usual estimator for the mean of a multivariate normal distribution. *Proc. Third Berkeley Symp. Math. Stat. Prob.*, **1**, 197–206.

Stein, C. (1965). Inadmissibility of the usual estimator for the variance of a normal distribution with unknown mean. *Ann. Inst. Stat. Math.*, **16**, 155–156.

Stephens, M. A. (1970). Use of the Kolmogorov–Smirnov, Cramer–von Mises and related statistics without extensive tables. *J. R. Stat. Soc. B*, **32**, 115–122.

Stephens, M. A. (1974). EDF statistics for goodness-of-fit and some comparisons. *J. Am. Stat. Assoc.*, **69**, 730–737.

Stewart, G. W. (1973). *Introduction to Matrix Computations*. Academic Press: New York.

Steyn, H. S. (1978). On approximations for the central and noncentral distributions of the generalized variance. *J. Am. Stat. Assoc.*, **73**, 670–675.

Stigler, S. M. (1977). Do robust estimators work with *real* data (with discussion). *Ann. Stat.*, **5**, 1055–1098.

Stoddard, A. M. (1979). Standardization of measures prior to cluster analysis. *Biometrics*, **35**, 765–773.

Sturt, E. (1981). Computerized construction in Fortran of a discriminant function for categorical data. *Appl. Stat.*, **30**, 213–222.

Subbaiah, P., and Mudholkar, G. S. (1978). A comparison of two tests for the significance of a mean vector. *J. Am. Stat. Assoc.*, **73**, 414–418.

Subbaiah, P., and Mudholkar, G. S. (1981). On a multivariate analog of studentised range test. *J. Am. Stat. Assoc.*, **76**, 725–728.

Subrahmaniam, K. (1975). On the asymptotic distributions of some statistics used for testing $\Sigma_1 = \Sigma_2$. *Ann. Stat.*, **3**, 916–925.

Subrahmaniam, K., and Chinganda, E. F. (1978). Robustness of the linear discriminant function to nonnormality: Edgeworth series distribution. *J. Stat. Plan. Infer.*, **2**, 79–91.

Subrahmaniam, K., and Subrahmaniam, K. (1973). On the multivariate Behrens–Fisher problem. *Biometrika*, **60**, 107–111.

Subrahmaniam, K., and Subrahmaniam, K. (1975). On the confidence region comparison of some solutions for the multivariate Behrens–Fisher problem. *Commun. Stat.*, **4**, 57–67.

Subrahmaniam, K., and Subrahmaniam, K. (1976). On the performance of some statistics in discriminant analysis based on covariates. *J. Multivar. Anal.*, **6**, 330–337.

Sugiura, N. (1969). Asymptotic expansions of the distributions of the likelihood ratio criteria for covariance matrix. *Ann. Math. Stat.*, **40**, 2051–2063.

Sugiura, N. (1976a). Asymptotic expansions of the latent roots and the latent vectors of the Wishart and multivariate *F* matrices. *J. Multivar. Anal.*, **6**, 500–525.

Sugiura, N. (1976b). Asymptotic non-null distributions of the likelihood ratio criteria for the equality of several characteristic roots of a Wishart matrix. *In* S. Ikeda et al. (Eds.), *Essays in Probability and Statistics* (dedicated to J. Ogawa), pp. 253–264. Shinko Tsusho: Tokyo.

Sugiura, N., and Fujikoshi, Y. (1969). Asymptotic expansions of the non-null distributions of the likelihood ratio criteria for multivariate linear hypothesis and independence. *Ann. Math. Stat.*, **40**, 942–952.

Sugiyama, G. (1966). On the distribution of the largest latent root and corresponding latent vector for principal component analysis. *Ann. Math. Stat.*, **37**, 995–1001.

Susarla, V., and Walter, G. (1981). Estimation of a multivariate density function using delta sequences. *Ann. Stat.*, **9**, 347–355.

Symons, M. J. (1981). Clustering criteria and multivariate normal mixtures. *Biometrics*, **37**, 35–43.

Szatrowski, T. H. (1980). Necessary and sufficient conditions for explicit solutions in the multivariate normal estimation problem for patterned means and covariances. *Ann. Stat.*, **8**, 802–810.

Takahane, Y., Young, F. W., and de Leeuw, J. (1977). Nonmetric individual differences multidimensional scaling: an alternating least squares method with optimal scaling features. *Psychometrika*, **42**, 7–67.

Tang, W. Y., and Chang, W. C. (1972). Some comparisons of the method of moments and the method of likelihood in estimating parameters of a mixture of two normal densities. *J. Am. Stat. Assoc.*, **67**, 702–708.

Ten Berge, J. M. F. (1977). Orthogonal procrustes rotation for two or more matrices. *Psychometrika*, **42**, 267–276.

Thompson, G. H. (1934). Hotelling's method modified to give Spearman's *g*. *J. Educ. Psychol.*, **25**, 366–374.

Thompson, M. L. (1978a). Selection of variables in multiple regression: Part I. A review and evaluation. *Int. Stat. Rev.*, **46**, 1–19.

Thompson, M. L. (1978b). Selection of variables in multiple regression: Part II. Chosen procedures, computations and examples. *Int. Stat. Rev.*, **46**, 129–146.

Thompson, R. C., and Therianos, S. (1973). On the singular values of matrix products, I. *Scripta Math.*, **29**, 99–110.

Thomson, G. H. (1951). *The Factorial Analysis of Human Ability*. London University Press: London.

Thorndike, R. L. (1953). Who belongs in a family? *Psychometrika*, **18**, 267–276.

Tiku, M. L. (1967). Tables of the power of the *F*-test. *J. Am. Stat. Assoc.*, **62**, 525–539.

Tiku, M. L. (1972). More tables of the power of the *F*-test. *J. Am. Stat. Assoc.*, **67**, 709–710.

Tiku, M. L., and Singh, M. (1982). Robust statistics for testing mean vectors of multivariate distributions. *Commun. Stat. Theor. Methods A*, **11**, 985–1001.

Timm, N. H. (1975). *Multivariate Analysis with Applications in Education and Psychology*. Brooks-Cole: Monterey, Calif.

Timm, N. H. (1980). Multivariate analysis of variance of repeated measurements. *In* P. R. Krishnaiah (Ed.), *Handbook of Statistics*, Vol. 1, pp. 41–87. North-Holland: Amsterdam.

Titterington, D. M. (1976). Updating a diagnostic system using unconfirmed cases. *Appl. Stat.*, **25**, 238–247.

Titterington, D. M. (1977). Analysis of incomplete multivariate binary data by the kernel method. *Biometrika*, **64**, 455–460.

Titterington, D. M. (1978). Estimation of correlation coefficients by ellipsoidal trimming. *Appl. Stat.*, **27**, 227–234.

Titterington, D. M. (1980). A comparative study of kernel-based density estimates for categorical data. *Technometrics*, **22**, 259–270.

Titterington, D. M., Murray, G. D., Murray, L. S., Speigelhalter, D. J., Skene, A. M., Habbema, J. D. F., and Gelpke, D. J. (1981). Comparison of discrimination techniques applied to a complex data set of head injured patients. *J. R. Stat. Soc. A*, **144**, 145–175.

Torgerson, W. S. (1952). Multidimensional scaling: I–theory and method. *Psychometrika*, **17**, 401–419.

Torgerson, W. S., (1958). *Theory and Methods of Scaling*. Wiley: New York.

Toussaint, G. (1974). Bibliography on estimation of misclassification. *IEEE Trans. Inf. Theory*, **IT-20**, 472–479.

Tretter, M. J., and Walster, G. W. (1975). Central and noncentral distributions of Wilks' statistic in MANOVA as mixtures of incomplete beta functions. *Ann. Stat.*, **3**, 467–472.

Tso, M. K.-S. (1981). Reduced-rank regression and canonical analysis. *J. R. Stat. Soc. B*, **43**, 183–189.

Tsui, K.-W., Weerahandi, S., and Zidek, J. (1980). Inadmissibility of the best fully equivariant estimator of the generalized residual variance. *Ann. Stat.*, **8**, 1156–1159.

Tu, C.-T., and Han, C.-P., (1982). Discriminant analysis based on binary and continuous variables. *J. Am. Stat. Assoc.*, **77**, 447–454.

Tukey, J. W. (1957). On the comparative anatomy of transformations. *Ann. Math. Stat.*, **28**, 602–632.

Tukey, J. W. (1960). A survey of sampling from contaminated distributions. *In* I. Olkin (Ed.), *Contributions to Probability and Statistics*, pp. 448–485. Stanford University Press: Palo Alto, Calif.

Tukey, J. W., and McLaughlin, D. H. (1963). Less vulnerable confidence and significance procedures for location based on a single sample: trimming/winsorization 1. *Sankhya A*, **25**, 331–352.

Tyler, D. E. (1981). Asymptotic inference for eigenvectors. *Ann. Stat.*, **9**, 725–736.

Urbakh, V. Yu. (1971). Linear discriminant analysis: loss of discriminating power when a variate is omitted. *Biometrics*, **27**, 531–534.

Utts, J. M., and Hettmansperger, T. P. (1980). A robust class of tests and estimates for multivariate location. *J. Am. Stat. Assoc.*, **75**, 939–946.

Van de Geer, J. P. (1968). Fitting a quadratic function to a two-dimensional set of points. Unpublished research note, RN 004-68, Department of Data Theory for the Social Sciences, University of Leden, The Netherlands.

Van Driel, O. P. (1978). On various causes of improper solutions in maximum likelihood factor analysis. *Psychometrika*, **43**, 225–243.

Van Ness, J. (1979). On the effects of dimension in discriminant analysis for unequal covariance populations. *Technometrics*, **21**, 119–127.

Van Ness, J. W., and Simpson, C. (1976). On the effects of dimension in discriminant analysis. *Technometrics*, **18**, 175–187.

Vasicek, O. (1976). A test for normality based on sample entropy. *J. R. Stat. Soc. B*, **38**, 54–59.

Venables, W. N. (1974). Algorithm AS 77: Null distribution of the largest root statistic. *Appl. Stat.*, **23**, 458–465.

Venables, W. N. (1976). Some implications of the union–intersection principle for tests of sphericity. *J. Multivar. Anal.*, **6**, 185–190.

Wachter, K. (1975). *User's Guide and Tables to Probability Plotting for Large Scale Principal Component Analysis*. Research Report W-75-1, Department of Statistics, Harvard University, Cambridge, Mass.

Wachter, K. (1976a). Probability plotting points for principal components. *In* Hoaglin and Welsch (Eds.), *Proc. Ninth Interface Symp. Comput. Sci. Stat.*, pp. 299–308. Prindle, Weber and Schmidt: Boston.

Wachter, K. (1976b). *Asymptotic Distributions of Principal Component Variances: The k-Atom Case*. Research Report W-76-1, Department of Statistics, Harvard University, Cambridge, Mass.

Wachter, K. W. (1980). The limiting empirical measure of multiple discriminant ratios. *Ann. Stat.*, **8**, 937–957.

Wahl, P. W., and Kronmal, R. A. (1977). Discriminant functions when covariances are unequal and sample sizes are moderate. *Biometrics*, **33**, 479–484.

Wainer, H. (1981). Comment on a paper by Kleiner and Hartigan. *J. Am. Stat. Assoc.*, **76**, 272–275.

Wainer, H., and Schacht, S. (1978). Gapping. *Psychometrika*, **43**, 203–212.

Wakimoto, K., and Taguri, M. (1978). Constellation graphical method for representing multidimensional data. *Ann. Inst. Stat. Math.*, **30**, 97–104.

Wald, A., and Brookner, R. J. (1941). On the distribution of Wilks' statistic for testing the independence of several groups of variates. *Ann. Math. Stat.*, **12**, 358–372.

Wallace, C. S., and Boulton, D. M. (1968). An information measure for classification. *Comput. J.*, **11**, 185–194.

Walster, G. W., and Tretter, M. J. (1980). Exact noncentral distributions of Wilks' Λ and Wilks-Lawley U criteria as mixtures of incomplete beta functions: for three tests. *Ann. Stat.*, **8**, 1388–1390.

Wang, P. C. C. (Ed.) (1978). *Proceedings of the Symposium on Graphic Representation of Multivariate Data*. Academic Press: New York.

Wang, Y. H. (1980). Sequential estimation of the mean of a multinormal distribution. *J. Am. Stat. Assoc.*, **75**, 977–983.

Ward, J. H., Jr. (1963). Hierarchical grouping to optimize an objective function. *J. Am. Stat. Assoc.*, **58**, 236–244.

Waternaux, C. M. (1976). Asymptotic distribution of the sample roots for a nonnormal distribution. *Biometrika*, **63**, 639–645.

Watson, G. S. (1964). A note on maximum likelihood. *Sankhya A*, **26**, 303–304.

Watson, L., Williams, W. T., and Lance, G. N. (1966). Angiosperm taxonomy: a comparative study of some novel numerical techniques. *J. Linn. Soc. Bot.*, **59**, 491–501.

Wegman, E. J., and Carroll, R. J. (1977). A Monte Carlo study of robust estimators of location. *Commun. Stat. Theor. Methods A*, **6**, 795–812.

Weisberg, S., and Bingham, C. (1975). An approximate analysis of variance test for non-normality suitable for machine calculation. *J. Am. Stat. Assoc.*, **17**, 133–134.

Welch, B. L. (1939). Note on discriminant functions. *Biometrika*, **31**, 218–220.

Welch, B. L. (1947). The generalization of "Student's" problem when several different population variances are involved. *Biometrika*, **34**, 28–35.

Welsch, R. E., (1976). Graphics for data analysis. *Comput. Graphics*, **2**, 31–37.

White, J. W., and Gunst, R. F. (1979). Latent root regression: large sample analysis. *Technometrics*, **21**, 481–488.

Wilf, H. S. (1977). A method of coalitions in statistical discriminant analysis. *In* K. Enslein, A. Ralston, and H. S. Wilf (Eds.), *Mathematical Methods for Digital Computers*, Vol. 3, pp. 96–120. Wiley: New York.

Wilk, M. B., and Gnanadesikan, R. (1968). Probability plotting methods for the analysis of data. *Biometrika*, **55**, 1–17.

Wilk, M. B., Gnanadesikan, R., and Huyett, M. J. (1962a). Probability plots for the gamma distribution. *Technometrics*, **4**, 1–20.

Wilk, M. B., Gnanadesikan, R., and Huyett, M. J. (1962b). Estimation of parameters of the gamma distribution using order statistics. *Biometrika*, **49**, 525–545.

Wilkinson, J. H. (1965). *The Algebraic Eigenvalue Problem*. Oxford University Press: London.

Wilkinson, J. H. (1967). The solution of ill-conditioned linear equations. *In* A. Ralston and H. S. Wilf (Eds.), *Mathematical Methods for Digital Computers*, Vol. 2, pp. 65–93. Wiley: New York.

Wilkinson, J. H., and Reinsch, C. (1971). *Handbook for Automatic Computation*. Vol. 2. *Linear Algebra*. Springer-Verlag: Berlin.

Wilks, S. S. (1932). Certain generalisations in the analysis of variance. *Biometrika*, **24**, 471–494.

Wilks, S. S. (1946). Sample criteria for testing equality of means, equality of variances and equality of covariances in a normal multivariate distribution. *Ann. Math. Stat.*, **17**, 257–281.

Wilks, S. S. (1963). Multivariate statistical outliers. *Sankhya A*, **25**, 407–426.

Williams, E. J. (1970). Comparing means of correlated variates. *Biometrika*, **57**, 459–461.

Williams, J. S. (1978). A definition for the common-factor analysis model and the elimination of problems of factor score undeterminacy. *Psychometrika*, **43**, 293–306.

Williams, J. S. (1979). A synthetic basis for a comprehensive factor-analysis theory. *Biometrics*, **35**, 719–733.

Williams, W. T., and Lance, G. N. (1977). Hierarchical classificatory methods. *In* K. Enslein, A. Ralston, and H. S. Wilf (Eds.), *Statistical Methods for Digital Computers*, Vol. 3, pp. 269–295. Wiley: New York.

Williams, W. T., Lambert, J. M., and Lance, G. N. (1966). Multivariate methods in plant ecology. V. Similarity analyses and information analysis. *J. Ecol.*, **54**, 427–445.

Williams, W. T., Clifford, H. T., and Lance, G. N. (1971a). Group-size dependence: a rationale for choice between numerical classifications. *Comput. J.*, **14**, 157–162.

Williams, W. T., Lance, G. N., Dale, M. B., and Clifford, H. T. (1971b). Controversy concerning the criteria for taxonometric strategies. *Comput. J.*, **14**, 162–165.

Winer, B. J. (1962). *Statistical Principles in Experimental Design*. McGraw-Hill: New York.

Wish, M., Deutsch, M., and Biener, L. (1970). Differences in conceptual structures of nations: an exploratory study. *J. Person. Social Psychol.*, **16**, 361–373.

Wishart, D. (1969a). An algorithm for hierarchical classifications. *Biometrics*, **25**, 165–170.

Wishart, D. (1969b). Mode analysis. *In* A. J. Cole (Ed.), *Numerical Taxonomy*, pp. 282–308. Academic Press: New York.

Wishart, D. (1978a). *Clustan User Manual*, 3rd ed., Report No. 47, Program Library Unit, Edinburgh University, Edinburgh.

Wishart, D. (1978b). Treatment of missing values in cluster analysis. *In* L. C. A. Corsten and J. Hermans (Eds.), *Compstat 1978*, pp. 281–287. Physica-Verlag: Vienna.

Wishart, J. (1938). Growth rate determination in nutrition studies with the bacon pig, and their analysis. *Biometrika*, **30**, 16–28.

Wold, S. (1978). Cross-validatory estimation of the number of components in factor and principal components models. *Technometrics*, **20**, 397–405.

Wolde-Tsadik, G., and Yu, M. C. (1979). Concordance in variable-subset discriminant analysis. *Biometrics*, **35**, 641–644.

Wolfe, J. H. (1970). Pattern clustering by multivariate mixture analysis. *Multivar. Behav. Res.*, **5**, 329–350.

Wooding, R. A. (1956). The multivariate distribution of complex normal variables. *Biometrika*, **43**, 212–215.

Woolson, R. F., and Leeper, J. D. (1980). Growth curve analysis of complete and incomplete longitudinal data. *Commun. Stat. Theor. Methods A*, **9**, 1491–1513.

Worsley, K. J. (1977). A non-parametric extension of a cluster analysis method by Scott and Knott. *Biometrics*, **33**, 532–535.

Yao, Y. (1965). An approximate degrees of freedom solution to the multivariate Behrens–Fisher problem. *Biometrika*, **52**, 139–147.

Yerushalmy, J. et al. (1965). Birth weight and gestation as indices of "immaturity." *Am. J. Dis. Child.*, **109**, 43–57.

Yochmowitz, M. G., and Cornell, R. G. (1978). Stepwise tests for multiplicative components of interaction. *Technometrics*, **20**, 79–84.

Yohai, V. J., and Garcia Ben, M. S. (1980). Canonical variables as optimal predictors. *Ann. Stat.*, **8**, 865–869.

Young, F. W., Takahane, Y., and Lewyckyj, R. (1978). Three notes on ALSCAL. *Psychometrika*, **43**, 433–435.

Young, F. W., Takahane, Y., and Lewyckyj, R. (1980). ALSCAL: a multidimensional scaling package with several individual differences options. *Am. Stat.*, **34**, 117–118.

Zadeh, L. A. (1977). Fuzzy sets and their application to pattern classification and clustering analysis. *In* J. Van Ryzin (Ed.), *Classification and Clustering*, pp. 251–299. Academic Press: New York.

Zahn, D. A. (1975a). Modifications of and revised critical values for the half-normal plot. *Technometrics*, **17**, 189–200.

Zahn, D. A. (1975b). An empirical study of the half-normal plot. *Technometrics*, **17**, 201–211.

Zellner, A. (1962). An efficient method of estimating seemingly unrelated regressions and tests for aggregation bias. *J. Am. Stat. Assoc.*, **57**, 348–368.

Zellner, A. (1963). Estimators for seemingly unrelated regression equations: some exact finite sample results. *J. Am. Stat. Assoc.*, **58**, 977–992.

Zerbe, G. O. (1979a). Randomization analysis of the completely randomized design extended to growth curves. *J. Am. Stat. Assoc.*, **74**, 215–221.

Zerbe, G. O. (1979b). Randomization analysis of randomized blocks design extended to growth and response curves. *Commun. Stat. Theor. Methods A*, **8**, 191–205.

Zerbe, G. O., and Jones, R. H. (1980). On application of growth curve techniques to time series data. *J. Am. Stat. Assoc.*, **75**, 507–509.

Zerbe, G. O., and Walker, S. H. (1977). A randomization test for comparison of growth curves with different polynomial design matrices. *Biometrics*, **33**, 653–657.

Index

Applied Probability and Statistics (Continued)

DRAPER and SMITH • Applied Regression Analysis, *Second Edition*
DUNN • Basic Statistics: A Primer for the Biomedical Sciences, *Second Edition*
DUNN and CLARK • Applied Statistics: Analysis of Variance and Regression
ELANDT-JOHNSON and JOHNSON • Survival Models and Data Analysis
FLEISS • Statistical Methods for Rates and Proportions, *Second Edition*
FOX • Linear Statistical Models and Related Methods
FRANKEN, KÖNIG, ARNDT, and SCHMIDT • Queues and Point Processes
GALAMBOS • The Asymptotic Theory of Extreme Order Statistics
GIBBONS, OLKIN, and SOBEL • Selecting and Ordering Populations: A New Statistical Methodology
GNANADESIKAN • Methods for Statistical Data Analysis of Multivariate Observations
GOLDBERGER • Econometric Theory
GOLDSTEIN and DILLON • Discrete Discriminant Analysis
GREENBERG and WEBSTER • Advanced Econometrics: A Bridge to the Literature
GROSS and CLARK • Survival Distributions: Reliability Applications in the Biomedical Sciences
GROSS and HARRIS • Fundamentals of Queueing Theory
GUPTA and PANCHAPAKESAN • Multiple Decision Procedures: Theory and Methodology of Selecting and Ranking Populations
GUTTMAN, WILKS, and HUNTER • Introductory Engineering Statistics, *Third Edition*
HAHN and SHAPIRO • Statistical Models in Engineering
HALD • Statistical Tables and Formulas
HALD • Statistical Theory with Engineering Applications
HAND • Discrimination and Classification
HILDEBRAND, LAING, and ROSENTHAL • Prediction Analysis of Cross Classifications
HOAGLIN, MOSTELLER, and TUKEY • Understanding Robust and Exploratory Data Analysis
HOEL • Elementary Statistics, *Fourth Edition*
HOEL and JESSEN • Basic Statistics for Business and Economics, *Third Edition*
HOGG and KLUGMAN • Loss Distributions
HOLLANDER and WOLFE • Nonparametric Statistical Methods
IMAN and CONOVER • Modern Business Statistics
JAGERS • Branching Processes with Biological Applications
JESSEN • Statistical Survey Techniques
JOHNSON and KOTZ • Distributions in Statistics
Discrete Distributions
Continuous Univariate Distributions—1
Continuous Univariate Distributions—2
Continuous Multivariate Distributions
JOHNSON and KOTZ • Urn Models and Their Application: An Approach to Modern Discrete Probability Theory
JOHNSON and LEONE • Statistics and Experimental Design in Engineering and the Physical Sciences, Volumes I and II, *Second Edition*
JUDGE, HILL, GRIFFITHS, LÜTKEPOHL and LEE • Introduction to the Theory and Practice of Econometrics
JUDGE, GRIFFITHS, HILL and LEE • The Theory and Practice of Econometrics
KALBFLEISCH and PRENTICE • The Statistical Analysis of Failure Time Data